Dear Dale:

Thanks for working with me on the chapter.

Regards

Yoga

Accidental Injury

Narayan Yoganandan • Alan M. Nahum
John W. Melvin
The Medical College of Wisconsin Inc
on behalf of Narayan Yoganandan
Editors

Accidental Injury

Biomechanics and Prevention

Third Edition

Editors
Narayan Yoganandan, PhD, FAAAM,
　FAIMBE, FASME, FSAE
Chief Editor
Professor of Neurosurgery and
　Chair, Biomedical Engineering
Professor of Orthopaedic Surgery
Department of Neurosurgery
Medical College of Wisconsin and
　Zablocki Veterans Affairs
　Medical Center
Milwaukee, WI, USA

Adjunct Professor
Department of Biomedical Engineering
Marquette University
Milwaukee, WI, USA

John W. Melvin, PhD (Deceased)
President, Tandelta Inc.
Ann Arbor, Michigan, USA

Alan M. Nahum, MD, FACS
Professor Emeritus of Surgery
School of Medicine
University of California at San Diego
San Diego, CA, USA

The Medical College of Wisconsin Inc
　on behalf of Narayan Yoganandan
Milwaukee
Wisconsin, USA

ISBN 978-1-4939-1731-0　　ISBN 978-1-4939-1732-7 (eBook)
DOI 10.1007/978-1-4939-1732-7
Springer New York Heidelberg Dordrecht London

Library of Congress Control Number: 2014953845

© Springer Science+Business Media New York 1993, 2002, 2015
This work is subject to copyright. All rights are reserved by the Publisher, whether the whole or part of the material is concerned, specifically the rights of translation, reprinting, reuse of illustrations, recitation, broadcasting, reproduction on microfilms or in any other physical way, and transmission or information storage and retrieval, electronic adaptation, computer software, or by similar or dissimilar methodology now known or hereafter developed. Exempted from this legal reservation are brief excerpts in connection with reviews or scholarly analysis or material supplied specifically for the purpose of being entered and executed on a computer system, for exclusive use by the purchaser of the work. Duplication of this publication or parts thereof is permitted only under the provisions of the Copyright Law of the Publisher's location, in its current version, and permission for use must always be obtained from Springer. Permissions for use may be obtained through RightsLink at the Copyright Clearance Center. Violations are liable to prosecution under the respective Copyright Law.
The use of general descriptive names, registered names, trademarks, service marks, etc. in this publication does not imply, even in the absence of a specific statement, that such names are exempt from the relevant protective laws and regulations and therefore free for general use.
While the advice and information in this book are believed to be true and accurate at the date of publication, neither the authors nor the editors nor the publisher can accept any legal responsibility for any errors or omissions that may be made. The publisher makes no warranty, express or implied, with respect to the material contained herein.

Printed on acid-free paper

Springer is part of Springer Science+Business Media (www.springer.com)

To John W. Melvin (1938–2014)

Foreword

This new edition of *Accidental Injury* is a reference and a milestone on the quest for preventing and/or reducing accidental human injuries and deaths associated with vehicular accidents. In many cases this information can be applied to non-vehicular accidental injuries. The mixture of contributors and subject matter is a testament to the comprehensive and multidisciplinary investigation necessary to fully understand and activate preventive measures.

Injury biomechanics has a long and colorful history with pioneers such as H. Yamada who performed laboratory impact experiments with post-mortem human anatomical specimens and Col. John Stapp who designed and performed acceleration experiments on himself.

Dr. William Haddon, the first head of the National Highway Traffic Safety Administration, enumerated the principles of public health policy which were essential to his mission. This included the need to prevent injury even if the accident could not be prevented. This has fostered the development of injury biomechanics. This area of investigation has required the compilation of precise statistical data on accidental injuries and deaths and the mechanisms and relevant circumstances surrounding the injuries. With this information in hand researchers have studied human tolerance using human surrogates, volunteers and computer simulations to explain more accurately the mechanism of injuries to the human body. This book summarizes current research and conclusions in many areas applicable to injury prevention and is essential knowledge for all those interested in reducing and eliminating many of the preventable deaths and injuries.

San Diego, CA, USA   Alan M. Nahum, MD, FACS

Preface

The book covers the biomechanical and prevention aspects of accidental injuries of the human body in 29 chapters. Research efforts focused on injuries, injury mechanisms and human tolerance are included. Experimental studies on all body regions including their anatomy; collection, analysis and scoring of injuries; computational modeling including finite element and stochastic techniques; motor vehicle safety standards; aviation studies; and ballistic environment research are presented.

Well-known multidisciplinary authors, ranging from practicing physicians in neurosurgery, orthopaedic surgery and trauma surgery, to biomedical engineers have contributed to this book. Many have attained unique statuses such as fellows of multiple scientific organizations including the American Institute of Medical and Biological Engineering, American Society of Mechanical Engineers, Association for the Advancement of Automotive Medicine and Society of Automotive Engineers; Presidents of International Societies; members of the National Academy of Engineers; and members of the Editorial Boards of scientific international clinical and bioengineering peer-review journals. Their collective experience of over 1,000 years, spanning from academia to industry to private organizations, are reflected in this book. I offer my sincere thanks for their timely contributions.

I would like to place on record my deep sense of appreciation and gratitude to all individuals who have helped shape my professional career. While I was a student at the Indian Institute of Science, Bangalore, India, I was fortunate to learn the "art" of conducting research under the guidance and supervision of Professor and former Deputy Director of the Institute, Asuri Sridharan, Ph.D. His mentoring during the budding days of my graduate studies generated lifelong curiosity to pursue academic research. Although his guidance was in structural mechanics and geotechnical engineering, principles continue to remain the same. I owe a debt of gratitude to my Professor.

The Department of Neurosurgery has been an intellectual home for me. I would like to thank Dennis J. Maiman, M.D., Ph.D., Chair, Department of Neurosurgery, Medical College of Wisconsin, for cultivating an excellent research atmosphere spanning decades and extending his unselfish support to researchers like me in our institutions. I thank Frank A. Pintar, PhD, Professor and Vice Chair of Neurosurgery Research in our department, my colleague and a trusted friend for over 30 years for all his valuable and timely suggestions. The encouragement and assistance of Brice Osinski, MBA,

Department Administrator, is also acknowledged. My thanks are in order to Anne Brown, Associate General Counsel; Sara Cohen, General Counsel; and Marjorie Spencer, Chief Financial Officer, Medical College of Wisconsin, for their assistance in contractual matters. I thank Gregory Baer and his staff at Springer Inc., for timely publication. Thanks are to Ms. Sumathy Thanigaivelu, Project Coordinator for typesetting this work and Mr. Bharath Krishnamoorthy, Project Manager, SPi Technologies India Pvt. Ltd., Pondicherry, India, for handling the production of the third edition of the book on behalf of Springer Inc.

Jan Schiebenes, Administrative Assistant III, Department of Neurosurgery, Medical College of Wisconsin, deserves special acknowledgment for her tireless work that included weekends and "countless" e-mail communications. While electronic media has helped administrative and editorial aspects compared to my experiences with two books, an effort of this nature still demands diligence and perseverance. She took personal pride and always had a smile on her face in all her work. On behalf of all authors, I offer my heartfelt thanks.

I thank my wife Malini and daughter Asha for their understanding of this profession and help in bringing this work to fruition. My brother Prasad S. Narayan, sister-in-law Ashwini and nephew Nikil Prasad deserve a special note of thanks for their support and taking care of my parents and my family, especially during this period. My parents are always a source of inspiration and hard work, and I am fortunate to have their support to explore my academic curiosities and expand my knowledge. Blessings of the Lord, parents, teachers and individuals such as those mentioned above, are what I need.

Milwaukee, WI, USA   Narayan Yoganandan, PhD, FAAAM,
                                                          FAIMBE, FASME, FSAE

Contents

1	**Introduction to and Applications of Injury Biomechanics**........ Albert I. King	1
2	**Automotive Field Data in Injury Biomechanics**...................... Hampton C. Gabler, Ashley A. Weaver, and Joel D. Stitzel	33
3	**Medical Imaging and Injury Scaling in Trauma Biomechanics**... Jacob R. Peschman and Karen Brasel	51
4	**Anthropomorphic Test Devices and Injury Risk Assessments**.. Harold J. Mertz and Annette L. Irwin	83
5	**Restraint System Biomechanics**... Richard W. Kent and Jason Forman	113
6	**Mathematical Models, Computer Aided Design, and Occupant Safety**.. King H. Yang and Clifford C. Chou	143
7	**Computational Analysis of Bone Fracture** .... Daniel P. Nicolella and Todd L. Bredbenner	183
8	**Skull and Facial Bone Injury Biomechanics**............................. Cameron R. Bass and Narayan Yoganandan	203
9	**Biomechanics of Brain Injury: A Historical Perspective** .... John W. Melvin and Narayan Yoganandan	221
10	**Biomechanics of Brain Injury: Looking to the Future**............. David F. Meaney	247
11	**Neck Injury Biomechanics** .... Roger W. Nightingale, Barry S. Myers, and Narayan Yoganandan	259
12	**Upper Extremity Injury Biomechanics**...................................... Mei Wang, Raj D. Rao, Narayan Yoganandan, and Frank A. Pintar	309

| 13 | **Thorax Injury Biomechanics** | 331 |

John M. Cavanaugh and Narayan Yoganandan

| 14 | **Impact and Injury Response of the Abdomen** | 373 |

Warren N. Hardy, Meghan K. Howes, Andrew R. Kemper, and Stephen W. Rouhana

| 15 | **Thoracic Spine Injury Biomechanics** | 435 |

Mike W.J. Arun, Dennis J. Maiman, and Narayan Yoganandan

| 16 | **Lumbar Spine Injury Biomechanics** | 451 |

Brian D. Stemper, Frank A. Pintar, and Jamie L. Baisden

| 17 | **Knee, Thigh, and Hip Injury Biomechanics** | 471 |

Jonathan D. Rupp

| 18 | **Leg, Foot, and Ankle Injury Biomechanics** | 499 |

Robert S. Salzar, W. Brent Lievers, Ann M. Bailey, and Jeff R. Crandall

| 19 | **Pain Biomechanics** | 549 |

Nathan D. Crosby, Jenell R. Smith, and Beth A. Winkelstein

| 20 | **Thoracolumbar Pain: Neural Mechanisms and Biomechanics** | 581 |

John M. Cavanaugh, Chaoyang Chen, and Srinivasu Kallakuri

| 21 | **Role of Muscles in Accidental Injury** | 611 |

Gunter P. Siegmund, Dennis D. Chimich, and Benjamin S. Elkin

| 22 | **Pediatric Biomechanics** | 643 |

Kristy B. Arbogast and Matthew R. Maltese

| 23 | **Best Practice Recommendations for Protecting Child Occupants** | 697 |

Kathleen D. Klinich and Miriam A. Manary

| 24 | **Pedestrian Injury Biomechanics and Protection** | 721 |

Ciaran Knut Simms, Denis Wood, and Rikard Fredriksson

| 25 | **Design and Testing of Sports Helmets: Biomechanical and Practical Considerations** | 755 |

James A. Newman

| 26 | **Normalization and Scaling for Human Response Corridors and Development of Injury Risk Curves** | 769 |

Audrey Petitjean, Xavier Trosseille, Narayan Yoganandan, and Frank A. Pintar

27	**Injury Criteria and Motor Vehicle Regulations**.....................	793
	Priyaranjan Prasad	
28	**Civil Aviation Crash Injury Protection**......................................	811
	Richard L. DeWeese, David M. Moorcroft, and Joseph A. Pellettiere	
29	**Ballistic Injury Biomechanics**...	829
	Cynthia Bir	

Index.. 841

Contributors

Kristy B. Arbogast, Ph.D. Children's Hospital of Philadelphia, University of Pennsylvania, Philadelphia, PA, USA

Mike W.J. Arun, Ph.D. Neuroscience Research Lab, Department of Neurosurgery, Medical College of Wisconsin, Milwaukee, WI, USA

Ann M. Bailey, B.S. Center for Applied Biomechanics, University of Virginia, Charlottesville, VA, USA

Jamie L. Baisden, M.D. Department of Neurosurgery, Medical College of Wisconsin, Milwaukee, WI, USA

Cameron R. Bass, Ph.D. Department of Biomedical Engineering, Duke University, Durham, NC, USA

Cynthia Bir, Ph.D. Keck School of Medicine, University of Southern California, Los Angeles, CA, USA

Karen Brasel, M.D., M.P.H. Medical College of Wisconsin, Milwaukee, WI, USA

Todd L. Bredbenner, Ph.D. Southwest Research Institute, San Antonio, TX, USA

John M. Cavanaugh, M.D. Department of Biomedical Engineering, Wayne State University, Detroit, MI, USA

Chaoyang Chen, M.D. Department of Biomedical Engineering, Wayne State University, Detroit, MI, USA

Dennis D. Chimich, Ph.D., P.Eng. MEA Forensic Engineers & Scientists, Richmond, BC, Canada

Jeff R. Crandall, Ph.D. Center for Applied Biomechanics, University of Virginia, Charlottesville, VA, USA

Clifford C. Chou, Ph.D. Department of Mechanical Engineering, Wayne State University, Detroit, MI, USA

Nathan D. Crosby, Ph.D. Department of Bioengineering, University of Pennsylvania, Philadelphia, PA, USA

Richard L. DeWeese, B.S. Federal Aviation Administration, Civil Aerospace Medical Institute, Oklahoma City, OK, USA

Benjamin S. Elkin, Ph.D. MEA Forensic Engineers & Scientists, Richmond, BC, Canada

Jason Forman, Ph.D. University of Virginia, Charlottesville, VA, USA

Rikard Fredriksson, Ph.D. Autoliv Research, Vargarda, Sweden

Hampton C. Gabler, Ph.D. Center for Injury Biomechanics, Virginia Tech-Wake Forest University, Blacksburg, VA, USA

Warren N. Hardy, Ph.D. Center for Injury Biomechanics, Virginia Tech – Wake Forest University, Blacksburg, VA, USA

Meghan K. Howes, Ph.D. Center for Injury Biomechanics, Wake Forest University, Blacksburg, VA, USA

Annette L. Irwin, Ph.D. General Motors Company, Troy, MI, USA

Srinivasu Kallakuri, Ph.D. Department of Biomedical Engineering, Wayne State University, Detroit, MI, USA

Andrew R. Kemper, Ph.D. Center for Injury Biomechanics, Virginia Tech – Wake Forest University, Blacksburg, VA, USA

Richard W. Kent, Ph.D. University of Virginia, Charlottesville, VA, USA

Albert I. King, Ph.D. Department of Biomedical Engineering, Wayne State University, Detroit, MI, USA

Kathleen D. Klinich, Ph.D. Transportation Research Institute, University of Michigan, Ann Arbor, MI, USA

W. Brent Lievers, Ph.D., P.Eng. Bharti School of Engineering, Laurentian University, Sudbury, ON, Canada

Dennis J. Maiman, M.D., Ph.D. Department of Neurosurgery, Medical College of Wisconsin, Milwaukee, WI, USA

Matthew R. Maltese, Ph.D. Children's Hospital of Philadelphia, University of Pennsylvania, Philadelphia, PA, USA

Miriam A. Manary, M.Eng. Transportation Research Institute, University of Michigan, Ann Arbor, MI, USA

David F. Meaney, Ph.D. Departments of Bioengineering and Neurosurgery, University of Pennsylvania, Philadelphia, PA, USA

John W. Melvin, Ph.D. Tandelta Inc., Ann Arbor, MI, USA

Harold J. Mertz, Ph.D. General Motors Corporation, Harper Woods, MI, USA

David M. Moorcroft, M.S. Federal Aviation Administration, Civil Aerospace Medical Institute, Oklahoma City, OK, USA

Barry S. Myers, M.D., Ph.D. Duke University, Durham, NC, USA

Alan M. Nahum, M.D., F.A.C.S. School of Medicine, University of California at San Diego, San Diego, CA, USA

James A. Newman, Ph.D. Newman Biomechanical Engineering Consulting Inc., Edmonton, AB, Canada

Daniel P. Nicolella, Ph.D. Southwest Research Institute, San Antonio, TX, USA

Roger Nightingale, Ph.D. Duke University, Durham, NC, USA

Joseph A. Pellettiere, Ph.D. Federal Aviation Administration, Washington, DC, USA

Jacob R. Peschman, M.D. Department of Surgery, Medical College of Wisconsin, Milwaukee, WI, USA

Audrey Petitjean, M.Sc. CEESAR, Nanterre, France

Frank A. Pintar, Ph.D. Department of Neurosurgery, Medical College of Wisconsin, Milwaukee, WI, USA

Priyaranjan Prasad, Ph.D. Prasad Consulting, LLC., Plymouth, MI, USA

Raj Rao, M.D. Department of Orthopaedic Surgery, Medical College of Wisconsin, Milwaukee, WI, USA

Stephen W. Rouhana, Ph.D. Ford Motor Company, Dearborn, MI, USA

Jonathan D. Rupp, Ph.D. University of Michigan Transportation Research Institute and the Departments of Emergency Medicine and Biomedical Engineering, University of Michigan, Ann Arbor, MI, USA

Robert S. Salzar, Ph.D. Center for Applied Biomechanics, University of Virginia, Charlottesville, VA, USA

Gunter P. Siegmund, Ph.D. Injury Biomechanics, MEA Forensic Engineers & Scientists, Richmond, BC, Canada

Ciaran Knut Simms, Ph.D. Department of Mechanical and Manufacturing Engineering, Trinity College, Dublin, Ireland

Jenell R. Smith, Ph.D. Department of Bioengineering, University of Pennsylvania, Philadelphia, PA, USA

Brian D. Stemper, Ph.D. Department of Neurosurgery, Medical College of Wisconsin, Milwaukee, WI, USA

Joel D. Stitzel, Ph.D. Biomedical Engineering, Wake Forest University, Center for Injury Biomechanics, Virginia Tech-Wake Forest University, Winston-Salem, NC, USA

Xavier Trosseille, Ph.D. LAB PSA Peugeot Citroen RENAULT, Nanterre, France

Mei Wang, Ph.D. Department of Orthopaedic Surgery Research, Medical College of Wisconsin, Milwaukee, WI, USA

Ashley A. Weaver, Ph.D. Biomedical Engineering, Wake Forest University, Center for Injury Biomechanics, Virginia Tech-Wake Forest University, Winston-Salem, NC, USA

Beth A. Winkelstein, Ph.D. Departments of Bioengineering and Neurosurgery, University of Pennsylvania, Philadelphia, PA, USA

Denis Wood, Ph.D. Denis Wood Associates, Dublin, Ireland

King H. Yang, Ph.D. Bioengineering Center, Wayne State University, Detroit, MI, USA

Narayan Yoganandan, Ph.D. Department of Neurosurgery, Medical College of Wisconsin, Milwaukee, WI, USA

Introduction to and Applications of Injury Biomechanics

Albert I. King

Abstract

The four aims of impact biomechanics are (1) Identification and explanation of injury mechanisms (2) Quantification of mechanical response of body components to impact (3) Determination of tolerance levels to impact and (4) Assessment of safety devices and techniques to evaluate prevention systems. These are briefly described followed by a discussion of the methods used to study injury biomechanics. The rest of the chapter is devoted to injury mechanisms which need to be understood if preventative measures are to be developed and implemented. The mechanisms covered are brain injury and acute subdural hematoma, neck pain due to whiplash, aortic rupture, spinal injury due to vertical acceleration, disc rupture, hip fracture in the elderly, ankle (pilon) fracture and foot fracture. It is concluded that much misinformation exists in the literature regarding these injuries and the statement in Lancet (Issue 9773): "The most entrenched conflict of interest in medicine is a disinclination to reverse a previous opinion" is indeed true.

1.1 Introduction

Injury biomechanics is a huge field and covers many areas of study, involving many parts of the human body. Obviously, it is not possible to cover the subject in a single chapter. This invitation by the Editors asking me to write the first chapter was for me to introduce the reader to this field of endeavor and to encourage him/her to seek out more detailed information from journal papers and books on this subject.

There is no question that we all want to avoid being injured. Severe injuries can be fatal or at least life changing. The fact that we are getting older is another reason why we need to be more careful in our activities, whether we are driving, walking, or even sitting. Generally, the main symptom of injury is pain which is a warning that something bad has happened and we should try to get out of that situation or not to repeat what we have been doing. Of course, when an impact

A.I. King, Ph.D. (✉)
Department of Biomedical Engineering, Wayne State University, Detroit, MI, USA
e-mail: king@rrb.eng.wayne.edu

occurs, this is not possible. So we have to do the next best thing – to completely avoid the impact or to ameliorate the impact so that the injury would not be as severe. This brings up the question of injury prevention or injury mitigation. So the study of injury biomechanics boils down to a concerted effort to prevent injury.

Of course, there are many ways one can get injured. The two principal categories are intentional and unintentional injury. The former is difficult to prevent because there are so many ways someone can inflict injury on another person or oneself. However, unintentional injuries are preventable and, for biomechanicians, the focus of injury prevention is on unintentional injuries.

There are many forms of unintentional injuries. Perennially, motor vehicle crashes appear to be the leading cause of unintentional injury deaths [1]. In 2010, this was followed by unintentional poisoning and unintentional falls. Sports-related injuries do not rank high as one of the leading causes of nonfatal unintentional injuries in the US but are headline grabbers because of such issues as concussion in American football.

In the automotive safety arena, a very effective method of injury prevention is active safety which aims to avoid the crash. By the use of electronic sensors and other devices, the vehicle can be made to slow down or the driver can be warned to take evasive action to avoid the collision. This is not always possible if there is no time to react or the momentums of the vehicles involved is too large to overcome. An inebriated pedestrian stepping out from behind a parked truck just as a car is passing by would probably not benefit from any currently available active safety systems. Similarly, if a vehicle is suddenly and deliberately made to swerve and cross the centerline of a busy two-lane highway, it will most likely hit an oncoming vehicle before the active safety system of either vehicle can take effective action. That is, despite what the computer can do, crashes will still occur and passive safety in its many forms will still be necessary to protect the occupant. There are also opportunities to combine active and passive safety for more effective injury prevention. One such example on whiplash prevention is discussed in this chapter.

1.2 Definition of Injury Biomechanics

Biomechanics is the application of the principles of mechanics to biological systems and injury biomechanics is a subfield that studies the effect of mechanical impact on biological systems, in particular, the human body. Although the field is made up of a relatively small number of researchers and professionals, it was the first area of biomechanics to be studied. Gurdjian and Lissner [2] published the first paper on head injury in 1944 and initiated skull fracture mechanism research in 1939 at Wayne State University.

1.3 Motivation to Study Injury Biomechanics

The principal motivation to study injury biomechanics is, or at least, it should be that of saving lives and reducing suffering due to injury. Safety professionals have always said that injury is preventable if we only take the necessary steps to protect the body. For the elderly who tend to fall frequently, the obvious countermeasure is to instruct them to walk carefully and watch where they are going. Of course, this does not cover falls due to dizziness, heart and cardiovascular problems or fracture. In the case of automotive related injuries, the Federal government and the automotive industry have worked on improving automotive safety over the last half century. Traffic deaths as measured by the number of fatalities per 100,000 miles driven has been going down steadily with the introduction of belt restraints, padding of interior surfaces of cars and the airbag. These passive safety features are found in a modern vehicle. They constitute the use of environmental control to protect the occupant and do not need occupant participation with the exception of the wearing of the belt restraints. In the US, there are seat belt laws in place in all 50 states and the wearing rate has been going up steadily. Nationally, the average rate in 2010 was 85 % with the rate exceeding 90 % in 15 states and the District of Columbia [3]. Biomechanically, seat belt use is an essential component of passive safety and all occupants need to be belted when traveling in cars, regardless of the length of the trip.

In sports, the motivation is to reduce brain and joint injuries, especially in contact sports. Much remains to be done in the area of concussion prevention in American football. The football helmet in use currently was designed to prevent skull fracture because the standards governing helmet performance were drawn up for that purpose. There have been many attempts to reduce the incidence of concussion but there is no concerted effort on the part of any national sports organization or the Federal government to initiate the design and fabrication of an anti-concussion helmet. In particular, a computer based design which takes into account the response of the human brain to impact would be advisable because the helmet can be hit in almost every direction and reliance on experimental impact data alone would not be adequate.

1.4 The Four Aims of Injury Biomechanics

Research in injury biomechanics evolved from ad hoc crude experiments and back of the envelope calculations to well organized studies funded by the Federal government and large corporations. The principal aims of this research are [4]:

1. Identification and explanation of injury mechanisms
2. Quantification of mechanical response of body components to impact
3. Determination of tolerance levels to impact
4. Assessment of safety devices and techniques to evaluate prevention systems

1.4.1 Identification and Explanation of Injury Mechanisms

As the saying goes, "you cannot prevent an injury unless you know the cause." In this case, it means the mechanism whereby the injury is caused. In the section below, several lesser known or somewhat controversial mechanisms are discussed. Some mechanisms are obvious. For example, if a bicyclist falls and hits his head on the ground,

suffering a skull fracture, the mechanism is impact of the head with a hard surface and the method of prevention would be to wear a helmet. Other injury mechanisms are not immediately obvious and it is necessary to perform detailed studies to ensure that exact cause can be identified before a fix is introduced. A perfect example of this is the headrest mounted atop of automotive seat backs. The headrest was meant to prevent neck pain after a rearend collision (whiplash). It has been installed in cars sold in the US since 1986 but the whiplash problem did not go away. Several mechanisms of injury are discussed in Sect. 1.6 below.

1.4.2 Quantification of Mechanical Response

In engineering, the standard procedure for classifying a material is to define its mechanical response to an applied load. The same can be done for biological materials. For bone and, we can apply a compressive load and obtain its force –deflection response. Similarly, for ligaments, we can apply a tensile load and obtain the same response. However, for more complex body regions, such as the chest, we need to load it in different directions and locations to obtain its response to frontal and lateral loads for the upper, mid and lower thorax. For organs that are within the body, the task is more challenging, especially if we want to test it in situ. Accurate quantification of brain motion inside the skull due to a blunt impact was not achieved until beginning of this century [5].

Mechanical response can take many forms. The traditional engineering approach is to define the response in terms of stress versus strain or load versus deflection. However, for dynamic response, we can define a force-time response or a displacement-time response. Figure 1.1 shows the displacement response of the brain. Targets within the cadaveric brain make motions in the shape of a figure 8 when the head is impacted. The force–deflection response of the thorax is shown in Fig. 1.2. The response data are not as clean cut and predictable as with inanimate manufactured materials. Variations among individuals are large and are affected by age and gender. Reliability of the response data depends heavily

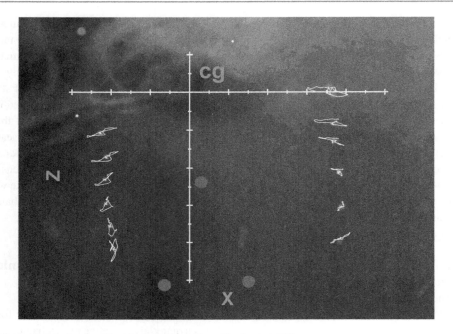

Fig. 1.1 Brain motion due to a blunt impact (Taken from Hardy et al. [5])

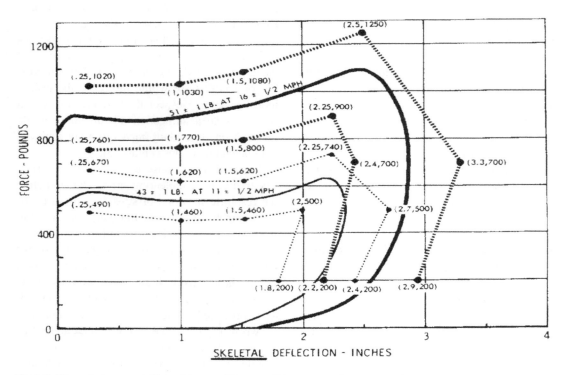

Fig. 1.2 Thoracic response to frontal impact. The two solid curves are for responses by two different impactors

1 Introduction to and Applications of Injury Biomechanics

Fig. 1.3 Photograph of a Hybrid III dummy. (Courtesy of Humanetics Innovative Solutions, Inc.

on the number of specimens tested, especially if a response corridor is to be drawn.

To design a human surrogate (crash dummy) for impact testing, a large amount of impact response data is needed so that the dummy can be as human-like as possible. The current dummy used universally by automakers is the Hybrid III. It is shown in Fig. 1.3. Not all mechanical responses of the Hybrid III are human-like because it is very difficult to have a device that is simultaneously biofidelic and behaves in a repeatable and reproducible manner during testing. In general, the dummy is stiffer than the human and its torso is less flexible.

1.4.3 Determination of Human Tolerance to Impact

In the design of safety systems for vehicles of any kind, it is necessary to know how much acceleration or force the body or any part of the body can take before serious injury is incurred. This is the study of human tolerance to impact. To the perennial question: How many g's can I take? The answer is another question: How badly do you want to get hurt? There are several levels of injury and the safety design can focus on one of those levels for a given impact severity. The injury levels are:

1. The "Ouch" level
2. Minor injury
3. Moderate injury
4. Serious injury
5. Critical injury
6. Fatal injury

The "Ouch" level is used when testing human volunteers. If the test causes pain in any way, the test should not proceed beyond that level and the volunteer can withdraw from the program. Minor and moderate injuries are real injuries which may require a visit to the emergency room but should not require prolonged hospitalization. At the serious injury level, the healthy individual will require hospitalization but the injuries are not life threatening. They may be for the elderly and the infirm. For a vehicle to be affordable, the design should result in injuries between the moderate and the serious. Critical injuries are life threatening to the healthy individual and should be avoided. Of course, no design should result in fatal injuries.

Like response, tolerance is also highly variable among subjects and is also dependent on age and gender. Because of this variability, absolute tolerance values are not very meaningful. Instead, a probabilistic approach is taken and, for a given level of injury, the tolerance values are expressed as probability of injury. For example, for head injury at a serious injury level, the probability of injury in terms of angular acceleration is shown Fig. 1.4. In this case, the angular acceleration for 50 % probability of a minor traumatic brain injury is 5,500 rad/s^2. Other parameters can be used as injury predictors, using this Logist analysis. Statistical parameters can be computed to determine which parameter is the best predictor of a particular injury.

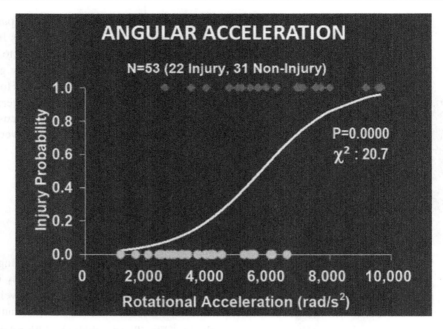

Fig. 1.4 Logist plot of mTBI versus rotational (angular) acceleration. For a 505 probability of injury the rotational acceleration is predicted to be approximately 5,500 rad/s²

1.4.4 Assessment of Safety Devices and Techniques

The knowledge gained in the three areas of study described above can be used to assess safety devices in vehicles and other systems. The two principal tools are the crash dummy (Hybrid III) and computer models. For the last half century, most of the assessment has been experimental, using the crash dummy. This assessment is made as the design is progressing and calls for many crash tests that are costly and time consuming. The automotive industry has begun to realize that computer models are now sufficiently reliable to be used as an assessment tool. With the aid of models, the designer can skip much of the crash testing and do a final test of the system at the end of the development process. There is now a consortium of model developers, in the US, funded by the automotive manufacturers to develop a single total human body model that can be used in place of the Hybrid III. This is a lofty goal that may not be achievable because all kinds of models can be developed quickly and at low cost whereas a repeatable and reproducible crash dummy takes years of work to come to fruition. Nevertheless, this modeling effort is a laudable goal that will save the industry time and money.

1.5 Methods to Study Injury Biomechanics

Impact biomechanics began as an experimental discipline, much like many other fields of study. From observation, it progressed to testing and organized experimentation. The study began with injuries to particular body region and was gradually extended to the whole body. For example, Gurdjian and Lissner [6] concentrated their efforts on head injury and performed experiments on human skulls initially before venturing to test anesthetized animals. Evans and Lissner [7] also initiated studies on the thoracolumbar spine by testing embalmed whole cadavers on a specially designed vertical accelerator that was housed in an elevator shaft of an eight-story building. Mertz and Patrick [8] studied human neck response using a horizontal sled. Later on, the response and tolerance of the knee was also studied using horizontal sleds.

Mathematical models were developed in conjunction with the experimental studies, usually after some experimental data have become available to test the viability of the model. As computers became faster and capable of storing large amounts of data, the models became more complex and more detailed. A record of model development over half a century can be found in Yang et al. [9]. It goes without saying that most researchers have more faith in experimental data than in the predictions of mathematical or computer models. As a result, it is traditional to expect the model developers to validate their models against appropriate experimental data. The policy of the Stapp Crash Journal is that every model published in the journal must be validated. As a result, improvements in model predictions have been rapid and they are becoming reliable predictors of response and injury. In fact, because of the ready availability of models of many body regions, it is now possible to model the experiment and determine what can be expected to occur in an experiment, where to make the critical measurements and what levels of force or acceleration are to be expected in a given series of experiments. For example, in a current project on blast-related brain injury, the modelers are being asked to determine where to expect the maximum pressures in the brain and what pressure levels will be reached before any testing is done. These predictions will assist with the placement of pressure transducers in the brain to measure the peak pressures.

1.6 Injury Mechanisms from Head to Toe: Application to Design

1.6.1 Head Injury Mechanisms

1.6.1.1 Brain Injury Mechanisms

The existence of two injury mechanisms for the brain is well known. Gurdjian et al. [10] proposed that linear acceleration producing pressure waves within the brain was a mechanism for brain concussion. On the other hand, Ommaya et al. [11] were of the opinion that angular acceleration was the prime cause of concussion. This debate became quite heated for a time in the 1970s but has since died down somewhat when it became apparent that both forms of acceleration usually increased in a monotonic fashion with increased impact severity. However, the injury mechanisms due to these two forms of acceleration are quite different. Based on the work of Hardy et al. [5], we can say that angular acceleration is responsible for the relative motion of the brain within the skull during a blunt impact and that the resulting diffuse axonal injury is due to this motion. Although axonal stretch has not been measured directly, it can be deduced that this relative motion of the brain is essentially the mechanism that can stretch the axons. As for linear acceleration, it causes a pressure wave to be generated, starting as a compressive wave at the site of impact and becoming a reflected tensile wave as it is reflected from the skull at the opposite or contrecoup side. Gurdjian et al. [12] invented the fluid percussion method of causing concussion in dogs after they discovered the existence of a transient pressure wave traversing the brain. Fluid percussion tests that varied in duration from less than 1 ms to 46 ms and in magnitude from 34.5 to 345 kPa (5–50 psi), resulted in concussion in experimental animals. Unfortunately, no biomarkers were identified to describe the injury. Thus, we know that pressure is an injury mechanism but we do not know what cells are injured or what parts of the cell are damaged by pressure. Ongoing research to study the effects of blast overpressure on military personnel may soon reveal one or more mechanisms of injury at the cellular and/or the molecular level. Preliminary data on rodents [13] indicate that blast overpressure is causing the glial cells to go into apoptosis with possible deleterious effects on the neurons they support and that there was no axonal injury associated with blast exposure. This finding is reasonable in that there was very little head motion during a pure (primary) blast.

1.6.1.2 Injury Mechanisms for Acute Subdural Hematoma

The accepted injury mechanism for acute subdural hematoma (ASDH) is bridging vein rupture.

However, from an engineering point of view, the acute formation of a hematoma from a ruptured vein violates the principles of fluid mechanics. The mechanism proposed in this book chapter is a hypothesis with no data to support its veracity. The reader is asked to consider the logic of the hypothesis and decide if it has more merit than the accepted mechanism.

The physiopathogenesis of ASDH formation has been a subject of debate since the early thinking of an organized space between the arachnoid and dura was detailed by Key and Retzius in the late 1800s [14]. This group described the structures of the meninges and experimentally determined that substances injected into the presumed subdural space did not mix with other substances within the tissue. Early researchers believed that fluids within the alleged space could move between compartments of the brain [15]. Thus, authors of this time period believed and offered evidence that a fluid-filled space existed between the dura and arachnoid [15–19]. As Weed continued his studies, he determined that the structures were fused together in embryos, but could be separated in mature animals [19]. These early investigators injected fluids into the subdural area and visualize the distinctive compartmentalization of these fluids. Microscopically, layers of unique cells between the dura and arachnoid tissue were recognized and these cells were thought to produce a fluid which appeared to be present within the 'space'. Leary [20] concluded that the inner dura was lined with fibroblasts and that the cells lining the outer arachnoid were dissimilar. Thus, investigators began examining the dura and arachnoid as two exclusively separate identities.

The Dura Mater. The dura mater appears to be a thick layer of fibroblasts and extracellular collagen [21]. The cells look large and flattened and the collagen is abundant and somewhat organized. Haines [14] summarized the dura-arachnoid organization. The dura is characterized as having an inner and outer portion. The periosteal dura is adherent to the inner skull and the meningeal layer of the dura, contains a specialized layer that Nabeshima et al. [22] named the dural-border cell layer. This layer appears to be continuous with the dural aspect of the arachnoid and the histological aspects of this dural-border cell layer have brought much interest to researchers [22–25]. This amorphous layer appears to have flattened cell processes, varying sizes of extracellular spaces and little collagenous material. The amorphous structure possibly makes this an area of weakness within the tissue. A cross-section of the meninges and cell layers is shown in Fig. 1.5. If an ASDH is to form, the bleed needs to occur in the border cell layers.

The Arachnoids. The arachnoid portion of the meninges also consists of two distinct areas, the arachnoid barrier cell layer, which is attached to the dural-border cell layer, and the arachnoid trabeculae, which is closely attached to the pia mater. Both the cells and the extracellular material are dissimilar as compared to the dura mater. The cells are larger, more densely packed, having numerous mitochondria and filaments within their cytoplasm making the layer distinctive [22, 23, 26]. This closely packed structure of the arachnoid barrier cell layer excludes the presence of extracellular space, making it distinctive from the attached dural-border cell layer. Existing literature supports this idea. The description of the layers above has been verified [14, 27, 28] and testing has shown that the 'space' is not pre-existing. However, the junction between the dural and arachnoid border cells would be an area of weakness in cases of brain impact injury because the loosely organized dural border cell layer is attached to the more rigid arachnoid barrier cell layer. In fact, there is evidence that the space is easily created by a mechanical separation [29–31]. Since the biomechanical properties of the border cell layers have not been investigated, the adhesive properties of the layers in radial traction or in shear need to be quantified. These properties are crucial to the understanding of the formation of ASDH because of the close association of the bridging vein and cortical arteries within these layers. Only when this mechanism is established will preventative and clinical strategies be able to be discovered and tested. This will ultimately decrease morbidity and mortality rates associated with these types of brain injuries.

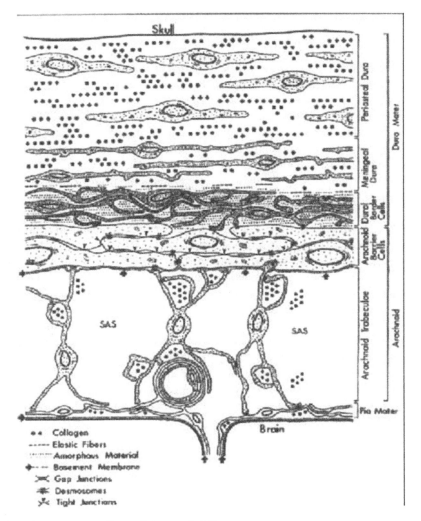

Fig. 1.5 Meningeal structure according to Haines et al. [14]

On the other hand, neurosurgeons are often of the opinion that the dura is attached to the skull and the arachnoid goes with the brain. Thus, even if there is no space in the subdural layer in the young, an actual space maybe created in the elderly should their brain shrink because not all of that space can be accommodated in the CSF layer. Since this is still controversial, we need to consider the mechanism of ASDH with no subdural space as well as in the presence of a subdural space occupied by CSF.

Anatomy of Cortical Vessels. The cortical vessels consist of bridging veins and cortical arteries and veins. The bridging veins traverse the dural/arachnoid complex. Their rupture has been traditionally considered responsible for ASDH and they have been studied extensively by researchers [32, 33]. The number of veins and their range of diameters have all been documented. Yamashima and Friede [34] provided a detailed description of the vessel wall as it traverses the dura/arachnoid complex in a straight course with no tortuosity to allow for the possible displacement of the brain. The cranial end is firmly attached to the rigid dura while the cerebral end is attached to the movable hemisphere. Leary [20] found that the thickness of the bridging vein walls varied remarkably in the subdural portion, the thinnest part measuring 10 μm with a range of 10–600 μm. In the subarachnoid portion, the walls have a more consistent thickness of 50–200 μm. The collagen fibers in the subdural portion were loosely woven with a pattern that was more resistant to distension while less resistant to traction. That is, bridging veins are vulnerable to leakage in the subdural region. In fact, Yamashima and Friede [34] speculated that the bridging vein can rupture in the dura/arachnoid complex due to a physiological increase in venous pressure or due to cardiac resuscitation as well as due to a head impact. Trotter [35] regarded the rupture of the bridging vein as the cause of chronic subdural hematoma. However, another bleed source is the cortical artery traversing the dura/arachnoid complex. Information on the size, distribution and number of cortical vessels is sparse. Cortical arteries are found in the CSF layer. They run along the surface for a short distance and penetrate the pia to enter the cerebral cortex. However, some of the arteries running under the arachnoid can extend branches into the subdural layer. There is even evidence of a cortical artery forming a kink (knuckle) in the subdural space, as shown in Fig. 1.6 [36]. When the dura separates from the arachnoid, the vessel wall of the knuckle is torn off and bleeding from this tear results in an ASDH.

Acute Subdural Hematomas. Subdural hematoma (SDH) is a clinical condition due to a quickly clotting blood collection amid the dura and arachnoid membranes. ASDH's are most frequently the result of an acute head injury, however they can sometimes occur spontaneously in the elderly. The mechanism behind the separation of the arachnoid from the dura has yet to be determined. ASDH's usually transpire when the brain is subjected to a high energy, short duration force from trauma. It is thought that this shearing force will tear the bridging veins and as a consequence ASDH will form. However, epidemiological studies have shown that injuries other than bridging vein rupture accounted for a significant portion of ASDH cases. Thus, the need to determine the mechanism behind the injury is vital before any effective preventative and therapeutic strategies can be attempted and implemented. Finding the pathogenic mechanism through a more open-minded approach will lead to new innovative treatments for this disabling condition.

Epidemiology. Traumatic ASDH's are among the most lethal of all head injuries, carrying the highest

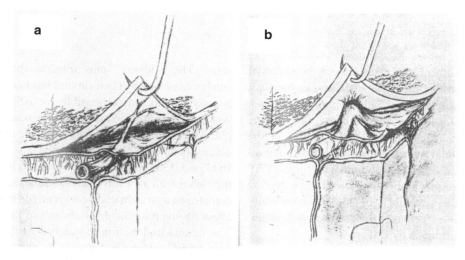

Fig. 1.6 (**a**) Bridging cortical artery connected to the dura. (**b**) Adherence of cortical arterial knuckle to dura and arachnoid (Bongioanni et al. [36])

risk to the patient, with a mortality rate of greater than 50 % in most studies. ASDH kills or severely disables more head injured patients than any other complication of cranial trauma. The main pathological factor involved is ischemic neuronal damage that results from cerebral vascular damage, raising the intracranial pressure. ASDH was found in patients who were involved in motor vehicle crashes, falls and assaults [37]. It is also found in boxers [38]. According to Gennarelli and Thibault [39], ASDH is the most important cause of death in severely head injured patients due to high incidence (30 %), high mortality (60 %) and head injury severity (2/3 with Glasgow Coma scores of 3–5). They also found that the cause of ASDH by falls or assaults was 72 % while that due to motor vehicle crashes was only 24 %. Maxeiner [40] attributed the source to bleeding in ASDH cases to extensive brain surface damage (contusion) and ruptured superficial cerebral vessels, including bridging veins and small arteries of the cortex. However, he also indicated in another publication [41] that rupture of the bridging veins did not lead to the formation of ASDH's. In fact, Maxeiner and Wolff [42] showed that there was an equal probability of ASDH caused by bridging vein rupture and by cortical artery rupture. Moreover, Shenkin [43] reviewed 39 consecutive cases of ASDH and found that there was a high incidence of cortical artery rupture (61.5 %). Bleeds of venous origin constituted 25.6 % of the cases and cerebral contusions were the cause in 7.7 % of the cases. The elderly were found to be more susceptible to ASDH [44, 45]. Since there can be brain shrinkage with age resulting in stretching of the bridging veins, the high incidence among the elderly can be explained by bridging vein rupture. However, the simple rupture of the bridging vein should not lead to ASDH formation unless additional mechanical factors are present, as explained below. Thus, clear mechanisms of ASDH formation need to be formulated before we can claim to understand why there is a high incidence of ASDH among the elderly. Karnath [46] found that ASDH usually occurs in younger adults while chronic SDH usually occurs in older individuals between 60 and 70 years of age. Finally, although the literature is silent in terms of a detailed injury mechanism, there is an implication that ASDH occurs when there is head contact with a rigid surface. However, in their experiments on subhuman primates, Gennarelli and Thibault [39] applied a pure angular acceleration to the head without direct impact to cause ASDH in these animals. More than one injury mechanism is in play in the formation of ASDH.

In terms of the locations of ASDH, not all ASDH's occur along the superior sagittal sinus into which the bridging vein empties the venous blood. Obviously, non-bridging vein related ASDH's are caused by bleeds from other sources, such as cortical vessels and brain contusion or laceration [47]. We will now consider the mechanism of ASDH formation from cortical bleeds for reasons stated in the section below.

Biomechanical Mechanisms for the Formation of ASDH

Based on current thinking, ASDH can arise from one of three sources, the first being the cortical arteries and veins. Laceration or rupture of these vessels can occur with penetrating injuries. Secondly, closed head injuries resulting in large contusions can cause similar bleeding into the adjacent subdural area. Thirdly and the most common type of ASDH is thought to occur from tearing of the veins that bridge the subdural area as they travel from the surface of the brain to the various dural sinuses. This last mechanism assumes rupture of the bridging vein in the subdural space since if it ruptured below the arachnoid, the result would be a subarachnoid hematoma. Ultimate strain to failure of the bridging veins and possibly other tissue components is inversely related to the strain rate [48]. Thus, the threshold for injury decreases as the strain rate or acceleration is increased. Gennarelli and Thibault [39] opined that ASDH is due to the rupture bridging veins during angular acceleration of the head associated with rapid onset rates (high strain rate). They contend that nothing needs to strike the head in order for ASDH to occur. That is, although impact to the head is certainly the most common cause of ASDH, it is the angular acceleration induced by the impact and not the head contact that causes ASDH. Examples include violent non-cranial impacts of football players and motorcycle riders. A rapid head movement is sufficient to exceed the bridging vein tolerance. This group also demonstrated that sensitivity of the ultimate strain of bridging veins to strain rate and provided acceleration tolerance data for subdural hematoma in primates [39]. Lee and Haut [49] reported the insensitivity of tensile failure properties of human cerebral bridging veins to strain rate. However, data scattering drew concerns regarding this finding. To confirm their data, this group performed similar experiments on the carotid arteries and jugular veins of ferrets [50]. These vessels were stretched longitudinally in vitro at either a low (0.2–2.0 s^{-1}) or high rate (200 s^{-1}). The ultimate stretch ratios and loads were found to be independent of strain rate in all the vessels tested. Therefore, the results appeared to support their previous finding on human bridging veins. Maxeiner [40] tested the hypothesis that subdural hematomas are less frequent in acceleration injuries in traffic accidents compared to falls or assaults. They reported that this hypothesis did not hold true in the same way for bridging vein ruptures. Ruptures of these vessels without subdural bleeding were only seldom mentioned in the literature. However, if no subdural hematomas were present, no one would look for these structures. They predicted that the frequency of bridging vein lesions in severe head injuries was likely underestimated in the clinical as well as in the postmortem literature, hypothesizing that a rapid increase of intracranial pressure after the accident produced a collapse of the cerebral circulation which is probably responsible for the absence of the subdural hematomas in the presented cases. Most recently, Monson et al. [51] examined the mechanical behavior of human cerebral blood vessels. This group determined that cerebral arteries were noticeably stiffer than the cerebral veins. Pang et al. [52] studied the morphological properties of pig cerebral bridging vein. They demonstrated that there is a narrow cuff at the junction of the cerebral bridging veins and the superior sagittal sinus. This finding could play an important role in maintaining intracranial pressure (ICP) and ASDH formation. Collectively, all these reports implicate that failure of the bridging vein structure is associated with a large number of subdural hematomas.

However, how the failure of bridging veins causes the venous blood to form an ASDH is difficult to explain based on the principles of fluid mechanics. Since the path to the sagittal sinus is still open and the creation of a subdural space constitutes resistance to fluid flow, it is not clear how this can happen in the absence of other mechanical factors. This principle is demonstrated in Fig. 1.7. We see that for the blood to form a space, and thus an ASDH, in the border cell layer, it must overcome the adhesive resistance between the dural and arachnoid border cells and enlarge the space until it becomes symptomatic and visible on scans. On the other hand, the flow of venous blood into the superior sagittal sinus is the path of least resistance because there is no need to push apart solid boundaries (cell layers) and the formation of an ASDH in the subdural layer by venous blood would violate this very basic principle of fluid mechanics – namely, flow will proceed in the direction of least resistance.

If we consider the possible pre-existence of a subdural space filled with CSF, then a rupture of the bridging vein in the subdural space will allow the blood to enter this space. However, it is not likely that this flow is strong enough to progress into an ASDH because the ICP in the CSF is essentially the same as the pressure in the veins.

To summarize findings to date, ASDH is found in victims of falls, assaults and motor vehicle crashes who sustain a direct impact to the head against a rigid surface. The head undergoes both linear and angular acceleration. However, Gennarelli and Thibault [39] have demonstrated the formation of ASDH with extremely high pure angular acceleration in subhuman primates and found ruptured bridging veins in these animals. Similarly, Depreitere et al. [53] used a reverse injection method to demonstrate the feasibility of ASDH formation after a direct impact that resulted in high angular accelerations. They may have indeed created the observed SDH by reverse injection which is unnatural and not representative of a venous overpressure. Nevertheless, the

1 Introduction to and Applications of Injury Biomechanics

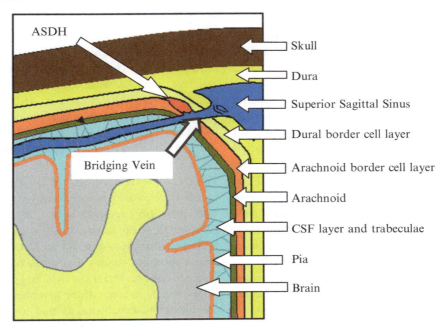

Fig. 1.7 ASDH formation due to bridging vein rupture is not possible in the subdural layer, based on principles of fluid mechanics

circumstantial evidence of associating bridging vein rupture with ASDH is extremely strong. However, as mentioned above, why would the normal flow of blood into the superior sagittal sinus be diverted into the subdural space to form an ASDH against a much higher resistance or against the ICP in the CSF? It violates the basic principle of fluid flow in that the flow will go in the direction of least resistance. We can hypothesize a series of mechanisms that can explain the observed phenomenon and provide a logical explanation for the formation of ASDH. However, before we list the various hypotheses, it is necessary to reiterate the assumption that the bridging vein rupture we are concerned with needs to occur in the dural/arachnoid complex because if it occurred below the arachnoid, a subarachnoid hematoma (SAH) would be the result. It is also necessary to reiterate the finding that the dural and arachnoid border cells can be easily separated although the amount of force needed to achieve this separation has never been measured. It is also interesting to note that a subarachnoid hematoma will form in the CSF space but a SDH will not form in the subdural space from a ruptured bridging vein. The explanation is rather simple. A torn bridging vein below the arachnoid should open the CSF to the sagittal superior sinus and there will be leakage of CSF into the sinus. As a result, venous blood from the torn bridging vein will fill the CSF space to maintain the volume of the CSF layer. Thus, a subarachnoid hematoma can easily form when the bridging vein is ruptured below the arachnoid.

Hypothesis I – Almost all ASDH's from non-penetrating impacts are due to the rupture of cortical arteries in the dural/arachnoid complex, not including the bridging veins.

We postulate that border cell layers need to either separate radially or deform in shear to a sufficient extent to tear these cortical vessels and that the tearing of the cortical arteries will generate enough pressure in the subdural layer to cause the separation to progress, resulting in an ASDH. If this hypothesis can be validated, it can provide a logical explanation for the formation of ASDH.

Hypothesis II – Radial separation of the border cell layers can occur when there is skull in-bending and skull rebound due to a direct impact.

Not all direct impacts produce high angular accelerations but they are more likely to cause skull deformation in the form of in-bending and subsequent rebound. This in-bending can cause delamination of the border cell layer sandwiched between two layers with dissimilar material properties and the rebound applies tension to the dura to cause separation of the border cells and rupture of the cortical vessels within the dural/arachnoid complex. Large deformations can occur in the temporal area of the adult skull as well as in the skulls of children. This mechanism can also explain the formation of ASDH at sites remote from the site of impact because the skull is a closed container which will increase in diameter in one direction when the orthogonal diameter is decreased by impact. This hypothesis can be tested using an animal model as well as a computer model.

Hypothesis III – Shearing deformation of the border cell layer can rupture the cortical vessels when the head is subjected to high levels of angular acceleration.

Results of the brain motion mapping study reported by Hardy et al. [5] reveal that most of the relative motion of the brain with respect to the skull occurs near the center of the brain, above the brain stem. Relative motion of the brain surface is small but it is not non-existent under angular accelerations below 10,000 rad/s². It is conceivable that under very high angular accelerations, the border cell layers will undergo shear deformation that is large enough to rupture the cortical vessels. We also postulate that the relative motion does not occur in the cerebral spinal fluid (CSF) layer because the trabeculae in the CSF layer have a higher shear resistance as compared to the border cells. A study to measure the shear resistance of the pia-arachnoid junction was completed by Jin et al. [54]. This hypothesis can again be tested using either an animal model or a computer model.

Hypothesis IV – Even though the bridging vein in the border cell layer may be ruptured, it does not result in the formation of ASDH.

The prevailing understanding of ASDH formation due to bridging vein rupture is inconsistent with all known principles of fluid mechanics.

Yet, bridging vein rupture was found in patients with ASDH, in cadaveric subjects and animal subjects. This hypothesis can be tested by rupturing the bridging vein of an animal in the border cell layer with no head impact to demonstrate that the rupture will not cause an ASDH to form and progress. The reverse injection method used by Depreitere et al. [53] forces the contrast medium out of the ruptured vessel into the border cell layer which can then be easily separated. There is no reversal of flow of this type in vivo unless there is a large increase in venous pressure. It is also necessary to test the amount of pressure needed to reverse the flow and cause an ASDH to form. However, for an ASDH to progress, a sustained venous over pressure is needed for a period of several hours, which is physiologically unlikely to occur. If the hypothesis is valid, the inevitable conclusion is that many of the observed bridging vein ruptures were due to the autopsy or surgical procedure used to identify the ASDH or that there is cortical bleeding associated with the observed rupture that was not detected. In addition, SDH associated with a pure bridging vein rupture will likely result in a chronic SDH, from occasional increases in venous pressure, such as in a valsalva maneuver.

The reader is encouraged to do research in this area to confirm these hypotheses or to prove them wrong.

1.6.2 Neck Injury Mechanisms

1.6.2.1 The Cause of Neck Pain After Whiplash

Neck pain, resulting from being the victim of a low speed rearend impact (whiplash), is the most commonly reported injury or syndrome [55–57]. The societal cost of this syndrome has been found to be high in North America and in Europe [58–60]. Some researchers have expressed the opinion that complaints of pain and disability are related to the amount of compensation a victim can expect to receive. In Lithuania, Schrader et al. [61] found that whiplash victims all returned to work 2 weeks after the event because there is no compensation for injury. This statistic is cited as a basis for this opinion. Ferrari and Russell [62] deny the severity of the problem while Radanov [63] and Bogduk [64] are of the opinion that chronic symptoms are real and are present in 10–20 % of Bogduk's patients 6 months after whiplash.

In terms of prevention of neck pain following whiplash, the current headrest does not seem to be effective, as evidenced by the continuing interest in the subject in the medical literature. From the biomechanical point of view, an injury cannot be prevented unless the causes or mechanisms are known. For whiplash, several hypotheses have been proposed and until the biomechanics community can come to an agreement as to the true cause of whiplash, prevention of the syndrome would be difficult. The discussion on neck pain and whiplash hypotheses provided below is taken from Deng et al. [65].

To understand the neck pain incurred due to rear-end impacts, current knowledge about spinal pain is reviewed first. The brain interprets acute pain from the signals it receives from a network of pain-sensing nerve endings or nociceptors that are found throughout the body. They are quite numerous in the skin, muscles and tendons but are not normally found in articular cartilage of synovial joints. Nociceptors have a high threshold. That is, a large stimulus is needed to get them to fire and to signal the brain that there is potential or real hazard to that region of the body. This means that the tissue in which the nociceptor is imbedded needs to be deformed considerably before the nociceptor is set off. It is also known that when the soft tissue in the intervertebral joint is inflamed due to degenerative processes, the threshold of nociceptors is reduced and much less deformation is necessary to cause pain.

In normal intervertebral discs, nociceptors are rather sparse and focused on the peripheral regions. They may not even be sensitive to mechanical loading. However, in a degenerated disc, nociceptors become more numerous and are sensitive to mechanical loading. Elsewhere in the intervertebral joint, nociceptors are found in the facet capsules, spinal ligaments, tendons and muscles. In the lumbar spine facet capsule, Cavanaugh et al. [66] were able to show that facet capsules are a source of low back pain. Bogduk and Marsland [67] and Barnsley et al. [68] have clinical evidence of cervical facet pain after whiplash.

There is also another type of pain related to the spine. It is a radicular or radiating type of pain felt in the extremities as a result of degeneration in the intervertebral joint. It is well known that numbness and tingling in the arms are due to pressure on the nerve roots exiting the spinal cord. The pressure can be from a bulging or herniated disc or from bony overgrowth narrowing the foramen through which the nerve exits. The reason for radicular pain is still not entirely clear but Ozaktay et al. [69] were able to show that a toxic chemical found in the nucleus pulposus can inflame the nerve root and produce this referred or phantom pain. If the dorsal root ganglion is affected by this chemical, the pain can be quite severe. However, it is important to recognize that mechanical or chemical effects on the nerve root cause radiculopathy or pain that appears to be located in the extremities and not localized pain in the area of the intervertebral joint. There are at least five different hypotheses, which offer explanations for the source of neck pain following whiplash. These hypotheses are reviewed briefly below.

MacNab [70] proposed an early hypothesis, identifying neck hyperextension as the injury mechanism. This was based on primate studies in which the headrest was not used and head and neck extensions exceeded 90°. However, current automotive head restraints were not effective in reducing the frequency of this problem. As a result, other hypotheses were proposed.

A popular theory is muscle injury from hyperextension, during which the anterior cervical muscles undergo an eccentric contraction. That is, the muscles are contracting as they are being stretched. It has been shown by Garrett et al. [71] that muscles can only be injured during eccentric contraction. However, in the case of these anterior muscles, they tend to hurt for few days after the impact but are not a chronic problem. The difficulty appears to be located in the back of the neck where the extensor muscles are located. These muscles undergo concentric contraction and are not likely to be injured. Tencer et al. [72]

proposed that the injury to the extensor muscles occurred during rebound of the head and neck as they undergo eccentric contraction. The mechanism is biomechanically consistent with the results of Garrett et al. [71] but the hypothesis flies in the face of many more frontal impacts in which there is hyperflexion due to severe crashes and there are not a large number of complaints of neck ache from these crash victims. More recent data by Azar et al. [73, 74] show that stretching of the facet joint capsule can induce muscle spasms and pain. That is, muscular involvement is secondary to facet capsular stretch.

Ono et al. [75] and Yoganandan et al. [76] both proposed a facet joint impingement injury mechanism. Specifically, Ono et al. [75] theorized that a portion of the facet capsule can be trapped between the facet joint surfaces and pinched, causing pain. There is no biomechanical evidence that the capsule is loose enough to be trapped between the facet joint and even it was trapped, evidence is lacking to show that nociceptors are present in the trapped portion of the capsule and are indeed set off by the pressure. The proposition that compression of the facet surfaces can produce pain is also untenable since cartilage is devoid of nociceptors and there is no neurophysiological evidence that the nociceptors in the subchondral bone can be made to fire by this presumed compression.

Aldman [77] found that there was a pressure increase in the spinal canal during whiplash causing pressure on the nerve roots. Subsequent studies by Svensson et al. [78] revealed that this pressure can affect the dorsal root ganglion (DRG) causing it to send pain signals to the brain. Stimulation of or pressure on the nerve roots and DRG will produce radicular pain and not pain in the neck. Furthermore, a generalized increase in spinal canal pressure cannot selectively affect the nerve roots and DRG in the lower cervical spine where most of the problems appear to reside. The originators of this hypothesis also failed to provide objective evidence regarding DRG dysfunction due to transient pressures. Because of these inconsistencies, this hypothesis is not likely to be valid. While political support is keeping this hypothesis viable it has very little scientific merit.

Yang and Begeman [79] proposed the shear hypothesis which attributes the pain to the facet capsule which can be stretched during whiplash. The motion of the torso precedes that of the neck and in order for the head to remain attached to the torso, a shear force is generated at each cervical level and transmitted up the cervical spine until it reaches the occipital condyles where the force can act on the head to cause it to move forward. This shearing action causes relative motion between adjacent vertebrae which can be most pronounced at the lower cervical levels where the facet angle is less steep. It remains to be shown that facet stretch is large at these lower cervical levels and that there is indeed sufficient stretch to cause the nociceptors to fire and thus cause pain.

In support of the shear hypothesis, neurophysiological studies were performed by Lu et al. [80] to show that nociceptors in the goat facet joint capsule could be made to fire if the capsule was stretched both statically and dynamically. The capsular strains reached were found to be as high as 60 %. These strains are consistent with the cadaveric studies by Deng et al. [65] who found large capsular strains as high as 92 % and who also found relative sliding and rotation between adjacent vertebrae, confirming the shear hypothesis.

With the aid of active safety technology, we now can eliminate the whiplash problem by designing the seat back and headrest to work together during a rear impact. The idea is to eliminate shear forces in the neck. We will need a collision sensor on the rear bumper to provide the signal for an impending rearend collision. This signal will activate the headrest and bring it forward to contact the head of the occupant prior to collision. During the collision, the head and torso need to be moved forward at the same rate to eliminate shear between the cervical vertebrae. It is likely that such a seat is already available but it has not been fine tuned to accomplish the zero shear objective. The stiffnesses of the headrest and seat back need to be selected to match those of the head and torso so as to minimize relative motion between the head and torso. If this last step can be accomplished, we will have a seat that can prevent whiplash associated disorders.

The conclusion is that unless and until the biomechanics community accepts a single hypothesis for whiplash related pain, no effective preventive measure can be implemented, even though a solution is at hand.

1.6.3 Injury Mechanisms of the Thorax

1.6.3.1 Traumatic Rupture of the Aorta

Of all the traumatic injuries to the thorax, traumatic rupture of the aorta (TRA) due to blunt impact has the highest fatality rate, even among victims who made it to the hospital. When a frank rupture occurs at the crash scene, it is very difficult to save the victim. TRA related to automobile crashes has been known to auto safety professionals and surgeons for a long time [81]. The injury was originally found to be due to chest impact with the steering wheel at barrier impact speeds in excess of 45 mph. Recently however, TRA was found to occur in side impact after belt and airbag restraints were introduced to mitigate the effects of a frontal collision. One Canadian statistic indicates that the fatality rate due to TRA in the early 1990s was 21 % of all lateral impact cases.

One of the first known attempts at reproducing TRA experimentally was the work of Roberts et al. [82] who impacted anesthetized dogs in the sternum and achieved aortic rupture. A flash x-ray system was used to track the movement of the aorta but it is not clear how many ruptures were obtained out of the six animals they impacted. It can be confirmed that the rupture was transverse. Subsequent efforts to reproduce this injury in cadavers were largely unsuccessful [83] until Hardy et al. [84] managed to cause TRA in seven of the eight cadavers tested. They were tested in an inverted position with the lower extremities removed and were held up for pendulum impact by a clamp that went around the vertebral column. The peri-isthmus region of the aorta was targeted with radio-opaque markers and was visualized during impact by means of a biplanar high speed x-ray unit. Video images of aortic motion and deformation were recorded at 1,000 frames per second. The impact speeds varied from 6.8 to 12.2 m/s and the impact modes took form of shoveling, side impact, submarining and combined. These modes are described in detail [84] and they all were able cause TRA in the cadaver. Most of the tears were transverse with two oblique tears. The failure of the aorta was tensile in nature and the failure strains were estimated to vary from 4.1 % to 35.6 % near the site of failure. It was concluded that the failure mechanism was tension in the aorta that developed when it moved away from the impactor.

1.6.4 Injury Mechanisms of the Thoracolumbar Spine

1.6.4.1 Effect of Bending During Vertical Acceleration

Although vertical (caudocephalad) accelerations are rare in automotive collisions, they were of concern to the Air Force during the development of ejection seats and are now a concern to the Army because of the effects of blast from improvised explosive devices (IED's) on military personnel riding in vehicles. It is perhaps important to review the biomechanics of spinal injury during ejection from a disabled jet aircraft to enable the reader to see the similarities and differences between seat ejection and vehicle acceleration due to blast.

The injury mechanism for anterior wedge fractures in the thoracolumbar spine of pilots who eject from disabled aircraft was studied using the Wayne State vertical accelerator which was housed in an elevator shaft of an 8-story building. The acceleration pulse simulated that of an ejection seat and is shown in Fig. 1.8. Vulcan et al. [85] first discovered the mechanism which was later confirmed by King et al. [86]. In order to explain how this particular injury occurs in the thoracolumbar spine, it is necessary to show that there are two load paths in the spine for transmitting vertical load from the head down to the pelvis. Prasad et al. [87] went to extraordinary lengths to demonstrate the existence of a facet load in the lumbar spine in addition to the load carried by the intervertebral disc. Cadaveric studies were carried out on a vertical accelerator at

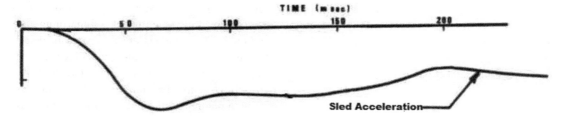

Fig. 1.8 Ejection seat pulse

Fig. 1.9 Intervertebral load cell (*IVL*)

Wayne State University. An intervertebral load cell (IVL) was inserted into the lumbar spine just above a lumbar disc, usually the L3-4 disc. This is shown in Fig. 1.9. The IVL was inserted by removing the inferior portion of a lumbar vertebra, using a double-bladed saw and it did not substantially change the stiffness of the spine. The IVL measured the load borne by the disc or the intervertebral body force. The total load on the spine was estimated as the product of the mass above the IVL and the sled acceleration assuming that dynamic overshoot was present in the acceleration pulse itself. Thus, the total load had the same shape as the vertical sled acceleration. It was found that the measured disc load was different from the estimated total load on the spine and the inevitable conclusion was that this difference could only be attributed to the facets. As shown in Fig. 1.10, the intervertebral or disc load (IVL) was less than the total load at the beginning of the acceleration pulse, implying that the facets were in compression and carrying some of the inertial load. However, as the acceleration progressed, the torso began to flex forward due to the anterior eccentricity of the body center of mass and this flexion moment caused the disc load to exceed the total load, with the facets going into tension. Of course, this tensile load is borne by the posterior ligaments of the vertebral column and not by the facet capsules alone. Figure 1.11 shows how the disc load can exceed the total load. By replacing the flexion moment by a couple, it is seen that disc compression is increased and the posterior spine goes into tension. Therefore, we see anterior wedge fractures in pilots who eject.

In the course of conducting this research, hundreds of cadaveric tests were run. Initially, the acceleration increased rapidly to a pre-set level which was maintained for about 200 ms. The rate of change of acceleration or jerk could be varied. We also attached strain gages to the sides of lumbar vertebral bodies as an indicator of disc load. Although no deliberate attempt was made to correlate the measured strain to jerk, it was found that the strain increased with acceleration but did not change when jerk was varied over a large range. That is, there is no change in dynamic overshoot in vertebral strain when jerk was increased.

With respect to the Dynamic Response Index (DRI) used by the Air Force for ejection seat assessment, recall that it was based entirely upon the response of a spring-mass system subjected to a base excitation, as shown in Fig. 1.12. This model was one of the first used to represent the spine subjected to vertical acceleration and is, anatomically, not representative of the human spine and torso. It is one dimensional and is only capable of predicting the dynamic overshoot for different jerk rates. Patrick ran whole body tests on cadavers in the vertical accelerator and

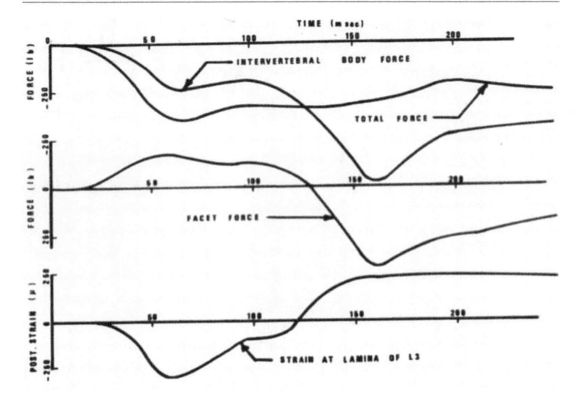

Fig. 1.10 Facet load

attached strain gages to the walls of lumbar vertebrae. The input pulse was trapezoidal as shown in Fig. 1.8 and the rate of rise of acceleration (jerk rate) could be varied from 50 to 2,500 g/s. He measured the vertebral strains at 4 different jerk rates and found inconsistent overshoots of the strain data. This is because the strain gage on the anterior aspect of a vertebral body senses not only compression but also bending. In subsequent testing on the same accelerator, we found a slight evidence of dynamic overshoot in the intervertebral load. The increase in intervertebral or disc load was less than 16 % when the rate of onset increased from 460 to 760 g/s. [88] for seat ejection and the injuries to the spinal column are also expected to be much more severe. Smith and Begeman [89] simulated such blasts on a horizontal sled by testing the seated cadaver lying on its back. They demonstrated the feasibility of using a horizontal sled to simulate a vertical acceleration caused by an IED. The seat was mounted on a deceleration sled which was stopped before the seat hit the concrete barrier. By controlling the thickness of the polymer used between the seat and the barrier, high decelerations were achieved. The seat had an added feature of simulating floor deformation just prior to the impact of the seat against the barrier. A high speed impactor under the feet of the dummy could be deployed using compressed gas. Velocities of up to 21 m/s and accelerations in excess of 1,400 g could be achieved. Thus, massive injuries to the lower extremities can be reproduced on this sled, simulating the real world injuries reported by Ramasamy et al. [90, 91]. Similarly, at the thoracolumbar junction, far more severe injuries than those seen in pilot ejection are expected to occur. Thus, simple linear models like the DRI are obviously inadequate to handle these highly non-linear situations and even finite element models have to be carefully developed to simulate structural collapse of the spine during

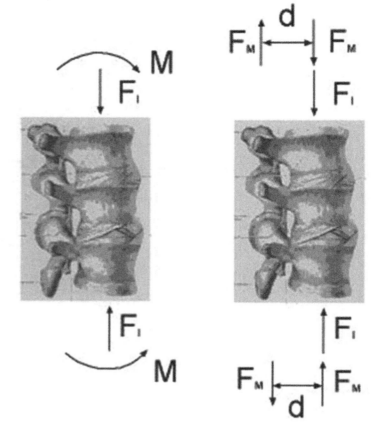

Fig. 1.11 Demonstration of how the IVL can be larger than the total load when the torso flexes forward. F_I is the total load and M is the flexion moment, as shown on the *left*. On the *right*, the moment, M, is replaced by a couple, $F_M \times d$, where d is the moment arm and F_M is the additional compressive force on the disc due to couple and it is also the tensile load imposed on the posterior structure of the vertebral column. Thus, the IVL can be larger than the total load

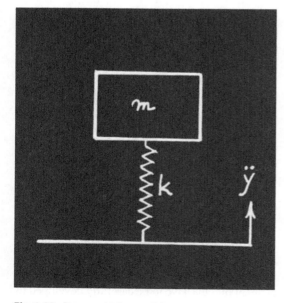

Fig. 1.12 Base excitation model

the impact. The study of military injuries opens a new area of blunt impact research for which the basic principles impact biomechanics are still valid but the events are in a different time and acceleration domain. Impact durations are much shorter and impact accelerations are much higher. Surrogates to be used in military research need to be frangible to reveal injury locations and severity and they need to be much more humanlike than the currently available automotive crash dummies or computer models.

1.6.4.2 Spinal Compression During Horizontal Acceleration/Deceleration

To say that a compressive force can be generated in the spine when a belted automotive occupant is involved in a frontal crash sounds somewhat incredible. This phenomenon was predicted by a

Fig. 1.13 Spine load due to horizontal deceleration

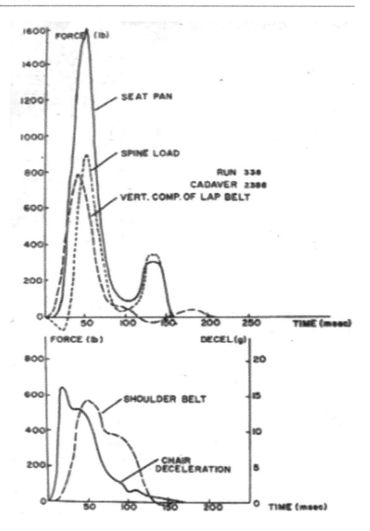

two dimensional spinal model developed by Prasad and King [92]. It was verified experimentally by Begeman et al. [93] who subjected restrained cadavers to frontal decelerations on a sled. They instrumented the seat pan with a load cell which measured both the vertical component of the lapbelt load and the transmitted spinal load due to the shoulder belt, as shown in Fig. 1.13. Several thoracolumbar fractures were found in the cadaveric spines, as evidence of the existence of a compressive load in the lumbar spine. This load is due to the straightening of the kyphotic thoracic spine when the chest is restrained from moving forward during a frontal crash. There is also evidence from real world crashes that such injuries are sustained by three-point belted occupants involved in frontal crashes [94–97].

It should also be mentioned that during a rear-end collision, the occupant of the struck vehicle also sustains the same compressive spinal load due to the seat back pushing on the kyphotic thoracic spine. In this case, the compression in the cervical spine is contributory to the neck pain due to whiplash. The compression is generated as soon as the seat back pushes on the torso to move it forward when the car is struck and it tends to loosen the ligaments and tendons attached to the cervical vertebrae, making it easier for the vertebrae to slide with respect to each other and allowing the facet capsules to stretch.

1.6.4.3 Acute Disc Rupture

The question of an acute disc rupture occurring as the result of a single loading event on the spine is controversial because many motor vehicle crash victims complain of low back pain (LBP) and/or neck pain after the crash and were eventually diagnosed with one or more ruptured discs. Although some of them have a history of being treated for spinal pain, a small percentage of these victims claim that they have never had a history of back or neck pain. The treating physician inevitably opines that the rupture is causally related to crash because his/her diagnosis depends heavily on the history provided by patient. Additionally, payment for the office visit and all subsequent treatments by the automotive insurance company is only made if the treatment provided is related to the incident. This indeed is a major conflict of interest as far as the opinions of the treating physician are concerned.

In the Second Edition of this book, King [98] described the etiology of low back pain and the reasons why disc rupture cannot occur as a result of a single impact. This is summarized below for completeness.

Since LBP is a very common complaint inasmuch as eight out of ten persons will have at least one such attack in their lifetime, it becomes extremely difficult to pinpoint the cause of any particular attack or to relate it to any given incident. Cavanugh et al. [99] have shown that the disc is not the only cause of LBP and that the posterior structures of the spine, including the facet joint capsules and the muscle and tendons around the intervertebral joint can be a source of LBP. Work by Yamashita et al. [100] focused on the source pain in terms of the presence and activation of pain sensing nerve endings or nociceptors. There are three basic requirements to sense pain:

1. The presence of nociceptors in the soft tissue being stretched or deformed
2. Adequate deformation of the soft tissue containing the nociceptors
3. Firing of the nociceptors in response to the deformation.

The first requirement is self-evident and needs no elaboration. As for the morphological distribution of nerve endings in the facet capsule, Yamashita et al. [100] have found these endings in the capsular tissue, extending to the border of nearby ligaments and tendons. In terms of the human facet capsule, Cavanaugh et al. [66] reported that the human capsule can undergo large deformations, particularly during spinal extension. In some cases, the static stretch in the superior lateral corner of the capsule is observable by the naked eye. Quantitatively, the stretch can vary over a large range of strain, depending on the geometry of the facets. Table 1.1 shows data provided in [66] in the form of percent per unit moment (in N-m). The peak applied flexion

Table 1.1 Maximum tensile facet capsule stretch data (%/Nm)

	Extension tests				Flexion tests			
Cadaver #	X-axis	Y-axis	Z-axis	Resultant	X-axis	Y-axis	Z-axis	Resultant
400	7.1	8.7	5.3	12.4	0.6	0.2	0.1	0.6
464	4.8	5.5	8.2	11	0.1	0.4	0.3	0.5
807	3.3	2.5	6.4	7.6	0.4	0.6	0.3	0.8
329	7.4	6.3	0.2	9.7	0.8	2.6	1.9	3.3
455	1.3	3.9	1.3	4.3	0.3	1.3	0.3	1.4
490	0.8	2.5	4.7	5.4	2.2	4.5	1.2	5.2
2	0.8	7.4	7	10.2	1.3	3.5	6.4	7.4
117	0.8	6.6	6.6	9.4	0.9	7.1	1.3	7.3
34	12.4	21	21.1	32.2	2	0.3	3	3.6
SD	4	5.6	6	7.8	0.7	2.4	2	2.6
CV	93.8	78.4	88.7		77.5	104.5	122.8	
AVG	4.3	7.2	6.8	11.4	1	2.3	1.6	3.3

and extension moments were 24 and 18 N-m respectively. With respect to the third requirement, Avramov et al. [101] demonstrated that a mechanical load of sufficient magnitude on the lumbar spine of anesthetized New Zealand white rabbits could activate the high threshold pain fibers. This confirms the previously reported facet syndrome by Mooney and Robertson [102] and others. In fact, facet pain may be more common form of LBP than what the clinicians diagnose. Thus, LBP is not necessarily disc pain.

Returning now to the issue of acute disc rupture, we have dispelled the myth that LBP felt after a crash is always disc pain and that the victim suffered a ruptured disc as a result. A review of the literature on disc rupture reveals that Henzel et al. [103] indicated that early researchers, such as, Ruff [104], Brown et al. [105] and Roaf [106] observed that the vertebral body always broke before the adjacent disc incurred visible damage. In particular, the work of Brinckmann [107] should be described in some detail. He showed that a severely weakened lumbar disc with posterior elements removed could not be ruptured or even made to bulge when loaded in compression to 1 kN. Additional loads causing fracture of the vertebral body did not result in herniation or excessive bulging. There are three reports of disc rupture due to a single loading event in the literature. Farfan et al. [108] applied torsional loads to intact lumbar motion segments without any compressive preload and was able to cause posterior and anterior disc ruptures after a rotation averaging 22.9° for normal discs and 15.2° for abnormal discs. They also tested facet joints and facet capsules to failure and found that the average angle at which they failed was 14° and 12° respectively. This meant that if the facets were allowed to slide over each other or fracture, resulting in a large rotation, then the rupture can occur. Normally, in the presence of a preload on the facets, a single torsional load that does not disrupt the facets or tear the capsules does not cause rupture. The second report is by Adams and Hutton [109] who caused spontaneous rupture of several discs by compressing the spine while it was hyperflexed both laterally and sagittally. If the disc did not rupture on the first try, it was flexed 1° or 2° more and loaded again with the same load. The average angle of flexion needed to cause rupture was 12.9°, implying that the lumbar spine was flexed a total of 64°. The average applied load was 5.45 kN (1,225 lb). This situation is again not representative of a realistic loading condition, as it is extremely rare that a large compressive force could be applied to a spine that is virtually doubled over. Moreover, the herniation occurred between the disc and the endplate due to extreme tension on the posterior aspect of the disc. There was no rupture of the annulus. The third report is by Adams et al. [110] who purported to show the scan of a ruptured disc due to compressive load with moderate flexion. However, there was no scan of the specimen prior to load application. The rupture could have been pre-existing. These reports, in fact, tend to reinforce the point of view that a single loading event is unable to cause disc rupture.

On the contrary, repeated loading of the disc can produce rupture but the number of loading cycles is very high. Yang et al. [111] produced disc ruptures (extrusion of nuclear material to the outside of the disc) by the application of repetitive torsional loads combined with compression and flexion, with the facets removed. The rupture occurred after about 20,000 cycles of torsional loading. This herniation is shown in Fig. 1.14 in which the extruded nucleus can be seen on the right side of the figure. Gordon et al. [112] caused nuclear extrusion in 4 of 14 specimens that were

Fig. 1.14 Herniated disc

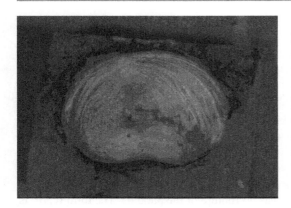

Fig. 1.15 Cross-section of a herniated disc

subjected to an average of 36,750 cycles of combined axial load, flexion and torsion. Other studies involving repetitive loading resulted in extrusion of the nucleus into the endplate [113–115]. In one specimen, Hardy et al. [113] was able to cause an annular rupture after loading it in compression for 1.29 million cycles. Figure 1.15 shows the tortuous path taken by the nucleus material before it can exit the disc. The annular layers do not rupture in a straight line and nucleus material is too viscous to move through this maze under an impulsive load. Disc ruptures are the result of a slow and degenerative process that takes a long time to develop. In other words, discs do not rupture like a balloon, allowing all of the material in the center to come out at once and the pain felt after an accident or fall is not necessarily discogenic since the facet capsules are also a source of LBP The same principles apply to the cervical discs which are morphologically the same as the lumbar discs.

1.6.5 Hip Fractures in the Elderly

Fracture of the hip occurs in the elderly due to bone loss. This fracture is characterized by the failure of the neck of the femur or fracture of the greater trochanter. It is more common in women than in men [116–118]. Men have a greater cross-sectional area of bone than women, based on quantitative CT data and women have a lower bone volumetric density than men.

The association between hip fracture and falls has been documented [119–121] and is attributed to the increased risk of falling among the elderly. However, the etiology of injury is unclear. A search of the literature showed that falls are blamed for hip fracture and it is accepted almost universally that "Grandma fell and broke her hip". A biomechanical argument is provided in this section to demonstrate that the fall is due to hip fracture. That is, a confirmation of spontaneous hip fractures, as proposed by Freeman et al. [122], Sloan and Holloway [123] and Smith [124], is provided based on the work of Yang et al. [125] and data from automotive related injuries to the hip/pelvis in side impact.

Yang et al. [125] hypothesized that hip fracture was due to a strong contraction of either the iliopsoas (inserted into the lesser trochanter) or the gluteus medius (inserted into the greater trochanter). Muscular contraction was responsible for the fracture, resulting in a fall. An experiment was performed to simulate muscular loading through the lesser or greater trochanter to test this hypothesis. They found that lesser trochanteric loading produced femoral neck fractures while greater trochanteric loading produced sub- or inter-trochanteric fractures. The loads to fracture averaged 4.5 times body weight (BW) which approximates the estimated muscle load generated in the iliopsoas during normal walking (4 X BM). Thus, these rotator muscles are strong enough to cause a hip fracture in osteoporotic bone, especially if there is a misstep or accidental firing of the fibers.

The results of Yang et al. [125] only suggest that spontaneous hip fractures can occur and the literature indicates that such fractures do occur [126–128]. To demonstrate that they are the principal mechanism of injury, it is necessary to look at the biomechanics of the fall. It has been reported by Parkkari et al. [129] that 76 % of the hip fracture patients reported that they fell to the side. This was confirmed by data from Lotz and Hayes [120] showing that 78 % of the fractures occurred as a result of falls to the side. The implication is that a side impact to the greater trochanter was the cause of hip fracture. However, data from side impact crashes in which the near-side occupant was impacted by the door, before side door airbags were available, did not sustain hip fractures and instead had pubic rami fractures. Additionally, Smith [124] loaded the proximal femur in many directions but failed to duplicate a femoral neck fracture. Thus, the torsional loads due to the rotator muscles are a likely cause of hip fracture and the fall is due to the fracture.

There is an epidemiologic study that can be done to further validate the Yang hypothesis. Since a femoral neck fracture will cause a fall to the side of the fractured femur while a greater trochanteric fracture will cause a fall to the contralateral side, a detailed accident analysis of falls among the elderly, including the type of fracture sustained, will provide further proof that the fall is due to the fracture.

1.6.6 The Pilon Fracture Mechanism

Pilon fractures are comminuted fractures of the distal tibia that are frequently the result of an offset frontal collision in which there is footwell intrusion. The injury is particularly bothersome because the fractures extend into the region of the capsule of the ankle joint and treatment to restore normal ankle function is difficult. As to the mechanism of this fracture, it was necessary to reproduce the injury in a cadaver so that the cause can be identified. Most attempts to create this injury were unsuccessful because the pilon fracture force needed would cause calcaneal fracture, resulting in a lower force being transmitted to the ankle [130–132]. One of the successful attempts was made by Kitagawa et al. [133]. In addition to impacting the plantar surface of the foot, they also simulated the action of the calf muscles as though the driver was doing maximal braking just prior to impact. In this way, a lower impact force to the foot could be used to avoid a calcaneal fracture and the sum of the impact force and the muscle force acting on the ankle was sufficient to cause a pilon fracture. The experimental setup is shown in Fig. 1.16. It was estimated that the force generated in the calf muscles was about 2 kN and a special tendon catcher was used to grip the Achilles tendon so that a force of this magnitude could be applied through the tendon. An energy absorber limited

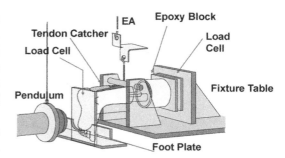

Fig. 1.16 Test setup used by Kitagawa et al. [133] to produce pilon fractures

the force to 2 kN. Five pilon fractures were produced out of the 16 specimens tested. There were 11 calcaneal fractures and in one specimen there was no fracture.

These results show that it was possible to reproduce this injury in a cadaveric specimen but they do not explain the mechanism of injury. Kitagawa et al. [133] developed a finite element model simulating the impact and calculated the stresses developed in the ankle joint. As shown in Fig. 1.17, it is seen that when the talus is jammed into the distal tibia, the medial malleolus is pushed outward, causing the development of a high tensile stress on the medial undersurface of the distal tibia. This stress can initiate a crack which then propagates into the distal tibia, causing the typical comminuted fracture.

1.6.7 Mechanism of Mid-foot Fracture

Mid foot fractures involve the distal row of the tarsal bones and the metatarsal bones. The most famous injury is the Lisfranc injury, named after Jacques Lisfranc de St. Martin, a military surgeon who amputated the foot of a cavalryman in Napoleon's army. The soldier fell from his horse but his foot was caught in the stirrup causing it to twist and fracture. (*Nouvelle méthode opératoire pour l'amputation partielle du pied dans son articulation tarso-metatarsienne.* Paris, Gabon, 1815). The current definition of a Lisfranc injury is fracture of the foot accompanied by lateral displacement and/or dislocation of one or more of the tarsometatarsal joints.

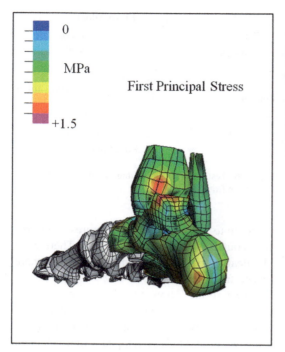

Fig. 1.17 Finite element model of the ankle showing high tensile stresses on the underside of the medial malleolus

Fig. 1.18 Plantar flexed position of the foot

In frontal offset collisions where there is footwell and brake pedal intrusion, foot fractures are also likely to occur (references). In particular, it was noted by Crandall et al. [134] that shorter drivers tend to sustain foot fractures more frequently than those with normal stature. This statement hinted at the mechanism of injury inasmuch as the toes may need to be plantar flexed for the injury to occur. In fact, Wiley [135] had proposed this mechanism based on his clinical experience. However, biomechanical tests on foot and ankle injuries were performed with the foot in its normal plantar configuration [136] and attempts to reproduced [132] his injury did not take advantage of the work of Wiley [135]. Impacts were carried out with the toes fully extended and these attempts were largely unsuccessful in creating Lisfranc injuries in the cadaver. Smith et al. [137] began by impacting cadaveric specimens consisting of the foot and lower leg on the plantar surface of the foot, with the toes extended, to simulate brake pedal intrusion. At an impactor speed of 16 m/s, four Lisfranc injuries were created out of the 13 specimens tested. Lower speeds did not produce the injury. Since the impact speed of 16 m/s is too high for footwell intrusion, it became necessary to simulate the braking action of a short driver who needs to stretch the foot out to reach the pedals, as shown in Fig. 1.18. When the foot was tested in the plantar flexed configuration or with the toes plantar flexed, 27 feet out of 41 sustained Lisfranc injuries at speeds varying from 1 to 16 m/s. Thus, the injury is caused by the transmission of an axial load to the Lisfranc joints by the metatarsal bones.

1.7 Discussion

Although the research community in injury biomechanics is relatively small compared to other specialties in biomedical engineering, these researchers have accomplished many of the studies needed to design a safe automobile. There is now a good understanding of most of the mechanism of injury so that designs can be implemented to prevent the injuries. A few mechanisms remain unresolved. Part of the reason is contained in a cryptic statement on the cover of Lancet, (Issue No. 9773) It states that **"The most entrenched conflict of interest in medicine is a disinclination to reverse a previous opinion"**. This statement is valid for the mechanism of the formation of an acute subdural hematoma, the causes of low back pain and the reason for hip fractures among the elderly.

In terms of response data, the need for such data at high strain rates of different tissues is expanding rapidly and there are not enough researchers or funds to keep up with this demand. In fact, the development of accurate models is severely hampered by the lack of response data. The situation is now even more critical in the field of blast-related injuries where the impact severity levels are orders of magnitude higher than those encountered in automotive crashes and the durations are at least an order of magnitude shorter. Response data for these high loads and high strain rates are currently unavailable.

Human tolerance levels to blunt impact for automotive safety design are now generally available and accepted by the community. These are called Injury Assessment Reference Values (IARV) and can be found in Mertz et al. [138]. The same cannot be said for military injuries where new tolerance levels need to be established.

Assessment of safety devices and techniques should turn gradually from the use of crash dummies to the use of computer models. There is no question that this will save the designer much time and money and will allow for the evaluation of safety systems that cannot be easily tested in the laboratory. For example, complex rollover impacts can be simulated to yield results that are difficult to obtain experimentally. Multiple crashes and multiple directional impacts can be simulated by a model but are almost impossible to stage in the laboratory or at the proving grounds.

1.8 Conclusions

Since heavy emphasis has been placed on injury mechanisms in this chapter, the conclusion will deal largely with this topic.

1. A lot of misinformation exists with respect to injuries to the spine. It ranges for the causes of neck pain due to whiplash and of the low back to the prevalence of acute disc rupture. More research needs to be undertaken to clarify these murky areas that lead to unnecessary claims for compensation and unverified medical opinions.
2. Long held medical beliefs or myths in the causes for acute subdural hematoma and hip fracture need to be proven wrong by more research. The difficulty is that the people who review research proposals are doing exactly what appeared on the cover of Issue No. 9773 of Lancet. They are disinclined to reverse a previous opinion and tend to disapprove any proposal that will reverse this opinion.
3. Greater attention needs to be paid to foot and ankle injuries. They are not life threatening but they restrict mobility and are disabling injuries.

References

1. Centers for Disease Control (CDC) (2012) National Center for Injury Prevention and Control: WISQARs leading causes of death reports, national and regional. Webapp.cdc.gov/sasweb/ncipc/leadingcaus10_us.html
2. Gurdjian ES, Lissner HR (1944) Mechanism of head injury as studied by the cathode ray oscilloscope, preliminary report. J Neurosurg 1:393–399
3. US Department of Transportation (2010) Seat belt used in 2010 – use rates in the states and territories, DOT HS811-493. US Government Printing Office, Washington, DC
4. Viano DC, King AI, Melvin JW, Weber K (1989) Injury biomechanics research: an essential element in the prevention of trauma. J Biomech 22(5):403–417
5. Hardy WN, Foster CD, Mason MJ, Yang KH, King AI, Tashman S (2001) Investigation of head injury mechanisms using neutral density technology and high-speed biplanar X-ray. Stapp Car Crash J 45: 337–368
6. Gurdjian ES, Lissner HR (1945) Deformation of the skull in head injury; a study with the stresscoat technique. Surg Gynecol Obstet 81:679–687
7. Evans FG, Lissner HR (1963) Studies on the energy absorbing capacity of human lumbar intervertebral discs. In: Proceedings of the 7th Stapp car crash conference, Los Angeles, pp 386–402
8. Mertz HJ, Patick LM (1967) Investigation of the kinematics and kinetics of whiplash. Paper presented at the proceedings of the 11th Stapp car crash conference, Anaheim
9. Yang KH, Hu J, White NA, King AI, Chou CC, Prasad P (2006) Development of numerical models for injury biomechanics research: a review of 50 years of publications in the Stapp car crash conference. Stapp Car Crash J 50:429–490
10. Gurdjian ES, Lissner HR, Latimer FR, Haddad BF, Webster JE (1953) Quantitative determination of acceleration and intracranial pressure in experimental head injury; preliminary report. Neurology 3(6): 417–423

11. Ommaya AK, Grubb RL Jr, Naumann RA (1971) Coup and contre-coup injury: observations on the mechanics of visible brain injuries in the rhesus monkey. J Neurosurg 35(5):503–516. doi:10.3171/jns.1971.35.5.0503
12. Gurdjian ES, Webster JE, Lissner HR (1955) Observations on the mechanism of brain concussion, contusion, and laceration. Surg Gynecol Obstet 101(6):680–690
13. Vandevord PJ, Bolander R, Sajja VS, Hay K, Bir CA (2012) Mild neurotrauma indicates a range-specific pressure response to low level shock wave exposure. Ann Biomed Eng 40(1):227–236. doi:10.1007/s10439-011-0420-4
14. Haines DE, Harkey HL, Al-Mefty O (1993) The "subdural" space: a new look at an outdated concept. Neurosurgery 32:111–120
15. Weed LH (1917) An anatomical consideration of the cerebro-spinal fluids. Anat Rec 12:461–496
16. Cushing H (1914) Studies on the cerebro-spinal fluid: I. Introduction. J Med Res 31(1):1–19
17. Penfield (1923) The cranial subdural space. Anat Rec 28:173–175
18. Weed LH (1920) The cells of the arachnoid. Johns Hopkins Hosp Bull 31:343–357
19. Weed LH (1938) Meninges and cerebrospinal fluid. J Anat 72(Pt 2):181–215
20. Leary T (1939) Subdural or intradural hemorrhages? Arch Path Lab Med 28:808–820
21. Allen DJ, DiDio LJ (1977) Scanning and transmission electron microscopy of the encephalic meninges in dogs. J Submicrosc Cytol 9:1–22
22. Nabeshima S, Reese TS, Landis DM, Brightman MW (1975) Junctions in the meninges and marginal glia. J Comp Neurol 164(2):127–169. doi:10.1002/cne.901640202
23. Alcolado R, Weller RO, Parrish EP, Garrod D (1988) The cranial arachnoid and pia mater in man: anatomical and ultrastructural observations. Neuropathol Appl Neurobiol 14(1):1–17
24. Rascol MM, Izard JY (1976) The subdural neurothelium of the cranial meninges in man. Anat Rec 186(3):429–436. doi:10.1002/ar.1091860308
25. Yamashima T, Yamamoto S (1984) How do vessels proliferate in the capsule of a chronic subdural hematoma? Neurosurgery 155:672–678
26. Schachenmayr W, Friede RL (1979) Fine structure of arachnoid cysts. J Neuropathol Exp Neurol 38(4):434–446
27. Frederickson RG (1991) The subdural space interpreted as a cellular layer of meninges. Anat Rec 230(1):38–51. doi:10.1002/ar.1092300105
28. Friede RL, Schachenmayr W (1978) The origin of subdural neomembranes. II. Fine structural of neomembranes. Am J Pathol 92(1):69–84
29. Orlin JR, Osen KK, Hovig T (1991) Subdural compartment in pig: a morphologic study with blood and horseradish peroxidase infused subdurally. Anat Rec 230(1):22–37. doi:10.1002/ar.1092300104
30. Reina MA, De Leon Casasola O, Lopez A, De Andres JA, Mora M, Fernandez A (2002) The origin of the spinal subdural space: ultrastructure findings. Anesth Analg 94(4):991–995, table of contents
31. Yamashima T (2000) The inner membrane of chronic subdural hematomas: pathology and pathophysiology. Neurosurg Clin N Am 11(3):413–424
32. Andrews BT, Dujovny M, Mirchandani HG, Ausman JI (1989) Microsurgical anatomy of the venous drainage into the superior sagittal sinus. Neurosurgery 24(4):514–520
33. Ehrlich E, Maxeiner H, Lange J (2003) Postmortem radiological investigation of bridging vein ruptures. Leg Med (Tokyo) 5(Suppl 1):S225–S227
34. Yamashima T, Friede RL (1984) Why do bridging veins rupture into the virtual subdural space? J Neurol Neurosurg Psychiatry 47:121–127
35. Trotter W (1914) Chronic subdural haemorrhage of traumatic origin, and its relation to pachymeningitis haemorrhagica interna. Br J Surg 2:271–291
36. Bongioanni F, Ramadan A, Kostli A, Berney J (1991) Acute subdural hematoma of arteriolar origin. Traumatic or spontaneous? Neurochirurgie 37(1):26–31
37. Wilberger JE Jr, Harris M, Diamond DL (1991) Acute subdural hematoma: morbidity, mortality, and operative timing. J Neurosurg 74(2):212–218. doi:10.3171/jns.1991.74.2.0212
38. Guterman A, Smith RW (1987) Neurological sequelae of boxing. Sports Med 4(3):194–210
39. Gennarelli TA, Thibault LE (1982) Biomechanics of acute subdural hematoma. J Trauma 22(8):680–686
40. Maxeiner H (1997) Detection of ruptured cerebral bridging veins at autopsy. Forensic Sci Int 89(1–2):103–110
41. Maxeiner H, Spies C, Irnich B, Brock M (1999) Rupture of several parasagittal bridging veins without subdural bleeding. J Trauma 47(3):606–610
42. Maxeiner H, Wolff M (2002) Pure subdural hematomas: a postmortem analysis of their form and bleeding points. Neurosurgery 50(3):503–508; discussion 508–509
43. Shenkin HA (1982) Acute subdural hematoma. Review of 39 consecutive cases with high incidence of cortical artery rupture. J Neurosurg 57(2):254–257. doi:10.3171/jns.1982.57.2.0254
44. Howard MA 3rd, Gross AS, Dacey RG Jr, Winn HR (1989) Acute subdural hematomas: an age-dependent clinical entity. J Neurosurg 71(6):858–863. doi:10.3171/jns.1989.71.6.0858
45. Maxeiner H (1991) Arterial misplacement of a central venous catheter with a fatal cerebral embolism. Anaesthesist 40(8):452–455
46. Karnath B (2004) Subdural hematoma. Presentation and management in older adults. Geriatrics 59(7):18–23
47. Tandon PN (2001) Acute subdural haematoma: a reappraisal. Neurol India 49(1):3–10
48. Lowenhielm P (1974) Dynamic properties of the parasagittal bridging veins. Z Rechtsmed 74(1):55–62
49. Lee MC, Haut RC (1989) Insensitivity of tensile failure properties of human bridging veins to strain rate:

implications in biomechanics of subdural hematoma. J Biomech 22(6–7):537–542
50. Lee MC, Haut RC (1992) Strain rate effects on tensile failure properties of the common carotid artery and jugular veins of ferrets. J Biomech 25(8):925–927
51. Monson KL, Goldsmith W, Barbaro NM, Manley GT (2003) Axial mechanical properties of fresh human cerebral blood vessels. J Biomech Eng 125(2):288–294
52. Pang Q, Wang C, Hu Y, Xu G, Zhang L, Hao X, Zhang Q, Gregerson H (2001) Experimental study of the morphology of cerebral bridging vein. Chin Med Sci J 16(1):19–22
53. Depreitere B, Van Lierde C, Sloten JV, Van Audekercke R, Van der Perre G, Plets C, Goffin J (2006) Mechanics of acute subdural hematomas resulting from bridging vein rupture. J Neurosurg 104(6):950–956. doi:10.3171/jns.2006.104.6.950
54. Jin X, Yang KH, King AI (2011) Mechanical properties of bovine pia-arachnoid complex in shear. J Biomech 44(3):467–474. doi:10.1016/j.jbiomech.2010.09.035
55. Deans GT, Magalliard JN, Kerr M, Rutherford WH (1987) Neck sprain–a major cause of disability following car accidents. Injury 18(1):10–12
56. Squires B, Gargan MF, Bannister GC (1996) Soft-tissue injuries of the cervical spine. 15-year follow-up. J Bone Joint Surg Br 78(6):955–957
57. States JD, Balcerak JD, Williams JS, Morris AT, Babcock W, Polvino R, Dawley RE (1972) Injury frequency and head restraint effectiveness in rear-end impact accidents. In: Proceedings of the 16th Stapp car crash conference, Detroit, pp 228–257
58. Castro WH, Schilgen M, Meyer S, Weber M, Peuker C, Wortler K (1997) Do "whiplash injuries" occur in low-speed rear impacts? Eur Spine J 6(6):366–375
59. Kleinberger M (1993) Application of finite element techniques to the study of cervical spine mechanics. Paper presented at the proceedings of the 37th Stapp car crash conference, San Antonio
60. Lubin S, Sehmer J (1993) Are automobile head restraints used effectively? Can Fam Physician 39:1584–1588
61. Schrader H, Obelieniene D, Bovim G, Surkiene D, Mickeviciene D, Miseviciene I, Sand T (1996) Natural evolution of late whiplash syndrome outside the medicolegal context. Lancet 347(9010):1207–1211
62. Ferrari R, Russell AS (1999) Epidemiology of whiplash: an international dilemma. Ann Rheum Dis 58(1):1–5
63. Radanov BP, Sturzenegger M, Di Stefano G (1995) Long-term outcome after whiplash injury. A 2-year follow-up considering features of injury mechanism and somatic, radiologic, and psychosocial findings. Medicine (Baltimore) 74(5):281–297
64. Bogduk N (2000) Epidemiology of whiplash. Ann Rheum Dis 59(5):394–395; author reply 395–396
65. Deng B, Begeman PC, Yang KH, Tashman S, King AI (2000) Kinematics of human cadaver cervical spine during low speed rear-end impacts. Stapp Car Crash J 44:171–188
66. Cavanaugh JM, Ozaktay AC, Yamashita T, Avramov A, Getchell TV, King AI (1997) Mechanisms of low back pain: a neurophysiologic and neuroanatomic study. Clin Orthop Relat Res 335:166–180
67. Bogduk N, Marsland A (1988) The cervical zygapophysial joints as a source of neck pain. Spine (Phila Pa 1976) 13(6):610–617
68. Barnsley L, Lord SM, Wallis BJ, Bogduk N (1995) The prevalence of chronic cervical zygapophysial joint pain after whiplash. Spine (Phila Pa 1976) 20(1):20–25; discussion 26
69. Ozaktay AC, Cavanaugh JM, Blagoev DC, King AI (1995) Phospholipase A2-induced electrophysiologic and histologic changes in rabbit dorsal lumbar spine tissues. Spine (Phila Pa 1976) 20(24):2659–2668
70. MacNab I (1965) Whiplash injuries of the neck. In: Proceedings of the 8th annual AAAM conference. Rochester, MN
71. Garrett WE, Seaber AV, Best TM, Glisson RR, Nikolaou PK, Taylor DC (1997) Muscle strain injury: basic science and clinical application. Kappa delta paper. In: Proceedings of the 43rd annual meeting of Orthopaedic Research Society, San Francisco
72. Tencer A, Mirza S, Martin D, Goodwin V, Sackett R, Schaefer J (1999) Development of a retro-fit anti-whiplash seat cushion based on studies of drivers and human volunteers. In: Proceedings of the 9th injury prevention through biomechanics symposium, Wayne State University, Detroit, pp 39–45
73. Azar NR, Kallakuri S, Chen C, Cavanaugh JM (2011) Muscular response to physiologic tensile stretch of the caprine c5/6 facet joint capsule: dynamic recruitment thresholds and latencies. Stapp Car Crash J 55:441–460
74. Azar NR, Kallakuri S, Chen C, Lu Y, Cavanaugh JM (2009) Strain and load thresholds for cervical muscle recruitment in response to quasi-static stretch of the caprine C5-C6 joint capsule. J Electromyogr Kinesiol 19:e387–e394
75. Ono K, Kaneoka K, Wittek A, Kajzer J (1997) Cervical injury mechanism based on the analysis of human cervical vertebral motion and head-neck-torso kinematics during low speed rear impacts. In: Proceedings of the 41st Stapp car crash conference, Orlando, pp 339–356
76. Yoganandan N, Pintar FA, Cusick JF, Sun E, Eppinger R (1998) Whiplash injury mechanisms. In: Whiplash'98 symposium, Phoenix, p 23
77. Aldman B (1986) An analytical approach to the impact biomechanics of head and neck. In: Proceedings of the 30th annual AAAM conference, pp 439–454
78. Svensson MY, Aldman B, Hansson HA, Lovsund P, Seeman T, Suneson A, Oertengren T (1993) Pressure effects in the spinal canal during whiplash extension motion: A possible cause of injury to the cervical spinal ganglia. In: Proceedings of the 1993 IRCOBI conference, Eindhoven

79. Yang KH, Begeman PC (1996) A proposed role for facet joints in neck pain in low to moderate speed rear end impacts. Part I: Biomechanics. In: Proceedings of the 6th injury prevention through biomechanics symposium, Wayne State University, Detroit, pp 59–63
80. Lu Y, Chen C, Kallakuri S, Patwardhan A, Cavanaugh JM (2005) Neural response of cervical facet joint capsule to stretch: a potential whiplash pain mechanism. Stapp Car Crash J 49:49–65
81. Strassman G (1947) Traumatic rupture of the aorta. Am Heart J 33(4):508–515
82. Roberts VL, Jackson FR, Berkas EM (1966) Heart motion due to blunt trauma to the thorax. In: Proceedings of the 10th Stapp car crash conference, Detroit, pp 242–248
83. Viano DC (2011) Chest impact experiments aimed at producing aortic rupture. Clin Anat 24(3):339–349. doi:10.1002/ca.21110
84. Hardy WN, Shah CS, Mason MJ, Kopacz JM, Yang KH, King AI, Van Ee CA, Bishop JL, Banglmaier RF, Bey MJ, Morgan RM, Digges KH (2008) Mechanisms of traumatic rupture of the aorta and associated peri-isthmic motion and deformation. Stapp Car Crash J 52:233–265
85. Vulcan AP, King AI, Nakamura GS (1970) Effects of bending on the vertebral column during +Gz acceleration. Aerosp Med 41(3):294–300
86. King AI, Prasad P, Ewing CL (1975) Mechanism of spinal injury due to caudocephalad acceleration. Orthop Clin North Am 6(1):19–31
87. Prasad P, King AI, Ewing CL (1974) The role of articular facets during +Gz acceleration. J Appl Mech 41:321–326
88. Patrick LM (1961) Caudo-cephalad static and dynamic injuries to the vertebrae. In: Proceedings of the 5th Stapp car crash conference, Minneapolis, pp 171–181
89. Smith BR, Begeman P (2012) The development of a new injury criteria for the spine and tibia. In: Advanced technologies and new frontiers in military injury biomechanics symposium, Arlington
90. Ramasamy A, Hill AM, Masouros S, Gibb I, Bull AM, Clasper JC (2011) Blast-related fracture patterns: a forensic biomechanical approach. J R Soc Interface 8(58):689–698. doi:10.1098/rsif.2010.0476
91. Ramasamy A, Masouros SD, Newell N, Hill AM, Proud WG, Brown KA, Bull AM, Clasper JC (2011) In-vehicle extremity injuries from improvised explosive devices: current and future foci. Philos Trans R Soc Lond B Biol Sci 366(1562):160–170. doi:10.1098/rstb.2010.0219
92. Prasad P, King AI (1974) An experimentally validated dynamic model of the spine. J Appl Mech 41:546–550
93. Begeman PC, King AI, Prasad P (1973) Spinal loads resulting from -Gx acceleration. In: Proceedings of the 17th staff car crash conference, Oklahoma City, pp 343–360
94. Ball ST, Vaccaro AR, Albert TJ, Cotler JM (2000) Injuries of the thoracolumbar spine associated with restraint use in head-on motor vehicle accidents. J Spinal Disord 13(4):297–304
95. Huelke DF, Mackay GM, Morris A (1995) Vertebral column injuries and lap-shoulder belts. J Trauma 38(4):547–556
96. Robertson A, Branfoot T, Barlow IF, Giannoudis PV (2002) Spinal injury patterns resulting from car and motorcycle accidents. Spine (Phila Pa 1976) 27(24):2825–2830. doi:10.1097/01.BRS.0000035686.45726.0E
97. Rutherford WH (1985) The medical effects of seat-belt legislation in the United Kingdom: a critical review of the findings. Arch Emerg Med 2(4):221–223
98. King AI (2001) Injury to the thoracolumbar spine and pelvis. In: Melvin J, Nahum A (eds) Accidental injury: biomechanics and prevention, 2nd edn. Springer, New York
99. Cavanaugh JM, Ozaktay AC, Yamashita HT, King AI (1996) Lumbar facet pain: biomechanics, neuroanatomy and neurophysiology. J Biomech 29(9):1117–1129
100. Yamashita T, Cavanaugh JM, el-Bohy AA, Getchell TV, King AI (1990) Mechanosensitive afferent units in the lumbar facet joint. J Bone Joint Surg Am 72(6):865–870
101. Avramov AI, Cavannaugh JC, Ozaktay CA, Getchell TV, King AI (1992) The effects of controlled mechanical loading on Group II, III, and IV afferents from the lumbar facet joint and surrounding tissue: an in vitro study. J Bone Joint Surg 74A:1464–1471
102. Mooney V, Robertson J (1976) The facet syndrome. Clin Orthop Relat Res 115:149–156
103. Henzel JH, Mohr GC, von Gierke HE (1968) Reappraisal of biodynamic implications of human ejections. Aerosp Med 39(3):231–240
104. Ruff S (1950) Brief acceleration less than one second. In: German aviation medicine, world war II, vol I. US Government Printing Office, Washington, DC, pp 584–597
105. Brown T, Hansen R, Yorra A (1957) Some mechanical tests on the lumbo-sacral spine with particular reference to the intervertebral discs. J Bone Joint Surg 39A:1135–1164
106. Roaf R (1960) A study of the mechanics of spinal injuries. J Bone Joint Surg 42B:810–823
107. Brinckmann P (1986) Injury of the annulus fibrosus and disc protrusions. An in vitro investigation on human lumbar discs. Spine (Phila Pa 1976) 11(2):149–153
108. Farfan HF, Cossette JW, Robertson GH, Wells RV, Kraus H (1970) The effects of torsion on the lumbar intervertebral joints: the role of torsion in the production of disc degeneration. J Bone Joint Surg Am 52(3):468–497
109. Adams MA, Hutton WC (1982) Prolapsed intervertebral disc. A hyperflexion injury 1981 Volvo Award in Basic Science. Spine (Phila Pa 1976) 7(3):184–191
110. Adams MA (2004) Biomechanics of back pain. Acupunct Med 22(4):178–188
111. Yang KH, Byrd AJI, Kish VL, Radin EL (1988) Annulus fibrosus tears – an experimental mode. Orthop Trans 12:86–87

112. Gordon SJ, Yang KH, Mayer PJ, Mace AH Jr, Kish VL, Radin EL (1991) Mechanism of disc rupture. A preliminary report. Spine (Phila Pa 1976) 16(4):450–456
113. Hardy WG, Lissner HR, Webster JE, Gurdjian ES (1959) Repeated loading tests on the lumbar spine: a preliminary report. Surg Forum 9:690–695
114. Liu YK, Goel VK, Dejong A, Njus G, Nishiyama K, Buckwalter J (1985) Torsional fatigue of the lumbar intervertebral joints. Spine (Phila Pa 1976) 10(10):894–900
115. Liu YK, Njus G, Buckwalter J, Wakano K (1983) Fatigue response of lumbar intervertebral joints under axial cyclic loading. Spine (Phila Pa 1976) 8(8):857–865
116. Cawthon PM (2011) Gender differences in osteoporosis and fractures. Clin Orthop Relat Res 469(7):1900–1905. doi:10.1007/s11999-011-1780-7
117. Khosla S, Melton LJ 3rd, Riggs BL (1999) Osteoporosis: gender differences and similarities. Lupus 8(5):393–396
118. Riggs BL, Melton Iii LJ 3rd, Robb RA, Camp JJ, Atkinson EJ, Peterson JM, Rouleau PA, McCollough CH, Bouxsein ML, Khosla S (2004) Population-based study of age and sex differences in bone volumetric density, size, geometry, and structure at different skeletal sites. J Bone Miner Res 19(12):1945–1954. doi:10.1359/JBMR.040916
119. Cummings SR, Kelsey JL, Nevitt MC, O'Dowd KJ (1985) Epidemiology of osteoporosis and osteoporotic fractures. Epidemiol Rev 7:178–208
120. Lotz JC, Hayes WC (1990) The use of quantitative computed tomography to estimate risk of fracture of the hip from falls. J Bone Joint Surg Am 72(5):689–700
121. Melton LJI (1988) Epidemiology of fractures. In: Osteoporosis, etiology, diagnosis and management. Raven Press, New York, pp 111–131
122. Freeman MA, Todd RC, Pirie CJ (1974) The role of fatigue in the pathogenesis of senile femoral neck fractures. J Bone Joint Surg Br 56-B(4):698–702
123. Sloan J, Holloway G (1981) Fractured neck of the femur: the cause of the fall? Injury 13(3):230–232
124. Smith LD (1953) Hip fractures; the role of muscle contraction or intrinsic forces in the causation of fractures of the femoral neck. J Bone Joint Surg Am 35-A(2):367–383
125. Yang KH, Shen KL, Demetropoulos CK, King AI, Kolodziej P, Levine RS, Fitzgerald RH Jr (1996) The relationship between loading conditions and fracture patterns of the proximal femur. J Biomech Eng 118(4):575–578
126. Alffram PA (1964) An epidemiologic study of cervical and trochanteric fractures of the femur in an urban population. Analysis of 1,664 cases with special reference to etiologic factors. Acta Orthop Scand Suppl 65(Suppl 65):61–109
127. Kelly JP (1954) Fractures complicating electroconvulsive therapy and chronic epilepsy. J Bone Joint Surg Br 36-B(1):70–79
128. Phillips R, Williams JF, Melick RA (1975) Predictions of the strength of the neck of the femur from its radiological appearance. Biomed Eng 10:367–372
129. Parkkari J, Kannus P, Palvanen M, Natri A, Vainio J, Aho H, Vuori I, Jarvinen M (1999) Majority of hip fractures occur as a result of a fall and impact on the greater trochanter of the femur: a prospective controlled hip fracture study with 206 consecutive patients. Calcif Tissue Int 65(3):183–187
130. Begeman P, Paravasthu N (1997) Static and dynamic compression loading of the lower leg. In: Proceedings of the 7th injury prevention through biomechanics symposium, Wayne State University, Detroit
131. Klopp GS, Crandall JR, Hall GW, Pilkey WD, Hurwitz SR, Kuppa S (1997) Mechanisms of injury and injury criteria for the human foot and ankle in dynamic axial impacts to the foot. In: Proceedings of the IRCOBI conference, Hanover, pp 73–86
132. Yoganandan N, Pintar F, Boynton M, Begeman P, Prasad P, Kuppa S, Morgan R, Eppinger R (1996) Dynamic axial tolerance of the human foot-ankle complex. In: Proceedings of the 40th Stapp car crash conference, Albuquerque, pp 207–218
133. Kitagawa Y, Ichikawa H, King AI, Levine RS (1998) A severe ankle and foot injury in frontal crashes and its mechanism. In: Proceedings of the 42nd Stapp conference, Tempe, pp 1–12
134. Crandall JR, Martin PG, Bass CR, Pilkey WD, Dischinger PC, Burgess AR, O'Quinn TD, Schmidhauser CB (1996) Foot and ankle injury: the roles of driver anthropometry, footwear, and pedal controls. In: Proceedings of the 40th AAAM conference, Vancouver, pp 1–18
135. Wiley JJ (1971) The mechanism of tarso-metatarsal joint injuries. J Bone Joint Surg Br 53(3):474–482
136. Crandall JR, Portier L, Petit P, Hall GW, Bass CR, Klopp GS, Hurwitz SR, Pilkey WD, Trosseille X, Tarriere C, Lassau PJ (1996) Biomechanical response and physical properties of the leg, foot, and ankle. In: Proceedings of the 40th Stapp car crash conference, Albuquerque, pp 173–192
137. Smith BR, Begeman PC, Leland R, Meehan R, Levine RS, Yang KH, King AI (2005) A mechanism of injury to the forefoot in car crashes. Traffic Inj Prev6(2):156–169.doi:10.1080/15389580590931635
138. Mertz HJ, Irwin AL, Prasad P (2003) Biomechanical and scaling bases for frontal and side impact injury assessment reference values. Stapp Car Crash J 47:155–188

Automotive Field Data in Injury Biomechanics

Hampton C. Gabler, Ashley A. Weaver, and Joel D. Stitzel

Abstract

A crucial component to understanding the biomechanics of injury is the study of real world injuries. These studies are essential to characterize the incidence and characteristics of impact injuries, to establish impact test configurations, and to evaluate the effectiveness of injury intervention measures. This chapter will describe the data sources for these evaluations, metrics of performance for injury, and representative applications, i.e., the generation of injury risk curves, measurement of societal costs of injuries, and estimating the mortality associated with specific injuries.

2.1 Injury Metrics

The nature of impact injuries can be described with several different injury metrics. This section describes each of the more widely used injury metrics.

H.C. Gabler, Ph.D. (✉)
Biomedical Engineering, Virginia Tech, Center for Injury Biomechanics, Virginia Tech-Wake Forest University, Blacksburg, VA, USA
e-mail: gabler@vt.edu

A.A. Weaver, Ph.D. • J.D. Stitzel, Ph.D.
Biomedical Engineering, Wake Forest University, Center for Injury Biomechanics, Virginia Tech-Wake Forest University, Winston-Salem, NC, USA
e-mail: asweaver@wakehealth.edu; jstitzel@wakehealth.edu

2.1.1 Abbreviated Injury Scale (AIS)

The AIS is an injury coding lexicon established by the Association for the Advancement of Automotive Medicine [1–3]. It is the most advanced trauma-specific, anatomically-based coding lexicon and was first conceived as a system to define the type and severity of injuries arising from motor vehicle crashes (MVCs). AIS is currently in its sixth revision. To calculate AIS scores, medical records of traumatic incidents are transcribed into specific codes that capture individual injuries. AIS is a proprietary classification system, meaning it requires specialized training for coding personnel. Therefore, AIS is not captured at every hospital.

The AIS code consists of two numerical components. The first component is a six-digit injury descriptor code ("pre-dot"), which is

Table 2.1 Abbreviated Injury Scale

AIS score	Injury severity
1	Minor
2	Moderate
3	Serious
4	Severe
5	Critical
6	Maximal (currently untreatable)

Equation 2.1. Injury Severity Score (ISS)

$$\text{ISS} = \left(\text{MaxAISseverity}_{bodyregion1}\right)^2 \\ + \left(\text{MaxAISseverity}_{bodyregion2}\right)^2 \\ + \left(\text{MaxAISseverity}_{bodyregion3}\right)^2 \quad (2.1)$$

unique to each traumatic injury. Pre-dots classify the injury by region, type of anatomic structure, specific structure, and level. The second component of the AIS code is a measure of injury severity in terms of threat to life. Each injury incurred by a person or subject is coded on a six point scale which ranges from 1 for minor injuries to 6 for maximal (currently untreatable) injuries as shown in Table 2.1. The AIS system is based on the assessment of threat to life, which was developed by a consensus of trauma surgeons, for an extensive compendium of injuries. The AIS system is widely used in in-depth crash investigation databases including NASS/CDS and CIREN. AIS is periodically updated and revised to encompass the most recent data needs for research in trauma and emergency medicine. The most recent version of AIS is AIS 2005 (Update 2008) [2].

The maximum AIS (MAIS) can be used as an AIS measure of the overall severity of a patient's injuries. The Injury Severity Score (ISS) is another metric of overall injury severity and is calculated using AIS severity scores for body regions [4]. The highest AIS severity scores in each of the three most severely injured body regions are squared and summed together (Eq. 2.1). The six body regions used in the ISS calculation are: (1) head and neck, (2) face, (3) chest, (4) abdomen, (5) extremities, and (6) external. The ISS scores range from 1 to 75. If any of the three AIS severity scores is a 6, the score is automatically set at 75. A patient with an ISS greater than 15 is used by many sources to designate a "major trauma" or "seriously injured" patient that needs treatment at a Level I or II trauma center.

The New Injury Severity Score (NISS) is a modified version of ISS in which the three highest AIS severity scores (regardless of body region) are squared and summed together [5]. NISS scores range from 1 to 75 and if any of the three AIS severity scores is a 6, the score is automatically set at 75. An example of the ISS and NISS calculation for a subject with a particular set of injuries is provided in Fig. 2.1. While ISS is a more widely used overall injury severity measure, NISS has been shown to be superior in predicting post-injury multiple organ failure and length of hospitalization and intensive care unit admission in patients with multiple orthopedic injuries [6, 7].

2.1.2 KABCO Scale

The KABCO scale is a five point scale commonly used by law enforcement in the US to code injury severity for police accident reports. As shown in Table 2.2, the KABCO scale ranges from 'K' for killed to 'O' for property damage only, i.e. not injured. Police officers would typically provide a code for each person involved in a crash. Crash databases which rely on police accident reports typically code injury severity using the KABCO scale. This would include FARS, GES, and state accident databases. Because KABCO is not a medically based scale and is coded by non-medical personnel, KABCO has been found to not accurately conform to medically-based systems such as AIS [9].

2.1.3 International Classification of Diseases Version 9 (ICD-9)

The International Classification of Diseases version 9 (ICD-9) injury coding lexicon is used commonly in the US [10]. ICD-9 codes are nominal,

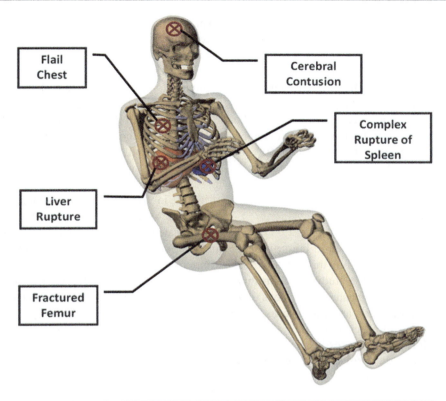

Fig. 2.1 Example ISS and NISS calculation for injuries illustrated on a full body model [8]

Region	Injury Description	AIS	ISS: Square Top Three Injuries By Body Region	NISS: Square Top Three Overall Injuries
Head & Neck	Cerebral Contusion	3	9	
Face	No Injury	0		
Chest	Flail Chest	4	16	16
Abdomen	Liver Rupture	4		16
	Complex Rupture of Spleen	5	25	25
Extremity	Fractured Femur	3		
External	No Injury	0		
Sum of Squared Top Three Injuries			ISS: 50	NISS: 57

Table 2.2 KABCO injury scale

KABCO code	Injury severity
K	Killed
A	Incapacitated
B	Moderate injury
C	Complaint of pain
O	Property damage only

meaning they are unordered, qualitative categories not ranked by severity. ICD is not trauma-specific, but rather is a general, all-purpose diagnosis taxonomy for all health conditions. It is over 110 years old and is currently in its tenth revision (ICD-10), though in the US the nineth revision (ICD-9) is most commonly used. Codes exist for over 10,000 medical conditions, about 2,000 of which are physical injuries (the block of ICD-9 codes from 800.0 to 959.9 encompasses all traumatic injuries). ICD-9 codes are used by all hospitals in the US, primarily to classify diagnoses for administrative purposes, such as billing and event reporting. Databases such as NTDB and

NIS use the ICD-9 coding lexicon for injuries. MVC cases can be designated with the ICD-9 external cause codes (ecodes). Ecodes 810–819 with a post-dot decimal of 0 or 1 are used to designate motor vehicle traffic accidents involving drivers and passengers. Mapping between the AIS and ICD-9 injury code lexicons can be accomplished using a robust one-to-one mapping algorithm developed by Barnard et al. (2013) or ICDMAP software [11, 12].

2.2 Data Sources

The sources of field data used in injury biomechanics can be grouped into three categories: (1) fatal crash data used in studies of mortality, (2) databases of in-depth crash investigations used for studies of both crash and injury causation, and (3) databases of all crashes, fatal and nonfatal, used for producing estimates of crash exposure. This section provides summaries of a selected subset of the primary field data sources used in injury biomechanics.

2.2.1 The Fatality Analysis Reporting System (FARS)

FARS is a comprehensive census of all traffic related fatalities in the United States. FARS data is available from 1975 to the present and includes records of approximately 30,000–40,000 fatalities which occur on U.S. highways each year. FARS contains records of traffic fatalities in all vehicle types and crash modes, i.e. cars, light trucks, heavy trucks, bicyclists, motorcyclists, and pedestrians. In injury biomechanics, FARS is useful for determining the characteristics of fatal crashes.

FARS has been maintained and operated by the National Highway Traffic Safety Administration (NHTSA) since 1975 [13]. For a case to be included in this database, it must involve a motor vehicle traveling on a primarily public roadway and death of an individual involved within 30 days of the incident. Each incident is characterized by the collection of approximately 175 data elements split among an accident table, a vehicle table, and a person table. In 2010, FARS was expanded and restructured to include additional crash data in tables on crash events, contributing factors, driver's vision, violations, pre-crash maneuvers, distraction, impairment, safety equipment, pedestrian and bicyclist crash circumstances.

In the U.S., all states must collect and provide NHTSA with records of all traffic related fatalities on their highways. FARS data can be obtained by downloading any of the published files from the NHTSA web site. The files are available in SAS, DBF, and sequential ASCII file formats. FARS data is assembled by FARS analysts located in each state who supplement each jurisdiction's police accident report with data from death certificates, coroner/medical examiner reports, driver history data from the state department of motor vehicles, and toxicology from police or medical facilities. The injury severity suffered by each person is coded using the KABCO scale. Note that FARS includes fatalities within 30 days of the MVC and the KABCO scale is an assessment of occupant injury severity on scene, so the KABCO assessment can vary ("Killed", "Incapacitated", etc.) for FARS occupants.

2.2.2 The National Automotive Sampling System/ Crashworthiness Data System (NASS/CDS)

NASS/CDS provides a detailed record of a national sample of approximately 5,000 towaway crashes investigated each year by NHTSA at 24 locations throughout the United States [14]. This database includes a probability sample of thousands of minor, serious, and fatal crashes involving cars, light trucks, vans and sport utility vehicles. Compared with the FARS database, the data collected in NASS/CDS is much more detailed and includes approximately 400 data elements. Each case provides detailed information describing the collision including vehicle characteristics, severity of each injury, crash configuration, occupant contact sources inside the vehicle, and occupant restraint performance. Trained crash investigators obtain data from

Table 2.3 NASS/CDS 2000–2011 case counts per year (Excluding NASS/CDS 2009–2011 cases with model year vehicles greater than 10 years old)

Year	Case count
2000	4,307
2001	4,090
2002	4,589
2003	4,754
2004	5,597
2005	4,481
2006	4,940
2007	4,963
2008	5,167
2009	4,425
2010	3,989
2011	3,401
Total	54,703

crash sites, studying evidence such as skid marks, fluid spills, broken glass, and bent guard rails. They locate the vehicles involved, photograph them, measure the crash damage, and identify interior locations that were struck by the occupants. These researchers follow up on their on-site investigations by interviewing crash victims and reviewing medical records to determine the nature and severity of injuries. By applying weighting factors to NASS/CDS data, a US representative population can be analyzed [15]. Injury severity is coded using the AIS scale.

NHTSA requirements for NASS/CDS crash investigations changed in 2009 and many variables (including all injury data) are not collected for model year vehicles greater than 10 years old. Model year vehicles greater than 10 years old accounted for approximately one-third of the unweighted NASS/CDS 2009–2011 cases. Excluding these NASS/CDS 2009–2011 cases with model year vehicles greater than 10 years old, the NASS/CDS 2000–2011 dataset contains 54,703 cases, 94,283 vehicles, 115,159 occupants, and 303,230 injuries (Table 2.3).

2.2.3 NASS General Estimates System (GES)

GES is a comprehensive database containing information on approximately 60,000 randomly sampled police reported accidents each year. GES can be analyzed to determine the number of occupants who were involved - both injured and uninjured. Because GES contains all crashes, GES analyses are frequently conducted to provide a measure of exposure when computing injury or fatality risk in particular crash modes. Cases from GES are assigned weights that can be used to estimate the number of similar accidents that may have taken place that year that were not sampled. GES data was first made available by NHTSA in 1988. Beginning in 2009, FARS and GES were restructured to standardize the definitions of common variables [16].

The function of NASS/GES is to present a representative sample of all police-reported motor vehicle accidents in the United States [17]. Criteria for selection include the involvement of a motor vehicle traveling on a public road and a crash that results in property damage, occupant injury, a fatality, or any combination of these outcomes. The accident reports are sampled from approximately 400 police jurisdictions in 60 primary sampling units across the United States. Each area is selected with the intent of providing a spectrum of regions that is indicative of geographical, roadway, population, and traffic characteristics in the entire country. Complexity of collected crash information is analogous to that of the FARS database and includes approximately 130 data elements. Injury severity is coded for each person using the KABCO scale. In 2012, NHTSA began an effort to modernize NASS which will include improving the IT infrastructure of NASS and FARS, updating NASS data collected, and reexamining the NASS sample design [18].

2.2.4 National Motor Vehicle Crash Causation Survey (NMVCCS)

NMVCCS is a nationally representative dataset that includes a probability sample of over 6,900 U.S. crashes involving 13,300 vehicles which occurred from July 3, 2005 to December 31, 2007 [19]. NMVCCS is an in-depth crash investigation database similar to NASS/CDS. However unlike NASS/CDS, NMVCCS focuses on crash causation rather than crash outcomes. In order to be included in NMVCCS, Emergency Medical Services (EMS) must have been activated and an investigator must have been at the scene of the crash before it was cleared. This allowed investigators to interview occupants, witnesses, and first responders in order to determine crash causation factors not available in traditional crash databases which focus on injury. The NMVCCS database was developed by NHTSA as a means to obtain insight into the primary pre-crash circumstances and behaviors that are associated with driver crash risk in light vehicles.

During data collection, NMVCCS relied on special arrangements between crash investigators, EMS, and police agencies as well as constant monitoring of crash occurrences with the aid of police scanners to allow for immediate crash-site investigations and on-site driver interviews. To further ensure the accuracy of the data and inhibit the loss of critical information, the NMVCCS protocol required that a responding officer was on-scene at the time of the crash investigation and a particular focus was placed on driver interviews. This provided an opportunity to collect evidence and conduct interviews with the involved parties immediately after the crash regarding the pre-crash events and behaviors. Injury severity is coded for each person using the KABCO scale.

2.2.5 State Crash Data

In the U.S., many states maintain databases containing digital records of all police accident reports. The datasets are large and are used by many injury researchers to obtain exposure data which is used to compute injury or fatality risk. State data is similar in function to GES. An example is the NJCRASH database of all police reported crashes in New Jersey. NJCRASH contains electronic summaries of the approximately 300,000 police reported crashes that occur each year in New Jersey. NJCRASH can be downloaded from the New Jersey Department of Transportation web site.

State data can be obtained in a number of ways: (1) directly from the states, (2) from NHTSA through their state data system, and (3) through the Highway Safety Information Systems (HSIS). HSIS is a collection of state data from seven states (California, Illinois, Maine, Minnesota, North Carolina, Ohio, and Washington State) which is maintained by Federal Highway Administration (FHWA). Additional historical state data is also available through HSIS for Michigan and Utah.

The advantage of using state data over GES is that state data is a census of crashes and avoids the analytical difficulties of using a probability sample which is the basis for GES. The difficulty with state data is that (1) it can be difficult to obtain from the states, (2) reliance on a single state may not be representative of other states in the U.S., (3) there is lack of standardization, and (4) typically injury is only coded using the KABCO scale. One approach to achieve regional balance is to combine the state data of several states for an analysis. However, the lack of standardization can make this challenging. The lack of standardization may be resolved in the future. Many states are now adopting the Model Minimum Uniform Crash Criteria Guideline (MMUCC) which provides standardized definitions for many critical police-reported crash characteristics [20].

2.2.6 Crash Outcome Data Evaluation System (CODES)

One promising method of combining the benefits of large state accident databases with the detailed medical coding of injuries practiced in CIREN and NASS/CDS is to link databases of police accident reports with associated medical records. NHTSA founded an effort known as Crash Outcome Data Evaluation System (CODES) which includes the efforts of numerous state Departments of Transportation for linking vehicle, medical, and insurance related sources [21, 22]. CODES links police accident reports, EMS records, emergency department records, and hospital discharge records to build a comprehensive set of crash records with associated medically-based accounts of injury. CODES was implemented by states on a voluntary basis with NHTSA sponsorship. The number of states in the system varied over the course of the CODES program. In 1996, there were seven CODES states (Hawaii, Maine, Missouri, New York, Pennsylvania, Utah, and Wisconsin). In 2009, there were 16 CODES states (Connecticut, Delaware, Georgia, Illinois, Indiana, Kentucky, Maryland, Massachusetts, Minnesota, Missouri, Nebraska, New York, Ohio, South Carolina, Utah, and Virginia). Although the CODES program has now ended, many states still maintain these linked crash-medical records data programs.

2.2.7 Special Crash Investigations (SCI)

Since 1972, the Special Crash Investigations (SCI) Program by the National Center for Statistics and Analysis (NCSA) has provided NHTSA with detailed crash data collected for the purpose of studying new and rapidly changing vehicle technologies or special crash circumstances. Approximately 200 SCI cases are collected annually from NHTSA's Auto Safety Hotline, the Department of Transportation's National Crash Alert System, NHTSA's regional offices, automotive manufacturers, other government agencies, law enforcement agencies, engineers, and medical personnel. Case selection is based on the program manager's discretion and may include investigation of new emerging technologies such as the safety performance of alternative fueled vehicles, child safety restraints, airbag systems, adapted vehicles, safety belts, vehicle-pedestrian interactions, and potential safety defects. SCI crashes have been used widely in the past to study airbag systems and school bus crashes [23, 24]. Data collected on SCI cases may include police reports, insurance crash reports, and reports from professional crash investigation teams. Hundreds of data elements are collected describing the vehicle, scene, occupants, safety systems, injuries, and injury mechanisms. SCI cases can be queried using NHTSA's electronic case viewer [25].

2.2.8 Crash Injury Research and Engineering Network (CIREN)

CIREN is an in-depth crash investigation system sponsored by NHTSA and select automakers. CIREN investigates approximately 300 cases each year which are collected by six level 1 trauma centers. CIREN teams are multi-disciplinary teams which combine the expertise of trauma physicians, crash investigators, biomechanics and automotive engineers, and emergency medical personnel to determine the detailed injury mechanisms associated with a crash.

CIREN cases provide extraordinary detail on the injuries suffered by each crash victim. Injuries are coded using the AIS scale, and members of the CIREN network have access to radiology and other medical data which support enhanced diagnosis of injury mechanisms. CIREN is however a clinical sample of persons who enter a level 1 trauma center, and, unlike NASS/CDS, is not nationally representative of the U.S. population. Methods have been developed by Stitzel et al. (2007), Yu et al. (2008) and Elliott et al. (2010) to assess CIREN and NASS similarity and allow national estimates of injury outcomes to be estimated using CIREN [26–28].

Summaries of CIREN cases can be reviewed using the NHTSA online CIREN case viewer [29]. CIREN is a sample of seriously injured MVC occupants. MVC occupants admitted to a level 1 trauma center at an enrolling CIREN center are eligible for enrollment in the database if they meet certain inclusion criteria. Case occupants 13 years of age and older must sustain an AIS 3 or higher injury, with limited exceptions for AIS 2 injuries in multiple body regions or injuries of special interest to NHTSA. The model year of the vehicle of the case occupant must also be within 6 years of the time of enrollment unless NHTSA approves enrollment of an older vehicle. Cases of catastrophic impacts and occupant ejection are also excluded because the injury causation scenarios are difficult to define and often do not yield information that could be used in countermeasure development [30, 31]. Through the CIREN program, detailed vehicle, crash, and medical information is collected for each patient and presented at a multidisciplinary case review. The goal of each case review is to bring together engineering and medical knowledge to assess the crash mechanics, biomechanics, and clinical aspects of an injury. All cases undergo a case review with medical, engineering, and crash reconstruction specialists to determine injury causation [32].

2.2.9 Pedestrian Crash Data Study (PCDS)

Table 2.4 NTDB-RDS version 7.1 admission counts per year

Year	Admission count
2002	324,907
2003	356,577
2004	342,881
2005	430,667
2006	471,213
Total	1,926,245
MVC total	529,362

PCDS is a 5-year compilation of pedestrian crash data collected from six major United States cities from 1994 to 1998 [33, 34]. The database focused on late model year vehicles that struck pedestrians. The PCDS contains over 500 cases with detailed information describing the collision including injury severity, vehicle characteristics, and crash configuration. The U.S. National Highway Traffic Safety Administration conducted the study to better define the problem of pedestrian safety and to compare current data with previously conducted pedestrian reports to determine any modifications in trends over the years. PCDS is invaluable for the analysis of pedestrian crashes, both fatal and non-fatal. Each crash was investigated in detail, and provided information unavailable through FARS, NASS/GES, or NASS/CDS, including detailed descriptions of injuries. Although this is an older dataset, PCDS is still widely used for pedestrian studies in the U.S. At the time of this report, there was no more current in-depth crash investigation data on U.S. pedestrian crashes. PCDS has been used to understand the fatality and injury risk of pedestrian crashes [35, 36].

2.2.10 National Trauma Data Bank (NTDB)

The National Trauma Data Bank (NTDB) is the largest aggregation of trauma registry data ever assembled [37]. It is supported by the American College of Surgeons (ACS) and provides information about patients, injuries, and treatments. NTDB collects trauma registry data from participating trauma centers on an annual basis. All hospitals with trauma registries are encouraged to participate and all institutions that contribute are either ACS-verified, state-designated, or self-designated as trauma centers. The criteria for participation in the database is based on the reporting institution's trauma accreditation and hospital or emergency department admission. Data submitted to NTDB are rigorously examined using both the NATIONAL TRACS system (Digital Innovation, Inc., Forest Hill, MD) institutionally and an additional logical checks system created and enforced by NTDB administrators. Data regarding patient demographics, injury severity, and injury origin is collected as well as descriptive accounts of each traumatic incident. Patients who are dead upon arrival are excluded from the NTDB. NTDB contains injuries coded using both the ICD-9 and AIS coding lexicons.

The NTDB Research Data System (RDS) contains all records sent to NTDB for each admission year. NTDB-RDS version 7.1 contains 1,926,245 cases from 2002 to 2006 admission years (Table 2.4). ICD-9 ecodes 810–819 with a post-dot decimal of 0 or 1 which correspond to motor vehicle traffic accidents involving drivers and passengers were used to designate MVC cases, resulting in 529,362 cases. No crash data is available from NTDB aside from ICD-9 ecodes indicating MVC as the mechanism of injury, but NTDB serves as a large dataset of AIS and ICD-9 coded injuries for studying traumatic injury.

2.2.11 National Inpatient Sample (NIS)

The National Inpatient Sample (NIS) is a database that contains hospital discharge data from approximately eight million hospital stays each year. NIS is supported by HCUP, the Healthcare Cost and Utilization Project [38]. This is a Federal-State-Industry partnership sponsored by the Agency for Healthcare Research and Quality, and HCUP data is used to inform decision making at the national, state, and community levels. NIS is a unique and powerful database of hospital inpatient stays. Researchers and policymakers use the NIS to identify, track, and analyze national trends in health care utilization, access, charges, quality, and outcomes. NIS is drawn from states participating in HCUP. For 2010, the participating HCUP states comprised 96 % of the US population. In 2010, NIS contained all discharge data from 1,051 hospitals located in 45 states, approximating a 20 % stratified sample of US community hospitals. NIS contains patient injury information coded with the ICD-9 lexicon. The NIS database contains discharge and hospital weighting factors which can be applied to generate population samples.

NIS 1998–2007 contains 76,879,567 cases and 401,222 of these were classified as MVC cases using ICD-9 ecodes 810–819 with a post-dot decimal of 0 or 1 which correspond to motor vehicle traffic accidents involving drivers and passengers (Table 2.5). When discharge or hospital weighting factors are applied to the MVC cases, the weighted counts total approximately two million MVC cases. No crash data is available from NIS aside from ICD-9 ecodes indicating MVC as the mechanism of injury, but NIS serves as a large population US sample of ICD-9 coded injuries for studying traumatic injury.

Table 2.5 NIS 1998–2007 discharge counts per year

Year	Discharge count
1998	6,827,350
1999	7,198,929
2000	7,450,992
2001	7,452,727
2002	7,853,982
2003	7,977,728
2004	8,004,571
2005	7,995,048
2006	8,074,825
2007	8,043,415
Total	76,879,567
MVC total	401,222
Discharge-weighted MVC total	1,976,266
Hospital-weighted MVC total	1,952,889

2.2.12 Pre-hospital Databases (North Carolina PreMIS)

Pre-hospital data, including Emergency Medical Services (EMS) records, can be useful to the study of injury biomechanics and patient triage. EMS data is collected and maintained at the state level and data collection and storage methods vary from state to state. The North Carolina (NC) EMS Performance Improvement Center (EMSPIC) maintains one of the most comprehensive databases of pre-hospital data in the nation [39]. The EMSPIC develops, maintains, and supports the Prehospital Medical Information System (PreMIS) database to collect comprehensive data on pre-hospitalization trauma services.

NC has had statewide mandatory compliance since 2008 ensuring that all EMS call reports are submitted to PreMIS. PreMIS is in use by over 700 NC EMS agencies and 40,000 EMS personnel. Within NC, the PreMIS system collects an estimated 1,000,000 EMS call reports per year, containing well over 200 data elements per record. The EMSPIC also implemented PreMIS in South Carolina and West Virginia and it is in use by over 100,000 EMS personnel in these states. Over 2,000,000 EMS records are collected annually through daily data submission by every EMS agency encompassing the three states.

The PreMIS database contains extensive pre-hospital data that has been linked by EMSPIC to police reports and hospital databases with state trauma registry and hospital discharge information. Similar to CODES, this linked crash-medical data can be used to study patient triage and evaluate patient treatment and outcome following the transfer of the patient from EMS care to the hospital. A summary of the PreMIS data elements is provided in Table 2.6.

Table 2.6 Summary of PreMIS data elements

Pre- hospital data	Unit/agency information
	Unit/call information
	Times
	Scene
	Situation
Patient	Patient
Injury	Billing
	Situation/trauma
Assessment	Assessment/vital signs
	Assessment/injury
	Assessment/exam
Procedures	Intervention/procedure
Patient outcome, hospital, disposition	Disposition
	Outcome and linkage
Medical history	Medical history
Other	Situation/CPR

2.2.13 Event Data Recorder Data

Widespread deployment of Event Data Recorders (EDRs), sometimes called "black boxes," promise a new and unique glimpse of the events that occur during a highway traffic collision. NHTSA maintains a database of EDR records downloaded from more than 9,000 NASS/CDS crashes of U.S. cars, pickup trucks, vans, and SUVs. The EDR in a colliding vehicle can provide a comprehensive snapshot of the entire crash event: pre-crash, crash, and post-crash. By carefully collecting and analyzing the details provided by the growing number of EDR-equipped vehicles, the injury biomechanics research community has an unprecedented opportunity to understand the interaction of the vehicle-road-driver system as experienced in thousands of U.S. highway crashes each year [40].

Figures 2.2, 2.3, and 2.4 present an example of EDR data downloaded for a 2003 Pontiac Trans Am which collided with an impact attenuator. As seen in Fig. 2.3, the EDR recorded vehicle speed, brake status, engine RPM, and throttle position

Fig. 2.2 Event data recorders are a promising method for measurement of impact response in vehicle crashes, e.g. as this 2003 Pontiac Trans Am collision with impact attenuator [NASS/CDS Case 2003-50-066]

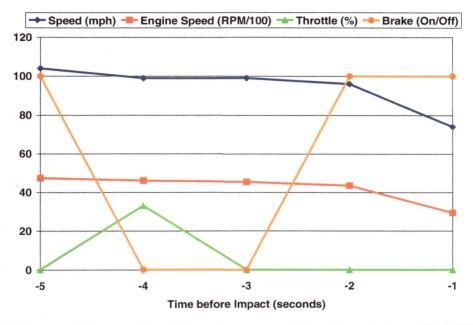

Fig. 2.3 EDR pre-crash data showed that the Pontiac Trans Am speed = 72 mph 1 s before impact [NASS/CDS Case 2003-50-066]

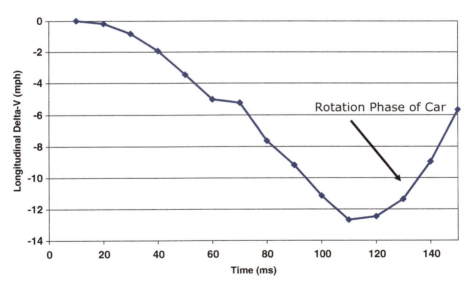

Fig. 2.4 EDR recorded longitudinal delta-V vs. time of Trans Am during the impact. [NASS/CDS Case 2003-50-066]

for 5 s before impact. Newer EDRs also record yaw rate, roll angle, and steering wheel input. The EDR recorded the longitudinal delta-V for 150 ms during the crash event.

2.3 Applications of Real World Crash Data

Crash data has been used in a wide number of applications in injury biomechanics. Applications include the development of injury metrics, e.g. delta-V, determining the effectiveness of impact injury countermeasures, e.g. safety belts or airbags, determining the relative incidence of particular types of injury, and the development of laboratory test procedures which are representative of the real world. Here we describe three selected applications of automotive crash data: the estimation of mortality associated with injuries, measurement of societal costs of injuries, and the generation of injury risk curves.

2.3.1 Measuring the Mortality Associated with Injuries

The mortality associated with injuries can also be measured using mortality risk ratios (MRRs), the probabilistic complement of survival risk ratios (SRR) first proposed by Osler et al. [41]. An injury's MRR is both lexicon- and database-specific. MRRs and SRRs have been calculated for both of the major trauma coding systems, AIS and ICD-9, and several data sources including the NTDB [42–46]. The MRR, a measure of the proportion of people who died that sustained a given injury, is reported to be among the more powerful discriminators of mortality following trauma [43, 47]. The SRR is a measure of the proportion of people who survived that sustained a given injury. MRRs and SRRs can be used as an alternative to AIS-based metrics for estimating patient mortality. The International Classification of Diseases Injury Severity Score (ICISS), derived as the product of all the SRRs of a patient's ICD-9 codes, is a time-tested approach that has been found to be a better discriminator of mortality compared to several AIS-based metrics such as the ISS, NISS, and MAIS [42, 43, 47]. Kilgo et al. (2003) showed that a similar metric to ICISS, known as the Trauma Registry AIS Score (TRAIS) and computed using the product of all the SRRs of a patient's AIS codes, represented the best AIS-based score for predicting mortality [42]. Higher injury severity will be indicated by lower ICISS and TRAIS scores (product of lower SRRS). Patients with a greater number of injuries will also tend to have lower ICISS and TRAIS scores. ICISS and TRAIS have also been shown to discriminate for mortality better when only the SRR from the patient's worst injury is included in the calculation as opposed to SRRs from all injuries [42].

Other overall injury severity measures include the Trauma – Injury Severity Score (TRISS) and the Revised Trauma Score (RTS) that account for age or physiological measures [48, 49]. The RTS is a physiologic scoring system based on the initial vital signs of the patient: Glasgow Coma Scale (GCS), systolic blood pressure (SBP), and respiratory rate (RR) [48]. Equation 2.2 is used to calculate RTS. GCS is heavily weighted in the calculation to compensate for major head injury without multisystem injury or major physiological changes. RTS scores range from 0 to 7.8408, with lower scores indicating higher injury severity. $RTS<4$ is the proposed threshold to identify patients needing treatment at a trauma center. TRISS determines the probability of survival (P_s) of a patient from the ISS and RTS using Eq. 2.3 [49, 50].

Equation 2.2. Revised Trauma Score (RTS)

$$RTS = 0.9368\ GCS + 0.7326\ SBP + 0.2908\ RR \quad (2.2)$$

Equation 2.3. Trauma-Injury Severity Score (TRISS)

$$P_s = \frac{1}{1+e^{-[b0+b1(RTS)+b2(ISS)+b3(AgeIndex)]}} \quad (2.3)$$

where [b0, b1, b2, b3] = [−0.4499, 0.8085, −0.0835, −1.7430] for blunt trauma or patients younger than 15; [b0, b1, b2, b3] = [−2.5355, 0.9934, −0.0651, −1.1360] for penetrating trauma; AgeIndex is 0 for patients 0–54 years and 1 for patients 55+ years.

2.3.2 Measuring Societal Cost with Harm

The Harm metric is a means of measuring the societal cost of traffic crashes. The Harm metric is frequently used in the evaluation of impact injury countermeasures. The Harm metric was first developed by Malliaris et al. (1982) as a means of balancing number of injuries with the severity or cost of an injury [51]. Severity is measured using the Abbreviated Injury Scale (AIS) which describes the relative threat to life of an injury [1, 2]. AIS levels range from 0 for no injury to 6 for unsurvivable, or fatal, injuries. Using the Malliaris Harm metric, each AIS level has a prescribed societal cost [51]. This societal cost includes both medical costs and indirect costs such as loss of wages. For each injured person, the Harm is the societal cost which corresponds to their maximum AIS injury level.

This original Harm metric was a remarkable new method of injury assessment, but had two weaknesses. First, societal cost is not a function exclusively of AIS level. The societal cost of injury varies by body region as well as by injury severity. For example, an AIS 3 head injury has a higher societal cost than an AIS 3 leg injury. Second, the original Harm metric assigned a cost to only the injury of highest severity. This approach can underestimate the total societal cost of a person who suffers multiple injuries as multiple injuries can aggravate the total threat to a crash victim's life.

Fildes et al. (1992) developed an improved Harm metric which addressed these two issues [52, 53]. The improved method assigns a societal cost to each injury, and sums these costs to estimate a total societal cost of injury.

$$Harm = \sum_{i=1}^{Num\ Injuries} Cost_i\,(bodyregion, AIS)$$

In the above equation, $Cost_i$, the societal cost of an injury i as defined by Fildes et al. (1992) was used as a measure of societal cost [52, 53]. $Cost_i$ is a function of the injury severity as measured by the AIS scale, and the body region which has been injured. The cost components include not only treatment and rehabilitation costs but also all other costs to society such as loss of wages and productivity, medical and emergency service infrastructure costs, legal, and insurance costs, family and associated losses and allowances for pain and suffering.

Gabler et al. (2005) developed a variation of the Fildes method for computation of Harm [54]. In some cases, there may be multiple injuries to a single body region. In this methodology, the maximum injury to a single body region is used when assigning costs as costs are typically assigned to treat a single body region not individual injuries of that body region. The costs used for the Fildes Harm metric were normalized to cost of a fatality and are presented in Table 2.7.

Table 2.7 Average cost per injury (Normalized to the cost of a fatal injury)

Body Region	Injury Severity						Unknown
	Minor (AIS=1)	Moderate (AIS=2)	Serious (AIS=3)	Severe (AIS=4)	Critical (AIS=5)	Maximum (AIS=6)	
External	0.0045	0.0250	0.0698	0.1135	0.1646	1.0000	0.0045
Head	0.0063	0.0295	0.1213	0.2796	0.9877	1.0000	0.0045
Face	0.0063	0.0295	0.1213	0.1601	0.3277	1.0000	0.0045
Neck	0.0063	0.0295	0.1213	0.1601	0.3277	1.0000	0.0045
Chest	0.0045	0.0250	0.0698	0.1135	0.1646	1.0000	0.0045
Abdomen	0.0045	0.0250	0.0698	0.1135	0.1646	1.0000	0.0045
Pelvis	0.0045	0.0250	0.0698	0.1135	0.1646	1.0000	0.0045
Spine	0.0045	0.0250	0.1631	1.4054	1.6804	1.0000	0.0045
Upper extremity	0.0063	0.0433	0.1026				0.0045
Lower extremity	0.0045	0.0433	0.1303	0.1926	0.3277		0.0045

2.3.3 Injury Risk Estimation

Previous studies have identified important parameters for estimating occupant injury [55–65]. These studies have utilized multiple databases such as NASS/CDS, CIREN, FARS, GES, census data, Federal Highway Administration (FHWA) statistics, and EDR results. The authors of these studies also used different types of statistical analysis such as ordered probit methods and regression analysis. Some of the most important parameters for estimating occupant injury were crash configuration, delta-v, occupant age, ejection status, safety equipment used, intrusion, occupant seating position, location of crash (rural or urban), impact type (wide or narrow), and size difference (mismatch) between the impacting vehicles.

Advanced automatic crash notification (AACN) algorithms utilize vehicle telemetry data and additional data to estimate injury risk. The OnStar AACN algorithm and its performance in predicting occupant injury was published recently [64]. The model variables (crash direction, delta-v, multiple impacts, belt use, vehicle type, age, and gender) and criteria defining serious injury (ISS 15+) were based on recommendations by the CDC National Expert Panel on Field Triage [66]. A multivariate logistic regression model was applied at the vehicle-level to predict the probability that a crash involved vehicle will contain one or more occupants with serious injuries. The model was trained with NASS–CDS 1999–2008 planar, non-rollover crashes involving model year 2000 or newer cars, light trucks, and vans. A bootstrapping technique was used to validate the model with the same subset of NASS/CDS. The key model predictors were delta-v, belt use, crash direction, and age (55+ years), with additional limited predictive power for multiple impacts and gender predictors.

BMW currently employs an AACN algorithm called URGENCY to aid emergency personnel in making triage decisions [55–63]. The literature describes numerous versions of the URGENCY algorithm. URGENCY uses multivariate logistic regression to predict the probability of MAIS 3+ injury. Prediction of injury risk is for individual occupants which is in contrast to OnStar which reports injury risk at the vehicle level. Model variables differ between the different versions of URGENCY and the different datasets used to train and validate URGENCY vary between studies. Many of the model variables URGENCY uses to predict risk of MAIS 3+ injury are not obtainable by current EDR technology or vehicle sensors (examples: age, gender, height, and partial/complete ejection of the occupant).

Augenstein et al. (2001) applied a version of URGENCY to frontal crashes with belt and frontal airbag restrained injured occupants [55]. NASS/CDS 1988–1995 data was analyzed to predict MAIS 3+ injury. Model variables included delta-v, maximum crush, single vehicle crash, vehicle curb weight, as well as occupant age, gender, and entrapment.

One of the most recent publications on URGENCY investigated MAIS 3+ injury risk thresholds and their effect on sensitivity, specificity, and positive predictive value (PPV) [59]. NASS/CDS 1998–2007 was used for training and NASS/CDS 2008–2009 was used for validation of URGENCY with the following model variables: delta-v, multiple impacts, rollover, belt use, and age. Risk thresholds were explored with the goal of achieving PPV values greater than 20 %, based on cited recommendations by the CDC expert panel that overtriage should not be greater than 80 % [59, 66]. Bahouth et al. (2012) recommends risk thresholds of 10 % for frontal crashes and 5 % for near side and far side crashes for the injury risk curves developed.

Injury risk functions have also been developed from EDR data. Gabauer and Gabler (2008) developed injury risk curves using the data from 108 EDRs downloaded in NASS/CDS cases to develop serious injury risk curves for three competing injury metrics: delta-V, occupant impact velocity (OIV), and acceleration severity index (ASI) [67]. The goal was to compare the relative ability of each of these injury metrics to discriminate between serious and non-serious occupant injury in real world frontal collisions. Stigson et al. (2012) used multiple logistic regression models to determine the EDR variables most predictive of MAIS 2+ injury risk for frontal crashes

[68]. Models were selected that avoided collinear predictor variables and had the lowest Akaike's information criterion (AIC) indicating a high likelihood and a lower number of variables. In total, 31 models were evaluated and the best model was achieved with delta-v and peak acceleration*age predictor variables (AUROC = 0.92).

References

1. Association for the Advancement of Automotive Medicine (2001) The abbreviated injury scale: 1990 revision, Update 98
2. Association for the Advancement of Automotive Medicine (2008) Abbreviated injury scale 2005 (Update 2008)
3. Gennarelli TA, Wodzin E (2006) AIS 2005: a contemporary injury scale. Injury 37(12):1083–1091. doi:10.1016/j.injury.2006.07.009, S0020-1383(06)00419-0 [pii]
4. Baker SP, O'Neill B, Haddon W Jr, Long WB (1974) The injury severity score: a method for describing patients with multiple injuries and evaluating emergency care. J Trauma 14(3):187–196
5. Osler T, Baker SP, Long W (1997) A modification of the injury severity score that both improves accuracy and simplifies scoring. J Trauma 43(6):922–925, discussion 925–926
6. Balogh Z, Offner PJ, Moore EE, Biffl WL (2000) NISS predicts postinjury multiple organ failure better than the ISS. J Trauma 48(4):624–627, discussion 627–628
7. Balogh ZJ, Varga E, Tomka J, Suveges G, Toth L, Simonka JA (2003) The new injury severity score is a better predictor of extended hospitalization and intensive care unit admission than the injury severity score in patients with multiple orthopaedic injuries. J Orthop Trauma 17(7):508–512
8. Gayzik FS, Moreno DP, Geer CP, Wuertzer SD, Martin RS, Stitzel JD (2011) Development of a full body CAD dataset for computational modeling: a multi-modality approach. Ann Biomed Eng 39(10):2568–2583. doi:10.1007/s10439-011-0359-5
9. National Center for Health Statistics (NCHS), Centers for Medicare & Medicaid Services (CMS) (2008) The international classification of diseases, vol 6th edn, 9th revision
10. Compton CP (2005) Injury severity codes: a comparison of police injury codes and medical outcomes as determined by NASS CDS Investigators. J Safety Res 36(5):483–484. doi:10.1016/j.jsr.2005.10.008, S0022-4375(05)00081-2 [pii]
11. Johns Hopkins University and Tri-Analytics Inc (2007) ICDMAP-90. Bloomberg School of Public Health, Center for Injury Research and Policy, Baltimore
12. Barnard RT, Loftis KL, Martin RS, Stitzel JD (2013) Development of a robust mapping between AIS 2+ and ICD-9 injury codes. Accid Anal Prev 52:133–143. doi:10.1016/j.aap.2012.11.030, S0001-4575(12)00423-X [pii]
13. National Highway Traffic Safety Administration (2012) Fatality analysis reporting system (FARS) analytical users manual: 1975–2011. Report DOT HS 811 693, Dec 2012
14. Radja G (2012) National automotive sampling system – crashworthiness data system, 2011 analytical user's manual. Report DOT HS 811 675, Washington, DC
15. Zhang F, Chen C-L (2013) NASS-CDS: sample design and weights. Report no. DOT HS 811 807, Washington, DC
16. National Highway Traffic Safety Administration (2011) 2010 FARS/NASS GES standardization. Report number DOT HS 811 564, Washington, DC
17. National Highway Traffic Safety Administration (2013) National automotive sampling system (NASS) general estimates system (GES) analytical users manual 1988–2011. Report DOT HS 811 704
18. National Highway Traffic Safety Administration (2012) Data modernization project: better data, safer roads. http://www.nhtsa.gov/Data/DataMod/DataMod
19. Bellis E, Page J (2008) National motor vehicle crash causation survey (NMVCCS) SAS analytical users manual. Report DOT HS 811 053
20. Federal Highway Administration (2012) Model minimum uniform crash criteria guideline (MMUCC), 4th edn. Report DOT HS 811 631, June 2012
21. National Highway Traffic Safety Administration (2010) The crash outcome data evaluation system (CODES) and applications to improve traffic safety decision-making. NHTSA technical report DOT HS 811 181, Apr 2010
22. Johnson SW, Walker J (1996) The crash outcome data evaluation system (CODES). NHTSA final report DOT HS 808 338
23. Winston FK, Reed R (1996) Air bags and children: results of national highway traffic safety administration special investigation into actual crashes. In: Women's travel issues second national conference, Baltimore
24. Elias JC, Sullivan LK, McCray LB (2001) Large school bus safety restraint evaluation. In: National Highway Traffic Safety Administration (ed) 17th international technical conference on the enhanced safety of vehicles
25. National Highway Traffic Safety Administration (2013) Special crash investigations. http://www-nass.nhtsa.dot.gov/BIN/logon.exe/airmislogon
26. Elliott MR, Resler A, Flannagan CA, Rupp JD (2010) Appropriate analysis of CIREN data: using NASS-CDS to reduce bias in estimation of injury risk factors in passenger vehicle crashes. Accid Anal Prev 42(2):530–539. doi:10.1016/j.aap.2009.09.019, S0001-4575(09)00247-4 [pii]

27. Stitzel JD, Kilgo P, Schmotzer B, Gabler HC, Meredith JW (2007) A population-based comparison of CIREN and NASS cases using similarity scoring. Ann Proc Assoc Adv Automot Med 51:395–417
28. Yu MM, Danelson KA, Stitzel JD (2008) Categorical similarity comparison of CIREN and NASS. Biomed Sci Instrum 44:304–309
29. National Highway Traffic Safety Administration (2013) CIREN data. http://www.nhtsa.gov/Research/Crash+Injury+Research+(CIREN)/Data
30. National Highway Traffic Safety Administration (2006) Crash injury research engineering network coding manual, Version 1.6
31. National Highway Traffic Safety Administration (2010) Crash injury research engineering network coding manual, version 2.0
32. Schneider LW, Rupp JD, Scarboro M, Pintar F, Arbogast KB, Rudd RW, Sochor MR, Stitzel J, Sherwood C, Macwilliams JB, Halloway D, Ridella S, Eppinger R (2011) BioTab–a new method for analyzing and documenting injury causation in motor-vehicle crashes. Traffic Inj Prev 12(3):256–265. doi:10.1080/15389588.2011.560500, 938347404 [pii]
33. National Highway Traffic Safety Administration (1996) 1996 Pedestrian crash data study data collection, coding, and editing manual
34. Chidester AB, Isenberg RA (2001) Final report – the Pedestrian crash data study. In: Paper presented at the proceedings of the seventeenth international enhanced safety vehicle conference, Amsterdam
35. Lefler DE, Gabler HC (2004) The fatality and injury risk of light truck impacts with pedestrians in the United States. Accid Anal Prev 36(2):295–304, S0001457503000071 [pii]
36. Roudsari BS, Mock CN, Kaufman R (2005) An evaluation of the association between vehicle type and the source and severity of pedestrian injuries. Traffic Inj Prev6(2):185–192.doi:10.1080/15389580590931680, G5L6613042205666 [pii]
37. American College of Surgeons (2007) National trauma data bank – research data system, vol RDS 7.1. American College of Surgeons Committee on Trauma, Chicago
38. (1998–2007) Agency for Healthcare Research and Quality, Rockville. www.hcup-us.ahrq.gov/databases.jsp
39. EMS Performance Improvement Center (2013) EMS performance improvement center. http://www.emspic.org/
40. Gabler HC, Hinch J, Steiner J (2008) *Event Data Recorders: A Decade of Innovation*, SAE International, Warrendale, PA (2008)
41. Osler T, Rutledge R, Deis J, Bedrick E (1996) ICISS: an international classification of disease-9 based injury severity score. J Trauma 41(3):380–386;discussion 386–388
42. Kilgo PD, Osler TM, Meredith W (2003) The worst injury predicts mortality outcome the best: rethinking the role of multiple injuries in trauma outcome scoring. J Trauma 55(4):599–606. doi:10.1097/01.TA.0000085721.47738.BD; discussion 606–597
43. Meredith JW, Evans G, Kilgo PD, MacKenzie E, Osler T, McGwin G, Cohn S, Esposito T, Gennarelli T, Hawkins M, Lucas C, Mock C, Rotondo M, Rue L, Champion HR (2002) A comparison of the abilities of nine scoring algorithms in predicting mortality. J Trauma 53(4):621–628. doi:10.1097/01.TA.0000032120.91608.52; discussion 628–629
44. Weaver A, Barnard R, Kilgo P, Martin R, Stitzel J (2013) Mortality-based quantification of injury severity for frequently occurring motor vehicle crash injuries. Assoc Adv Automot Med Accept 57:235–246
45. Kilgo PD, Weaver AA, Barnard RT, Love TP, Stitzel JD (2013) Comparison of injury mortality risk in motor vehicle crash versus other etiologies. Accid Anal Prev (in review)
46. Meredith JW, Kilgo PD, Osler T (2003) A fresh set of survival risk ratios derived from incidents in the National Trauma Data Bank from which the ICISS may be calculated. J Trauma 55(5):924–932. doi:10.1097/01.TA.0000085645.62482.87
47. Sacco WJ, MacKenzie EJ, Champion HR, Davis EG, Buckman RF (1999) Comparison of alternative methods for assessing injury severity based on anatomic descriptors. J Trauma 47(3):441–446; discussion 446–447
48. Champion HR, Sacco WJ, Copes WS, Gann DS, Gennarelli TA, Flanagan ME (1989) A revision of the Trauma Score. J Trauma 29(5):623–629
49. Boyd CR, Tolson MA, Copes WS (1987) Evaluating trauma care: the TRISS method. Trauma Score and the Injury Severity Score. J Trauma 27(4):370–378
50. Brohi K (2007) TRISS: Trauma-Injury Severity Score. http://www.trauma.org/index.php/main/article/387/
51. Malliaris AC, Hitchcock R, Hedlund J (1982) A search for priorities in crash protection. In: International congress and exposition, vol SAE 820242. SAE, Warrendale
52. Fildes BN, Lane JC, Lenard J, Vulcan AP (1994) Passenger cars and occupant injury: side impact crashes. Report CR 134, Canberra
53. Fildes BN, Monash University. Accident Research Centre, Australia. Federal Office of Road Safety (1992) Feasibility of occupant protection measures. Federal Office of Road Safety
54. Gabler HC, Digges K, Fildes BN, Sparke L (2005) Side impact injury risk for belted far side passenger vehicle occupants. SAE transactions, journal of passenger car – mechanical systems, vol 114, section 6, paper no. 2005-01-0287
55. Augenstein J, Digges K, Ogata S, Perdeck E, Stratton J (2001) Development and validation of the URGENCY algorithm to predict compelling injuries. In: Paper presented at the enhanced safety of vehicles
56. Augenstein J, Perdeck E, Bahouth GT, Digges KH, Borchers N, Baur P (2005) Injury identification: priorities for data transmitted. In: Paper presented at the enhanced safety of vehicles

57. Augenstein J, Perdeck E, Stratton J, Digges K, Bahouth G (2003) Characteristics of crashes that increase the risk of serious injuries. Ann Proc Assoc Adv Automot Med 47:561–576
58. Augenstein J, Perdeck E, Stratton J, Digges K, Steps J, Bahouth G (2002) Validation of the urgency algorithm for near-side crashes. Ann Proc Assoc Adv Automot Med 46:305–314
59. Bahouth G, Digges K, Schulman C (2012) Influence of injury risk thresholds on the performance of an algorithm to predict crashes with serious injuries. Ann Adv Automot Med 56:223–230
60. Bahouth GT, Digges KH, Bedewi NE, Kuznetsov A, Augenstein JS, Perdeck E (2004) Devleopment of URGENCY 2.1 for the prediction of crash injury severity. Top Emerg Med 26(2):157–165
61. Champion H, Augenstein J, Blatt A, Cushing B, Digges K, Flanigan M, Hunt R, Lombardo L, Siegal J (2005) New tools to reduce deaths and disabilities by improving emergency care: URGENCY software, occult injury warnings, and air medical services database. In: Paper presented at the enhanced safety of vehicle conference, Washington, DC
62. Malliaris AC, Digges KH, DeBlois JH (1997) Relationships between crash casualties and crash attributes. In: Paper presented at the SAE international congress & exposition, Detroit
63. Rauscher S, Messner G, Baur P, Augenstein J, Digges K, Perdeck E, Bahouth G, Pieske O (2009) Enhanced automatic collision notification system – improved rescue care due to injury prediction – first field experience. In: Paper presented at the enhanced safety of vehicle conference, Stuttgart
64. Kononen DW, Flannagan CA, Wang SC (2011) Identification and validation of a logistic regression model for predicting serious injuries associated with motor vehicle crashes. Accid Anal Prev 43(1):112–122. doi:10.1016/j.aap.2010.07.018, S0001-4575(10)00206-X [pii]
65. Kusano KD, Gabler HC (2013) Comparison of logistic regression and ensemble machine learning algorithms injury risk models for advanced automated crash notification algorithms. In: Proceedings of the 2013 road safety and simulation international conference, Rome
66. Centers for Disease Control and Prevention (2008) Recommendations from the expert panel: advanced automatic collision notification and triage of the injured patient. Centers for Disease Control and Prevention, Atlanta
67. Gabauer DJ, Gabler HC (2008) Comparison of roadside crash injury metrics using event data recorders. Accid Anal Prev 40(2):548–558. doi:10.1016/j.aap.2007.08.011, S0001-4575(07)00139-X [pii]
68. Stigson H, Kullgren A, Rosen E (2012) Injury risk functions in frontal impacts using data from crash pulse recorders. Ann Adv Automot Med 56:267–276

Medical Imaging and Injury Scaling in Trauma Biomechanics

Jacob R. Peschman and Karen Brasel

Abstract

Victims of trauma present unique diagnostic and management challenges to healthcare providers. The role of medical imaging in the evaluation of this patient population has evolved over the last several decades to help provide fast, accurate information to clinicians. The ability to identify important clinical signs to help identify patients in need of screening for different potential injuries can still be a challenge. Utilizing evidence based patient and mechanism specific factors to recognize those at risk of life threatening and potentially life threatening injuries will help guide the decision making process in choosing the most appropriate imaging modality. Additionally, improvements in imaging techniques has allowed for the establishment of image based injury grading systems to expand the role of non operative management of solid organ injuries. These advancements are not without downsides as radiation exposure, contrast exposure, and cost are also on the rise. Therefore, expanding provider familiarity with available imaging modalities, injury patterns, and injury severity is key in providing trauma patients with the best possible care.

3.1 Background

When a patient presents to the Emergency Department or a hospital's Trauma Bay after suffering a traumatic injury, approaching their resuscitation and evaluation in an organized manner as part of a trauma team can improve their outcome. Evaluation of a trauma patient begins with a Primary Survey, evaluating the ABC's of Airway, Breathing, Circulation, Disability, and Exposure of the patient to identify life threatening injuries. Immediately following, a head to toe Secondary Examination is performed to further identify the extent of their injuries. The use of imaging in the evaluation of trauma patients has become a key part of providing appropriate immediate and long term post injury care. X-ray imaging of the Chest and Pelvis and an ultrasound FAST exam, focused assessment sonogram in trauma, are medical

J.R. Peschman, M.D. (✉) • K. Brasel, M.D., M.P.H
Division of Trauma & Critical Care, Department of Surgery, Medical College of Wisconsin,
Milwaukee, WI, USA
e-mail: jpeschma@mcw.edu; kbrasel@mcw.edu

imaging tools that can be used as adjuncts by providers during this process. Once the primary and secondary surveys are complete, the typical next stop for a trauma patient is to the imaging department for further selective diagnostic and screening imaging assessments using X-rays, Computed Tomography (CT), Ultrasound, Magnetic Resonance Imaging (MRI), or Angiography based on their injury mechanism and potential injuries identified during their initial resuscitation. Since its introduction in the 1980s CT has rapidly become the work horse in trauma imaging as will become evident throughout this chapter. Appropriately identifying injuries helps to guide the remainder of the patient's care, whether that is discharge home, a trip to an operating room, or admission to the hospital. This accurate assessment, and the ability to grade injuries radiographically, has also expanded the role of nonoperative management of many less severe injuries, such as the spleen that had previously warranted immediate operative removal.

Injury Score Scaling and Organ Injury Scores developed by the American Association for the Surgery for Trauma (AAST) first published in 1988 and available on their website www.AAST.org are commonly used clinical tools that have allowed practitioners to study and develop management algorithms for different injuries [1]. Originally developed using findings at the time of operation, they now are based primarily on CT findings, specifically with contrast enhanced multidetector helical CT imaging. Additionally, the Injury Severity Score (ISS) introduced by Baker and colleagues in the late 1960s has helped allow advancements in trauma research by tracking, comparing and standardizing the severity of multiply injured patients [2]. The ISS score is derived from assignments of an Abbreviated Injury Scale score, a 7 part number system most recently updated in 1990 that ultimately assigns a 1–6 point value, (1 being minor, 6 unsurvivable) to different injuries that a patient may have suffered in six different body regions; Head or Neck, Face, Chest, Abdomen, Extremities, and External. The regions AIS is the highest value in that region. The ISS is calculated using the squared sum of the 3 AIS regions with the highest assigned values as shown below.

$$ISS = (\text{Highest AIS region A})^2 \\ + (\text{Highest AIS region B})^2 \\ + (\text{Highest AIS region C})^2$$

For ISS calculation only the highest value within each region can be used. For example, in a patient with two thoracic injuries assigned a 3 and 4, an extremity injury assigned a 2, and an abdominal injury assigned a 2, the ISS would not include both the 3 and 4 from the thoracic region even though they are both higher than the extremity and abdomen regions values of 2. The ISS would therefore be 24 ($4^2 + 2^2 + 2^2$), not 29 ($4^2 + 3^2 + 2^2$).

The remainder of this chapter will further investigate the imaging of a trauma patient by focusing on the different anatomic region.

3.2 Head

Head injuries are frequently encountered following blunt trauma and often lead to long term disability. Though injuries to the scalp and calvarium can require intervention, the true morbidity and mortality from head trauma occur as a result of intracranial injuries or traumatic brain injuries (TBI). Therefore, early recognition of these injuries is necessary to provide appropriate treatment to patients. One of the first true clinical assessments of intracranial injury is the calculation of the patient's Glasgow Coma Scale or GCS score during the "D" portion of the Primary Survey [3]. The GCS score is a well validated clinical scoring system developed in the 1970s that assesses a patient's eye opening, motor response, and verbal response on a 3–15 scale [4]. A GCS of less than 8 is typically categorized as a severe brain injury, 9–12 as a moderate, and 13–15 as minor (Table 3.1).

GCS also plays a key point in early decision making for head injuries. In patients presenting with a normal GCS of 15 *and* without loss of consciousness, amnesia to the event, physical evidence of head trauma, headache, vomiting, use of anticoagulants, neurologic deficits or use or drugs or alcohol, dedicated intracranial imaging may not be needed [5, 6].

Table 3.1 Glasgow coma scale

Patient assessment	Score
Eye opening (E)	1–4
Spontaneously	4
To speech	2
To pain	3
None	1
Best motor response (M)	1–6
Follows commands	6
Localizes to pain	5
Withdraws from pain	4
Decorticate posturing (Flexion)	3
Decerebrate posturing (Extension)	2
None	1
Verbal response (V)	1–5
Oriented	5
Confused conversation	4
Inappropriate words	3
Incomprehnsible sounds	2
None	1

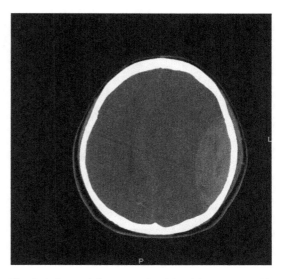

Fig. 3.1 Large left temporal epidural hematoma with ventricular compression and significant midline shift requiring early neurosurgical evaluation for potential evacuation

When further investigation of a potential head injury is indicated in the early stages of trauma evaluation, axial imaging with non contrast enhanced CT is the modality of choice due to its speed and ability to identify pathologies that require intervention. MRI, PET, fMRI and other emerging imaging techniques are often too slow and not as readily available and therefore play limited role early but can be used to assess progression of certain types of traumatic brain injury [7]. Injuries to the brain are typically a result of the acceleration and deceleration forces on tissues. Epidural hematomas, subdural hematomas, intraparenchymal injuries, and skull fractures are the most common early injuries following trauma. Diffuse axonal injuries also develop though the onset is more insidious as they develop later due to fluid shifts along injured axons. These injuries are life threatening when they cause an increase in intracranial pressure potentially leading to brainstem herniation. CT findings of midline shift >5 mm, ventricular collapse due to epidural or subdural hematomas, and depressed skull fractures require immediate neurosurgical evaluation as operative methods to reduce intracranial pressure or to place devices to allow pressure monitoring to guide pharmaceutical therapies are often needed [8, 9].

Epidural hematomas typically form as the result of lacerations of arteries between the dura and the skull due to movement of the brain within the skull or injury by skull fractures. Most commonly they develop in the region that was struck during the inciting events rather than the opposite side of the brain due to a contra coup injury. They are evident on CT scans as having the classically described "Biconvex Lens" shape which forms as the blood collecting in the extradural space pushes the dura towards the brain (Fig. 3.1).

Traumatic subdural hematomas typically form as the result of shearing or tearing of bridging veins between the dura and the arachnoid matter. As they are a result of venous bleeding they often develop more gradually than epidural hematomas, though they still can become large enough within the fixed space of the skull to cause dangerous increases in intracranial pressure. Outcomes are typically worse with traumatic subdural hematomas due to the associated shearing forces exerted on axons resulting in diffuse axonal injuries (see below). They are evident on CT scans as a "Crescent" shaped collection of blood which pushes the dura toward the skull as it collects outside of the parenchyma of the brain (Fig. 3.2).

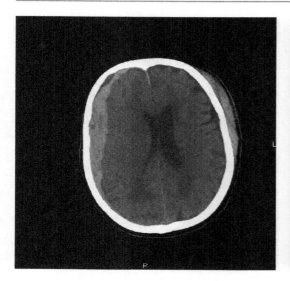

Fig. 3.2 Right fronto-temporal subdural hematoma with midline shift. Some ventricular compression is noted

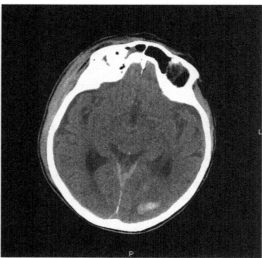

Fig. 3.3 Left occipital intraparenchymal hemorrhage several hours following motor vehicle collision

Intraparenchymal injuries (i.e. subarachnoid hemorrhages) typically form as a result of injury to the brain parenchyma itself and can often "blossom" from contusions to brain matter. This can result in their formation in areas 180° from the site the head was struck resulting in a contra coup injury. They can appear as indistinct higher density areas within the brain matter itself, with larger or older intraparenchymal injuries having the appearance of distinct collections of blood within the brain matter (Fig. 3.3).

Depressed skull fractures form as a result of direct traumatic forces to the head. Fragments of skull push into the brain, often directly injuring brain tissue or causing injury to intracranial arteries or veins as previously described. Though skull fractures can be identified on plain X-ray imaging, the presence of a skull fracture necessitates further evaluation of intracranial structures. Therefore, if the clinician has a high suspicion for skull fracture, CT imaging is the modality of choice. There is little to no role for isolated plain films for head injuries if CT imaging is available. Depressed skull fractures can increase intracranial pressures directly and in combination with the associated bleeding, often requiring early surgical intervention (Fig. 3.4).

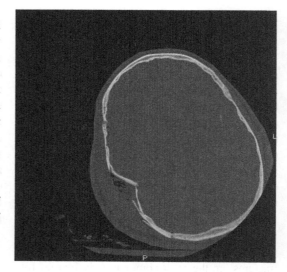

Fig. 3.4 Open right temporal comminuted depressed skull fracture with air in the subcutaneous tissue

Diffuse axonal injuries (DAI) form as a result of shearing and straining forces exerted on axons, often within areas of the brain with differing densities such as the grey and white matter interface. These types of injuries are can be associated with severe morbidity and mortality and present in up to 50 % of traumatic brain injuries. Early recognition is important as non surgical techniques to reduce swelling and optimize oxygen delivery should be employed early in the course of treatment. CT imaging is not as sensitive as conventional Magnetic Resonance Imaging (cMRI) for DAI, though a definitive diagnosis historically has been made at autopsy. Generalized cytotoxic brain edema as characterized by loss of the grey-white differentiation and scattered petechial hemorrhages can be indicators of severe DAI and seen early using CT. cMRI like CT focuses on structural abnormalities. In DAI this can be higher resolution detection of petechial hemorrhages and vastly improved detection of non hemorrhagic lesions due to edema following shear injuries [10].

Other imaging modalities find limited use in trauma. Functional imaging studies such as Positron Emission Tomography (PET) and functional MRI (fMRI) have limited clinical utility, with use primarily as tools to assess response to therapies aimed at improving cerebral perfusion. They are still primarily limited to research settings due to their cost and slow image acquisition which is often not tolerated by critically ill trauma patients. CT angiography can be used if there is concern for extension of blunt injuries to extracranial vessels which will be described in more detail later. CTAs as well as nuclear medicine brain flow do serve a role late in TBI patients as a confirmative test for brain death as evidenced by a lack of intracranial blood flow.

3.2.1 Injury Grading

The Abbreviated Injury Scale Score as related to head injuries may not relate as closely to the clinical outcome as in some other injuries. Most often injuries are based on the CT findings on initial presentation. The long term morbidity of head injuries in trauma patients is well established, though in much of the published data, the presence of a head injury and a severe ISS are often evaluated separately during multivariate logistic analyses. Injuries to brain structures or vasculature are all attributed AIS scores of 3–5 translating to a 9–25 point contribution to the overall ISS. Even clinical evidence of a TBI in the absence of anatomic injury such a clinically diagnosed concussion can carry a significant AIS score of 2–4.

3.3 Spine

In the AIS system, injuries to the spine and vertebrae are scored by the body region through which that segment of the spine traverses (i.e. cervical spine within neck, thoracic spine within chest, and lumbar spine within abdomen). In regards to imaging of these regions and principles for evaluation, injuries to the spine and spinal cord follow similar patterns and will be presented together here. Vertebral injuries in any area of the spinal column are primarily clinically significant in relationship to how they affect the spinal cord, usually defined by whether or not they are stable or unstable. Evaluation of spine stability remains somewhat controversial with one of the more widely accepted systems being that proposed by Denis in the 1980s which divides the spine into three vertical columns, the posterior, middle, and anterior columns [11]. Instability occurs with injury to 2 of 3 columns, making accurate evaluation of the middle column which includes the posterior 1/3 of the vertebral body and posterior longitudinal ligament, key in this determination. Advances in imaging continue to play a key role in this evaluation as multidetector CT has been shown to be an improvement over conventional radiographs. Flexion/Extension radiographs, along with MRI though, are better for ligamentous and soft tissue injury, and are used primarily as an adjunct to CT.

Additionally, despite the potential for severe neurologic deficits especially in the cervical spine, imaging evaluation can at times be temporarily deferred. This allows appropriate spinal precautions including a cervical collar and flat bed rest to be maintained, while other life threatening injuries (i.e. exsanguination, respiratory compromise) are addressed. Injuries to the spine are also often associated with other predictable injuries which should help guide clinical decision making when deciding when to further image the patient.

Cervical spine injuries are most common, comprising over half of all injuries to the spinal cord, and are often associated with other injuries to the spine as 10 % of patients with fractures of their cervical spine will have a non contiguous spine fracture. Up to 25 % of patients with spine injuries also will have an associated TBI, which though often mild, warrants a prudent intracranial evaluation. Injuries to the spine are usually due to the effects of blunt acceleration and deceleration forces on the spinal column. As its rate of being most frequently injured implies, the cervical spine is at greatest risk due to the lack of extraneous support that the thoracolumbar spine has from the rib cage and abdomen. For this reason the cervical spine requires special attention and will be addressed first.

3.3.1 Screening

Screening evaluation of the cervical spine should be considered in essentially all patients who suffer from blunt trauma. This includes clinical evaluation as well as typically radiographic evaluation. Current recommendations from the Eastern Association for the Surgery of Trauma (EAST) Practice Management Guidelines Committee are to perform initial evaluation with CT imaging for any patient with a high risk mechanism of injury which include motor vehicle crash (MVC) at >35 mph, fall from >3 ft, any neurologic deficit, when the mechanism includes an axial load on the spine, in those 65 years of age or older, or when a neurologic deficit is present [12]. In the absence of any of these CT is still indicated if the patient has any pain, tenderness, or altered mental status during the attempt to clinically clear the patient's cervical spine. If the CT is felt to be negative after being evaluated by a trained surgeon or radiologist, accuracies of >99 % for excluding injury are often reported, with missed injuries usually of no clinical significance. Even with normal imaging, clinical evaluation should be performed prior to cervical collar removal. In patients who have pain with active range of motion despite a normal CT scan, the practitioner can choose to perform an MRI or repeat the clinical exam in 7–10 days prior to removing the collar. In patients with abnormal cervical spine CT scans, immediate consultation with a Neurosurgeon or Orthopedic Surgeon is indicated.

As with cervical spine screening, identifying unstable thoracolumbar fractures are important due to potential spinal cord compromise, so screening imaging should be considered in many cases of blunt traumatic injury [13]. As stated before, there is a 10 % rate of non-contiguous vertebral fractures with the first detected not always being the most clinically significant. Also as with the cervical spine, EAST Guidelines recommend CT as the modality of choice for screening evaluation of the thoracolumbar spine [14]. There are several reasons for this. X-ray has consistently been found to have a relatively low sensitivity for detecting injury ranging from 25 % to 75 % compared to >95 % for CT. Additionally, patients who would require imaging of their thoracic or lumbar spine based on mechanism such as fall >10 ft, high speed MVC, motor pedestrian crashes, or those who report back pain or tenderness often also meet criteria to have further thoracic or abdominal imaging. Protocols for Chest, Abdomen, and Pelvis CT scans for trauma now can include dedicated reformats of the thoracic and lumbar spine without need for additional time or radiation exposure.

3.3.2 Unique Cervical Spine Injuries

Other unique fractures of the cervical spine include atlanto-occipital instability, C1, and C2 fractures. Occipital condyle fractures are often seen of CT imaging but frequently carry little clinical significance. Dens fractures are more often associated with spinal instability. They are classified as Type 1, involving just the tip of the dens, Type 2, involving the base (most common) and Type 3, involving the C2 body. Injury to the spinal cord at these levels can cause devastating neurologic outcomes including quadriplegia and death. Therefore early consultation with a Neurosurgeon or Orthopedic Surgeon is necessary to help determine the stability of the affected area [15]. MRI may be needed to evaluate

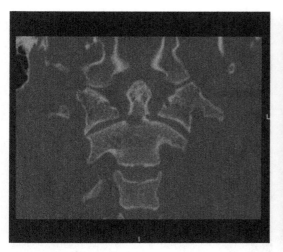

Fig. 3.5 Type 2 dens fracture with distraction ultimately requiring halo placement for stabilization

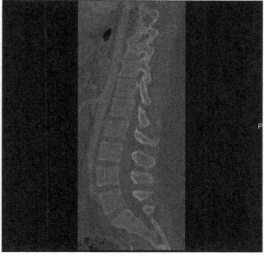

Fig. 3.6 Mildly displaced fracture of the anterior/superior endplate of L2 following a 20 ft fall

ligamentous injury, and stabilization with either an aspen collar, halo vest device, or surgical fixation may be needed early in the course (Fig. 3.5).

3.3.3 Compression and Burst Fractures

Compression and Burst fractures occur as the result of axial load forces on the spinal cord. The most common compression fractures are anterior wedge compression fractures which occur as a result of hyper flexion as the upper body and torso continue travel forward while the lower body is restrained during deceleration. Wedge compression fractures typically only involve the anterior column and therefore are less likely to cause spinal instability. Burst fractures are more severe as they can involve both the anterior and middle columns and can have retropulsion into the spinal canal causing direct cord compromise. They are significant based on the effect they have on alignment, vertebral body height loss, percentage of canal compromise (increasing risk of cord damage) and risk of injury to the posterior longitudinal ligament leading to instability [16, 17]. Fractures in the lumbar region are often of greater concern due to the increased axial load on these vertebrae when standing, the altered directions of the forces due to the natural lordosis of the region and lack of stabi-

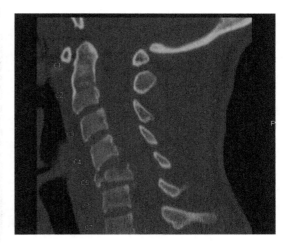

Fig. 3.7 Burst fracture of C5 following motor vehicle crash with retropulsion

lization by the rib cage that the thoracic vertebrae have (Figs. 3.6 and 3.7).

3.3.4 Flexion Distractions and Fracture Dislocations of the Spine

Flexion distraction injuries as described by Denis are unstable fracture patterns due to injuries of the middle and posterior columns usually due to

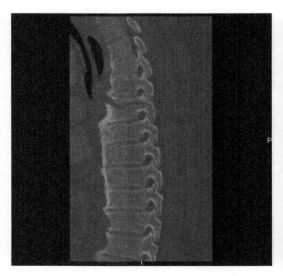

Fig. 3.8 Chance fracture at T5 with moderate displacement but no canal compromise in a pedestrian struck by a car

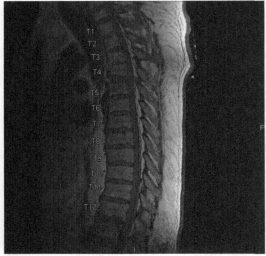

Fig. 3.9 T1 weighted MRI images of the patient in Fig. 3.8 with a T5 chance showing the focus on fat containing structures compared to bony CT windows

hyperflexion injuries with vertebral compression while the most serious fracture dislocations usually result in jury to all three columns as a result of hyperextension or hyperflexion. Chance Fractures are similar to flexion distraction injuries however a fracture line in the axial plane of the vertebral body and spinous process forms rather than a compression fracture. These bony fractures are seen best on CT imaging though, there is a low threshold for MRI to evaluate for cord injury when any potential for neurologic deficit exists (Figs. 3.8 and 3.9).

3.3.5 Transverse Process Fractures

Isolated transverse process (TP) fractures thoracolumbar spine are typically not clinically significant from a spinal stability standpoint and usually do not require special precautions or surgical evaluation. However, in the cervical spine, TP fractures will often include the transverse foramen. This injury pattern does not carry major significance to the spinal column or spinal cord but does warrant investigation of the vertebral artery with CT angiography (will be discussed further later in this chapter) (Fig. 3.10).

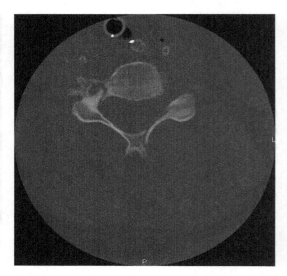

Fig. 3.10 Fracture through the right C4 transverse foramina which prompted a CTA showing occlusion of the right vertebral artery from C3–C6. A left lamina fracture can also be seen

3.3.6 Spinal Cord Trauma

Injuries to the ligamentous structures and spinal cord are not well seen on CT. Ligamentous injuries are extremely rare in the thoracolumbar spine

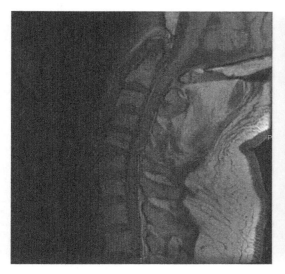

Fig. 3.11 T1 weighted MRI images showing epidural hematoma extending from C5 up to C2. This was not evident on CT but due to unexplained upper extremity weakness, and MRI was obtained

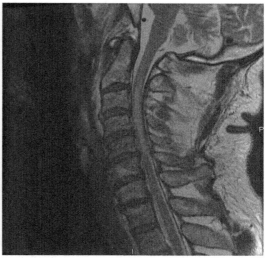

Fig. 3.12 Brightness within the cord in T2 weighted images making water bright shows cord edema in a patient with symptoms of incomplete cervical paraplegia

in the absence of bony injury however persistent midline pain in the cervical spine despite a normal CT scan can warrant MRI prior to removal of a c-collar as a final method of excluding ligamentous injury. Fractures causing narrowing of the canal, clinical neurologic deficits, or unstable spine injuries seen on initial CT imaging can guide the clinician to pursue MRI imaging for evaluation for spinal cord infarcts, hemorrhages, contusions and epidural hematomas not seen on CT (Figs. 3.11 and 3.12).

3.3.7 Injury Grading

As stated at the beginning of this section, injuries to the spinal column are assigned AIS scores in the area through which that portion traverses. Injuries to the brachial plexus are also included with the cervical spine. Cord injuries are assigned considerably higher values, often 4–5, regardless of the presence of a fracture based on the associated morbidity. Higher injuries (above C3) are also scored higher. Dedicated spinal cord imaging, usually MRI, is needed to assign these scores radio graphically unless CT shows obvious evidence of major cord injury. However, as imaging purely for the purposes of AIS scoring is not recommended, clinical evidence of a cord injury can be sufficient to assign an AIS score. Scores for fractures and dislocations are assigned based on imaging, however their inclusion in the patient's ISS will only occur in the setting of no damage to the cord as the scores range from 1 to 3.

3.4 Neck

Imaging of the neck following trauma has traditionally focused primarily on the cervical spine and spinal cord as previously discussed. Major blunt injuries to aerodigestive structures are extremely uncommon and will typically be associated with external findings that will help guide further management. However, vascular injuries of the neck, especially if missed, can lead to delayed stroke or death devastating outcomes to otherwise young previously healthy patients and therefore cannot be overlooked. Rates of blunt carotid injuries of 0.3–1 % have been quoted in patients suffering from blunt trauma, therefore appropriate selected screening to identify at risk

Table 3.2 Blunt carotid injury grading scale

Grade	Criteria
I	Luminal irregularity or dissection with <25 % luminal narrowing
II	Dissection or intramural hematoma with >25 % luminal narrowing, thrombus, or raised flap
III	Pseudoaneurysm
IV	Occlusion
V	Transection with free extravasation

patients should be part of the evaluation algorithm for all trauma patients. Additionally, any patient with cervical transverse processes fractures that include the transverse foramen places the patient at risk for vertebral artery injury. Prior to the widespread availability of CT angiography (CTA), traditional angiograms were needed to evaluate for blunt carotid and vertebral injuries. Though highly sensitive and specific, and still the gold standard for diagnosis, they are also time consuming and expensive. CTA can be performed as a screening tool in a cost effective manner with appropriate selection criteria based on mechanism, exam findings, and associated injuries [18]. In addition to transverse foramen fractures, basilar skull fractures involving the carotid canal should also warrant CTA. Neurologic deficits on exam not consistent with other injuries, a "seat belt sign" or other abrasion to the anterior neck, severe hyperextentsion/rotation/flexion of the cervical spine especially when associated with midface or mandibular fractures and near hanging injuries also warrant further examination for vertebral or carotid injuries (Table 3.2).

If an irregularity is discovered, the Blunt Carotid Injury Grading Scale developed by the Denver Health Center has become the major system for injury scaling [19]. The injury scale originally used for the extracranial carotids are now also used to grade vertebral artery injuries. The grading system has important clinical applicability in prediction of stroke rates as well as guiding clinical treatment. Grade 5 injuries need urgent intervention and have the highest associated mortality. Grade 4 injuries have high stroke rates though often no intervention is needed and the goal of anticoagulation is to prevent propagation more than improving flow in the already occluded system. Similarly, though no intervention other

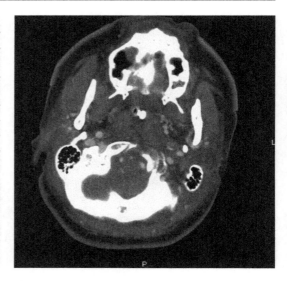

Fig. 3.13 Left internal carotid dissection with thrombus in over 50 % of the lumen on axial images of the CTA of the neck following a motor vehicle crash

than anticoagulation is immediately warranted for most grade 1 and 2 injuries they can progress to pseudoaneursyms (grade 3) which do require further intervention, though typically not urgently. Additionally, with the high resolution multidetector CT scanners providing much finer detail than has been available in the past, many times potential grade 1 injuries cannot be excluded. It is for this reason that follow up CTA is needed in grade 1 or 2 injuries 7–10 days following injury to either select patients who no longer require anticoagulation if the scan has normalized or need further intervention if progression has occurred. This is where traditional angiography still plays a role in the management of vertebral and carotid injuries, as the angiogram provides more definitive diagnostic ability as well as the potential for intervention (Figs. 3.13 and 3.14).

3.4.1 Injury Grading

The AIS grading for extracranial carotid and vertebral injuries are scored in the neck section. Though the AIS system does not perfectly correlate with the Blunt Carotid Injury Grading Scale CTA or angiographic findings do help with injury scaling. Also, as with spinal cord trauma, the presence of

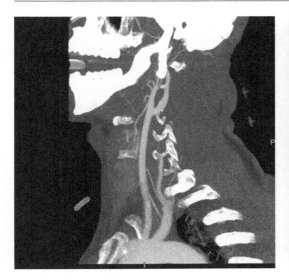

Fig. 3.14 Reconstructions of the CTA of the neck from the same patient presented in Fig. 3.10, showing an abrupt cutoff of the right vertebral artery at C6

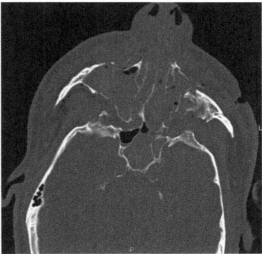

Fig. 3.15 Complex midface fractures with Lefort II and III components after being kicked in the face by a deer

neurologic deficits also can increase the AIS score for a specific arterial injury. Venous injuries, also included in the system are of more concern in non accidental and penetrating trauma. Lastly, though not described in detail here, injury to other structures of the neck including the thyroid and salivary glands, laryngeal and pharyngeal structures, and hyoid bone are also scored in this section.

3.5 Face

Imaging of the face is typically guided by physical exam findings consistent with facial trauma. Swelling, ecchymosis, and deep lacerations can all indicate injury to underlying structures. In general, imaging focuses on bony structures of the face. If injury to cranial nerves or branches are suspected based on physical exam findings, the mechanism of injury will often play the largest role in deciding if operative exploration (for lacerations) or a watchful waiting approach (for blunt contusions) is taken. Bony injury involving a specific foramen can additionally guide the decision making process. As with many other systems, CT scans are the primary imaging modality. However, using specific programs, 3D reconstructions can be generated using standard CT technology which can greatly aid surgeons in determining the need for facial reconstruction. The fracture patterns of the face are complex and beyond the focus of this chapter, however the principles for using imaging to help decide when intervention is warranted is based on determining stability and cosmesis which requires consultation with a surgeon with training in facial trauma. Likewise, injuries to the orbital walls which lead to intrusion into the orbit or cause retro bulbar hematomas may indicate the need for immediate ophthalmologic consultation. Dedicated dental imaging, when needed, continues to rely on plain film X-ray, though dedicated imaging machines are necessary. The ordering practitioner should keep in mind that in many institutions, consultation with Oromaxillofacial Surgeons or Dentists may be needed to interpret these images as some radiologists will not read these types of films (Figs. 3.15, 3.16, and 3.17).

3.5.1 Injury Scaling

Injuries included in the Face category of the AIS, including the eye and ear, are heavily biased towards bony fractures. The only nerve included in this area is the intraorbital portion of the optic nerve. Injuries to the globe itself, though often morbid due to impairment in vision, still receive fairly low AIS scores of 1 or 2. Fractures to the Mandible, Maxilla, and Orbit also carry reasonably low scores despite the fact they can require multiple operations and carry long term consequences in altering the cosmetic appearance of the face and in affect the psychologic recovery of the patient. Complex fractures of the maxilla that involve deeper bony structures including the pterygoid or sphenoid bones can lead to an overall higher ISS score as they are coded in both the facial and head categories (assuming no other intracranial injuries are present).

3.6 Chest

Imaging of the chest occurs earlier in the evaluation of trauma patients than any other body region as a chest X-ray is considered such an important tool it is an adjunct to both Primary and Secondary Surveys. Current recommendations from ATLS favor liberal use of chest X-rays as they can detect several potentially life threatening injuries such as a large hemothorax, simple pneumothorax (tension pneumothorax being a clinical diagnosis), tracheobronchial tree injuries, traumatic aortic disruptions (widening of the mediastinum), blunt esophageal injuries, and diaphragmatic injuries [20]. Detection of many of these processes are important early in the evaluation of a trauma patient as they can trigger immediate intervention such as tube thoracostomy (i.e. chest tube) placement prior to continuing the Secondary Survey as well as help guide the decision making process as to what additional imaging may be needed (Figs. 3.18, 3.19, and 3.20).

The other imaging modality used early in the resuscitation process as an adjunct to the Primary Survey is ultrasound. A FAST exam can be used to identify life threatening injuries very early in the resuscitation. The use of abdominal FAST will be discussed in greater detail later. The ultrasound imaging view obtained during a FAST exam to evaluate for thoracic trauma is a pericardial view with the goal primarily being to identify fluid, presumably blood in the setting of trauma, within the pericardial sac which could be causing cardiac tamponade. Beyond evaluation for tamponade, FAST in the thoracic cavity can

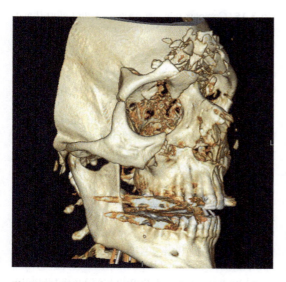

Fig. 3.16 3D reconstructions of the facial fractures in the same patient as Fig. 3.12 further showing the comminuted nature and midface depression

Fig. 3.17 Panorex image illustrating the detail in dental hardware and alignment in a patient with bilateral mandibular fractures

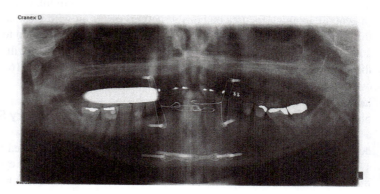

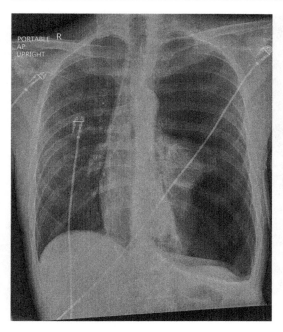

Fig. 3.18 Large left pneumothorax with mediastinal shift

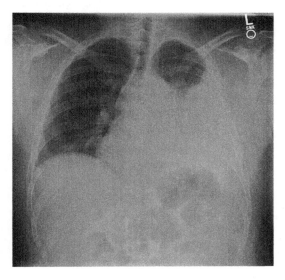

Fig. 3.19 Large left hemothorax identified 3 days after motor vehicle crash. A left seventh rib fracture is also identified

also detect hemothorax and pneumothorax. Hypotension can come from a multitude of places in the trauma patient; however, early recognition of tamponade physiology is critical. Physical examination identifying the classic Beck's triad of muffled heart tones, hypotension and jugular venous distension can be difficult in the initial assessment of a critically ill patient. Ultrasound provides the Emergency Medicine physician or Trauma Surgeon the opportunity for a rapid assessment of the pericardial space, with an accuracy of up to 90 % when performed by an experienced provider.

3.6.1 Heart

In the vast majority of patients with blunt cardiac injury, the injury does not pose significant danger, and those with the most significant injuries often do not survive long enough for them to be recognized. Screening for blunt cardiac injury begins with obtaining an Electrocardiogram (ECG) on arrival to the hospital in patients with a suspected injury and abnormality noted on telemetry [21]. Significant chest wall trauma should be an early warning sign, however studies have shown sternal fractures alone are infrequently associated with blunt cardiac injury [22]. The most common ECG findings include unexplained sinus tachycardia, frequent premature ventricular contractions, arrhythmias, heart block, or signs of ischemia. Though FAST exam in the hands of an experienced operator can help identify some of the more significant valvular disruptions and pericardial bleeding, ECG remains the screening modality of choice, with a negative predictive value (NPV) of >95 %. Cardiac enzymes including Troponin I have been evaluated as an additional measure, but have been found to be more indicative of cardiac ischemia than an actual cardiac injury [23]. If tested and elevated, concomitant myocardial infarction should be appropriately excluded, as elderly trauma patient can be at increased risk of MI as an inciting event or as a result of trauma. In patients with ECG abnormalities concerning for a blunt cardiac injury or otherwise unexplainable hemodynamic compromise, formal 2D Echocardiogram transthoracic or trans esophageal should be performed as a diagnostic test [24]. Screening with echocardiogram is not recommended due to the cost and low yield in low risk patients. Lastly, the role of CT and MRI in

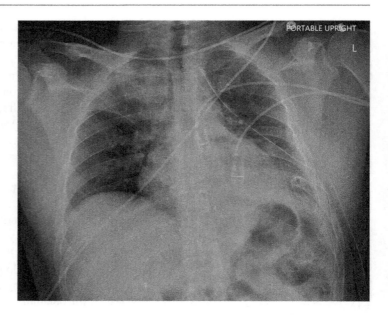

Fig. 3.20 Post operative chest X-ray of the patient in Fig. 3.20 showing marked improvement of the hemothorax

evaluation has long been limited by their speed and poor ability to detect abnormalities in a beating heart. As the technology improves especially with MDCT and the advent of ECG gated image acquisition, their role may expand. Presently, the utility of ECG gated imaging is limited to case reports and series and the feature is not widely available or part of most trauma imaging protocols.

3.6.2 Chest Wall

Chest wall injuries, fractures of the ribs are among the most common injuries following blunt trauma, occurring in up to 70 % of injured patients, with injury to the underlying lung occurring frequently, even without bony fracture [25]. In selected young otherwise healthy patients with minimal associated injuries, non-displaced rib fractures with minimal symptoms can be insignificant enough to allow discharge directly from the emergency department. However, often times rib fractures carry a significant morbidity and mortality, especially in older patients, and therefore identifying them to develop an appropriate plan of care and rule out injuries to the lung is important. Fractures to >3 ribs has been suggested as potential criterion for transfer to a trauma center, and rib fractures in patients over the age of 65 are associated with a significantly increased morbidity and mortality. Chest wall imaging begins with the initial Chest X-ray during the Primary or Secondary Survey. Significant hemothoraces or pneumothoraces may require intervention at that time. A normal supine chest X-ray in an otherwise hemodynamically normal patient without a concerning mechanism of injury can be enough information to exclude the need for further CT imaging.

Though less frequently performed with the wide availability of and sensitivity of CT, dedicated rib X-rays are superior to supine chest X-rays to identify rib fractures in some patients that will either help to guide directed pain control therapies such as intercostal nerve injections, identify elderly patients with multiple fractures requiring admission for respiratory monitoring, or lower rib fractures that may require further intra-abdominal imaging [26]. However, CT has become by for the next step in imaging of the chest. CT is highly sensitive for detecting rib fractures. CT scans with or without contrast can be windowed to provide highly detailed bony evaluation that can identify non displaced fractures that would be missed on X-ray imaging. Rib fractures are typically managed non operatively and require

Fig. 3.21 Initial chest X-ray from a patient run over by a car revealing bilateral hemothoraces, subcutaneous emphysema, bilateral rib fractures and a right clavicle fracture

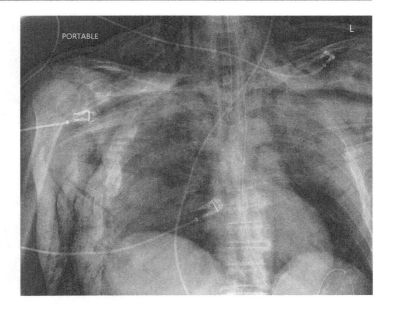

appropriate respiratory therapy and pain control. Some controversy still exists in the developing field of rib stabilization. The additional value to CT imaging can come in the ability to perform 3D reconstructions for choosing patients that may benefits from operative intervention and for operative planning. Flail chest, when multiple contiguous ribs are fractures in two or more locations, and significantly displaced fractures have been proposed criteria and are better visualized on 3D imaging (Figs. 3.21, 3.22, and 3.23).

3.6.3 Pulmonary

Injuries to the lung occur commonly with thoracic trauma. The presence of rib fractures, subcutaneous emphysema, pulmonary contusions, hemothorax, or pneumothorax on chest X-ray are among the most common findings. CT imaging provides detailed imaging of the lung parenchyma with negative windowing, resulting in only air to appear black compared to other forms where fatty tissue can have a very similar appearance. With that said however, chest X-ray remains more predictive of clinically significant lung injuries, especially pulmonary contusions [27]. A hemothorax or pneumothorax seen on CT but not identified on

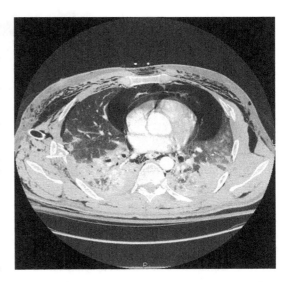

Fig. 3.22 Lung windowed chest CT in the same patient as Fig. 3.21. The comminuted right rib fractures are again seen and bilateral pulmonary contusions are more evident. A chest tube is now present, entering from the right side

chest X-ray is referred to as occult. Rates of detection vary depending on the frequency with which CTs are obtained, whether the imaging is interpreted by a radiologist, emergency medicine physician, or trauma surgeon, but are typically reported as occurring in 5–15 % of patients with occult pneumothorax and up to 20–30 % for

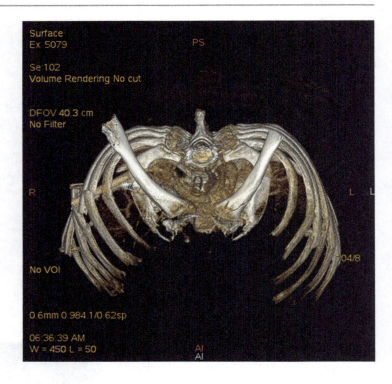

Fig. 3.23 3D reconstruction images generated from the CT scan in Fig. 3.22

occult hemothorax [28]. Often time subtle findings such as a deep sulcus sign, an elongated costophrenic angle, on CXR can result in initial under recognition of a pneumothorax that will become more evident on CT [29]. At present, identification of an occult pneumothorax is not an indication for the placement of a chest tube or even follow up imaging if found in isolation. Management of occult hemothorax is less well defined and follow up imaging in 1–2 days may be helpful in identifying patients that would benefit from a chest tube or Video Assisted Thoracic Surgery (VATS) for hematoma evacuation to prevent formation of an empyema or trapped lung. For these indications follow up imaging with a chest X-ray can be useful, however non contrast CT can be performed to help distinguish pulmonary contusion from hemothorax if intervention is being considered (Figs. 3.24 and 3.25).

FAST and ultrasound is now being proposed as a potential screening tool for detection of pneumothorax in trauma patient and is highly sensitive and specific, with rates up to 89 % and 99 % respectively, being reported [30]. Ultrasound detects the interface, or lack thereof, between the parietal pleura and lung pleura. A normal interface in a D-mode of detection can reveal a "waves on the beach" appearance while, lack of this finding indicates air between the surfaces and a pneumothorax by definition [31]. Much like CT findings of occult pneumothorax however, the clinical significance of pneumothoraces detected in the manner are unknown.

3.6.4 Mediastinum and Thoracic Aorta

Imaging of the mediastinum focuses primarily on the great vessels of the chest, including the Superior Vena Cava, Inferior Vena Cava and particularly the Thoracic Aorta. Evaluating these structures requires some form of a contrast medium. As discussed earlier chest X-ray remains the primary screening tool for most thoracic trauma, with >90 % sensitivity in detecting mediastinal bleeding. In relationship to major vascular injuries, a normal supine chest X-ray is

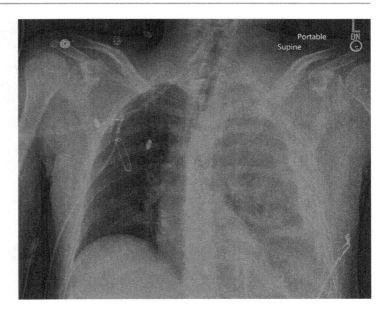

Fig. 3.24 Chest X-ray in a patient 5 days after a 15 ft fall showing a delayed left hemothorax. An X-ray 2 days prior had been normal

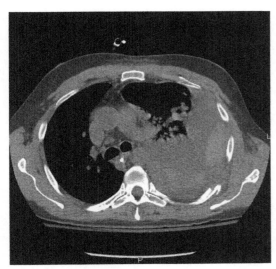

Fig. 3.25 Non contrast CT in the patient in Fig. 3.24 confirming the presence of a large heterogenous hemothorax with clot and a posterior rib fracture. He ultimately underwent a VATS procedure for clot evacuation

defined as one where the mediastinum is less than 8 cm as measured directly above the aortic knob. In patients with a low-risk mechanism, further imaging is not necessary. A low risk mechanism is often defined as a restrained (including three points and airbag deployment) person at <35 mph in a motor vehicle crash which excludes many of the patients who will

present after trauma [32]. In all others contrast enhanced chest CT should at least be considered as the next step in imaging to evaluate primarily for injury to the great vessels [33]. Findings of a widened mediastinum, loss of the aortic knob, left sided hemothorax, and first rib fractures all warrant further evaluation [34, 35]. Traumatic thoracic aortic injuries occur due to rapid deceleration and often tearing at tethering points such as the ligamentum arteriosus, the remnant of the fetal ductus arteriosus. These injuries have a very high associated morbidity and mortality, with historical reports of thoracic aortic trauma accounting for 20 % of all blunt chest trauma mortalities, with 85 % of these occurring at the scene [36]. Survivors have high rates of death if these injuries are not identified early, therefore screening practices are considered warranted even though the yield is often low. Prior to the broad availability of CT, angiography or trans esophageal echocardiogram were the imaging modalities necessary to identify these injuries. Current use of contrast enhanced MDCT have replaced these, and is now the recommended imaging choice as it allows for rapid detection of images. The use of intravenous iodine based contrast agents is necessary for this type of imaging, and institution specific protocols have been developed in major trauma centers to help

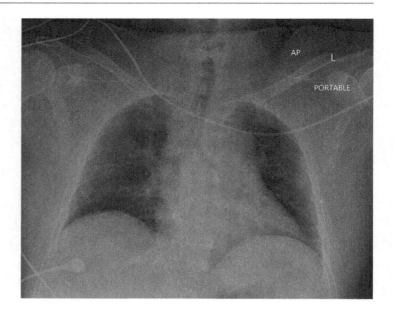

Fig. 3.26 Chest X-ray with findings concerning for aortic injury including widened mediastinum and loss of the aortic knob

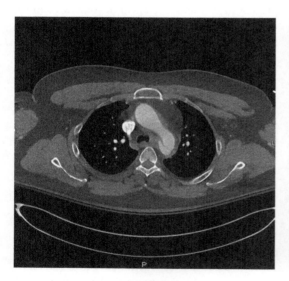

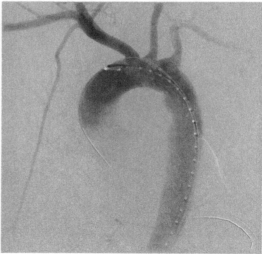

Fig. 3.27 Contrast enhanced CT of the chest in the patient in Fig. 3.26 showing a traumatic descending aortic arch dissection with surrounding hematoma

Fig. 3.28 Post stent placement angiogram in the patient in Figs. 3.26 and 3.27

determine the rates, volumes, and timing of the contrast delivery. Angiography now plays its primary role in the management of thoracic aortic injuries as dissections and disruptions can be managed using endovascular techniques [37] (Figs. 3.26, 3.27, and 3.28).

3.6.5 Diaphragm

Injuries to the diaphragm following blunt trauma are extremely rare and occur as a result of radial tears through the diaphragm. Large injuries can be seen on chest X-ray if intra-abdominal contents herniate into the chest cavity, however this is even more rare and more commonly detected with left sided injuries as the liver prevents herniation on the right. Placement of a nasal or oral gastric tube prior to chest X-ray can aid in the diagnosis if it is seen to terminate in the thoracic cavity. More subtle findings can be misinterpreted as an enlarged gastric bubble, elevated diaphragm, loculated hemopneumothorax, or subpulmonary hematoma. Helical CT has greatly improved the detection rates of these. With thin sliced images <3–5 mm between cuts, discontinuity of the diaphragm musculature, a rim sign (curling of the muscle around viscera) and visceral herniation can more easily be seen [38]. Another descriptive feature that has been identified is a "dependent viscera sign" which occurs when the posterior diaphragm no longer supports the abdominal viscera with the patient lying supine and they fall and lay directly along the posterior ribs [39]. If not identified on imaging, many are found intraoperatively, and all require some form of operative repair to prevent eventual formation of a symptomatic diaphragmatic hernia.

3.6.6 Injury Grading

The AIS grading of intrathoracic injuries can be difficult as some require both clinical and radiologic factors to assess their value. Hemothoraces, for example, can code for three points if simple and five if associated with tension physiology or air embolus. Likewise, pulmonary contusions are assigned a score of three if unilateral and four if bilateral independent of their radiographic size on CT and clinical impact. While these can almost be considered incidental findings, pericardial tamponade is attributed only three points and be a cause of death and thoracic spine injuries requiring operative stabilization often also only receiving a three. For this reason, trauma registrars should have discussions with the clinical providers when assigning some of these scores until adjustments in updated editions are developed.

3.7 Abdomen

Imaging of the abdomen also can occur early in the evaluative process. As with chest X-ray, a supine film of the pelvis should also be considered in the trauma patient with suspected multisystem injuries. The utility of the pelvic X-ray is in determining if a pelvic fracture is present that could cause vascular disruption with blood loss to explain hypotension such as open book or vertical shear injuries. Outside of this purpose, the yield of a plain abdominal X-ray for identifying injury to intra-abdominal organs is actually quite low. FAST exam does have an important role in the rapid assessment for intra-abdominal injury. In addition to the pericardial view previously discussed, three views are obtained in the abdomen; on the right of the hepatorenal fossa, on the left of the splenorenal fossa, and suprapubic to evaluate the pelvis or rectouterine pouch. FAST will detect intra-abdominal fluid, indicating either bleeding or hollow viscus rupture. It is a highly specific test for this purpose, reportedly >99 % with experienced users [40]. However, a negative FAST exam does not exclude an intra-abdominal injury, and it is recommended that the exam ideally be repeated in approximately 30 min if no other evaluation is planned as this increased the sensitivity from approximately 30 % to nearly 70 % to rule out the presence of fluid [41]. FAST also is limited in that it does not provide information as to the source of the fluid, whether it be bleeding from a solid organ injury or enteric contents from a hollow viscus.

In hemodynamically normal blunt trauma patients or those initially unstable but responsive to resuscitative efforts, the CT scan is usually the next stop after completion of the secondary survey. Unstable patients should not be sent for this evaluation as their presence in the scanner itself limits providers ability to adequately care for them and can compound an already life threatening situation. For appropriate patients, the decision to obtain abdominal imaging remains a moving target with many different criteria having been proposed to help decide who will benefit from evaluation while attempting to reduce the cost, radiation and contrast exposure to patients who have a very low likelihood of having injury. In patients with a high risk mechanism and GCS <9 or other evidence of neurologic injury, a CT should be highly considered as examination of the patient is unreliable. Those with abdominal or pelvic pain or tenderness, a "seatbelt sign" (abrasion due to vehicle restraint along lower abdomen) or gross hematuria should proceed with CT imaging. Other imaging or laboratory findings including lower rib fractures on chest X-ray, Pelvic fractures, or unexplained acidosis also warrant further work up. Using algorithms based on physiologic and lab parameters rather than purely based on mechanism can reduce the need for CT. However, even those outside of these protocols a mechanism of injury that includes a deceleration of >30 mph, a fall from more than standing height, or patient age >75 years should trigger the clinician to use sound clinical judgment when considering whether or not to obtain a CT scan [42]. From that point, management and additional imaging will typically proceed based on the CT findings.

3.7.1 Solid Organs

Intra-abdominal solid organ injuries include those to the Spleen, Liver, Pancreas, and Kidney. At present, the current management strategy of the majority of blunt solid organ injuries is non-operative with good rates of success. Studies have found correlation between the AAST OIS grade of the injuries and these success rates. At present >95 % of splenic injuries are successfully managed non operatively, with Grade 1 injuries approaching 100 % and the lowest success rates (though still 40 %) seen in Grade 4 and 5 injuries and those with active bleeding [43]. Angiography plays no real role in early screening or diagnosis, and the role of interventional radiology in management of solid organ injuries, especially to the spleen, is still controversial. Those that do have access consider IR as an adjunct to non-operative therapy as it may save a patient a trip to the operating room, and preservation of even just the short gastric arteries perfusion of the kidney can save the patient from becoming asplenic (Table 3.3).

Additional controversy exists in splenic injury patients non operatively managed is follow up imaging. Again, CT is the modality of choice and therefore weighing the risks and benefits can be patient specific [44]. One concern is the development of splenic pseudoaneurysms, with rates as high as 7 % reported, including Grade 1 and 2 injuries, when reimaged even as early as 24–48 h after injury [45]. These patients can require intervention due to risk of delayed rupture, with the role of angiography and embolization more widely accepted. Repeat imaging can be recommended in patients who plan to return to contact sports, putting them at higher risk of re-injury [46] (Figs. 3.29 and 3.30).

Table 3.3 AAST grading scale for spleen injuries

Grade	Criteria	
	Hematoma	Laceration
I	Subcapsular: nonexpanding, <10 % surface area	Capsular tear, nonbleeding, <1 cm deep
II	Subcapsular: nonexpanding, 10–50 % surface area	Capsular tear, active bleeding, 1–3 cm deep, no trabecular vessel involvement
	Intraparenchymal: nonexpanding <2 cm in diameter	
III	Subcapsular: >50 % surface area or expanding	>3 cm deep or trabecular vessel involvement
	Ruptured capsule with active bleeding	
	Intraparenchymal: >2 cm or expanding	
IV	Ruptured intraparenchymal with active bleeding	Segmental or hilar vessel involvement with >25 devascularization
V	Completely shattered	Hilar vascular injury that devascularizes spleen

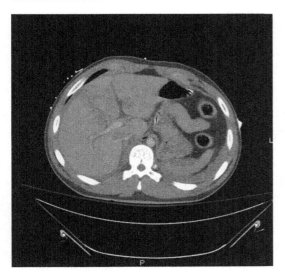

Fig. 3.29 Axial contrast enhanced CT image showing a grade IV liver laceration and a grade II splenic injury following motor vehicle crash

Table 3.4 AAST grading scale for liver injuries

	Criteria	
Grade	Hematoma	Laceration or other pattern
I	Subcapsular: nonexpanding, <10 % surface area	Capsular tear, nonbleeding, <1 cm deep
II	Subcapsular: nonexpanding, 10–50 % surface area,	Capsular tear, active bleeding, 1–3 cm deep, <10 cm Long
III	Subcapsular: >50 % surface area or expanding	>3 cm deep
	Ruptured capsule with active bleeding	
	Intraparenchymal: >10 cm	
IV	Ruptured intraparenchymal with active bleeding	Parenchymal disruption involving 25–50 % of a hepatic lobe
V	Completely shattered	Juxtahepatic venous injury (Vena cava, major hepatic vein)
VI	Hepatic avulsion	Hepatic avulsion

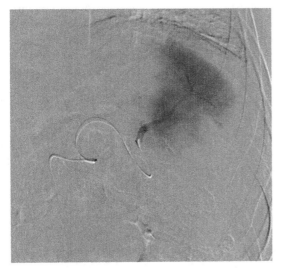

Fig. 3.30 Pre embolization angiogram in a patient with a delayed splenic bleed in the superior pole

Liver injuries have followed a similar non operative management pattern to spleen injuries, again with success rates >90 % [47, 48]. Higher OIS grade, evidence of active extravasation, and large volumes of hemoperitoneum are associated with increased rates of intervention, and in liver injuries, the role of angiography and embolization is expanded. While splenectomy is a reasonably straightforward operation that can be performed by most trained surgeon, surgical management of damage to the liver can be difficult. The primary approach is packing of the liver during damage control surgery. Major resections and hepatorrhaphy is often fraught with complications. Angiography has developed as a useful adjunct in patients requiring ongoing resuscitation. Uncontrolled bleeding from the arterial supply to the liver is more common due to the higher pressures compared to the venous or portal system, and catheter directed therapeutic interventions can be useful in identify these injuries. Need for intervention during hepatic angiography come with reported rates from 40 % to as high as 100 % [49]. Additionally, in patients who develop late complications following hepatic injury including hemobilia, arteriovenous fistulas, or persistent bile leaks, the role of interventional radiology to provide a non-operative therapy is invaluable (Table 3.4, Figs. 3.31 and 3.32).

Isolated pancreas injuries due to blunt trauma are extremely uncommon, with rates of less than 1.5 % [50]. Outcomes following major pancreatic injury requiring operative intervention, especially of the proximal pancreas, can be poor. The primary concern with pancreatic injury is

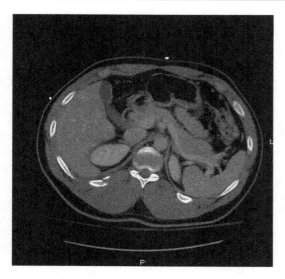

Fig. 3.31 Axial CT showing a subtle grade II liver laceration just anterior to the right kidney in segment 6/7 with associated posterior rib fractures of 10–12

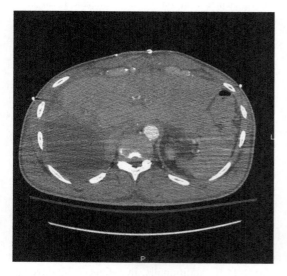

Fig. 3.32 Large biloma lateral to the liver 9 days following a non anatomic resection of the liver after a grade IV injury suffered in a motor vehicle crash. This was subsequently drained percutaneously under ultrasound guidance

disruption of the duct, and can be indicated by peripancreatic fluid on CT. Follow up imaging is warranted in the setting of suspected pancreatic injuries as they may not be immediately evident on CT imaging in the first 8 h after injury and normal serum amylase early on does not exclude

injury [51]. Additionally, late complications can develop including pseudocyst formation or evidence of ongoing duct leak. Pancreatic protocols for CT scans can be used that time image acquisition based on the delivery of contrast to better delineate the surround vascular structures. Additionally, MRCP and ERCP provide better imaging of the duct system, with ERCP allowing therapeutic options as well though often fraught with complications including development of strictures [52, 53]. CT guided drain placement, similar to the thought process used in managing complications of pancreatitis have been employed in the setting of pancreatic trauma with good outcomes.

Renovascular injuries and injury to the kidney parenchyma can be identified and graded similar to other solid organs on CT imaging and nonoperative management is again utilized with >95 % success. Injuries to the renal pelvis, ureter, and bladder may not be as evident on the standard contrast abdominal CT scan used for most trauma patients, even if delayed bladder images are obtained. Patients with gross hematuria or major pelvic fractures should be considered at high risk for injuries to these systems and warrant further evaluation. Contrast directed into these structures either via CT cystogram, Intravenous Pyelogram (IVP), or Retrograde Urethrogram (RUG) is the modalities of choice [54]. In patients with blood at the meatus of their urethra, evaluation for a urethral injury is MANDATORY prior to attempts at insertion of a Foley catheter. Insertion of an 8 F Foley into the urethra and gentle injection of contrast with a portable X-ray in the trauma bay or ED has been described for that purpose. Early consultation with a Urologist is also highly recommended (Table 3.5, Figs. 3.33 and 3.34).

3.7.2 Hollow Viscus

In the setting of blunt trauma, the most commonly injured hollow viscus is the small intestine due to the fact that it is subject to deceleration forces within the abdomen causing subsequent injury to tethered points including the ligament of Trietz and Ileocecal Valve with rates of 5 % in patients suffering blunt trauma quoted. Though uncommon,

Table 3.5 AAST grading scale for kidney injuries

Grade	Criteria			
	Contusion	Hematoma	Laceration	Vascular injury
I	Microscopic or gross hematuria Normal urologic studies	Subcapsular, nonexpanding	No parenchymal laceration	None
II	Not applicable	Nonexpanding perirenal, retroperitoneal	<1 cm deep without urinary extravasation	None
III	Not applicable	Not applicable	>1 cm deep without urinary extravasation	None
IV	Not applicable	Not applicable	Parenchymal laceration extending through cortex, medulla, and collecting system	Main renal artery or vein injury with contained hemorrhage
V	Not applicable	Not applicable	Completely shattered	Avulsion of renal hilum

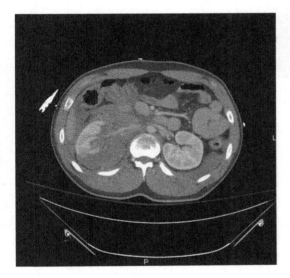

Fig. 3.33 CT revealing a grade IV right kidney injury in the same patient as Fig. 3.29

missed small bowel injuries have a high associated morbidity and mortality, and with the emergence of more non operative management of liver and spleen injuries, though low, the rates may be go up [55, 56]. CT traditionally had not been considered a particularly good method for detecting blunt small bowel injuries. However, with helical images provider small cuts, findings of extraluminal air or moderate (on 4–5 consecutive 5 mm cuts) or large (>6 consecutive cuts) amounts of free fluid in the absence of solid organ injury can have up to a 90 % positive predictive value in identifying clinically significant small bowel injuries [57, 58]. The use of oral contrast may also help in diagnosis if it is seen out of the lumen of the bowel, however, most protocols for trauma imaging at this point due not incorporate its use. However, it is still recommended that patients with potential for small bowel injury either undergo Diagnostic Peritoneal Lavage (DPL) or be admitted to undergo serial abdominal examinations after an oral food tolerance challenge for 24 h to exclude unrecognized injury.

3.7.3 Vascular

Injuries to the abdominal aorta and Inferior Vena Cava (IVC) are rare in blunt trauma, though retrohepatic IVC injuries carry an extremely high mortality. The primary concerns for abdominal vascular injuries that require intervention are mesenteric vessels and pelvic vessels. The diagnosis of clinically significant mesenteric injury using CT scan is difficult with active extravasation and significant bowel wall thickening being the primary finding. Mesenteric hematomas can also be seen but are rarely clinically significant.

Bleeding pelvic vessels can be an important clinical problem. As previously discussed, obtaining a pelvic X-Ray early in the resuscitation process of a hemodynamically unstable patient can be a helpful clinical decision making tool to

Fig. 3.34 Standard trauma CT with delayed bladder images. Though not the ideal imaging study, a bladder injury is evident explaining the gross hematuria noted in the foley catheter

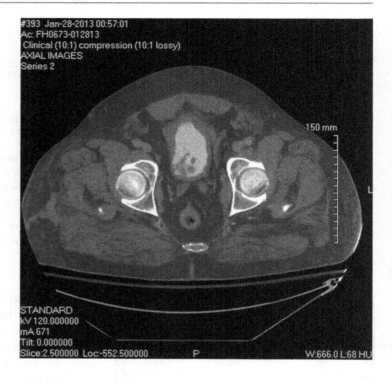

identify high risk pelvic fractures. If identified the first step should be placement of a pelvic binder or tightly bound sheet. Algorithms for the management and imaging of pelvic trauma have been proposed by several groups including the Western Trauma Association [59]. The algorithms typically include several points where imaging test help direct patient care. For hemodynamically stable patients, CT imaging is the modality of choice with evidence of a pelvic blush with no other blood losing injuries directing the patient as a potential candidate for Angiography and embolization. Hemodynamically unstable patients, should not undergo CT and therefore using either FAST or DPL to detect intra-abdominal fluid is often used to determine if they should go to the OR if positive or IR if negative for potentially therapeutic purposes (Fig. 3.35).

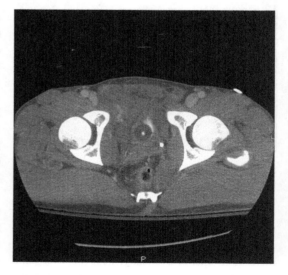

Fig. 3.35 CT showing a hematoma along the right pelvic side wall associated with a trauma pelvic diastasis and sacral fracture with a blush indicating ongoing extravasation

3.7.4 Incidentalomas

AS CT technology has improved and image resolution to detect submillimeter abnormalities, the term "Incidentaloma" has been coined in reference to incidental findings on a CT obtained to evaluate a completely unrelated process. Rates of these findings have been reported between 10 % and 50 % with the most common being pulmonary nodules (7 %) followed in decreasing order by

3 Medical Imaging and Injury Scaling in Trauma Biomechanics

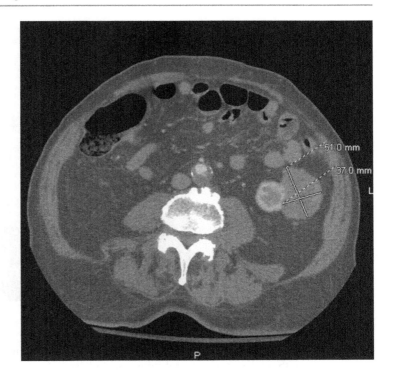

Fig. 3.36 Left renal exophytic mass concerning for malignancy in an 80 year old man who underwent CT after a fall. He had no injuries identified and the incidental findings of this scan were explained to the patient with results sent to his Primary Care Physician

mediastinal lymphadenopathy (6 %) adrenal lesions and hepatic lesions (4 %), and thyroid nodules (3 %) [60, 61]. Malignancy rates in these lesions are not insignificant, up to 30 % in some types. The key is ensuring appropriate follow up with a primary care provider as most of these lesions do not require immediate attention. Unfortunately follow up is already challenging in the trauma population, with the risks of inadequate follow up of these incidental findings truly unknown (Fig. 3.36).

3.7.5 Injury Grading

The AIS for abdominal injuries including solid organs vary widely. Solid organ injuries grades of the Liver, Kidney, and Spleen are correlated to the point assignments in AIS which is why documentation of the grade of injury either by the radiologist of the treating physician is helpful for appropriate coding and injury assignment by trauma registrars.

3.8 Extremity

Except for pelvic fractures and their role in vascular injury as previously discussed, the vast majority of orthopedic extremity injuries are not life threatening and therefore can become a secondary priority or overlooked during the primary and secondary surveys. Any trauma patient admitted to the hospital should have a Tertiary Survey, an additional head to toe physical exam, to identify and overlooked injuries or new areas of pain approximately 24 h after discharge. The rate of injury identification on tertiary survey is approximately 10 %, usually finding fractures in the distal extremities. The other role of the tertiary survey is to review all the final reads of imaging tests that have already been obtained to ensure no injuries were missed when the initial preliminary or provider interpretations occurred immediately after they were obtained.

Unlike many of the other injuries suffered by blunt trauma patients, where screening for serious though uncommon injuries is justified, imaging of

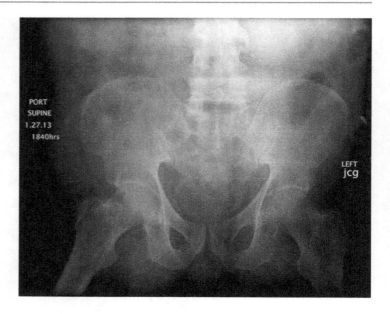

Fig. 3.37 Portable pelvic X-ray obtained in the same patient as Fig. 3.34 revealing superior and inferior pubic rami fractures consistent with a lateral compression injury

the extremities should be directed as a diagnostic test for areas with evidence of external injury or pain. Most patients will report pain in various joints or areas after blunt trauma due to soft tissue injuries and contusions making physical examination and appropriate part of the decision making process in patients who can participate in the exam. The remainder of this section will present imaging of pelvic fractures and some of the concepts that should be used to guide extremity trauma imaging. An inclusive discussion of all orthopedic injuries and patterns of injuries to the extremities and their management is beyond the scope of this chapter.

3.8.1 Pelvis

There are four commonly identified patterns of pelvic injury based on the type of force exerted on it during blunt trauma. The most common is a lateral compression fracture, typically occurring when the force is exerted along the sides of the patient (T-Bone mechanism of crash). These forces often cause fractures of the hip as well as superior and inferior rami fractures. These are not considered high risk for vascular injuries as they tend to make the pelvic space smaller, and not shear it open resulting in tearing forces in pelvic vessels. These are the most common fractures with frequency of 60–70 %. Imaging with plain films including inlet and outlet to assess the pelvic ring and Judet views (45° oblique views) to assess for acetabular fractures within the hip joint are usually sufficient, and the less often require operative stabilization (Fig. 3.37).

Anterior-posterior compression fractures are most commonly associated with the unstable, open book type fractures and account for 15–20 % of pelvic fractures. Forces directed in these directions (head on collision) can cause widening of the pubic symphysis and outward rotation related fractures along the sacroiliac joints. These do cause shearing forces on the pelvic vasculature and opening of the pelvic space. When identified, these patients can benefit from pelvic binder placement. Imaging with inlet, outlet and Judet films can be obtained first. 3D pelvic reconstructions can be requested at the time of abdominal CT to evaluate for other injuries, or obtained later to aid in operative planning. This has also led to increasing use of intra-operative fluoroscopy and CT imaging in the OR at the time of operation to insure proper alignment when percutaneous pinning techniques are used.

Fig. 3.38 Portable X-ray showing a complex fracture with open book and vertical shear components following MVC. This patient would benefit from a binder if hemodynamically unstable

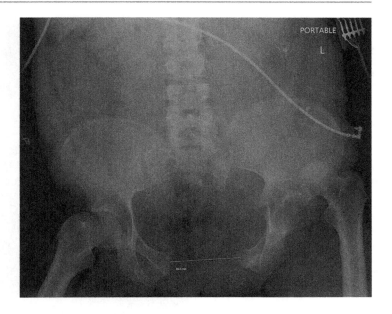

Vertical shear injuries occur due to high energy forces directed upward (fall or jump from height landing on feet or head on motorcycle crash) disrupting the ligamentous connections at the pubis and SI joints. These injuries are highly unstable and can result in significant pelvic bleeding often unresponsive to a pelvic binder. They occur least frequently, 5–15 % of the time. These warrant early orthopedic consultation. The final injury pattern is a complex pattern that occur as any combination of the other three types and carry with them similar stability and vascular injuries concerns. These injuries commonly require CT imaging with reconstruction for planning for complex operative stabilization (Figs. 3.38 and 3.39).

3.8.2 Extremities

Imaging of the extremities should be performed with X-Rays to assess for injury based on patient symptoms or deficits. When imaging an extremity for trauma, they can be thought of as joints and long bones when ordering X-rays. If the patient is complaining of long bone pain or has outward evidence of trauma, imaging the joint proximally and distally to the injury should be performed. This is especially true of radial/ulnar fractures and tibia/fibula fractures as the interosseous membranes between them can transmit forces to the adjacent joints resulting in injury. In patients who have fallen or jumped from >10 ft and landed on their feet these transmitted forces can extend up into the lower vertebrae of the spine and cause compression fractures, so identification of lower extremity injuries may warrant further imaging of the spine. If there is an open laceration overlying a suspected fracture, imaging is needed urgently as these fractures should be washed out and addressed in the operating room by an orthopedic surgeon expeditious due to the risk of development of osteomyelitis [62] (Figs. 3.40 and 3.41).

The roles of CT scans and MRI in extremity trauma have expanded over time. CT is now being used not infrequently in extremity trauma when intraarticluar involvement is suspected. Comminuted fractures of the acetabulum, distal femur, tibial plateau, and ankle are now being evaluated with CT scans, with or without 3D reconstructions as a means of operative planning. CT has also found a niche in evaluation of scapular fractures and scapulothoracic dissociation. MRI has limited use early after trauma due to the soft tissue swelling that occurs which decreases the resolution of the exam [63]. However, it is the best test to

Fig. 3.39 3D reconstruction of CT imaging of the patient in Fig. 3.38. A left acetabular fracture becomes more evident on CT as well

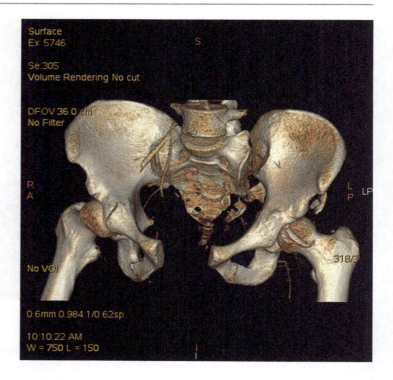

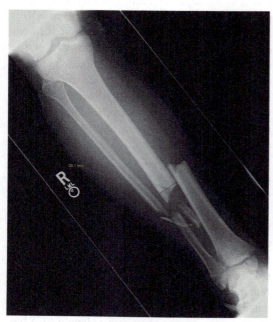

Fig. 3.40 X-ray imaging of comminuted right tibia and fibula fractures in a patient hit by a motor vehicle. There was an open wound in the medial calf classifying this as an open fracture

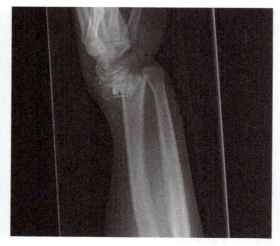

Fig. 3.41 X-ray showing a comminuted right distal radius fracture and ulnar styloid fracture. Additional imaging of the hand and wrist was obtained without evidence of carpal bone fractures and the patient was taken for fixation

evaluate for ligamentous joint injury, especially in the absence of operative bony injury when intraoperative evaluation will not occur. Therefore patients with suspected ligamentous injuries, such as to the

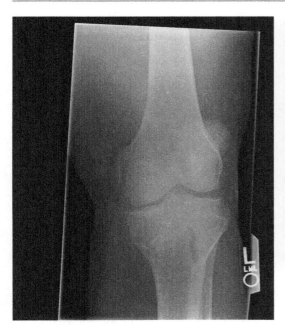

Fig. 3.42 X-ray showing a comminuted fracture of the proximal left tibia after motor vehicle crash. Additional views raised concern for further intra articular involvement

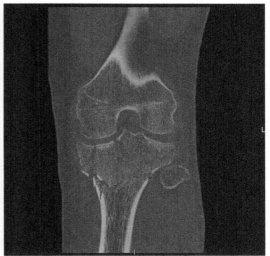

Fig. 3.43 Non contrast coronal CT view with bone windowing of the left knee of the patient in Fig. 3.42 showing the extent of the joint involvement. Distal femur fracture was excluded

Anterior Cruciate Ligament or Rotator Cuff, will often be discharged with weight bearing restrictions and planned follow up in several weeks. If the exam or symptoms remain concerning, MRI can be performed on an elective basis at that time when the swelling has subsided (Figs. 3.42 and 3.43).

3.8.3 Injury Grading

Assignment of AIS scores to the Extremity region includes injuries to the bony pelvis and upper and lower extremities. Major injuries to the soft tissue, such as degloving injuries, and neurovascular structures are included in this region. The values assigned to bony fractures increase sequentially for open fractures and fractures with associated neurovascular injuries.

3.9 External

The external AIS category refers to any injury to the skin and subcutaneous tissues in the absence of injury to underlying structures. These do not typically contribute significantly to the overall ISS except in the case of burn patients without any or with limited other injuries. There is little role for specific imaging of injuries isolated to the skin though abrasions, contusions, and other superficial injuries are helpful in directing imaging of specific underling structures.

3.10 Conclusion

Due to the complex nature of the patterns of injuries seen in blunt trauma and high associated morbidity and mortality of many missed injuries the role of imaging will only continue to grow. Since the first uses of CT in the evaluation of trauma patients in the 1980s, it has rapidly advanced to become a part of the standard of care for evaluating the patients and the basis of many practice management guidelines. Unfortunately, this is not without its downside. While CT provide fast, reliable, high resolution imaging that is improving in quality almost every year, its use has become so prevalent that new issues are evolving in deciding when it should be used. Contrast induced nephropathy is the third leading cause of acute kidney injury in the United States [64].

Especially at risk are the increasing number of elderly trauma patients, more likely to have underlying renal dysfunction, and whom most imaging protocols err on the side of more liberal use of CT scans than not. For younger patients, the risks of exposure to ionizing radiation are real, and the exposure rates have been rising with wider adoption of CT use in many settings, especially the emergency department and trauma [65, 66]. The "Biologic Effects of Ionizing Radiation" committee have published age and sex based models related to cancer risks from radiation exposure and have estimated that, for example 30 year old females receiving a typical chest CT yields 10 new malignancies per 10,000 scans [67]. For these reasons, improving technology will be a major focus for years to come. Faster data acquisition times from non-radiating modalities such as MRI, image gating and improved detectors allowing less radiation with CTs are just the beginning [68]. Additionally further uniting of the IR and surgical realms are leading to hybrid operating rooms where advanced endovascular procedures can be performed during initial exploratory laparotomies to ideally reducing cost and delays in care. Imaging and trauma will forever be intertwined, and will continue to grow to provide the best care for these trauma patients.

References

1. Moore EE, Cogbill TH, Malangoni M, Jurkovich GJ, Champion HR (2013) Scaling system for organ specific injuries. http://www.aast.org/Library/TraumaTools/InjuryScoringScales.aspx. Accessed 11 Feb 2013
2. Baker SP, O'Neill B, Haddon W Jr, Long WB (1974) The injury severity score: a method for describing patients with multiple injuries and evaluating emergency care. J Trauma 14(3):187–196
3. American College of Surgeons Committee on Trauma (2008) Head trauma. In: ATLS, 8th edn. American College of Surgeons, Chicago, pp 131–155
4. Teasdale G, Jennett B (1976) Assessment and prognosis of coma after head injury. Acta Neurochir (Wien) 34(1–4):45–55
5. Stiell IG, Clement CM, Rowe BH, Schull MJ, Brison R, Cass D et al (2005) Comparison of the Canadian CT head rule and the New Orleans criteria in patients with minor head injury. JAMA 294(12):1511–1518
6. Stiell IG, Lesiuk H, Wells GA, McKnight RD, Brison R, Clement C et al (2001) The Canadian CT head rule study for patients with minor head injury: rationale, objectives, and methodology for phase I (derivation). Ann Emerg Med 38(2):160–169
7. Duckworth JL, Stevens RD (2010) Imaging brain trauma. Curr Opin Crit Care 16(2):92–97
8. Jagoda AS et al (2008) Clinical policy: neuro imaging and decision making in adult mild traumatic brain injury in the acute setting. Ann Emerg Med 52(6):714–748
9. Rosenfeld JV, Maas AI, Bragge P, Morganti-Kossmann MC, Manley GT, Gruen RL (2012) Early management of severe traumatic brain injury. Lancet 380(9847):1088–1098
10. Li XY, Feng DF (2009) Diffuse axonal injury: novel insights into detection and treatment. J Clin Neurosci 16(5):614–619
11. Denis F (1984) Spinal instability as defined by the three-column spine concept in acute spinal trauma. Clin Orthop Relat Res 189:65–76
12. Como JJ, Diaz JJ, Dunham CM, Chiu WC, Duane TM, Capella JM et al (2009) Practice management guidelines for identification of cervical spine injuries following trauma: update from the eastern association for the surgery of trauma practice management guidelines committee. J Trauma 67(3):651–659
13. Inaba K, DuBose JJ, Barmparas G, Barbarino R, Reddy S, Talving P et al (2011) Clinical examination is insufficient to rule out thoracolumbar spine injuries. J Trauma 70(1):174–179
14. Sixta S, Moore FO, Ditillo MF, Fox AD, Garcia AJ, Holena D et al (2012) Screening for thoracolumbar spinal injuries in blunt trauma: an eastern association for the surgery of trauma practice management guideline. J Trauma Acute Care Surg 73(5 Suppl 4):S326–S332
15. Longo UG, Denaro L, Campi S, Maffulli N, Denaro V (2010) Upper cervical spine injuries: indications and limits of the conservative management in halo vest. A systematic review of efficacy and safety. Injury 41(11):1127–1135
16. Izzo R, Guarnieri G, Guglielmi G, Muto M (2013) Biomechanics of the spine part I: spinal stability. Eur J Radiol 82(1):118–126
17. Izzo R, Guarnieri G, Guglielmi G, Muto M (2013) Biomechanics of the spine part II: spinal instability. Eur J Radiol 82(1):127–138
18. Kaye D, Brasel KJ, Neideen T, Weigelt JA (2011) Screening for blunt cerebrovascular injuries is cost-effective. J Trauma 70(5):1051–1056; discussion 1056–1057
19. Biffl WL, Moore EE, Offner PJ, Brega KE, Franciose RJ, Burch JM (1999) Blunt carotid arterial injuries: implications of a new grading scale. J Trauma 47(5):845–853
20. American College of Surgeons Committee on Trauma (2008) Chest trauma. In: ATLS, 8th edn. American College of Surgeons, Chicago, pp 96–100

21. Clancy K, Velopulos C, Bilaniuk JW, Collier B, Crowley W, Kurek S et al (2012) Screening for blunt cardiac injury: an eastern association for the surgery of trauma practice management guideline. J Trauma Acute Care Surg 73(5 Suppl 4):S301–S306
22. Athanassiadi K, Gerazounis M, Moustardas M, Metaxas E (2002) Sternal fractures: retrospective analysis of 100 cases. World J Surg 26(10):1243–1246
23. Malbranque G, Serfaty JM, Himbert D, Steg PG, Laissy JP (2011) Myocardial infarction after blunt chest trauma: usefulness of cardiac ECG-gated CT and MRI for positive and aetiologic diagnosis. Emerg Radiol 18(3):271–274
24. Garcia-Fernandez MA, Lopez-Perez JM, Perez-Castellano N, Quero LF, Virgos-Lamela A, Otero-Ferreiro A et al (1998) Role of transesophageal echocardiography in the assessment of patients with blunt chest trauma: correlation of echocardiographic findings with the electrocardiogram and creatine kinase monoclonal antibody measurements. Am Heart J 135(3):476–481
25. Livingston DH, Shogan B, John P, Lavery RF (2008) CT diagnosis of rib fractures and the prediction of acute respiratory failure. J Trauma 64(4):905–911
26. Bhavnagri SJ, Mohammed TL (2009) When and how to image a suspected broken rib. Cleve Clin J Med 76(5):309–314
27. Moore MA, Wallace EC, Westra SJ (2011) Chest trauma in children: current imaging guidelines and techniques. Radiol Clin North Am 49(5):949–968
28. Ball CG, Kirkpatrick AW, Laupland KB, Fox DI, Nicolaou S, Anderson IB et al (2005) Incidence, risk factors, and outcomes for occult pneumothoraces in victims of major trauma. J Trauma 59(4):917–924; discussion 924–925
29. Ball CG, Kirkpatrick AW, Laupland KB, Fox DL, Litvinchuk S, Dyer DM et al (2005) Factors related to the failure of radiographic recognition of occult post-traumatic pneumothoraces. Am J Surg 189(5):541–546; discussion 546
30. Wilkerson RG, Stone MB (2010) Sensitivity of bedside ultrasound and supine anteroposterior chest radiographs for the identification of pneumothorax after blunt trauma. Acad Emerg Med 17(1):11–17
31. Ding W, Shen Y, Yang J, He X, Zhang M (2011) Diagnosis of pneumothorax by radiography and ultrasonography: a meta-analysis. Chest 140(4):859–866
32. Streck CJ Jr, Jewett BM, Wahlquist AH, Gutierrez PS, Russell WS (2012) Evaluation for intra-abdominal injury in children after blunt torso trauma: can we reduce unnecessary abdominal computed tomography by utilizing a clinical prediction model? J Trauma Acute Care Surg 73(2):371–376; discussion 376
33. Demehri S, Rybicki FJ, Desjardins B, Fan CM, Flamm SD, Francois CJ et al (2012) ACR appropriateness criteria((R)) blunt chest trauma–suspected aortic injury. Emerg Radiol 19(4):287–292
34. Gundry SR, Williams S, Burney RE et al (1983) Indications for aortography: radiography after blunt chest trauma; a reassessment of the radiographic findings associated with traumatic rupture of the aorta. Invest Radiol 18:230–237
35. Gupta A, Jamshidi M, Rubin JR (1997) Traumatic first rib fracture: is angiography necessary? A review of 730 cases. Cardiovasc Surg 5(1):48–53
36. Parmley LF, Mattingly TW, Manion TW et al (1958) Nonpenetrating traumatic injury of the aorta. Circulation XVII:1086
37. Demetriades D, Velmahos GC, Scalea TM, Jurkovich GJ, Karmy-Jones R, Teixeira PG et al (2008) Diagnosis and treatment of blunt thoracic aortic injuries: changing perspectives. J Trauma 64(6):1415–1418; discussion 1418–1419
38. Killeen KL, Mirvis SE, Shanmuganathan K (1999) Helical CT of diaphragmatic rupture caused by blunt trauma. AJR Am J Roentgenol 173(6):1611–1616
39. Bergin D, Ennis R, Keogh C, Fenlon HM, Murray JG (2001) The "dependent viscera" sign in CT diagnosis of blunt traumatic diaphragmatic rupture. AJR Am J Roentgenol 177(5):1137–1140
40. American College of Surgeons Committee on Trauma (2008) Abdominal and pelvic trauma. In: ATLS, 8th edn. American College of Surgeons, Chicago, pp 118–121
41. Blackbourne LH, Soffer D, McKenney M, Amortegui J, Schulman CI, Crookes B et al (2004) Secondary ultrasound examination increases the sensitivity of the FAST exam in blunt trauma. J Trauma 57(5):934–938
42. Brasel KJ, Nirula R (2005) What mechanism justifies abdominal evaluation in motor vehicle crashes? J Trauma 59(5):1057–1061
43. Caddeddu M, Garnett A, Al-Anezi K et al (2006) Management of spleen injuries in the adult trauma population: a ten year experience. Can J Surg 49:386–390
44. Allins A, Ho T, Nguyen TH et al (1996) Limited value of routine follow-up CT scans in nonoperative management of blunt liver and spleen injuries. Am Surg 62:883–886
45. Thaemert BC, Cogbill TC, Lambert PJ (1997) Nonoperative management of splenic injury: are follow-up computed tomographic scans of any value? J Trauma 43:748–751
46. Lyass S, Sela T, Lebensart PD et al (2001) Follow-up imaging studies of blunt splenic injury: do they influence management? Isr Med Assoc J 3:731–733
47. Christmas AB, Wilson AK, Manning B et al (2005) Selective management of blunt hepatic injuries including non-operative management is a safe and effective strategy. Surgery 138:606–610; discussion 610–611
48. Velmahos GC, Toutouzas K, Radin R et al (2003) High success with non-operative management of blunt hepatic trauma: the liver is a sturdy organ. Arch Surg 138:475–480; discussion 480–481
49. Johnson JW, Gracias VH, Gupta R et al (2002) Hepatic angiography in patients undergoing damage control laparotomy. J Trauma 52:1102–1106

50. Phelan HA, Velmahos GC, Jurkovich GJ, Friese RS, Minei JP, Menaker JA et al (2009) An evaluation of multidetector computed tomography in detecting pancreatic injury: results of a multicenter AAST study. J Trauma 66(3):641–646; discussion 646–647
51. Buccimazza I, Thomson SR, Anderson F et al (2006) Isolated main pancreatic duct injuries. Spectrum and management. Am J Surg 191:448–452
52. Holalkere NS, Soto J (2012) Imaging of miscellaneous pancreatic pathology (trauma, transplant, infections, and deposition). Radiol Clin North Am 50(3):515–528
53. Lin B, Liu N, Fang J et al (2006) Long-term results of endoscopic stent in the management of blunt major pancreatic duct injury. Surg Endosc 20(10):1551–1555
54. Avery LL, Scheinfeld MH (2012) Imaging of male pelvic trauma. Radiol Clin North Am 50(6):1201–1217
55. Joseph DK, Kunac A, Kinler RL, Staff I, Butler KL (2013) Diagnosing blunt hollow viscus injury: is computed tomography the answer? Am J Surg 205(4):414–418
56. Miller PR, Croce MA, Bee TK, Malhotra AK, Fabian TC (2002) Associated injuries in blunt solid organ trauma: implications for missed injury in nonoperative management. J Trauma 53(2):238–242; discussion 242–244
57. Brasel KJ, Olson CJ, Stafford RE, Johnson TJ (1998) Incidence and significance of free fluid on abdominal computed tomographic scan in blunt trauma. J Trauma 44(5):889–892
58. LeBedis CA, Anderson SW, Soto JA (2012) CT imaging of blunt traumatic bowel and mesenteric injuries. Radiol Clin North Am 50(1):123–136
59. Davis JW, Moore FA, McIntyre RC Jr, Cocanour CS, Moore EE, West MA (2008) Western trauma association critical decisions in trauma: management of pelvic fracture with hemodynamic instability. J Trauma 65(5):1012–1015
60. Barett T, Schierling M, Zhou C et al (2009) Prevalence of incidental findings in trauma patients detected by computed tomography imaging. Am J Emerg Med 27:428–435
61. Paluska TR, Sise MJ, Sack DI et al (2007) Incidental CT findings in trauma patients: incidence and implications for care of the injured. J Trauma 62(1):157–161
62. Scalea TM, DuBose J, Moore EE, West M, Moore FA, McIntyre R et al (2012) Western trauma association critical decisions in trauma: management of the mangled extremity. J Trauma Acute Care Surg 72(1):86–93
63. van Dijk CN, de Leeuw PA (2007) Imaging from an orthopaedic point of view. What the orthopaedic surgeon expects from the radiologist? Eur J Radiol 62(1):2–5
64. Venkataraman R (2008) Can we prevent acute kidney injury? Crit Care Med 36:S166–S171
65. Ahmadinia K, Smucker JB, Nash CL, Vallier HA (2012) Radiation exposure has increased in trauma patients over time. J Trauma Acute Care Surg 72(2):410–415
66. Tien H, Tremblay L, Rizoli S et al (2007) Radiation exposure from diagnostic imaging in severely injured trauma patients. J Trauma 62:151–156
67. Committee to Assess Health Risks from Exposure to Low Levels of Ionizing Radiation, National Research Council (2005) Health risks from exposure to low levels of ionizing radiation: BEIR VII—phase 2. National Academies Press, Washington, DC
68. Tepper B, Brice JH, Hobgood CD (2012) Evaluation of radiation exposure to pediatric trauma patients. J Emerg Med 44(3):646–652

Anthropomorphic Test Devices and Injury Risk Assessments

Harold J. Mertz and Annette L. Irwin

Abstract

Anthropomorphic test devices (ATDs), commonly referred to as dummies, are mechanical surrogates of the human that are used by the automotive industry to evaluate the occupant protection potential of various types of restraint systems in simulated collisions of new vehicle designs. Dummies are classified according to size, age, sex and impact direction. There are adult male and female dummies of different sizes, and child dummies that represent different ages. These dummies are used to assess occupant protection in frontal, side, rear and rollover collision simulations. The midsize adult male dummy is the most utilized size in automotive restraint testing. It approximates the median height and weight of the 50th percentile adult male population. The heights and weights of the small female and large male adult dummies are approximately those of the 5th percentile female and the 95th percentile male, respectively. The adult dummies are referred to as small female, midsize male and large male to avoid debates about whether the dimensions are consistent with the latest published percentile classifications. Child dummies have the median heights and weights of children of the specific age groups that they represent without regard to sex. They are referred to by age.

Current ATDs are designed to be biofidelic; that is, they mimic pertinent human physical characteristics such as size, shape, mass, stiffness, and energy absorption and dissipation, so that their mechanical responses simu-

H.J. Mertz, Ph.D. (retired) (✉)
General Motors Corporation,
Harper Woods, MI, USA
e-mail: bud.mertz@yahoo.com

A.L. Irwin, Ph.D.
General Motors Company, Warren, MI, USA
e-mail: Annette.L.Irwin@gm.com

late corresponding human responses of trajectory, velocity, acceleration, deformation, and articulation when the dummies are exposed to prescribed simulated collision conditions. They are instrumented with transducers that measure accelerations, deformations, and loads of various body parts. Analyses of these measurements are used to assess the efficacy of restraint system designs. Limits placed on these dummy measurements are called Injury Assessment Reference Values (IARVs). The restraint system design goal is for the dummy's responses to be at or below their corresponding IARVs for all the test conditions being evaluated.

This chapter will discuss (1) the pertinent characteristics of various dummies that have been used by the auto industry over the last 50 plus years, (2) the IARVs for the various measurements made with these dummies, and (3) the Injury Risk Curves that served as the bases for many of the IARVs.

4.1 ATDs for Frontal Crash Testing

4.1.1 Early ATDs

The first crash test dummy used by the domestic automobile industry for restraint system testing was Sierra Sam, a 95th percentile adult male dummy that was developed by Sierra Engineering in 1949 for ejection seat testing by the U.S. Air Force. In the mid-1960s, Alderson Research Laboratories produced their Very Important People (VIP) family of dummies which consisted of a small female, a midsize male and a large male dummy. In the same time period, Sierra Engineering produced a midsize male dummy, Sierra Stan and a small female dummy, Sierra Susie. While these dummies mimicked human shape and weight and were quite durable, they had a number of deficiencies. First, their design specifications were not sufficient to assure that two dummies of the same size made by a given manufacturer or by different manufacturers would give the same responses when tested under the same conditions. In addition, the dummies lacked humanlike stiffness in the important areas such as the head, neck, thorax, and knee, and they were not extensively instrumented to measure responses that could be associated with all the pertinent injury concerns.

4.1.2 Hybrid II Dummy Family

The most used of the early frontal impact dummies was the Hybrid II midsize adult male dummy, which was developed by General Motors (GM) in 1972 to assess the integrity of lap/shoulder belt systems [1–3]. This dummy mimicked the size, shape, mass, and ranges of arm and leg motion of the 50th percentile adult male. It was instrumented to measure the orthogonal linear accelerations of the center of gravity of its head and a point prescribed in its "thoracic spine." Its femurs were instrumented to measure axial-shaft loading. The dummy was quite durable and gave repeatable responses (coefficients of variation of 10 % or less) when subjected to repeat tests. In 1973, the GM Hybrid II became the first dummy specified in the Federal Motor Vehicle Safety Standard 208 (FMVSS 208) for compliance testing of vehicles equipped with passive restraints [4]. It remained as a compliance dummy until September 1997.

In addition to the midsize male, Humanoid Systems developed Hybrid II-type small adult female and large adult male dummies by scaling the shapes and features of the midsize male [1–3]. These dummies had instrumentation capabilities similar to those of the Hybrid II midsize adult male. Also, a 3-year-old and a 6-year-old child dummy were developed. These two child dummies, along with the three sizes of adult dummies, became known as the Hybrid II Dummy Family.

The Hybrid II dummy family had two major deficiencies that limited their usefulness in assessing the efficacy of restraint systems: (1) they lacked humanlike response stiffness for their heads, necks, thoraxes, and knees, and (2) they were sparsely instrumented.

4.1.3 Hybrid III Dummy Family

In 1974, GM initiated a research program to define and develop a biofidelic midsize adult male dummy, called Hybrid III, to replace the GM Hybrid II dummy. The Hybrid III [6, 8] was designed to mimic human responses for forehead impacts, fore and aft neck bending, distributed sternal impacts, and knee impacts. The instrumentation of the dummy is quite extensive and is noted in Table 4.1 [3]. A deformable face [6, 7] and a deformable abdomen [6, 8] are available that can be used to assess the risk of facial bone fracture and abdominal injury due to lap belt submarining, respectively. Since its creation in 1976, the Hybrid III midsize male dummy has undergone design changes to improve the biofidelity of its hips [9] and ankle joints [6].

Because of its excellent biofidelity and measurement capability, GM petitioned the National Highway Traffic Safety Administration (NHTSA) in 1983 to allow the use of the Hybrid III midsize adult male dummy as an alternative test device to the Hybrid II for FMVSS 208 compliance testing of passive restraints. Its use was allowed in 1986. In 1990, GM filed a second petition requesting that the Hybrid II dummy be deleted from FMVSS 208 compliance testing. NHTSA deleted the Hybrid II in September 1997, making the Hybrid III the only midsize adult male dummy specified for regulatory frontal restraint evaluation

Table 4.1 Instrumentation for Hybrid III family of dummies [3]

	3 year	6 year	10 year	Small female	Midsize male	Large male
Head						
Accel. (A_x, A_y, A_z)	Yes	Yes	Yes	Yes	Yes	Yes
Neck						
Head/C1 (F_x, F_y, F_z, M_x, M_y, M_z)	Yes	Yes	Yes	Yes	Yes	Yes
C7/T1 (F_x, F_y, F_z, M_x, M_y, M_z)	Yes	Yes	Yes	Yes	Yes	Yes
Shoulder						
Clavicle force (F_x, F_z)	Yes	No	Yes	Yes	Yes	No
Thorax						
Spine accel. (A_x, A_y, A_z)	Yes	Yes	Yes	Yes	Yes	Yes
Spine force (F_x, F_y, F_z, M_x, M_y)	No	No	No	Yes	Yes	Yes
Sternum defl. (δ_x)	Yes	Yes	Yes	Yes	Yes	Yes
Sternum accel. (A_x)	Yes	Yes	Yes	Yes	Yes	No
Abdomen						
Lumbar (F_x, F_y, F_z, M_x, M_y, M_z)	Yes	Yes	Yes	Yes	Yes	Yes
Pelvis						
Accel. (A_x, A_y, A_z)	Yes	Yes	Yes	Yes	Yes	Yes
ASIS (F_x)	Yes	Yes	Yes	F_x, M_y	Load Bolt	Yes
Pubic loads (F_x, F_z)	Yes	No	No	No	No	No
Lower extremities						
Femur (F_x, F_y, F_z, M_x, M_y, M_z)	No	Yes	Yes	Yes	Yes	Yes
Tibia-Femur displ. (δ_x)	No	No	No	Yes	Yes	Yes
Knee clevis (F_z)	No	No	No	Yes	Yes	Yes
Tibia (F_x, F_y, F_z, M_x, M_y, M_z)	No	No	No	Yes	Yes	Yes

Table 4.2 Key dimensions and weights for various sizes of dummies [11, 12, 14]

	Infant dummies			Child dummies			Adult dummies		
	6-month	12-month	18-month	3-year	6-year	10-year	Small female	Mid male	Large male
	Key dimensions – mm								
Erect sitting height	439	480	505	546	635	719	812	907	971
Buttock to knee	170	198	221	284	381	455	521	589	638
Knee to floor	125[a]	155[a]	173[a]	221	358	432	464	544	595
Shoulder to elbow	130	150	160	193	234	287	305	366	382
Elbow to finger tip	175	198	213	254	310	366	400	465	502
Standing height	671	747	813	953	1,168	1,374	1,513	1,751	1,864
	Weight – kg								
Head	2.11	2.47	2.73	3.05	3.47	3.66	3.67	4.54	4.94
Neck	0.30	0.32	0.32	0.40	0.50	0.65	0.77	1.54	2.04
Torso	4.02	4.99	5.75	7.48	10.74	16.66	24.09	40.23	52.89
Upper extremity	0.95	1.18	1.31	1.79	1.87	2.96	4.81	8.53	10.91
Lower extremity	0.42	0.74	1.09	1.79	4.28	8.47	13.38	23.36	31.72
Total	7.80	9.70	11.20	14.51	20.86	32.40	46.72	78.20	102.5

[a]Corrected from prior publication

throughout the world. Because the Hybrid III is specified in worldwide regulations, all design changes have to be approved by the various regulatory bodies.

In 1987, the Centers for Disease Control (CDC) awarded a grant to Richard Stalnaker of the Ohio State University (OSU) to develop a Hybrid III-based dummy family. To support this effort, the Mechanical Human Simulation Subcommittee of the Human Biomechanics and Simulation Standards Committee of the Society of Automotive Engineers (SAE) formed a task group of biomechanics, test dummy, transducer, and restraint system experts. They defined the specifications for a small adult (5th percentile) female dummy, a large adult (95th percentile) male dummy, and a 6-year-old child dummy having the same level of biofidelity and measurement capacity as the Hybrid III midsize adult male dummy. Key body segment lengths and weights were defined based on anthropometry data for the U.S. population, Table 4.2 [59]. Biofidelity response requirements for the head, neck, thorax, and knee of each size of dummy were scaled from the respective biofidelity requirements of the Hybrid III midsize adult male dummy [10, 11]. The instrumentation for each dummy is listed in Table 4.1. The dummies became commercially available in 1991.

In 1992, the SAE Hybrid III Dummy Family Task Group initiated a program to develop a Hybrid III 3-year-old child dummy. Again, this dummy was designed to have the same level of biofidelity and measurement capacity as the other Hybrid III-type dummies, except for the knee impact requirement and the leg instrumentation [11]. These items were omitted from the design requirements since knee impact is an unlikely event for properly restrained 3-year-old child. Since this dummy was to be designed to replace the GM 3-year-old airbag dummy [12] for evaluating unrestrained child interactions with deploying passenger airbags as well as to be used to assess the efficacy of child restraints, its sternum was instrumented to measure its response to the punch-out forces of deploying passenger airbags. Its instrumentation is summarized in Table 4.1. The dummy became commercially available in 1997.

At its June 15, 2000 meeting, the SAE Hybrid III Dummy Family Task Group initiated a program to develop a Hybrid III 10-year-old child dummy [13]. The dummy was designed to have the same level of biofidelity and measurement capacity as the Hybrid III 6-year-old child dummy. The dummy was designed so it could assume a "normal" or a "slouched" seating posture. Its instrumentation is given in Table 4.1. The dummy was approved for production build at the July 26, 2001 Task Group meeting.

The SAE Hybrid III Dummy Task Group has documented the designs of the Hybrid III small female, large male, 3-year-old, 6-year-old, and 10-year-old dummies so that they could be incorporated into Part 572 of the U.S. Transportation Regulations, replacing their respective Hybrid II dummies. The dummies are well defined, durable and give highly repeatable and reproducible results. To date, all but the large male dummy have been incorporated into Part 572. No large male dummy is specified in Part 572 since the FMVSS 208 Passive Restraint Regulation does not require testing with that size dummy.

4.1.4 CRABI Infant Dummies

In 1990, the Child Restraint Air Bag Interactions (CRABI) Task Group of the SAE Human Mechanical Simulation Subcommittee was convened. The specific purpose of the CRABI Task Group was to develop instrumented 6-month-old, 12-month-old, and 18-month-old infant dummies to be used to assess the injury potential associated with the interactions of deploying passenger airbags with rearward-facing child restraints if they are placed in the front seat of vehicles. These three dummies are called CRABI infant dummies. Prototypes of these dummies were available in 1991. The sizes and weights of the dummies were based on anthropometry studies of the U.S. child population, Table 4.2. Biofidelity response requirements were defined for the head and neck [11]. The instrumentation used with these dummies is given in Table 4.3. The designs of these dummies were documented by the SAE so that they could be incorporated into Part 572 of the U.S. Transportation Regulations. To date, only the CRABI 12-month-old has been incorporated into Part 572.

Table 4.3 Instrumentation of CRABI dummies [3, 11]

	6 month	12 month	18 month
Head			
Accel. (A_x, A_y, A_z)	Yes	Yes	Yes
Neck			
Head/C1 (F_x, F_y, F_z, M_x, M_y, M_z)	Yes	Yes	Yes
C7/T1 (F_x, F_y, F_z, M_x, M_y, M_z)	Yes	Yes	Yes
Shoulder			
Force (F_x, F_z)	No	Yes	Yes
Thorax			
Spine accel. (A_x, A_y, A_z)	Yes	Yes	Yes
Abdomen			
Lumbar (F_x, F_y, F_z, M_x, M_y, M_z)	Yes	Yes	Yes
Pelvis			
Accel. (A_x, A_y, A_z)	Yes	Yes	Yes
Pubic loads (F_x, F_z)	No	Yes	Yes

4.1.5 THOR

In the early 1990s, NHTSA initiated a program to develop an advanced mid-size adult male, frontal impact dummy. Under a NHTSA contract, UMTRI (University of Michigan Transportation Research Institute) developed the TAD-50 M (Trauma Assessment Device) which was a Hybrid III midsize male dummy with modifications made to its thorax, abdomen and pelvis [14–17]. Simultaneously, VRTC (NHTSA's Vehicle Research and Test Center) was working on new neck and lower extremities designs. The dummy parts from these two projects were merged into an advanced, midsize adult male dummy which NHTSA called THOR (Test Device for Human Occupant Restraint). Early versions of THOR included the THOR alpha (2001), the THOR-FT (2004) and the THOR-NT (2005). The distinguishing features of THOR are its two-segment thoracic spine, a rib cage shaped like a human's, and passive muscle force simulations for the neck and the ankle structures [18, 19].

In 2004, the SAE THOR Evaluation Task Group was formed with the goal of identifying and recommending dummy enhancements [20]. The task group was disbanded in 2013.

Table 4.4 Key dimensions and weights for Q dummies [22]

	Q0	Q1	Q1.5	Q3	Q6
	Key dimensions – mm				
Chest depth	90	117	–	145.5	141
Shoulder width	141	227	227	259	305
Hip width	98	191	194	200	223
Buttock to popliteus	–	161	185	253	299
Sitting height	354	479	499	544	601
Standing height	–	740	800	985	1,143
	Weights – kg				
Head & neck	1.1	2.41	2.8	3.17	3.94
Torso	1.5	4.46	5.04	6.40	9.62
Upper extremity	0.25	0.89	1.20	1.48	2.49
Lower extremity	0.55	1.82	2.06	3.54	6.90
Total	3.4	9.6	11.1	14.6	22.9

Table 4.5 Instrumentation for Q0, Q1, Q1.5, Q3 and Q6 dummies [22]

	Q0	Q1 & Q1.5	Q3 & Q6
Head			
Accel. (A_x, A_y, A_z)	Yes	Yes	Yes
Angular velocity (ω_x, ω_y, ω_z)	No	Optional	Optional
Neck			
Head/C1 (F_x, F_y, F_z, M_x, M_y, M_z)	Yes	Yes	Yes
C7/T1 (F_x, F_y, F_z, M_x, M_y, M_z)	Yes	Yes	Yes
Thorax			
Spine Accel. (A_x, A_y, A_z)	Yes	Yes	Yes
Ribcage Accel. (A_x, A_y)	No	Optional	Optional
Ribcage deflection (δ_x or δ_y)	No	Yes	Yes
Abdomen			
Lumbar (F_x, F_y, F_z, M_x, M_y, M_z)	No	Yes	Yes
Abdomen pressure sensor	No	No	Experimental
Pelvis			
Accel. (A_x, A_y, A_z)	Yes	Yes	Yes

Recent modifications, including the mod kit, metric THOR, and SD3 shoulder, have not been thoroughly evaluated by the safety community.

4.1.6 Q Series of Infant and Child Dummies

In 1993, the Netherlands Organization for Applied Technical Research (TNO), was funded by the European Union (EU) to develop a new family of omni-directional child dummies called the Q-series [21–23]. These dummies were to replace the TNO P-Series (Pinocchio-Series), which have been specified in the European regulations since 1981. An international Child Dummy Working Group was formed of experts from research institutes and from child restraint system (CRS) and dummy manufacturers to define design targets for frontal impact responses of the head, neck, thorax, abdomen and lower extremities and for lateral impact responses of the shoulder and pelvis of the Q dummies. Commercially available Q dummies are the Q0, Q1, Q1.5, Q3 and Q6, with anthropometry representing a newborn, 12-month, 18-month, 3-year and 6-year-old, respectively. Their key dimensions and weights are given in Table 4.4. The instrumentation used with these dummies is given in Table 4.5. The Q10, representing a 10-year-old child, is under development by the European EPOCh (Enabling Protection for Older Children) consortium.

4.2 ATDs for Side Impact Testing

4.2.1 SID and SID-HIII

The first Side Impact Dummy (SID) was developed in 1979 by UMTRI under contract with NHTSA [24–27]. SID was based on the Hybrid II midsize male dummy, but its chest structure consists of a hydraulic shock absorber that links five interconnected steel ribs to the spine. SID has no arm or shoulder structure.

Major deficiencies of the SID are the lack of a shoulder load path, no elasticity in the thoracic compliance, and a very heavy rib mass. SID lacks instrumentation to measure neck, shoulder, and abdominal loads and rib deflection [28]. Due to these deficiencies, SID does not provide appropriate data to assess side impact protection [28–36]. Despite these deficiencies, SID was specified by NHTSA as the only dummy to be used to evaluate compliance to the side impact rule, FMVSS 214, which was implemented in 1990. SID was phased-out of FMVSS 214 as a compliance dummy in 2012.

Table 4.6 Instrumentation for small female and mid-size male side impact dummies [25, 42, 44, 48, 52, 53]

	SID-IIs	WorldSID small female	SID-HIII	BioSID	ES-2 ES-2re	WorldSID midsize male
Head						
Accel. (A_x, A_y, A_z)	Yes	Yes	Yes	Yes	Yes	Yes
Neck						
Head/C1 (F_x, F_y, F_z, M_x, M_y, M_z)	Yes	Yes	Yes	Yes	Yes	Yes
C7/T1 (F_x, F_y, F_z, M_x, M_y, M_z)	Yes	Yes	No	Yes	Yes	Yes
Shoulder						
Force (F_x, F_y, F_z)	Yes	Yes	No	Yes	Yes	Yes
Deflection (δ_y)	Yes	Yes	No	Yes	No	Yes
Upper extremities						
Upper arm accel. (A_x, A_y, A_z)	Yes	Yes	No	Yes	No	Yes
Upper arm (F_x, F_y, F_z, M_x, M_y, M_z)	Yes	Yes	No	No	No	Yes
Forearm accel. (A_x, A_y, A_z)	Yes	Yes	No	No	No	Yes
Forearm (F_x, F_y, F_z, M_x, M_y, M_z)	Yes	Yes	No	No	No	Yes
Thorax						
Spine accel. (A_x, A_y, A_z)	Yes	Yes	Yes	Yes	Yes	Yes
Rib defl. (δ_y)	Yes	Yes	No	Yes	Yes	Yes
Rib accel. (A_y)	Yes	Yes	Yes	Yes	Yes	Yes
Back plate force (F_x, F_y, M_x, M_y)	No	No	No	No	Yes	No
Abdomen						
Force (F_y)	No	No	No	No	Yes	No
Rib defl. (δ_y)	Yes	Yes	No	Yes	No	Yes
Lumbar (F_x, F_y, F_z, M_x, M_y, M_z)	Yes	F_y, F_z, M_x, M_z	Yes	Yes	Yes	Yes
Pelvis						
Accel. (A_x, A_y, A_z)	Yes	Yes	Yes	Yes	Yes	Yes
Iliac wing (F_y)	Yes	Yes	No	Yes	No	Yes
Sacrum (F_y)	No	No	No	Yes	No	No
Acetabulum (F_y)	Yes	No	No	Yes	No	Yes
Pubic (F_y)	Yes	Yes	No	Yes	Yes	Yes
Lower extremities						
Femoral neck (F_x, F_y, F_z)	No	Yes	No	No	No	Yes
Femur (F_x, F_y, F_z, M_x, M_y, M_z)	Yes	Yes	Yes	Yes	Yes	Yes
Tibia (F_x, F_y, F_z, M_x, M_y, M_z)	Yes	Yes	No	Yes	No	Yes

SID-HIII is the SID dummy with its Hybrid II head and neck replaced with Hybrid III head and neck. This improves the biofidelity of its head and neck response and provides for measurement of neck loads as indicated in Table 4.6. In 1998, NHTSA specified SID-HIII as the dummy to be used to evaluate side impact head airbags in the side impact pole test of FMVSS 201.

4.2.2 EUROSID and EUROSID-1

In 1986, several European laboratories, Association Peugeot-Renault (APR), the French National Institute for Research in Transportation and Safety (INRETS), the Transport Research Laboratory (TRL) in the United Kingdom, and TNO in the Netherlands, formed an ad-hoc group under the auspices of the European Experimental Vehicle Committee (EEVC) to develop a midsize male side impact dummy called EUROSID [37–39]. Between 1987 and 1989, production prototypes of EUROSID were evaluated worldwide by governments, the car industry, the International Standards Organization (ISO), and the SAE [29, 34–36].

Based on these evaluations, the dummy's biofidelity, durability, and instrumentation were improved and its name was changed to EUROSID-1 [40]. The final specification for EUROSID-1 was established by EEVC in April 1989 when it was specified for side impact compliance testing by the European Side Impact Regulation ECE R95. It is also specified in the Australian and Japanese Side Impact Regulations. EUROSID-1 instrumentation is listed in Table 4.6.

Major deficiencies of the EUROSID-1 are its marginal biofidelity [30], the flat-topping that can occur on its rib deflection measurement due to binding of the piston rod used to control its rib motion [41] and the lack of neck, iliac, and acetabulum load measurements [28].

4.2.3 BioSID

BioSID (Biofidelic Side Impact Dummy) was developed by SAE for side impact testing in 1989 following international evaluations of SID and EUROSID, which indicated the need for a more biofidelic dummy with additional measurement capability [29, 34, 35]. BioSID was the first dummy designed to the ISO impact response biofidelity guidelines for the head, neck, shoulder, thorax, abdomen, and pelvis [42, 43]. BioSID instrumentation is listed in Table 4.6.

The design of the dummy is well documented and quite durable, resulting in excellent repeatability and reproducibility. BioSID has been commercially available since 1990. While it is an acceptable test device for assessing side impact protection [28, 30, 36], BioSID has never been adopted into a regulation.

4.2.4 ES-2 and ES-2re

EEVC Working Group 12 coordinated an international task force to compile and address the deficiencies with the EUROSID-1 dummy. The ES-2 resulted from that effort [44] and was intended for use as an interim harmonized side impact dummy, until the more advanced WorldSID could be completed. In addition to the EUROSID-1 instrumentation, the ES-2 has upper neck, lower neck, clavicle, back plate, T12, and femur load cells, as noted in Table 4.6. Needle-bearing rib guides were added to reduce rib binding which was a major problem with EUROSID-1 design. The ES-2 demonstrated a slight improvement in biofidelity compared to EUROSID-1 [45]. In 2003, contractors of the 1959 United Nations Agreement on Passive Safety adopted a proposal by GRSP, the government experts on passive safety, to add ES-2 to ECE R95.

The ES-2 was further modified by NHTSA with the addition of rib extensions that provide a smooth surface to transition from the rearward end of the ribs to the lateral edge of the dummy's backplate, thus preventing the penetration of seatback foam into the gap that exists in the ES-2 [46]. The modified dummy is called ES-2re. As noted in Table 4.6, it has the same instrumentation as the ES-2. Biofidelity of the ES-2 and ES-2re are similar. The ES-2re has been incorporated into Part 572 (Subpart U) and could be used to demonstrate compliance to FMVSS 214 beginning in 2007, when advanced credits could be earned toward the phase-in period for the new rule.

4.2.5 SID-IIs

The SID-IIs is a second generation (II), small (s) SID. SID-IIs has the anthropometry of a small female adult dummy, Table 4.2, but its total weight is less than the Hybrid III small female. SID-IIs has no arm on the non-impacted side and either a stub arm or an instrumented full arm on the impacted side. The dummy was developed to fill the need for a small female test device to evaluate side impact protection countermeasures, notably airbags [48, 49].

The Small Size Advanced Side Impact Dummy Task Group of OSRP first met in January 1994 to define the general characteristics of the SID-IIs. The biofidelity response targets were scaled from ISO targets [48, 50] that were defined for the midsize adult male. Many of the parts from the Hybrid III small adult female were incorporated into the design. The thorax, abdomen, and pelvis structures are scaled versions of the BioSID design. The SID-IIs is extensively instrumented, as noted in Table 4.6. This includes the use of an instrumented arm [51] that has been incorporated into the design. Using the ISO rating scheme [28, 31–33] the SID-IIs biofidelity rating is good and its instrumentation rating is excellent. SID-IIs has been incorporated into Part 572 (Subpart V). Along with the ES-2re, the SID-IIs could be used for FMVSS 214 compliance testing beginning in 2007.

4.2.6 WorldSID

The WorldSID Task Group was formed in 1997 under Working Group 5 of ISO TC22/SC12. Its tri-chair structure had equal funding and representation from Asia-Pacific, Europe, and North America. The first prototype dummy became available for testing in 2001 [52]. However, it took 10 more years to develop a production dummy due to a lengthy development process and an extensive evaluation performed by dummy and instrumentation manufacturers, auto companies and suppliers, governments, and research institutions.

The dummy's head and the neck are designed to be biofidelic both laterally and fore-aft. Unlike prior side impact dummies, WorldSID's thorax and abdomen have sagittal plane symmetry, with its inner and outer rib bands attaching to a central spine plate such that the bands are horizontal when the dummy is seated in a vehicle. The upper and lower extremities are unique to WorldSID, with both a full and half arm available and a molded shoe.

WorldSID has the best biofidelity of all midsize male side impact dummies according to both the ISO biofidelity rating [50] and BioRank [47]. WorldSID can be instrumented with more than 170 channels, some of which are listed in Table 4.6.

4.2.7 WorldSID Small Female

Development of the WorldSID 5th female dummy began in 2004 with European Commission 6th Framework project APROSYS [53]. Unlike the midsize male WorldSID, development and

Table 4.7 Key dimensions and weights for Q3s [57]

	Key dimensions – mm
Chest depth	151
Shoulder width	247
Hip width	202
Buttock to knee	305
Sitting height	556
Standing height	986
	Weights – kg
Head & neck	3.12
Torso	5.78
Upper extremity	1.41
Lower extremity	3.55
Q3s suit	0.40
Total (with suit)	14.26

evaluation of the female dummy was funded entirely by Europe. Much of the female dummy is scaled from the male design, but differences include the use of 2-dimensional IR-TRACC units throughout the WorldSID small female [54]. Some of the instrumentation available for the WorldSID small female is listed in Table 4.6. The revision 1 of the WorldSID small female met its design goals of good biofidelity, both overall and for each body region. Dummy modifications are being coordinated by the WorldSID 5th Technical Evaluation Group.

4.2.8 Q3s

The Q3s is a 3-year-old side impact dummy, with improved durability and side impact biofidelity compared to the Q3. Early work in developing the Q3s from the Q3 was performed by First Technology Safety Systems [55, 56]. NHTSA and Humanetics Innovative Solutions completed the redesign with significant modifications to the neck, shoulder, thorax, and hip joint. The key dimensions and weights of Q3s are given in Table 4.7 [57]. Instrumentation available for the Q3s includes the instrumentation for the Q3, listed in Table 4.5, plus lateral deflection of the shoulder (δ_y), shoulder forces (F_x, F_y, F_z), and lateral pubic force (F_y). The Q3s is proposed for use in evaluating child restraints in a side impact sled test proposed for FMVSS 213 [58].

4.3 ATDs for Rear Impact Testing

4.3.1 Hybrid III Family of Dummies

One of the biomechanical response features of the Hybrid III midsize adult male dummy is that its neck responses mimic the average male's forward, rearward and lateral neck bending responses [5, 7]. The necks of the other members of the Hybrid III Family of dummies also have forward, rearward and lateral bending responses that mimic human responses for their relative sizes since their neck designs and response requirements were scaled from the Hybrid III midsize male's responses [10, 11, 13]. All the Hybrid III family of dummies can be used in rear impact testing. Instrumentation for the Hybrid III dummy family is listed in Table 4.1.

4.3.2 BioRID II and RID2 Dummies

In the late 1990s, two rear impact dummies, BioRID II [60–62] and RID2 [63, 64] were developed by two different European groups for assessing the potential of seat/head-restraint designs to reduce (eliminate) the occurrence of neck pain to occupants involved in low-severity rear end collisions.

The BioRID II (Biofidelic Rear Impact Dummy II) is a midsize male rear impact dummy developed by a Swedish consortium. It has a spine consisting of 24 articulating segments; 7 cervical, 12 thoracic and 5 lumbar. Articulations of the segments are constrained to the sagittal plane. Cables are used to control the position and motion of the head and neck vertebrae. The cable tensions are controlled by a rotary spring/damper mechanism. The complexity of its spine design and neck cabling mechanism poses concerns with repeatability and reproducibility of its test results. BioRID 1 and BioRID P3 were prototypes of BioRID II, the production dummy. BioRID II instrumentation is given in Table 4.8.

In 1997, the European Whiplash Project was started with the goal of developing passive restraint systems to reduce the occurrence of neck injuries in rear end impacts. One of the project's objectives was to develop a Rear Impact Dummy (RID) to be used to assess the efficacy of such restraint systems. The resulting dummy was called RID2 which was completed in 2000. The RID2 is based on the Hybrid III midsize male dummy, but uses the THOR chest, the EUROSID-1 lumbar spine and a modified TRID neck [63, 64]. The TRID neck structure was modified to allow neck twist and lateral neck bending. In addition, its sagittal plane bending stiffness was modified to allow 15° of fore/aft head rotation without any bending resistance at the nodding joint. RID2 instrumentation is listed in Table 4.8.

Table 4.8 Instrumentation for RID 2 and BioRID II [60, 63]

	RID 2	BioRID II
Head		
Accel. (A_x, A_y, A_z)	Yes	Yes
Angular rate (ω_x, ω_y, ω_z)	Yes	Yes
Skull cap load (F_x, F_y, F_z)	Yes	Yes
Skull cap contact switch	No	Yes
Tilt sensor	No	Yes
Neck		
Head/C1 (F_x, F_y, F_z, M_x, M_y, M_z)	Yes	Yes
C7/T (F_x, F_y, F_z, M_x, M_y, M_z)	Yes	F_x, F_z, M_y
Thorax		
T1 accel. (A_x, A_y, A_z)	Yes	Yes
T1 angular rate (ω_x, ω_y, ω_z)	No	Yes
T1 loads (F_x, F_z, M_y)	No	Yes
T1 tilt sensor	No	Yes
T8 accel. (A_x, A_y, A_z)	No	Yes
T8 angular rate (ω_x, ω_y, ω_z)	No	Yes
T12 accel.	Yes	No
Abdomen		
Lumbar accel. (A_x, A_y, A_z)	No	No
Lumbar angular rate (ω_x, ω_y, ω_z)	No	No
Lumbar loads (F_x, F_y, F_z, M_x, M_y, M_z)	Yes	Yes
Pelvis		
Accel. (A_x, A_y, A_z)	Yes	Yes
Angular rate (ω_x, ω_y, ω_z)	No	Yes
Tilt sensor	No	Yes
Lower extremities		
Femur load (F_x, F_y, F_z, M_x, M_y, M_z)	No	Yes

In 2003, the Occupant Safety Research Partnership (OSRP) of the United States Council for Automotive Research (USCAR) evaluated the biofidelity of the rearward bending responses of the Hybrid III midsize male, BioRID II and RID2 by comparing their responses to those of a tensed volunteer and two flaccid cadavers in a series of low and medium severity rear impact collision simulation sled tests [65]. The Hybrid III responses were the closest to those of the volunteer. Its neck flexibility was greater than the tensed volunteer's, but less than the flaccid cadavers' which is consistent with its neck's design objective. The necks of both the BioRID II and the RID2 were much more flexible. The flexibility of the BioRID II neck was comparable to the flaccid cadavers'. The RID2 neck was more flexible than the cadavers' which is a major concern.

Currently, work is being done on a Global Technical Regulation involving rear impact testing which proposes using BioRID II as the test dummy. Due to ongoing concerns over the repeatability and reproducibility of BioRID II's test results, the ISO working group for Anthropomorphic Test Devices (TC22/SC12/WG5) no longer recommends BioRID II for use in low-severity rear impact tests [66].

4.4 Injury Assessment Reference Values

In the late 1970s, GM began using their newly developed Hybrid III midsize male dummy to evaluate the restraint performances of their second generation air bag system designs [67]. They prescribed limits for the various Hybrid III midsize male response measurements which they called Injury Assessment Reference Values (IARVs). Each IARV was chosen so that if the value was not exceeded, then the corresponding injury would be unlikely to occur. Unlikely was defined as less than a 5 % chance of the injury occurring [68]. When GM petitioned NHTSA in 1983 to allow the use of the Hybrid III midsize male dummy in FMVSS 208 passive restraint system testing, they made the dummy's IARVs public [69].

The Hybrid III midsize male IARVs were scaled for size and material strength differences [2, 3, 68–76] to give corresponding IARVs for the other sizes of Hybrid III dummies. The IARVs for the head, torso and extremities of the Hybrid III family of dummies are given in Table 4.9. Neck IARVs for the Hybrid III family of dummies are given in Table 4.10 for in-position tests with 80 % muscle tone and Table 4.11 for out-of-position airbag tests with no muscle tone [73, 74]. It is not known if any of these neck IARVs address the occurrence of neck pain that might occur in minor rear end collisions. While theories on the causes of neck pain and proposed criteria have been published, laboratory tests have not identified the mechanism of neck pain. Time dependent IARV curves are given in Figs. 4.1, 4.2, 4.3, 4.4, and 4.5.

Similar type of scaling was used to prescribe IARVs for the CRABI family of dummies [68, 71, 73, 77] which are given in Table 4.12. The IARVs for the neck of the CRABI family of dummies are given in Table 4.13 for in-position tests and Table 4.14 for out-of-position airbag tests.

IARVs have been prescribed for the SID-IIs and BioSID side impact dummies [2, 3, 70, 73] and these IARVs are listed in Table 4.15. Neck IARVs are given in Table 4.10 for in-position tests and Table 4.11 for out-of-position tests with these dummies. Time dependent IARV curves for the necks of SID-IIs and BioSID are given in Figs. 4.1, 4.2, 4.3, and 4.4. In Tables 4.10 and 4.11 and in Figs. 4.1, 4.2, 4.3, and 4.4, use the small adult female values for SID-IIs and the midsize adult male values for BioSID.

4.5 Injury Risk Curves

The 5 % risk value of an injury risk curve could be used as an IARV. However, there are a number of precautions that one must be aware of when using published injury risk curves to prescribe IARVs. First, the sample of specimens used to formulate a risk curve may not be representative of the size and strength of the population that the dummy represents. In most cases, the test samples consist mostly of older cadavers whose

Table 4.9 Injury assessment reference values for measurements made with the Hybrid III family of dummies [73]

	3 year	6 year	10 year	Small female	Midsize male	Large male
Head						
Pk. 15 ms HIC	568	723	741	779	700	670
Pk. CG Accel. (G)	175	189	189	193	180	175
Upper extremity						
Pk. upr. arm resultant moment (Nm)	–	–	–	130	214	308
Pk. forearm resultant moment (Nm)	–	–	–	44	90	110
Thorax						
Pk. stern. defl. (mm)						
Shoulder belt	28	31	36	41	50	55
No shoulder belt	–	–	40.4	39.0	47.7	52.8
Pk. defl. rate (m/s)	8.0	8.5	8.4	8.3[a]	8.3[a]	8.3[a]
Pk. T4 accel (G)	92	93	82	73	60	54
Lower extremity						
Femur-time curve refer to Fig. 4.5	–					
Tibia-femur displ. (mm)	–	–	–	12	15	17
Pk. medial or lateral knee clevis (N)	–	–	–	2,550	4,000	4,910
Pk. tibia compr. (N)	–	–	–	5,100	8,000	9,820
Tibia index, TI	–	–	–	1	1	1
TI intercepts						
M_c (Nm)				114	225	306
F_c (N)				22,900	35,900	44,100
Pk. ankle moment (Nm)	–	–	–	114	225	306

[a]Corrected from prior publication

material strengths are much less than the average of the population, and whose size variations have not been normalized to the size of the dummy. Second, the data are usually censored, that is, no subject's measured value is the subject's injury threshold value which is defined as the lowest value that will produce the specified injury to the given subject. If failure does not occur in the test, the specimen's test value is defined as "right-censored" since its threshold value is greater than (to the right of) the measured value by some unknown amount. If failure occurs in the test, the test value is defined as "left-censored" since its threshold value is less than (to the left of) the measured value by some unknown amount.

The two most important measured values from a set of censored data are the lowest value associated with injury, X_{WEAK}, and the highest value associated with non-injury, X_{STRONG}. If additional testing is needed, the "weakest" specimens should be tested such that their values are less than X_{WEAK}. The "strongest" specimens should be tested such that their values are greater than X_{STRONG}. Such testing will solidify the bounds (X_{WEAK}, X_{STRONG}) for the range of threshold strengths of the population.

A number of statistical methods have been used to develop injury risk curves from censored data sets. The Maximum Likelihood (ML) Method [78] evaluates thousands of sets of possible threshold values based on a given set of censored data and chooses the best fit for a prescribed type of distribution function. A major concern with the ML selection process is that the distribution of the test sample is not usually representative of the population that the dummy represents. As a result, the ML Method can choose a distribution function that gives a significant injury risk with no stimulus [79] as shown by the ML Normal curve of Fig. 4.6. To avoid such selections, Nakahira et al. [80] proposed to use only distribution functions that required zero injury risks at zero stimuli, such as Log Normal or 2-parameter Weibull. However, no constraint

Table 4.10 Neck injury assessment reference values for peak loads (N) and moments (Nm) for Hybrid III family of dummies, SID-IIs and BioSID in in-position tests [73]

	3 year	6 year	10 year	Small female	Midsize male	Large male
Upper neck						
Shear, F_x, F_y	1,070	1,410	1,710	1,950	3,100	3,740
Tension, $+F_z$	1,430	1,890	2,290	2,620	4,170	5,030
Compression, $-F_z$	1,380	1,820	2,200	2,520	4,000	4,830
Lat. moment, M_x	32	45	59	72	144[a]	190
Flex. moment, $+M_y$	42	60	78	95	190	252
Ext. moment, $-M_y$	21	30	40	49	97[a]	128
Twist, M_z	21	30	40	49	97[a]	128
Upper Neck N_{ij}						
Combined N_{ij}	1	1	1	1	1	1
N_{ij} intercepts						
F_T	2,330	3,080	3,710	4,260	6,780	8,180
F_C	2,130	2,820	3,390	3,900	6,200	7,480
M_F	67.0	96.0	125	153	305	405
M_E	29.3	42.0	54.8	66.9	133	177
Lower neck						
Shear, F_x, F_y	1,070	1,410	1,710	1,950	3,100	3,740
Tension, $+F_z$	1,430	1,890	2,290	2,620	4,170	5,030
Compression, $-F_z$	1,380	1,820	2,200	2,520	4,000	4,830
Lat. moment, M_x	63	90	118	144	287[a]	380
Flex. moment, $+M_y$	84	120	156	190	380	504
Ext. moment, $-M_y$	42	60	80	98	194[a]	256
Twist, M_z	21	30	40	49	97[a]	128
Lower Neck N_{ij}						
Combined N_{ij}	1	1	1	1	1	1
N_{ij} intercepts						
F_T	2,330	3,080	3,710	4,260	6,780	8,180
F_C	2,130	2,820	3,390	3,900	6,200	7,480
M_F	134	192	250	306	610	810
M_E	58.6	84.0	110	134	266	354

[a]Corrected from prior publication

is placed on the upper end (strongest) and the selection process can choose distribution functions that allow overly strong people and too many weak people as shown by the ML Log Normal and Weibull curves of Fig. 4.6.

The Mertz-Weber [75] and the Certainty Method [79] are two methods that use only the data points that are bounded by X_{WEAK} and X_{STRONG}, including these two points. However, the Certainty Method, like the Maximum Likelihood Method, has a problem if the test sample distribution is not consistent with the distribution of the desired population. The best fit to a biased data set will not be the distribution of the desired population. The Mertz-Weber Method, which is a modified Median Rank Technique [81], does not have this problem since it only fits a symmetric distribution, usually a Normal distribution, to the boundary points, X_{WEAK} and X_{STRONG} and the resulting distribution function is consistent with the data, as shown in Fig. 4.6.

Figure 4.6 shows all the censored data points that were used to develop the 15 ms HIC risk curve (Mertz-Weber Normal) for skull fracture, Fig. 4.7. Association Peugeot-Renault (APR) conducted 20 tests where the skull thickness was measured for each cadaver [79, 82]. As shown on Fig. 4.6, the APR cadaver with the smallest skull

Table 4.11 Neck injury assessment reference values for peak loads (N) and moments (Nm) for Hybrid III family of dummies, SID-IIs and BioSID in out-of-position tests [73]

	3 year	6 year	10 year	Small female	Midsize male	Large male
Upper neck						
Shear, F_x, F_y	1,070	1,410	1,710	1,950	3,100	3,740
Tension, $+F_z$	1,130	1,490	1,800	2,070	3,290	3,970
Compression, $-F_z$	1,380	1,820	2,200	2,520	4,000	4,830
Lat. moment, M_x	30	42	55	67	134	178
Flex. moment, $+M_y$	42	60	78	95	190	252
Ext. moment, $-M_y$	17	24	32	39	78	103
Twist, M_z	17	24	32	39	78	103[a]
Upper neck N_{ij}						
Combined N_{ij}	1	1	1	1	1	1
N_{ij} intercepts						
F_T	2,130	2,820	3,390	3,900	6,200	7,480
F_C	2,130	2,820	3,390	3,900	6,200	7,480
M_F	67.0	96.0	125	153	305	405
M_E	26.8	38.4	50.1	61.2	122	162
Lower neck						
Shear, F_x, F_y	1,070	1,410	1,710	1,950	3,100	3,740
Tension, $+F_z$	1,130	1,490	1,800	2,070	3,290	3,970
Compression, $-F_z$	1,380	1,820	2,200	2,520	4,000	4,830
Lat. Moment, M_x	59	84	110	134	268	355
Flex. Moment, $+M_y$	84	120	156	190	380	504
Ext. Moment, $-M_y$	34	48	64	78	156	206
Twist, M_z	17	24	32	39	78	103[a]
Lower neck N_{ij}						
Combined N_{ij}	1	1	1	1	1	1
N_{ij} intercepts						
F_T	2,130	2,820	3,390	3,900	6,200	7,480
F_C	2,130	2,820	3,390	3,900	6,200	7,480
M_F	134	192	250	306	610	810
M_E	53.6	76.8	100	122	244	324

[a]Corrected from prior publication

thickness (HIC = 517) is located near X_{WEAK} and their cadaver with the greatest skull thickness (HIC = 2,351) is located near X_{STRONG}. Based only on the APR sample of 20 cadavers and using the Mertz-Weber Method, the Median Rank Technique [81] fracture risks for these two cadavers are 3 % and 97 %, respectively. Using all the data, the Mertz-Weber Normal curve gives risks of 2 % and 98 %, respectively, which are nearly identical to the risk values calculated using only the APR sample.

The Injury Risk Curves [68, 73, 74, 83] associated with many of the IARVs listed in Tables (4.9, 4.10, 4.11, 4.12, 4.13, 4.14, and 4.15) are shown in Figs. 4.7, 4.8, 4.9, 4.10, 4.11, 4.12, 4.13, 4.14, 4.15, 4.16, and 4.17. Figures 4.7 and 4.8 are risk curves for skull fracture and AIS ≥ 4 brain injury, based on 15 ms HIC values for the adult population. Note that their mean values (1,500 and 1,434) and standard deviations (488 and 430) are quite similar. Figure 4.9 is the skull fracture risk curve based on peak head acceleration for the adult population. Risk curves for a given size of occupant could not be estimated since head size and weight data for the test subjects were not given. Normalized AIS ≥ 3 neck injury risk curves for tension, extension moment and N_{TE} measured at the Occipital Condyles for various occupant

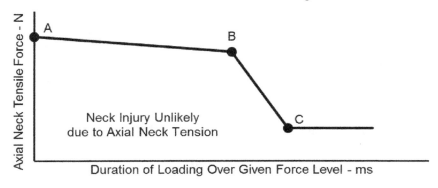

Point	6 Month Old		12 Month Old		18 Month Old	
	Time ms	Force N	Time ms	Force N	Time ms	Force N
A	0	930	0	990	0	1080
B	20	820	20	870	21	950
C	26	250	26	260	27	290

Point	3 Year Old		6 Year Old		10 Year Old	
	Time ms	Force N	Time ms	Force N	Time ms	Force N
A	0	1430	0	1890	0	2290
B	22	1260	24	1670	26	2000
C	29	380	31	500	34	610

Point	Small Adult Female		Mid-Size Adult Male		Large Male	
	Time ms	Force N	Time ms	Force N	Time ms	Force N
A	0	2620	0	4170	0	5030
B	28	2310	35	3670	38	4420
C	36	690	45	1100	49	1330

Fig. 4.1 Injury assessment reference curve for axial neck tension loading for the CRABI and Hybrid III dummy families, SID-IIs and BioSID for in-position tests

sizes are given in Figs. 4.10, 4.11, and 4.12. The normalizing constants given in the table below each graph are the IARV values for each size of occupant [84]. The injury risks associated with the IARVs are 3 % for neck tension, 5 % for neck extension moment and 2 % for N_{TE}.

Figures 4.13 and 4.14 are the injury risk curves for AIS ≥ 3 and AIS ≥ 4 thorax injury based on normalized sternal deflections for various size occupants for anterior, distributed (airbag) thorax loading. The normalizing values are the dummies' IARVs and correspond to a 5 % injury risk. Note that there are no AIS ≥ 3 risk curves for children since AIS = 3 injuries are based on rib fractures. Due to the elastic properties of their ribs, child ribs seldom fracture. Figure 4.15 is the normalized risk curve for AIS ≥ 4 thorax injury based on the rate of sternal deflection. The normalizing values are the IARVs for the dummy sizes listed and represent a 5 % injury risk. This curve was developed to assess OOP airbag interactions with the chest [75].

Figure 4.16 is the AIS ≥ 3 thorax injury risk to adults due the shoulder belt loading based on

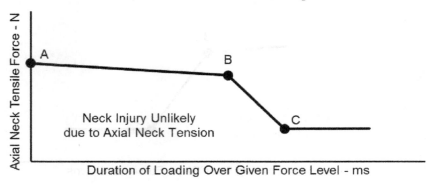

Fig. 4.2 Injury assessment reference curve for axial neck tension loading for the CRABI and Hybrid III dummy families, SID-IIs and BioSID for out-of-position airbag tests

Point	6 Month Old		12 Month Old		18 Month Old	
	Time ms	Force N	Time ms	Force N	Time ms	Force N
A	0	730	0	780	0	850
B	20	640	20	680	21	750
C	26	250	26	260	27	290

Point	3 Year Old		6 Year Old		10 Year Old	
	Time ms	Force N	Time ms	Force N	Time ms	Force N
A	0	1130	0	1490	0	1800
B	22	990	24	1310	26	1570
C	29	380	31	500	34	610

Point	Small Adult Female		Mid-Size Adult Male		Large Male	
	Time ms	Force N	Time ms	Force N	Time ms	Force N
A	0	2070	0	3290	0	3970
B	28	1810	35	2880	38	3480
C	36	690	45	1100	49	1330

sternal deflection. Figure 4.17 is the risk of tibia shaft fracture to male adults based on tibia bending moment. Risk curves for a given size of occupant could not be estimated since test subject size data were not given.

4.6 Future ATD and IARV Needs

In the 1960s, many new occupant protection technologies, such as high penetration resistance windshield glass; energy absorbing steering columns and knee restraints; padded headers, instrument panels and roof rails; and lap/shoulder belts, were developed and installed in cars. To assess the efficacies of these restraint technologies, component and sled tests were developed as well as their associated test devices [85–89]. In addition, research projects were conducted to determine human impact tolerance levels that were needed to define design limits (IARVs) for these restraint systems [88, 90–92].

In the 1970s and early 1980s, the automobile manufacturers began working on developing frontal airbags for drivers and front seat passengers. The main concern with airbags was deploying

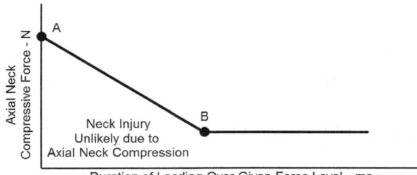

Point	6 Month Old		12 Month Old		18 Month Old	
	Time ms	Force N	Time ms	Force N	Time ms	Force N
A	0	890	0	960	0	1,040
B	17	250	18	260	18	290

Point	3 Year Old		6 Year Old		10 Year Old	
	Time ms	Force N	Time ms	Force N	Time ms	Force N
A	0	1,380	0	1,820	0	2,200
B	19	380	21	500	23	610

Point	Small Adult Female		Mid-Size Adult Male		Large Male	
	Time ms	Force N	Time ms	Force N	Time ms	Force N
A	0	2,520	0	4,000	0	4,830
B	24	690	30	1,100	33	1,330

Fig. 4.3 Injury assessment reference curve for axial neck compression loading for the CRABI and Hybrid III dummy families, SID-IIs and BioSID

the bag fast enough to provide restraint to an unbelted occupant in the FMVSS 208 frontal barrier test, but slow enough to not injure an occupant who may be in the path of the deploying bag [93]. This out-of-position (OOP) injury concern required that research be conducted to determine IARVs and to develop dummies (Hybrid III midsize male dummy and 3 year-old airbag dummy) to measure the OOP interactions [5, 7, 12]. In the late 1980s, the US Congress mandated that front seat airbags be installed in all cars. To address the OOP injury concern, SAE Task Groups developed the Hybrid III and CRABI families of dummies [11, 10] and issued two SAE Information Reports [94, 95] describing how to conduct various OOP tests for airbags using these dummies and their corresponding IARVs.

In the late 1990s, automobile manufacturers started working on side airbags. The NHTSA Administrator, Dr. Martinez, requested that the Alliance of Automobile Manufacturers (Alliance), the Association of International Automobile Manufacturers (AIAM), the Automotive Occupant Restraints Council (AORC), and the Insurance Institute for Highway Safety (IIHS) form a Technical Working Group (TWG) to define how to manage the OOP injury concerns of side airbags. On August 8, 2000, the TWG issued a report that recommended test procedures for assessing the OOP injury risk from deploying side airbags [96].

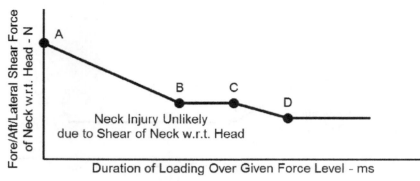

Point	6 Month Old		12 Month Old		18 Month Old	
	Time ms	Force N	Time ms	Force N	Time ms	Force N
A	0	690	0	740	0	810
B	14	330	15	360	15	390
C	20	330	20	360	21	390
D	26	250	26	260	27	290

Point	3 Year Old		6 Year Old		10 Year Old	
	Time ms	Force N	Time ms	Force N	Time ms	Force N
A	0	1,070	0	1,410	0	1,710
B	16	520	17	680	19	830
C	22	520	24	680	26	830
D	29	380	31	500	34	610

Point	Small Adult Female		Mid-Size Adult Male		Large Male	
	Time ms	Force N	Time ms	Force N	Time ms	Force N
A	0	1,950	0	3,100	0	3,740
B	20	950	25	1,500	27	1,810
C	28	950	35	1,500	38	1,810
D	36	690	45	1,100	49	1,330

Fig. 4.4 Injury assessment reference curve for fore/aft/lateral shear force of neck with respect to head for the CRABI and Hybrid III dummy families, SID-IIs and BioSID

The prescribed tests were to be conducted with the Hybrid III 3- and 6-year old, the Hybrid III small female and the SID-IIs dummies. Their IARVs were specified as design limits that were not to be exceeded. No new dummies, transducers or IARVs were needed.

Current automobiles are equipped with frontal and side airbags, force-limiting belt systems and energy absorbing knee restraints. For Model Years (MY) 2000–2007, field accident data indicated that there was a 61 % reduction in fatalities in frontal accidents of these vehicles [97]. However, the data also indicated that a quarter of the fatalities (29/122 = 0.24) occurred in corner or oblique impacts that involved only a small portion of the front structure of the car. To address this concern, IIHS has begun to evaluate vehicles in 64 km/h small overlap, rigid barrier tests [98]. To obtain acceptable performance in this test, some vehicles' front structures may have to be

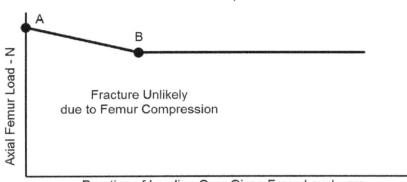

Fig. 4.5 Injury assessment reference curve for knee loading for the Hybrid III dummy family

Table 4.12 Injury assessment reference values for measurements made with CRABI family of dummies [73]

	CRABI 6 month	CRABI 12 month	CRABI 18 month
Head			
Peak 15 ms HIC	377	389	440
Peak CG Accel. (G)	156	154	160
Thorax			
Peak spine accel. (G)	88	87	89
Pelvis			
Peak pubic loads (N)	–	1,540	1,690

modified to prevent toe pan and A-pillar intrusion on the impacted side, and in some vehicles the airbag coverage area may have to be enlarged. No new dummies, transducers or IARVs were needed to assess vehicle performance.

Another observation from the frontal accident data for the 2000–2007MY vehicles is that there were 13 fatalities to elderly occupants (75 or older) some in minor severity crashes [97]. Since the elderly cannot tolerate the same level of restraint loading as younger adults, the load limiting values of various energy absorbing restraints (shoulder belts, knee restraints) may have to be reduced [99]. This would require setting the IARVs for chest deflection at a lower risk level. However, with a lower restraint load, the severity of accident where protection is provided will be reduced. This trade-off between protecting the elderly in the more frequent mild-to-moderate frontal crashes or protecting younger and stronger adults in infrequent severe frontal crashes needs to be resolved in the future.

Available field accident data do not contain enough side impact data of vehicles equipped with a full array of side airbags to assess their

Table 4.13 Neck injury assessment reference values for loads and moments for CRABI family of dummies in in-position tests [73]

	CRABI 6 month	CRABI 12 month	CRABI 18 month
Upper neck			
Peak shear, F_x, F_y (N)	690	740	810
Peak tension, $+F_z$ (N)	930	990	1,080
Peak compr., $-F_z$ (N)	890	960	1,040
Pk. lat. moment, M_x (Nm)	19	21	22
Pk. flex. moment, $+M_y$ (Nm)	25	27	29
Pk. ext. moment, $-M_y$ (Nm)	13	14	15
Pk. twist, M_z (Nm)	13	14	15
Upper neck N_{ij}			
Combined N_{ij}	1	1	1
N_{ij} intercepts			
F_T	1,510	1,610	1,760
F_C	1,380	1,470	1,610
M_F	39.5	42.5	46.8
M_E	17.3	18.6	20.4
Lower neck			
Peak shear, F_x, F_y (N)	690	740	810
Peak tension, $+F_z$ (N)	930	990	1,080
Peak compr., $-F_z$ (N)	890	960	1,040
Pk. lat. moment, M_x (Nm)	38	41	44
Pk. flex. moment, $+M_y$ (Nm)	50	54	58
Pk. ext. moment, $-M_y$ (Nm)	26	28	30
Pk. twist, M_z (Nm)	13	14	15
Lower neck N_{ij}			
Combined N_{ij}	1	1	1
N_{ij} intercepts			
F_T	1,510	1,610	1,760
F_C	1,380	1,470	1,610
M_F	79.0	85.0	93.6
M_E	34.6	37.2	40.8

Table 4.14 Neck injury assessment reference values for loads and moments for CRABI family of dummies in out-of-position tests [73]

	CRABI 6 month	CRABI 12 month	CRABI 18 month
Upper neck			
Peak shear, F_x, F_y (N)	690	740	810
Peak tension, $+F_z$ (N)	730	780	850
Peak compr., $-F_z$ (N)	890	960	1,040
Pk. lat. moment, M_x (Nm)	18	19	21
Pk. flex. moment, $+M_y$ (Nm)	25	27	29
Pk. ext. moment, $-M_y$ (Nm)	10	11	12
Pk. twist, M_z (Nm)	10	11	12
Upper neck N_{ij}			
Combined N_{ij}	1	1	1
N_{ij} intercepts			
F_T (N)	1,380	1,470	1,610
F_C (N)	1,380	1,470	1,610
M_F (Nm)	39.5	42.5	46.8
M_E (Nm)	15.8	17.0	18.7
Lower neck			
Peak shear, F_x, F_y (N)	690	740	810
Peak tension, $+F_z$ (N)	730	780	850
Peak compr., $-F_z$ (N)	890	960	1,040
Pk. lat. moment, M_x (Nm)	35	38	41
Pk. flex. moment, $+M_y$ (Nm)	50	54	58
Pk. ext. moment, $-M_y$ (Nm)	20	22	24
Pk. twist, M_z (Nm)	10	11	12
Lower neck N_{ij}			
Combined N_{ij}	1	1	1
N_{ij} intercepts			
F_T (N)	1,380	1,470	1,610
F_C (N)	1,380	1,470	1,610
M_F (Nm)	79.0	85.0	93.6
M_E (Nm)	31.6	34.0	37.4

effectiveness in reducing fatalities. The main difference between side and frontal impact protection is the space available to absorb the crash forces. This space is less in side crashes than in frontal crashes because the door crush space is much smaller than the front-end crush space. However, test data of vehicles equipped with frontal and side airbags show that side airbags will provide similar head and chest protection as frontal airbags for the test conditions used in their development.

In summary, there does not appear to be a need to change the ATDs that were used in tests to develop the current frontal and side impact restraint systems. Switching to either THOR or WorldSID dummy technology would be an added expense to the car manufacturers with no benefit in restraint system improvements being realized.

Table 4.15 Injury assessment reference values for measurements made with side impact dummies [3, 73]

	SID-IIs	BioSID
Head		
Peak 15 ms HIC	779	700
Peak CG accel. (G)	193	180
Neck		
Refer to Table 4.10 for in-position tests		
Refer to Table 4.11 for out-of-position tests		
Shoulder		
Peak lateral force (N)	2,920[a]	4,000
Peak lateral deflection (mm)	61	75
Upper extremity		
Upper arm peak resultant moment (Nm)	130	214
Forearm peak resultant moment (Nm)	44	90
Thorax		
Peak rib to spine defection (mm)	34	42
Peak deflection rate (m/s)	8.3[a]	8.3[a]
Peak T4 accel (G)	73	60
Abdomen		
Peak rib to spine deflection (mm)	32	39
Peak deflection rate (m/s)	8.3[a]	8.3[a]
Internal load (N)	1,830	2,500
Pelvis		
Peak pubic load (N)	4,390	6,000
Peak iliac crest load (N)	4,390	6,000
Peak sacrum load (N)	4,390	6,000

[a]Corrected from prior publication

For rear end collisions, the cause of neck pain is not known, nor is it known if it even exists for some people who claim that it does. Consequently, it is not known whether BioRID II and/or new IARVs are needed. It should be noted that a human volunteer tolerated a 44 mph simulated rear impact with no neck injury or neck pain [100]. Alternative approaches such as warning the occupants of impending rear impacts may allow enough time for occupants to brace, minimizing their rearward head rotation and hyperextension of their necks.

Child restraints systems (CRS) have not progressed as far as adult restraints. There are three major concerns with CRS. First, there is no forward restraint for the child's head in forward facing CRS in a frontal crash. Second, older children use adult belts which are not optimized for their small body structure. Third, there are no airbags or EA belts for children. If CRS were designed to have the same level of restraint as the current adult systems, the existing child dummies (Hybrid III, Q and CRABI) and their corresponding IARVs may be adequate.

In the future, as it has been in the past, the need for improvements in dummy technology and additional IARVs will be dependent on whether or not the current ATD technologies can be used to evaluate future restraint systems.

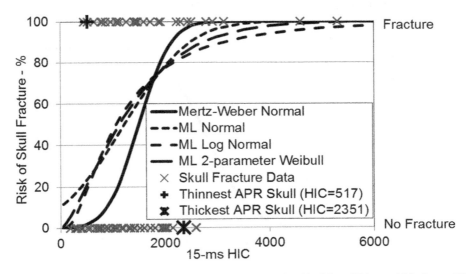

Fig. 4.6 Skull fracture risk curves based on 15 ms HIC values determined by Mertz-Weber and Maximum Likelihood methods

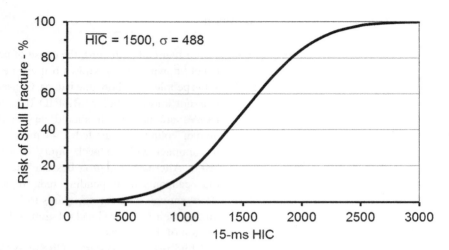

Fig. 4.7 Risk of skull fracture based on 15 ms HIC for adult population

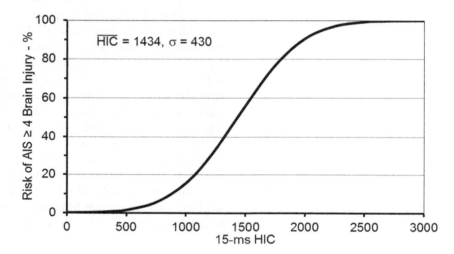

Fig. 4.8 Risk of AIS ≥ 4 brain injury based on 15 ms HIC for adult population

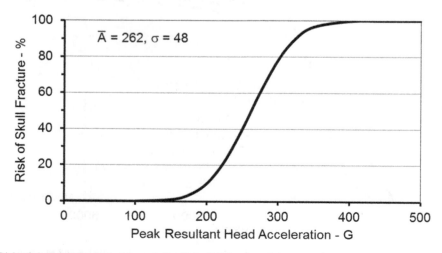

Fig. 4.9 Risk of skull fracture based on peak head acceleration for adult population

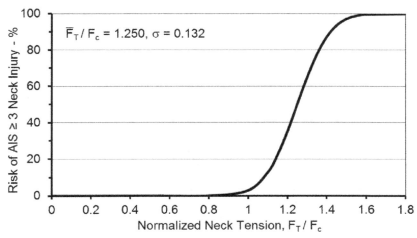

Dummy	Critical Values for Neck Tension, F_c - N	
	In-Position Tests	Out-of-Position Tests
CRABI 12 Month Old	990	780
Hybrid III 3 Year Old	1430	1130
Hybrid III 6 Year Old	1890	1490
Hybrid III 10 Year Old	2290	1800
Hybrid III Small Female	2620	2070
Hybrid III Midsize Male	4170	3290
Hybrid III Large Male	5030	3970

Fig. 4.10 Risk of AIS $\geq$ 3 neck injury based on peak normalized neck tension measured at the occipital condyles for in-position or OOP testing

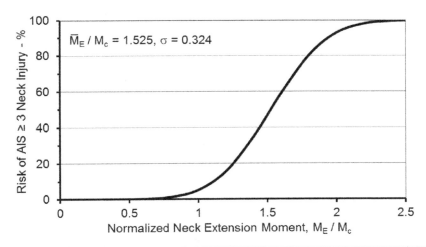

Dummy	Critical Values for Neck Extension Moment, M_c - Nm	
	In-Position Tests	Out-of-Position Tests
CRABI 12 Month Old	14	11
Hybrid III 3 Year Old	21	17
Hybrid III 6 Year Old	30	24
Hybrid III 10 Year Old	40	32
Hybrid III Small Female	49	39
Hybrid III Midsize Male	97	78
Hybrid III Large Male	128	103

Fig. 4.11 Risk of AIS $\geq$ 3 neck injury based on peak normalized neck extension moment measured at the occipital condyles for in-position or OOP testing

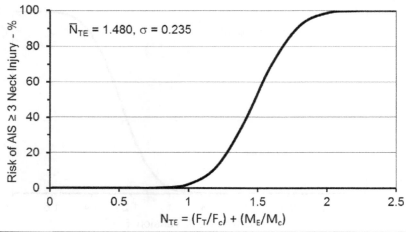

Dummy	Critical Values for In-Position Tests		Critical Values for Out-of-Position Tests	
	Neck Tension, F_c - N	Neck Extension Moment, M_c - Nm	Neck Tension, F_c - N	Neck Extension Moment, M_c - Nm
CRABI 12 Month Old	1610	18.6	1470	17.0
Hybrid III 3 Year Old	2330	29.3	2130	26.8
Hybrid III 6 Year Old	3080	42.0	2820	38.4
Hybrid III 10 Year Old	3710	54.8	3390	50.1
Hybrid III Small Female	4260	66.9	3900	61.2
Hybrid III Midsize Male	6780	133	6200	122
Hybrid III Large Male	8180	177	7480	162

Fig. 4.12 Risk of AIS ≥ 3 neck injury based on peak normalized Nte measured at the occipital condyles for in-position or OOP testing

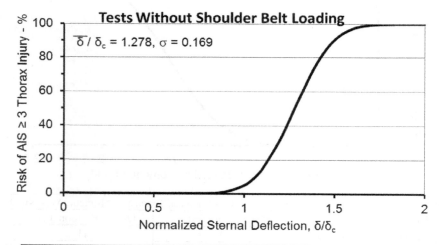

Dummy	Critical Values for Sternal Deflection, δ_c - mm
Hybrid III 10 Year Old	40.4
Hybrid III Small Female	39.0
Hybrid III Midsize Male	47.7
Hybrid III Large Male	52.8

Fig. 4.13 Risk of AIS ≥ 3 thorax injury due to distributed, anterior-thoracic loading based on peak normalized sternal deflection

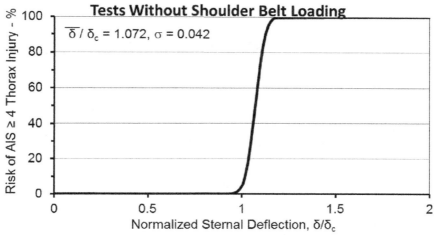

Dummy	Critical Values for Sternal Deflection, δ_c - mm
Hybrid III 3 Year Old	35.8
Hybrid III 6 Year Old	39.7
Hybrid III 10 Year Old	46.5
Hybrid III Small Female	52.5
Hybrid III Midsize Male	64.3
Hybrid III Large Male	71.1

Fig. 4.14 Risk of AIS ≥ 4 thorax injury due to distributed, anterior-thoracic loading based on peak normalized sternal deflection

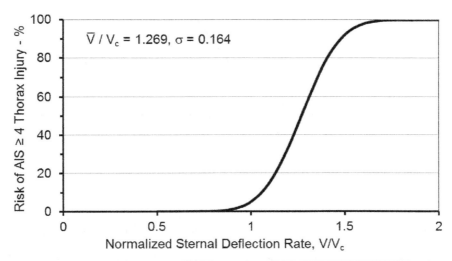

Dummy	Critical Values for Sternal Deflection Rate, δ_c - mm
Hybrid III 3 Year Old	8.0
Hybrid III 6 Year Old	8.5
Hybrid III 10 Year Old	8.4
Hybrid III Small Female	8.3
Hybrid III Midsize Male	8.3
Hybrid III Large Male	8.3

Fig. 4.15 Risk of AIS ≥ 4 thorax injury based on peak normalized sternal deflection rate

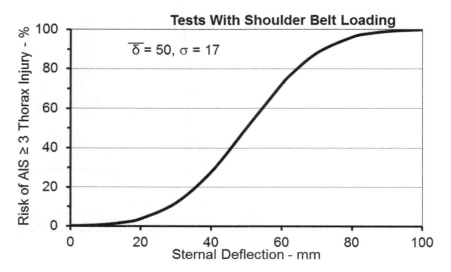

Fig. 4.16 Risk of AIS ≥ 3 thorax injury based on sternal deflection for adult shoulder belt loading

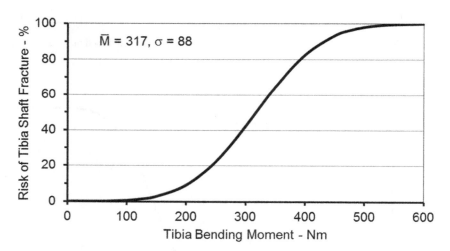

Fig. 4.17 Risk of tibia shaft fracture based on tibia bending moment for male adults

References

1. Mertz HJ (1985) Anthropomorphic models. In: Nahum AM, Melvin JW (eds) The biomechanics of trauma. Appleton, Norwalk, pp 31–60
2. Mertz HJ (1993) Anthropomorphic test devices. In: Nahum AM, Melvin JW (eds) Accidental injury: biomechanics and prevention, 1st edn. Springer, New York, pp 66–84
3. Mertz HJ (2002) Anthropomorphic test devices. In: Nahum AM, Melvin JW (eds) Accidental injury: biomechanics and prevention, 2nd edn. Springer, New York, pp 72–88
4. Anthropomorphic Test Dummy (1973) Code of Federal Regulations, Title 49, Chapter V, part 572, subpart B. Federal Register 38(142)
5. Foster JK, Kortge JO, Wolanin MJ (1977) Hybrid III – a biomechanically-based crash test dummy. In: 21st Stapp car crash conference. SAE, Warrendale. Paper 770938
6. Backaitis SH, Mertz HJ (eds) (1994) Hybrid III: the first human like crash test dummy. SAE PT-44, Warrendale
7. Melvin JW, Shee TR (1989) Facial injury assessment techniques. In: 12th ESV conference
8. Rouhana SW, Jedrzejczak EA, McCleary JD (1990) Assessing submarining and abdominal injury risk in

the Hybrid III family of dummies: part II-development of the small female frangible abdomen. In: 34th Stapp car crash conference. SAE, Warrendale. Paper 902317
9. Klinich KD, Beebe MS, Backaitis SH (1995) Evaluation of a proposed Hybrid III hip modification. In: 39th Stapp car crash conference. SAE, Warrendale. Paper 952730
10. Mertz HJ, Irwin AL, Melvin JW, Stalnaker RL, Beebe MS (1989) Size, weight and biomechanical impact response requirements for adult size small female and large dummies. SAE paper no 890756
11. Irwin AL, Mertz HJ (1997) Biomechanical bases for the CRABI and Hybrid III child dummies. In: 41st Stapp car crash conference. SAE, Warrendale. Paper 973317
12. Wolanin MJ, Mertz HJ, Nyznyk RS, Vincent JH (1982) Description and basis of a three-year old child dummy for evaluating passenger inflatable restraint concepts (Republished as SAE 826040. automatic occupant protection systems. SP736. 1988) In: 9th ESV conference
13. Mertz HJ, Jarrett K, Moss S, Salloum M, Zhao Y (2001) The Hybrid III 10-year-old dummy. Stapp Car Crash J 45:319–328
14. Shams T, Rangarajan N, Higuchi K, Keller J, Haffner M (1996) Performance of the TAD-50M in vehicle barrier tests and comparison with Hybrid III. In: Paper 96-S10-O-07 presented at the 15th ESV conference. Melbourne, 13–16 May 1996
15. Nusholtz GS, Xu L, Venkatesh A, Athey J, Balser J, Daniel RP, Hultman RW, Kirkish S, Mertz HJ, Parks D, Prater J, Rouhana SW, Scherer R (1997) Comparison of the pre-prototype NHTSA advanced dummy to the Hybrid III. SAE technical paper 971141
16. Schneider LW, Ricci LL, Beebe MS, King AI, Rouhana SW, Neathery RF (1994) Design and development of an advanced ATD thorax system for frontal crash environments Final Report on NHTSA contract DTNH22-83-C-07005
17. Rangarajan N, White R, Shams T, Beach D, Fullerton J, Haffner MP, Eppinger RE, Pritz H, Rhule D, Dalmotas DJ, Fournier E (1998) Design and performance of the THOR advanced frontal crash test dummy thorax and abdomen assemblies. In: Paper 98-S9-O-12 presented at the 16th ESV conference. Windsor, 31 May – 4 June 1998
18. Xu L, Jensen J, Byrnes K, Kim A, Venkatesh A, Davis KL, Hultman RW, Kostyniuk G, Marshall ME, Mertz H, Nusholtz G, Rouhana S, Scherer R (2000) Comparative performance evaluation of THOR and Hybrid III. SAE 2000-01-0161
19. Haffner M, Eppinger R, Rangarajan N, Shams T, Artis M, Beach D (2001) Foundations and elements of the NHTSA THOR alpha ATD design. In: Paper 458 presented at the 17th ESV conference. Amsterdam, 4–7 June 2001
20. Ridella SA, Parent DP (2011) Modifications to improve the durability, usability and biofidelity of the THOR-NT dummy. In: Paper 11-0312 presented at the 22nd ESV conference. Washington, DC, 13–16 June 2011
21. van Ratingen MR, Twisk D, Schrooten M, Beusenberg MC, Barnes A, Platten G (1997) Biomechanically based design and performance targets for a 3-year old child crash dummy for frontal and side impact. In: Child occupant protection 2nd symposium proceedings. SAE, Warrendale. Paper 973316
22. de Jager K, van Ratingen M, Lesire P, Guillemot H, Pastor C, Schnottale B, Tegera G, Lepretre JP (2005) Assessing new child dummies and criteria for child occupant protection in frontal impact. In: Paper 05-0157 presented at the 19th ESV conference. Washington, DC. 6–9 June 2005
23. Bellais P, Alonzo F, Chevalier MC, Lesire P, Leopold F, Trosseille X, Johannsen H (2012) Abdominal twin pressure sensors for the assessment of abdominal injuries in Q dummies: in-dummy evaluation and performance in accident re-constructions. Stapp Car Crash J 56:387–410
24. Melvin JW, Robbins DH, Benson JE (1979) Experimental application of advanced thoracic instrumentation techniques to anthropomorphic test devices. In: Paper presented at the 7th ESV conference, Paris
25. Donnelly BR (1982) Assembly manual for the NHTSA side impact dummy. CALSPAN report no. 70733V-I; Contract no. DTNH22-82-C-07366
26. Eppinger RH, Marcus JH, Morgan RM (1984) Development of dummy and injury index for NHTSA's thoracic side impact protection research program. In: Paper 840885 presented at the SAE government/industry meeting and exposition, Washington, DC
27. Morgan RM, Marcus JH, Eppinger RH (1986) Side impact – the biofidelity of NHTSA's proposed ATD and efficacy of TTI. In: 30th Stapp car crash conference. SAE, Warrendale. Paper 861877
28. Mertz HJ (1990) Rating of measurement capacity of side impact dummies ISO/TC22/SCI2/WG5 Document N281
29. The biofidelity test results on SID and EUROSID. ISO/TC22/SC12/WG5 Document N213
30. Resolution No.1. ISO/TC22/SC12/WG5 Document N298 (1990)
31. A method to calculate a single, weighted biofidelity value for a side impact dummy (1990). ISO/TC22/SC12/WG5 Document N253
32. Proposed weighting factors for rating the impact response biofidelity of various side impact dummies (1990). ISO/TC22/SC12/WG5 Document N278
33. Ballot results on biofidelity weighting factors for side impact dummies (1990). ISO/TC22/SC12/WG5 Document N284
34. Bendjellal F, Tarriere C, Brun-Cassan F, Foret-Bruno JY, Caillibot P, Gillet D (1988) Comparative evaluation of the biofidelity of EUROSID and SID side impact dummies. In: 32nd Stapp car crash conference. SAE, Warrendale. Paper 881717

35. Irwin AL, Pricopio LA, Mertz HJ, Balser JS, Chkoreff WM (1989) Comparison of the EUROSID and SID impact responses to the response corridors of the International Standards Organization, SAE 890604
36. Mertz HJ, Irwin AL (1990) Biofidelity ratings of SID, EUROSID and BIOSID ISO/TC22/SC12/WG5 Document N288
37. Neilson L, Lowne R, Tarriere C, Bendjellal F, Gillet D, Maltha J, Cesari D, Bouquet R (1985) The EUROSID side impact dummy. In: Paper presented at the 10th ESV conference
38. Lowne RW, Neilson JD (1987) The development and certification of EUROSID. In: 11th ESV conference
39. Janssen EG, Vermissen ACM (1988) Biofidelity of the European side impact dummy-EUROSID. In: 32nd Stapp car crash conference. SAE, Warrendale. Paper 881716
40. The EUROSID-1 (1995) Technical information from the TNO Crash Safety Research Center, Delft
41. EUROSID-1 Developments (1997) ISO/TC22/SC12/WG5 Document N533
42. Beebe MS (1990) What is BIOSID? SAE, Warrendale, Paper 900377
43. Beebe MS (1991) BIOSID update and calibration requirements. SAE, Warrendale, Paper 910319
44. van Ratingen MR (2001) Development and evaluation of the ES-2 side impact dummy. In: Paper 336 presented at the 17th ESV conference, Amsterdam, 4–7 June 2001
45. Byrnes K, Abramczyk J, Berliner J, Irwin A, Jensen J, Kowsika M, Mertz HJ, Rouhana SW, Scherer R, Shi Y, Sutterfield A, Xu L, Tylko S, Dalmotas D (2002) ES-2 dummy biomechanical responses. Stapp Car Crash J 46:353–396
46. Sutterfield A, Pecoraro K, Rouhana SW, Xu L, Abramczyk J, Berliner J, Irwin A, Jensen J, Mertz HJ, Nusholtz G, Pietsch H, Scherer R, Tylko S (2005) Evaluation of the ES-2re dummy in biofidelity, component, and full vehicle crash tests. Stapp Car Crash J 49:481–508
47. Rhule H, Moorhouse K, Donnelly B, Stricklin J (2009) Comparison of WorldSID and ES-2re biofidelity using an updated biofidelity ranking system. In: Paper 09-0563 presented at the 21st ESV conference, Stuttgart, 15–18 June 2009
48. Daniel RP, Irwin A, Athey J, Balser J, Eichbrecht P, Hultman RW, Kirkish S, Kneisly A, Mertz H, Nusholtz G, Rouhana S, Scherer R, Salloum M, Smrcka J (1995) Technical specifications of the SID-IIs dummy. In: 39th Stapp car crash conference. SAE, Warrendale. Paper 952735
49. Kirkish SL, Hultman RW, Scherer R, Daniel R, Rouhana S, Nusholtz G, Athey J, Balser J, Irwin A, Mertz H, Kneisly A, Eichbrecht P, Salloum M (1996) Status of prove-out testing of the SID-IIs alpha-prototype. In: Paper presented at the 15th ESV conference, Melbourne, 13–16 May 1996
50. Road vehicle-anthropomorphic side impact dummy-lateral impact response requirements to assess the biofidelity of the dummy. (1998). ISO/TR9790. ANSI, New York
51. Saul RA, Backaitis SH, Beebe MS, Ore LS (1996) Hybrid III dummy instrumentation and assessment of arm injuries during airbag deployment. In: 40th Stapp car crash conference. SAE, Warrendale. Paper 962417
52. Scherer R, Cesari D, Uchimura T, Kostyniuk G, Page M, Asakawa K, Hautmann E, Bortenschlager K, Sakurai M, Harigae T (2001) Design and evaluation of the WorldSID prototype dummy. In: Paper 409 presented at the 17th ESV conference, Amsterdam, 4–7 June 2001
53. Been B, Meijer R, Bermound F, Bortenschlager K, Hynd D, Martinez L, Ferichola G (2007) WorldSID small female side impact dummy specifications and prototype evaluation. In: Paper 07-0311 presented at the 20th ESV conference, Lyon, 18–21 June 2007
54. Eggers A, Schnottale B, Been B, Waagmeester K, Hynd D, Carroll J, Martinez L (2009) Biofidelity of the WorldSID small female revision 1 dummy. In: Paper 09-0420 presented at the 21st ESV conference, Stuttgart, 15–18 June 2009
55. Carlson M, Burleigh M, Barnes A, Waagmeester K, van Ratingen M (2009) Q3s 3 year old side impact dummy development. In: Paper 07-0205 presented at the 20th ESV conference, Lyon, 18–21 June 2007
56. Wang ZJ, Yao C, Jacuzzi E, Marudhamuthu K (2009) Development and improvement of Q3s – a three year old child side impact dummy. In: Paper 09-0358 presented at the 21st ESV conference, Stuttgart, 15–18 June 2009
57. Federal Register (2013) Anthropomorphic test devices; Q3s 3-year-old child side impact test dummy; Proposed Rule. 49 CFR Part 572, vol 78(225), 21 Nov 2013
58. Federal Register (2014) Federal motor vehicle safety standards; Child restraint systems – Side impact protection; Proposed Rule. 49 CFR Part 571, vol 79(18), 28 Jan 2014
59. Irwin AL, Mertz HJ, Elhagediab AM, Moss S (2002) Guidelines for assessing the biofidelity of side impact dummies of various sizes and ages. Stapp Car Crash J 46:297–319
60. Davidsson J, Svensson MY, Flogard A, Haland Y, Jakobsson L, Lovsund P, Wiklund K (1998) BioRID I – a new biofidelic rear impact dummy. In: IRCOBI
61. Philippens M, Cappon H, van Ratingen M, Wismans J, Svensson M, Sirey F, Ono K, Nishimoto N, Matsuoka F (2002) Comparison of the rear impact biofidelity of BioRID II and RID2. Stapp Car Crash J 46:461–476
62. Svensson MY, Louscind PA (1992) Dummy for rear-end collisions: development and validation of a new dummy neck. In: IRCOBI
63. Cappon H, Philippens M, van Ratingen M, Wismans J (2001) Development and evaluation of a new rear-impact crash dummy: the RID 2. Stapp Car Crash J 45:225–238

64. Thunnissen JG, van Ratingen MR, Beusenberg MC, Janssen EO (1996) A dummy neck for low severity rear impacts. In: Paper 96-Sl0-0-12 presented at the 15th ESV conference, Melbourne, 13–16 May 1996
65. Kim A, Anderson KF, Berliner J, Hassan J, Jensen J, Mertz HJ, Pietsch H, Rao A, Scherer R, Sutterfield A (2003) A biofidelity evaluation of the BioRID II, Hybrid III and RID2 for use in rear impacts. Stapp Car Crash J 47:489–523
66. ISO TC22/SC12/WG5 (2012) Adult dummies recommended by WG5. Document N1012
67. Nyquist GW, Begeman PC, King AI, Mertz HJ (1979) Correlation of field injuries and GM Hybrid III dummy responses for lap-shoulder belt restraint. In: ASME winter meeting (79-WA/BIO-2)
68. Mertz HJ, Prasad P, Irwin AL (1997) Injury risk curves for children and adults in frontal and rear collisions. In: 41st Stapp car crash conference. SAE, Warrendale. Paper 973318
69. Mertz HJ (1984) Injury assessment reference values used to evaluate Hybrid III response measurements. NHTSA docket 74-14. Notice 32. Enclosure 2 of Attachment I of Part III of General Motors Submission USG2284
70. Anthropomorphic dummies for crash and escape system testing (1996) AGARD-AR-330 ISBN 921039-3
71. Proposal for dummy response limits for FMVSS 208 compliance testing. Attachment C of AAMA Submission S98-13 (1998). NHTSA Docket 98-4405. Notice 3
72. Mertz HJ, Horsch JD, Horn G, Lowne RW (1991) Hybrid III sternal deflections associated with thoracic injury severities of occupants restrained with force-limiting shoulder belts. SAE 910812
73. Mertz HJ, Irwin AL, Prasad P (2003) Biomechanical and scaling bases for frontal and side impact injury assessment reference values. Stapp Car Crash J 47:155–188
74. Mertz HJ, Prasad P (2000) Improved neck injury risk curves for tension and extension moment measurements of crash dummies. Stapp Car Crash J 44:59–75
75. Mertz HJ, Weber DA (1982) Interpretations of the impact responses of a three-year old child dummy relative to child injury potential. (Republished as SAE 826048, Automatic Occupant Protection Systems, SP-736, 1988). In: Paper presented in the 9th ESV conference
76. Prasad P, Daniel RP (1984) A biomechanical analysis of head, neck, and torso injuries to child surrogates due to sudden torso acceleration. In: 28th Stapp car crash conference. SAE, Warrendale. Paper 841656
77. Melvin JW (1995) Injury assessment reference values for the CRABI 6-month dummy in a rear-facing infant restraint with airbag deployment. SAE 950872
78. Golub C, Sheth NJ (1956) Analysis of sensitivity experiments when levels of the stimulus cannot be controlled. J Am Stat Assoc 51:257–265
79. Mertz HJ, Prasad P, Nusholtz G (1996) Head injury risk assessment for forehead impacts, SAE 960099, Warrendale
80. Nakahira Y, Furukawa K, Niimi H, Ishihara T, Miki K, Matsuoka F (2000) A combined evaluation method and a modified maximum likelihood method for injury risk curves. In: IRCOBI
81. Lipson C, Sheth HJ (1973) Statistical design and analysis of engineering experiments. McGraw-Hill, New York
82. Prasad P, Mertz HJ (1985) The position of the United States delegation to the ISO working group 6 on the use of HIC in the automotive environment. SAE 851246
83. Mertz HJ (2002) Injury risk assessment based on dummy responses. In: Nahum AM, Melvin JW (eds) Accidental injury: biomechanics and prevention, 2nd edn. Springer, New York, pp 89–102
84. Mertz HJ (1984) A procedure for normalizing impact response data. SAE 840884, Washington, DC
85. Fredericks RH (1965) SAE test procedures for steering wheels. In: 9th Stapp car crash conference, University of Minnesota, Minneapolis. Paper 650962
86. Hansen AM (1965) SAE test procedure for instrument panels. In: 9th Stapp car crash conference, University of Minnesota, Minneapolis. Paper 650963
87. Patrick LM, Kroell CK, Mertz HJ (1966) Impact dynamics of unrestrained, lap belted, and lap and diagonal chest belted vehicle occupants. In: 10th stapp car crash conference. SAE, Warrendale. Paper 660788
88. Rieser RG, Michaels GE (1965) Factors in the development and evaluation of safer glazing. In: 9th Stapp car crash conference, University of Minnesota, Minneapolis. Paper 650959
89. Skeels PC, Falzon RG (1962) A new laboratory device for simulating vehicle crash conditions In: 6th Stapp car crash conference. AAAM, Des Plaines. Paper 1962-12-0014
90. Daniel RP, Patrick LM (1965) Instrument panel impact study In: 9th Stapp car crash conference, University of Minnesota, Minneapolis. Paper 650958
91. Gadd CW (1966) Use of a weighted impulse criterion for evaluating injury hazard. In: 10th Stapp car crash conference. SAE, Warrendale. Paper 660793
92. Patrick LM, Kroell CK, Mertz HJ (1965) Forces on the human body in simulated crashes. In: 9th Stapp car crash conference, University of Minnesota, Minneapolis. Paper 650961
93. Mertz HJ, Marquardt JF (1985) Small car air cushion performance considerations. SAE 851199. Warrendale
94. SAE International (1990) Guidelines for evaluating out-of-position vehicle occupant interactions with deploying frontal airbags. SAE J1980, Warrendale
95. SAE International (1993) Guidelines for evaluating child restraint system interactions with deploying airbags. SAE J2189, Warrendale
96. Lund AK (2000) Recommended procedures for evaluating occupant injury risk from deploying side airbags. IIHS (1st Revision published in 2003)

97. Bean JD, Kahane CJ, Mynatt M, Rudd RW, Rush CJ, Wiacek C (2009) Fatalities in frontal cashes despite seat belts and air bags – review of all CDS cases – model and calendar years 2000–2007 – 122 fatalities. NHTSA technical report, DOT HS 811 202
98. Mueller BC, Sherwood CP, Arbelaez RA, Zuby DS, Nolan JM (2011) Comparison of Hybrid III and THOR dummies in paired small overlap tests. Stapp Car Crash J 55:379–409
99. Mertz HJ, Dalmotas DJ (2007) Effects of shoulder belt limit forces on adult thoracic protection in frontal collisions. Stapp Car Crash J 51:361–380
100. Mertz HJ, Patick LM (1967) Investigation of the kinematics and kinetics of whiplash. In: 11th Stapp car crash conference. SAE, Warrendale. Paper 670919

Restraint System Biomechanics

Richard W. Kent and Jason Forman

Abstract

The fundamentals of occupant protection in a crash involve vehicle crashworthiness and occupant restraint. Crashworthiness refers to implementing a strong occupant compartment that resists intrusion and to designing crushable front and rear structures that deform and perform work to dissipate the kinetic energy of the crash. This combination provides controlled vehicle deceleration and survival space in the occupant compartment. Occupant restraint refers to the use of lap-shoulder belts, airbags, and other systems to provide ride-down of the vehicle deceleration, containment on the seat, and distribution of forces on the pelvis, shoulder, chest, and other designated anatomical structures to decelerate the occupant. The effectiveness of current restraint designs for protecting the occupant and reducing the risk of serious injury and death in a crash are well documented. Lap-shoulder belts are approximately 40 % effective in preventing death, with the highest effectiveness of nearly 80 % in rollovers and the lowest of approximately 30 % in near-side impacts [1]. In frontal crashes, the addition of the airbag to the three-point belt system raises the effectiveness level to approximately 50 %. Despite this impressive safety record, belt system performance is continually being refined. For example, recent papers have discussed the development of four-point harnesses for use in production vehicles [2, 3], and devices such as pretensioners and belt load

R.W. Kent, Ph.D. (✉) • J. Forman, Ph.D.
Mechanical and Aerospace Engineering,
University of Virginia, Charlottesville, VA, USA
e-mail: Rwk3c@virginia.edu; Jlf3m@virginia.edu

limiters [4] are becoming common features in contemporary vehicles. A pretensioner, which is usually pyrotechnic, triggers from the deceleration characteristic imparted to the vehicle by an impact. The charge winds several centimeters of belt around its storage spool, thereby removing slack in the system, generating a decelerating force on the occupant, and reducing the time required to generate force (tension) in the belt (i.e., the response time). Load-limiters typically involve an element within the belt retractor that yields when a pre-determined belt tension level is reached. This tension may range from 2 to 6 kN, and in the most advanced systems can be transient and programmable. These refinements introduce significant changes in the shape of the restraining force profile (Fig. 5.1), which have been shown to enhance belt performance both in the laboratory and on the road [6–12]. As this technology continues to develop, the characteristics of the restraining force applied to the occupant are becoming increasingly controllable. Continued improvement in restraint performance may be possible by implementing active control as an integral part of the restraint system. Steps have been taken to develop active systems (sometimes referred to as "smart" restraints) that adapt based on various inputs. For example, dual-stage pretensioners may modulate the magnitude of belt retraction based on the severity of the collision. Other "smart" aspects of restraint systems have been discussed by several researchers [4, 13–16].

This chapter reviews the biomechanics of restraints. The discussion includes a description of occupant kinematics for belted and unbelted occupants in frontal impacts with and without an airbag. The synergistic integration of restraint system components is a particular focus. For example, the airbag's role in facilitating force-limiting belts is discussed. Non-frontal impacts are then discussed, followed by some special restraint design considerations, such as the occupant's age and body habitus.

5.1 General Considerations

The fundamentals of occupant protection in a crash involve vehicle crashworthiness and occupant restraint. Crashworthiness refers to implementing a strong occupant compartment that resists intrusion and to designing crushable front and rear structures that deform and perform work to dissipate the kinetic energy of the crash. This combination provides controlled vehicle deceleration and survival space in the occupant compartment. Occupant restraint refers to the use of lap-shoulder belts, airbags, and other systems to provide ride-down of the vehicle deceleration, containment on the seat, and distribution of forces on the pelvis, shoulder, chest, and other designated anatomical structures to decelerate the occupant. The effectiveness of current restraint designs for protecting the occupant and reducing the risk of serious injury and death in a crash are well documented. Lap-shoulder belts are approximately 40 % effective in preventing death, with the highest effectiveness of nearly 80 % in rollovers and the lowest of approximately 30 % in near-side impacts [1]. In frontal crashes, the addition of the airbag to the three-point belt system raises the effectiveness level to approximately 50 %. Despite this impressive safety record, belt system performance is continually being refined. For example, recent papers have discussed the development of four-point harnesses for use in production vehicles [2, 3], and devices such as pretensioners and belt load

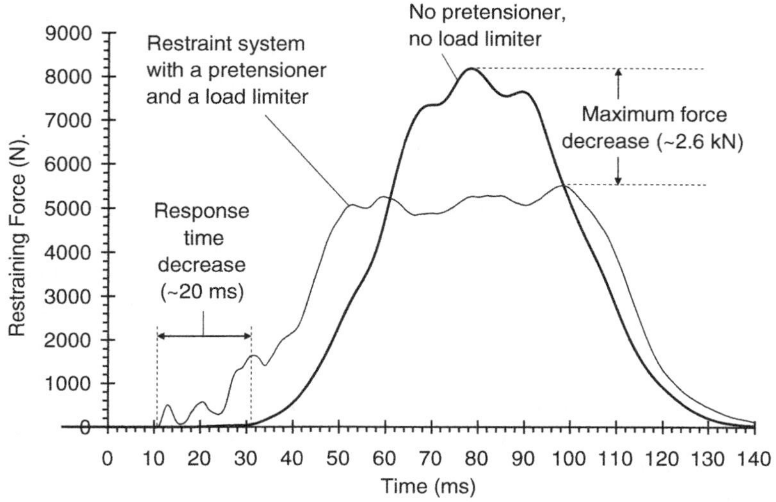

Fig. 5.1 Impact test data from Kent et al. [5] illustrating the effect of a pretensioner and load limiter. The total forward motion of the occupant relative to the vehicle is similar for the two systems, despite the significant reduction in force applied to the occupant. New restraint technology is creating the potential for even greater control and flexibility in the shape of the restraining force profile

limiters [4] are becoming common features in contemporary vehicles. A pretensioner, which is usually pyrotechnic, triggers from the deceleration characteristic imparted to the vehicle by an impact. The charge winds several centimeters of belt around its storage spool, thereby removing slack in the system, generating a decelerating force on the occupant, and reducing the time required to generate force (tension) in the belt (i.e., the response time). Load-limiters typically involve an element within the belt retractor that yields when a pre-determined belt tension level is reached. This tension may range from 2 to 6 kN, and in the most advanced systems can be transient and programmable. These refinements introduce significant changes in the shape of the restraining force profile (Fig. 5.1), which have been shown to enhance belt performance both in the laboratory and on the road [6–12]. As this technology continues to develop, the characteristics of the restraining force applied to the occupant are becoming increasingly controllable. Continued improvement in restraint performance may be possible by implementing active control as an integral part of the restraint system. Steps have been taken to develop active systems (sometimes referred to as "smart" restraints) that adapt based on various inputs. For example, dual-stage pretensioners may modulate the magnitude of belt retraction based on the severity of the collision. Other "smart" aspects of restraint systems have been discussed by several researchers [4, 13–16].

This chapter reviews the biomechanics of restraints. The discussion includes a description of occupant kinematics for belted and unbelted occupants in frontal impacts with and without an airbag. The synergistic integration of restraint system components is a particular focus. For example, the airbag's role in facilitating force-limiting belts is discussed. Non-frontal impacts are then discussed, followed by some special restraint design considerations, such as the occupant's age and body habitus.

5.2 Introduction

Airbags and safety belts serve complementary functions for occupant protection since a system of technologies is needed to provide maximum safety in the wide variety of real-world crashes [17]. For example, safety belts can prevent fatal injuries in many crash modes, but they must be worn by the occupant to be effective.

Conversely, airbags require no action by the occupant, but they do not provide protection in all impact scenarios. Even as many thousands of lives are being saved by belt use and supplemental airbags, however, serious injury and fatalities still occur in severe crashes. Likewise, any restraint system can have a negative effect in certain instances. Seat belts can cause injurious thoracic, abdominal and, more rarely, neck loading, particularly if they are misused. In particular, injury can result from or be exacerbated by placement of the shoulder harness under the arm and wearing of the lap-belt high on the abdomen with poor seating posture. An airbag can also cause injury if the occupant is in the path of deployment. Therefore, the airbag design requires a balance between a long fill time to reduce the risk of inflation injury and a rapid inflation to fill the space between the occupant and the interior. The design, implementation, and legislation of these restraint systems therefore require consideration of a complex web of biomechanical and other tradeoffs.

5.3 Seat Belt Biomechanics and General Restraint Maxims

The fundamental principles of crash energy management are most easily illustrated with a collinear frontal impact between a moving vehicle and a fixed rigid barrier. The kinetic energy of the vehicle must be dissipated through the process of work. Most of that work is done by the deforming structure of the vehicle. Likewise, the kinetic energy of the occupant must be dissipated through the application of forces to the occupant. In broad terms, anything that applies these forces can be considered to be part of the restraint system, including the seat, knee bolster, steering wheel, seatbelt, and airbag. The vehicle structural mechanics and the restraint mechanics are inextricably linked as the vehicle's deceleration during an impact is dictated in large part by its structure and the vehicle deceleration can be thought of as an input that drives the restraint system's development of forces on the occupant.

These two components of the vehicle's occupant protection system (structure and restraints) are designed to maximize the dissipation of the occupant's kinetic energy while minimizing the risk of injury from the restraining forces.

These fundamental concepts of occupant restraint can be illustrated if the speeds of two particles are plotted as a function of time. Consider the vehicle as a particle and the occupant inside the vehicle as another. Assume that the motion is one-dimensional and that the vehicle and the occupant are both translating at a speed of 48 km/h. If the vehicle strikes, for example, a rigid fixed barrier, it will rapidly undergo a change in speed from 48 km/h to 0 km/h. For a contemporary vehicle front end, this will occur over approximately 100 ms, though this time varies depending on the nature of the vehicle's structure. If the front end of the vehicle crushes 60 cm during this impact, the vehicle will decelerate at an average of approximately 15 g. All else equal, if it crushes less it will have a greater average deceleration, and vice-versa. The occupant's speed over the ground will remain constant until the occupant is acted upon by a force, but the deceleration of the vehicle will build relative speed between the vehicle and the occupant and the occupant will begin to translate relative to the vehicle. The force acting on an unrestrained occupant will be virtually zero until the occupant strikes the interior of the vehicle. Assuming that the occupant is 60 cm from the interior at the start of the impact, the occupant will strike the vehicle interior after approximately 100 ms, and the relative speed between the occupant and the vehicle will have become close to the pre-impact speed of the vehicle over the ground. The occupant will then rapidly undergo a change in speed from its pre-impact speed to zero. This scenario is shown in the top plot of Fig. 5.2.

Contrast this scenario with the case of a restrained occupant illustrated in the bottom plot of Fig. 5.2. In this case, again the vehicle will rapidly decelerate upon impact and the occupant's speed over the ground will remain constant until the occupant is acted upon by a force. If that force is generated by a restraint system, it can occur earlier and in a more controlled manner

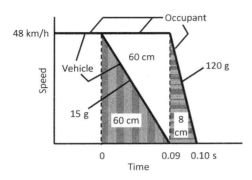

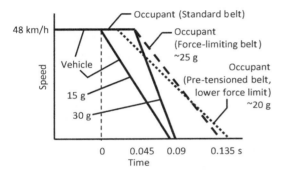

Fig. 5.2 Particle speed plotted as a function of time to illustrate the fundamental behavior of an unrestrained (*top*) and restrained (*bottom*) occupant in a frontal impact

than the forces that are generated by a strike with the vehicle interior. This will generate a more gradual change in the occupant's speed during the impact. The restraint system will begin to generate force on the occupant a few milliseconds after impact (pre-tensioners and occupant positioning systems reduce this time) and apply that force over the entire available distance between the occupant and the vehicle interior. With a well-designed, contemporary restraint, the occupant's deceleration during the impact may actually be less than the deceleration of the vehicle since the occupant has a greater distance available over which to decelerate (the vehicle decelerates over the distance that it crushes while the occupant decelerates over that distance plus the initial space between the occupant and the vehicle interior). The peak force acting on an unrestrained occupant can be several multiples, even orders of magnitude, greater than the peak force acting on a restrained occupant.

These fundamental concepts, combined with general considerations of human anatomy and injury mechanics, suggest several general restraint maxims:

Restraint Maxims (Mechanical)
- The time over which restraint force is applied should be maximized. This minimizes the magnitude of that force and hence the associated risk of injury.
- The distance traveled by the occupant over the ground as restraining forces are applied should be maximized. This minimizes the magnitude of the applied force and maximizes the work performed to dissipate the kinetic energy of the occupant. This distance is limited by the available distance between the occupant and the vehicle interior and by the distance available for vehicle crush.
- The restraining force should be maximized below an injurious level and should be applied as early as possible. This maximizes the work performed to dissipate the kinetic energy of the occupant and minimizes the required distance between the occupant and the vehicle interior.

Restraint Maxims (Anatomical)
- Body articulations, local deformation, rate of deformation, and acceleration of the occupant should be minimized to minimize the risk of injury.[1]
- The restraining force should be applied over the largest possible area. Load distribution reduces body deformation and surface pressure for a given magnitude of force.
- The magnitude of restraining forces applied to a body region should not exceed the force tolerance of that region. This generally results in restraining forces being applied primarily to the bony structures of the head, shoulder, upper thorax, pelvis, and femur as these are most able to bear load without injury.

---

[1] These kinematic factors are not, however, all equal in terms of priority or injury risk. As discussed in this chapter, load distribution and the locations of application influence the relationship between applied force and each of these potentially injurious kinematics. Furthermore, whole-body acceleration is tolerable to very high magnitudes if the body does not deform [18, 19] and the role of deformation rate is negligible in many injurious loading scenarios [20–22].

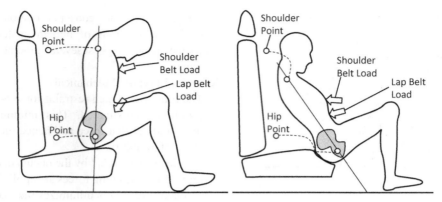

Fig. 5.3 Fundamental kinematics of "high quality" (*left*) and "low quality" (*right*) belt restraint in a frontal crash [23]

A series of papers in the mid-1970s described motion sequence and restraint design criteria to focus restraining loads on biomechanically favorable anatomical locations and to minimize relative displacement between individual body parts [23–25]. These criteria have formed an important biomechanical foundation for restraint design, and remain key aspects of contemporary restraint design and evaluation [3]. The key concepts propounded in these studies were the trajectories of the hip and shoulder, which should conform to the following:

1. Vertical downward motion of the hip should be avoided. A "gentle upward motion" is allowable if it does not exceed 40 mm.
   This minimizes pelvic rotation and reduces the tendency for the lap-belt to slide off the ileum and directly load the abdomen.
2. The torso should be pitched forward to a vertical orientation or beyond when the shoulder belt force and the forward excursion of the chest reach their peak values.
   Forward rotation of the upper torso, to slightly greater than 90° upright posture, directs a major portion of the upper torso restraint into the shoulder (Fig. 5.3). The second criterion was not discussed by the original authors for the case of a primarily trapezoidal shoulder belt force-time history that has a long period of essentially constant force, such as that generated by some force-limiting belt systems. Regardless, it seems reasonable to assert that the recommended temporal relationship between torso pitch and forward excursion remains valid regardless of whether a load limiter is used.
   Finally, though not discussed explicitly in the Adomeit studies, it seems axiomatic that, all else being equal, a third criterion could be stated:
3. A restraint that allows less excursion, especially of the head, is preferable.

In a frontal impact, a snug-fitting lap-shoulder belt ties the occupant to the passenger compartment and allows the occupant to "ride-down" the crash as the vehicle front-end crushes. This coupling and ride-down decelerate the occupant more gradually than is possible with energy-absorbing interiors. Even with shoulder belt restraint, however, there is forward excursion of the torso and, in particular, movement of the head and neck toward the steering wheel or windshield. Figure 5.4 shows three time-points in a frontal crash of a lap-shoulder belted rear-seat occupant and highlights the forward lean of the upper torso, the flexion of the neck, and the excursion of the head. While this kinematic is more favorable than the consequence of an unrestrained occupant striking the interior surfaces at high speed, there is still the potential for head and face contact with the vehicle interior and for inertial injuries of the neck. In fact, significant neck injuries have been generated in cadavers with restraint mechanics that fail the Adomeit criteria [26]. Figure 5.5 shows the same type of crash sequence as Fig. 5.4 with a combined

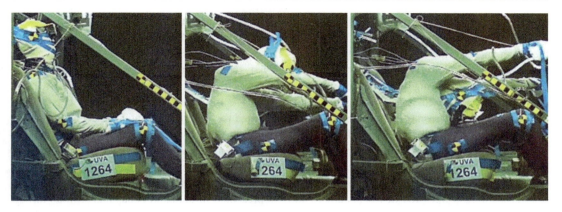

Fig. 5.4 Kinematics of a belted occupant without an airbag in a frontal collision [26]. Note the neck loading via head inertia and the potential for head/face contact with the vehicle interior

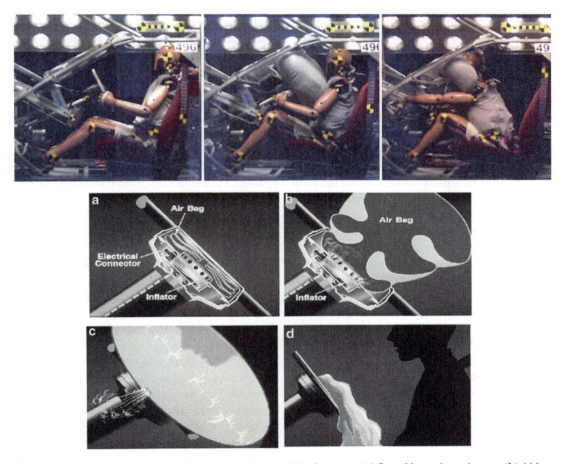

Fig. 5.5 Driver interaction with an airbag in a frontal collision. Inertial neck loading and the potential for head/face contact are mitigated and work is performed to dissipate kinetic energy. (**a**) Stored bag prior to impact. (**b**) Airbag inflation. (**c**) Kinetic energy dissipated via venting. (**d**) Airbag deflated after venting

driver airbag and lap-shoulder belt restraint. Early in the crash, sensors detect the severity of the crash and activate the inflation of the airbag if the collision severity is above a set threshold. This causes a rapid filling of the bag as it deploys out of the steering wheel hub. The bag then loads the head, neck, and torso to restrain the upper body in ways not possible with only the shoulder belt. Vent holes in the back of the airbag relieve pressure and absorb energy such that, even after occupant rebound, the airbag continues to deflate. This type of restraint provides a benefit to the occupant via several mechanisms, as described in the following section.

5.4 Inflatable Restraint Biomechanics

In the early days of airbag development, belt use was low and showed no signs of increasing in the near term [27]. Low belt use rate was the original impetus behind the development of airbag systems since they overcome the primary weakness of belt systems: to be effective the occupant must fasten the belts in advance of the crash. Using a pyrotechnic device to generate nitrogen gas, a bag can be rapidly inflated during the early phase of vehicle frontal crush without action by the occupant. The bag then "fills" some of the space between the occupant and the interior, which couples the occupant to the passenger compartment and achieves the safety benefits of ride-down and load distribution. Energy dissipation is achieved via venting since a non-vented bag acts essentially as an elastic spring once it is inflated [28].

For unbelted occupants, the airbag's principle benefit is load distribution and attenuation on the thorax. Despite low belt use being the original motivation, however, the airbag also provides a safety benefit for a belted occupant, albeit via a different mechanism. For belted occupants, the principle benefits of an airbag are due to mitigation of head contact and neck loading. The primary mechanisms by which airbags protect occupants are outlined below.

5.4.1 Head and Face Contact Mitigation

First, the airbag provides primary restraint to the head and face so that contact with the steering wheel is mitigated. This benefit has been documented thoroughly in frontal impact tests. This is not to say that an airbag precludes any risk of head or face contract, especially in the absence of concomitant belt use. Crandall et al. [29] and others have observed that head and face injuries are over-represented in unbelted occupants with an airbag compared to belted occupants with no airbag. Berg et al. [30] presented a series of seven cadaver tests, where an airbag restraint with no belt was used. They noted head impacts into the upper region of the windscreen or the roof in all tests. Depending on the size of the occupant, the severity of the collision, the geometry of the steering column and the vehicle interior, and the characteristics of the airbag, it is possible for an occupant to translate vertically over the airbag and sustain head or face loading via the steering wheel or the windshield/windshield header (Fig. 5.6). This injury mechanism emphasizes, again, the importance of the belt as a primary restraint and the airbag as a supplemental restraint. With a properly positioned belt restraint, the whole-body motion shown in Fig. 5.6 does not occur.

5.4.2 Load Sharing, Force Distribution, and Work

Second, the airbag distributes forces on the chest and shares loading with the belt system. Numerous studies have shown that cadavers restrained by a belt and an airbag sustain lower levels of injury than cadavers restrained by a belt alone or by an airbag alone [32] [7, 21, 33]. Shoulder injuries related to belt loading also seem to be reduced with the addition of an airbag restraint [34] (Fig. 5.7). These injury reductions are primarily due to the load sharing and force distributing effects of an airbag. It is well established that a person's tolerance to a force applied on the thorax is highly dependent upon the area

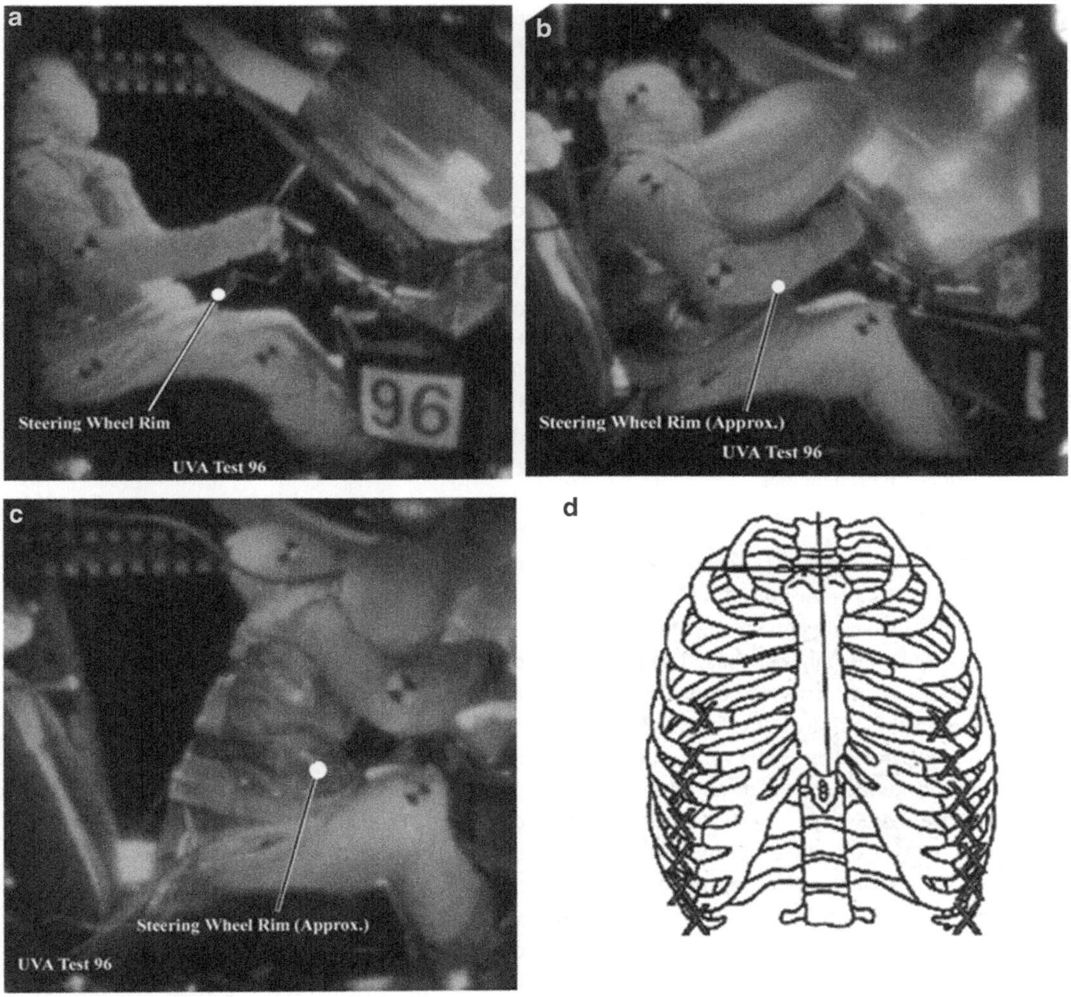

Fig. 5.6 Unbelted cadaver translating vertically over the airbag and sustaining a substantial head strike on the header (~175 g acceleration spike recorded at the first thoracic vertebra). Also note the steering wheel loading on the inferior anterior thorax, which resulted in multiple bilateral rib fractures, as shown. Adapted from Kent et al. [31]. (**a**) time = 0 ms. (**b**) time = 68 ms. (**c**) 122 ms. (max. chest deflection) (note head strike). (**d**) 14 rib fractures (7 displaced)

over which that force is applied. For example, Patrick et al. [35] performed a series of sled tests with embalmed cadavers impacting padded load cells anteriorly. They found that a 3.3 kN hub load to the sternum resulted in minor trauma, while similar injuries required approximately 8.8 kN if the load was distributed over the shoulders and chest. Bierman et al. [36] tested volunteers with a drop device that loaded a four-point belt harness 490 cm² in area. Painful reactions and some minor injuries occurred when loads exceeded 8.9 kN. When the load area was increased to 1,006 cm², loads up to 13.3 kN were sustained without pain or injury. The potential exists, therefore, to increase substantially the work done on an occupant and to perform this work at a non-injurious force level if the force can be distributed over the entire anterior thorax.

This work can be increased even further, without increasing thoracic injury risk, if the occupant

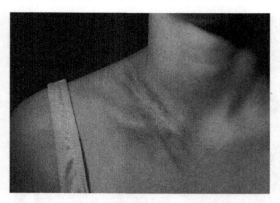

Fig. 5.7 Abrasion from shoulder belt loading. Belt injuries include fractured or dislocated clavicle, rib fractures, or others. The presence of an airbag can reduce the frequency and severity of belt-related injuries

is tensed during the impact. In a series of human volunteer sled tests performed at 48 km/h, loading through the steering wheel via tensed arms and through the floorpan via tensed legs had a pronounced effect in terms of controlling occupant kinematics and minimizing restraining loads through the belt and airbag. The role of arm bracing has been studied by Horsch and Culver [37] and shows a considerable ability to restrain the upper torso away from the steering wheel when the driver is alert of the pending crash. Similar effects can be seen with bracing by volunteers in simulated crashes [38]. However, the proportion of supplemental restraint by bracing decreases with increasing crash severity, simply because the occupant's strength becomes negligible compared to the inertial forces exerted on the upper body. In these cases, the airbag plays a greater role. Nonetheless, comparison of the tensed human volunteer in Fig. 5.8 with the atonic cadaver in Fig. 5.6 indicates the substantial effect of musculature in a crash as severe as 48 km/h. Armstrong et al. [40] estimated that as much as 55 % of an occupant's kinetic energy in a tolerable frontal impact may be dissipated via the work done by "propriotonic" restraint.

The load-sharing characteristics of airbags have led to the observation that thoracic acceleration, a widely used injury indicator, may not reflect an airbag's benefit to the occupant. By distributing forces on the chest, the total force on the chest can be higher than the force generated by belt loading, thereby resulting in higher chest acceleration. Furthermore, the fundamental of upper torso lean is an important attribute of belt restraint and loading through the bony structures of the body. Forces concentrated on the shoulder or upper thorax are more tolerable than forces concentrated lower on the thorax or abdomen. With this background, it is necessary to evaluate chest acceleration critically when comparing belt-dominated loading with airbag-dominated loading since a higher level of force, and therefore acceleration, can be tolerated if the force is distributed or if the torso is pitched forward. Grosch [41] found that the expected benefit due to a supplemental airbag was not reflected by the acceleration levels measured in laboratory tests and concluded that it is necessary to consider other indicators of injury risk. Others have observed that the chest acceleration peak due to bag slap during deployment can be the global maximum, though the magnitude of this peak (approximately 50 g) is insufficient to cause acceleration-induced injury [42]. An acceleration peak such as this, having a different cause than a comparable sled test with belt loading, can lead to misinterpretation of the relative benefits of different restraint conditions. In the Grosch test series, chest acceleration and chest deflection exhibited contradictory trends, with acceleration increasing when an airbag was added, while deflection decreased. Further complicating the assessment is the fact that this phasing is different for drivers and passengers. Driver-side airbags typically deploy rearward nearly far enough to load the occupant prior to any forward translation of the driver in the vehicle. The result of the driver's proximity to the deploying bag is that the airbag loading begins sooner after the bag deploys (approximately 50 ms after impact in a 48-km/h full-frontal sled test) than it does for the passenger. The driver's chest acceleration therefore typically exhibits a single peak, which occurs during combined belt and airbag loading. A passenger-side occupant, on the other hand, sustains primarily belt loading over the first approximately 75 ms after impact, which results in peak chest acceleration due primarily to belt loading (Fig. 5.9).

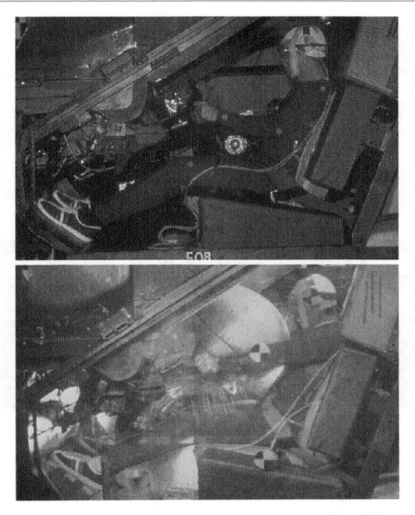

Fig. 5.8 Human volunteer frontal sled test (48 km/h) showing airbag system feasibility and pronounced effect of musculature, which reduces restraint loading relative to a dummy or cadaver test (From Holloman Air Force Base tests of the GM driver air cushion described in Smith et al. [39])

The later airbag loading then typically generates a second chest acceleration peak. Depending on the characteristics of the belt, airbag, vehicle interior, and occupant, a passenger's global acceleration maxima may occur either under belt loading or under combined loading, while a normally seated driver's maximum acceleration occurs under combined loading. This phenomenon was observed by Vezin et al. [43] and its importance for thoracic injury prediction in the laboratory is discussed in more detail by Kent et al. [31]. The combinatorial nature of belt and airbag loading also necessitates appropriate phasing of belt and airbag loading so that the combined effect does not exceed the acceleration tolerance of the human body, though cases exceeding the acceleration tolerance of a properly restrained occupant in an otherwise survivable frontal crash are assumed to be extremely rare (by inference from Melvin et al., 1998) [44]. Biomechanics researchers are currently working on the development of injury criteria that apply to combined belt and airbag systems. These criteria incorporate load distribution and load phasing within the functional form of the criterion, through multipoint measurement methods, or through the integration of the criterion within an injury risk function [5, 45].

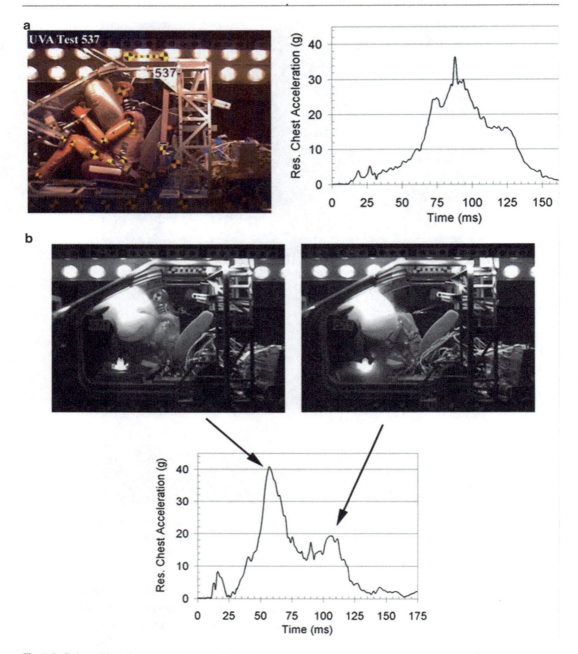

Fig. 5.9 Driver-side and passenger-side occupant kinematics and chest acceleration in a 30-mph frontal sled test with a force-limiting belt and airbag restraint system [31]. (**a**) Driver-side test showing Hybrid III 50th male at time of maximum chest acceleration. (**b**) Passenger-side showing Hybrid III 50th male at times of peaks in chest acceleration

While the airbag does share loading with the seat belt and thereby reduces the level of concentrated belt force exerted on the chest, it is typically an insufficient restraint by itself. As shown in Fig. 5.6, injurious thoracic loading can occur from steering wheel contact, even with an airbag restraint. This behavior was observed by Smith et al. [39] in volunteer tests and produced injuries in cadaver tests conducted by Yoganandan et al. [33]. Because airbags neither remain inflated nor provide lateral restraint, seat belts are needed to adequately control occupant kinematics over the range of crash types, including rollovers and side impacts.

5.4.3 Restraint System Design Flexibility

Third, an airbag facilitates the use of force-limiting belts, and allows for the use of lower force limits than would be possible without a supplemental restraint. While numerous experimental studies have documented the benefits in the laboratory, the value of lower belt force limits (e.g., 4 kN vs. 6 kN) has also been observed in the field [8]. By distributing loads, mitigating head and face contact, and sharing loading with the belt system, an airbag allows for lower levels of belt force to be applied to the occupant. This has two primary benefits. First, a force-limiting belt can be designed to yield before injurious belt forces are developed (Fig. 5.10). Second, a force-limiting belt promotes forward rotation of the torso, which generates a "higher quality" biomechanical restraint condition (Figs. 5.3 and 5.11) [23]. While a reduction of belt force is associated with a lower risk of belt-induced thoracic injury, there is a limit on how low the shoulder belt force should be reduced without compromising protection in crashes involving long-duration deceleration or multiple impacts. If the belt has yielded in a previous collision, there can be a loss of upper torso control and (with a sliding latch plate design) lap control in subsequent collisions. Since the airbag will deflate after the primary impact, it cannot be relied upon to provide restraint in secondary impacts. In this case, the belt restraint system must be available and functioning to provide retention and restraint of the occupant.

The discussion above describes the mechanisms by which airbags supplement seat belts in frontal collisions, and also illustrates how some benefit can be realized by unbelted occupants. The discussion also implies, however, the biomechanical tradeoffs associated with an inflatable restraint. First, occupant loading should occur as early as possible in a crash, so that the maximum ride-down benefit may be achieved. This emphasizes early deployment of a relatively large airbag and also the use of belt pretensioning with early activation and fast retraction to initiate early belt restraint. Second, the magnitude of restraint loading applied to the occupant should be the maximum tolerable so that injurious contact with interior components can be avoided. These requirements emphasize an airbag with relatively high internal pressure so that large forces can be developed. In contrast, the work performed on the occupant should occur over as much distance as possible prior to contact with the vehicle interior in order to minimize the magnitude of force that must be applied. Depending on the crash severity and occupant size, this emphasizes an airbag with lower internal pressures. Furthermore, occupants of different mass and geometry must be considered. It is a substantial design challenge to optimize the airbag system's performance given the range of conditions in which it must perform, including a single crash that may have both belted and unbelted occupants.

5.5 Near-Side Collisions

Side-impact collisions can be classified based on the seating position of the occupant of interest as either near-side or far-side collisions. Near side collisions occur when the study occupant is seated on the struck side of the vehicle. Near side collisions place occupants in close proximity to the point of collision, often with less than ½ meter between the occupant and the point of external vehicle contact [46]. This limits the time and space available to provide restraint, and is reflected in the risk of severe injury or death relative to other collision types [47–49]. Approximately 35 % of car occupant fatalities occur in side-impact collisions [50]; approximately 65 % of those are near-side collisions. Injury risk is associated with impact severity and is there correlated with the presence of intrusion into the vehicle [51, 52]. Serious injuries most commonly occur in the head, chest, thorax, abdomen, and pelvis [53–63]. Fatal cases often

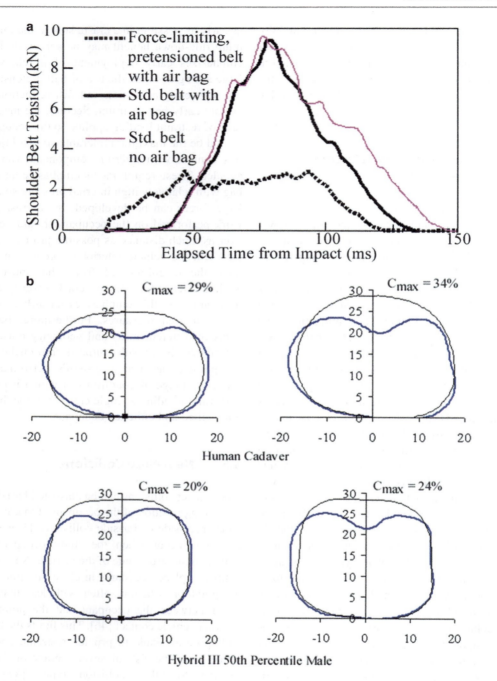

Fig. 5.10 Shoulder belt tension and thoracic cross-sectional deformation in a 48 km/h frontal impact with three different restraint conditions. (**a**) Shoulder belt tension. Note that the airbag reduces the magnitude and the duration of belt loading with a standard belt. Note also the substantial reduction in belt force that can be achieved with a force-limiting belt and airbag (the belt is also pretensioned in this case). (**b**) Thoracic profiles at maximum sternal deflection (C_{max}) with 4-kN force-limiting belt (*left*) and standard belt (*right*) with airbag. Note the increased C_{max} (expressed as a percent of undeformed chest depth) and decreased radius of curvature for the standard belt (units are cm)

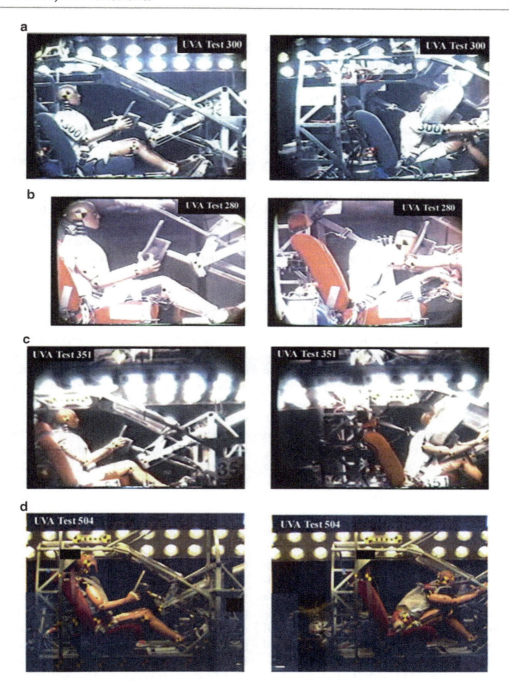

Fig. 5.11 57-km/h frontal impact restraint scenarios (50th percentile male) illustrating the benefit of an airbag. Comparing (**a**) and (**b**) shows the bag's benefit for mitigating head/face contact and for load sharing. The inadequacy in this test of a 4-kN force-limiting belt without an airbag is shown in (**d**). (**a**) Standard belt with an airbag, at impact (*left*) and maximum occupant excursion (*right*). (**b**) Standard belt with no airbag, at impact (*left*) and maximum occupant excursion (*right*). (**c**) 4-kN force-limiting belt with an airbag, at impact (*left*) and maximum occupant excursion (*right*). Note the increased occupant excursion and torso pitch relative to (**a**). (**d**) 3.5-kN force-limiting belt with no airbag, at impact (*left*) and maximum occupant excursion (*right*). Note that this is a different dummy (THOR) than the Hybrid III used in (**a**), (**b**), and (**c**)

include injury to the brain, aorta, heart, and spleen [53]. Brain injuries in near-side occupants most often resulted from striking either the pillar, the vehicle interior, or the striking vehicle (e.g., through the window) [53, 63].

The last decade has seen the widespread adoption of airbags for near-side occupant protection. Among other considerations, side airbags seek to fill the space between the body and the struck-side vehicle interior, distributing load and reducing the risk of directly impacting striking objects or the vehicle interior. These airbags take several different forms, both in shape and in the targeted body structure. Most systems can be classified as targeting either the torso, the head, or a combination of both [50, 64, 65]. Some combination systems can consist of a single airbag covering both the torso and the head. More commonly (at least in newer model years) combination systems consist of separate torso and head bags [66]. Torso airbags are mounted either in the seat back or the door [67], and vary widely in shape to engage the shoulder, lower chest, or to extend superiorly towards the head. Head protection airbags consist of either an inflatable tubular structure or an inflating curtain which extends from the roof rail down to the bottom of the window [50, 66].

Side airbags are generally protective against fatality and serious injury. Torso-only airbag systems are approximately 11–26 % effective at reducing fatalities in near-side collisions (compared to no airbag) [50, 68, 69]. Systems that include head protection are approximately 24–37 % effective at reducing fatalities [50, 69]. Kahane [50] concluded that adopting a combination of improved side structures and padding, a torso airbag, and a head curtain airbag, the fatality risk of drivers and right front passengers in all side-impact collisions (including far-side) could be reduced by 30 % for four-door cars and 42 % for two-door cars. Although data on non-fatal injuries are limited, McGwin et al. [70] suggested that side airbag availability is approximately 75 % effective in reducing the risk of any (AIS1+) head injury, and 68 % effective in reducing the risk of any thorax injury. As with other protective systems, however, the effectiveness of side airbags is closely related to the impact severity – decreasing at high impact velocities [46].

5.6 Far-Side Collisions

In far-side collisions (side-impact collisions where the occupant is opposite the struck side of the vehicle) occupant restraint occurs primarily from the seatbelt with some interaction with the seat cushion, seat back, and center armrest or console if present. The geometry of the shoulder belt – with outboard mounting of the upper shoulder belt anchor and the belt passing over the outboard shoulder – creates the potential for the belt to slip off of the shoulder [71]. This, in turn, creates the potential for greater lateral motion of the head and thorax, and greater loading of the upper abdomen and lower thorax by the lower portion of the shoulder belt. Head and thorax injuries, in general, account for over one half of all serious injuries in far-side impacts [72, 73]. Also of concern are abdominal injuries, especially to the liver and spleen [49, 74]. Much of the injury harm in far-side collisions comes from the head or thorax striking the door or window of the struck-side of the vehicle [71, 72, 75, 76], often in concert with intrusion [75]. Belt slip off of the shoulder has been attributed as a contributing factor in head injuries [77], strikes of the head and thorax against the intruding struck-side interior [72], increased loading of the lower chest and abdomen by the shoulder belt, and potential impact interaction between the struck-side and far-side vehicle occupants. Belt-slip off of the shoulder may also contribute to increased occupant excursion in oblique-far-side collisions [78].

Several studies have suggested a strong sensitivity between belt geometry and the presence of pretensioning on the risk of belt slip off of the shoulder in far-side collisions. Douglas et al. [79] presented a sensitivity study using volunteers, a Hybrid III dummy, and a MADYMO human model validated under far-side impact. That study found that both moving the D-ring rearward and adding a pretensioner tended to decrease the risk of belt slip off of the shoulder. Recent cadaveric tests have produced similar results, indicating a sensitivity to both D-ring position and pretensioning in both low-acceleration (6.6 g vehicle acceleration) and higher-acceleration (14 g) environments [80]. That study showed a decrease in

5 Restraint System Biomechanics

 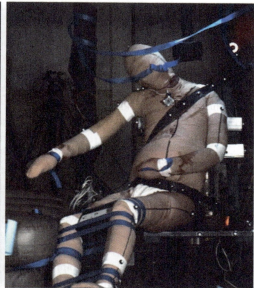

No pretensioner — With pretensioner

Fig. 5.12 Video captures at the time of maximum lateral head excursion in 15 km/h sled tests investigating shoulder-belt retension in a far-side collision. *Left*: Standard (not pre-tensioned) three-point belt. *Right*: same cadaver, belt with a (nominal) 1.8 kN retractor pretensioner. Notice the reduction in shoulder belt slip and lateral head motion with the pretensioning belt [80]

lateral head excursion when belt slip off of the shoulder was prevented through pretensioning, an improved D-ring position, or both (e.g., Fig. 5.12). That study also found that even when the belt did slip off of the shoulder, the addition of a pretensioner still tended to decrease lateral excursion of the head.

Other restraint concepts have been explored to prevent belt rollout (slip off of the shoulder) and reduce lateral head excursion in far-side collisions. Kallieris and Schmidt [81] performed a series of full-scale, far-side crash tests (50 km/h impacting speed; 60° and 90° impacting direction) with cadavers restrained by a reversed three-point belt system (i.e., the belt passed over the inboard shoulder). The authors concluded that the reversed belt system reduced the lateral excursion of the torso and head, however most of their cadaver exhibited AIS 1 injuries to the neck. Bostrom and Haland [2] explored the concept of adding a supplemental second shoulder belt (passing over the inboard shoulder and anchored to the seatback) to an existing three-point belt system. Bostrom et al. [72] also investigated the potential efficacy of incorporating an inboard airbag to support the torso (a "side support airbag" or "SSA"). That study suggested that the SSA may reduce AIS 3+ injuries in far-side collisions by as much as 57 %. Pintar et al. [82, 83] explored various combinations of generic side-support countermeasures in experiments with ATDs and cadavers, also suggesting that inboard torso and shoulder support may reduce lateral head excursion. Similarly, since a large portion of far-side injuries result from striking the struck-side door or window, some studies have suggested a potential benefit for far-side occupants from side-curtain airbags deployed on the struck-side of the vehicle [72]. Kahane [50] estimated a potential 24 % fatality reduction for far-side occupants when torso + head curtain side impact airbags are present on the struck side of the vehicle.

5.7 Rollover

Rollovers account for fewer than 10 % of all tow-away collisions, but are responsible for 20–33 % of fatal crashes and 15–25 % of crashes with serious injury [84–87]. Moderate-to-severe injuries in rollovers most often occur in the head, spine (including neck), and thorax [88–91]. There are several potential reasons for this disparity in risk compared to other collision modes. Some studies have suggested that rollovers tend to occur at greater vehicle speeds than normally observed in the entire population of tow-away collisions [86, 92]. Rollovers also carry a risk of ejection or partial ejection from the vehicle, especially for unbelted occupants [93]. In 2000, approximately 50 % of cases of fatally injured, unbelted occupants in rollovers in the U.S. were completely ejected from the vehicle [85]. Less than 10 % of the cases of fatal injuries to belted occupants involved complete ejection from the vehicle. Even when ejection is prevented, however, rollover collisions represent a chaotic environment containing multiple potential sources of injury. Head strikes against the vehicle roof, the side interior are common sources of head and neck injury for belted occupants in rollovers [86, 91]. Head injuries in belted occupants are also occasionally attributed to strikes against exterior structures through partial ejection [86]. Three-point seatbelts, when used, are approximately 64.2–77.9 % effective at preventing serious injury and 43.8 (far-side occupants) to 82.5 % (near-side occupants) effective at preventing fatality in rollover collisions [94].

A major goal of restraints in rollovers is to control the occupant kinematics to reduce the risk and severity of strikes to the vehicle interior (e.g., the vehicle roof and window) and to exterior structures (e.g., through partial ejection). Rollovers are complex events, with several stages, or phases of the collision resulting in various types of motion of the occupant [95], all depending on the specific characteristics of the collision. For example, in a tripped rollover the torso and head of the occupant may initially move laterally towards the tripped side of the vehicle. This may be followed by vertical occupant motion toward the roof as the vehicle begins to roll. This may be followed by lateral motion of the occupant in the opposite direction, depending on the nature and severity of the role. Eventually, the occupants will move towards the side of the vehicle (including the roof) that interacts with the ground to arrest the overall vehicle motion [95]. In some cases, this is also accompanied by intrusion of the vehicle interior toward the occupant.

Because of the prevalence of head and neck injuries attributed to head strikes against the roof, vertical head excursion toward the roof is a commonly studied topic. Moffat and James [96] described a number of factors that may affect vertical occupant excursions, related to the amount of headspace available. Several studies have investigated head excursion under three-point belt restraint using either quasi-static or dynamic volunteer, dummy, or cadaver roll-tests in laboratory settings [97–100]. Moffat et al. [100] concluded that vertical head excursion could be reduced with a lap-belt with a steep angle and with short webbing length, and that the shoulder belt portion of a three-point belt can reduce head excursion by limiting forward rotation of the torso. Other studies have investigated the potential efficacy of belt restraints with seat-integrated anchors, seeking to reduce excursions through shorter webbing lengths and more consistent control of belt geometry and fit [101, 102]. Other studies have investigated occupant kinematics under other phases of the rollover event, studying different directions of motion [103, 104]. Shoulder-belt retention issues similar to those observed in far-side lateral collisions have been observed in some laboratory-based roll studies [86, 104]. Due to the complex inertial environments presented, rollover collisions have also presented a concern for collision-sensing mechanical belt locks in retractors [105–110], prompting increased investigation of backup locking mechanisms (such as the webbing payout lock) in systems that do not include pretensioners [109].

Various supplemental belt technologies have been investigated to reduce occupant excursions, the most notable of which include pretensioning devices. Increasing the tension in the belt restraint system (in the lap belt in particular) has the potential to limit the motion of the occupant during a rollover [100, 111, 112]. Several studies have demonstrated that pyro-mechanical pretensioners may reduce occupant excursions during simulated rollover events [98, 102, 113]. Others have studied the potential efficacy of electro-mechanical retractor pretensioning devices [102, 114, 115]. Unlike pyrotechnic-based devices, electro-mechanical based pretensioners are reversible and, thus, can be disengaged and re-used after deployed. These are also usually much less aggressive than pyro-mechanical pretensioners, generating pretensioning forces on the order of 140 N, compared to forces between 1 and 2 kN with pyrotechnic pretensioners [102, 114, 116]. As a result of the reversibility and decreased aggressivity, electro-mechanical pretensioners can be deployed very early in the collision event, even when the occurrence of a collision is anticipated but not certain. For example, an electro-mechanical pretensioner may be programmed to deploy when the Electronic Stability Control system is activated, detecting an erratic vehicle dynamic that may result in a collision [102]. Electro-mechanical pretensioners may also be used in concert with pyro-mechanical pretensioners, where the electro-mechanical device removes belt slack and locks the belt early in the event, followed by the pyrotechnic device increasing the belt tension when a collision is certain. As a result, electro-mechanical pretensioners are often referred to as pre-pretensioners. Data from Sword et al. [102] suggests that the use of an electro-mechanical pre-pretensioner (at the initiation of rollover) in concert with a pyrotechnic pretensioner may reduce occupant lateral and vertical excursions relative to a pyrotechnic pretensioner alone.

Even when seatbelts are worn, there exists the possibility of partial ejection where a body structure strikes, and passes through, the window, increasing the risk of striking objects exterior to the vehicle. Some have investigated the efficacy of rollover-activated curtain airbags in reducing partial ejections and injury risk, most notably of the head [117, 118]. Rollover-activated side airbags (a.k.a. rollover curtains) are distinct from traditional side-curtain airbags designed solely for use in side-impact collisions. Because rollovers take place over substantially a substantially longer time-frame than planar side-impact crashes, rollover curtains must maintain their internal pressure longer than traditional side curtains [117, 118]. Takahashi et al. [118] also concluded that the volume of rollover curtains should be greater than traditional side airbags. Because rollover curtains also fill the role of side curtain airbags in planar lateral collisions, they must be designed to fill the protection, deployment, and sensing functions required in both types of collisions [118]. In a matched-pair field data study, Padmanaban and Fitzgerald [117] found that rollover curtains (as standard equipment) were approximately 23 % effective in reducing fatalities for all belted, front seat, outboard occupants in rollovers.

5.8 Special Considerations

5.8.1 Rear Seat Occupants

Several recent studies have suggested that the adoption of advanced restraint technologies in the front seat of vehicles has contributed to a disparity in safety between front seat and rear seat occupants [119]. Kuppa et al. [120] reported that the fatality risk for rear seat occupants was greater than front seat occupants after approximately age 50. Similarly, Bilston et al. [121] found that the rear seat carried a higher risk of AIS 3+ injury compared to the front seat for restrained occupants over the age of 50 years. This was exacerbated in newer model year vehicles (1997–2007), where the front seat presented a lower injury risk for all restrained occupants over the age of 16. Similar trends have been suggested by other studies [119, 122]. Sahraei et al. [123] found that in model years 1990–1999, the rear seat was safer than the front seat for all belted occupants over the age of 25. This trend was switched for model years 2000–2009, where the risk of fatality for all belted occupants over the age of 25 was greater for rear seat occupants

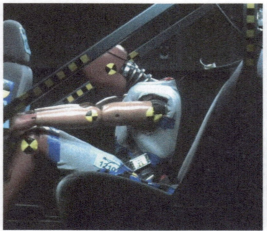 

Fig. 5.13 Video captures at the time of maximum forward head excursion in 48 km/h frontal-impact sled tests with a Hybrid III 50th percentile male ATD in an environment simulating the rear seat of a mid-sized sedan. *Left*: Standard (not force-limited, not pretensioned) belt. *Right*: Belt with a retractor pretensioner and a force-limiter. The force-limiting, pretensioning system reduced shoulder belt force by 50 % and chest deflection by 30 % with little increase in forward excursion of the head [116]

than front seat occupants. It should be noted, however, that the rear seat has consistently been shown to be the safest seating position for occupants for at least age 15 or younger [119–121, 123, 124].

Unlike in the front seat, rear seat restraints have yet to see the widespread adoption of technologies such as pretensioners and force limiters. The rear seat environment presents some challenges different from the front seat environment. Unlike in the front seat, frontal-impact airbags have not been adopted widely in the rear seat. As a result, rear-seat occupant restraint in frontal collisions must be managed almost entirely by the seatbelt and the seat. While the airbag acts as a complimentary load path in front seat restraint systems, helping to both distribute load and manage occupant forward excursion in concert with force limiting, no analogous alternative load path exists for rear seat occupants. Thus, care must be taken when incorporating force-limiters into rear seat restraints to ensure that occupant forward motion is properly managed to limit the risk of striking forward objects. Some experimental studies have suggested that three-point belt restraint systems incorporating a pretensioner and a progressive force-limiter may reduce the risk of chest and neck injury in rear seat occupants in frontal collisions up to 48 km/h compared to a standard (not pretensioning or force-limiting) belt with little-to-no increase in occupant forward excursion [116, 125] (Fig. 5.13). Pretensioning in general had the added effect of reducing forward motion of the pelvis, which may be beneficial in the fleet where large variations in rear seat seatpan and cushion design may result in substantial variations in pelvis restraint. Other rear seat restraint options recently investigated have included belt-integrated airbags [126, 127], which have shown promise through an increase in load distribution and through a pretensioning effect during inflation.

5.8.2 Obesity

The obesity rate has increased dramatically in the U.S. and throughout much of the world. From 1980 to 2000, the prevalence of obesity in Americans increased from 14.4 % to 30.5 % [128]. In 2005–2006, approximately 72 million Americans, over one third of the adult U.S.

population, were obese [129]. This trend may be exacerbated in the coming years by a trend towards increasing obesity in younger people. Between the time periods of 1988–1994 and 2003–2004, the largest increases in the prevalence of abdominal obesity in U.S. adults occurred in the age range of 20–29 years [130].

Obesity presents several challenges in automotive restraint, stemming from both biomechanical and patient outcome considerations. Obesity is often associated with increased comorbidity and complications resulting from blunt impact trauma [131]. In studies of patients admitted to level I trauma centers, Choban et al. [132] and Neville et al. [133] both observed increased mortality in obese patients (compared to non-obese patients) despite similar injury severities. Choban et al. [132] observed an increase in complications (mainly pulmonary) and length of stay in the obese group. Neville et al. [133] and Boulanger et al. [131] also observed that the majority of obese patients admitted to level I trauma centers with blunt trauma retained those injuries in motor vehicle collisions. This increase in co-morbidity may affect an overall increase in automobile collision fatality risk as a function of obesity. In studies of NASS-CDS data, several studies have observed increased fatality risks in obese occupants (defined by body-mass index (BMI) >30 kg/m^2) compared to non-obese occupants [134–136].

Obesity may also affect the distribution of body regions injured in automobile collisions. In their studies of blunt trauma patients admitted to level I trauma centers, Boulanger et al. [131] and Neville et al. [133] both observed that obese patients were more likely than non-obese patients to exhibit lower extremity injuries. Boulanger et al. [131] also observed that obese patients were more likely to exhibit rib fractures than non-obese patients. In a study of NASS-CDS data, Mock et al. [134] observed that increased BMI tended to increase the prevalence of AIS 3+ chest injuries. Jakobsson and Lindman [137] observed an increase in AIS 2+ chest and lower extremity injuries in obese occupants in Volvo's automobile collision database compiled in Sweden, but did not observe those trends in data from NASS-CDS (1998–2002).

Obesity (and body habitus in general) causes vehicle occupants to interact with restraints differently than 50th percentile occupants. Greater body mass requires greater restraining force to be applied to the body to decelerate it in a collision. Body mass is distributed throughout the body differently in the obese than in the non-obese [138], affecting occupant kinematics in a collision. The physical, geometric relationship between an automobile occupant and the restraints and interior of a vehicle also change with obesity and with body size and shape. Reed et al. [139] found that an increase of 10 kg/m^2 in BMI was associated with a lap belt positioned 43 mm further anterior of the anterior-superior iliac spines and 21 mm more superior. This increase in BMI was also associated with 130 mm of additional lap belt webbing and 60 mm of additional shoulder belt webbing. In a study of driver position and clearance with vehicle interior structures, Bove et al. [140] found that the minimum distance between the steering wheel and the driver decreased with increasing BMI. Obese occupants are also more likely to not use their seatbelts than non-obese occupants [141], often citing discomfort.

In addition to these belt fit and occupant positioning issues, obesity is associated with thicker layers of subcutaneous soft tissue, which affects the interaction of a restraint system (most notably belt systems) with the skeletal structure of the body. Belt restraint systems are designed to engage the naturally strong structures of the body. The lap belt is designed to load the relatively strong structure of the anterior superior iliac spines of the pelvis; the shoulder belt is designed to load the relatively stiff structures of the clavicle and the upper ribcage. Increases in soft tissue depth at the thigh, abdomen, and chest limit the ability of the belt to engage the bony structures of the pelvis and the shoulder [125, 142]. In a series of frontal impact sled tests with obese and non-obese cadaveric subjects, Forman et al. [125] and Kent et al. [142] found that the lap belt tended to slip in to the abdomen of the obese subjects due to a combination of relatively extreme magnitudes of soft tissue depth coupled with a non-optimal initial lap belt position.

Consistent with the belt fit studies of Reed et al. [139], the extreme tissue depth in the abdomen and thighs of those cadavers prevented lap belt positioning low enough on the lap to engage the bony structures of the pelvis regardless of the belt donning procedure employed. One consequence was that the obese subjects exhibited very limited pelvis restraint through the lap belt. Where the non-obese subjects exhibited limited forward pelvis motion resulting in forward pitch of the torso (through a shoulder belt force limiter), the obese subjects tended to move forward in a linear fashion, with no forward torso pitch and extreme forward motion of the pelvis and lower extremities (Fig. 5.14). While no knee bolster was included in those studies, the increased subject mass, the increased forward knee motion, and the presumed decrease in the restraint of the pelvis by the lap belt all would tend to increase the forces generated in the lower extremity resulting from interaction with a front seat or knee bolster. This is consistent with studies that have observed a relationship between obesity and an increase in lower extremity injury risk in automobile collisions [137]. The reduction in forward torso pitch may also tend to increase loading of the lower chest by the shoulder belt, also consistent with the set of field data studies which have suggested an effect of obesity on the prevalence of chest injuries [131, 134].

5.8.3 Aging

As the world population ages, elderly vehicle occupant protection is gaining as a research priority. By 2030, 25 % of the U.S. population will be age 65 or older [143] and the average age of the U.S. population is projected to increase through 2100 (2000 U.S. Census). Aging is not just a U.S. domestic public health issue – for example, China will have 285 million people over age 60 by 2025 [144]. People are also tending to drive later in life. In 2000, 14 % of new car buyers in the U.S. were over the age of 60 [145]. That percentage is expected to increase.

Older occupants differ in several respects from young or even middle-aged occupants in terms of both crash exposure and outcomes [146]. Older occupants tend to sustain greater injuries for a given crash severity [144, 147, 148]. Injury patterns also tend to shift with age, from a greater prevalence of head injury in younger occupants to chest injury in older occupants [147–149]. These macro-scale differences in population outcomes reflect independent aspects of aging. First, older people are more fragile than younger: they tend to sustain a greater level of injury for a given magnitude of loading. This is reflected, for example, in a higher probability of injury at a given AIS level for a specified crash condition (delta-V, restraint use, etc.). In the chest in particular, increased fragility is observed as a decrease in deformation tolerance of the ribcage prior to the occurrence of fracture [150, 151], attributed in part to decreases in the rib cortical bone thickness and failure strain and changes in ribcage geometry [152, 153]. In addition, older people are more frail than younger: they tend to have worse outcomes for the same injury. In general, morbidity, mortality, and treatment costs are higher for older people than for younger people with the same injury [154–159]. These changes and fragility and frailty combine to affect changes in injury and fatality outcome for older occupants.

Older people also tend to have different exposure patterns than younger. For example, NHTSA [160] found that older drivers have slightly higher belt usage than younger and Viano and Ridella [161] concluded that older drivers are overrepresented in lateral impacts at intersections. Compared to younger drivers, older drivers involved in fatal collisions are less likely to be drunk, ejected from the vehicle, involved in a single-vehicle collision, or involved in a collision of such severity as to be considered "unsurvivable" [149]. Older drivers involved in fatal collisions are more likely, however, to be belted, have pre-existing health issues, or to have delayed death occurring after the collision.

Among older vehicle occupants, the increased fragility, frailty, and differences in exposure present unique challenges in restraint design and occupant protection. Kent et al. [149] estimated that as many as 50 % of fatally injured older drivers die from crashes that would be survived by

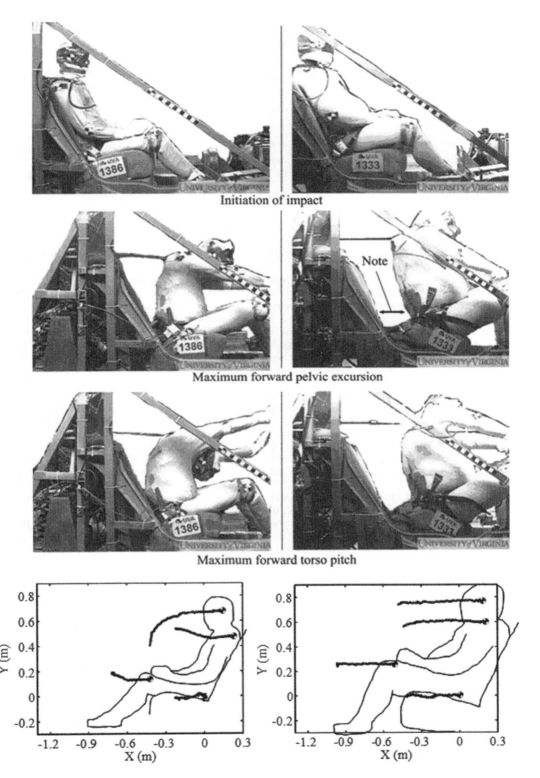

Fig. 5.14 Kinematics of obese (*right*, 151 kg mass, 182 cm stature, 45 kg/m^2 BMI) and non-obese (*left*, 69 kg mass, 175 cm stature, 23 kg/m^2 BMI) cadavers in 48 km/h, frontal impact sled tests with a pre-tensioned, force-limited belt. Also shown are pre-impact body contours and impact trajectories of the left head, shoulder, knee, and hip up to the time of maximum forward head excursion (drawn to-scale). Note the lap belt routed over the extended abdomen of the obese subject; its extreme forward pelvis, knee, and shoulder motion; and the pronounced differences in torso recline [125, 142]

younger drivers. While disconcerting, this may be an indication that a substantial number of elderly fatality cases are within the potentially-survivable design space. Research in improved elderly occupant protection is ongoing, and is a priority for several government, industry, and academic bodies [2, 3, 162–165].

References

1. Viano D (1991) Effectiveness of safety belts and air bags in preventing fatal injury. In: Paper #910901, Society of Automotive Engineers, Warrendale
2. Bostrom O, Haland Y (2005) Benefits of a 3 + 2 point belt system and an inboard torso-side support in frontal, far side and rollover crashes. Int J Veh Safety 1:181–199
3. Rouhana SW, Bedewi P, Kankanala S, Prasad P, Zwolinski J, Meduvsky A, Rupp J, Jeffreys T, Schneider L (2003) Biomechanics of 4-point seat belt systems in frontal impacts. Stapp Car Crash J 47:367–399
4. Viano D (2003) Lap-shoulder belts: some historical aspects. In: Viano D (ed) Seat belts: the development of an essential safety feature, PT-92. Society of Automotive Engineers, Warrendale, PA
5. Kent R, Patrie J, Benson N (2003) The hybrid III dummy as a discriminator of injurious and non-injurious restraint loading. Ann Proc Assoc Adv Automot Med 47:51–75
6. Adomeit D, Balser W (1987) Items of an engineering program on an advanced web-clamp device. In: SAE World Congress, paper #870328, Society of Automotive Engineers, Warrendale
7. Crandall JR, Bass CR, Pilkey WD, Miller HJ, Sikorski J, Wilkins M (1997) Thoracic response and injury with belt, driver side airbag, and force limited belt restraint systems. Int J Crashworthiness 2(1): 119–132
8. Foret-Bruno JY, Trosseille X, Page Y, Huere JF, Le Coz JY, Bendjellal F, Diboine A, Phalempin T, Villeforceix D, Baudrit P, Guillemot H, Coltat JC (2001) Comparison of thoracic injury risk in frontal car crashes for occupant restrained without belt load limiters and those restrained with 6 kN and 4 kN belt load limiters. Stapp Car Crash J 45:205–224
9. Foret-Bruno J-Y, Trosseille X, Le Croz J-Y, Bendjellal F, Steyer C, Phalempin T, Villeforceix D, Dandres P, Got C (1998) Thoracic injury risk in frontal car crashes with occupant restrained with belt load limiter. In: SAE paper #983166. Stapp car crash conference
10. Haland Y, Skanberg T (1989) A mechanical buckle pretensioner to improve a three point seat belt. In: 12th Technical conference on the enhanced safety of vehicles. Paper #896134
11. Kent R, Lessley D, Shaw G, Crandall J (2003) The utility of hybrid III and THOR chest deflection for discriminating between standard and force-limiting belt systems. Stapp Car Crash J 47:267–297
12. Petitjean A, Lebarbe M, Potier P, Trosseille X, Lassau JP (2002) Laboratory reconstructions of real world frontal crash configurations using the hybrid III and THOR dummies and PMHS. Stapp Car Crash J 46:27–54
13. Andrews S (1995) Occupant sensing in smart restraint systems. In: 39th annual proceedings, association for the advancement of automotive medicine, pp 543–555
14. Bernat A (1995) "Smart" safety belts for injury reduction. In: 39th annual conference proceedings, association for the advancement of automotive medicine, pp 567–576
15. Johannessen HG, Mackay GM (1995) Why "intelligent" automotive occupant restraint systems? In: 39th annual conference proceedings, association for the advancement of automotive medicine, pp 519–526
16. Miller HJ (1995) Injury reduction with smart restraint systems. In: 39th annual proceedings, association for the advancement of automotive medicine, pp 527–541
17. Viano DC (1988) Limits and challenges of crash protection. Accid Anal Prev 20(6):421–429
18. Forman J, Stacey S, Evans J, Kent R (2008) Posterior acceleration as a mechanism of blunt traumatic injury of the aorta. J Biomech 41(6):1359–1364
19. Hardy WN, Shah CS, Mason MJ, Kopacz JM, Yang KH, King AI, Van Ee CA, Bishop JL, Banglmaier RF, Bey MJ, Morgan RM, Digges KH (2008) Mechanisms of traumatic rupture of the aorta and associated peri-isthmic motion and deformation. Stapp Car Crash J 52:233–265
20. Kent R, Stacey S, Kindig M, Woods W, Evans J, Rouhana SW, Higuchi K, Tanji H, St LS, Arbogast KB (2008) Biomechanical response of the pediatric abdomen, part 2: injuries and their correlation with engineering parameters. Stapp Car Crash J 52:135–166
21. Kent RW, Crandall JR, Bolton J, Prasad P, Nusholtz G, Mertz H (2001) The influence of superficial soft tissues and restraint condition on thoracic skeletal injury prediction. Stapp Car Crash J 45:183–204
22. Lau I, Viano D (1986) The viscous criterion – bases and applications of an injury severity index for soft tissues. In: Proceedings of the 30th stapp car crash conference society of automotive engineers, pp 123–142
23. Adomeit D (1977) Evaluation methods for the biomechanical quality of restraint systems during frontal impact. In: Proceedings of the 21st stapp car crash conference. SAE, Warrendale, pp 911–932
24. Adomeit D (1979) Seat design – a significant factor for safety belt effectiveness. In: Proceedings of the 23rd stapp car crash conference. SAE, Warrendale, pp 39–68
25. Adomeit D, Heger A (1975) Motion sequence criteria and design proposals for restraint devices in order

to avoid unfavorable biomechanic conditions and submarining. Stapp Car Crash Conf 19:139–165
26. Forman J, Lopez-Valdes F, Lessley D, Kindig M, Kent R, Ridella S, Bostrom O (2009) Rear seat occupant safety: an investigation of a progressive force-limiting, pretensioning 3-point belt system using adult PMHS in frontal sled tests. Stapp Car Crash J 53:49–74
27. NHTSA (1984) FMVSS 208 regulatory impact analysis. National Highway Traffic Safety Administration. U.S. Department of Transportation, Washington, DC
28. Patrick LM, Nyquist GW (1972) Air bag effects on the out-of-position child. Paper #720442, Society of Automotive Engineers, Warrendale
29. Crandall J, Kuhlmann T, Martin P, Pilkey W, Neeman T (1994) Differing patterns of head and facial injury with airbag and/or belt restrained drivers in frontal collisions. In: Proceedings of the advances in occupant restraint technologies: joint AAAM-IRCOBI special session. Lyon
30. Berg F, Schmitt B, Epple J, Mattern R, Kallieris D (1998) Results of full-scale crash tests, stationary tests and sled tests to analyse the effects of airbags on passengers with or without seat belts in the standard sitting position and in out-of-position situations. Paper #98-S5-O-10. In: Proceedings of the international research council on the biomechanics of impact (IRCOBI)
31. Kent RW, Crandall JR, Bolton JR, Duma SM (2000) Driver and right-front passenger restraint system interaction, injury potential, and thoracic injury prediction. Ann Proc Assoc Adv Automot Med 44: 261–282
32. Kallieris D, Mellander HG, Barz J, Mattern R (1982) Comparison between frontal impact tests with cadavers and dummies in a simulated true car restrained environment. In: Proceedings of the 26th stapp car crash conference, pp 353–367
33. Yoganandan N, Pintar FA, Skrade D, Chmiel W, Reinartz JM, Sances A (1993) Thoracic biomechanics with air bag restraint. In: Stapp car crash conference, pp 133–144. doi:SAE Paper Number 933121
34. Werner J, Sorenson W (1994) Survey of air bag involved accidents an analysis of collision characteristics. Paper #940802, Society of Automotive Engineers, Warrendale
35. Patrick L, Kroell C, Mertz H (1965) Forces on the human body in simulated crashes. In: 9th proceedings, stapp car crash conference, pp 237–259
36. Bierman HR, Wilder RM, Hellems HK (1946) The physiological effects of compressive forces on the torso. Report #8, Naval Medical Research Institute X-630, Bethesda
37. Horsch J, Culver C (1983) The role of steering wheel structure in the performance of energy absorbing steering systems. Paper #831607. Warrendale
38. Hendler E, O'Rourke J, Schulman M, et al. (1974) Effect of head and body position and muscular tensing on response to impact. Paper #741184. Society of Automotive Engineers, Warrendale
39. Smith GR, Gullash EC, Baker RG (1974) Human volunteer and anthropometric dummy tests of the general motors driver air cushion system. Paper #740578, Society of Automotive Engineers, Warrendale
40. Armstrong R, Waters H, Stapp J (1968) Human muscular restraint during sled deceleration. Paper #680793. Society of Automotive Engineers, Warrendale
41. Grosch L (1985) Chest injury criteria for combined restraint systems. Paper #851247. Society of Automotive Engineers, Warrendale
42. Enouen S, Guenther D, Saul R, MacLaughlin T (1984) Comparison of models simulating occupant response with air bags. Paper number 840451, Society of Automotive Engineers, Warrendale
43. Vezin P, Bruyere-Garnier K, Bermond F, Verriest JP (2002) Comparison of hybrid III, Thor-alpha and PMHS response in frontal sled tests. Stapp Car Crash J 46:1–26
44. Melvin JW, Baron KJ, Little WC, Gideon TW (1998) Biomechanical analysis of Indy race car crashes. Stapp Car Crash J 42:247–266
45. Petitjean A, Baudrit P, Trosseille X (2003) Thoracic injury criterion for frontal crash applicable to all restraint systems. Stapp Car Crash J 47:323–348
46. Sunnevang C, Rosen E, Bostrom O, Lechelt U (2010) Thoracic injury risk as a function of crash severity – car-to-car side impact tests with WorldSID compared to real-life crashes. Ann Adv Automot Med 54:159–168
47. Bertrand S, Cuny S, Petit P, Trosseille X, Page Y, Guillemot H, Drazetic P (2008) Traumatic rupture of thoracic aorta in real-world motor vehicle crashes. Traffic Inj Prev 9(2):153–161
48. Haland Y, Lovsund P, Nygren A (1993) Life-threatening and disabling injuries in car-to-car side impacts–implications for development of protective systems. Accid Anal Prev 25(2):199–205
49. Yoganandan N, Pintar FA, Gennarelli TA, Maltese MR (2000) Patterns of abdominal injuries in frontal and side impacts. In: 44th annual proceedings of the association for the advancement of automotive medicine, pp 17–36
50. Kahane CJ (2007) An evaluation of side impact protection: FMVSS 214 TTI(d) improvements and side airbags. DOT HS 810 748. Department of Transportation National Highway Traffic Safety Administration. Washington, DC
51. Newgard CD, Lewis RJ, Kraus JF, McConnell KJ (2005) Seat position and the risk of serious thoracoabdominal injury in lateral motor vehicle crashes. Accid Anal Prev 37(4):668–674
52. Schiff MA, Tencer AF, Mack CD (2008) Risk factors for pelvic fractures in lateral impact motor vehicle crashes. Accid Anal Prev 40(1):387–391
53. Augenstein J, Bowen J, Perdeck E, Singer M, Stratton J, Horton T, Rao A (2000a) Injury patterns in near-side collisions. SAE World Congress, SAE technical paper #2000-01-0634

54. Augenstein J, Perdecl E, Bowen J, Stratton J, Singer M, Horton T, Rao A, Digges K, Malliaris A, Steps J (1999) Injuries in near side collisions. In: 43rd annual proceedings, association for the advancement of mutomotive medicine, pp 139–158
55. Bohman K, Rosen E, Sunnevang C, Bostrom O (2009) Rear seat occupant thorax protection in near side impacts. In: 53rd Proceedings of the association for the advancement of automotoive medicine, pp 3–12
56. Huelke DF (1990) Near side passenger car impacts – CDC, AIS & body areas injured (NASS data). SAE World Congress. SAE technical paper #900374
57. O'Connor JV, Kufera JA, Kerns TJ, Stein DM, Ho S, Dischinger PC, Scalea TM (2009) Crash and occupant predictors of pulmonary contusion. J Trauma 66(4):1091–1095
58. Pintar FA, Maiman DJ, Yoganandan N (2007) Injury patterns in side pole crashes. Ann Proc Assoc Adv Automot Med 51:419–433
59. Rouhana SW, Foster ME (1985) Lateral impact – an analysis of the statistics in the NCSS. SAE World Congress, SAE technical paper 851727
60. Ryb GE, Dischinger PC, Braver ER, Burch CA, Ho SM, Kufera JA (2009) Expected differences and unexpected commonalities in mortality, injury severity, and injury patterns between near versus far occupants of side impact crashes. J Trauma 66(2):499–503
61. Scarboro M, Rudd R, Sochor M (2007) Nearside occupants in low delta-V side impact crashes: analysis of injury and vehicle damage patterns. In: International technical conference on the enhanced safety of vehicles (ESV), Paper #07-0225
62. Stein DM, O'Connor JV, Kufera JA, Ho SM, Dischinger PC, Copeland CE, Scalea TM (2006) Risk factors associated with pelvic fractures sustained in motor vehicle collisions involving newer vehicles. J Trauma 61(1):21–30
63. Welsh R, Morris A, Hassan A (2007) Struck-side crashes involving post-regulatory European passenger cars: crash characteristics and injury outcomes. Int J Veh Safety 2(1):103–115
64. Yoganandan N, Pintar FA (2006) Chest injuries and occupant positioning in lateral impacts with side airbags. In: Proceedings of the BIO2006, paper #BIO2006-157524, 2006 summer bioengineering conference, ASME
65. Yoganandan N, Pintar FA, Gennarelli T (2005) Evaluation of side impact injuries in vehicles equipped with side airbags. In: Proceedings of the 2005 international research council on the biomechanics of impact, Prague
66. Yoganandan N, Pintar FA, Stemper BD, Gennarelli TA, Weigelt JA (2007) Biomechanics of side impact: injury criteria, aging occupants, and airbag technology. J Biomech 40(2):227–243
67. Hallman JJ, Brasel KJ, Yoganandan N, Pintar FA (2009) Splenic trauma as an adverse effect of torso-protecting side airbags: biomechanical and case evidence. Ann Adv Automot Med 53:13–24
68. Braver ER, Kyrychenko SY (2004) Efficacy of side air bags in reducing driver deaths in driver-side collisions. Am J Epidemiol 159(6):556–564
69. McCartt AT, Kyrychenko SY (2007) Efficacy of side airbags in reducing driver deaths in driver-side car and SUV collisions. Traffic Inj Prev 8(2):162–170
70. McGwin G Jr, Metzger J, Rue LW III (2004) The influence of side airbags on the risk of head and thoracic injury after motor vehicle collisions. J Trauma 56(3):512–516
71. Stolinski R, Grzebieta R, Fildes B (1998) Side impact protection-occupants in the far-side seat. Int J Crashworthiness 3(2):93–122
72. Bostrom O, Gabler HC, Digges K, Fildes B, Sunnevang C (2008) Injury reduction opportunities of far side impact countermeasures. Ann Adv Automot Med 52:289–300
73. Gabler HC, Fitzharris M, Scully J, Fildes BN, Digges K, Sparke L (2005) Far side impact injury risk for belted occupants in Australia and the United States. In: Proceedings of the international technical conference on the enhanced safety of vehicles, paper #05-0420
74. Augenstein J, Perdeck E, Martin P, Bowen J, Stratton J, Horton T, Singer M, Digges K, Steps J (2000) Injuries to restrained occupants in far-side crashes. Ann Proc Assoc Adv Automot Med 44:57–66
75. Bostrom O, Fildes B, Morris A, Sparke L, Smith S, Judd R (2003) A cost effective far side crash simulation. Int J Crashworthiness 8(3):307–313
76. Fildes BN, Lane JC, Lenard J, Vulcan AP (1991) Passenger cars and occupant injury: side impact crashes. Monash University Accident Research Centre, CR95, Monash, Australia
77. Mackay GM, Hill J, Parkin S, Munns JAR (1993) Restrained occupants on the nonstruck side in lateral collisions. Accid Anal Prev 25(2):147–152
78. Tornvall FV, Svensson MY, Davidsson J, Flogard A, Kallieris D, Haland Y (2005) Frontal impact dummy kinematics in oblique frontal collisions: evaluation against post mortem human subject test data. Traffic Inj Prev 6(4):340–350
79. Douglas CA, Fildes BN, Gibson TJ, Bostrom O, Pintar FA (2007) Factors influencing occupant-to-seat belt interaction in far-side crashes. Ann Proc Assoc Adv Automot Med 51:319–339
80. Kent R, Lopez-Valdes F, Forman J, Seacrist T, Higuchi K, Arbogast K (2012) PMHS experiments to evaluate shoulder belt retention in oblique impacts. JSAE Annual Congress. Paper #20125172
81. Kallieris D, Schmidt G (1990) Neck response and injury assessment using cadavers and the US-SID for far-side lateral impacts of rear seat occupants with inboard anchored shoulder belts. In: Proceedings of the 34th stapp car crash conference, Orlando
82. Pintar FA, Yoganandan N, Stemper BD, Bostrom O, Rouhana SW, Digges KH, Fildes BN (2007) Comparison of PMHS, WorldSID, and THOR-NT responses in simulated far side impact. Stapp Car Crash J 51:313–360

83. Pintar FA, Yoganandan N, Stemper BD, Bostrom O, Rouhana SW, Smith S, Sparke L, Fildes BN, Digges KH (2006) Worldsid assessment of far side impact countermeasures. Ann Proc Assoc Adv Automot Med 50:199–219
84. Bedewi PG, Godrick DA, Digges KH, Bahouth GT (2003) An investigation of occupant injury in rollover: NASS-CDS analysis of injury severity and source by rollover attributes. In: International technical conference on the enhanced safety of vehicles (ESV) paper #419
85. Deutermann W (2002) Characteristics of fatal rollover crashes. DOT HS 809 438, U.S. Department of Transportation National Highway Traffic Safety Administration, Washington, DC
86. Parenteau CS, Thomas P, Lenard J (2001) U.S. and U.K. field rollover characteristics. SAE World Congress, SAE technical paper #2001-01-0167
87. Viano DC, Parenteau CS (2004) Rollover crash sensing and safety overview. SAE World Congress, SAE technical paper 2004-01-0342
88. Cuerden R, Cookson R, Richards D (2009) Car rollover mechanisms and injury outcome. In: Proceedings of the international technical conference on the enhanced safety of vehicles, paper #09-0481
89. Digges K, Eigen AM (2007) Injuries in rollover by crash severity. In: Proceedings of the international technical conference on the enhanced safety of vehicles, paper #07-0236
90. Frechede B, McIntosh AS, Grzebieta R, Bambach MR (2011) Characteristics of single vehicle rollover fatalities in three Australian states (2000–2007). Accid Anal Prev 43(3):804–812
91. Hu J, Lee JB, Yang KH, King AI (2005) Injury patterns and sources of non-ejected occupants in tripover crashes: a survey of NASS-CDS database from 1997 to 2002. Ann Proc Assoc Adv Automot Med 49:119–132
92. Digges KH, Malliaris AC, Ommaya AK, McLean AJ (1991) Characterization of rollover casualties. In: Proceedings of the international research council on the biomechanics of impact, pp 309–319
93. Otte D, Pape C, Krettek C (2005) Kinematics and injury pattern in rollover accidents of cars in German road traffic – an in-depth-analysis by GIDAS. Int J Crashworthiness 10(1):75–86
94. Viano DC, Parenteau CS, Edwards ML (2007) Rollover injury: effects of near- and far-seating position, belt use, and number of quarter rolls. Traffic Inj Prev 8(4):382–392
95. McIntosh AS, Frechede B, Grzebieta RH, Bambach MR (2010) Biomechanical considerations for a dynamic rollover crash test. In: Proceedings of the international crashworthiness conference
96. Moffat EA, James MB (2005) Headroom, roof crush, and belted excursion in rollovers. SAE World Congress, SAE technical paper #2005-01-0942
97. Bahling GS, Bundorf RT, Kaspzyk GS, Moffat EA, Orlowski KF, Stocke JE (1990) Rollover and drop tests – the influence of roof strength on injury mechanics using belted dummies. SAE World Congress, SAE technical paper #902314
98. Meyer SE, Davis M, Chng D, Herbst B, Forrest S (2000) Three-point restraint system design considerations for reducing vertical occupant excursion in rollover environments. SAE World Congress, SAE technical paper #2000-01-0605
99. Meyer SE, Herbst B, Forrest S, Syson SR, Sances A Jr, Kumaresan S (2002) Restraints and occupant kinematics in vehicular rollovers. Biomed Sci Instrum 38:465–469
100. Moffat EA, Cooper ER, Croteau JJ, Parenteau C, Toglia A (1997) Head excursion of seat belted cadaver, volunteers and Hybrid III ATD in a dynamic/static rollover fixture. SAE World Congress, SAE technical paper #973347
101. Padmanaban J, Burnett RA (2008) Seat integrated and conventional restraints: a study of crash injury/fatality rates in rollovers. Ann Adv Automot Med 52:267–280
102. Sword ML, Louden AE (2009) NHTSA research on improved restraints in rollovers. In: Proceedings of the international technical conference on the enhanced safety of vehicles, paper #09-0483
103. Adamec J, Praxl N, Miehling T, Muggenthaler H, Schonpflug M (2005) The occupant kinematics in the first phase of a rollover accident experiment and simulation. In: International IRCOBI conference on the biomechanics of impacts, pp 145–156
104. Kress T, Rink R, Sewell P (2004) Occupant kinematics and restraint effectiveness during a quarter-turn rollover in a heavy truck. SAE World Congress, SAE technical paper #2004-01-0327
105. Klima ME, Toomey DE, Weber MJ (2005) Seat belt retractor performance evaluation in rollover crashes. SAE World Congress. SAE technical paper #2005-01-1702
106. Meyer SE, Forrest S, Hock D, Herbst B, Sances A (2001) Test methods for evaluating seatbelt retractor response in multiplanar acceleration environments. ASME-PUBLICATIONS-BED 50:915–916
107. Meyer SE, Hock D, Forrest S, Herbst B, Sances A Jr, Kumaresan S (2003) Motor vehicle seat belt restraint system analysis during rollover. Biomed Sci Instrum 39:229–240
108. Meyer SE, Hock D, Herbst B, Forrest S (2001b) Dynamic analysis of ELR retractor spoolout. SAE World Congress, SAE technical paper #2001-01-3312
109. Meyer SE, Hock DA, Forrest SM, Herbst BR (2009) Webbing sensitivity as a means for limiting occupant excursion in rollovers. In: Proceedings of the international technical conference on the enhanced safety of vehicles, paper #09-0501
110. Renfroe DA (1996) Rollover ejection while wearing lap and shoulder harness: the role of the retractor. SAE World Congress, SAE technical paper #960096
111. Arndt MW (1998) Testing of seats and seat belts for rollover protection systems in motor vehicles. SAE World Congress, SAE technical paper #982295

112. Parenteau C (2006) A comparison of volunteers and dummy upper torso kinematics with and without shoulder belt slack in a low speed side/pre-roll environment. Traffic Inj Prev 7(2):155–163
113. Newberry W, Lai W, Carhart M, Richards D, Brown J, Raasch C (2006) Modeling the effects of seat belt pretensioners on occupant kinematics during rollover. SAE World Congress, SAE technical paper #2006-01-0246
114. McCoy RW, Balavich KM (2005) Analysis of a prototype electric retractor, a seat belt pre-tensioning device and dummy lateral motion prior to vehicle rollover. SAE World Congress. SAE technical paper #2005-01-0945
115. McCoy RW, Chou CC (2007) A study of kinematics of occupants restrained with seat belt systems in component rollover tests. SAE World Congress. SAE technical paper #2007-01-0709
116. Forman J, Michaelson J, Kent R, Kuppa S, Bostrom O (2008) Occupant restraint in the rear seat:ATD responses to standard and pre-tensioning, force-limiting belt restraints. Ann Proc Assoc Adv Automot Med 52:141–154
117. Padmanaban J, Fitzgerald M (2012) Effectiveness of rollover-activated side curtain airbags in reducing fatalities in rollovers. In: Proceedings of the international research council on the biomechanics of impact, pp 76–90
118. Takahashi H, Iyoda M, Aga M, Sekizuka M, Kozuru Y, Ishimoto S (2003) Development of rollover curtain shield airbag system. In: Proceeding of the international technical conference on the enhanced safety of vehicles, paper #548
119. Kent R, Forman J, Parent D, Kuppa S (2007) Rear seat occupant protection in frontal crashes and its feasibility. In: 20th Proceedings of the international technical conference on the enhanced safety of vehicles (ESV). Paper #07-0386. Lyon
120. Kuppa S, Saunders J, Fessahaie O (2005) Rear seat occupant protection in frontal crashes. In: 19th proceedings of the international technical conference on the enhanced safety of vehicles. Paper #05-0212. Washington, DC
121. Bilston LE, Du W, Brown J (2010) A matched-cohort analysis of belted front and rear seat occupants in newer and older model vehicles shows that gains in front occupant safety have outpaced gains for rear seat occupants. Accid Anal Prev 42(6):1974–1977
122. Esfahani E, Digges K (2009) Trend of rear occupant protection in frontal crashes over model years of vehicles. SAE World Congress. SAE technical paper #2009-01-0377
123. Sahraei E, Digges K, Marzougui D (2010) Reduced protection for belted occupants in rear seats relative to front seats of new model year vehicles. Ann Adv Automot Med 54:149–158
124. Smith KM, Cummings P (2006) Passenger seating position and the risk of passenger death in traffic crashes: a matched cohort study. Inj Prev 12(2):83–86
125. Forman J, Lopez-Valdes FJ, Lessley D, Kindig M, Kent R, Bostrom O (2009) The effect of obesity on the restraint of automobile occupants. Ann Proc Assoc Adv Automot Med 53:25–40
126. Forman JL, Lopez-Valdes FJ, Dennis N, Kent RW, Tanji H, Higuchi K (2010) An inflatable belt system in the rear seat occupant environment: investigating feasibility and benefit in frontal impact sled tests with a 50(th) percentile male ATD. Ann Adv Automot Med 54:111–126
127. Sundararajan S, Rouhana SW, Board D, DeSmet E, Prasad P, Rupp JD, Miller CS, Schneider LW (2011) Biomechanical assessment of a rear-seat inflatable seatbelt in frontal impacts. Stapp Car Crash J 55:161–197
128. Flegal KM, Carroll MD, Ogden CL, Johnson CL (2002) Prevalence and trends in obesity among US adults, 1999–2000. JAMA 288(14):1723–1727
129. Ogden CL, Carroll MD, McDowell MA (2007) Obesity among adults in the United States – no statistically significant change since 2003–2004. Data brief number 1, Centers for Disease Control, Atlanta, Georgia, p 8
130. Li C, Ford ES, McGuire LC, Mokdad AH (2007) Increasing trends in waist circumference and abdominal obesity among US adults. Obesity (Silver Spring) 15(1):216–224
131. Boulanger BR, Milzman D, Mitchell K, Rodriguez A (1992) Body habitus as a predictor of injury pattern after blunt trauma. J Trauma 33(2):228–232
132. Choban PS, Weireter LJ Jr, Maynes C (1991) Obesity and increased mortality in blunt trauma. J Trauma 31(9):1253–1257
133. Neville AL, Brown CV, Weng J, Demetriades D, Velmahos GC (2004) Obesity is an independent risk factor of mortality in severely injured blunt trauma patients. Arch Surg 139(9):983–987
134. Mock CN, Grossman DC, Kaufman RP, Mack CD, Rivara FP (2002) The relationship between body weight and risk of death and serious injury in motor vehicle crashes. Accid Anal Prev 34(2):221–228
135. Tagliaferri F, Compagnone C, Yoganandan N, Gennarelli TA (2009) Traumatic brain injury after frontal crashes: relationship with body mass index. J Trauma 66(3):727–729
136. Viano DC, Parenteau CS, Edwards ML (2008) Crash injury risks for obese occupants using a matched-pair analysis. Traffic Inj Prev 9(1):59–64
137. Jakobsson L, Lindman M (2005) Does BMI (Body Mass Index) influence the occupant injury risk pattern in car crashes? In: Proceedings of the 2005 international research council on the biomechanics of impact, Prague
138. Friess M, Corner BD (2004) From XS to XL: statistical modeling of human body shape using 3D scans. SAE World Congress. SAE technical paper #2004-01-2183
139. Reed MP, Ebert-Hamilton SM, Rupp JD (2012) Effects of obesity on seat belt fit. Traffic Inj Prev 13(4):364–372
140. Bove RTJ, Fisher JL, Ciccarelli L, Cargill RSI, Moore TLA (2006) The effects of anthropometry on

driver position and clearance measures. SAE World Congress, SAE technical paper #2006-01-0454
141. Hunt DK, Lowenstein SR, Badgett RG, Steiner JF (1995) Safety belt nonuse by internal medicine patients: a missed opportunity in clinical preventive medicine. Am J Med 98(4):343–348
142. Kent RW, Forman JL, Bostrom O (2010) Is there really a "cushion effect"?: a biomechanical investigation of crash injury mechanisms in the obese. Obesity (Silver Spring) 18(4):749–753
143. OECD (2001) Ageing and transport – mobility needs and safety issues. Organization for Economic Co-operation and Development, Paris
144. Kent R, Funk J, Crandall J (2003) How future trends in societal aging, air bag availability, seat belt use, and fleet composition will affect serious injury risk and occurrence in the United States. Traffic Inj Prev 4(1):24–32
145. Alonso-Zaldivar R (2000) Auto makers retool to fit an aging US Safety: study puts focus on protecting the growing population of older drivers. LA Times July 31
146. Islam S, Mannering F (2006) Driver aging and its effect on male and female single-vehicle accident injuries: some additional evidence. J Safety Res 37(3):267–276
147. Morris A, Welsh R, Frampton R, Charlton J, Fildes B (2002) An overview of requirements for the crash protection of older drivers. In: 46th proceedings, association for the advancement of automotive medicine, pp 141–156
148. Morris A, Welsh R, Hassan A (2003) Requirements for the crash protection of older vehicle passengers. In: 47th proceedings, association for the advancement of automotive medicine, pp 165–180
149. Kent R, Henary B, Matsuoka F (2005a) On the fatal crash experience of older drivers. In: 49th proceedings, association for the advancement of automotive medicine. Annual Advancement of Automotive Medicine
150. Kent R, Patrie J, Poteau F, Matsuoka F, Mullen C (2003d) Development of an age-dependent thoracic injury criterion for frontal impact restraint loading. In: Proceedings of the conference on the enhanced safety of vehicles. Paper #72
151. Laituri TR, Prasad P, Sullivan K, Frankstein M, Thomas RS (2005) Derivation and evaluation of a provisional, age-dependent, AIS3+ thoracic risk curve for belted adults in frontal impacts. Society of Automotive Engineers. Paper #2005-01-0297
152. Kent R, Lee SH, Darvish K, Wang S, Poster CS, Lange AW, Brede C, Lange D, Matsuoka F (2005) Structural and material changes in the aging thorax and their role in crash protection for older occupants. Stapp Car Crash J 49:231–249
153. Stein I, Granik G (1976) Rib structure and bending strength: an autopsy study. Calcif Tissue Res 20:61–73
154. Bulger EM, Arneson MA, Mock CN, Jurkovich GJ (2000) Rib fractures in the elderly. J Trauma 48(6):1040–1046
155. Evans L (2001) Age and fatality risk from similar severity impacts. J Traffic Med 29:10–19
156. Martinez R, Sharieff G, Hooper J (1994) Three-point restraints as a risk factor for chest injury in the elderly. J Trauma 37(6):980–984
157. Miller TR, Lestina DC, Spicer RS (1998) Highway crash costs in the United States by driver age, blood alcohol level, victim age, and restraint use. Accid Anal Prev 30(2):137–150
158. Miltner E, Salwender HJ (1995) Influencing factors on the injury severity of restrained front seat occupants in car-to-car head-on collisions. Accid Anal Prev 27(2):143–150
159. Peek-Asa C, Dean BB, Halbert RJ (1998) Traffic-related injury hospitalizations among California elderly, 1994. Accid Anal Prev 30(3):389–395
160. Cerelli E (1998) Research note: crash data and rates for age-sex groups of drivers, 1996. National Highway Traffic Safety Administration, U.S. Department of Transportation, Washington, DC
161. Viano D, Ridella S (1996) Significance of intersection crashes for older drivers. SAE World Congress, paper #960457
162. National Highway Traffic Safety Administration (1993) Traffic safety plan for older drivers. U. S. Department of Transportation, Washington, DC
163. El-Jawahri RE, Laituri TR, Ruan JS, Rouhana SW, Barbat SD (2010) Development and validation of age-dependent FE human models of a mid-sized male thorax. Stapp Car Crash J 54:407–430
164. IRTAD (2009) Road safety 2009 annual report. International Traffic Safety Data and Analysis Group
165. Kent R, Trowbridge M, Lopez-Valdes FJ, Ordoyo RH, Segui-Gomez M (2009) How many people are injured and killed as a result of aging? Frailty, fragility, and the elderly risk-exposure tradeoff assessed via a risk saturation model. Ann Adv Automot Med 53:41–50

Mathematical Models, Computer Aided Design, and Occupant Safety

King H. Yang and Clifford C. Chou

Abstract

Imagine what it would be like if, in the near future, we were able to take all the crash test dummies out of all the cars in all crash test labs around the world, because we no longer needed them. Imagine how quick, inexpensive, and accurate crash testing would become if we could marry the mathematical data representing all the adult male, adult female, and child vehicle occupants, and run all the crash tests as simulations. If you can imagine this scenario, you can imagine that all future cars will be capable of providing individualized restraints based on information of each occupant instead of affording safety only for a handful of anthropomorphic test devices (ATDs), also known as crash dummies.

6.1 Introduction

Since the invention of the Electronic Numerical Integrator and Computer (ENIAC) more than 60 years ago, harnessing the high-speed computational power provided by the digital computer has been the key to advancing all fields of computer aided engineering (CAE). Taking advantage of a computer, the first mathematical model constructed for crash safety evaluation can be traced back to McHenry [1] who calculated the dynamic response of a vehicular occupant involved in a collision. Limited by computational power, early-stage simulation models in the 1960s and 1970s tended to be very simple and were based on solving simultaneous differential equations.

In mid-stage around 1980s to 2000s, software packages based on rigid body dynamics became the main workhorses for occupant and restraint system simulations while software based on finite element (FE) methods was the main solver used for vehicular structure modeling. An initial review of several gross-motion simulators was made by King, Chou [2]. Prasad [3] further contrasted the capabilities of various occupant simulation tools while Prasad and Chou [4] reviewed

K.H. Yang, Ph.D. (✉)
Bioengineering Center, Wayne State University, Detroit, MI, USA
e-mail: Aa0007@wayne.edu

C.C. Chou, Ph.D.
Department of Mechanical Engineering, Wayne State University, Detroit, MI, USA
e-mail: Cc2chou@sbcglobal.net

mathematical occupant simulation models. Some simple one-dimensional and two-dimensional frontal and side impact models were reviewed in the previous editions of this book by Prasad and Chou [5, 6].

As computational power continues to rise and with no ending in sight, FE models have become the tool of choice for most automotive safety CAE analyses in recent years. Integrated occupant, restraint, and vehicle models involving millions of elements are now routinely employed in supporting design of production vehicles that meet various regulations prescribed by a number of consumer and governmental agencies around the globe. Without mathematical models, it would be a monumental task attempting to optimize the structural integrity and passive safety designs through crash testing for active and passive safety protections.

Aside from saving the high cost associated with crash testing, mathematical models offer a number of "theoretical" advantages. Unlike early- or mid-stage mathematical models where differential equations are needed to describe and tackle the problem, complex geometry can be accurately represented in modern-day FE models to better characterize the issue on hand. As new materials are developed to more efficiently absorb crash energy, unconventional materials (such as foam) can be precisely modeled. In terms of biological tissues where material properties are nonlinear, loading rate dependent, and viscoelastic, proper constitutive laws can be implemented in software to mimic behaviors of these tissues. Additionally, mathematical models can handle complicated loading conditions with ease and commercially available software is available to manage a large number of Design of Computer Experiments (DOCE) to identify the trend of changing design parameters and determine optimized design specifications. More importantly, mathematical models are the only ethical way to study the effect of active muscle and post fracture responses. They can be used to calculate stress and strain within each composition material to better estimate the risk of injury. Once a complete picture of tissue-level injury threshold is identified, specific injury (such as diffuse axonal injury, skull fracture, liver laceration, and knee ligament rupture) can be predicted instead of using the current region-based injury criteria (such as the head injury criterion and viscous criteria for chest injury) and associated injury assessment reference value (IARV).

Just as Rome was not built in a day, mathematical models developed for automotive safety research involved numerous efforts, researchers, and spanned several decades. Readers are referred to Chaps. 6 and 7 of the previous editions, the first and the second edition, respectively of this book for more detailed reviews of research works prior to 2002 [5, 6]. A number of concluding remarks made in those articles are becoming reality over the last 20 years. For example, the statement that "the use of mathematical models to explore injury criteria and countermeasures will be feasible" is currently being practiced routinely in injury biomechanics research. Additionally, readers are referred to a review article of nearly 200 mathematical models published in Stapp conference proceedings [7] for acquiring appreciations on how mathematical models evolved over the years. Due to large volume of models available, this chapter can be considered as Part II of these publications. At the onset of this chapter, some proposed model validation methods are presented to emphasize the need for high quality model validation. This is followed by a very brief review of rigid body dynamics models published since 1993 for long duration events such as rollovers or human modeling. Issues related to FE model development and generalized procedures used to address a variety of safety regulations are presented next. Discussion on occupant modeling, including ATD (anthropomorphic test device) and human modeling, is emphasized in greater detail because more researchers are moving towards this field. Finally, recommendations are drawn at the end to highlight critical future steps.

6.2 Importance of Model Validation

Mathematical models reviewed in this chapter are intentionally limited because there are a number of other review articles already available. Most models have been developed for various purposes for:
- ATD and human occupant kinematics study
- Restraint systems, such as airbag and belts development
- Addressing regulatory and non-regulatory requirements
- Injury biomechanics study
- Injury mechanism and causation studies
- Component level test methodology development
- Prototype and product development cycle reduction
- Others

In order to constitute numerical models as reliable and predictive, they need to be assessed through a series of rigorous validation processes. The extent of model validation against experimental data needs to be assessed **quantitatively** rather than qualitative descriptions (e.g. well validated, in good or favor agreement) seen in most publications reviewed in this study. Several government agencies, such as the United States Department of Defense (DOD) [8], American Institute of Aeronautics and Astronautics (AIAA) [9], American Society of Mechanical Engineers (ASME) [10], and Advanced Simulation and Computing (ASC) of the United States of Department of Energy (DOE) [11], have investigated fundamental concepts/methodologies for validation of large-scale numerical models. Unfortunately, there are no universally accepted quantification methods to determine the degree of model validation against experimental observations currently. This problem is due in part to the large variations and uncertainties seen in experimental data and many experimental studies do not provide detailed test conditions used while conducting experiments.

Generally speaking, experimental data are recorded digitally in the form of time histories associated with different measurement units such as g's, kN (or lbs), mm (inches), etc. Additionally, multiple responses are generated in experiments at different spatial and temporal locations. Many publications used graphical plots comparing experimental data and simulated results for subjective engineering judgments of how well the models are and extent of validation. Such graphical comparisons cannot be considered **quantitative** when determining the reliability and predictive capability of numerical models. Based on a limited review of validation methods used in the open literature, the approaches in the process of model validation can be largely divided into two groups: simplistic and rigorous approaches.

6.2.1 Simplistic Approaches

In these approaches, metrics, indices or parameters are defined to compare between data obtained experimentally and from simulations. Jovanovski [12] proposed the Normalized Integral Square Error (NISE) method for quantitatively evaluating the goodness of fit between model-predicted response time histories with experimental data. The NISE method, based on cross-correlation coefficients, considers phase shift, amplitude difference, and shape difference, thus allowing examination of discrepancies and similarities between the test data and simulation results. The smaller the differences, the better the model will generate accurate predictions. In 1997, Prasad, [13] based on criteria formulated but un-finalized by the SAE Human Biomechanics and Simulation Sub-committee, also proposed an eight-level scale, with Level Eight being the best match, to judge the degree of validation, especially for head models. Using the Prasad's descriptive scale, most of the models scored very low, between Levels 3 and 5, implying that this scale may be too difficult to be achieved at the high levels. This pointed out that continued effort should be directed toward developing validation and verification methodologies for improving validity and quantitative report for numerical models.

In 2010, Deb et al. [14] proposed a Gross Correlation Index (GCI) by including metrics such as mean load, peak load, and energy absorption for correlation assessment of an FE model of a top-hat section component. Their GCI is defined as:

$$GCI = 1 - \left[\frac{1}{3} \left\{ \frac{(P-P_{Test})^2}{P_{Test}^2} + \frac{(M-M_{Test})^2}{M_{Test}^2} + \frac{(E-E_{Test})^2}{E_{Test}^2} \right\} \right]^{\frac{1}{2}}$$

where:

P, P_{Test} = predicted and experimental values respectively of peak load,

M, M_{Test} = predicted and experimental values respectively of mean load, and

E, E_{Test} = predicted and experimental values respectively of energy absorption.

When GCI = 0.0, it means there is no correlation at all. A GCI of one (1.0) implies a perfect correlation of numerical model prediction to experimental results. Therefore, a higher GCI indicates a better correlation. Deb et al. [14] was able to achieve a GCI of 0.94 of their FE model in predicting a double top-hat section column component under axial compressive loadings. Again, Haorongbam et al. [15] used the same GCI for assessing FE models in off-set axial crushing on single and double hat sections. A Modified Gross Correlation Index (MGCI) was adopted by Zhu et al. [16] who used optimization-based methodologies to obtain a set of optimal material input properties of die cast AM60B magnesium alloy, which yielded simulated results for both slow-speed and high-speed axial crushing as well as quasi-static four-point bending of thin-walled double top-hat beam components. By selecting the peak load, displacement at the peak load, and energy absorption as metrics, the correlation between the simulated results and test data was made using the MGCI with different weighting factors for evaluating the goodness-of-fit or the degree of agreement. This MGCI reads as:

$$GCI = 1 - \left[a \cdot \frac{(F_c - F_m)^2}{F_m^2} + b \cdot \frac{(D_c - D_m)^2}{D_m^2} + c \cdot \frac{(E_c - E_m)^2}{E_m^2} \right]^{\frac{1}{2}} \quad (6.1)$$

where

F_c, F_m: simulated and experimental values of peak load, respectively,

D_c, D_m: simulated and experimental values of the displacement at the peak load,

E_c, E_m: simulated and experimental values of energy absorption, respectively;

a, b and c are weighting factors of the parameters F, D and E, respectively, and $a + b + c = 1$.

The weighting factors a, b, and c can be judiciously chosen depending on importance of their respective metrics.

Results indicate that material constants generated from both these procedures replicate experimentally obtained force–deflection curves and fracture patterns well with a GCI of 0.91. The GCI can be generalized to take any number of parameters into consideration in the following form:

$$GCI = 1 - \left[\frac{1}{n} \left\{ \frac{(P_1 - P_{1test})^2}{P_{1test}^2} + \frac{(P_2 - P_{2test})^2}{P_{2test}^2} + \frac{(P_3 - P_{3test})^2}{P_{3test}^2} + \ldots + \frac{(P_n - P_{ntest})^2}{P_{ntest}^2} \right\} \right]^{\frac{1}{2}}$$

where

P_1, P_{1test} = predicted and experimental values respectively of parameter 1,

P_2, P_{2test} = predicted and experimental values respectively of parameter 2,

P_3, P_{3test} = predicted and experimental values respectively of parameter 3,

etc.

P_n, P_{ntest} = predicted and experimental values respectively of parameter n

Note: P_1, P_2, P_3, ….P_n are to-be-determined parameters of a given material type. It is believed that this generalized GCI can be applicable to practical assessment of quality of CAE generated results against corresponding experimental data.

A remark should be made here that ISO-TR9790 Procedures [17] was specifically developed for evaluation of biomechanical responses of WorldSID prototype dummy, Cesari, Compigne et al. [18], and subsequently adopted by Ruan, El-Jawahri et al. [19] for assessing the biofidelity of the human body FE model in side impact simulations.

6.2.2 Rigorous Approaches

By taking test uncertainty and multi responses into consideration, this set of approaches used more rigorous mathematical treatment with statistical tools and probabilistic approaches for model validation of the dynamic system. When developing a rigorous approach to provide qualitative model validation under uncertainty, statistical hypothesis testing is frequently involved. Oberkampf and Barons [20] published a comprehensive state-of-the-art review of statistical hypothesis testing approaches that can be divided into classical hypothesis testing and Bayesian approach. The Bayesian approach focuses on model acceptance whereas classical hypothesis testing focuses on statistical model rejection. Recently, the Bayesian approach has become widely applied, but a review of those applications is beyond the scope of this chapter. However, Jiang et al. [21] applied Bayesian probabilistic principal component approach for model validation of a child restraint system which can be served as an example of this approach.

Recently, Kokkolaras et al. [22] presented an overview of their most recent and ongoing research efforts in developing a comprehensive frame-work for simulation-based design evaluation of vehicle system. It is the hope that such an approach can be streamlined to become more user-friendly for easy application by modelers in the future. Based on methods described above, it can be seen that the simplistic approach is much easier to apply for faster assessment of model reliability and predictive capabilities.

6.3 MADYMO-Based Rollover and Human Models

Both CAL (Cornell Aeronautics Laboratory) 3D and MADYMO (*MA*thematical *DY*namic *MO*del) 3D are the most sophisticated mathematical occupant simulation models, and are widely used in automotive and aerospace industries. These two simulation tools were basically extensions from their respective 2D versions. Currently, MADYMO 3D developed by TNO is marketed through TASS Safety in Netherlands, while CAL3D developed by CAL is maintained by the Air Force Research Laboratory. Since these two packages were previously reviewed in detail by Prasad and Chou [6], and nowadays, MADYMO 3D is widely used in automotive industry, this chapter will discuss only applications of this crash victim simulator (CVS) in rollover and whole body human models.

MADYMO 3D is a whole-body kinematics simulator, also known as gross-motion simulator, which is a class of mathematical models formulated to describe a vehicle occupant in three-dimensional (3D) motion in a crash environment. The occupant is generally assumed to be a set of rigid bodies with prescribed masses and moments of inertia, linked by various types of joints in open- and closed-loop systems known as "tree structures." Any number of rigid bodies can be used to describe the occupant, resulting in simple to complex models of the occupant. The governing equations of motion of such a collection of rigid bodies are derived automatically in closed form using the Lagrangian or Newtonian approach. A gross-motion simulator, in general, consists of a body dynamic sub-model, a contact sub-model, a restraint system sub-model, and an injury criteria sub-model. In addition, MADYMO has an excellent dummy database consisting of various types, sizes of adult and child dummies. These subsystems and dummy database form the framework for versatile applications to analysis of frontal, side, rollover, and pedestrian impacts, as well as various restraint systems. In the late 90s, FE capability had been added to version 5.0 and its efficiency was improved in version 5.3. This added FE feature allowed MADYMO to expand its modeling capability, thus enabling application further for simulating more complex airbag deployment/interaction with occupants and deformable structural components.

6.3.1 MADYMO-Based Rollover Models

The rollover condition is reviewed because relatively few studies have been conducted due to the need to simulate long duration impacts. Blum

[23] explored the feasibility of using MADYMO to simulate rollovers in various conditions. Aljundi et al. [24] gave a brief description of rollover impact simulation using MADYMO Package. Yaniv [25] developed a MADYMO model and validated it against test results for restrained occupant with inflatable tubular structure (ITS). Their model was then exercised to evaluate the effectiveness of ITS in preventing occupant ejection during rollover events. Sharma [26] used the model to help develop a rollover component test methodology for evaluating restraint systems under a Notional Highway Traffic Safety Administration (NHTSA) contract. Renfroe et al. [27] presented the MADYMO modeling of vehicle rollovers and resulting occupant kinematics. Chou and Wu [28] used MADYMO models to simulate different rollover test procedures.

Furthermore, Takagi et al. [29] adopted the MADYMO methodology to simulate occupant behaviors in various rollover initiation types. MADYMO models, in general, give fairly good predictions of vehicle kinematics at the initial and airborne phases during a rollover, and can be applied to: (a) helping establish threshold(s) for rollover sensor system development, and (b) guiding and determine the initial conditions for rollover test procedure development. However, the rigid-body approach cannot predict vehicle structural crush during the contact phases and its effect on occupant kinematics. For predictive structural model development, test data from numerous rollover events and development of FE rollover models are needed.

Rollover models of varying degrees of complexity based on rigid-body and FE assumptions are initially reviewed by Chou et al. [30] and updated by Chou et al. [31]. The analytical studies and model simulations are becoming useful methods for determining the influence of vehicle parameters on vehicle responses in rollovers. MADYMO-based models for simulating vehicle kinematics prescribed in SAE J2114, side curb trip, critical sliding velocity, and corkscrew ramp tests were developed and reported by Chou [32] and Chou and Wu [33]. The rigid-body based MADYMO models are easier to develop and run to provide trend analysis and design direction for rollover testing, sensing systems, and restraint system development. Recently, Gopal et al. [34] used MADYMO and PC-Crash as analytical tools for simulation and testing of a suite of field relevant rollovers.

The CAE methodologies can be developed along with the test methodologies. Experimental data obtained from testing are generally used for developing rollover CAE models that replicate vehicle motion under similar test conditions. Analyses of simulated results provide valuable feedback to help improve and/or enhance the test procedures. Testing with improved/enhanced procedures can provide additional and more accurate data for continued model refinements. MADYMO-based CAE tools can provide high quality models with better simulated and/or predicted results. Generally speaking, MADYMO rollover models consist of sprung and unsprung masses, suspension systems and tires, using characteristics extracted from an ADAMS-based vehicle handling model. Uses of the MADYMO-based models to support rollover testing, rollover sensing algorithm development and rollover protection system development, and many issues associated with rollover CAE simulations were given by Chou [32]. Reviews of rigid-body-based mathematical rollover models indicated that such analytical tools are good for:

- Providing test conditions for roll and non-rollover events for a given test mode
- Selecting test vehicle configurations for sensor development testing
- Supporting countermeasures development
- Trend analysis.

However, the quality of signals generated from existing rigid-body-math-based models is not yet mature enough for use in rollover sensor algorithm and calibration [34]. Development of rollover models is a continually improving process, which requires experimental data for validation, refinement, and enhancement. With the advancement of computer technology, CAE methodology continues to grow in the use of FE analysis for rollover modeling to study vehicle structural deformation and occupant kinematics interacting with the restraint system and vehicle interior. In addition, an assessment tool for evaluating quality of CAE-generated sensor signals is needed. An example of such a tool that can be mentioned was developed by Le and Chou [35] for CAE generated rollover signal analysis.

In recent years, one area in the advancement of rollover CAE methodology is MADYMO with new suspension capability. The MADYMO program [36] is primarily used to study kinematics/responses of occupants within vehicles as the consequences of impact to these vehicles. Airbags and belt systems can be modeled to investigate how the occupants interact with them. Since MADYMO has the capability in modeling the behavior of tires with the Magic Formula (MF) –Tire model and allows the use of actuators to apply forces or displacements at defined points in a mechanical system, it is therefore feasible to develop a model for studying vehicular dynamics with input characteristics of vehicles. The aforementioned functionalities of MADYMO thus allow engineers to study the vehicle handling capabilities and the rollover crash characteristics with one single model. According to a review article by Chou et al. [31], a trend in modeling approach is to use one software package to model laboratory-based rollover tests using prescribed motion of the vehicular body from either experimental data or output from the vehicular dynamics program. Using a single software package could reduce model development time/cost without needs to translate model output from a vehicular dynamics software into another occupant simulation package.

The feasibility of such a single-model approach was investigated by McCoy et al. [37], who developed a MADYMO model of a mid-size sport utility vehicle, including not only a detailed suspension system translated from the model of a typical SUV from ADAMS vehicle dynamic software package, but also tire characteristics that incorporated the Delft-Tire MF description. The model was completed by adding an actuator that applied a translation motion to the steering rack, which in turn displaced the wheels. A steering wheel and column was also implemented, making it possible to provide input by turning the steering wheel. This model was correlated by simulating a vehicle suspension kinematics and compliance test. The correlated model was then used to simulate a J-turn vehicular dynamics test maneuver, a roll and non-roll ditch test, corkscrew ramp, and a lateral trip-over test. Results indicated that MADYMO was able to reasonably predict the vehicular and occupant responses in these types of applications and was potentially suitable for helping set up a suite of vehicular configurations and test conditions for rollover sensor testing. A sensitivity study was also conducted using the new suspension system for the laterally tripped non-roll event.

6.3.2 MADYMO-Based Human Models

In the early 90s, the advent of dynamic side impact crash test requirements in FMVSS 214 added impetus to the importance of developing human models for predictive human injuries due to insufficient biofidelity of existing side impact dummies. This sort of development was made possible with the advancement of FE technology and injury studies through biomechanics research in the past. There are MADYMO-based, ATB (Articulate Total Body)-based, and FE-based human models currently in the literature. Procedures used to develop FE human models are discussed in Sect. 6.6.2. In this section, only MADYMO-based human models are reviewed.

Huang, King et al. [38] developed a MADYMO 3D whole body model in side impact configurations to predict injury parameters such as TTI (Thoracic Trauma Index), chest compression, viscous criterion (VC), and average spinal acceleration (ASA). The model consists of rigid body segments connected by inextensible joints. The inertial properties used for each of the body segments were based on a 50th percentile male with a body mass of 76 kg. However, the stiffness characteristics of the torso are representative of the human and not an anthropomorphic dummy. Since major emphasis of Huang's study is on the torso, four segments representing the shoulder, thorax, abdomen, and pelvis were used. Simulation of the soft tissue covering the rib cage was accomplished by using a viscoelastic mini-model. This human model was validated against 17 cadaveric sled tests and 44 pendulum tests on 10 parameters. The average difference between the model outputs and average experimental results was 12 %. This model was used to study the effect of padding on injury parameters for side impact protection.

Happee et al. [39] developed a MADYMO-based human model consisting of a 6-DOF (Degree-of-Freedom) multi-segment spine and neck model with lumped joint resistance for ligament and muscle, and a seven-flexible-body thorax. This model was set up for omnidirectional impact applications, and then to predict human responses (such as kinematics, accelerations, loads, and chest deflection) as in side impact and out-of-position frontal impacts. An anthropometric RAMSIS database, which consists of body height, corpulence, and torso height (relative) of various seated civilian population, was used to develop a MADYMO-based 50th percentile male RAMSIS model for this study. This human model was validated against available volunteer and cadaver data under various test conditions in frontal impacts (volunteers, blunt thorax impact, belted/unbelted) and lateral impacts (thorax). This human model was then applied to study airbag/occupant interaction in out-of-position conditions under which the occupant was also restrained by belts that were simulated with finite elements. The authors reported that crash safety designs based on a real human body model of varying anthropometry are expected in the future.

Van der Horst et al. [40] developed a human head-neck model to study the injury mechanism, with emphasis on effects of muscle response. They improved the detailed head-neck model developed by De Jager et al. [41] by the addition of neck muscles. The model was composed of nine rigid bodies representing the head, the seven cervical vertebrae, and the first thoracic vertebrae. Ellipsoids represented the skull and the vertebrae with spinous and transverse processes, and articular facet surfaces. The bodies were connected through 3D linear viscoelastic disks, 2D nonlinear viscoelastic ligaments, frictionless facet joints, and contractile Hill-type muscles. Muscle geometry was also improved by including more muscles that were divided into a number of segments, thus allowing the muscles to curve around the vertebrae during neck bending. Therefore, the model was called the curved muscle neck (CMN) (Fig. 6.1) and validated against human volunteer data obtained from Hyge experiments for frontal impact at 15 g's and side impacts at 7 g's with and without muscle activation. Simulation results indicated that active muscle behavior seemed essential to accurately describe the human head-neck response to impacts.

Early in the 1970s, a number of rigid-body-based pedestrian models had been developed for simulations of pedestrian kinematics upon impact by vehicles; for example, Padgaonkar [42] using CAL3D, Wijk et al. [43] and Janssen and Wismans [44] using MADYMO. In these mod-

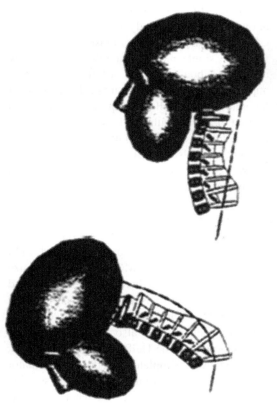

Fig. 6.1 Path of a multi-segment muscle in initial and flexed potions: curved muscle neck model [40] (Reprinted with permission of the Stapp Association)

els, their associated joint properties and body contact characteristics were based on experimental data of early dummies with limited biofidelity, thus not being able to predict accurately the pedestrian kinematics upon car impact.

Since the 1990s, more rigid-body-based models with improved biofidelity for pedestrian safety research has been developed. Ishikawa et al. [45] reported development of a MADYMO 3D pedestrian model consisting of 15 body segments, using geometrical and biomechanical properties taken from anthropomorphic and biomechanical data available in the literature. This model also laid out the underlying basis for various models appeared later in the open literature. Examples that can be mentioned are HONDA model by Yoshida et al. [46], Japan Automobile Research Institute (JARI) model by Mizuno and Ishikawa [47], and a model by Anderson and McLean [48].

Using anatomical knee joint structures, Yang et al. [49] developed a pedestrian model with a human-like knee joints and breakable legs for simulating pedestrian-car interaction, thus enabling study of tibia and fibula fractures. This model was validated against Post Mortem Human Subject (PMHS) test data in overall pedestrian kinematics, body segment responses and injury predictions. As an outgrowth of this model via the scale method, a series of child pedestrian models were developed by Liu and Yang [50]. Additionally, TNO [51] developed a more detailed 50th percentile human pedestrian model consisting of 52 body segments and 64 contact ellipsoids, of which the simulated responses were correlated with those PMHS test data favorably. However, the model still requires further refinements for the knee and shoulder body regions, according to the authors.

6.4 CAE Tools and Critical Aspects in FE Modeling

Developing models for solvers based on rigid body dynamics such as MADYMO was reasonably straightforward. On the other hand, developing an FE model typically requires the use of

Table 6.1 Exemplary FE pre- and post-processing software packages

Software title	Company	Location
Hypermesh	Altair	Troy, MI, USA
FX + Modeler	Midas	London, UK
ANSA	Beta CAE Systems	Thessaloniki, Greece
FEMB	Engineering Tech. Associates	Troy, MI, USA
Patran	MSC Software	Santa Ana, CA, USA

several software packages. To facilitate user friendliness for creating a simulation model and allowing ease of analyzing simulation results, most mathematical simulation packages have developed pre- and post-processor capabilities. Aside from these solver orientated packages, there are also software packages dedicated to pre- and post-processing works with numerous features for mesh development. Some exemplary software packages are summarized in Table 6.1. In Sect. 6.6.2, additional software needed to process medical images to generate human or animal models will be further discussed.

6.4.1 Mesh Convergence in FE Model

An important, but frequently overlooked, aspect in developing FE models is mesh convergence. Mesh convergence refers to how small the element size should be in an FE model to ensure that simulation results are unaffected by changing the size of the mesh. There are three main reasons for this oversight. First, an FE model with high mesh density may not be solvable when computing resources are limited. This issue is no longer a critical one as newer computers are capable of handling a large quantity of random access memory. Second, developing FE models with a different mesh density requires significant effort. Unless each refinement represents a division of one 3-D element into eight elements (that is, dividing each edge of an element into two), substantial laborious work is involved when refining a mesh. While this issue currently persists, it is less critical now because there are software packages which allow users to parameterize the mesh so that little effort is needed to adjust the mesh density (e.g. Mao et al. [52]). Still, refinement of the mesh using such automatic meshing software usually has limitations if the parameterized surface is poorly formulated and there is no guarantee that the refined mesh will be of high quality. Third, many research groups have in their databases a number of numerical models available and hence have a tendency to take an old model that was previously published for use in a new loading condition without testing for convergence to ensure that the mesh density is sufficient to solve the new problem.

To check for convergence, strains or stresses in several regions of interest are computed and plotted as a function of mesh density. If simulation results from two FE models with different mesh densities differ within a few percentage points of each other, then mesh convergence has been achieved. Otherwise, continued refinement of the mesh should be carried out and the FE simulation repeated. Typically, differences in strains or stresses in two consecutive refinements will decrease as the mesh is refined. Eventually, the difference will be sufficiently small so that convergence is deemed to be achieved. In some cases, such as impact of soft tissue by a very small diameter indenter or pendulum, convergence is very difficult to achieve due to the large deformation confined to a small region. In this case, a report must be generated to indicate how far away the mesh is from full convergence. Although some advanced FE solution methods are, in theory, not affected by mesh size, testing for convergence is a recommended practice in any laboratory/institution in the event that the software does not live up to its expectation.

6.4.2 Material Model Verifications

FE models need to be validated at the material and sub-system levels first prior to a system level model can be reliably used to predict the system behaviors. Engineering materials, such as mild steel and aluminum, are well described in literature. Their constitutive laws and associated material properties are well investigated and published. On the other hand, newly developed alloy (such as magnesium alloy for example) designed to reduce vehicular weight while maintaining high material strength is less explored. Similarly, constitutive laws and associated properties for biological tissues are not readily available and require a great number of assumptions in choosing the material properties.

Many of the new materials cannot be fully described with linear elastic models. Some FE software packages now offer a large number of material libraries for users to choose from. Because developing suitable constitutive laws for new materials requires much greater efforts, alternative material laws can be used to reasonably replicate the behaviors of such material while new or refined constitute laws are being developed. Zhu et al. [16] successfully modeled a novel die case magnesium alloy AM60B by investigating the possibilities of using MATs 24, 88, 99, and 107, all with varying degree of strain rate effect and failure simulation capabilities, available in LS-DYNA. Design optimization procedures were used to compare simulation results with physical tests including coupon testing from quasi-static to 800 s^{-1} strain rate, and four-point bending and crush testing of thin-walled structural components. A goodness fit procedure using GCI as described in Sect. 6.2 was used to judge which material law provided the best fit for all test conditions. Iterative procedures were continued until new material laws are validated.

In engineering materials, specimens are abundant and can be machined into specific sizes for testing under various loading conditions to identify/determine the properties. Biological tissues are not easily obtained and exhibited greater variations due to different age, gender, ethnicity, among other reasons. These tissues are very compliant in nature and hence difficult to control the exact geometry of the test specimen. Partially due to inaccurate geometry utilized in determining material properties, extremely large variations were observed in literature for biological tissues. As a result, selecting any parameters to represent biological tissues seems acceptable, although it is fundamentally erroneous. To account for specimen-to-specimen variations in geometry by removing the geometric effect in physical testing, Guan et al. [53] utilized specimen-specific FE models to identify material properties through reverse engineering procedures. It was concluded that properties obtained through this set of rigorous procedures can better represent the material behavior. Figure 6.2 shows the procedures deemed appropriate when investigating very compliant biological tissues for implementation as material properties in numerical models. Currently, laser or micro CT or MRI scanning is in place to obtain specimen-specific geometry. Specimen-specific FE models can then be developed before reverse engineering methods can be applied through optimization procedures to identify material properties.

6.4.3 Procedures in Meeting Safety Regulations

Aside from governmental regulations, such as FMVSS (Federal Motor Vehicle Safety Standard), there are also consumer-based organizations (such as NCAP organizations throughout the world and IIHS) prescribing the minimum safety requirements each vehicle should possess. To meet all these regulatory and non-regulatory requirements is a monumental task for safety engineers. With the advancement of CAE tools used in design of new product, more and more cars are meeting all regulatory standards and non-regulatory requirements with good rating. This section describes a basic set of procedures that can be used in CAE process to help meet these standards and/or test protocol requirements.

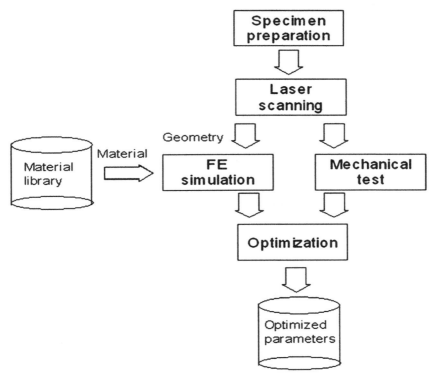

Fig. 6.2 Procedures to identify accurate material properties of very compliant biological tissues. High resolution laser or micro CT, micro MRI scanning is used to identify specimen-specific geometry while reverse engineering method is used in conjunction with optimization procedures to determine high fidelity properties

At the onset of designing a safe vehicle, engineers are to gather the information related to intended design boundaries such as target buyer, price range, safety rating, corporate average fuel economy (CAFE), etc. The following three-stage CAE processes can be used in conjunction with physical testing to achieve the desired design goals. The first stage is to complete a CAE material library which includes commonly used material models and associated constants. In the second stage, critical structural load bearing components related to crash energy management are meshed with coarse mesh FE model to obtain the target crash pulses. Experiments are conducted on either a body-in-white prototype or a previously developed vehicle platform to check the fidelity of the coarse mesh model. Once the energy absorption characteristics fit the desired goal, a detailed FE model of the entire vehicle can be created and simulated. Iterative procedures between experimental validation and refinement of the FE model are needed to account for any simplifications assumed in the FE model. Figure 6.3 shows the three stages of this design process via CAE approaches.

The objective of the first stage is to create (and maintain) a material library for CAE related activities. For each material needed to manufacture a vehicle, a material model should be developed with proper material laws (such as linear elastic, viscoelastic, etc.). Specimens are then cut into the desired shape and dimensions, and tested under different test modes and loading rates. Results from these experiments are compared with the FE material model. The model will be corrected if there is any mismatch between the model-predicted and experimentally obtained results. This set of procedures will continue until all materials needed to manufacture a car are available. Figure 6.4 shows this set of iterative procedures in establishing the material library. In some CAE software packages, such as the virtual proving ground (eta/VPG) developed by ETA, a library consisting of many commonly used materials is available but users are recommended for additional checking to account for different manufacturing processes.

In the second CAE stage, it is critical to meet ideal crash pulse targeted for crash energy management. In the early stage of designing a safe

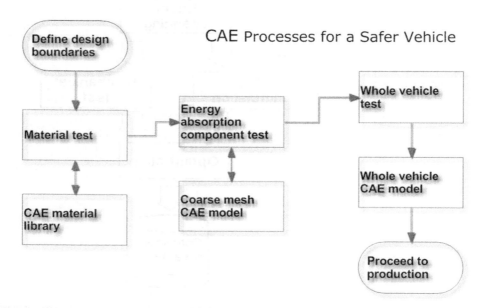

Fig. 6.3 Overall schemes of using CAE tools to meet regulatory and non-regulatory requirements

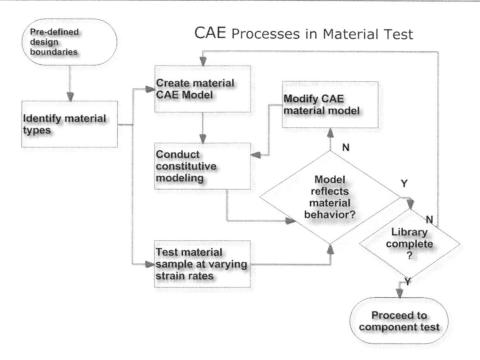

Fig. 6.4 Steps recommended for setting up a material library for CAE applications

car, the important issue is to set the design specifications for all major structures. A common practice is to develop a FE model consisting of critical energy absorption components to set up the load path in order to manage crash energy absorption for crashworthiness study. If the model-predicted crash pulse is far from the targeted idealized one (i.e., equivalent square wave, or crash pulse of an image car), the CAE model is modified and the processes repeated until idealized crash pulses are achieved. At this time, a prototype consists of these components are built (or modified from a previous production car) and tested at different impact modes to determine the fidelity of the FE model. Again, iterative procedures are needed to correct/refine the FE model until its predictions fit experimental results meeting certain validation criteria established consensually by CAE engineers. Successful completion of the coarse mesh model at this stage can be used to design the entire vehicle.

Figure 6.5 shows the iterative procedures at this stage of the design.

In the third stage of the design, occupant models are integrated into detailed FE vehicle model to determine if occupant responses meet the intended design safety rating target. Modifications will be made until the occupant responses meet the targeted star rating. At this stage, the FE model, when validated with predictive confidence, may also be used to generate signals for various crash modes required for the algorithm development for airbag sensing system among other things. Based on the satisfied FE model, a prototype vehicle will be developed and tested according to regulatory safety standards. Variations between test results and model predictions need to be resolved and solutions documented as "lessons learned" for future reference in modeling and hand-on-training of future safety engineers. Figure 6.6 shows this set of CAE processes.

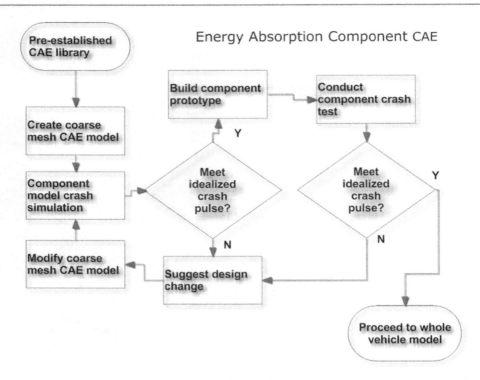

Fig. 6.5 Schemes for setting up a model consisting of critical energy absorption components for basic crash energy management

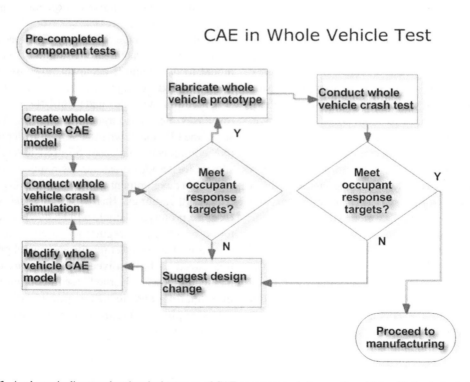

Fig. 6.6 A schematic diagram showing the last stage of CAE processes to design a safer vehicle with the desired star rating

6.5 FE-Based Models

As mentioned previously, FE-based models are the primary choice of CAE simulation tools recently because they can be used for both structure and occupant modeling and the speed of computer has been improved to an acceptable level. Additionally, it is more cost-effective to use only one software package so that engineers need to be trained to use one package only. In this section, exemplar CAE methodologies are presented for frontal impact, side impact, and rollover analyses.

6.5.1 Frontal Impact

Frontal impact tests may include the following crash test modes in accordance with regulatory, non-regulatory and sensor development requirements:
- Full rigid barrier impact
- Offset rigid barrier or deformable barrier impact
- 30° right/left angular impacts
- Center pole impact
- Small offset impact (IIHS test protocol)
- Over-ride and under-ride impacts

CAE analyses can be carried out using both rigid-body and FE-based methods for occupant and structural analyses depending on complexity of the problems at hand.

Figure 6.7 shows a schematic diagram of a frontal crash event, which can be considered as a three-impact event. In the first (primary) impact, the vehicle strikes a barrier or another car, causing the front-end to crush. Kinetic energy of the vehicle is expended in deforming the vehicle's front structure. The design of the front-end, rear, or side to crumple in a collision and absorb crash energy is called crash energy management or crashworthiness of a vehicle.

The second (secondary) impact occurs when the occupant continues to move forward as a free-flight mass and strikes the vehicle interior or interacts with or loads the restraint system. Some of the kinetic energy is expended in deforming the vehicle interior or the restraint system, and in compressing the occupant's torso. The remaining kinetic energy is dissipated as the occupant decelerates with the vehicle.

The first and the second impacts are frequently discussed in frontal impact analyses. Now, a third impact can be added concerning with brain impact against skull or internal organs within human skeleton, when the occupant impacts the vehicle interiors. This phase of impact analysis needs to be addressed using human models.

The crash event shown in Fig. 6.7 results in vehicle response and occupant response as exhibited in Fig. 6.8. Analysis of results of these two responses contributes to the evaluation of vehicle performance and occupant safety. Referring to Fig. 6.8, the vehicle response is primarily due to collapse of energy absorbing members of front-end structures, interaction and stack-up of rigid structural components (i.e., engine, transmission, steering box/gear, etc.), depending on packaging of power-train and accessories and available crush space. Depending on the impact velocity and crash energy management, intrusion of the engine into the occupant compartment may or may not occur. Such vehicle responses, often referred to as the crash pulse, can be recorded by accelerometers mounted at the un-deformed occupant compartment locations. In the meantime, the responses of occupant are measured for each body region (i.e., head, neck, torso, pelvis, leg, lower extremity) with instrumented dummy as occupant moves within the vehicle compartment. These measurements are in form of time histories of acceleration, force, displacement, etc. Dummy injury performance numbers can be calculated for HIC (Head Injury Criterion), Nij (neck), Chest G's, chest deflection, femur load, and tibia index, etc.

Based on results of both the vehicle and occupant responses, crash energy management of the vehicle and compliance with regulations (i.e., safety restraint system performance to meet mandated regulation) can be assessed. Traditionally, such assessment is carried out by crash testing of prototypes in laboratory environments. With the advancement of computer and CAE software, numerical methods help greatly in shorting the product development cycle and cost savings.

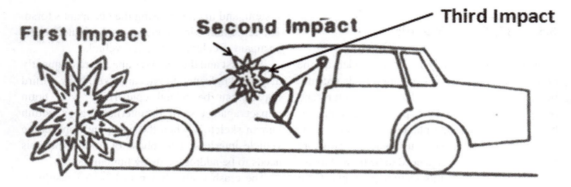

Fig. 6.7 Schematic of a frontal crash event

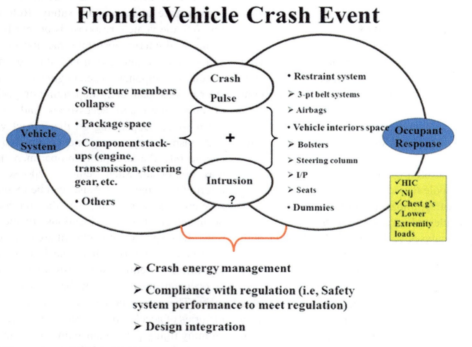

Fig. 6.8 Components of a frontal vehicle crash event

The crash pulse obtained from physical crash tests described above can be used as a forcing function for direct input into MADYMO occupant models to study the occupant kinematics for assessing potential occupant injury risk. Occupant response data obtained from crash tests also provide useful information for validating the occupant CAE model. In addition, a crash pulse can be idealized to become an equivalent square wave or two-step function for target pulse setting for future vehicle development.

Along with the line of numerical analysis, early stage frontal impact simulations can be performed using a classical lumped-mass-spring (LMS) method as shown in Fig. 6.9a, where a vehicle front end is represented by a five-mass model in addition to a barrier mass in a 90° frontal impact with a fixed rigid barrier test configuration. The masses are connected by non-linear spring and damper elements to account for energy absorbing characteristics of structural members, such as bumper, radiator support and fore and aft portions

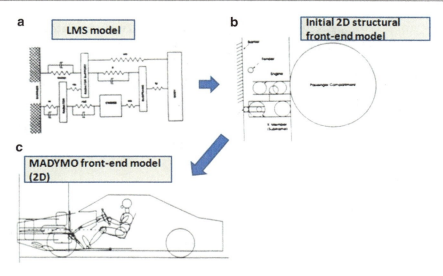

Fig. 6.9 Linkage model using MADYMO

of upper and lower rails, etc. The load-deflections of these non-linear spring characteristics can be obtained from static crusher test or from FE analysis of corresponding structural components. LMS models are in general considered as the simplest model, whose parameters, i.e., spring characteristics, relative positions between rigid components, can be further utilized for developing an integrated front-end model with an occupant model as shown conceptually in Fig. 6.9b using MADYMO, and can be further developed to become a better representation of an integrated structure and occupant model as shown in Fig. 6.9c [54]. Such an integrated model allows simultaneous simulations of the vehicle and occupant responses, without inputting the crash pulse, which is actually generated in the model. However, both the LMS and MADYMO-type models mentioned above are semi-empirical (or semi-analytical) approach, which required structural characteristics are a priori. More complex numerical simulation of full vehicle in frontal impact can resort to FE method approach.

Knowing that most certification tests of new production vehicles are run at high impact velocity of 35 mph (56 km/h) as specified in FMVSS 208, most FE models, referred to as crash models here, are developed for crash analysis at this speed. However, such crash models may not yield good prediction when performed at low impact speeds, say below 20 mph (32 km/h) as reported by Chou et al. [55]. Crash tests conducted at low velocities normally provide data for airbag sensing system calibration and algorithm developments. FE models developed for those purposes are referred to as "sensor models" accordingly to Chou et al. [55]. Figure 6.10 shows a FE vehicle model modified based on a crash model for developing a sensor model, which consisted of roughly 200,000 elements, and was validated for a 14 mph rigid barrier impact, a 19 mph center pole impact, and a 9.3 mph Thacham 40 % offset rigid barrier impact. Some model generated signals were used for developing a velocity-based algorithm for crash sensing system with tunnel accelerometer sensors and one or two front crash sensors [56]. Once such model is developed, one model can then be used for simulating the different frontal crash test modes as shown in Fig. 6.11 by changing impact mode, test weight and impact velocity for the same carline.

6.5.2 Side Impact

The significant crash event in a typical FMVSS 214 moving deformable barrier (MDB) side impact test is explained by analyzing the velocity

This model was correlated with data from the following tests:

• 14 mph rigid barrier impact

• 19 mph pole impact

• 9.3 mph Thatcham 40% offset rigid barrier

Number of Parts	330
Shell (triangular and quads) Elements	184,616
Solid Elements	272
Nodes	194,513

Fig. 6.10 A finite element model for frontal crash impact analysis [55] (Reproduced by permission of Inderscience Publishers from IJ Vehicle Safety (2007) Vol. 2, No.3, pp. 241–260)

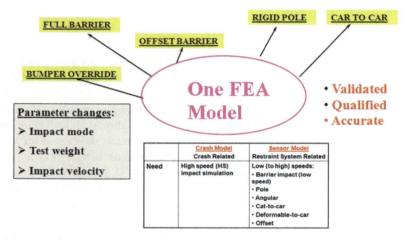

Fig. 6.11 One FEA model for simulating several different frontal crash test modes [55] (Reproduced by permission of Inderscience Publishers from IJ Vehicle Safety (2007) Vol. 2, No.3, pp. 241–260)

profiles shown in Fig. 6.12. The profiles are obtained by the numerical integration of accelerometer data taken from the following locations:
- Center of gravity (CG) of the MDB
- Non-impacted right-hand rocker of the target vehicle
- Door inner panel at the armrest
- Side Impact Dummy (SID), EuroSID, ES2-re, and SIDIIs pelvis

The following momenta exchanges are taking place during a dynamic side impact test according to FMVSS 214 test procedure with MDB:

1. The primary momentum exchange taking place between the MDB and the impacted vehicle. During this event, the rigid body motion of the impacted vehicle increases while the MDB velocity decreases until at some point in time, both the MDB and the impacted vehicle achieve a common velocity.

2. The momentum exchange taking place between the MDB and the door of the impacted vehicle. The door quickly attains the high velocity of the MDB.

3. Finally, the momentum exchange taking place as the intruding door comes into contact with the stationary dummy. The dummy pelvis, hit by a fast-intruding door, is quickly accelerated in the lateral direction. Door-to-dummy interaction forces can be reviewed by looking at the

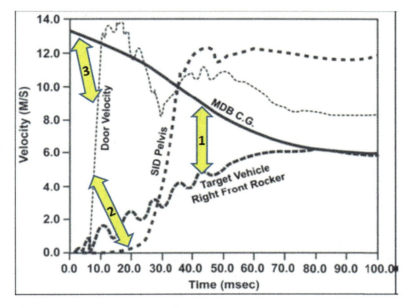

Fig. 6.12 Typical velocity profile in side impact [57]

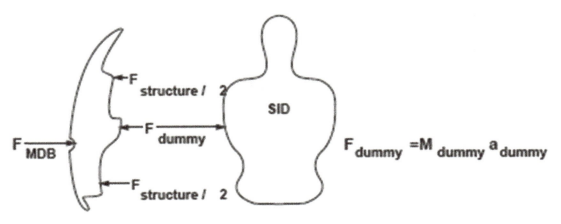

Fig. 6.13 Door and dummy "free-body" diagrams [57]

door and the dummy "free-body" diagram of Fig. 6.13 with the force components acting on the door shown:

F_{MDB} is the punch-through force of the MDB acting on the door, $F_{structure}$ is the body side structural resistance of the impacted vehicle that resists door intrusion, and F_{dummy} is the door-to-dummy interaction force, which also is the reaction force acting on the dummy.

Using Newton's Second Law, the rate of change in the door linear momentum is equal to the summation of forces acting on the door free-body diagram.

$$-\frac{d}{dt}[M_{door}V_{door}] = \sum F$$
$$= F_{MDB} - F_{structure} - F_{dummy}$$

and solving for F_{dummy}, one obtains the following:

$$F_{dummy} = F_{MDB} + \frac{d}{dt}[M_{door}V_{door}] - F_{structure}$$

Thus, it is noted that:
- Decreasing the MDB punch-through force (F_{MDB}) would decrease the force acting on the dummy (*Fdummy*).
- Decreasing the rate of change in the linear momentum of the door by making the door lighter or decreasing the door intrusion velocity would decrease the force acting on the dummy.
- Increasing the impacted vehicle body side structural resistance (*Fstructure*) would also decrease the force acting on the dummy.

These explain how countermeasures can be made by limiting the force on the dummy while strengthening the side door structure and using padding for cushioning the occupant, detailed of such effects is given in describing fundamental principles for vehicle/occupant system analysis in side impacts by Chou [57].

To develop countermeasures using simple CVS (Crash Victim Simulator), Padgaonkar and Prasad [42] developed a CAL3D model to simulate a perpendicular side impact as shown in Fig. 6.14. This model used ellipsoid features of CVS for generating interaction forces between structural and occupant body regions, and validated with experimental data with excellent agreement as shown in Fig. 6.15. Consequently, the model was used to study both the effects of side structure upgrade to reduce the velocity of the intruding door and padding for occupant protection via parametric study as shown in Fig. 6.16, where the results showed that a thicker padding provided a better protection. This model demonstrates that, though simple, can still provide trend analysis extremely accurate and useful.

Using MADYMO 3D capability, Low et al. [58] developed an integrated side impact simulation model as shown in Fig. 6.17, where hyper-ellipsoids

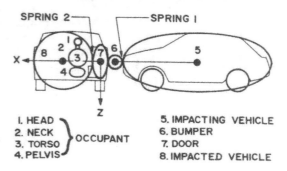

Fig. 6.14 An integrated CAL3D side impact model (1D math model) [42]

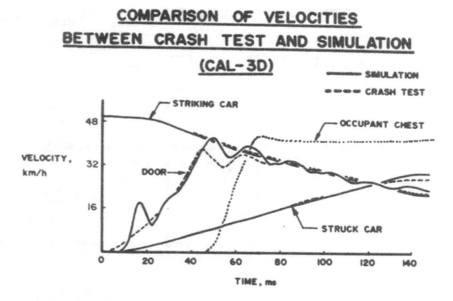

Fig. 6.15 Comparison between simulated results and test data [42]

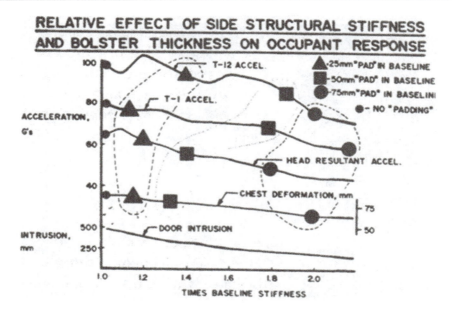

Fig. 6.16 Effects of side structure upgrade and padding [42]

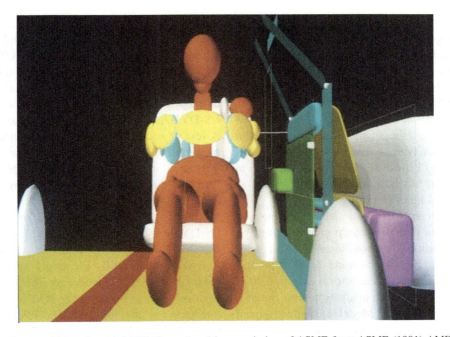

Fig. 6.17 Target vehicle sub-model [58] (Reproduced by permission of ASME from ASME (1991) AMD-Vol.126/BED-Vol. 19, pp. 155–168)

and planes were used for contact simulations. Upper and lower ribs of SID were also modeled to monitor their respective responses in terms of acceleration. Rib forces were obtained from static testing of SID rib cage, while door and MDB characteristics were obtained experimentally from static crusher tests. This model yielded good acceleration responses of upper rib and

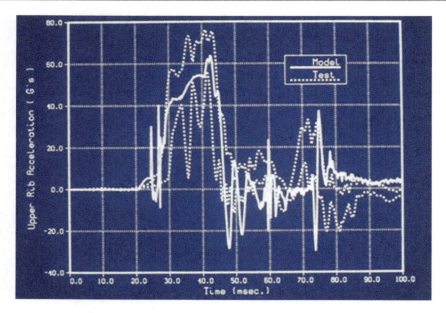

Fig. 6.18 Comparison of upper rib acceleration: model vs. test [58] (Reproduced by permission of ASME from ASME (1991) AMD-Vol.126/BED-Vol. 19, pp. 155–168)

pelvis (Figs. 6.18 and 6.19), which were then used for calculating TTI (Thorax Trauma Index) when SID was specified as an ATD in then FMVSS 214 dynamic test. It should be remarked that the ES2-re and SIDIIS are currently used in the updated FMVSS 214 test instead of SID. However, the methodology is still valid immaterial of type of dummy used. The model developed by Low et al. [58] was used to help in development of deployable door trim system as reported by Lim et al. [59].

Initial attempts of side impact modeling was started in the mid-1980s as primarily driven by safety regulation when non-linear FE crash codes just became available. Nowadays, FE methodology using FE methods as shown in Fig. 6.20 is routinely used by CAE analysts among automotive OEMs worldwide. Model size and associated degree of complexity were also dramatically increased since then. The turn-around time of side impact analysis was shortened due to advancement of technologies in both computer hardware and software development as previously mentioned.

Many types of side impact ATDs are specified in different safety standards, test protocols worldwide among government and safety research communities. Side impact ATDs such as SID, BioSID, EuroSID, ES2-re, SIDIIs, 5th and 50th WorldSIDs, exist, thus FE side impact dummy models were developed using LS-DYNA, RADIOSS, and Pam-Crash codes. A few of such side impact dummy models are mentioned here as references to those who are interested in this area: an FE model of the EuroSID dummy using Pam-Crash by Ruckert et al. [60]; a SID FE model using RADIOSS by Midoun et al. [61]; A SID FE model using LS-DYNA by Kirkpatrick et al. [62]; LS-DYNA BioSID FE model by Fountain et al. [63]; A RADIOSS EuroSID FE model by Pal et al. [64]; LS-DYNA SID model by Shkolnikov and Bhaisod [65]; a BioSID using RADIOSS model by Khan et al. [66]; SIDIIs FE model in RADIOSS by Chai et al. [67]; SIDIIs model in Pam-Crash by Kobayashi et al. [68]; WorldSID 5th and 50th dummy models by Liu et al. [69]. It is remarked here that the 5th WorldSID was developed by scale down from the WorldSID 50th dummy model. The prototype of the WorldSID 5th dummy hardware was then built as reported by Wang et al. [70].

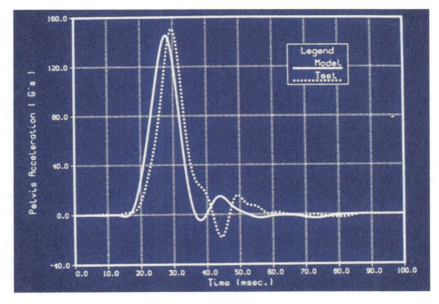

Fig. 6.19 Comparison of pelvis acceleration: model vs. test [58] (Reproduced by permission of ASME from ASME (1991) AMD-Vol.126/BED-Vol. 19, pp. 155–168)

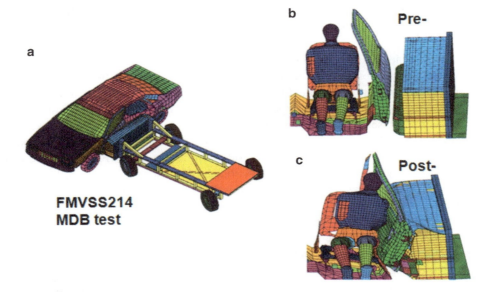

Fig. 6.20 A side FE model for analysis of FMVSS 214 side impact with MDB

The process used in development of the aforementioned ATD models generally follows the steps given below:
- Create geometry of each body part (i.e., head, neck, torso, pelvis, etc.) by either using CAD dummy data or scanning physical dummy parts, and generating models through meshing
- Parts can be treated as components (such as head, neck, torso, etc.) or sub-system assembly

(such as head/neck, upper leg/lower leg/lower extremities, etc.), and finally the complete dummy model

- Component and sub-system level models need to be validated using calibration of dummy parts, such as drop test for head, pendulum tests for neck and chest etc.
- Subsystem models can also be validated using sled test data.
- The complete (or total system) dummy model is validated using sled test and/or crash test data

To conclude the side impact analysis for this section, three examples of applying CAE methodology to help in developing side impact component test methodologies are given for: (a) selection of structural component for inclusion in the test setup targeted for side impact sensor response, (b) calculation of energy absorption for developing energy absorbing devices used in the door mounting fixture [71], and (c) use of FE model for side impact sled system simulation. These are subsequently described below.

(a) Selection of component(s) for inclusion in the test setup:

Most test methodologies reviewed by Chou et al. [72] were developed for aiding in restraint systems development. A new test methodology was developed for testing a component or subsystem for side impact sensor calibration and potential adaptation for FMVSS 214 pole impact simulations in conjunction with side impact airbag (SAB) restraint system development. To develop such test methodology for side impact sensor application, one must:

- Identify side impact modes for sensor development and firing time requirements, and
- Analyze side impact sensor signals and determine the pulse length in time (msec) that needs to be simulated.

Results from the above analysis of signals obtained at the sensor locations from side impact test data are used for CAE simulation to decide the followings:

- Identify side structural engagement in both pole and FMVSS 214 MDB impact modes,
- Use CAE to: (i) determine influential components for inclusion in component/subsystem level testing and (ii) study test setup configuration for both non-deployment and must-deployment condition.
- Effect of frame on the responses of sensors at both satellite and RCM (Restraint Control Module) locations

Aekbote et al. [73] developed a new sled-to-sled side impact component test methodology by following the process outlined above to derive targeted velocity profiles for bullet and door sub-sleds using a decelerator to achieve the required gap closure between the door and the occupant. A honeycomb was also used to maintain gap closure between the MDB represented by a moving bullet sled and the door sled once engaged. This concept study led to a physical sled test setup for development of a side airbag (SAB) for successful implementing into productions.

(b) Calculation of energy for energy absorbing devices:

Clements [71] developed a subsystem test method that allowed an undeformed door to a mounting fixture with energy absorbing devices to support various rigid elements representing A-, B- and/or C-pillars to which the door was mounted. These energy devices used to represent energy absorbed by the side structure through pillar areas during a full-scale side impact. The amounts of energy dissipated at different locations at the pillars at different levels were calculated using a full scale FE model of a side impact test mode (Fig. 6.21). He demonstrated this approach to both IIHS and EuroNCAP test modes.

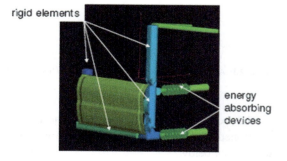

Fig. 6.21 CAE analysis for developing energy absorbing devices [71]

(c) Use of FE model for side impact sled system simulation:

In 2006, Teng et al. [74] reported the development of a FE model consisting of a sled rail sub-model, a door sled submodel, a seat sled submodel and a FE dummy sub-model to simulate a side impact sled system using BASIS System of TNO for side impact sensing system development.

Although a FE model is useful for side impact component test setup simulation, it should be remarked that the FE model must be fully validated to ensure timely deployment of side airbag system prior to such an application. Quality check and accuracy of FE model are still an important part of model development now and in the future.

6.5.3 Rollover

The limitation of the rigid-body-based rollover models is obviously due to its incapability in simulating the vehicular structural deformations. To study a rollover event with complex component deformations can be resorted to the use of FE models. FE methods became the most up-to-date advanced technology available to aid in automotive crash analysis in both frontal and side impacts as mentioned previously. However, to date, applications of FE models to rollover simulations are still limited. A fairly extensive review of math-based models for rollover simulations was reported by Chou et al. [31]. Hu et al. [74] developed a FE model of a sports utility vehicle (SUV) based on the non-linear FE code, LS-DYNA, and validated the simulated results against experimental data. Chou et al. [75] used this model to demonstrate how a FE model could be used in developing a methodology to determine the vehicle-to-ground contact load (or force) when a vehicle hits the ground during an SAE J2114 laboratory-based rollover test. When using FE models for rollover simulation, one should bear in mind not to use a fine-meshed model. Should one intends to use a detailed front impact analysis model for rollover simulations, the number of finite elements need to be reduced to about 100,000 to 200,000 elements in order to save computer run time.

The FE SUV model developed by Hu et al. [74] consisted of several sub-models, including a vehicle structure sub-model (Fig. 6.22a), four tire sub-models from eta/VPG (Fig. 6.22b), and two suspension sub-models (Fig. 6.22c and d). These sub-models were first individually developed and then integrated together. The full vehicle model consisted of about 100,000 elements, and used piecewise linear plasticity material for most structural members. The powertrain components were defined as rigid material, and lumped masses added to the model to account for all non-structural components. A Mooney-Rivlin type rubber material was defined for the tire, and an airbag model was defined to simulate the tire pressure. Several validations have been conducted to each sub-model. The tire and suspension sub-models were also validated against previously published experimental data by Lee et al. [76] and Zhang et al. [77], respectively.

The structural sub-model was validated against data from a quasi-static roof crush test (i.e. FMVSS 216 Roof Crush Strength Test) as shown in Fig. 6.23. The full vehicle model was also validated by three dynamics laboratory-based rollover tests, i.e., an SAE J2114 dolly test, a curb-trip test, and a corkscrew test by Hu et al. [74]. The simulated vehicle motion and comparisons between experimental and simulated results of SAE J2114 dolly test are shown in Fig. 6.24.

Such FE models were used for determining vehicle-to-ground contact forces when using vehicle kinematics data obtained from photogrammetric analysis of crash videos [75]. Currently, vehicle-to-ground contact forces cannot be measured during a full scale rollover impact test in laboratory environment.

6.5.4 Restraint System Modeling

Currently, two types of restraint systems, namely a belt restraint and an airbag restraint system, are mandatory standard safety features equipped in production vehicles, although EA steering column and wheel and knee bolsters can also be considered as part of such systems. The belt restraint system is a safety device that requires

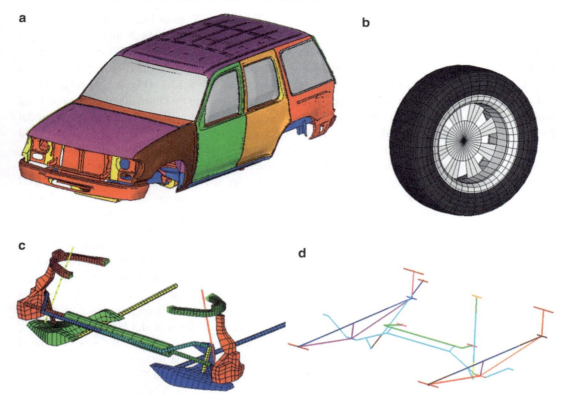

Fig. 6.22 FE vehicle model consists of four sub-models [74] (Reproduced by permission of ASME from IMECE'07 (2007) Paper No. IMECE2007-44083)

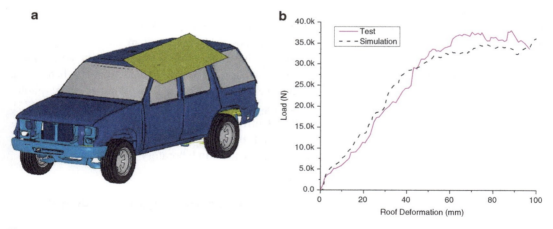

Fig. 6.23 Validation of the SUV model against test data based on FMVSS 216 [74] (Reproduced by permission of ASME from IMECE'07 (2007) Paper No. IMECE2007-44083)

action of the user to buckle up into operational position. The most well-known belt restraint system is the continuous 3-point belt restraint system that was developed by Volvo in late 1960s. This system, when placed, provides a shoulder belt portion and a lap belt portion for effective restraining the torso and the pelvis of occupant, respectively, during the vehicular deceleration in

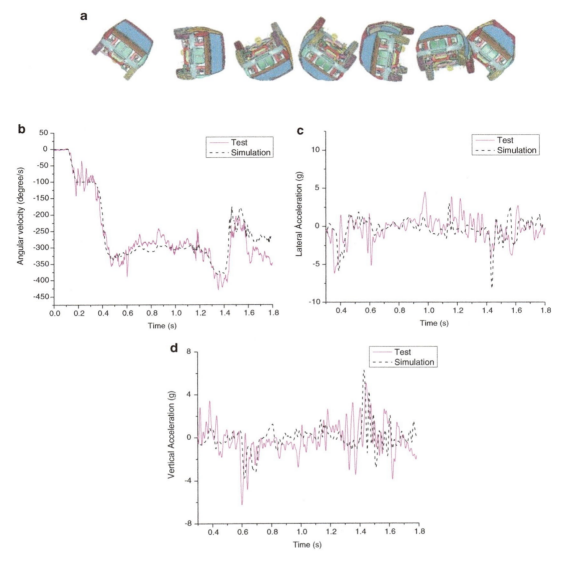

Fig. 6.24 Model predicted vehicle motion and comparison between experimental and simulated results of SAE J2114 dolly test [74] (Reproduced by permission of ASME from IMECE'07 (2007) Paper No. IMECE2007-44083)

a crash event. The belt restraint system is also considered as a primary restraint system for occupant safety protection in frontal impacts and rollovers. The airbag restraint system was originally developed for driver occupant protection in frontal impacts, but now also covers the passenger airbag for frontal impact, side airbag (SAB) for side impact and curtain bag for side and rollover impact protections. The airbag systems for the driver and passenger protection in frontal impact are considered to be a supplemental restraint system, because they provide additional restraint besides the belt restraint system. The airbag restraint systems are categorized as a passive restraint system as they will automatically deploy when needed, thus not requiring user's action. An excellent article on occupant restraint systems pertaining to some fundamental laws, concepts and applications can be found in the work by Eppinger [78].

Development of a restraint system for occupant protection requires knowledge of vehicular structural performance in a specific crash mode. Take the frontal impact for instance; the roles of the front end structures are to absorb the vehicle crash energy through structural deformation, but need to attenuate the impact load to tolerance levels of occupant deceleration. Therefore, a restraint system needs to work with vehicle structural performance by knowing crash pulse, vehicle/occupant kinematics and human tolerance, and hence should be developed by taking these into consideration. Generally speaking, the stiffer the vehicular front end structure, the higher impact force the occupant will experience, thus placing a greater demand on the restraint system for occupant protection.

Both rigid-body- and FE-based models discussed in this chapter have capabilities in simulating such systems. Some general discussion is given below.

Various types of belt restraint systems can be modeled:

(a) Simple belts: These are basically springs. However, Kelvin elements can be used to model simple belts.
(b) Advanced belts: These are a more realistic belt model, accounting for the spatial geometry of a belt system, such as a continuous 3-point belt restraint system. Locations of belt anchorage points can be specified in the inertial space coordinates. Slip between belt segments is possible, e.g., as between the lap belt portion and shoulder harness portion of a three-point belt system. For belt systems with a retractor. The "film spool effect" can be simulated if the data are available, e.g. applied belt load versus length of belt spooled out of the retractor. The capability of simulating acceleration-sensitive retractor locking mechanism was made in rigid-body-based model. Multiple belts can be done within the same system and belt contact with surfaces done by prescribed reference points.
(c) The pre-tensioner can be modeled for the retractor with various triggering criteria.
(d) The web grabber model is available.
(e) The FE belt model using triangular membrane elements is available. Sliding of belts over occupant body surface can be simulated.

Airbag restraint systems are most frequently represented by 3D model for both 2D and 3D simulations. The 3D airbag simulation can be made with a FE module in rigid-body-based model, such as MADYMO, where the airbag fiber is modeled using triangular membrane elements. The FE airbag model takes the inertia bag effects and bag slap into consideration and determines forces acting on the occupant more accurately. Multiple inflators for each airbag and multiple sensors for each inflator are allowed. Both the driver-side and passenger-side airbags can be simulated. Effects of vent holes and porosity can be studied. Various gas properties are predefined in the program to allow users to select the gas mixture flowing out of the inflator. Lastly, Various types of airbag folding can be done with most non-linear FE crash codes, such as LS-DYNA, RADIOOS, and Pam-Crash.

Out-of-position (OOP) occupant-airbag interaction is an important aspect regarding the development of low risk deployment of airbag restraint systems for protection of small size occupant seated in the front-seat position. Cheng et al. [79] developed a methodology by coupling the rigid-body-based ATB (Articulated Total Body) occupant model with the LS-DYNA FE airbag model that simulates OOP occupant-airbag interaction. This model included the airbag, 5th percentile female driver occupant and major vehicle interior components such as the steering-wheel/column, instrument panel, windshield, toe-board and seat which includes seat cushion, seat back and head rest. The 5th percentile female dummy, which was scaled down from the 50th percentile dummy model, was unbelted in the simulations of (a) sled tests with a half-sine deceleration pulse at 30 mph and (b) static deployment tests for (i) chin above the airbag and (ii) sternum on the airbag module positions. AIRBAG_WANG_NEFSKE_JETTING in LS-DYNA was used to define the thermodynamic behavior of the gas and jetting effect of the inflator.

Another FE numerical simulation of OOP front passenger injuries in frontal crashes was presented by Yamagishi et al. [80] using Pam-Crash code with both the 50th male and 5th female dummy models developed by then First Technology Safety System (FTSS) to study static test conditions for OOP of driver and passenger sides as specified in FMVSS 208. Some critical design factors that influence side impact injury performance numbers are considered including:

- Airbag force that depends on gas flow and pressure in the bag and bag material type/properties
- Airbag deployment kinematics which is affected the out-flow of gas, bag pressure and back support by the steering wheel ot instrument panel, and
- Steering wheel/column and instrument panel support force, which is largely dependent of structural component and bag material properties.

In this study, influential design factors listed above were adopted with the help of an FE airbag model which took gas flow into account. Comparison of simulated airbag deployment behavior appears to be in good agreement with experimental counterpart.

6.6 Occupant FE Models

Obviously, any safety simulation is aimed to improve an occupant's safety in the event of a crash. Initially, CVS-based (namely, rigid-body-based-) dummy model is the emphasis of such developments. As computational speed advances, FE models become CAE engineers' choice for the reason that structure and occupant responses can be analyzed in one simulation. Most recently, researchers realized that the FE dummy model has all the inherited shortcomings of a physical dummy and thus FE-based human models have become the points of interest for most automotive safety research institutions. In this section, FE-based ATD (i.e., dummy) models are discussed first followed by more in-depth descriptions of developing human models.

6.6.1 FE-Based ATD Modeling

Compliance with the FMVSS regulations is a critical first step in making a vehicle safer. By the 1980s, CVS programs like MADYMO and CAL3D had made significant progresses in ATD modeling. For this reason, dynamic coupling between the structure analysis using FE codes and occupant response using CVS code was the main research focus at this period. In the 1990s, developing FE dummy models became an emphasis of numerous research efforts so that all analyses can be done in one software package.

Limited by computational power, early FE dummy models were highly simplified or included only a section of the dummy. Midoun et al. [81] developed a whole body FE Hybrid III model using shell elements with interior supported by spring elements and rigid bodies connected by joints (Fig. 6.25). In essence, this model was developed by simply transferring a MADYMO Hybrid III into an FE code. Lasry et al. [82] developed a FE model of a Hybrid III dummy chest with detailed internal structure, such as the

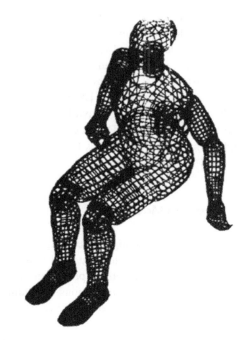

Fig. 6.25 The whole body Hybrid III shell model developed by Midoun et al. [81] (Reprinted with permission of the Stapp Association)

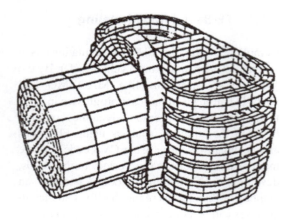

Fig. 6.26 Detailed Hybrid III chest model with chest foam developed by Lasry et al. [82] (Reprinted with permission of the Stapp Association)

rib, rib energy absorber, and chest foam, and validated it against standard pendulum impact calibration test data (Fig. 6.26). The next year, Yang and Le [83] developed a detailed Hybrid III neck model and validated it against standard calibration test results. Since then, FE ATD models have included detailed geometry with composition identical to physical ATDs. Section 6.5.2 provides a list of recent side impact dummy models for readers to review on their own the procedures used to develop and validate the models.

Aside from model developers at various research institutions, readers can find ATD models either provided free by LSTC (Fig. 6.27) (http://models.lstc.com/pages/available.htm) or license fully calibrated ATD models (Fig. 6.28) through Humanetics (http://www.humaneticsatd.com/virtual-models/finite-element).

6.6.2 Human Modeling

Human modeling technologies are being applied to areas where human models are used to aid in the design of the motor vehicle passenger compartment and aircraft via simulations, including seats, controls, displays, and occupant restraints. ATDs are mechanical devices that can be simulated quite accurately. However, the lack of biofidelity associated with mechanical ATDs is also reflected in their counterpart mathematical models.

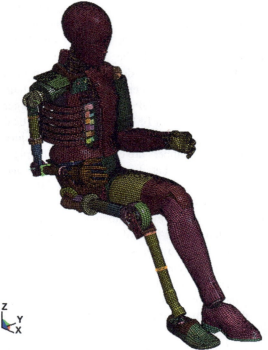

Fig. 6.27 The alpha version of Hybrid III FE model provided by LSTC

ATDs designed for safety evaluations are intentionally made with limited degrees-of-freedom in favor of high repeatability. As a result, available ATDs are directionally specific, such as frontal or side impact ATDs, instead of omnidirectional. ATDs are made of mostly steel, foam, and rubber. While the total mass of a dummy can be made the same as human, large difference in density between steel and bone makes it difficult to have the same distribution patterns for the mass inertial. Also, vital internal organs are not included in existing ATDs for evaluations of organ injury. Additionally, ATDs use force, deflection and acceleration to predict the risk of injury. That is, they will not break under typical crash scenarios. However, this design makes it inaccurate for predicting post fracture or rupture behaviors once a part failure is initiated. Lastly, protection provided for non-standard sizes may not be fully effective.

Ideally, a safe vehicle should be designed for its breathing human occupant, rather than just

Fig. 6.28 The Hybrid III model available for licensing through Humanetics

ATDs that needed to satisfy various regulations. Because testing human at the injury level is not feasible and there are ethical concerns in the use of animals, numerical human model is the best alternative in making vehicle safe for its human occupants. In general, developing a human component or whole body model involves five separate and yet integrated techniques: (1) an imaging technique to acquire geometry, (2) a numerical technique to smooth/idealize geometry, (3) a FE model construction technique to make high quality mesh, (4) a physical testing technique to obtain properties at tissue, sub-system, or system level, and (5) model validation and application.

Unlike ATDs, with geometry being well defined and computer aided design (CAD) files available for constructing the FE models, detailed geometry of a human is very complex and requires a combination of multiple tools to segment different tissues. Aside from physical dissection, 3D medical scanning tools such as CT and MRI machines are used to acquire the detailed anatomy. Processing these images into FE mesh is usually the first step in developing a human component model.

Biomedical imaging is an essential component in many fields of modern day medicine. Radiologists routinely analyze MRI and CT scans for tumors. Freeware to view the data is frequently provided with images. However, analysis of these images requires sophisticated computational and visualization tools. The Center for Information Technology (CIT) of the National Institutes of Health (NIH) provides a general-purpose image processing and visualization program called Medical Image Processing, Analysis, and Visualization (MIPAV). The same tool can be used to segment different anatomical structures of interest for FE mesh development. Aside from this freeware, commercially available imaging processing software such as MIMICS (Materialise, Leuven, Belgium) and ScanIP (Simpleware, Exeter, United Kingdom) come with special module(s) to create FE meshes.

Once the geometry is acquired, basic geometric scaling and curve fitting are generally needed. Unlike constructing a specimen-specific FE model, many models are designated to represent a group of population (such as 50th percentile male). Because anatomy acquired from one or a couple of representative subjects cannot be used to represent the group, scaling some key geometry according to the population average is a common practice. Although the computer is very powerful now, it is still insufficient when modeling the complex human structures. Thus, curve fitting techniques are frequently used to simplify the structure of interest. In some cases where subject-specific model is needed, one can morph a well-developed and well-validated FE model into subject-specific geometry using the techniques developed by Hu et al. [84] to calculate the magnitude of brain shift for surgical tumor removal planning while saving precious time in the model development.

As previously mentioned in Sect. 6.4, a number of CAE pre-processing programs are available to develop the FE model. Because all FE solvers are practically based on isoparametric formulation, it is critical that the quality of mesh needs to be as close as possible to the original idealized element used to derive the element stiffness matrix. For a 3D element, the idealized base element is a square cube, that is, an element of the same length in all edges with all internal angles as 90°. Similarly, a 2D element should be as close to a square as possible. Unfortunately, human anatomy is very complex and hence, unrealistic to mesh it with such idealized elements. Yang and King [86] recommended that any FE mesh should have a minimum Jacobian of 0.7, all internal angles between 45° and 135°, a maximum aspect ratio of 3, and a maximum skew angle of 30°. Additionally, the mesh quality goal for warpage angle, taper, and element length to thickness ratio shall be less than 20°, 0.5°, and 3°, respectively, for 2-D elements and the warpage angle and tetra collapse shall be less than 5° and 0.5°, respectively, for 3-D elements. Additionally, the block technique, such as that reported by Mao et al. [52], is recommended for generating high quality meshes of different mesh densities to allow for convergence study. Lastly, parametric study and Designs of Computer Experiments (DOCE) methods are frequently used to test out the model to decide which parameter(s) is the most influential to model responses.

Physical testing involves both experiments to identify material properties as well as to obtain impact responses for model validations. Typical material tests include uniaxial or bi-axial tension and compression, three-point, and four-point bending tests while typical subsystem or whole body tests involved pendulum or sled. For hard tissues, strain gauges can be used to identify tissue level responses. On the other hand, photographic method is preferred in measuring soft tissue deformation. In the case that the test specimen is of irregular shape, laser scanning [85] or micro CT scanning technique [53] is used to acquire specimen-specific geometry so that reverse engineering method can be used to identify material properties.

Once all aforementioned steps are completed, the model needs to be rigorously validated before it can be used to predict responses under conditions which cannot be easily conducted using experimental methods. A problem persists in validating human component or whole body model is a lack of high quality experimental data. Past experiments are not intended to generate data for model validation; hence not all critical information is presented in published studies. Also, large biological variations make response corridors wide and multiple combinations of material properties can fit the model into the corridors. Because there is no good way to circumvent this problem, any FE model needs to be validated against multiple test conditions under different speeds in order to increase the confidence level in using the model. Figure 6.29 shows the interrelationships among these five sets of technique discussed above to generate a human model.

The prevalence of head and brain injuries has promoted the development of a great deal of human head models in a number of institutions. Yang and King [86] reviewed these models and identified some common shortcomings associated with the head model development. Aside from head models, Yang et al. [7] reviewed a number of component and whole body models published in the 50 years span of Stapp Car Crash Conferences and those models will not be repeated here. Considering the potentials on using human models to improve automotive safety, many consortiums are formed for the developments of human models for their members. Figure 6.30 shows the three whole body human models, developed respectively at the Global Human Body Models Consortium (GHBMC), European Commission, and Japan Automobile Manufacturers Association, Inc. (JAMA), through consortium efforts. There are other entities, such as the Total Human Model for Safety (THUMS) developed by Toyota Motor (Nagoya, Japan) and H-Model developed by ESI (Paris, France), that have human models available for commercial licensing. These models tend to be equipped with scaling and positioning tools for easy manipulation of the model.

In addition to rigid-body-based modeling for pedestrian safety described in Sect. 6.3.2, developments of FE human models for pedestrian

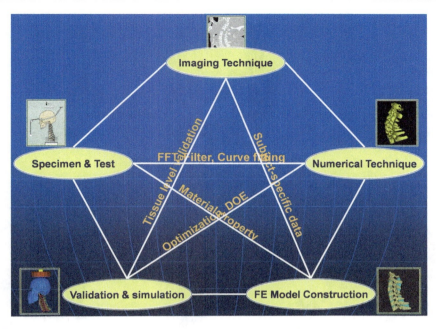

Fig. 6.29 Five interrelated sets of techniques needed to develop human model

safety research have also been progressing well in the past. Some examples that can be mentioned included:

(a) Takahashi et al. [87] modified the lower limb model of the H-Dummy™ model with new features to simulate bone fractures and ligament ruptures, and integrated into its upper body model to obtain a full scale model of a pedestrian in standing posture. In order to use such model for pedestrian accident investigations Dokko et al. [88] improved this model further by replacement of the thorax model with more detailed thoracic and cervical spines, and validated the simulated responses with cadaver test results.

(b) Maeno and Hasegawa [89] modified the lower extremities of the THUMS human model to develop an improved FE pedestrian model, which was validated against data obtained from cadaver tests. The correlation was used to assess the capabilities of this pedestrian model in replicating overall kinematics and predicting pedestrian lower extremity injuries as observed in real world scenarios.

(c) Under the HUMOS and HUMOS2 projects in Europe, Robin [90] and Verzin and Verriest [91] developed a series of FE human models representing a broad range of European population for car occupant and pedestrian simulations. The pedestrian models were particularly validated at both the organ and whole body levels in overall pedestrian kinematics during impact to provide capabilities in predicting injuries with greater accuracy.

(d) Regarding to the child pedestrian model, Okamoto et al. [92] developed a 6-year-old model with detailed lower extremities using geometry data obtained from the MRI scan, thus allowing reproduction of specific fracture patterns of child pedestrian injuries resulted from car-pedestrian accidents.

Pertaining to car-to-pedestrian tests, both POLAR-II pedestrian dummy and the Chalmers-Autoliv pedestrian dummy were used in many research activities. Therefore, various FE pedestrian dummy models were developed in lieu of physical dummies. Shin et al. [93] developed an FE POLAR-II model, while Yao et al. [94] developed FE pedestrian model for the Chalmers-Autoliv pedestrian dummy consisting of 68 components and 82,000 elements representing the head, neck, thorax, pelvis, thigh, feet, jacket

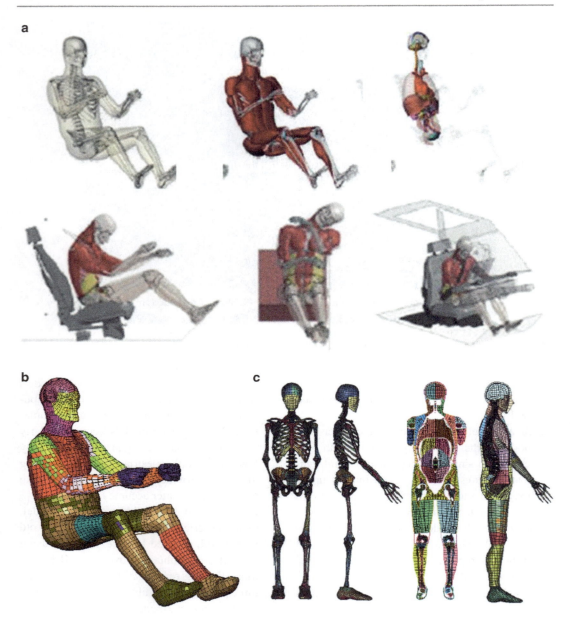

Fig. 6.30 Several available human models

and shoes, and was validated against results from a full-scale pedestrian impact at 30 km/h. The FE POLAR_II model was used to investigate the effect of vehicle front end designs on the kinematics of a pedestrian upon impact [95, 96]. Yao et al. [94] applied the C-A pedestrian FE model to designs of concept countermeasures, i.e., a pedestrian airbag pop-up hood system [98],

which was sensed by a bumper-pedestrian contact sensor system as described by Huang et al. [97]. The pedestrian airbag and pop-up system has been implemented into Volvo V40 as reported in a case study in O'Brian [99]. However, it is interesting to note that one development and evaluation of a kinematic hood for pedestrian protection was done by using Pam-Crash to establish the required pop-up space and multi-body simulation method to determine the force needed to deploy the hood [98]. It should be remarked here that all the aforementioned researches in pedestrian safety have focused almost exclusively on aspects of pedestrian injuries during primary impact of pedestrians with vehicle without addressing the injuries caused by the second impact of pedestrians with ground. Recently, Gupta and Yang [100] studied effects of vehicle front-end profiles on pedestrian secondary head impact to ground, and proposed potential design directions towards mitigating/reducing pedestrian injuries under such circumstances.

Most human models include bulk passive muscles that act as a cushion during impact as well as resistance to stretching of these muscles. Only a handful of modeling studies discuss the effect of active muscles. One common problem in modeling the muscle lies in the fact that multiple muscles are commonly used to achieve a desired motion of any segment. Thus, this is a statically indeterminate problem and optimization procedures are needed to determine force generated by each muscle group. Unfortunately, what was the strategy (such as minimal energy consumption, minimal change of sustaining injury, etc.) used by muscle for a certain task is far from completion. Also, most researchers used Hill-type muscle characteristics to simulate muscle actions but the response curves for all muscles are not available. Muscle modeling will be the topic of research for many years to come.

6.7 Remarks and Conclusions

Rather than providing an exhaustive description of existing rigid-body and FE-based numerical simulation models, this chapter discussed only a group of selective models that have been or can be used for dummy and human occupant and vehicular structural responses, and airbag supplemental restraint system simulations in frontal, side, rollover and pedestrian impacts. Some remarks will be made which will then be followed by conclusions:

With increasing computational capabilities and new sophisticated and realistic numerical models becoming available, these models should be properly validated and applied to a variety of impact conditions in order to render them as the potential future tools to understand human response in a wide range of crash configurations. The capability of rigid-body-based models has been witnessed in developing effective countermeasures for occupant safety protection systems among automobile OEMs. The ability of the human models to estimate the stresses and strains in the various parts of the human body and material failure criterion need further advance so that they can ultimately lead to the development of more refined injury criteria and a continuum of occupant sizes.

Future development of a well-validated human model requires:

1. More material property data on human tissues. These biomaterials exhibit anisotropic behaviors under tensile/compressive loadings, and have time-dependent properties under dynamic impacts.
2. Better experimental testing methods to reduce and/or minimize biological variations at high speed test conditions.
3. More detailed and accurate body parts/organ models to allow thorough investigation of various internal organ interactions, and
4. Muscle modeling

A few means of model validation were discussed. Simple criteria such as Normalized Integral Square Error (NISE), and Global Correlation Index (GCI) were described, and can be easily or readily applicable when needed. Statistical approach methodology is rigorously and mathematically treated, and needs to be further explored. Validation process of such approach needs to be streamlined to make it more user-friendly. Metrics and/or criteria for validating need to be established as guidelines for general practice or for specific application. It should remark that a task force of ISO is currently undertaking development of validation and verification methodologies for quantitative assessing model quality, thus improving model validity.

To further refine the human model, more thorough biomechanical studies are needed at the component, sub-systems, and whole body levels. These data collected are needed to validate current/existing models not only in their prediction of bony kinematics and tissue deformation locally, but also for the global validation and accurate prediction of future whole body human models.

Finally, predictive capability of numerical techniques can be further improved by reconstruction of a sufficient number of real-world crashes using validated dummy and human models integrated with a detailed car model. This allows comparison among occupant kinematics and injury metrics from the dummy model, human-model-predicted tissue and organ failures, and injuries sustained by the crash victims. This process will lead to further improvements in numerical models and anthropometric test devices. In addition, FE human models must be validated to ensure that they are capable of robust simulation of occupant response for a continuous distribution of human anthropometry under various impact conditions in real-world crashes.

In summary, mathematical models using rigid-body-based and nonlinear FE techniques have been developed in large numbers for understanding of response, injury mechanisms, and injury mitigation of occupants in accidents. The application of these models has a substantial influence on advances in the automotive and aircraft safety fields. Such numerical models, if rigorously and fully validated, will have the potential of becoming a valuable tool in the evaluation of innovative restraint systems that can further reduce injuries and fatalities in vehicle accidents and aid in the improved design of the next generation of vehicles with advanced technologies for the years to come. Specifically, efforts should be directed towards improving human model capabilities in the following areas:

- Standardization of material properties: Complete constitutive models for human soft tissues must be developed to better simulate the complete behavior of tissues including their nonlinear, strain-rate sensitive properties up to speeds experienced in crashes. Associated tissue test methodologies may also need to be addressed. Additionally, material databases for metallic (such as steel, aluminum, advanced high strength, and lightweight alloys, etc.), composites, and foam materials need to be created.
- Material property sensitivity study: Parametric studies related to tissue material properties are required to investigate the effect of various tissue response sensitivities. Reverse engineering and optimization approaches are promising to be adopted to help study of complex material properties of human models.
- Additional validation of human models: These can be accomplished by validating human models at either the component or whole-body level against more data of experimental tests carried out at various impact speeds, thus leading to assurance of their predictive accuracy and robustness in a variety of impact conditions such as frontal, side, rear, oblique and rollover crashes, as well as car-to-pedestrian impact.
- Accident reconstruction with human models: Human models are currently used in simulation of regulatory-related frontal and rollover test modes. This task of reconstructing injuries sustained by the crash victims in real world accident scenarios definitely help in validating the models.
- Guidelines for human modeling: Best practice procedures need to be established for modeling of the human.
- Model assessment tool development: There is a need to develop assessment tools for evaluating quality of signals and responses generated by numerical models. Such tools help greatly in the validating process of numerical models under development.
- Rollover crash ATD: Currently, ATDs exist for frontal, side, rear, and pedestrian impact testing for assessing potential dummy occupant injuries. There is a need to develop a physical dummy for rollover testing.

References

1. McHenry RR (1963) Analysis of the dynamics of automobile passenger-restraint systems. In Proceedings of the 7th Stapp Car Crash Conference. Los Angles, CA, USA, paper no.1963-12-0017
2. King AI, Chou CC (1976) Mathematical modelling, simulation and experimental testing of biomechanical system crash response. J Biomech 9(5):301–317
3. Prasad P (1984) An overview of major occupant simulation models. SAE 84085. In: Mathematical simulation of occupant and vehicle kinematics. SAE Publication, Warrendale, p 146.
4. Prasad P, Chou CC (1989) A review of mathematical occupant simulation models. In: Crashworthiness and occupant protection in transportation systems, AMD-Vol. 106/BED, Vol 13. The American Society of Mechanical Engineers, New York, pp 96–112
5. Prasad P, Chou CC (1993) A review of mathematical occupant simulation models, Chapter 6. In: Accident injury – biomechanics and prevention. Springer, New York, pp 102–150
6. Prasad P, Chou CC (2002) A review of mathematical occupant simulation models. Chapter 7. In: Accident injury – biomechanics and prevention. Springer, New York, pp 121–186
7. Yang KH, Hu J, White NA, King AI, Chou CC, Prasad P (2006) Development of numerical models for injury biomechanics research: a review of 50 years of publications in the Stapp Car Crash Conference. Stapp Car Crash J 50:429–490
8. DOD (1996) Verification, validation, and acceleration (VV&A) recommended practice guide. Alexandria, http://vva.dmso.mil/.
9. AIAA (1998) Guide for the verification and validation of computational fluid dynamics simulations. American Institute of Aeronatics and Astronatics, AIAA-G-077, Reston
10. ASME (2006). Guide for verification and validation in computational solid mechanics. American Society of Mechanical Engineers, ASME V&V 10, New York
11. DOE (2008) Advanced simulation and computing (ASC) program plan. Office of Advanced Simulation & Computing, DOE/NNASA NA-114 http://www.sandia.gov/NNSA/ASC/pubs/pubs.html
12. Jovannovski J (1981) Crash data analysis and model validation using correlation techniques. SAE 810471
13. Prasad P (1997) Occupant simulation models: experiment and practice. In: Computation of transportation systems: structural impact and occupant protection. Kluwer, The Netherlands, pp 209–219
14. Deb A, Haorongbam B, Chou CC (2010) Efficient approximate methods for predicting behaviors of steel hat section under impact axial loading. SAE 2010-01-1015
15. Haorongbam B, Deb A, Chou CC (2013) Numerical prediction of dynamic progressive buckling behaviors of single-hat double-hat steel components under axial loading. SAE 2013-01-0458
16. Zhu F, Chou CC, Yang KH, Chen X, Wagner D, Bilkhu S (2012) Obtaining material parameters for die cast AM60B magnesium alloy using optimization techniques. Int J Vehicle Safety 6(2):178–190
17. ISO TR9790, ISO/TC22/SC120/WG5 (2000) Road vehicles – anthropomorphic side impact dummy – lateral impact response requirements to assess the biofidelity of the dummy
18. Cesari D, Compigne S, Scherer R, Xu L, Takahashi N, Page M, Asakawa K, Kostyniuk G, Hautmann E, Bortenschlager K, Sakurai M, Harigae T (2001) WorldSID prototype dummy biomechanical responses. Stapp Car Crash J 45:285–318
19. Ruan JS, El-Jawahri R, Rouhana SW, Barbat S, Prasad P (2006) Analysis and evaluation of the biofidelity of the human body finite element model in lateral impact simulations according to ISO-TR9790 procedures. Stapp Car Crash J 50:491–507
20. Oberkampf WL, Barone MF (2006) Measures of agreement between computation and experiment. Validation metrics. J Comput Phys 217(1):5–36
21. Jiang X, Yang RJ, Barbat S, Weerappuli P (2009) Bayesian probabilistic PCA approach for model validation of dynamic systems. SAE 2009-01-1404
22. Kokkolaras J, Hulbert G, Papalambros P, Mourelatos Z, Yang RJ, Brudnak M, Gorsich D (2013) Towards a comprehensive framework for simulation-based design validation of vehicle systems. Int J Vehicle Design 61(1/2/3/4):233–248
23. Blum P (1997) Passenger vehicle rollover – model development, test method design and sensor algorithm assessment. Master's thesis at Chalmers University of technology, Mechanical Engineering Department of Injury Prevention, Chalmers University of Technology, 41296 Goteborg
24. Aljundi B, Skidmore M, Poeze E, Alaats P (1997) Rollover impact. In Proceedings of the 35th annual symposium. SAFE Association, Phoenix, pp. 395–96, Sept 8–10
25. Yaniv G, Duffy S, Summers S (1998) Rollover ejection mitigation using an inflatable tubular structures. 16th Enhanced Safety Vehicles (ESV). Winsor, June 1–4, paper no. 98-S8-W18
26. Sharma D (1997) Status of research on restraint systems for rollover protection. Presented by G. Rains of Vehcile Research Transportation Center (VRTC) in Government/Industry meeting at Washington, DC
27. Renfroe DA, Partain J, Lafferty J (1998) Modeling of vehicle rollover and evaluation of occupant injury potential using MADYMO. SAE 980021
28. Chou CC, Wu F (2001) MADYMO-based rollover simulations. 17th international technical conference on the enhanced safety vehicles, Amsterdam, June 4–7
29. Takagi H, Maruyama A, Dix J, Kawaguchi K (2003) MADYMO modeling methods of rollover event and

occupant behavior in each rollover initiation type. 18th international technical conference on the enhanced safety of vehicles, Nagoya, paper no. 236
30. Chou CC, Wu F, Gu L, Wu SR (1998) A review of mathematical models for rollover simulation. In: Crashworthiness, occupant protection and biomechanics in transportation systems, AMD-Vol. 230/BED-Vol 41, pp. 223–239. Presented at the 1998 ASME Winter Annual Meeting, California
31. Chou CC, Wagner CD, Yang KH, King AI, Hu J, Hope K, Arepally S (2008) A review of math-based CAE tools for rollover simulations. Int J Vehicle Safety 3(3):236–275
32. Chou CC (2003) CAE methodology for rollover simulation – current status and future trends, presented at the SAE Government/Industry Meeting, May 12–14
33. Chou CC, Wu F (2005) Development of MADYMO-based model for simulation of laboratory rollover test modes. 19th ESV, June 6–9, Washington, DC, paper no. 05–0347
34. Gopal M, Baron K, Shah M (2004) Simulation and testing of a suite of field relevant rollovers. SAE 2004 world congress and exhibition, Detroit, Michigan, SAE 2004-01-0335
35. Le J, Chou CC (2007) Assessment tool development for rollover SAE signal analysis. SAE 2007-01-0681
36. TNO – MADYMO, V6.0 User's Manual 3-D. TNO Road-Vehicles Research Institute (2003)
37. McCoy RW, Chou CC, van de Velde R, Twisk D, van Schie C (2007) Vehicle and rollover sensor test modeling. SAE 2007-01-0686
38. Huang Y, King AI, Cavanaugh JM (1994) A MADYMO model of near-side human occupants in side impacts. J Biomech Eng 116(2):228–235
39. Happee R, Morsink P, Wismans J (1999) Mathematical human body modeling for impact simulation. SAE 1999-01-1909
40. Van der Horst MJ, Thunnissen JG, Happer R, van Haastere RM (1997) The influence of muscle activity on head-neck response during impact. SAE 973346. 41st Stapp Car Crash Conference, pp 487–507
41. De Jager M, Sauren A, Thunissen J, Wismans J (1997) A global and a detailed mathematical model for head-neck dynamics. In: Proceedings of the 35th SAFE symposium, Phoenix
42. Padgaonkar AJ, Prasad P (1982) A Mathematical analysis of side impact using the CAL3D simulation model. In: Proceedings of the 9th international ESV conference, Nagoya, Japan
43. Wijk J, Wismans J, Wittrebrood L (1983) MADYMO pedestrian simulations. SAE 830060
44. Janssen EG, Wismans J (1986) Experimental and mathematical simulation of pedestrian-vehicle and cyclist-vehicle accidents. In: Proceedings of 10th international tech conference on experimental safety vehicles, Oxford, UK
45. Ishikawa H, Kajzer J, Schroeder G (1993) Computer simulation of impact response of the human body in car-pedestrian accidents. SAE 933129. In: Proceedings of 37th Stapp Car Crash Conference, San Antonio, Texas, USA
46. Yoshida S, Matsuhashi T, Matsuoka Y (1995) Simulation of car-pedestrian accident for evaluating car structure. In: Proceedings of 16th international technical conference on the enhanced safety of vehicles, Ontario
47. Mizuno Y, Ishikawa H (2001) Summary of IHRA Pedestrian Safety WG activities-proposed test methods to evaluate pedestrian protection afforded by passenger cars. In: Proceedings of 17th technical conference on the enhanced safety of vehicles, Amsterdam
48. Anderson R, Mclean J (2001) Vehicle design and speed and pedestrian injury: Australia's involvement in the Harmonization Research Activities. In: Proceedings of 2001 road safety conference, Australia
49. Yang JK, Lovsund P, Cavallero C, Bonnoit J (2000) A Human-body 3D mathematical model for simulation of car-pedestrian impacts. J Crash Prev Inj Cont 2(2):131–149
50. Liu XJ, Yang JK (2002) Development of child pedestrian mathematical models and evaluation with accident reconstruction. Traffic Inj Prev 4(4):337–344
51. TNO - MADYMO Database Manual, Version 6.2, TNO Road-Vehicles Research institute (2004)
52. Mao H, Gao H, Cao L, Genthikatti VV, Yang KH (2013) Development of high-quality hexahedral human brain meshes using feature-based multi-block approach. Comput Methods Biomech Biomed Eng 16(3):271–279. doi:10.1080/10255842.2011.617005
53. Guan F, Han X, Mao H, Wagner C, Yeni YN, Yang KH (2011) Application of optimization methodology and specimen-specific finite element models for investigating material properties of rat skull. Ann Biomed Eng 39(1):85–95. doi:10.1007/s10439-010-0125-0
54. Chou CC, Neriya S, Low TC, Prasad P (1993) MADYMO2D/3D vehicle structural/occupant simulation models. AMD Vol. 169/BED Vol. 25, Crashworthiness and occupant protection in transportation systems, ASME winter annual meeting, San Francisco, CA, USA, pp 207–222
55. Chou CC, Chen P, Le J (2007) Development of a unified model for sensor and crash analyses. Int J Vehicle Safety 2(3):241–260
56. Stutzler F-J, Chou CC, Le J, Chen P (2003) Development of CAE-based crash sensing algorithm and system calibration. SAE 2003-01-0509
57. Chou CC (2004) Chapter 4: Fundamental principles for vehicle/occupant system analysis In: P Prasad and J Belwafa (ed) Vehicle crashworthiness and occupant safety. American Iron and Steel Institute (AISI), Southfield, MI. The book can be downloaded through the link: http://www.autosteel.org/~/media/Files/Autosteel/Research/Safety/safety_book.pdf
58. Low TC, Prasad P, Lim GG, Chou CC, Sundararajan S (1991) An integrated three-dimensional side

impact model, ASME AMD-Vol. 126/BED-Vol 19, pp 155–168
59. Lim GG, Prasad P, Chou CC, Walker LA, Sundararajan S, Fletcher GD, Chicola JA (1996) Development of deployable door trim system. Automotive body interior & safety systems, IBEC (International engineering body conference), Detroit, MI, USA, Oct 1–3
60. Ruckert J, Marcault P, Lasry D, Haug E, Cesari D, Bermond F, Bouquet R (1992) A finite element model of the EuroSID dummy. SAE 922528
61. Midoun DE, Abramoski E, Rao MK, Kalidindi R (1993) Development of a finite element based model of the side impact dummy. SAE 930444
62. Kirkpatrick SW, Homies B, Hollowell WT, Gabler HC, Trella T (1993) Finite element modeling of the Side Impact Dummy (SID). SAE 930104
63. Fountain M, Altamore P, Skarakis J, Spiess O (1994) Mathematical modeling of the BioSID dummy. SAE 942226
64. Pal C, Ichikawa H, Sagawa K (1997) Development and improvement of finite element side impact dummy (EuroSID) model based on experimental verifications. SAE 971041
65. Shkolnikov MB, Bhaisod D (1997) LS-DYNA3D finite element model of side impact dummy SID. SAE 971525
66. Khan A, Subbian T, O'Conner C (1997) Finite element model development of the BioSID. SAE 971140
67. Chai L, Subbian T, Khan A, Barabt S, O'Conner C, McCoy R, Prasad P (1999) Finite element model development of SIDIIs. SAE 99SC06
68. Kobayashi M, Matsuoka Y, Matsumoto T (2003) Validation of SIDIIs dummy FE-model and study of relation between design parameter and injury. SAE 2003-01-2820
69. Liu Y, Zhu F, Wang ZH, van Ratingen M (2007) Development of advanced finite element models of WorldSID 5th and 50th – the next generation side impact dummies. SAE 2007-01-0891
70. Wang ZJ, Been BW, Barness AS, Burleigh MJ, Schmidt A, Dotinga M, van Ratingen MR (2007) WorldSID 5th percentile prototype dummy development. SAE 2007-01-0701
71. Clements D (2005) Sub-system testing method for evaluation of the protective potential of door structures during side impact. Crash Expo, Novi, Oct 26
72. Chou CC, Aekbote K, Le J (2007) A review of side impact component test methodology. Int J Vehicle Safety 2(1/2):141–184
73. Aekbote K, Sobick J, Zhao L, Abramczyk J, Maltarich M, Stiyer M, Bailey T (2007) A dynamic sled-to-sled methodology for simulating dummy responses in side impact. SAE 2007-01-0710
74. Hu J, Yang KH, Chou CC, King AI (2007a) Development of a finite element model for simulation of rollover crashes. Proceedings of the ASME mechanical engineering congress and exposition, Crashworthiness, occupant protection and biomechanics in transportation systems. In: Proceedings of IMECE'07 2007 ASME international mechanical engineering congress Nov 10–16, 2007, Seattle, Washington. Paper no IMECE2007-44083
75. Chou CC, Hu J, Yang KH, King AI (2008) A Method for determining the vehicle-to-ground contact load during laboratory-based rollover tests. SAE 2008-01-0351
76. Lee C-R, Kim J-W, Hallquist JO, Zhang Y, Farahani AD (1997) Validation of a FEA tire model for vehicle dynamic analysis and full-vehicle real-time proving ground simulations. Paper presented at the SAE International Congress and Exposition, Detroit, Michigan
77. Zhang Y, Xiao P, Palmer T, Farahani A (1998) Vehicle chassis/suspension dynamics analysis~Finite element model versus rigid body model. Paper presented at the SAE international congress and exposition, Detroit, Michigan
78. Eppinger R (2002) Occupant restraint systems. In: Accidental injury – biomechanics and prevention, 2nd edn. Springer, New York, Chapter 8, pp 187–197
79. Cheng Z, Rizer AL, Pellettiere JA (2003) Modeling and simulation of OOP occupant-airbag interaction. SAE 2003-01-0510
80. Yamagishi M, Jun Iyama J, Araki T, Natori S (2012) Numerical simulation of out-of-position front passenger injuries in frontal crashes using an accurate finite element model of the cockpit module. SAE 2012-01-0552
81. Midoun DE, Rao MK, Kalidindi B (1991) Dummy models for crash simulation in finite element programs. SAE 912912. Stapp conf proceedings, San Diego, pp 351–367, Nov 1–3
82. Lasry D, Hoffmann R, Hong P, Yang KH (1991) Mathematical modeling of the hybrid III dummy chest with chest foam. SAE 912892. Stapp conference proceedings, San Diego, pp 65–71, Nov 1–3
83. Yang KH, Le J (1992) Finite element modeling of Hybrid III head-neck complex. Stapp Conf Proc 36:219–233
84. Hu J (2007) Numerical investigation of neck injury mechanism in rollover crashes and a systematic approach of improving rollover neck protection. PhD dissertation, Wayne State University, November
85. Zhu F, Jin X, Guan F, Zhang L, Mao H, Yang KH, King AI (2010) Identifying the properties of ultrasoft materials using a new methodology of combined specimen-specific finite element model and optimization techniques. Mater Design 31:4704–4712
86. Yang KH, King AI (2011) Modeling of the brain for injury simulation and prevention. In: Miller K (ed) Biomechanics of the brain. Springer, New York. ISBN 1441999965
87. Takahashi Y, Kikuchi Y, Mori F, Konosu A (2003) Advanced FE lower limb model for pedestrian. 18th international technical conference on the enhanced safety of vehicles, Nagoya
88. Dokko Y, Anderson R, Manavix J, Blumburgs P, Mclean J, Zhang L, Yang KH, King AI (2003) Validation of the human head FE model against pedestrian accident and its tentative application to

the examination of existing tolerance curve. In: Proceedings of the 18th ESV, Nagoya
89. Maeno T, Hasegawa J (2001) Development of a finite element model of a total human model for safety (THUMS) and applications to car-pedestrian impacts. In: Proceedings of 17th ESV, Amsterdam
90. Robin S (2001) HUMOS: human model for safety – a joint effort towards the development of refined human-like car occupant models. In: Proceedings of the 17th ESV, Amsterdam
91. Vezin P, Verriest JP (2005) Development of a set of numerical human models for safety. In: Proceedings of 19th ESV, Washington, DC
92. Okamoto M, Takahashi Y, Mori F, Hitosugi M, Madeley J, Ivarsson J, Crandall J (2003) Development of finite element model for child pedestrian protection. In: Proceedings of 18th ESV, Nagoya
93. Shin J, Untaroiu C, Kerrigan J, Crandall J, Subi D, Takahashi Y, Akiyama A, Kikuchi Y, Longitano D (2007) Investigating pedestrian kinematics with the POLAR-II finite element model. SAE 2007-01-0756
94. Yao JF, Yang JK, Fredriksson R (2011) Development of a pedestrian dummy FE model for the design of pedestrian friendly vehicles. Int J of Vehicle Design 57(2/3):254–274
95. Untaroiu C, Shin J, Crandall J, Fredriksson R, Bostrum O, Takahashi Y, Akiyama A, Okamoto M, Kikuchi Y (2009) Development and validation of pedestrian sedan bucks using finite element simulations; applications in study the influence of vehicle automatic braking on the kinematics of the pedestrian involved in vehicle collisions. In: Proceedings of 21st ESV, Stuttgart
96. Untaroiu C, Shin J, Iversson J, Crandall J, Takahashi Y, Akiyama A, Kikuchi Y (2007) Pedestrian kinematics investigation with finite element dummy models based on anthropometry scaling method. In: Proceedings of 20th international technical conference on the enhanced safety of vehicles, Lyon
97. Huang SN, Yang JK, Fredriksson R (2008) Performance analysis of a bumper-pedestrian contact sensor system by using FE models. Int J Crashworthiness 13(2):149–157
98. Krenn M, Mlekusch B, Wilfling C, Dobida F, Deutscher E (2003) Development and evaluation of a kinematic hood for pedestrian protection. SAE 2003-01-0897
99. O'Brian J (2012) Pedestrian zone. Crash Test Technology International, pp 10–16, Sept
100. Gupta V, Yang HK (2013) Effect of vehicle front end profiles leading to pedestrian secondary head impact to ground. Stapp Car Crash J 57:139–155

Computational Analysis of Bone Fracture

Daniel P. Nicolella and Todd L. Bredbenner

Abstract

Age-related non-traumatic fractures are a major public health problem with fracture of the proximal femur resulting in significant patient mortality. Currently, the clinical gold standard for assessing an individual's fracture risk is dual-energy x-ray absorptiometry (DEXA). Although DXA-based bone mineral density (BMD) measurements have been shown to correlate with fracture risk, BMD distributions describing normal individuals and those who have suffered a hip fracture contain significant overlap, thereby reducing the specificity of BMD based fracture risk categorization. Since bone strength results from a combination of bone mass, bone micro-architecture and material properties, and overall bone geometry, two-dimensional imaging based diagnostic approaches such as DEXA are limited in their ability to specifically predict bone strength. Engineering models of skeletal structures that combine descriptions of three dimensional bone geometry, bone mass distribution, and bone material behavior into a high fidelity simulation have been shown to predict bone strength with an improved accuracy compared to DEXA alone, albeit with some limitations. Specifically, voxel based finite element models derived from QCT image data are based on correlations between predicted structural stiffness and structural failure and therefore are generally only valid for simple uniaxial loading; it is unlikely this approach would be valid for more complex applied loads. Furthermore, these models generally use simplified bone material descriptions that do not capture the demonstrated non-linear damaging behavior of bone and do not model bone material

D.P. Nicolella, Ph.D. (✉) • T.L. Bredbenner, Ph.D.
Musculoskeletal Biomechanics Section,
Material Engineering Division, Mechanical
Engineering Division, Southwest Research Institute,
San Antonio, TX, USA
e-mail: daniel.nicolella@swri.org;
Todd.bredbenner@swri.org

failure. Another major limitation is that these models require three dimensional QCT image data, which can be expensive and limited in its availability, and they are not parametric – they are specific to an individual. Finally, all models used to date are deterministic and do not account for uncertainty and/or variability in bone structure and/or material properties, and cannot be used to predict the *probability* of bone failure. Each of these limitations can be addressed using advanced computational methods such as statistical shape and bone density modeling, non-linear continuum damage mechanics modeling, and probabilistic analysis methods with the goal of improving the identification of individuals who are at a greater risk of fracture and focusing resources on those areas where treatment can be directed to significantly improve an individual's fracture risk. Ultimately, this approach will significantly improve the ability to clinically quantify the risk of fracture in an individual and allow treatment to be administered in a timely manner.

7.1   Skeletal Fragility Is a Major Social and Economic Problem

The problem of increased risk of skeletal fractures due to bone mass loss in aging or disease is a major clinical problem with associated estimated health care costs of nearly $17 billion [1] in the US. It has been estimated that 40–46 % of all women over the age of 50 and 13–22 % of all men over the age of 50 will suffer a fracture [2]. With the number of persons aged 60 years or older projected to more than triple by the year 2050 [3], the aging of the general population will lead to a significant increase in the at-risk population for fractures. As such, the number of worldwide fractures will likely increase from 1.26 million, as estimated in 1990, to 2.6 million by 2025 [4]. Not withstanding the economic burden, non-vertebral fractures are a significant cause of morbidity and mortality in the aging population [5–7]. Thus, concerted efforts are needed to not only identify those at risk of bone fractures, but also to identify treatment strategies that can maintain the health of the skeleton with age.

7.2   Clinical Imaging Based Fracture Risk Assessment Is Non-specific

BMD measurements are widely used for determining bone strength, especially in women, and can account for up to 70 % of bone strength. In simple terms, the increase in fracture risk is due to a loss of load-carrying material leading to higher tissue-level stresses in the skeleton at any given load. On this basis, one could argue that relative fracture risk could be simply described in a relationship with bone density [8, 9], perhaps including the effects of the anisotropic microstructure of the bone [10, 11]. This general view has driven a great deal of work involving the direct use of imaging techniques to predict the relative fracture risk. Attempts to directly predict fracture risk based on bone mass have been made using x-ray, ultrasound, dual energy x-ray absorptiometry (DXA), computed tomography (CT), high-resolution CT, and magnetic resonance imaging (MRI) techniques [12–22]. While some correlations of fracture risk with the various imaging and biochemical data have been demonstrated, they remain nonspecific and have low sensitivity [23–25] and, as a result, fall well short of predictive ability for individual subjects. Moreover, estimates of in vitro bone strength using BMD alone show only moderate correlative strength [26, 27] with correlation coefficients in the range of 0.6–0.7. These correlation models generally have no physical meaning since they do not describe the physics of the fracture process and, as such, it is not possible to use these methods to gain insight into the mechanisms of skeletal fragility that ultimately would lead to the development of improved treatment strategies. Further, correlation models require information from hundreds or thousands of subjects gathered over many years in order to achieve the statistical power required to draw conclusions regarding the consequences of bone loss and treatment effectiveness [28, 29]. Consequently, there is a critical clinical need for the ability to accurately and confidently predict the risk of fracture using tools that are readily available to clinicians. Accordingly, a more accurate overall determination of bone strength is still needed.

7.3 Bone Strength Is a Function of Bone Geometry, BMD, and Bone Quality

The shape of a structure significantly influences the deformations and stresses resulting from external loads and therefore the structure's likelihood of failure. It follows that simple descriptions of the shape of the proximal femur have been shown to improve predictions of fracture risk in combination with DXA-based BMD measurements. When measures of BMD are used in combination with descriptions of geometry such as neck angle, cortical thickness, etc. the predictive strength of these models generally increases thus indicating the importance of structural information in fracture risk [16, 17, 30–36]. Hip axis length, but not femoral neck width or the neck-shaft angle as measured from DXA scans predicted significant increases femoral neck and trochanter fracture [37]. Further, measures of proximal femur morphology from conventional radiographs predict hip fractures as well as BMD measurements [38] and improved correlation results when used in combination with BMD measures [31]. A two dimensional description of the proximal femur derived from an active shape model in combination with BMD measured at Ward's triangle prospectively classified 90 % of patients who had suffered a hip fracture [39]. Correlation models that combine BMD measurements with maximum principal strain predictions computed from two-dimensional finite element analysis (FEA) models and subject's height and neck shaft angle were able to correctly identify 82 % of subjects who previously reported a femoral fracture [40]. Although, these investigations demonstrate the bone geometry in an important determinant of fracture risk alone or in combination with BMD, the predictions are based on correlations and ultimately suffer from the same shortcomings of correlations based on BMD alone.

7.4 Engineering Models Can Improve the Assessment of Fracture Risk

There is a great deal of evidence that effective predictive models of structural strength of skeletal structures under general loading histories will require a much more detailed and realistic model of the mechanics of the bone structure [41–44], and, further, that engineering models based on imaging data greatly improve estimates of bone strength and fracture risk for an individual [26, 40, 45–51]. FEA models [52, 53] have been used in many biomechanics investigations including efforts to improve predictions of in vivo fracture loads. Automated meshing of QCT-data has been used to predict femoral fracture load. This method demonstrated accuracy similar to DXA-based methods [46, 54] using linear FEA analyses, and improved correlations using non-linear material models [55–57]. However, these analyses predict significantly greater structural stiffness compared to experimental results bringing into question their ability to accurately identify mechanisms of structural deformation and failure. Furthermore, the use of "voxel" finite elements results in a non-realistic surface representation of the bone (partial volume effect) and does not allow accurate modeling of failure at the surface, ostensibly where the largest stresses occur and thus where failure initiates. In fact, the surface elements are not used in these analyses for femur failure load predictions [54]. FEA models of CT-derived vertebral bodies and CT-derived proximal femora were also able to predict compressive failure loads in experimental tests based on correlations of structural failure load to structural stiffness [26, 58, 59]. Similar studies with vertebrae demonstrated that FEA models produced from QCT that included trabecular property variation along with an effective modulus assigned to a vertebral shell of constant thickness successfully explained 81 % of the variation in vertebral compressive stiffness and strength [60]. In these analyses, the FEA model is used to predict structural stiffness, which is used as a surrogate for structural strength, and does not model bone failure explicitly. While improved correlations to in-vitro failures loads were obtained, these approaches fail to discern the underlying mechanisms of skeletal fragility.

There is a great deal of evidence that effective predictive models of structural strength of skeletal structures under general loading histories will require a much more detailed and realistic model of the mechanics of the bone structure [41–44], and, further, that engineering models based on imaging data greatly improve estimates of bone strength and fracture risk for an individual [26, 40, 45–51]. FEA models [52, 53] have been used in many biomechanics investigations including efforts to improve predictions of in vivo fracture loads. Automated meshing of QCT-data has been used to predict femoral fracture load. This method demonstrated accuracy similar to DXA-based methods [46, 54] using linear FEA analyses, and improved correlations using non-linear material models [55–57]. However, these analyses predict significantly greater structural stiffness compared to experimental results bringing into question their ability to accurately identify mechanisms of structural deformation and failure. Furthermore, the use of "voxel" finite elements results in a non-realistic surface representation of the bone (partial volume effect) and does not allow accurate modeling of failure at the surface, ostensibly where the largest stresses occur and thus where failure initiates. In fact, the surface elements are not used in these analyses for femur failure load predictions [54]. FEA models of CT-derived vertebral bodies and CT-derived proximal femora were also able to predict compressive failure loads in experimental tests based on correlations of structural failure load to structural stiffness [26, 58, 59]. Similar studies with vertebrae demonstrated that FEA models produced from QCT that included trabecular property variation along with an effective modulus assigned to a vertebral shell of constant thickness successfully explained 81 % of the variation in vertebral compressive stiffness and strength [60]. In these analyses, the FEA model is used to predict structural stiffness, which is used as a surrogate for structural strength, and does not model bone failure explicitly. While improved correlations to in-vitro failures loads were obtained, these approaches fail to discern the underlying mechanisms of skeletal fragility.

Improvements in material constitutive models used in FEA models have also resulted in improved strength predictions. Nonlinear material behavior in the form of isotropic perfect plasticity has been incorporated into FEA models for simulations of uniaxial compression of midsagittal sections of vertebral bodies and one-legged stance and fall-type loading of femora. These models predicted yield loads that correlated with measured values in similar experimental tests [47, 50, 61]. These computational studies demonstrate the added predictive value of models that incorporate material behavior, but for the most part, they used simplistic models of trabecular bone behavior in light of the current knowledge of the elastic and inelastic (damaging and post-yield) behavior of trabecular bone [62–65]. It is our contention that such models will ultimately be of limited value in understanding fracture risk due to bone mass loss, because they fail to incorporate the effects of damage accumulation and other nonlinear material behavior in the structural response and failure of trabecular and cortical bone and whole bone structures. These more sophisticated models predict proximal femur load much better than correlative analyses incorporating simple measures of geometry and BMD. Inclusion of the abductor muscle force in the five structural models that were examined decreased the computed proximal femoral strength by only 0.5–1.3 % (mean, 0.9 %) and had virtually no effect on the predicted fracture locations [57].

7.5 Statistical Shape Modeling Efficiently and Compactly Describes Complex Biological Shapes

Statistical shape models, introduced over ten years ago in the image processing community [66], have been successfully used to simplify image processing tasks such as image segmentation, registration, object recognition, and diagnosis [67–73]. The basic principle is to reduce the shape dimensionality of the object of interest from a large set of highly correlated variables (typically a large set of surface vertices) to a compact set of independent and uncorrelated variables. Statistical shape modeling also provides for a method of parameterizing complex anatomical shapes and constraining the set possible shapes using a priori knowledge of a training population.

These methods are based on the use of prior knowledge of the statistical variability of the shape of a structure. One requirement of shape modeling of anatomical structures is that the model should resemble a large portion of the individual shape instances within a specific population or subpopulation. In statistical shape modeling, this is accomplished by statistically analyzing a set of shapes representing the structure of interest within a given population [74]. The result of this analysis is the projection of a large number of degrees of freedom representing the complex shape of the structure to a simplified orthogonal shape space using spectral decomposition (or principal components analysis). The complex shape of the structure of interest within the subject population is then described as the mean shape vector plus a linear combination of the principal modes of variation within the set [75]. A major challenge when constructing statistical shape models is ensuring point to point correspondence across the set of subjects or specimens of interest [66, 76–79]. Point to point correspondence ensures that specific points on each surface represent the same anatomical structure or substructure with optimal point-to point correspondence resulting in a compact statistical shape model where the variability in shape instances is described by a minimum number of modes of variation [76].

Typical finite element models of skeletal structures involve a heterogeneous distribution of bone material properties defined by local bone tissue BMD measurements resulting in a large number of material property variables defined over the volume of the structure. While this distribution can be approximated by mapping QCT derived bone tissue BMD to individual elements within a model [80], significant uncertainty exists in the definition of material properties from bone density measures as demonstrated by the correlation coefficients in the bone density bone material property regression equations. Furthermore, the spatial distributions of BMD within the bone (and therefore bone material properties) exhibit significant variability within and between subjects. Random fields provide a systematic way of describing the spatial variability and uncertainties observed in real structures. They may be used to describe spatial variations in material properties in a non-homogeneous medium, or to express geometry variations within and between structures [81].

7.6 Statistical Shape and Bone Density Model of the Proximal Human Femur

A statistical shape and density model (SSDM) based parametric finite element model describing a set of human femurs was developed and used to analyze the effect of shape and spatial bone density distribution variations as well as uncertainty in the density-material property relationship on the strength of the proximal human femur. This model uses a combination of optimal correspondence based statistical shape modeling and parametric finite element model generation to compactly describe the geometry and spatial bone density distribution of the proximal femur.

Briefly, seven fresh-frozen human female right femurs (69.9 ± 8.8 years old) were obtained from willed body programs (Anatomical Board of the State of Texas, San Antonio, TX;

Anatomical Gifts Registry, Hanover, MD). The femurs were submerged in distilled water in a plastic bin along with a density calibration standard (Mindways, Austin, TX) and scanned using a clinical helical CT scanner (Lightspeed Ultra, GE Healthcare, Chalfont St. Giles, UK). CT data was reconstructed with $0.488 \times 0.488 \times 1.25$ mm voxels. An iterative thresholding scheme was used to extract the bones from the imaging data and triangulated surfaces were defined to describe the outer cortical boundary for each femur (Amira v.4.0, Mercury Computing, Chelmsford, MA). One of the exterior femur surfaces was arbitrarily chosen from the set as the template surface, and iterative closest point analyses were performed to register the remaining surfaces to the template surface (MATLAB R2006b, The Mathworks, Inc., Natick, MA). A finite element mesh was defined using commercial mesh generation software (TrueGrid, v2.2, XYZ Scientific Applications, Inc., Livermore, CA) and projected to the template femur surface to produce a volumetric FE mesh of the template femur, consisting of 11,184 8-node hexahedral continuum elements. Mesh distortion was corrected using iterative elliptical smoothing (TrueGrid). A set of landmarks was defined by identifying 100 specific attachment points of the computational FE mesh on the template femur surface. These landmarks were mapped on to the remaining femur surfaces using a modified closest point transform [82]. Point-to-point correspondence was obtained between landmarks on all femurs using a hierarchical optimization approach consisting of coarse landmark repositioning followed by levels of refinement [76]. Parametric descriptions of the femurs surfaces were developed by mapping the triangulated surfaces to a unit sphere [83]. A kernel-based method was used to perturb the landmark positions in parametric space using constrained nonlinear optimization to minimize the variance between corresponding sets of landmarks (MATLAB) [76]. The computational mesh was then tied to each set of landmarks and projected onto the corresponding femur surfaces to generate a total of seven individual volumetric FE meshes, each consisting of 11,184 8-node hexahedral elements with mesh-to-mesh correspondence between femurs. Image intensity values corresponding to the spatial location of each node in the meshes were determined from the QCT data for each femur (MATLAB), and nodal intensity values were converted to apparent ash density using a linear relationship determined between known K_2HPO_4 densities in the calibration phantom and corresponding image intensities [84] along with a relationship between K_2HPO_4 density and apparent ash density [85].

The finite element mesh and corresponding nodal bone density values are described by a shape and density parameter vector as

$$\boldsymbol{p}_i = \left[\left(v_{1x}, v_{1y}, v_{1z}, v_{1d} \right), \ldots, \left(v_{jx}, v_{jy}, v_{jz}, v_{jd} \right) \right] \quad (7.1)$$

where $v_{j(xyz)}$ are the three dimensional coordinates of the nodes in the finite element mesh, v_{jd} is the corresponding bone density at that node, $j = 1, \ldots$, number of nodes in the finite element mesh, and $i = 1 \ldots n = 7$ denote each femur in the training set [76]. The mean finite element model of all femurs in the training set is then defined as:

$$\bar{\boldsymbol{p}} = \frac{1}{n} \sum_{i=1}^{n} \boldsymbol{p}_i \quad (7.2)$$

and the correlation between individual FE meshes is given by the empirical covariance matrix [86].

$$\boldsymbol{S} = \frac{1}{n} \sum_{i=1}^{n} \left(\boldsymbol{p}_i - \bar{\boldsymbol{p}} \right) \left(\boldsymbol{p}_i - \bar{\boldsymbol{p}} \right)^T \quad (7.3)$$

Principal components analysis (PCA) of $\boldsymbol{S}$ results in a set of eigenvalues (λ_i) and eigenvectors (q_i) [72] that are the principal directions of a shape and bone density space centered at $\bar{\boldsymbol{p}}$. Each eigenvalue defines the variance of the finite element mesh and corresponding bone density values from the mean in the direction of the corresponding eigenvector. The proportion of the total variance described by each eigenvector is equal to its corresponding eigenvalue normalized by the total variance (sum of all eigenvalues). The finite element mesh with associated bone density for each bone in the training set is now described as a statistical shape and density model (SSDM):

$$\boldsymbol{p}_i = \bar{\boldsymbol{p}} + \sum_{j=1}^{m} \lambda_j \boldsymbol{q}_j \quad (7.4)$$

where λ_j are eigenvalues corresponding to the q_i eigenvectors.

Multiple finite element models were developed using:

$$p_v = \bar{p} + \sum_{j=1}^{m_M} c_j \sqrt{\lambda_j} q_j \quad (7.5)$$

where m_M is the number of major eigenvalues (the eigenmodes that describe greater than 75 % of the total training set variability), p_v is a vector containing coordinates and apparent bone density value for all nodes in the FE model, and deviation from the average femur was determined as the sum of the products of a set of scalars, c_j, and SSDM standard deviations, $\sqrt{\lambda}$, along the q_i direction [87]. FE models were developed for the mean geometry and bone density distribution as well as one each investigating the effects of ±1 standard deviation (SD) for each major independent eigenmode (Fig. 7.1).

In this analysis, elastic- plastic, linear hardening material behavior was assumed. Isotropic elastic moduli and ultimate stress values were determined as functions of ash density for all bone elements using empirical relationships [88]. Post-yield hardening moduli were set as 10 % of the elastic moduli. The models were oriented in a fall position [89, 90], the nodes representing the distal extent of the model at the greater trochanter were fixed, and a uniform displacement was applied through a flexible cup that contacted the femoral head. Models were solved using LS-DYNA (LSTC, Livermore, CA) on a Linux cluster and the vertical reaction force at the greater trochanter was recorded.

Principal components analysis of the statistical shape and density model demonstrated that 78 % of the variation in the set of 7 human proximal femur finite element models was captured by the first three independent modes of variation (eigenmodes) (Fig. 7.1). The finite element model representing the mean − 1 SD variation in the second eigenmode resulted in an increase in neck angle length, decrease in neck-shaft angle, increase in neck diameter, and an increase in cortex width (Fig. 7.2), all of which have been shown to decrease fracture risk [17, 30–36]. This independent mode of variation (eigenmode) resulted in the greatest predicted maximum load in a simulated fall configuration (Fig. 7.2). Thus, this modeling approach directly relates variations in bone geometry and bone density to bone strength through a high fidelity predictive engineering model. The advantage of the approach described herein is that a physics based predictive finite element model is used to directly assess bone strength, rather than a statistical correlation based on epidemiological data of fracture incidents, bone density measures, and geometry in a given population.

A set of verification experiments was performed to verify the performance of the SSDM based finite element model predictions of femur strength. The femurs used to create the SSDM were sectioned transversely at a distance of four inches below the lesser trochanter. The distal portion of the proximal shaft was cleaned and embedded in PMMA within custom cylindrical mounting cups (Fig. 7.3). Using a custom fixture, each femur was mounted in a servo-hydraulic materials test machine (Instron, Canton, MA) in the fall loading configuration [89, 90]. Displacement was applied to the head of the femur through a rubber cup at a rate of 9.94 mm/s to a total displacement of 35 mm. The average maximum load recorded for the set of seven femurs was 1821.4 N +/− 1065.4 N similar to but slightly lower than the average SSDM FE model predictions of 1852.0 N+/− 1129.0 N.

7.7 Probabilistic Modeling Accounts for Biological Variability and Uncertainty in Data Parameters

All of the applications of engineering models to predict bone strength fail to account for the inherent variability in biological materials and structures as well as uncertainty in contained in our ability to directly measure these properties in-vivo. In many structural systems, there is a great deal of uncertainty associated with the environment in which the structure is required to function, the material properties of the system components, and in the structure itself (i.e. geometry).

Fig. 7.1 *Top*: the SSDM describes over 75 % of the variability in femur shape and BMD distribution using the mean and first three principal components. *Middle*: clear differences are seen in the proximal femur geometry and BMD distribution (as indicated by the fringe plot). *Bottom*: finite element models efficiently created using the SSDM were used to predict proximal femur strength in the fall loading condition

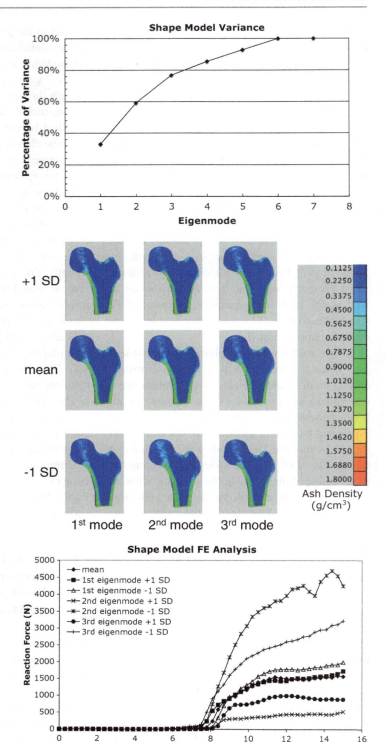

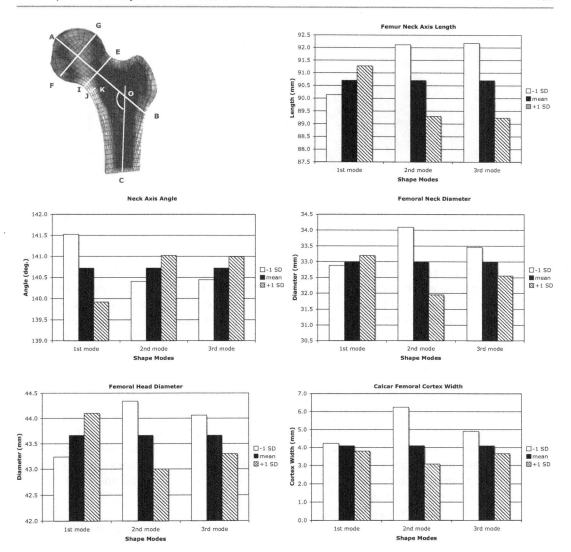

Fig. 7.2 Measures of proximal femur geometry determined for finite element models based on the statistical shape model of the proximal femur. The principal modes of variation from the mean represented by the shape model eigenvalues and eigenvectors generally have no direct physical meaning. However, by tracking the locations of specific points in the FE models that represent common measures of proximal femur geometry, the physical significance of each eigenmode can be discerned. *Top Left*: cross sectional view of a proximal femur FE model indicating definitions of typical proximal femur geometry measures. *A–B* femoral neck axis length, *AOC* neck-shaft angle, *E–I* femoral neck diameter, *G–F* femoral head diameter, *J–K* calcar femoral cortex width; *Top right*: proximal femur neck axis length; *Middle Left*: Proximal femur neck axis angle; *Middle Right*: proximal femur neck axis diameter; *Bottom Left*: femoral head diameter; *Bottom Right*: calcar femoral cortex width

This variability or uncertainty has a direct effect on the ability to predict the structural response of the system and therefore its reliability [91–93]. Biological systems are a prime example: uncertainty and variability exist in the physical and mechanical properties of bone tissue and geometry of the bone, ligaments, cartilage, as well as uncertainty in joint and muscle loads. Since these quantities are used as input variables to finite element models, the predicted model response or

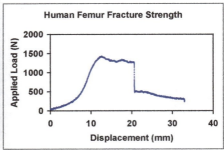

Fig. 7.3 *Left*: femur experimental test configuration. The human femur was loaded in a simulated fall configuration. *Right*: the measure load versus displacement response of the proximal femur. The load configuration resulted in an intertrochanteric fracture

performance will also be uncertain. Deterministic analyses, the class to which current applications of FEA to predict fracture risk belong, cannot quantify failure probabilities [92, 94, 95] because they do not include statistical information that describes the structure and material behavior. The recognition that predicting the probability of failure requires the use of probabilistic analysis techniques is a major motivation for this work.

Probabilistic analyses incorporate statistical information describing the materials, structural configuration, and environment from the outset whereas deterministic analyses discard this data. Modeling material behavior, structural geometry, and environmental loads and boundary conditions using random variables captures this statistical information. Using an appropriate performance function such as a simple strength-stress function, the probability of failure of a structure is quantified, rather than using a "factor of safety" approach. For example, it is more meaningful to say that a system has a 15 % probability of failing rather than saying the computed stress is 43 % less than the strength of the material or the system has a factor of safety of 1.75 [95–97]. Furthermore, probabilistic analyses identify those random variables that affect the probability of failure the most, thereby identifying where resources can be directed to have the most significant impact on improving either the diagnosis of risk or reducing the risk of failure directly [95–97]. Since deterministic analyses do not incorporate statistical data into the analysis, the effect of variability or uncertainty cannot be quantified. For example, the probability of failure of a structure depends not only on the mean value of the material properties, structural geometry, and environmental variables, but also on their variance and type of distribution these variables follow. In a simple stress-strength performance function, when the variance of the strength of the material is reduced, a deterministic analysis will result in an unchanged failure prediction whereas a probabilistic analysis quantifies the resulting decrease in probability of failure [95–97].

7.8 Bone Damage and Its Effects

Bone damage is known to accumulate as a normal process in living bone [98]. It is believed that the small amounts of normally occurring damage are repaired by ordinary bone turnover processes. However, if left un-repaired, or if accumulation outpaces the normal repair process, bone damage can lead to fatigue failures at stresses below the monotonic yield or failure strength during highly strenuous activities such as running or marching [99, 100]. Laboratory experiments, both in vitro and in vivo, have demonstrated that microdamage is produced by cyclic loading at physiological levels of stress or strain [101, 102].

The actual effects of damage accumulation on the mechanical properties of bone are fairly complex. Fatigue loading has been associated with modulus degradation [98], and a decrease in fracture toughness [100, 103]. When damage is accumulated in compression and the bone is tested in tension, the impact strength is reduced by an average of 40 %, although damage in tension does not reduce the impact strength in compression [104]. It was found that damage induced in machined cortical bone in tension, compression or torsion produced varying levels of stiffness change in all three loading modes [105]. The results of these initial experiments suggest that the variable effects are consistent with the microstructure of the cortical bone.

In addition to damage accumulation due to fatigue loading, damage accumulation in bone is associated with post-yield loading. It has been have postulated that inelastic behavior is primarily associated with damage accumulation, and that both inelastic strain accumulation and viscous property changes reflect the formation of new internal surfaces and voids [106–109]. There is some evidence that damage associated with inelastic strains (creep or low cycle fatigue damage) and damage associated with sub-yield strains (high cycle fatigue) are different mechanisms [110]. However, no physical basis for such a distinction has yet been identified.

Recent results suggest that the inelastic behavior and the failure of cancellous bone involve damage accumulation processes [63, 111, 112]. In cortical bone, which has been studied much more extensively than cancellous bone, the damage accumulation rate is an extremely nonlinear function of stress amplitude. In simple loading such as tension, compression or torsion, it obeys a power law relationship with stress amplitude of the form $R = A\sigma^n$, where R is a measure of damage accumulation, A is a proportionality parameter which is usually dependent on accumulated damage level, and n is a constant typically greater than 10 [109, 110]. Sufficient work has been done with cancellous bone to establish that damage is also an inherent process in cancellous bone [111, 113] and gives evidence of a similar nonlinear rate of damage accumulation [114, 115].

7.9 Continuum Damage Mechanics Modeling of Bone

Continuum damage mechanics (CDM) models, particularly when incorporated into FEA models, offer the potential for characterizing the damage accumulation process and the risk of fracture in real skeletal structures. This approach is exemplified by the work of Zysset and co-workers [115], who implemented a quite general damage model in application to cancellous bone. This work can potentially accommodate all the major features that have seen in bone behavior including plastic or viscoplastic flow and damage accumulation. Damage accumulation is incorporated around the central assumption that plastic flow and damage accumulation are intrinsically related. Zysset and Curnier incorporated this material model into a three-dimensional continuum model of anisotropic trabecular bone. Although the general model they proposed had the capability to include greater complexity, they applied the model using an assumption of isotropic damage and time independent plastic flow.

Based on evidence from our own studies [116] and others that time-dependence [117] and damage anisotropy are significant in trabecular bone and a lack of specific experimental basis for assuming an intrinsic link between plastic flow and damage, we developed a unified constitutive model to describe the elastic and inelastic (viscoelastic, plastic, and damage) behavior of human vertebral trabecular bone [116]. Material parameters were defined in terms of measures of trabecular density and architecture. Reasonable predictions of experimental specimen behavior were obtained and the ability to predict damage measures, such as evolving and accumulated modulus degradation, was achieved despite the complicated nature of the material model and limited amount of data for multiaxial material characterization. FEA simulations of specimen-level experiments support the basic assumptions of the model and demonstrated that it is possible to obtain estimates of all model parameters based on experimental data and the literature. This constitutive model was also used to investigate the effects of uniform or focal mass loss on the behavior of finite element models of idealized vertebrae, demonstrating that reduction in bone mineral density leads to an increase in the effect of accumulated damage, especially for focal loss.

7.10 Probabilistic Continuum Damage Modeling of Cortical Bone

7.10.1 Uniaxial Material Modeling

To investigate the applicability of using probabilistic methods in combination with CDM material models to predict bone failure, an initial probabilistic damage model for cortical bone was developed based on isotropic elasto-visco-plastic damage continuum mechanics (CDM) model [118]. Primary model features include the following: (a) Inelastic strains are assumed to result from either plastic flow or damage accumulation. The onset of damage accumulation and plastic flow is governed by two independent threshold surfaces in stress space. The yield criterion and the damage threshold are described by closed, convex surfaces in stress space, both of which can change by a linear hardening rule. Plastic hardening incorporates both isotropic and kinematic hardening. (b) Damage is defined by two scalar damage variables, one associated with the shear stresses and the other associated with a volumetric damage that only accumulates under positive (tensile) volumetric stress. The model incorporates nine material constants including two elastic moduli, three plasticity parameters, three damage parameters, and a time constant. A uniaxial version of the model was implemented in MATLAB for simulating axial loading. An extensive series of simulations were carried out to investigate the roles of the material parameters vis-à-vis known general behavior of bone (from experimental uniaxial test data). Parameter fits from these simulations were used to generate random variable descriptions of the CDM material model internal variables. Probabilistic calculations were then performed using the general-purpose probabilistic analysis software NESSUS (NESSUS v.7.5 Southwest Research Institute, San Antonio, TX).

The initial analysis investigated the probabilistic behavior of the model when driven to a specific strain level by computing the probability of bone failure at that strain level. Probability of bone failure was defined as the probability of reaching a specific reduction in modulus (resulting from the damage process). Modulus reduction at failure for each experimental data set was computed and used as a random variable describing strength. The performance function for this analysis was defined as $g = R - S$ where R is the strength random variable (modulus reduction limit at failure) and S is the model computed response (computed reduction in modulus). The probability of failure is defined as $p[g \leq 0]$.

The parameter estimation produced a set of material model random variables that best described the behavior of the material under uniaxial loading. The average reduction in modulus compiled from the curve fits to the experimental results was 52 % (±18 %) The predicted probability of failure for a set of give strain levels is given in Fig. 7.4 as a function of strain rate. At a strain level of 1.5 %, the probabilistic model predicts a probability of bone failure of 62 % for the slowest strain rate to 98 % for the highest strain rate. Slower strain rate simulations resulted in lower probabilities of failure for a given strain level. The uniaxial implementation of the CDM model accurately predicts the experimental

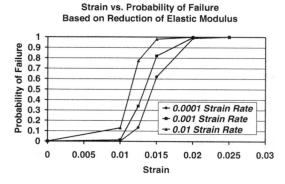

Fig. 7.4 Probability of bone failure at a given strain level. Failure is based on accumulated damage (modulus reduction)

behavior of cortical bone uniaxially loaded to failure in tension. Furthermore, using a probabilistic modeling approach, this type of material model can be used to predict the probability of failure given a set of uncertain material model parameters and boundary conditions.

7.10.2 Probabilistic Analysis of Cortical Bone Failure using FEA

To further investigate the utility of probabilistic methods combined with continuum damage mechanics models of bone behavior to predict bone failure, we recently modeled the tensile uniaxial load to failure behavior of cortical bone an existing elastic, viscoplastic, CDM material model available in the commercial finite element analysis code LS-DYNA (LS-DYNA v.960, LSTC, Livermore CA). This CDM model differed from the one described above in that, since it was developed to model the behavior of metals, it does not include viscoelastic effects. Furthermore, damage is dependent upon plastic strain. A nonlinear optimization procedure was implemented to determine the material parameters such the computed stress–strain behavior agreed with the experimental behavior (Fig. 7.5). This procedure was performed independently using archival data from five specimens [119] resulting in five sets of material parameters. Random variable descriptions of each material model parameter were determined from this set of material parameters and a probabilistic analysis simulating the uniaxial tensile loading was performed using the maximum stress attained during the simulation as the performance function. Thus, the result of probabilistic analysis is a predicted cumulative distribution function (CDF) of the maximum stress attained during a uniaxial tension test to failure. To quantify the predictive accuracy of the probabilistic analysis, the CDF of maximum attained stress compiled from the results of 60 independent cortical bone tension tests, in which each specimen was loaded to failure, was constructed and compared to the predicted CDF from the probabilistic analysis (Fig. 7.6). The CDF of the experimental data quantifies the variability associated with the tensile behavior of cortical bone.

Good agreement is shown between the predicted CDF and CDF based on the measured experimental data. Based on a Kolmogorov-Smirnov goodness of fit test, there is no statistical difference between the experimental data CDF and the computationally predicted CDF. The probabilistic model slightly under predicts the mean value and over predicts the standard deviation of the maximum tensile stress.

A deterministic analysis would only predict the mean value of the maximum stress giving no indication of the variability that can be expected. In contrast, the probabilistic analysis not only accurately predicts the mean value it also accurately predicts the variation in the response.

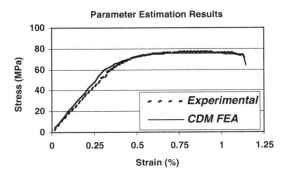

Fig. 7.5 CDM material model parameter estimation results. The results from fits of five independent tensile tests were used to determine the random variable descriptions of each material model parameter

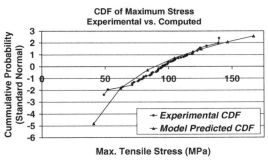

Fig. 7.6 Experimental versus predicted CDF of the maximum stress attained during a probabilistically simulated tensile test of cortical bone

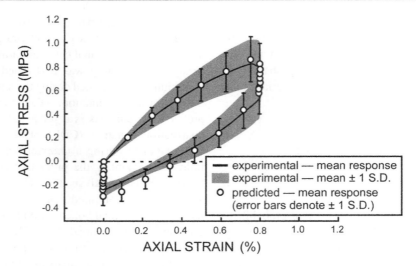

Fig. 7.7 Probabilistic trabecular bone CDM material model response

A deterministic analysis can only provide a binary answer when assessing a structure: either the stress is less than the strength in which case the structure does not fail, or the stress is equal to or greater than the strength wherein the structure will fail. A probabilistic analysis, alternatively, will quantify the probability that a stress will be greater than the material strength. Thus, this straightforward application demonstrates the ability of probabilistic analysis methods to successfully model uncertainty and variability in the tensile behavior of cortical bone.

7.10.3 Probabilistic Analysis of Trabecular Bone Damaging Behavior

In order to investigate variability in the evolving response of human vertebral trabecular bone, material model parameters were estimated individually for 10 experimental specimens using genetic algorithm based optimization combined with a phenomenological constitutive model (above and [120, 121]). The constitutive model and resulting material parameter distributions were implemented using MATLAB within a probabilistic analysis framework (NESSUS, SwRI) and the mean and variation of the experimental specimen response were successfully determined (Fig. 7.7).

7.11 Computational Model Verification and Validation (V&V)

There is a critical need to quantify the level of credibility that can be associated with computational model predictions. Model verification and validation (V&V) is a methodology for the development of models that can be used to make engineering predictions with quantified confidence. Verification is the process of determining that a model implementation accurately represents the developer's conceptual description of the model and the solution to the model. Validation is the process of determining the degree to which a model is an accurate representation of the real world from the perspective of the intended uses of the model. In short, verification deals with the mathematics associated with the model, whereas validation deals with the physics associated with the model.

Uncertainties exist in the outputs of computational simulations due to inherent and/or subjective uncertainties in the model inputs or in the form of the model itself. Likewise, measurements taken from physical experiments to validate these simulation results will also contain uncertainties. While the experimental measurements are taken as the reference for the validation, the validation process does not presume the experiment to be more accurate than the simulation. Rather, the validation process seeks to quantify the accuracy of the model considering all uncertainties in the model and the experiment [96, 122].

7.12 Summary

In summary, the long-term goal of our computational methods applied to injury research should be to quantitatively predict the risk or ***probability*** of skeletal fracture. To this end, we seek to understand the origins of skeletal fragility from both a tissue and structural basis and to investigate the limits of computational modeling within the context of biological variability and the current ability to determine material and structural properties from clinical image data. The methods described herein is based the combination of image-derived parametric statistical shape and density based modeling with non-linear continuum damage mechanics bone material modeling and multi-level predictive model verification and validation (V&V), all within a probabilistic analysis framework. The paradigm described here is different from previous and current applications of image derived engineering models in several important ways: current engineering models are deterministic and therefore cannot quantitatively predict the probability of fracture since they are unable to account for the large variations in material properties, geometry and boundary conditions associated with biological systems; uncertainty and variability are almost entirely unaccounted for in current engineering modeling applications to skeletal fragility research. In contrast, these new methods employ probabilistic methods that will enable us to account for uncertainty and variability in geometry and biological tissue properties from the outset. The effect of variability in geometry is investigated by employing a parametric description of bone geometry and BMD using statistical shape and density modeling. The structural failure process is also modeled using a continuum damage mechanics material model of both cortical and cancellous bone that is validated at the tissue level. This approach results in probabilistic sensitivity factors that will relate variability and uncertainty in modeling parameters to predictions of the probability of bone failure.

References

1. Burge R, Dawson-Hughes B, Solomon D, Wong J, King A, Tosteson A (2006) Incidence and economic burden of osteoporosis-related fractures in the United States, 2005–2025. J Bone Miner Res 22:465–475
2. Kanis J, Johnell O, Oden A, Sembo I, Redlund-Johnell I, Dawson A, De Laet C, Jonsson B (2000) Long-term risk of osteoporotic fracture in Malmo. Osteoporos Int 11(8):669–674
3. UN (1999) World population prospects: the 1998 revision. Department of Economic and Social Affairs, Population Division, United Nations, New York
4. Gullberg B, Johnell O, Kanis J (1997) World-wide projections for hip fracture. Osteoporos Int 7(5): 407–413
5. Melton L (2003) Adverse outcomes of osteoporotic fractures in the general population. J Bone Miner Res 18(6):1139–1141
6. Kanis J, Oden A, Johnell O, De Laet C, Jonsson B, Oglesby A (2003) The components of excess mortality after hip fracture. Bone 32(5):468–473
7. Johnell O, Kanis J, Oden A, Sernbo I, Redlund-Johnell I, Petterson C, De Laet C, Jonsson B (2004) Mortality after osteoporotic fractures. Osteoporos Int 15(1):38–42
8. Ebbesen E, Thomsen J, Beck-Nielsen H, Nepper-Rasmussen H, Mosekilde L (1999) Age- and gender-related differences in vertebral bone mass, density, and strength. J Bone Miner Res 14(8):1394–1403
9. Testi D, Viceconti M, Baruffaldi F, Cappello A (1999) Risk of fracture in elderly patients: a new predictive index based on bone mineral density and finite element analysis. Comput Methods Programs Biomed 60(1):23–33
10. Luo G, Kaufman J, Chiabrera A, Bianco B, Kinney J, Haupt D, Ryaby J, Siffert R (1999) Computational methods for ultrasonic bone assessment. Ultrasound Med Biol 25(5):823–830
11. Pietruszczak S, Inglis D, Pande G (1999) A fabric-dependent fracture criterion for bone. J Biomech 32(10):1071–1079
12. Magland JF, Wald MJ, Wehrli FW (2009) Spin-echo micro-MRI of trabecular bone using improved 3D fast large-angle spin-echo (FLASE). Magn Reson Med 61(5):1114–1121

13. Majumdar S, Genant HK (1997) High resolution magnetic resonance imaging of trabecular structure. Eur Radiol 7(Suppl 2):S51–S55
14. Majumdar S, Genant HK, Grampp S, Newitt DC, Truong VH, Lin JC, Mathur A (1997) Correlation of trabecular bone structure with age, bone mineral density, and osteoporotic status: in vivo studies in the distal radius using high resolution magnetic resonance imaging. J Bone Miner Res 12(1):111–118
15. Wehrli FW, Ladinsky GA, Jones C, Benito M, Magland J, Vasilic B, Popescu AM, Zemel B, Cucchiara AJ, Wright AC, Song HK, Sana PK, Peachey H, Snyder PJ (2008) In vivo magnetic resonance detects rapid remodeling changes in the topology of the trabecular bone network after menopause and the protective effect of estradiol. J Bone Miner Res 23(5):730–740
16. Gnudi S, Malavolta N, Testi D, Viceconti M (2004) Differences in proximal femur geometry distinguish vertebral from femoral neck fractures in osteoporotic women. Br J Radiol 77(915):219–223
17. Gnudi S, Ripamonti C, Gualtieri G, Malavolta N (1999) Geometry of proximal femur in the prediction of hip fracture in osteoporotic women. Br J Radiol 72(860):729–733
18. Gnudi S, Ripamonti C, Lisi L, Fini M, Giardino R, Giavaresi G (2002) Proximal femur geometry to detect and distinguish femoral neck fractures from trochanteric fractures in postmenopausal women. Osteoporos Int 13(1):69–73
19. Genant H, Gordon C, Jiang Y, Link T, Hans D, Majumdar S, Lang T (2000) Advanced imaging of the macrostructure and microstructure of bone. Horm Res 54(Suppl 1):24–30
20. Cheng XG, Lowet G, Boonen S, Nicholson PH, Brys P, Nijs J, Dequeker J (1997) Assessment of the strength of proximal femur in vitro: relationship to femoral bone mineral density and femoral geometry. Bone 20(3):213–218
21. Nicholson P, Muller R, Lowet G, Cheng X, Hildebrand T, Ruegsegger P, van der Perre G, Dequeker J, Boonen S (1998) Do quantitative ultrasound measurements reflect structure independently of density in human vertebral cancellous bone? Bone 23(5):425–431
22. Bauer D, Ewing S, Cauley J, Ensrud K, Cummings S, Orwoll E (2007) Quantitative ultrasound predicts hip and non-spine fracture in men: the MrOS study. Osteoporos Int 18:771–777
23. Kanis J (2002) Assessing the risk of vertebral osteoporosis. Singapore Med J 43(2):100–105
24. Kanis J (2002) Diagnosis of osteoporosis and assessment of fracture risk. Lancet 359(9321):1929–1936
25. Kanis J, Johnell O, De Laet C, Jonsson B, Oden A, Ogelsby A (2002) International variations in hip fracture probabilities: implications for risk assessment. J Bone Miner Res 17(7):1237–1244
26. Cody DD, Gross GJ, Hou FJ, Spencer HJ, Goldstein SA, Fyhrie DP (1999) Femoral strength is better predicted by finite element models than QCT and DXA. J Biomech 32(10):1013–1020
27. Lochmuller E, Miller P, Burklein D, Wehr U, Rambeck W, Eckstein F (2000) In situ femoral dual-energy X-ray absorptiometry related to ash weight, bone size and density, and its relationship with mechanical failure loads of the proximal femur. Osteoporos Int 11(4):361–367
28. Black D, Cummings S, Karpf D, Cauley J, Thompson D, Nevitt M, Bauer D, Genant H, Haskell W, Marcus R, Ott S, Torner J, Quandt S, Reiss T, Ensrud K (1996) Randomised trial of effect of alendronate on risk of fracture in women with existing vertebral fractures. Fracture intervention trial research group. Lancet JID – 2985213R 348(9041):1535–1541
29. Roschger P, Rinnerthaler S, Yates J, Rodan G, Fratzl P, Klaushofer K (2001) Alendronate increases degree and uniformity of mineralization in cancellous bone and decreases the porosity in cortical bone of osteoporotic women. Bone JID – 8504048 29(2):185–19
30. Pulkkinen P, Eckstein F, Lochmüller EM, Kuhn V, Jämsä T (2006) Association of geometric factors and failure load level with the distribution of cervical vs. trochanteric hip fractures. J Bone Miner Res 21(6):895–901
31. Pulkkinen P, Partanen J, Jalovaara P, Jämsä T (2004) Combination of bone mineral density and upper femur geometry improves the prediction of hip fracture. Osteoporos Int 15(4):274–280
32. Patron MS, Duthie RA, Sutherland AG (2006) Proximal femur geometry and hip fractures. Acta Orthop Belg 72(1):51–54
33. Specker B, Binkley T (2005) High parity is associated with increased bone size and strength. Osteoporos Int 16(12):1969–1974
34. El Kaissi S, Pasco JA, Henry MJ, Panahi S, Nicholson JG, Nicholson GC, Kotowicz MA (2005) Femoral neck geometry and hip fracture risk: the Geelong osteoporosis study. Osteoporos Int 16(10):1299–1303
35. Crabtree NJ, Kroger H, Martin A, Pols HA, Lorenc R, Nijs J, Stepan JJ, Falch JA, Miazgowski T, Grazio S, Raptou P, Adams J, Collings A, Khaw KT, Rushton N, Lunt M, Dixon AK, Reeve J (2002) Improving risk assessment: hip geometry, bone mineral distribution and bone strength in hip fracture cases and controls. The EPOS study European prospective osteoporosis study. Osteoporos Int 13(1):48–54
36. Center JR, Nguyen TV, Pocock NA, Noakes KA, Kelly PJ, Eisman JA, Sambrook PN (1998) Femoral neck axis length, height loss and risk of hip fracture in males and females. Osteoporos Int 8(1):75–81
37. Faulkner K, Cummings S, Black D, Palermo L, Gluer C, Genant H (1993) Simple measurement of femoral geometry predicts hip fracture: the study of osteoporotic fractures. J Bone Miner Res 8(10):1211–1217

38. Gluer C, Cummings S, Pressman A, Li J, Gluer K, Faulkner K, Grampp S, Genant H (1994) Prediction of hip fractures from pelvic radiographs: the study of osteoporotic fractures. The Study of Osteoporotic Fractures Research Group. J Bone Miner Res 9(5):671–677
39. Gregory J, Testi D, Stewart A, Undrill P, Reid D, Aspden R (2004) A method for assessment of the shape of the proximal femur and its relationship to osteoporotic hip fracture. Osteoporos Int 15(1):5–11
40. Testi D, Viceconti M, Cappello A, Gnudi S (2002) Prediction of hip fracture can be significantly improved by a single biomedical indicator. Ann Biomed Eng 30(6):801–807
41. Hipp JA, Jansujwicz A, Simmons CA, Snyder BD (1996) Trabecular bone morphology from micromagnetic resonance imaging. J Bone Miner Res 11(2):286–297
42. Müller R, Hildebrand T, Hauselmann HJ, Ruegsegger P (1996) In vivo reproducibility of three-dimensional structural properties of noninvasive bone biopsies using 3D-pQCT. J Bone Miner Res 11(11): 1745–1750
43. Odgaard A, Andersen K, Ullerup R, Frich LH, Melsen F (1994) Three-dimensional reconstruction of entire vertebral bodies. Bone 15(3):335–342
44. Turner CH, Takano Y, Hirano T (1996) Reductions in bone strength after fluoride treatment are not reflected in tissue-level acoustic measurements [see comments]. Bone 19(6):603–607
45. Keyak JH (2001) Improved prediction of proximal femoral fracture load using nonlinear finite element models. Med Eng Phys 23(3):165–173
46. Keyak JH, Meagher JM, Skinner HB, Mote CD Jr (1990) Automated three-dimensional finite element modelling of bone: a new method. J Biomed Eng 12(5):389–397
47. Lotz JC, Cheal EJ, Hayes WC (1991) Fracture prediction for the proximal femur using finite element models: part II–nonlinear analysis. J Biomech Eng 113(4):361–365
48. Lotz JC, Cheal EJ, Hayes WC (1991) Fracture prediction for the proximal femur using finite element models: part I–linear analysis. J Biomech Eng 113(4):353–360
49. Lotz JC, Cheal EJ, Hayes WC (1995) Stress distributions within the proximal femur during gait and falls: implications for osteoporotic fracture. Osteoporos Int 5(4):252–261
50. Bessho M, Ohnishi I, Matsuyama J, Matsumoto T, Imai K, Nakamura K (2007) Prediction of strength and strain of the proximal femur by a CT-based finite element method. J Biomech 40(8):1745–1753
51. Bessho M, Ohnishi I, Okazaki H, Sato W, Kominami H, Matsunaga S, Nakamura K (2004) Prediction of the strength and fracture location of the femoral neck by CT-based finite-element method: a preliminary study on patients with hip fracture. J Orthop Sci 9(6):545–550
52. Cook RD (1995) Finite element modeling for stress analysis. Wiley, New York
53. Crisfield MA (1991) Non-linear finite element analysis of solids and structures, vol 1. Wiley, New York
54. Keyak JH, Rossi SA, Jones KA, Skinner HB (1998) Prediction of femoral fracture load using automated finite element modeling. J Biomech 31(2):125–133
55. Keyak JH (2000) Nonlinear finite element modeling to evaluate the failure load of the proximal femur. J Orthop Res 18(2):337
56. Keyak JH (2001) Improved prediction of proximal femoral fracture load using nonlinear finite element models. Med Eng Phys 23(3):165–173
57. Keyak JH, Kaneko TS, Tehranzadeh J, Skinner HB (2005) Predicting proximal femoral strength using structural engineering models. Clin Orthop Relat Res 437:219–228
58. Crawford RP, Cann CE, Keaveny TM (2003) Finite element models predict in vitro vertebral body compressive strength better than quantitative computed tomography. Bone 33:744–750
59. Crawford RP, Rosenberg WS, Keaveny TM (2003) Quantitative computed tomography-based finite element models of the human lumbar vertebral body: effect of element size on stiffness, damage, and fracture strength predictions. J Biomech Eng 125(4): 434–438
60. Liebschner MA, Kopperdahl DL, Rosenberg WS, Keaveny TM (2003) Finite element modeling of the human thoracolumbar spine. Spine 28(6):559–565
61. Silva MJ, Keaveny TM, Hayes WC (1998) Computed tomography-based finite element analysis predicts failure loads and fracture patterns for vertebral sections. J Orthop Res 16(3):300–308
62. Bowman SM, Guo XE, Cheng DW, Keaveny TM, Gibson LJ, Hayes WC, McMahon TA (1998) Creep contributes to the fatigue behavior of bovine trabecular bone. J Biomech Eng 120:647–654
63. Keaveny TM, Wachtel EF, Guo XE, Hayes WC (1994) Mechanical behavior of damaged trabecular bone. J Biomech 27(11):1309–1318
64. Kopperdahl DL, Keaveny TM (1998) Yield strain behavior of trabecular bone. J Biomech 31(7): 601–608
65. Kopperdahl DL, Keaveny TM (1995) Compressive and tensile elastic and failure properties of human vertebral trabecular bone. In: Hochmuth RM, Langrana NA, Hefzy MS (eds) 1995 Bioengineering Conference, 1995. BED. ASME, Beaver Creek, pp 357–358
66. Cootes TF, Hill A, Taylor CJ, Haslam J (1994) Use of active shape models for locating structure in medical images. Image Vis Comput 12(6):355–365
67. Benameur S, Mignotte M, Labelle H, De Guise JA (2005) A hierarchical statistical modeling approach for the unsupervised 3-D biplanar reconstruction of the scoliotic spine. IEEE Trans Biomed Eng 52(12): 2041–2057

68. Dornaika F, Ahlberg J (2006) Fitting 3D face models for tracking and active appearance model training. Image Vis Comput 24(9):1010–1024
69. Ferrarini L, Palm WM, Olofsen H, van Buchem MA, Reiber JHC, Admiraal-Behloul F (2006) Shape differences of the brain ventricles in Alzheimer's disease. Neuroimage 32(3):1060–1069
70. Koikkalainen J, Hirvonen J, Nyman M, Lotjonen J, Hietala J, Ruotsalainen U (2007) Shape variability of the human striatum–Effects of age and gender. Neuroimage 34(1):85–93
71. Babalola KO, Cootes TF, Patenaude B, Rao A, Jenkinson M (2006) Comparing the similarity of statistical shape models using the Bhattacharya metric. In: Medical image computing and computer-assisted intervention – Miccai 2006, Pt 1, vol 4190. Lecture notes in computer science. pp 142–150
72. Rueckert D, Frangi AF, Schnabel JA (2003) Automatic construction of 3-D statistical deformation models of the brain using nonrigid registration. Ieee Trans Med Imaging 22(8):1014–1025
73. Shan ZY, Parra C, Ji Q, Jain J, Reddick WE (2006) A knowledge-guided active model method of cortical structure segmentation on pediatric MR images. J Magn Reson Imaging 24(4):779–789
74. Lorenz C, Krahnstover N (2000) Generation of point-based 3D statistical shape models for anatomical objects. Comput Vis Image Underst 77(2):175–191
75. Cootes TF, Taylor CJ, Cooper DH, Graham J (1995) Active shape models – their training and application. Comput Vis Image Underst 61(1):38–59
76. Davies RH, Twining CJ, Cootes TF, Waterton JC, Taylor CJ (2002) A minimum description length approach to statistical shape modeling. IEEE Trans Med Imaging 21(5):525–537
77. Davies RH, Twining CJ, Cootes TF, Waterton JC, Taylor CJ (2002) 3D statistical shape models using direct optimisation of description length. In: Computer vision – ECCV 2002 Pt III, vol 2352, pp 3–20
78. Styner MA, Rajamani KT, Nolte L-P, Zsemlye G, Székely G, Taylor CJ, Davies RH (2003) Evaluation of 3D correspondence methods for model building. Inf Process Med Imaging 18:63–75
79. Walker KN, Cootes TF, Taylor CJ (2003) Determining Correspondences for Statistical Models of Appearance. Computer Vision-ECCV 2000. Berlin, Heidelberg: Springer, Berlin, Heidelberg, pp 829–843
80. Viceconti M, Zannoni C, Testi D, Cappello A (1999) A new method for the automatic mesh generation of bone segments from CT data. J Med Eng Technol 23(2):77–81
81. Pepin JE, Thacker BH, Rodriguez EA, Riha DS (2002) A probabilistic analysis of a nonlinear structure using random fields to quantify geometric shape uncertainties. In: Proceedings of the AIAA/ASME/ASCE/AHS/ASC 43rd structures, structural dynamics and materials conference. AIAA, Denver, 22 Apr 2002
82. Mauch S, Breen D (2000) A fast algorithm for computing the closest point and distance function. Unpublished technical report. California Institute of Technology, September. http://www.acm.caltech.edu/seanm/projects/cpt/cpt.pdf
83. Brechbuhler C, Gerig G, Kubler O (1995) Parametrization of closed surfaces for 3-D shape-description. Comput Vis Image Underst 61(2):154–170
84. Taddei F, Pancanti A, Viceconti M (2004) An improved method for the automatic mapping of computed tomography numbers onto finite element models. Med Eng Phys 26(1):61–69
85. Les CM, Keyak JH, Stover SM, Taylor KT, Kaneps AJ (1994) Estimation of material properties in the equine metacarpus with use of quantitative computed tomography. J Orthop Res 12(6):822–833
86. Rajamani K, Nolte L, Styner M (2004) A novel approach to anatomical structure morphing for intra-operative visualization. In: Medical image computing and computer-assisted intervention – MICCAI 2004, Pt 2, Proceedings, vol 3217, pp 478–485
87. Bredbenner TL, Bartels KA, Havill LM, Nicolella DP (2007) Probabilistic shape-based finite element modeling of baboon femurs. In: ASME, Amelia Island
88. Keller TS (1994) Predicting the compressive mechanical behavior of bone. J Biomech 27(9):1159–1168
89. Courtney AC, Wachtel EF, Myers ER, Hayes WC (1994) Effects of loading rate on strength of the proximal femur. Calcif Tissue Int 55(1):53–58
90. Courtney AC, Wachtel EF, Myers ER, Hayes WC (1995) Age-related reductions in the strength of the femur tested in a fall-loading configuration. J Bone Joint Surg Am 77(3):387–395
91. Wu Y, Millwater H, Cruse T (1990) An advanced probabilistic structural analysis method for implicit performance functions. AIAA J 28(9):1663–1669
92. Haldar A, Mahadevan S (2000) Probability, reliability, and statistical methods in engineering design. Wiley, New York
93. Ang AHS, Tang WH (1975) Probability concepts in engineering planning and design: Vol. 1. John Wiley & Sons, New York
94. Nicolella DP, Francis WL, Bonivtch AR, Thacker BH, Paskoff GR, Shender BS Development, verification, and validation of a parametric cervical spine injury prediction model. In: Collection of technical papers – AIAA/ASME/ASCE/AHS/ASC structures, structural dynamics and materials conference, 2006. pp 3977–3985
95. Nicolella DP, Thacker BH, Katoozian H, Davy DT (2001) Probabilistic risk analysis of a cemented hip implant. J Math Model Sci Comput 13(1–2):98–108
96. Nicolella D, Francis W, Bonivtch A, Thacker B, Paskoff G, Shender B (2006) Development, verification, and validation of a parametric cervical spine

injury prediction model. In: Collection of technical papers – AIAA/ASME/ASCE/AHS/ASC structures, structural dynamics and materials conference, vol 6, pp 3977–3985

97. Nicolella DP, Thacker BH, Katoozian H, Davy DT (2006) The effect of three-dimensional shape optimization on the probabilistic response of a cemented femoral hip prosthesis. J Biomech 39(7):1265–1278
98. Burr D, Turner C, Naick P, Forwood M, Ambrosius W, Hasan M, Pidaparti R (1998) Does microdamage accumulation affect the mechanical properties of bone? J Biomech 31(4):337–345
99. Norman TL, Wang Z (1997) Microdamage of human cortical bone: incidence and morphology in long bones. Bone 20(4):375–379
100. Norman TL, Yeni YN, Brown CU, Wang Z (1998) Influence of microdamage on fracture toughness of the human femur and tibia. Bone 23(3):303–306
101. Burr DB, Martin RB, Schaffler MB, Radin EL (1985) Bone remodeling in response to in vivo fatigue microdamage. J Biomech 18:189–200
102. Schaffler M, Radin E, Burr D (1989) Mechanical and morphological effects of strain rate on fatigue of compact bone. Bone 10(3):207–214
103. Chan KS, Nicolella DP (2012) Micromechanical modeling of R-curve behaviors in human cortical bone. J Mech Behav Biomed Mater 16:136–152. doi:10.1016/j.jmbbm.2012.09.009
104. Reilly GC, Currey JD (2000) The effects of damage and microcracking on the impact strength of bone. J Biomech 33(3):337–343
105. Jepsen K, Bensusan J, Davy DT (2001) Intermodal effects of damage on mechanical properties of human cortical bone. Trans Orthop Res Soc, San Francisco, California, Feb 25–28
106. Fondrk MT (1989) An experimental and analytical investigation into the nonlinear constitutive equations of cortical bone (Doctoral dissertation, Case Western Reserve University)
107. Fondrk M, Bahniuk E, Davy D (1999) Inelastic strain accumulation in cortical bone during rapid transient tensile loading. J Biomech Eng 121(6):616–621
108. Fondrk M, Bahniuk E, Davy D (1999) A damage model for nonlinear tensile behavior of cortical bone. J Biomech Eng 121(5):533–541
109. Fondrk M, Bahniuk E, Davy DT, Michaels C (1988) Some viscoplastic characteristics of bovine and human cortical bone. J Biomech 21(8):623–630
110. Carter DR, Caler WE (1985) A cumulative damage model for bone fracture. J Orthop Res 3(1):84–90
111. Fyhrie DP, Lang SM, Hoshaw SJ, Schaffler MB, Kuo RF (1995) Human vertebral cancellous bone surface distribution. Bone 17(3):287–291
112. Wachtel EF, Keaveny TM (1997) Dependence of trabecular damage on mechanical strain. J Orthop Res 15(5):781–787
113. Burr DB, Forwood MR, Fyhrie DP, Martin RB, Schaffler MB, Turner CH (1997) Bone microdamage and skeletal fragility in osteoporotic and stress fractures. J Bone Miner Res 12(1):6–15
114. Courtney AC, Keaveny TM (1994) Post-yield behavior and damage in bovine cortical bone. In: Proceedings of the 29th annual meeting. Society for Biomaterials, Boston
115. Zysset PK, Curnier A (1996) A 3D damage model for trabecular bone based on fabric tensors. J Biomech 29(12):1549–1558
116. Bredbenner T (2003) Damage modeling of vertebral trabecular bone
117. Deligianni D, Maris A, Missirlis Y (1994) Stress relaxation behaviour of trabecular bone specimens. J Biomech 27(12):1469–1476
118. Zhu Y, Cescotto S, Habraken A (1992) A fully coupled elastoplastic damage modeling and fracture criteria in metalforming processes. J Mater Process Technol 32(1–2):197–204
119. Wilson J, Jepsen K, Bensusan J, Davy D (1998) Simple mechanical measures as predictors of tensile failure in human cortical bone. Trans Orthop Res Soc 23:963
120. Bredbenner TL, Nicolella DP, Davy DT (2006) Modeling damage in human vertebral trabecular bone under experimental loading. In: Annual conference and exposition. Society for Experimental Mechanics, St. Louis, 4–7 June 2006
121. Bredbenner TL, Nicolella DP, Davy DT (2006) Modeling damage in human vertebral trabecular bone under experimental loading. Experimental Mechanics: (submitted)
122. Francis WL, Eliason TD, Thacker BH, Paskoff GR, Shender BS, Nicolella DP (2012) Implementation and validation of probabilistic models of the anterior longitudinal ligament and posterior longitudinal ligament of the cervical spine. Comput Methods Biomech Biomed Engin. doi:10.1080/10255842.2012.726353

Skull and Facial Bone Injury Biomechanics

Cameron R. Bass and Narayan Yoganandan

Abstract

The skull and facial bones are biomechanically complex. This chapter outlines existing research on the impact and fracture response of the skull and facial bones including the calvarium, maxilla, zygomatic bone, nasal bone, orbit, and mandible. The chapter begins with a discussion of biomechanically relevant anatomy and a discussion of injury severity scales for skull and facial fracture. Discussion includes a detailed description of the available research on impact tolerance of skull and facial bones. These studies show that impact and fracture response is strongly dependent on location, bony impact geometry, and contact area. The chapter concludes with an outline of the available, but limited, information on injury risk functions for skull and facial bones. These risk functions are restricted to skull fracture and maxilla/zygoma injury assessments.

8.1 Introduction

The skull and facial bones have manifest biomechanical complexities. Impact and fracture response is generally dependent on location, bony geometry, contact surface and contact area. This is especially true for the facial bones, and care must be taken in assessing biomechanical results from impactors, especially flat impactors on curved bony surfaces. Much of the study of injury biomechanics in the skull and facial bones arises in the context of automobile crashes with contact interactions of the head and interior of vehicles, falls and assaults. Jaegar found in a US national survey of emergency departments that skull fractures occurred in approximately 16 % of severe TBI cases and were predominantly associated with motor vehicles and bicyclists/pedestrians (29 %), assaults (23 %) and falls (39 %) [1]. Similarly, basilar skull fractures often occurred as a consequence of motor vehicle crashes, either hit pedestrians or occupants, or falls [2, 3].

Mandible fractures are common in blunt trauma and rank second, after nasal bone fractures, in

C.R. Bass, Ph.D. (✉)
Department of Biomedical Engineering,
Duke University, Durham, NC, USA
e-mail: Dale.bass@duke.edu

N. Yoganandan, Ph.D.
Department of Neurosurgery, Medical College
of Wisconsin, Milwaukee, WI, USA
e-mail: yoga@mcw.edu

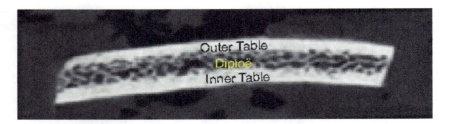

Fig. 8.1 High resolution CT image of a diploic human skull showing inner table, outer table and diploe

facial bones fractured [4, 5]. In a review of 5 years of cases at a major level one trauma center, Haug et al. found that the ratio of mandible fractures to zygomatic and maxillary fractures is 6:2:1 [6]. They further found that motor vehicle crashes and assaults were the most common cause of facial fractures (e.g. Greene 1997). In a study of 882 patients with facial fractures, the incidence of comorbid cranial fractures was found to be 4.4 % though there appears to be a stronger correlation of facial fracture with cranial injuries when concussion is included [7]. Approximately half of mandible fractures occur from assaults and gunshot wounds and over a third from motor vehicle crashes, either occupants or pedestrians. The distribution of injuries across the mandible in civilian trauma [5] is qualitatively similar to that seen in the peacetime US military [8]. Parasymphyseal (34 %), angle (15 %) and body (21 %) fractures are common with condylar complex injuries (21 % condyle, subcondyle) (Fig. 8.1). Injuries to the ramus are less common owing to its usual location outside of the direct loading path (parasymphysis, body) and reaction support (condyle, subcondyle).

8.2 Anatomy

From a biomechanical perspective, the human skull bone lies inferior to the scalp, a connective tissue [9]. It is generally composed of three layers: the outer and inner tables sandwiching a diploe layer (Fig. 8.1). Structurally, the diploe is soft, with material properties similar to the cancellous bone of the human vertebral column, and the two tables are relatively rigid, comparable to the cortical bone of the vertebrae or long bones [10]. The bones of the calvarium, the frontal, occipital, parietal, temporal, sphenoid and ethmoid (not shown) bones are regionally connected by interdigitated sutures, knit since childhood with adult structural properties similar to the nearby diploic skull (Fig. 8.2). The geometry of the cranial vault is complex and three-dimensional. Generally, the cranial vault is symmetric about the mid-sagittal plane. The temporal region is concave medially and the parietal region is convex. This changing geometry including thickness variations, often challenges, investigators to accurately, load or impact one region without engaging the other. Despite geometrical differences, skull bones are constitutionally similar. Dura separates the inner table from the brain.

The upper facial skeleton (Fig. 8.2) is comprised of a complex system of interconnected bones that articulate with the skull [11, 12]. The principal bones studied include the maxilla, zygomatic bone, nasal bone and orbit. The maxilla is in the mid-face and supports both the orbital floor and the upper mandible. Dentition of the upper mandible resides in the alveolar process. The zygomatic bone, or cheek bone, articulates with the maxilla and forms the prominence of the cheeks, known as the malar eminence. The zygomatic arch extends posteriorly where it articulates with the temporal bone. The frontal bone forms the forehead, and it is connected with the nasal bones, maxilla, and the zygomatic bone. The orbit is a generally conical cavity comprised of several bony structures. The frontal bone extends to the superior wall of the orbit while the maxilla extends to the orbital floor. The zygoma extends to the orbit lateral wall. The lacrimal, ethmoid and sphenoid bones form the remainder of the orbit. The medial and inferior walls are extremely thin. Fractures of the bones of the orbit walls are usually produced from two sources or

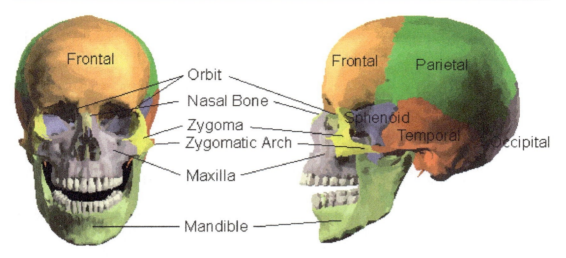

Fig. 8.2 Bones of the skull on the *left* and facial bones including the orbit on the *right*

combinations of these two. The first is fractures of the maxilla, zygoma or frontal bone that continue into the orbit. The second is, indirect contact from the globe of the eye, also known as blowout fractures. These fractures may result in subsequent herniation of the globe.

The mandible (or lower jaw) is biomechanically complex [13]. The seat of the lower dentition, the mandible is mechanically articulated (hinged) at the condyle with muscular insertions at the coronoid process (Fig. 8.3). Though the mandible is the strongest bone in the face, the articular dislocation force is not large, especially in the absence of muscular control [14–16]. The major functions of the mandible in humans are eating and modulation of speech. The maximum bite force is generally large, 200–400 N in adult males [17, 18] and approximately 100–200 N in adult females [18].

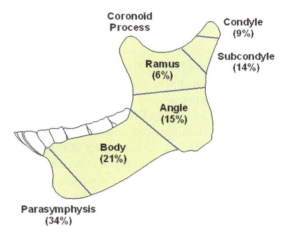

Fig. 8.3 Anatomy of the mandible and typical fracture locations

8.3 Injury Classification/Severity Scales

Several injury severity scales have been devised for grading skull/facial fractures. These include the all-encompassing Injury Severity Score based on the AIS (Abbreviated Injury Scale, AIS), a threat to life scale with limited granularity for facial fractures [19, 20]. The LeFort Classification grades maxilla fractures into three categories based on the extent of the fracture [21]. LeFort I is a horizontal fracture of the alveolar process. LeFort II is fracture where the body of the maxilla is separated a from the facial skeleton and is pyramidal in shape, and LeFort III is a complete separation of the maxilla from the skull. This system, developed for military blunt trauma, is appropriate for large area fracture but is generally inadequate for localized injuries as it was developed in the early 1900s for military applications – typically blunt trauma to the face over a wide area.

Zingg et al. developed an injury classification based on facial fracture patterns and locations. Incomplete fractures are characterized as Type

A, fracture of the zygomatic arch is classified as A1, lateral orbital wall fracture is A2 and infraorbital rim fracture is A3 [22]. Complete monofragment fractures are Type B in which the zygoma is completely separated at all three sites. Multifragment fractures are Type C in which the zygoma is separated and the body is comminuted. This classification scheme does not generally address the severity of fracture patterns. Donat et al. defined a system that hypothesizes six notional horizontal beams and five notional vertical buttresses supporting the facial bones [23]. Fractures are described using disruption of support segments. This method does not well represent the structural support of the facial bones and is not strictly a severity rating, but is useful as a descriptive tool for physicians. Rhee et al. developed a classification similar to the AIS for facial bones, ranging in severity from 1 (no fracture) to 6 (displaced zygomaticomaxillary fracture with significant concomitant fractures) based on subjective measures [24]. Unlike the AIS, it is applicable to facial fracture alone, allowing finer distinctions between similar injuries.

There are a number of isolated mandible fracture severity classifications. Joos et al. developed a mandibular fracture score ranging from 0 to 15 based on preoperative factors (location, displacement, fracture complexity and systemic factors) and intraoperative factors, (difficulty of reduction, undefined occlusion and difficulty of soft tissue coverage [25]. This score was found to be well correlated with complication rates (binary classification) with R^2 value of 0.88. Zhang et al. proposed the Maxillofacial Injury Severity Score (MFISS) based on AIS as the product of the three highest AIS scores in the maxillofacial region combined with three functional parameters [26]. This score showed significant correlation between the MFISS score and days of hospitalization, but the correlation coefficients were quite low (e.g., 0.26 for cost). More recently, Shetty et al. developed the UCLA Mandible Injury Severity Score [27]. This scale is based on a multiple level scale of fracture characterizations. Constituent scaled fracture variables in this composite score include fracture (F) type, fracture location (L), occlusion (O), soft tissue involvement (S), presence of infection (I) and interfragmentary displacement (D). Ranging in value between 0 and 21, MISS shows a statistically significant correlation with sensory nerve deficits, hospitalization and pain, but no significant correlation with postoperative complications.

8.4 Biomechanical Studies

8.4.1 Skull (Cranial Vault)

Gurdjian et al. dropped intact heads from cadavers onto a solid steel slab [28]. Strain sensitive lacquer was used to identify the pathology [29]. Linear fractures occurred due to tensile stresses from local skull bending. Energies ranged from 650 to 1,230 Nm. Velocities ranged from 4.6 to 6.3 m/s. Fracture limits for dry skulls were lower than for intact heads and comparison of data with static tests demonstrated loading rate effects [30]. To study the effects of contact area on bone tolerance, Nahum et al. delivered impacts using an impacting mass to the temporo-parietal junction of ten PMHS heads [31]. Average fracture forces were lower in females (3,123 N ± 623) in males (3,944 N ± 1,287). Minimum thresholds of 2,450 N for males and 2,000 N for females were suggested for clinically significant fractures. In a follow up study, Schneider et al. conducted additional temporo-parietal impact studies using six intact and nine isolated head-C7 PMHS [32]. Intact PMHS were supine and isolated specimens were supported by wedges of polyurethane padding. Weights were dropped at velocities from 3 to 6 m/s. Injuries ranged from none to severe comminuted fractures. Comparison of the temporo-parietal data with frontal bone tolerance (Fig. 8.4) from tests conducted showed that peak forces for the two regions fall within the range of each other.

Hodgson and co-workers investigated the biomechanics of frontal, lateral, occipital and facial regions of embalmed PMHS [33–35]. Results were reported from 35 drop tests conducted with seven male PMHS [34]. Two pairs of bi-axial accelerometers were corded to record transverse and antero-posterior (AP) skull accelerations. A load cell mounted under the rigid impact sur-

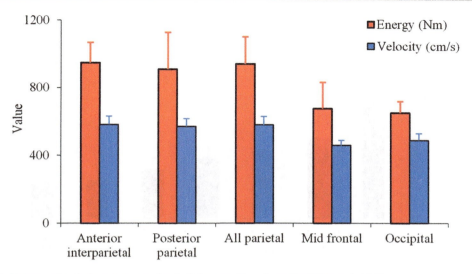

Fig. 8.4 Energy and velocity parameters for skull fracture (Data from Gurdijian et al. 1943)

face recorded the forces. The head was restrained by a cord from rotating into the preferred position during free fall. The PMHS were raised to the desired height and dropped (1.6 to 4.7 m/s) to contact a rigid flat plate. One specimen was dropped on the left side four times and on the right side nine times without fracture. All other six specimens showed linear fractures. Peak impact forces ranged from 5,560 to 17,792 N (mean: 10,151 ± 4,928). Peak AP accelerations ranged from 190 to 325 g (mean: 267 ± 71). Pulse times ranged from 2.5 to 6 ms (mean: 4.6 ± 1.5). These values were lower than those from frontal impact tests. A comparison of side impact force (Fig. 8.5) and acceleration (Fig. 8.6) data with their companion study on rear and frontal bones indicated the skull to be the strongest in the posterior, followed by the lateral and frontal regions [36].

McElhaney et al. reported data from sub-failure quasi-static tests conducted using intact PMHS heads [37]. The head was positioned between two steel platens. Force–deflection curves from loading to the temporal and frontal regions were reported from 12 tests. From the non-linear force–deflection responses, bilinear approximations were made, and the stiffness in the second region ranged from 700 to 1,750 N/mm for the temporal and 1,400 to 3,500 N/mm for the frontal regions. Significant overlap existed (Fig. 8.7) in the response between the two anatomical sites. Further, both sites responded with forces ranging from 4,000 to 8,000 N. Viano et al. investigated frontal bone fracture at high velocities to 69 m/s finding peak forces ranging from 2,450 to 4,080 N without producing fracture [38]. This is consistent with the results of McElhaney, but it is unknown whether fracture force tolerances for the frontal bone are higher at higher velocities.

Stalnaker et al. conducted left lateral impacts on PMHS seated in the upright position using a pneumatic piston [39]. Three padded and two rigid impacts were conducted on one male and four female PMHS. Peak forces and accelerations ranged from 4.21 to 9.59 kN (mean: 6.1 ± 2.3) and 125 to 532 g (mean: 247 ± 169). All three drops with padded surfaces did not result in bony pathology with peak forces ranging from 4.2 to 4.8 kN and peak accelerations from 125 to 179 g. However, both rigid drop specimens sustained AIS 3 fractures: temporal and occipital bones in one case at a peak force of 7.15 kN and acceleration of 262 g, and a comminuted fracture of the temporal bone at a peak force and acceleration of 9.6 kN and 532 g in another case. Impact velocities were 7.2 and 6.8 m/s for these two impacts. Padded impacts resulted in lower forces and accelerations than did rigid impacts, with

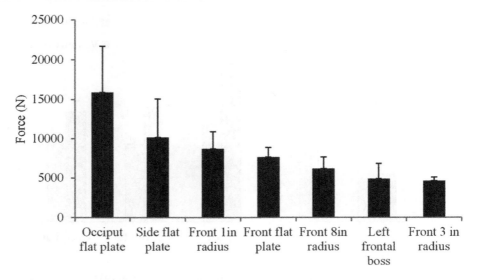

Fig. 8.5 A comparison of peak side impact forces with data from rear and frontal bones (Data from Hodgson et al. 1972)

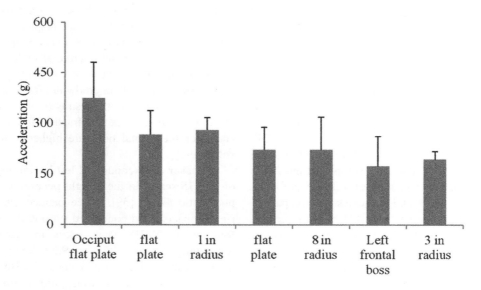

Fig. 8.6 A comparison of peak side impact acceleration with data from rear and frontal bones (Data from Hodgson et al. 1971)

lower pulse durations. Impact tolerance to the side head was found to be approximately 5 kN.

The temporo-parietal responses were determined from five intact PMHS subjected to 19 rigid and padded impacts [40]. Three drop heights were used: 1.83 m according to the United States Federal Motor Vehicle Safety Standards, FMVSS 218, 2.5 m according to the French standard and 3 m. Two rigid impact tests produced peak forces of 12.2 and 12.5 kN, and three tests with padded impact produced peak forces ranging from 5.0 to 10.1 kN. Although no fractures were identified in padded impacts, both specimens in rigid impacts sustained skull fractures. In one rigid impact, HIC exceeded 7,000 with frontal and exceeded 1,700 with right temporal accelerometer data.

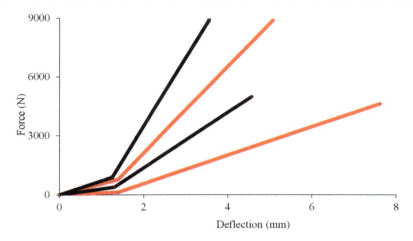

Fig. 8.7 Force–deflection responses for the side (red) and frontal (black) regions of the human head showing the range of responses bounded by the two respective curves (Redrawn from Ref. [37])

The use of the contralateral accelerometer produced minimal variations in HIC while the use of computed center of gravity (CG) accelerations increased the HIC by a factor of 2.5; the force increased by approximately 25 %. The HIC computed from the accelerometer placed on the right side did not exhibit the same tendency (high or low) when compared with the HIC from sensors placed at other anatomic locations or computed at the CG. However, HIC computed from the CG accelerations were always lower than those computed from the frontal accelerometer.

Impact tests using the frontal bone as the site of loading (7.7 m/s) produced no fractures in six specimens with peak forces from 5.6 to 14.0 kN. HIC computed using the CG acceleration data were 1,000 and 2,800 for two specimens (forces 5.6 and 6.0 kN). HIC computed using rear accelerometer signals ranged from 1,022 to 3,870. The force data matched very well with such data for the side loading suggesting the overlap phenomenon between the fracture thresholds for these sites. However, such clear conclusions could not be drawn for fracture tests because of differences in impact velocities. In a later study, these investigators reported poor correlations ($R<0.2$) between age, cranial vault dimensions and fracture force [41]. Higher correlations were found between vault weight and mineralization ($R=0.74$). Skull bone condition factor, defined using the thickness of the skull, skull diameter, skullcap mineralization and head mass, correlated well with mechanical tests performed on skullcap fragments.

Nahum et al. conducted tests using one embalmed (four impacts) and five unembalmed PMHS (one impact to each cadaver) using a rigid mass at velocities from 6.5 to 7.5 m/s for the embalmed cadaver and from 7.0 to 10.2 m/s for unembalmed specimens [42]. Peak head accelerations correlated well with HIC (334–1,466 for embalmed and 1,340–5,246 for unembalmed cadaver tests). The time interval maximizing HIC was considerably lower for unembalmed than for embalmed specimens. Peak head accelerations linearly correlated ($R^2=0.86–0.99$) with positive pressures in the right frontal area near the region of impact ($R^2=0.99$).

Allsop et al. conducted impact tests on 31 isolated unembalmed PMHS heads [43]. The specimens were impacted using flat rectangular (at 4.3 m/s) or circular (at 2.7 m/s) plates. The impactors were fixed to a drop tower. Two locations on the temporo-parietal regions were selected for the circular impactor. For the rectangular impactor, the parietal region was selected as the impact site. The mean fracture for rectangular plate impacts was 12,390 N (±3,654). The average fracture force for both impact sites with the circular impactor was 5195 N (±1010).

Rectangular plate tests responded with stiffness ranging from 1,600 to 6,430 N/mm (mean: 4,168 ± 1,626). Circular plate tests resulted in stiffness values ranging from 700 to 4,760 N/mm (mean: 1,800 ± 881). The contact area of the impactor significantly affected peak forces. Hodgson et al. and Yoganandan et al. advanced similar conclusions for facial bone structures in experimental studies in the 1970s and 1980s [11, 12, 34, 44, 45].

McIntosh et al. conducted lateral impacts to 11 unembalmed PMHS using an aluminum impactor [46]. Tests at velocities from 3.9 to 6.1 m/s were conducted without padding; tests at velocities from 2.8 to 3.8 m/s were conducted with padding in front of the impactor. Skull fractures occurred in three undamped impacts at >4.0 m/s and AIS ranged from 2 to 4. Peak biomechanical parameters were lower in specimens with no injury (AIS = 0) compared to injured specimens. Head mass was scaled according to procedures outlined by Reynolds [47]. Because of the limited sample size, data from occipital impacts (four fracture and four non fracture) was grouped with side impact data, and risk curves were developed using a logistic regression analysis. At the 200 Hz filter level, a HIC value of 800 represented 50 % injury risk.

Yoganandan et al. conducted studies by fixing the inferior end of the intact PMHS head using a custom-designed device and applying static and dynamic loads with the piston of the electro-hydraulic testing apparatus [48]. Failure forces were 5,292 and 5,915 for the two parietal loading specimens and 6,182 N for the temporally loaded specimen. Temporal and parietal fractures were reported. Failure deflections for these specimens were 8.9, 7.8 and 15.4 mm, and stiffness was 695, 1,143 and 487 N/mm. The parietal region was stronger than the temporal region, although the temporal bone was more compliant. Force tolerance increased by approximately a factor of two under dynamic loading, while the stiffness increase was higher (Fig. 8.8). While x-rays and computed tomography scans identified skull fractures, the precise location and direction of the impact on the skull were not apparent in these images. The authors concluded that, based on retrospective imaging, it may not be appropriate to extrapolate the anatomical region that sustained the external insult. Because the experimental design included monitoring forces and deflection, actual force-deflection plots were provided.

McElhaney et al. conducted a study to investigate three mechanisms of basilar skull fractures using human cadaver specimens [49]. These include transmission of loading from the mandible through the temporomandibular joint (TMJ) in a anteroposterior (AP) direction, direct TMJ loading with neck tension simulating inertial loading and direct superior-inferior cranial vault impact [50]. They found no evidence that direct transmission in the AP direction through the TMJ results in basilar skull fracture, but did produce basilar skull fracture with applied neck tension with loading vertically through the TMJ with force levels ranging from 2,590 to 4,880 N. Of the 11 tests with loading applied vertically by impacts to the cranial vault, 7 tests produced cervical spinal fractures while only one produced a basilar skull fracture in concert with a cervical spinal fracture.

8.4.2 Facial Bones

Many of the investigations of facial fracture are associated with automobile safety, aimed at assessing injury from contact with the steering column, dashboard, windshield or air bag. For example, a Society for Automotive Engineers (SAE) task force summarized available data to 1986 for facial fracture forces [51]. The focus of these and more recent investigations has been generally direct impact on the maxilla, zygoma and blowout fractures of the orbit.

Schneider et al. conducted 13 maxillary impacts and 27 zygomatic impacts on 17 human cadavers using a circular disc impactor with a surface area of 6.5 cm^2 with various velocities and masses [32]. Eleven of the maxilla impacts produced fractures. The peak force to fracture ranged from 625 to 1,980 N. Seventeen of the zygoma impacts produced fractures with the peak force ranging from 930 to 2,850 N. A series

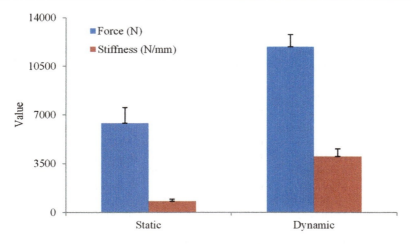

Fig. 8.8 Comparison of force and stiffness at two different rates of loading from isolated specimens (Adapted from Ref. [48])

of 113 impact experiments to various facial bones were performed by Nahum using both male and female cadavers that were both embalmed and unembalmed [52, 53]. Maxilla and zygoma were found to have approximately 25 % of the impact strength of the frontal bone for a 6.5 cm impactor (Table 8.1). Karlan and Cassisi studied fractures of the body of the zygoma, treating the structure as a flattened pyramid with multipoint support [54]. Stanley and Nowak emphasized the importance of impacts to the buttressing structures of the zygoma and maxilla in creating more severe facial fracture. Nyquist et al. reported 11 impacts to the nasal bones of human cadavers [55]. They found an association of impact kinetic energy with fracture severity. Impact forces greater than 3 kN were associated with fractures of the maxilla, zygoma and frontal bone. Impact kinetic energy varied from 241 to 815 J.

Allsop et al. reported 30 impacts on 15 human cadavers. Fifteen impacts were applied to the midface region (either the maxilla or zygoma) with a 14.5 kg semicircular impactor that was 23 cm long [56]. This configuration represents a steering wheel rim. Impactor drop heights ranged from 30.5 to 61 cm. Allsop et al. used acoustic emission sensors to identify the timing of fracture initiation (cracks in the bone create characteristic ultrasonic waves that are detected by the acoustic sensor). They measured initiation of fracture from 900 to 2,400 N force in the zygomatic region. Zygoma impact stiffness ranged from 90 to 230 N/mm. For the maxilla, fracture loads varied from 1,000 N to 1,800 N, with impact stiffness from 80 to 250 N/mm.

Yoganandan et al. conducted quasistatic and dynamic tests using 22 human cadaver heads impacted with a steering wheel assembly [11]. Zygomatic region fracture loads varied from 1,200 N to 1,400 N for the quasistatic tests. The dynamic impact velocities varied from 2.7 m/s to 7.0 m/s. These impacts produced peak forces from 1,300 N to 4,600 N with impacts often producing maxilla fractures. In a later study, Yoganandan reported the Weibul-based probability function (Fig. 8.9) for fracture [44, 57]. Bisplinghoff et al. reported 92 tests on 14 human cadaver facial bones to develop injury criteria for facial bones [58]. Test velocities ranged from 1 to 6 m/s using a 5 cm^2 indentor. The 50 % probability of globe rupture from a 4.5 mm BB impact was 107 N. They reported only 50 % risk values for the maxilla (1,150 N), nasal bone (340 N), zygomatic bone (1,360 N), frontal bone (2,670 N) and mandible (1,780 N).

Welbourne et al. reported eight impacts to the sub-nasal maxilla of human cadaver heads with a horizontal steel bar [59]. Four of these impacts produced fracture, only one of which was considered extensive by the authors. The peak

Table 8.1 Summary of experimental studies of facial fracture

Author (year)	Number of specimens	Impact location	Impactor geometry	Impactor area (cm^2)	Initial energy (J)	Velocity (m/s)	Peak force (N)
Schneider (1972)	13	Maxilla	Circular	6.5	4.9–26.3	3–5.4	625–1,980
	27	Zygoma			6.4–46.7	3–5.2	930–2,850
Nahum (1975a)	9	Maxilla	Circular	6.5	4.9–26.3	3–5.4	625–1,980
	16	Zygoma	Circular	6.5	12.6–86.1	2.2–7.5	610–3,470
	9	Zygomatic arch	Circular	6.5	6.4–46.7	3.0–5.4	930–1,940
	21	Frontal bone	Circular	6.5	25.3–133.8	3.1–7.1	2,670–9,880
	8	Lateral mandible	Flat	25 (2.5×10 cm)	56.9–68.4	5.4–6.0	820–3,400
	9	Mandible symphysis (AP)	Circular	6.5	37.3–46.7	4.9–5.4	1,900–4,100
Nyquist (1986)	11	Nasal bone	Cylinder	5	240–815	2.8–7.1	2,000–4,500
Allsop (1988)	6	Maxilla	Cylinder	2 cm diameter bar	43–87	2.4–3.5	1,000–1,800
	8	Zygoma					900–2,400
Yoganandan (1988)	9 quasistatic	Zygoma	Steering wheel			0.0025	1,200–1,400
	15 dynamic					2.0–6.9	1,300–4,600
Welbourne (1989)	8	Maxilla	Cylinder	5	15.1–85.9	1.35–3.22	788
Garza (1993)	12	Zygoma	Cylinder	NA	29.4–58.8	2.2–3.1	1,100–2,800
McElhaney (1995)	6	Mandible	Flat platen	127	11.4–119	6.08–7.15	4,500–6,700
	6	Basilar skull through TM/neck tension		NA	4.3–14	Quasistatic	2,590–4,880
	11	Basilar skull through vault		182	NA	2.4–3.5	1,290–4,290
Rhee (2001)	14	Zygoma	Rounded rod end	3.0–13.9	NA	2.0–6.9	1,400–4,600
Davis (2004)	11	Maxilla, zygoma, orbit	Rounded cylinder	13.8	57.3–81.2	4.2–5	1,100–4,400
Viano (2002)	4	Frontal bone	Cylinder	10.8	14–84	33–69	2,450–4,080
	5	Zygoma			10–50	28–41	1,520–2,300
	6	Mandible			16–28	36–47	1,460–3,710
Bisplinghoff (2008)	14	Nasal bone, maxilla, zygoma	Bevelled cylinder	5	1.6–57.2	1–6	NA
Green (1990)	22	Orbit	Rounded cylinder	1.2	2.08–3.28	NA	NA
Waterhouse (1999)	47	Orbit	Rod end	4.5		1.1–2.3	NA
Warwar (2000)	21	Orbit floor	Rod end	0.38–0.50	29–127 mJ	NA	NA

force for the impacts producing fracture varied from 516 N to 1,254 N, while peak force for the non-fracture tests varied from 660 N to 1,362 N. Garza et al. studied rigid fixation of the zygomatic complex following injurious blows [60]. The fracture force of the intact zygoma was found to range from 1,100 to 2,800 N, a very similar range of failure forces to that of Nahum [52]. Rhee et al. reported impacts on the zygoma of 14 unembalmed cadaver heads with resulting fracture force ranging from approximately 1,500 N to 4,500 N [24]. Severe fractures were

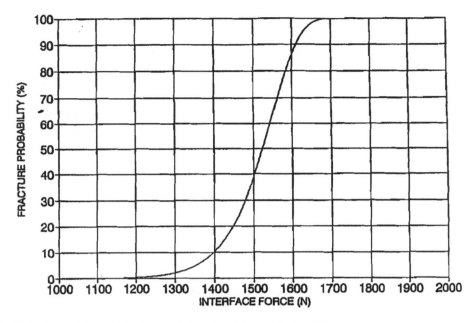

Fig. 8.9 Weibull probability distribution for facial injury (From Refs. [44, 57])

correlated with velocities above 3.5 m/s. Rhee et al. found that velocity of impact was better correlated with their injury severity than was contact area, skin thickness or impact force [24].

Davis assessed the directional dependence of force and fracture response of maxilla and zygoma fractures with an indentor that represented the eyepiece of military night vision goggles [61]. Tests included four dynamic impacts into the orbits under airbag loading and 22 impacts on 11 human cadavers at velocities of 4.2–5 m/s. These include directional impacts to the maxilla and zygoma at 20- and 40-deg relative to the head coronal plane in the flexion/extension (pitch) axis for the head. Davis used acoustic sensing techniques to identify fracture initiation. These tests produced force at acoustic emission onset from 166 to 2,345 N and peak forces from 1,470 to 4,423 N. This study found that fracture initiation force generally increased with angle and developed an injury risk function for force to fracture initiation for direct impacts to the facial bones. Viano et al. studied zygoma impacts at velocities to 41 m/s from a rigid cylindrical impactor [38]. Peak forces ranged from 1,520 to 2,300 N, similar to those found by Allsop et al., Yoganandan et al., and Davis (at substantially lower impact velocities) [11, 56, 61].

Fractures of the thin sidewalls of the orbit, or orbital blowout fractures, are produced by two general mechanisms. The first, a buckling mechanism, fractures the interior orbital surface from impacts to the frontal bone, maxilla or zygoma. The second mechanism, or the hydraulic theory, arises when an impact to the globe indirectly loads the wall of the orbit. Green et al. reported tests on primates in vivo [62]. These impacts consistently produced hydraulic blowout fractures with input energies above 2 J. Fractures were associated with globe rupture in five of the 16 tests. Waterhouse et al. reported tests using 47 cadaver orbits [63]. They found that the hydraulic and buckling mechanisms result in different detailed fracture patterns. Hydraulic fractures were generally larger than buckling fractures, included the medial orbit wall or roof and herniated the orbital contents in contrast with buckling fractures. Warwar et al. also studied both blowout fracture mechanisms with impacts to 21 human cadaveric orbits [64]. They reported impactor energies between 29 and 127 mJ are necessary to fracture orbit floors by direct impact. These are consistent with theoretical models of buckling/hydraulic mechanisms predicting fracture energies from 68 to 71 mJ [64]. The large variation in fracture energies in the tests may be attributed to differences in local contact area and location of impact. In a clinical study, Burm et al. reported that medial orbital fractures occur twice as often as inferior fractures [65]. They also found that combined fractures occurred in about one-third of the observed orbital blowout fractures, suggesting impacts to both globe and orbital rim.

8.4.2.1 Mandible

There is a long history of biomechanics experimentation on the mandible. Messerer determined that approximately 2,000 N was required to fracture the subcondylar region from symphyseal loading directed through the line connecting the condyles [30]. He found that the rami spread approximately 1 cm from each other before fracture. Huelke and Patrick investigated the biomechanics of mandible impacts with strain gauges finding large tensile and compressive strains in the subcondylar regions from chin impacts [66]. Hodgson used a large area impactor to determine the blunt impact tolerance of the mandible in embalmed cadavers for impacts to the condyles [35]. Peak fracture forces ranged from 1,600 to 2,600 N and fracture injuries ranged from 35 to 65 J.

Owing to the embalming, it is not clear how the tolerance values translate to unembalmed tissue. Da Fonseca impacted the mandibles of 15 unembalmed cadavers using a large quadrangular club of 64 cm^2 area applied to the symphyseal region with three mandible configurations: symphyseal impact with mouth open and mouth closed and an impact aimed at the angle region [67]. Though no force values were reported, he found that the condylar/subcondylar region was more likely to be fractured in the mouth open configuration and that it is difficult to dislocate the condyle into the middle cranial fossa.

Nahum investigated the biomechanics of mandible fracture, assessing fracture in 18 impact experiments with male and female cadavers that were both embalmed and unembalmed [53]. The mandible fracture region is shown in Fig. 8.10 and is compared with other facial bone tolerance values proposed by Nahum. He found that both the AP and lateral mandible were generally stronger than the maxilla and zygoma for impacts with 6.5 cm^2 area, but were substantially weaker than the frontal bone. McElhaney et al. impacted the mandible of six unembalmed cadavers, generally producing condylar fractures and occasionally symphyseal fractures [49]. The mean fracture force of 5,270 ± 930 N is substantially larger than the mean tolerance of Nahum at 2,840 N, a differ-

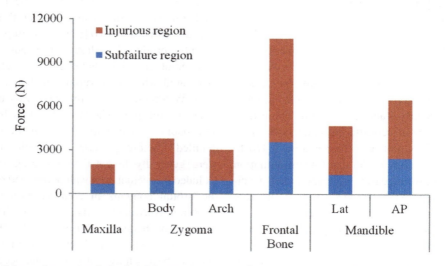

Fig. 8.10 Injury tolerance bands for facial fractures with 6.5 cm^2 impactor (Data from Nahum 1975)

ence that may be attributable to the large difference in contact loading area between the two studies [53]. Viano et al. studied high velocity impacts (42 ± 10 m/s) to the mandible in six unembalmed cadavers [38]. Subfracture biomechanical force displacement response of the mandible in 15 unembalmed cadavers was investigated by Craig et al. [68]. They found a bilinear stiffening response under direct symphyseal loading of 475 ± 200 kN/m to 600 N and 2,380 ± 500 kN/m from 600 to 3,250 N.

8.5 Injury Criteria

8.5.1 Skull (Calvarium) Fracture

The Society of Automotive Engineers specifies peak force data [69]. Peak linear acceleration has been adopted by agencies such as the Snell and Canadian standards for helmets [70, 71]. Peak linear acceleration and dwell times are suggested for motorcycle helmet standards by the US government Federal Motor Vehicle Safety Standard 218 (FMVSS 218). The Gadd severity index (Eq. 8.1) uses an integral of the head CG resultant acceleration response [72]. This index is used in football helmet standards (National Operating Committee on Standards for Athletic Equipment, NOCSAE, 1997). Other indices such as the rotational acceleration and head injury power have been proposed, although standards have not been promulgated [73].

The HIC replaced the severity index [74]. It is also based on the integral of the resultant acceleration at the CG of the head (Eq. 8.2) and remains the most widely used metric for frontal impacts around the world [75]. The criterion uses time-averaged weighted acceleration data and represents the kinetic energy transfer over a selected period. Based on the argument that HIC depends on the impacting boundary condition, another index, termed skull fracture correlate (SFC, Eq. 8.3), has been proposed recently for frontal impacts [76]. The skull fracture correlate uses the resultant acceleration at the center of gravity of the head in its determination [76, 77, 78].

$$SI = \int \left[a(t)\right]^{2.5} dt \quad (8.1)$$

$$HIC = Max \left[\frac{1}{(t_2 - t_1)} \int_{t_1}^{t_2} a(t) dt\right]^{2.5} (t_2 - t_1) \quad (8.2)$$

$$SFC = \left[\frac{\int_{t_1}^{t_2} a(t) dt}{(t_2 - t_1)}\right] \text{ or } \Delta V_{HIC} / \Delta t_{HIC} \quad (8.3)$$

Where, a(t) represents the resultant acceleration at the center of gravity of the head, and t_1 and t_2 denote the time interval that maximized the HIC criterion. The original limit for the HIC in the FMVSS-208 frontal impact standards was 1000; this has been changed in recent years to reflect advances in impact biomechanics (this topic is covered in a later chapter). Yoganandan and associates conducted temporo-parietal impact tests using isolated PMHS head specimens [76, 78, 79]. The biomechanical responses of the human head (translational CG accelerations, rotational accelerations and HIC) under lateral impact to the parietal-temporal region were investigated. Free drop tests were conducted at impact velocities ranging from 2.4 to 7.7 m/s with 40 and 90 durometer flat padding and 90 durometer cylinder targets. Specimens were isolated from PMHS subjects at the level of occipital condyles, and the intracranial contents were replaced with brain simulant. Three tri-axial accelerometers were used at the anterior, posterior and vertex of the specimen, and a pyramid nine accelerometer package (pNAP) was used at the contra-lateral site. Biomechanical responses were computed by transforming accelerations measured at each location to the head CG. The results indicated significant "hoop effect" from skull deformation. Translational head CG accelerations were accurately measured by transforming the pNAP, the vertex accelerations or the average of anterior/posterior acceleration to the CG. The material stiffness and structural rigidity of the padding changed the biomechanical responses of the head with stiffer padding resulting in higher head accelerations. The average fracture impact velocity was the highest (7.0 m/s) in 40D tests, followed by 90D-cylinder (5.64 m/s) and lowest (5.29 m/s) in 90D tests. Skull fractures occurred after HIC values reached 2,200, 2,996 and 2,353: resultant rotational head accelerations were 20.1, 31.1 and 34.1 krad/s/s in 40D, 90D and 90D-cylinder tests. At the same impact velocity before skull fracture, peak translational accelerations in the 90D tests were approximately two times greater than in the 40D tests and rotational accelerations were more than 2.5 times. Threshold of HIC for the temporo-parietal region was more than two-to-three times greater than the frontal bone, indicating differences between the two bones of the skull. Rotational head accelerations up to 42.1 krad/s/s were obtained from this data before skull fracture, indicating possible severe brain injury without skull fracture in lateral head impact. Combined analyses of frontal and lateral impact data from these and other PMHS tests in literature resulted in injury risk curves (Fig. 8.11) as estimated by the logistic regression [76, 77, 78]. A value of SFC=150 g corresponds to 42 % mean probability of fracture for frontal impact, 33 % for lateral impact and 37 % for the combined dataset.

8.5.1.1 Facial Fracture

Using acoustic data to identify the force at initiation of fracture, the data of Davis may be used to produce an injury risk function for facial fracture of the maxilla or zygoma [61]. This assessment is based on a survival analysis with a Weibull distribution treating the acoustic initiation data as statistically uncensored (cf. Kent 2004). In this form, the probability of initiation of fracture for the 13.8 cm² impactor is

$$P_{fracture}(F) = 1 - \exp\left[-\left(\frac{F}{B}\right)^\alpha\right]$$

where F is the input force in N, index alpha=2.27 is the Weibull shape parameter and B=887.7 N is the Weibull scale parameter (Fig. 8.12). The mean force for fracture initiation is 755±120 N. This injury risk function for fracture initiation is consistent with facial fracture data associated with peak impact force from Schneier et al., Nyquist et al., Yoganandan et al., Wellbourne et al., Rhee et al., and Davis [11, 24, 32, 55, 59, 61].

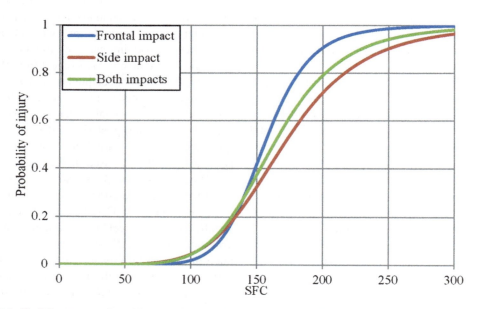

Fig. 8.11 Skull fracture correlate risk curves for frontal, side and combined frontal and side impacts (Data from Refs. [77, 78])

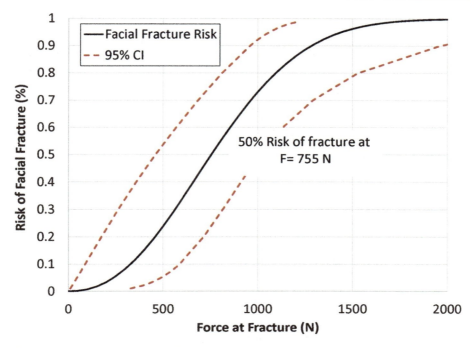

Fig. 8.12 Weibull injury risk function and 95 % confidence intervals for fractures to the maxilla and zygoma [61]

8.6 Summary

An attempt has been made to synthesize the biomechanics of skull and facial bones in this chapter. Anatomy, fractures and biomechanical metrics such as forces and injury risk functions are covered with a primary focus on human cadaver models. Automotive studies have been a major contributor to the current understanding of human tolerance in this area. This has been only possible through decades of research using human cadavers, embalmed and unembalmed, intact and isolated head specimens and using impacting surfaces ranging from rigid plates to circular diameter to steering wheels. Probability curves reported in this chapter, while not completely specific to the individual bones of the skull and face, have been used to derive injury thresholds such as HIC, GSI, and SFC and advance safety in vehicle and athletic-related environments. More studies are needed to further examine the role of skull and facial injury biomechanics in other environments such as those occurring in military events (example, underbody blast), and the present information should be of value in pursuing these types of goals.

Acknowledgments This material is the result of work supported with resources and the use of facilities at the Zablocki VA Medical Center, Milwaukee, Wisconsin and the Medical College of Wisconsin. Narayan Yoganandan is a part-time employee of the Zablocki VA Medical Center, Milwaukee, Wisconsin. Any views expressed in this chapter are those of the authors and not necessarily representative of the funding organizations.

References

1. Jager TE, Weiss HB, Coben JH, Pepe PE (2000) Traumatic brain injuries evaluated in U.S. emergency departments, 1992–1994. Acad Emerg Med 7(2): 134–140
2. Chee CP, Ali A (1991) Basal skull fractures. A prospective study of 100 consecutive admissions. ANZ J Surg 61(8):597–602
3. Liu-Shindo M, Hawkins DB (1989) Basilar skull fractures in children. Int J Pediatr Otorhinolaryngol 17(2): 109–117

4. Hwang K, You SH (2010) Analysis of facial bone fractures: an 11-year study of 2,094 patients. Indian J Plast Surg 43(1):42–48. doi:10.4103/0970-0358.63959
5. King RE, Scianna JM, Petruzzelli GJ (2004) Mandible fracture patterns: a suburban trauma center experience. Am J Otolaryngol 25(5):301–307
6. Haug RH, Prather J, Indresano AT (1990) An epidemiologic survey of facial fractures and concomitant injuries. J Oral Maxillofac Surg 48(9):926–932
7. Pappachan B, Alexander M (2006) Correlating facial fractures and cranial injuries. J Oral Maxillofac Surg 64(7):1023–1029. doi:10.1016/j.joms.2006.03.021
8. Boole JR, Holtel M, Amoroso P, Yore M (2001) 5196 mandible fractures among 4381 active duty army soldiers, 1980 to 1998. Laryngoscope 111(10):1691–1696. doi:10.1097/00005537-200110000-00004
9. Williams PL (1995) Gray's anatomy. Churchill Livingstone, New York
10. Agur AMR, Lee ML (1991) Grant's atlas of anatomy, 9th edn. Williams and Wilkins, Baltimore
11. Yoganandan N, Pintar FA, Sances AJ, Harris G, Chintapalli K, Myklebust JB, Schmaltz D, Reinartz J, Kalbfleisch J, Larson SJ (1989) Steering wheel induced facial trauma. SAE Trans 97(4):1104–1128
12. Yoganandan N, Sances A Jr, Pintar FA, Larson S, Hemmy D, Maiman D, Haughton V (1991) Traumatic facial injuries with steering wheel loading. J Trauma 31(5):699–710
13. Osborn JW (1996) Features of human jaw design which maximize the bite force. J Biomech 29(5): 589–595
14. Bellman MH, Babu KV (1978) Jaw dislocation during anaesthesia. Anaesthesia 33(9):844
15. Patel A (1979) Jaw dislocation during anaesthesia. Anaesthesia 34(4):376
16. Sosis M, Lazar S (1987) Jaw dislocation during general anaesthesia. Can J Anaesth 34(4):407–408. doi:10.1007/BF03010145
17. Hellsing G (1980) On the regulation of interincisor bite force in man. J Oral Rehabil 7(5):403–411
18. Paphangkorakit J, Osborn JW (1997) The effect of pressure on a maximum incisal bite force in man. Arch Oral Biol 42(1):11–17
19. AIS (2005) The abbreviated injury scale. American Association for Automotive Medicine, Arlington Heights
20. Baker SP, O'Neill B, Haddon W Jr, Long WB (1974) The injury severity score: a method for describing patients with multiple injuries and evaluating emergency care. J Trauma 14(3):187–196
21. LeFort R (1901) Etude Experimental sur les Fractures de la Machoire Superieure. Revue di Chirurgie (Paris) 23:208–227
22. Zingg M, Laedrach K, Chen J, Chowdhury K, Vuillemin T, Sutter F, Raveh J (1992) Classification and treatment of zygomatic fractures: a review of 1,025 cases. J Oral Maxillofac Surg 50(8):778–790
23. Donat TL, Endress C, Mathog RH (1998) Facial fracture classification according to skeletal support mechanisms. Arch Otolaryngol Head Neck Surg 124(12): 1306–1314
24. Rhee JS, Posey L, Yoganandan N, Pintar F (2001) Experimental trauma to the malar eminence: fracture biomechanics and injury patterns. Otolaryngol Head Neck Surg 125(4):351–355. doi:10.1067/mhn.2001. 118692
25. Joos U, Meyer U, Tkotz T, Weingart D (1999) Use of a mandibular fracture score to predict the development of complications. Int J Oral Maxillofac Surg 57(1): 2–5; discussion 5–7
26. Zhang J, Zhang Y, El-Maaytah M, Ma L, Liu L, Zhou LD (2006) Maxillofacial injury severity score: proposal of a new scoring system. Int J Oral Maxillofac Surg 35(2):109–114. doi:10.1016/j.ijom.2005.06.019
27. Shetty V, Atchison K, Der-Matirosian C, Wang J, Belin TR (2007) The mandible injury severity score: development and validity. Int J Oral Maxillofac Surg 65(4):663–670. doi:10.1016/j.joms.2006.03.051
28. Gurdjian ES, Webster JE, Lissner HR (1949) Studies on skull fracture with particular reference to engineering factors. Am J Surg 78:736–742
29. Gurdjian ES, Webster JE (1946) Deformation of the skull in head injury studied by stresscoat technique. Surg Gynecol Obstet 83:219–233
30. Messerer O (1880) Uber Elasticitat und Festigkeit der Menschlichen Knochen. Cotta, Stuttgart
31. Nahum A, Gatts J, Gadd C, Danforth J (1968) Impact tolerance of the skull and face. In: Stapp car crash conference, 22–23 Oct 1968, Detroit, p 302
32. Schneider DC, Nahum AM (1972) Impact studies of facial bones and skull. In: Stapp car crash conference, 1972, Detroit, p 186
33. Hodgson V, Thomas L (1973) Breaking strength of the human skull vs impact surface curvature, DOT-HS-801-002. Washington, DC
34. Hodgson VR, Thomas LM (1971) Breaking strength of the human skull vs. impact surface curvature. National Technical Information Service Accession Number 00224104.
35. Hodgson VR (1967) Tolerance of the facial bones to impact. Am J Anat 120:113–122
36. Yoganandan N, Pintar F (2004) Biomechanics of temporo-parietal skull fracture. Clin Biomech 19(3): 225–239
37. McElhaney JH, Stalnaker RL, Roberts VL (1972) Biomechanical aspects of head injury. In: King W, Mertz H (eds) Human impact injury tolerance. Plenum, New York, pp 85–112
38. Viano DC, Bir C, Walilko T, Sherman D (2004) Ballistic impact to the forehead, zygoma, and mandible: comparison of human and frangible dummy face biomechanics. J Trauma 56(6):1305–1311
39. Stalnaker R, Melvin J, Nuscholtz G, Alem N, Benson J (1977) Head impact response. In: Stapp car crash conference, New Orleans, 19–21 Oct 1977, pp 305–335
40. Got C, Patel A, Fayon A, Tarriere C, Walfisch G (1978) Results of experimental head impacts on cadavers: the various data obtained and their relations to some measured physical paramenters. In: Stapp car crash conference, Ann Arbor, 24–26 Oct 1978

41. Got C, Guillon F, Patel A, Mack P, Brun-Cassan F, Fayon A, Tarriere C, Hureau J (1983) Morphological and biomechanical study of 126 human skulls used in experimental impacts, in realtion with the observed injuries. In: Stapp car crash conference, Bron
42. Nahum A, Ward C, Raasch F, Adams S, Schneider D (1980) Experimental studies of side impact to the human head. In: Stapp car crash conference, Troy, 15–17 Oct 1980
43. Allsop D, Perl T, Warner C (1991) Force/deflection and fracture characteristics of the temporo-parietal region of the human head. In: Stapp Car Crash conference, San Diego, 18–20 Nov 1991
44. Yoganandan N, Pintar FA, Reinartz J, Sances A Jr (1993) Human facial tolerance to steering wheel impact: a biomechanical study. J Saf Res 24(2):77–85
45. Yoganandan N, Pintar FA, Sances A Jr (1991) Biodynamics of steering wheel induced facial trauma. J Saf Res 22:179–190
46. McIntosh AS, Svensson NL, Kallieris D, Mattern R, Krabbel G, Ikels K (1996) Head impact tolerance in side impacts. In: Enhanced safety of vehicles, Melbourne
47. Reynolds HM, Clauser CE, McConnville J, Chandler R, Young JW (1975) Mass distribution properties of the male cadaver. In: Society of automotive engineers, 19–21 Oct 1975, pp 305–335. SAE Technical Paper 750424, doi:4271/750424
48. Yoganandan N, Pintar FA, Sances AJ, Walsh PR, Ewing CL, Thomas DL, Snyder RG (1995) Biomechanics of skull fracture. J Neurotrauma 12(4):659–668
49. McElhaney JH, Hopper RH Jr, Nightingale RW, Myers BS (1995) Mechanisms of basilar skull fracture. J Neurotrauma 12(4):669–678
50. Cooter RD, David DJ (1990) Motorcyclist craniofacial injury patterns. In: International motorcycle safety conference, Orlando, FL
51. Society of Automotive Engineers (1986) Vehicle occupant restraint systems and components standards manual. Society of Automotive Engineers, Warrrendale
52. Nahum AM (1975) The biomechanics of facial bone fracture. Laryngoscope 85(1):140–156. doi:10.1288/00005537-197501000-00011
53. Nahum AM (1975) The biomechanics of maxillofacial trauma. Clin Plast Surg 2(1):59–64
54. Karlan MS, Cassisi NJ (1979) Fractures of the zygoma. A geometric, biomechanical, and surgical analysis. Arch Otolaryngol 105(6):320–327
55. Nyquist G, Cavanaugh J, Goldberg S, King A (1986) Facial impact tolerance and response, SAE paper number 861896. In: 30th Stapp car crash conference, SAE Technical Paper 861906. doi:4271/86196
56. Allsop D, Warner C, Mille M, Schneider D, Nahum A (1988) Facial impact repsonse – a comparison of the hybrid III dummy and human cadaver. In: Stapp car crash conference, Atlanta, 17–19 Oct 1988
57. Yoganandan N, Haffner M, Pintar FA (1996) Facial injury: a review of biomechanical studies and test procedures for facial injury assessment. J Biomech 29(7):985–986
58. Bisplinghoff J, Cormier J, Duma S, Kennedy E, Depinet P, Brozoski F (2008) Development and validation of eye injury and facial fracture criteria for the focus headform. In: Proceedings of the 26th army science conference, Orlando, 1–4 Dec 2008
59. Welbourne E, Ramet M, Zarebski M (1989) A comparison of human facial fracture tolerance with the performance of a surrogate test device. In: 12th experimental safety vehicles conference, Göteborg, Sweden
60. Garza JR, Baratta RV, Odinet K, Metzinger S, Bailey D, Best R, Whitworth R, Trail ML (1993) Impact tolerances of the rigidly fixated maxillofacial skeleton. Ann Plast Surg 30(3):212–216
61. Davis M (2002) Facial fracture and eye injury tolerance from night vision Goggle loading. University of Virginia, Charlottesville
62. Green RP Jr, Peters DR, Shore JW, Fanton JW, Davis H (1990) Force necessary to fracture the orbital floor. Ophthal Plast Reconstr Surg 6(3):211–217
63. Waterhouse N, Lyne J, Urdang M, Garey L (1999) An investigation into the mechanism of orbital blowout fractures. Br J Plast Surg 52(8):607–612. doi:10.1054/bjps.1999.3194
64. Warwar RE, Bullock JD, Ballal DR, Ballal RD (2000) Mechanisms of orbital floor fractures: a clinical, experimental, and theoretical study. Ophthal Plast Reconstr Surg 16(3):188–200
65. Burm JS, Chung CH, Oh SJ (1999) Pure orbital blowout fracture: new concepts and importance of medial orbital blowout fracture. Plast Reconstr Surg 103(7):1839–1849
66. Huelke DF, Patrick LM (1964) Mechanics in the production of mandibular fractures: strain-gauge measurements of impacts to the chin. J Dent Res 43:437–446
67. da Fonseca GD (1974) Experimental study on fractures of the mandibular condylar process (mandibular condylar process fractures). Int J Oral Surg 3(3):89–101
68. Craig M, Bir C, Viano D, Tashman S (2008) Biomechanical response of the human mandible to impacts of the chin. J Biomech 41(14):2972–2980. doi:10.1016/j.jbiomech.2008.07.020
69. Society of Automotive Engineers (1980) Human tolerance to impact conditions as related to motor vehcile design-J885. SAE, Warrendale
70. Canadian Standards Association (1985) Standard CAN3-D230-M85, protective headgear in motor vehicle applications
71. Snell (1995) Snell Standard for protective headgear for use in bicycles
72. Gadd CW (1968) Use of a weighted-impulse criterion for estimating injury hazard In: Stapp car crash conference, New York, Nov 1968, pp 164–174
73. Newman JA, Shewchenko N, Welbourne E (2000) A proposed new biomechanical head injury assessment function – the maximum power index. Stapp Car Crash J 44:215–247

74. Versace J (1971) A review of the severity index. In: Stapp car crash conference, Coronado, Nov 1971, pp 771–796
75. NHTSA (2002) Code of federal regulations, title 49, part 571. FMVSS – Federal Motor Vehicle Safety Standards. National Highway Traffic Safety Administration, Washington, DC
76. Vander Vorst M, Stuhmiller J, Ho K, Yoganandan N, Pintar F (2003) Statistically and biomechanically based criterion for impact-induced skull fracture. Annu Proc Assoc Adv Automot Med 47: 363–381
77. Chan P, Lu X, Rigby P, Zhnag J, Yoganandan N, Pintar FA (2007) Development of a generalized linear skull fracture criterion. In: ESV, 2007, Lyon, France
78. Vander Vorst M, Chan P, Zhang J, Yoganandan N, Pintar F (2004) A new biomechanically-based criterion for lateral skull fracture. Annu Proc Assoc Adv Automot Med 48:181–195
79. Zhang J, Yoganandan N, Pintar FA (2009) Dynamic biomechanics of the human head in lateral impacts. Annu Proc Assoc Adv Automot Med Sci Conf 53: 249–256
80. Greene, D., Raven, R., Carvalho, G., & Maas, C. S. (1997). Epidemiology of facial injury in blunt assault: determinants of incidence and outcome in 802 patients. Archives of Otolaryngology–Head & Neck Surgery, 123(9), 923–928
81. Kent, R. W., & Funk, J. R. (2004). Data Censoring and Parametric Distribution Assignment in the Development of Injury Risk Functions from Biochemical Data(No. 2004-01-0317). SAE Technical Paper
82. Gurdjian, E. S., & Webster, J. E. (1943). Experimental head injury with special reference to the mechanical factors in acute trauma. Surg. Gynec. Obstet, 76, 623–634
83. Hodgson, V. R., & Thomas, L. M. (1972). Breaking strength of the human skull vs. impact surface curvature (No. Final Rept), National Technical Information Service Accession Number 00224104

Biomechanics of Brain Injury: A Historical Perspective

John W. Melvin* and Narayan Yoganandan

Abstract

The brain may be the organ most critical to protect from trauma, because anatomic injuries to its structures are currently nonreversible, and consequences of injury can be devastating. The experimental study of brain-injury mechanisms is unparalleled, because effects of trauma to the organs responsible for control and function of the body are the objects of the study. Injury of the central nervous system results not only from local primary effects, but from effects on physiologic homeostasis that may lead to secondary injury. The brain controls the flow of information, including autonomic control as well as sensory perception and motor function. The brain is the source of intentional actions, and it functions in time to store, retrieve, and process information. The state of self-awareness or consciousness is the highest level of brain function in man.

This chapter addresses the historical state of knowledge in the biomechanics of brain injury. It is organized to provide a summary of head and brain anatomy and of clinical head-injury issues, followed by sections on experimental brain-injury models and bio-mechanical mechanisms of brain-injury and head-injury criteria. It serves as a preface to the following chapter on the future of brain injury biomechanics.

9.1 Anatomy of the Head

This section is a brief summary of head anatomy, starting with the scalp and skull, followed by more detailed descriptions of the structures internal to the skull. It is intended to provide a basic description for use in the following sections of the chapter. More complete descriptions can be found in standard anatomy texts.

J.W. Melvin, Ph.D. *Deceased

N. Yoganandan, Ph.D. (✉)
Department of Neurosurgery, Medical College of Wisconsin, Milwaukee, WI, USA
e-mail: yoga@mcw.edu

9.1.1 Scalp

The scalp is 5–7 mm (0.20–0.28 in.) thick and consists of three layers: the hair-bearing skin (cutaneous layer), a subcutaneous connective-tissue layer, and a muscle and fascial layer. Beneath the scalp there is a loose connective-tissue layer plus the fibrous membrane that covers bone (periosteum). The thickness, firmness, and mobility of the outer three layers of scalp as well as the rounded contour of the cranium function as protective features. When a traction force is applied to the scalp, its outer three layers moved together as one.

9.1.2 Skull

The skull is the most complex structure of the skeleton. This bony network is neatly molded around and fitted to the brain, eyes, ears, nose, and teeth. The thickness of the skull varies between 4 and 7 mm (0.16 and 0.28 in.) to snugly accommodate and provide protection to these components. The skull is composed of eight bones that form the brain case, 14 bones that form the face, as well as the teeth. Excluding the face, the cranial vault is formed by the ethmoid, sphenoid, frontal, two temporal, two parietal, and occipital bones. The inner surface of the cranial vault is concave and relatively smooth. The base of the braincase is an irregular plate of bone containing depressions and ridges plus small holes (foramen) for arteries, veins, and nerves, as well as the large hold (the foramen magnum) that is the transition area between the spinal cord and the brainstem.

9.1.3 Meninges

Three membranes know as the meninges protect and support the brain and spinal cord. One function of the meninges is to isolate the brain and spinal cord from the surrounding bones. The meninges consist primarily of connective tissue, and they also form part of the walls of blood vessels and the sheaths of nerves as they enter the brain and as they emerge from the skull.

The meninges consist of three layers: the dura mater, the arachnoid, and the pia mater. The dura mater is a tough, fibrous membrane that surrounds the spinal cord and, in the skull, is divided into two layers. The outer cranial, or periosteal layer, lines the inner bony surface of the calvarium. The inner layer, or meningeal layer, covers the brain. In the braincase, the two layers of dura mater are fused except where they separate to form venous sinuses, which drain blood from the brain. Folds of the meningeal layer form the falx cerebri, which projects into the longitudinal fissure between the right and left cerebral hemispheres; and the tentorium cerebelli, a shelf on which the posterior cerebral hemispheres are supported.

The arachnoid mater is a delicate spider-web-like membrane that occupies the narrow subdural space. The pia mater is a thin membrane of fine connective tissue invested with numerous small blood vessels. It is separated from the arachnoid by the subarachnoid space. The pia mater covers the surface of the brain, dipping well into its fissures.

The subarachnoid space and the ventricles of the brain are filled with a colorless fluid (cerebrospinal fluid, or CSF), that provides some nutrients for the brain and cushions the brain from mechanical shock. Cerebrospinal fluid is formed in the lateral and third ventricles by the choroid plexus and passes via the cerebral aqueduct into the fourth ventricle. From this site the fluid circulates in the sub-arachnoid spaces surrounding both the brain and the spinal cord. The majority of the CSF is passively returned to the venous system via the arachnoid villi. The specific gravity of cerebrospinal fluid is about 1.008 in the adult, which is approximately that of blood plasma. About 140 ml of CSF constantly circulates and surrounds the brain on all sides, so it serves as a buffer and helps to support the brain's weight. Since the subarachnoid space of the brain is directly continuous with that of the spinal cord, the spinal cord is suspended in a tube of CSF. For normal movement, a shrinkage or expansion of the brain is quickly balanced by an increase or decrease of CSF.

9.1.4 Central Nervous System

The central nervous system (CNS) consists of the brain and the spinal cord. At a microscopic level, the CNS is primarily a network of neurons and supportive tissue functionally arranged into areas that are gray or white in color. Gray matter is composed primarily of nerve-cell bodies concentrated in locations on the surface of the brain and deep within the brain; white matter is composed of myelinated nerve-cell processes (axons) that largely form tracts to connect parts of the central nervous system to each other.

The brain can be divided structurally and functionally into five parts: cerebrum, cerebellum, midbrain, pons and medulla oblongata. In addition, it has four ventricles (CSF cisterns with exits), three membranes (meninges), two glands (pituitary and pineal), 12 pairs of cranial nerves, and the cranial arteries and veins. The average length of the brain is about 165 mm (6.5 in.), and its greatest transverse diameter is about 140 mm (5.5 in.). Because of size differences, its average weight is 1.36 kg (3.0 lb) for the male and a little less for the female. The specific gravity of the brain averages 1.036, and it is gelatinous in consistency. The brain constitutes 98 % of the weight of the central nervous system and represents about 2 % of the weight of the body.

9.1.5 Cerebrum

The cerebrum makes up seven-eighths of the brain and is divided into right and left cerebral hemispheres. These are incompletely separated by a deep midline cleft called the longitudinal cerebral fissure. The falx cerebelli process projects downward into this fissure. Beneath the longitudinal cerebral fissure, the two cerebral hemispheres are connected by a mass of white matter called the corpus callosum. Within each cerebral hemisphere is a cistern for cerebrospinal fluid called the lateral ventricle. The surface of each hemisphere is composed of gray matter and referred to as the cerebral cortex. In man, the cerebral cortex is arranged in a series of folds or convolutions. The ridge of the fold is referred to as a gyrus whereas the valley of the fold is called a sulcus. Each cerebral hemisphere is further subdivided into four lobes by fissures, each lobe being named by its association to the nearest cranial bone. Thus, the four lobes are the frontal, parietal, temporal, and occipital lobes.

The interior of each cerebral hemisphere is composed of white matter arranged in tracts that serve to connect one part of cerebral hemisphere with another, to connect the cerebral hemispheres to each other, and to connect the cerebral hemispheres to the other parts of the central nervous system. In addition, within these interior areas of white matter are concentrations of gray matter called nuclei.

9.1.6 Midbrain

The midbrain connects the cerebral hemispheres above to the pons below. Anteriorly the midbrain is composed of two stalks that consist of fibers passing to and from the cerebral hemispheres above. The midbrain also contains gray matter nuclei. Within the midbrain is a narrow canal, the cerebral aqueduct, which connects to the third ventricle above the fourth ventricle below.

9.1.7 Pons

The pons lies below the midbrain, in front of the cerebellum, and above the medulla oblongata. It is composed of white matter nerve fibers connecting the cerebellar hemispheres. Lying deep within its white matter are areas of gray matter that are nuclei for some of the cranial nerves.

9.1.8 Medulla Oblongata

The medulla oblongata appears continuous with the pons above and the spinal cord below. In the lower part of the medulla oblongata, motor fibers cross from one side to the other so that fibers from the right cerebral cortex pass to the left side of the body. Some sensory fibers passing upward toward the cerebral cortex also cross from one side to the other in the medulla oblongata. The medulla oblongata also contains areas of gray matter within its white matter. These are nuclei for cranial nerves and relay stations for sensory fibers passing upward from the spinal cord.

9.1.9 Cerebellum

The cerebellum lies behind the pons and the medulla oblongata. Its two hemispheres are joined at the midline by a narrow, striplike structure called the vermis. The outer cortex of the cerebellar hemispheres is gray matter; the inner cortex is white matter. The outer surface of the cerebellum forms into narrow folds separated by deep fissures. Nerve fibers enter the cerebellum in three pairs of stalks that connect the cerebellar hemispheres to the midbrain, pons, and medulla oblongata.

9.2 Traumatic Brain Injury from Clinical Experience

9.2.1 Introduction

Data compiled by the Department of Health and Human Services (1989) indicate that someone receives a head injury every 15 s in the United States. This places the total number of injuries at over two million per year with half a million severe enough to require hospital admission. This study further indicates that 75,000–100,000 die each year as a result of traumatic brain injury, and it is the leading cause of death and disability in children and young adults.

9.2.2 Types of Brain Injuries

Clinically brain injuries can be classified in two broad categories: diffuse injuries and focal injuries. The diffuse injuries consist of brain swelling, concussion, and diffuse axonal injury (DAI). The focal injuries consist of epidural hematomas (EDH), subdural hematomas (SDH), intracerebral hematomas (ICH), and contusions (coup and contrecoup). Studies have attempted to describe the incidence and sequelae of the above brain injuries [1–3]. The results of these studies vary somewhat, however, mainly due to the criterion of selection of the patients in the studies. For example, the patients in the Chapon, et al. study were all critically injured with immediate unconsciousness without a lucid interval after accidents involving automobiles (occupants and pedestrian) and two-wheelers [1]. Only 7 % of their sample contained injuries from falls and sports. Gennarelli's study contained 48 % patients whose injuries were due to falls and assaults while the other 52 % were from automobile accidents (occupants and pedestrians) and data indicated that the average injury was serious to severe [2]. This difference in the population considered may explain the different results in the two series, since Gennarelli's study shows a marked difference in the type of brain injuries sustained by automobile accident victims as compared to assault and fall victims. In this study, it was found that three out of four automobile/pedestrian brain injuries were of the diffuse type, and one out of four was of the focal type. The assault and fall victims had one out of four diffuse injuries and three out of four focal injuries.

Gennarelli's study points out that acute subdural hematoma and diffuse axonal injury were the two most important causes of death. These two lesions together accounted for more head injury deaths than all other lesions combined. The injuries most often associated with a good or moderate recovery were cerebral concussions and cortical contusion.

9.2.3 Diffuse Injuries

Diffuse brain injuries form a spectrum of injuries ranging from mild concussion to diffuse white matter injuries. In the mildest forms, there is mainly physiological disruption of brain function and, at the most severe end, physiological and anatomical disruptions of the brain occur.

Mild concussion does not involve loss of consciousness. Confusion, disorientation, and a brief duration of post-traumatic and retrograde amnesia may be present. This is the most common form of diffuse brain injury, it is completely reversible, and due to its mildness it may not be brought to medical attention. According to Scott, minor concussions form approximately 10 % of all minor-to-serious injuries involving the brain, skull, and spinal column [4].

The classical cerebral concussion involves the immediate loss of consciousness following injury. Clinically, the loss of consciousness should be less than 24 h and is reversible. Post-traumatic and retrograde amnesia is present, and the duration of amnesia is a good indicator of the severity of concussion. Thirty-six percent of these cases involve no lesions of the brain. The remainder may be associated with cortical contusions (10 %), vault fracture (10 %), basilar fracture (7 %), depressed fracture (3 %), and multiple lesions (36 %) [5]. Hence, the clinical outcome of patients with this type of symptom depends on the associated brain injuries. In general, 95 % of the patients have good recovery at the end of 1 month. Close to 2 % of the patients might have severe deficit, and 2 % might have moderate deficit [5].

The injury considered to be the transition between pure physiological dysfunction of the brain to anatomical disruption of the brain generally involves immediate loss of consciousness lasting for over 24 h. This injury is also called diffuse brain injury. It involves occasional decerebrate posturing, amnesia lasting for days, mild-to-moderate memory deficit, and mild motor deficits. At the end of 1 month, only 21 % of these cases have good recovery. Fifty percent of the cases end up with moderate to severe deficits, 21 % of the cases have vegetative survival, and 7 % are fatal [5]. The incidence of diffuse injury among severely injured patients is 13 % [5] to 22 % [1].

Diffuse axonal injury (DAI) is associated with mechanical disruption of many axons in the cerebral hemispheres and subcortical white matter. Axonal disruptions extend below the midbrain and into the brainstem to a variable degree [6]. These are differentiated from the less-severe diffuse injuries by the presence and persistence of abnormal brainstem signs. Microscopic examination of the brain discloses axonal tearing throughout the white matter of both cerebral hemispheres. It also involves degeneration of long white-matter tracts extending into the brainstem. High-resolution CT scans may show small hemorrhages of the corpus callosum, superior cerebellar peduncle, or periventricular region. These hemorrhages are quite small and may often be missed on CT scans.

Diffuse axonal injury involves immediate loss of consciousness lasting for days to weeks. Decerebrate posturing, with severe memory and motor deficits, is present. Post-traumatic amnesia may last for weeks. At the end of 1 month, 55 % of the patients are likely to have died, 3 % might have vegetative survival, and 9 % would have severe deficit [5].

Brain swelling, or an increase in intravascular blood within the brain, may be superimposed on diffuse brain injuries, adding to the effects of the primary injury by increased intracranial pressure [5]. Penn and Clasen report from 4 % to 16 % of all head-injured patients and 28 % of pediatric head-injured patients have brain swelling [7]. Tarriere has reported the incidence of edema in 21 % of CT-scanned head-injured subjects [3]. It should be noted that in general brain swelling and edema are not the same but are often used interchangeably in literature. Cerebral edema is a special situation in which the brain substance is expanded because of an increase in tissue fluid [7]. The course of treatment for the two may be different. According to Penn and Clasen, the mortality rate due to brain swelling among adults is 33–50 % and 6 % in children [7].

9.2.4 Focal Injuries

Acute subdural hematoma (ASDH) has three sources: direct lacerations of cortical veins and arteries by penetrating wounds, large-contusion bleeding into the subdural space, and tearing of veins that bridge the subdural space as they travel from the brain's surface to the various dural sinuses. The last mechanism is the most common in the production of ASHD [5]. Gennarelli and Thibault report a high incidence (30 %) of ASDH among severely head-injured patients, with a 60 % mortality rate. From various earlier studies, Cooper reports the incidence rates of ASDH to be between 5 % and 13 % of all severe head injuries [8]. According to Cooper, acute subdural hematomas generally coexist with severe injury to the cerebral parenchyma, leading to poorer outcome when compared to chronic subdural and extradural hematomas that generally do not coexist with injuries to the cerebral parenchyma. The mortality rate in most studies is greater than 30 % and greater than 50 % in some.

Extradural hematoma (EDH) is an infrequently occurring sequel to head trauma (0.2–6 %). It occurs as a result of trauma to the skull and the underlying meningeal vessels and is not due to brain injury [8]. Usually skull fracture is found, but EDH may occur in the absence of fracture. From 50 % to 68 % of the patients have no significant intracranial pathology [8]. The remainder of the patients may have subdural hematoma and cerebral contusions associated with the EDH. These associated lesions influence the outcome of the EDH. The mortality rate from various studies ranges from 15 % to 43 %. This rate is greatly influenced by age, presence of intradural lesions, the time from injury to the appearance of symptoms, level of consciousness, and neurological deficit [8].

Cerebral contusion is the most frequently found lesion following head injury. It consists of heterogeneous areas of necrosis, pulping, infarction, hemorrhage, and edema. Some studies have shown the occurrence of contusions in 89 % of the brains examined post-mortem [8]. CT-scan studies have shown incidence rates from 21 % to 40 %. A recent CT-scan study by Tarriere shows an incidence of 31 % for contusions alone and 55 % associated with other lesions [3]. In contrast, Gennarelli's study [5] shows an incidence of contusions in only 13 % of the patients studied.

Contusions generally occur at the site of impact (coup contusions) and at remote sites from the impact (contrecoup contusions). The contrecoup lesions are more significant than the coup lesions [8]. They occur predominantly at the frontal and temporal poles, which are impacted against the irregular bony floor of the frontal and middle fossae. Contusions of the corpus callosum and basal ganglia have also been reported [8]. Contusions are most often multiple and are frequently associated with other lesions (cerebral hemorrhage, SDH and EDH). Contusions are frequently associated with skull fracture (60–80 %). In some cases they appear to be more severe when a fracture is present than when it is absent [8]. Mortality from contusions is reported to range from 25 % to 60 % [8]. Adults over 50 years of age fare worse than children. Patients in coma have a generally poor outcome.

Intracerebral hematomas (ICH) are well-defined homogeneous collections of blood within the cerebral parenchyma and can be distinguished from hemorrhagic contusions (mixture of blood and contused and edematous cerebral parenchyma) by CT scans. They are most commonly caused by sudden acceleration/deceleration of the head. Other causes are penetrating wounds and blows to the head producing depressed fractures below which ICH develop. Hemorrhages begin superficially and extend deeply into the white matter. In one-third of the cases they extend as far as the lateral ventricle. Some cases of the hematomas extending into the corpus callosum and the brainstem have been reported [8].

According to Cooper, the incidence of ICH has been underestimated in the past. With the advent of the CT scan, better estimates are now available [8]. Recent studies show the incidence to be between 4 % and 23 %. Gennarelli shows an incidence rate of 4 % and Chapon et al. (1983) show an incidence rate of 8 % in severely injured patients [1, 5].

Depending on the study, the mortality for traumatic ICH has been as high as 72 % and as low as 6 %. The outcome is affected considerably by the presence or absence of consciousness of the patient [8].

9.2.5 Skull Fractures

Clinically the presence or absence of linear skull fracture does not have much significance for the course of brain injury, although the subject is still controversial. This controversy continues because some studies show that, when dangerous complications develop after an initial mild injury, they are associated with skull fracture [9]. According to Cooper, failure to detect fractures would have little effect on the management of the patient [10]. Cooper further cites a study in which half the patients with extradural hematomas had skull fracture and half did not. Another study showed the occurrence of subdural hematomas in twice the number of patients without skull fracture as in those with skull fractures.

Gennarelli found the incidence of fracture to be similar across all severities of brain injuries in 434 patients [11]. There was a slight tendency for the severity of fracture to increase as the severity of brain injuries increased. However, this tendency was not statistically significant. Similar conclusions were drawn by Chapon et al. [1]. Cooper (1982b) states that there is no consistent correlation between simple linear skull fractures and neural injury [10].

Comminuted skull fractures result from severe impacts and are likely to be associated with neural injury. Cooper further states that clinically depressed fractures are significant where bone fragments are depressed to a depth greater than the thickness of the skull. The incidence of depressed skull fracture is 20 per 1,000,000 persons per year, with an associated mortality rate of 11 % related more to the central nervous system (CNS) injury than the depression itself. From 5 % to 7 % have coexisting, intracranial hematoma and 12 % have involvement of an underlying venous sinus [10]. Clinically, basal skull fractures are significant because the dura may be torn adjacent to the fracture site, placing the CNS in contact with the contaminated paranasal sinuses. The patient will be predisposed to meningitis [12].

9.3 Experimental Brain-Injury Models

Mechanical injury to the brain can occur by a variety of mechanisms. General categories include (1) direct contusion of the brain from skull deformation or fracture; (2) brain contusion from movement against rough interior surfaces of the skull; (3) reduced blood flow due to infarction or pressure; (4) indirect (contrecoup) contusion of the brain opposite the side of the impact; (5) tissue stresses produced by motion of the brain hemispheres relative to the skull and each other; and (6) subdural hematoma produced by rupture of bridging vessels between the brain and dura mater.

The latter three mechanisms are hypothesized to be involved in both impact and nonimpact (inertial acceleration) head injury [13].

Regardless of the experimental-outcome variable to be studied by the neurotraumatologist, the model chose by the researcher will produce the injury by mimicking one or more of the mechanisms described earlier. Experimental models of brain injury have used different techniques and a variety of species, with the common objective to develop a reproducible model of trauma that exhibits anatomical, physiological, and functional responses similar to those described clinically. Despite these efforts, there is not a single "ideal" experimental injury model; rather the goals and objectives of the research must dictate what is required of the model.

To this end there exists a variety of models. However, there is no entirely satisfactory experimental model that succeeds in producing the complete spectrum of brain injury seen clinically and yet is sufficiently well controlled and quantifiable to be a useful model for experimental studies. Experimental injury models that can be characterized biomechanically are required for physical and analytical modeling of tissue deformation. These modeling efforts will be useful for correlation of experimental results and injury response with human-injury initiation and response. Assuming that neural and vascular tissues are injured by induced local stresses and strains, the technique used to load or deform the neural tissue at the contact surface is unimportant if the pattern of injury seen is comparable to that observed in clinical brain injury. One can argue that since the structural changes, such as contusion, hemorrhage, or axonal injury, are distributed in patterns similar to those observed clinically, the local mechanics of the injury at a tissue level are comparable even if the gross-level input is different [14, 15].

This section will review specific experimental models of brain injury developed to study biomechanical and physiological mechanisms of clinical CNS trauma. The experimental brain injury models will be divided into three general categories: (1) head impact; (2) head acceleration; and (3) direct brain deformation. Examples will be reviewed and discussed for each model category. Finally, the use, value, and appropriate validation of physical and analytical models will be discussed in relation to development of predictive criteria for functional neural injury in humans. The section focuses on experimental models of nonpenetrating mechanical brain injury and will discuss the appropriateness and applications of those models for investigation of injury biomechanics, pathophysiology, and histopathology.

9.3.1 Method Development

The development and evolution of any experimental model of traumatic neural injury depends in part on the investigator's objectives. Differing scientific objectives and goals may result in different, but equally appropriate, models. A goal of injury prevention through improved understanding of injury biomechanics poses somewhat different constraints and requirements for the design of the model than, for example, testing efficacy of pharmacologic interventions on specific physiologic aspects of the injury response. However, all models must mimic to some extent the mechanical mechanisms of brain trauma observed in real world trauma to assure comparability of mechanisms. Despite differing experimental objectives, there are certain requirements that are common to any successful experimental model for traumatic brain injury: (1) The mechanical input used to produce the injury must be controlled, reproducible and quantifiable; (2) The resultant injury must be reproducible and quantifiable and produce functional deficit and neuropathology that emulate human brain injury; (3) The injury outcome, whether specified in anatomical, physiological, or behavioral terms, must cover a range of injury severities; (4) The level of mechanical input used to produce the injury should be predictive of the outcome severity.

The primary objective for any experimental model of brain injury is to create specific pathophysiologic outcomes in a reproducible manner. The importance of reproducible, intermediate-severity injury levels should be emphasized, because valuable data are likely to be found at transition- or threshold-level regions. Whether one is evaluating the relationship of injury biomechanics to outcome severity or the efficacy of a particular treatment regimen, the range of outcome modification due to altered mechanical input or treatment is likely to be greater at the intermediate-severity level than at either supramaximal or trivial injury-severity levels.

Although the mechanical input to the brain mimics, at least in part, the biomechanics of human trauma, some simplifications are helpful and indeed necessary to enable interpretation of mechanisms of injury at a tissue level. For example, direct brain deformation reproduces only some of the dynamics of closed head-impact trauma, but can be characterized biomechanically. Provided that the injury technique is designed to be relevant to hypothesized biomechanics and pathophysiology or clinical injury, the critical factor becomes the production of functional and anatomic sequelae comparable to those observed in human neurologic injury.

A simplified model, once characterized biomechanically and physiologically, can be combined with analytic and physical models of the neural, vascular, and skeletal structures to define injury-tolerance criteria at the tissue level. A reproducible, well-defined mechanical input simplifies to a degree the interpretation of the pathophysiologic, biomechanical, and biochemical mechanisms of the resultant injury.

9.4 Models of Brain Injury

9.4.1 Head-Impact Models

Nonpenetrating head-impact models can be grouped according to whether the resulting head motion is constrained to a single plane or whether the head is allowed to move freely in an unconstrained manner. These two types of head impact models were first studied in primates and later in non-primate species. Controlled nonpenetrating head impact was pioneered by Denny-Brown and Russell in primates [16, 17].

In early experiments performed by these researchers, the head motion resulting from the impact to the head was unconstrained. Their experiments laid the groundwork for the use of controlled impact to the unrestrained head as a common experimental technique because it was observed that when the head was fixed, cerebral concussion was less likely to result. The neurological and physiological indicators of injury produced by direct impact to the head were evaluated in fully anesthetized animals. Since a surgical plane of anesthesia is required to assure absence of discomfort to the animal during the experimental impact procedures, immediate assessment of the injury severity is limited to brainstem reflexes and systemic physiologic changes.

Gurdjian and associates used an impact piston with a 1-kg mass to strike the head of anesthetized primates at a predetermined location on the skull. Physiologic parameters, impact force, and head acceleration were monitored in an effort to determine thresholds for concussion and coma, coup and contrecoup contusions, and relative brain motions [18–20]. Later experiments performed by Ommaya and others used a molded protective skull cap to prevent fracture. Slight variations of this technique have been used to evaluate changes in intracranial pressure, behavior, systemic physiology, cerebral metabolism and histopathology following impact injury to the primate head [21–28].

Head motion in these primate studies was constrained only by the neck, resulting in complex and poorly characterized three-dimensional head movement. Severity was determined by selection of impactor velocity, impactor mass, interface material, and impactor contact area. However, like human brain injury the potential injury mechanisms remained complex. While a wide range of biomechanical responses exhibiting clinical relevance were likely produced, including local skull deformation, development of pressure gradients within the brain tissue, and relative motion between the brain and skull, because of the dynamics of the resultant head movement, it was not technically feasible to quantify or separate the various biomechanical components of the injury event for more detailed analysis of injury mechanisms. Although some studies attempted to evaluate the effects of rotational and translational head acceleration separately, careful biomechanical analysis of the experiments found that impact to the unrestrained head always produced both rotational and translational acceleration.

The various direct head-impact methods are successful at producing either fatality or short-duration unconsciousness but generally suffer from a high degree of variability in the response. This variability may result from a lack of control over the precise conditions of the impact, an absence of control of head dynamic response, and inconsistent impactor-skull interface parameters. As a result, the expected variability due to biological differences is confounded by mechanical variability of the injury event, necessitating a large number of experiments to obtain a representative sample of pathophysiologic response. In addition, the uncontrolled injury biomechanics make analysis of brain-injury biomechanics or testing of therapeutic efficacy difficult at best.

A number of studies have therefore constrained, at least partially, the head motion during and after impact in an effort to increase the reproducibility of injury outcome [29–31]. Full constraint was not obtained, but the head motion was typically confined to a single plane. An analysis of the pathophysiology and relation of impact biomechanics to outcome in these studies indicates that the partial control of head motion does not represent a significant improvement over the gradation and reproducibility of impact to the unconstrained head.

Closed head impact has been applied to nonprimates, including rat and cat, with limited success [32–39]. Similar to early experiments involving primates, biomechanical interpretation of the test results was initially impeded by the unpredictable head movement resulting from the impact. This was due in part to unspecified level of head support or restraint. As a result, biomechanical analysis of head and brain dynamics following head impact was inconsistent.

Head-impact studies performed in cats, using a captive bolt pistol to deliver the impact, have been most extensively developed by Tornheim [36–39]. Oblique impacts delivered to the coronal suture cause a reproducible contusion with associated edema and skull fracture. The captive-bolt head-impact model has been used successfully to test efficacy of anti-edema drugs. Subsequent experiments have used an oblique lateral impact, with the head resting on a collapsible hex-cell support [36, 38]. This model is useful for studies of cerebral contusion and resultant edema, though reliable gradation of impact severity is still needed.

In general, whole head impact has not been successful at producing clinically relevant brain injury in the rat. In most cases, early attempts using direct head impact to the rat resulted in either no identifiable injury or immediate convulsions associated with apnea. This finding highlights a common problem shared by early whole head-impact models in subprimates, particularly the rat. That problem is the inability to produce a graded, reproducible range of injury outcomes by varying the biomechanical parameters of the impact. In some rat impact models high-level input parameters resulted in either fatality or a brief period of unconsciousness, usually measured by absence of the righting reflex [32, 40-43].

Skull fracture commonly occurs at impact-severity levels insufficient to produce prolonged coma in the rat; moreover, the occurrence of skull fracture is generally not well correlated with injury severity. Convulsive activity, a common result of direct head impact in the rat, may affect overall outcome and must be considered in interpretation of cerebral metabolic, neurophysiologic and electroencephalographic changes which follow injury. The histopathological evidence of neural injury is typically restricted to the lower brainstem in contrast to findings in human brain injury. Finally, the injury response curve is extremely steep; only a slight change in impact parameters is needed to produce an injury spectrum from minor to fatal.

Goldman and collaborators have reported results obtained using a newly developed closed head-injury impact model of mild to moderate head injury in the rat [44]. Moderate concussion in this model is characterized by 4-10 min of unconsciousness in the absence of skull fractures or brain contusions. This new experimental brain-injury model utilizes a pendulum to deliver a controlled impact to the intact skull. The skull and maxilla rest on an energy absorbing rubber mat to reduce the likelihood of a skull fracture. The pendulum impact surface incorporates a load cell that enables measurement of the amplitude and duration of force produced by the impact. This model reproducibly produces brain swelling accompanied by increases in intracranial pressure lasting more than 3 days post-injury. Significant morphologic changes include edema and neuronal death.

The model developed by Goldman is significant and unique in producing a reproducible range of mild- to moderate-severity injury based on pendulum stroke, impact load, and animal body weight, in the absence of skull fracture [44]. Repeatability of experimental outcome is dependent on brain mass and is affected by skull strength. A key step in the success of this model was development of an impact force/bodyweight nomogram that compensates for differences in skull strength over the body-weight range (330-430 g) so that similar cerebrovascular responses are achieved in animals at the extremes of body weight. Well-controlled animal husbandry conditions are critical for the reproducibility achieved by these investigators. Goldman reported variances in estimates of regional cerebral permeability that were well within the range encountered in other studies [45]. Physiologic and morphologic changes parallel many aspects of human head trauma, particularly elevation of intracranial pressure. These observations suggest this may be a useful and relatively inexpensive tool for investigating the mechanisms and therapeutics of brain trauma, and especially edema and elevated intracranial pressure.

9.4.2 Head Acceleration Models

Various head acceleration models have been developed, some of which tightly control the motions of the head and others of which allow free head motion. Free head acceleration is attained through an abrupt deceleration of a moving frame to which the rest of the body is firmly affixed. This produces a whiplash head motion and can result in concussion [26, 28]. These studies have been performed in large nonhuman primates since the anatomic relationships between brain, brainstem, and spinal cord are most similar to man, as is the ratio of brain mass to head mass.

In any whole head acceleration model it is the geometry and mass distribution that will determine the brain dynamics and resultant injury. A significant difference between the head-impact and head-acceleration models is the relative absence of skull fracture in the acceleration models; however, in real situations the range of induced accelerations that produce injury experimentally is only achieved in man when head impact occurs.

Since unconstrained head dynamics make analysis of the injury biomechanics difficult and contribute to outcome variability, the majority of studies using head acceleration techniques have employed some type of helmet and linkage to control the head motions [6, 46]. Such a system controls the path direction, path length, and duration of acceleration so that the head motion is comparable between experiments with a comparable input. In addition, controlling the path of head movement during acceleration reduces the incidence of skull fracture and local skull deformation which are common occurrences in impact techniques.

The most extensive series of experiments studying acceleration induced brain injury was initiated by Ommaya and Gennarelli and concluded by Gennarelli and Thibault. The investigators were able to produce graded anatomic and functional injury severities, including prolonged traumatic unconsciousness (>30 min) in nonhuman primates [6]. This series of experiments built upon earlier head-acceleration studies that had indicated that loss of consciousness was more readily produced by high levels of angular acceleration than high levels of translational acceleration [46, 47, 48]. Consequently, a device and control linkage was designed to allow high levels of angular acceleration to be delivered to the primate head without skull fracture [6].

The technique used a pneumatic thrust column, helmet, and pivoting linkage to apply 60″ of angular rotation within 11–22 ms, attaining peak angular accelerations of $1-2 \times 10^5$ rad/s². The acceleration pulse was biphasic, with a relatively long, ramp-like acceleration phase followed by an abrupt deceleration phase, with injury presumed to occur during the deceleration phase. In addition to prolonged traumatic coma, these animals exhibited diffuse axonal injury in the subcortical white matter, comparable to pathology observed in clinical brain injury [6, 49]. Histopathological evaluation of the brains shows a high degree of comparability to clinical brain injury pathology, substantiating the relevance of the technique [50–52]. The spectrum of injury obtainable using this technique ranges from mild, subconcussive brain injury with only sparse histologic evidence of axonal damage, through prolonged (hours to days) traumatic coma with extensive axonal injury throughout the white matter and brainstem, to immediately fatal injury.

Biomechanical analysis has been partially successful in relating the magnitude of acceleration input to the severity of injury outcome in this model, although interpretation is confounded by the two-phase acceleration-deceleration pulse, which results in nearly comparable acceleration and deceleration force peaks and the potential for neural and vascular injury during either or both pulses. Nevertheless, the need to use nonhuman primates for the technique (because of geometrical considerations) slows the progress of research since it is not possible to perform the large number of experiments necessary to fully use the experimental power of the technique. The injury device is also complex and very costly to duplicate. It requires collaboration of medical scientists and biomechanical engineers for implementation and meaningful data interpretation. Finally, due to the biphasic acceleration-deceleration pulse, conclusions regarding the biomechanics and timing of neural and vascular tissue injury must be drawn cautiously.

In 1993, Meany et al. reported on the application of this device to the miniature pig [53]. The use of a large nonprimate animal model allows more experiments to be performed and, as discussed in the last section of this chapter, using mathematical modeling of the brain response as an adjunct in interpreting the experimental results significantly resolves the concerns expressed above.

Nelson and his colleagues developed a controlled nonimpact head acceleration model in the cat, using repetitive accelerations [54–56]. Repeated acceleration/deceleration is delivered at 1,200–1,400 oscillations/min and combined with induced post-traumatic hypoxia or ischemia. Injury severities were categorized into three outcomes: immediate fatality, delayed mortality, and extended coma. Although clinical brain injury may involve a period of hypoxia or relative cerebral ischemia, the biomechanics of brain neural and vascular tissue response to rapid repetitive accelerations is unknown. The outcomes observed, however, are clearly relevant to clinical brain injury. This technique induces a consistent range of injury, with a reproducible percentage distribution in each of the three outcome categories, but does not allow prediction of the injury severity that is likely to result from a specific exposure. Hence, testing for efficacy of pharmacologic intervention is limited to observations regarding changes in the percent immediate fatality, delayed mortality, and prolonged coma, requiring larger sample sizes and impairing interpretation on a mechanistic basis.

9.4.3 Direct Brain Deformation Models

Nonpenetrating focal deformation of cortical tissue has been employed as an alternative to head-impact and induced-acceleration models of brain injury. From a theoretical point of view, since brain tissue is injured by the induced stresses and pressures within the cerebrum and brainstem, if a direct cortical deformation results in the same pattern of intracerebral forces within the brain as that resulting from an acceleration or impact model, then the injuries induced are directly comparable.

The extent to which this can be successfully achieved remains an open question; however, since the functional changes and histopathology are in many aspects comparable to clinical brain injury, these models have become widely used. Further, since the brain deformation is under direct experimental control, it is unnecessary to restrict studies to primate species on the basis of brain geometry, as has sometimes been required for clinical relevance in head-impact or acceleration techniques. Theoretically, direct brain-deformation models should allow better control over the mechanics of the injury input and therefore pose some potential advantages for biomechanical analysis of the injury event and relation of brain-tissue deformation to injury outcome. However, these advantages remain potential since in almost every instance, little or no analysis of the brain-tissue deformation has been performed.

Early experimental work produced a direct focal cortical compression using either weight-drop devices or gas-pressure jets [19, 57, 58]. Although different mechanical inputs were used to produce cortical deformation, each model type displayed both localized cortical contusions and distributed cerebral metabolic effects (Dail et al. 1981). In both models the pathologic changes in neural and vascular elements were predominantly observed in the impact region. For this reason direct focal cortical compression is a useful model in studies in which a desired histopathological outcome is cortical contusion, and for testing the efficacy of therapies aimed at reducing cortical contusion size. However, widespread diffuse injuries of clinical relevance are not produced, nor are long-term function consequences of importance to clinical brain injury. From a mechanistic viewpoint this may be due to damping of the mechanical input by the dura and cortical tissue, effectively preventing propagation of the input throughout the remainder of the brain. The absence of biomechanical data on brain deformations in these injury models prevents any further conclusions.

The fluid percussion method is the most widely accepted direct brain-deformation technique for producing experimental brain trauma, including reactive axonal injury, in subprimate species such as cat and rat [59–68]. In this technique, a brief fluid pressure pulse is applied to the dural surface of the brain through a craniotomy.

The procedure has been applied to both cats and rats to generate a range of brain injury from mild to severe [60, 61, 65, 66, 68, 69]. The pressure pulse is delivered via a fluid column to the intact dura of the brain and results in functional changes accompanied by subcortical axonal damage and brainstem pathology. By increasing or decreasing the magnitude of the pulse pressure (1.0–4.0 atm, with relatively constant duration of 20 ms), a graded, reproducible range of injury severities can be produced.

The fluid-percussion technique reproduces some of the features associated with moderate head injury in man, characterized clinically by transient unconsciousness with prolonged alteration of mental status, a low incidence of hematomas, and prolonged neuropsychological deficits [59, 70]. A key aspect of human head injury that is not reproduced in midline fluid percussion is the cortical contusion; this likely reflects the distributed nature of cortical loading in the technique. Recently, McIntosh and colleagues have shown that the fluid percussion technique applied through a lateral craniotomy produces cortical contusion, a histological endpoint previously unreported using this technique [64, 71].

While the fluid percussion technique may be suitable for studying the physiology and pharmacology of moderate-severity head injury, it is not well suited for the study of the biomechanics of trauma. Experimental evidence taken from high-speed radiographic movies of the fluid-percussion event demonstrates a consistent, though complex, movement of fluid within the cranial cavity in the rat [14, 60]. Identical experiments performed in the ferret showed a consistent, though less complex, wave pattern of fluid flow in the epidural space; however, the pattern observed was completely different from that observed in the rat. This would indicate that the pattern of cerebral deformation resulting from fluid percussion is species dependent (Lighthall and Dixon, unpublished observation). The interaction of the fluid pulse with the cranial contents does not lend itself to accurate biomechanical analysis of tissue deformations that produce the injury.

Although the mechanics of the fluid-percussion injury model have been investigated, they are not yet well defined, and are arguably different from closed head injury in man. However, the pathophysiologic changes offer a reasonable model in which to study mechanisms for brainstem injury and secondary cerebrometabolic abnormalities. The ability to produce graded brainstem and subcortical axonal injury continues to make the fluid-percussion technique a useful model for those aspects of clinical brain injury.

A direct cortical-impact model was developed in 1988 that allows independent control of brain contact velocity and level of brain deformation. This technique was originally developed to study spinal-cord injury [72] and later modified to be used for experimental brain-injury research by Lighthall [73]. Initially characterized in the laboratory ferret, the cortical impact model has now been applied with success to the rat [74]. The controlled impact is delivered using a stroke-constrained pneumatic impactor. The impact device and impact procedures have been described elsewhere [73]. Briefly stated, the device consists of a stroke-constrained stainless-steel pneumatic cylinder with a 5.0 cm stroke. The cylinder is mounted in a vertical position on a crossbar that can be adjusted in the vertical axis. Impact velocity can be adjusted between 0.5 and 10 m/s by varying the air pressure driving the pneumatic cylinder. Depth of impact is controlled by vertical adjustment of the crossbar that holds the cylinder.

Results with this model indicate that cortical impact produces acute vascular and neuronal pathology that is similar to observations in currently accepted experimental models and, most importantly, in clinical brain injury. Controlled impacts using the pneumatic impactor produce a range of injury severities that are a function of contact velocity, level of deformation, and site of impact. This model is unique in its ability to produce graded cortical contusion, subcortical injury and, in high-severity impacts, brainstem contusion. Impacts performed at 4.3 m/s or 8.0 m/s, with $\cong 10$ % compression (2.5 mm), produce extensive axonal injury at 3 and 7 days post-injury using both velocity/compression combinations [15]. Regions displaying axonal injury were the subcortical white matter, internal capsule, thalamic relay nuclei, midbrain, pons, and medulla. Axonal injury was also evident in the white matter of the cerebellar folia and the region of the deep cerebellar nuclei.

Because diffuse axonal injury was observed, this would imply, based on clinical and experimental observation, that rapid mechanical deformation of the brain can produce behavioral suppression and/or functional changes similar to coma. Behavioral assessment showed functional coma lasting up to 36 h following 8.0 m/s impacts, with impaired movement and control of the extremities over the duration of the post-injury monitoring time. The spectrum of anatomic injury and systemic physiologic responses closely resembled aspects of closed head injury seen clinically.

This procedure complements and improves upon existing techniques by allowing independent control of contact velocity and level of deformation of the brain to facilitate biomechanical and analytic modeling of brain trauma. Graded cortical contusions and subcortical injury are produced by precisely controlled brain deformations, thereby allowing questions to be addressed regarding the influence of contact velocity and level of deformation on the anatomic and functional severity of brain injury.

These cortical impact experiments confirm that direct brain deformation models of experimental injury can produce many aspects of traumatic brain injury in humans and can be used to investigate mechanisms of axonal damage and prolonged behavioral suppression.

More recently, variations on the direct cortical impact method have been developed. Meany et al. (1994) modified the technique by adding a second, contralateral, craniotomy site to more easily produce axonal damage in the rat cerebral cortex. This method was also used by Lighthall's group in the ferret. Another variation on the method, intended to study focal lesions in the cortex, was reported by Schreiber et al. [75]. A special coupling device placed on a single craniotomy site allowed a vacuum pulse to be applied to the surface of the brain with dura mater removed.

9.5 Biomechanical Mechanisms of Brain Injury and Injury Criteria

Impact injury to the brain can be caused by forces applied to the head and by the resulting abrupt motions imparted to the head. These forces can either be external to the body and act directly on the head and/or internal to the body and act on the head through the head/neck junction. Forces transmitted to the head from the neck are a combination of muscle forces and cervical spinal forces. In the case of involuntary loading due to impact, these forces are the result of external loads applied to other body regions, such as vehicle restraint forces applied to the torso, and are sometimes referred to as forces due to "whiplash."

Application of external forces to the head can result in local injury to the scalp, bones of the skull and/or brain tissue due to the effects of concentrated load. In the case of the brain, local injury at the site of impact can occur due to penetration of the skull by the striking surface or due to local deflection of the skull without skull fracture. Externally applied head impact forces can also deform the skull globally, which can cause pressure and deformation gradients throughout the brain.

Motions of the head associated with impacts to the head and torso quite often result in severe accelerations and large attendant velocity changes that, typically, are both translational and rotational in nature. Such abrupt motions result in inertial or body forces being developed in the brain tissue that in turn result in stresses and deformations throughout the brain. Thus, the state of deformation of the tissue in a region of the brain in a head undergoing impact loading will depend on; (1) its location relative to the point of force application, (2) the nature of the distribution of the force, and (3) the nature of the motion of the head due to the forces acting on the head. In addition, abrupt head motion can also result in relative motion of the brain or parts of the brain with respect to the skull. Such relative motions can deform brain tissue due to impingement upon irregular skull surfaces or interaction with meningeal membranes and can stretch the connecting blood vessels between the surface of the brain and the skull. Finally, one last mechanism of deformation of brain tissue is that of local stretching of the brain-stem and spinal cord due to motions produced at the head/neck junction. This motion can occur as a result of either head impact or head motion due to torso loading.

If the magnitudes of the deformations and stresses induced in the tissues by any of the above mechanisms are sufficiently great, the tissues will fail, either in the physiological or the mechanical sense, and injury will occur. It is evident from the above discussion that there are many possible combinations of loading conditions that could result in injury to the tissues of the head. Because of the multiplicity of conditions, the history of head injury biomechanics has produced a wide variety of proposed mechanical factors that have been presented as being responsible for producing injury. Ideally, a biomechanical understanding of brain injury mechanisms requires a description of the mechanical states that occur in the tissue during an impact and of the resulting dysfunction in the physiological processes at the tissue level. The resulting tissue stresses and strains must be related to the dysfunction in order to develop truly predictive injury criteria. It is our view that this is best done through finite element modeling of the brain response to impact.

The sections that follow discuss the major studies that have addressed one or another aspect of brain injury biomechanics and their associated attempts at developing biomechanical injury criteria. It should be noted that the resulting injury criteria are all based on mechanical inputs to the head, rather than the more desirable mechanical responses of the brain itself.

9.5.1 Early Studies

The earliest studies of the biomechanical factors associated with brain injury were those of Scott (1940), Denny-Brown and Russell (1941), Holbourn (1943), Walker, et al. (1944), Gurdjian and Lissner (1944), and Pudenz and Sheldon (1946) [16, 76–80]. The early studies with experimental animals were severely limited by the lack of suitable dynamic measurement methods available for characterizing the impact severity and the responses such as intracranial pressure. Holbourn did no actual experiments on animals but, rather, used simple photoelastic models of the head to demonstrate his theories. He hypothesized that translational acceleration of the head would not produce significant deformations in the brain due to the incompressible nature of confined brain tissue. He thus concluded that shearing deformations, which produce no volume change, caused by rotational acceleration could develop the shear strains throughout the brain required to produce the diffuse effects needed for concussive brain injuries. Gurdjian and Lissner, on the other hand, attributed intracranial damage to deformation of the skull and changes in intracranial pressure brought about by skull deformation and acceleration of the head due to a blow to the head. Later, Gurdjian, et al. also recognized that movement of the brain relative to the skull and resulting injuries can be caused by head rotation [19].

The head-impact work of Pudenz and Sheldon used anesthetized primates that had transparent plastic calvarias replacing the top of the skull. This surgical modification allowed viewing of the surface of the brain with high-speed photography during the impact. The authors demonstrated very dramatic motions of the brain and, undoubtedly, gave apparent credence to Holbourn's hypothesis in the minds of some researchers. However, careful reading of Pudenz and Sheldon indicates that the authors were unable to eliminate air bubbles under the plastic calvarium and that they understood that this would accentuate the brain motion effects. Since inclusion of compressible gas bubbles can eliminate the constraining effect of brain tissue incompressibility on translational acceleration-induced brain motions, the results of their study cannot be relied upon as indicators of the degree of brain motion during impact. Similar comments apply to the extent of observed internal brain motions reported in the radiographic head impact study of Shatsky et al. [31]. The effect of slight compressibility on head-impact brain deformations was discussed through finite element brain modeling by Lee et al. [81].

9.5.2 Rotational Brain-Injury Studies

Holbourn's hypothesis, that rotational acceleration is the primary means by which diffuse brain injury is produced, makes no differentiation between direct impact and indirect loading in terms of the outcome if the rotational acceleration levels are the same in both exposures. This led some subsequent researchers to explore the concept in experiments with monkeys in which abrupt rotational motion was imparted without direct impact loading to the head. Ommaya, et al. demonstrated that abrupt rotation with impact could affect sensory responses in the monkey and that a cervical collar that reduced the rotational acceleration in occipital impacts also raised the impact severity needed to reach the threshold for concussion [27]. This work demonstrated that the angular accelerations necessary to produce concussion by direct impact were approximately half those needed by indirect or inertial loading and, thus, did not support Holbourn's hypothesis. Ommaya and Hirsch [26] subsequently revised the rotational theory to state that "Approximately 50 % of the potential for brain injury during impact to the unprotected movable head is directly proportional to the amount of head rotation...the remaining potential for brain injury will be directly proportional to the contact phenomena of impact (e.g., skull distortion)." The authors reported on experimental head impact and whiplash injury studies with three different sizes of subhuman primate species in support of their hypothesis. The data from the experiments were compared with values predicted from scaling considerations based on dimensional analysis. The authors predicted that a level of head rotational acceleration during whiplash in excess of 1,800 rad/s^2 would have a 50 % probability of resulting in cerebral concussion in man.

Gennarelli, Thibault, and co-workers, in a series of studies, continued the investigation into the relative roles of translational and rotational accelerations in causing brain injuries by using experimental devices that could produce large translational and rotational head accelerations independently and without deforming the skull of subhuman primate subjects. The nature of the test apparatus produced an acceleration followed by a deceleration since it employed a limited stroke motion. While such a start-stop motion is typical of angular motions due to impact, it is not typical of translational motions in which an initial velocity is usually involved. In an early study using these techniques, the effects of translational and angular acceleration on the brain were studies using squirrel monkeys [48]. Two groups of animals were subjected to similar acceleration loading. The first group received translational accelerations and the second group received angular accelerations in a 45° arc about a center of rotation in the cervical spine. The physiological and pathological results were quite different in each of the groups. Pure translation did not produce diffuse injury although focal lesions were produced. It was only when rotation was added to translation that diffuse injury types were seen. At the highest acceleration exposures, it was impossible to induce a cerebral concussion in animals subjected to pure translational acceleration, while the combination of angular acceleration and translational acceleration easily produced concussion for the same tangential acceleration exposure level measured at the top of the head.

Later studies in this series reported by Abel, et al. and Gennarelli and Thibault applied accelerations in the sagittal plane by subjecting the heads of rhesus monkeys to controlled rotation about a lateral axis in the lower cervical spine [6, 82]. The Abel, et al. study focused on the production of subdural hematomas and expressed the result of these tests in terms of a translational component, tangential acceleration, which is the product of the initial angular acceleration and the radius of curvature of the applied motion [82]. The test results were plotted as functions of the two parameters with maximum tangential acceleration shown as a series of lines of constant magnitude. The line denoting a maximum tangential acceleration of 714G was found to be the apparent boundary separating those cases in which subdural hematoma occurred from those in which it did not.

In the later study, Gennarelli and Thibault (1982) chose to use the angular deceleration phase of the enforced motion, rather than the initial acceleration phase, as the mechanical parameter with which to relate to the occurrence of brain injuries [5]. They plotted the test results as functions of the peak angular deceleration and the associated pulse duration of the stopping phase of the motion. They found the occurrence of cerebral concussion to generally follow a decreasing angular deceleration level as pulse duration increased. Diffuse brain injuries occurred at higher levels of angular deceleration and generally followed the same trend as that for concussion. Plotting the subdural hematoma data on the same graph produced the unusual result that the magnitude of the angular deceleration necessary to produce subdural hematoma increased for increasing pulse duration. They attributed this effect to rate sensitivity of the failure strength of the bridging veins.

Lee and Haut have shown that the bridging veins are not rate sensitive [83]. However, Lee, et al. used a two-dimensional finite element model representation of the rhesus monkey brain to simulate these experiments to better understand the mechanisms of traumatic subdural hematoma and to estimate its threshold of occurrence in the rhesus monkey [81]. The brain was treated as an isotropic homogeneous elastic material with and without structural damping and the skull was treated as a rigid shell. The complete acceleration and deceleration time history of the enforced motion was applied to the model. During both phases of the motion, high shear strains occurred at the vertex where the parasagittal veins are located.

The conclusions of the model analysis indicated that subdural hematoma may have occurred during the acceleration phase in the primate experiments. Because the experiments used a fixed total angle of rotation, the relationship between maximum angular acceleration and its pulse duration is fixed. Thus, if the angular deceleration and its pulse duration were increased in the experiments, the initial angular acceleration would also be very high and its pulse duration would, conversely be shortened.

This finding may explain the apparent anomaly of the Gennarelli and Thibault analysis of the subdural hematoma data and supports the Abel, et al. analysis using the initial acceleration data. Lee, et al. were able to replot the experimental data as a function of both tangential and angular accelerations to show their combined effect on bridging vein deformation and estimate tolerance thresholds for subdural hematoma in the rhesus monkey as a linear combination of both types of acceleration.

Further analysis of the injuries produced in the rhesus monkey experiments led Gennarelli, et al. to conclude that the extent of axonal injury, duration of coma, and outcome of this type of experimental injury correlate better with coronal (lateral)-plane rotational impacts rather than sagittal (frontal)-plane rotational impacts [6, 84]. In an effort to understand the implications of the injury data in terms of estimating equivalent exposures for the human, they duplicated the lateral rotation tests using physical models of skull-brain structures of the baboon and human. The skulls were prepared as coronal hemisections that allowed visual determination of shear deformations in a silicone gel simulation of brain material [85]. The falx cerebri membrane was simulated in both models. Since no measurement of brain bib mechanical response, such as pressure or displacement within the head, was made in any of the experimental animals, there was no direct way to verify the reality of the response of the physical animal model. Instead, the authors chose to compare the overall deformation pattern with the pattern of diffuse axonal injury found in the animals. This method allowed them to estimate an empirically derived value of approximately 10 % for critical shear strain associated with the onset of severe diffuse axonal injury in primates. This estimate of critical shear strain was then used to determine a corresponding coronal plane rotational acceleration value for the human head model. This resulted in an estimate of 16,000 rad/s^2 rotational acceleration for the threshold of severe diffuse axonal injury in man.

The above study indicates the difficulty in developing comprehensive injury-prediction criteria for functional brain injury. As noted above, the lack of biomechanical measurement in the experimental animal brains during the test required estimations of brain responses through completely separate physical-model tests. The very high level of angular and translational accelerations necessary to produce closed head injury in the small brains of animals makes such measurements difficult. Additionally, the shapes of the brains of animals typically used for brain-injury studies make them more resistant to impact injury and also make it very difficult to scale the results to man, since structural scaling laws assume geometric similitude [86].

9.5.3 Translational Motion Studies

The early emphasis on skull deformations and intracranial pressure gradients as sources of brain injury led Gurdjian, Lissner, and their co-workers to focus their brain injury studies on direct blows to the head and the resulting translational accelerations associated with such blows. In impact experiments in which blows were delivered to the fixed or free heads of dogs, Gurdjian, et al. demonstrated that cellular damage in the upper spinal cord occurred as frequently with the head fixed as with the head free to move [18]. They attributed the damage to shear stresses at the craniospinal junction as a result of pressure gradients in the brain. These researchers used an indirect approach to study human concussion by impact testing embalmed cadaver heads.

The rationale for using a nonphysiological test subject was based on the clinical observation that concussion is present in 80 % of patients with simple linear skull fractures. Even though most concussions do not involve skull fractures, the easily observed occurrence of a skull fracture was felt to be an endpoint that could be studied in the cadaver. Thus, by impacting the foreheads of embalmed cadaver head against rigid surfaces at different impact energies, it would be possible to determine the levels of head acceleration associated with the onset of linear skull fractures. By extension, then, the results could be used to infer the onset of concussion for the case of rigid head impacts to the front of the head. The translational head acceleration in the anterior-posterior direction was measured in the tests with a uniaxial accelerometer attached to the back of the skull of the cadaver.

This work was summarized in terms of acceleration level and impulse time duration by Lissner, et al. [87]. The relationship between acceleration level and impulse duration was presented by a series of six data points that indicated a decreasing tolerable level of acceleration as duration increased. This relationship became known as the Wayne State Tolerance Curve (WSTC), for the affiliation of the researchers, and has become the foundation upon which most currently accepted indexes of head injury tolerance are based. The original data only covered a time duration range of 1–6 ms and, of course, only addressed the production of linear skull fractures in embalmed cadaver heads. The curve was later extended to durations above 6 ms with comparative animal and cadaver impact data and with human volunteer restraint system sled test data. The volunteer information consisted of whole body deceleration without head impact at very long (100 ms and greater) durations. The WSTC was subsequently used by Gadd to develop the weighted impulse criterion that eventually became the Gadd Severity Index (GSI) [88, 89]. In 1972, the National Highway Traffic Safety Administration proposed a modification of the GSI that has become known as the Head Injury Criterion (HIC) currently used to assess head injury potential in automobile crash test dummies. It is based on the resultant translational acceleration rather than the frontal axis acceleration of the original WSTC. A complete discussion of the development of the current HIC and its predecessors, the GSI and WSTC, is contained in SAE Information Report, J885-APR84.

The WSTC has been criticized on various grounds since its inception. The paucity of data points, questionable instrumentation techniques, lack of documentation regarding the scaling of animal data used in its extension to longer durations, and uncertainty of definition of the acceleration levels have all been questioned. From a biomechanical standpoint the main criticism of the WSTC is that, just like the work of those advocating a rotational acceleration-induced mechanism of brain injury, there has been no direct demonstration of functional brain damage in an experiment in which biomechanical parameters sufficient to determine a failure mechanism in the tissue were measured. The assumption behind the WSTC work is that translational acceleration produces pressure gradients in the region of the brainstem that result in shear-strain-induced injury. The extension of this hypothesis to human brain injury remains to be verified. Ono, et al. conducted an extensive series of experiments with monkeys and demonstrated that cerebral concussion can be produced by impacts that produce pure translational acceleration of the skull [90]. Using additional experiments with cadaver skulls and scaling of the animal data, the authors developed a tolerance curve for the threshold of cerebral concussion in humans. The curve was called the Japan Head Tolerance Curve (JHTC). The difference between the JHTC and the WSTC in the 1–10 ms duration range was shown to be negligible, while minor differences exist in the longer duration region of the curves. The Ono, et al. data show that the threshold for human skull fracture is slightly higher than that for cerebral concussion.

In both the above discussions, researchers attempted to predict various kinds of brain injury from one specific head impact input parameter, such as rotational or translational acceleration, with little regard for the effects of the other parameter. In the vast majority of head impact situations it can be expected that both rotational and translational accelerations are present and combine to cause the brain responses that produce the injury in the tissue. As discussed at the beginning of this section, the global patterns of stress and deformation in the brain during an impact are complex. Accordingly, comprehensive brain-injury prediction will require a complete description of tissue mechanical states throughout the brain for any combination of mechanical inputs. This state of knowledge has yet to be realized and will depend on continued research in brain injury and the development of sophisticated analytical models for the prediction of brain mechanical response to impact.

9.5.4 Recent Developments in Brain-Injury Biomechanics Research

Standardization of experimental protocols dictates and encourages the use of a simple, single, controlled mechanical input to produce a neural injury. Several advantages accrue. If the mechanical input is designed to be quantifiable and graded, then various relationships can be made between the tissue deformation parameters, including applied force, the amount of deformation and its time-history, and the resultant pathology and functional changes. Such analysis will ultimately lead to enhanced understanding of the interaction between the physical input, the severity of the physiologic injury response, and the functional outcome.

The restriction to one quantifiable mechanical input variable facilitates biomechanical analysis of experimental CNS injury. Parallel analysis, utilizing physical and analytical modeling of tissue deformation then allows correlation to the more complex dynamics of human CNS injury, especially involving the brain, through derivation of tissue biomechanical parameters that produce transient neurological changes, coma, or fatality.

In all cases, physiological responses to the mechanical injury must be considered in the design of the experimental model and must be taken into account when ascribing hypothesized modalities of treatment. In addition, anesthetic interaction with the desired physiological outcome must be considered. While a surgical plane of anesthesia is mandatory during the experimental impact procedures, anesthesia will mask certain aspects of neurological outcome and behavioral assessment. Ethically there is no way to avoid this aspect of experimental CNS trauma research; however, consistent use of specific standardized anesthetic protocols between laboratories using the same experimental technique will enhance comparability of experimental results.

Techniques for producing experimental brain injury are sometimes not compatible with detailed biomechanical analysis required for development of an injury criterion. The cortical-impact technique described here in general has provided the ability to reproducibly generate graded levels of functional-injury severity in order to investigate mechanisms of clinical injury, and to conduct initial evaluations of potential therapeutic interventions. Species differences in brain mass and geometry and the nature of the interaction between the physical and physiological factors dictate first that certain simplifications will add to the usefulness of an experimental model and second that analytical or mathematical methods such as finite element models are necessary to make extrapolations from experimental studies to humans [91]. The approach of using a finite element model of the human brain will allow information from impact tests with dummies to be analyzed to predict the potential for brain injury. The method characterizes the brain as a deformable body bounded by the skull. The complete head motion from an impact is enforced on the skull by the accelerations. The contact surface of the skull-brain boundary transfers the motion from the skull to the brain resulting in stresses and strains in the brain tissue. Critical sites of potential injury may be indicated by maximum pressure, stress, strain, or displacement relative to the boundary. These potential injury sites must then be assessed using tissue-level injury criteria that have been validated using anatomical and functional data from experimental physiologic models.

The development of tissue-level injury criteria for the CNS can best be achieved by using a three component approach consisting of (1) a simplified, biomechanically characterized physiological model that produces clinically relevant injury; (2) a physical model constructed from material that incorporates geometry, mass, and physical material properties matching the physiological model; and (3) a comprehensive finite-element model of the brain. Analytical modeling of the tissue response leading to injury requires that mechanical input to the tissue be quantifiable and reproducible. Therefore simplification of the input is required in designing a model that can be used to address the mechanics of neural and vascular injury at a tissue level. Validation of analytical finite-element models using experimental data requires correlation of the mechanical responses of the system in terms of pressures, displacements, and local accelerations of the important regions of the brain. This approach can lead to tissue-level injury criteria, which, after physiological and anatomical validation of the analytical model, can be applied to the analysis of neural injury risk for impact and acceleration trauma in humans.

The considerations outlined above have been implemented in various ways by a number of research studies. Ueno et al. report on the development of a finite-element model of double-craniotomy cortical impact studies using the ferret animal model [86]. The goal of the work is to determine the mechanical variable best associated with the threshold of diffuse axonal injury observed in the animal experiments. Another example is the study by Shreiber et al. on mechanical injury to the blood-brain barrier [75]. They used a cortical deformation method, applying a dynamic vacuum pulse to the surface of the rat brain. Nine unique loading conditions were exercised by varying pulse durations in levels of 25, 50, and 100 ms and pulse vacuum levels of 2, 3, and 4 psi. A finite-element model was developed and compared to experimental injury maps demonstrating blood-brain barrier breakdown. Statistical analysis of the data suggested that the blood-brain barrier is most sensitive to maximum principle logarithmic strain, and that breakdown occurs above a strain of 0.188 ± 0.0324.

Meany et al. performed experiments producing severe diffuse axonal injury using controlled, nonimpact head accelerations applied to the heads of miniature pigs [92, 93]. Zhou et al. developed a three-dimensional finite-element model of the pig head for biomechanical analysis of the experiments [94]. The model results showed that regions of high normal or shear strains were not in the regions where DAI was found, primarily because a distinction was not made between white and gray matter. This discrepancy led to the development of a two-dimensional finite-element model by Zhou, et al. that included accurate anatomic and material description. The model consisted of a three-layered skull, dura mater, CSF, white matter, gray matter, and ventricles [95]. The two-dimensional model demonstrated that the regions where DAI was found in the experiments were regions of high shear stress. However, other regions of high shear stress occurred where no DAI was seen, indicating that further investigation was needed. In support of improving the model, Arbogast and Margulies measured the dynamic material response of the pig brainstem and cerebrum [96]. Complex shear moduli were calculated over a range of frequencies from 20 to 200 HZ at two engineering strain amplitudes of 2.5 % and 5.0 %. At the 2.5 % strain level, no statistical regional difference was evident. At the 5 % strain level the difference in complex shear moduli between regions was statistically significant, with the brainstem region modulus being 80 % greater than the cerebrum.

Miller et al. incorporated these improved mechanical property characterizations into two different two-dimensional finite-element models of the pig brain [97]. Their goal was to study the ability of the different modeling approaches to predict specific forms of traumatic brain injury. They found that a model using a sliding frictional interface between the cerebral cortex and the dura mater gave better topographic distributions of axonal injury than did one that characterized that interface as a low shear modulus, incompressible solid. Injury indices based on either maximum principal strain or von Mises stress predicted comparable patterns of axonal and macroscopic cortical contusions, while negative pressure was a poor predictor of both forms of injury.

Prediction of the occurrence of axonal injury requires a tissue level injury criterion that directly relates a mechanical parameter to the injury outcome. The above studies were aimed at defining such a criterion through the use of statistical methods that relate mechanical states computed in a particular brain region to injury seen to occur experimentally in the same region. This approach is necessary because it is generally not possible to test functioning neural tissue under simplified stress and deformation states, as is customarily done with engineering materials. Gennarelli, et al. reported on a method that provides an exception to this experimental shortcoming. Using dynamic uniaxial elongation of the guinea pig optic nerve, they were able to produce in vivo axonal injury under simple loading conditions [98]. The architecture of the tissue is simple, allowing relatively straightforward characterization of axonal deformation, although some undulation in the neural fibers must be taken into account [99]. Because the simple loading conditions are easily controlled and quantified, and the functional injury outcome can be determined using standard electrophysiological techniques, the experiment has a unique potential to determine a strain-based tissue-level threshold for functional impairment to CNS white matter. Bain and Meaney report on a detailed analysis of such experiments [100]. They determined three strain-based thresholds for electrophysiological impairment to the optic nerve tissue. A liberal threshold strain of 0.28 minimized false-positive rate, a conservative threshold strain of 0.13 minimized the false-negative rate, and an optimal threshold strain of 0.18 balanced the specificity and sensitivity measures.

Success appears to be imminent in establishing tissue-level injury criteria for neural tissue from animal studies. However, there remain two important steps before the goal of predicting head-injury potential from head impact data can be achieved. First, an analytic model of the human head must be validated in terms of the model's ability to describe the detailed mechanical states in the critical regions of the brain during an impact. Second, the ability of the combination of the human model and the injury criteria to predict human injuries must be evaluated.

The most accurate physical model of the human head is the approximation given by the cadaver head. However, great care must be given to restoring, as closely as possible, the mechanical state of the soft tissues of the head through fluid filling and repressurization of the vascular and CSF systems and minimizing the post-mortem time to test-time period. Such work is difficult and time-consuming. Wayne State University (WSU) researchers, Hardy et al. have developed these techniques and have applied them, along with new methods for measurement of head-impact induced internal brain motions through the use of neutral-density accelerometers and targets for embedding in brain tissue in combination with high-speed cineradiography techniques [101]. This work has great promise in providing the basis for evaluating the ability of analytical models to predict states of strain in regions of the brain under realistic impact conditions.

9.6 Summary

The biomechanics of brain injury continues to be the most critical subject in injury biomechanics. Many studies, described in this chapter, have made significant contributions to the understanding of general injury mechanisms and mechanical factors that affect them. Definitive injury criteria for the assessment of head injury potential in the wide variety of loading conditions which exist in modern life, such as car crashes, sports impacts and warfare, remain an unfulfilled goal. The work presented in this chapter provides a historical background for the following chapter on the future of brain injury biomechanics.

Acknowledgment This material is the result of work supported with resources and the use of facilities at the Zablocki VA Medical Center, Milwaukee, Wisconsin and the Medical College of Wisconsin. Narayan Yoganandan is a part-time employee of the Zablocki VA Medical Center, Milwaukee, Wisconsin. Any views expressed in this chapter are those of the authors and not necessarily representative of the funding organizations.

References

1. Chapon A, Verriest JP, Dedoyan J, Trauchessec R, Artru R (1983) Research on brain vulnerability from real accidents. ISO document no. ISO/TC22SC12/GT6/N139
2. Gennarelli TA (1981) Mechanistic approach to head injuries: clinical and experimental studies of the important types of injury. In: Ommaya AK (ed) Head and neck injury criteria: a consensus workshop. U.S. Department of Transportation, National Highway Traffic Safety Administration, Washington, DC, pp 20–25
3. Tarriere C (1981) Investigation of brain injuries using the C.T. Scanner. In: Ommaya AK (ed) Head and neck injury criteria: a consensus workshop. U.S. Department of Transportation, National Highway Traffic Safety Administration, Washington, DC, pp 39–49
4. Scott WE (1981) Epidemiology of head and neck trauma in victims of motor vehicle accidents. In: Ommaya AK (ed) Head and neck criteria: a consensus workshop. U.S. Department of Transportation, National Highway Traffic Safety Administration, Washington, DC, pp 3–6
5. Gennarelli TA, Thibault LE (1982) Biomechanics of acute subdural hematoma. J Trauma 22(8):680–686
6. Gennarelli TA, Thibault LE, Adams JH, Graham DI, Thompson CJ, Marcincin RP (1982) Diffuse axonal injury and traumatic coma in the primate. Ann Neurol 12(6):564–574. doi:10.1002/ana.410120611
7. Penn RD, Clasen RA (1982) Traumatic brain swelling and edema. In: Cooper PR (ed) Head injury. Williams and Wilkins, Baltimore/London, pp 233–256
8. Cooper PR (1982) Post-traumatic intracranial mass lesions. In: Cooper PR (ed) Head injury. Williams and Wilkins, Baltimore/London, pp 185–232
9. Jennett B (1976) Some medicolegal aspects of the management of acute head injury. Br Med J 1(6022):1383–1385
10. Cooper PR (1982) Skull fracture and traumatic cerebrospinal fluid fistulas. In: Cooper PR (ed) Head injury. Williams and Wilkins, Baltimore/London, pp 65–82
11. Gennarelli TA (1980) Analysis of head injury severity by AIS-80. In: 24th annual conference of the American association of automotive medicine, AAAM, Morton Grove, pp 147–155
12. Landesman S, Cooper PR (1982) Infectious complications of head injury. In: Cooper PR (ed) Head injury. Williams and Wilkins, Baltimore/London, pp 343–363
13. Ommaya AK (1985) Biomechanics of head injury: experimental aspects. In: Nahum AM, Melvin JW (eds) The biomechanics of trauma. Appleton-Century-Crofts, Norwalk, pp 245–269
14. Lighthall JW, Dixon CE, Anderson TE (1989) Experimental models of brain injury. J Neurotrauma 6(2):83–97
15. Lighthall JW, Goshgarian HG, Pinderski CR (1990) Characterization of axonal injury produced by controlled cortical impact. J Neurotrauma 7(2):65–76
16. Denny-Brown D (1945) Cerebral concussion. Physiol Rev 25:296–325
17. Denny-Brown D, Russell WR (1941) Experimental cerebral concussion. Brain 64:93–164
18. Gurdjian ES, Roberts VL, Thomas LM (1966) Tolerance curves of acceleration and intracranial pressure and protective index in experimental head injury. J Trauma 6(5):600–604
19. Gurdjian ES, Webster JE, Lissner HR (1955) Observations on the mechanism of brain concussion, contusion, and laceration. Surg Gynecol Obstet 101(6):680–690
20. Hodgson VR, Thomas LM, Gurdjian ES, Fernando OU, Greenber SW, Chason JL (1969) Advances in understanding of experimental concussion mecha-

nisms. In: 13th Stapp car crash conference, vol 13. Society of Automotive Engineers, Detroit, MI, pp 18–37
21. Foltz EL, Schmidt RP (1956) The role of the reticular formation in the coma of head injury. J Neurosurg 13(2):145–154. doi:10.3171/jns.1956.13.2.0145
22. Lewis HP, McLaurin RL (1972) Cerebral blood flow and its responsiveness to arterial PCO2: alterations before and after experimental head injury. Surg Forum 23(10):413–415
23. Martins AN, Doyle TF (1977) Blood flow and oxygen consumption of the focally traumatized monkey brain. J Neurosurg 47(3):346–352. doi:10.3171/jns.1977.47.3.0346
24. McCullough D, Nelson KM, Ommaya AK (1971) The acute effects of experimental head injury on the vertebrobasilar circulation: angiographic observations. J Trauma 11(5):422–428
25. Ommaya AK, Grubb RL Jr, Naumann RA (1971) Coup and contre-coup injury: observations on the mechanics of visible brain injuries in the rhesus monkey. J Neurosurg 35(5):503–516. doi:10.3171/jns.1971.35.5.0503
26. Ommaya AK, Hirsch AE (1971) Tolerances for cerebral concussion from head impact and whiplash in primates. J Biomech 4(1):13–21
27. Ommaya AK, Hirsch AE, Flamm ES, Mahone RH (1966) Cerebral concussion in the monkey: an experimental model. Science 153(3732):211–212
28. Ommaya AK, Hirsch AE, Martinez JL (1966) The role of whiplash in cerebral concussion. 10th Stapp car crash conference, vol 10. Society of Automotive Engineers, pp 314–324
29. Gosch HH, Gooding E, Schneider RC (1970) The lexan calvarium for the study of cerebral responses to acute trauma. J Trauma 10(5):370–376
30. Langfitt TW, Tannanbaum HM, Kassell NF (1966) The etiology of acute brain swelling following experimental head injury. J Neurosurg 24(1):47–56. doi:10.3171/jns.1966.24.1.0047
31. Shatsky SA, Evans DE, Miller F, Martins AN (1974) High-speed angiography of experimental head injury. J Neurosurg 41(5):523–530. doi:10.3171/jns.1974.41.5.0523
32. Bakay L, Lee JC, Lee GC, Peng JR (1977) Experimental cerebral concussion. Part 1: an electron microscopic study. J Neurosurg 47(4):525–531. doi:10.3171/jns.1977.47.4.0525
33. Bean JW, Beckman DL (1969) Centrogenic pulmonary pathology in mechanical head injury. J Appl Physiol 27(6):807–812
34. Beckman DL, Bean JW (1970) Pulmonary pressure-volume changes attending head injury. J Appl Physiol 29(5):631–636
35. Hall ED (1985) High-dose glucocorticoid treatment improves neurological recovery in head-injured mice. J Neurosurg 62(6):882–887. doi:10.3171/jns.1985.62.6.0882
36. Tornheim PA, Liwnicz BH, Hirsch CS, Brown DL, McLaurin RL (1983) Acute responses to blunt head trauma. Experimental model and gross pathology. J Neurosurg 59(3):431–438. doi:10.3171/jns.1983.59.3.0431
37. Tornheim PA, McLaurin RL (1981) Acute changes in regional brain water content following experimental closed head injury. J Neurosurg 55(3):407–413. doi:10.3171/jns.1981.55.3.0407
38. Tornheim PA, McLaurin RL, Sawaya R (1979) Effect of furosemide on experimental traumatic cerebral edema. Neurosurgery 4(1):48–52
39. Tornheim PA, McLaurin RL, Thorpe JF (1976) The edema of cerebral contusion. Surg Neurol 5(3):171–175
40. Bergren DR, Beckman DL (1975) Pulmonary surface tension and head injury. J Trauma 15(4):336–338
41. Govons SR, Govons RB, VanHuss WD, Heusner WW (1972) Brain concussion in the rat. Exp Neurol 34(1):121–128
42. Nilsson B, Ponten U, Voigt G (1978) Experimental head injury in the rat. Part I: mechanics, pathophysiology, and morphology in an impact acceleration trauma. J Neurosurg 47:241–251
43. Ommaya AK, Geller A, Parsons LC (1971) The effect of experimental head injury on one-trial learning in rats. Int J Neurosci 1(6):371–378
44. Goldman H, Hodgson V, Morehead M, Hazlett J, Murphy S (1991) Cerebrovascular changes in a rat model of moderate closed-head injury. J Neurotrauma 8(2):129–144
45. Rapoport SI, Fredericks WR, Ohno K, Pettigrew KD (1980) Quantitative aspects of reversible osmotic opening of the blood-brain barrier. Am J Physiol 238(5):R421–R431
46. Unterharnscheidt FJ (1969) Pathomorphology of experimental head injury due to rotational acceleration. Acta Neuropathol 12:200–204
47. Gennarelli TA, Thibault LE (1983) Experimental production of prolonged traumatic coma in the primate. In: Villiani R (ed) Advances in neurotraumatology. Excerpta Medica, Amsterdam, pp 31–33
48. Ommaya AK, Gennarelli TA (1974) Cerebral concussion and traumatic unconsciousness. Correlation of experimental and clinical observations of blunt head injuries. Brain 97(4):633–654
49. Adams JH, Doyle DI (1984) Diffuse brain damage in non-missile head injury. In: Anthony PP, MacSween RNM (eds) Recent advances in histopathology. Churchill Livingstone, Edinburgh, pp 241–257
50. Adams JH, Doyle D, Ford I, Gennarelli TA, Graham DI, McLellan DR (1989) Diffuse axonal injury in head injury: definition, diagnosis and grading. Histopathology 15(1):49–59
51. Adams JH, Doyle D, Graham DI, Lawrence AE, McLellan DR (1986) Gliding contusions in non-missile head injury in humans. Arch Pathol Lab Med 110(6):485–488
52. Adams JH, Graham DI, Gennarelli TA (1985) Contemporary neuropathological considerations regarding brain damage in head injury. In: Becker

DP, Povlishock JT (eds) Central nervous system trauma status report. Sponsored by NIH, NINCDS, Bethesda, pp 143–452
53. Meaney DF, Smith DH, Ross DT, Gennarelli TA (1993) Diffuse axonal injury in the miniature pig: biomechanical development and injury threshold. In: Crashworthiness and occupant protection in transportation systems, ASME, AMD-vol 169/BED-vol 25. New York, NY, pp 169–175
54. Barron KD, Auen EL, Dentinger MP, Nelson L, Bourke R (1980) Reversible astroglial swelling in a trauma-hypoxia brain injury in cat. J Neuropathal Exp Neurol 39:340
55. Nelson LR, Auen EL, Bourke RS, Barron KD (1979) A new head injury model for evaluation of treatment modalities. Neurosci Abstr 5:516
56. Nelson LR, Auen EL, Bourke RS et al (1982) A comparison of animal head injury models developed for treatment modality evaluation. In: Grossman RG, Gildenber PL (eds) Head injury: basic and clinical aspects. Raven, New York, pp 117–128
57. Feeney DM, Boyeson MG, Linn RT, Murray HM, Dail WG (1981) Responses to cortical injury: I. Methodology and local effects of contusions in the rat. Brain Res 211(1):67–77
58. Meyer JS (1956) Studies of cerebral circulation in brain injury. III. Cerebral contusion, laceration and brain stem injury. Electroencephalogr Clin Neurophysiol 8(1):107–116
59. Clifton GL, Lyeth BG, Jenkins LW, Taft WC, DeLorenzo RJ, Hayes RL (1989) Effect of D, alpha-tocopheryl succinate and polyethylene glycol on performance tests after fluid percussion brain injury. J Neurotrauma 6(2):71–81
60. Dixon CE, Lighthall JW, Anderson TE (1988) Physiologic, histopathologic, and cineradiographic characterization of a new fluid-percussion model of experimental brain injury in the rat. J Neurotrauma 5(2):91–104
61. Dixon CE, Lyeth BG, Povlishock JT, Findling RL, Hamm RJ, Marmarou A, Young HF, Hayes RL (1987) A fluid percussion model of experimental brain injury in the rat. J Neurosurg 67(1):110–119. doi:10.3171/jns.1987.67.1.0110
62. Hayes RL, Stalhammar D, Povlishock JT, Allen AM, Galinat BJ, Becker DP, Stonnington HH (1987) A new model of concussive brain injury in the cat produced by extradural fluid volume loading: II. Physiological and neuropathological observations. Brain Inj 1(1):93–112
63. McIntosh TK, Faden AI, Bendall MR, Vink R (1987) Traumatic brain injury in the rat: alterations in brain lactate and pH as characterized by 1H and 31P nuclear magnetic resonance. J Neurochem 49(5): 1530–1540
64. McIntosh TK, Vink R, Noble L, Yamakami I, Fernyak S, Soares H, Faden AL (1989) Traumatic brain injury in the rat: characterization of a lateral fluid-percussion model. Neuroscience 28(1):233–244

65. Povlishock JT, Becker DP, Cheng CL, Vaughan GW (1983) Axonal change in minor head injury. J Neuropathol Exp Neurol 42(3):225–242
66. Povlishock JT, Becker DP, Sullivan HG, Miller JD (1978) Vascular permeability alterations to horseradish peroxidase in experimental brain injury. Brain Res 153(2):223–239
67. Stalhammar D, Galinat BJ, Allen AM, Becker DP, Stonnington HH, Hayes RL (1987) A new model of concussive brain injury in the cat produced by extradural fluid volume loading: I. Biomechanical properties. Brain Inj 1(1):73–91
68. Sullivan HG, Martinez J, Becker DP, Miller JD, Griffith R, Wist AO (1976) Fluid-percussion model of mechanical brain injury in the cat. J Neurosurg 45(5):521–534
69. Lewelt W, Jenkins LW, Miller JD (1980) Autoregulation of cerebral blood flow after experimental fluid percussion injury of the brain. J Neurosurg 53(4):500–511. doi:10.3171/jns.1980.53.4.0500
70. Rimel RW, Giordani B, Barth JT, Jane JA (1982) Moderate head injury: completing the clinical spectrum of brain trauma. Neurosurgery 11(3):344–351
71. Cortez SC, McIntosh TK, Noble LJ (1989) Experimental fluid percussion brain injury: vascular disruption and neuronal and glial alterations. Brain Res 482(2):271–282
72. Anderson TE (1982) A controlled pneumatic technique for experimental spinal cord contusion. J Neurosci Methods 6(4):327–333
73. Lighthall JW (1988) Controlled cortical impact: a new experimental brain injury model. J Neurotrauma 5(1):1–15
74. Dixon CE, Clifton GL, Lighthall JW, Yaghmai AA, Hayes RL (1991) A controlled cortical impact model of traumatic brain injury in the rat. J Neurosci Methods 39(3):253–262
75. Shreiber DI, Bain AC, Meaney DF (1997) In vivo thresholds for mechanical injury to the blood–brain barrier. In: 41st Stapp car crash conference, vol 41. Society of Automotive Engineers, Lake Buena Vista, FL, pp 277–291
76. Gurdjian ES, Lissner HR (1944) Mechanism of head injury as studied by the cathode ray oscilloscope, preliminary report. J Neurosurg 1:393–399
77. Holbourn AH (1943) Mechanics of head injury. Lancet 2:438–441
78. Pudenz RH, Shelden CH (1946) The lucite calvarium; a method for direct observation of the brain; cranial trauma and brain movement. J Neurosurg 3(6):487–505
79. Scott WW (1940) Physiology of concussion. Arch Neurol Psychiatry 43:270–283
80. Walker AE, Kollros JJ, Case TJ (1944) The physiological basis of concussion. J Neurosurg 1:103–116
81. Lee MC, Melvin JW, Ueno K (1987) Finite element analysis of traumatic subdural hematoma. In: 31st

Stapp car crash conference, vol 31. Society of Automotive Engineers, New Orleans, LA, pp 67–77
82. Abel JM, Gennarelli TA, Segawa H (1978) Incidence and severity of cerebral concussion in the rhesus monkey following sagittal plan angular acceleration. In: 22nd Stapp car crash conference, vol 22. Society of Automotive Engineers, Ann Arbor, MI, pp 35–53
83. Lee MC, Haut RC (1989) Insensitivity of tensile failure properties of human bridging veins to strain rate: implications in biomechanics of subdural hematoma. J Biomech 22(6–7):537–542
84. Gennarelli TA, Thibault LE, Tomei G, Wiser R, Graham D, Adams J (1987) Directional dependence of axonal brain injury due to centroidal and noncentroidal acceleration. In: 31st Stapp car crash conference, vol 31. Society of Automotive Engineers, Warrendale, pp 49–53
85. Margulies SS, Thibault LE, Gennarelli TA (1990) Physical model simulations of brain injury in the primate. J Biomech 23(8):823–836
86. Ueno K, Melvin JW, Li L, Lighthall JW (1995) Development of tissue level brain injury criteria by finite element analysis. J Neurotrauma 12(4):695–706
87. Lissner HR, Lebow M, Evans FG (1960) Experimental studies on the relation between acceleration and intracranial pressure changes in man. Surg Gynecol Obstet 111:329–338
88. Gadd CW (1961) Criteria for injury potential. In: Impact acceleration stress symposium, national research council publication no. 977, National Academy of Sciences, Washington DC, pp 141–144
89. Gadd CW (1966) Use of a weighted impulse criterion for estimating injury hazard. In: 10th Stapp Car Crash conference. Society of Automotive Engineers, Holloman Air Force Base, NM, pp 164–174
90. Ono K, Kikuchi A, Nakamura M, Kobayashi H, Nakamura N (1980) Human head tolerance to sagittal impact reliable estimation deduced from experimental head injury using sub-human primates and human cadaver skulls. SAE 801303. In: Proceeding of 24th Stapp car crash conference, Troy, MI
91. Ueno K, Melvin JW, Lundquist E, Lee MC (1989) Two-dimensional finite element analysis of human brain impact responses: application of a scaling law. In: Crashworthiness and occupant protection in transportation systems, AMD-Vol 106. The American Society of Mechanical Engineers, New York, pp 123–124
92. Meaney DF, Smith DH, Shreiber DI, Bain AC, Miller RT, Ross DT, Gennarelli TA (1995) Biomechanical analysis of experimental diffuse axonal injury. J Neurotrauma 12(4):689–694
93. Meaney DF, Thibault LE, Brasko J, Ross DT, Gennarelli TA (1993) Significance of impact velocity in the production of axonal injury in the rat cerebral cortex using rigid indentation. J Neurotrauma 9(3):393
94. Zhou C, Khalil TB, King AI (1994a) Comparison between the impact responses of the human brain and the porcine brain by finite element analysis. In: Proceedings of the 4th injury prevention through biomechanics symposium, vol 4. Wayne State University, Detroit, pp 143–158
95. Zhou C, Khalil TB, King AI (1994b) Shear stress distribution in the porcine brain due to rotational impact. In: 38th Stapp car crash conference, vol 38. Society of Automotive Engineers, pp 133–143
96. Arbogast KB, Margulies SS (1997) Regional differences in mechanical properties of the porcine central nervous system. In: 41st Stapp car crash conference SAE, vol 41. pp 293–300
97. Miller RT, Margulies SS, Leoni M et al (1998) Finite element approaches for predicting injury in an experimental model of severe diffuse axonal injury. In: 42nd Stapp car crash conference, vol 42. Society of Automotive Engineers, Tempe, AZ, pp 155–167
98. Gennarelli TA, Thibault LE, Tipperman R, Tomei G, Sergot R, Brown M, Maxwell WL, Graham DI, Adams JH, Irvine A et al (1989) Axonal injury in the optic nerve: a model simulating diffuse axonal injury in the brain. J Neurosurg 71(2):244–253. doi:10.3171/jns.1989.71.2.0244
99. Bain AC, Biliar KL, Schreiber DI, McIntosh TK, Meaney DF (1996) In vivo thresholds for traumatic axonal damage. In: AGARD specialists meeting, Mescalero
100. Bain AC, Meaney DF (1999) Thresholds for electrophysiological impairment to in vivo white matter. In: Proceedings of the 9th injury prevention through biomechanics symposium, vol 9. Wayne State University, Detroit, MI, pp 67–78
101. Hardy WN, Foster CD, Mason MJ, Yang KH, King AI, Tashman S (2001) Investigation of head injury mechanisms using neutral density technology and high-speed biplanar X-ray. Stapp Car Crash J 45:337–368

Biomechanics of Brain Injury: Looking to the Future

David F. Meaney

Abstract

In this chapter, we review the state of the field in understanding how the cellular components of the brain and spinal cord respond to the biomechanical loading that occurs at the moment of traumatic injury. Several other recent reviews can be collected and reviewed in their own right for this purpose (Kumaria A, Tolias CM, Br J Neurosurg 22(2):200–206, 2008; Morrison B 3rd, Elkin BS, Dolle JP, Yarmush ML, Annu Rev Biomed Eng 13:91–126, 2011; Chen YC, Smith DH, Meaney DF, J Neurotrauma 26(6):861–876, 2009; LaPlaca MC, Simon CM, Prado GR, Cullen DK, Prog Brain Res 161:13–26, 2007). Rather, we intend to provide a broad overview of the basic principles that led to our current understanding of how cells in the nervous system respond to mechanical force. We also point out critical emerging areas in this discipline as we move from molecules, genes, and cells to circuit, behavior and degenerative disease. Our holistic objective is bringing a mechanistic understanding of how mechanotransmission in the CNS can shape the neurobehavioral response of the organism after traumatic CNS injury.

10.1 Introduction

Biomechanics is critical to understanding both human physiology and pathophysiology. Several application domains exist for biomechanics – growth regulation, skeletal development, organism morphogenesis, and regeneration biology.

D.F. Meaney, Ph.D. (✉)
Departments of Bioengineering and Neurosurgery,
University of Pennsylvania, Philadelphia, PA, USA
e-mail: dmeaney@seas.upenn.edu

Impact biomechanics is one of the core sciences used to understand how traumatic loading in the central nervous system results in a change of physiological function. The events and loading scenarios that distinguish physiology from pathophysiology is especially important, as early pathophysiological changes to the CNS can strongly influence circuit plasticity and survival. A key challenge in impact biomechanics is linking the timescale of the problems that occur initially, with the biological consequences that occur later, when formulating modern tolerance criteria.

Brain injury biomechanics has three rather unique characteristics in the field of biomechanics. First, the mechanical event is nearly always considered as a single event that lasts from the microsecond to millisecond timescale, rather than a series of repetitive loading cycles that lasts from minutes to years. Indeed, the brain and spinal cord are considered 'mechanically protected' organs and do not have a clear constant level of mechanical stimulation. Second, traumatic loading is probably the fastest event studied in biomechanics, especially considering the very recent work on blast injury biomechanics. The very brief nature of the event – the acceleration/deceleration event in a blunt impact event is typically delivered in less than 50 ms, while the mechanical loading in a blast event lasts only a few milliseconds – present enormous technical challenges because it means that traditional biomechanical tools to study the forces, displacements, deformations, and stresses in tissue/cellular structures are often lacking. Third, the brain circuitry diagram remains to be fully explored and this large missing piece of information can greatly impact the functional consequences of any applied mechanical force.

Simplified laboratory models in animals, cell culture and even in silico representations provide the main methodological tools currently in use to solve these unexplored areas of brain injury biomechanics. The most pressing issue for any of these models is to establish how each model connects to the clinical condition of human head injury, for it is this connection that will largely establish the meaningful relevance of information generated from these models on modern tolerance criteria. This connection to the clinical picture of brain injury will become increasingly sophisticated as we develop even more fundamental knowledge on how the brain is organized, structured, and how the brain functions.

We define mechanotransmission as a key process in this new understanding of brain injury biomechanics. Mechanotransmission is formally defined as the conversion of mechanical input into the resulting biochemical cascades, and this concept has transformed biomechanics from a study of structure–property relationships into a more integrated structure-function-property triad across length scales in the central nervous system (CNS). At the most reductionist level, this means that we are now beginning to explore how some cells respond directly to the mechanical load with a set of biochemical signatures, even now extending these into genomic signatures. Not surprisingly, as the mechanical load is adjusted to either individual cells or clusters of cells, one may find that some cell types respond while others do not.

10.2 The Complex and Critical Aspects of Mechanical Loading to the Brain During Injury

Over 50 years of research indicates that three characteristics – direction, duration, and type of head motion – are the primary determinants for how the brain moves and deforms within the skull during impact or non-impact (e.g., whiplash) conditions. The most prominent effect of acceleration direction is that it strongly influences brain regions affected during an injury and, thus, the resulting neurological sequelae. The timing or duration of acceleration also contributes to the points within the brain that are maximally affected during impact; generally, cortical regions are affected the most for extremely brief head motions, while the deeper brain structures are influenced more following longer duration accelerations. Third, and perhaps most importantly, the type of acceleration – rotational and/or translational – clearly contributes to the mechanics of loading within the intracerebral tissues.

The past two decades has shown a tremendous growth in the use of computational models to further dissect the role of these three primary loading characteristics. The development of these models, initially developed well after experimental data was collected to describe the in situ biomechanics of loading, has now quickly outpaced the rate of experimental data collection. With increasing frequency, the first exploration of new types of impact/injury loading conditions (e.g., traumatic blast loading) is done virtually with computational techniques (recent studies: [5–14]; reviews: [15–17]). These computational approaches are extremely critical because they provide a mechanical road map for the rational design of in vitro models of CNS trauma that, in turn, will provide critical biological response information at the millimeter and micron scale. In the next section, we review the salient features of the traumatic mechanical loading that need to be reproduced in any in vitro representation of CNS trauma.

10.3 Re-creating the Salient Traumatic Loading Scenarios In Vitro

The primary starting point for any in vitro model of traumatic brain injury is to establish some fidelity to the 'real world' scenario of injures that can span the mild (e.g., concussion) to the more complex and severe (depressed skull fracture and intracranial hematoma). Establishing this model fidelity may be more difficult than it first appears because nearly every clinical head injury occurs in a biomechanically unique situation. As a result, these in vitro models often reduce the complex input into a series of well-prescribed, controllable, and repeatable inputs at the cellular and tissue scale. A desirable trait of these models is to bring the complexity and variability of biology under the more prescribed conditions of an engineer: namely, modulating both stress and strain input independently and monitoring the resulting cellular, molecular and physiological changes to the circuit. Three main factors drive the design of nearly every model system interrogating the biomechanical input-output relationships for cellular and organotypic tissue systems.

The first decision faced by an investigator is the preparation used to interrogate this functional input-output relationship. An ideal and common platform is dissociated cell monolayers, since they offer the opportunity to directly constrain the input (mechanical) and immediately observe the output (biological). Cell lines derived from oncogenic cell populations were the first types of dissociated cells used over two decades ago, principally because there were no available techniques to isolate primary CNS cells from adult or embryonic animals. Beginning in the mid-1990s, though, the commercial availability of feeding media supplements made the isolation of a large number of diverse primary cell types possible, including primary neurons, microglia, astrocytes, and oligodendrocytes. These technologies have developed enough recently so that one may even prepare living slices of brain tissue (absent vascular perfusion), thereby preserving the architecture of the brain at the time of biomechanical testing. These preparations – termed organotypic cultures – represent the most fidelity to the living, in vivo environment. However, even these preparations lack an intact vasculature. Recent work showing how photopolymer and natural protein networks can provide an intact vasculature provides hope that even this technical hurdle of vascularized tissue in vitro is not too far away from solution. To date, though, no investigator has provided an in vitro model that contains an active vascular system to mimic the potential interaction of vascular perfusion on outcome of tissue injured in vitro.

The second major decision facing an investigator is defining the mechanical input. A very pragmatic part of selecting the mechanical input is driven by the output that is measured. For example, one may only want to measure the survival or degeneration of neurons, astrocytes and other cell types in the network for hours to days following the mechanical injury. If this is the only objective, the mechanical input device can be highly customized and the optical path for viewing the cell preparations during and immediately following the injury is not critical. Many questions, though, are critically reliant on how the cellular/tissue system responds immediately following the initial mechanical stimulation; examples include fluorescence-based ion indicators to indirectly assess ion flux immediately after injury, release of neurotransmitters, or changes in the membrane potential [18–21]. For these measures, testing systems often mount directly to a microscope [22–26]. Microscope mounting frequently means that the mechanical testing system is relatively simple, reliable and therefore may have less fidelity to the complex loading that occurs within the in vivo brain. However, the advantages of drawing relationships between these acute events and longer term assays of cellular and organelle viability make these simple models the most widely used. To date, the most versatile of the in vitro models provide the flexibility to challenge the in vitro preparations with more 'real world' mechanical inputs [27]. However, the practical constraints of exploring this parametric space – there are literally millions of combinations that one could explore – limit the utility of these more flexible architectures.

One often overlooked aspect is that the in vivo brain does not experience a uniform mechanical deformation during any traumatic brain injury. Several past experimental and computational studies show that, for the in vivo loading scenario, some brain regions are deformed far more than neighboring regions. Therefore, an additional design element in any in vitro model of traumatic brain injury is to select *how much* of the cellular population will be injured, and the mechanical uniformity of this injury. Many studies apply the injury to the entire cellular populations; a far smaller number apply the injury only to a subregion of the culture (see reviews: [1–3]). These in vitro preparations, since they contain several different cell types, are useful surrogates of the integrated response approximating small regions in the brain. Recently, the ability to mechanically stimulate single individual cells has appeared, opening up new opportunities to explore how single cells can influence the behavior of small neuronal ensembles. Therefore, a closer simulation of the in vivo environment can be targeted by studying how spatially defined patterns of mechanical stimulation can affect the broad behavior of cellular networks.

10.4 Simulating the Critical Mechanical Inputs from the In Vivo Environment – How Do We Know the Mechanical State for Individual Cells in the Brain?

The high water content of brain tissue, combined with its relatively disorganized extracellular matrix, means that brain tissue is one of the softest tissues on the entire human body. Characterized by a high bulk modulus and low shear modulus, brain is often considered weakest under shear loading. Although impact and the subsequent acceleration of the head can produce transient pressures within the cranial vault, several studies have shown that the effects of these pressures that occur during impact are relatively modest. As such, most of the critical inputs that are recreated in in vitro models are linked to the primary shear deformation that occurs throughout the brain at the moment of injury. A common confusion appears, though, when designing how shear deformation can be represented with in vitro systems. This confusion often centers on how macroscopic shear exactly translates to deformations of cellular structures within the complex CNS tissue.

Examples of the tissue-cellular transformation in representative brain tissue samples can help illustrate how easily these concepts can be confused. In a simple shear cube deformation, a dendrite or axon that is oriented along the diagonal of this cube can either compress or extend, based on the initial orientation of the dendrite/axon. In comparison, if the dendrite/axons align along one edge of the cube, or is oriented in the plane of the cube, it would experience no deformation during simple shear.

The correct prediction of how bulk or macroscopic mechanical loading causes deformation of cellular structures – illustrated in the simple example of the previous paragraph – is an enormously significant problem in CNS mechanics. It is a problem that is highly influenced by the boundary conditions – i.e., how does the cellular, vascular, and extracellular structure couple together? It is also influenced greatly by the initial conditions – e.g., does the axon or dendrite traverse a normal, straight course through the tissue or does it have a meandering, tortuous path through the tissue? Indeed, a key knowledge gap now being filled is building structurally accurate models of CNS tissue that can be used to understand how an applied mechanical loading at the macroscopic scale – i.e., a sample of excised tissue, or a small volume of brain tissue in vivo- results in the deformation of cellular and subcellular structures [28–32]. Even simple models of composite material behavior show that cellular/subcellular deformation is very complex, depends critically on the mechanical coupling among components, and may vary substantially across different brain regions subject to the same macroscopic loading because of differences in tissue microstructure. Experimental platforms are also showing how subcellular structures are deformed uniquely in response to the applied mechanical loading [33].

With a continuing focus on the role that deformation plays in the subsequent injury sequelae to the brain, several technologies have emerged to study this phenomenon in vitro. In a crossover of technology from the vascular hemodynamics field, fluid flowing quickly past the model layer of cells will induce a deformation sufficiently high enough to activate glutamate receptors and, at higher levels, cause an immediate increase in plasma membrane permeability and changes in network dynamics [34, 35]. Alternatively, cells in two-dimensional culture can be moved rapidly and then stopped very suddenly [36]. In this model, the different densities of subcellular organelles and nuclear components will cause in intracellular deformation. In both of these techniques, it is difficult to establish the in vivo biomechanical fidelity of these models. Rather, the techniques are useful to study input/output responses not possible with other methods.

By far the most common method to study input-output relationships for cellular populations is stretching flexible substrates, where the cells of interest are adhered directly to the substrate [22–26]. Many studies evaluate the effects of deforming the cell population in two perpendicular directions simultaneously (termed 'biaxial'), while a less frequently used method deforms the substrate primarily along one axis ('uniaxial'). The uniaxial configuration is most readily comparable to shear deformation in vivo, as the principal directions of stretch are identical to a pure shear deformation. The biaxial method is more complex, because it involves a mixture of planar expansion and pure shear. Some differences appear when comparing these techniques directly: the uniaxial model of deformation requires higher peak deformations to cause immediate alterations in plasmalemma permeability, while biaxial deformation offers a symmetric deformation field that can apply a more homogeneous input to dissociated or organotypic culture networks.

Applying deformations to cells in vitro is a relatively straightforward technique, and has been the study of several past laboratories. However, the application of a defined stress, both the direction and magnitude of the stress, is a much more difficult technical challenge. One must first define the stress level that is appropriate. Next, one should consider how to focally apply these controlled stresses without affecting neighboring cells. Third, the material properties of the cellular and subcellular constituents may mean the loading regime will differ uniquely among individual cells, as well as across regions within individual cells. It is perhaps for this reason that models to accurately apply stresses at the micron and submicron scale are lacking still in the literature.

10.5 The Critical Components of the Early Response Are Mechanosensors

We define mechanosensors as the critical receptors/channels/organelles on neurons and glia that respond directly to applied mechanical stimulation. Like many sensors, the biological entities contain a 'dynamic range' – they often have a critical threshold for activation, and this threshold often changes when the force is applied dynamically, rather than statically. Much like how a typical instrumentation transducer will show a dynamic operating range over a range of frequencies, these mechanosensors are often equipped with a molecular architecture that makes them tuned for a particular part of the mechanical force spectrum. Models for the mechanically-influenced binding of ligands to receptors – a 'mechanokinetics' of receptor activation – have been proposed for some ligand-receptor interactions [37], but none have been applied to channels and receptors expressed in neurons and glia.

Analogous to developing measures of how an agonist binds to a receptor over a range of ligand concentrations, we expect mechanical sensors will possess a minimum force threshold (concentration) for their activation, and a saturating force (saturating agonist) for their maximal activation levels. Likewise, mechanosensors may respond differently to competing mechanical inputs-for example, some receptors may be sensitive only to the pressure applied, while other receptors will be sensitive to a certain type of deformation (i.e. tension), and still yet others may change their sensitivity to deformation based on the applied input pressure.

Despite the potential importance of mechanosensors, we know very little about the mechanoactivation profiles for most receptors expressed in the brain. Evidence shows that the NMDA receptor can respond or 'sense' mechanical forces that are applied during traumatic injury [38–40]. One important physiological consequence of this mechanical loading is the loss of magnesium block in the NMDA receptor, a feature that normally limits the ion flux across the receptor at normal resting membrane potentials [39]. Moreover, evidence shows that the NMDAR agonist affinity is altered after mechanical stretch. Linking the NMDAR to the cytoskeleton appears key for this mechanosensing property, and a single serine residue of the NR2B subunit of the receptor appears to substantially influence its mechanical force sensing ability [41]. With the role of the NR2B subunit of the NMDAR for neuronal death, the unique mechanosensing ability of the NR2B subunit appears uniquely positioned to control the survivability of neurons in the circuit after mechanical injury [42].

The α-subunit of the voltage gated sodium channel is a second mechanosensor [43–45]. Following activation of the sensor, changes in calcium homeostasis can appear and lead to the proteolytic cleavage of this channel, setting up a scenario where the activation of a channel may lead to its direct modulation. One direct consequence of altering sodium channel structure after injury is an effect on the network dynamics, suppressing the activity of the network until compensatory changes in the network can help restore a balance in activity. Intriguingly, this breakdown of the channel by proteases is influenced by a part of the cell cytoskeleton (microtubule), suggesting that the mechanical structure of the neuron confers some level of protection against degenerative changes that can be caused by mechanical trauma. This rapid change (<15 min) in channel integrity is contrasted with a later stage degradation phase (4–6 h) that precedes any evidence of neuronal death [46]. However, it is not yet clear if this later phase of degradation can be blocked to rescue neurons in the network, or if this later degradation is part of a larger irreversible sequence of events that will lead to the removal of the neuron from the injured network.

Other receptors are known to either activate rapidly and/or change their properties after mechanical injury, but there is no conclusive evidence yet to term these as primary mechanosensors. For example, AMPA receptors lose one important physiological property early after mechanical loading; they express a loss in the desensitization of the receptor [47]. The reason(s) for this loss in desensitization of the receptor are not completely known, but it may owe more to the upstream mechanoactivation of the NMDAR than the molecular structure of the receptor and any postsynaptic tethering to the cytoskeleton [48]. In this light, AMPA receptor modulation is a secondary effector for the primary mechanosensor (NMDAR). In addition, changes in GABA receptor currents appear after injury, but they also depend directly on NMDAR activation [49]. Similarly, focal mechanical stimulation in glial cells can elicit broad intercellular waves that are mediated by release of ATP and glutamate [50]. Although the mechanism(s) responsible for intercellular wave transmission is well known, specific mechanoactivated receptors in astrocytes are not well described for this phenomenon. Thus, although glial cells may have primary mechanosensors, this is inevitably complicated with the stimulated release of agonists, that can in turn activate other receptors.

As many candidate mechanosensors play important roles in cellular and neural circuit physiology, the activation of these mechanosensors can have broad impact on function.

However, only scant evidence appears for mechanosensing properties of other glutamate receptors and channels in neurons. Future knowledge of sensors that are associated with specific neuronal subtypes is also needed, as well as knowing the differential role(s) for mechanosensors in adjusting the balance between inhibitory and excitatory circuitry in the brain after trauma. To this end, glial cell types may also provide a critical modulatory role that can influence the rebuilding of network function for days to weeks following initial injury.

10.6 Do These Mechanosensors Cause Functional Changes?

Across length scales – from individual synapses, neurons, microcircuits and cell clusters to whole brain regions – mechanosensors very likely play a profound role in the recovery of circuits after injury. The most straightforward effect on function is when neurons and glia within an intact circuit degenerate, thereby degrading the information capacity of the circuit. Although the mechanical tolerance for neuronal survival exists for several testing systems, and these estimates span both dissociated and organotypic slice culture preparations [51–53], no published work extends how the function of the system is altered with this degeneration. Relatedly, new evidence shows the spontaneous electrical activity of the network can be impaired with mechanical stimulation [35, 54], yet there is no consensus yet on how these thresholds for activity impairments correspond to the thresholds for neuronal death. Several changes in the neural circuit state can now be traced back to mechanoactivation of the NMDA receptor, including a slow, neuronal depolarization that can lead to altered network excitability, the loss of AMPAR desensitization that can also change excitability, and the insertion of new glutamate receptor subtypes that appear over several hours (see [3]). At this time, though, little data exists on how the function of individual neurons within the network contributes to the overall network dynamics of the circuit.

The recovering neural circuit is an evolving entity, with alterations occurring at both the synaptic and connectivity level. New network mapping methods estimate the connection map among neuronal ensembles, and these methods show the clear development of circuit connections that naturally occur during neural development [55]. Optogenetic tools provide an important complement to these mapping methods, since they provide a resolution to activate or inhibit the activity of single/multiple network nodes within the network population [56], providing an opportunity to draw functional relationships and identify neurons driving rhythms in networks, or shaping the complex patterns that emerge in these networks. It is reasonable to propose that the patterns of recovery within an intact neural circuit are influenced by the neurons degenerating within the circuit, and recent observations show that the persisting patterns are certainly influenced by neuronal loss [54]. One future area of investigation could study a potentially important complement to pattern generation within a network – the potential to synchronize a network. Network theory is now showing that highly connected nodes can influence the synchronization of a network, but the relative mechanical tolerance of a mechanical circuit to synchronization would provide important insights into the lost plasticity of the network after injury.

10.7 Can We Use *In Silico* Simulations as a Complement to In Vitro Systems?

The fundamental building block of neurotransmission – the release of a glutamate vesicle from the presynaptic terminal and its subsequent binding to glutamate receptors on the postsynaptic surface – is a well-studied process in neuroscience. In the past decade, this largely experimental work has now turned computational. Glutamate binding to its cognate receptors is a stochastic phenomenon at the single synaptic scale and is influenced heavily by the amount, location, temporal timing (e.g., single, multivesicular) and the composition of the postsynaptic surface [57–59]. Many, if not all, of these features are likely altered in the traumatically injured brain. Already, there is evidence that the loss of the magnesium block can provide an important functional consequence for the NMDA receptor in the traumatically injured network [42]. The evolution of the synapse and its glutamate receptor composition within hours of injury can clearly disrupt the timing of the signaling within circuits. In parallel with these postsynaptic models of neurotransmission, relatively recent models of the presynaptic and postsynaptic signaling networks show how changes at the pre/post-synaptic terminal can also contribute to the precise timing and duration of signaling network activation in the postsynaptic zone [59, 60, 61–63].

With the integration of multiple synaptic inputs onto individual neurons, the next level of analysis involves how these individual dendritic inputs, regardless of their noisy behavior, will influence the activation signal propagation throughout neuronal networks. A large number of simple models of different neuronal types can be developed and solved analytically, using approximations of the neural connectivity that appears in vivo. These computational models can become increasingly complex by incorporating features that one measures at the individual dendritic and synaptic scale, building models of networks with high anatomic fidelity. The use of these more elegant models to understand posttraumatic signaling, though, is not available in the literature. One recent analysis using a simple, reductionist model from our group revealed that a key feature of injury is the loss in ability to synchronize signals among neuronal ensembles [64]. This ability to synchronize was critically dependent on the activation of calpain, a common calcium activated neutral protease capable of cleaving the α-subunit of the sodium channel to disrupt the timing of networks [54]. Expanding the use of these simple models, adapted to include the stochastic properties of receptor activation at the single spine, represents great promise for the future to ultimately decode how single cell scale changes can influence the behavior of broader neuronal networks.

10.8 Looking into the Future: Multiscale Questions and the Key for the Future

Using in vitro models of neurotrauma to examine the multiscale behavior of function in the brain after trauma eventually lead us back to the beginning. Namely, can we use these models to understand how the neurobehavior of an animal or human is altered after trauma? This multiscale question is certainly not unique to neurotrauma; this is one of the penultimate questions broadly across neuroscience. We are moving towards this level of understanding in the traumatically injured brain with a newfound appreciation of how events at the synaptic, dendritic, single neuron, and the neuronal/glial ensemble levels integrate to affect this overlying behavior. Moreover, we may soon realize that the coordinated activation of ensembles across major pathways in the brain may provide one of the keys in recovery after injury. As we look forward to the future, it is likely that our information at the in vitro scale will feed into important insights into how the consequences of traumatic mechanical loading condition will help synchronize the activity of neural circuits in vivo, and also how the cellular/molecular mechanotransmission process will influence the capacity of these circuits to relay information across this frequency spectrum. In concert with understanding how these circuits are synchronized to optimize information transfer, we will be in a much better position to understand how behaviors can change after traumatic injury, as well as understanding effective therapies to improve outcome after these injuries.

References

1. Kumaria A, Tolias CM (2008) In vitro models of neurotrauma. Br J Neurosurg 22(2):200–206. doi:10.1080/02688690701772413
2. Morrison B 3rd, Elkin BS, Dolle JP, Yarmush ML (2011) In vitro models of traumatic brain injury. Annu Rev Biomed Eng 13:91–126. doi:10.1146/annurev-bioeng-071910-124706
3. Chen YC, Smith DH, Meaney DF (2009) In-vitro approaches for studying blast-induced traumatic brain

injury. J Neurotrauma 26(6):861–876. doi:10.1089/neu.2008.0645

4. LaPlaca MC, Simon CM, Prado GR, Cullen DK (2007) CNS injury biomechanics and experimental models. Prog Brain Res 161:13–26. doi:10.1016/S0079-6123(06)61002-9

5. Sundaramurthy A, Alai A, Ganpule S, Holmberg A, Plougonven E, Chandra N (2012) Blast-induced biomechanical loading of the rat: an experimental and anatomically accurate computational blast injury model. J Neurotrauma. doi:10.1089/neu.2012.2413

6. Cloots RJ, van Dommelen JA, Kleiven S, Geers MG (2012) Multi-scale mechanics of traumatic brain injury: predicting axonal strains from head loads. Biomech Model Mechanobiol. doi:10.1007/s10237-012-0387-6

7. Panzer MB, Myers BS, Capehart BP, Bass CR (2012) Development of a finite element model for blast brain injury and the effects of CSF cavitation. Ann Biomed Eng. doi:10.1007/s10439-012-0519-2

8. Coats B, Eucker SA, Sullivan S, Margulies SS (2012) Finite element model predictions of intracranial hemorrhage from non-impact, rapid head rotations in the piglet. Int J Dev Neurosci 30(3):191–200. doi:10.1016/j.ijdevneu.2011.12.009

9. Lamy M, Baumgartner D, Willinger R, Yoganandan N, Stemper BD (2011) Study of mild traumatic brain injuries using experiments and finite element modeling. Ann Adv Automot Med. Annual Scientific Conference Association for the Advancement of Automotive Medicine Association for the Advancement of Automotive Medicine Scientific Conference 55:125–135

10. Chatelin S, Deck C, Renard F, Kremer S, Heinrich C, Armspach JP, Willinger R (2011) Computation of axonal elongation in head trauma finite element simulation. J Mech Behav Biomed Mater 4(8):1905–1919. doi:10.1016/j.jmbbm.2011.06.007

11. Kimpara H, Iwamoto M (2012) Mild traumatic brain injury predictors based on angular accelerations during impacts. Ann Biomed Eng 40(1):114–126. doi:10.1007/s10439-011-0414-2

12. Zhu F, Mao H, Dal Cengio Leonardi A, Wagner C, Chou C, Jin X, Bir C, Vandevord P, Yang KH, King AI (2010) Development of an FE model of the rat head subjected to air shock loading. Stapp Car Crash J 54:211–225

13. Nyein MK, Jason AM, Yu L, Pita CM, Joannopoulos JD, Moore DF, Radovitzky RA (2010) In silico investigation of intracranial blast mitigation with relevance to military traumatic brain injury. Proc Natl Acad Sci U S A 107(48):20703–20708. doi:10.1073/pnas.1014786107

14. Ho J, Kleiven S (2009) Can sulci protect the brain from traumatic injury? J Biomech 42(13):2074–2080. doi:10.1016/j.jbiomech.2009.06.051

15. King AI, Ruan JS, Zhou C, Hardy WN, Khalil TB (1995) Recent advances in biomechanics of brain injury research: a review. J Neurotrauma 12(4):651–658

16. Voo K, Kumaresan S, Pintar FA, Yoganandan N, Sances A Jr (1996) Finite-element models of the human head. Med Biol Eng Comput 34(5):375–381

17. Cohen AS, Pfister BJ, Schwarzbach E, Grady MS, Goforth PB, Satin LS (2007) Injury-induced alterations in CNS electrophysiology. Prog Brain Res 161:143–169. doi:10.1016/S0079-6123(06)61010-8

18. Lusardi TA, Wolf JA, Putt ME, Smith DH, Meaney DF (2004) Effect of acute calcium influx after mechanical stretch injury in vitro on the viability of hippocampal neurons. J Neurotrauma 21(1):61–72. doi:10.1089/089771504772695959

19. Geddes DM, LaPlaca MC, Cargill RS 2nd (2003) Susceptibility of hippocampal neurons to mechanically induced injury. Exp Neurol 184(1):420–427

20. Tavalin SJ, Ellis EF, Satin LS (1995) Mechanical perturbation of cultured cortical neurons reveals a stretch-induced delayed depolarization. J Neurophysiol 74(6):2767–2773

21. LaPlaca MC, Thibault LE (1998) Dynamic mechanical deformation of neurons triggers an acute calcium response and cell injury involving the N-methyl-D-aspartate glutamate receptor. J Neurosci Res 52(2):220–229

22. Cargill RS 2nd, Thibault LE (1996) Acute alterations in [Ca2+]i in NG108-15 cells subjected to high strain rate deformation and chemical hypoxia: an in vitro model for neural trauma. J Neurotrauma 13(7):395–407

23. McKinney JS, Willoughby KA, Liang S, Ellis EF (1996) Stretch-induced injury of cultured neuronal, glial, and endothelial cells. Effect of polyethylene glycol-conjugated superoxide dismutase. Stroke 27(5):934–940

24. Smith DH, Wolf JA, Lusardi TA, Lee VM, Meaney DF (1999) High tolerance and delayed elastic response of cultured axons to dynamic stretch injury. J Neurosci 19(11):4263–4269

25. Lusardi TA, Rangan J, Sun D, Smith DH, Meaney DF (2004) A device to study the initiation and propagation of calcium transients in cultured neurons after mechanical stretch. Ann Biomed Eng 32(11):1546–1558

26. Morrison B 3rd, Meaney DF, McIntosh TK (1998) Mechanical characterization of an in vitro device designed to quantitatively injure living brain tissue. Ann Biomed Eng 26(3):381–390

27. Morrison B 3rd, Cater HL, Benham CD, Sundstrom LE (2006) An in vitro model of traumatic brain injury utilising two-dimensional stretch of organotypic hippocampal slice cultures. J Neurosci Methods 150(2):192–201. doi:10.1016/j.jneumeth.2005.06.014

28. Meaney DF (2003) Relationship between structural modeling and hyperelastic material behavior: application to CNS white matter. Biomech Model Mechanobiol 1(4):279–293. doi:10.1007/s10237-002-0020-1

29. Karami G, Grundman N, Abolfathi N, Naik A, Ziejewski M (2009) A micromechanical hyperelastic modeling of brain white matter under large deformation. J Mech Behav Biomed Mater 2(3):243–254. doi:10.1016/j.jmbbm.2008.08.003

30. Bain AC, Shreiber DI, Meaney DF (2003) Modeling of microstructural kinematics during simple elongation of central nervous system tissue. J Biomech Eng 125(6):798–804

31. Pan Y, Shreiber DI, Pelegri AA (2011) A transition model for finite element simulation of kinematics of central nervous system white matter. IEEE Trans Biomed Eng 58(12):3443–3446. doi:10.1109/TBME.2011.2163189
32. Cohen TS, Smith AW, Massouros PG, Bayly PV, Shen AQ, Genin GM (2008) Inelastic behavior in repeated shearing of bovine white matter. J Biomech Eng 130(4):044504. doi:10.1115/1.2939290
33. LaPlaca MC, Cullen DK, McLoughlin JJ, Cargill RS 2nd (2005) High rate shear strain of three-dimensional neural cell cultures: a new in vitro traumatic brain injury model. J Biomech 38(5):1093–1105. doi:10.1016/j.jbiomech.2004.05.032
34. LaPlaca MC, Thibault LE (1997) An in vitro traumatic injury model to examine the response of neurons to a hydrodynamically-induced deformation. Ann Biomed Eng 25(4):665–677
35. Prado GR, Ross JD, DeWeerth SP, LaPlaca MC (2005) Mechanical trauma induces immediate changes in neuronal network activity. J Neural Eng 2(4):148–158. doi:10.1088/1741-2560/2/4/011
36. Murphy EJ, Horrocks LA (1993) A model for compression trauma: pressure-induced injury in cell cultures. J Neurotrauma 10(4):431–444
37. Bell GI (1978) Models for the specific adhesion of cells to cells. Science 200(4342):618–627
38. Paoletti P, Ascher P (1994) Mechanosensitivity of NMDA receptors in cultured mouse central neurons. Neuron 13(3):645–655
39. Zhang L, Rzigalinski BA, Ellis EF, Satin LS (1996) Reduction of voltage-dependent Mg2+ blockade of NMDA current in mechanically injured neurons. Science 274(5294):1921–1923
40. Kloda A, Lua L, Hall R, Adams DJ, Martinac B (2007) Liposome reconstitution and modulation of recombinant N-methyl-D-aspartate receptor channels by membrane stretch. Proc Natl Acad Sci U S A 104(5):1540–1545. doi:10.1073/pnas.0609649104
41. Singh P, Doshi S, Spaethling JM, Hockenberry AJ, Patel TP, Geddes-Klein DM, Lynch DR, Meaney DF (2012) N-methyl-D-aspartate receptor mechanosensitivity is governed by C terminus of NR2B subunit. J Biol Chem 287(6):4348–4359. doi:10.1074/jbc.M111.253740
42. DeRidder MN, Simon MJ, Siman R, Auberson YP, Raghupathi R, Meaney DF (2006) Traumatic mechanical injury to the hippocampus in vitro causes regional caspase-3 and calpain activation that is influenced by NMDA receptor subunit composition. Neurobiol Dis 22(1):165–176. doi:10.1016/j.nbd.2005.10.011
43. von Reyn CR, Spaethling JM, Mesfin MN, Ma M, Neumar RW, Smith DH, Siman R, Meaney DF (2009) Calpain mediates proteolysis of the voltage-gated sodium channel alpha-subunit. J Neurosci 29(33): 10350–10356. doi:10.1523/JNEUROSCI.2339-09.2009
44. Iwata A, Stys PK, Wolf JA, Chen XH, Taylor AG, Meaney DF, Smith DH (2004) Traumatic axonal injury induces proteolytic cleavage of the voltage-gated sodium channels modulated by tetrodotoxin and protease inhibitors. J Neurosci 24(19):4605–4613. doi:10.1523/JNEUROSCI.0515-03.2004
45. Wolf JA, Stys PK, Lusardi T, Meaney D, Smith DH (2001) Traumatic axonal injury induces calcium influx modulated by tetrodotoxin-sensitive sodium channels. J Neurosci 21(6):1923–1930
46. von Reyn CR, Mott RE, Siman R, Smith DH, Meaney DF (2012) Mechanisms of calpain mediated proteolysis of voltage gated sodium channel alpha-subunits following in vitro dynamic stretch injury. J Neurochem 121(5):793–805.doi:10.1111/j.1471-4159.2012.07735.x
47. Goforth PB, Ellis EF, Satin LS (1999) Enhancement of AMPA-mediated current after traumatic injury in cortical neurons. J Neurosci 19(17):7367–7374
48. Goforth PB, Ellis EF, Satin LS (2004) Mechanical injury modulates AMPA receptor kinetics via an NMDA receptor-dependent pathway. J Neurotrauma 21(6):719–732. doi:10.1089/0897715041269704
49. Kao CQ, Goforth PB, Ellis EF, Satin LS (2004) Potentiation of GABA(A) currents after mechanical injury of cortical neurons. J Neurotrauma 21(3):259–270. doi:10.1089/089771504322972059
50. Charles AC, Merrill JE, Dirksen ER, Sanderson MJ (1991) Intercellular signaling in glial cells: calcium waves and oscillations in response to mechanical stimulation and glutamate. Neuron 6(6):983–992
51. Geddes DM, Cargill RS 2nd (2001) An in vitro model of neural trauma: device characterization and calcium response to mechanical stretch. J Biomech Eng 123(3):247–255
52. Elkin BS, Morrison B 3rd (2007) Region-specific tolerance criteria for the living brain. Stapp Car Crash J 51:127–138
53. Cater HL, Sundstrom LE, Morrison B 3rd (2006) Temporal development of hippocampal cell death is dependent on tissue strain but not strain rate. J Biomech 39(15):2810–2818. doi:10.1016/j.jbiomech.2005.09.023
54. Patel TP, Ventre SC, Meaney DF (2012) Dynamic changes in neural circuit topology following mild mechanical injury in vitro. Ann Biomed Eng 40(1):23–36. doi:10.1007/s10439-011-0390-6
55. Soriano J, Rodriguez Martinez M, Tlusty T, Moses E (2008) Development of input connections in neural cultures. Proc Natl Acad Sci U S A 105(37):13758–13763. doi:10.1073/pnas.0707492105
56. Bernstein JG, Garrity PA, Boyden ES (2012) Optogenetics and thermogenetics: technologies for controlling the activity of targeted cells within intact neural circuits. Curr Opin Neurobiol 22(1):61–71. doi:10.1016/j.conb.2011.10.023
57. Franks KM, Bartol TM Jr, Sejnowski TJ (2002) A Monte Carlo model reveals independent signaling at central glutamatergic synapses. Biophys J 83(5):2333–2348. doi:10.1016/S0006-3495(02)75248-X
58. Franks KM, Stevens CF, Sejnowski TJ (2003) Independent sources of quantal variability at single glutamatergic synapses. J Neurosci 23(8):3186–3195
59. Santucci DM, Raghavachari S (2008) The effects of NR2 subunit-dependent NMDA receptor kinetics on

synaptic transmission and CaMKII activation. PLoS Comput Biol 4(10):e1000208. doi:10.1371/journal.pcbi.1000208
60. Faas GC, Raghavachari S, Lisman JE, Mody I (2011) Calmodulin as a direct detector of Ca2+ signals. Nat Neurosci 14(3):301–304. doi:10.1038/nn.2746
61. Nadkarni S, Bartol TM, Sejnowski TJ, Levine H (2010) Modelling vesicular release at hippocampal synapses. PLoS Comput Biol 6(11):e1000983. doi:10.1371/journal.pcbi.1000983
62. Volman V, Levine H, Ben-Jacob E, Sejnowski TJ (2009) Locally balanced dendritic integration by short-term synaptic plasticity and active dendritic conductances. J Neurophysiol 102(6):3234–3250. doi:10.1152/jn.00260.2009
63. Keller DX, Franks KM, Bartol TM Jr, Sejnowski TJ (2008) Calmodulin activation by calcium transients in the postsynaptic density of dendritic spines. PLoS One 3(4):e2045. doi:10.1371/journal.pone.0002045
64. Volpicelli-Daley LA, Luk KC, Patel TP, Tanik SA, Riddle DM, Stieber A, Meaney DF, Trojanowski JQ, Lee VM (2011) Exogenous alpha-synuclein fibrils induce Lewy body pathology leading to synaptic dysfunction and neuron death. Neuron 72(1):57–71. doi:10.1016/j.neuron.2011.08.033

Neck Injury Biomechanics

11

Roger W. Nightingale, Barry S. Myers, and Narayan Yoganandan

Abstract

This chapter summarizes the research on the biomechanical responses of the neck in a form that will be useful in design of protective systems and in the development of societal strategies to reduce the number of cervical spine injuries. Neck injuries are described in detail and injury mechanisms are explained. Accidents that involve neck injuries are analyzed. Real-life neck injuries are presented and laboratory results are synthesized to provide a rational basis for understanding neck injury. In order to maintain a reasonable scope, we restrict our discussion to severe fractures – those with a high probability of spinal cord injury. Low AIS injuries, such as whiplash, sprains and strains are left to other sources.

Most of what we know about catastrophic neck injury comes from two sources: clinical studies including accident reconstruction, and cadaver studies. In the last 30 years, the field of biomechanics has made significant gains in understanding the complex dynamics that accompany both head-impact neck injury and inertial neck injury. Cervical injury mechanisms involve a wide range of loading modes with various combinations of force and concurrent bending moment. Using the historical concepts of excessive head motion as a cause of neck injury frequently leads to paradoxical and incorrect conclusions regarding neck injury mechanism. In contrast, when the neck is viewed as a mechanical structure, whose injury is a consequence of the deformations of a segmented beam column

---

R.W. Nightingale, Ph.D. (✉)
B.S. Myers, M.D., Ph.D.
Biomedical Engineering, Duke University,
Durham, NC, USA
e-mail: rwn@duke.edu; bsm@duke.edu

N. Yoganandan, Ph.D.
Department of Neurosurgery, Medical College
of Wisconsin, Milwaukee, WI, USA
e-mail: yoga@mcw.edu

bounded by two large masses, the neck's mechanical behavior and the resulting traumatic injuries can be more readily explained. That is, by considering the effects of buckling and examining the forces and moments that act at the level of injury in the spine at the instant of injury, the mechanisms of each injury become clear and rational.

From a mechanical and structural point of view, the cervical spine is a very complex mechanism. The human neck contains vital neurologic, vascular, and respiratory structures as well as the cervical vertebrae, musculature, and spinal cord. Although injury statistics generally attribute only 2–4 % of serious trauma to the neck, any neck injury can have debilitating if not life-threatening consequences. Permanent paralysis is a particularly devastating and costly injury. When it is a consequence of accidental trauma, frequently a young productive member of society is transformed into a highly dependent member. The advent of high-speed land and air transportation has made us increasingly aware of the serious consequences that can result from a structural failure of the neck. Also, as more people pursue leisure-time activities, the potential for serious neck injuries increases. Football, diving, gymnastics, skiing, hang gliding, mountain climbing, surfing and cycling are examples of activities with ample kinetic energy to expose the neck to a risk of serious injury. As a result, a variety of devices have been developed that offer a measure of protection to the head and neck from mechanical trauma. Head and seat restraints, motorcycle and football helmets, energy-absorbing pads and collars, and gymnastic mats are a few examples of head and neck protective devices. Unfortunately, the design of many of these has proceeded with insufficient biomechanical input owing historically to a lack of data. However, as a result of research beginning in the late 1980s supported in large part by the NHTSA and the CDC, the lack of data or a robust mechanistic understanding of the neck is no longer a limitation in design.

With the publication of the first edition of this text, the view of the neck as a slender, viscoelastic, segmented beam-column had been formulated and significant science on compression-flexion and compression-extension had been conducted. The publication of the second edition in 2002 saw significant advances in our understanding of cervical spine injury, especially in bending and in tension. The bending results in particular have necessitated a re-assessment of the role of this load in cervical spine injury mechanism and tolerance. With this evolving understanding came new injury criteria integrating bending and tension compression into a single metric (Nij) and harmonizing basic-science research with field epidemiology that culminated in new FMVSS 208 safety standards. Although the general descriptions of injuries and mechanisms at the segment level have remained relatively unchanged, there has been a significant shift in our understanding of the importance of compression and tension as the external loads. In this third edition, we incorporate these changes and focus on the mechanical factors influencing injury. There is less emphasis on historically valuable classifications of injury based on clinical experience, which often suffer from highly anecdotal evidence, confusing language, and in many cases, incorrect assertions of how the neck gets hurt.

11.1 Incidence

The National Head and Spinal Cord Injury Survey estimated the occurrence of spinal cord injury with quadriplegia in the United States at 5 per 100,000 or in excess of 10,000 cases each year. According to the National Spinal Cord Injury Database, which catalogues data from approximately 15 % of these new cases, cervical lesions were documented in 54 % of discharged patients with 51 % in the lower cervical spine: 14.7 % at C4, 15.3 % at C5, and 10.6 % at C6 (www.nscisc.uab.edu, 2012). Despite the relative low frequency of these injuries, the large direct medical costs and lost productivity result in annual costs that are conservatively estimated at $97 billion [1].

While injuries to the cervical spine can result from almost any activity, the literature suggests that automobile accidents, sports, gunshot wounds, and falls are the circumstances most often identified. Unfortunately, injuries due to violence (mainly gunshot wounds) have increased dramatically over the last three decades (www.nscisc.uab.edu, 2012).

Automobile and aircraft accidents undoubtedly produce extensive injuries because of the speeds involved and the associated energy that must be dissipated in a crash. According to Huelke et al., 56 % of all spinal cord injuries are the result of highway accidents, with 67 % of those involved being vehicle occupants [2]. Pedestrians and motorcyclists were also significantly involved in the injury statistics. Other studies of automobile and highway-related cervical spine injuries include those of Alker et al., Mertz et al., Schutt and Dohan, Sims et al., Thorson, Tonga et al., Voight and Wilfert, Yoganandan et al., and Yule.

Automobile restraint systems have also been associated with spinal cord in juries. These are frequently described as noncontact injuries produced in crashes with the upper torso belt restraining the torso [3–11]. It should be realized, however, that the reduction in injury by restraint systems far exceeds these relatively uncommon injuries.

Sports and leisure activities account for a significant portion of injuries to the cervical spine. In 1978, Shield et al. analyzed 10 years of data on 152 cervical spinal cord injuries caused by sports participation that were treated at the Rancho Los Amigos Hospital [12]. Considerable regional variation in sports-related cervical injury exists primarily due to the degree of participation in the various activities. However, these injuries can be generally classified as most commonly resulting from contact sports, of which football is the most well-known, and falls and dives from height. The incidence of cervical injuries in the game of football is described by Torg and Otis using the National Football Injury Register [13–18].

The development and use of improved head protection in american football resulted in a significant increase in cervical injury in the 1960s. Because of the increased security of facial and head protection, players, unfortunately, began using the head as a method for tackling other players. Awareness of this problem, and the subsequent use of "heads-up" tackling together with penalizing spearing, resulted in a significant drop in neck injury in football.

Schneider, in his book on football injuries as well as in his numerous papers, related trauma due to the impingement of the rear of the helmet shell on the neck [19]. Subsequent authors have described the role of the helmet in producing neck injuries and have questioned the mechanism proposed by Schneider. These authors, such as Hodgson and Thomas, Mertz et al., and Virgin, have attempted, without success, to verify Schneider's experiments [20–22].

Swimming and diving accidents have also been identified as causing significant numbers of fractures and dislocations of the cervical spine. Kewairamani and Taylor found that 18 % of all spinal cord injuries in their series were related to diving accidents [23]. Albrand and Walter published curves that related depth in feet and the head velocity to the height of a diver above the water [24]. McElhaney et al. also provided experimental data relative to body velocity and the depth of the water [25]. This series of accidents included not only springboard diving but water slides as well; an early work on water slides is Gabrielsen and McElhaney [26]. McElhaney et al. suggested that a head velocity of approximately 10 ft/s (3.05 m/s) with a following body is sufficient to cause compression fractures of the cervical spine, most frequently at the level of C5, the fifth cervical vertebra. As this is equivalent to a vertical drop height of less than 0.5 m, we realize that many activities, including activities of daily living, contain the potential for neck injury. Reid and Saboe provide a proportionate distribution of

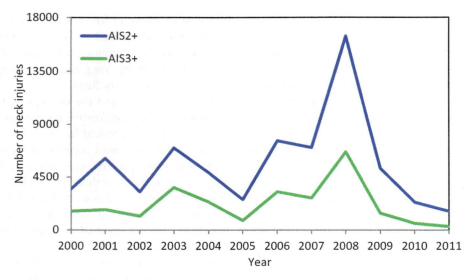

Fig. 11.1 The number of moderate and severe neck injuries in motor vehicle crashes is at historical lows

these injuries in specific sports [27]. Care must be exercised in interpreting data of this type because the incidence is determined more by the number of participants and exposures than by other factors.

As shown in Fig. 11.1, neck injuries in motor vehicle crashes continue to occur despite improvements in vehicle technologies, public awareness of safety, and regulatory efforts. An analysis was conducted using the US National Automotive Sampling System (NASS-CDS) with the following inclusion criteria: frontal impacts with occupants 16 years or older, front seat outboard occupants, passenger cars, light trucks and vans; AIS 1 through 6 (unknown excluded); and years 2000–2011. Rollovers and ejections were excluded.

11.2 Anatomy

The complete spine is a structure composed of 7 cervical, 12 thoracic, 5 lumbar, 5 sacral and 4 coccygeal vertebrae. Each vertebra is composed of a cylindrical vertebral body connected to a series of bony elements collectively referred to as the posterior elements. The posterior elements include the pedicles, lamina, spinous and transverse processes, and the superior and inferior facet joint surfaces. This structure is also known as the neural arch. It provides mechanical protection for the spinal cord and contributes to the stability and kinematics of the vertebral column.

The cervical spine is comprised of seven vertebrae that form eight motion segments between the base of the skull and the first thoracic vertebra, T1 (Fig. 11.2). In the neutral upright posture, the cervical spine has slight curvature with a posterior concavity that is referred to as the normal lordosis. The vertebrae are numbered such that the upper most vertebra is denoted C1, also called the atlas, and the lowermost vertebra is C7. The motion segments are similarly labeled C1 through C8. The C1 motion segment denotes the articulation of the base of the skull with the first cervical vertebra and C8 denotes the articulation of the C7 vertebra with T1. Because this nomenclature scheme gives an identical name to the motion segment superior to the vertebra of the same name, e.g., motion segment C4 is superior to the C4 vertebra, an alternate labeling scheme is often used for clarification. The motion segment and intervertebral disc between two vertebrae can be described with reference to both surrounding vertebrae. For example, the C4 motion segment is also called the C3–C4 motion segment. These vertebrae and motion segments can be structurally grouped morphologically and mechanically

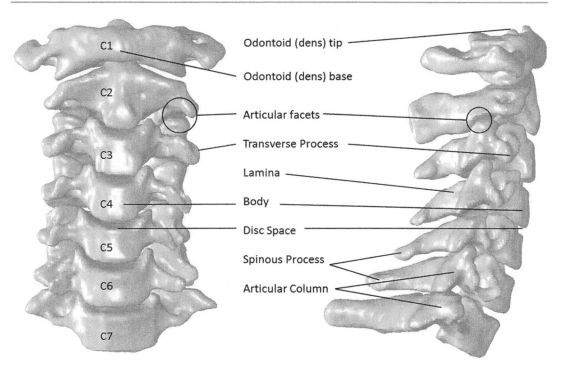

Fig. 11.2 The anatomy of the cervical spine (anatomical reconstruction from the NIH visible human project). The slight curvature that is visible on the *right* is called the lordosis

into two regions, the upper cervical spine and the lower cervical spine.

The vertebrae of the upper cervical spine (or UCS) are anatomically distinct and consist of three bony elements: the occiput (the base of the skull), the atlas (C1) and the axis (C2) (Fig. 11.3). They produce two joints – the occipito-atlantal joint (C1 or O-C1) and the atlantoaxial joint (C2 or C1–C2). Unlike the other vertebrae, the atlas has no vertebral body. The atlas is a bony ring with enlarged facets on the lateral portions of the ring, and is divided into the anterior and posterior arches. The anterior arch forms a synovial articulation with the second cervical vertebra, the axis, and the posterior arch provides protection for the spinal cord and brainstem. Like the vertebrae in the lower cervical spine, the axis is composed of a vertebral body and a posterior bony arch. But it also has an additional element called the odontoid process (the dens) that is embryologically derived from the fusion to C2 of what would have been the C1 vertebral body. The odontoid process points superiorly from the C2 vertebral body, and its anterior surface articulates with the posterior portion of the anterior arch of the atlas. The lateral portions of the axis contain flattened, enlarged articular facet surfaces and lateral to these surfaces are the transverse processes.

The O-C1 joint, which allows large flexion-extension rotations, is formed by the superior facets of the atlas and two bony protuberances on the base of the skull called the occipital condyles. The C1–C2 joint, which allows large axial head rotations, is composed of three synovial articulations between the atlas and the axis. The superior facet surfaces of the axis form two synovial articulations with the inferior facet surfaces of the atlas. The third synovial joint is formed by the articulation of the anterior arch of the atlas with the odontoid process of the axis.

The ligamentous structures of the occipitoatlantoaxial complex include continuations of the lower cervical ligaments and an additional set of structures unique to the upper cervical spine. The transverse ligament is a stout horizontal ligament that connects the medial portions of the C1 lateral

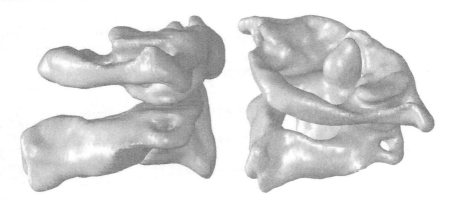

Fig. 11.3 Saggital and oblique views of the upper cervical spine. C1 (the Atlas) is a ring shaped vertebra with no body. C2 (Axis) has a bony process called the odontoid or dens about which C1 can rotate

masses and constrains the odontoid process posteriorly. The transverse ligament is the horizontal portion of the cruciate ligament. The vertical portion of this ligament attaches to the anterior inferior aspect of foramen magnum of the base of the skull and the posterior aspect of the C2 vertebral body. The apical ligament is a midline structure that connects the apex of the odontoid to the base of the skull anterior to the cruciate ligament. The alar ligaments originate on the posterior lateral aspect of the odontoid process and ascend laterally to insert directly to the base of the skull. The posterior longitudinal ligament, which runs the entire length of the spinal column, inserts on the base of the skull posterior to the transverse ligament and is named the tectorial membrane. The anterior longitudinal ligament (ALL) inserts on the base of the skull. The anterior atlantoepistrophical and atlantooccipital ligaments lie posterior to the ALL and connect the C2 vertebral body to the atlas and the atlas to the base of the skull, respectively. The lower cervical flaval ligaments insert on the base of the skull and are denoted the posterior atlanto-occipital membrane. Finally, the nuchal ligament, which is formed by the midline fusion of the paravertebral muscular fascia, inserts on the occipital protuberance, which is small bump on the back of the head just above the neck.

The lower cervical spine (or LCS) contains vertebrae C3 through C7 and motion segments C3 through C8. The vertebrae, which increase in absolute size from superior to inferior, are geometrically similar with a roughly cylindrical vertebral body and posterior bony arch. The vertebral bodies are connected to one another by a fibrocartilaginous intervertebral disc. The disc is composed of a central fluid-like nucleus pulposus bounded by a laminar set of spirally wound fibrous sheets called the annulus fibrosus. The anterior surfaces of the vertebral bodies are connected from the sacrum to the base of skull by the ALL (Fig. 11.4). Similarly, the posterior longitudinal ligament (PLL) connects the posterior surface of the vertebral bodies and forms the anterior surface of the spinal canal. The pedicles run posterolaterally from the posterior of the vertebral bodies and terminate in the articular column (or pars interarticularis). The articular column denotes a bony region bounded superiorly by the superior facet surface, inferiorly by the inferior facet surface, laterally by the transverse process, and medially by the spinal canal. Lateral to these elements are the transverse processes, which contain the vertebral artery, the major blood supply for the brainstem and the posterior portions of the brain. This artery is contained in the foramen transversarium of the transverse process. The lamina project posteromedially from the pars interarticularis and meet at the midline. The lamina form the posterior lateral surface of the spinal canal. Protruding from the midline fusion of the two lamina on each vertebra is the spinous process. The spinous processes are the palpable protuberances on the dorsal surface of the neck and back.

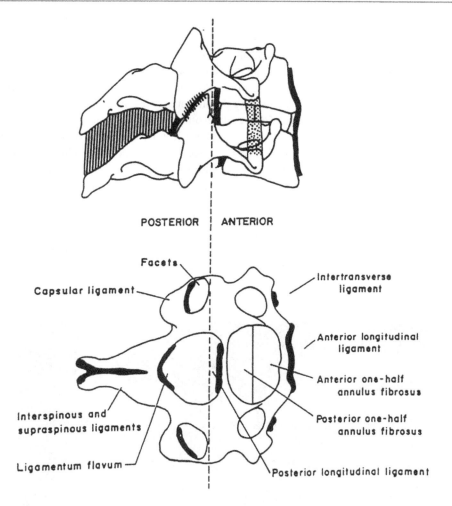

Fig. 11.4 A schematic of the cervical spine ligaments

The interspinous and supraspinous ligaments connect the spinous processes of adjacent vertebra. The flaval, yellow, ligaments similarly connect adjacent lamina from the sacrum to the base of the skull. The facet joints are formed by the inferior facet surface of the superior vertebra and the superior facet surface of the inferior vertebra. These are synovial joints, which are wrapped in a capsular ligament with paraspinal muscular insertions. The spinal canal of the vertebrae is the location of the spinal cord. Nerve roots exit between adjacent vertebrae through the intervertebral foramen. This foramen is composed of the inferior vertebral notch of the superior vertebra and the superior vertebral notch of the inferior vertebra.

11.3 Fractures and Dislocations

In this section, we review the more common fracture patterns and their association with spinal cord injuries (SCIs). The purpose is to present the clinical picture, without regard to the mechanism of injury, so that the reader has some context for the relationship between cervical spinal fracture (AIS 3) and spinal cord injury (AIS 5+). Understanding the dynamics of fracture, the stresses created on the spinal cord during fracture, and the potential for SCI is an exceptionally complex biomechanical problem. Making a robust prediction of specific fractures is well outside the scope of our current understanding and computational abilities. Fortunately, however, it is possible to mechanistically understand the probability of SCI by examining the extent to which the spinal cord is impinged during a typical fracture and to validate that understanding against clinical studies that provide SCI frequency for particular fractures.

Broadly speaking, lower cervical spinal (LCS) cord injuries are caused by burst fractures, dislocations, and fracture-dislocations. Cord injuries are much less common with simple fractures (i.e. vertebral compression fractures and other minor fractures). Spinal cord injuries are also associated with fractures that are unique to the upper cervical spine (UCS), so these are discussed separately.

11.3.1 Lower Cervical Spine

Burst fractures typically include both the anterior and posterior cortex of the vertebral body (Fig. 11.5). They occur in the lower cervical spine at all vertebral levels from C3 to T1 and are responsible for 30 % of spinal cord injuries [28]. These lesions may consist of destruction of the cancellous bony centrum with loss of disk height and/or vertebral end-plate and multi-part fractures of the vertebral body. One variation of burst fractures is a midsagittal cleavage fracture. The most common sites of these fractures in the lower cervical spine are C4, C5, and C6 [25]. Because the vertebrae are a closed bony ring, complete fracture through the anterior and posterior cortex

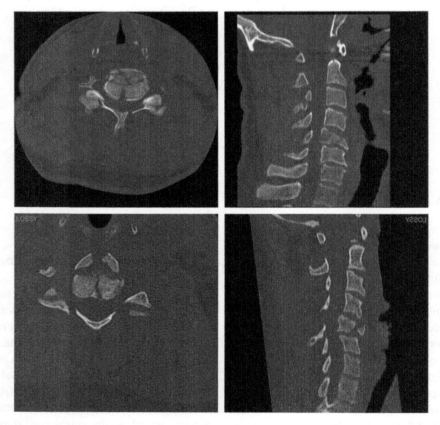

Fig. 11.5 Burst fractures of C5 from a mountain bike crash (*top*) and a shallow pool dive (*bottom*). Note the retropulsion of bone fragments into the spinal canal and the anterior "tear drop" fragment on the sagittal reconstructions (*right*)

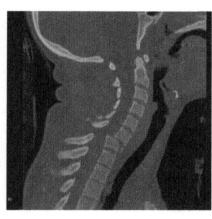 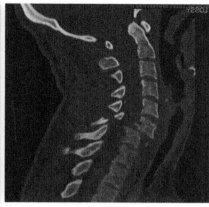

Fig. 11.6 Bilateral facet dislocations of C6–C7 from rollover crashes. A "teardrop" fragment is observed on the *left*. On the *right*, there is an associated lamina fracture of C6

of the vertebral body is often accompanied by fractures to the lamina, and disruption of the facet joints. They are grossly unstable injuries and bony fragments, often trapezoidal in shape, are displaced posteriorly and impinge on the spinal cord. The risk of spinal cord injury from these fractures is very high in the reported literature (deficit on admission >90 %, complete lesions 40–75 %), and tends to increase with the severity of the fracture [29].

A cervical spine dislocation typically refers to an injury in which the superior vertebral body is displaced (subluxed) relative to its subjacent vertebra. This displacement, which is most often in the anterior direction, results in disruption and frequently dislocation of the facet joints with subsequent locking of the surfaces in a tooth-to-tooth fashion secondary to muscle spasm (Fig. 11.6). If both facets are dislocated, it is called a bilateral facet dislocation (or BFD). The lesion occurs at all levels of the lower cervical spine, but predominates between C5 and C7 [30–32]. A pure dislocation without any fracture is relatively unusual and accounts for only 5 % of spinal cord injuries. BFDs are more commonly associated with fractures of the facets and often of the superior-anterior lip of the inferior body, or the anterior-inferior lip of the superior body. This injury results in greater than 50 % anterolisthesis and a significant reduction in the anteroposterior diameter of the neural canal and is therefore usually associated with spinal cord injury. Braakman and Vinken report permanent, partial, or complete spinal cord injury in 27 of 35 patients with bilateral dislocations [33]. Other studies have found that permanent quadriplegia occurs in over 50 % of these injuries [31, 33, 34]. Bohlman found that over one half of patients with tetraparesis had sustained some form of bilateral facet fracture dislocation.

Unilateral facet dislocation (UFD) refers to the anterosuperior displacement of only one of the two facets over the facet of the inferior vertebra with subsequent locking in a tooth-to-tooth fashion. The most common sites of this injury are the C5, C6, and C7 motion segments [33]. Clinically, the lesion is frequently asymptomatic, or associated with radicular pain on the side of the injury (25 of 39 patients). Spinal cord injury is relatively un-common in this type of injury [33, 35].

Fracture dislocations are often characterized by comminution of vertebral bodies and gross displacements of the vertebrae (Fig. 11.7). The injured vertebrae may not be contiguous and concurrent injuries to the upper cervical spine may be seen. A typical clinical presentation might involve displaced C6–C7 burst fracture with posterior element fracture and concomitant fracture of C2.

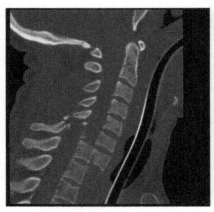 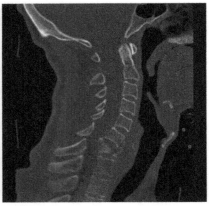

Fig. 11.7 (*Left*) A severe fracture dislocation of C7-T1 from a rollover crash with full body antero-lithesis and downward translation of the C7 body. (*Right*) A fracture dislocation of C6–C7 with a burst fracture of the C7 vertebral body from a lacrosse accident

11.3.2 Upper Cervical Spine

Because of its unique anatomy and function, the upper cervical spine fractures in idiosyncratic ways. The clinical literature indicates that many of these injuries do not have neurological involvement, but this reflects a clinical epidemiologic bias. Because damage to the spinal cord at this level is rarely survived, upper cervical spine fractures may be under-represented in the clinical literature because the victims are not transported to emergency rooms [36].

One of the more frequently described injuries to the upper cervical spine is a multipart fracture of the atlas (Fig. 11.8). While these fractures are frequently described as Jefferson fractures, the Jefferson fracture properly refers to a particular four part fracture of the atlas [37]. Fatalities and instability from this group of injuries are common [38, 39].

Traumatic spondylolisthesis of the axis, or Hangman's Fracture, describes a fracture of the elongated pars interarticularis of the posterior arch of the second cervical vertebra (Fig. 11.8). It has been historically attributed to hyperextension and distraction (tension and rearward rotation of the head), which can result from blows to the face and chin or as its name would suggest, from judicial hanging [2, 38, 40–42]. Rupture of the C2–C3 intervertebral disc accompanies the pars fracture and creates dramatic instability in the more serious forms of the injury. Automobile crashes have replaced hangings as the most common cause of these often fatal injuries [43].

Another important class of C2 injury is odontoid fracture (Fig. 11.9). These injuries are frequently mechanically and clinically unstable and are of considerable importance because of the potential for lethal spinal cord impingement, and the technical difficulty associated with surgical stabilization of the injury. They have been classified clinically by Anderson and D'Alonzo into three groups or types: Type I, avulsion fractures of the superior-most tip of the odontoid; Type II, fractures through the body of the odontoid; and Type III, fractures through the base of the odontoid that extend into the axis vertebral body [44]. Clinically, Type I injuries are stable, and Type III injuries tend to heal with appropriate immobilization. Type II injuries, however, have a high incidence of nonunion, and surgical repair is common. Odontoid fractures are biomechanically important because most of the strongest ligaments attaching the head to the neck insert or originate on the dens. Therefore, these fractures can lead to atlanto-axial (C1–C2) dislocation. An atlantoaxial dislocation can also occur due to transverse ligament rupture; however, a traumatic etiology for this problem is controversial as the transverse ligament is thought to be stronger than

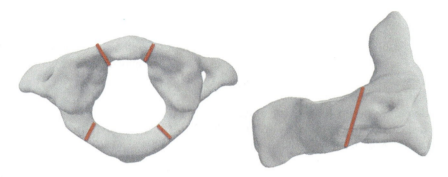

Fig. 11.8 Schematic of the classic four-part Jefferson fracture (*left*) and the Hangman's fracture (*right*)

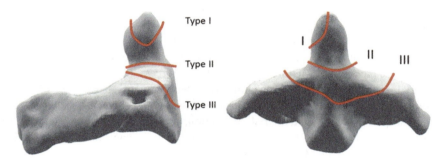

Fig. 11.9 Schematic of the three classes of odontoid (or dens) fracture (Adapted from Anderson and D'Alonzo)

the odontoid process [45]. Experimental work by Fielding et al., based on shear loads applied directly to the atlas, suggests that the transverse ligament can fail in postero-anterior shear loading prior to odontoid fracture, indicating that a traumatic transverse ligament injury is possible [46]. A traumatic tearing of the transverse ligament of the axis does occur in patients with preexisting soft tissue laxity or collagen vascular disease (e.g., Down's Syndrome or rheumatoid arthritis). In these patients the trauma producing the lesion is minor and the injury is described as a spontaneous atlantoaxial dislocation.

Montane et al. report on four cases of occipitoatlantal dislocation resulting from rapid decelerations in motor vehicle accidents in which the victims survived past the immediate event [47]. The prevalence of this type of injury is debated. Because of the nature of the injury and the method of dissection at autopsy, occipitoatlantal dislocations are frequently undetected and were, therefore, thought to be uncommon. However, according to Buckolz and Burkhead the lesion is a common cause of death in motor vehicle accidents where it occurred in 9 of 26 automotive fatalities with cervical spine injury, making it the most common cervical injury in their study of 112 total automotive fatalities [48].

Another noteworthy upper cervical spine injury is the basilar skull fracture. These are very serious and often fatal injuries that came to attention in the last decade with some high profile fatalities of race car drivers. Reviewed in detail by Myers and Nightingale, basilar skull fractures (BSFs) can occur from a variety of mechanisms, but when the loads on the skull base arise from compression or tension of the cervical spine, the fractures have distinctive characteristics, including the "ring" fracture [49]. These injuries are often immediately fatal owing to rupture of the main blood vessels of the head, and direct injury to brain and brain stem. Clinical presentation can include significant craniofacial bleeding, or bruising around the eyes and ears.

11.3.3 Other Fractures

Clay Shoveler's fracture is an isolated fracture of the spinous process [42]. It is named for its historical occurrence in people while shoveling clay; the clay sticks to the shovel resulting in a whiplash motion of the head with avulsion of the spinous process. Clinically, the lesion is of little significance, as it is associated with neither neurologic deficit nor spinal instability.

Teardrop, or limbus, fractures refer to the bony avulsion of either the antero-inferior or anterosuperior portion of the vertebral body in the lower cervical spine. These fragments were designated as teardrop by Schneider and Kahn because of their water droplet appearance in the lateral view on X-ray, and because of the strong emotional component frequently associated with such lesions. Because the overall structural integrity of the cervical spine varies considerably with this presentation, a simple relationship between teardrop fractures, spinal stability, and the risk of spinal cord injury does not exist [50]. However, the risk for spinal cord injury due to retropulsed bony fragments associated with the teardrop fracture must always be considered when the teardrop fracture is seen as part of a burst fracture. There has been extensive discussion of the etiology of teardrop fractures over the years based largely, but not exclusively, on subjective evaluation of patient injury data [41, 42, 50, 51, 52, 53, 54]. Review of the literature reveals a wide variety of mechanisms through which teardrop fragments may be formed and suggests that the size and geometry of the fragment depends as much on host factors, (e.g., local bone strength, ligamentous strength, and the presence of degeneration) as on the type of applied loading. Thus, generalizations about the injury based solely on the presence of a teardrop fracture are not appropriate. The clinical importance of these lesions stems from the recognition of their association with other structural injuries to the spine, and the associated instability and neurologic injury that they can produce. It should be noted that the teardrop fracture was designated as such in an era of plain radiographs, which were not were not able to capture the vertical body fracture planes with the fidelity seen on modern CT and MRI [28]. Therefore, the clinical implications of the teardrop fragment may not be as important in a modern radiological setting.

11.3.4 Clinical Classification

Because our knowledge of cervical spine trauma has evolved from a wide variety of institutions and sources, several authors have attempted to classify the injuries so that there can be agreement on what investigators are referring to when describing a particular injury. These include Allen et al., Babcock, Dolen, Harris et al., Melvin et al., Moffatt et al., Myers and Winkelstein, Portnoy et al., and White and Panjabi [32, 38, 42, 51, 55, 56, 57, 58]. The bases for these classifications has included retrospective reviews of patient data, whole and segmental cadaveric experimentation, and retrospective evaluations of various known injury environments [14, 25, 36, 55]. While most of the above classification schemes are very thoughtful, they have been necessarily speculative and often based on anecdotal or inferential evidence. Injury classifications are further complicated by the terminology used to determine the mechanism. The global motion of the head relative to the torso is often different from, and mistaken for, the motions of the cervical spine at the injured level. For example, the observation of a flexion motion of the head relative to the torso may actually be concurrent with local (i.e. a single motion segment) extension in the cervical spine and may produce an extension injury. Further, experimental studies have shown that small changes of the impact location or the initial position of the head (less than 1.0 cm) can change the injury from compression-flexion to compression-extension in cadaveric studies [59, 60]. In addition, the observed motions of the head may occur after the injury has occurred and thus reflect the motions of an unstable head and spine rather than the true injury mechanism [61].

Understanding whether the classifications are based on global motions of the head (as in [42]), or local deformations of the motion segment [55] is key to an understanding the validity of the injury classification. But all too often in both the clinic and in litigation, the assumption is that the classification refers to motions of the head, not the spine at the site of injury.

11.4 Experimental Studies

Experimental investigations using whole cadavers, whole cervical spines, and cervical spine segments have been invaluable in developing kinematic relationships as well as suggesting injury mechanisms. The primary limitation on such research is the paucity of cadaveric tissue for the study of injury and the lack of active musculature. This lack of tissue restricts the size of most studies and has been, perhaps, the greatest limitation on advancing our understanding of biomechanics. Despite these limitations, the neck injury literature includes far more studies than we can reasonably summarize in this chapter. This section reviews the studies that have had the most impact in shaping our current understanding of cervical spine injury mechanisms. The reader is referred to Myers and Winkelstein for a more comprehensive review of the literature [58].

11.4.1 Compression

Most cervical spine injuries occur when the neck is forced to manage the kinetic energy of the moving torso. This situation is best exemplified by diving injuries, but it also occurs in other athletic/recreational activities as well as in motor vehicle crashes and household accidents. For this reason, much of the research on neck injury mechanisms has been focused on compression. There are a number of experimental studies on the compressive responses of cervical spine segments, but they are of limited utility in understanding injury because of the complex dynamics associated with compressive impact. For this reason, we will limit our review of compression to whole cadaver tests and whole cervical spine tests.

11.4.1.1 Whole Cadavers

Culver et al. performed 11 superior-inferior head impacts of unembalmed cadavers using a padded impactor [62]. No fractures were observed at the peak head forces below 5.7 kN. Mean head impact force producing cervical injury was 7.58 ± 0.94 kN. Most of the injuries produced were in the posterior elements, which the authors attributed to "compressive arching". These authors were among the first to suggest that neck position, particularly the degree of initial lordosis, influenced the injury mechanism.

Hodgson and Thomas performed a series of quasi-static and impact tests using embalmed cadavers, with one side of the spine exposed to allow for high-speed imaging [20]. The authors concluded that increased constraint of the head in the padded impact surface increased the likelihood of injury to the neck. They also concluded that compression from crown impacts produces exaggerated regional cervical spine flexion and increased the likelihood of fracture or dislocation between C4 and C7. It was one of the first studies to note that global compression of the head and neck could produce local flexion and concomitant shear and described the complex neck dynamics that can lead to a wide variety of compressive injuries to a number of cervical levels.

Nusholtz et al. performed impacts to the head vertex of 12 unembalmed cadavers with a guided 56 kg impactor [63]. Compressive extension cervical injury occurred with the necks buckling in extension or flexion. Head impact loads ranged from 1.8 to 11.1 kN. Nusholtz et al. also performed free-fall drop tests using eight unembalmed cadavers [60]. Drop heights that varied from 0.8 to 1.8 m produced extensive and varied cervical and thoracic fractures and dislocations. Head impact forces ranged from 3.2 to 10.8 kN in these tests. Nusholtz et al. drew a variety of conclusions from these cadaver tests that still remain quite valid. (1) The initial orientation of the spine is a critical factor in influencing the type of dynamic response and injuries produced. (2) Descriptive motion of the head relative to the torso is not a good indicator of neck damage. (3) Energy-absorbing materials are effective methods for reducing peak impact force but do not necessarily reduce the amount of energy transferred to the head, neck, and torso or the damage produced in a cadaver model.

Yoganandan et al. conducted a study to evaluate the effect of head restraints on spinal injuries with vertical impact [64]. Sixteen human male cadavers were suspended head down and dropped vertically from heights of 0.9 to 1.5 m. In 8 of the 16 specimens, the head was restrained to simulate the effect of muscle tone, producing preflexion of the cervical spine in its initial position. Head-impact forces ranged from 3.0 to 7.1 kN in the unrestrained and from 9.8 to 14.7 kN in the restrained specimens. On average, the peak load in the C5 or C6 vertebral body was 70 % lower than the peak head load. Cervical vertebral body damage was observed most commonly when the cadavers remained in contact with the impacting surface without substantial rotation or rebound. This caused the neck to be the major element in stopping the torso motion.

One important factor, not well controlled in all of these whole cadaver impact studies, is the degree of head constraint. When the head is free to rotate on the occipital condyles, bending of the neck occurs with minimal compression. When head rotation is constrained, bending of the neck is restricted and the system becomes much stiffer. The factors that influence head motion are the stiffness and conformability of the impacting surface and the cadaver head. Another factor that cannot be controlled in whole cadaver experiments is the extent to which failure progresses. A certain amount of kinetic energy is available in these tests and failures progress until this energy is dissipated. Finally, it is technically difficult to record the internal forces without violating the integrity of the structure, and because the dynamics are complex, the head impact loads are very different from the cervical spine loads [64, 65]. It is these factors, as well as lack of muscle action, that makes it difficult to determine injury tolerance and mechanisms from whole cadaver tests.

11.4.1.2 Whole Cervical Spines

Experiments on whole cervical spines offer the advantage of measuring the cervical spine loads in situ while preserving the geometry and biomechanics. For this reason they have been the most important experiments for establishing cervical spine injury kinetics and tolerance. One of the first such studies is from Bauze and Ardran, [66]. In 1978, Bauze and Ardran recognized that the biomechanics of the bilateral facet dislocation (BFD) were not well understood and undertook the first systematic approach to producing the injury. This was a very significant study, the implications of which would not become clear until decades later. They were able to produce bilateral facet dislocations in 6 of 14 cadaver cervical spines using an apparatus that applied compression while restricting head rotation. They were the first to describe the resulting buckling deformations and hypothesized that it occurred during compressive impacts. The primary limitations of the study were that it was static, and that a non-physiologic stress riser existed at the point of dislocation to control lower cervical bending. In addition, the loads were applied in discrete steps, and only the maximum compressive load for all specimens was reported (1,422 N).

Myers et al. tested 18 cadaver spines as beam-columns in compression using three different mechanical end conditions for the head: fully constrained, rotationally constrained, and unconstrained [67]. Somewhat to the surprise of the authors, the neck could not be broken in the unconstrained mode, which produced a large amount of flexion bending (>90°) with relatively little compressive force (<300 N). Bilateral facet dislocations could be consistently produced in the lower-cervical spine with a rotationally constrained head end condition. Bilateral facet dislocations occurred with an average of 1,720 N of axial force and 2.9 cm of compressive displacement. The study also examined the fully constrained end condition and found it to be substantially stiffer than the other two. In this mode of loading, wedge and compression fractures were produced in the mid-cervical spine. The average axial force and displacements at failure were 4,810 N and 1.4 cm respectively. Increasing the constraints on head motion resulted in a stiffer response and a greater axial load to failure, but there was no statistically

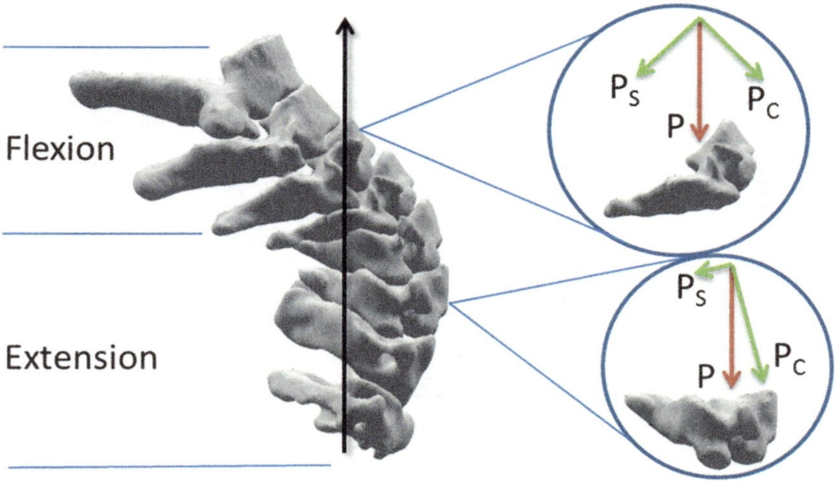

Fig. 11.10 Compressive buckling leads to multiple possible injury mechanisms. C1 through C4 are primarily in compression with extension. C5 through C7 are undergoing shear (s), compression (c) and flexion. Which vertebra fails will depend on timing and subject specific vertebral strengths and geometries

significant difference in the energy to failure between the fully constrained specimens and the rotationally constrained specimens. This suggests that the mode of failure shifted from primarily compressive, to a combined loading (compression-flexion) with significant internal bending.

Yoganandan et al. were the first to examine the dynamics of isolated whole cervical spines in compressive impact [68]. They quantified the resulting complex kinematics and suggested that buckling instability may play a role. They also were the first to observe the decoupling between the head response and the cervical spine response.

Pintar et al., studied the responses of whole head and neck preparations to compressive impacts [69]. In these experiments, the necks were straightened (the lordosis was removed) with between 15 and 30° of head flexion and the crown of the head was impacted with a hydraulic ram. The primary variable was the eccentricity of the occipital condyles over the T1 vertebra, which varied from 0.5 cm posteriorly to 2.5 cm anteriorly. They found that compressive injuries in the mid-cervical column were most likely to occur when the condyles were aligned vertically within 1 cm of T1. However, there was no single biomechanical metric that was predictive of injury. They produced a striking variety of injury patterns and concluded that *"for essentially the same loading vector, (crown impact to the head) a variety of different injury mechanisms can be produced depending on the configuration of the spine prior to loading. This implies that without sufficient external data (e.g. scalp laceration) an assumed mechanism of injury from retrospective evaluation of radiographs may not be correct."*

Nightingale et al. studied a variety of head constraints and impact angles in compressive drop tests on cadavers [61, 65, 70]. Compressive failure loads were 1.06 ± 0.27 kN for females with a mean age of 65, and 2.24 ± 0.57 kN for males with a mean age of 62. This difference was statistically significant and indicates that sex should be controlled for in biomechanical testing. The injuries produced included nearly the full spectrum of clinically relevant fractures: burst fractures, Hangman's fractures, odontoid fractures, bilateral facet dislocations, posterior element fractures and anterior avulsion fractures. The authors also observed first and second order buckling of the cervical spine, which gave rise to complex kinematics that can explain mid-cervical fracture patterns (Fig. 11.10). Another significant finding was the speed with which the fractures occurred – in rigid impacts, the spine failed

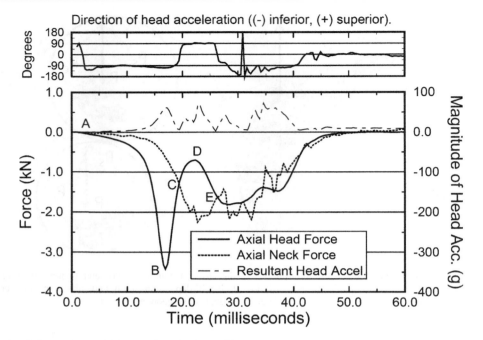

Fig. 11.11 In accordance with Newton's second law, the difference between the neck force and the head force is M*A of the head. At B the peak head force is always greater than the peak neck force as the head is stopped by the impact surface. The direction of the head acceleration reverses at C as the head rebounds towards the neck. The difference between the curves from C to E is the portion of the neck load due to head rebound

within 8 ms, and in padded impacts, less than 25 ms. Head rebound was another kinetic observation, leading to the conclusion that impact surface should absorb as much of the kinetic energy of the head as possible so as not to return it to the cervical spine (Fig. 11.11). The study had been designed with the thought that angled impacts (contacts fore or aft of the head vertex) would allow the head and neck to rotate and escape the path of the following torso, but because of the head inertia this frequently did not occur and the cervical column fractured.

Toomey et al., performed drop tests very similar to those on Nightingale et al., except that they were designed to apply combined compression and lateral bending [71]. It was thought that this configuration might be more relevant to rollover injuries. The fracture loads and timing were not affected by the lateral orientation, nor was the expression of the first order-buckling mode. The authors concluded that the asymmetric loading produces similar kinetics and injury patterns to sagittal plane loading.

More recently, Saari et al., conducted impact tests on whole human cervical spines with high-speed cine-radiography [72]. The injuries produced were primarily compressive-extension fractures of the cervical spine, with multiple "teardrop" avulsions of the anterior vertebral body, all from extension. As with prior impact studies, very little head motion was observed at the time of injury. One specimen had a C1–C2 fracture dislocation with C1 anterior ring fracture and odontoid fracture, which was observed to spontaneously reduce on the cine-radiography. Maximum cord compression occurred between 8 and 15 ms after impact, which supports the notion that cord injury occurs rapidly following head impact.

Ivancic et al., subjected five cadaver spine specimens attached to a Hybrid III dummy head to high-energy horizontal impacts to simulate athletic neck injuries [73]. Catastrophic neck injuries were produced at many levels of the cervical spine. Using the ATD load cell inserted into the occipitocervical junction, the authors observed similar forces and times indicating that inertial forces of the individual vertebrae are small relative to the reactions in the neck and validating the use of T1 load cells as a means to study the neck without violating the craniocervical junction.

11.4.1.3 Discussion

The human neck is fairly strong in compression, but it is also very stiff in that direction and not nearly strong enough to manage the kinetic energy of the torso. Head first impacts from low drop heights reliably produce a variety of cervical spine fractures which arise from various buckling modes in the cervical spine [61, 64, 65, 67, 68, 69, 70, 74, 75].

Differences in axial load to failure have been shown to depend on the injury produced and, therefore, on the degree of constraint imposed by the contact surface. From a tolerance standpoint, these injuries should be evaluated separately. As an example, cadaver experiments by Maiman et al., Myers et al., and Pintar et al. show that compression-flexion and compression-extension injuries occur with smaller axial loads than pure compressive injuries [67, 76, 77]. Bilateral facet dislocations reported by Myers occurred at $1,720 \pm 1,230$ N, and flexion injuries reported by Maiman occurred at approximately 2,000 N. In contrast, compression injuries reported in these studies occurred at $4,810 \pm 1,290$ N and $5,970 \pm 1,049$ N, respectively. Given the importance of the musculature in resisting flexion and extension in the lower cervical spine, the above values should be considered as a lower bound for human tolerance in compression-flexion and compression-extension. In contrast, the compressive loads associated with pure compression injuries are a reasonable estimate of cervical compressive tolerance.

Much of the scatter in the existing compressive tolerance data for the cervical spine is due to differences in experimental technique. Since neck responses are very sensitive to end conditions, inertia, and orientations, it is important to use data from studies that employ similar experimental setups. A synthesis of data that meet the inclusion criteria for a dynamic cervical spine tolerance are presented in Nightingale et al. [70]. The results of studies from two institutions were combined to formulate a tolerance for males and females. Averaging the means of the two studies resulted in the following tolerance values: 1.68 kN for females, and 3.03 kN for males. The tolerance values were scaled to account for the age of the cadaver specimens and the authors suggest a cervical spine tolerance for the young human male in the range of 3.64–3.94 kN. The tolerance is lower for a neutrally positioned spine and higher for a straightened spine; however, these are a reasonable first-order representation of the mean resultant force required for a catastrophic cervical spine injury in any given impact situation. The threshold velocity required to produce catastrophic neck injuries is often quoted to be 3 m/s, but catastrophic injuries can occur at even lower velocities during those circumstances in which the neck geometry and initial torso velocity maximize injury potential (i.e. a forward of vertex impact with a straightened spine and the torso velocity perpendicular to the impact surface).

A number of studies have reported on the total compressive displacements of the cervical spine at the time of injury and this metric has been suggested as a measure of tolerance [69]. Compressive impact tests by Pintar et al. with straightened spines resulted in injuries at a mean compressive displacement of 18 ± 3 mm. Myers et al. reported a mean of 14 ± 4 mm in neutrally positioned (lordotic) spines [67]. Displacements can also be estimated from the Nightingale et al., 1997 from the product of injury time and impact velocity for rigid impacts. This results in displacements of 13 ± 7 mm.

Both the neutrally positioned and the straightened cervical spines are stiff structures. As indicated above, the cervical spine generates a great deal of load very quickly and without much compressive displacement. Pintar et al. (straightened spine) produced neck fracture between 1 and 7 ms after impact. Nightingale et al. (lordotic spines) produced injury at an average of 5 ms for rigid impacts and 18 milliseconds for padded impacts. It appears that a burst fracture is more likely in the straight configuration than in the lordotic configuration. A bilateral facet dislocation is more likely in the lordotic configuration than in the straightened. But either injury can happen in either configuration depending on the initial positions, the rate of loading, and chance. The complex deformations resulting from cervical spine buckling explain, in part, why the complete spectrum of cervical spine injuries can be produced in what appear to be very similar compressive impacts [78].

Because cervical spine fracture happens so quickly, there is not enough time for much head motion to occur before the neck breaks. Head rotations greater than 20° have been shown to occur well after neck injury (as much as 80 ms after) and maximum head rotations occur at least 150 ms after injury [61]. In addition, the direction of head rotation (flexion or extension) is not correlated with the neck injury mechanism. The reason for this is that the inertia of head resists changes in linear and angular momentum. This means that it takes both force and time to change the direction and rotation of the head – much more time, it turns out, than it takes to break the neck.

The mechanism of the bilateral facet dislocation (BFD) was long described with biomechanically ambiguous phrases like "forced" flexion or "hyperflexion" and was thought to occur as a result of motion of the head that forces the chin towards the chest. This "hyperflexion" hypothesis was satisfying to lay intuition and became dogmatic in the orthopedic literature [38, 42, 52, 79]. But as early as 1960, difficulties were reported in producing this "flexion" injury in experiments on cadavers [80, 81]. In 1978, Bauze and Ardran were able to produce the injury using an apparatus that applied compression while restricting head rotation [66]. They called the buckling deformation "ducking" and hypothesized that it occurred during compressive impacts. In 1980, Hodgson and Thomas demonstrated these ducking modes of deformation and concluded that compression from crown impacts produces exaggerated regional cervical spine flexion and increased the likelihood of fracture or dislocation between C4 and C7 [20]. Since the earlier work highlighted the importance of mechanical end-conditions on injury type, Myers et al. performed a controlled parametric study to examine the effects of end-conditions and determined that the neck could not be broken in flexion bending, but could be consistently dislocated in compression with constrained head rotation – an end condition similar to the one applied by Bauze and Ardran. To date, the 1991 Myers study is the only one to consistently produce the elusive BFD and it definitively precluded forced flexion of the head as a mechanism. In this same time frame, Dr. Torg was hypothesizing that compressive buckling in a straightened neck was the BFD mechanism. This was based on his review of injuries and film of football players [17, 82]. In fact, Dr. Torg and his colleagues produced the buckling kinematics and failure in a skeleton model [13]. In the mid 1990s, Nightingale et al. applied varied head constraints in compressive impact tests on cadavers and were able to produce some bilateral facet dislocations. The mechanism for the bilateral facet dislocation is now thought to be combined loading at the segment level secondary to neck compression and buckling (Fig. 11.12). There are one or two anecdotal cases (e.g. a full-nelson hold, performed until muscles fatigued) where static flexion-tension loads have been reported to cause injury but these are a rare exception [83]. The role of "hyperflexion" as a mechanism for BFD is still being debated in the interesting and unusual circumstance of the rugby scrum [83, 84].

11.4.2 Tension

Compared to compression, pure tensile injuries of the cervical spine are relatively rare. The development of active automotive restraint systems and increased seat belt use over the last three decades has markedly reduced the overall fatality and injury rates in automotive accidents. However, these technologies have given rise to a number of reported severe and fatal cervical spine injuries due to tensile mechanisms. These occur primarily during noncontact head decelerations and out-of-position airbag deployments. There is also some concern for inertial injuries to rear-seated occupants [85]. Inertial injuries typically have a significant component of flexion or extension and will be discussed further in

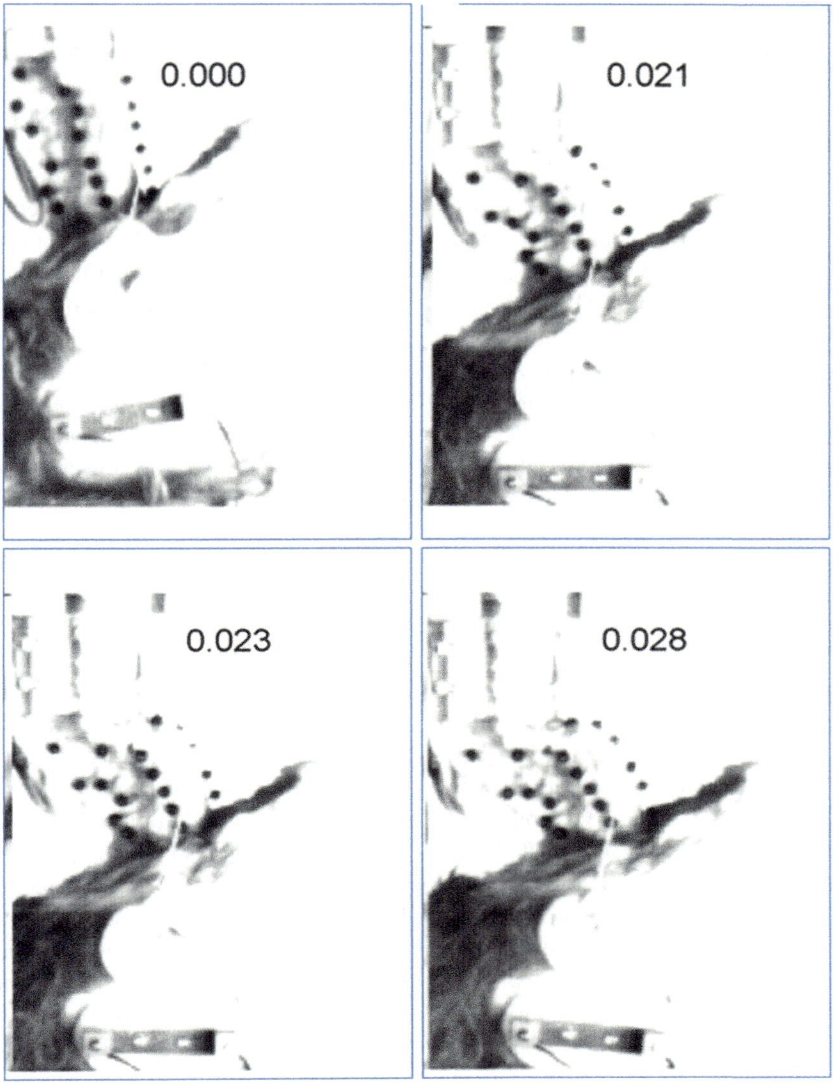

Fig. 11.12 Injury sequence showing first mode buckling and the initiation of a C6–C7 bilateral facet dislocation at 21 ms with full displacement at 28 ms. Note how little head rotation or translation occurs during this time period. The small amount of head rotation that does occur is in extension

Sect. 11.4.4. The research reviewed in this section is most relevant to out-of-position airbag deployments, but is also important in inertial loading.

Tensile experiments on the cervical spine have been performed by relatively few investigators. In 1971, Mertz and Patrick proposed a tensile tolerance of 1,135 N based on a single volunteer test [86]. This noninjurious static strength was offered as a lower bound for the corresponding dynamic strength. Sances et al. reported results obtained from tests on human ligamentous spines under tensile displacement [87]. Three specimens were tested and failed at a loading rate of 0.13 cm/min with failure forces ranging from 1,446 to 1,940 N and total distractions of 1.14–3.0 cm.

Yoganandan et al. reported on the experiments of Sances et al. and added the results of additional tests [88]. The first set of tests was performed on 20 isolated discs from eight cadavers (C2–C3, C3–C4, C4–C5, C5–C6, C7–T1). The average failure load was 569±54 N at a displacement of 11.1±1.1 mm. A second set of experiments reported by the authors was conducted on four isolated ligamentous spine preparations from skull to T3. Failure loads ranged from 800 to 1,900 N with a mean of 1,555±459 N at a mean displacement of 27.1±4.7 mm. All four specimens disarticulated at the C6–C7 level. The authors also performed a series of tests on three whole cadavers using a shoulder yoke and cable setup to apply tensile loads to the head. Only upper cervical spinal injuries were observed, including odontoid fractures and a basilar skull fracture. The mean failure load was 3,373±464 N at an average distraction of 21.3±1.9 mm. However, the head end condition was not well described in these tests.

In 2000, Van Ee at al published a systematic study of cervical spine stiffness and tolerance in tension [89]. Structural tests were conducted on six whole cervical spines of mixed sex. The specimens were separated into motion segments for component and failure testing. The strength of the upper cervical spine was 2,400±270 N and was significantly stronger that the lower cervical spine (1,780±230 N). The injuries in the lower cervical spine were all complete joint disruptions without bony fractures. The upper cervical spine injuries were O–C1 or C1–C2 dislocations and one Type II odontoid fracture. The strength and stiffness results were used in a computational model that accounted for the contributions of the cervical musculature. The muscle increased the tensile tolerance to 4,200 for full activation of all neck muscles.

Chancey et al., used optimization to estimate physiologically appropriate muscle activation levels for relaxed (sufficient to hold the head upright in equilibrium) and tensed (maximal activation without head motion) neck muscles [90]. Head and neck tension simulations under these activations showed that the muscles significantly affect tensile tolerance. Even though the ligamentous cervical spine will fail at a tensile load of 1.8 kN, the muscles increase the overall neck tolerance to between 3.1 and 3.7 kN for the relaxed and tensed states, respectively.

More recently, Dibb et al., performed a series of experiments on whole and segmented adult male cadavers [91]. Twenty male cervical spines were loaded in tension with a variety of end conditions and loading points (the anterior-posterior position of a vertical load relative to the occipital condyles). Cervical spine stiffness was found to be sensitive to both end condition and the loading point. Loading points that are eccentric from the condyles allow more head rotation and a more compliant response. As in prior studies, the upper cervical spine was found to be stronger in tension (2.1±0.3 kN) than the lower cervical spine (1.6±0.3 kN and 1.4±0.3 kN for C4–C5 and C6–C7 respectively). The tensile strength of the upper cervical spine increased with a more anterior eccentric axial load, which is contrary to elementary beam theory. The authors hypothesized that this is due to the unique anatomy of the upper cervical spine which includes robust ligaments located anterior to the condyles. With loading aligned with those ligaments, they act in parallel and strength is increased over more posterior loading. In addition to O–C1 and C1–C2 dislocations, failures of the upper cervical spine included type III odontoid fractures (n=6), basilar skull fractures (n=2), occipital condyle fractures, and vertebral body fractures. Failure of the C4–C5 and C6–C7 also produced complete joints disruptions with some vertebral body and spinous process fractures originating near the fixation. No Hangman's fractures were produced, but this may be due to the fixation of C2 and C3.

11.4.2.1 Discussion

Interest in the tensile tolerance of the neck was primarily sparked by a string of severe and fatal injuries to children and adults in low speed frontal crashes with airbag deployments. In 1991, a NHTSA Special Crash Investigation was initiated in response and it was determined that most of the victims were either not wearing belts or were very close to the airbag when it deployed. The mechanism of injury was thought to be tension and extension in the neck caused by interaction with a deploying airbag under the chin. The resulting injuries were basilar skull fractures, occipito-atlantal dislocations, and C1–C2 injuries [92–97].

One of the most interesting findings of the ensuing research on tension was that the upper cervical spine is stronger than the lower cervical spine. However this is not consistent with the field studies above in which upper cervical spine injuries are much more common. As a segmented column, one would expect that the weakest point in the neck would be the site of failure, but the findings of the NHTSA investigation showed a large number of upper cervical spine injuries in both adults and children. Chancey et al., found that this discrepancy is most likely due to the effects of the active musculature [90]. The muscles of the neck share tensile loads with the cervical spine by providing a parallel load path. Such load sharing increases the overall strength and stability of the neck and provides greater protection to the caudal motion segments because of the larger size and number of muscles in the lower cervical spine. So, here is a situation where the cadaver, without the benefit of a biofidelic computational model of the muscles, is not a good surrogate for the living human and this has implications for both tension experiments and the deceleration experiments discussed in Sect. 11.4.4.1.

Finally, the epidemiology and the experimental research show that tensile neck injuries in the lower cervical spine are rare. Since tension produces primarily upper cervical spine injuries, further discussion of these will be left to Sect. 11.4.5.

11.4.3 Torsion

Torsion has been thought to be a principal cause of both upper and lower cervical injury and dislocation; specifically unilateral facet dislocations and rotatary atlantoaxial dislocations. Types of atlantoaxial subluxations include unilateral, as well as bilateral, anterior and posterior subluxations [98]. These injuries refer to the dislocation of one or both of the atlas facet surfaces on the axis facet joint surface and are thought to represent combinations of shear and torsion [38].

Cadaver studies have shown that torsional loading of the head can produce rotary atlantoaxial dislocation, with or without tearing of the alar ligaments [99, 100]. Estimates of the lower bound of torsional tolerance are available from Myers et al. [101]. Axial torque of 17.2 ± 5.1 Nm produced upper cervical injury in whole cervical spines (occiput to T1). Goel et al. report a slightly lower value of 13.6 ± 4.5 Nm in isolated upper cervical spine segments; however, the rates of loading in these experiments were considerably lower than in the Myers study ($4°/s$ vs. $500°/s$). Extrapolating piece-wise linear torsional stiffness data from volunteer decelerations reported by Wismans and Spenny to the $114 \pm 6.3°$ of rotation required to produce injury in the cadaver, we obtain an estimate of human torsional tolerance of approximately 28 N-m [102].

11.4.3.1 Discussion

Myers et al. showed that the lower cervical spine is stronger in torsion than the atlantoaxial joint, mitigating torsion as a primary cause of lower cervical injury [67]. The notion that torsion can predispose the lower cervical spine to injury from other types of loading (i.e., flexion), while a common belief, is clearly not true for catastrophic injury (though it does play a role in whiplash) [103].

The etiology of unilateral facet dislocations has been the subject of interest and controversy. Huelke et al. and Harris et al. reported that unilateral dislocation was the result of combined flexion and rotation [2, 42]. Gosch et al. reported that in purely anterior to posterior deceleration sled tests of primates, axial rotation was observed and unilateral facet dislocation produced [104]. Roaf reported that torsion was required to produce lower cervical ligamentous injury and dislocation and that hyperflexion alone could not produce ligamentous injury [80]. Rogers and White and Panjabi noted that unilateral dislocation resulted from an exaggeration of the normal coupling of lower cervical lateral bending and rotation but did not describe the loads required to produce the injury [38, 105]. Braakman and Vinken suggested that unilateral facet dislocation was the result of combined flexion and rotation

[33]. Torg and Otis stated that the lesion was the result of compression and buckling [13, 14, 17]. Bauze and Ardran produced unilateral facet dislocations in a number of specimens loaded in compression and flexion [66]. Their apparatus applied no torque to the specimen, and yet lower cervical rotation was observed prior to unilateral dislocation. Myers et al. demonstrated that while unilateral facet dislocation could be produced by direct torsional loading of the lower cervical spine, the lesion could not be the result of torsional loading of the head because of the comparative weakness of the atlantoaxial joint [100]. The authors currently believe that the lesion is produced by a mechanism similar to the bilateral facet dislocation, with the difference being due to bending out of the plane of symmetry, or the presence of pre-existing facet tropism of other structural asymmetry. Another possibility is that the unilateral facet dislocation is a partially reduced bilateral facet dislocation, which might account for the cohort of these lesions that have complete spinal cord injury [33].

11.4.4 Sagittal Plane Bending

Cervical spine bending has long been thought to play a major role in injury mechanisms. However, recent research calls this into question, particularly in flexion. The research on neck bending falls into two broad categories. The first is deceleration tests designed to create inertial loads similar to those in fore and aft impacts. These produce either flexion or extension with a large component of tension; therefore, they are not pure bending loads. The second category is isolated cervical spine tests in which the bending moment is pure. Until the turn of the millennium, while many orthopaedic studies has examined the kinematics of the cervical spine when subjected to pure moments, there had been very little research on tolerance to moment in the second category.

11.4.4.1 Inertial Loads

One of the most frequently cited studies is the work of Mertz and Patrick [86, 106]. They summarized data from forward and rearward sled decelerations of one human volunteer (Patrick) and four cadavers with weights added to the head at, above, or below the center of gravity of the head. Despite having extremely limited data, these authors were able to define corridors for the kinematic response and recommended noninjurious tolerance values in dynamic and static loading situations that proved useful in motor vehicle safety standards for many years. Reporting data from both volunteer and cadaver decelerations, the authors determined a lower bound for the risk for neck injury based on the bending moment estimated at the occipital condyles for flexion and extension loading. For extension, the authors report noninjurious loading of 35 ft-lb (47.3 Nm) with ligamentous injury expected above 42 ft-lb (56.7 Nm). For flexion, the authors report initiation of pain at 44.0 ft-lb (59.4 Nm), a maximum voluntary loading of 65 ft-lb (87.8 Nm), and the risk for structural injury above 140 ft-lb (189 Nm) in the cadaver following contact of the chin on the chest. The largest deceleration applied to a cadaver produced calculated moment of 189 Nm with no evidence of injury detected by x-rays. This limit was suggested as the upper bound of ligamentous spine tolerance in flexion, despite the lack of injury. The authors note however, that muscular injury may occur at loading below the latter value.

Bass et al. tested 36 head and neck preparations in frontal decelerations impacts to develop a predictive Neck Injury Index [107]. The tests produced AIS 3+ injuries in 11 of the specimens, with four injuries that were judged to have the potential for cord damage. Most of the injuries were to the soft tissues, with the osseous injuries being relatively minor. Upright orientation of the head and neck produced significantly less tension and shear (1,251 and 1,376 N, respectively) than orientations run with 30° of forward flexion (2,146 and 3,747 N, respectively) despite being run at significantly lower velocities. The authors suggest that this was because the neck was loaded in a stiffer mode when flexed forward (more tensile than bending). There was no significant difference in peak flexion moments.

Pintar et al. [108] subjected five cadavers to belted frontal impacts ranging in severity from 13 to 57 km/h and resulting in sled peak accelerations from 10 to 20 Gs. The instrumentation included accelerometer packs on the head and T1 as well as belt load sensors. Chin-to chest contact occurred in all tests, but after the peak calculated forces. The injuries produced in four of the five specimens were dislocations of the C7-T1 (n=2) and dislocations of C6–C7 and T1–T2. The four specimens with cervical spine fractures also had multiple rib fractures and assorted fractures of the manubrium, clavicle and sternum. In the high velocity tests, the mean axial force and moment at the base of the neck were 1,424 N and 192 N-m respectively. At the occipital condyles they were 1,570 N and 41 N-m. The resultant force in both locations was not reported, but was substantially higher than the z component.

11.4.4.2 Bending Loads

The clinical finding of a spinal cord injury without radiographic abnormality (SCIWORA) by Taylor and Blackwood prompted interest in extension as a causative agent [109]. In a cadaver study by Taylor, which combined myelography with manual loading of the head in flexion and extension, demonstrated posterior canal defects in "forced" hyperextension. These defects were thought to be the result of bulging of the flaval ligaments into the neural canal and were produced without overt structural damage to the spine. The author postulated, based on this observation, that extension could produce spinal cord injury without trauma to the cervical spine. Further evaluation and the recognition of disc bulging during loading reinforced this as a possible mechanism of SCIWORA [110, 111]. The mechanism has also been observed in the more recent compressive impact testing of Saari et al. [72].

In a study by Gadd et al., four whole cadavers were loaded manually using levers coupled to the head to apply combined shear and bending (flexion-extension or lateral) [112]. Rotation in extension greater than 80° or 60° in lateral bending produced audible sounds in two of the four specimens and resulted in decreased stiffness following reapplication of the same load. Defining these events as minor injury, mean mid-cervical moment at injury was 24.3 ± 12 Nm.

The end condition study by Myers et al. in 1991 applied compressive displacements to whole cervical spines with a free rostral end condition [113]. This was not pure moment loading; however, the tests produced primarily flexion because of the limited ability of the cervical spine to react compression with an unconstrained head. The mean peak axial load was 289 ± 81 N with a mean axial displacement of 8.6 ± 1.2 cm and a mean head rotation of $96 \pm 7°$. None of the specimens were injured despite the large, anatomically impossible head rotations.

Nightingale et al. examined the flexibility and strength of 52 female motion segments in both flexion and extension [114]. This was followed in 2007 by a similar study of 41 male cervical spines [115]. Their flexibility results were similar to prior studies [116–118] and they were the first to provide statistically meaningful tolerance data in pure bending (Table 11.1). They found that the upper cervical spine was stronger than the lower cervical spine. For both males and females, the lower cervical spine failures were complete joint disruptions (tearing of all the ligaments and the disc) with a few associated avulsion fractures of the vertebral bodies. The most common injuries to the upper cervical spine were odontoid fractures, which occurred primarily in extension, but were also observed in flexion.

Wheeldon et al., 2006 tested whole lower cervical spines in pure flexion and extension [119]. The segments were from C2 to T1 and the range of motion for a 2.5 N-m moment was near 50° for flexion and near 30° for extension. The study was unusual in that the mean age of the donors was

Table 11.1 Mean bending tolerance of the cervical spine [114, 115]

	Upper C-spine		Lower C-spine	
	Moment (N-m)	Angle (deg.)	Moment (N-m)	Angle (deg.)
Flexion				
Male	39.0	58.7	20.5	14.0
Female	23.7	56.2	17.4	19.3
Extension				
Male	49.5	42.4	17.1	15.1
Female	43.3	50.2	21.2	20.5

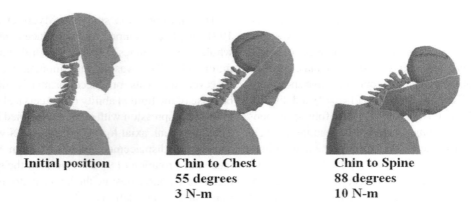

Fig. 11.13 The C6–C7 moment as a function of head flexion angle. At chin to chest contact, the neck moment is only 15 % of the lower cervical spine bending tolerance. If the chin were able to contact the spine, this would only be 50 % of tolerance

only 33 years old. By comparing their results with prior studies, the authors concluded that younger specimens were significantly less stiff in bending than older specimens. This was also the only study to use whole lower cervical spines and the large rotations encountered even without the O-C1 joint illustrate the technical difficulty in applying a pure follower moment to these very flexible structures.

11.4.4.3 Discussion

Bending studies on isolated cervical spine have shown that the neck is remarkably flexible in bending and its toughness in this mode of loading is underestimated. Researchers who have tried to fail the whole spine in this mode have not succeeded [113] and they have had to resort to segment tests. Even applying pure moments to isolated upper cervical spine segments is technically difficult and requires very specialized test equipment. Summing the angles at failure for the upper and lower cervical spine gives an estimate of the degree of head flexion required for a pure bending injury of the cervical spine. This value is greater than 120° and is clearly nonphysiologic as shown in Fig. 11.13. It can be stated with confidence that failure of the cervical spine in pure flexion is an anatomical impossibility, validating the hypothesis suggested decades ago by Roaf [66]. This may also be true of extension, although further modeling and experimental work is necessary to state this with similar confidence.

11.4.5 The Upper Cervical Spine (UCS)

As documented in the prior sections of this chapter, upper cervical spine injuries can be produced in almost all loading modalities. However, there was a time when they were attributed to very specific loads. In this section, we will review the historical hypotheses and the subsequent testing that have led to our current understanding of how these injuries occur.

Much of the tolerance data for horizontal shear stems from efforts to understand the mechanisms of occipitoatlantoaxial injuries. Whether or not these injuries are actually due to shear is unclear. The shear loads required to produce transverse ligament failure and odontoid fracture in isolated segments have been reported by a number of authors. Fielding et al. observed transverse ligament rupture at 824 N when the atlas is driven anterior relative to the axis in the isolated cadaver upper cervical spine [46]. Doherty et al. produced odontoid fractures at $1,510 \pm 420$ N by applying posterior and laterally directed loads directly to the odontoid process [120]. Based on the volunteer data in flexion, Mertz and Patrick suggest a lower bound of tolerance of 847 N [86]. However, direct loading of the cervical spine, as done in these studies is extremely rare. Although this mode of loading can be readily applied to isolated upper cervical spine motion segments in a laboratory setting, it is much more unusual with a head and full cervical spine as in the real world. Shear loads in the cervical spine are almost always the result of anteroposterior horizontal head loading. Therefore, they are usually accompanied by large flexion or extension bending moments. Moreover, as the head flexes or extends the shear loads resolve into tensile loading and injury occurs in a tension-bending mode. This, together with the observation of compression-bending modes as the most common cause of neck injury, prompted the NHTSA to upgrade its standards (FMVSS 208) in 2001 to use the Nij, a compression-bending injury criterion.

A 1956 clinical review of 51 patients with dens fracture suggested that extreme flexion, extension, or rotation may avulse the odontoid process [121]. The authors felt that tension in the alar ligaments was responsible for the avulsion. Rogers presented five cases of non-displaced fractures of the odontoid and suggested the mechanism to be flexion and rotation because of an associated crushing of the superior articular process of C2 on one side [105]. He indicated that posterior fracture dislocation was produced in two cases by hyperextension and in one of these cases *"the vertebral arch of the atlas was fractured, evidently by compression against the second cervical vertebra."* Anterior fracture dislocation of the dens occurred twice and he suggested that this was produced by a shearing force directed anteriorly, the dens being carried forward with the atlas by the transverse ligament. Cheng et al., Clemens and Burow, and Maiman and Yoganandan also reported that odontoid fractures occurred or could occur under flexion and extension [122–124]. These bending tests also produced occipito-atlantal dislocation in flexion and extension.

Buckolz and Burkhead postulated, based on the presence of submental lacerations, that the injury mechanism for occipito-atlantal dislocations is hyperextension [48]. Yoganandan et al. report an association of this injury with passenger ejections during motor vehicle accidents [36]. The other known cause of this injury is high-speed decelerations associated with military aircraft crashes into water, with inertial tensile loading of the neck by the mass of the head and helmet. Estimates based on these events suggest that decelerations must exceed 100 g to produce the injury in the healthy and muscular military pilot population. Notably, other authors have postulated that occipito-atlantal dislocations may be the result of a variety of head loads; however, it is likely that all mechanisms lead to the final common path of tensile loading of the spinal ligaments inserting to the occiput.

Recent studies have shown that pure tension, pure flexion, and pure extension can all result in C1–C2 or O–C1 dislocations secondary to a dens fracture [89, 91, 114, 115]. The reason for this is elucidated by a study of the O–C2 anatomy [125]. The robust ligaments of the O–C2 complex insert on the dorsum of the dens (C2) and originate on the occipital bone, forming a strong fibrous chord between C2 and the skull, which is effectively responsible for holding the head onto the cervical spine. Like tension, flexion and extension bending result in large tensile loads on the odontoid and on the ligaments between the C2 and the occiput. Flexion moments are reacted by compression of the chin on the sternum and by tension in the dens (Fig. 11.14). Extension moments are reacted by compression in the stacked posterior elements and tension in the dens. Failure in either mode of bending can occur in the dens or in the occipito-atlanto-axial ligamentous complex.

It is possible that the occipito-atlantal dislocation and the atlanto-axial dislocation occur by identical loading mechanisms that result in two different structural failures along the same load path. In the occipito-atlantal dislocation, tensile stresses on the alar ligaments cause them to fail. This results in rapid failure of the remaining, less robust, occipito-cervical ligaments, and continued motion of the head with subsequent cord or brainstem injury. The atlanto-axial injury occurs when the same tensile stresses avulse the dens from the body of C2 (a Type II or Type III dens fracture) causing C1 to separate from C2. Which of the two failures occurs may be related, in part, to the age of the victim [126, 127]. In elderly osteoporotic individuals, the dens fracture may be more likely to occur whereas in the younger individual, the bone may be stronger than the

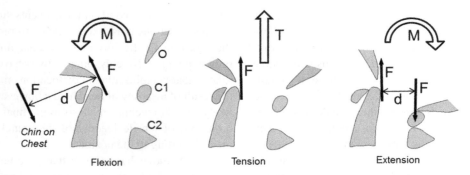

Fig. 11.14 Flexion, tension, and extension loading all produce tension in the odontoid. The flexion moment is reacted by compression between chin and chest and by tension in the odontoid ligamentous complex. Extension is reacted by compressive stacking of the posterior elements and tension in the odontoid ligamentous complex. For a given moment, larger forces (F) would be expected in extension than in flexion because of the differences is eccentricity (d)

ligamentous structures, resulting in the occipto-atlantal dislocation.

An increasingly important tensile injury of the upper cervical spine is the basilar skull fracture. The National Center for Statistical Analysis (NCSA) has been conducting Special Crash Investigations since 1991 on all airbag injuries in low to moderate severity crashes (www.nhtsa.dot.gov/people/ncsa/sci.html). Study of these data shows a large number of basilar skull fractures in both adults and children. Basilar skull fractures to out-of-position occupants during airbag deployments are clearly tensile injuries [92–94, 96, 97, 128, 129]. Basilar skull fractures have also been reported in race car drivers sustaining high G frontal decelerations with four- or five-point torso restraints. Improvements in head-supports, vehicle crashworthiness, and racetrack crashworthiness have dramatically reduced the incidence of these injuries. The lesion has been produced experimentally by Hopper et al. [130]. In five of six specimens tested, classic basilar skull ring fractures were produced at an average tensile load of $4,300 \pm 350$ N.

Although the above discussion has focused on bending and tension, it is important to recognize that many, if not most, upper cervical spine injuries occur from compression. Upper cervical injuries have also been reported with near vertex head impact testing in most compressive impact studies [60, 63, 64, 69, 72, 131]. It is also interesting to note that the elderly are particularly susceptible to UCS fractures. In fact, UCS fractures account for 65 % of all cervical spine injuries in ages greater than 70, while accounting for only 20 % of injuries in those less than 30 [126, 132]. The mechanism of odontoid fractures when the elderly fall warrants further investigation.

11.4.6 Lateral Bending and Lateral Shear

Injuries due to coronal plane motions from lateral bending or lateral shear are considerably less common than those due to loading in the sagittal plane. This is likely due to the relatively lower incidence of lateral loading and the flexibility of the neck in lateral bending. While out-of-plane loading was once thought to be required for many asymmetric injuries, current research has shown that that is not the case. During compressive impact, a lateral impact point on the head can result in symmetric neck injuries as the momentum of the torso is still in the midsagittal plane. In that regard, this out-of-midsagittal plane head contact, results in post-injury lateral head motion but may not influence the injury produced. By contrast, the research presented herein, which includes very high energies, does show that lateral loading can give rise to injuries in their own right including the peripheral nerves of the neck.

Experimental studies using whole cadavers in automotive side impact of far-side belted occupants at impact velocities of 50 km/h (30 mph) produced minor injury in only two of seven tests performed (Kallieris and Schmidt) [133]. In 2007, McIntosh et al. reported an analysis of data from 15 sled tests conducted in 1980s [134]. At delta-V ranging from 6.4 to 11.1 m/s, resultant neck force and moments were estimated from 1.4 to 6.4 kN and 93 to 358 Nm. Injuries ranged from hemorrhages "into the intervertebral discs" to "ligament tears," with the majority occurring between the C3 and C6 levels. There were no fractures and none of the injuries were greater than AIS 2. Yoganandan et al. conducted lateral sled tests using eight intact PMHS at varying velocities and with varying torso restraints [135]. The sled change in velocity ranged from 8.7 to 17.9 m/s. The neck injuries were between AIS 0 and AIS 3 depending on velocity and restraint condition. For all groups of specimens, mean axial forces ranged from 1,348 to 2,257 N, shear forces ranged from 438 to 970 N, lateral shear forces ranged from 240 to 658 N; lateral bending moments ranged from 17.4 to 61.5 Nm, torsion moments ranged from 12.9 to 27.2 Nm, and sagittal moments ranged from 6.4 to 26.3 Nm. In a follow up study, Yoganandan et al. conducted tests using intact head-neck complexes and intact PMHS and applied pure lateral bending moments by connecting the occiput to the piston of an electro-hydraulic testing device via a custom apparatus. Specimens were loaded incrementally at velocities ranging from 2 to 7 m/s. The results from four PMHS repeated tests indicated that PMHS head-neck complexes can resist a lateral bending moment of 75 Nm at the occipital condyles without neck injury. Ear to shoulder contact occurred in all cases, alleviating the moment on the cervical spine.

11.5 Important Factors in Spine Injury Mechanics and Mechanisms

Mechanically, the spine is best thought of as a segmented beam column composed of nonlinear viscoelastic joints (the discs and ligaments) connecting discrete, modestly deformable masses (the vertebrae). The motions of the vertebrae are coupled and the connective tissues can withstand large strains. For many years, neck dynamics in impact were poorly understood and the relevant mechanical factors were frequently ignored or mischaracterized. Over the last 20 years these factors have been examined using cadaveric specimens and computational models, resulting in a significant increase in our understanding of neck dynamics. Among these factors are the musculature, buckling, loading rate, head motion, initial position (lordosis and preflexion), anatomical variation and asymmetry, head end condition, impact attitude and head inertia, and the role of impact surface properties.

11.5.1 Cervical Musculature

Muscles impose loads on the cervical spine through reflex, passive, and stimulated mechanisms. Foust et al. [136] and Schneider et al. reported muscle reflex times to be about 50–65 ms for volunteers exposed to head loading. These reflex times are considerably longer than the 5–18 ms required to produce injury in compression head [61, 65], suggesting that muscle reflexes may be neglected in the head impacts that cause compressive injuries. Szabo and Welcher recorded surface electromyogram (EMG) measurements in volunteers during low-speed rear-impact collisions and found the onset of myoelectric activity to be between 100 and 125 ms following the start of vehicle acceleration [137]. Siegmund et al. conducted a similar study and found onset times from 70 to 80 ms and peak EMG at 110 ms depending on the subjects awareness levels [138, 139]. Recognizing that muscle force lags the initiation of myoelectric activity further mitigates the role of muscle reflexes in neck injury dynamics. However, the role of passive and pre-activation (pre-impact tensing) is not well characterized and may influence the behavior of the spine. Moreover, some degree of pre-injury muscle stimulation is always present in the cervical spine to control the head weight. Studies of muscle mechanics show that elongation of stimulated muscle significantly increases muscle forces, which increases their importance in tensile and bending modes of neck loading.

The effects of muscle on cervical injury depend on the specific mode of loading. For impacts in which the dominant mode of loading is compressive, the role of the neck musculature is minimized. This is because they go slack according to the age old mechanical axiom, "you can't push on a rope". However, even in this mode of loading, the musculature, particularly the deep intersegmental muscles, influences the flexural rigidity of the ligamentous spine and, therefore, its buckling behavior (and the associated injury mechanisms). For cases of tensile loading of the spine, the muscle fibers represent a robust parallel load path for the spine, and the significance of neck musculature in tension has been quantified [89, 140, 141]. Neck musculature also influences each of the bending modes of the cervical spine: flexion-extension, lateral bending, and torsion. Comparison of volunteer and cadaveric sled decelerations show larger head excursions and rotations in the cadaver than human [86, 106, 133, 142], owing to an absence of the effects of stimulated muscle and decreased passive muscle forces in the cadaver [143].

The effects of muscles on head and neck dynamics during severe neck loading can only be fully understood using computational models. This is because the cadaver is a poor surrogate for live muscle [143]. A modeling study of pure tension from Chancey et al. shows that, depending on the level of muscle activation, compressive preloads in the cervical spine can range from 120 N for a relaxed state to 1,400 N for a tensed state [90]. And between the active and passive muscle responses, the muscles account for between 35 % and 65 % of the total neck force in tension, depending on the activation state and the spinal level.

For inertial loading in frontal impacts, the effects of muscles are similarly profound. Simulations using FE and multi-body dynamics models have shown that the most important modeling parameters dictating head kinematics are the cervical muscle properties, activations, and activation timing [144–148]. These studies have also demonstrated large compressive loads in the cervical joints during frontal impact due to the action of the extensor muscles – a finding that is somewhat counterintuitive. Compressive neck injuries in belted frontal impacts are unusual and typically attributed to contact with the A-pillar, or to airbag interaction – the possibility of a muscular mechanism bears further investigation. The role of muscle activity on cervical spine injury mechanisms in frontal impacts is complex and requires further study.

11.5.2 Buckling and Buckled Deformation

Injuries to the cervical spine are often multiple, noncontiguous, and the result of various mechanisms at different levels within the spine [149, 150]. Concomitant flexion and extension injuries have been observed in clinical studies and produced in experimental studies [61, 65, 150, 151]. The existence of concomitant injuries with seemingly disparate mechanisms can be explained by cervical spine buckling. The first order buckling mode of the cervical spine can also explain many of the most common cervical spine injuries and patterns.

Torg et al. wrote extensively that buckling of the cervical spine causes neck injury [13, 14, 15, 17]. This is not strictly correct because buckling is a structural failure not a material failure (a neck injury). As such, buckling does not cause injury directly, but it does influence cervical spine injury patterns. Because the buckle occurs prior to injury and the cervical spine has post-buckled stability, the spine can continue to resist load after buckling [61, 65, 67]. Once the critical buckling load of the cervical spine is reached, it buckles with a rapid transition from a predominantly compressive mode of deformation to a combined mode of compression and bending. It should be noted that the static critical load for the human cervical spine is less than the weight of the head [78, 152], which points to the importance of muscles in maintaining cervical spine stability. The buckling mode that is

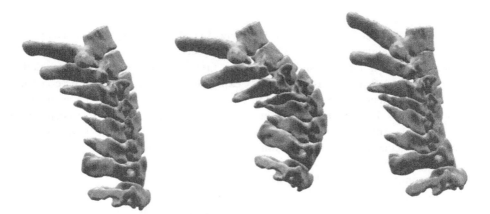

Fig. 11.15 Computer model of the effect of rate and preflexion on buckling modes. A neutrally positioned neck (*left*), first order buckling (*center*), and second order buckling (*right*). Failure in the straight, neutral, or second order configurations tends to produce vertebral body fractures. The first order mode produces a broad spectrum of injuries, including BFDs and posterior element fractures

expressed in a dynamically loaded beam-column depends on the rate of loading, the inertial characteristics of the vertebrae, and the viscoelastic characteristics of the joints. The complex interactions of the above factors can result in the second order modes have been observed in cervical spine impact studies and in modeling studies (Fig. 11.15).

While buckling itself is not an injury, the post-buckled deformation pattern of local compression with either flexion or extension bending explains the mechanism of injuries that are observed clinically. The deformation pattern of the spine is a first mode buckle for a fixed-pinned column (Fig. 11.10). This deformation exhibits regions of extension in the upper cervical spine and flexion in the lower cervical spine. As a result, the neck load vector passes behind the upper cervical spine causing upper cervical compression-extension, and in front of the lower cervical spine causing lower cervical compression-flexion. Thus, with continued increases in load, the failures in the post buckled spine may be either compression-extension or compression-flexion. Therefore, buckling plays a critical role in determining which types of injuries are produced and the peak forces required to cause them. It also provides an additional mechanistic explanation for the absence of a relationship between head motion and compression-bending neck injury.

11.5.3 Loading Rate

The cervical spine and its constituents exhibit viscoelastic behavior. Load relaxation, the decrease in load with constant deformation, is a hallmark of these viscoelastic effects. Differences in load that would be observed between high rates and quasi-static tests can be as much as 70 % [59].

In addition to having a significant component of viscoelasticity, the spine also demonstrates a second component of time dependence – preconditioning. Following large periods (hours) of inactivity, additional fluid is absorbed by the soft tissues of the spine, placing it in a stiffened state, the fully re-equilibrated state. During activity fluid is extruded from the structure and the stiffness decreases. Eventually, a steady state is reached in which subsequent activity does not produce a change in stiffness. This is the mechanically stabilized or preconditioned state. The cyclic modulus is defined as the peak load over the peak deflection during sinusoidal loading, and for the mechanically stabilized state it is from 45 % to 55 % of the fully re-equilibrated state. Given that most individuals are continually active, it has always been assumed that all cadaveric structural testing should be performed in the mechanically stabilized state.

Given the viscoelastic nature of the constituents of the cervical spine, it is not surprising that its reported structural and failure properties are sensitive to loading rate. An increase in loading velocity has been reported to increase the stiffness, failure load, and energy at a failure for a variety of spinal tissues. For example, the intervertebral disc exhibits increased stiffness with rate of loading in compression [153]. Bone exhibits increased stiffness and ultimate load with increased compression rate [154, 155]. Similar behavior has been associated with the spinal ligaments in tension [156]. Chang et al. reported increased stiffness of the cervical spine with increased rate of loading [157]. Indeed, McElhaney et al. reported that varying the loading rate by a factor of 500 resulted in a change in spinal stiffness from 1,285 to 2,250 N/cm for a single specimen [59]. A significant positive correlation between failure load, age, and loading rate for the cervical spine in nearly pure compressive loading has been reported by Pintar et al. [158]. Their linear regression model predicts forces to failure for a 30-year-old male ranging from 2 kN at 2 m/s to 6 kN at 6 m/s.

In addition to viscoelastic effects, loading rates during impact (durations of approximately 20 ms or less) will include inertial effects of the head. Measured head impact forces must then be interpreted with the realization that force measured at the point of head contact is considerably larger than the force transmitted to the neck because the neck force is reduced by the product of head mass and the head acceleration. This effect is evident in the studies that have simultaneously measured the head impact load and the neck load. Yoganandan et al. [64] reported peak head loads of 5.9 ± 3.0 kN and peak neck loads of 1.7 ± 0.57 kN. Pintar et al. reported similar data, with mean peak impact loads of 11.78 ± 5.7 kN, while the corresponding neck loads were 3.5 ± 1.95 kN [75]. Nightingale et al. report a mean head impact load of 5.8 ± 2.5 kN and a resultant neck failure load of 1.9 ± 0.7 kN [70]. Head inertial forces explain why head injury or head impulse is a poor predictor of neck injury risk (when compared to neck force and moment).

Finally, as discussed in Sect. 11.5.2, loading rate affects the buckling modes that are expressed in the cervical spine. At low impact velocities, the first order mode is the most likely to be expressed and injuries would be consistent with the deformations in Fig. 11.15. At higher velocities, cervical spine failure can occur in a higher order mode, or even before there is time for neck buckling to occur at all.

11.5.4 Head Motion

Early clinical studies hypothesized a quasi-static mechanistic approach that assumed that neck injury was caused by exaggerated motions of the head that move the cervical spine beyond its normal range of motion [51, 159]. For example, cervical flexion injuries were thought to occur in accidents in which head flexion was observed. In the same context, extension head motions were believed to be associated with extension mechanisms of injuries. However, it is quite clear from the pure moment experiments that the cervical spine is very tough in bending. Experimental studies have shown that compression is required to produce many of the injuries that have been attributed to bending. In the full cadaver tests by Nusholtz et al., head motion did not serve as a good indicator of neck injury mechanism. Bauze and Ardran, and Myers et al. were able to produce compression-flexion injuries quasi-statically without any head rotation. Nightingale et al. provided a mechanistic foundation for this observation, showing that cervical spine injury occurred within 18 milliseconds, while the large head motions did not occur until much later [61, 65]. The observed head motions were not the cause of injury - they were the dynamic consequence of the inertial characteristics of the head, the loads by the contact surface, and the reactions from an unstable cervical spine. However, the segmental behavior within the spine and the position and direction of the load vector at the instant of injury provides an excellent explanation of the cervical injury mechanism (see Sect. 11.5.2).

Similar assertions about the role of head motion on tensile injures have been put forward by Babcock, Allen et al., and Kazarian [51, 55, 159]. Unlike compression, in tensile modes, head motions are typically more descriptive of the associated neck injury mechanism. However, because some fractures, have multiple mechanisms (e.g. Hangman's fractures), the injury is still not definitive of the head motion and/or injury mechanism.

11.5.5 Lordosis and Preflexion

The initial curvature of the spine (its neutrally positioned lordosis) has received considerable attention as a mechanical determinant for neck injury risk. Specifically, removal of this initial lordotic curvature of the spine prior to impact (preflexion), has been thought to be a requisite condition for neck injury, to increase the injury risk, and to alter the mechanism of injury. Theory dictates that a column with an initial curvature will have a lower failure force compared to a straight column because bending stresses are an increasingly large percentage of the total stress in the materials with increasing curvature. However, initially curved beam-columns are less stiff and can absorb energy through both bending and compression deformation modes. As a result, while peak force at failure may be greater in a straight column, the total energy absorbed by the neck and the impulse to failure may be decreased in a straightened spine during dynamic loading. This is evident in the end-condition study by Myers et al. [67]. Curved beam-columns, if sufficiently slender, have lower buckling loads than straight columns. On the basis of purely mechanical grounds, the effects of an initial curvature on injury potential are not immediately clear.

In early studies, Torg et al. suggested that preflexion was required for the production of cervical injuries [16, 160]. However, a number of biomechanical investigations have produced clinically observed cervical fractures by impacting cadavers with unconstrained, naturally curved cervical spines [60, 61, 63, 70, 161], proving that preflexion is not a requirement for neck injury.

Torg et al's studies of video from American football were among the first to report the impulsive loading of the torso on the head that gives rise to compressive impact conditions and hence the wide range of compression bending injuries. However, the limits of standard video frame rates and the pads and helmets worn by the players made it impossible for them to accurately discern small changes in curvature of the neck just prior to and during impact.

Although preflexion of the cervical spine is not requisite for injury, it alters the types of injuries produced. Biomechanical investigations suggest that cervical spine preflexion yields a greater incidence of lower cervical compression and burst fractures than neutrally positioned spines [69, 75, 162, 163]. It appears that straightening of the cervical spine and applying forces along the spinal axis with the load vector positioned anteriorly allows transmission of the load to the middle and lower regions of the cervical spine without necessarily producing upper cervical injury [69, 75]. It also appears that preflexion increases the flexural rigidity of the spine owing to its nonlinear response, thereby decreasing its propensity to buckle. As a result, the neck fails as a column, and cervical injuries that result due to bending moments, including posterior element fractures, soft tissue injuries, and facet dislocation, are less commonly observed. While burst and wedge compression fractures are most commonly produced in the presence of preflexion, production of these injuries in the midcervical spine can also occur in neutrally positioned spines during to head impact [61, 65].

The above findings are explained by the increase in compressive stiffness that results from preflexion. This increase in stiffness can be illustrated with a computational model. Figure 11.16 shows the effects of preflexion on the peak loads in the cervical spine for a constant input energy of 16 kg at 3 m/s. The stiffer preflexed configuration must manage the impact energy over a shorter time, which results in much higher peak forces. Unfortunately; however, the neck has extremely limited ability to manage energy and accordingly is at risk for injury from falls from 0.5 m.

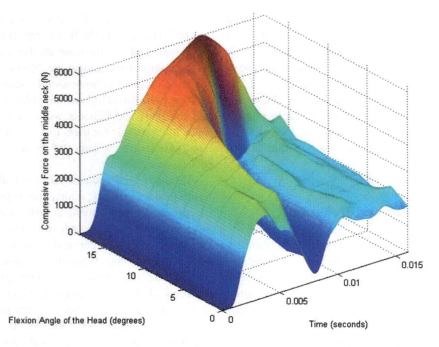

Fig. 11.16 Compressive force in the midcervical spine as a function of pre-flexion. In the neutral posture (0°), the cervical spine is prone to buckling and does not develop large compressive loads. But with neck flexion of 20° (10 from head and 10 from torso) the cervical spine is straightened and the peak loads nearly double

11.5.6 Anatomical Variation and Asymmetry

There is an understandable tendency to observe a fracture pattern that is more severe on one side and assume that an asymmetrical load (outside the midsagittal plane) was the cause. Unfortunately, it is not that simple. In all of the impact studies from The University of Michigan, Duke University, and the Medical College of Wisconsin, many asymmetrical injuries were produced in the most carefully controlled symmetrical conditions [20, 60, 63, 64, 70, 75, 164]. Asymmetric injury likely occurs due to normal variations in our anatomy where one side is a little weaker than the other and failures tend to propagate on whichever side fails first. This is one of the reasons it is difficult to deduce the loading mechanism from the injury pattern.

Toomey et al. conducted one of the few rigorous studies on compressive impact with a lateral component [71]. The hypothesis was that such an impact would reliably produce asymmetric cervical spine fractures. While asymmetric fractures were produced, the authors concluded that their injury patterns were not significantly different from those seen in symmetric impacts.

Foster et al. examined all of the compressive impact literature to see if injury patterns in rollover crashes could be correlated with conditions in impact experiments [35]. They observed a large number of unilateral fractures, but few at the AIS 3+ level (cord contusion with recovery) and none at the AIS 4+ level (permanent neurological deficit). They concluded that rollover injuries were primarily compressive in nature and the catastrophic injuries were similar to cadaver tests. But rollover injuries differed from cadaver tests in some of the more minor structures fractured. For instance, transverse process fractures are significantly underrepresented in the cadaver literature, most likely because they are caused by active musculature.

11.5.7 End Condition

Like any beam-column, the internal loads of the cervical spine are governed by the imposed end conditions of the head-neck-torso system. The approximation of the torso as a fixed end condition has been used by several investigators in cadaver tests and typically results in the production of realistic cervical injuries [65, 67, 69, 75, 165, 166]. The rationale for this approximation is that the large rotary inertia of the torso precludes significant torso rotations in the short time in which neck injuries occur. While this appears reasonable for neck injuries due to compressive impact, it is likely less valid for the longer duration dynamic events like noncontact frontal decelerations. The head geometry, mass, and interaction with a contact surface confound the simple definition of the upper cervical end condition, and as a result, investigators have imposed a variety of different end conditions on the head in an effort to better understand the importance of this factor on neck injury.

Mechanical analysis based on the effective mass of the torso and the energy absorption of the neck reveals that the cervical spine is capable of managing impacts equivalent to vertical drops of less than 0.50 m when the neck is called upon to stop the torso [25]. Unfortunately, most impact situations have considerably larger impact energies. Put in its simplest terms, because of the large mass of the torso, the head and neck must escape from the path of the moving torso or the neck will be at risk for injury.

The hypothesis that constraints on head motion increase the neck's risk for injury has been suggested by a number of investigators. Roaf was unable to produce lower cervical ligamentous injury in unconstrained flexion [80]. Hodgson and Thomas opined that restriction of motion of the atlantoaxial joint greatly increased the risk of injury [20]. Bauze and Ardran produced bilateral facet dislocation in the cadaver by constraining the rotation of the head and inserting a peg in the neural foramen [66]. Yoganandan et al. noted that head constraint increased the measured axial load and the number of injuries in whole cadaver impacts [64]. Subsequent studies have also found increased injury risk with increases in head constraint [61, 65, 67, 131, 167]. For example, Myers et al. performed cadaveric studies of isolated whole cervical spines to study the effect of imposed end condition on the observed axial and flexural stiffness of the cervical spine [67]. Indeed, stiffness increased as much as 12-fold with increasing constraint. Full constraint, which might occur when the head stops in the contact surface and the trajectory of the torso is collinear with the axis of the neck, resulted in compression and compressive-flexion types of injury in six of six specimens. Mean axial load to failure was large (4,810 N); however axial displacement was small and energy absorbed to failure was small compared to real-world impact energies. Rotational constraint, as might occur when the head pockets in the contact surface with the torso moving posteriorly, produces an S-shape to the deforming spine and bilateral facet dislocations in the lower cervical spine in six of six specimens studied. The mean axial load for bilateral facet dislocation was 1,720 N. As with full constraint, the energy absorbed to failure was small compared with the energy available during a typical impact, and the axial deflection to failure was also small.

11.5.8 Impact Attitude and Head Inertia

While increased constraint and increased contact time may increase neck injury risk, they are not requisite for neck injury and their influence is relatively small compared to other biomechanical factors (like alignment of the head, neck, torso, and impact surface). The effects of head inertia and the impact surface attitude (or equivalently, the direction of the velocity of the body relative to the impact surface) are among the most important determinants of neck injury risk. Head inertial loads alone have been shown by Nightingale et al. to constrain cervical spine motion and produce neck injury in the absence of any constraints imposed by the impact surface [65, 70]. The attitude of the impact surface has a profound effect on the risk for neck injury. Attitudes in which the cervical spine is oriented more closely to perpendicular to the impact surface place the neck at greater risk for injury than those in which the orientation of the spine is less perpendicular to the impact surface [60, 61, 63, 65]. Recognizing that the neutrally positioned cervical spine has a flexion angle of approximately 25° from horizontal at T1, it is not surprising that impacts to the head vertex and slightly anterior to the head vertex have a higher frequency and severity of neck injuries than impacts to the posterior of the head. For the anterior and vertex impacts, the neck loading vector has only a small component of force in the direction required to accelerate the head and neck out of the path of the following torso [131]. In contrast, for impacts posterior to the head vertex, the head has a component of velocity in the anterior direction that is immediately increased by the neck loading vector. As a result, the head and neck are more likely to escape the torso loading in impacts posterior to the crown of the head. If the impact surface attitude is sufficiently far enough forward that the face hits the contact surface, the probability of neck injury is again decreased with the head and neck escaping through an extension mechanism. This observation is supported by an analysis of shallow water diving accidents reported by McElhaney et al. in which divers suffering facial injuries did not suffer neck injuries, and divers with neck injuries did not suffer facial injuries [25].

The highest risk for injury appears to be a conic region within 15° of the axis of the neck as shown in Fig. 11.17. Within this zone, neck fracture is almost certain at velocities greater than 3 m/s. The head inertia makes it very difficult for the head and neck to be directed out of the path of the following torso prior to injury. Beyond this cone, there is still a large zone of vulnerability depending on many factors including velocity, impact surface, and age. Within 30°, injury appears to be more likely than not at velocities of 3 m/s [70]. But in general, the less aligned the force vector is with the cervical spine, the lower the risk for injury.

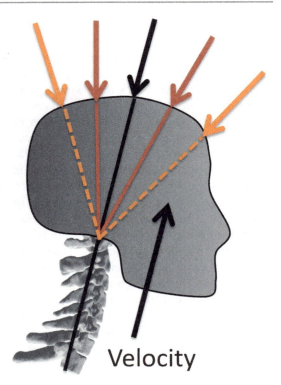

Fig. 11.17 For an impact velocity greater than 3 m/s, if the axis of the cervical spine in aligned with the contact reaction (*black*), injury is a near certainty. This is also true for reaction vectors within 15° of the cervical spine axis (*red*). Vectors within 30° (*dashed*) have a very high risk of cervical spine injury. The direction of the contact reaction is dependent on the both the velocity vector and the orientation and friction of the impacted surface

11.5.9 Impact Surface Stiffness and Padding Characteristics

Because of the obvious benefits in reduction of head injury risk and the high frequency of head injuries as compared to neck injuries, padding is often added to a potential impact surfaces. However, the addition of compliant pads to the impact surface may increase the risk of injury to the cervical spine. The mechanism by which this increase in risk occurs is more involved than the end condition-head constraint mechanism often cited in the literature (so-called pocketing). In impacts with a compliant surface, the head pockets into the pad, and its motion parallel to the impact surface is opposed by both pressure and surface friction effects. In other words, the deformed padding applies forces to the head that may oppose the escape velocity of the head.

Padding, by virtue of a change in stiffness, also delays the escape of the head from the impact surface. Specifically, by lowering the stiffness, the magnitude of the head force is diminished and the contact time is increased. As a result, the motion of the head in the escape direction, which occurs due to impact surface reactions in the escape direction, is delayed. Nightingale et al. have shown that for impacts near the vertex of the head there is a significant increase in the frequency and severity of neck injuries sustained in padded impacts when compared with rigid impacts [168]. This finding is particularly clear in the cases where the head is impacted at 15° posterior to the vertex. In this impact orientation, the head and neck have an initial velocity component directed anteriorly in an escape direction. For impacts into a frictionless rigid surface, the impact surface and neck apply forces to the head that increase the velocity in the escape direction and flex the head, allowing the head and neck to escape out of the way of the following torso. However, in padded impacts the deformed padding applies forces directed posteriorly and increases the time the head spends in the contact surface. As a result, the neck sees a significantly larger torso impulse and suffers injuries. It should be recognized that the materials used in these experiments were considerably more compliant than those used currently in automotive and other safety environments. However, it should also be apparent that the addition of surface padding cannot be considered as a method of protecting the neck from injury.

More recent studies have identified impact surface friction as a contributor to neck injury risk in compressive impacts [169–171]. Indeed, increases in the coefficient of friction between the head and impact surface greater than or equal to 0.5 were found to constrain the head translation, prohibiting motion in the escape direction [170]. Increases in the coefficient of friction were reported to increase both the neck forces and occiput-C1 moments. The mechanism for increased risk for neck injury is due to frictional forces constraining the head and preventing escape from the following torso. Moreover, in a parametric analysis including pad thickness and surface friction, increases in the coefficient of friction had the most profound effect on increasing neck forces and moments [171]. This study further reported that for increases in pad thickness, head forces were decreased with no change in neck injury risk. In that regard, the historically cited pocketing mechanism relates more to friction than the depth of the padding. Work is ongoing to evaluate the ability of "slippery" helmets to reduce the risk of both traumatic brain injury and cervical spine injury [172, 173]. For brain, there may be no net benefit to such a design because friction can increase or decrease head accelerations depending on the direction of head motion relative to the impact surface [174]. But this has yet to be evaluated for the cervical spine.

11.6 Activity-Related Injuries and Case Studies

Studies of accidents involving serious cervical injuries provide information about etiology, injury mechanisms, and human tolerance. The mechanisms of falls and diving accidents are frequently simple enough that many of the kinematics can be estimated by calculations or computer-based simulations. Some sports activities, especially American football, are videotaped and the body motions involved in the injury are preserved for analysis. Statistical data of automobile injuries provides normative data to calibrate and validate injury tolerance criteria against field epidemiology and exposure, but are often too complex for individual analysis without a great effort. Recognizing that neck injuries occur in the domain of milliseconds, even these well-documented events may fail to capture the actual time of injury and yield erroneous conclusions regarding mechanism. Moreover the annual incidence data reported herein does not account for the degree of participation and therefore the relative risks of the activities. For example, hang gliding, which is a relatively uncommon activity, may have a higher relative risk than other activities whose participation and incidence are larger.

11.6.1 Automobile Accidents

Auto accidents remain the most common cause of fractures and dislocations of the cervical spine. The impacted structures, force directions, and velocities are extremely varied and result in a wide range of cervical spine injuries. Because of the large numbers involved, it is often possible to draw correlations between the head impact site and the type and level of cervical fracture. Frequently, there are facial or scalp lacerations that can be associated with permanent structural deformations or imprints in the vehicle. However, the vehicle motions and occupant kinematics are usually too complex to allow the detailed analysis required to estimate the impact forces, velocities, and accelerations. Thus, these accidents do not provide much neck tolerance data.

Yoganandan et al. (Table 11.2), in a study of the epidemiology and injury biomechanics of cervical injuries in 103 motor vehicle accidents (MVAs), concluded:

- Cervical spine injuries from MVAs concentrate at the occiput-axis in the upper cervical region and primarily around C5–C6 in the lower cervical region.
- The craniocervical junction and upper regions of the cervical spine (O-C1-C2) are the most common sites of trauma in fatal spinal injury from MVAs.
- Among survivors with spinal injury from MVAs, the lower cervical spine is the more frequently injured region than the upper cervical spine.

Table 11.2 Injury score by spinal level in MVCs

Spinal level	AIS injury level				
	2	3	4	5	6
C1					1
C2	1	8	1		1
C3		4		1	
C4		4	4	2	
C5	1	5	11	15	
C6		10	9	17	
C7		4	3	2	

- A strong association exists between craniofacial injury and serious cervical spine trauma; the overall beneficial role of belt restraint in the reduction of serious cervical spine injuries is most likely due to the prevention of head/face impact with vehicle contact.

Regardless of the circumstance of the accident, in most instances the victim is injured because of head contact with some object and neck load developing because of the inertia of the torso. That said, evidence of head injury (superficial abrasion, laceration, or extracranial hematoma) is usually absent or not-detected owing to the presence of hair and the under-reporting of minor, non-clinically meaningful injuries. The position of the head and neck, the impact site, the nature of the impacted surface, and the direction of the cervical spine loading determines the resulting cervical fracture. Many combinations of head-neck position, impact site, and cervical spine loading can occur, and research has shown that the relationship between impact, the point of impact, the head motion, and neck injury is not simple (Sect. 11.5, & [70]). In a few instances the head is not impacted and neck injury occurs entirely because of the head inertia. These however are uncommon and suggest a lack of preimpact awareness, underlying weakness, or large accelerations [2, 7, 175, 176].

Case 1

Shear et al., 1988 reported on a 39-year-old woman who sustained an immediate and complete motor and sensory quadriplegia at C5 in a motor vehicle accident in which the vehicle she was the passenger in was struck broadside on the passenger side [151]. In addition to her spinal injuries, she suffered a closed head injury [Glasgow Coma Scale (GCS) 7], facial lacerations, bilateral forearm fractures, and a fractured left acetabulum. (The GCS score ranges from 3 to 15 with decreasing severity; a severe brain injury is scored GCS 8 or less). Radiographs revealed a burst fracture of the fifth cervical vertebral body with a 25 % forward subluxation of C4 on C5, in addition to a fracture of the posterior arch of C1.

Fig. 11.18 Motorcycle crash illustrating the kinematics associated with the thoracic fractures that are common for motorcycle crash victims

The cervical spine injuries were managed with initial skeletal traction, followed by a posterior fusion of C3 to C6. The patient's complete sensory and motor paralysis remained unchanged.

This case illustrates the common occurrence of multiple noncontiguous cervical injuries. The mechanisms are upper cervical compression-extension and lower cervical compression-flexion. These differing mechanisms occur during one impact and are a direct result of the buckling of the spine that precedes, but does not cause, the cervical injuries.

11.6.2 Motorcycle Accidents

A study of spine injuries in motorcycle accidents by Kupferschmid et al. reported on 266 motorcycle accidents [177]. Thirteen cases of thoracic spine fractures, four cases of cervical spine fractures, and two cases of lumbar spine fractures were identified. The association of cervical and thoracic injuries (i.e., multiple noncontiguous fractures) is a common finding in motorcycle accidents. This is likely the result of the high kinetic energy of the victim and the kinematics. Similar kinematics and injuries are also seen in bucked horseback riders and cyclists.

The typical accident situation associated with this injury involves the victim somersaulting over the handlebars of the motorcycle and landing on his/her head, shoulders, or upper back (Fig. 11.18). Thus, the victim may experience significant impact to the head or the upper thorax. Depending on the vault height, timing, and orientation, they can experience head, neck or thoracic spinal injuries.

In many motorcycle, bicycle and equestrian accidents, the helmet provides a unique opportunity to improve our understanding of human tolerance and recreate the injury mechanism. Specifically, in many helmets, a clearly discernible crush area in the polystyrene helmet liner and a characteristic imprinting of the helmet shell at the point of impact are observed. Since the liner provides a permanent record of the impact pressure on the head, researchers have been studying methods of interpreting it. As an example, impact tests on exemplar helmets with loads of 6–8 kN replicate the damage seen on helmets following an accident. Because of its proximity to the injury, it is commonly thought that helmets should play a role in neck injury prevention. Unfortunately, this is not the case. While the helmet decreases the head acceleration it cannot prevent inertial loading by the torso.

11.6.3 Football

Injury in football receives considerable attention despite being a hallmark in which characterization of the injury biomechanics and appropriate intervention have dramatically decreased injury rates. When football helmets with rigid shell and energy absorbing liners were introduced, coaches and players assumed that improved head and neck protection would allow a new style of blocking and tackling using the head to spear the opposition. The incidence of serious cervical spine injuries increased dramatically. Torg et al. and Schneider correctly analyzed the cause (compression owing to near vertex head impact) of this increase and assisted in initiating a rule change prohibiting the use of head-first blocking and tackling [16, 19]. The result was a threefold reduction in the numbers of quadriplegic injuries in the late 1970s and a subsequent continued reduction by sixfold as a result of coaching and trainer education [178]. Put into its proper context, in 2009, diving, cycling and all-terrain vehicles accounted for approximately 2,200 quadriplegic injuries whereas football accounted for 136 injuries [179].

Case 2

While carrying the ball, a high school football halfback received a neck injury when two tacklers held his arms at his side as he fell with his neck flexed and his head bent downward [19]. A third opponent struck the group so that the ball carrier fell forward with the neck so completely flexed that both shoulders rested directly on the ground. On impact with the ground, the halfback had an immediate incomplete paralysis of all four extremities, with a total loss of sensation for a few minutes after injury in his lower extremities. Between the time of injury and admission to his community hospital, the patient gradually recovered all motor function and only noted pain and tingling radiating along the inner aspect of his forearms. Five weeks later, a neurosurgical examination revealed some right triceps paresis and weakness of his interosseus muscles. X-rays of the cervical spine demonstrated bilateral facet dislocation of C6 on C7 with marked vertebral subluxation. A fracture of the lateral mass of C7 and possible C6 vertebral fracture was demonstrated.

This injury includes a facet dislocation and compression fractures of the contiguous vertebrae. Historically, this mechanism has been described as a "forced flexion". However, the presence of significant components of compression and the results of impact neck injury research show that this is not the case. Indeed, this injury likely occurred in the first 20 ms following impact, before the head rotated significantly. The compression bending (flexion) injury is a caused by flexion in the lower cervical spine that occurred as a result of the neck buckling. It also illustrates that while complete spinal cord injuries typically remain permanent, partial injuries can have significant recovery.

Case 3

A 20-year-old college football player attempted to tackle a charging ball carrier. Just prior to this attempt, a teammate hit the ball carrier at an angle, causing him to tumble forward (Fig. 11.19). The tackler struck the back of the now upside-down ball carrier with his head, neck, and torso in line. A severe fracture dislocation resulted with complete neurologic deficit below the C4–C5 level.

An analysis of the game film showed a closing velocity between the tackler's head and the ball carrier's back of 4.27 m/s. This is the equivalent of a three-foot fall and illustrates how fragile the neck is when the head, neck, and torso are aligned, and the impact is near the vertex of the head - the state was ideal for a catastrophic injury. Eighteen inch fall heights with only a portion of the torso following have been shown to reliably produce neck injuries in cadaver studies. In this regard, one need not look for "high energy" events to have sufficient energy to cause injury. This particular player's situation was compounded by a strap placed by the equipment manager between the face-mask

Fig. 11.19 The indicated player, who struck the lower back of the inverted player, sustained a fracture dislocation of C4–C5 resulting in complete neurological deficit

and shoulder pads to reduce the risk of an extension injury, which instead maintained a spearing alignment of the body and a near vertex head impact.

11.6.4 Diving

Diving remains a leading cause of cervical spine injury and has been studied and published extensively [180]. Unlike other activities, diving provides a unique opportunity for reconstruction using volunteers who can replicate observed kinematics in deeper water. Estimated depth of water where the victim struck the bottom in open water was derived from statements of witnesses (Fig. 11.20). In 18 % of the accidents the impact depth was 2 ft or less and in 85 % of accidents the water was 4 ft deep or less. McElhaney et al reconstructed 76 diving accidents by observing anthropometrically similar volunteers performing specific dives in deep water [25]. These included shallow water dives from the edge of the pool, springboard dives, and head-first entry from a water slide. Dive kinematics where quantified using 200 frame/s film and a grid placed behind the diver (Fig. 11.21). In this way, a lower bound for critical head impact velocity resulting in cervical injury was developed. Based on these studies, in a clear dive, hydrodynamic drag generally does not start to slow the diver's in-water speed to values less than that at water entry until a significant portion of the diver's body is immersed. On the order of one-half to three-fourths of the body must be in the water when executing a clean dive before deceleration commences. Indeed, under the action of gravity, the diver's in-water head speed increases for the first meter of travel (Fig. 11.22).

For dives from the edge of the pool, fall heights ranged from 116 to 219 cm. Estimated head impact speeds ranged from 3.11 to 6.55 m/s. Springboard and platform dives had a wide range of fall heights and water depths; however, impact velocities remained between 3.8 and 8.1 m/s. Head-first entry from a water slide has the potential for neck injury when, instead of skimming

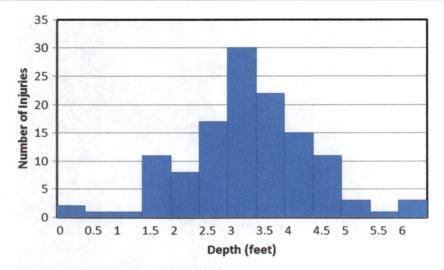

Fig. 11.20 Histogram of injuries by water depth. Fifty-five percent of cervical injuries occur between 2.5 and 4 ft of water

Fig. 11.21 Experimental technique for reconstructing diving accidents

across the surface, the head and hands are lowered and a snap roll or tumble occurs. Head impact velocities for the snap roll mode range from 3.57 to 4.94 m/s.

It is clear from the number and severity of the accidents presented here that diving or head-first sliding into shallow water is potentially dangerous and should be actively discouraged. The snap-roll motion probably occurred in many of these accidents. Keeping the head and hands up and the back arched is critical in shallow water diving.

Neck injuries observed in diving are highly similar, clustering in the lower cervical spine as in other activities (Table 11.3). Interestingly, there is a negative association between facial trauma and neck injury suggesting that if the face hits the impact surface, the neck moves into extension and

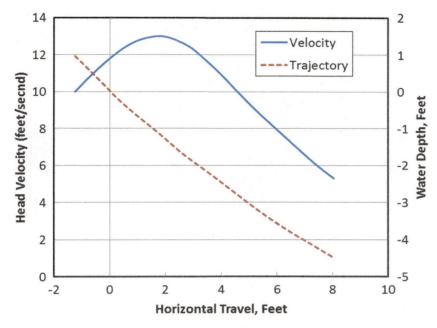

Fig. 11.22 The head trajectory and head velocity for a typical dive from the edge of a pool, showing the increase in head velocity following entry into the water. Water level is at depth=0 (Reproduced from [25])

the torso does not load the neck thereby avoiding injury. Moderate to severe head injuries are also rare in diving. As head impact occurred in most all of these accidents, these data indicate that the velocities required to produce head injuries are higher than those required to produce catastrophic neck injuries. Taken together, these simulations suggest that compression-bending injuries can occur with head impacts in a diving mode at velocities greater than 10 ft/s (3.05 m/s), where the head is suddenly stopped and the neck must stop the torso.

Case 4

A young man was participating in a late night pool party when he decided to dive into approximately 75 cm of water. This case was somewhat unique in that the entire dive sequence was capture by video camera. Based on photogrammetry, he struck the bottom at approximately 7 m/s while his abdomen and lower extremities were still above the surface (Fig. 11.23). The resulting injury is shown in Fig. 11.5 and was clinically described as a comminuted compression fracture

Table 11.3 Level of injury in diving accidents occurring in pools and the natural environment

Level	Pool	Natural	Total
C1	3	1	4
C1–2	3	0	3
C2	1	0	1
C2–3	1	1	2
C3	1	1	2
C3–4	6	0	6
C4	6	2	8
C4–5	41	6	47
C5	106	48	154
C5–6	113	42	155
C6	18	4	22
C6–7	33	12	45
C7	3	3	6
Thoracic	5	2	7
Total	340	122	462

of C5 with disruption of the posterior longitudinal ligament. There was a superior occipital laceration on his head and a small occipital bone fracture near the foramen magnum. He was immediately quadriplegic and was rescued from the pool by friends. The location of the scalp

Fig. 11.23 Shallow water pool diving accident captured on video

laceration suggests that he had his head down at the time of impact and the video shows that his dive was becoming increasingly vertical as he hit the bottom. This resulted in a classic burst fracture in a preflexed (straightened) spine. Alcohol was a factor, as it is in nearly half of all diving related cervical spine injuries [30], which suggests that many of these injuries result from impaired judgment.

Case 5

A middle-aged male dove head first on a toy known as a Slip & Slide. This is a plastic sheet that is spread on the ground and sprinkled with water. Children are encouraged to dive onto this slippery surface and slide on their chest. In this case, the victim struck his face and forehead with sufficient force to produce a posterior dislocation of C5 on C6 and a midsagittal splitting of the body of C6. Analysis of videotape showed that he did not strike the top of his head nor did his head flex or extend significantly. He suffered incomplete sensory and motor loss. This case also illustrates the relationship between the head impact and neck injury and illustrates the limitations of conventional imaging in providing detailed kinematic information regarding the injury mechanism. At 30 Hz, or 60 Hz in split frame, the duration between frames is longer than the time required for the injury to occur. As in prior cases, the injury is the result of compressive forces, buckling, and compression-bending failure of vertebra and motion segment.

11.7 Summary

The summary of the literature presented in this chapter shows that much has been written about the biomechanical aspects of cervical injury. Most of what we know comes from two sources: clinical studies including accident reconstruction, and cadaver studies. In the last 30 years, the field of biomechanics has made significant gains in understanding the complex dynamics that accompany both head-impact neck injury and inertial neck injury. Cervical injury mechanisms involve a wide range of loading modes with various combinations of force and concurrent bending moment. Using the historical concepts of excessive head motion as a cause of neck injury frequently leads to paradoxical and incorrect observations regarding neck injury mechanism. In contrast, when the neck is viewed as a mechanical structure, whose injury is a consequence of the deformations of a segmented beam column bounded by two large masses, the neck's mechanical behavior and the resulting traumatic injuries can be readily explained. That is, by considering the effects of buckling and examining the forces and moments that act at the level of injury in the spine at the instant of injury, the mechanisms of each injury become clear and rational.

Contrary to current clinical thinking and many of the injury classification schemes in the literature, neck injury is primarily driven by tension and compression. The neck acts very much like a stout cable or green wood. It is tough and flexible in bending and is most easily damaged in compression and tension: the modes in which it is most stiff. It is clear from the most current science that the cervical spine injuries with the most devastating neurological outcomes can be explained or understood within the context of the deformations that result from compression, compressive buckling, and tension.

In compression, the fractures occur while the cervical spine is either straightened or lordotic. Regardless, compressive forces transmitted from the head and torso are resolved as primarily compressive forces at the vertebral levels. Burst fracture and Jefferson fractures result from compressive stresses in the vertebra and occur more frequently when the cervical spine is in a straightened configuration.

Compressive buckling results in a characteristic deformation pattern (Fig. 11.10) that can give rise to a wide spectrum of fractures. The inflexion point (transition region from flexion to extension) likely varies with time, impact energy and individual anatomy. First order buckling typically produces compressive-flexion related injuries in the lowest parts of the spine and compressive-extension injuries further up, with axial compression fracture possible at the inflexion point. Experimental and modeling observations of second order buckling modes at higher rates or straightened configurations give rise to the possibility of even more complex injury kinematics.

Tension is a fairly straightforward mechanism that is of particular concern in the upper cervical spine. It can cause a wide variety of C2 fractures as well as ligament disruption and dislocation at O–C1 and C1–C2. However, many of the C2 fractures seen in tension can also occur in compression or following compressive buckling.

There has been a historical desire to define unique injury mechanisms for all manner of cervical spine fractures. For instance, "compressive-extension" for lamina fractures, "distractive-flexion" for bilateral facet dislocations, and "vertical-compression" for comminuted fractures of the vertebral body. However, it is clear from the literature, that even under the most controlled circumstances, all of these injuries patterns can be produced from what is essentially the same loading mechanism. For this reason, without corroborating evidence such as a scalp laceration or independent basis for knowing the victim's kinematics, the loading mechanism cannot be definitively deduced from a retrospective evaluation of the radiology alone. The mechanistic injury classifications are quite useful for describing what is happening at the involved joints in the cervical spine, but are quite useless in describing the system level inputs (i.e., head and torso kinetics) that produced the loads.

In the clinical setting, the bilateral facet dislocation is almost universally misunderstood to result from "hyperflexion". However this was a hypothesis that had never been validated. To our knowledge, there has never been a published biomechanical study that has produced this injury in a whole cadaver or whole cervical spine in flexion. This is because "hyperflexion" is not possible within the anatomical range of motion of the neck. The mechanism for a BFD is either post compressive buckling or combined tension and bending – both of these conditions can completely disrupt the joint. The two modes can often be distinguished on the basis of smaller associated facet fractures. Regardless, the result is a grossly unstable segment that has sufficient mobility to allow the superior facets of the joint to "lock" over the inferior facets. This locking is not typically maintained by mechanical engagement – it is maintained by muscle activation and spasm. In the cadaver, the injury can be effortlessly reduced and reproduced by hand owing to an absence of muscle forces. But in the living victim, it can only be reduced by applying sufficient tension to overcome the muscles. Ironically, standard clinical practice already treats the BFD with the opposite load that typically produces it – namely traction. Perhaps for this reason, correctly classifying this injury as compression may not lead to significant changes in clinical practice.

Although applications of injury biomechanics research by the present authors and others to automotive safety standards were not covered in this chapter, it should be noted that the United States Federal Motor Vehicle Safety Standards, No. 208, was upgraded in 2001. The neck injury criteria (Nij), was introduced for the first time for crashworthiness. The criteria, which combines loads based on treating the neck as a slender beam-column, was based on research findings described in this chapter and elsewhere in the book and was introduced to the US public in 1998. The initial introduction was based on research conducted in the late 1970s to 1990s. The formulated interaction criteria were validated with crashworthiness tests. The authors of this chapter played a role in the process, in concert with personnel from the National Highway Traffic Safety Administration. Combining the axial force with the sagittal bending moment, applied to different dummy populations appears to have produced a positive result in terms of decreasing neck injuries, as evidenced by an initial analysis of recent NASS-CDS databases (Fig. 11.1). Other fields have also received attention for using combined criteria. With continuing technological advances in dummy design, instrumentation, and biofidelity, and with newer dummies envisaged for special applications such as rollovers, the approaches used in recent decades of combining experimental models with numerical simulations, together with correlation of injuries with clinical cases, will continue to drive the safety engineering field in mitigating neck injuries in motor vehicle and other environments.

Work remains to be done. Improvements in computational models of neck dynamics have been dramatic and continue. These models, however, are only as good as the data on which they were developed. There is still a need for experimental work, particularly on the role of the muscles and in characterizing the cervical spine in lateral bending. The concepts of head pocketing and the constraints imposed by the impact surface have been studied and have identified surface friction as a source of constraints that can increase injury risk. However, head inertia also provides a large constraining force (forces that oppose motion in the direction that would reduce neck loading) at injury and this may defy attempts to redirect the head quickly enough to prevent neck injury. Future research to develop injury prevention strategies that capitalize on these results will be required.

Acknowledgments This material is the result of work supported with resources and the use of facilities at the Zablocki VA Medical Center, Milwaukee, Wisconsin and the Medical College of Wisconsin. Narayan Yoganandan is a part-time employee of the Zablocki VA Medical Center, Milwaukee, Wisconsin. Any views expressed in this chapter are those of the authors and not necessarily representative of the funding organizations.

References

1. Berkowitz M (1998) Spinal cord injury: an analysis of medical and social costs. Demos Medical Publishing, New York
2. Huelke DF, Moffat EA, Mendelsohn RA, Melvin JW (1979) Cervical fractures and fracture dislocations – an overview. In: Proceedings of the 23rd Stapp Car Crash Conference, vol 790131. pp 462–468
3. Burke DC (1973) Spinal cord injuries and seat belts. Med J Aust 2:801–806
4. Epstein BS, Epstein JA, Jones MD (1978) Lap-sash three point seat belt fractures of the cervical spine. Spine 3(3):189–193
5. Gögler H, Athanasiadis S, Adomeit D (1979) Fatal cervical dislocation related to wearing a seat belt: a case report. Injury 10(3):196–200
6. Horsch JD, Schneider DC, Kroell CK, Raasch F (1979) Response of belt restrained subjects in simulated lateral impact
7. Huelke DF, Mackay GM, Morris A, Bradford M (1992) Non-head impact cervical spine injuries in frontal car crashes to lap-shoulder belted occupants
8. Nyquist GW, Begeman PC, King AI, Mertz HJ (1980) Correlation of field injuries and GM Hybrid III dummy responses for Lap-Shoulder Belt Restraint. J Biomech Eng 102:103–109
9. Marsh J, Scott R, Melvin J (1975) Injury patterns by restraint usage in 1973 and 1974 passenger cars.

In: 19th annual conference of the American association for automotive medicine
10. Schmidt G, Kallieris D, Barz J, Mattern R, Klaiber J (1975) Neck and thoray tolerance levels of belt-protected occupants in head-on collisions, vol SAE# 751149
11. Taylor TK, Nade S, Bannister JH (1976) Seat belt fractures of the cervical spine. J Bone Joint Surg Br 58(3):328–331
12. Shields C Jr, Fox J, Stauffer E (1978) Cervical cord injuries in sports. Phys Sports Med 6(9):71–76
13. Otis J, Burstein A, Torg J (1991) Mechanisms and pathomechanics of athletic injuries to the cervical spine. Athletic injuries to the head, neck, and face. Mosby–Year Book Inc., St. Louis, pp 438–456
14. Torg JS (1985) Epidemiology, pathomechanics, and prevention of athletic injuries to the cervical spine. Med Sci Sports Exerc 17(3):295–303
15. Torg JS, Sennett B, Pavlov H, Leventhal MR, Glasgow SG (1993) Spear tackler's spine an entity precluding participation in tackle football and collision activities that expose the cervical spine to axial energy inputs. Am J Sports Med 21(5): 640–649
16. Torg JS, Truex R Jr, Quedenfeld T, Burstein AH, Spealman A, Nichols C (1979) The national football head and neck injury registry: report and conclusions 1978. JAMA 241:1477–1479
17. Torg JS, Vegso JJ, O'Neill MJ, Sennett B (1990) The epidemiologic, pathologic, biomechanical, and cinematographic analysis of football-induced cervical spine trauma. Am J Sports Med 18(1):50–57
18. Torg JS, Vegso JJ, Sennett B, Das M (1985) The national football head and neck injury registry. 14-year report on cervical quadriplegia, 1971 through 1984. JAMA 254(24):3439–3443
19. Schneider RC (1973) Head and neck injuries in football: mechanisms, treatment, and prevention. Williams & Wilkins, Baltimore
20. Hodgson VR, Thomas LM (1980) Mechanisms of cervical spine injury during impact to the protected head. In: Proceedings of the 24th Stapp Car Crash Conference, pp 17–42
21. Mertz HJ, Hodgson VR, Thomas LM, Nyquist GW (1978) An assessment of compressive neck loads under injury-producing conditions. Phys Sportsmed 6(11):95–106
22. Virgin H (1980) Cineradiographic study of football helmets and the cervical spine. Am J Sports Med 8(5):310–317
23. Kewalramani LS, Taylor RG (1975) Injuries to the cervical spine from diving accidents. J Trauma Acute Care Surg 15(2):130–142
24. Albrand O, Walter J (1975) Underwater deceleration curves in relation to injuries from diving. Surg Neurol 4(5):461
25. McElhaney JH, Snyder RG, States JD, Gabrielsen MA (1979) Biomechanical analysis of swimming pool injuries. Soc Automot Eng 790137:47–53
26. Gabrielsen MA (1990) Diving injuries: the etiology of 486 case studies with recommendations for needed action. NOVA University Press, Ft. Lauderdale
27. Reid D, Saboe L (1989) Spine fractures in winter sports. Sports Med 7(6):393–399
28. Sekhon LHS, Fehlings MG (2001) Epidemiology, demographics, and pathophysiology of acute spinal cord injury. Spine 26(24S):S2–S12
29. Kiwerski J (1991) The influence of the mechanism of cervical spine injury on the degree of the spinal cord lesion. Spinal Cord 29(8):531–536
30. Bailes JE, Herman JM, Quigley MR, Cerullo LJ, Meyer PR Jr (1990) Diving injuries of the cervical spine. Surg Neurol 34(3):155–158
31. Bedbrook G (1979) Spinal injuries with tetraplegia and paraplegia. J Bone Joint Surg Br 61(3):267
32. Portnoy HD, McElhaney HJ, Melvin JW, Croissant PD (1972) Mechanism of cervical spine injury in auto accidents. In: Proceedings of the 15th annual conference of the American association for automotive medicine, pp 58–83
33. Braakman R, Vinken PJ (1967) Unilateral facet interlocking in the lower cervical spine. J Bone Joint Surg 49B(2):249–257
34. Bohlman H (1979) Acute fractures and dislocations of the cervical spine. An analysis of three hundred hospitalized patients and review of the literature. J Bone Joint Surg 61(8):1119
35. Foster J, Kerrigan JR, Nightingale RW, Funk JR, Cormier J, Bose D, Sochor M, Ridella SA, Ash J, Crandall JR (2012) Analysis of cervical spine injuries and mechanisms for CIREN rollover crashes. In: Proceedings of the 2012 IRCOBI Conference, Dublin
36. Yoganandan N, Haffner M, Maiman DJ, Nichols H, Pintar FA, Jentzen J, Weinshel SS, Larson SJ, Sances A (1989) Epidemiology and injury biomechanics of motor vehicle related trauma to the human spine. In: Proceedings of the 33rd Stapp Car Crash conference, 1989
37. Jefferson G (1920) Fracture of the atlas vertebra. Br J Surg 7:407
38. White AA, Panjabi MM (1990) Clinical biomechanics of the spine. JB Lippincott Company, Philadelphia
39. Jefferson G (1927) Remarks on fractures of the first cervical vertebra. Br Med J 2(3473):153
40. Rayes M, Mittal M, Rengachary SS, Mittal S (2011) Hangman's fracture: a historical and biomechanical perspective. J Neurosurg Spine 14(2): 198–208
41. Braakman R, Penning L (1971) Causes of spinal lesions. In: Injuries of the cervical spine. Excerpta Medica, Amsterdam
42. Harris J Jr, Edeiken-Monroe B, Kopaniky D (1986) A practical classification of acute cervical spine injuries. Orthop Clin North Am 17(1):15
43. Levine AM, Edwards CC (1985) The management of traumatic spondylolisthesis of the axis. J Bone Joint Surg 67A(2):217–226

44. Anderson LD, D'Alonzo RT (1974) Fractures of the odontoid process of the axis. J Bone Joint Surg 56(8):1663–1674
45. Werne S (1957) Studies in spontaneous atlas dislocation. Acta Orthop Scand Suppl 23:1
46. Fielding JW, Cochran GVANB, LAWSING JF III, Hohl M (1974) Tears of the transverse ligament of the atlas a clinical and biomechanical study. J Bone Joint Surg Am 56(8):1683–1691
47. Montane I, Eismont FJ, Green BA (1991) Traumatic occipitoatlantal dislocation. Spine 16(2):112
48. Bucholz RW, Burkhead WZ (1979) The pathological anatomy of fatal atlanto-occipital dislocations. J Bone Joint Surg Am 61(2):248–250
49. Myers BS, Nightingale RW (1999) Review: the dynamics of near vertex head impact and its role in injury prevention and the complex clinical presentation of basicranial and cervical spine injury. Traffic Inj Prev 1(1):67–82
50. Torg JS, Pavlov H (1991) Axial load teardrop fracture. In: Athletic injuries to the head, neck and face. Lea & Febiger, Philadelphia
51. Babcock JL (1976) Cervical spine injuries. Diagn Classif Arch Surg 111(6):646–651
52. Norton WL (1962) Fractures and dislocations of the cervical spine. J Bone Joint Surg Am 44(1):115–139
53. Schneider RC, Kahn EA (1956) Chronic neurological sequelae of acute trauma to the spine and spinal cord the significance of the acute-flexion or tear-drop fracture-dislocation of the cervical spine. J Bone Joint Surg 38(5):985–997
54. Whitley JE, Forsyth H (1960) The classification of cervical spine injuries. Am J Roentgenol Radium Ther Nucl Med 83:633
55. Allen BL, Ferguson RL, Lehmann TR, O'Brien RP (1982) A mechanistic classification of closed indirect fractures and dislocations of the lower cervical spine. Spine 7(1):1–27
56. Dolan K (1977) Cervical spine injuries below the axis. Radiol Clin North Am 15(2):247–259
57. Moffatt C, Advani S, Lin CJ (1971) Analytical and experimental investigations of human spine flexure
58. Myers BS, Winkelstein BA (1995) Epidemiology, classification, mechanism, and tolerance of human cervical spine injuries. Crit Rev Biomed Eng 23(5–6):307–409
59. McElhaney JH, Paver JG, McCrackin HJ, Maxwell GM (1983) Cervical spine compression responses. In: Proceedings of the 27th Stapp Car Crash Conference, vol 831615. pp 163–177
60. Nusholtz GS, Huelke DE, Lux P, Alem NM, Montalvo F (1983) Cervical spine injury mechanisms. In: Proceedings of the 27th Stapp Car Crash Conference, pp 179–197
61. Nightingale RW, McElhaney JH, Richardson WJ, Best TM, Myers BS (1996) Experimental impact injury to the cervical spine: relating motion of the head and the mechanism of injury. J Bone Joint Surg Am 78(3):412–421
62. Culver RH, Bender M, Melvin JW (1978) Mechanisms, tolerances, and responses obtained under dynamic superior-inferior head impact. UM-HSRI-78-21
63. Nusholtz GS, Melvin JW, Huelke DF, Alem NM, Blank JG (1981) Response of the cervical spine to superior-inferior head impact. In: Proceedings of the 25th Stapp Car Crash Conference, pp 197–237
64. Yoganandan N, Sances A, Maiman DJ, Myklebust JB, Pech P, Larson SJ (1986) Experimental spinal injuries with vertical impact. Spine 11(9):855–860
65. Nightingale RW, McElhaney JH, Richardson WJ, Myers BS (1996) Dynamic responses of the head and cervical spine to axial impact loading. J Biomech 29(3):307–318
66. Bauze RJ, Ardran GM (1978) Experimental production of forward dislocation of the human cervical spine. J Bone Joint Surg 60-B:239–245
67. Myers BS, McElhaney JH, Richardson WJ, Nightingale RW, Doherty BJ (1991) The influence of end condition on human cervical spine injury mechanisms. In: Proceedings of the 35th Stapp Car Crash Conference, 912915, pp 391–400
68. Yoganandan N, Pintar FA, Sances A Jr, Reinartz J, Larson SJ (1991) Strength and kinematic response of dynamic cervical spine injuries. Spine 16(10):S511–S517
69. Pintar FA, Yoganandan N, Pesigan M, Voo L, Cusick JF, Maiman DJ, Sances A Jr (1995) Dynamic characteristics of the human cervical spine. In: Proceedings of the 39th Stapp Car Crash Conference, vol 95722. pp 195–202
70. Nightingale RW, McElhaney JH, Camacho DL, Winkelstein BA, Myers BS (1997) The dynamic responses of the cervical spine: the role of buckling, end conditions, and tolerance in compressive impacts. In: 41st Stapp Car Crash Conference Proceedings 41 (973344), pp 451–471
71. Toomey DE, Mason MJ, Hardy WN, Yang KH, Kopacz JM, Van Ee C (2009) Exploring the role of lateral bending postures and asymmetric loading on cervical spine compression responses. In: Proceedings of the ASME 2009 international mechanical engineering Congress & exposition, ASME, Lake Buena Vista
72. Saari A, Itshayek E, Cripton P (2011) Cervical spinal cord deformation during simulated head-first impact injuries. J Biomech 44(14):2565–2571
73. Ivancic PC (2012) Head-first impact with head protrusion causes noncontiguous injuries of the cadaveric cervical spine. Clin J Sport Med 22(5):390–396
74. Nightingale RW, Camacho DL, Armstrong AJ, Robinette JJ, Myers BS (1997) Cervical spine buckling: the effects of vertebral mass and loading rate. In: 1997 Advances in bioengineering, ASME IMECE, pp 231–232
75. Pintar FA, Sances A, Yoganandan N, Reinartz J, Maiman DJ, Unger G, Cusick JF, Larson SJ (1990) Biodynamics of the total human cadaveric cervical spine. In: Proceedings 34th Stapp Car Crash Conference, vol 902309. pp 55–72

76. Maiman DJ Jr, Sances A, Myklebust JB, Larson SJ, Houterman C, Chilbert M, El-Ghatit AZ (1983) Compression injuries of the cervical spine: a biomechanical analysis. Neurosurgery 13(3):254–260
77. Pintar FA, Yoganandan N, Pesigan M, Reinartz J Jr, Sances A, Cusick JF (1995) Cervical vertebral strain measurements under axial and eccentric loading. J Biomech Eng 117(4):474–478
78. Nightingale RW, Camacho DL, Armstrong AJ, Robinette JJ, Myers BS (2000) Inertial properties and loading rates affect buckling modes and injury mechanisms in the cervical spine. J Biomech 32(2):191–198
79. Beatson T (1963) Fractures and dislocations of the cervical spine. J Bone Joint Surg Br 45(1):21
80. Roaf R (1960) A study of the mechanics of spinal injuries. J Bone Joint Surg Br 42(4):810–823
81. Roaf R (1963) Lateral flexion injuries of the cervical spine. J Bone Joint Surg Br 45:36–38
82. Torg JS, Sennett B, Vegso JJ, Pavlov H (1991) Axial loading injuries to the middle cervical spine segment. Am J Sports Med 19(1):6–20
83. Kuster D, Gibson A, Abboud R, Drew T (2012) Mechanisms of cervical spine injury in Rugby Union: a systematic review of the literature. Br J Sports Med 46:550–4
84. Dennison CR, Macri EM, Cripton PA (2012) Mechanisms of cervical spine injury in rugby union: is it premature to abandon hyperflexion as the main mechanism underpinning injury? Br J Sports Med 46(8):545–549
85. Michaelson J, Forman J, Kent R, Kuppa S (2008) Rear seat occupant safety: kinematics and injury of PMHS restrained by a standard 3-point belt in frontal crashes. Stapp Car Crash J 52:295
86. Mertz HJ, Patrick LM (1971) Strength and response of the human neck. In: Proceedings 15th Stapp Car Crash Conference, . vol 710855. pp 207–254
87. Sances A, Myklebust J, Cusick JF, Weber R, Houterman C, Larson SJ, Walsh P, Chilbert M, Prieto T, Zyvoloski M, Ewing C, Thomas D, Saltzberg B (1981) Experimental studies of brain and neck injury. In: Proceedings of the 25th Stapp Car Crash Conference, vol 811032. pp 149–194
88. Yoganandan N, Pintar FA, Maiman DJ, Cusick JF, Sances A Jr, Walsh PR (1996) Human head-neck biomechanics under axial tension. Med Eng Phys 18(4):289–294
89. Van Ee CA, Nightingale RW, Camacho DL, Chancey VC, Knaub KE, Sun EA, Myers BS (2000) Tensile properties of the human muscular and ligamentous cervical spine. Stapp Car Crash J 44:85–102
90. Chancey VC, Nightingale RW, Van Ee CA, Knaub KE, Myers BS (2003) Improved estimation of human neck tensile tolerance: reducing the range of reported tolerance using anthropometrically correct muscles and optimizing physiologic initial conditions. Stapp Car Crash J 47:135–154
91. Dibb AT, Nightingale RW, Chancey VC, Van Ee AC, Luck JF, Fronheiser LE, Myers BS (2009) The human cervical spine in tension: strength, stiffness and the effects of end conditions. J Biomech Eng 131(8)
92. Kleinberger M, Summers L (1997) Mechanisms of injuries for adults and children resulting from airbag interaction. In: AAAM proceedings of the 41st annual conference
93. Maxeiner H, Hahn M (1997) Airbag-induced lethal cervical trauma. J Trauma 42(6):1148–1151
94. Perez J, Palmatier T (1996) Air bag-related fatality in a short, forward-positioned driver. Ann Emerg Med 28(6):722–724
95. Willis BK, Smith JL, Falkner LD, Vernon DD, Walker ML (1996) Fatal air bag mediated craniocervical trauma in a child. Pediatr Neurosurg 24(6):323–327
96. Winston FK, Reed R (1996) Airbags and children: results of a national highway traffic safety administration special investigation into actual crashes. In: 40th Stapp Car Crash Conference Proceedings P-305, vol 962438. pp 383–389
97. Brown DK, Roe EJ, Henry TE (1995) A fatality associated with the deployment of an automobile airbag. J Trauma 39(6):1204–1206
98. Moore KR, Frank EH (1995) Traumatic atlantoaxial rotatory subluxation and dislocation. Spine 20:1928–1928
99. Goel VK, Winterbottom JM, Schulte KR, Chang H, Gilbertson LG, Pudgil AG, Gwon JK (1990) Ligamentous laxity across C0-C1-C2 complex. Axial torque-rotation characteristics until failure. Spine 15(10):990–996
100. Myers BS, McElhaney JH, Doherty BJ, Paver JG, Gray L (1991) The role of torsion in cervical spine trauma. Spine 16(8):870–874
101. Myers BS, McElhaney JH, Doherty BJ, Paver JG, Ladd TP, Nightingale RW, Gray L (1989) Responses of the human cervical spine to torsion. In: Proceedings of the 33rd Stapp Car Crash Conference, vol 892437. pp 215–222
102. Wismans J, Spenny CH (1983) Performance requirements for mechanical necks in lateral flexion. In: Proceedings of the 27th Stapp Car Crash Conference, vol 831613
103. Winkelstein BA, Nightingale RW, Myers BS (2000) A biomechanical investigation of the cervical facet capsule and its role in whiplash injury. Spine 25(10):1238–1246
104. Gosch HH, Gooding E, Schneider RC (1972) An experimental study of cervical spine and cord injuries. J Trauma 12(7):570–576
105. Rogers WA (1957) Fractures and dislocations of the cervical spine an end-result study. J Bone Joint Surg 39(2):341–376
106. Mertz HJ, Patrick LM (1967) Investigation of the kinematics and kinetics of whiplash. In: Proceedings of the 11th Stapp Car Crash Conference 11, 670919
107. Bass C, Donnellan L, Salzar R, Lucas S, Folk B, Davis M, Rafaels K, Planchak C, Meyerhoff K, Ziemba A (2006) A new neck injury criterion in combined vertical/frontal crashes with head supported mass.

In: 2006 international IRCOBI conference on the biomechanics of impact
108. Pintar FA, Yoganandan N, Maiman DJ (2010) Lower cervical spine loading in frontal sled tests using inverse dynamics: potential applications for lower neck injury criteria. Stapp Car Crash J 54:133
109. Taylor AR (1951) The mechanism of injury to the spinal cord in the neck without damage to the vertebral column. J Bone Joint Surg Br 33(4):543–547
110. Barnes R (1948) Paraplegia in cervical spine injuries. J Bone Joint Surg Br 30(2):234–244
111. Schneider RC, Thompson JM, Bebin J (1958) The syndrome of acute central cervical spinal cord injury. Br Med J 21(3):216
112. Gadd CW, Culver CC, Nahum AM (1971) A study of responses and tolerances of the neck. In: Proceedings of the 15th Stapp Car Crash Conference, vol 710856
113. Myers BS, McElhaney JH, Richardson WJ, Nightingale RW, Doherty BJ (1991) The influence of end condition on human cervical spine injury mechanisms. Soc Automot Eng J Passeng Cars 6(100):2040–2048
114. Nightingale RW, Winkelstein BA, Knaub KE, Myers BS (2002) Comparative bending strengths and structural properties of the upper and lower cervical spine. J Biomech 35(6):725–732
115. Nightingale RW, Chancey VC, Ottaviano D, Luck JF, Tran LN, Prange MT, Myers BS (2007) Flexion and extension structrural properties and strengths for male cervical spine segments. J Biomech 40(3):535–542
116. Goel VK, Clark CR, Gallaes K, Liu YK (1988) Moment-rotation relationships of the ligamentous occipito-atlanto-axial complex. J Biomech 21(8):673–680
117. Panjabi M, Dvorak J, Crisco JJ 3rd, Oda T, Hilibrand A, Grob D (1991) Flexion, extension, and lateral bending of the upper cervical spine in response to alar ligament transections. J Spinal Disord 4(2):157–167
118. Voo LM, Pintar FA, Yoganandan N, Liu YK (1998) Static and dynamic bending responses of the human cervical spine. J Biomech Eng 120(6):693–696
119. Wheeldon JA, Pintar F, Knowles S, Yoganandan N (2006) Experimental flexion/extension data corridors for validation of finite element models of the young, normal cervical spine. J Biomech 39:375–380
120. Doherty BJ, Heggeness MH, Esses SI (1993) A biomechanical study of odontoid fracture and fracture fixation. Spine 18(2):178–184
121. Blockey N, Purser D (1956) Fractures of the odontoid process of the axis. J Bone Joint Surg Br 38(4):794–817
122. Cheng R, Yang KH, Levine RS, King AI, Morgan R (1982) Injuries to the cervical spine caused by a distributed frontal load to the chest. In: Proceedings of the 26th Stapp Car Crash Conference, vol 821155
123. Clemens HJ, Burow K (1972) Experimental investigation on injury mechanism of cervical spine at frontal and rear-front vehicle impacts. In: Proceedings of the 16th Stapp Car Crash Conference, vol 720960
124. Maiman D, Yoganandan N (1989) Biomechanics of cervical spine trauma. In: Congress of neurological surgeons, Atlanta
125. Nightingale RW, Winkelstein BA, Van Ee CA, Myers BS (1998) Injury mechanisms in the pediatric cervical spine during out-of-position airbag deployments. In: 42nd annual proceedings: association for the advancement of automotive medicine, pp 153–164
126. Ryan MD, Henderson JJ (1992) The epidemiology of fractures and fracture-dislocations of the cervical spine. Injury 23(1):38–40
127. Ngo B, Hoffman J, Mower W (2000) Cervical spine injury in the very elderly. Emerg Radiol 7(5):287–291
128. Hollands CM, Winston FK, Stafford PW, Shochat SJ (1996) Severe head injury caused by airbag deployment. J Trauma 41(5):920–922
129. Willis BK, Smith JL, Falkner LD, Vernon DD, Walker ML (1996) Fatal air bag mediated craniocervical trauma in a child. Pediatr Neurosurg 24(6):323–327
130. Hopper RH, McElhaney JH, Myers BS (1994) Mandibular and basilar skull fracture tolerance. In: Proceedings of the 38th Stapp Car Crash Conference, vol 942213. pp 123–131
131. Nightingale RW, Richardson WJ, Myers BS (1997) The effects of padded surfaces on the risk for cervical spine injury. Spine 22(10):2380–2387
132. Mower W, Hoffman JR, Zucker M (2000) Odontoid fractures following blunt trauma. Emerg Radiol 7(1):3–6
133. Kallieris D, Schmidt G (1990) Neck response and injury assessment using cadavers and the US-SID for far-side lateral impacts of rear seat occupants with inboard anchored shoulder Belts. In: Proceedings of the Stapp Car Crash Conference
134. McIntosh AS, Kallieris D, Frechede B (2007) Neck injury tolerance under inertial loads in side impacts. Accid Anal Prev 39(2):326–333
135. Yoganandan N, Pintar FA, Maiman DJ, Philippens M, Wismans J (2009) Neck forces and moments and head accelerations in side impact. Traffic Inj Prev 10(1):51–57
136. Foust DR, Chaffin DB, RG RGS, Baum JK (1973) Cervical range of motion and dynamic response and strength of cervical muscles. In: Proceedings of the 17th Stapp Car Crash Conference, pp 285–308
137. Szabo TJ, Welcher JB (1996) Human subject kinematics and electromyographic activity during low speed rear impacts. In: Proceedings of the 40th Stapp Car Crash Conference
138. Siegmund GP, Sanderson DJ, Myers BS, Inglis JT (2003) Awareness affects the response of human subjects exposed to a single whiplash-like perturbation. Spine 28(7):671–679
139. Siegmund GP, Sanderson DJ, Myers BS, Timothy Inglis J (2003) Rapid neck muscle adaptation alters the head kinematics of aware and unaware subjects undergoing multiple whiplash-like perturbations. J Biomech 36(4):473–482
140. Valsamis MP (1994) Pathology of trauma. Neurosurg Clin N Am 5(1):175–183

141. Dibb AT, Nightingale RW, Chancey VC, Fronheiser LE, Tran LN, Ottaviano D, Myers BS (2006) Comparative structural neck responses of the THOR-NT, Hybrid III, and human in tension and bending. Stapp Car Crash J 50:567–583
142. Kallieris D, Mattern R, Miltner E, Schmidt G, Stein K (1991) Considerations for a neck injury criterion. In: Proceedings of the 35th Stapp Car Crash Conference, vol 912916
143. Van Ee CA, Chasse AL, Myers BS (1998) The effect of postmortem time and freezer storage on the mechanical properties of skeletal muscle. In: Proceedings of the 42nd Stapp Car Crash Conference, vol submitted
144. De Jager M, Sauren A, Thunnissen J, Wismans J (1996) A global and a detailed mathematical model for head-neck dynamics. In: Proceedings of the 40th Stapp Car Crash Conference
145. van der Horst MJ, Thunnessen JGM, Happee R, van Haaster RMHP, Wismans JSHM (1997) The influence of muscle activity on head-neck response during impact. In: Proceedings of the 41st Stapp Car Crash Conference, vol 973346. pp 487–507
146. Brolin K, Halldin P, Leijonhufvud I (2005) The effect of muscle activation on neck response. Traffic Inj Prev 6(1):67–76
147. Dibb AT, Cox CA, Nightingale RW, Luck JF, Cutcliffe HC, Myers BS, Arbogast K, Bass CR (2013) Importance of muscle activations for biofidelic pediatric neck response in computational models. In: Paper presented at the 23rd international technical conference on the enhanced safety of vehicles, Seoul
148. van der Horst M (2002) Human head neck response in frontal, lateral and rear end impact loading: modelling and validation. Eindhoven University of Technology, Eindhoven
149. Bohlman H (1973) Correlative pathology and biomechanics of craniospinal injuries: a post mortem study of fifty fatal cases and a clinical review of 300 treated cases. In: Proceedings of the international research council on the biomechanics of injury conference, 1973. International Research Council on Biomechanics of Injury
150. Levine AM, Edwards CC (1991) Fractures of the atlas. J Bone Joint Surg 73(5):680–691
151. Shear P, Hugenholtz H, Richard MT, Russel NA, Peterson EW, Benoit BG, DaSilva VF (1988) Multiple noncontiguous fractures of the cervical spine. J Trauma 28(5):655–659
152. Panjabi MM, Cholewicki J, Nibu K, Grauer J, Babat LB, Dvorak J (1998) Critical load of the human cervical spine. Clin Biomech 13(1):11–17
153. Casper R (1980) The viscoelastic behavior of the human intervertebral disc. Phd Thesis
154. Kaplan F, Hayes W, Keaveny T, Boskey A, Einhorn T, Iannotti J (1994) Form and function of bone. In: Orthopaedic basic science. pp 127–185
155. Kazarian L, Graves GA Jr (1977) Compressive strength characteristics of the human vertebral centrum. Spine 2(1):1
156. Yoganandan N, Pintar F, Butler J, Reinartz J, Sances A Jr, Larson SJ (1989) Dynamic response of human cervical spine ligaments. Spine 14(10):1102–1110
157. Chang H, Gilbertson LG, Goel VK, Winterbottom JM, Clark CR, Patwardhan A (1992) Dynamic response of the occipito–atlanto–axial (C0-C1-C2) complex in right axial rotation. J Orthop Res 10(3):446–453
158. Pintar FA, Yoganandan N, Voo LM (1998) Effect of age and loading rate on human cervical spine injury threshold. Spine 23(18):1957–1962
159. Kazarian L (1981) Injuries to the human spinal column: biomechanics and injury classification. Exerc Sport Sci Rev 9:297–352
160. Torg J, Sennett B, Vegso J (1987) Spinal injury at the level of the third and fourth cervical vertebrae resulting from the axial loading mechanism: an analysis and classification. Clin Sports Med 6(1):159
161. Nahum A, Gatts J, Gadd C, Danforth J (1968) Impact tolerance of the skull and face. In: SAE transactions #680785. pp 631–645
162. Paine RR, Godfrey LR (1997) The scaling of skeletal microanatomy in non-human primates. J Zool 241:803–821
163. Carter JW, Mirza SK, Tencer AF, Ching RP (2000) Canal geometry changes associated with axial compressive cervical spine fracture. Spine 25(1):46
164. Alem NM, Nusholtz GS, Melvin JW (1984) Head and neck response to axial impacts. In: Proceedings of the 28th Stapp Car Crash Conference, 1984. vol 841667. pp 275–288
165. Pintar FA, Yoganandan N, Sances A, Reinartz J, Larson SJ, Harris G (1989) Kinematic and anatomical analysis of the human cervical spinal column under axial loading. In: Proceedings of 33rd Stapp Car Crash Conference, 1989. vol 892436. pp 987–1010
166. Yoganandan N Jr, Sances A, Pintar F (1989) Biomechanical evaluation of axial compressive responses of human cadaveric and manikin necks. J Biomech Eng 111(1):250–255
167. McElhaney JH, Doherty BJ, Paver JG, Myers BS, Gray L (1988) Combined bending and axial loading responses of the human cervical spine. In: Proceeding of 32nd Stapp Car Crash Conference, 1988. vol 881709. pp 21–28
168. Nightingale RW, Myers BS (1997) The effects of padding on the risk for cervical spine injury. In: 9th annual meeting of the federation of spine associations, 1997
169. Hu J, Yang KH, Chou CC, King AI (2008) A numerical investigation of factors affecting cervical spine injuries during rollover crashes. Spine 33(23):2529–2535
170. Camacho DLA, Nightingale RW, Myers BS (1999) Surface friction in near-vertex head and neck impact increases the risk of injury. J Biomech 3(32):293–301
171. Camacho DLA, Nightingale RW, Myers BS (2001) The influence of surface padding properties on head and neck injury risk. J Biomech Eng 123:432–439

172. Aare M, Halldin P (2003) A new laboratory rig for evaluating helmets subject to oblique impacts. Traffic Inj Prev 4(3):240–248
173. Oxland TR, Bhatnagar T, Choo AM, Dvorak MF, Tetzlaff W, Cripton PA (2011) Biomechanical aspects of spinal cord injury. Neural Tissue Biomech 3:159–180
174. Finan JD, Nightingale RW, Myers BS (2008) Experimental investigation of a helmet modification designed to mitigate head and neck injury. Traffic Inj Prev 9(5):483–488
175. Hadley M, Sonntag V, Rakate H, Murphy A (1989) The infant whiplash-shake injury syndrome: a clinical and pathological study. Neurosurgery 24(4):536–540
176. Huelke DF, Mendelsohn RA, States JD, Melvin JW (1978) Cervical fractures and fracture-dislocations sustained without head impact. J Trauma 18(7):533–538
177. Kupferschmid JP, Weaver ML, Raves JJ, Diamond DL (1989) Thoracic spine injuries in victims of motorcycle accidents. J Trauma 29(5):593
178. Heck JF, Clarke KS, Peterson TR, Torg JS, Weis MP (2004) National Athletic Trainers' Association position statement: head-down contact and spearing in tackle football. J Athl Train 39(1):101
179. CPSC (2009) NEISS: National Electronic Injury Surveillance System. US Consumer Product Safety Commission
180. Gabrielsen MA, McElhaney J, O'Brien RF (2001) Diving injuries: research findings and recommendations for reducing catastrophic injuries. Informa HealthCare

Upper Extremity Injury Biomechanics

12

Mei Wang, Raj D. Rao, Narayan Yoganandan, and Frank A. Pintar

Abstract

The scapula, clavicle, humerus, radius and ulna, and bones in the hand are joined by distinct soft tissues and joints in the human. As is true in the other components of the body, these structures are complex and have unique biomechanical characteristics. The purpose of this chapter is to present some of the basic anatomy of this region with a focus on the shoulder and its complex and the forearm. Most of the injuries to this region are high energy injuries. Experimental studies using post mortem human subject (PMHS) delineating the tolerance are described. A considerable majority of tolerance literature due to impact loading is from the automotive area, similar to the other regions. Studies using component models such as isolated forearm and intact PMHS models are described from injuries and injury biomechanics perspectives. Biomechanical testing using component models provide specific loading response information of individual bone and joint, while whole-body PMHS studies facilitate development of injury criteria and understanding of the dynamic interaction between linked components. The chapter concludes with a brief discussion on field injuries and the role of the shoulder in affecting the kinematics, loading and injuries to the thorax, abdomen and pelvis are discussed, with a focus on side impacts. Where possible, injury tolerance information is provided in the form of probability curves.

M. Wang, Ph.D. (✉) • R.D. Rao, M.D.
Department of Orthopaedic Surgery, Medical College of Wisconsin, Milwaukee, WI, USA
e-mail: meiwang@mcw.edu; rrao@mcw.edu

N. Yoganandan, Ph.D. • F.A. Pintar, Ph.D.
Department of Neurosurgery, Medical College of Wisconsin, Milwaukee, WI, USA
e-mail: yoga@mcw.edu; fpintar@mcw.edu

12.1 Biomechanically Relevant Anatomy

12.1.1 Bones

The shoulder girdle consists of three bones, the clavicle, scapula, and humerus (Fig. 12.1). The clavicle is a horizontally double-curved long bone that extends from the sternum to the scapula in front of the thorax. The clavicle has a rounded medial end and a flattened lateral end. The medial part of the clavicle is curved anteriorly, and the lateral part is curved posteriorly. The clavicle forms the front of the shoulder girdle and is palpable along its entire length. In the anatomic position, the clavicle is angled approximately 20° posterior to the coronal plane. The main function of the clavicle is to act as a strut to hold the shoulder joint away from the thoracic cage, allowing the upper extremity the freedom of movement. Because of its vulnerable position and relative thinness, the clavicle fractures more frequently than many other bones in the body. When it does, the shoulder loses support in its anatomic position.

The scapula is located on the posterolateral thoracic wall behind the second through seventh ribs, forming the back portion of the shoulder girdle. The flat and triangular body of the scapular has three processes called the acromion, spine, and coracoid process. The horizontally oriented scapular spine separates the smaller supraspinous fossa from the infraspinous fossa. The scapular spine and fossae provide attachment sites for muscles that move the arm. The spine ends in the large flat acromion that articulates with the clavicle and is an attachment site for chest and arm muscles. Below the acromion is the glenoid fossa that acts as a shallow socket for the head of the humerus. The coracoid process is a thick curved structure that projects over the glenoid fossa and directed anteriorly. The coracoid process serves as an attachment for ligaments and muscles of the arm and chest. In the anatomic position, the scapula is deviated 35° anterior to the frontal plane.

The humerus is the longest and largest bone of the upper limb. The shaft of the humerus is cylindrical in its proximal half and flattened in the anterior-posterior direction in the distal half. The head of the humerus articulates with the glenoid fossa of the scapula to form the glenohumeral joint. There are two protuberances on the proximal humerus, the greater and lesser tubercles. They serve as attachment sites for rotator cuff muscles. Retroversion of the humeral head is roughly 30° posterior to the medial-lateral axis at the elbow.

Two bones form the forearm (Fig. 12.2). The longer one of the two and located on the medial aspect of the forearm is the ulna. The proximal end of the ulna has a posterior projection called olecranon, which forms the prominence of the elbow, and an anterior projection called coronoid process. The olecranon and coronoid process together form the trochlear notch that articulate with the humerus. Radius is the bone located on

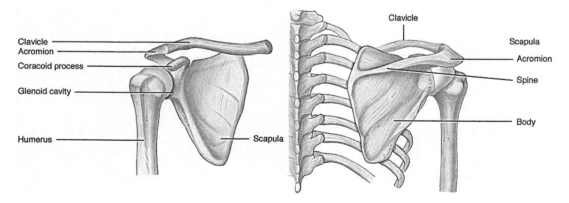

Fig. 12.1 Anterior (L) and posterior (R) views of bone in the shoulder joint

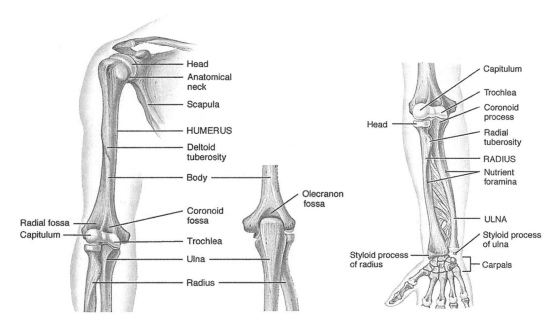

Fig. 12.2 Bones of the upper extremity

the lateral aspect of the forearm. The proximal end of the radius has a disc-shaped head that articulates with the humerus. Distally, the shaft of the radius widens at the wrist, with a styloid process pointing laterally.

12.1.2 Joints

The shoulder girdle is comprised of three synovial joints and one gliding mechanism involving the sternum, clavicle, scapula, ribs, and humerus. These articulations, namely the sternoclavicular, acromioclavicular, and glenohumeral joints and the scapulothoracic gliding mechanism, function in a coordinated manner to allow complex arcs of motion of the upper limb. The most proximal joint within the shoulder girdle is the sternoclavicular joint, which connects the appendicular skeleton to the axial skeleton. The sternoclavicular joint is formed by the medial end of the clavicle articulating with the manubrium of the sternum just above the first rib. Approximately half of the medial end of the clavicle protrudes above the shallow sternal socket. The sternal surface is saddle shaped concave anteroposteriorly and convex downwardly [1]. In between the articular surfaces is a strong, nearly circular fibrocartilaginous disc, which is attached superiorly to the upper medial end of the clavicle. The disc functions as a sliding hinge that facilitates a greater range of motion at this joint and stabilizes the joint against medially directed forces transmitted through the clavicle.

The lateral end of the clavicle articulates with the acromion process, forming the acromioclavicular joint. This articulation forms the roof of the shoulder. The acromioclavicular joint has oblique surfaces that allow the acromion, and thus the scapula, to glide forward or backward over the clavicle. This movement of the scapula keeps the glenoid fossa continuously facing the humeral head. The acromioclavicular joint is not as mobile as the large glenohumeral joint but has an important function in contributing to total arm movement, especially when the shoulder is overhead or across the chest (adducted). The acromioclavicular joint also facilitates force transmission between the clavicle and the acromion.

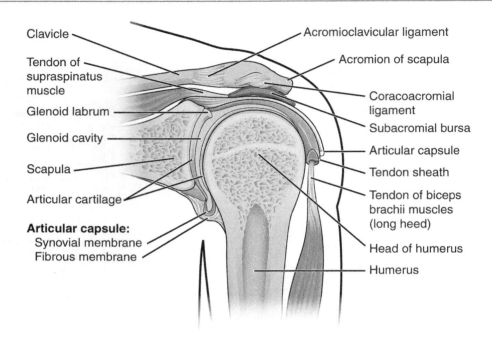

Fig. 12.3 Coronal section through right glenohumeral joint (posterior view)

The articulation that forms between the glenoid fossa of the scapular and the head of the humerus is the glenohumeral joint (Fig. 12.3). The glenohumeral joint is a multi-axial ball-and-socket joint. The joint surfaces are asymmetrical and incongruent. The surface area of the pear shaped glenoid fossa is one-third to one-fourth that of the humeral head; as a result of the mismatch, only a portion of the humeral head can be in articulation in any position of the joint. Along the margin of the glenoid fossa attaches a rim of fibrocartilaginous tissue caked the glenoid labrum. The labrum is thought to deepen the articular cavity, protect the edges of the bone, and assist in lubrication of the joint [2]. Typical healthy glenohumeral range of motion is 120° of abduction, 120–180° of flexion, 45–55° of extension, 75–85° of internal rotation, and 60–70° of external rotation. This high mobility afforded by the glenohumeral joint is at the expense of joint stability.

The scapulothoracic gliding mechanism is not a true joint but is the riding of the concave anterior surface of the scapula along the convex posterolateral surface of the thoracic cage (ribs 2–7). The thorax and scapula are separated by the subscapularis and serratus anterior muscles, which glide over each other during movements of the scapula. The scapula is held in close approximation to the chest wall by muscular attachments. In movements of the shoulder complex, the scapula can be protracted, retracted, elevated, depressed, and rotated about a variable axis perpendicular to its flat surface. However, fibrous adhesions can sometimes occur following a shoulder injury, particularly if the joint has been immobilized for a long period of time.

The elbow joint is a modified hinge joint composed of two articulations, where the humerus connects to the radius and ulna. The humeroradial joint is formed by the head of the radius articulates with capitulum of the humerus. The humeroulnar joint is formed by the trochlear notch of the ulna articulate with the trochlea of the humerus (Fig. 12.4). The radius and ulna also articulates with one another at the proximal and distal ends, called radioulnar joints, and through a strong fibrotic band known as the syndesmosis in between the two end joints.

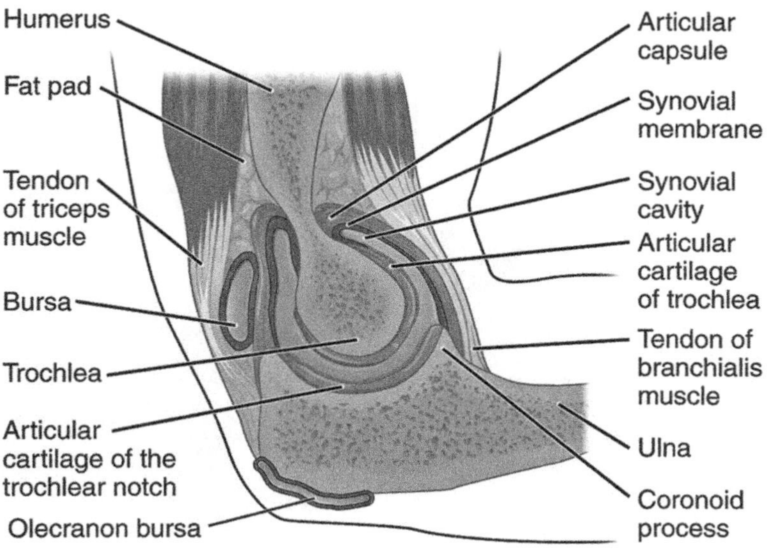

Fig. 12.4 Sagittal section through right elbow (lateral view)

12.1.3 Ligaments

Because the articular surfaces of the glenohumeral joint contribute little to joint stability, joint capsule reinforced by a group of ligaments connecting the humerus to the glenoid become the main source of stability. The coracohumeral ligament is one of the most important ligamentous structures in the shoulder complex [3]. This ligament extends from the coracoid process of the scapula to the greater tubercle of the humerus. The downward pull of gravity on the arm is counteracted largely by the superior capsule and the coracohumeral ligament. Also because it is located anterior to the axis of axial rotation of the humerus, the coracohumeral ligament restricts lateral rotation and extension.

The three glenohumeral ligaments lie on the anterior aspect of the joint (Fig. 12.5). The superior glenohumeral ligament originates from the supraglenoid tubercle and inserts on the humerus near the lesser tuberosity. The superior glenohumeral ligament, together with the coracohumeral ligament and the supraspinatus muscle, assists in preventing downward displacement of the humeral head. The middle glenohumeral ligament originates from the supraglenoid tubercle and anterosuperior region of the labrum, passes laterally, gradually enlarges, and attaches to the anterior aspect of the anatomical neck of the humerus. The middle glenohumeral ligament limits lateral rotation up to 90° of abduction. The inferior glenohumeral ligament is a hammock-like structure with anchor points on the anterior and posterior sides of the glenoid. The ligament originates from the anterior, inferior, and posterior margins of the glenoid labrum and inserts laterally to the inferior aspects of the humeral neck. The superior band strengthens the capsule anteriorly and supports the joint most effectively in the middle ranges of abduction. The inferior band supports the joint most effectively in the upper ranges of abduction and also prevents anteroinferior subluxation and dislocation [4].

The coracoacromial ligament is a strong triangular ligament originated from the lateral border of the coracoid process and inserted to the top of the acromion process. Together with the acromion and the coracoid processes, the coracoacromial ligament forms an important protective arch over the glenohumeral joint and a restraining socket for the humeral head. This protects the joint from trauma from above and preventing dislocation of the humeral head superiorly.

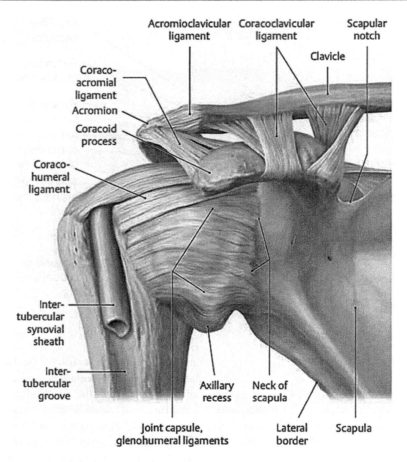

Fig. 12.5 Major ligaments of the shoulder (posterior view)

The clavicle is held in place by strong ligaments at both ends. Around the sternoclavicular joint, the anterior and posterior sternoclavicular ligaments are found anterior and posterior to the joint; the interclavicular ligament links the ends of the two clavicles to each other and to the superior surface of the manubrium of the sternum; and the costoclavicular ligament that is positioned laterally to the joint and links the proximal end of the clavicle to the first rib and its related costal cartilage. On the lateral side, the acromioclavicular ligament strengthens the acromioclavicular joint and binds the clavicle to the scapula. The coracoclavicular ligament consists of two parts: trapezoid and conoid. These two components, functionally and anatomically distinct, are united at their corresponding borders. These ligaments suspend the scapula from the clavicle and transmit the force of the superior fibers of the trapezius to the scapula.

There are two strong ligament complex at the elbow joint, the medial collateral ligaments (MCL) on the ulna side and the lateral collateral ligaments (LCL) on the radius side (Fig. 12.6). The MCL complex consists of three ligaments: the anterior oblique, posterior oblique, and transverse ligaments, originated from the anteroinferior surface of the medial epicondyle. The MCL complex is a primary medial stabilizer of the flexed elbow joint. The LCL complex is made up of four ligaments: the lateral ulnar collateral ligament, radial collateral ligament, annular ligament, and accessory lateral collateral ligament, originated along the inferior surface of the lateral epicondyle. The LCL complex is the primary stabilizer of the elbow for varus and external rotation forces.

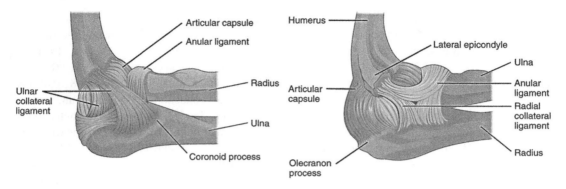

Fig. 12.6 Ligaments of the elbow (*Left:* medial view, *Right:* lateral view)

12.1.4 Muscles

The shoulder joint is surrounded by small muscles (rotator cuff) that stabilize the glenohumeral joint and large muscles that power its strong movements. The rotator cuff is a conjoined tendon of four muscles, the supraspinatus, infraspinatus, teres minor, and subscapularis. The rotator cuff muscles predominantly controls the dynamic stability of the glenohumeral joint but also contribute significantly to shoulder movement. All these muscles except subscapular attach to greater tuberosity of head of humerus, while subscapular muscle tendon attaches to lesser tuberosity. The supraspinatus is an initiator of abduction, and active throughout the range of shoulder abduction. It is the most common site for tear. The infraspinatus and teres minor lie below the scapular spine and are external rotators of the shoulder. The infraspinatus is mainly active with the arm in neutral and the teres minor contributes more when the arm is in 90° of abduction. The subscapularis is the main internal rotator of the shoulder. It is the largest and strongest rotator cuff muscle, providing 53 % of total strength.

Several muscles are responsible for elevation of the arm. The deltoid muscle is the only shoulder elevator if the supraspinatus is torn and dysfunctional. It comprises anterior, middle and posterior portions which are more active depending on the direction of arm elevation. The long head of biceps passes over the humeral head curving in two planes, and is recognized as providing a small degree of stability to the glenohumeral joint. This is predominantly with abduction and external rotation of the arm in the scapular plane. The scapular muscles that control upward rotation and protraction of the scapulothoracic joint are the serratus anterior and trapezius muscles. The latissimus dorsi and the sternocostal head of the pectoralis major are the largest adductor and extensor muscles of the shoulder. The teres major, long head of the triceps, posterior deltoid, infraspinatus, and teres minor are also primary muscles for shoulder adduction and extension. The primary muscles that internally rotate the humerus are the subscapularis, anterior deltoid, pectoralis major, latissimus dorsi, and teres major. External rotators of the humerus are the infraspinatus, teres minor, and posterior deltoid.

At the elbow joint, the primary flexor muscles include the biceps brachii, brachialis, and brachioradialis. The triceps brachii and anconeus muscles are the elbow extensors. The muscles that are responsible for forearm supination are the biceps brachii and supinator muscles. The pronator muscles of the forearm are the pronator quadratus located in the distal portion of the forearm and pronator teres on the proximal portion of the forearm.

12.2 Component Biomechanics and Tolerance

12.2.1 Clavicle

The clavicle is the most commonly fractured bone in the human body. The majority of clavicular fractures (75–80 %) occur within the middle third of the bone. In males, the annual incidence was highest under 20 years of age, decreasing in each subsequent cohort until the seventh decade. In females, the incidence was more constant, but relatively frequent in teenagers and the elderly. The common mechanism of clavicle fracture, regardless of site, is a direct blow on the shoulder such as occurs during a fall or road traffic accident [5].

Several studies have investigated the impact forces involved in clavicle fracture. Duprey et al. conducted dynamic axial compression tests until failure on 18 clavicles using a drop tower system (6 kg impactor mass), at four impact speeds between 1 and 2.5 m/s [6]. A schematic of the system is shown (Fig. 12.7). Clavicle fractures occurred at an average velocity of 1.41 ± 0.4 m/s, with an average fracture force of 1.48 ± 0.46 kN and average deflection of 5.4 ± 1.1 mm. Failure force of female clavicles was significantly lower than the male, and at the slowest impact speed (1 m/s) only female clavicles ruptured. Effect from bone mineral density (BMD) on failure force was not detected due to small samples. In a comprehensive imaging study, Harrington et al. investigated the cross-sectional geometric properties of 15 adult clavicles from five females and ten males [7]. Cross-sectional area, anatomic moments of inertia, and porosity along the length of the clavicle were measured. Mechanical properties, including axial, flexural, and torsional rigidities and critical buckling force were derived using beam theory based on the geometric data. The clavicle was found to be weakest in the central third of its length. Analysis of Euler buckling predicted a

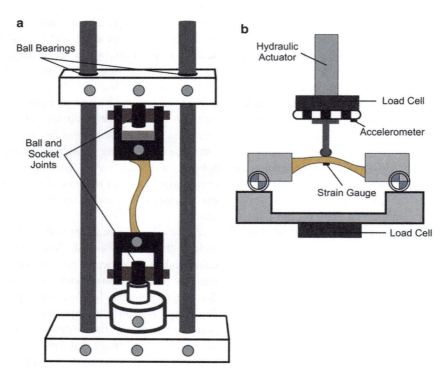

Fig. 12.7 Schematic showing the axial loading with drop weight (*left*) and three-point bending (*right*) using an actuator used in [6, 8]

minimum critical force for buckling during axial loading of approximately two to three body weights for an average adult, estimation in support of findings from an early study [6].

Mechanical response of the clavicle was reported by Kemper et al. [8]. Ten cadaver clavicles were tested in three–point bending, with impact delivered by a servo hydraulic actuator at a velocity of 0.15 m/s. A schematic of the system is shown (Fig. 12.7). The average failure load found was 732±175 Nm and the average failure moment was 28.3±7.8 Nm. Most clavicles fractured in the mid-shaft region in the form of a transverse, oblique, or comminuted-wedge fracture.

12.2.2 Humerus

Early tolerance studies of the humerus were conducted using three-point bending loading technique and maximum bending moments were reported. In one of the earliest studies conducted in 1859, according to a 1995 citation moments for males and females were 115 and 73 Nm [9, 10]. Greater moments of 151 and 85 Nm were reported for males and females in the 1880 study, according to the cited 1995 reference [11]. In a more recent study reported in 1970, the average fracture load for the humerus was determined to be 1,330 N and this was associated with the quasi-static load applied in the antero-posterior (AP) direction [12]. Studies conducted in the 1990s have determined the effect of loading direction (AP versus lateral-medial, ML) and rate (218 and 0.635 mm/s) on the shear forces, AP and lateral bending moments corresponding to the ML and AP loadings [13]. These authors subjected 24 specimens from one female and 13 male PMHS with age ranging from 44 to 75 years, to three-point bending. The overall moments and shear forces for all combinations of variables were 155±44.6 Nm (±denotes standard deviation) and 1.96±0.67 kN. These authors normalized the data to fifth and fiftieth percentile anthropometries using the impulse-momentum approach [14]. The summary of data is shown in Table 12.1. Because of the lack of statistical difference (p>0.05) between the two loading directions and rates, the authors suggested the following magnitudes to represent the tolerance of the humerus: fiftieth percentile: 2.5 kN and 230 Nm; and fifth percentile: 1.7 kN and 130 Nm (Table 12.1). Fractures transverse to the longitudinal axis were common and this was followed by angled and spiral/irregular fractures.

A later study used 9.48-kg dropping mass, similar to those used in the clavicle studies described above, described earlier, to impact 12 female humerii under the three-point bending configuration at a velocity of 3.6 m/s [15]. The peak failure moment was 154±27 Nm and the normalized peak moment corresponding to the fifth percentile anthropometry was 128±19 Nm [16]. The impactor force is shown (Fig. 12.8, the dashed and dotted curves are not the subject matter of this chapter).

Table 12.1 Summary of humerus tolerance from three-point bending tests (refer to text for details)

Standard for normalization	Injury metric	Lateral-medial loading	Antero-posterior loading
Fiftieth percentile	Force (kN)	2.68±0.69	2.17±0.28
	Moment (Nm)	236±57.7	226±77.3
	2.5 kN and 230 Nm for both loadings		
Fifth percentile	Force (kN)	1.86±0.48	1.51±0.20
	Moment (Nm)	137±33.6	131±44.9
	1.7 kN and 130 Nm for both loading		

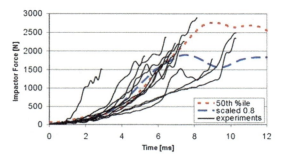

Fig. 12.8 Force recorded from the impacting mass accounting for inertial compensation from the 12 humerii tests, from Ref. [17] (Reprinted with permission from the Stapp Association)

12.2.3 Forearm and Elbow

Like the humerus, early human forearm tests were done using quasi-static loading. For example, a 1970 study found that the failure of ulna and radius occurs at 630 and 520 N [12]. Dynamic tests were conducted to determine effects of pronation and supination tolerance and potential application to airbag loading in motor vehicle crashes [18]. Two drop tests were conducted at 4.6 m/s with a 9-kg impacting mass. In the first, the forearm was pronated with the distal third of the ulna oriented towards the impacting mass (simulating 'natural driving posture'), and in the second, the specimen was fully supinated position with the anterior surface of the wrist facing the mass. The peak failure loads and moments for the supinated and pronated forearms were 2.1 and 1.5 kN, and 115 and 80 Nm, respectively. Ulna fractures preceded radius fractures in both specimens.

In a later study, Pintar et al. evaluated the response of the PMHS forearms at dynamic rates using a material testing device [19]. They subjected 30 forearm specimens (12 female and 18 male) with age ranging from 41 to 89 years to three-point bending from 15 PMHS (Fig. 12.9).

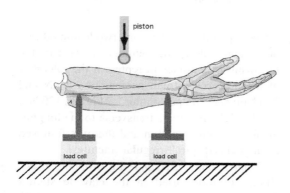

Fig. 12.9 Dynamic three-point bending loading studies on isolated forearm (From Ref. [19])

Loading velocities were 3.3 and 7.6 m/s in the two groups of equally divided specimens. Thirteen had butterfly fractures of the ulna (10 from male specimens). Out of the 16 comminuted fracture specimens, 11 were from the high rate of loading. There were no statistical ($p > 0.05$) differences on the fracture location based on the two loading rates and no statistical differences were apparent with regard to the anatomical aspect ($p > 0.05$). However, peak force and bending moments were significantly ($p < 0.05$) lower in female than male specimens. These results clearly indicated that the tolerances of female forearms to three-point bending are lower than males (Table 12.2). Furthermore, compared to the quasi-static tests, these results showed a dynamic factor of 1.7 to account for the increased rate of loading. This factor was found to be in-line with similar data for other bones. For example, the dynamic rise factor is reported to be 1.68 and 1.66 for the tibia and femur bones [20, 21]. Based on this research, using the mean moment of 94 Nm for all specimen as the basis, Pintar et al. proposed a tolerance of 47 Nm for a smaller size occupant [19]. Fractures from this study are shown (Fig. 12.10).

In another study, 9.48-kg dropping mass was used to impact ten forearms (with intact elbow joint) under the three-point bending configuration (Fig. 12.11) at a velocity of 4.4 m/s [15]. Three matched-pair tests with supinated and pronated initial positions showed that the peak failure moments was 21 % greater in the former (92 ± 5 and 75 ± 7 Nm) posture. Tests with additional six pronated specimens indicated the failure moment to be 70 ± 13 Nm. The normalized peak moment corresponding to the fifth percentile anthropometry was 58 ± 12 Nm when the pronated data were normalized using the equal stress equal velocity approach [16]. Although oblique, transverse and

Table 12.2 Summary of forearm tolerance data based on velocities and sex (From Ref. [19])

Parameter	Velocity (m/s)		Sex		All tests
	3.3	7.6	Male	Female	
Sample size	15	15	18	12	30
Force (N)	1,860±918	2,083±813	2,368±812	1,377±534	1,971±860
Moment (Nm)	89±44	99±39	113±39	66±25	94±41
Forearm mass	1.39±0.51	1.42±0.52	1.63±0.44	1.07±0.42	1.41±0.51

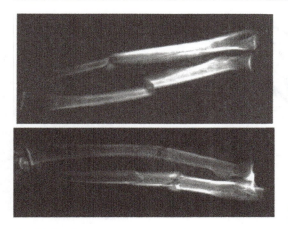

Fig. 12.10 Top x-ray: Fracture in the ulna and radius, with butterfly-type fragment on the ulna. Bottom x-ray: fractures of both long bones with comminuted fractures of the ulna (From Ref. [19])

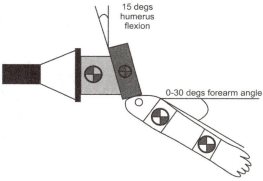

Fig. 12.12 Schematic of elbow joint loading using a horizontal piston at various angles (From Ref. [24]) (Reprinted with permission from the Stapp Association)

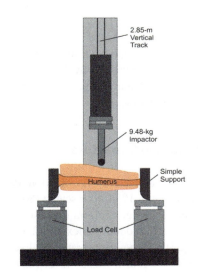

Fig. 12.11 Schematic of three-point bending loading studies using a dropping mass on isolated humerus (From Ref. [15])

butterfly fractures of both bones were reported for both postures, ulna and radius were sequentially loaded in the pronated and simultaneously loaded in the supinated postures. It should be noted that PMHS studies have been conducted with isolated forearms and intact specimens with deploying airbag as the load delivering device [15, 18, 22, 23]. Some of these studies are described later.

In addition to determining the bending tolerance of the forearm under three-point loading for frontal impact and airbag deployment applications, studies have been done with a focus on side impact loading in automotive environments [24]. The experimental design forced the elbow into the compressive mode, and this was achieved by fixing the distal humerus, removing the surrounding flesh, and delivering the load to the humeral with a horizontal piston. A schematic is shown (Fig. 12.12). In effect, the forearm acted like a cantilever with the humerus end sustaining the load, resulting in rotation of the humerus and forearm. The humerus was positioned with a 15-deg initial angle and the forearm was positioned at 15-, 20- and 30-deg with respect to the horizontal. A total of 40 tests were done with 18 PMHS specimens. Injuries included chondral and osteochondral fractures of the proximal radius head and coronoid. The peak elbow force was normalized to the 5th percentile female according to equal stress equal velocity approach and risk curves were developed [16]. The loading angle relative to the forearm longitudinal axis was determined to be a significant ($p<0.05$) indicator of elbow injury, with the higher risk associated with a higher angle. The peak force was found to be a significant indicator ($p<0.04$) of elbow fracture, with a force of 2,715 N to be associated with 50 % risk, regardless of the angle. The authors found that the combination of angle and force resulted in a more significant risk function for fracture ($p<0.01$). A 30-deg, the force at 50 % risk was 1,780 N. Risk curves are shown (Fig. 12.13).

In addition to determining the bending tolerance of the forearm, for helicopter applications,

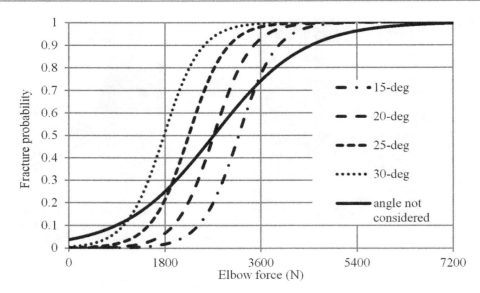

Fig. 12.13 Injury risk curves for elbow force (From Ref. [24]) (Reprinted with permission from the Stapp Association)

three-point bending experiments were conducted to 24 isolated female PMHS forearms with intact elbow joint [23]. The impact load was delivered with a dropping mass of 9.75 kg (2.4 or 4.2 m/s) at the fixed end, humerus bone. Injuries ranged from minor cartilage to dislocations and humerus fractures. The mean loading onset rate was 31 ± 30 and 44 ± 27 Nm/ms at the high low velocities and these were statistically insignificant ($p > 0.05$). Peak bending moments at the elbow joint ranged from 42 to 146 Nm. This metric was found to be significant ($p < 0.05$) descriptor of injury and joint dislocation. Statistical analyses of these data revealed that 56 and 93 Nm represent the tolerance for any injury and elbow joint dislocation at the 50 % risk level. These results indicated that the hyperextension moment required to induce dislocation is greater than the moment to sustain fracture. Risk curves are shown (Fig. 12.14).

12.3 Studies Using Intact Post Mortem Human Subjects (Pmhs)

Experiments using intact PMHS have been conducted to better understand the mechanical response and injury tolerance of the human shoulder and upper extremity under dynamic impact. A variety of impact delivery systems are involved in these studies, including guided impactors such as the pendulum and sled equipment.

12.3.1 Guided Impactors

In a study conducted by Association Peugeot-Renault, France in 1984, four seated PMHS were tested using a 23-kg, 150-mm diameter cylinder to impact the shoulder complex at 4.2 to 4.6 m/s (cited in [25]). Three impacts were delivered laterally while the fourth was at 15-deg anterior of lateral. The force and acceleration of the impacting mass, acceleration of the thoracic spine, and the deflection of the shoulder with respect to the thoracic spine were recorded. The cadaver was autopsied for fractures of the ribs, clavicle, or scapula. Shoulder deflection with respect to the thoracic spine was also calculated. Based on the resulting normalized force time responses and corresponding biofidelity corridors, the study suggested boundary of maximum shoulder deflection with respect to the thoracic spine as 34 to 41 mm (10 % deviation from the average value of 37.5 mm). The group reported little difference between oblique and lateral loading in pendulum impact loading scenario.

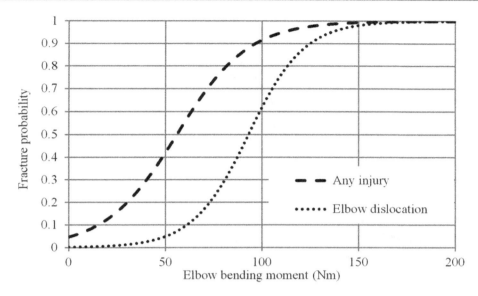

Fig. 12.14 Injury risk curves for elbow moment (From Ref. [23]) (Reprinted with permission from the Stapp Association)

In piston impactor test, only the selected anatomical region is impacted, which allows to study precisely its response at a variety of velocities and impact directions. A padded pneumatic impacting ram was employed by Bolte et al. on 11 PMHS to study lateral impact response of the glenohumeral joint. Impact velocities ranged from 3.5 to 7 m/s. The most traumatic injury found in the study was fracture of the distal end of the clavicle, while the frequent injury documented at autopsy was looseness of the sternoclavicular joint. Results from regression analysis showed that the best predictors of an AIS level 2 injury were the displacement measured between the impacted acromion and the sternum and the rate at which this deflection occurred. The study concluded that a shoulder deflection higher than 47 mm led to a 50 % probability of an AIS 2 injury. In 2003, the same group of authors conducted additional impact pure lateral impact tests (n = 4), 15° oblique anteriorly (n = 4), and 30° oblique anteriorly (n = 6), andcombined with previous data to study the protective role of the shoulder join tot the thorax. The shoulder girdle was found less stiff during oblique impacts, allowing for the shoulder to displace posteriorly and medially, potentially transferring load to the upper thoracic cage.

A study 2004 study reported lateral impact tests on seven PMHS [26]. The right shoulder was impacted Three times with different impact orientations (lateral direction and ±15-deg from lateral) at a low velocity of 1.5 m/s. The left shoulder was laterally impacted at a higher velocity, between 3 and 6 m/s to generate injuries. The impact was delivered by a 23 kg-mass rigid plate. Shoulder and thoracic bone structures were instrumented with tri-axial accelerometers and markers for three-dimensional high speed cameras. Autopsies were performed and injuries were coded using AIS. No injuries occurred from low velocity impacts. The injury threshold was identified to be around 4 m/s. In oblique impacts, slight modifications in the shoulder responses were observed, with higher force peaks obtained for the +15-deg impacts in comparison to the 0- and -15-deg impacts. Acromion-to-sternum deflection peak values were closely associated with clavicle fracture. Deflection analysis showed that fractures occurred for a low acromion-sternum deflection of 25 mm, proving again that padding affects impact outcomes. Injury risk curves for forces and deflections are presented (Figs. 12.15, 12.16, and 12.17).

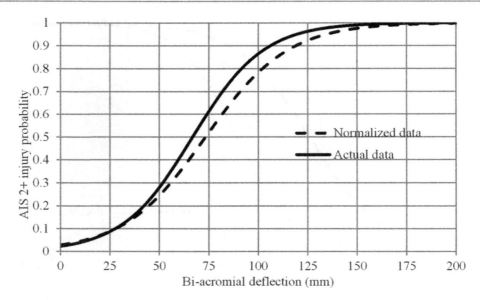

Fig. 12.15 Injury risk curves based on actual and normalized data at the AIS 2+ levels (Adapted from Ref. [26]) (Reprinted with permission from the Stapp Association)

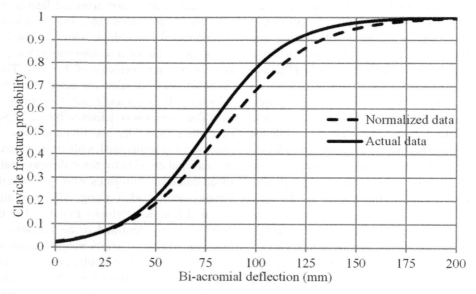

Fig. 12.16 Clavicle injury risk curves based on actual and normalized data (Adapted from Ref. [26]) (Reprinted with permission from the Stapp Association)

12.3.2 Sled

In a study reported in 1993, shoulder and thorax behaviors during lateral impacts were studied from seven PMHS tests and the following results were based on three tests due to experimental issues [27]. Impact velocities ranged from 6.7 to 9.1 m/s for rigid and the velocity was 9 m/s for padded impacts. The PMHS were instrumented with accelerometers and markers. High speed

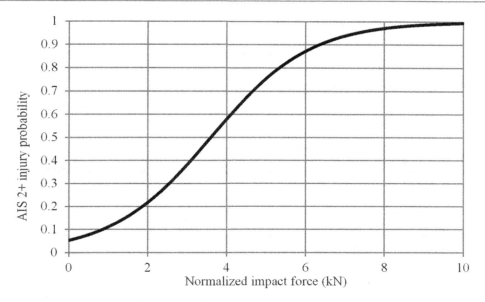

Fig. 12.17 Injury risk curve for forces to fracture (Adapted from Ref. [26]) (Reprinted with permission from the Stapp Association)

cameras were used to obtain the three-dimensional positions of skeletal landmarks within the shoulder area during impact. Time history of the displacements at various locations on the impacted shoulder relative to the twelfth thoracic vertebra and non-impacted shoulder during the impact was used to estimate the response of a fiftieth percentile male using the impulse-momentum approach [14]. The injuries found for the tests with a rigid wall were acromioclavicular dislocations and acromion fractures. With a padded wall, acromioclavicular dislocations, clavicle and rib fractures were found. The analysis revealed that the PMHS rotated during impact and considerably during rebound, confirmed the forward deflection of the sternum due to side impact, and non-impacted rib structures deflected laterally with respect to the spinal column.

A reanalysis of the 17 lateral sled impact tests conducted earlier indicated that the sternum acceleration and shoulder deflection are lower with padded impacts [28]. Common injuries from rigid wall impacts included acromioclavicular joint separations and fractures to the acromion process of the scapula. Common injuries in padded wall impacts included clavicle fractures and acromioclavicular joint separations. Padding was beneficial in reducing the occurrence of acromion fracture, but not clavicle fracture. Logistic regression analysis revealed that the best predictor for the assessment of MAIS 2 shoulder injury risks was the maximum shoulder deflection (Fig. 12.18) or a combination of the maximum shoulder deflection and the average spine (Fig. 12.19). Based on the maximum deflection criterion, a 50 % probability of a MAIS 2 injury to the shoulder corresponds to a shoulder deflection of 106 mm between T1 and the shoulder edge.

12.3.3 Airbag Deployment

A variety of upper extremity injuries have been reported as a result of airbag deployment. Female occupants are at higher risk of these injuries due to their small stature, bone structure and low bone density. Two different injury mechanisms have been proposed [29]. One involves the compression of the distal humerus head onto the distal trochlear notch and proximal head of the radius. This occurs early in the deployment event as the side air bag loads the posterior humerus and forces the forearm forward, resulting the

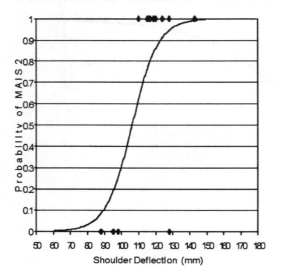

Fig. 12.18 Logist probability distribution of MAIS 2 to the T1 to shoulder edge deflection (From Ref. [28]) (Reprinted with permission from the Stapp Association)

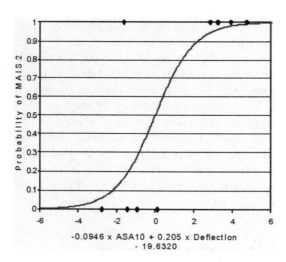

Fig. 12.19 Logist probability distribution of MAIS 2 to the shoulder versus combination of ASA10 and deflection (From Ref. [28]) (Reprinted with permission from the Stapp Association)

chondral and osteochondral fractures of the distal trochlear notch and proximal head of the radius. The second is related to proximal trochlear notch injuries, which occurs later in the deployment as the elbow becomes fully extended, the proximal trochlear notch is forced into the olecranon fossa of the distal humerus. In tests with small female cadavers, chondral and osteochondral fractures were found in the elbow joint in seven of the twelve cadaver tests subjected to upper extremity loading from a deploying seat-mounted side-impact airbag. These authors also determined the tolerance from isolated tests, described earlier (Fig. 12.20).

12.4 Field Data with a Focus on Side Impacts

The human shoulder acts as a load path and influences the kinematics and load-sharing of other components in motor vehicle crashes [30]. Recent field-based studies reported an analysis of injury patterns in survivors from nearside impacts with and without shoulder injuries (SI), using the National Automotive Sampling System (NASS) and Crash Injury Research Engineering Network (CIREN) databases [31, 32]. One study focused on narrow object impacts and the other included all crashes. In both studies, the vehicle model years were 2000 or newer. Left or right side impacts with the highest ranked collision deformation code were used to identify nearside crashes. All occupants were in the front outboard seat, restrained and contained and rollovers were excluded. Only occupants in passenger cars, light trucks and vans seated on the side of the impact were considered. Medical records and crash information were analyzed. The damage class was determined based on the crush profiles and the assigned vehicle collision deformation code. Based on the principal direction of force (PDOF), impacts were classified into anterior oblique, pure lateral and posterior lateral impacts. Data were obtained for the following categories: matched-vehicle crashes representing two vehicles of the same mass/type; mismatched vehicle crashes wherein the impacted vehicle was different from the other in terms of mass/type; vehicle-to-fixed narrow object crashes; and other fixed object crashes. The change in velocity (ΔV), abbreviated injury score (AIS), injury severity score (ISS), maximum AIS (MAIS) data were obtained along with the MAIS of different body regions [33]. AIS 2+ injuries to body regions

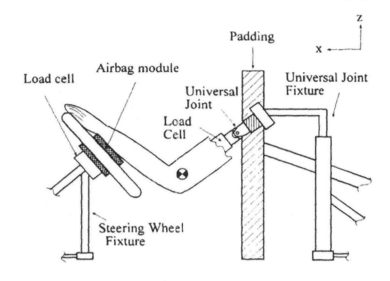

Fig. 12.20 Upper extremity/airbag test fixture – side view (From Ref. [18]) (Reprinted with permission from the Stapp Association)

were analyzed based on the presence or absence of SI. Procedures incorporated in the analysis of NASS were similar to CIREN, with the exception that data were used for the years 2009 and 2010. Data were analyzed based on the impacted vehicle match/mismatch/pole, ΔV and ISS. Analyses were done by combining left and right nearside occupants into two groups.

12.4.1 Occupants and Injury Types

CIREN had 61 and 276 occupants and NASS had 49 and 162 with and without SI. Males represented 48 % of SI occupants in both databases, compared to 42 and 43 % of non SI occupants in CIREN and NASS. Isolated fractures of the clavicle, scapula and humerus, and AC joint dislocations were found in 59, 15, 15, and 2 % of the occupants in CIREN and 45, 16, 6, and 10 % in NASS. The scapula was associated with clavicle, AC joint, acromion and humerus injuries in 10 % in both databases. Clavicle was associated with AC joint and humerus in 10 % of NASS. Occupants without and with SI accounted for 27 and 28 % in matched vehicle, 52 and 59 % in mismatched vehicle, and 17 and 11 % in vehicle-to- narrow object crashes in CIREN.

12.4.2 Vehicle Factors

Passenger cars accounted for 87 and 71 % of the case vehicles with SI in CIREN and NASS. For non-SI occupants, passenger cars accounted for 79 and 70 % in the CIREN and NASS databases. In NASS, occupants without and with SI groups were 32 and 29 % for matched, 43 and 53 % for mismatched, and 18 % each for vehicle-to-narrow object crashes. The nearside occupant was in the right front passenger seat in 30 % of crashes with SI in CIREN and 35 % in NASS, while the occupant was in the right front seat in 24 % of non-SI crashes in CIREN and 29 % in NASS.

12.4.3 Impact Types and Velocities

The majority of crashes in both databases were in the anterior oblique category, accounting for 59 and 49 % of without and with SI occupants in CIREN, and 59 % each in the SI and non-SI groups in NASS. Pure lateral impacts consisted of 38 and 41 % of without and with SI occupants in CIREN and 26 and 20 % in NASS. The posterior oblique loadings in both databases for both groups were below ten percent. The highest ΔV (40 km/h+) accounted for a large proportion of SI occupants: 44 and 18 % in CIREN and NASS. These data for non-SI occupants were 31 and 23 %, respectively. All crashes in CIREN database were coded with known ΔV. The unknown ΔV group accounted for 43 % in NASS with SI and 28 % without SI. Approximately two-thirds of all crashes occurred at <40 km/h. At the 50 % cumulative frequency level, ΔV for SI and non-SI occupants were: 36 and 32 km/h in CIREN, and 29 and 32 km/h in NASS.

12.4.4 Injury Scores

MAIS 3 accounted for 47 and 39 % of non-SI and SI occupants in CIREN and 35 and 20 % in NASS. The trend was opposite for MAIS 4 level injuries: non-SI and SI occupants represented 33 and 48 % in CIREN, and 10 and 22 % in NASS. MAIS 5 occupants were below 11 % each group in CIREN database and 9 and 8 % for the non-SI and SI groups in NASS. Occupants without and with SI in the ISS 4–13 group accounted for 20 and 10 % in CIREN and 58 and 47 % in NASS. These data at the ISS 14–24 level in CIREN were 43 and 33 %, and in NASS were 22 and 24 %. Occupants without and with SI in the ISS 25+ group were 37 and 57 % in CIREN and 20 and 29 % in NASS.

12.4.5 Injuries to Other Body Regions

Head injuries in the without and with SI groups were found in 50 and 51 % of the occupants in CIREN, and 46 and 41 % in NASS. In contrast, skull fractures (with or without associated brain trauma) in the without and with SI groups occurred in 11 and 13 % in CIREN, and 10 and 22 % NASS. Regarding thoracic fractures to the ribs and sternum, both databases showed higher rates in the SI group: 69 and 41 % in CIREN and NASS than the 49 and 30 % without SI, respectively. Regarding abdominal injuries, in the SI group, CIREN and NASS showed 49 and 20 %, and these data for the non-SI group were 39 and 20 %, respectively. Regarding pelvic injuries, CIREN showed similar rates between the SI and non-SI occupants (62 and 61 %), while NASS showed an increase with SI at 31 % compared to 23 % without SI. In general, torso injuries appeared to increase with the presence of SI in both databases.

12.4.6 Implications of Injuries

The SC joint was not traumatized in both databases, indicating that SI in nearside crashes does not involve the ribcage-sternum-clavicle complex. Although the sternum complex is considered a part of the thorax body region in the AIS coding manual, this junction is an integral part of the shoulder complex. Separation of the anatomical structures like the manibrium in the coding process adds to the understanding of injuries to this complex. Scapula and humerus bone fractures were the most frequently injured shoulder components in both databases. This finding parallels other studies wherein both databases have identified similar results to other body regions [34]. Scapula and clavicle fractures were often associated with injuries to other parts of the shoulder complex, particularly the humerus and AC joint. These results suggest that these bones act as a media to transfer the side impact energy to the other components of the shoulder complex (and its inferior anatomical structures). They are the most frequent load paths for nearside occupants. Similar trends (example, types of collisions, confinement of median DV within a small range, decrease in MAIS 3 injuries for SI occupants and increase in MASI 4 injuries, scapula and humerus fractures) between the two databases for occupants with and without SI indicate that it is appropriate to use these databases to explore opportunities for advancing safety.

Because dummies continue to be designed with capabilities to measure kinematics and forces in the shoulder (e.g., a sensor is housed in the ipsilateral shoulder rib of the mid-size WorldSID dummy for recording deflections), it may be necessary to gather such field data to assist in developing biofidelic dummies [35–37]. As recent post mortem human surrogates (PMHS) sled tests have been conducted with a modular and scalable load-wall to accommodate subject-specific anthropometry and determine region-specific (example, shoulder) injury metrics, more inclusive field data will be of value for dummy and PMHS injury assessments [38].

Skull fractures increased with SI in both databases. Acknowledging that skull fractures occur from contact loading, these results suggest that SI resulting from shoulder contact loading exposes the cranium for trauma [39, 40]. Thorax bone injuries also increased with SI in both databases, suggesting a related load-sharing/load path between the two regions: shoulder and thorax. In an analysis of Indy car crashes, Melvin et al. postulated that shoulder loading may be protective of "internal organ damage in the chest" [30]. A study reported PMHS sled tests at 15 km/h [41]. One PMHS sustained 16 rib fractures and SI, while the other two had no injury, and the authors stated: "the collected response data suggest that the shoulder injury may have contributed to rib fractures in the injured subject." Acknowledging the lower change in velocity used in sled tests compared to the present field data, and as indicated in their biomechanical results, the shared load path between the shoulder and thorax may render additional forces to the inferior regions of the shoulder in the presence of SI. The increase in torso injuries with SI appears to support this biomechanical postulate. Additional studies are needed to quantify the role of the shoulder to influence traumas to the other body regions, including the head. Sled tests may be an option.

Anterior oblique impacts accounted for a larger share than pure lateral and posterior oblique crashes in both databases, with and without SI. The lowest percentage for the posterior oblique loading indicates that these impacts are relatively infrequent, but this observation is tempered by the inclusion criteria. However, the recognition of the anterior oblique vector and its relatively greater importance than the pure lateral condition, underscores the importance of oblique loading in side impact crashworthiness. This finding is further reinforced when one considers that current side impact dummies are based only on pure lateral impacts [42–44]. Shoulder deflection metrics can be measured with the WorldSID dummy. Recognizing the importance of shoulder injuries and implications of load paths as discussed above, controlled matched-pair tests should be conducted with PMHS to investigate injury biomechanics and assess the biofidelity of this dummy. This includes anterior oblique full-scale vehicle and sled tests with anthropometry-specific modular and scalable load-walls to isolate region-specific deflections and forces for deriving injury criteria for these conditions. As injury metrics can be measured in the WorldSID shoulder, it would be necessary to clearly define the role of this body region in side impacts and develop an injury criterion. The possibility of the shoulder acting as a load path and affecting the kinematics, loads, energy transfer and injury assessments to the other body regions should be more thoroughly examined in future research.

12.5 Summary

This complex anatomy of the upper extremity presents challenges when studying injury biomechanics of the various individual components. The floating shoulder joint that contributes to the normal physiologic range of motion is stabilized nearly entirely by active muscles, and a multi-articulation elbow joint coordinates motions of three large bones of the upper extremity, the radius, ulna and humerus. Current knowledge on injury tolerance and risk prediction of the upper extremity are largely based on experimental studies using PMHS under loads delivered by different systems with applications to automobile crashes. Although early studies were focused on frontal impacts and airbag deployments, recent field injury data on side impacts covered in this chapter shed light on the interaction of the shoulder complex due to contact loading during the event. These studies complement findings from laboratory tests and provide insight into injury mechanisms of different body regions. The role of the shoulder complex in modulating the kinematics, internal-load paths and injuries to the other body regions including the thorax, abdomen and pelvis are presented in this brief chapter.

Acknowledgments This material is the result of work supported with resources and the use of facilities at the Zablocki VA Medical Center, Milwaukee, Wisconsin and the Medical College of Wisconsin. Narayan Yoganandan is a part-time employee of the Zablocki VA Medical Center, Milwaukee, Wisconsin. Any views expressed in this chapter are those of the authors and not necessarily representative of the funding organizations.

References

1. Dempster WT (1965) Mechanisms of shoulder movement. Arch Phys Med Reha 46:49–70
2. Moore KL (1980) Glenohumeral joint. In: Clinically oriented anatomy. Williams & Wilkins, Baltimore, pp 814–819
3. Basmajian JV (1969) Recent advances in the functional anatomy of the upper limb. Am J Phys Med 48(4):165–177
4. Burkart AC, Debski RE (2002) Anatomy and function of the glenohumeral ligaments in anterior shoulder instability. Clin Orthop Relat Res 400:32–39
5. Robinson CM (1998) Fractures of the clavicle in the adult. Epidemiology and classification. J Bone Joint Surg Br 80(3):476–484
6. Duprey S, Bruyere K, Verriest J-P (2008) Influence of geometrical personalization on the simulation of clavicle fractures. J Biomech 41(1):200–207
7. Harrington MA Jr, Keller TS, Seiler JG 3rd, Weikert DR, Moeljanto E, Schwartz HS (1993) Geometric properties and the predicted mechanical behavior of adult human clavicles. J Biomech 26(4–5):417–426
8. Kemper A, Stitzel J, Gabler C, Duma S, Matsuoka F (2006) Biomechanical response of the human clavicle subjected to dynamic bending. Biomed Sci Instrum 42:231–236
9. Weber C (1859) Chirurgische Erfahrungen und Untersuchungen, pp. 171–174. (cited in Melvin 1995)
10. Melvin JW (1995) Injury assessment reference values for the CRABI 6-month infant dummy in a rear-facing infant restraint with airbag deployment. SAE Congress, Detroit
11. Messerer O (1880) U X ber ElasticitaX t und Festigkeit der menschlichen Knochen. Stuttgart: J. G. Cotta'schen Buchhandlung. (cited in Melvin, 1995)
12. Yamada H (1970) In: Evans FG (ed) Strength of biological materials. Williams and Wilkins, Baltimore
13. Kirkish SL, Begeman PC, Paravasthu NS (1996) Proposed provisional reference values for the humerus for evaluation of injury potential. Paper no. 962416. Society of Automotive Engineers, Warrendale
14. Mertz HJ (1984) A procedure for normalizing impact response data. Society for Automotive Engineers, Warrendale
15. Duma S, Schreiber P, McMaster J, Crandall J, Bass C, Pilkey W (1998) Dynamic injury tolerances for long bones of the female upper extremity. In: Proceedings of the 1998 International IRCOBI Conference on the Biomechanics of Impact. Sept. 16–18, 1998; Göteberg, Sweden. pp 189–201
16. Eppinger RH, Marcus JH, Morgan RM (1984) Development of dummy and injury index for NHTSA's thoracic side impact protection research program. Paper no. 840885. Society of Automotive Engineers, Warrendale
17. van Rooij L, Bours R, van Hoof J, Mihm JJ, Ridella SA, Bass CR, Crandall JR (2003) The development, validation and application of a finite element upper extremity model subjected to air bag loading. Stapp Car Crash J 47:55–78
18. Bass CR, Duma SM, Crandall JR, Morris R, Martin P, PIlkey WD (1997) The interaction of air bags with upper extremities. Paper no. 973324. Society of Automotive Engineers, Warrendale
19. Pintar FA, Yoganandan N (2002) Dynamic bending tolerance of the human forearm. Traffic Inj Prev 3:48–52
20. Mather BS (1967) A method of studying the mechanical properties of long bones. J Surg Res 7(5): 226–230
21. Mather BS (1967) Correlations between strength and other properties of long bones. J Trauma 7(5): 633–638
22. Hardy WN, Schneider LW, Rouhana SW (2001) Prediction of airbag-induced forearm fractures and airbag aggressivity. Stapp Car Crash J 45:511–534
23. Duma SM, Hansen GA, Kennedy EA, Rath AL, McNally C, Kemper AR, Smith EP, Brolinson PG, Stitzel JD, Davis MB, Bass CR, Brozoski FT, McEntire BJ, Alem NM, Crowley JS (2004) Upper extremity interaction with a helicopter side airbag: injury criteria for dynamic hyperextension of the female elbow joint. Stapp Car Crash J 48:155–176
24. Duma SM, Boggess BM, Crandall JR, Mac Mahon CB (2002) Fracture tolerance of the small female elbow joint in compression: the effect of load angle relative to the long axis of the forearm. Stapp Car Crash J 46:195–210
25. Bolte JH, Hines MH, McFadden JD, Saul RA (2000) Shoulder response characteristics and injury due to lateral glenohumeral joint impacts. Stapp Car Crash J 44:261–280
26. Compigne S, Caire Y, Quesnel T, Verries JP (2004) Non-injurious and injurious impact response of the human shoulder three-dimensional analysis of kinematics and determination of injury threshold. Stapp Car Crash J 48:89–123
27. Irwin A, Walilko T, Cavanaugh J, Zhu Y, King A (1993) Displacement responses of the shoulder and thorax in lateral sled impacts. In: Stapp car crash conference, San Antonio, pp 166–173
28. Koh SW, Cavanaugh JM, Zhu J (2001) Injury and response of the shoulder in lateral sled tests. Stapp Car Crash J 45:101–142
29. Duma SM, Crandall JR, Hurwitz SR, PIlkey WD (1998) Small female upper extremity Interaction with a deploying side air bag. Society of Automotive Engineers, Warrendale. Paper no. 983148

30. Melvin JW, Baron KJ, Little WC, Gideon TW, Pierce J (1998) Biomechanical analysis of Indy race car crashes. Society of Automotive Engineers, Warrendale. Paper no. 983161
31. Yoganandan N, Stadter GW, Halloway DE, Pintar FA (2013) Injury patterns to other body regions and load vectors in nearside impact occupants with and without shoulder injuries. Ann Adv Autom Med. Annual Scientific Conference Association for the Advancement of Automotive Medicine Association for the Advancement of Automotive Medicine 57: 133–144
32. Stadter GW, Yoganandan N, Halloway DE, Pintar FA Analysis of nearside narrow object impacts with and without shoulder injuries in real-world crashes. In: IRCOBI, Gotenborg, Sweden, 13–15 Sept 2013
33. AIS (1990) The abbreviated injury scale, 1998 update. American Association for Automotive Medicine, Arlington Heights
34. Yoganandan N, Pintar FA (2005) Odontoid fracture in motor vehicle environments. Accid Anal Prev 37(3):505–514. doi:10.1016/j.aap.2005.01.002
35. Yoganandan N, Humm JR, Pintar FA, Brasel K (2011) Region-specific deflection responses of WorldSID and ES2-re devices in pure lateral and oblique side impacts. Stapp Car Crash J 55:351–378
36. Yoganandan N, Humm JR, Pintar FA (2012) Modular and scalable load-wall sled buck for pure-lateral and oblique side impact tests. J Biomech 45(8):1546–1549. doi:10.1016/j.jbiomech.2012.03.002
37. Yoganandan N, Humm JR, Pintar FA, Maiman DJ (2013) Determination of peak deflections from human surrogates using chestbands in side impact tests. Med Eng Phys. doi:10.1016/j.medengphy.2012.12.012
38. Yoganandan N, Humm JR, Pintar FA, Brasel K (2012) Deflection responses post mortem human surroagtes in pure lateral and oblique side impacts. Stapp Car Crash J 55:351–378
39. Yoganandan N, Pintar FA, Sances A Jr, Walsh PR, Ewing CL, Thomas DJ, Snyder RG (1995) Biomechanics of skull fracture. J Neurotrauma 12(4):659–668
40. Yoganandan N, Pintar FA (2004) Biomechanics of temporo-parietal skull fracture. Clin Biomech (Bristol, Avon) 19(3):225–239. doi:10.1016/j.clinbiomech.2003.12.014
41. Lessley D, Shaw G, Parent D, Arregui-Dalmases C, Kindig M, Riley P, Purtsezov S, Sochor M, Gochenour T, Bolton J, Subit D, Crandall J, Takayama S, Ono K, Kamiji K, Yasuki T (2010) Whole-body response to pure lateral impact. Stapp Car Crash J 54:289–336
42. Kuppa S, Eppinger RH, McKoy F, Nguyen T, Pintar FA, Yoganandan N (2003) Development of side impact thoracic injury criteria and their application to the modified ES-2 dummy with Rib extensions (ES-2re). Stapp Car Crash J 47:189–210
43. Maltese MR, Eppinger RH, Rhule HH, Donnelly BR, Pintar FA, Yoganandan N (2002) Response corridors of human surrogates in lateral impacts. Stapp Car Crash J 46:321–351
44. Yoganandan N, Pintar FA, Stemper BD, Gennarelli TA, Weigelt JA (2007) Biomechanics of side impact: injury criteria, aging occupants, and airbag technology. J Biomech 40(2):227–243. doi:10.1016/j.jbiomech.2006.01.002

Thorax Injury Biomechanics

John M. Cavanaugh and Narayan Yoganandan

Abstract

This chapter provides a broad overview of the biomechanics of the human thorax with a focus on impact loading. The introduction to the thoracic anatomy emphasizes its skeletal structures, organs and soft tissues, and this is followed by a brief description of injury quantification using the Abbreviated Injury Score and Injury Severity Score. Injury mechanisms are reviewed from a historical perspective. They include rib fractures and flail chest, lung contusions, hemo- and pneumo-thoraces, and heart and aorta injuries. The next two sections are focused on frontal and side impact injury aspects. In the frontal impact, different methods used in loading are described, including blunt, belt and airbag loading, along with biomechanical response data and injury mechanisms. Various candidates for injury criteria are also described. Pure lateral and oblique side impacts are then described using a similar method, albeit separately. The oblique loading modality discovered in modern vehicular environments is relatively new. Oblique impacts are shown to be more injurious than pure lateral impacts. Injury criteria associated with this vector needs further research as studies are beginning to appear in published literature. The reader is then exposed to a brief assessment and development of dummy thorax although dummy-related topic is covered elsewhere in the book.

13.1 Introduction and Epidemiology

Chest injury ranks second only to head injury in overall number of fatalities and serious injuries in motor vehicle crashes in the United States. The U.S. Centers for Disease Control (CDC) reported that injuries to the torso are the second major cause of death by specific body region next to the head and neck [1]. During a motor vehicle impact,

J.M. Cavanaugh, M.D. (✉)
Department of Biomedical Engineering,
Wayne State University, Detroit, MI, USA
e-mail: jmc@wayne.edu

N. Yoganandan, Ph.D.
Department of Neurosurgery, Medical College
of Wisconsin, Milwaukee, WI, USA
e-mail: yoga@mcw.edu

the thorax can contact various components of the automobile interior, including restraint systems. Contacts include unrestrained driver or passenger with steering wheel or instrument panel, and contact with active or passive restraints, including three point lap/shoulder belts, two-point shoulder belts, knee bolsters, and air bags. Injury to the thorax commonly occurs in frontal and side impacts and in oblique directions intermediate to these two.

Nirula and Pintar analyzed the National Automotive Sampling System (NASS) databases from 1993 to 2001 and the Crash Injury Research and Engineering Network (CIREN) databases from 1996 to 2004 [2]. The incidence of severe chest injury (AIS 3 and greater) in NASS and CIREN were 5.5 % and 33 %, respectively. The steering wheel, door panel, armrest and seat were identified as contact points associated with an increased risk of severe chest injury. The door panel and arm rest were consistently a frequent cause of severe injury.

In a study of motor vehicle crashes in the UK, Morris et al. [3] examined vehicle crash injury data to determine to determine the relative injury risk of occupants of different age groups. For all occupants, the body region most prone to injury in frontal impact crashes was the chest. Older and middle-aged occupants were at greater risk of sustaining Maximum AIS 3+ (MAIS 3+) chest injuries. In frontal impacts, the vast majority of chest injuries were caused by the restraint system, whereas other interior vehicle components accounted for only 4 % of the injuries. A significant portion of middle-aged and older passengers were female. A seat-belt pre-tensioner was found to have a general effect of reducing the risk of MAIS 3+ chest injury to all age groups.

13.1.1 Thorax Anatomy

A brief review of the anatomy of the thorax is presented below to provide a basis for understanding anatomical regions that can be injured in blunt chest trauma. The thorax is that region of the body between the head and the neck and includes consists of the rib cage and the underlying soft tissue organs. The thorax is bounded inferiorly by the diaphragm, a thin muscular sheet which separates the thoracic contents from the abdominal contents.

13.1.1.1 Rib Cage

The rib cage consists of 12 pairs of ribs connected to the sternum anteriorly and the thoracic vertebrae posteriorly (Fig. 13.1). Posteriorly, each of the 12 ribs is joined to its corresponding thoracic vertebra (T1–T12) at the costal facet joints. Anteriorly, each of the first seven ribs joins the sternum at the chondro-sternal junctions. Ribs 8–10 join the bottom of the sternum by cartilaginous attachments to the seventh rib. Ribs 11 and 12, the floating ribs, have no anterior attachment to the sternum or other skeletal structures and are sometimes called floating ribs.

13.1.1.2 Lungs and Mediastinum

The left lung consists of two lobes, the upper (superior) and lower. The right lung consists of three lobes, the upper, middle and lower lobes (Fig. 13.2). The lungs are covered in a serous membrane called the visceral pleura. The parietal pleura lines the inner surface of the chest wall, covers the diaphragm and encloses the structures in the middle of the thorax. The central region of the thoracic cavity is called the mediastinum (Fig. 13.2). The mediastinum is bordered in front by the sternum and in back by thoracic vertebrae and contains the heart, the great vessels entering and leaving the heart, thymus gland, esophagus and lower portion of the trachea, thoracic duct and thoracic lymph nodes, and nerves passing into and through the thorax, including the vagus and phrenic nerves.

13.1.1.3 Heart and Great Vessels

The heart (Fig. 13.3) is a hollow muscular organ that lies in the lower part of the thoracic cavity in the middle mediastinum. It is roughly the size of a man's fist, weighing 300 g in the adult male and 250 g in the female. The heart is divided into four chambers, left and right atria, and left and right ventricles. The right atrium receives the returning deoxygenated blood from all body tissues except the lungs. This returning blood enters the right

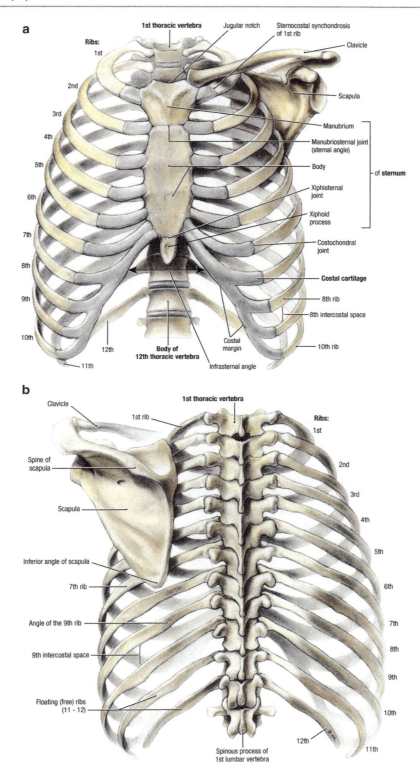

Fig. 13.1 (a) and (b) Illustration of the skeletal anatomy of the thorax. (a) Anterior (*front*) view. (b) Posterior (*back*) view (Reproduced with permission from Grant's Atlas of Anatomy, 13th edition, Lippincott, Williams and Wilkins, 2013)

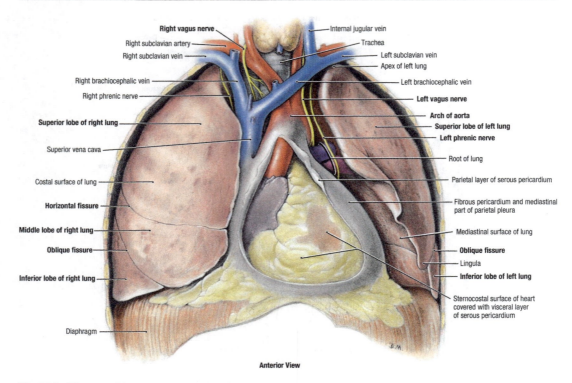

Fig. 13.2 Diagram of the lungs. The right lung has upper, middle and lower lobes. The left lung has upper and lower lobes. The mediastinum, in the central region of the thorax, contains the heart and great vessels and other structures as illustrated (Reproduced with permission from Grant's Atlas of Anatomy, 13th edition, Lippincott, Williams and Wilkins, 2013)

atrium, through the superior and inferior venae cavae. The right ventricle pumps the deoxygenated blood though four pulmonary veins. From the left atrium, the oxygenated blood goes to the thick-walled left ventricle, which pumps it out to through the aorta to all parts of the body except the lungs.

13.2 Injury Scaling Methods

13.2.1 AIS

The standard method for classifying the level of injury to a body region or organ is the Abbreviated Injury Scale (AIS) of the Association for the Advancement of Automotive Medicine (AAAM), periodically updated by the Committee on Injury Scaling. The last update was in 2008 [4]. The numerical rating system ranges from 0 (no injury) to 6 (maximum, virtually unsurvivable). The higher the AIS level, the higher the mortality or threat to life. The scale does not quantify long term disability or medical and societal costs of injury. Typical AIS injuries to the rib cage and to thoracic soft tissues are shown in Table 13.1.

Regarding rib fractures, one rib fracture is AIS1, two fractures is AIS2 and 3 or more fractures is AIS3. Flail chest can range from AIS3 to AIS5. A flail chest is an unstable chest wall in which a portion of the rib cage does not rise on inspiration because of loss of structural rib cage integrity. Unilateral flail chest is rated AIS 3 if 3–5 ribs are involved and AIS 4 if more than five ribs are involved. Bilateral flail chest is rated AIS 5. In cadaveric studies Walfish et al. considered flail chest as two fractures each of four consecutive ribs [5]. At an international meeting in Paris it was proposed that two fractures of three or more

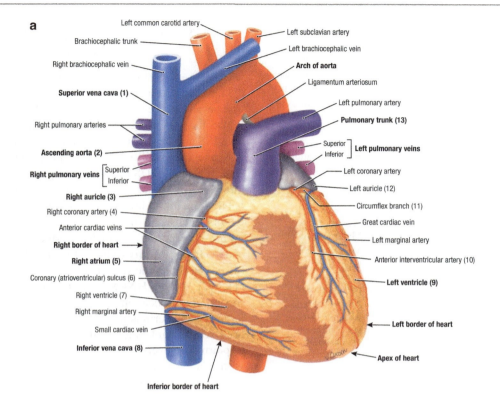
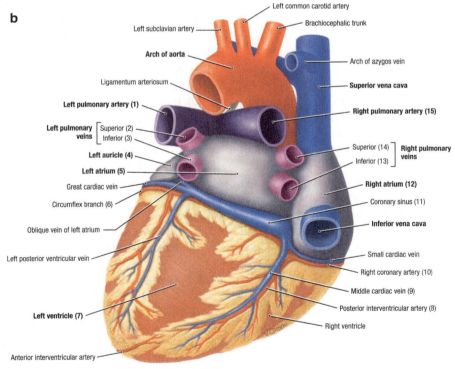

Fig. 13.3 (a) and (b). Diagram of the heart showing right and left atrium, right and left ventricles and great vessels. (a) Anterior view. (b) Posterior view (Reproduced with permission from Grant's Atlas of Anatomy, 13th edition, Lippincott, Williams and Wilkins, 2013)

Table 13.1 Skeletal and soft tissue injuries to the thorax (Ranked by AIS)

AIS	Skeletal injury	AIS	Soft tissue injury
1	1 rib fracture	1	Contusion of bronchus distal to main stem
2	2 rib fractures Sternum fracture	2	Partial thickness Bronchus tear distal to main stem
3	3 or more fractures Unilateral flail chest with 3–5 flail ribs	3	Lung contusion NFS Heart laceration, no perforation
4	Unilateral flail chest with >5 flail ribs	4	Bilateral lung laceration NFS Minor aortic laceration Major heart contusion
5	Bilateral flail chest	5	Major aortic laceration Tension pneumothorax
6		6	Aortic laceration with hemorrhage not confined to mediastinum

consecutive ribs be defined as flail chest [6]. The Walfisch criterion or nine or more rib fractures to the thorax was considered flail chest in a study by Viano [7].

13.2.2 ISS and PODS

The Injury Severity Scale (ISS) is a measure of the probability of survival. It is obtained by summing the square of the highest AIS in each of three body regions and was developed to account for the effect of injuries to multiple body regions. The ISS uses six body regions: head and neck, face, chest, abdomen and pelvis, extremities and bony pelvis, and external [8]. The Probability of Death Score (PODS) and PODSa (which accounts for subject age) were developed for the same purpose [9]. The equation for PODS is as follows:

$$PODS = e^x / (1 + e^x)$$

$$\text{for PODS} \quad x = 2.2(AIS_1) + 0.9(AIS_2) - 11.25 + C$$

$$\text{for PODSa} \quad x = 2.7(AIS_1) + 1.0(AIS_2) + 0.06 \times AGE - 15.4 + C$$

where $C = -0.764$ for cars, and AIS_1 and AIS_2 are the first and second highest AIS.

The advantages of PODS over ISS are that PODS has a better goodness-of-fit to real data than ISS and has an inherent definition to health outcome (probability of death) [10]. Whereas AIS, ISS and PODS quantify injury in terms of threat to life, other concepts have been developed to quantify injury in terms of additional factors, such as quality of life or societal cost of injury. These methodologies include HARM and the Injury Priority Rating (IPR). HARM establishes an economic value to each level of injury [11, 12]. Using HARM, chest injury ranked second only to head injury in overall number of fatalities and serious injuries [11]. Hirsch et al. used AIS to rank injuries by impairment [13]. Impairment severity ranged from one to four, with four being most severe. Six different aspects of impairment (mobility, cognitive/psychological, cosmetic, sensory, pain, and daily living) for three different durations (0-1 year, 1-5 years, > 5 years) were thus ranked. Carsten and O'Day developed the Injury Priority Rating (IPR) to evaluate overall impairment [14]. Carsten modified this and developed the Multi-Injury Priority Rating (MIPR) to express overall injury impairment [15]. The IPR provides for post-accident survival and can distinguish the impairment due to injuries with the same AIS value. Marcus extended impairment studies to more recent data and reviewed IPR and impairment methodologies [16].

13.3 Thoracic Injury and Mechanisms

13.3.1 General

In earlier work, Cohen reported that for the thorax and abdomen, most of the contact for the driver was to the steering assembly, while for the passenger, it was the instrument panel [17]. The NASS data base from 1979 to 1984 was reviewed by Haffner for passenger cars in non-ejection and non-rollover cases and published by Schneider et al. [18]. For drivers and passengers, skeletal injury represented the highest percentage of thoracic AIS, followed by pulmonary/lung injuries for the driver and then liver and heart injuries. The liver and spleen are located under the diaphragm, but are bounded by the lower lateral rib cage. The rim of the steering wheel or the instrument panel can produce soft tissue injuries to these components. Although arterial injuries accounted for 6–8 % of AIS >2, they accounted for 27–30 % of estimated HARM. With the increased use of seat belts and the requirement for frontal air bags, these injury statistics have changed. More current data are reviewed in subsequent sections.

13.3.2 Rib Fractures and Flail Chest

Rib fracture and flail chest occur with blunt impact to the rib cage. Rib fractures are the most frequent type of AIS3+ chest injuries [19]. It is probable that ribs fail in bending, with failure occurring on the tensile side of the ribs. Stalnaker and Mohan, and Melvin et al. concluded that maximum compression of the chest is the determining factor for rib fracture [20, 21]. In their review of injury and response data from cadaveric studies, rib fractures were more frequent with chest deflections over 3 in., while none occurred at deflections less than 2.3 in. Thus, it appears that the number of rib fractures depends on the magnitude of rib deflection, rather than the rate of deflection. However, the amount of force depends on the rate at which the force is applied due to the viscous nature of the thorax. Thus, for a given loading rate, force appears to be related to the number of rib fractures. Eppinger developed a relationship between the number of rib fractures and upper torso shoulder belt force, age and weight of cadaveric subject [22].

In 2008, Trosseille et al. showed a promising methodology to examine rib fracture mechanisms, using strain gauges glued on the ribs of PMHS [23]. In 2009, they published the results of static airbag tests performed on 50th percentile male PMHS at different distances and angles (pure lateral and 30° forward oblique direction) [24]. To complete the study, Leport et al. carried out eight PMHS lateral and oblique impactor tests with the same methodology [19]. The rib cages were instrumented with more than 100 strain gauges on the ribs, cartilage and sternum. A 23.4-kg impactor was propelled at 4.3 or 6.7 m/s. The forces applied to the PMHS at 4.3 m/s ranged from 1.6 to 1.9 kN and the injuries varied from 4 to 13 rib fractures.

Kindig et al. [25] identified rib-level differences in fracture characteristics for individual ribs subjected to anterior-posterior loading. Twenty-seven individual ribs from levels 2 to 10 from 3 postmortem human subjects (two females and one male) were tested in anterior-posterior loading at a quasistatic (2 mm/s) loading rate. Rib 2 was 3–4 times stiffer than rib 3, whereas all other ribs were comparable in stiffness to rib 3. Fracture forces, fracture displacement, and work to fracture showed no clear rib-level differences. The data were compared to a similar study at dynamic loading rates (1.43–1.85 m/s). The quasistatic tests exhibited lower peak force and greater normalized fracture displacement than the dynamic tests, though the work was comparable between the 2 studies [25].

13.3.3 Lung Contusions

Unlike rib fractures, lung contusion is rate dependent. Fung and Yen determined that it was a velocity dependent phenomenon [26]. At high velocities, a compression or pressure wave is transmitted through the chest wall to the lung tissue, causing damage to the capillary bed of the alveoli. Lacerations of lung tissue can also occur at sites of rib fractures. Gayzik et al. induced pulmonary contusion in male Sprague Dawley rats (n = 24) through direct impact to the right lung at 5.0 m/s [27]. Force vs. deflection data were collected and used for model validation and optimization. CT scans were taken at 24 h, 48 h, 1 week, and 1 month post contusion and used to quantify the volume of pathologic lung tissue. The best fit injury measure at the 24-h time point was the product of maximum principal strain and strain rate, with a value of 28.5 s^{-1}. The maximum principal strain and the maximum principal strain rate also characterized pathology well, with the threshold for lung injury at 24 h post-impact being 15.4 % and 304 s^{-1}, respectively.

13.3.4 Hemothorax and Pneumothorax

Hemothorax is bleeding into the pleural space. Blood vessels can be lacerated by broken ribs, leading to hemothoax. If a hole is created in the pleural sac between the lungs and the rib cage, a pneumothorax occurs. The puncture or laceration can be in the parietal pleura lining the inside of the rib cage or in the visceral pleura surrounding the lungs. The pneumothorax is often the result of rib fracture. Thus, rib fracture, hemothorax and pneumothorax appear to be deflection dependent, while lung contusion appears to be velocity dependent.

13.3.5 Heart Injuries

During high-speed thoracic impact the heart can be subject to contusion, laceration or cardiac arrest. Lasky et al. in a review of 67 cases of frontal and side impact collisions by the Vehicle Trauma Research Group at UCLA found that in frontal collisions, energy absorbing steering wheel hubs and columns resulted in reduced cardiovascular injury [28]. Side impact collisions were shown to produce a large proportion of the cardiovascular injury. Contusion appears to be due to compression and the velocity of compression while cardiac laceration may be due to high magnitudes of compression over the sternum. At high rates of loading, the heart may undergo fibrillation or arrest. High speed blunt impacts (15–20 m/s) appear to interrupt the electromechanical transduction of the heart wall. Cooper et al. reported on 38 mid-sternal impacts to the chest of pigs by a cylindrical mass of 0.14–0.38 kg and 3.7 and 10 cm in diameter at velocities of 20–74 m/s [29]. Acute ventricular fibrillation appeared to be associated with mid-sternal blows during the T-wave of the electrocardiogram (ECG). Kroell et al. reported 11 cases of ventricular fibrillation (VFB) in impacts to the mid-sternum of 23 anesthetized swine [30]. Eight of the VFB's were immediate and five of these were in impacts which occurred during the T-wave of the ECG.

13.4 Effects of Age on Injury Tolerance

Zhou et al. studied the effects of age on thoracic injury tolerance by analyzing the mechanical properties of human bones and soft tissues and examining results from literature on thoracic impact tests to PMHS [31]. The work divided the age range into three groups: 16–35, 36–65, and 66–85 years. The reduction ratios with age for thoracic injury tolerance of AIS 3 ranged from 1 to 0.47 to 0.28 for belt loading, for these three age groups; the tolerance decreased from 1.0 to 0.84 to 0.79 for blunt frontal impacts and from 1 to 0.80 to 0.73 for side impact loads. Tolerance under concentrated loading such as seat belts appeared to be governed more by the properties of the rib cage and tolerance under blunt loading appeared to be governed mainly by the properties soft tissues. In computer simulations by Kent et al. the structural and material changes played approximately equal roles in the force–deflection response of the thorax [32]. Changing the rib angle to be more perpendicular to the spine as in the older person increased the effective thoracic stiffness, while the "old" material properties and the thin cortical shell decreased the effective stiffness. All three effects tended to decrease chest deflection tolerance for rib fractures. The findings indicated an older person's thorax, relative to a younger, does not necessarily deform more in response to an applied force but the tolerable sternal deflection level is much less.

Mertz and Dalmatos assessed the effectiveness of 3-point restraint systems using 1988–2005 NASS data for 3-point belted, front outboard-seated, adult occupants in passenger vehicles equipped with airbags and involved in frontal, tow away collisions [33]. These data showed that (i) half of the occupants with AIS ≥ 3 chest injuries were in collisions with a ΔV ≤ 40 km/h; (ii) for older occupants (50+ years), half experienced their chest injuries at ΔV ≤ 34 km/h; and (iii) the chest injury rate for the older occupants was more than double that of the younger occupants. Analysis indicated that 2.5 kN shoulder belt limit load would substantially reduce shoulder belt-induced AIS ≥ 3 chest injuries in 99 % of frontal collisions to all adult, front outboard seated occupants whose normalized bone strengths were greater than 0.4.

Cormier [34] assessed the influence of body mass index (BMI) on thoracic injury potential. The data for this study were obtained from the National Automotive Sampling System-Crashworthiness Data System (NASS-CDS) database for years 1993–2005. Obese occupants had a 26 % and 33 % higher risk of AIS 2+ and AIS 3+ thoracic injury when compared to lean occupants. The increased risk of AIS 3+ injury due to obesity was slightly higher for older occupants, but the influence of age was greater than that of obesity. The increase in injury potential was higher for unbelted obese occupants than unbelted. Non-parametric and parametric risk curves were developed to estimate the risk of thoracic injury based on occupant BMI, belt use and delta-V. Overall, increase in thoracic injury risk due to obesity was more prominent in males and older occupants and for occupants sustaining AIS 3+ thoracic injuries [34].

Koppel et al. [35] studied data from the Australian Transport Accident Commission insurance claims database for two groups of drivers: aged 41–55 years (middle-aged drivers) and aged 65 years and older (older drivers). The majority of crashes involved a collision with another vehicle (70.0 % of middle-aged drivers and 68.7 % of older drivers). Older drivers sustained a significantly higher proportion of injuries to the thorax (30.9 % compared to 18.5 % of middle-aged drivers). Conversely, a significantly higher proportion of middle-aged drivers sustained injury to the neck (30.6 % compared to 12.1 % of older drivers) [35].

13.5 Aortic Trauma

13.5.1 Epidemiology

Previous studies showed that traumatic rupture of the aorta (TRA) accounted for 10–25 % of all deaths from MVCs [36–39]. Previous studies estimated that 7,500–8,000 cases of blunt aortic injury occurred annually in the US and Canada [40, 41]. Eighty to eighty-five percent of the victims of TRA died at the scene of the accident [42]. In a prospective study using 50 trauma centers in North America Fabian et al. reported 274 (199 men, 75 women, mean age 38.7 years) blunt aortic injury cases studied over 2.5 years, of which 81 % were caused by automobile crashes [43]. Overall mortality was 31 %, with 63 % of deaths attributable to the aortic rupture itself. Motor vehicle crashes were responsible for 222 of the 274 cases with 72 % of these being head-on and 24 % from side impacts. The other causes of TRA were motorcycle crashes, auto-pedestrian impacts, falls and other. Forty-six % had multiple rib fractures, 38 % pulmonary contusion, 31 % pelvic injury and 51 % closed head injury. In a study of 63,763 patients with blunt trauma over a 5-year period, 8.1 % sustained pelvic fractures and 441 (1.6 %) had aortic ruptures. Eighty-six had both aortic rupture and pelvic fracture. The incidence of aortic rupture in the pelvic fracture cases was more than twice that in the overall blunt trauma population. Mechanisms of injury were predominantly high-speed motor vehicle accidents followed by falls and being struck as a pedestrian [38].

Bertrand et al. [44] analyzed crash data collected from 1998–2006 as part of the Co-operative Crash Injury Study (CCIS) to assess frontal, near-side, and far-side injury risks related to TRA. The database included 15,074 occupants with detailed autopsy reports. The influences of gender, age, Equivalent Test Speed (ETS), compartment intrusion, and restraint system on TRA were analyzed. TRA occurred in 1.2 % of all occupants but accounted for 21.4 % of all fatalities. The incidence of TRA was higher in side impacts (2.4 %) than in frontal impacts

(1.1 %). TRA injury risk increased with ETS, intrusion, and age and decreased with the absence of intrusion regardless of the impact direction. It also decreased for belted occupants in frontal impacts. Multiple rib fractures were the most common injuries associated with TRA (79.1 %). TRA victims with uninjured or slightly injured (AIS 1) rib cage were significantly younger (p<0.0001) than other TRA victims. Whatever the impact type, the TRA victims sustained mostly bilateral rib fractures. Fractures concerned mainly the 2nd up to the 7th ribs of TRA victims. The typical TRA involving partial or complete aorta transection within the peri-isthmic region was independent of age and impact type. The high frequency of bilateral rib cage fractures observed in TRA victims and the significant influence of intrusion on TRA occurrence emphasized that the aortic injury mechanism mainly involves a severe direct chest impact or compression [44].

13.5.2 Anatomical Location of Aortic Injury

Sites of aortic laceration include the aortic isthmus, root, and aortic insertion into the diaphragm. Ninety percent of thoracic aortic injuries occur in the region of the aortic isthmus, just distal to the origin of the left subclavian artery [45]. The isthmus is a portion of the aorta between the ligamentum-arteriosum and left subclavian artery (Fig. 13.3) that develops from the fetal ductus arteriosus. The aortic isthmus averages only about 1.5 cm in length in adults and the region of the proximal descending aorta is often referred to loosely as the region of the aortic isthmus [45]. Clinically, thoracic aortic injury involves the ascending aorta in only 5 % of cases. At autopsy, it was the site of injury in 20–25 % of cases. This area of the aorta is associated with grave complications such as valve rapture and coronary artery laceration that are threats to life.

13.5.3 Aortic Injury Mechanisms

Three general mechanisms for TRA have been proposed: (1) traction or shear forces generated between relatively mobile portions of the vessel and points of fixation, (2) direct compression over the vertebral column, (3) and excessive sudden increases in intraluminal pressure [45]. Hossack et al. in an Australian study pointed out that 12.7 % of fatal road crash victims had a rupture of the aorta [46]. He also noted that the recurrent laryngeal nerve hooks around the aorta just near the ligamentum-arteriosum and that forcible traction on this nerve may exert pressure at the site of aortic tears. Viano [47] noted that inertial loading of the blood-filled heart can cause the heart to displace in the chest cavity and stretch points of attachment of the aortic arch such as the superior arteries or the ligamentum-arteriosum [47]. This may occur if the heart is displaced vertically, laterally or obliquely. Aortic lacerations generally have a transverse orientation. Mohan and Melvin showed that the descending mid-thoracic aorta failed in the transverse rather than longitudinal direction under uniform biaxial stretch [48]. Lasky et al. postulated that large fluid pressures resulting in a water hammer effect may be important in causing vessel rupture [28].

Newman and Rastogi observed that in all 12 cases of aortic rupture, the impact was not directly frontal, suggesting that a transverse component of impact is required to produce aortic wall laceration between fixed and mobile sections of the aorta [49]. In cadaveric aide impact sled tests performed by Cavanaugh et al. the impacts were to the left side of the PMHS at 6.7 and 9 m/s with one test at 10.4 m/s [50, 51]. In these tests, the sudden deceleration of the sled caused the unembalmed PMHS to slide across a Teflon™ seat into a side wall instrumented with load cells. The arterial system of the torso was pressurized to 2 psi and the venous system to 1 psi. After the impact, each PMHS had a necropsy performed by a board-certified pathologist. In five of 17 tests, a tear of the aortic arch was noted. All tears were in the transverse direction.

The authors hypothesized that the more mobile heart and aortic arch translated laterally, producing tears at the top of the more firmly anchored descending aorta.

Shah et al. determined mechanical properties of planar aorta tissue at high strain rates [52]. Cruciate tissue samples from various regions of 12 PMHS thoracic aortas were subjected to equibiaxial stretch at two nominal speed levels using a new biaxial tissue-testing device. The aorta tissue exhibited nonlinear behavior. In a series of component-level tests, the response of the intact thoracic aorta to longitudinal stretch was obtained using seven aorta specimens. The aorta failed within the peri-isthmic region with transverse tears and the intima failed before the media or adventitia. Complete transection was observed at 92 N axial load and 0.221 axial strain.

Hardy et al. investigated TRA mechanisms in PMHS in four quasi-static and one dynamic tests [53]. The quasi-static tests included anterior, superior, and lateral displacement of the heart and aortic arch in the mediastinum, resulting in partial tears to complete transection. All injuries occurred within the peri-isthmic region. The average failure load and stretch were 148 N and 30 %, respectively, for the quasi-static tests. The results indicated that intraluminal pressure and whole-body acceleration are not required for TRA to occur and that the role of the ligamentum-arteriosum is likely limited. The studies indicated that tethering of the descending thoracic aorta by the parietal pleura was a principal aspect of this injury.

Hardy et al. investigated the mechanisms of TRA in eight unembalmed PMHS which were inverted and tested in various dynamic blunt loading modes [54]. Impacts were conducted using a 32-kg impactor with a 152-mm face. High-speed biplane x-rays of radiopaque markers on the aorta were used to visualize aortic motion. Clinically relevant TRA was observed in seven of the tests. Peak average longitudinal Lagrangian strain was 0.644 and the average peak strain for all tests was 0.208 +/− 0.216. Peak intraluminal pressure was 165 kPa. Longitudinal stretch of the aorta was found to be a principal component of injury causation. Stretch of the aorta was generated by thoracic deformation, which was required for injury to occur. Atherosclerosis further promoted injury.

13.6 Biomechanics of Frontal Impact

Many biomechanical tests have been performed on PMHS under controlled laboratory conditions to measure biomechanical responses (forces, accelerations, deformations, pressures) and obtain details of resulting injury through necropsy of the body after impact. Pendulum and sled tests have been performed to ascertain these data for frontal and lateral impacts. This data has been used to develop frontal and side impact dummies and to develop injury criteria. A description of these tests follows.

13.6.1 Biomechanical Response of the Thorax in Frontal Impact

13.6.1.1 Pendulum Impacts to the Sternum

Some of the most extensive testing analyzed in the 1970s involved six-inch diameter rigid pendulum impacts to the sternum of unembalmed PMHS (Fig. 13.4). The data were presented by Kroell et al., Nahum et al., and Stalnaker et al. [55–60]. These data have also been analyzed by Lobdell et al. and Neathery [61, 62]. In the Kroell tests, a six-inch diameter impactor contacted the sternum at the level of the interspace between the fourth and fifth ribs. Figure 13.5 shows force-deflection curves for 4.02–5.23 and 6.71–7.38 m/s in the Kroell et al. tests.

The total chest deflection including the flesh overlying the sternum was included by Kroell et al. [55, 56]. Neathery developed corridors of skeletal chest deflection, based on these data [62]. Lobdell et al. showed that lower impactor masses resulted in lower deflections [61]. Patrick developed force-deflection curves under conditions similar to the Kroell unrestrained back conditions, but used himself as a volunteer and a padded (2.4 cm Rubatex R310V) six-inch diameter striker at velocities of 2.4–4.6 m/s [63]. Forces in the tensed condition were slightly greater than in the relaxed condition for the same impact velocity. Apparent initial stiffness were: 79 N/mm at 2.4 m/s for tensed, 57 N/mm at 2.4 m/s for relaxed and 250 N/mm at 4.6 m/s for tensed conditions. Peak skeletal deflections were 44–46 mm (16–17 % deflection of the rib cage).

The idealized force-deflection curves derived from these responses can be divided into a loading and an unloading phase (Fig. 13.6), with the loading phase having three components [64]. An initial rapid rise or apparent initial stiffness (A) is due in large part to the viscous properties of the thorax. The force plateau (B) is also due to a viscous response. At maximum deflection (C), the impactor and subject are moving at a common velocity and the forces are due to inertial forces caused by whole-body acceleration and the elastic forces due to tissue compression. The unloading portion of the curve (D), is due to unloading of the compressed tissues, and follows the elastic non-linear unloading of the thorax seen in quasi-static tests.

13.6.1.2 Belt Loading

Beginning in the early 1960s, three-point restraint systems have been installed in some vehicles. However, in the 1987 model year all vehicle manufacturers had to certify that their cars would meet the 50th percentile, adult male protection requirements in the 48 km/h frontal, rigid-barrier test specified in FMVSS 208. Manufacturers

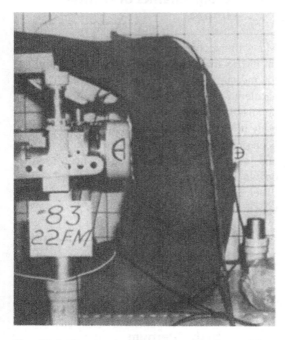

Fig. 13.4 Photograph showing a frontal pendulum impact applied to the sternum (From [55], reproduced by permission of the Stapp Association)

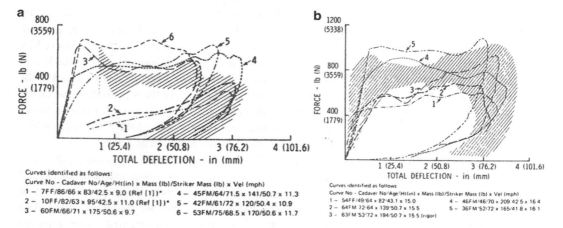

Fig. 13.5 (**a**) and (**b**) Force-deflection plots generated by Kroell et al. [56] for frontal impacts to the sternum. (**a**) 4.02–5.23 m/s impacts. *Shaded area* represents a corridor of three tests at 5.14 m/s with 19.3 kg mass. (**b**) 6.71–7.38 m/s impacts. *Shaded area* represents a corridor of seven tests at 6.7-7.4 m/s with 23.1 kg mass. The impactor had a 6 in. diameter flat, rigid surface and a mass of 19.3 or 23.1 kg (Reproduced by permission of the Stapp Association)

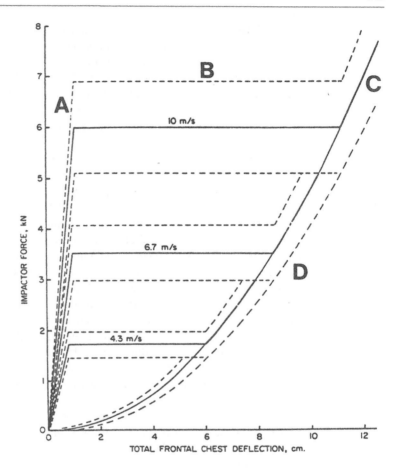

Fig. 13.6 Plots of pendulum impact force-deflection response of the thorax illustrating the various components of the curve. *A*. Rise. *B*. Plateau. *C*. Maximum deflection *D*. Unloading. See text for equations used to generate various portions of these plots. Based on pendulum impacts using 6 in. diameter rigid flat disc and 23.4 kg impact mass [162]

began installing 3-point restraints with force-limiting shoulder belts and frontal airbags for the driver and right front passenger [33].

Fayon et al. calculated the resultant normal force to the thorax from the geometry and tension forces at belt anchorages and determined that sternal thoracic stiffness was 166.4 N/mm [65]. Walfisch came up with slightly smaller values of 70–116 N/mm (mean 119.4 N/mm) [5]. L'Abbe et al. conducted static and dynamic tests and belt loading and consequent chest deflection in human volunteers [66]. The dynamic loads were up to 3,600 N and produced mid-sternal stiffness of 137 N/mm, right 7th rib stiffness of 123 N/mm and 200 N/mm at left clavicle. The Hybrid III dummy tended to produce stiffer responses than relaxed volunteers, with better agreement to tensed volunteers, particularly at mid-sternum. Backaitis and St.-Laurent in similar tests showed greatest deflections at the seventh rib in volunteers and least in the shoulder, while Hybrid III showed maximum deflections at the shoulder [67].

13.6.1.3 Two-Point Belt Loading Plus Knee Bolster

Cheng et al. performed eleven PMHS tests with a VW two-point belt and knee bolster in the right-front passenger position [68]. The belt crossed from the right shoulder to the left side. There were more rib fractures on the left in seven cadavers. The number of rib fractures ranged from 0 to 14, with seven subjects having seven or more rib fractures. Peak upper shoulder belt load was 5.1–7.6 kN for 48 km/h impacts with peak sled deceleration of 22 g. The peak belt loads ranged from 3.6 to 9.7 kN for four runs with peak sled deceleration of 35 g. The ratio of lower to upper belt peak loads was approximately 0.9.

Kent et al. tested 15 PMHS in single and double diagonal belts and distributed, and hub loading on the anterior thorax [69]. Subjects were positioned supine on a table and a hydraulic master–slave cylinder with a high-speed materials testing machine provided controlled chest deflection at rates similar to sled tests. Non-injurious tests were followed by a final injurious test (40 % chest deflection). The distributed loading condition generated the stiffest response (3.33 kN at 4.6 cm), followed by the double diagonal belt condition (3.18 kN at 4.6 cm), single diagonal belt (2.28 kN at 4.6 cm) and hub (1.14 kN at 4.6 cm). Corridors were developed to a deflection level of 20 % of the 50th percentile male's external chest depth. Kindig et al. studied of force-displacement and kinematic responses under a highly-localized loading condition using three PMHS (ages 44, 61, and 63 years) ribcages [25]. In general, bilateral loading produced an approximately symmetric deformation pattern, while unilateral loading resulted in approximately twice the resultant deformation on the ipsilateral side compared to the contralateral side.

Forman et al. [70] investigated a 3-point restraint system in the rear seat with a force-limiting (FL) and pretensioning (PT) shoulder belt in a series of 48 km/h frontal impact sled tests (20 g, 80 ms sled acceleration pulse) performed with PMHS [70]. The results of these tests were compared to matched PMHS tests, published in 2008, performed in the same environment with a standard 3-point belt restraint. The FL+PT restraint system resulted in significant ($p<0.05$) decreases in peak shoulder belt tension (average ± standard deviation: 4.4 ± 0.13 kN with the FL+PT belt, 7.8 ± 0.6 kN with the standard belt) and 3 ms-resultant, mid-spine acceleration (FL+PT: 34 ± 3.8 g; standard belt: 44 ± 1.4 g). The FL+PT tests also produced more forward torso rotation caused by decreased forward excursion of the pelvis and increased payout out of the shoulder belt by the force-limiter.

13.6.1.4 Quasi-Static Tests

As three point belts and air bags are more frequently used, lower rate loading will become more important in frontal impact. The distributed loading to the ribs due to the airbag, and rib and clavicle loading due to the shoulder belt make the biomechanical response of the ribs and clavicle (in addition to the sternum) increasingly important. Thus, in addition to the dynamic pendulum data from sternal impacts, quasi-static chest loading data is also important. These data are described below.

Stalnaker et al. loaded the sternum of volunteers and PMHS with a 153-mm diameter rigid plate with the subject's back against a rigid wall [60]. Stiffness averaged 12.2 N/mm for unembalmed PMHS, 40.2 N/mm for relaxed and 114 N/mm tensed volunteers. Lobdell et al. showed much less stiffness under similar loading conditions: 7 N/mm for relaxed volunteers and 23.7 N/mm for the tensed subject [61].

Tsitlik et al. [71] measured force versus deflection due to loading the sternum of eleven patients with a 48×64 mm rubber thumper [71]. The average stiffness was 9.1 N/mm (range: 5.25–15.9 N/mm) for deflections of 30–61 mm. At WSU, Cavanaugh et al. loaded the sternum and rib cage of three unembalmed PMHS with 25-mm strokes at quasi-static loading rates (described in [72]). The thorax stiffness were as follows: upper and mid-sternum, 8.6–12.3 N/mm; lower sternum, 5.7–11.4 N/mm; 2nd rib, 5.6–7.3 N/mm; 5th rib, 5.1–8.4 N/mm and 7th rib, 3.4–5.2 N/mm. Weisfeldt loaded the chest of volunteers at 60 Hz with a 22×64 mm pad [73]. The mean stiffness was 6.3 N/mm. On one subject loading the chest with a 153-mm diameter pad resulted in a stiffness of 21 N/mm. Melvin et al. after reviewing the quasi–static load–deflection literature, concluded that up to deflections of 41 mm the thorax has an approximate linear stiffness of 26.2 N/mm and for deflections greater than 76 mm, the linear stiffness is 120 N/mm [74]. Melvin also modeled these force–deflection relationships with a quadratic equation:

$$F = Kd^2$$

where $K = 47.6$ N/mm^2.

Fayon et al. demonstrated stiffness of 17.5–26.3 N/mm due sternal deflections up to 25 mm for static belt loading to supine volunteers [65]. This compared to 8.8–17.5 N/mm for disc loading up to 38 mm. At the 2nd and 9th ribs the estimated stiffness were 17.5–35 N/mm and

8.8–17.5 N/mm. Melvin and Weber estimated an apparent stiffness from the static belt tests of L'Abbe et al. [64, 66] (1982). These results were 67.5 N/mm at mid-sternum, 40 N/mm at right 7th rib, and 95 N/mm at left clavicle for deflections up to 10 mm and estimated normal forces up to 667 N. Melvin has suggested that these data may be significantly higher than that of Fayon et al. because spinal curvature may have reduced stiffness in the latter study. Arbogast et al. compared thoracic force-deflections measured during adult cardiopulmonary resuscitation (CPR) to that measured during hub-based loading of adult PMHS [75]. A load cell and accelerometer was integrated into a monitor-defibrillator to measure chest compression and applied force during live human CPR. Chest response during CPR demonstrated more hysteresis than the PMHS, indicating the viscoelastic nature of the thorax is more pronounced when tissues are naturally perfused and substantial tissue autolysis has not begun.

13.6.2 Injury Tolerance of the Thorax in Frontal Impact

Much of the early work on frontal impact tolerance was reviewed by Mertz and Kroell [76]. Some of this data and later data are summarized below. These data have been used in the development of the Hybrid III frontal impact dummy [77].

13.6.2.1 Acceleration Criteria

Human tolerance for severe chest injury is often stated as the peak spinal acceleration (sustained for 3 ms or longer) not to exceed 60 g in a frontal crash. The Hybrid III dummy is used for assessment of frontal impact crashworthiness per Federal Motor Vehicle Safety Standard 208 (Code of Federal Regulations, Title 49, Part 571.208, 1986). Early acceleration tolerance data was obtained by Stapp, who demonstrated the human tolerance to rocket sled acceleration when belt restraints were worn [78, 79]. Thoracic accelerations up to 40 g for 100 ms or less were tolerated. In one subject a maximum of 45 g was tolerated which resulted in a calculated pressure of 252 kPa under the harness. Peaks of 30 G reached at a rate of 1,000 g/s were not tolerated.

Eiband in an analysis of the Stapp data showed that the acceleration tolerance decreased as duration of exposure increased [80]. Spinal acceleration is a general indicator of the overall severity of whole body impact, but is sensitive to changing impact conditions [81]. Mertz and Gadd studied the tolerance and response of a 40 year old male stunt man who took 16 dives from heights of 27–57 ft onto a thick mattress [82]. For each dive the stunt man executed three quarters of a turn and landed supine on the mattress. In ten dives chest acceleration was measured. The authors concluded that 50 g chest acceleration for a pulse less than 100 ms was within voluntary range of healthy adult males. A 60-g acceleration with duration less than 100 ms was recommended as the tolerance level until other the availability of data. This finding was used in the development of the FMVSS 208 chest injury criterion.

13.6.2.2 Force Criteria

Bierman et al. tested young male volunteers with a drop-weight device that loaded a lap/double-shoulder belt harness (490 cm^2 in area) [83]. Painful reactions and minor injury occurred when loads exceeded 8.9 kN. When load area was increased to 1,006 cm^2, loads of 8.0–13.3 kN were sustained without injury. Patrick et al. performed a series of sled tests with embalmed PMHS to simulate the response of an unrestrained occupant [84]. The head, chest and knees impacted padded load cells. These studies were used as the basis for the design of the energy absorbing steering column. Gadd and Patrick, and Patrick et al. (1969) used a prototype energy absorbing steering column in unrestrained cadaveric sled tests [85]. A 3.3 kN hub load to the sternum and 8.8 kN distributed load to the shoulders and chest resulted in only minor trauma in centered impacts. The reactive force is due to the inertial, elastic and viscous components of the torso.

13.6.2.3 Energy Criterion

Eppinger and Marcus examined the results of 82 laboratory impact tests and concluded that severity of injury to the thorax was proportional to the amount of specific energy that the thorax must absorb [86]. The severity of injury was found to be inversely proportional to the impacted area and the duration of time over which the energy was transferred.

13.6.2.4 Compression Criteria

Kroell et al. analyzed a large number of blunt thoracic impact experiments (Fig. 13.4) and determined that chest compression correlated well with AIS (r=0.730) while maximum plateau force did not (r=0.524) [55, 56]. The linear equation relating AIS to compression was:

$$AIS = -3.78 + 19.56 C$$

where C was chest deformation divided by chest depth, resulting in C=0.3 (30 %) for AIS 2 and 0.4 (40 %) for AIS 4. Thus for the 230 mm chest depth of the 50th percentile male, 30 % compression or 69 mm deflection predicts AIS 2 and 40 % compression or 92 mm deflection predicts AIS 4.

The integral of spinal acceleration which is a measure of the velocity of deformation correlated almost as well as compression to injury severity [59]. In 5–7 m/s sternal impacts to PMHS, compressions >20 % regularly produced rib fractures. Compressions of 40 % produced flail chest. Neathery et al. extrapolated the 40 % tolerance for severe injury to the Hybrid III frontal impact dummy to obtain maximum allowable compression of 75 mm [62]. Viano demonstrated that severe injury to internal organs occurred at an average maximum compression (C_{max}) of 40 % and recommended C_{max} of 32 % to maintain enough rib cage stability to protect internal organs [87]. The Federal Motor Vehicle Standards 208 (Code of Federal Regulations, 571.208), allowed a maximum 76 mm chest deflection in the 50th percentile Hybrid III dummy in frontal impact crashworthiness testing of automobiles. This was reduced to 63 mm in the FMVSS 208 standard upgrade, commencing in model year 2003, and including the entire vehicle fleet by model year 2006. The upgrade also included the 5th percentile Hybrid III dummy with a maximum allowable chest deflection of 52 mm [88].

13.6.2.5 Viscous Criterion

Soft tissue injury is compression and rate dependent [81]. Lau and Viano in impact studies over the liver of anesthetized rabbits, found that when maximum compression (C_{max}) was held to 16 %, liver injury increased as impact velocity increased from 5 to 20 m/s [89]. In frontal thoracic impacts to anesthetized rabbits at 5, 10 and 18 m/s, severity of lung injury was found to increase with C_{max} at each level of velocity [90]. The alveolar region was more sensitive to the rate of loading than regions of vascular junctions. Data from 123 frontal impacts to anesthetized rabbits were used to define the viscous tolerance [91]. This lead to the development of the viscous criterion: VC_{max}, the maximum product of velocity of deformation and compression is an effective predictor of injury risk, and is a measure of the energy dissipated by the viscous elements of the thorax [92].

Kroell et al. verified the validity of the viscous criterion in blunt thoracic frontal impacts to anesthetized swine [30, 93]. In these studies, 23 swine (53.3 kg average mass) were impacted at 15 and 30 m/s with a 4.9 kg striker mass with a 150 mm diameter striker plate [30]. According to logistic analysis, VC_{max} and $V_{max}C_{max}$ were good predictors of the probability of heart rupture and thoracic MAIS>3, while C_{max} was not. Lau and Viano concluded that the viscous criterion is the best indicator for soft tissue injury to many body regions for velocities of deformation of 3–30 m/s [81]. The velocity of impact of automobile occupants to various parts of the automobile interior is in this 3–30 m/s range, which is intermediate to the high velocity pressure waves of a pure blast (in which injury occurs with little compression) and the pure crushing injuries of quasi-static loading (in which injury is due to compression alone). In an analysis of the 39 unembalmed PMHS sternal impacts (average age 62 years) performed by Kroell and others, VC_{max} of 1.3 m/s was the value for 50 % probability of thoracic AIS>3, and VC_{max} of 1.0 m/s for a 25 % probability of AIS>3, based on Probit analysis [81, 92].

13.6.2.6 Combined Compression and Acceleration Criterion

In 1998 the NHTSA proposed Combined Thoracic Index (CTI) as an injury criterion for frontal impact that combined chest compression and acceleration responses, with the aim of addressing both air bag and belt loading [94].

$$CTI = A_{max} / A_{int} + D_{max} / D_{int}$$

Where, A_{max} = Maximum observed acceleration
A_{int} = Maximum allowable intercept value of acceleration
D_{max} = Maximum observed deflection
D_{int} = Maximum allowable intercept value of deflection

The principles of CTI are explained by the differences in loading of the thorax by belt versus bag systems. For a given load, a belt system would apply greater pressure along its contact area than an air bag system, which has a larger contact area. The chest is more vulnerable to injury under the more concentrated belt loading. With the combined belt/bag system, the predominant loading could range from stiff belt/soft bag which would produce predominantly a line load, to soft belt/stiff bag which would produce a more distributed load. CTI was proposed as a criterion that reflects both the conditions and the various combinations in between. Peak chest acceleration is a measure of the magnitude of total forces applied to the torso, in proportion to the mass of the torso. Chest deflection is an indication of the belt loading contributing to the restraint system effect. The greater the deflection per unit of acceleration, the more the relative contribution of the belt system [94]. Figure 13.7a, b show the cross plot of chest deflection to chest accelerations for different AIS levels and the proposed regulation limits for CTI.

CTI was developed based on 71 PMHS tests reported by Kuppa et al. [95]. Five different restraint systems were used: 3-point belt, 2-point belt/knee bolster, 3-point belt/air bag, air bag/knee bolster and air bag/lap belt. Impact velocities ranged from 23 to 57 km/h. Peak g's in the deceleration profiles ranged from 15 to 25 g's. Instrumentation included tri-axial accelerometers at T1 and chestbands wrapped around the chest at the location of the 4th and 8th ribs. The 3 ms clip T1 resultant acceleration, maximum chest deflection (D_{max}), V_{max} and VC_{max} were analyzed statistically as injury functions using logistic regression analysis of the probability of thoracic AIS of 3 or greater. Univariate and multivariate analyses of the linear combination of these responses were performed. The combination with the best predictive value was $-6.43 + 0.076As + 13.68D_{max}$. The probability of AIS 3 or greater injury using this function is shown in Fig. 13.8. CTI was not incorporated into FMVSS 208 after review and comment to the Notice of Proposed Rulemaking (NPRM). However, stricter criteria for chest deflection were incorporated into FMVSS 208 commencing in model year 2003 as described above under Compression Criteria [88].

13.6.3 Biomechanics of Belt Loading and Combined Restraint Loading

13.6.3.1 Three-Point and Force-Limiting Belt Loading

In a Canadian study of injury to 121 belted occupants, Dalmotas found that shoulder/chest injuries constituted 23 % of AIS 3 and greater injuries [96]. For those drivers who did not contact the steering assembly, injuries were skeletal fractures of clavicle, sternum and ribs in the belt line. No intra-thoracic or abdominal injuries were attributed to belt loading. Patrick and Anderson reported injury results in crash investigations of drivers wearing three-point belts in Volvos [97]. Fourteen of 169 (8.3 %) occupants had rib fractures. Barrier equivalent velocity (BEV) ranged from 1 to 24 m/s. At 0 to 4 m/s, 29 of 32 occupants had no injury. There were 14 rib cage injuries in the 1–24 m/s BEV. Schmidt et al. reported thoracic injury data in a test series with 49 fresh PMHS [98]. Tests were conducted at a crash velocity of 50 km/h and a stopping distance of about 40 cm. Three-point retractor belts were used on 30 and two-point belts plus knee-bar were used on 19 PMHS. Age range was 12–82 years.

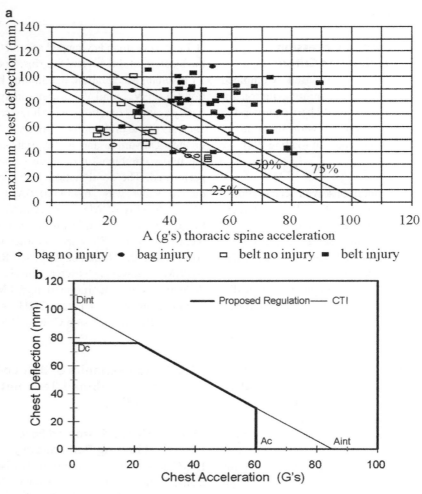

Fig. 13.7 (a) Lines of equal probability of AIS 3 or greater chest injury using the linear combination of maximum deflection and spinal acceleration. (b) Injury criteria for mid-sized adult male Hybrid III dummy (From [94])

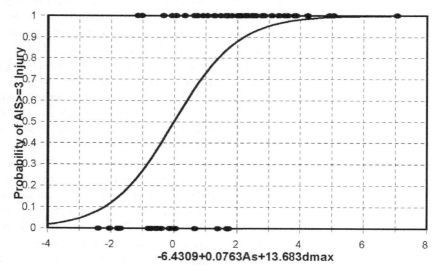

Fig. 13.8 The probability of chest injury in frontal impact using the linear combination of d_{max} and A as a risk factor (From [94])

The fracture patterns followed the belt line and there was an approximate linear relationship between the number of rib fractures and age, with more than 20 fractures occurring in several cases. Injuries to internal organs of chest and abdomen were also documented.

Horsch et al. studied the APR and Volvo real world crash studies relating shoulder belt loading to chest injuries [99]. The analysis of the data indicated that belt loading of the Hybrid III dummy chest resulting in 40 mm chest compression represented about a 25 % risk of AIS 3 or greater thoracic injury. Injury due to belt loading appeared to be driven by chest compression rather than VC_{max}. In APR and Volvo test data rib fractures increased with age but more so in the APR data base. Crashes severe enough to result in more than 40 mm Hybrid III dummy chest compression were infrequent. Cesari and Bouquet used a diagonal shoulder belt to dynamically load the chest of unembalmed PMHS and dummies [100]. Using two chestbands and an array of linear displacement transducers, the authors determined that the dummy thorax is almost twice as stiff as that of the cadaver. The human thorax had complex deformations with maximum occurring away from the sternum at the lower region of the rib cage.

In 1995 the first generation programmed restraint system (PRS) was introduced in France with seat belt force threshold of 6 kN. Thirty-seven frontal accident cases involving this restraint system were investigated by Bendjellal et al. [101] Analysis of this data showed only two cases were related to AIS 3 level injury but the authors concluded that reducing the shoulder belt force to 4 kN was a necessary further step and required redesign of the air bag. The new system was called PRS II. The PRS II uses a pyrotechnic buckle pretensioner with a 4 kN belt load limiter and air bag deployment to the sides and from the top to bottom in order to reduce risk in out-of-position situations. Foret-Bruno et al. sought to establish a thoracic injury risk function for belted occupants as a function of age and loads applied to the shoulder level [102]. Fifty-percent probability of AIS 3+ occurred at a belt load of 6.9 kN. The authors concluded that a belt load limit of 4 kN combined with a specially designed air bag could protect 95 % of those involved in frontal impacts from AIS 3+ chest injuries. Foret-Bruno et al. compared thoracic injury risks for two occupant populations: belted occupants involved in accidents in which the vehicle was not equipped with a load limiter (378 cases with pyrotechnic pretensioners), and belted occupants involved in accidents in which the vehicles were equipped with 4 or 6 kN load limiters and pyrotechnic pretensioners (347 cases) [103]. A 4 kN load limit resulted in a reduction of thoracic injury risk for all AIS levels, compared to other belt configurations. There was 50–60 % reduction for AIS 2+, 75–85 % for AIS 3+ and a complete absence of AIS 4+ with a 4 kN load limiter.

In a PMHS study Petitjean et al. reported the same trends as real-world accident studies. Some of the criteria proposed in the literature did not show a better protection of the 4 kN load limiter belt with airbag restraint, in particular thoracic deflection maxima for THOR and Hybrid III dummies [104]. More thoracic deflections such as in the THOR dummy may allow for more accurate analysis of the loading pattern and therefore of injury risk. Rouhana et al. investigated the biomechanical behavior of 4-point seat belt systems through MADYMO modeling, dummy tests and post mortem human subject tests [105]. A 3-point belt with an extra shoulder belt that crisscrossed the chest (X4) appeared to add constraint to the torso and increased chest deflection and injury risk. Harness style shoulder belts (V4) loaded the body in a different biomechanical manner than 3-point and X4 belts. The V4 belt appeared to shift load to the clavicles and pelvis and to reduce traction of the shoulder belt across the chest, resulting in a reduction in chest deflection by a factor of two. This was associated with a 5- to 500-fold reduction in thoracic injury risk, depending on whether the 4-point belts applied concentrated or distributed loading. In four of six PMHS restrained by V4 belts during 40 km/h sled tests, chest compression was zero or negative and rib fractures were nearly eliminated. Submarining was not observed in any PMHS. The authors pointed out there were still issues to be resolved before 4-point belts can be considered for production vehicles. These issues included potential effects on hard and soft neck tissues, interaction with inboard shoulder belts in far-side impacts and potential effects on the fetus of latch/buckle junctions at the centerline of pregnant occupants.

Shaw et al. evaluated the response of restrained PMHS in 40 km/h frontal sled tests [106]. Eight male PMHS were restrained on a rigid planar seat by a custom 3-point shoulder and lap belt. The seat pan was horizontal, the open cable-design seat back was intended to support the back of the torso and allow optical motions analyses, knees of the subjects were initially in contact with the aluminum knee-bolster, and no airbag was used. The snug lap belt, stiff and channeled knee bolster, and stiff footrest minimized pelvic and lower extremity kinematics while allowing motion of the torso due to belt interaction. A video motion tracking system measured three-dimensional trajectories of multiple skeletal sites on the torso allowing quantification of ribcage deformation. Anterior and superior displacements of the lower ribcage may have contributed to sternal fractures occurring early in the event, at displacement levels below those typically considered injurious, suggesting that fracture risk is not fully described by traditional definitions of chest deformation.

13.6.3.2 Combined Restraint Systems

As described above, there has been a progression of changes in the design of restraint systems for front seat ocupants. Peititjean et al. pointed out that after initial limitation of forces in the shoulder belt to 6 kN, these forces are now usually limited to 4 kN, with airbags intentionally designed to absorb the surplus of energy [107]. Evaluation of the performance of force-limiting belts with airbags can be limited when using the Hybrid III dummy and applied to all restraint systems. Peititjean et al. proposed a new criterion for combined restraint configurations (belt, airbag only or airbag and belt) based on the measurement of the shoulder belt forces and of the central deflection and directly applicable to the Hybrid III dummy [107]. The use of shoulder belt forces allowed the separation of the belt and airbag contributions to chest deflection. A weighted chest injury criterion was developed, taking into account the different risks associated with a belt and an airbag for the same deflection.

Kent et al. reported that an inflatable air-belt generated lower head, neck, and thoracic injury responses and PMHS trauma than non-inflatable rear-seat restraints using standard 3-point belt and also a pre-tensioned shoulder belt with a progressive load limiter [108].

13.6.3.3 Highly Restrained Occupants: Analysis of Indy Car Crashes

Melvin et al. described the results of the GM Motorsports Safety Technology Research Program to investigate Indianapolis-type race car crashes [109]. A total of 202 cases had peak decelerations above 20 g. The mean peak rigid body chassis deceleration was 53 g. The mean total velocity change was 45.3 km/h. There were 143 cases of side impact, with 41 cases above 60 g and 7 cases above 100 g. There was no incidence of serious torso injury in the 202 cases. The data has called into question the use of spinal acceleration alone in injury assessment, particularly in side impact. Factors contributing to occupant protection included lack of intrusion, uniform support of the body, tight wide double shoulder belts, and load paths from the seat/chassis through the shoulder and pelvis which may bypass the chest.

13.6.3.4 Airbag Loading and Air Bag Versus Belt Loading

The concern for injury to the child and out-of-position occupant is not new [110, 111]. Situations of concern include the shorter stature older subject who sits close to the steering wheel, the subject leaning over during an impact or the subject brought closer to the steering wheel during a relatively minor first impact who is then impacted by the airbag during a major second impact. Bag slap from high velocity impacts (15 m/s) may injure the lungs and heart. It is hypothesized that these injuries could be due to stress waves. At WSU Cheng et al. conducted a series of frontal impact sled tests with a pressurized air bag [68]. Peak bag pressure was

93–139 kPa. Chest AIS was 0–2, but overall MAIS was 2–6 due to cervical spine injury.

A detailed study of the effect of airbag deployment on the Hybrid III dummy was performed for the bag opposite head, neck and thorax by Horsch et al. [112]. Highest response amplitudes occurred with the body part directly against the air bag module. Additional tests were performed with the module opposite the sternum of anesthetized swine. Severe to critical injuries were seen in all tests where the swine's torso covered the module. Injuries included heart contusions and perforations. Injury was related to internal bag pressure. As long as bag volume was greater than gas volume generated, there would be minimal pressure and thus minimal force to the subject. There are two main times when available volume is less than generated volume: "punch out" during initial pressurization of the module and "membrane-force" phases when the subject is in the path of the bag and the force is due to bag pressure on the subject. This includes tension forces from bag wrap-around. Hybrid III dummy test conditions in which $VC_{max} < 1$ m/s, resulted in no chest injury to the swine, and Hybrid III dummy dummy tests in which $VC_{max} > 1$ m/s resulted in severe chest injury to swine. Chest compression velocities were as high as 14 m/s.

Fourteen unembalmed PMHS were subjected to deceleration sled tests at velocities of 9–13 m/s at the Medical College of Wisconsin [113]. When three-point belts were used, there was considerable local compression compared to cases of knee bolster and lap belt use. Morgan et al. analyzed these and other data from 63 cadaver frontal impact sled tests run at 23–50 km/h at three different test centers [114]. Restraint systems included 3-point belt, airbag plus lap belt, airbag plus knee bolster, airbag plus 3-point belt and 2-point belt plus bolster. For the same level of mechanical performance, belt restraint systems had a higher associated injury rate than air bag restraint systems. The authors pointed out that this would introduce an injury performance bias. The bias would either under-represent the performance of airbag-like restraints or over-represent the performance of belt-like restraint systems. A dichotomous process was introduced which determines whether the combined restraint system was more belt-like or bag-like. Appropriate injury predictive criteria were then applied. The best separators of injury probability were a linear combination of normalized central chest compression, 3-ms clip resultant T1 spine acceleration and age of the subject [94, 114].

Kallieris et al. performed frontal impact tests at 48 km/h using PMHS instrumented with accelerometers and two chest bands [115]. Tests with belts of 6 % and 16 % elongation showed localized deformation of the chest along the shoulder belt path. With airbag only, forces were distributed more evenly. The goal was then to obtain the chest injury mitigation of the airbag and the overall restraint of the belt. Using a belt with a force limiter of 4 kN, Hybrid III dummy tests showed more bag-like compression of the chest with the belt effect only slightly pronounced. In comparable cadaver tests, only AIS 1 injuries were observed in the thorax in the 60–65 year age range.

Lebarbéet et al. [116] performed tests on nine PMHS on a static test bench using a driver airbag module described by Petit et al. [117]. The steering wheel was replaced by a plate to increase the loading generated by the airbag. Three loading configurations were performed: membrane only, punch-out only, and both types combined. Both pure punch-out and pure membrane loading resulted in thoracic injuries. However, the rib fracture locations appeared to differ from one type of loading to the other. Also, for the same initial distance between the airbag module and thorax, the injuries were more severe in the combined effect tests than in the pure punch-out or pure membrane conditions.

13.6.4 Biomechanics of Lateral Impact

Side impact is a most serious automotive injury problem, second only to frontal impact in terms of injury and fatality in the United States. Each year, about 8,000 automobile occupants are killed and thousands more injured due to side impact. In a review of fatality data by Viano et al. it was found that 31.8 % of passenger car fatalities occur in crashes with the principal direction of force lateral to the vehicle [7]. Of those, 2/3 of the fatalities are due to multi-vehicle crashes and the remainder involved the impact of a single vehicle with a fixed object. Multi-vehicle crashes frequently involve the older victim. Per Viano et al., 76 % of side impact victims were 50 years or older and 28 % were 70 years or older; and in single vehicle frontal crashes, 26 % of fatalities were over age 50 years and 8 % over age 70 years [7].

In a study of CIREN side impact crashed by Tencer et al. [118], in side impact crashes with the lower border of the door leading, 81 % of occupants sustained pelvic injury, 42 % suffered rib fractures, and the rate of organ injury was 0.84. With the upper border of the door leading, 46 % of occupants sustained pelvic injury, 71 % sustained rib fracture, and the rate of organ injuries increased to 1.13 [118]. The differences in the groups with respect to pelvic injury were significant at p=0.01, rib fracture, p=0.10, and organ injury, p=0.001. MADYMO modeling indicated that when the bumper of the striking vehicle overroad the door beam, the upper part of the door led the intrusion into the passenger compartment. With the upper border of the door leading, more severe chest and organ injuries resulted [118].

13.6.4.1 Drop Tests
Stalnaker et al. analyzed force-deflection data of the struck-side half-thorax in a series of 15 lateral drop tests from a height of 1–3 m onto an unpadded or padded force plate using unembalmed PMHS [60]. A corridor of normalized force versus relative deflection (%) of the half-thorax was formulated. It was proposed that this be a corridor for the development of a side impact dummy. Compression of 35 % was the value for AIS of 3 or less. Tarriere et al. analyzed this and additional data (sixteen 1–2 m drop tests in which force and deflection were measured and nine 3 m drop tests in which force was measured but not deflection) [119] Maximum normalized force for AIS 0 was 7.40 kN and for AIS 3 10.20 kN. Compression of approximately 30 % of the whole chest width was the tolerance for AIS 3 or less. Compression of 35 % for the struck side half-thorax was the tolerance for AIS 3 or less. Maximum 3-ms lateral acceleration of the T4 vertebra did not show a close relation to the number of rib fractures, but the maximum 3-ms acceleration averaged 49 g in nine subjects with AIS 3 and 60 g in 11 subjects with AIS 4 and 5.

Sacreste et al. analyzed bone condition factor (BCF) in 62 cadavers in an attempt to reduce the scatter in injury severity in cadaveric side impact tests [120]. Using rib samples from these cadavers, eight parameters were measured: ash mass/total mass, ash mass per unit length, shear strength, bending strength, initial slope of force-deflection curve, shear energy, maximum bending stress and Young's modulus. These data were examined in a factorial analysis and BCF, an indicator of bone resistance which integrates these factors, was formulated using each of these parameters. The linear relationship of BCF to subject age had a 0.60 correlation coefficient.

13.6.4.2 Sled Tests
In earlier side impact research an extensive series of side impact sled tests were sponsored by the NHTSA and performed at the University of Heidelberg. Unembalmed PMHS were placed on a seat of low coefficient of friction and 0.6–0.9 m from the impacted wall. The sled was accelerated slowly and suddenly decelerated, so that the PMHS slid across the seat at the same speed as the sled and impacted the padded or unpadded sidewall. In the earliest tests the subjects were instrumented with the 12 accelerometer array developed by Robbins et al. and Eppinger et al. [22, 121]. This array measures accelerations at the ribs, sternum, and thoracic vertebrae. Forces were not measured in the earlier tests. In the first series of these tests, Kallieris et al. concluded that acceleration responses were identical but injury varied greatly even for the same response, leading to the conclusion that the injury function must have physical descriptors of the population as well as kinematic parameters [122]. Eppinger et al. analyzed 30 side impact cadaver tests, 27 of them being the Heidelberg sled tests [123]. The 12 accelerometer array was used and the analysis concentrated on the lateral responses of the fourth rib and the twelfth thoracic vertebrae. Injury could be successfully partitioned by using the variables of sub-

ject age vs. struck-side fourth rib acceleration or rib relative velocity. Marcus et al. analyzed data from eleven Heidelberg sled tests, most of which included force measurements on the impacted wall [124]. A normalized thoracic AIS of AIS − 0.025 (AGE-45) was proposed to normalize the injury to a 45-year-old subject. Eppinger et al. further analyzed the Heidelberg sled data and proposed the Thoracic Trauma Index (TTI), which summed an age factor and the average of fourth struck-side rib and lateral T12-y accelerations scaled for mass [125]. The equation was as follows:

$$TTI = 1.4 \times AGE + 0.5 \times (RIB_y + T12y) \times (MASS/MASSstd)$$

Morgan et al. analyzed Heidelberg data and ForschungsvereinigungAutomobiltechnik (FAT) side impact test series in which a moving deformable barrier struck an Opel Kadett car body in which a cadaver was seated [126]. The authors proposed a revised TTI which utilized the maximum of either 4th or 8th struck-side rib acceleration for RIB_y in the equation above. Figure 13.9a is a plot of maximum hard thorax AIS vs. TTI for left-sided impact tests and Fig. 13.9b a plot of probability of AIS 4 versus TTI for left- and right-sided impacts. The "hard thorax" included those structures in the upper abdomen bounded by the lower portion of the rib cage [123]. Thus, the organs of the hard thorax included the liver and spleen. The side impact dummy (SID) was developed to measure upper and lower lateral rib accelerations and thoracic spine accelerations so that TTI can be measured to assess side impact crashworthiness [126]. The SID dummy was used as the test-measuring device in assessing side impact crashworthiness of automobiles in FMVSS 214 (Code of Federal Regulations, Title 49, Part 571.214, 1990).

Cavanaugh et al. performed 12 sled tests with unembalmed PMHS in which the side-wall was divided into shoulder, thoracic, abdominal and pelvic beams (Fig. 13.10), and found that compression and velocity times compression (viscous criterion) were more predictive of thoracic injury than acceleration and force-based criteria [50]. Five additional tests were performed and the data from 17 tests were published [51]. In the analysis of this data, a new injury criterion was developed and proposed, Average Spine Acceleration (ASA). ASA was obtained by integrating the T12y acceleration pulse to obtain velocity and taking the slope of this velocity curve between specified points (15 % and 85 % of peak spine velocity, Fig. 13.11). Using logistic regression

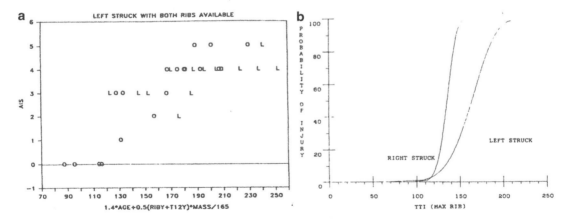

Fig. 13.9 (a) Plot of maximum AIS to the thorax versus the Thoracic Trauma Index (*TTI*) for left-sided impacts. O's are primarily from the Heidelberg sled test series and L's from the Forschungsvereinigung Automobiltechnik (*FAT*) test series in which a moving deformable barrier struck an Opel Kadett car body in which a cadaver subject was seated (From [126]). (b) Probability curve of AIS 4 or greater to the thorax versus TTI (From [126], reproduced by permission of the Stapp Association)

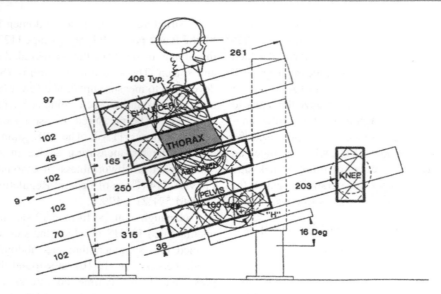

Fig. 13.10 Diagram of instrumented side-wall in WSU/CDC side-impact sled test series (From [51], reproduced by permission of the Stapp Association)

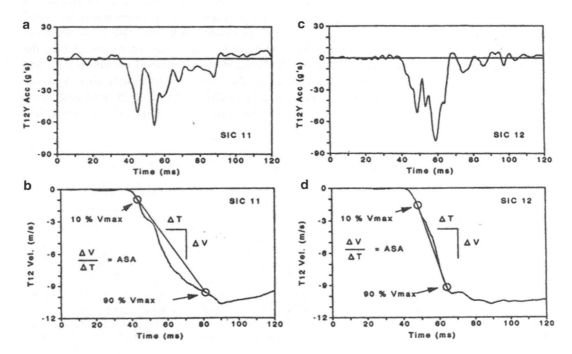

Fig. 13.11 T12-y acceleration and Average Spine Acceleration (*ASA*) for two cadaver tests, WSU/CDC SIC 11, using four inch thickness of soft paper honeycomb, and WSU/CDC SIC 12, using four inch thickness of stiff paper honeycomb. (**a**) T12-y acceleration for SIC 11. (**b**) T12-y velocity and ASA for SIC 11. (**c**) T12-y acceleration for SIC 12. (**d**) T12-y velocity and ASA for SIC 12. ASA for SIC 11 was 23.3 g's and maximum thoracic AIS was 2. ASA for SIC 12 was 46.3 g's and maximum thoracic AIS was 5 (From [51], reproduced by permission of the Stapp Association)

analysis, ASA was shown to have better predictive value than peak acceleration in the 17 sled test runs at WSU and in a combined data set of 58 WSU and NHTSA sled tests. In the WSU data set, ASA had better predictive value than any other function analyzed. In the combined data set ASA had almost identical predictive value to TTI in ascertaining the probability of AIS 4 or greater chest injuries. The efficacy of the viscous criterion was also borne out after analyzing additional five padded wall tests. In general, the padded tests in which VC_{max} was kept below 1 m/s resulted in AIS of 2 or less to the thorax and padded tests in which VC_{max} was > 1 m/s resulted in AIS of 4 or 5 to the thorax [51].

Cavanaugh et al. tested analyzed data from 37 SID side impact sled tests with rigid and padded walls with 14 different padding configurations [127]. SID ASA predicted that padding greater than 20 psi crush strength is harmful (ASA > 40 g). SID TTI predicted that padding greater than 20 psi crush strength is beneficial (TTI < 85 g's). In SID padded tests, ASA was able to discriminate the harmful effects of stiff padding compared to soft padding while SID TTI did not. This was largely attributed to the concentrated chest mass of SID, which resulted in the ability to crush stiff padding and produce low rib accelerations.

Irwin et al. performed three-dimensional film analysis of seven cadavers in WSU side impact sled tests [128]. The cadavers rotated slightly during impact and significantly during rebound. The study confirmed forward deflection of the sternum in response to a lateral impact. The non-impacted ribs deflected laterally inward with respect to the spine. Five of seven cadavers in this analysis experienced considerable damage to the thoracic skeleton. Displacements of the struck-side half chest at the T5 level were 75–110 mm. It was proposed that shoulder motion included winging of the scapula with rib deformation occurring without rib fracture and a scapula that did not wing during cases of multiple rib fracture at the scapula.

Pintar et al. reported the results of 26 sled tests conducted at the Medical College of Wisconsin (Fig. 13.12) and the NHTSA Vehicle Research and Test Center [129]. Twenty-four and 32 km/h tests were run in which the PMHS impacted a rigid wall, a padded wall or a wall with a 12-cm pelvic load plate offset. The padding was 10-cm of Ethafoam LC200 with a compressive stiffness of 103 kPa. Two or three chest bands were used on the rib cage to ascertain C_{max} and VC_{max}. TTI, C_{max}, VC_{max} and ASA were evaluated as injury criteria. Of these, TTI had the best predictive value

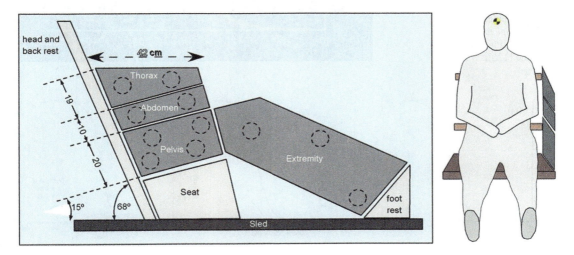

Fig. 13.12 Instrumented side wall in MCW side impact sled tests shown on the *left* and the orientation of the surrogate with respect to the wall (pure lateral loading) shown in the *right* (From Pintar et al. 1998)

in this test series (chi-square = 12.561, p = 0.0004). The product TTI*C_{max} had better predictive value (chi-square = 15.685, p = 0.0001) than either criterion individually. Figure 13.13 shows the good match of TTI between sled and full-scale vehicle with PMHS tests [126].

13.6.4.3 Impactor Tests

Viano performed a series of impactor tests with unembalmed PMHS at the WSU [7]. The pendulum mass was 23.4 kg and direction of impact 30° anterior to lateral. Logistic regression analysis was performed (Fig. 13.14). Peak compression and VC_{max} were found to be good predictors of thoracic injury, and better predictors than peak T8–y and T12–y accelerations. For maximum AIS of 4 or greater, peak impactor force was also a good injury function. The force-deflection and force-time corridors for 4.4, 6.5 and 9.5 m/s impacts were also generated.

The BioSID side impact dummy was developed by an SAE task force in response to the perceived need for a dummy that could measure thoracic compression and rate of compression and also the perceived need for a dummy more biofidelic than SID [130]. Viano et al. conducted side impact pendulum tests on Eurosid I and Biosid to assess biofidelity of the thorax, abdomen and pelvis and injury tolerance levels at impact speeds of 4.5, 6.7 and 9.4 m/s [131]. Test conditions duplicated those of

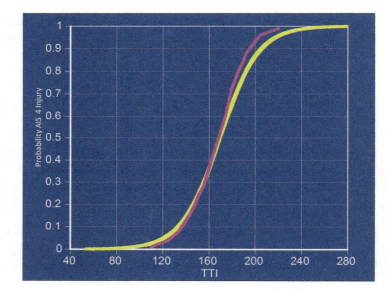

Fig. 13.13 Comparison of the Thoracic Trauma Index (*TTI*) for AIS 4 injury probability between sled (*yellow* curve) and full-scale vehicle with PMHS (*magenta* curve) tests (From Pintar et al. 1998 and [126])

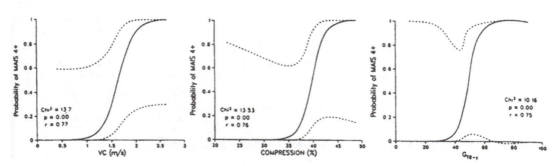

Fig. 13.14 Curve showing the probability of thoracic AIS 4 or greater as a function of VCmax (From [86], reproduced by permission of the Stapp Association). These data were derived from unembalmed cadaver impacts run at University of California, San Diego in the late 1960s and early 1970s (See Kroell et al. [55, 56] and Nahum et al. [57, 59] for details)

PMHS impacts, using a 23.4-kg impactor mass [7]. Overall Biosid had better biofidelity than Eurosid 1 for all three body regions. Both dummies showed better biofidelity in lower speed impacts. Eurosid I and BioSID injury tolerance levels were proposed. For the destructive tests, two test conditions were evaluated by performing high-energy lateral impacts, 23.4-kg at 12 m/s, with a pneumatic impactor. Only one destructive test was performed per PMHS with the arm placed at either 45° or parallel with the thorax. The highest average peak forces, peak rib deflections, and peak rib strains were observed when only the ribs were impacted and lowest when the shoulder was impacted. The study showed that in low-energy side impacts both the arm and shoulder reduce impactor force, rib deflection, and rib strain. In high-energy side impacts, the position of the arm has a considerable effect on both the total number and distribution of rib fractures. Higher average peak forces, peak rib deflections, and rib strains were observed when the arm was placed parallel with the thorax versus 45°.

13.6.4.4 The Side Impact Punch

Using a CAL3D simulation program, Deng showed that there are critical differences between free-flight impacts and velocity-pulse impacts [132]. Typical laboratory lateral impact studies are free-flight impacts represented by a moving mass into a stationary subject (pendulum impacts) or a moving subject into a stationary mass (sled tests). Car-to-car side impacts are velocity pulse impacts in which the impacting car imparts a velocity pulse to the struck door, which after traversing the space between struck door and occupant strikes the occupant with little decrease in door velocity because of the large amount of energy behind the impact. Deng's velocity pulse simulation showed that: (1) a stiffer side door structure can reduce injury, (2) increasing the spacing between occupant and door is beneficial, and (3) the use of padding on the inner door panel reduces occupant acceleration but increases occupant deformation. Thus, TTI predicts that padding on the inner door panel is beneficial in side-impact and C_{max} and VC_{max} predict that padding is harmful in side impact. This conclusion has been borne out in full-scale side impacts analyzed by Campbell et al. in which three inches of ARCEL 512™ padding on the inner door panel of a mid-size four-door passenger car decreased TTI in the SID and increased VC_{max} in the BIOSID [133].

This hypothesis was also investigated in a laboratory study at WSU [134]. In BioSID tests it was found that the addition of padding decreased rib accelerations. Chest compression increased with the introduction of stiff padding. In four unpadded cadaver tests the chest compression (C_{max}) was 27.4–36.4 %. In two tests with four inches of Dytherm (2 pound per cubic foot density), C_{max} was 52.1 and 59.7 %. C_{max}, stored energy criterion and peak T12-y correlated best to the number of rib fractures. Lau et al. pointed out that energy absorbing materials like padding can reduce peak accelerations but also prolong occupant contact in the side impact punch [135]. That padding can increase energy transfer to the dummy chest is reflected by higher chest deflection and lower TTI(d). A carefully chosen pad may reduce both TTI(d) and deflection.

13.6.4.5 Airbag Tests

Trosseille et al. performed 17 side airbag tests on PMHS at different severities and angles. The subjects were instrumented with accelerometers on the spine and strain gauges on the ribs [24]. They were loaded by an unfolded airbag at different distances in pure lateral or 30° forward. The airbag forces ranged from 1,680 N to 6,300 N and injuries resulted in up to 9 separated fractured ribs. Statistical analysis showed the effect of the loading angle on the injury outcome.

13.6.5 Biomechanics of Oblique Lateral impact

While the above discussed side impact studies represent advancements in our understanding of injuries, injury mechanisms, and injury metrics, focus was on pure lateral loading. Recent descriptive and epidemiological studies have indicated that oblique loading in side impacts occurs more often than pure lateral loadings and furthermore, injuries are more severe in real world crashes. Pintar et al. identified 53 cases (ΔV 15–58 km/h) in the CIREN database which demonstrated unique injury patterns: unilateral thoracic injuries with oblique side impacts and narrow object (pole/tree) crashes [136]. Unilateral chest traumas were hypothesized to be due to oblique loading transferred to the nearside occupant from door intrusion and resulting in antero-lateral chest loading instead of direct lateral loading through the arm or shoulder seen in pure lateral crashes. Upon examination of the NASS-CDS data for the years 2003–2006, it was found that pure lateral crashes, right or left side (3 or 9 o'clock position) accounted for fewer cases (raw and weighted) than oblique crashes. Impacts at 10, 20 and 30° forward from 90 and 270° were approximately 2.9, 2.5 and 1.7 times of pure lateral crashes [137]. The recent introduction of 75° pole test into the FMVSS-214 Standards induces antero-lateral oblique loading to the occupant [138]. Recognizing that oblique impacts were not examined in any laboratory experiment and or computational modeling, these field data suggested the need to examine oblique lateral crashes [139]. Oblique (anterolateral) chest loading was inferred from vehicle deformation images along with door intrusion into the occupant space. An oblique load to the chest results in a different injury mechanism to the rib cage due to difference in arm position and direct exposure of the rib cage to the load with little shoulder protection, and internal organs may sustain more severe exposure as the lungs and heart are aligned with the impact vector. The authors developed a classification scheme to assess maximum damage location with respect to the center of the wheelbase (Fig. 13.15), identified as class 1–4 impacts. Class-2 impacts produced almost an equal distribution of severe injuries between pelvis, chest, and head body regions and were attributed to induce oblique loading.

Pintar et al. conducted two full-scale mid-size sedan vehicle-to-pole crash tests at 32 km/h at in-house laboratories of the Medical College of Wisconsin: one with PMHS and the other with ES-2 surrogates [136]. Test designs were based on results and classifications from above field data analyses to induce maximum head and chest loads to nearside-positioned surrogates. The principal direction of force was 15° anterior from pure left lateral impact. The surrogates were instrumented with pyramid nine-accelerometer package to obtain head angular accelerations; T1,

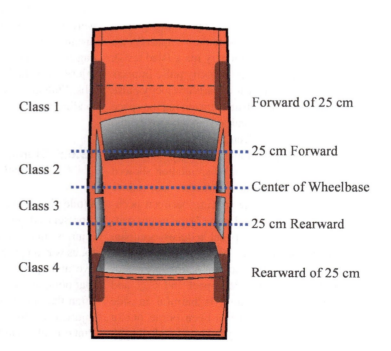

Fig. 13.15 Classification scheme to define location of maximum damage from real-world crashes (from [136])

T12 and sacrum accelerometers to obtain local kinematics; and chestband for obtaining chest deflection profiles [140]. The intruding door velocities in the front and rear were 5–10 m/s, compared to the 8.9 m/s impact velocity of the vehicle, implying that chest is subjected to oblique loading at higher rates, which may be a cause for increased propensity of severe trauma in a narrow object side impact compared to vehicle-to-vehicle side crashes. This finding paralleled literature: twice the risk of AIS = 3+ injuries in nearside crashes with narrow objects [141, 142]. Chest deflection profile showed antero-lateral loading of the thorax in PMHS. Both surrogates struck the door approximately at the same time. The selected crash configuration induced oblique loading based chest deflection profiles. A comparison of the response between the two surrogates is shown (Fig. 13.16). Similar evaluations have not been made between the PMHS and other

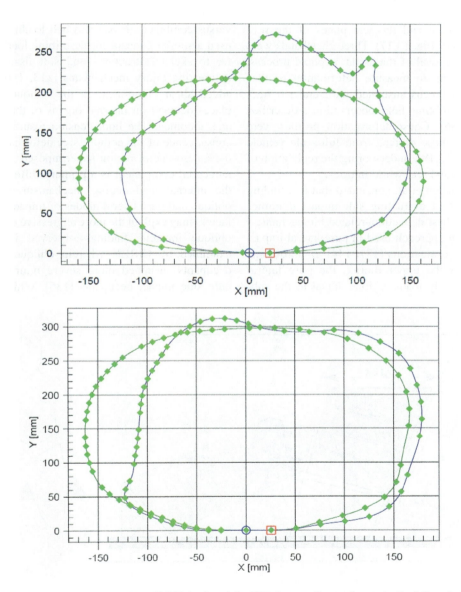

Fig. 13.16 Chestband contours of the PMHS (*top*) and the ES2 dummy (*botttom*) tests in the full-scale laboratory vehicle pole tests (From [136])

dummies including WorldSID mid-size and small female, and SID-IIs and with different class of vehicles to understand the full-scale vehicle test-induced biomechanics, including injuries and injury tolerances under oblique loading.

Unembalmed PMHS were placed on a bench seat fixed to the platform of a sled, configured with an impacting load-wall. The four-plate configuration consisted of the upper plate contacting the mid-thorax, middle plate the abdomen, lower plate the pelvis, and extremity plate the lower extremities. To simulate an oblique lateral impact, the abdominal and thoracic plates of the wall were angled (Fig. 13.17). Three chestbands were fixed at the level of the axilla, xyphoid process, and tenth rib to measure deformation profiles [143]. Specimens sustained unilateral rib fractures, resembling field observations, described above [136]. Chest deformation profiles were similar to those derived from full-scale vehicle tests (Fig. 13.18), underscoring the replication of these tests and real-world injuries.

Yoganandan et al. concluded that this finding may assist in evaluating side-impact dummies based on chest deflection–related biomechanical metrics, an approach used in pure lateral impact biofidelity evaluations [137, 144]. In an oblique impact, at the upper thorax, the pure lateral vector directly loads regions dorsal to the subclavian artery while an oblique vector applies impact forces to ventral arterial regions engaging the common carotid artery and brachio-cephalic vein. The former vector introduces postero-anterior load transfer to these tissues, in contrast to the antero-posterior load transfer by the oblique vector [145]. In other words, an obliquely oriented load vector induces antero-lateral compression of the rib cage on the struck side and the impact force is thus transferred via a combined shear and compression mechanism. Frontal impact-induced chest injuries with belt-only versus combined air bag and belt loadings have used a similar concept for describing load transfer to skeletal structures and soft tissues and delineating injury mechanisms [113, 146]. The added shear component in the oblique vector places demand on internal organs of the thorax and abdomen. The hoop tension resulting as a consequence of the compressive deformation on the antero-lateral region superimposed with the tangential component is a primary difference in the internal load-sharing mechanism between oblique and pure lateral modes of impact. These factors may explain the more aggressive nature of oblique impacts, findings observed in cases examined by CIREN wherein oblique impact occupants sustained more severe injuries than pure side impact occupants [136]. While these

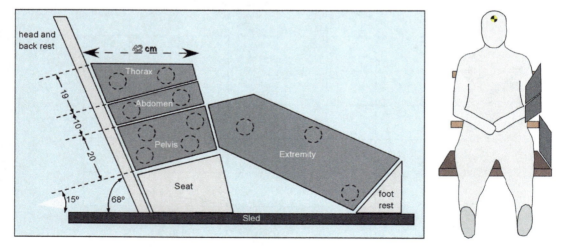

Fig. 13.17 Instrumented side wall in MCW oblique side impact sled tests shown on the *left* and the orientation of the surrogate with respect to the wall (obliquel loading) shown in the *right* (From [143])

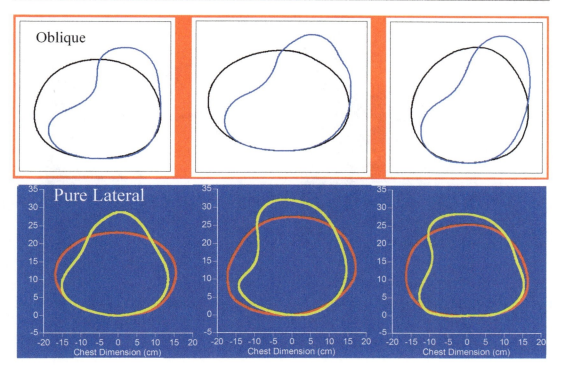

Fig. 13.18 Comparison of deformation contours at the top, middle and lower chest levels for oblique (*top row*) and pure lateral (*bottom row*) tests

studies offered a hypothesis for injuries depending on load vector, as the load-wall was non-specific to subject anthropometry, and modern dummies incorporated sensors to determine parameters such as deflections at different levels of the chest and abdomen, it was deemed necessary to redesign the load-wall to obtain such data. Hence, this early oblique side impact research laid a foundation to obtain region-specific force and deflections and injury evaluations.

The angled non-anthropometry-specific load-wall was redesigned to accommodate individual specimen demographics (stature and seating height), resulting in a modular, scalable and anthropometry-specific load-wall design [147, 148]. Five metal plates formed the STAPP load-wall: shoulder, thorax, abdomen, and superior (corresponding to the location of the ipsilateral iliac crest) and inferior pelvis (corresponding to the location of the ipsilateral greater trochanter) plates. All plates were modular to accommodate variations in PMHS anthropometry and their positions were adjustable (superior-inferior and left-to-right) to ensure contact of each plate with the intended specific body region. All five STAPP plates were rigidly attached to a vertical fixture, securely connected to the platform of the acceleration sled equipment. The vertical height and lateral positioning of the fixture were adjusted based on the surrogate anthropometry (Fig. 13.19). The shoulder, thoracic and abdominal plates were oriented such that the left antero-lateral portion of the surrogate chest contacted the oblique load-wall. The lateral positions of the shoulder, thorax and abdomen plates were adjusted based on specimen anthropometry to ensure initial simultaneous contact of each body region at the time of impact.

While upper thorax deflections were approximately the same in the pure lateral and oblique two groups of tests, lower thorax and abdomen deflections from oblique tests were significantly greater ($p < 0.05$) than those obtained from pure lateral tests (Fig. 13.20). This finding implies that the loading of the thorax, abdomen and pelvis of the

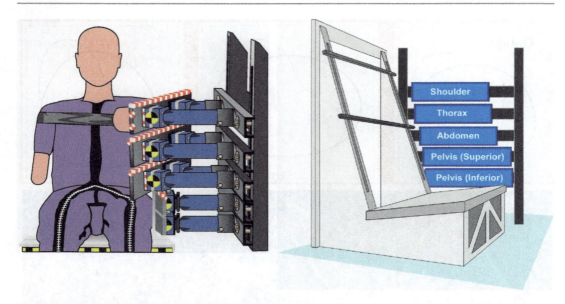

Fig. 13.19 Schematic of the surrogate (WorldSID) with positioning for modular oblique load-wall tests (*left*) and schematic of the segmented and anthropometry-specific load-wall for oblique sled test (From [147])

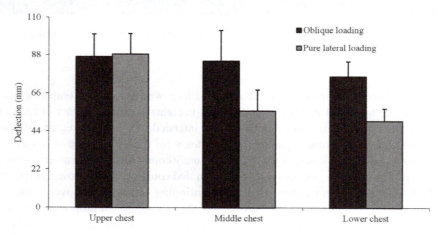

Fig. 13.20 Comparison of maximum chest deflections (*left*, *middle* and *lower*) between oblique and pure lateral tests (From [143])

occupant in antero-lateral oblique side impacts results in the thorax region absorbing greater share of the external insult than the abdomen. Angulations in oblique tests were different ($p<0.05$) than those from pure lateral tests for all regions.

Injuries in modular and non-modular non-anthropometry-specific oblique load-wall tests included fractures and organ traumas. The mean AIS for oblique tests were greater than pure lateral tests (3.4 for modular and 2.8 for non-modular oblique versus 2.0 for pure lateral), suggesting that oblique tests result in more severe injuries than the pure lateral vector at the same input energy, findings in-line with field observations: oblique crashes susceptible to more trauma than pure lateral crashes.

To explore the role of obliqueness, it was recognized that the load-wall used in the described two groups of tests had two distinct features: constant dimension, regardless of subject demographics and regional geometry, and the shoulder contact issue. The invariant load-wall dimension was deemed inadequate to accurately delineate region-specific biomechanics: in some specimens the thorax plate may have spanned the abdomen and vice versa. The shoulder contact issue was also deemed less than representative of modern motor vehicle environments. For example, current vehicle fleet in the United States includes higher beltlines, increasing the propensity to engage the shoulder even in mid-size occupants.

Because shoulder contact was absent in some previous sled experiments, the modular scalable anthropometry-specific load-wall design offered a method to determine the role of shoulder acting as a load-path in side impacts [109, 122, 136, 149]. This is relevant as modern dummies have the ability to determine shoulder-related metrics in side impacts [147, 150]. Further, current vehicle fleet in the United States includes higher beltlines, increasing the propensity to engage the shoulder even in mid-/small-size occupants. The load-wall design has the flexibility to conduct pure lateral and left or right side oblique tests, although only left side oblique impacts were conducted [151]. Likewise, the design can be used to conduct postero-lateral oblique tests, a topic of potential importance to rear seat occupants and side airbags.

Hallman et al. characterized the responses of the thorax and abdomen due to postero-lateral loading from seat-mounted torso side airbag, simulating real-world scenarios [152]. PMHS were subjected to stationary airbag deployment or dynamic lateral impact using sled equipment with close-proximity airbag boundary condition, according to procedures described above for pure lateral impact. For stationary airbag deployments, a rigid wall was attached to the seat assembly such that it corresponded to a level just inferior to the shoulder complex. The torso side airbag was mounted approximately 130 mm away from the wall. Subjects were positioned adjacent to the wall with the unmodified folded airbag approximately 1.0 cm from the posterolateral thorax between T6 and L1 levels. For dynamic impacts, a four-plate force-instrumented load-wall configuration was located on the left end of the seat assembly (Fig. 13.21). Adjustments were made to ensure that the superior edge of the thorax plate corresponded to a level just inferior to the shoulder complex, inducing no shoulder engagement. The unmodified side airbag was mounted posterior to the subject at a distance of 150 mm from the wall. The subjects were positioned 400 mm from the load-wall, and the side airbag was activated when the outboard edges of the module and subject torso were coincident in the frontal plane to ensure the interaction of the deploying airbag with the postero-lateral region of PMHS thorax and abdomen. One of the three specimens sustained spleen laceration along with rib fractures, the other sustained only skeletal trauma, and the third specimen sustained no injury in stationary tests. In contrast, all four

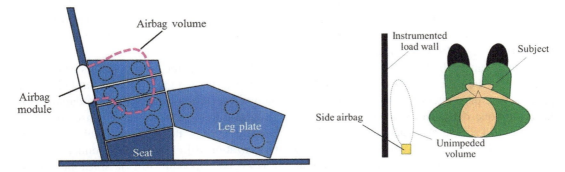

Fig. 13.21 Schematic of the test setup for posterior-oblique tests with side airbag. *Left* shows the load-wall and approximate airbag position and *right* shows the overhead view of the PMHS with the load-wall and side airbag (From [152])

specimens in sled tests sustained skeletal injuries and three out of four had organ injuries, renal and spleen traumas. Pleural laceration occurred with flail chest in a specimen.

Chestband contour analyses indicated that the deformation direction was transient during sled impact, progressing from 122-deg at the onset of deformation to 90-deg following peak deflection. Angles from stationary subjects progressed from 141 to 120°. Peak normalized deflections, rate, and VC_{max} ranged from 0.07 to 0.17, 3.7 to 12.7 m/s, and 0.3 to 0.6 m/s with stationary airbag; and 0.17 to 0.30, 7.4 to 18.3 m/s, and 0.7 to 3.0 m/s dynamic impacts. Peak deflections occurred at between 99 and 135 and 98 and 125-deg for stationary and dynamic conditions. Because of deflection angle transience and localized injury response, both postero-lateral and lateral injury metrics were suggested for postero-lateral impacts, and it was recommended that the biomechanical response to side airbag interaction may be augmented by peak normalized deflection or VC_{max} at 130-deg angulation.

Shaw et al. performed two low energy impacts on each of seven subjects at 2.5 m/s, with one lateral impact and one oblique impact to opposite sides of each PMHS [153]. Data were obtained from a chestband placed on the thorax at the level of impact to measure thoracic deflection. Force versus deflection response corridors were generated and results compared to existing data for oblique and lateral thoracic impacts. The lateral forces were greater than the oblique forces, and oblique deflection was greater than lateral deflection for equal energy impacts.

Rhule et al. performed oblique and lateral thoracic impacts to PMHS at velocities of 4.5 and 5.5 m/s to assess lateral versus oblique responses [154]. Twelve PMHS were impacted by a 23-kg pneumatic ram with a 152.4 mm × 304.8 mm rectangular face plate at the level of the xyphoid process in a pure lateral or 30-deg anterior-to-lateral oblique direction. Normalized responses demonstrated similar characteristics for both lateral and oblique impacts, indicating that it may be reasonable to combine lateral and oblique responses together at these higher speeds to define characteristic PMHS response as was done by ISO. However, less chest compression may be required to obtain serious thoracic injury in oblique impacts as compared to lateral impacts at these speeds.

13.7 Assessment and Development of Dummy Thorax

A prototype thorax system was developed for frontal crash dummies [72]. The dummy was produced with a more humanlike rib cage than previous dummies including lower ribs over the regions of the liver and spleen. The thoracic spine has a flexible link at T7. The shoulder was designed with a humanlike clavicle connecting the sternum and lateral aspects of the shoulders. A modified version of the GM frangible abdomen was used. The dummy has enhanced chest deflection instrumentation for three-dimensional measurement of chest displacement at three injury regions.

Daniel et al. reported on the technical specifications of the SID-IIs side impact dummy [155]. This dummy has the anthropometry of the 5th percentile adult female with a mass of 43.5 kg and has over 100 available data channels. The size is equivalent to an average 12-13 year old adolescent. The response targets for this dummy were derived from responses of cadavers and volunteers subjected to lateral pendulum impacts, drop tests and sled impacts. The thorax has individual ribs that measure displacement and the viscous response. A goal was to achieve rib deflection of 75 mm without binding, bottoming or permanent deformation. Muscle tone of 500 N was added. The half thorax dimension for VC was 142 mm.

The 5th percentile female SID-IIs was further developed, termed the SID-IIs Beta[+] [156]. The biomechanical response of the new dummy was calculated using a weighted response procedure developed by the International Standards Organization [6]. The overall biofidelity rating was 7.0, corresponding to a classification of good by the ISO. Changes were made to the head, shoulder/arm, abdomen and pelvis. The dummy has been tested worldwide and is capable of measuring over 100 data channels for injury assessment of the head, neck, arm, thorax, abdomen, pelvis and leg. Anthropometry is based on the Hybrid III 5th female dummy [157].

An international consortium has supervised the development of the WorldSID dummy with the goal of world harmonization of a side impact dummy [158]. The biofidelity rating of the WorldSID prototype was calculated using the weighted biomechanical test response procedure developed by ISO [6]. Based on this evaluation of the dummy biofidelity, the WorldSID prototype dummy showed a biofidelity rating of 6.15, corresponding to an ISO biofidelity rating of "fair". The dummy showed good repeatability with a global coefficient of variation of 3.30 % for the pendulum and rigid sled tests.

The results of a large side impact tests series sponsored by the NHTSA were reported by Kuppa et al. [159]. Forty-two side impact PMHS sled tests were conducted at 24 and 32 km/h impact speeds into rigid and padded walls. The PMHS were instrumented with accelerometers on the ribs and spine and chest bands around the thorax and abdomen to characterize their mechanical response during the impact. Load cells at the wall measured the impact force at the level of the thorax, abdomen, pelvis, and lower extremities. The resulting injuries were determined through detailed autopsy and radiography. Full and half thorax deflections were computed from the chest band data. The cadaver test data was analyzed using ANOVA and logistic regression. Rib fractures with or without hemo/pneumo thorax or flail chest were the most common injury, chest injury severity ranging from AIS=0 to 5. Subject age influenced injury outcome while gender and mass had little or no influence. Existing side impact injury criteria were evaluated, including the Thoracic Trauma Index (TTI), Average Spinal Acceleration (ASA), full and half thorax deflections, chest velocity and viscous criterion, and contact force. The results indicated that maximum normalized average half thorax deflection was the best predictor of $AIS \geq 3$ and $AIS \geq 4$ thoracic injury. TTI and upper spine accelerations were also good predictors of thoracic injury. Sixteen side impact sled tests were also conducted with the modified ES-2 dummy with rib extensions (ES-2re) under similar impact conditions as the cadaver tests. The rib extensions were added to the original ES-2 dummy ribs to prevent the "seat grabbing" action of the back plate that was observed in side impact vehicle crash tests with the ES-2. A separate analysis was conducted using the injury response and subject characteristics from the cadaver tests and the impact response measurements of the ES-2re dummy in sled tests under similar conditions as the cadaver tests, resulting in thoracic injury criteria that could be directly applied to the ES-2re dummy. Maximum rib deflection and ASA of the ES-2re were the best predictors of thoracic injury. A 50 % risk of $AIS \geq 3$ thoracic injury for a 45 year old corresponds to 44 mm (standard error range: 32–54 mm) of ES-2re maximum rib deflection and ASA of 46 g (standard error range: 34–58 g).

Kuppa [159] reported a further analysis using the injury response and subject characteristics from the cadaver tests and the impact response measurements of the SID-IIS small female dummy in sled tests under similar conditions as the cadaver tests, resulting in thoracic injury criteria that could be applied to the SID-IIs dummy. Logistic regression analyses indicated that the best predictors of thoracic injury were lower spine acceleration, ASA, and peak and average rib deflection. Average rib deflection (Davg) was the mean of the peak deflection of the three thoracic ribs. Maximum deflection (dmax) was the maximum of the peak deflection of the three thoracic ribs. Thoracic injury risk curves were developed using logistic regression. Age of the subject was included in all the models and the risk functions were normalized to a 56 year old [159].

Vezin et al. reported on two series of nine frontal sled tests conducted to evaluate the response of Hybrid III and Thor-alpha dummies [160]. The first series was conducted at 50 km/h with airbag and 4 kN force-limited shoulder belt and the second series at 30 km/h and only a 4 kN force-limited shoulder belt. In each series, three replicate tests were conducted with each dummy and compared with three PMHS. The data provided by the same instrumentation located at the same position were compared to assess the biofidelity of both dummies. The goodtest-to-test repeatability for each dummy permitted to compare the mean value of each recorded parameter. Based on the cadaver response, the results showed that the Thor-alpha provides responses more similar to those of PMHS than the Hybrid III dummy. The flexible joints in the thoracic spine, the sternum design and the more humanlike ribcage give more similar accelerations than the Hybrid III as compared to those of the PMHS. Nevertheless, some parts need improvement to better follow the behavior of the human subject. The head-neck complex, the chest, the shoulder and the pelvis of the Thor-alpha have a more humanlike behavior but some differences remain. The distribution of the deceleration between the components is sometimes different compared to those of the cadaver, even if the resultants are similar. The dummies, particularly the Hybrid III, were less sensitive to the change in restraint systems and tests conditions than the cadavers. Yaguchi et al. [161] conducted component tests to evaluate the biofidelity of the THOR-NT as well as the Hybrid III. Responses on the head and femur of both the THOR-NT and Hybrid III were within the PMHS response corridors. For other body parts each component of THOR-NT did not yield results that satisfied all the PMHS response corridors but the responses of THOR-NT were closer to the corridors than those of the Hybrid III [161].

13.8 Summary

Forces, accelerations and deflections of human cadavers have been measured in controlled laboratory tests to ascertain the biomechanical response of the thorax under a variety of test conditions. Injury criteria have been developed to relate biomechanical response to probability of injury for these various test conditions.

The intent of this chapter was to introduce the reader to the anatomical aspects of the human thorax and provide chronological developments in the area of chest injury biomechanics with a focus on injuries specific to automotive environments. Frontal and side impacts were described due to their strong association with thoracic trauma in the field, development of restraint system/technologies specific to these crash modes and updated regulatory efforts for these crash modes. The PMHS model, regardless of the loading method, has resulted in more skeletal than organ trauma. However, field data continue to involve both skeletal and soft tissue injuries in both crash modalities. Despite this lack of full one-to-one correlation between the laboratory and real-world conditions, injury metrics and criteria that have been developed based on PMHS testing have been for the most part effective. Even though much has been determined regarding thoracic injury tolerance over many decades, in the last 10 years new and important findings have been made. These include the following: Data on the injury tolerance of the older occupant has been obtained in both field studies and laboratory tests. The benefit of pretensioners and load limiters for the older occupant has been determined. A much greater understanding of injury mechanisms to the aorta has been obtained. Longitudinal stretch and the tethering of the descending aorta have been determined to be important in the production of aortic tears. In side impact the effects of oblique loading on thoracic response and injury has been characterized as well as the effects of side airbag loading to the torso. Under oblique loading mid and lower chest deflections were shown to be greater than in previous studies of pure lateral impact. Recent recognitions of the oblique loading mode in side impacts and improvements in computational modeling with the potential emergence of advanced frontal and side impact dummies (from Hybrid III to THOR, and from SID to ES2-re and WorldSID, and others) and instrumentation systems such as sensors to determine the timing of skeletal fracture and their use in stochastic modeling continue to challenge injury biomechanics researchers and improve occupant safety in these environments. Although not discussed in this chapter due to lack of similar levels of research, another topic is the protection of occupants in military environments such as underbody blast. The attention and knowledge gained in the automotive area, in the opinion of the authors of this chapter, are valuable in advancing this emerging area.

Acknowledgments This material is the result of work supported with resources and the use of facilities at the Zablocki VA Medical Center, Milwaukee, Wisconsin and the Medical College of Wisconsin. Narayan Yoganandan is a part-time employee of the Zablocki VA Medical Center, Milwaukee, Wisconsin. Any views expressed in this chapter are those of the authors and not necessarily representative of the funding organizations.

References

1. Services UDoHaH (2007) Injury in the United States: 2007. Chartbook, Atlanta
2. Nirula R, Pintar FA (2008) Identification of vehicle components associated with severe thoracic injury in motor vehicle crashes: a CIREN and NASS analysis. Accid Anal Prev 40(1):137–141. doi:10.1016/j.aap.2007.04.013
3. Morris A, Welsh R (2003) Requirements for the crash protection of older vehicle passengers. Annu Proc Assoc Adv Automot Med 47:165–180
4. Abbreviated Injury Scale (AIS) 2005–2008 Revision (2008). American Association for Automotive Medicine, Morton Grove
5. Walfisch G, Fayon A, Lestrelin D et al. (1985) Predictive functions for thoracic injuries to belt wearers in frontal collisions and their conversion into protection criteria. In: Proceedings of the 29th Stapp Car Crash Conference, pp 49–68
6. ISO/TC22/SC12/WG6 (1993). In: Special seminar on injury criteria in side impact. Paris
7. Viano DC (1989) Biomechanical responses and injuries in blunt lateral impact. In: Proceedings of the 33rd Stapp Car Crash Conference, pp 113–142
8. Baker SP, O'Neill B, Haddon W Jr, Long WB (1974) The injury severity score: a method for describing patients with multiple injuries and evaluating emergency care. J Trauma 14(3):187–196
9. Somers RL (1983) The probability of death score: an improvement of the injury severity score. Accid Anal Prev 15:247–257
10. Somers RL (1982) New ways to use the 1980 abbreviated injury scale. Odense University Hospital, Accident Analysis Group, Laboratory for Public Health and Health Economics, Odense
11. Malliaris AC (1985) Harm causation and ranking in car crashes. vol SAE 850090
12. Malliaris AC, Hitchock R, Hedlund J (1982) A search for priorities in crash protection. SAE 820242
13. Hirsch A, Eppinger RH, Shams T, Nguten T, Levine RS, Mackenzie J, Marks M, Ommaya A (1983) Impairment scaling from the abbreviated injury scale. Washington, DC: National Highway Traffic Safety Administration
14. Carsten O, O'Day J (1984) Injury priority analysis. NHTSA
15. Carsten O (1986) Relationship of accident type to occupant injuries. General Motors Research Laboratories
16. Marcus JH, Blodgett R (1988) Priorities of automobile crash safety based on impairment. Proceedings of the 11th international technical conference on experimental safety vehicles
17. Cohen DS (1987) The safety problem for passengers in frontal impacts: analysis of accidents, laboratory and model simulation data. In: Proceedings of the 11th international technical conference on experimental safety vehicles, Washington, DC
18. Schneider LW, King AI, Beebe MS (1988) Design requirements and specifications; thorax abdomen development task. Interim report: trauma assessment device development program
19. Leport T, Baudrit P, Potier P, Trosseille X, Lecuyer E, Vallancien G (2011) Study of rib fracture mechanisms based on the rib strain profiles in side and forward oblique impact. Stapp Car Crash J 55:199–250
20. Melvin JW, Mohan D, Stalnaker RL (1975) Occupant injury assessment criteria
21. Stalnaker RL, Mohan D (1974) Human chest impact protection. In: Proceedings of 3rd international conference on occupant protection, New York, pp 384–393
22. Eppinger RH, Augustyn K, Robbins DH (1978) Development of a promising universal thoracic trauma prediction methodology. In: Proceedings of the 22nd Stapp Car Crash Conference, pp 211–268
23. Trosseille X, Baudrit P, Leport T, Vallancien G (2008) Rib cage strain pattern as a function of chest loading configuration. Stapp Car Crash J 52:205–231
24. Trosseille X, Baudrit P, Leport T, Petitjean A, Potier P, Vallancien G (2009) The effect of angle on the chest injury outcome in side loading. Stapp Car Crash J 53:403–419
25. Kindig MW, Lau AG, Forman JL, Kent RW (2010) Structural response of cadaveric ribcages under a localized loading: stiffness and kinematic trends. Stapp Car Crash J 54:337–380
26. Fung YC, Yen MR (1984) Experimental investigation of lung injury mechanisms. Topical Report. U.S. Army Medical Research and Development Command
27. Gayzik FS, Hoth JJ, Daly M, Meredith JW, Stitzel JD (2007) A finite element-based injury metric for pulmonary contusion: investigation of candidate metrics through correlation with computed tomography. Stapp Car Crash J 51:189–209
28. Lasky II, Siegel AW, Nahum AM (1968) Automotive cardio thoracic injuries: a medical engineering analysis, vol SAE 680052. Automotive Engineering Congress, Detroit
29. Cooper GJ, Pearce BP, Stainer MC, Maynard RL (1982) The biomechanical response of the thorax to nonpenetrating impact with particular reference to cardiac injuries. J Trauma 22(12):994–1008

30. Kroell CK, Allen SD, Warner CY, Perl TR (1986) Interrelationship of velocity and chest compression in blunt thoracic impact to swine II. In: Proceedings of the 30th Stapp Car Crash Conference, pp 99–121
31. Zhou Q, Rouhana SW, Melvin JW (1996) Age effects on thoracic injury tolerance. In: Proceedings of the 40th Stapp Car Crash Conference, pp 137–148
32. Kent R, Lee SH, Darvish K, Wang S, Poster CS, Lange AW, Brede C, Lange D, Matsuoka F (2005) Structural and material changes in the aging thorax and their role in crash protection for older occupants. Stapp Car Crash J 49:231–249
33. Mertz HJ, Dalmotas DJ (2007) Effects of shoulder belt limit forces on adult thoracic protection in frontal collisions. Stapp Car Crash J 51:361–380
34. Cormier JM (2008) The influence of body mass index on thoracic injuries in frontal impacts. Accid Anal Prev 40(2):610–615. doi:10.1016/j.aap.2007.08.016
35. Koppel S, Bohensky M, Langford J, Taranto D (2011) Older drivers, crashes and injuries. Traffic Inj Prev 12(5):459–467
36. Greendyke RM (1966) Traumatic rupture of aorta; special reference to automobile accidents. JAMA 195(7):527–530
37. Hurley ET (1986) Trauma management, vol III
38. Ochsner MG Jr, Champion HR, Chambers RJ, Harviel JD (1989) Pelvic fracture as an indicator of increased risk of thoracic aortic rupture. J Trauma 29(10):1376–1379
39. Parmley LF, Mattingly TW, Manion WC, Jahnke EJ Jr (1958) Nonpenetrating traumatic injury of the aorta. Circulation 17(6):1086–1101
40. Jackson DH (1984) Of TRAs and ROCs. Chest 85(5):585–587
41. Mattox KL (1989) Fact and fiction about management of aortic transection. Ann Thorac Surg 48(1):1–2
42. Smith RS, Chang FC (1986) Traumatic rupture of the aorta: still a lethal injury. Am J Surg 152(6):660–663
43. Fabian TC, Richardson JD, Croce MA, Smith JS Jr, Rodman G Jr, Kearney PA, Flynn W, Ney AL, Cone JB, Luchette FA, Wisner DH, Scholten DJ, Beaver BL, Conn AK, Coscia R, Hoyt DB, Morris JA Jr, Harviel JD, Peitzman AB, Bynoe RP, Diamond DL, Wall M, Gates JD, Asensio JA, Enderson BL et al (1997) Prospective study of blunt aortic injury: multicenter trial of the American association for the surgery of trauma. J Trauma 42(3):374–380; discussion 380–373
44. Bertrand S, Cuny S, Petit P, Trosseille X, Page Y, Guillemot H, Drazetic P (2008) Traumatic rupture of thoracic aorta in real-world motor vehicle crashes. Traffic Inj Prev 9(2):153–161
45. Creasy JD, Chiles C, Routh WD, Dyer RB (1997) Overview of traumatic injury of the thoracic aorta. Radiographics 17(1):27–45
46. Hossack DW (1980) Rupture of the aorta in road crash victims. Aust N Z J Surg 50(2):136–137
47. Viano DC (1983) Biomechanics of non-penetrating aortic trauma: a review. In: Proceedings of the 27th Stapp Car Crash Conference, pp 109–114
48. Mohan D, Melvin JW (1983) Failure properties of passive human aortic tissue. II–Biaxial tension tests. J Biomech 16(1):31–44
49. Newman RJ, Rastogi S (1984) Rupture of the thoracic aorta and its relationship to road traffic accident characteristics. Injury 15(5):296–299
50. Cavanaugh JM, Walilko TJ, Malhotra A, Zhu Y, King AI (1990) Biomechanical response and injury tolerance of the thorax in twelve sled side impacts. In: Proceedings of the 34th Stapp Car Crash Conference, pp 23–38
51. Cavanaugh JM, Zhu Y, Huang Y, King AI (1993) Injury and response of thorax in side impact cadaveric tests. In: Proceedings of the 37th Stapp Car Crash Conference, Society of Automotive Engineers, San Antonio, pp 199–221
52. Shah CS, Hardy WN, Mason MJ, Yang KH, Van Ee CA, Morgan R, Digges K (2006) Dynamic biaxial tissue properties of the human cadaver aorta. Stapp Car Crash J 50:217–246
53. Hardy WN, Shah CS, Kopacz JM, Yang KH, Van Ee CA, Morgan R, Digges K (2006) Study of potential mechanisms of traumatic rupture of the aorta using insitu experiments. Stapp Car Crash J 50:247–266
54. Hardy WN, Shah CS, Mason MJ, Kopacz JM, Yang KH, King AI, Van Ee CA, Bishop JL, Banglmaier RF, Bey MJ, Morgan RM, Digges KH (2008) Mechanisms of traumatic rupture of the aorta and associated peri-isthmic motion and deformation. Stapp Car Crash J 52:233–265
55. Kroell CK, Schneider DC, Nahum AM (1971) Impact tolerance and response to the human thorax. In: Proceedings of the 15th Stapp Car Crash Conference, pp 84–134
56. Kroell CK, Schneider DC, Nahum AM (1974) Impact tolerance and response to the human thorax II. In: Proceedings of the 18th Stapp Car Crash Conference, pp 383–457
57. Nahum AM, Gadd CW, Schneider DC, Kroell CK (1970) Deflection of the human thorax under sternal impact. In: 1970 international automobile safety conference compendium, pp 797–807
58. Nahum AM, Gadd CW, Schneider DC, Kroell CK (1971) The biomechanical basis for chest impact protection. I. Force-deflection characteristics of the thorax. J Trauma 11(10):874–882
59. Nahum AM, Schneider DC, Kroell CK (1975) Cadaver skeletal response to blunt thoracic impact. In: Proceedings of the 19th Stapp Car Crash Conference, pp 259–293
60. Stalnaker RL, McElhaney JH, Roberts VL, Trollope ML (1973) Human torso response to blunt trauma. In: King WF, Mertz H (eds) Human impact response measurement and simulation. Plenum, New York, pp 181–199
61. Lobdell TE, Kroell CK, Schneider DC, Hering WE, Nahum AM (1973) Impact response of the human thorax. In: King WF, Mertz H (eds) Human impact response measurement and simulation. Plenum, New York, pp 201–245

62. Neathery RF (1974) Analysis of chest impact response data and scaled performance specifications. In: Proceedings of the 18th Stapp Crash conference, pp 459–493
63. Patrick LM (1981) Impact force deflection of the human thorax. In: Proceedings of the 25th Stapp Car Crash Conference, pp 471–496
64. Melvin JW, Weber K (1985) Review of biomechanical response and injury in the automotive environment. U.S. Department of Transportation, National Highway Traffic Safety Administration, Washington, DC
65. Fayon A, Tarriere C, Walfisch G, Got C, Patel A (1975) Thorax of three point belt wearers during a crash (experiments with cadavers). In: Proceedings of the 19th Stapp Car Crash Conference, pp 195–223
66. L'Abbe RJ, Dainty DA, Newman JA (1982) An experimental analysis of thoracic deflection response to belt loading. In: Proceedings of the 7th international IRCOBI conference on the biomechanics of impacts, Bron, pp 184–194
67. Backaitis SH, St. Laurent A (1986) Chest deflection characteristics of volunteers and HYBRID III dummies. In: Proceedings of the 30th Stapp Car Crash Conference, pp 157–166
68. Cheng R, Yang KH, Levine RS, King AI, Morgan R (1982) Injuries to the cervical spine caused by a distributed frontal load to the chest. In: Proceedings of the 26th Stapp Car Crash Conference, pp 1–40
69. Kent R, Lessley D, Sherwood C (2004) Thoracic response to dynamic, non-impact loading from a hub, distributed belt, diagonal belt, and double diagonal belts. Stapp Car Crash J 48:495–519
70. Forman J, Lopez-Valdes F, Lessley D, Kindig M, Kent R, Ridella S, Bostrom O (2009) Rear seat occupant safety: an investigation of a progressive force-limiting, pretensioning 3-point belt system using adult PMHS in frontal sled tests. Stapp Car Crash J 53:49–74
71. Tsitlik JE, Weisfeldt ML, Chandra N, Effron MB, Halperin HR, Levin HR (1983) Elastic properties of the human chest during cardiopulmonary resuscitation. Crit Care Med 11(9):685–692
72. Schneider LW, Haffner MP, Eppinger RH et al. (1992) Development of an advanced ATD thorax system for improved injury assessment in frontal crash environments. In: Proceedings of the 36th Stapp Car Crash Conference, pp 129–156
73. Weisfeldt ML (1979) Compliance characteristics of the human chest during cardiopulmonary resuscitation
74. Melvin JW, King AI, Alem NM (1985) AATD system technical characteristics, design concepts, and trauma assessment criteria. U.S. Department of Transportation, National Highway Traffic Safety Administration, Washington, DC
75. Arbogast KB, Maltese MR, Nadkarni VM, Steen PA, Nysaether JB (2006) Anterior-posterior thoracic force-deflection characteristics measured during cardiopulmonary resuscitation: comparison to post-mortem human subject data. Stapp Car Crash J 50:131–145
76. Mertz HJ, Kroell CK (1970) Tolerance of thorax and abdomen. In: Thomas CC (ed) Impact injury and crash protection. Charles C. Thomas, Springfield, pp 372–401
77. Foster JK, Kortge JO, Wolanin MJ (1977) Hybrid III – a biomechanically-based crash test dummy. In: Proceedings of the 21st Stapp Car Crash Conference, Warrendale
78. Stapp JP (1951) Human exposure to linear decelerations, Part 2. The forward facing position and the development of a crash harness. Wright-Patterson AFB, Dayton
79. Stapp JP (1970) Voluntary human tolerance levels. In: Gurdjian ES, Lange W, Patrick LM, Thomas LM (eds) Impact injury and crash protection. Charles C. Thomas, Springfield, pp 308–349
80. Eiband AM (1959) Human tolerance to rapidly applied acceleration. A survey of the literature. National Aeronautics and Space Administration, Washington, DC
81. Lau IV, Viano DC (1986) The viscous criterion – bases and applications of an injury severity index for soft tissues. In: Proceedings of 30th Stapp Car Crash Conference, pp 123–142
82. Mertz HJ, Gadd CW (1971) Thoracic tolerance to whole-body deceleration. In: Proceedings of the 15th Stapp Crash conference, pp 135–152
83. Bierman HR, Wilder RM, Hellems HK (1946) The physiological effects of compressive forces on the torso. Naval Medical Research Institute Project X 630, Bethesda, p Report #8
84. Patrick LM, Kroell CK, Mertz HJ (1965) Forces on the human body in simulated crashes. In: Proceedings of the 9th Stapp Car Crash Conference, pp 237–259
85. Gadd CW, Patrick LM (1968) Systems versus laboratory impact tests for estimating injury hazard, vol SAE 680053. SAE, New York
86. Eppinger rH, Marcus JH (1985) Prediction of injury in blunt frontal impact. In: Proceedings of the10th international conference on experimental safety vehicles, Oxford, UK, pp 90–104
87. Viano DC (1978) Evaluation of biomechanical response and potential injury from thoracic impact. Aviat Space Environ Med 49(1 Pt. 2):125–135
88. Eppinger R, Sun E, Bandak F, Haffner M, Khaewpong N, Maltese M, Kuppa S, Nguyen T, Takhounts E, Tannous R (1999) Development of improved injury criteria for the assessment of advanced automotive restraint systems–II. National Highway Traffic Safety Administration, pp 1–70
89. Lau VK, Viano DC (1981) Influence of impact velocity on the severity of nonpenetrating hepatic injury. J Trauma 21(2):115–123
90. Lau VK, Viano DC (1981) Influence of impact velocity and chest compression on experimental pulmonary injury severity in rabbits. J Trauma 21(12):1022–1028

91. Viano DC, Lau VK (1983) Role of impact velocity and chest compression in thoracic injury. Aviat Space Environ Med 54(1):16–21
92. Viano DC, Lau IV (1985) Thoracic impact: a viscous tolerance criterion. In: Proceedings of the Tenth international technical conference on experimental safety vehicles, Oxford, pp 104–114
93. Kroell CK, Pope ME, Viano DC, Warner CY, Allen SD (1981) Interrelationship of velocity and chest compression in blunt thoracic impact. In: Proceedings of the 25th Stapp Car Crash Conference, pp 549–579
94. Kleinberger M, Sun E, Eppinger R, Kuppa S, Saul R (1998) Development of improved injury criteria for the assessment of advanced automotive restraint systems. NHTSA, Washington, DC
95. Kuppa SM, Eppinger RH (1998) Development of an improved thoracic injury criterion. In: Proceedings of the 42nd Stapp Car Crash Conference, pp 139–154
96. Dalmatos DJ (1980) Mechanism of injury to vehicle occupants restrained by three point seat belts. In: Proceedings of the 24th Stapp Car Crash Conference, pp 439–476
97. Patrick LM, Bohlin NI, Anderson A (1974) Three point harness accident and laboratory data comparison. In: Proceedings of the 18th Stapp Car Crash Conference, Ann Arbor, MI, pp 201–282
98. Schmidt G, Kallieris D, Barz J, Mattern R (1974) Results of 49 cadaver tests simulating frontal collision of front seat passengers. In: Proceedings of the 18th Stapp Car Crash Conference, Ann Arbor, MI, pp 283–291
99. Horsch JD, Melvin JW, Viano DC, Mertz HJ (1991) Thoracic injury assessment of belt restraint systems based on Hybrid III chest compression. In: Proceedings of the 35th Stapp Car Crash Conference, Dan Diego, CA, pp 85–108
100. Cesari D, Bouquet R (1994) Comparison of Hybrid III and human cadaver thorax deformations loaded by a thoracic belt. In: Proceedings of the 38th Stapp Car Crash Conference, Ft. Lauderdale, FL, pp 65–76
101. Bendjellal F, Walfisch G, Steyer C et al. (1997) The programmed restraint system – a lesson from accidentology. In: Proceedings of the 41st Stapp Car Crash Conference, Lake Buena Vista, FL, pp 249–264
102. Foret-Bruno JY, Trosseille X, Le Coz JY et al. (1998) Thoracic injury risk in frontal car crashes with occupant restrained with belt load limiter, vol SAE Paper No. 983166, Warrendale, PA
103. Foret-Bruno JY, Trosseille X, Page Y, Huere JF, Le Coz JY, Bendjellal F, Diboine A, Phalempin T, Villeforceix D, Baudrit P, Guillemot H, Coltat JC (2001) Comparison of thoracic injury risk in frontal car crashes for occupant restrained without belt load limiters and those restrained with 6 kN and 4 kN belt load limiters. Stapp Car Crash J 45:205–224
104. Petitjean A, Lebarbe M, Potier P, Trosseille X, Lassau JP (2002) Laboratory reconstructions of real world frontal crash configurations using the Hybrid III and THOR dummies and PMHS. Stapp Car Crash J 46:27–54
105. Rouhana SW, Bedewi PG, Kankanala SV, Prasad P, Zwolinski JJ, Meduvsky AG, Rupp JD, Jeffreys TA, Schneider LW (2003) Biomechanics of 4-point seat belt systems in frontal impacts. Stapp Car Crash J 47:367–399
106. Shaw G, Parent D, Purtsezov S, Lessley D, Crandall J, Kent R, Guillemot H, Ridella SA, Takhounts E, Martin P (2009) Impact response of restrained PMHS in frontal sled tests: skeletal deformation patterns under seat belt loading. Stapp Car Crash J 53:1–48
107. Petitjean A, Baudrit P, Trosseille X (2003) Thoracic injury criterion for frontal crash applicable to all restraint systems. Stapp Car Crash J 47:323–348
108. Kent R, Lopez-Valdes FJ, Dennis NJ, Lessley D, Forman J, Higuchi K, Tanji H, Ato T, Kameyoshi H, Arbogast K (2011) Assessment of a three-point restraint system with a pre-tensioned lap belt and an inflatable, force-limited shoulder belt. Stapp Car Crash J 55:141–159
109. Melvin JW, Baron KJ, Little WC, Gideon TW, Pierce J (1998) Biomechanical analysis of Indy race car crashes. In: Proceedings of the 42nd Stapp Car Crash Conference, Tempe, AZ, pp 247–268
110. Aldman B, Anderson A, Saxmark O (1974) Possible effects of air bag inflation on a standing child. In: Proceedings of the 18th conference of the American association for automotive medicine, Tornoto, Canada
111. Takeda H, Kobayashi S (1980) Injuries to children from airbag deployment. In: Proceedings of the 8th international technical conference on experimental safety vehicles, Wolfsburg, Germany
112. Horsch JD, Lau IV, Andrzejak DV, Viano DC, Melvin JW, Pearson J, Cok D, Miller G (1990) Assessment of air bag deployment loads. In: Proceedings of the 34th Stapp Car Crash Conference, Orlando, FL, pp 267–288
113. Yoganandan N, Pintar FA, Skrade D et al. (1993) Thoracic biomechanics with air bag restraint. In: Proceedings of the 37th Stapp Car Crash Conference, San Antonio, TX, pp 133–144
114. Morgan RM, Eppinger RH, Haffner MP et al. (1994) Thoracic trauma assessment formulations for restrained drivers in simulated frontal impacts. In: Proceedings of the 38th Stapp Car Crash Conference, Ft. Lauderdale, FL, pp 15–34
115. Kallieris D, Rizzetti A, Mattern R et al. (1995) On the synergism of the driver air bag and the three point belt in frontal collisions. In: Proceedings of the 39th Stapp Car Crash Conference, San Diego, CA, pp 389–402
116. Lebarbe M, Potier P, Baudrit P, Petit P, Trosseille X, Vallancien G (2005) Thoracic injury investigation using PMHS in frontal airbag out-of-position situations. Stapp Car Crash J 49:323–342

117. Petit P, Trosseille X, Baudrit P, Gopal M (2003) Finite element simulation study of a frontal driver airbag deployment for out-of-position situations. Stapp Car Crash J 47:211–241
118. Tencer AF, Kaufman R, Huber P, Mock C (2005) The role of door orientation on occupant injury in a nearside impact: a CIREN, MADYMO modeling and experimental study. Traffic Inj Prev 6(4):372–378. doi:10.1080/15389580500256813
119. Tarriere C, Walfisch G, Fayon A et al. (1979) Synthesis of human tolerances obtained from lateral impact simulations. In: Proceedings of the 7th international technical conference on experimental safety vehicles, Paris, France, pp 359–373
120. Sacreste J, Brun-Cassan F, Fayon A, Tarriere C, Got C, Patel A (1982) Proposal for a thorax tolerance level in side impacts based on 62 tests performed with cadavers having known bone condition. In: Proceedings of the 26th Stapp Car Crash Conference, Ann Arbor, MI, pp 155–171
121. Robbins DH, Melvin JW, Stalnaker RL (1976) The prediction of thoracic impact injuries. In: Proceedings of the 20th Stapp Car Crash Conference, Dearborn, MI, pp 699–729
122. Kallieris D, Mattern R, Schmidt G, Eppinger RH (1981) Quantification of side impact responses and injuries. In: Proceedings of the 25th Stapp Car Crash Conference, Society of Automotive Engineers, San Francisco, 28–30 Sept 1981, pp 329–368
123. Eppinger RH, Morgan RM (1982) Side impact data analysis. In: Ninth international conference on experimental safety vehicles, Kyoto, Japan
124. Marcus JH, Morgan RM, Eppinger RH, Kallieris D, Mattern R, Schmidt G (1983) Human response to injury from lateral impact. In: Proceedings of the 27th Stapp Crash conference, San Diego, CA, pp 419–432
125. Eppinger RH, Marcus JH (1984) Morgan RM Development of dummy and injury index for NHTSA's thoracic side impact protection research program. Government/Industry Meeting and Exposition, Washington, DC
126. Morgan RM, Marcus JH, Eppinger RH (1986) Side impact – the biofidelity of NHTSA's proposed ATD and efficacy of TTI. In: Proceedings of the 30th Stapp Car Crash Conference, San Diego, CA
127. Cavanaugh JM, Walilko TJ, Walbridge A et al. (1994) An evaluation of TTI and ASA in SID side impact sled tests. In: Proceedings of the 38th Stapp Car Crash Conference, Ft. Lauderdale, FL, pp 292–308
128. Irwin AI, Walilko TJ, Cavanaugh JM et al. (1994) Displacement responses of the shoulder and thorax in lateral sled impacts. In: Proceedings of the 37th Stapp Car Crash Conference, San Antonio, TX, pp 165–174
129. Pintar FA, Yoganandan N, Hines MH, Maltese MR, McFadden J, Saul R, Eppinger RH, Khaewpong N, Kleinberger M (1997) Chestband analysis of human tolerance to side impact. In: Proceedings of the 41st Stapp Car Crash Conference, Lake Buena Vista, pp 63–74
130. Beebe MS (1990) What is BIOSID? SAE technical paper 900377
131. Viano DC, Fan A, Ueno K, et al. Biofidelity and injury assessment in Eurosid I and Biosid. In: Proceedings of the 39th Stapp Car Crash Conference, San Diego, CA, pp 307–326
132. Deng YC (1988) Design considerations for occupant protection in side impact – a modeling approach. In: Proceedings of the 32nd Stapp Car Crash Conference, Atlanta, GA, pp 71–79
133. Campbell KL, Wasko RJ, Hensen SE (1990) Analysis of side impact test data comparing SID and BIOSID. In: Proceedings of the 34th Stapp Car Crash Conference, Orlando, FL, pp 185–205
134. Chung J, Cavanaugh JM, Koh SW, King AI (1999) Thoracic Injury mechanisms and biomechanical responses in lateral velocity pulse impacts. In: Proceedings of the 43rd Stapp Car Crash Conference, San Diego, CA
135. Lau IV, Capp JP, Obermeyer JA (1991) A comparison of frontal and side impact: crash dynamics, countermeasures and subsystem tests. In: Proceedings of the 35th Stapp Car Crash Conference, San Diego, CA, pp 109–124
136. Pintar FA, Maiman DJ, Yoganandan N (2007) Injury patterns in side pole crashes. Annu Proc Assoc Adv Automot Med 51:419–433
137. Yoganandan N, Pintar FA (2008) Deflections from two types of human surrogates in oblique side impacts. Ann Adv Automot Med Annu Sci Conf Assoc Adv Automot Med 52:301–313
138. FMVSS-214 (2008) FMVSS 214: 49Code of Federal Regulations: 571.214, vol 55. US Government Printing Office, Washington, DC
139. Yoganandan N, Pintar FA, Stemper BD, Gennarelli TA, Weigelt JA (2007) Biomechanics of side impact: injury criteria, aging occupants, and airbag technology. J Biomech 40(2):227–243. doi:10.1016/j.jbiomech.2006.01.002
140. Yoganandan N, Zhang J, Pintar FA, King Liu Y (2006) Lightweight low-profile nine-accelerometer package to obtain head angular accelerations in short-duration impacts. J Biomech 39(7):1347–1354. doi:10.1016/j.jbiomech.2005.03.016
141. Zaouk A, Eigen A, Digges K (2001) Occupant injury patterns in side crashes. In: SAE World Congress, Detroit, pp 77–81
142. Augenstein J, Perdeck E, Stratton J, Digges K, Bahouth G (2003) Characteristics of crashes that increase the risk of serious injuries. Annu Proc Assoc Adv Automot Med 47:561–576
143. Yoganandan N, Pintar FA, Gennarelli TA, Martin PG, Ridella SA (2008) Chest deflections and injuries in oblique lateral impacts. Traffic Inj Prev 9(2):162–167. doi:10.1080/15389580701775942
144. Rhule HH, Maltese MR, Donnelly BR, Eppinger RH, Brunner JK, Bolte JH (2002) Development of a

145. Yoganandan N, Pintar FA, Maltese MR (2001) Biomechanics of abdominal injuries. Crit Rev Biomed Eng 29(2):173–246
146. Yoganandan N, Morgan RM, Eppinger RH, Pintar FA, Sances A Jr, Williams A (1996) Mechanisms of thoracic injury in frontal impact. J Biomech Eng 118(4):595–597
147. Yoganandan N, Humm JR, Pintar FA, Brasel K (2011) Region-specific deflection responses of WorldSID and ES2-re devices in pure lateral and oblique side impacts. Stapp Car Crash J 55:351–378
148. Yoganandan N, Humm JR, Pintar FA (2012) Modular and scalable load-wall sled buck for pure-lateral and oblique side impact tests. J Biomech 45(8):1546–1549. doi:10.1016/j.jbiomech.2012.03.002
149. Maltese MR, Eppinger RH, Rhule HH, Donnelly BR, Pintar FA, Yoganandan N (2002) Response corridors of human surrogates in lateral impacts. Stapp Car Crash J 46:321–351
150. Scherer R, Bortenschlager K, Akiyama A, Tylko S, Hartlieb M, Harigae T (2009) WorldSID production dummy biomechanical responses. In: Experimental safety of vehicles, Stuttgart, 18–22 June 2009
151. Yoganandan N, Humm JR, Brasel K, Pintar FA (2012) Thoraco-abdominal deflection responses of post mortem human surrogates in side impacts. Stapp Car Crash J 56:49–64
152. Hallman JJ, Yoganandan N, Pintar FA (2010) Biomechanical and injury response to posterolateral loading from torso side airbags. Stapp Car Crash J 54:227–257
153. Shaw JM, Herriott RG, McFadden JD, Donnelly BR, Bolte JH (2006) Oblique and lateral impact response of the PMHS thorax. Stapp Car Crash J 50:147–167
154. Rhule H, Suntay B, Herriott R, Amenson T, Stricklin J, Bolte JH (2011) Response of PMHS to high- and low-speed oblique and lateral pneumatic ram impacts. Stapp Car Crash J 55:281–315
155. Daniel RP, Irwin AI, Athey J et al. (1995) Technical specifications of the Sid-IIs dummy. In: Proceedings of the 39th Stapp Car Crash Conference. pp 359–388
156. Scherer RD, Kirkish SL, McCleary JP (1998) Sid IIs Beta + prototype dummy biomechanical responses. In: Proceedings of the 42nd Stapp Car Crash Conference. Tempe, AZ, pp 89–114
157. http://www.humaneticsatd.com/crash-test-dummies/side-impact/sid-iis
158. Cesari D, Compigne S, Scherer R, Xu L, Takahashi N, Page M, Asakawa K, Kostyniuk G, Hautmann E, Bortenschlager K, Sakurai M, Harigae T (2001) WorldSID prototype dummy biomechanical responses. Stapp Car Crash J 45:285–318
159. Kuppa S, Eppinger RH, McKoy F, Nguyen T, Pintar FA, Yoganandan N (2003) Development of side impact thoracic injury criteria and their application to the modified ES-2 dummy with rib extensions (ES-2re). Stapp Car Crash J 47:189–210
160. Vezin P, Bruyere-Garnier K, Bermond F, Verriest JP (2002) Comparison of Hybrid III, Thor-alpha and PMHS response in frontal sled tests. Stapp Car Crash J 46:1–26
161. Yaguchi M, Ono K, Masuda M (2008) Biofidelic responses of the THOR-NT and Hybrid III based on component tests
162. Melvin, J (1988) The Engineering Design, Development, Testing and Evaluation of an Advanced Anthropometric Test Device, Phase 1:Concept Definition. DOT HS 807 224. February 1988

Impact and Injury Response of the Abdomen

14

Warren N. Hardy, Meghan K. Howes, Andrew R. Kemper, and Stephen W. Rouhana

Abstract

The impact and injury response of the abdomen has been studied using a variety of testing modes on a variety of scales. Much of the research has focused on the whole-body level response of the abdomen to simplified inputs simulating interaction with steering components or restraint systems. These studies generally strive to provide the load vs. penetration response of the abdomen, with some correlation of injury to potential injury metrics, and they attempt to enhance our understanding of possible injury mechanisms. On the structure scale, investigators have examined the response of organ systems (or portions thereof) to impact or quasi-static tension or compression. On the tissue scale, dog bone or plug samples of solid organs have been tested in tension or compression, and cruciate samples of hollow organs have been tested under high-rate equibiaxial stretch.

This chapter provides a retrospective look at, and analysis of, a number of the cadaver and animal tests of the abdomen that are available in the literature. It begins with a review of pertinent anatomy, and discusses some of the injury mechanisms and metrics that investigators have studied, which have contributed to the design and implementation of injury mitigation tools such as improved physical surrogates. The discussion includes some of the organ and tissue testing that supports the continuing development of computational models, and concludes with thoughts on issues important to future study of abdominal biomechanics.

W.N. Hardy, Ph.D. (✉) • M.K. Howes, Ph.D.
A.R. Kemper, Ph.D.
Center for Injury Biomechanics, Virginia Tech – Wake Forest University, Blacksburg, VA, USA
e-mail: whardy@vt.edu; mkhowes@vt.edu; akemper@vt.edu

S.W. Rouhana, Ph.D.
Ford Motor Company, Dearborn, MI, USA
e-mail: srouhana@ford.com

14.1 Anatomy

This review addresses abdominal anatomy relevant to the biomechanics of crash-induced abdominal injury. The abdominal cavity consists of the peritoneal cavity and the abdominal viscera and is bounded superiorly by the diaphragm and extends into the pelvis. The abdomen is traditionally divided into nine regions: the right and left hypochondriac, right and left lumbar, right and left iliac (or inguinal), and the epigastric, umbilical, and hypogastric (or suprapubic) regions, as shown in Fig. 14.1. The nine regions are separated by four planes: the subcostal and transtubercular planes extending horizontally and two vertical midclavicular planes [1]. Four abdominal quadrants are defined by the median plane and the transumbilical plane, as shown in Fig. 14.1 as dashed lines.

The upper abdomen is afforded protection from blunt trauma by the lower rib cage. An additional level of protection is provided to the abdominal viscera by the anterolateral abdominal wall consisting of skin, subcutaneous tissue, musculature, fascia, and parietal peritoneum [1]. The parietal peritoneum is a serous membrane lining the interior abdominal wall, and the visceral peritoneum surrounds the abdominal viscera. The thin space between the parietal and visceral peritoneum forms the peritoneal cavity. The kidneys, located posterior to the parietal peritoneum and bordering the posterior abdominal wall, are therefore classified as retroperitoneal.

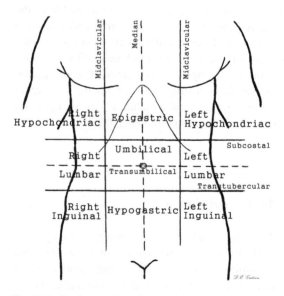

Fig. 14.1 Regions and quadrants of the human abdomen

Mesentery consists of two layers of peritoneum either extending between organs or connecting underlying viscera to the abdominal wall [1]. Mesentery provides vascular and neural connections to the abdominal organs and is essential to the mobility of the viscera within the abdominal cavity. The length of the tethering mesentery permits or restricts the relative motion of organs. Subdivisions of mesentery include the peritoneal ligaments extending between organs and omentum that connects the stomach to the surrounding viscera. The lesser omentum consists of the gastrohepatic and gastroduodenal ligaments and extends between the lesser curvature of the stomach and the liver. The greater omentum spans from the greater curvature of the stomach to cover the majority of the abdominal viscera and folds over to connect to the transverse colon. The sheet-like, fatty omentum separates the viscera from the parietal peritoneum and the abdominal wall, adding to the protection of the abdominal organs during blunt impact. The anatomical features of the abdomen are depicted in Fig. 14.2.

14.1.1 Hollow or Membranous Abdominal Organs

The hollow abdominal organs include the stomach, small intestine, large intestine, and gallbladder. The urinary bladder, a hollow, muscular organ, is commonly included in this classification. For the purposes of material testing, the diaphragm, mesentery, and vasculature are also categorized with the hollow abdominal organs as membranous soft tissues with similar biomechanical test protocols. The most common crash-induced injuries of the hollow abdominal organs are serosal tears, or lacerations, and perforations, or rupture-type injuries [2–5]. Therefore, the wall structure and the biomechanical response of the intact wall are essential to understanding the injury mechanisms and failure tolerances for accurately predicting hollow abdominal organ injury in blunt trauma.

Together the esophagus, stomach, and the small and large intestines make up the alimentary canal, or gastrointestinal tract, and perform the various aspects of digestion [6]. In general,

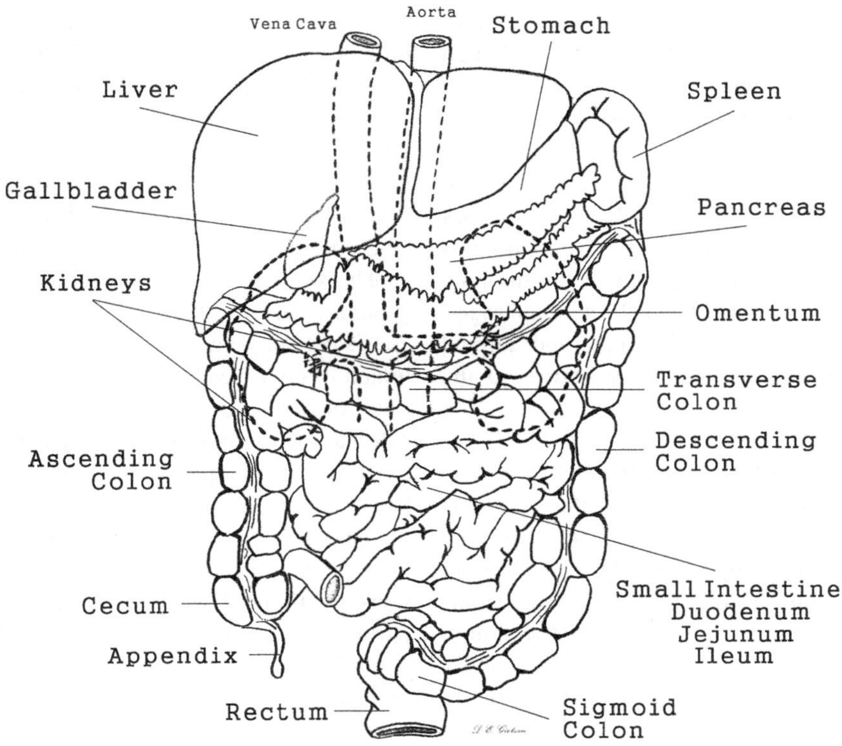

14 Impact and Injury Response of the Abdomen

Fig. 14.2 Gross anatomy of the abdominal cavity

the wall structure of the alimentary canal has a consistent layered architecture throughout its length. The four wall layers, described from the lumen to the outer layer, consist of the mucosa, submucosa, muscularis externa, and serosa or adventitia. The mucosa is organized in three concentric layers: an epithelial lining, a layer of connective tissue known as the lamina propria, and a smooth muscle layer, the muscularis mucosae. Nervous, vascular, and lymphatic connections reach the mucosa through the connective tissue of the submucosal layer. The muscularis externa is comprised of an inner circular smooth muscle layer bound in tight helices and a loosely bound outer longitudinal smooth muscle layer. The outer layer of the wall structure varies throughout the gastrointestinal tract. Structures suspended by peritoneum have an outer layer of serosa consisting of connective tissue and mesothelium, and structures attached to the abdominal or pelvic wall have an outer layer of connective tissue known as adventitia. Variations in the wall structure of specific organs are described in the respective sections.

14.1.1.1 The Stomach

The stomach is continuous superiorly with the esophagus and inferiorly with the proximal duodenum of the small intestine [1]. The organ itself is a hollow reservoir with four regions: the cardia, adjacent to the esophageal opening, the fundus, the corpus or body of the stomach, and the pyloric antrum. The lesser curvature extends from the cardia to the pyloric region on the medial side, and the greater curvature forms the lateral and inferior borders of the stomach. The inner mucosal layer of the stomach forms longitudinal folds, or rugae, that permit extension of the stomach during filling [7]. Rugae are less prominent in the fundus and upper region of the corpus, where the stomach wall generally appears more thin and smooth.

The stomach can accommodate a volume of up to 1.0 or 1.5 l of food or fluid [8]. Although the position and shape of the stomach change with the stages of digestion, the stomach generally extends from the level of the 5th left rib to the level of the 9th costal cartilage or the first lumbar vertebra [1]. The pyloric region crosses the midline of the body, where it is continuous with the duodenum to the right of the midline. The stomach is situated beneath the diaphragm, adjacent to the spleen, left kidney and adrenal gland, and pancreas, and lies posterior to the transverse colon.

14.1.1.2 The Small Intestine

The small intestine is divided into three regions: the duodenum, the jejunum, and the ileum [1]. The duodenum is approximately 23–28 cm in length, extending from the pylorus of the stomach to the duodenojejunal flexure, and wrapping around the head of the pancreas. The remainder of the cadaveric intestine is 6–7 m in length, extending from the duodenojejunal junction to the ileocecal junction. The jejunum and ileum are continuous but can be distinguished visually by a decrease in diameter and a change in wall thickness at the ileum. With the exception of the duodenum, the small intestine is extremely mobile within the abdominal cavity. Mesentery connects the jejunum and ileum to the posterior abdominal wall and provides the blood and lymphatic supply to the small intestine.

The wall structure of the small intestine is continuous throughout its length and consistent with the remainder of the gastrointestinal tract. As the three main functions of the intestine include mixing, propulsion, and absorption, mucosal folds appear as a result of contractions of the muscularis mucosae during mixing and propulsion, and the folds exist to increase absorption rates [8].

14.1.1.3 The Large Intestine

The cecum, colon, rectum, and anal canal make up the large intestine [1]. The colon is divided into four anatomic regions arcing around the border of the abdomen: the ascending, transverse, descending, and sigmoid colon. Both the ascending and descending colon are retroperitoneal, which provides some protection from blunt impact. The transverse colon is fairly mobile extending between the ascending and descending colon, with the curvatures between the regions known as the right and left colic flexure, respectively. The sigmoid colon curves from the descending colon, adjacent to the iliac fossa, to the rectum.

The wall structure of the colon matches the general structure of the alimentary canal and is consistent throughout the length [7]. From the lumen to the outer layer, the wall structure consists of the mucosa, submucosa, muscularis externa, and serosa. The distinguishing characteristics of the large intestine are teniae coli and haustra. Teniae coli are thin bands of muscle fibers extending axially throughout the large intestine to the rectum, and haustra coli are sections of tissue between the bands of teniae. During digestion, the proximal portion of the colon functions for absorption and the distal portion functions for storage of feces [8]. The contents of the colon, therefore, tend to be more fluid in the proximal portion and more solid in the distal half of the colon.

14.1.1.4 The Gallbladder

The gallbladder is a hollow reservoir that stores and concentrates bile produced in the liver and transported to the gallbladder through the common hepatic duct. In humans, the gallbladder can distend to accommodate approximately 50 ml [7]. The structure is surrounded by peritoneum, which fixes the body and neck of the gallbladder to the liver on the visceral surface. Concentrated bile is secreted from the gallbladder through the common bile duct to the duodenum of the small intestine. The wall structure of the gallbladder does not include a muscularis mucosae or submucosa layer, and it is the contractions of the smooth muscle cells in the muscularis externa of the lamina propria that empty the contents of the gallbladder into the cystic duct.

14.1.1.5 The Urinary Bladder

The urinary bladder is a round, hollow organ with thick, muscular walls. The bladder is located within the lesser pelvis when empty and extends toward the umbilicus during filling [1]. Two ureters open into the bladder, and the urethra extends to the external urethral orifice. The trigone is the triangular region formed by the three orifices of the bladder and has a consistent thickness [7]. The remainder of the bladder wall consists of thick, muscular folds arranged to direct urine toward the urethra during contraction.

14.1.1.6 The Uterus

The uterus is a thick-walled hollow organ positioned within the female lower pelvis between the bladder and the rectum. The uterus is divided into two regions: the body and the cervix [1]. The wall of the body consists of three layers: the outer perimetrium, a thick, muscular layer known as the myometrium, and the internal mucosal lining called the endometrium. The histology of the uterus varies throughout the menstrual cycle, and the position and structure alter considerably during pregnancy.

14.1.1.7 The Diaphragm

The diaphragm is a sheet of muscle and tendon that forms the superior border of the abdominal cavity [1]. The structure is mobile in the central region and fixed on the outer margins to the rib cage and lumbar spine with ligamentous attachments. The central tendon spans the central arc of the diaphragm and is surrounded by three regions of muscle fibers extending radially: the sternal and costal fibers, and the lumbar crura. The diaphragm arcs to form right and left domes. Three major diaphragmatic openings connect the thorax and the abdomen: the caval opening, the esophageal hiatus, and the aortic hiatus.

14.1.2 Solid Abdominal Organs

The solid abdominal organs include the liver, spleen, kidneys, and pancreas, and are frequent contributors to increased injury severity in automobile collisions. Injury distributions indicate that the liver and the spleen are the most frequently injured abdominal organs in motor vehicle collisions [9–11].

14.1.2.1 The Liver

The liver is the largest internal organ in the body, weighing up to 1,500 g and occupying the majority of the right upper quadrant of the abdomen [1]. The rib cage offers partial protection to the liver in blunt impact; however, blunt trauma can still occur and fractured ribs have the potential to result in penetrating trauma, or lacerations. Although damage to the liver is commonly described as "laceration", in the absence of a true penetrating wound or broken rib and resulting cut or puncture, this damage is better described as "fracture", which is a term frequently encountered in the biomechanics community in reference to solid organ disruption. Depending upon the extent of blunt trauma to the liver, the injury can be described as a "rupture" or "burst", indicating extreme organ disruption, which is indicative of the highly vascular/fluid-filled nature of the liver.

The anterior, superior surface of the liver is known as the diaphragmatic surface due to the location inferior to the diaphragm [1]. The posterior, inferior surface of the liver is known as the visceral surface and is adjacent to the stomach, duodenum, colon, and right kidney. The ligaments of the liver are reflections of the peritoneum. The diaphragmatic surface is separated into right and left lobes by the falciform ligament connecting to the anterior abdominal wall. The coronary ligament suspends the liver by tethering it to the diaphragm. The layers of the coronary ligament give rise to the right and left triangular ligaments. The posterior surface of the liver is divided into four lobes: the left, right, caudate, and quadrate lobes.

14.1.2.2 The Spleen

The spleen is a lymphatic organ responsible for the breakdown of red blood cells [1]. The spleen is positioned in the left upper quadrant of the abdomen, adjacent to the left colic flexure. The diaphragmatic surface of the spleen is anterior and inferior to the diaphragm, and the medial hilum, the location of the splenic artery and vein, borders the tail of the pancreas. The gastrosplenic ligament and the splenorenal ligament connect the spleen to the greater curvature of the stomach and to the left kidney, respectively. The spleen is generally up to 12 cm in length and 7 cm wide and does not extend below the rib cage, which offers some protection from blunt impact. The parenchyma of the spleen is comprised of white and red pulp and is protected by a thin, but tough, capsule. The mode of splenic injury familiar to most is rupture.

14.1.2.3 The Kidneys

The kidneys, positioned in the retroperitoneal space, are encased in fat and therefore remain well protected from blunt impact injury. The kidneys generally extend from T12 to L3, with the right kidney positioned lower than the left due to the expanse of the liver [1]. The suprarenal glands are adjacent to the kidneys superiorly. The kidneys are responsible for urine production. The renal artery, renal vein, and ureter enter and exit the kidney at the renal hilum. The kidney is comprised of renal lobes, which consist of renal pyramids (part of the renal medulla) and their associated cortices. Other major constituents of the kidneys include the cortex, renal pelvis, and the calyces.

14.1.2.4 The Pancreas

The pancreas is a digestive gland located in the retroperitoneal space and situated between the spleen and duodenum on the left and right sides and posterior to the transverse colon [1]. From the medial to lateral edge, the pancreas is classified into four regions: the head, neck, body, and tail. The length of the pancreas extends across the vertebral column, potentially facilitating compression injuries. Although relatively rare in occurrence, the pancreas can rupture and result in the release of pancreatic enzymes into the abdominal cavity. Pancreatic injury usually occurs in conjunction with other injuries, and can be associated with bleeding.

14.1.3 Vasculature

The major vessels within the abdominal region are positioned adjacent to the vertebral column and are retroperitoneal in location. The arteries are efferent vessels, resistant to blood pressure with a muscular and elastic composition. The veins are afferent, thin-walled vessels with a high expansion capacity for containing blood volume.

14.1.3.1 The Arteries

The aorta enters into the abdominal cavity through the aortic hiatus, located at the posterior aspect of the diaphragm at the level of the T12 vertebra [1]. The abdominal aorta extends anterior to the vertebral column to L4, where the aortic bifurcation into the common iliac arteries occurs at the left of the midline. The left and right common iliac arteries branch into the internal and external iliac arteries before exiting the abdominal region. The branches of the abdominal aorta are paired, extending into visceral or parietal branches to the left and right of the midline, or unpaired, dividing into the celiac trunk, superior and inferior mesenteric arteries, and the median sacral artery.

The superior mesenteric artery branches from the anterior surface of the abdominal aorta and extends through layers of mesentery to supply the majority of the intestines [12]. Its divisions reach the length of the jejunum and ileum, cecum, ascending colon, and the proximal portion of the transverse colon. The inferior mesenteric artery branches from the inferior region of the abdominal aorta and supplies the remainder of the colon and the proximal length of the rectum.

14.1.3.2 The Veins

Blood from the abdominal viscera is transported back to the heart via the portal venous system entering into the inferior vena cava (IVC) through the hepatic veins [1]. The common iliac veins combine to form the IVC at the level of L5, posterior to the right common iliac artery. The IVC extends to the right of the vertebral column and aorta, entering into the thorax through the caval opening in the diaphragm. In addition to draining the hepatic portal system through the hepatic veins, the other tributaries to the IVC include the iliac, lumbar, renal, suprarenal, and phrenic veins [12].

14.2 Field Accident Data

Nahum et al. (1970) analyzed data from the accident files of the UCLA Trauma Research Group [13]. The files contained data on collisions of 604 vehicles produced in the 1960–1969 model years. There were 972 occupants involved in the collisions, of which 20 % were lap belted and 1 % were shoulder belted. The abdomen was injured in 3.5 % of all occupants. The authors calculated injury rates for the abdomen-pelvis area in association with specific contact points. In the 1960–1966 vehicles 19.8 % of the drivers received abdominal injuries which were associated with the steering system (55 %), side interior (22 %), "impact" (9 %), seatbelts (3.6 %), and other contacts. In the 1967–1969 vehicles, after the introduction of energy-absorbing steering systems, the abdominal injury rate decreased to 16.5 % of the drivers. The steering system association decreased dramatically to 21 %, side interior increased (33 %), "impact" (11 %), seatbelts (11 %), and the rest from other contacts. The severity of the abdominal injuries was moderate to fatal for 73 % of the 1960–1966 group, and 52 % of the 1967–1969 group.

Nearly one in ten (9.8 %) passengers had abdominal injuries in the 1960–1966 group, and 15.3 % in the 1967–1969 group. For the 1960–1966 group, these injuries had contact associated with the lower instrument panel (38 %), the side interior (23 %), "impact" (15 %), ejection (8 %), the A-pillar (8 %), the seatbelt (8 %), and other contacts. In the 1967–1969 group, the injuries were associated with contact to the lower instrument panel (30 %), the seatbelt (15 %), the upper instrument panel (10 %), side interior (10 %), and other contacts.

Rear seat occupants saw a reduction in abdominal injury rate from 12.5 % to 2.2 % when comparing the 1960–1966 and 1967–1969 groups, respectively. The associated contacts were: the seat back (57 %), side interior (29 %), and side glass (14 %) in the 1960–1966 group; and the seat belt (100 %) in the 1967–1969 group. Note, there was a change in abdominal injury severity from 71 % moderate to fatal injury in the 1960–1967 group, to 100 % minor injury in the 1967–1969 group.

In 1979, Danner and Langwieder [14] and Danner et al. [15] analyzed data from 2,545 pedestrian collisions with passenger cars investigated by German motor vehicle insurers. They found that 19.8 % of all pedestrian injuries, and 6.5 % of all serious pedestrian injuries (AIS 3+) were to the abdomen (only the head and tibia had greater frequency). Children and the elderly appeared to be overrepresented in terms of serious abdominal injury, with 22.7 % of all the serious injuries to children in the abdominal region (only head and femur had more), and 24.8 % of all serious injuries to the elderly in the abdominal region (only head and tibia had more).

Ricci (1980) analyzed data from the National Crash Severity Study (NCSS), which reported information for collisions in which the most severe injury occurred to an occupant of a towed passenger car [16]. The study dates were January 1977 through March 1979, and included a stratified sampling scheme in seven different areas of the US. The NCSS includes data on approximately 25,000 actual vehicle occupants, and over 106,000 occupants if weighted data is used. Ricci found that 3.8 % of all injuries in the NCSS data base were to the abdominal region. However, when severity was taken into account, injury to the abdomen was overrepresented, accounting for 8.3 % of all serious (AIS 3+), 29.9 % of all severe (AIS 4+), and 30.7 % of all critical (AIS 5+) injuries.

Hobbs (1980) reported on a 2-year in-depth investigation of collisions involving approximately 2,500 occupants in the United Kingdom [17]. He found that while less than 1 % of belted occupants were fatally injured, 8.4 % of all unbelted occupants were fatally injured. Half of 1 % (0.5 %) of the belted occupants, and 1.7 % of the unbelted occupants received serious (AIS 3+) abdominal injuries in pure frontal and oblique frontal impacts. For injury of all severities, contact was associated with the seatbelt, instrument panel, and side door for belted occupants, and the steering system, instrument panel, and side door for the unbelted occupants. For rollovers, none (0 %) of the belted occupants and 3 % of the unbelted occupants sustained serious abdominal injuries.

Galer et al. (1985) studied 261 collisions in the UK which involved 297 vehicles and 506 occupants [18]. They analyzed the data for belted occupants, and found that 14.6 % of all injuries to drivers and 17.8 % of all injuries to front seat passengers were to the abdomen/pelvis (they did not examine the abdomen alone). These represented 6.4 % of the nonminor (AIS 2–6) injuries to drivers and 18.1 % of those to front seat passengers. The main contacts associated with minor injury (AIS 1) to the abdomen/pelvis of vehicle drivers were seatbelt webbing (59 %) and the side door (12 %), and those associated with nonminor injury were the steering system (30 %) and side door (70 %). For front seat passengers seatbelt webbing was the principal contact for minor injury (91 %), while the side door (50 %) and other vehicle contacts (50 %) were associated with nonminor injury to the abdomen/pelvis.

Rouhana and Foster (1985) analyzed the NCSS data (the same data set as Ricci [16]) specifically for side impact collisions [19]. They found that 15.6 % of all serious (AIS 3+) injuries, and 24.2 % of all severe (AIS 4+) injuries were abdominal. The main contact points associated with serious abdominal injury in left side impacts were the side interior (39 %), the armrest (30 %), and the steering system (18 %). The main contacts associated with serious abdominal injury in right side impacts were the glove compartment (39 %), side interior (28 %), and the armrest (28 %). The curious appearance of the glove compartment as an injury factor in side impacts reflects the large percentage of struck cars with forward velocity at the time of impact. The occupants on the side opposite the impact continue moving forward in those cases as described by Newton's First Law.

Using the concept of "severity weighted frequency" of injury, the abdominal organs seriously injured for drivers in left side impacts were ordered as kidneys (4.5 %), liver (4.0 %), spleen (3.7 %), digestive (2.0 %), and urogenital (0.8 %). For passengers in right side impacts, the injuries were ordered as liver (7.0 %), spleen (3.2 %), kidneys (2.5 %), urogenital (2.5 %), and digestive (1.2 %). The near doubling of the proportion of liver injuries in right side impacts when compared to left side impacts makes sense from anatomical considerations.

Bondy (1980) also analyzed the statistics in the NCSS database [9]. This analysis showed that in frontal impact the order of serious abdominal injury was liver (39 %), spleen (25 %), digestive (16 %), kidney (14 %), and urogenital (2.6 %). The contact points associated with serious abdominal injury were the steering system (51 %), side interior/armrest (26 %), instrument panel/glove compartment (17 %), front seat back (5 %), and belt webbing (1 %). It is interesting to note that the restraint system was not associated with injury to any of the solid abdominal organs.

Mackay et al. (1993) studied the 1983–1989 side impact crash experience in the Birmingham region of the United Kingdom [20]. This study provides an interesting contrast to that by Rouhana and Foster [19] because, while the database used by Rouhana and Foster contained mainly unbelted occupants, the study by Mackay et al. contained only belted occupants. They found that 41 of the 180 occupants (23 %) in the study received an injury to the abdomen and the vast majority (71 %) of those injuries were associated with seatbelt loading. However, they found only five cases of abdominal injury with AIS 3+. Thus, as belt use rates in the US increase, one would expect the injury pattern for farside occupants in lateral impacts in the US to change from more serious injuries associated with interior structures in front of the occupant to mainly less serious injuries associated with the seat belt. In fact, using the National Automotive Sampling System - Crashworthiness Data System (NASS-CDS) database from 1993 to 2002, Gabler et al. (2005) found that in farside impacts with belted occupants, only 6.8 % sustained AIS 3+ abdominal injuries and 86 % of AIS 2+ abdominal injuries were attributed to contact with the seatbelt or belt buckle [21].

Klinich and Burton (1993) analyzed data from the National Automotive Sampling System (NASS, formerly National Accident Sampling System) for the years 1988–1990 [22]. They studied injury to older children, defined as ages 6–12 years old. These data indicate that the frequency of serious injury (AIS 3+) to the pelvis and abdomen ranked fourth after head, lower extremities and thorax. They showed that while 62.4 % of restrained older children and only 36.4 % of unrestrained older children remain uninjured, the percentage of injuries to the abdomen and pelvis are 10.8 % and 2.6 %, respectively. They did not break this down further to examine injury severity which is typically lower when restrained.

Khaewpong et al. (1995) performed a study of 103 cases of restrained children who were admitted to the level I trauma center at the Children's National Medical Center (CNMC) [23]. They studied the correctness and appropriateness of the restraint and the injuries sustained by the children. There were 12 cases of abdominal injury with maximum AIS 3+ to these restrained children. While the sample size was very small, several conclusions could be drawn relative to abdominal injury. Nearly 90 % of the abdominal injuries in this group were associated with contact with the restraint system. However, all of the children who sustained abdominal injury were using the restraint system either incorrectly, inappropriately, or both incorrectly and inappropriately.

Another very interesting observation was noted but not discussed in this paper. When the authors compared their results from the CNMC to results from a similar analysis of NASS data, they found that 52 % of the abdominal injuries to restrained children in the NASS data were to children in convertible child seats. They did not go into detail relative to injury or collision severity. Abdominal injury to children in seats with crotch straps seems unlikely to be more than AIS 1, so a more in-depth look at this data appears warranted.

Elhagediab and Rouhana (1998) reviewed abdominal injuries in non-rollover frontal impacts to non-ejected drivers and right front passengers [10]. They used the NASS database for the years 1988–1994. They found that abdominal injuries constituted 8 % of all injuries of AIS 3+, 16.5 % of all injuries of AIS 4+, and 20.5 % of all injuries of AIS 5+. The organs injured most frequently were the liver (38 %), spleen (23 %) and digestive system (17 %). The objects within the vehicle that were most often associated with the abdominal injuries were the steering wheel (68 %), belt (17 %) and airbag/other interior objects (14 %). Of note was the fact that only 10 % of the vehicles in the study contained airbags. In addition, right front passengers appeared to be more at risk of abdominal injury when restrained by lap-shoulder belts (12.5 %) than unrestrained (10.7 %). This statement, however, does not reflect the head, neck and thoracic injuries prevented by the lap-shoulder belt.

Yoganandan et al. (2000) examined abdominal injury frequency and severity in frontal, nearside, and farside impacts using the NASS-CDS database for the years 1993–1998 [24]. For a combined dataset of belted and unbelted drivers and right front passengers, the distribution of organ injury frequency for each crash mode varied when comparing AIS 2+ and AIS 3+ injuries. The organs injured most frequently in frontal crashes were the liver (39 %), spleen (29 %), and digestive organs (11 %) for AIS 2+ and the liver (35 %), spleen (28 %), and arteries (14 %) for AIS 3+. For nearside crashes, the spleen (49 %), kidneys (20 %), and liver (16 %) were injured most frequently for AIS 2+, and the spleen (47 %), diaphragm (17 %), and liver (13 %) were the most frequently injured organs for AIS 3+. The injury frequency for farside crashes was kidney (38 %), liver (32 %), and digestive (15 %) for AIS 2+ and liver (50 %), spleen (13 %), and kidney (12 %) for AIS 3+. The spleen was the most frequently injured organ in left side impacts, whereas the kidneys and liver were injured more frequently in right side impacts. The kidneys are protected by fat encasements and their retroperitoneal position; therefore, renal injuries tended to be less severe and comprised a greater percentage of AIS 2+ injuries in side impacts. Similarly, injury to the digestive tract is typically less severe on the threat-to-life scale, and therefore these injuries comprised a greater percentage when considering AIS 2+ abdominal injuries in frontal and farside impacts.

Lee and Yang (2002) characterized abdominal injury patterns in a similar dataset, using the NASS database from 1993 to 1997 [25]. The study focused on injury patterns for the solid abdominal organs, finding these injuries to account for 35 % of all abdominal injuries, with 32 % of these injuries AIS 3+. Lacerations were the most common injury type for the solid organs. Liver injuries were most frequently attributed to contact with components in the front of the occupant compartment, such as the steering assembly. Spleen injuries were most commonly attributed to contact with the left side of the occupant compartment.

Lamielle et al. (2006) analyzed abdominal injury patterns in frontal crashes occurring in France from 1970 to 2005 [26]. The likelihood of AIS 3+ abdominal organ injury increased with increasing crash severity. Injury risk was reduced with the use of seatbelts with retractors. Occupants with AIS 3+ abdominal injuries were significantly older than occupants without an AIS 3+ abdominal injury. This study identified belted rear seat occupants as more likely to sustain abdominal injury compared to belted front seat occupants. For belted occupants, hollow abdominal organ injury occurred more frequently (59 %) than solid organ injury, whereas solid abdominal organ injury occurred more frequently for unbelted occupants (77 %). The organs injured most frequently (AIS 3+) were the spleen (22 %), jejunum (16 %), and the liver (16 %).

Klinich et al. (2010) used the NASS-CDS (1998–2008) and Crash Injury Research and Engineering Network (CIREN) databases to assess abdominal injury in frontal, farside, and nearside crashes for a population of occupants in which 85 % were belted and 68 % were travelling in airbag-equipped vehicles [11]. The analysis emphasized characterizing injury for occupants utilizing current safety technologies. For all crash modes, belted occupants had a significant reduction in risk of AIS 2+ and AIS 3+ abdominal organ injuries compared to unbelted occupants. No difference in injury incidence was found in frontal crashes with and without airbag deployment. Right front passengers in nearside impacts had the highest risk of sustaining AIS 2+ abdominal organ injury. The percentage of occupants sustaining AIS 2+ abdominal organ injury, as well as the percentage of occupants with rib fracture, increased with increasing crash severity for all crash modes. Controlling for crash severity, the authors found that the likelihood of sustaining AIS 2+ and AIS 3+ injuries to the solid abdominal organs (liver, spleen, kidneys) increased considerably with the presence of AIS 2+ rib fractures. No association was found between abdominal organ injury incidence and occupant age.

Frampton et al. (2012) investigated abdominal injury risk factors for belted front and rear seat occupants in frontal crashes using crash data from the UK Co-operative Crash Injury Study (CCIS) from 1998 to 2010 [27]. The authors found an increased risk of AIS 2+ and AIS 3+ abdominal injuries for rear seat passengers compared to right front passengers and drivers. These results were consistent with the results of Lamielle et al. [26]. Abdominal injury risk increased with age for both severity levels; however, the injury incidence was also high for occupants less than 19 years of age. The most frequently injured organs for drivers were the liver and spleen, and for right front passengers, the most frequently injured organs were the liver, jejunum-ileum, and spleen. Rear seat passengers sustained a different injury pattern with the jejunum-ileum, mesentery, and the colon injured more frequently than the solid organs, though liver and spleen injury incidence remained high for rear seat occupants.

The association between abdominal organ injury and multiple rib fractures was also assessed. The majority of liver, spleen, and mesentery injuries were sustained in association with two or more rib fractures, whereas the majority of the colon, duodenum, and jejunum-ileum injuries were associated with one or no rib fractures. The incidence of rib fractures did not differ significantly between seating positions while the distributions of organ injury varied considerably for front and rear seat occupants.

To further the understanding of clinical data, Campbell et al. (1994) compared the Injury Impairment Scale (IIS) scores with physicians recorded estimates of impairment for 7,502 patients in the United Kingdom Major Trauma Outcome Study [28]. The IIS is an estimate of the impairment a patient will experience 1 year post-injury. The injuries had been sustained in motor vehicle crashes for 19.9 % (1,483) of the patients in the study. They found a strong, statistically significant correlation between IIS and physician's estimates. They also observed that 99.5 % of all patients with abdominal injuries had an IIS score of 0 (no impairment) and only 1 patient was expected to have a permanent impairment. In contrast, about 8 % of all patients with head injury were expected to have permanent major impairment. Notably, these data reflected patients who had sustained a single injury to the head, abdomen or lower limbs. Information on impairment adds another dimension beyond the traditional and more immediate AIS threat-to-life scale.

These field accident studies have shown that blunt abdominal trauma is a common result of motor vehicle related collisions. As injury severity increases, abdominal organ injuries account for a greater percentage of all injuries. In general, there is a preponderance of abdominal injuries to the solid organs compared to hollow organs. The presence of AIS 2+ rib fractures is correlated with an increased risk of AIS 2+ abdominal organ injury. When age is accounted for, children and the elderly are overrepresented in terms of incidence of serious abdominal injuries. The order of organs injured is significantly different when comparing side and frontal impacts.

The contact points associated with injury to unbelted occupants include many of the structures within the vehicle (steering system, side door, instrument panel, etc.). While seatbelt use is associated with an increase in minor abdominal injury compared to the unbelted occupant, there is a clear decline in serious injury to the abdomen when seatbelts are used.

14.3 Clinical Data

Clinical studies of patients with blunt abdominal injury have identified Motor Vehicle Collision - Related incidents to account for 50–90 % of cases [29–32]. Abdominal injury distribution by organ varies in frequency and order depending on the inclusion criteria for the clinical studies. Historically, however, the solid abdominal organs (liver, spleen, kidneys) have been the most frequent contributors to blunt abdominal injury as well as injury severity. Although the hollow abdominal organs are injured less frequently in blunt trauma cases, high morbidity and mortality are associated with these injuries if not diagnosed [33].

Increases in the morbidity and mortality of abdominal injuries resulting from blunt trauma are often attributed to delayed diagnosis resulting from lacking or masked symptoms [33, 34]. However, improved evaluation techniques have considerably facilitated the diagnosis and treatment of abdominal injury, beginning with the introduction of diagnostic peritoneal lavage (DPL) as a simple and highly sensitive method compared to physical examination [35]. Further diagnostic advances including computed tomography (CT) and ultrasonography (US) have improved the ability to detect injuries with the advantage of being non-invasive [36–39]. While CT requires increased time for evaluation, for stable patients it can provide more specific information than DPL regarding which organs are injured, injury severity, and need for surgery. Traditional US and the focused abdominal sonography for trauma (FAST) screening exam have largely replaced DPL due to speed and simplicity [40]; however, US can be limited in the ability to detect free fluid and can fail to provide direct visualization of solid parenchymal or hollow visceral injuries [33, 39]. Despite limitations, these techniques have improved the evaluation of intra-abdominal injury.

Improved diagnostic methods, as well as multi-modality approaches, have helped physicians to develop nonoperative trauma management protocols, which appear to be quite effective, for stable patients with suspected abdominal trauma [41, 42]. Nonoperative management of liver injuries has shown increased effectiveness compared to splenic or kidney injuries [43]. Diaphragmatic injuries and hollow organ injuries continue to be a concern [44]. Due to the associated morbidity, reducing nontherapeutic laparotomy for these and other injuries remains as a goal for the management of abdominal trauma.

14.4 Material Properties Testing

Material properties testing can provide valuable insight in to the local response of the organ system, how a tissue fails, and at what level (stress or strain) it fails. Equibiaxial stretch testing of membranous tissues provides the response characteristics of the components of the hollow organs using loading conditions and rates commensurate with those to be expected in a car crash. Dog bone, slug, or intact organ testing provides the response characteristics and material properties of solid organ structures. Common to all tissue testing are the issues of controlling and quantifying the initial and boundary conditions. Recreating the baseline (in-vivo or neutral) conditions of the tissue as a starting point is challenging. Depending upon the structure, this does not always correspond to a state of zero stress. For example, testing a tubular structure as a flat sample might result in the measurement of properties that are different than those that the tissue might exhibit in vivo.

Measurement of strain in a sample is certainly a challenge, but high-resolution, high-speed video facilitates the quantification of dot or speckle patterns. Measurement of stress is more difficult, as the continual change of tissue thickness must be quantified. Video and laser systems help in this regard, but often some assumptions must be made as to the distribution of load through a sample, and whether or not a tissue is incompressible.

14.4.1 Membranous Tissues

Limited material properties data are available in the literature for the complex structures of the digestive system. These data were obtained in uniaxial tension at quasi-static rates [45–48]. No historical data exist characterizing the dynamic biaxial material and failure properties of the human post-mortem hollow abdominal organs. The planar equibiaxial test configuration creates a two-dimensional nearly homogeneous strain field and incorporates the effects of tissue anisotropy in the response to multidirectional stretch. This experimental technique was applied to fresh human cadaver tissue including stomach, small intestine, and colon samples using a custom equibiaxial tissue-testing device. Mason et al. (2005) provided a detailed description of the components of the original device [49]. Therefore, the design and function of the device are explained only briefly.

Fig. 14.3 The equibiaxial tissue testing apparatus designed by Mason [49] and used by Shah [50] and Howes [51]

The dynamic equibiaxial testing device was designed to apply simultaneous equal stretch to a cruciate-shaped tissue sample in four directions, as shown in Fig. 14.3. Sample arms were gripped in four low-mass tissue clamps designed to reduce inertial effects and limit slip between the sample and clamp. Each clamp was mounted to a lightweight carriage riding on linear rails with low-friction bearings. Carriages were individually coupled to a central drive disk by rigid links. Rotation of the disk has been generated by a pneumatic cylinder coupled to the test device [50], as well as by a hydraulic actuating mechanism rigidly linked to the device and loaded by a falling mass [51]. Rotation of the disk produced identical motion of the four carriages away from the disk center and orthogonal to the adjacent carriages. Miniature load cells and accelerometers were mounted to each clamp and carriage assembly. Laser displacement sensors were mounted above and below the sample to obtain changing sample thickness throughout the duration of the test. A high-speed video camera was mounted with the lens parallel to the tissue sample surface to obtain optical data without interfering with the test.

Howes and Hardy (2012) conducted tissue testing using samples harvested from the abdominal cavities of fresh, never-frozen post-mortem human surrogates [51]. A CNC-machined stamp was applied to tissue sections from each

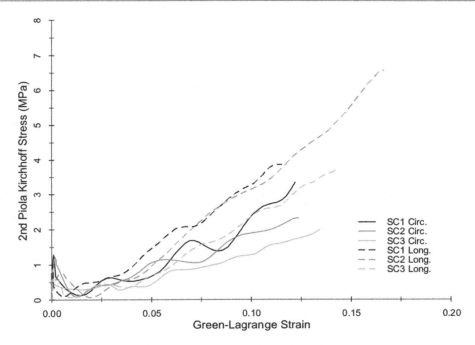

Fig. 14.4 Exemplar stress-strain responses for three cadaver colon samples undergoing high-rate equibiaxial stretch, from the research of Howes

excised organ of the digestive tract, producing cruciate-shaped samples with four arm branches converging to form a 10-mm central square region of interest (ROI). The stamp was aligned such that two co-linear arms were oriented with the visible longitudinal fiber direction of the tissue, or offset from the fiber direction by 22° [52]. The offset prevented failure across the arm branches of the sample [50], resulting in failure through the central ROI such that local strain calculated in the ROI was a direct measure of the failure strain. Samples were stretched until complete transection occurred.

High-speed video data were analyzed using tracking software to quantify the displacement of ink markers applied to the ROI of the sample. Average directional Green-Lagrange strain, maximum principal strain and strain rate, and maximum shear strain were calculated. Applied load in the ROI was calculated by scaling each of the inertia compensated measured loads, assuming a proportional transmission of load through each arm to the central region. The two compensated scaled loads in each direction were averaged to provide the load time histories in the x- and y-directions. True stress was calculated using the laser thickness data to determine changing cross-sectional area throughout the duration of each test. The 2nd Piola Kirchhoff stress was calculated using the applied load, initial cross-sectional area, and the inverse of the stretch ratio. Stress data were filtered using SAE J211 CFC 600 Hz [53]. All data were transformed to align with the material axes of the tissue and truncated at the time of tear initiation.

Considering all of the hollow organ tests, which were conducted at a maximum principal strain rate of 72 s^{-1} on average, the average failure strain (including both directions) was approximately 15 %, and the average failure stresses were on the order of 2 to 4 MPa. Figure 14.4 shows exemplar stress vs. strain responses for three colon samples. Directional differences in failure properties were observed throughout the length of the digestive tract with an overall trend toward an increased stiffness in the longitudinal direction for the intestines. This analysis quantified the material properties and failure thresholds of the hollow abdominal organs loaded in equibiaxial tension at impact-level rates, with application to the optimization of material model parameters to represent abdominal tissue response and to the advancement of finite element models of the human abdomen.

14.4.2 The Liver in Tension

There have been a number of studies that have investigated the tensile material properties of liver by performing tension tests on isolated liver samples. In the interest of brevity, the summary of previous studies provided in this section is limited to studies that have reported tensile failure properties of animal or human liver samples.

Uehara (1995) performed uniaxial tension tests on dog-bone shaped samples of porcine liver parenchyma at loading rates of 5, 20, 50, 200, and 500 mm/min [54]. The liver parenchyma samples were obtained from 7-month old pigs within 6 h of death (never frozen). Tests were performed at 21 °C and specimen hydration was maintained with Ringer's solution. It was reported that the failure stress and modulus increased with increased loading rate, while the extension ratio decreased with increased loading rate. The average failure stress, average modulus, and average extension ratio ranged from 126 to 205 kPa, 0.68 to 1.19 MPa, and 1.29 to 1.26 mm/mm between loading rates of 5 and 500 mm/min, respectively.

Hollenstein et al. (2006) performed uniaxial tension tests on the capsule of one bovine liver, which was tested within 8 h of death [55]. The liver was chilled with wet ice and wrapped in physiologic saline soaked cloth between the time of procurement and sample preparation. The capsule samples were separated from the underlying parenchyma by gently peeling the capsule off of the parenchyma using the fingers and surgical forceps. It was reported that the nominal failure stress for the bovine liver capsule was 9.2 MPa and nominal failure strain was 35.6 %.

Santago et al. (2009) evaluated the effect of temperature on the tensile material properties of bovine liver parenchyma at a strain rate of 0.07 s^{-1} [56]. Uniaxial tensile tests were conducted on fresh bovine liver parenchyma using dog-bone shaped samples. For this study, one bovine liver was received within 24 h of death. The organ was sealed in a plastic bag and chilled with wet ice between the time of procurement and arrival. Samples were allowed to hang under their own weight, i.e. 1 g of tension, during the clamping process to ensure that all samples had a minimal but consistent initial preload. The state of strain under this minimal preload was used to define the point of zero strain. It was reported that there were no statistically significant differences in failure stress or strain between specimens tested at 75 °F versus those tested at 98 °F. The average 2nd Piola Kirchhoff failure stress and Green-Lagrange failure strain for all samples in this study was 52.5 kPa and 0.25 strain respectively.

In a second study, Santago et al. (2009) evaluated the effect of freezing on the tensile material properties of bovine liver parenchyma at a strain rate of 0.07 s^{-1} [57]. As with the previous study [56], uniaxial tensile tests were performed on bovine liver parenchyma using dog-bone shaped samples. For this study, one bovine liver was received within 24 h of death. The organ was sealed in a plastic bag and chilled with wet ice between the time of procurement and arrival. Upon arrival, the liver was sectioned into two equal halves. One half was prepared and tested immediately after arrival. The second half was frozen for 26 days and then tested after thawing. The frozen half of the liver was thawed in a saline bath at 75 °F for 12 h. Again, specimens were allowed to hang under their own weight, i.e. 1 g of tension, during the clamping process to ensure that all specimens had a minimal but consistent initial preload. The state of strain under this minimal preload was used to define the point of zero strain. It was reported that freezing resulted in a significant reduction in Green-Lagrange failure strain, but there were not statistically significant differences in the 2nd Piola Kirchhoff failure stress.

Brunon et al. (2010) conducted quasi-static (0.001–0.01 s^{-1}) tensile failure tests on both fresh and previously frozen porcine and human liver capsule samples, which had the underlying parenchyma attached [58]. The porcine livers were obtained from adult pigs within 4–5 days after death. Fresh porcine samples were tested immediately, while frozen samples were stored in a freezer (24 h to a few days) until testing. Fresh, non-frozen human liver samples were tested within 5 days of death. The storage time between death and testing for frozen human liver samples was not reported. The mean ultimate true failure stress and Green-Lagrange failure strain for all fresh human hepatic capsule-parenchyma specimens were found to be 1.85 MPa and 32.6 %, and the mean ultimate true failure stress and Green-Lagrange failure strain for all fresh porcine hepatic capsule-parenchyma specimens were found to be 2.03 MPa and 43.3 %. A comparison between human and porcine tissues showed that the capsule characteristics were similar between porcine and human. However, freezing was found to significantly affect the failure properties of porcine liver capsule but not human liver capsule.

Kemper et al. (2010) performed uniaxial tension tests on dog-bone shaped samples of human liver parenchyma at loading rates of 0.01, 0.1, 1.0, and 10.0 s^{-1} (Fig. 14.5) [59]. Each organ was procured within 24 h of death, received within 36 h of death, and tested within 48 h of death to minimize the adverse effects of tissue degradation. The organs were immersed in Dulbecco's Modified Eagle Medium (DMEM), a tissue culture medium, and chilled with wet ice between the time of procurement and specimen preparation. Immediately prior to mounting the specimens on the experimental setup, the specimens were immersed in a bath of DMEM heated to 37 °C. In this study, specimens were allowed to hang under their own weight, i.e. 1 g of tension, during the clamping process to ensure that all specimens had a minimal but consistent initial preload. The state of strain under this minimal preload was used to define the point of zero strain. The results of this study showed that the response of human liver parenchyma is both non-linear and rate dependent in tensile loading (Fig. 14.6). Specifically, failure stress significantly increased with increased loading rate, while failure strain significantly decreased with increased loading rate. The average 2nd Piola Kirchhoff stresses for loading rates of 0.01, 0.1, 1, and 10 s^{-1} were reported to be 40.2, 46.8, 52.6 and 61.0 kPa, respectively.

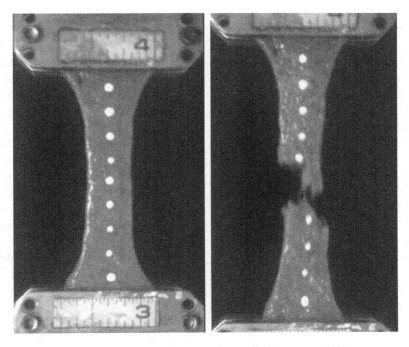

Fig. 14.5 Exemplar human liver parenchyma dog-bone sample used by Kemper et al. [59]

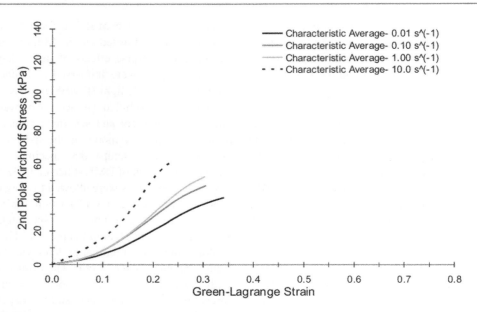

Fig. 14.6 Characteristic averages for tensile material response of human liver parenchyma dog-bone samples presented by Kemper et al. [59]

The average Green-Lagrange failure strains for loading rates of 0.01, 0.1, 1, and 10 s^{-1} were reported to be 0.34, 0.32, 0.30, and 0.24 strain, respectively.

Kemper et al. [59] also compared the results of the human liver parenchyma material response to the material response of bovine and porcine liver parenchyma presented by previous authors [54, 56, 57]. The comparison showed that the tensile failure stress of porcine liver reported by Uehara [54] was significantly larger than the tensile failure stress of human liver. In contrast, the comparison also showed that neither the tensile failure stress nor the tensile failure strain of human liver was significantly different from that of bovine liver reported by Santago et al. [56, 57]. It was noted that the differences between human, porcine, and bovine liver parenchyma could potentially be attributed to the differences in the microstructure of the liver parenchyma between each of these mammalian species. Specifically, the delineation of adjacent hepatic lobules in porcine liver parenchyma is different than that of human and bovine liver parenchyma. The hepatic lobules are a roughly hexagonal arrangement of hepatocytes radiating from a central vein to portal triads located at the vertices of the lobule. Each of the portal triads contains a bile duct, proper hepatic artery, and hepatic portal vein. In porcine liver, the lobules are easily recognized in histology sections because the portal triads are connected by relatively thick layers of collagenous septum [7, 60, 61]. Conversely, adjacent lobules are indistinctively separated from one another in human and bovine liver since there is no interlobular collagenous septum [60, 62, 63]. This accounts for the tougher nature of porcine liver compared to bovine liver [63]. Although collagen content was not quantified by Kemper et al. [59], previous studies have shown that the stiffness of the human liver increases with respect to the degree of liver fibrosis [64]. Since liver fibrosis is defined as the excessive accumulation of extracellular matrix proteins, primarily collagen, the stiffness of the liver can be considered proportional to the collagen content. Therefore, the considerably higher collagen content present in porcine liver compared to human and bovine liver could explain the significantly higher failure stress observed in porcine liver.

14.4.3 The Liver in Compression

There have been a number of studies that have investigated the compressive material properties of liver by performing compression tests on isolated samples of liver parenchyma. The majority of these studies have been limited to sub-failure loading on animal tissue [64–68]. However, this summary is limited to studies that have reported compressive failure properties of animal or human liver parenchyma.

Tamura et al. (2002) performed uniaxial unconfined compression tests on rectangular, previously frozen (~24 h), porcine liver parenchyma samples tested at loading rates of 0.005, 0.05, and 0.5 s^{-1} [69]. A series of pre-conditioning compression tests were performed on fresh and previously frozen porcine liver specimens from which it was determined that freezing had no considerable effects on the mechanical response of the tissue. However, it is possible that damage to the tissue during the pre-conditioning may have masked any potential changes in the compressive response resulting from the freezing process. The specimens were tested while immersed in a constant temperature (36 °C) bath of physiologic saline. The results of the study showed that the response of porcine liver parenchyma is both non-linear and rate dependent. Specifically, the average peak Cauchy stress increased with increased loading rate: 123.4, 135.2, and 162.5 kPa, respectively. The peak nominal strain, however, was not found to vary considerably with respect to loading rate: 0.432, 0.420, and 0.438 strain, respectively. It should be noted that the initial specimen height, i.e. point of zero strain, was defined as the height that corresponded to an arbitrary load of 0.6 N, and Cauchy stress was calculated by assuming incompressibility, i.e. constant volume.

Pervin et al. (2011) performed uniaxial unconfined compression tests on fresh "ring" shaped bovine liver parenchyma samples at loading rates between 0.01 and 3,000 s^{-1} [70]. The specimens were tested at room temperature within 3 h of death. Intermediate (1, 10, and 100 s^{-1}) and quasi-static (0.01 and 0.1 s^{-1}) unconfined compression experiments were performed using a hydraulically driven MTS. High strain rate (1,000, 2,000, and 3,000 s^{-1}) experiments were conducted using a Kolsky bar modified for soft material characterization. The point of zero strain was defined based on the measured initial specimen height, but the area of each sample was assumed to be equal to the ideal specimen area. The results of this study showed that the response of bovine liver parenchyma is both non-linear and rate dependent. It was reported that the average peak engineering stress increased with increased loading rate from 0.01 to 100 s^{-1}: 49.7, 62.3, 94.8, 117.2, and 131.1 kPa, respectively. The engineering peak strain (i.e. nominal strain) was not found to vary considerably with respect to loading rate until after 1.0 s^{-1}: 0.60, 0.60, 0.60, 0.59 and 0.55 strain, respectively. It was reported also that there were no considerable differences between samples oriented parallel to the exterior surface of the liver and samples oriented orthogonal to the exterior surface of the liver at intermediate and quasi-static loading rates, indicating that the bovine liver parenchyma is isotropic.

Kemper et al. (2013) performed uniaxial unconfined compression tests on cylindrical samples of human liver parenchyma at loading rates of 0.01, 0.1, 1, and 10 s^{-1} [71]. Each organ was procured within 24 h of death, received within 36 h of death, and tested within 48 h of death to minimize the adverse effects of tissue degradation. The organs were immersed in DMEM and chilled with wet ice between the time of procurement and specimen preparation. Immediately prior to testing, the specimens were immersed in a bath of DMEM heated to 37 °C. The point of zero strain was defined based on the initial specimen height, which was quantified using pre-test pictures. True stress was calculated based on the area of each cylindrical specimen, quantified using pre-test pictures, and assuming incompressibility, i.e. constant volume. The results of this study showed that the response of human liver parenchyma is both non-linear and rate dependent. Specifically, failure stress significantly increased with increased loading rate, while failure strain significantly decreased with increased loading rate. The average true failure stresses for loading rates of 0.01, 0.1, 1, and 10 s^{-1} were reported to be 63.7, 75.9, 103.4 and

110.5 kPa, respectively. The average true failure strains for loading rates of 0.01, 0.1, 1, and 10 s^{-1} were reported to be 0.94, 0.73, 0.61, and 0.60 strain, respectively.

14.4.4 The Spleen in Tension

There have been only two studies to the authors' knowledge that have investigated the tensile material properties of spleen by performing tension tests on isolated samples of spleen parenchyma or spleen capsule. The experimental methods and key results from these studies are summarized in the current section.

Uehara (1995) obtained spleen parenchyma samples from 7-month old pigs within 6 h of death (never frozen) [54]. The samples were tested in uniaxial tension using compression clamps at loading rates of 5, 20, 50, 200, and 500 mm/min. Tests were performed at 21 °C and specimen hydration was maintained with Ringer's solution. It was reported that the failure stress and modulus increased with increased loading rate, while the extension ratio decreased with increased loading rate. The average failure stress, average modulus, and average extension ratio ranged from 60 to 101 kPa, 164 to 272 kPa, and 1.59 to 1.51 mm/mm between loading rates of 5 and 500 mm/min, respectively.

Kemper et al. (2012) performed uniaxial tension tests on dog-bone shaped samples of human spleen parenchyma and spleen capsule at loading rates of 0.01, 0.1, 1, and 10 s^{-1} [72]. The capsule specimens were prepared with the underlying parenchyma attached (referred to as capsule/parenchyma specimens). This was done because the intimate connections between the capsule and parenchyma, specifically the collagenous trabeculae which extend from the capsule into the parenchyma, made it virtually impossible to isolate the capsule from the underlying parenchyma without causing damage. Each organ was procured within 24 h of death, received within 36 h of death, and tested within 48 h of death to minimize the adverse effects of tissue degradation. The organs were immersed in DMEM and chilled with wet ice between the time of procurement and specimen preparation. Immediately prior to mounting the specimens on the experimental setup, the specimens were immersed in a bath of DMEM heated to 37 °C. Similar to a previous study by Kemper et al. [59], specimens were allowed to hang under their own weight, i.e. 1 g of tension, during the clamping process to ensure that all specimens had a minimal but consistent initial preload. The state of strain under this minimal preload was used to define the point of zero strain. The results of this study showed that the responses of human spleen parenchyma specimens and spleen capsule/parenchyma specimens are both non-linear and rate dependent in tensile loading. Specifically, the failure stress significantly increased while the failure strain significantly decreased with increased loading rate for both parenchyma and capsule/parenchyma specimens. In addition, the failure stress of the capsule/parenchyma specimens was found to be significantly greater than that of the parenchyma specimens, while the failure strain was not significantly different. For the spleen parenchyma specimens, the average 2nd Piola Kirchhoff failure stresses for loading rates of 0.01, 0.1, 1, and 10 s^{-1} were reported to be 16.5, 23.9, 31.5 and 33.8 kPa, respectively. The average Green-Lagrange failure strains for loading rates of 0.01, 0.1, 1, and 10 s^{-1} were reported to be 0.26, 0.21, 0.19, and 0.18 strain, respectively. For the spleen capsule/parenchyma specimens, the average 2nd Piola Kirchhoff failure stresses for loading rates of 0.01, 0.1, 1, and 10 s^{-1} were reported to be 43.6, 46.1, 68.4 and 65.3 kPa, respectively. The average Green-Lagrange failure strains for loading rates of 0.01, 0.1, 1, and 10 s^{-1} were reported to be 0.23, 0.20, 0.19, and 0.17 strain, respectively.

14.4.5 The Spleen in Compression

There have been two studies that have investigated the compressive material properties of spleen by performing compression tests on isolated samples of spleen parenchyma. The experimental methods and key results from both of these studies are summarized herein.

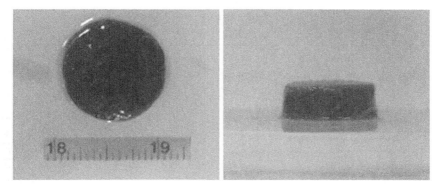

Fig. 14.7 Exemplar human spleen parenchyma compression sample used by Kemper et al. [73]

Tamura et al. (2002) performed uniaxial unconfined compression tests on rectangular, previously frozen (~24 h), porcine spleen parenchyma samples tested at loading rates of 0.005, 0.05, and 0.5 s^{-1}. [69]. The specimens were tested while immersed in a constant temperature (36 °C) bath of physiologic saline. The results of this study showed that the response of porcine spleen parenchyma is both non-linear and rate dependent. Specifically, it was reported that the average peak Cauchy stress increased with increased loading rate: 107.5, 114.6, and 146.3 kPa, respectively. The peak nominal strain, however, was not found to vary considerably with respect to loading rate: 0.825, 0.809, and 0.834 strain, respectively. It should be noted that the initial specimen height, i.e. point of zero strain, was defined as the height that corresponded to an arbitrary load of 0.6 N, and Cauchy stress was calculated by assuming incompressibility, i.e. constant volume.

Kemper et al. (2011) performed uniaxial unconfined compression tests on cylindrical samples of human spleen parenchyma at loading rates of 0.01, 0.1, 1, and 10 s^{-1} (Fig. 14.7) [73]. Each organ was procured within 24 h of death, received within 36 h of death, and tested within 48 h of death to minimize the adverse effects of tissue degradation. The organs were immersed in DMEM and chilled with wet ice between the time of procurement and specimen preparation. Immediately prior to testing, the specimens were immersed in a bath of DMEM heated to 37 °C. The point of zero strain was defined based on the initial specimen height, which was quantified using pre-test pictures. Engineering stress was calculated based on the initial area of each cylindrical specimen, which was quantified using pre-test pictures. The results of this study showed that the response of human spleen parenchyma is both non-linear and rate dependent. Specifically, failure stress increased with increased loading rate, while failure strain decreased with increased loading rate. In order to be consistent with the data published by Kemper et al. (2013), the engineering stress and engineering strain data presented by Kemper et al. (2011) have been converted to true stress, assuming incompressibility, and true strain (Fig. 14.8). The average true failure stresses for loading rates of 0.01, 0.1, 1, and 10 s^{-1} were reported to be 20.7, 25.3, 32.3 and 39.1 kPa, respectively. The average true failure strains for loading rates of 0.01, 0.1, 1, and 10 s^{-1} were reported to be 0.52, 0.51, 0.48, and 0.46 strain, respectively.

14.5 Surrogate Testing

Historically, investigators have turned to human surrogates to learn about the impact and injury response of the abdomen. This includes animals, animal cadavers, and human cadavers (also known as post-mortem human surrogates, or PMHS). Human volunteers are enlisted only for quasi-static, sub-injury testing, or posture and position studies. Differences in size, shape, anatomy, and physiology complicate the translation

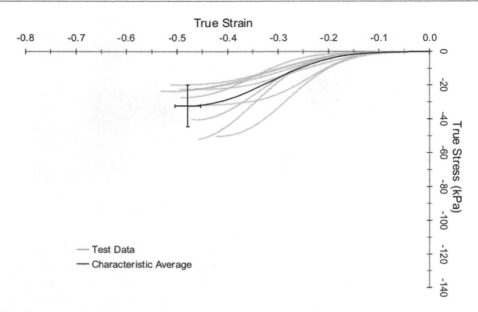

Fig. 14.8 Response of human spleen parenchyma in uniaxial compression at 1.0 s^{-1} presented by Kemper et al. [73]

of animal data to the human condition. Human cadaver testing is limited by post mortem changes. Cadavers have no muscle tone, no circulation, no respiration, and no physiological response or temperature. However, researchers attempt to make the cadaver respond more like the living by applying appropriate muscle/tendon tension when possible, or even inverting a specimen to counteract the effect of gravity, by perfusing and pressurizing the arterial and venous systems (sometimes with pulsatile flow), by ventilating the pulmonary system, and by heating the specimens. Regardless of the surrogate used, scaling the response data to the human is always an issue. Various scaling techniques have been proposed over the decades, but none has received universal acceptance, nor produced satisfying outcomes. This is particularly true when trying to assess pediatric response, for which there are precious few data.

Prior to the 1980s, little had been done to examine the response of the cadaver abdomen to impact. However, some animal testing had been conducted to investigate the response of specific organ systems as well as the abdomen as a whole [74–76]. Still today, the frontal crash regulatory dummy, the Hybrid III, does not have a biofidelic abdomen, and there are no abdominal injury criteria in the federal crash standards. Efforts began in the 1980s to develop improved crash dummies and parts thereof, such as the Advanced Anthropomorphic Test Device, or AATD, presented by Schneider et al. (1992), which fueled new interest in the response of the abdomen [77]. Several studies were conducted during that time, with some of the earliest being Rouhana et al. (1986), which studied the response of swine cadavers to seatbelt loading [78], and Cavanaugh et al. (1986), which studied the response of the cadaver abdomen to rigid-bar loading [79]. More recently, attention has focused on high-speed pretensioner loading under out-of-position conditions. Collectively, there is not an abundance of response data for the abdomen. And, some of the studies seem to provide conflicting results, although most of the discrepancies between datasets have been resolved [80]. Still, efforts continue to develop biofidelic abdominal inserts for crash dummies [81, 82], and corresponding injury metrics. This work extends in to the virtual world as well, with substantial effort being put toward the development of finite element representations of the human body.

Typically, researchers try to simulate interaction of a vehicle occupant with various aspects of the vehicle interior. Often, these interactions are simplified in an attempt to conduct more repeatable, and measurable, tests. Usually the tests are simplified in terms of occupant posture, seating system (or lack thereof), the type of interaction (e.g., rigid bars instead of steering rims), and the nature of the interaction (location and direction of applied load). For the abdomen, the specimen is usually positioned statically or even rigidly in a fixture and then impacted, avoiding the complications associated with sled testing. Studies have simulated interaction with the steering rim, steering column and dashboard components, restraint systems such as seatbelts and airbags, as well as passenger doors and armrests.

14.5.1 Rigid Impact

Steering rims are generally simulated using rigid bar indenters, both straight and curved. Other more blunt objects or surfaces within the occupant compartment, including steering column components, are often simulated using rigid disks of various diameters (152–305 mm). More penetrating interactions are simulated using smaller diameter probes (76 mm), which can be used for mapping the local, or regional, response of the abdomen to impact. These indenters and probes are typically fixed to heavy ballistic pendulums or linear impactors.

14.5.1.1 Frontal Impact

Stalnaker and Ulman (1985) compiled and analyzed data from three earlier studies [83]. These earlier studies were conducted by Beckman et al. (1971), Stalnaker et al. (1971), and Trollope et al. (1973), which all used rigid-bar and wedged-shaped impactors [74, 76, 84]. Tests were conducted using different primates ranging from 0.53 to 19.40 kg in free-back conditions and impact speeds ranging from 8.4 to 17.0 m/s. Three different impact locations were used, including the central regions of the upper, middle, and lower abdomen (mostly upper abdomen). All of the response data from the three different regions and six different impact surfaces was plotted on the same graph to define a single abdominal response corridor for 12.1 m/s. The force-deflection data were described in terms of a three-stage rise-plateau-rise response.

Cavanaugh et al. (1986) conducted abdominal impacts into 12 unembalmed cadavers using a 2.5-cm diameter, 38.1-cm long rigid bar attached to 31.5 and 63.5-kg linear impactors using impact speeds ranging from 4.9 to 13.0 m/s [79]. The bar was oriented with the long axis parallel to the width of the subject at the level of the third lumbar vertebra. Each subject was seated upright in a free-back condition, with the legs positioned straight out from the cadaver, with the torso held upright by straps under the arms that were released at the time of impactor contact. These tests were separated into two groups having average speeds of 6.1 +/− 1.5 m/s (five tests) and 10.4 +/− 1.1 m/s (seven tests). Four of the higher speed tests employed the 63.5-kg impactor. Importantly, abdominal response was found to be proportional to both impact speed and impactor mass, suggesting rate sensitivity of the response.

Viano et al. (1989) conducted impacts into unembalmed, repressurized free-back human cadavers using a 15.2-cm diameter rigid disk attached to a 23.4-kg pendulum and three speed ranges, averaging 4.5, 6.7, and 9.4 m/s [85]. Tests were performed at angles of 60° to the right or left of the midline through the center of gravity of the specimen, with the center of the impactor at 7.5 cm below the xiphoid process, covering approximately ribs 6–10. Deflection data was obtained by analysis of high-speed film. The subjects were suspended upright, with hands and arms overhead. The loading response showed a force plateau following the initial stiffness response, and the unloading response suggested action of restorative forces, which contrasts the sudden unloading reported by Cavanaugh et al. [79].

Nusholtz and Kaiker (1994) conducted impacts at the level of the second lumbar vertebra using six unembalmed, repressurized human cadavers [86]. The cadavers were positioned in a free-back seated posture with the knees bent and the legs hanging down. The impact apparatus consisted of an angled semicircular rigid-tube attached to an 18-kg pendulum, simulating the lower rim of a steering wheel. The bottom of the rim impacted equidistant from the inferior aspect of the tenth ribs and the iliac crests. A harness that was released at contact was used to keep the cadavers upright. Tests were conducted at nominal impact speeds of 6 and 10 m/s, varying from 3.9 to 10.8 m/s. In contrast to the Cavanaugh results, no rate sensitivity in the loading response was found.

Yoganandan et al. (1996) performed oblique lateral pendulum impacts similar to those of Viano et al. [85] on the right lower thorax (upper abdomen) of five human cadavers [87]. A 23.5-kg impactor with a 15.2-cm diameter, Ensolite-padded impact face was used, having velocity of 4.3 m/s. A chestband was used to monitor the contour of the cadavers during the impact. Force-deflection curves were presented, but the responses were not normalized for mass, which ranged from 56 to 82 kg. Two of the five curves showed large variability. The reasons for this variability were not addressed, but might include rotation of the test subjects, as there is no mention of an effort to strike through the center of mass of the subjects to prevent rotation, as was done by Viano et al. [85].

Of the aforementioned rigid impact studies, a number of them warranted further analysis to get a better understanding of the reasons for some of the seeming discrepancies between the reported responses in terms of shape and rate sensitivity. In 2001, Hardy, Rouhana, and Schneider [80] reanalyzed the existing body of rigid-bar abdominal-impact data by first digitizing the published results of the investigations of Stalnaker and Ulman [83], Cavanaugh et al. [79], Viano et al. [85], and Nusholtz and Kaiker [86]. For simplicity, only eight points were taken from each curve for this analysis. The Nusholtz cadaver data were split into high- and low-speed corridors, and equal-stress/equal-velocity scaling was applied. The Viano cadaver data were averaged within the 6.7- and 9.4-m/s ranges to yield two mean curves. Upper abdomen tests were eliminated from the Stalnaker primate data, and of the remaining tests, those conducted at 10 m/s, +/− 1.5 m/s were selected for analysis (seven tests). Again, equal-stress/equal-velocity scaling was applied. The data were averaged across all species to obtain the "human" response. The prescribed three loading stages were generated using 9.6 % and 27 % compression break points, and the data were normalized to an abdominal depth of 289 mm. Velocity scaling was used to generate a 6-m/s curve.

Because the Cavanaugh rigid-bar data showed a rate-dependent response, the Cavanaugh corridors were plotted for comparison to other reanalyzed datasets. The comparisons were simplified by digitizing five-to-eight points for each boundary of the corridors, which were piecewise continuous linear approximations of the continuous data. Similarly, for ease of comparison, the corridors were truncated at 9 kN for the 10-m/s corridor and 140 mm for the 6-m/s corridor. These comparisons are shown in Figs. 14.9 and 14.10. Although the various datasets show differences in curve characteristics, there is overall agreement in terms of stiffness for the 10-m/s corridor. For the 6-m/s corridor, there are differences in terms of curve shape and loading rate. This exercise suggests that the apparent rate effects, whether of a viscous or mass-recruitment nature, depend on the relative impactor-to-subject mass ratio, with higher ratios resulting in more rate sensitivity. The loading-phase characteristics depend on impactor shape, with larger relative surface area (distributed interface) resulting in a steeper rise to a plateau, and narrower bars (focused loading) resulting in a more gradual rise with no plateau. The degree of hysteresis during unloading is related to impact location and the nature of the impactor. Larger loading surfaces that result in contact with the ribcage, which produces restorative forces, are associated with less hysteresis than impacts that do not engage the ribcage. Focused loading that does not involve the ribcage generally results in a rapid drop in force for little reduction in penetration during the unloading phase of an abdominal impact.

After reanalyzing the available rigid-bar data and forming theories to explain the differences observed between datasets, Hardy et al. [80] conducted new testing of the cadaver abdomen in an

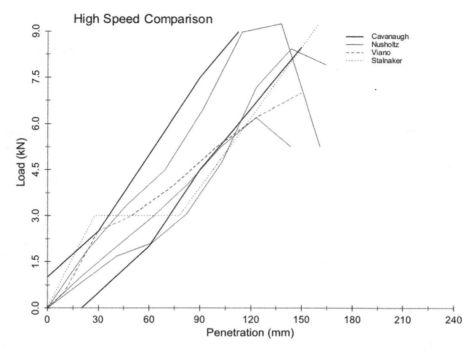

Fig. 14.9 Reanalysis of existing high-speed data (10 m/s) rigid-bar abdominal impact data showing reasonable agreement between studies [80] (Reprinted with permission from the Stapp Association)

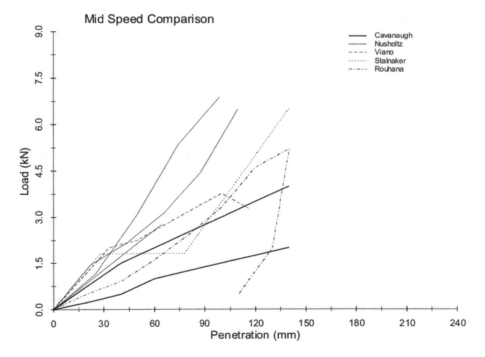

Fig. 14.10 Reanalysis of existing mid-speed data (6 m/s) showing differences in stiffness between the various studies [80] (Reprinted with permission from the Stapp Association)

effort to support their theories. In addition, a goal of their research was to examine the response of the human cadaver abdomen to various types of frontal impact. This included rigid-bar, seatbelt, and close-proximity airbag loading. Each of these loading modes is described in subsequent sections herein, beginning with the rigid-bar tests.

Hardy et al. [80] conducted both free- and fixed-back impacts of the cadaver abdomen, in two regions. A 48-kg ballistic pendulum was fitted with a 2.5-cm diameter, 46-cm long, rigid bar. The 48-kg mass was used because it was the median of the masses used by Cavanaugh and it was substantially larger than the effective mass of a typical abdomen. Each cadaver was positioned in a seated, upright, free-back posture with the legs positioned forward on a curved plastic skid, with the hands positioned in front of and above the head, as shown in Fig. 14.11. Accelerometer blocks were attached to T11 and L3, as were dual-marker target masts to facilitate kinematics analysis using high-speed (1 kfps) film. Prior to impact, the lungs of each cadaver were expanded using compressed air. Heated, 13.8-kPa, normal saline was used to perfuse the arterial and venous systems. Each cadaver was suspended prior to impact, and released at the moment of impact so as to be unconstrained during impact. A cargo net caught the cadavers after impact and prevented additional damage.

Nine tests were conducted using nine cadavers with average age, stature, and mass of 78 years, 170 cm, and 68 kg respectively. Six tests were conducted at the mid abdominal level (L3), with three of these in the 6-m/s range (6.3 m/s average) and three in the 9-m/s range (9.2 m/s average). Three additional tests were conducted at the upper-abdominal level, with two of these in the 6-m/s range and one in the 9-m/s range. For these tests, the rigid bar made initial contact with the epigastric region, approximately at the level of T11.

A fixed-back condition was also investigated. This configuration was designed to eliminate the motion of the spine and the effects of whole-body mass. This was in large part because the study was conducted in support of a new dummy abdomen that Rouhana et al. (2001) were developing [81]. Therefore, it was desired to investigate the role of spinal flexion and cadaver mass on abdominal response. A single large male cadaver (175 cm, 88 kg, 80 years) was used for all fixed-back tests. The specimen was held in a seated posture, fastened

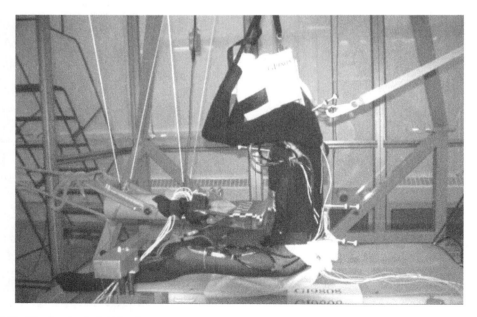

Fig. 14.11 The free-back rigid-bar impact apparatus and configuration used by Hardy et al. (2001) to investigate the response of the PMHS abdomen [80] (Reprinted with permission from the Stapp Association)

to a rigid seat back using "u" clamps installed around the spine from behind. To avoid damage to the specimen during repeated testing, the pendulum motion was arrested after 200 mm of abdominal penetration. Steel cables were used to tether the pendulum to an energy-absorption apparatus that consisted of steel rods. As tension developed in the cables, the rods were bent, which dissipated excess energy in the system. This specimen was impacted seven times using three different speed ranges: 3, 6, and 9-m/s. The order of the tests was designed to minimize changes in the response due to previous testing, as well as to disclose the changes that occurred.

This new set of free-back rigid-bar tests corroborated the Cavanaugh results, and reinforced the assertion that the response of the abdomen is rate sensitive. However, the observed rate sensitivity could be due to either mass recruitment or tissue viscosity, or both. The 48-kg impactor was important to obtaining this response. This is because the response of a rate-sensitive system is affected by the magnitude and rate of energy transfer, which are significantly influenced by the relationship between the effective mass of the system and the mass of the striking object.

Hardy et al. [80] also developed new response corridors (Figs. 14.12 and 14.13). The average dynamic stiffness of the mid-speed tests was 27 kN/m compared to 21 kN/m for Cavanaugh. The average dynamic stiffness of the high-speed tests was 63 kN/m compared to 70 kN/m for Cavanaugh. Differences in impact speed could account for the slight differences in stiffness between the results for the two studies. The upper abdomen tests also showed greater initial stiffness in both speed ranges (6 and 9 m/s) than did the mid abdomen tests. Further, there was less hysteresis in the upper abdomen response. In keeping with the original theory developed from examining existing data, these findings are thought to be due to greater involvement of the ribcage.

The loading responses to rigid-bar testing of the upper abdomen (high speed) and the fixed-back tests were similar to the responses reported by Stalnaker, who characterized the loading curve as a steep rise, followed by a gradual slope or plateau, and then another steep rise.

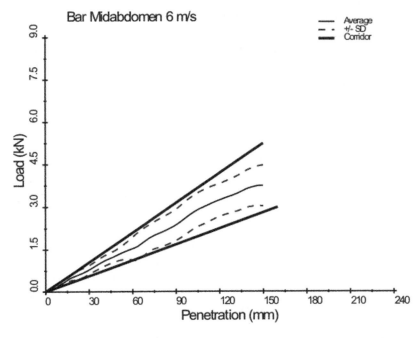

Fig. 14.12 The low-speed (6-m/s), midabdomen, rigid-bar impact response corridor developed by Hardy et al. (2001) for PMHS abdomen loading [80] (Reprinted with permission from the Stapp Association)

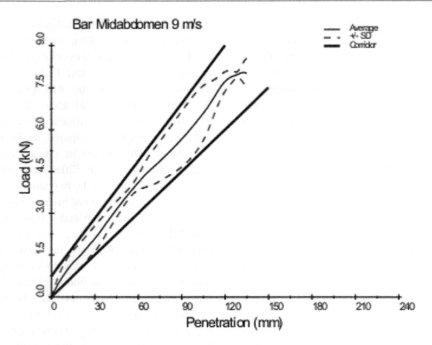

Fig. 14.13 The high-speed (9-m/s), midabdomen, rigid-bar impact response corridor developed by Hardy et al. (2001) for PMHS abdomen loading [80] (Reprinted with permission from the Stapp Association)

Greater rib involvement, as well as a greater degree of bottoming out against the spine, could have contributed to this rise-plateau-rise character. Melvin et al. (1988) suggested that for this type of response, the initial rise and subsequent plateau could be attributed to viscous characteristics, followed by inertial loading and tissue compression effects [88].

The fixed-back tests further also demonstrated the rate-dependent nature of the abdomen. The load-penetration responses of the fixed-back tests matched free-back results initially, but dropped below the corridors later in the impact. For the free-back tests, the reaction loads measured at the pendulum result from compressing abdominal tissue and accelerating the body. For the fixed-back tests, the reaction loads result from compressing abdominal tissue only. It is assumed that the whole-body motion of a free-back specimen would begin at a force level below that at which the fixed-back curves begin to fall below the corridors. Elimination of the inertial component of whole-body motion (i.e., removing a large mass-times-acceleration component) is likely the reason for the load-penetration responses for fixed-back tests falling below the corridors for free-back tests, in the absence of large compressive loads resulting from crushing tissues against the spine.

Howes et al. (2012) conducted tests using four PMHS that were subjected to four different fixed-back loading conditions [89]. The primary focus of this study was to quantify internal organ kinematics in blunt impacts to the thorax and abdomen. The secondary focus of this study was to gain an understanding of the structural interactions of the abdominal contents during impact to better understand injury mechanisms in motor vehicle crashes. High-speed biplane x-ray was used to image the motion of the thoracoabdominal contents in three dimensions. The quantification of relative internal organ kinematics during blunt impact can be used to identify potential crash-induced injury mechanisms for further investigation using finite element models and isolated experiments. For the abdomen (three cadavers), the loading modes consisted of a driver-side shoulder belt applied at a target speed of 3.0 m/s, a 114-mm diameter, 32.2-kg cylindrical probe impactor applied to the abdomen at 3.0 and 4.0 m/s without engaging the thorax, and a 25.4-mm diameter, 32.6-kg rigid-bar impact at 6.7 m/s centered at the mid-umbilicus. The two cadavers impacted in the abdominal region were subjected to multiple impacts of increasing severity. The two initial abdominal impacts were non-injurious, with the final test for each cadaver designed as a destructive test to induce a considerable amount of marker excursion for the instrumented organs. Tests were conducted in inverted, fixed-back configurations, as shown in Fig. 14.14. Tests were conducted using an inverted posture to adjust the initial position of the abdominal organs to obtain a more anatomically correct model [90], to counteract the effect of gravity.

The probe impacts resulted in anatomically cranial and posterior marker trajectories. The diaphragm moved in a circular pattern with a slight anterior-superior trajectory initially, followed by an arc in the left posterior to right anterior direction. The liver moved in a distinct arc to the right before continuing posterior and inferior from the initial position. The right lobe of the liver showed greater excursion than the left lobe, but the patterns were similar. Motion of the stomach and mesentery occurred in less defined patterns, first shifting superior and anterior before arcing toward the left, inferior, posterior region. The marker motion for an exemplar 3-m/s probe impact as measured by high-speed biplane x-ray is shown for the sagittal and transverse planes in Figs. 14.15 and 14.16, respectively.

Building upon the notion that inversion of a cadaver can place the thoracoabdominal contents closer to the in-vivo condition, Howes et al. (2013) investigated this hypothesis directly [91]. Biplane x-ray was used to image two cadavers in upright and inverted postures, and the three-dimensional variation in the relative abdominal organ position was quantified. The abdominal organs were instrumented using radiopaque markers. The specimens were ventilated and perfused, and residual air was removed from the abdominal cavity. Intuitive changes in organ position were observed due to the effect of gravity. When upright, the superior-inferior separation between the diaphragm and liver markers ranged from 95 to 169 mm, but when inverted, the organs shifted cranially and the separation fell to within 66 to 81 mm. These data were scaled and compared to the Global Human Body Models Consortium

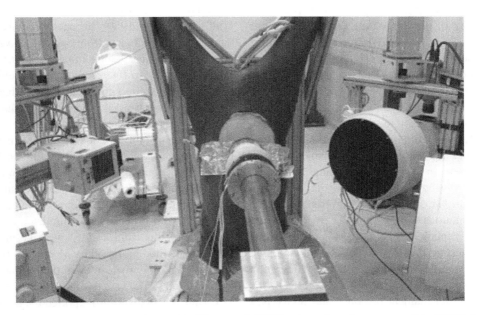

Fig. 14.14 The inverted cadaver position used by Howes et al. (2012) to investigate the response of the PMHS abdomen to probe impacts visualized by high-speed x-ray [89] (Reprinted with permission from the Stapp Association)

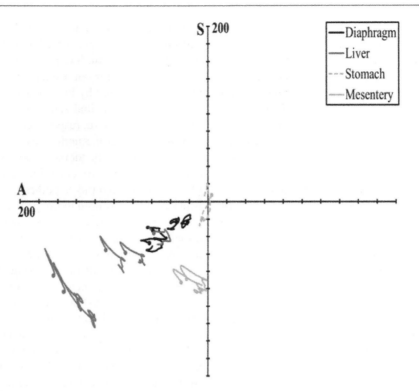

Fig. 14.15 A sagittal plane perspective of marker/organ motion for a 3-m/s probe impact of the PMHS abdomen from Howes et al. (2012), measured using high-speed biplane x-ray [89] (Reprinted with permission from the Stapp Association)

(GHBMC) model geometry and to the positional MRI data of Beillas et al. (2009) from nine human subjects in seated postures [92]. The overall shapes and relative positions of the inverted cadaver organs were deemed to compare better to both the human subjects and model geometry. These results reinforce the idea that interpretation of cadaver test results and comparison to finite element simulations is not straightforward. The ability of a model to have the same mechanical response and to predict the same injuries as observed during tests for which the abdominal viscera are in substantially different locations and orientations is very limited. This phenomenon can explain difficulties in using historical cadaver testing data for model development and validation. It is also an important consideration for investigators designing cadaver tests in the future: the mechanical response and injury patterns will be potentially more accurate for the inverted case, and the results will be more directly comparable to modeling efforts.

14.5.1.2 Side Impact

Walfisch et al. (1980) dropped unembalmed human cadaver subjects from one and two-meter heights to examine lateral impact response and injury [93]. The surface impacted was either a rigid or deformable simulated armrest that was struck by the right side of the subject at the level of the ninth rib. Contact speeds were 4.5 m/s for the 1 m drop, and 6.3 m/s for the 2 m drop. Deflection data was determined by film analysis, where deflection was defined as intrusion of the armrest relative to the spine, not the opposite side of the subject. Force data were obtained from load cells located beneath the simulated armrest. The data for the rigid armrest were normalized using the method proposed by Mertz [94] for lateral impact by the International Standards Organization [95].

Cavanaugh et al. (1996) reported the results of 16 side impact sled tests using a rigid or padded flat wall or a simulated armrest [96]. The setup was similar to that used at the University of

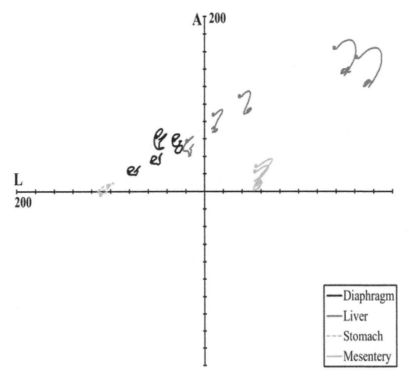

Fig. 14.16 A transverse plane perspective of marker/organ motion for a 3-m/s probe impact of the PMHS abdomen from Howes et al. (2012), measured using high-speed biplane x-ray [89] (Reprinted with permission from the Stapp Association)

Heidelberg [97] by Kallieris et al. (1981) in the tests which served as the basis for the Thoracic Trauma Index (TTI). The tests used a deceleration sled traveling 6.7 or 8.9 m/s that impacted a hydraulic snubber. During deceleration, the cadaver slid across a low friction surface such that the left side of the subject struck a rigid or padded wall that was instrumented with load cells. In some tests, the subject impacted a simulated armrest made of either soft or stiff paper honeycomb, nominally 69 or 138 kPa crush strength.

14.5.2 Restraint Testing

Various types of occupant restraints have been tested using animals and cadavers. These include primary seatbelt systems and supplemental passenger airbags. Both lap-belt only and lap-shoulder configurations have been tested, with some attention paid to load limiting, and considerable attention given to pretensioner loading. The 4-point restraint system developed by Rouhana et al. (2003) remains to be tested with respect to abdomen interaction [98]. Given repeatability and measurement issues, the impact response to airbags has been investigated using a surrogate airbag as well.

14.5.2.1 Seatbelt Tests

Although part of the recent effort of Howes [89] involved loading of the thorax and abdomen using a driver shoulder belt, the focus of that study was not interaction of the seatbelt with the abdomen. Several earlier studies have focused on this interaction using animal, animal cadaver, and human cadaver experimental models. Commonly, swine and swine cadaver models have been selected to investigate abdominal response to seatbelt loading. In particular, Rouhana et al. (1986) conducted 15 impacts on swine cadavers using a controlled-stroke MTS machine [78]. Also, Miller (1989) conducted 25 dynamic belt-loading tests into the abdomens of anesthetized swine [99]. These belt-loading tests were performed using a haversine displacement-time function with peak speeds near 3.7 and 6.3 m/s ("low" and "high" speeds). The animals were placed in a supine position in a V-shaped back support. An arch-shaped yoke was used to drive seatbelt webbing in to the abdomen at the level of the fourth lumbar vertebra. The load-penetration responses were characterized by a slightly convex ramp averaging 30 kN/m, followed by an essentially vertical unloading. Peak compression ranged from 6 % to 67 %. The swine cadaver and anesthetized swine data were compared to examine the scaling possibilities between human cadavers and living humans. Normalization was performed by Rouhana et al. (1989 and 1990) using equal-stress/equal-velocity scaling to account for differences between subject mass and anteroposterior dimension, and the data were separated in to low and high-speed corridors [100, 101]. These corridors were scaled by van Ratingen et al. (1997) to develop biofidelity seatbelt loading performance targets for the Q3, 3-year old child dummy [102]. These swine data were used by Rouhana to design the frangible abdomen for the Hybrid III dummy.

Hardy et al. (2001) conducted six seatbelt tests using three cadavers and a peak-loading rate of 3 m/s [80]. The cadavers were loaded about the mid abdominal region, approximately at the level of the umbilicus, as shown in Fig. 14.17. The webbing was placed flat against the anterior surface of the abdomen, and fastened to a pneumatic driving mechanism behind the cadaver. The geometry of the belt was designed to optimize belt/abdomen interaction, and to minimize "roping" of the belt. The seatbelt webbing was routed straight back from the sides of the cadaver. Peak loading was roughly 3 m/s using an approximately haversine penetration speed-time history. A seatbelt loading response corridor was generated for these data (Fig. 14.18). The initial stiffness was found to be 120 kN/m which is four times greater than the results of Miller [99], and probably results from differences in the way the tests were conducted as well as differences in the test subjects. Miller used a yoke device that resulted in the ventral-most portion of the abdomen being loaded initially, with an increasing section of

Fig. 14.17 The free-back seatbelt loading apparatus and configuration used by Hardy et al. (2001) to investigate the response of the PMHS abdomen [80] (Reprinted with permission from the Stapp Association)

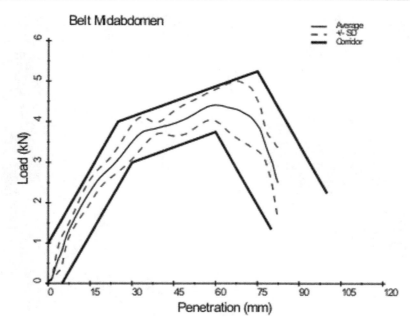

Fig. 14.18 The 3-m/s midabdomen seatbelt response corridor developed by Hardy et al. (2001) for PMHS abdomen loading [80] (Reprinted with permission from the Stapp Association)

abdomen being loaded as the yoke displacement progressed. The contact area increased as the penetration increased, but the sides of the abdomen were not particularly constrained. Hardy et al. [80] wrapped the webbing around the anterior and lateral aspects of cadaver abdomens. The contact area was quite large from the beginning of each test, and resulted in a more distributed and consistent application of load around the circumference of the abdomen, with the webbing restricting lateral motion of the sides of the abdomen. Both of these factors are thought to be responsible for the higher stiffness values found by Hardy.

Further, Hardy found the initial stiffness to be about the same for both the lower and midabdomen. The initial stiffness of the abdomen was about the same for both the free and fixed-back seatbelt tests as well. However, the fixed-back condition eventually resulted in greater stiffness than did the free-back condition. This is in contrast to rigid-bar tests, in which the stiffness eventually decreases for the fixed-back condition. Rigid-bar tests typically involve a free-flying mass and concentrated loading, while these seatbelt tests involved a driven pneumatic cylinder and distributed loading. It is possible that these differences can outweigh the influence of the missing whole-body acceleration during the fixed-back seatbelt tests.

Trosseille et al. (2002) presented the results from testing six cadavers in a study using high-power pretensioners, using seatbelt penetration rates from 8.2 to 11.7 m/s in an out-of-position condition [103]. The seatbelt webbing was routed above the iliac crests and around the abdomen, similar to the configuration used by Hardy et al. [80]. Abdominal compression was reported as ranging from 25 % to 32 %, but belt penetration and speed were measured as the travel of the lateral aspect of the belt and not at the umbilicus of the specimen. The load-penetration responses varied substantially, and were mapped onto the corridor developed by Hardy for 3-m/s belt loading using a simple lumped-parameter model. The model did not match the corridor well. No correlation was found between injury and measured response parameters.

Foster et al. (2006) measured the impact and injury response of the human cadaver abdomen

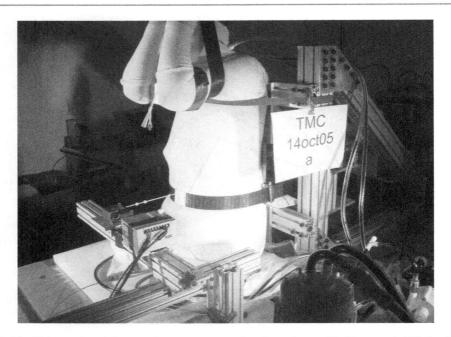

Fig. 14.19 The high-speed seatbelt pretensioner apparatus and configuration used by Foster et al. (2006) to investigate the response of the PMHS abdomen [104] (Reprinted with permission from the Stapp Association)

to seatbelt loading at high-speed using pyrotechnic pretensioners [104]. A specialized seatbelt loading apparatus equipped with pretensioners was developed for this task. The loading apparatus was designed to maximize belt/abdomen interaction in a controlled and measurable fashion. Load was measured using two low-mass seatbelt load cells. Eight subjects were tested under worst-case scenario, out-of-position conditions. Similar to Hardy et al. [80] and Trosseille et al. [103], a seatbelt was placed at the level of the mid-umbilicus and drawn back along the sides of the specimens. A seated, upright, fixed-back configuration was used as shown in Fig. 14.19. Penetration was measured by a laser tracking the anterior aspect of the abdomen and high-speed video. Three different pretensioner systems (A, B, and C) provided a spectrum of energy input to the cadaver, and therefore a spectrum of injury response. The B and C systems employed single pretensioners. The A system consisted of two B system pretensioners. The vascular systems of the subjects were perfused. Peak anterior abdominal loads due to the seatbelt ranged from 2.8 to 10.1 kN. Peak abdominal penetration ranged from 49 to 138 mm. Peak penetration speed ranged from 4.0 to 13.3 m/s. Three cadavers sustained liver injury: one AIS 2, and two AIS 3. Response corridors for the A and B system pretensioners were proposed. The response corridor for the A system is shown in Fig. 14.20.

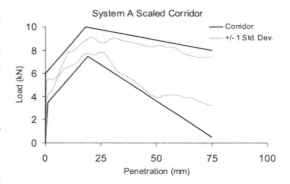

Fig. 14.20 The "A" system (highest penetration speed) pretensioner system response corridor developed by Foster et al. (2006) for PMHS abdomen loading [104] (Reprinted with permission from the Stapp Association)

During the initial phase of the pretensioner deployment, the seatbelt loads the lateral aspects of the cadaver abdomen. This results in medial motion of the sides of the cadaver, as the belt pushes each side toward the spine. This early lateral compression has the effect of increasing tension in the belt, and therefore the measured load, before penetration begins. This produces a very stiff initial response. The phenomenon is common to the studies of Hardy, Foster, Trosseille, and Lamielle [105], which is given further treatment subsequently. The piecewise continuous linear approximation corridor for Foster's A system is characterized by a 3,500 kN/m slope on the lower boundary for 1 mm followed by a general 222 kN/m slope for another 18 mm. The B system approximation is characterized by an initial 3,000 kN/m slope on the lower boundary followed by a local nadir in the response and then a gradual roll off. The upper boundary of both system corridors begins at 6 kN. In contrast, the seatbelt tests of Hardy produced a corridor with initial stiffness of 120 kN/m. In addition, differences between the motion of the seatbelt at the umbilicus and along the sides of the cadaver were observed, suggesting that Trosseille underestimated penetration and overestimated speed by measuring seatbelt deflection at the sides of the cadaver.

For Foster's tests, all of the accepted IARVs showed similar trends in transition from the no-injury to injury responses. This includes metrics developed by Rouhana. Rouhana et al. (1985) investigated the injury prediction ability of the product of maximum penetration speed (Vmax) and maximum compression (Cmax) [106]. The product Vmax*Cmax was termed the Abdominal Injury Criterion (AIC). The effect of work done by the abdomen during impact was investigated by Rouhana et al. (1987) by evaluating maximum load (Fmax) times maximum compression (Cmax) [107]. This product, Fmax*Cmax, was said to be related to the energy imparted to the abdomen during impact. Table 14.1 summarizes a comparison of the IARV levels determined by Foster [104] and common to Trosseille [103] and Hardy [80]. Most of the injury responses of Trosseille were AIS 0 to AIS 2, which corresponded to minor mesenteric tears. Only one test produced

Table 14.1 Comparison of IARV levels for seatbelt loading tests

Parameter range	Foster et al. [104]	Trosseille et al. [103]	Hardy et al. [80]
Fmax (kN)	2.8–10.1	6.1–10.3	3.1–6.1
Peak penetration (mm)	49–138	58–76	56–114
Cmax (%)	25–55	25–32	26–37
Penetration speed (m/s)	4.0–13.3	8.2–11.7	2.1–5.6
Peak V*C (m/s)	0.44–4.13	0.85–2.10	0.4–1.1
MAIS	0–3	0–5	0–4

AIS 5 tears of the liver, spleen, and omentum. Hardy obtained AIS 0 to AIS 4 injuries, which was due largely to rib fractures, which were not observed by Foster. Foster also found that pressure in the aorta above the abdominal bifurcation might be useful for predicting liver injury related to seatbelt loading. Currently, there is no one preferred or widely accepted metric for the prediction of abdominal injury, regardless of the loading mode.

More recently, Lamielle et al. (2008) tested eight cadavers under two different general conditions [105]. Lamielle used the same general configuration as Trosseille and Foster. The first was described as a "submarining" case, for which the penetration speed of the belt was lower (~4 m/s), but the peak compression was higher (~40 %). The second was described as an OOP (out-of-position) case, for which the penetration speed was higher (~8 m/s), but the peak compression was lower (~30 %). For both conditions, multiple contusions of the small bowel were noted. For the OOP case, two cadavers exhibited tears of the small bowel. However, it is important to note that the cadavers were frozen and thawed prior to testing, and were used for subsequent pelvis and thorax tests after the seatbelt tests. Further, during the subsequent tests, the specimens were perfused with a mixture of alcohol and India ink. Although it is not specified, it is assumed that the contusions were evaluated on the basis of observation of diffuse areas of ink (recorded as "infiltrated"). Given the frozen/thawed nature of the specimens originally, and the large amount of testing to which they were subjected and the associated time out of a cooler, it is not clear that they were not predisposed to small bowel tearing. Ignoring this possibility, this study suggests an OOP condition might be more likely to produce jejunum injury than a submarining condition. Overall, the cadavers sustained MAIS 2–3 abdomen injuries for the submarining condition and MAIS 2–4 injuries for the OOP. Nonetheless, the primary objective of the study was to measure the dynamic 3D deformation of the abdomen to aid in the development or validation of human computational models.

Rouhana et al. (2010) reviewed literature on pretensioners and provided consideration for a test to evaluate the risk of abdominal injury as a function of the speed and/or force with which a pretensioner pulls belt out of a belt system that is being worn by a vehicle occupant [108]. Abdominal injury risk curves were developed for AIS 2+ and AIS 3+ injuries based on published PMHS tests and with the help of the Ford Human Body Finite Element Model.

Very recently, Untaroiu et al. (2012) examined pretensioner loading of cadavers using a three-point seatbelt system. Pretensioners were used at the retractor and right anchor in a passenger seat configuration, or at the retractor and both left and right anchor locations [109]. Four cadavers having BMI ranging from 15.6 to 31.2 were positioned in production seats. Although scaled chest compression was limited to 10 %, two cadavers having higher BMI exhibited damage to the spleen. Seatbelt speed and Fmax*Cmax were found to best predict AIS 2+ and AIS 3+ injury, respectively.

14.5.2.2 Airbag Tests

In March 1997, the NHTSA issued an interim rule to FMVSS 208 that temporarily allowed vehicle manufacturers to conduct the unbelted-dummy portion of frontal crash testing using a generic 30-mph sled test. The prescribed 120-ms halfsine pulse effectively increased the time available to fully inflate and airbag compared to a 30-mph, rigid-barrier vehicle test. This facilitated the development of less aggressive depowered, or "sled certified" airbags. Augenstein et al. (2004) observed depowered airbags to be performing better than first generation airbags in-terms of reduction of airbag related fatalities. In an examination of 205 frontal impact cases, depowered airbags were found to have contributed to no fatalities for ΔV less than 20 mph for out-of-position adults, but first-generation airbags could have contributed to eight driver fatalities under out-of-position conditions. Kahane (2006) concluded that the overall fatality risk in frontal crashes of front seat child passengers (0–12 years) was 45 % lower for depowered airbags than first-generation airbags, while depowered airbags had "preserved the life-saving benefits of first generation airbags" in moderate to severe frontal collisions [110].

A laboratory study by Lau et al. (1993) examined the potential for injury from out-of-position interactions of thorax and abdomen of anesthetized swine with deploying first-generation driver airbags [111]. For the abdomen, they positioned a midline point on the abdomen that was 75 mm below the xiphoid process over the center of an airbag module/steering wheel unit. The abdomen was in contact with the airbag cover and the wheel was mounted rigidly to a bedplate during deployment. Splenic lacerations were the most frequent abdominal injury, often extending through the thickness of the spleen. Liver injuries were either stellate lacerations or junctional tears, typically with severe hemorrhage. While they did observe petechial hemorrhages in the omenta, they found no lacerations or contusions to the stomach or intestines.

Hardy et al. (2001) conducted out-of-position tests using passenger airbags to characterize the penetration-time history of the cadaver abdomen [80]. Two hollow carbon/composite shafts were inserted through the cadaver abdomen to measure anteroposterior abdomen compression during airbag loading. The cadavers were suspended such that the center of the vertically oriented face of the airbag module was aligned with the umbilicus (L3/L4). The response data were used to develop a surrogate airbag loading device capable of mimicking the conditions during the initial breakout phase of passenger airbag loading (low-mass, high-speed, distributed loading). The surrogate airbag device was therefore designed

Fig. 14.21 The surrogate airbag testing apparatus and configuration used by Hardy et al. (2001) to investigate the response of the PMHS abdomen [80] (Reprinted with permission from the Stapp Association)

to penetrate the cadaver abdomen at peak speed of approximately 13 m/s. The device includes a pneumatic firing mechanism that accelerates a lightweight (approximately 1 kg) aluminum impactor constructed of welded thin-wall tubing, as shown in Fig. 14.21. The face of the impactor is the sidewall of a 7.6-cm diameter tube that is 20 cm long. The loading by this device is characterized by an initial very fast, very steep rise, followed by a slightly oscillatory decay. The event occurs in less than 8 ms, but the most important part takes less than 2 ms. The peak loads range between 3.8 and 4.6 kN. A load-penetration corridor (Fig. 14.22) was developed for the cadavers' response to this device. This corridor essentially represents the impulse response of the abdomen. The initial stiffness is the most important part of this corridor. The lower bound of the initial stiffness is 500 kN/m, and the upper bound is 2,000 kN/m. The descending slope of the lower bound is −75 kN/m, while the two upper bound descending slopes are −100 kN/m followed by −50 kN/m. The airbag tests resulted in injuries to the ribs, colon, liver, mesentery, peritoneum, and diaphragm. Some of these injuries are commensurate with those observed from exposure to underwater blast events. Further, this is one of the few studies to generate tears of the colon in cadavers. The surrogate airbag tests resulted in similar but far fewer injuries, probably because only the initial breakout phase of loading is simulated by this device.

14.6 Injury Mechanisms, Criteria, and Tolerances

An injury mechanism from blunt trauma can be defined as a description of the cause of organ injury. That is, it is a description of how the state of stress or strain that produced injury was set up in the organ. An injury criterion is a mathematical relationship based on empirical observation that formally describes a relationship between some measurable physical parameter interacting with a test subject and the occurrence of injury which directly results from that interaction. Injury tolerance can be defined as the value of some known injury criterion which delineates, with a given statistical probability, a non-injurious event from an injurious one.

In general, there may be many physical parameters that have some correlation with the injury outcome of a set of experiments. But, often this is just because the physical parameter (e.g., acceleration) is a measure of the occurrence of the

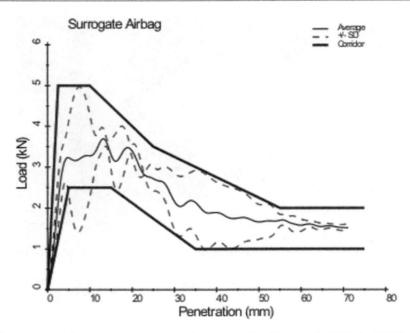

Fig. 14.22 The surrogate airbag impact response corridor developed by Hardy et al. (2001) for PMHS abdomen loading [80] (Reprinted with permission from the Stapp Association)

impact event. All physical measurements may increase as impact severity increases. This does not necessarily mean that there is a causal relationship between every physical parameter measured and the injury outcome. Similarly, mixing together multiple physical inputs in statistical analyses may improve correlation coefficients, but the resulting injury criteria may have no physical meaning or biomechanical basis. Caution is advised when performing, using, and interpreting these types of analyses.

In automotive safety studies, the injury criteria and tolerances are determined using human surrogates (human cadavers, anesthetized animals, etc.). The criteria become the basis for instrumentation in another human surrogate, the crash test dummy. Automotive safety engineers utilize crash test dummies in an attempt to measure the important physical parameters during a crash test of a vehicle under development. These measurements are then compared to the human tolerance values to interpret how well a design performs. The following discussion centers on the various criteria and mechanisms of abdominal injury from blunt trauma, first in general, and then as they apply to particular abdominal organs.

14.6.1 General Considerations for Injury Mechanism Data

Early studies of blunt abdominal trauma resulted in the hypothesis that the liver and spleen are less able to absorb energy from impact than the hollow organs because of their anatomic vulnerability to direct impact, "limited mobility", pedicle attachments, and proximity to the margins of the lower ribs [112–114]. Of course, the incompressible nature of the solid organs would predispose them to injury compared to the hollow organs.

Walker (1969) suggested that injury to the elderly or intoxicated can occur from a relatively minor blow because of weakened or relaxed muscles [115]. This has been supported by clinical observations of severe internal injury from apparently innocuous contact in the presence of a relaxed abdominal wall [116]. The pathological or normal physiological state of organs can also have a marked effect on the injury outcome in blunt trauma. For example, biomechanical experiments have shown that cirrhotic livers are stiffer and less extensible before failure in tension than non-cirrhotic livers [117]. Other considerations include liver fibrosis, cancer, splenomegaly associated with mononucleosis and other medical conditions, a full bladder, or presence of gas or chyme in the intestines (chyme is a mixture of partially digested food and digestive secretions) [116, 118, 119]. Previous surgical procedures can also be a predisposing factor to injury, although most experimental studies would not include such subjects, this may be the case for living humans in real crashes. The presence of adhesions in the abdominal cavity may or may not predispose the subject to injury.

The location of impact on the body can have a dramatic effect on the injury outcome [120, 121]. Baxter and Williams (1961) found a fivefold increase in hepatic injury and a twofold increase in renal injury when the impact location was the side as opposed to the front of their subjects [120].

Zhou et al. (1996) found an 80 % reduction in thoracic tolerance as a function of age [122]. Similarly, Yamada (1970) has shown that the tensile strength of the stomach, large and small intestines, kidneys, ureters, and urinary bladder decreases with age [46]. He noted 28 %, 42 %, 42 %, 17 %, 15 %, and 30 % reductions, respectively, when comparing strength at age 20 versus age 70. The ultimate tensile strength of the rectus abdominus muscle of a 75-year old is approximately 60 % of what it is for a 25-year old. On the younger end of the spectrum, Sturtz (1980) found that the "loadability" of the liver and spleen is higher at the age of 10 than at ages over 15 [123]. He also observed that the abdominal region is proportionately larger and that the liver is more exposed in children than in adults.

analysis techniques. In some cases, nonparametric approaches (e.g., Bayesian) are used. Much of the historical injury biomechanics data are double censored. For the injury cases, the peak parameters are known, but it is not known at what level injury occurred (left censored). For the non-injury cases, again the peak response parameters are known, but it is not known at what level injury might have occurred. Frequently, the highest value of a given parameter for which injury did not occur, and the lowest value of the same parameter for which injury did occur are reported in addition to any parametric or probabilistic analyses.

For example, in the high-speed pretensioner tests of Foster et al. (2006), the highest level of scaled seatbelt load for which injury did not occur was 7.9 kN and the lowest level for which injury occurred was 8.3 kN [104]. A transition from no injury to injury was observed between 112 and 124 mm for peak scaled penetration, between 38 % and 44 % for peak compression. The highest levels for which there was no injury for penetration speed, V*C, and Vmax*Cmax, were 7.5, 2.9, and 1.2 m/s, respectively. The lowest levels for which AIS 2+ injury occurred for penetration speed, V*C, and Vmax*Cmax, were 8.5, 1.8, and 4.1 m/s, respectively. The highest level of Fmax*Cmax for which injury did not occur was 2.8 kN and the lowest level for which injury occurred was 3.6 kN. The following is a discussion of potential metrics involved in the mechanism and prediction of abdominal injuries, and their associated values for different types of injury.

14.6.2 Injury Predictors

In general, researchers try to correlate measured physical response parameters with damage or injury outcomes in cadavers or animals, respectively. While there needs to be a biomechanical basis for such comparisons, the comparison does not have to be direct. That is, an input related to the parameter considered to be involved in the mechanism of injury can be used for injury risk prediction. Risk curves are often generated using nonlinear regression or survival

14.6.2.1 Acceleration

Acceleration of the body is a parameter that has long been examined for its role in the etiology of human organ injury [112, 124]. Eppinger et al. (1982), in the development of the Thoracic Trauma Index (TTI) for side impact, included the upper abdominal organs (liver, kidneys, and spleen) [125]. Using statistical techniques they developed a mathematical relation between thoracic AIS and a weighted sum of the input acceleration and the age of the cadaver subject tested. The accelerations used are those of the left upper or lower rib (whichever is larger), and the 12th thoracic vertebra. TTI has shown limited predictive ability for upper abdominal organs, but it does not apply to the middle or lower abdominal organs.

Mertz et al. (1982) and Mertz and Weber (1982) saw some association between acceleration and abdominal injury severity in anesthetized swine tested "out-of-position" with deploying airbags [126, 127]. When the injury data were paired with the lower spine peak resultant acceleration there appeared to be a reasonable correlation. However no statistical analysis data were presented for this series of tests.

In contrast, Horsch et al. (1985) and Lau et al. (1987) found "no correlation between abdominal injury and maximum lower spinal acceleration measured directly opposite to the impact site" [128, 129]. The lack of correlation between spinal acceleration and abdominal injury in this study is not surprising however, since the accelerometer was located on the spine of the subject, while the impact was to the front of the subject's abdomen. An accelerometer on the back of a subject will pick up only the whole-body acceleration from a frontal impact, while the local abdominal acceleration is expected to be related to injury.

Brun-Cassan et al. (1987) showed that "protection criteria based solely on thoracic acceleration measurements cannot account for the occurrence and severity of... abdominal injuries" [130]. More recently, Viano et al. (1989) found that spinal acceleration was not well correlated to abdominal injury in lateral pendulum impacts to human cadavers [85].

14.6.2.2 Compression

Crushing of organs can occur in blunt impact situations when the body surface deforms, and the organs interior to the impact site are compressed against an opposing surface. Crushing injury can occur at very low speeds. For example, crushing injury of the liver might occur in a low speed collision when an unbelted occupant strikes the instrument panel. During the event, the anterior surface (front) of the occupant is stopped by the instrument panel which causes local deformation. But, the posterior surface (back) of the occupant continues moving forward until enough force is built up to stop the entire body. Alternatively, an occupant in a collision with a lap belt improperly worn can sustain an abdominal injury when the lap belt stops the anterior abdominal wall, and the inertia of the rest of the body compresses the intestines between the belt/abdominal wall and the spinal column. The correlations of maximum compression with injury outcomes from the literature are summarized in Table 14.2

In a series of 45 lower abdominal impact experiments using anesthetized canine subjects, Williams and Sargent (1963) concluded that the mechanism of injury to the intestines in blunt impact is compression of the intestines against the spinal column [131]. These were fixed-back tests with the subjects supine. Melvin et al. (1973), in tests with surgically mobilized organs from anesthetized primates, showed that compressive strain (change in length per unit length), was related to the severity of kidney injury [75].

In experiments with primates and human cadavers, Stalnaker et al. (1973) found that abdominal compression was related to abdominal injury severity [132]. Different tolerance values were found for right versus left side impacts.

Rouhana et al. (1986) saw no correlation between compression and probability of either hepatic or renal injury in 117 experiments with rigid lateral impacts to anesthetized rabbits in a free-back condition [78]. However, in similar experiments with a crushable impactor there was a positive correlation for hepatic injury and compression [$r=0.97$; $p<0.03$], but not for renal injury. Similarly, Viano et al. (1989) showed poor correlation between compression and abdominal injury severity in lateral abdominal pendulum impacts of human cadavers [85].

In contrast, Miller (1989) found that maximum compression was well correlated with the severity of abdominal injury in 25 experiments using a seatbelt to load the lower abdomen of anesthetized porcine subjects [$r=0.69$; $p=0.0002$] [99]. These experiments were performed at relatively low speeds (mean = 3.6 m/s; range = 1.6 to 6.6 m/s).

Table 14.2 Correlations of maximum compression with injury

Author	(Year)	C_{max}*	Injury severity	Organ injured	Comments
Stalnaker et al.	(1973a)	60 % (L)	ESI>3	Upper abdomen	
		54 % (R)	ESI>3	Upper abdomen	
					N=96; primates and human cadavers; free side; scaled armrest; belt
Lau and Viano	(1981)	16 % at 12 m/s	AIS>3	Liver	Contusion
					N=26; *Oryctolagus cuniculus*; fixed back; compression=16 % for all experiments
Rouhana et al.	(1986)	29 %	AIS>3	Liver	ED_{50}
					N=214; *Oryctolagus cuniculus*; side impacts; 107 with force limiting Hexcel; free side; area=7.0 in.²
Miller	(1989)	37.8 %	AIS>3	Lower abdomen	ED_{25}
		48.4 %	AIS>3	Lower abdomen	ED_{50}
		48.3 %	AIS>4	Lower abdomen	ED_{25}
		54.2 %	AIS>4	Lower abdomen	ED_{50}
					N=25; *Sus scrofa*; lap belt impacts; fixed back; all energy into subject
Viano et al.	(1989)	43.7 %	AIS>4	Upper/midabdomen	ED_{25}
					N=14; unembalmed cadavers; rigid pendulum; side impact; free back; some energy into whole body motion
Talantikite et al.	(1993)	60 mm	AIS=4	Upper abdomen	AIS=4 threshold
					N=6; unembelmed cadavers; right side impact; 23.4 kg rigid pendulum impacts
Foster et al.	(2006)	44 %	AIS=2	Liver	Fractures through middle inferior and lateral inferior, quadrate lobe
		50 %	AIS=3	Liver	Minor fractures of cortex, diaphragmatic surface; burst disruption of parenchyma, visceral surface; transection of tip of right lobe
		55 %	AIS=3	Liver	Multiple fractures of cortex, diaphragmatic surface; disruption of parenchyma, visceral surface
					N=9; unembalmed cadavers; fixed-back seatbelt pretensioner loading, umbilicus
Lamielle et al.	(2008)	27 %	MAIS=3	Upper/midabdomen	Colon laceration (AIS=3); duodenum laceration (AIS=3); spleen laceration (AIS=2)
		30 %	MAIS=4	Upper/midabdomen	Spleen laceration (AIS=4); renal vessel laceration (AIS=4); colon laceration (AIS=3); liver capsular tear (AIS=2); jejunum-ileum laceration (AIS=2)
		31 %	MAIS=2	Jejunum-ileum	Laceration
		32 %	MAIS=3	Pancreas	Laceration
		41 %	MAIS=3	Upper abdomen	Pancreas contusion (AIS=3); liver capsular tear (AIS=2)
					N=8; unembalmed cadavers; fixed-back seatbelt loading in out-of-position or submarining configurations

*% of Entire A-P or lateral dimension of subject

Further, Talantikite et al. (1993) used a 23.4-kg mass driven at 5–7 m/s to study abdominal injury in six unembalmed human cadavers [133]. The best correlation for abdominal injury was obtained with deflection of the half abdomen (r = 0.85). They observed no AIS 4+ injuries below 60-mm deflection of the half abdomen and so proposed that value as a deflection limit.

14.6.2.3 Rate Effects and Viscous Injury

Rate of loading has long been recognized as a factor in injury outcome [134]. Many of the solid abdominal organs are fluid-filled. It is well known that fluid systems exhibit different mechanical characteristics under different rates of loading.

Impact studies by McElhaney et al. (1971), using live anesthetized primates, showed that small changes in impact speed had a profound effect on the injury level [135]. Melvin et al. (1973) also noted that the liver and the kidneys were both sensitive to rates of loading [75].

Lau and Viano (1981) held abdominal compression constant and varied the pre-impact speed in experiments with anesthetized rabbits [136]. They noted a significant increase in hepatic injury with increasing impact velocity.

Mertz et al. (1982) and Mertz and Weber (1982) in out-of-position airbag tests with anesthetized swine saw a sharp transition region for abdominal injury as a function of maximum rate of abdominal compression [126, 127]. The transition region began at a peak abdominal compression rate of 5.6 m/s, below which there were no AIS = 3 injuries. Between peak abdominal compression rates of 5.5 and 5.8 m/s there were some injuries above and below AIS = 3. Above 5.8 m/s there were no injuries less than AIS = 3. The data from this study were analyzed further by Mertz et al. (1997), examining injury risk as a function of the rate of abdominal compression for AIS 3+ and AIS 4+ injuries.

Rouhana et al. (1984, 1985) found that the product of maximum impact velocity (V) and maximum abdominal compression (C) was well correlated with the severity of abdominal injury from analysis of 117 abdominal impacts to anesthetized rabbits [106, 137]. They called this product, V*C, the Abdominal Injury Criterion (AIC).

Viano and Lau (1983) had previously found the same quantity was related to thoracic injury severity [138].

In their out-of-position airbag tests with anesthetized swine, Prasad and Daniel (1984) observed that the peak aortic blood pressure was well correlated with peak lower sternum velocity [139]. They used the peak aortic blood pressure as a measure of compression of the abdomen. Like Mertz and Weber [127], Prasad and Daniel [139] found that the maximum rate of abdominal compression was associated with the severity of injury. They observed the threshold of AIS 3 abdominal injury to be at 4.7 m/s which is slightly lower than that suggested by Mertz and Weber.

Citing the work by Rouhana et al. [137], Stalnaker and Ulman (1985) reexamined the data from 1973 studies on primates and human cadavers [83]. They concluded that the "V*C is a relevant parameter for predicting injury in sub-human primates", and that the "... values obtained from primates appear to be useful in predicting abdominal injury in man" [83]. When frontal and side impacts were considered separately, and each abdominal region was considered separately, the correlation of V*C with AIS 3+ injury was very high [r = 0.92 to 0.99]. Up to this time, the interrelationship between velocity and compression had used preimpact velocity, and maximum compression (Vmax*Cmax). In 1985, Viano and Lau extended the previous work by measuring the compression as a function of time, allowing the product of the velocity-time history and compression-time history to be compared with injury outcome [140]. Strong correlation was found between the maximum of the product, VC(t), and the resulting injury. This time-varying product was called the Viscous Tolerance Criterion for thoracic impact.

In a study of injury to 17 porcine subjects from steering system impacts, Horsch et al. (1985) showed good correlation of the abdominal injury outcome with the maximum of VC(t) [128]. They did not examine the relationship between Vmax*Cmax and injury. However, Kroell et al. (1986) examined the interrelationship between speed and chest compression in blunt impact to 23 porcine subjects [141]. Their results showed better correlation of Vmax*Cmax than VC(t)max for probability of heart rupture, and for probability of AIS 4+ injury.

The Viscous Criterion was further refined in 1986 by Lau and Viano and compared to other criteria [142]. When compared to Vmax*Cmax the authors noted that a time varying function gives one the ability to discriminate the timing of injury production which may help in the development of countermeasures. They also proposed that a time varying function could account for changes in speed caused by collapse of the surface being impacted. While this is true, the velocity environment of the subject (in this case a test dummy) will never be greater than the pre-impact velocity, and the probability that a structure will fail is also likely related to the impact velocity (since force, energy available, and impact velocity are very well correlated).

Much research has taken place since the Abdominal Injury Criterion and the Viscous Criterion were proposed. While some research has shown that there is little difference in the end result [141], other research has shown the utility of knowledge of the time of injury occurrence [143]. Therefore, the viscous response is probably the more desirable function to measure, but in cases where it cannot be measured, the AIC is a good substitute.

Rouhana showed that the interrelationship between velocity and compression can be observed by cross-plotting V and C [106, 137]. When lines of equal V*C were drawn they separated injury severities fairly well. This showed that when velocity of loading was very low, as in the case of seat belt loading to the abdomen, maximum compression was a better predictor of abdominal injury. When the speed of loading is very high, as in the case of airbags, maximum velocity is a better predictor of injury. In between low and high speed loading to the abdomen a combination of velocity and compression is a better predictor than either separately. These observations are corroborated by the studies of Mertz and Weber [127], Miller [99], and Prasad and Daniel [139]. More recent data by many researchers has provided supporting evidence of the correlation of abdominal injury with the Abdominal Injury Criterion and the Viscous Tolerance Criterion [85, 96, 99, 104, 107, 108, 133, 143–145]. The correlations of velocity, Vmax*Cmax, and [V*C]max with injury outcomes from the literature are summarized in Tables 14.3, 14.4, and 14.5, respectively.

14.6.2.4 Force

Force is defined as "that which changes the state of rest or motion in matter, measured by the rate of change of momentum" [146]. Newton's first law states that an object will remain at rest unless acted upon by an external force, and an object in motion will remain in motion at a constant velocity unless acted upon by an external force. When a vehicle is in a collision the state of motion of the occupants within the vehicle may be changed drastically depending on the severity of the collision. To do this, according to Newton's Laws, forces are exerted on the occupant. The actual etiology of these forces may be from inertia/acceleration, compression of elastic bodies, compression of viscous bodies, or other means. As such, impact force is expected to be well correlated with injury outcome. This has been demonstrated in multiple studies. The correlations of peak force with injury outcomes from the literature are summarized in Table 14.6.

In 85 impacts to primate and porcine subjects, Trollope et al. (1973) found the manner in which the force is applied to be important [76]. For intact animals, over 1.6 kN was needed to produce an Estimated Severity of Injury (ESI)=2 liver injury (similar to AIS=3 or 4), whereas, only 0.67 kN was necessary to produce injury of the same severity in a surgically exposed liver. Stalnaker [132, 147] found that the severity of abdominal injury was proportional to the logarithm of the impact force and time duration squared.

Gogler et al. (1977) studied impacts with minipigs at speeds of 9.8–16.9 m/s and saw a transition from AIS=3 abdominal injury to AIS=4 at 1.5 kN [148]. Peak force of 0.35 kN was associated with AIS 4+ abdominal injury in impacts to the abdomen of anesthetized rabbits conducted by Lau and Viano [136]. Leung et al. (1982) performed submarining tests with ten

Table 14.3 Correlations of velocity with injury

Author	(Year)	Velocity	Injury severity	Organ injured	Comments
McElhaney et al.	(1971)	11.3 m/s (L)	ESI>3	Mid abdomen	MM
		14.5 m/s (L)	ESI>3	Mid abdomen	PC
		9.8 m/s (R)	ESI>3	Mid abdomen	MM
		12.5 m/s (R)	ESI>3	Mid abdomen	PC
					N=13; primates (*Papio cynocephalus* (*PC*), *Macaca mulatta* (*MM*)); pneumatic impactor; free back/side; bar, belt, large flat block, wedge shape
Stalnaker et al.	(1973a)	7.3 m/s (L)	ESI>3	Upper abdomen	Primate
		6.1 m/s (R)	ESI>3	Upper abdomen	Primate
					N=96; primates and human cadavers; free side; scaled armrest; belt
Lau et al.	(1981)	8–10 m/s	AIS>3	Liver	Contusion
		12–14 m/s	AIS>4	Liver	Bursting injury
					N=26; *Oryctolagus cuniculus*; fixed back; compression=16 % for all experiments
Mertz and Weber	(1982)	5.7 m/s	AIS>3	Abdomen	ED50
					N=43; *Sus scrofa*; out-of-position airbag impacts
Prasad and Daniel	(1984)	4.7 m/s	AIS>3	Abdomen	Threshold
					N=15; *Sus scrofa*; out-of-position airbag impacts
Foster et al.	(2006)	8.5 m/s	AIS=3	Liver	Minor fractures of cortex, diaphragmatic surface; burst disruption of parenchyma, visceral surface; transection of tip of right lobe
		9.4 m/s	AIS=2	Liver	Fractures through middle inferior and lateral inferior, quadrate lobe
		13.3 m/s	AIS=3	Liver	Multiple fractures of cortex, diaphragmatic surface; disruption of parenchyma, visceral surface
					N=9; unembalmed cadavers; fixed-back seatbelt pretensioner loading, umbilicus
Lamielle et al.	(2008)	4.6 m/s	MAIS=3	Upper abdomen	Pancreas contusion (AIS=3); liver capsular tear (AIS=2)
		4.9 m/s	MAIS=4	Upper/midabdomen	Spleen laceration (AIS=4); renal vessel laceration (AIS=4); colon laceration (AIS=3); liver capsular tear (AIS=2); jejunum-ileum laceration (AIS=2)
		5.1 m/s	MAIS=3	Upper/midabdomen	Colon laceration (AIS=3); duodenum laceration (AIS=3); spleen laceration (AIS=2)
		5.8 m/s	MAIS=3	Pancreas	Laceration
		6.3 m/s	MAIS=2	Jejunum-ileum	Laceration
					N=8; unembalmed cadavers; fixed-back seatbelt loading in out-of-position or submarining configurations

Table 14.4 Correlations of $V_{max}*C_{max}$ with injury

Author	(Year)	$V_{max}C_{max}$	Injury severity	Organ injured	Comments
Rouhana et al.	(1984, 1985)	1.75 m/s (R)	AIS>3	Upper/midabdomen	ED_{25}
		2.71 m/s (R)	AIS>3	Upper/midabdomen	ED_{50}
		2.10 m/s (L)	AIS>3	Upper/midabdomen	ED_{25}
		3.31 m/s (L)	AIS>3	Upper/midabdomen	ED_{50}
					N=117; *Oryctolagus cuniculus*; side impacts; free side
Stalnaker and Ulman	(1985)	Frontal 3.0 m/s	AIS=3	Upper abdomen	
		Frontal 3.8 m/s	AIS=3	Midabdomen	
		Frontal 3.0 m/s	AIS=3	Lower abdomen	
		Right side 3.5 m/s	AIS=3	Abdomen	
		Left side 4.7 m/s	AIS=3	Abdomen	
					N=42; primates from previous studies; 6 different impactors; 4 different locations; from linear regression of VC with AIS
Rouhana et al.	(1986)	3.15 m/s (R & L)	AIS>3	Liver	ED_{50}
		5.5 m/s (R & L)	AIS>3	Kidney	ED_{50}
					N=214; *Oryctolagus cuniculus*; side impacts; 107 with force-limiting Hexcel; free side; area=7.0 in.^{2}
Rouhana	(1987)	0.75 m/s	AIS>3	Liver	
					N=8; human cadavers; right-side impacts; analysis of Walfisch data; probability of injury not stated
Foster et al.	(2006)	4.1 m/s	AIS=2	Liver	Fractures through middle inferior and lateral inferior, quadrate lobe
		4.3 m/s	AIS=3	Liver	Minor fractures of cortex, diaphragmatic surface; burst disruption of parenchyma, visceral surface; transection of tip of right lobe
		7.3 m/s	AIS=3	Liver	Multiple fractures of cortex, diaphragmatic surface; disruption of parenchyma, visceral surface
					N=9; unembalmed cadavers; fixed-back seatbelt pretensioner loading, umbilicus
Lamielle et al.	(2008)	1.45 m/s	MAIS=4	Upper/midabdomen	Spleen laceration (AIS=4); renal vessel laceration (AIS=4); colon laceration (AIS=3); liver capsular tear (AIS=2); jejunum-ileum laceration (AIS=2)
		1.83 m/s	MAIS=3	Pancreas	Laceration
		1.86 m/s	MAIS=3	Upper/midabdomen	Colon laceration (AIS=3); duodenum laceration (AIS=3); spleen laceration (AIS=2)
		1.87 m/s	MAIS=3	Upper abdomen	Pancreas contusion (AIS=3); liver capsular tear (AIS=2)
		1.92 m/s	MAIS=2	Jejunum-ileum	Laceration
					N=8; unembalmed cadavers; fixed-back seatbelt loading in out-of-position or submarining configurations

Table 14.5 Correlations of $[V*C]_{max}$ with injury

Author	(Year)	$[V*C]_{max}$	Injury severity	Organ injured	Comments
Lau and Viano	(1986)	1.20 m/s	AIS > 5	Liver	ED_{25}
		1.40 m/s	AIS > 5	Liver	ED_{50}
					N = 20; *Sus Scrofa*; chest/abdomen contact; steering wheel
Lau and Viano	(1988)	1.20 m/s	AIS > 4	Liver	Laceration ED_{25}
		1.24 m/s	AIS > 4	Liver	Laceration ED_{50}
					Same as Lau (1986) + 9 more subjects with "punch pulled"
Miller	(1989)	1.40 m/s	AIS > 4	Lower abdomen	ED_{25}
					N = 25; *Sus Scrofa*; lap belt impacts; fixed back; all energy into subject
Viano et al.	(1989)	1.98 m/s	AIS > 4	Upper/midabdomen	ED_{25}
					N = 14; unembalmed cadavers; rigid pendulum; side impact; free back some energy into whole body motion
Talantikite et al.	(1993)	1.98 m/s	AIS = 4	Upper abdomen	Threshold of AIS = 4
					N = 6; unembalmed cadavers; 23.4 kg rigid pendulum; side impact
Viano and Andrzejak	(1993)	1.16 m/s	MAIS = 3.6	Upper abdomen	N = 5; *Sus Scrofa*; stiff armrest; left side impacts
		0.33 m/s	MAIS = 2	Upper abdomen	N = 5; *Sus Scrofa*; soft armrest; left side impacts
Foster et al.	(2006)	1.8 m/s	AIS = 2	Liver	Fractures through middle inferior and lateral inferior, quadrate lobe
		2.0 m/s	AIS = 3	Liver	Minor fractures of cortex, diaphragmatic surface; burst disruption of parenchyma, visceral surface; transection of tip of right lobe
		4.1 m/s	AIS = 3	Liver	Multiple fractures of cortex, diaphragmatic surface; disruption of parenchyma, visceral surface
					N = 9; unembalmed cadavers; fixed-back seatbelt pretensioner loading, umbilicus
Lamielle et al.	(2008)	0.33 m/s	MAIS = 3	Upper/midabdomen	Colon laceration (AIS = 3); duodenum laceration (AIS = 3); spleen laceration (AIS = 2)
		0.42 m/s	MAIS = 4	Upper/midabdomen	Spleen laceration (AIS = 4); renal vessel laceration (AIS = 4); colon laceration (AIS = 3); liver capsular tear (AIS = 2); jejunum-ileum laceration (AIS = 2)
		0.75 m/s	MAIS = 3	Pancreas	Laceration
		0.78 m/s	MAIS = 2	Jejunum-ileum	Laceration
		1.25 m/s	MAIS = 3	Upper abdomen	Pancreas contusion (AIS = 3); liver capsular tear (AIS = 2)
					N = 8; unembalmed cadavers; fixed-back seatbelt loading in out-of-position or submarining configurations

Table 14.6 Correlations of peak force with injury

Author	(Year)	Peak force	Injury severity	Organ injured	Comments
Trollope et al.	(1973)	1.56 kN	ESI > 2	Liver	Intact animal
		0.67 kN	ESI > 2	Liver	Exposed liver
					N = 85 primates; N = 15 *Sus Scrofa*
Stalnaker et al.	(1973a)	3.11 kN (R & L)	ESI > 3	Upper abdomen	
					N = 96; primates and human cadavers; free side; scaled armrest; belt
Gogler et al.	(1977)	0.59–0.98 kN	N/A	Abdomen	Subacute shock – acute shock
		1.47 kN	AIS 3-4	Abdomen	AIS 3-4 transition
					N = 12; *Sus Scrofa*; projectile tests; frontal abdominal impacts; free back; 11.4–15.5 kg
Lau and Viano	(1981)	0.24 kN	AIS > 3	Liver	Contusion
					N = 26; *Oryctolagus cuniculus*; fixed back; compression = 16 % for all experiments
Rouhana et al.	(1986)	0.82 kN	AIS > 3	Liver	ED_{50}
		1.14 kN	AIS > 3	Kidney	ED_{50}
					N = 214; *Oryctolagus cuniculus*; side impacts; 107 with force limiting Hexcel; free side; area = 7.0 in.^{2}
Miller	(1989)	2.93 kN	AIS > 3	Lower abdomen	ED_{25}
		3.96 kN	AIS > 3	Lower abdomen	ED_{50}
		3.76 kN	AIS > 4	Lower abdomen	ED_{25}
		4.72 kN	AIS > 4	Lower abdomen	ED_{50}
					N = 25; *Sus Scrofa*; lap belt impacts; fixed back; all energy into subject
Viano et al.	(1989)	6.73 kN	AIS > 4	Upper/midabdomen	ED_{25}
					N = 14; unembalmed cadavers; rigid pendulum; side impact; free back; some energy into whole body motion
Cavanaugh	(1993)	3.0 kN	AIS = 3	Abdomen	AIS = 3 threshold
					N = 15; unembalmed cadavers; side impact; rigid and padded wall sled impacts
Talantikite et al.	(1993)	500 N	AIS = 3	Liver	Excised liver; drop tests
		4.4 kN	AIS = 4	Upper abdomen	AIS = 4 threshold
					N = 6; unembalmed cadavers; right side impact; 23.4 kg rigid pendulum impactors
Foster et al.	(2006)	8.3 kN	AIS = 2	Liver	Fractures through middle inferior and lateral inferior, quadrate lobe
		9.4 kN	AIS = 3	Liver	Minor fractures of cortex, diaphragmatic surface; burst disruption of parenchyma, visceral surface; transection of tip of right lobe
		9.7 kN	AIS = 3	Liver	Multiple fractures of cortex, diaphragmatic surface; disruption of parenchyma, visceral surface

(continued)

Table 14.6 (continued)

Author	(Year)	Peak force	Injury severity	Organ injured	Comments
					N=9; unembalmed cadavers; fixed-back seatbelt pretensioner loading, umbilicus
Lamielle et al.	(2008)	3.7 kN	MAIS=3	Upper abdomen	Pancreas contusion (AIS=3); liver capsular tear (AIS=2)
		4.6 kN	MAIS=2	Jejunum-ileum	Laceration
		4.6 kN	MAIS=3	Pancreas	Laceration
		4.8 kN	MAIS=4	Upper/midabdomen	Spleen laceration (AIS=4); renal vessel laceration (AIS=4); colon laceration (AIS=3); liver capsular tear (AIS=2); jejunum-ileum laceration (AIS=2)
		5.1 kN	MAIS=3	Upper/midabdomen	Colon laceration (AIS=3); duodenum laceration (AIS=3); spleen laceration (AIS=2)
					N=8; unembalmed cadavers; fixed-back seatbelt loading in out-of-position or submarining configurations

human cadavers [149]. They found a parabolic relationship between lap belt force and abdominal penetration in submarining. Likewise, they observed a parabolic relationship between lap belt force and severity of the resulting abdominal injuries. The authors performed simple trendline analyses by plotting belt force as the independent variable against abdominal AIS. This analysis showed the data could be well represented by a logarithmic fit with r=0.91. This analysis showed that the relationship between normalized lap belt force and abdominal AIS has merit if the test surrogate has the correct force-deflection response, i.e. biofidelity.

Rouhana et al. (1986) performed 214 experiments of lateral abdominal impacts to anesthetized rabbits using rigid and crushable impacting surfaces [78]. Their analysis showed that peak force correlated well with probability of AIS 3+ renal injury, but not with probability of hepatic injury. They were able to reduce the number of renal lacerations by a factor of 3 by use of a force-limiting material, but the material used had no effect on hepatic lacerations. They postulated that the renal lacerations occurred at the time of peak impact force. It is possible that the force-limiting did not reduce hepatic injury because the force limit was too high. Peak force was the best correlate with AIS 4+ abdominal injury of all the biomechanical measures compared in a study by Viano et al. (1989), including the Viscous Criterion, maximum compression, and acceleration measurements [85].

Miller (1989) showed that peak force was well correlated with the probability of AIS 3+ and AIS 4+ lower abdominal injury in belt loading in pigs [99]. Cavanaugh et al. (1993), in a series of lateral impact testing of human cadaver subjects using a side impact sled, noted that the peak force was not a good predictor of abdominal injury [144]. However, they observed a threshold for MAIS 3+ abdominal injuries to be approximately 3 kN.

Talantikite et al. (1993) examined the results of pendulum impacts to 25 excised human livers at speeds ranging from 1.5 to 4.1 m/s [133]. They found a threshold of AIS=3 injury for excised livers at a force of 500 N. For the whole-body portion of their study they found abdominal injury to be best correlated with deflection of the half abdomen, but they proposed a maximum tolerable force value of 4.4 kN because they observed no AIS 4 abdominal injuries below that level.

14.6.2.5 Impact Energy and Fmax*Cmax

Baxter and Williams (1961) performed a series of blunt abdominal impact experiments with anesthetized canine subjects [120]. They found that the number of abdominal injuries increased from 1.5 to 2.65 per subject as the impact energy increased from 271 to 407 J.

Mays (1966) dropped cadaver livers onto a concrete surface from varying heights in an attempt to reproduce clinically relevant bursting injuries [150]. Flaccid organs did not sustain bursting injuries unless they were repressurized before the test by filling them with barium and saline. Mays found that 36–46 J was associated with tears and superficial lacerations of the Glisson's membrane (AIS 4), 144–182 J was associated with deep fracture without vascular involvement (AIS 5), and that 386–488 J was associated with extensive pulpefication of the parenchymal tissue "as seen clinically in bursting injury", and "severe disruption of the tertiary divisions of the portal vein, hepatic artery, and bile ducts" (AIS 5). The Williams and Sargent (1963) experiments involving dropping weights onto the abdomen of anesthetized canine subjects showed that 542 J produced clinically relevant injuries [131]. The drop tests of Walfisch et al. (1980) showed that serious injury (AIS 3) was produced when both a force of 4.5 kN, and a compression of 28 % of the half abdomen (or 14 % of the whole abdomen) were simultaneously attained [93]. Rouhana (1987) reanalyzed the Walfisch data, and showed that the product of the maximum force and maximum abdominal compression (Fmax*Cmax) was well correlated with the probability of AIS 4+ abdominal injury, and proposed this measure as another injury criterion for abdominal injury [107].

That Fmax*Cmax might correlate well with probability of abdominal injury makes intuitive sense. It is related to the amount of work done on the subject during impact, and hence is related also to the amount of energy modulated by the body during impact (the integral of the load vs. penetration response). Tissue and organ disruption is certainly an energy dissipative process. Therefore, the larger the Fmax*Cmax product, the more energy dissipation and tissue destruction one might expect. However, except for when tissues are compressed maximally against rigid structures (e.g., the spine), Fmax and Cmax are not coincident in time. A reasonable approach would be to measure force and compression as a function of time during impact. One could then investigate the correlation of the maximum of $F(t)*C(t)$ with probability of injury in a manner analogous to the Viscous Criterion. The correlations of impact energy and Fmax*Cmax with injury outcomes from the literature are summarized in Tables 14.7 and 14.8.

Miller (1989) confirmed the predictive ability of Fmax*Cmax in seatbelt impacts to the lower abdomen of anesthetized porcine subjects [99]. In fact, Fmax*Cmax was better correlated than the Viscous Response with the probability of AIS 4+ injury to the lower abdomen.

14.6.2.6 Pressure

Many researchers have examined the pressure applied during impact as a way to account for impact surfaces of different shape and size. Williams and Sargent (1963) found that the average peritoneal pressure was greater than the

Table 14.7 Correlations of energy with injury

Author	(Year)	Energy (J)	Injury severity	Organ injured	Comments
Williams	(1963)	542	AIS > 3	Lower abdomen	N = 45; *Canidae*; 50 lb wt dropped 8′ onto a board over lower abdomen; 28 of 49 subjects had intestinal injury
Mays	(1966)	37–46	AIS = 4	Liver	Superficial lacerations
		144–182	AIS = 5	Liver	Deep lacerations, no vascular injury
		386–488	AIS > 5	Liver	Macerated
					N = 15; human cadaver livers; drop tests; livers injected with saline and barium. h = 2.1 to 35.7 m; M = 1.4 to 1.8 kg

Table 14.8 Correlations of $F_{max}*C_{max}$ with injury

Author	(Year)	$F_{max}*C_{max}$	Injury severity	Organ injured	Comments
Walfisch et al.	(1980)	4.5 kN 14 % Deflection (see Rouhana 1987)	AIS = 3	Liver	
					N = 8; human cadavers; drop tests; right side impacts only
Rouhana et al.	(1987)	0.63 kN	AIS > 3	Liver	4.5 kN; 14 % deflection whole abdomen
		0.88 kN	AIS > 3	Liver	4.5 kN; 19.5 % deflection whole abdomen
					N = 8; human cadavers; right side impacts; analysis of Walfisch data; probability of injury not stated
Miller	(1989)	1.33 kN	AIS > 3	Lower abdomen	ED_{25}
		1.96 kN	AIS > 3	Lower abdomen	ED_{50}
		2.00 kN	AIS > 4	Lower abdomen	ED_{25}
		2.67 kN	AIS > 4	Lower abdomen	ED_{50}
					N = 25; *Sus Scrofa*; lap belt impacts; fixed back; all energy into subject
Foster et al.	(2006)	3.63 kN	AIS = 2	Liver	Fractures through middle inferior and lateral inferior, quadrate lobe
		4.72 kN	AIS = 3	Liver	Minor fractures of cortex, diaphragmatic surface; burst disruption of parenchyma, visceral surface; transection of tip of right lobe
		5.35 kN	AIS = 3	Liver	Multiple fractures of cortex, diaphragmatic surface; disruption of parenchyma, visceral surface
					N = 9; unembalmed cadavers; fixed-back seatbelt pretensioner loading, umbilicus
Lamielle et al.	(2008)	1.36 kN	MAIS = 3	Upper/midabdomen	Colon laceration (AIS = 3); duodenum laceration (AIS = 3); spleen laceration (AIS = 2)
		1.40 kN	MAIS = 2	Jejunum-ileum	Laceration
		1.42 kN	MAIS = 4	Upper/midabdomen	Spleen laceration (AIS = 4); renal vessel laceration (AIS = 4); colon laceration (AIS = 3); liver capsular tear (AIS = 2); jejunum-ileum laceration (AIS = 2)
		1.46 kN	MAIS = 3	Pancreas	Laceration
		1.51 kN	MAIS = 3	Upper abdomen	Pancreas contusion (AIS = 3); liver capsular tear (AIS = 2)
					N = 8; unembalmed cadavers; fixed-back seatbelt loading in out-of-position or submarining configurations

intraluminal pressure in the intestine during impact [131]. In addition, the presence of air or water in the intestine did not affect the injury outcome. They concluded that intestinal rupture was not caused by intraluminal pressure exceeding the intraperitoneal pressure. The correlations of pressure with injury outcomes from the literature are summarized in Table 14.9.

Analysis by McElhaney et al. (1971) led to an estimate of 131 kPa for the tolerable pressure applied by "an armrest-like striker" to the midabdominal region [135]. This value was independent of the side of the body struck, and was associated with ESI 3 injury (similar to AIS 4 or 5). For a "flat rigid striker of larger cross-section than the animal struck", the tolerable value was 386 kPa [135].

In the early 1970s, Fazekas et al. [151–153] studied isolated cadaver organs and found that a pressure of 168.5 kPa was necessary to cause superficial fracture of the liver, while 319.8 kPa caused multiple ruptures. They also found that 44.0 kPa caused superficial fracture of the spleen.

In their work on surgically exposed livers and kidneys, Melvin et al. (1973) found that "moderate trauma (ESI 3+, similar to AIS 4 or 5) under dynamic loading occurs at a threshold stress level of approximately 310 kPa in the liver" [75]. They noted that kidney injury severity was not ordered by stress level. Analysis by Stalnaker et al. [132, 147] and Trollope et al. [76] gave conflicting tolerance numbers from the same data, but all showed a relationship between impact pressure and injury outcome.

In their series of side impact/cadaver drop tests, Walfisch et al. (1980) found that the average pressure on an armrest was a reliable indicator of injury severity (r = 0.93), and that pressure of 260 kPa was associated with AIS 3+ injury severity [93]. Lau and Viano (1981) saw the initiation of "hepatic surface injury" for pressure of 350 kPa in a study of belt loading to anesthetized canine subjects [136]. Rouhana et al. (1986) found that by using force-limiting honeycomb material beneath the impact surface, renal fractures were prevented when the crush strength was 231 kPa [78]. Liver injury was not affected by honeycomb of the same crush strength. This study showed that force-limiting materials could help mitigate abdominal injury in side impact although the crush strength of the honeycomb used (230–280 kPa) was too high to see elimination of injury. Miller (1989) found that peak pressure was well correlated to the probability of AIS 3+ injury and also correlated, although not as well, with the probability of AIS 4+ injury [99]. Viano and Andrzejak (1993) tested a series of armrests of different crush strengths using anesthetized swine as test subjects [145]. For the stiffer armrests, the average abdominal AIS was 3.6. For the softer armrests the average abdominal AIS was 2.2. This was further indication that force-limiting or pressure-limiting materials could help mitigate abdominal injury. Cavanaugh et al. (1996) ran an extensive series of side impact tests with unembalmed human cadavers, using force-limiting (and hence pressure-limiting) paper honeycomb [96]. Soft paper honeycomb (69 kPa) reduced abdominal injury in 9 m/s impacts to no internal organ injuries and just a few rib fractures in 1 of the 5 subjects tested. Since these tests were based on elderly human cadavers, the authors suggest that tolerable armrest crush strength is likely to be higher than 69 kPa for living subjects.

Prasad and Daniel (1984) used peak aortic blood pressure as a correlate to abdominal compression in a series of 15 "out-of-position" airbag tests with anesthetized swine [139]. They saw no injuries above AIS 2 for peak aortic blood pressure up to 53.3 kPa. In a revision of the pregnant dummy, Rupp et al. (2001) used pressure measured within a uterine insert to predict the likelihood of placental abruption [154]. Snedeker et al. (2005) conducted quasi-static compression tests of human kidney cortex [155]. These tests were conducted for comparison to results from dynamic compression tests of porcine whole kidneys as part of a modeling effort. The authors found the cortex of the kidney to fail at an average stress level of 116+/−28 kPa. Alonzo, et al. (2006) and Johannsen et al. (2012) proposed a set of injury criteria based on pressure and rate of pressure change for the Q series of child dummies in Europe [156, 157]. In the high-speed pretensioner tests of Foster et al. (2006), aortic

Table 14.9 Correlations of pressure with injury

Author	(Year)	Pressure	Injury severity	Organ injured	Comments
Williams and Sargent	(1963)	50 kPa	AIS>3	Lower abdomen	
					N=45; *Canidae*; 50 lb wt dropped 8 ft onto a board over lower abdomen; 28 of 49 subjects had intestinal injury
McElhaney et al.	(1971)	131 kPa (WS)	ESI>3	Midabdomen	
		386 kPa (LF)	ESI>3	Midabdomen	
					N=13; primates; pneumatic impactor; free back/side; bar, belt, large flat block (LF), wedge shape (WS)
Fazekas et al.	(1971, 1972)	44 kPa	AIS=4	Spleen	Rupture
		169 kPa	AIS=4	Liver	Superficial lacerations
		320 kPa	AIS=5	Liver	Multiple ruptures
					N=Unknown; human cadaver; isolated livers
Melvin et al.	(1973)	310 kPa	ESI>3	Liver	Exposed, perfused liver
					N=17 liver; N=6 kidney; *Macaca mulatta*; MTS type; fixed back; v=2.5 m/s
Trollope et al.	(1973)	600 kPa (CB)	ESI>3	Upper abdomen	
		152 kPa (LF)	ESI>3	Upper abdomen	
					N=85 primates; N=15 *Sus scrofa*; cylindrical bar (CB); large flat block (LF)
Stalnaker et al.	(1973a)	214 kPa	ESI>3	Upper abdomen	
					N=96; primates and human cadavers; free side; scaled armrest; belt
Stalnaker et al.	(1973b)	669 kPa (CB)	ESI>3	Upper abdomen	
		193 kPa (LF)	ESI>3	Upper abdomen	
					N=96; primates and *Sus scrofa*; cylindrical bar (CB); large flat block (LF); free side; scaled armrest
Walfisch et al.	(1980)	260 kPa	AIS>3	Liver	
					N=8; human cadavers; drop tests; right side impacts only
Lau and Viano	(1981)	67 kPa	AIS>3	Liver	
					N=26; *Oryctolagus cuniculus*; fixed back; Compression=16 % for all experiments
Rouhana et al.	(1984, 1985)	276 kPa	AIS>4	Kidney	ED_{50} for laceration
					N=117; *Oryctolagus cuniculus*; side impacts; free side
Rouhana et al.	(1986)	180 kPa	AIS>3	Liver	ED_{50}
		251 kPa	AIS>3	Kidney	ED_{50}
					N=214; *Oryctolagus cuniculus*; side impacts; 107 with force limiting Hexcel; free side; area=7.0 in.^{2}
Miller	(1989)	166 kPa	AIS>3	Lower abdomen	ED_{25}
		226 kPa	AIS>3	Lower abdomen	ED_{50}

(continued)

Table 14.9 (continued)

Author	(Year)	Pressure	Injury severity	Organ injured	Comments
		216 kPa	AIS > 4	Lower abdomen	ED_{25}
		270 kPa	AIS > 4	Lower abdomen	ED_{50}
					N = 25; *Sus scrofa*; lap belt impacts; fixed back; all energy into subjects
Foster et al.	(2006)	89 kPa	AIS = 2	Liver	Fractures through middle inferior and lateral inferior, quadrate lobe
		120 kPa	AIS = 3	Liver	Minor fractures of cortex, diaphragmatic surface; burst disruption of parenchyma, visceral surface; transection of tip of right lobe
		186 kPa	AIS = 3	Liver	Multiple fractures of cortex, diaphragmatic surface; disruption of parenchyma, visceral surface
					N = 9; unembalmed cadavers; fixed-back seatbelt pretensioner loading, umbilicus

pressure was the only parameter evaluated that exhibited overlap in the injury response for the dataset, but was still deemed predictive [104]. Aortic pressure as high as 103 kPa was measured in the non-injury cases, while pressure as low as 89 kPa was measured in the injury cases.

Sparks et al. (2007) investigated the relationship between internal pressure and liver injury severity [158]. Mays (1966) and Stein et al. (1983) asserted that rapid increases in internal pressure could be important to injury of fluid-filled solid abdominal organs [150, 159]. With this in mind, Sparks dropped a guided 23.4-kg impact plate on to isolated repressurized livers, positioned such that the visceral surface was in contact with a flat bench top. Clinically relevant injury patterns were observed. The best predictor of liver damage was found to be the product of peak tissue pressure (measured in the parenchyma) and peak rate of tissue pressure increase. Fifty-percent risk of AIS 3+ injury was associated with 1,370 kPa^2/ms. Tissue pressure alone was also found to be a good predictor, with 48 kPa corresponding to 50 % risk of AIS 3+ injury. A higher value of 64 kPa peak vascular pressure (measured at the midline of the liver) was found to correspond to the same risk.

14.6.2.7 Wave Phenomena

Cooper and Taylor (1989), and Cooper et al. (1991) suggest that stress and shear waves play a role in the production of injury in blast and impact loading of the body [160, 161]. They proposed wave phenomena as the manner in which injury occurs at locations that are remote from the impact site. The speed of deformation is the predominant factor in determining the magnitude of the wave created. Hardy et al. (2001) found abdominal injuries commensurate with those expected in response to an underwater blast during testing of passenger airbags in an out-of-position configuration using human cadavers [80].

Cooper suggested that displacements of external body surfaces may be propagated as a transverse wave of low velocity and long duration, i.e. a shear wave. Injury may result from: (a) differential motion of connected adjacent structures, (b) strain at the sites of attachments, or (c) collision of viscera with stiffer structures. These stress wave/shear wave theories are similar to those proposed by Von Gierke [162, 163] who based his postulates on a mechanical model of the human body exposed to impact. For higher-speed phenomena such as non-penetrating projectiles (50 m/s), Cooper suggested that stress waves emanate from the impact site, traveling at the speed of sound in the tissues. Injury then occurs at interfaces between unlike tissues or tissue/air boundaries (e.g., intestinal wall/intraluminal gas). The injury mechanisms proposed include: (a) stress wave induced compression and re-expansion of the stressed wall, (b) production of a pressure differential across the boundary, or (c) "spalling", where energy is released as the wave attempts to propagate from a dense to a much less dense medium (the reflected wave is tensile which may injure the tissue because many materials have lower strength in tension than compression) [160, 161].

14.6.3 Submarining Assessment and Anthropomorphic Test Device Abdomen Development

Submarining continues to be identified as being an important factor in AIS 3+ abdominal injuries [26]. However, an occupant properly restrained by a seatbelt system is likely to be spared serious abdominal injury, and, as stated by Garrett and Braunstein (1962), is "better off with a seatbelt than without one" [164]. Strictly speaking, in submarining the occupant is not properly restrained. Lap belt submarining occurs when the occupant's pelvis uncouples from the lap belt causing the occupant's abdomen, not their pelvis, to load the lap belt [100, 149]. The lap belt can be worn by an occupant such that it is not over the pelvis initially ("pre-submarined"), or the pelvis can move or slide under the lap belt such that the abdomen loads the belt during a crash. The work of Uriot et al. (2006) suggests the abdominal penetration rate associated with submarining to be on the order of 4 m/s [165].

Avoiding submarining is important to avoiding injury during a frontal crash. There is a direct relationship between occupant protection and occupant vehicle coupling [166]. Leung et al. (1982) attributed poor seatbelt geometry and slack to 74 % of submarining cases [149]. Miller (1989) stated that seatbelt designs should impart a constant load to an occupant throughout a crash event, which results in a reasonable excursion of the occupant. Initial stiffness of the system and seatbelt pretensioning can be adjusted to control occupant "ride down" characteristics [99]. Mueller and Linn (1998) state that when an occupant is coupled to the decelerating vehicle in the early stages of a crash event via a pretensioner, excursion may be reduced and submarining might be avoided [167]. Rouhana et al. (2010) proffered that limits to the amount of pretensioning may be desirable [108]. Nilson and Halaand (1995) describe how load limiting is an effective approach for crashes of greater severity, and that the risk of submarining can be reduced by this approach [168]. However, a method of assessing the occurrence of submarining is needed.

The Hybrid III crash dummy, which is the regulatory tool for current frontal crash standards, does not have a biofidelic abdomen. The Hybrid III abdominal insert essentially fills the space in the pelvis, and is made of soft foam covered in a urethane rubber skin. This abdomen does not fill the volume that would be occupied by the abdominal contents in a human, and does not respond to external loading like the human (or cadaver). As such, there are challenges using the Hybrid III dummy for abdominal loading applications, and it requires modifications to accurately assess the occurrence of submarining or to predict the potential for injury from submarining. The foam insert that comprises the standard Hybrid III abdomen is shown in Fig. 14.23.

In an effort to assess the occurrence of submarining, and therefore submarining injury risk, Rouhana designed the Frangible Abdomen for the Hybrid III dummy family [100, 101]. In the absence of belt-loading data for the human, the swine cadaver tests of Rouhana [78] and the anesthetized swine tests of Miller [99] were used

Fig. 14.23 The abdominal insert from the midsize male Hybrid III crash dummy

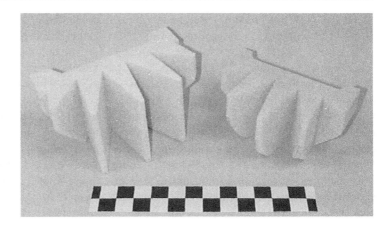

Fig. 14.24 The Frangible Abdomen developed by Rouhana et al. (1989) for use with the Hybrid III midsize male and small female crash dummies [100, 101]

to produce an abdomen insert that mimicked the human abdominal response at 3 m/s. In addition, these tests were compared to examine the scaling possibilities between human cadavers and living humans. The Frangible Abdomen was made from Styrofoam®. It has a curved ventral aspect with deep wedge-shaped pieces cut from it along the curvature. The remaining foam wedges that comprise the Frangible Abdomen have the appearance of teeth of a gear, and serve as an indicator of belt intrusion (via foam deformation), and to provide increasing resistance to belt intrusion as belt intrusion increases. Thus, the Frangible Abdomen had a reasonable degree of biofidelity, and was able to provide an indication of the occurrence and extent of submarining. Measurement of the depth of crush on the foam was correlated to injury risk for AIS 3+ and AIS 4+ abdominal injury in the 50th percentile and 5th percentile Hybrid III dummies with Frangible Abdomens. The Frangible Abdomens for the midsized male and small female dummies are shown in Fig. 14.24.

To improve upon the Frangible Abdomen, Rouhana et al. (2001) developed a Reusable Rate-Sensitive Abdomen (RRSA) [81]. The response data of Hardy [80] were used to tune the impact performance of the RRSA. A primary goal was for this abdomen to demonstrate good biofidelity under a variety of automotive-collision loading conditions. The resulting abdominal insert was designed for use within the Hybrid III midsized male crash dummy. The insert consists of a thick silicone shell filled with fluid (e.g., silicone gel, ultra-high viscosity liquid silicone, water, etc.). In addition to the viscosity of the fluid, the properties and thickness of the shell were adjusted to provide the desired response characteristics. This new dummy abdomen performs well compared to the human cadaver abdomen during rigid-bar, seatbelt, and airbag loading. One version of the RRSA with a prototype instrumentation system installed is shown in Fig. 14.25. This prototype instrumentation system represents an early attempt to use a resistive fluid and electrodes to measure abdominal deflection.

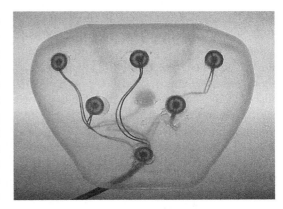

Fig. 14.25 The biofidelic Reusable Rate Sensitive Abdomen (RRSA) with a prototype deflection measurement system developed by Rouhana et al. (2001) for use with the Hybrid III midsize male crash dummy [81] (Reprinted with permission from the Stapp Association)

The AATD presented by Schneider et al. (1992) resulted from a large collaboration of universities, companies, and the National Highway Traffic Safety Administration [77]. This dummy became known as the TAD-50M (Trauma Assessment Device, 50th percentile male). As described in the original system technical characteristics specified by Melvin et al. (1985), the project had far reaching goals and aimed to improve all aspects of the Hybrid III [169]. This included response specifications for an abdominal insert. The new design elements had concentrated on the thorax, but by the project's end a modified version of the Frangible Abdomen was included. As dummy development continued beyond the TAD-50M, substantial portions of the TAD-50M found their way in to the THOR dummy. However, beginning with the first version of the THOR Alpha [170], there was an attempt to include an instrumented biofidelic abdomen, the lower section of which utilized two of the Double Gimbaled String Pots (DGSP) developed by Schneider et al. [77] for chest deflection measurement. The upper abdomen section used a single string pot to measure anteroposterior displacement. The THOR Alpha abdomen is shown in Fig. 14.26. The subsequent specifications for THOR (NT) called for two abdomen qualification tests [171]. The qualification tests were designed to compare the response of the THOR abdomen to the rigid-bar impact data of Cavanaugh [79] and Nusholtz [86], and the seatbelt and surrogate airbag loading profiles of Hardy [80]. The current version of this dummy is the THOR M.

The first attempt to construct a pregnant small female dummy was presented by Viano et al. (1996), with the idea in mind that such a device could be used to assess potential for harm to a fetus [172]. The dummy had an abdominal insert that represented the pregnant uterus, and the insert housed a representation of a fetus. However, the insert lacked biofidelity and an appropriate injury metric. The MAMA-2B developed by Rupp et al. (2001) aimed to address these shortcomings [154]. The abdominal insert was simplified, and tuned to mimic the available cadaver response data while possessing pregnant-like anthropometry. Building on the development of the RRSA, the MAMA-2B abdominal insert was designed as a fluid-filled silicon shell. The primary mechanism for fetal loss after a car crash is placental abruption, the potential for which was correlated to intra-abdominal pressure as the injury metric. The MAMA-2B abdominal insert is shown in Fig. 14.27 without the dummy jacket and with the dummy slightly reclined to expose the insert.

Elhagediab et al. (2006, 2007) extended the abdomen insert technology of the RRSA to the 6-year-old Hybrid III dummy [173, 82], which was done in conjunction with modification of the

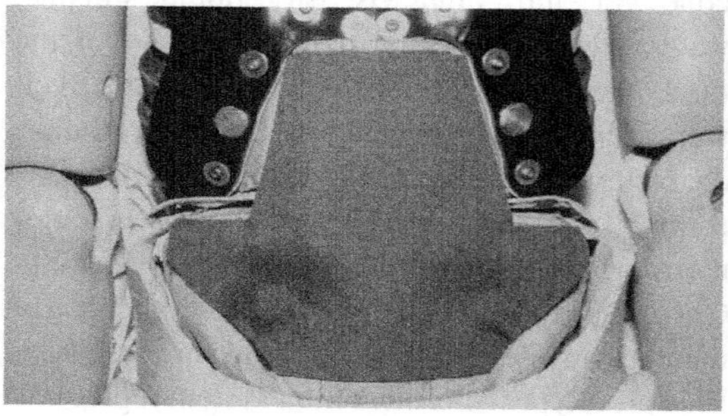

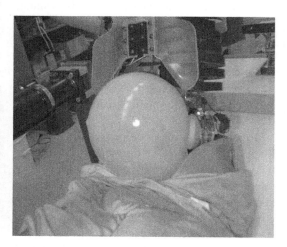

Fig. 14.26 The THOR Alpha abdomen (presented by Haffner et al. 2001), which has upper and lower sections instrumented to measure deflection [170]

Fig. 14.27 The MAMA-2B pregnant abdomen developed by Rupp et al. (2001) for the Hybrid III small female dummy (Courtesy of Dr. Jonathon Rupp)

Fig. 14.28 The prototype 6-year-old RRSA under development by Rouhana, shown in a modified (UMTRI) pelvis

pelvis described by Klinich et al. (2010), which was carried out by the University of Michigan Transportation Research Institute [174]. Arbogast et al. (2005), through their work at the Children's Hospital of Philadelphia, provided anthropometric values key to the development of the insert and pelvis [175]. The abdominal mechanical response of the 6-year-old from the porcine seatbelt tests of Kent et al. (2006) was used to fine tune the 6-year-old RRSA abdomen response [176]. The thickness of the insert's shell was determined via finite element modeling conducted at Ford Motor Company. An un-instrumented version of the 6-year-old RRSA positioned in the modified pelvis is shown in Fig. 14.28. Although various deflection-measurement systems were attempted throughout the development of the RRSA family, Gregory et al. (2012) introduced the first viable instrumentation concept for the 6-year old dummy [177]. Deflection is measured using differential signals between electrodes mounted within a conductive medium. The shell is filled with a low-conductivity (160 μS) saline solution. Ten stainless steel electrodes are mounted in the fluid-filled insert; six measurement electrodes are mounted to the inner ventral surface, one electrode is mounted to the inner surface of the dorsal aspect of the shell, and three fixed electrodes form a reference plane in the inferior/posterior portion of the insert. The measurement electrodes are mounted to the shell at locations where the quantification of deflection is desired for the assessment of belt loading. Three different high-frequency sinusoidal waveforms are applied to each fixed electrode in the reference plane. The potential difference between the fixed electrodes and the electrodes on the inner surface is measured using high input impedance circuitry. As the distance between the electrodes is varied, amplitude modulation of the sinusoidal carrier waveform that is proportional to the level of displacement occurs due to the change in resistance.

Faced with a lack of response data for the pediatric abdomen, Chamouard et al. (1996) conducted experiments with child volunteers to determine the quasi-static stiffness of the abdomen to belt loading [178]. These tests were done to judge the biofidelity of child dummies relative to belt interactions with the pelvis and thighs for appropriate submarining assessment. The children, aged about 3–10 years old, sat on a rigid plate with their backs supported by a wooden board that was perpendicular to the seat. A 47-mm wide lap belt, with a stretch characteristic of 10 % elongation at 1 t, was used to apply the load. Although not specified in the paper, the load appeared to be applied directly rearward. Six children all at least 10 months apart in age were tested. As the authors noted, properties of the dummy in this region are important to allow correct assessment of submarining performance of a restraint system.

Europe has adopted the Q dummy through its Child Advanced Safety Project for European Roads (CASPER) project [179]. The Q dummy contains an abdomen described as biofidelic and uses a set of twin pressure sensors for instrumentation. The Q dummy abdomen is instrumented with a set of pressure sensors [156] that give some left/right localization information through the use of two pressure vessels within the insert. A set of injury criteria based on pressure and rate of pressure change has been proposed [156, 157]. Beillas et al. (2012) tested the current instrumentation as well as an updated version, and correlated the responses with injury risk during accident reconstructions using sled testing for AIS 3+ injuries [180].

Additional research defining child, abdominal and pelvic responses is necessary to refine the current dummy and improve its capability to be a booster seat design tool. This applies specifically to the improvement and further development of the RRSA, and it's translation to the 10-year old dummy.

14.7 Summary

The abdomen has been gradually receiving increased attention from the injury prevention community. Prevention of injury to the abdomen and its contents notwithstanding, understanding the mechanical response of the abdomen to different restraint systems in various loading modes is important to understanding the response of the body as a whole. This is critical for design of better injury assessment tools such as crash dummies and computer models.

Of the research that has been done, the most plentiful pertains to the mechanical response of the midsized male. Rigid-bar and disk impact, seatbelt loading (including pretensioner rates, with and without submarining), and airbag interaction (including a surrogate airbag) data have been collected. What is less readily available is the correlation between injury (or damage in the cadaver) and injury metric. Not all studies have attempted this correlation. From those that have, there is no agreement as to which metric is superior. More data are needed in this regard, particularly with respect to hollow organ and vascular injuries, which are difficult to reproduce in the laboratory.

Also needed is more information about the impact and injury response of the small female abdomen and about the pediatric population in general. Future initiatives should focus on the local (the nine classic regions) response of the midsized male and small female abdomen, and on development of a satisfactory injury predictor. Further, the types of rigid bar, disk, and restraint system interactions that have been simulated in the laboratory should be repeated for the female. For children, the response of the abdomen to seatbelt systems is a very important area for continued study.

The hope and expectation is that the existing body of work and future efforts will influence federal regulations. That is, this information needs to be applied in the form of improved abdomen components in the current and future crash dummies. The Hybrid III does not have a biofidelic abdominal insert, and there is no compliance requirement related to the abdomen. The THOR dummy, the anticipated future frontal crash regulatory part, has an abdomen possessing some biofidelity and measurement capability in its current iteration. However, the most appropriate type of abdominal insert to use and the associated injury metric are still a subject of debate.

Some of the most recent work on the abdomen has focused on measuring the material properties of solid and hollow viscera, and understanding the interaction between organ systems during impact. These types of data have direct application to the development of computational models, as well as forwarding the understanding of injury mechanisms and injury tolerances. However, a lot of work remains before a virtual representation of the abdomen can replace other surrogates such as the cadaver. As they stand, finite element models are used largely for parametric studies to identify or examine phenomena worth investigating in the laboratory.

There exists a reasonable body of work dedicated to understanding and preventing injury to the abdomen, at least within the automobile environment. However, a complete solution remains elusive, and considerable work remains to improve safety and reduce the cost to society from these injuries.

References

1. Moore KL, Agur KMR (2002) Essential clinical anatomy, 2nd edn. Lippincott Williams & Wilkins, Philadelphia
2. Rutledge R, Thomason M, Oller D, Meredith W, Moylan J, Clancy T, Cunningham P, Baker C (1991)

The spectrum of abdominal injuries associated with the use of seat belts. J Trauma 31(6):820–826
3. Munns J, Richardson M, Hewett P (1995) A review of intestinal injury from blunt abdominal trauma. Aust NZ J Surg 65:4
4. Carrillo EH, Somberg LB, Ceballos CE, Martini MA, Ginzburg E, Sosa JL, Martin LC (1996) Blunt traumatic injuries to the colon and rectum. J Am Coll Surg 183:548–552
5. Hiroki S, Kikuchi T, Niwa H, Furuya Y, Morikane K, Naka S, Yasuhara H (2004) Characteristic features of abdominal organ injuries associated with gastric rupture in blunt abdominal trauma. Am J Surg 187(3): 394–397
6. Gartner LP, Hiatt JL (2006) Color atlas of histology, 4th edn. Lippincott Williams & Wilkins, Baltimore
7. Ross MH, Pawlina W (2011) Histology: a text and atlas: with correlated cell and molecular biology, 6th edn. Lippincott Williams & Wilkins, Baltimore
8. Guyton AC, Hall JE (2000) Textbook of medical physiology, 10th edn. WB Saunders, Philadelphia
9. Bondy N (1980) Abdominal injuries in the National Crash Severity Study. In: National Center for Statistics and Analysis collected technical studies, vol II, Accident data analysis of occupant injuries and crash characteristics. NHTSA, Washington, DC, pp 59–80
10. Elhagediab AM, Rouhana SW (1998) Patterns of abdominal injury in frontal automotive crashes. In: 16th International ESV conference proceedings, pp 327–337
11. Klinich KD, Flannagan CAC, Nicholson K, Schneider LW, Rupp JD (2010) Factors associated with abdominal injury in frontal, farside, and nearside crashes. Stapp Car Crash J 54:73–91
12. Saladin KS (2008) Human anatomy, 2nd edn. McGraw-Hill, New York
13. Nahum AM, Siegel AW, Brooks S (1970) The reduction of collision injuries: past, present and future. In: 14th Stapp car crash conference proceedings. SAE technical paper no. 700895
14. Danner M, Langwieder K (1979) Collision characteristics and injuries to pedestrians in real accidents. In: 7th International ESV conference proceedings
15. Danner M, Langwieder K, Wachter W (1979) Injuries to pedestrians in real accidents and their relation to collision and car characteristics. In: 23rd Stapp car crash conference proceedings, pp 161–198
16. Ricci L (1980) NCSS statistics: passenger cars. UM-HSRI report 80–36. Highway Safety Research Institute, University of Michigan
17. Hobbs CA (1980) Car occupant injury patterns and mechanisms. In: 8th International ESV conference proceedings, pp 755–768
18. Galer M, Clark S, Mackay GM, Ashton SJ (1985) The causes of injury in car accidents – an overview of a major study currently underway in Britain. In: 10th International ESV conference proceedings, pp 513–525
19. Rouhana SW, Foster ME (1985) Lateral impact – an analysis of the statistics in the NCSS. In: 29th Stapp car crash conference proceedings, pp 79–98, SAE technical paper no. 851727
20. Mackay GM, Hill J, Parkin S, Munns JAR (1993) Restrained occupants on the nonstruck side in lateral collisions. Accid Anal Prev 25(2):147–152
21. Gabler HC, Digges K, Fildes BN, Sparke L (2005) Side impact injury risk for belted far side passenger vehicle occupants. SAE paper no. 2005-01-0287
22. Klinich KD, Burton RW (1993) Injury patterns of older children in automotive accidents. In: 1st Child occupant protection symposium proceedings, vol SP-986, pp 17–24. SAE technical paper no. 933082
23. Khaewpong N, Nguyen TT, Bents FD, Eichelberger MR, Gotschall CS, Morrissey R (1995) Injury severity in restrained children in motor vehicle crashes. SAE technical paper no. 952711
24. Yoganandan N, Pintar FA, Gennarelli TA, Maltese MR (2000) Patterns of abdominal injuries in frontal and side impacts. Annu Proc Assoc Adv Automot Med 44:17–36
25. Lee JB, Yang KH (2002) Abdominal injury patterns in motor vehicle accidents: a survey of the NASS database from 1993 to 1997. Traffic Inj Prev 3:241–246
26. Lamielle S, Cuny S, Foret-Bruno JY, Petit P, Vezin P, Verriest JP, Guillemot H (2006) Abdominal injury patterns in real frontal crashes: influence of crash conditions, occupant seat and restraint systems. Annu Proc Assoc Adv Automot Med 50:109–124
27. Frampton R, Lenard J, Compigne S (2012) An in-depth study of abdominal injuries sustained by car occupants in frontal crashes. Ann Adv Automot Med 56:137–149
28. Campbell F, Woodford M, Yates DW (1994) A comparison of injury impairment scale scores and physician's estimates of impairment following injury to the head, abdomen and lower limbs. In: 38th Annual proceedings of the Association for the Advancement of Automotive Medicine
29. Perry JF (1965) A five-year survey of 152 acute abdominal injuries. J Trauma 5:53–61
30. Cox E (1984) Trauma – a 5-year analysis of 870 patients requiring celiotomy. Ann Surg 199(4): 467–474
31. Bergqvist D, Hedelin H, Lindblad B, Matzsch T (1985) Abdominal injuries in children: an analysis of 348 cases. Br J Accident Surg 16(4):217–220
32. Gupta S, Talwar S, Sharma RK, Gupta P, Goyal A, Prasad P (1996) Blunt trauma abdomen: a study of 63 cases. Indian J Med Sci 50(8):272–276
33. Broos P, Gutermann H (2002) Actual diagnostic strategies in blunt abdominal trauma. Eur J Trauma 28:64–74
34. Stuhlfaut JW, Anderson SW, Soto JA (2007) Blunt abdominal trauma: current imaging techniques and CT findings in patients with solid organ, bowel, and mesenteric injury. Semin Ultrasound CT MR 28: 115–129

35. Kearney PA (1989) Blunt trauma to the abdomen. Ann Emerg Med 18(12):1322–1325
36. Kinnunen J, Kivioja A, Poussa K, Laasonen EM (1994) Emergency CT in blunt abdominal trauma of multiple injury patients. Acta Radiol 35(4):319–322
37. Sherck J, Shatney C, Sensaki K, Selivanov V (1994) The accuracy of computed tomography in the diagnosis of blunt small-bowel perforation. Am J Surg 168:670–675
38. Lentz KA, McKenney MG, Nunez DB, Martin L (1996) Evaluating blunt abdominal trauma: role for ultrasonography. J Ultrasound Med 15:447–451
39. McGahan JP, Rose J, Coates TL, Wisner DH, Newberry P (1997) Use of ultrasonography in the patient with acute abdominal trauma. J Ultrasound Med 16:653–662
40. Dolich MO, McKenney MG, Varela JE, Compton RP, McKenney KL, Cohn SM (2001) 2,576 Ultrasounds for blunt abdominal trauma. J Trauma 50:108–112
41. Haller JA, Papa P, Drugas G, Colombani P (1994) Nonoperative management of solid organ injuries in children – is it safe? Ann Surg 219(6):625–631
42. Christmas AB, Wilson AK, Manning B, Franklin GA, Miller FB, Richardson JD, Rodriguez JL (2005) Selective management of blunt hepatic injuries including nonoperative management is a safe and effective strategy. Surgery 138(4):606–610
43. Velmahos GC, Toutouzas KG, Radin R, Chan L, Demetriades D (2003) Nonoperative treatment of blunt injury to solid abdominal organs. Arch Surg 138:844–851
44. Haan J, Kole K, Brunetti A, Kramer M, Scalea TM (2003) Nontherapeutic laparotomies revisited. Am Surg 69(7):562–565
45. Fung YC (1967) Elasticity of soft tissues in simple elongation. Am J Physiol 213:1532–1544
46. Yamada H (1970) In: Evans FG (ed) Strength of biological materials. Williams & Wilkins, Baltimore, Mechanical Properties of Respiratory and Digestive Organs and Tissues
47. Egorov VI, Schastlivtsev IV, Prut EV, Baranov AO, Turusov RA (2002) Mechanical properties of the human gastrointestinal tract. J Biomech 35:1417–1425
48. Zhao J, Liao D, Chen P, Kunwald P, Gregersen H (2008) Stomach stress and strain depend on location, direction and the layered structure. J Biomech 41:3441–3447
49. Mason MJ, Shah CS, Maddali M, Yang KH, Hardy WN, Van Ee CA, Digges K (2005) A new device for high-speed biaxial tissue testing: application to traumatic rupture of the aorta. SAE paper no. 2005-01-0741
50. Shah CS, Hardy WN, Mason MJ, Yang KH, Van Ee CA, Morgan R, Digges K (2006) Dynamic biaxial tissue properties of the human cadaver aorta. Stapp Car Crash J 50:217–246
51. Howes MK, Hardy WN (2012) Material properties of the post-mortem colon in high-rate equibiaxial elongation. Biomed Sci Instrum 48:171–178
52. Lanir Y, Lichtenstein O, Imanuel O (1996) Optimal design of biaxial tests for structural material characterization of flat tissues. J Biomech Eng 118:41–47
53. Society of Automotive Engineers (2007) 211/1 Instrumentation for impact test – part 1 – electronic instrumentation. SAE International, Warrendale
54. Uehara H (1995) A study on the mechanical properties of the kidney, liver, and spleen, by means of tensile stress test with variable strain velocity. J Kyoto Prefect Univ Med 104(1):439–451
55. Hollenstein M, Nava A, Valtorta D, Snedeker JG, Mazza E (2006) Mechanical characterization of the liver capsule and parenchyma. In: Szekely MHG (ed) ISBMS 2006. Springer, Berlin/Heidelberg, pp 150–158
56. Santago AC, Kemper AR, McNally C, Sparks JL, Duma SM (2009) The effect of temperature on the mechanical properties of bovine liver. Biomed Sci Instrum 45:376–381
57. Santago AC, Kemper AR, McNally C, Sparks J, Duma SM (2009) Freezing affects the mechanical properties of bovine liver. Biomed Sci Instrum 45:24–29
58. Brunon A, Bruyere-Garnier K, Coret M (2010) Mechanical characterization of liver capsule through uniaxial quasi-static tensile tests until failure. J Biomech 43:2221–2227
59. Kemper AR, Santago AC, Stitzel JD, Sparks JL, Duma SM (2010) Biomechanical response of human liver in tensile loading. Ann Adv Automot Med 50:15–26
60. Zhang S (1999) An atlas of histology. Springer, New York
61. Bacha W, Bacha L (2000) Color atlas of veterinary histology, 2nd edn. Lippincott Williams & Wilkins, Baltimore
62. Matthews J, Martin J (1971) Atlas of human histology and ultrastructure. Lea & Febiger, Philadelphia
63. Eurell J, Frappier B (2006) Dellmann's textbook of veterinary histology, 6th edn. Blackwell, Ames
64. Yeh W, Li P, Jeng Y, Hsu H, Kuo P, Li M, Yang P, Lee P (2002) Elastic modulus of human liver and correlation to pathology. Ultrasound Med Biol 28(4):467–474
65. Nasseri S, Bilston L, Tanner R (2003) Lubricated squeezing flow: a useful method for measuring the viscoelastic properties of soft tissue. Biorheology 40:545–551
66. Roan E, Vemaganti K (2007) The non-linear material properties of liver tissue determined from no-slip uniaxial compression experiments. J Biomech Eng 129:450–456
67. Chui C, Kobayashi E, Chen X, Hisada T, Sakuma I (2007) Transversely isotropic properties of porcine liver tissue: experiments and constitutive modeling. Med Biol Eng Comput 45:99–106

68. Gao Z, Lister K, Desai J (2010) Constitutive modeling of liver tissue: experiment and theory. Ann Biomed Eng 38(2):505–516
69. Tamura A, Omori K, Miki K, Lee J, Yang K, King A (2002) Mechanical characterization of porcine abdominal organs. Stapp Car Crash J 46:55–69
70. Pervin F, Chen W, Weerasooriya T (2011) Dynamic compressive response of bovine liver tissues. J Mech Behav Biomed 4(10):76–84
71. Kemper AR, Santago AC, Stitzel JD, Sparks JL, Duma SM (2013) Effect of strain rate on the material properties of human liver parenchyma in unconfined compression. J Biomech Eng 135(10):104503–8
72. Kemper AR, Santago AC, Stitzel JD, Sparks JL, Duma SM (2012) Biomechanical response of human spleen in tensile loading. J Biomechanics 45(2):348–55
73. Kemper AR, Santago AC, Sparks JL, Stitzel JD, Duma SM (2011) Multi-scale biomechanical characterization of human liver and spleen. In: 22nd International enhanced safety of vehicles conference proceedings. Paper no. 11–0195
74. Beckman DL, McElhaney JH, Roberts VL, Stalnaker RL (1971) Impact tolerance – abdominal injury. NTIS No PB204171
75. Melvin JW, Stalnaker RL, Roberts VL, Trollope ML (1973) Impact injury mechanisms in abdominal organs. In: 17th Stapp car crash conference proceedings, pp 115–126. SAE technical paper no. 730968
76. Trollope ML, Stalnaker RL, McElhaney JH, Frey CF (1973) The mechanism of injury in blunt abdominal trauma. J Trauma 13(11):962–970
77. Schneider LW, Haffner MP, Eppinger RH, Salloum MJ, Beebe MS, Rouhana SW, King AI, Hardy WN, Neathery RF (1992) Development of an advanced ATD thorax system for improved injury assessment in frontal crash environments. In: 36th Stapp car crash conference. SAE paper no. 922520
78. Rouhana SW, Ridella SA, Viano DC (1986) The effect of limiting impact force on abdominal injury: a preliminary study. In: 30th Stapp car crash conference proceedings, Warrendale, pp 65–79
79. Cavanaugh JM, Nyquist GW, Goldberg SJ, King AI (1986) Lower abdominal tolerance and response. In: 30th Stapp car crash conference proceedings, pp 41–63. SAE technical paper no. 861878
80. Hardy WN, Schneider LW, Rouhana SW (2001) Abdominal impact response to rigid-bar, seatbelt, and airbag loading. Stapp Car Crash J 45:1–41
81. Rouhana SW, Elhagediab AM, Walbridge A, Hardy WN, Schneider LW (2001) Development of a reusable rate-sensitive abdomen for the Hybrid III family of dummies. Stapp Car Crash J 45:1–27
82. Elhagediab AM, Hardy WN, Rouhana SW, Kent RW, Arbogast KB, Higuchi K (2007) Development of an instrumented rate-sensitive abdomen for the six year old Hybrid III dummy. JSAE annual congress, JSAE paper no. 20075245
83. Stalnaker RL, Ulman MS (1985) Abdominal trauma – review, response, and criteria. In: 29th Stapp car crash conference proceedings, pp 1–16. SAE technical paper no. 851720
84. Stalnaker RL, McElhaney JH, Snyder RG, Roberts VL (1971) Door crashworthiness criteria. NTIS no. PB203721
85. Viano DC, Lau IV, Asbury C, King AI, Begeman P (1989) Biomechanics of the human chest, abdomen, and pelvis in lateral impact. In: 33rd Annual proceedings of the Association for the Advancement of Automotive Medicine, Des Plaines, pp 367–382
86. Nusholtz GS, Kaiker PS (1994) Abdominal response to steering wheel loading. In: 14th International ESV conference proceedings, Washington, DC, pp 118–127. SAE technical paper no. 946024
87. Yoganandan N, Pintar FA, Kumaresan S, Sances A, Haffner M (1996) Response of human lower thorax to impact. In: 40th Annual proceedings of the Association for the Advancement of Automotive Medicine
88. Melvin JW, King AI, Alem NM (1988) AATD system technical characteristics, design concepts, and trauma assessment criteria. DOT-HS-807-224 Task E-F final report. US Department of Transportation, National Highway Traffic Safety Administration, Washington, DC
89. Howes MK, Gregory TS, Beillas P, Hardy WN (2012) Kinematics of the thoracoabdominal contents under various loading scenarios. Stapp Car Crash J 56:48
90. Hardy WN, Shah CS, Kopacz JM, Yang KH, Van Ee CA, Morgan R, Digges K (2006) Study of potential mechanisms of traumatic rupture of the aorta using in situ experiments. Stapp Car Crash J 50:19
91. Howes MK, Hardy WN, Beillas P (2013) The effects of cadaver orientation on the relative position of the abdominal organs. Ann Adv Automot Med 57:209–223
92. Beillas P, Lafon Y, Smith FW (2009) The effects of posture and subject-to-subject variations on the position, shape and volume of abdominal and thoracic organs. Stapp Car Crash J 53:127–154
93. Walfisch G, Fayon A, Tarriere C, Rosey JP, Guillon F, Got C, Patel A, Stalnaker RL (1980) Designing of a dummy's abdomen for detecting injuries in side impact collisions. In: 5th International IRCOBI conference proceedings, pp 149–164
94. Mertz HJ (1984) A procedure for normalizing impact response data. SAE technical paper no. 840884
95. International Standards Organization (1989) Road vehicles anthropomorphic side impact dummy – Part 5: lateral abdominal impact response requirements to assess biofidelity of dummy. Technical report, ISO TR 9790-5
96. Cavanaugh JM, Walilko T, Chung J, King AI (1996) Abdominal injury and response in side impact. In: 40th Stapp car crash conference proceedings, pp 1–16. SAE technical paper no. 962410
97. Kallieris D, Mattern R, Schmidt G, Eppinger RH (1981) Quantification of side impact responses and injuries. In: 25th Stapp car crash conference

proceedings, pp 329–366. SAE technical paper no. 811009
98. Rouhana SW, Schneider LW, Jeffreys TA, Rupp JD, Meduvsky AG, Zwolinski JJ, Prasad P, Kankanala SV, Bedewi PG (2003) Biomechanics of 4-point seat belt systems in frontal impacts. Stapp Car Crash J 47:33
99. Miller MA (1989) The biomechanical response of the lower abdomen to belt restraint loading. J Trauma 29(11):1571–1584
100. Rouhana SW, Viano DC, Jedrzejczak EA, McCleary JD (1989) Assessing submarining and abdominal injury risk in the Hybrid III family of dummies. In: 33rd Stapp car crash conference proceedings. SAE paper no. 892440
101. Rouhana SW, Jedrzejczak EA, McCleary JD (1990) Assessing submarining and abdominal injury risk in the Hybrid III family of dummies: Part II – development of the small female frangible abdomen. In: 34th Stapp car crash conference proceedings. SAE paper no. 902317
102. Van Ratingen MR, Twisk D, Schrooten M, Beusenberg MC, Barnes A, Platten G (1997) Biomechanically based design and performance targets for a 3-year old child crash dummy for frontal and side impact. In: 2nd Child occupant protection symposium proceedings, pp 243–260. SAE technical paper no. 973316
103. Trosseille X, Le-Coz J-Y, Potier P, Lassau J-P (2002) Abdominal response to high-speed seatbelt loading. Stapp Car Crash J 46:71–79
104. Foster CD, Hardy WN, Yang KH, King AI, Hashimoto S (2006) High-speed seatbelt pretensioner loading of the abdomen. Stapp Car Crash J 50:27–51
105. Lamielle S et al (2008) 3D deformation and dynamics of the human cadaver abdomen under seatbelt loading. Stapp Car Crash J 52:267–294
106. Rouhana SW, Lau IV, Ridella SA (1985) Influence of velocity and forced compression on the severity of abdominal injury in blunt, nonpenetrating lateral impact. J Trauma 25(6):490–500
107. Rouhana SW (1987) Abdominal injury prediction in lateral impact – an analysis of the biofidelity of the Euro-SID abdomen. In: 31st Stapp car crash conference proceedings, pp 95–104. SAE technical paper no. 872203
108. Rouhana SW, El-Jawahari R, Laituri TR (2010) Biomechanical considerations for abdominal loading by seat belt pretensioners. SAE paper no. 2010-22-0016
109. Untaroiu CD, Bose D, Lu YC, Riley P, Lessley D, Sochor M (2012) Effect of seat belt pretensioners on human abdomen and thorax: biomechanical response and risk of injuries. J Trauma Acute Care Surg 72(5):1304–1315.
110. Augenstein J, Perdeck E, Stratton J, Labiste L, Phillips J, Mackinnon J, Digges K, Morgan R, Bahouth G (2004) Using CIREN data to assess the performance of the second generation of air bags. SAE paper no. 2004-01-0842
111. Kahane CJ (2006) An evaluation of the 1998–1999 redesign of frontal airbag. Technical report no. DOT HS 810 085. NHTSA, Washington, DC
112. Lau IV, Horsch JD, Viano DC, Andrzejak DV (1993) Mechanism of injury from airbag deployment loads. Accid Anal Prev 25(1):29–45
113. Stapp JP (1971) Biodynamics of deceleration, impact, and blast. In: Randel HW (ed) Aerospace medicine, 2nd edn. Williams & Wilkins, Baltimore, pp 118–166
114. Widman WD (1969) Blunt trauma and the normal spleen; peacetime experience at a military hospital in Europe. Mil Med 134:25–35
115. Nusholtz GS, Melvin JW, Mueller G, MacKenzie JR, Burney R (1980) Thoraco-abdominal response and injury. In: 24th Stapp car crash conference proceedings, pp 187–228
116. Walker LG (1969) Mechanisms of injury. In: Martin JD Jr, Haynes CD, Hatcher CR, Smith RB III, Stone HH (eds) Trauma to the thorax and abdomen. Charles C. Thomas, Springfield, pp 47–58
117. Walt AJ, Wilson RF (1973) Blunt abdominal injuries: an overview. In: Biomechanics and its application to automotive design. SAE, New York
118. Yamanaka N, Okamoto E, Toyosaka A, Ohashi S, Tanaka N (1985) Consistency of human liver. J Surg Res 39(3):192–198
119. Clemedson C-J, Frankenberg L, Jonsson A, Pettersson H, Sundqvist A-B (1969) Dynamic response of thorax and abdomen of rabbits in partial and whole-body blast exposure. Am J Physiol 216(3): 615–620
120. Schmidt G (1979) The age as a factor influencing soft tissue injuries. In: 4th International IRCOBI conference proceedings, pp 143–150
121. Baxter CF, Williams RD (1961) Blunt abdominal trauma. J Trauma 1:241–247
122. Nusholtz GS, Kaiker PS, Huelke DF, Suggitt BR (1985) Thoraco-abdominal response to steering wheel impacts. In: 29th Stapp car crash conference proceedings, pp 221–245. SAE technical paper no. 851737
123. Zhou Q, Rouhana SW, Melvin JW (1996) Age effects on thoracic injury tolerance. In: 40th Stapp car crash conference proceedings, pp 137–148. SAE technical paper no. 962421
124. Sturtz G (1980) Biomechanical data of children. In: 24th Stapp car crash conference proceedings, pp 513–559
125. Eppinger RH, Marcus JH, Morgan RM (1984) Development of dummy and injury index for NHTSA's thoracic side impact protection research program. SAE technical paper no. 840885
126. Eppinger RH, Morgan RM, Marcus JH (1982) Side impact data analysis. In: 9th International ESV conference proceedings, pp 244–250
127. Mertz HJ, Driscoll GD, Lenox JB, Nyquist GW, Weber DA (1982) Responses of animals exposed to deployment of various passenger inflatable restraint system concepts for a variety of collision severities

and animal positions. In: 9th International ESV conference proceedings, pp 352–368.
128. Mertz HJ, Weber DA (1982) Interpretations of impact responses of a 3-year-old child dummy relative to child injury potential. In: 9th International ESV conference proceedings, pp 368–376.
129. Horsch JD, Lau IV, Viano DC, Andrzejak DV (1985) Mechanism of abdominal injury by steering wheel loading. In: 29th Stapp car crash conference proceedings, pp 69–78. SAE technical paper no. 851724
130. Lau IV, Horsch JD, Viano DC, Andrzejak DV (1987) Biomechanics of liver injury by steering wheel loading. J Trauma 27(3):225–235
131. Brun-Cassan F, Pincemaille Y, Mack P, Tarriere C (1987) Contribution and evaluation of criteria proposed for thorax abdomen protection in lateral impact. In: 11th International ESV conference proceedings, pp 289–301. SAE technical paper no. 876040
132. Williams RD, Sargent FT (1963) The mechanism of intestinal injury in trauma. J Trauma 3:288–294
133. Stalnaker RL, Roberts VL, McElhaney JH (1973a) Side impact tolerance to blunt trauma. In: 17th Stapp car crash conference proceedings, pp 377–408. SAE technical paper no. 730979
134. Talantikite Y, Brun-Cassan F, Lecoz J-Y, Tarriere C (1993) Abdominal injury protection in side impact – injury mechanisms and protection criteria. In: 1993 International IRCOBI conference proceedings, pp 131–144
135. Kroell CK, Pope ME, Viano DC, Warner CY, Allen SD (1981) Interrelationship of velocity and chest compression in blunt thoracic impact. In: 25th Stapp car crash conference proceedings, pp 547–580. SAE technical paper no. 811016
136. McElhaney JH, Stalnaker RL, Roberts VL, Snyder RG (1971) Door crashworthiness criteria. In: 15th Stapp car crash conference proceedings, pp 489–517. SAE technical paper no. 710864
137. Lau IV, Viano DC (1981) Influence of impact velocity on the severity of nonpenetrating hepatic injury. J Trauma 21(2):115–123
138. Mertz HJ, Prasad P, Irwin A (1997) Injury risk curves for children and adults in frontal and rear collisions. In: 41st Stapp car crash conference proceedings, pp. 13-30. SAE technical paper no. 973318
139. Rouhana SW, Lau IV, Ridella SA (1984) Influence of velocity and forced compression on the severity of abdominal injury in blunt, nonpenetrating lateral impact. GMR research publication no. 4763
140. Viano DC, Lau VK (1983) Role of impact velocity and chest compression in thoracic injury. Aviat Space Environ Med 54(1):16–21
141. Prasad P, Daniel RP (1984) A biomechanical analysis of head, neck, and torso injuries to child surrogates due to sudden torso acceleration. In: 28th Stapp car crash conference proceedings, pp 25–40. SAE technical paper no. 841656
142. Viano DC, Lau IV (1985) Thoracic impact: a viscous tolerance criteria. In: 10th International ESV conference proceedings
143. Kroell CK, Allen SD, Warner CY, Perl TR (1986) Interrelationship of velocity and chest compression in blunt thoracic impact to Swine II. In: 30th Stapp car crash conference proceedings, pp 99–121. SAE technical paper no. 861881
144. Lau IV, Viano DC (1986) The viscous criterion – bases and applications of an injury severity index for soft tissues. In: 30th Stapp car crash conference proceedings, pp 123–142. SAE technical paper no. 861882
145. Lau IV, Viano DC (1988) How and when blunt injury occurs – implications to frontal and side impact protection. In: 32nd Stapp car crash conference proceedings, pp 81–100. SAE technical paper no. 881714
146. Cavanaugh JM, Huang Y, Zhu Y, King AI (1993) Regional tolerance of the shoulder, thorax, abdomen and pelvis to padding in side impact. SAE technical paper no. 930435
147. Viano DC, Andrzejak DA (1993) Biomechanics of abdominal injuries by armrest loading. J Trauma 34(1):105–115
148. Weast RC, Astle MJ, Beyer WH (eds) (1984) CRC handbook of chemistry and physics, vol 919, 65th edn. CRC Press, Boca Raton
149. Stalnaker RL, McElhaney JH, Roberts VL, Trollope ML (1973) Human torso response to blunt trauma. In: King WF, Mertz HJ (eds) Human impact response: measurement and simulation. Plenum Press, New York
150. Gogler E, Best A, Braess H, Burst HE, Laschet G (1977) Biomechanical experiments with animals on abdominal tolerance levels. In: 21st Stapp car crash conference proceedings, pp 712–751. SAE technical paper no. 770931
151. Leung YC, Tarriere C, Lestrelin D (1982) Submarining injuries of 3 pt. belted occupants in frontal collisions – description, mechanisms and protection. In: 26th Stapp car crash conference proceedings, pp 173–205
152. Mays ET (1966) Bursting injuries of the liver. Arch Surg 93:92–106
153. Fazekas IG, Kosa F, Jobba G, Meszaros E (1971) Die Druckfestigkeit der menschlichen Leber mit besonderer Hinsicht auf die Verkehrsunfalle. Z Rechtsmed 68(4):207–224
154. Fazekas IG, Kosa F, Jobba G, Meszaros E (1971) Experimentelle Untersuchungen uber die Druckfestigkeit der menschlichen Niere. Zacchia 46: 294–301
155. Fazekas IG, Kosa F, Jobba G, Meszaros E (1972) Beitrage zur Druckfestigkeit der menschlichen Milz bei stumpfen Krafteinwirkungen. Arch Kriminol 149(5):158–174
156. Rupp JD, Klinich KD, Moss S, Zhou J, Pearlman MD, Schneider LW (2001) Development and testing of a prototype pregnant abdomen for the small-

female Hybrid III ATD. Stapp Car Crash J 45: 375–392

157. Snedeker JG, Barbezat M, Niederer P, Schmidlin FR, Farshad M (2005) Strain energy density as a rupture criterion for the kidney: impact tests on porcine organs, finite element simulation, and a baseline comparison between human and porcine tissues. J Biomech 38(5):993–1001

158. Alonzo F, Bermond F, Beillas P (2006) Child abdominal injuries in car restraint systems – an intra-abdominal pressure sensor for the Q-dummy family and proposed viscous injury criterion based on detailed accident analysis and their reconstructions. J Biomech 39(Supplement 1):S159

159. Johannsen H, Trosseille X, Lesire P, Beillas P (2012) Estimating Q-dummy injury criteria using the CASPER project results and scaling adult reference values. In: IRCOBI conference proceedings, pp 580–595

160. Sparks JL, Bolte JH, Dupaix RB, Jones KH, Steinberg SM, Herriott R, Stammen J, Donnelly B (2007) Using pressure to predict liver injury risk from blunt impact. Stapp Car Crash J 51:401–432

161. Stein PD, Sabbah HN, Hawkins ET, White HJ, Viano DC, Vostal JJ (1983) Hepatic and splenic injury in dogs caused by direct impact to the heart. J Trauma 23(5):395–404

162. Cooper GJ, Taylor DEM (1989) Biophysics of impact injury to the chest and abdomen. J R Army Med Corps 135:58–67

163. Cooper GJ, Townend DJ, Cater SR, Pearce BP (1991) The role of stress waves in thoracic visceral injury from blast loading: modifications of stress transmission by foams and high-density materials. J Biomech 24(5):273–285

164. von Gierke HE, Oestreicher HL, Franke EK, Parrack HO, von Wittern WW (1952) Physics of vibrations in living tissues. J Appl Physiol 4:886–900

165. von Gierke HE (1964) Biodynamic response of the human body. App Mech Rev 17(12):951–958

166. Garrett JW, Braunstein PW (1962) The seat belt syndrome. J Trauma 2:220–238

167. Uriot J, Baudrit P, Potier P, Trosseille X, Petit P, Guillemot H, Guérin L, Vallancien G (2006) Investigations on the belt-to-pelvis interaction in case of submarining. Stapp Car Crash J 50:53–73

168. Faidy JP (1995) The coupling phenomenon in the case of a frontal car crash. In: 39th Stapp car crash conference proceedings, pp 71–78

169. Mueller HE, Linn B (1998) Seat belt pretensioners. SAE technical paper no. 980557. Society of Automotive Engineers, Warrendale

170. Nilson G, Haaland Y (1995) An analytical method to assess the risk of the lap-belt slipping off the pelvis in frontal impacts. In: 39th Stapp car crash conference proceedings. SAE, Warrendale, pp 59–70

171. Melvin JW, King AI, Alem NM (1985) AATD system technical characteristics, design concepts, and trauma assessment criteria. Contract no. Dl" NH22-83-C-07005, Task E-F, Final report. The University of Michigan

172. Haffner M, Eppinger R, Rangarajan N, Shams T, Artis M, Beach D (2001) Foundations and elements of the NHTSA THOR ALPHA ATD design. In: 17th International ESV conference proceedings. Paper 458

173. Martin P, Shook L (2007) NHTSA'S THOR-NT database. In: 20th International ESV conference proceedings. Paper no. 07-0289-W

174. Viano D, Smrcka J, Jedrzejczak E, Deng B, Kempf P, Pearlman MD (1996) Belt and airbag testing with a pregnant Hybrid III female dummy. In: 15th International ESV conference proceedings. Paper no. 96-S1-O-03

175. Elhagediab AM, Hardy WN, Rouhana SW (2006) Advancements in the rate-sensitive abdomen for the Hybrid III family of dummies. J Biomech 39(Supplement 1):S158

176. Klinich KD, Reed MP, Manary MA, Orton NR (2010) Development and testing of a more realistic pelvis for the Hybrid III 6YO ATD. Traffic Inj Prev 11(6):7

177. Arbogast KB, Marigowda S, Higuchi K, Tanji H, Kent RW, Stacey S, Mattice J, Rouhana SW (2005) An experimental and epidemiological evaluation of abdominal injuries in children. JSAE Annual Congress. JSAE paper no. 20055409

178. Kent RW, Stacey S, Kindig M, Forman J, Woods W, Rouhana SW, Higuchi K, Tanji H, St. Lawrence S, Arbogast KB (2006) Biomechanical response of the pediatric abdomen, part 1: development of an experimental model and quantification of structural response to dynamic belt loading. Stapp Car Crash J 50:1–26

179. Gregory TS, Howes MK, Hardy WN (2012) Deflection measurement system for the Hybrid III six-year-old biofidelic abdomen. Paper and presentation. Biomed Sci Instrum 48:149–156

180. Chamouard F, Tarriere C, Baudrit P (1996) Protection of children on board vehicles – influence of pelvis design and thigh and abdomen stiffness on the submarining risk for dummies installed on a booster. In: 15th International ESV conference proceedings, pp 1063–1075. SAE technical paper no. 976088

181. Lesire P, Johannsen H, Willinger R, Longton A (2012) CASPER – improvement of child safety in cars. Procedia Soc Behav Sci 48:2654–2663

182. Beillas P et al (2012) Abdominal twin pressure sensors for the assessment of abdominal injuries in Q dummies: in-dummy evaluation and performance in accident reconstructions. Stapp Car Crash J 56

Thoracic Spine Injury Biomechanics

15

Mike W.J. Arun, Dennis J. Maiman, and Narayan Yoganandan

Abstract

Injury biomechanical studies on the human spinal column have been published in the 1970s and 1980s from different perspectives. Experimental studies in the 1970s and 1980s focused all three mobile regions: cervical, thoracic and lumbar spines [1–3]. In the 1980s and 1990s, computational modeling aspects of spinal injuries were published [4, 5]. Although these review articles provided data on the three regions of the human vertebral column, the presentation of the thoracic spine topic has been limited [6]. This chapter focuses only on the thoracic region and it describes the biomechanically relevant anatomy, followed by the determination of the biomechanical responses from component to mono- and multi-segment spines (one level functional and disc-body units, and more than one level spinal units) to intact human cadaver, termed post mortem human subject (PMHS) experiments. Field and clinical information are included along with radiological images to correlate some of the experimental model outputs. Biomechanical tolerances and injury mechanisms are given consideration in the brief chapter.

M.W.J. Arun, Ph.D. (✉)
Neuroscience Research Lab, Department of Neurosurgery, Medical College of Wisconsin, Milwaukee, WI, USA
e-mail: marun@mcw.edu

D.J. Maiman, M.D., Ph.D. • N. Yoganandan, Ph.D.
Department of Neurosurgery, Medical College of Wisconsin, Milwaukee, WI, USA
e-mail: dmaiman@mcw.edu; yoga@mcw.edu

15.1 Anatomy

15.1.1 General

The thoracic spine includes 12 contiguous thoracic vertebrae interconnected by 13 discs from the cervico-thoracic (C7–T1) junction to the thoraco-lumbar junction (T12–L1). While the cervical and lumbar columns are lordotic, the thoracic spine is kyphotic (backward curve) in curvature (Fig. 15.1).

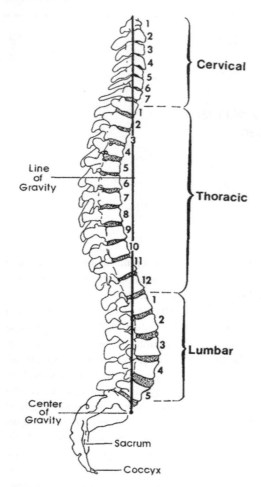

Fig. 15.1 Human vertebral column showing the line of gravity (Adapted from Ref. [1])

Variations in disc and vertebral body dimensions and the thickened thoracic laminae contribute to the formation and maintenance of this curvature, which may be altered by age-related changes.

15.1.2 Vertebrae

The vertebra consists of the load bearing anterior body and the posterior arch (Fig. 15.2). Its end plates consist of centrally roughened cortical bone and their purpose is to connect the intervertebral discs with the body. The sides of the body are somewhat concave, with almost similar antero-posterior and transverse diameters; the height at the ventral surface is normally 1–2 mm greater than its dorsal height, partially accounting for the thoracic curvature. The outer case of the vertebral body is composed of a thin layer of dense strong compact bone and the inner core is composed of trabecular bone aligned in vertical lamellae. This matrix resists external loads, primarily compression. The cross-sectional area of the body ranges from 800 to 1,055 mm^2 [2, 3].

The unique features of the thoracic spine are the ribs and their articulations. In addition to the facets on the pedicle (for spine articulation), articular surfaces exist for the ribs on each vertebra (Fig. 15.3). Rib facets are also seen on the transverse processes of all vertebrae, except T1 and T11–T12. The T11 and T12 vertebrae are unique, with their inferior articular facets oriented in sagittally. The vertebral arch is composed of laminae and bilateral pedicles, articulating surfaces (facets) and spinous processes (Fig. 15.2). The laminae are broad and heavily overlapped. The long and thin spinous processes project inferiorly and in the lower thoracic region, they become horizontal. While the vertebral bodies increase in size inferiorly, the transverse processes decrease in length. The thin but strong stout pedicles extend postero-laterally from the superior aspect of the body. Extending postero-medially from the pedicles, laminae fuse in the midline to form the posterior wall of the spinal canal. The articular processes arise from superior and inferior surfaces of the pedicle. Generally, the superior articular process projects cranially, with the articulating surface or facet on the posterior surface. The inferior articular process projects caudally, with the facet facing anteriorly.

A thin layer of hyaline cartilage lies on the surface of each facet with synovial joints in the articulated spine. The facets are almost coronal in orientation in the thoracic column (T1–T10). The facets face (vertical) sagittally at lower thoracic regions. The facets absorb compression loads, are important for spine stability during flexion, provide significant resistance to anterior translation from T1 to T10, change their properties in the lower thoracic region because of the anatomy and limit rotation and with less effect on translation [4, 5].

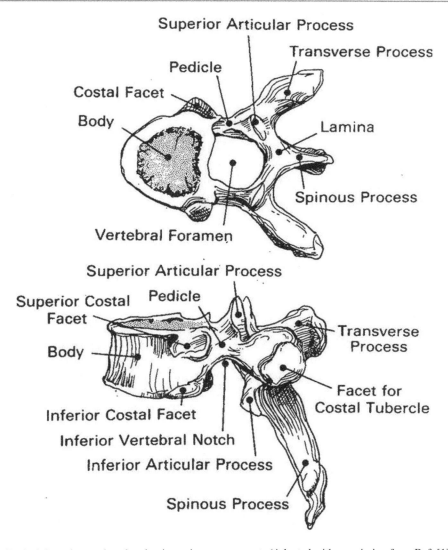

Fig. 15.2 Typical thoracic vertebra showing its various components (Adapted with permission from Ref. [1])

15.1.3 Intervertebral Discs

These fibro-cartilaginous structures attach to vertebral bodies centrally via the superior and inferior end plates, and peripherally via ligaments and fibers of the annulus fibrosus. The discs consist of the central nucleus surrounded by outer annulus. The nucleus pulposus is composed of a loose network of fibrous strands in a mucoprotein gel. The peripheral annulus fibrosus is composed of fibrous tissue arranged in concentric rings and the fiber alignment is oblique. Some fibers of the annulus blend into the anterior and posterior longitudinal ligaments. Others attach to the rim of the vertebral body. The annulus is deeper anteriorly than posteriorly with its fibers inserting into the cartilaginous endplates. The disc heights are narrower in height than the cervical and the lumbar spines. As a joining structural connective member between two vertebral bodies, the discs resist forces in all directions with the primary resistance under compression. Together with the facet joints and the ligament complex, the discs are responsible for all the load

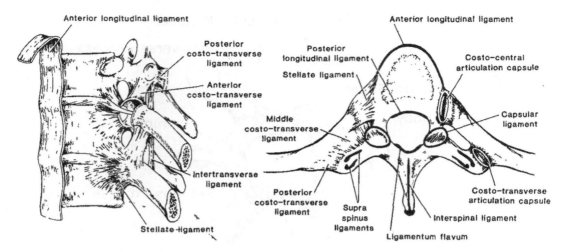

Fig. 15.3 Lateral and axial views of a thoracic vertebra showing its soft tissues (Adapted with permission from Ref. [1])

transfer between any two successive intervertebral joints and along the entire spinal column. Because of the almost vertical orientation of the facets, the discs carry the majority of the compressive loads placed on the trunk.

15.1.4 Ligaments

These are typically multilayered uni-axial structures; ligaments may connect adjacent vertebrae or extend over several segments. The elastin and collagen are the two main constituents. Several ligaments exist (Fig. 15.4). The anterior longitudinal ligament (ALL) is composed primarily of longitudinally arranged collagen fibers aligned in interdigitating layers and extend along the entire spinal column. The deepest of its three layers attaches to the edges of adjacent discs; the middle layer binds to the bodies and discs over three levels, and the superficial fibers extend four to five layers. Because of its anatomical location, its primary action is to control extension response of the spine. The posterior longitudinal ligament (PLL) is located on the posterior surface of the vertebral bodies. This ligament also consists of several layers, with the deep fibers extending only to adjacent vertebrae and the stronger superficial fibers spanning several levels. While the ligament closely adheres to the disc annulus, only marginal attachment exists to the vertebral body. It is thinner over the vertebral body than over the disc and thickest overall in the thoracic region. The function of the ligament is primarily to control flexion response of the spine. The ligamentum flavum (also called yellow ligament) is a broad, paired ligament that connects all spinal laminae. These ligaments arise from the anterior surface of the lower lamina and attach to the posterior border of the next lamina. Thus, they are discontinuous at mid-vertebral levels. They extend laterally to the joint capsules and become confluent. It has approximately 80 % elastin and is one of the most elastic tissues in the human body. The uniqueness of this ligament is that it is always under pretension to maintain the functionality of the spine. The capsular ligaments (or joint capsules) attach to the vertebra adjacent to the articular joints. The fibers are perpendicular to the plane of the facets [6]. They serve primarily to limit joint distraction and prevent excursion to a lesser degree. The interspinous ligaments connect adjacent spinous processes. The former attach from the base to the tip of each spinous process. These ligaments control flexion response of the spine.

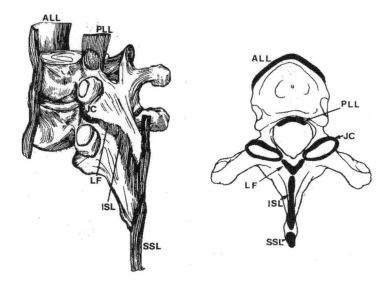

Fig. 15.4 Postero-lateral view on the left and axial views showing the ligaments (Adapted with permission from Ref. [7])

15.2 Biomechanical Models

In order to determine biomechanical properties of the spine, as it is composed of multiple components with differing load-bearing characteristics, a single experimental model is inadequate. Depending on the application, different models have been used by investigators. Models of individual components of the spine include individual vertebra/vertebral body models, isolated (in situ) ligament models, and disc models. Segmented models of the thoracic spine include single and multilevel functional units or motion segments and disc segments, along with longer columns spanning more than three levels between fixations. Intact PMHS models include the whole body. Contributions of the adjacent structures to the thoracic spinal column responses are automatically included in the structural response. Although micro-level models such as tests using small bone excised from vertebral bodies exist, these studies are not included in the gross response behaviors.

Each type of model has its own merits and demerits. The principal advantage of a component model lies in the relative ease of experimentation and provides fundamental data on the structural (force and deflection at failure for example) and material (stress and strain for example) properties of the component. This model achieves the objective of obtaining such properties independent of the connective tissues, although tests may be necessary at individual component (isolated ligament) level. However, the demerit is the in ability to predict injury because of the lack of connectivity, inherently present in vivo. Also, from a dynamic/injury perspective, inclusions of inertia effects and prediction of injury mechanisms are difficult from component models. Consequently, connective tissues are used to achieve these goals. While a functional unit consisting of vertebra-disc-vertebra structure can be fixed at the proximal and distal ends and loaded dynamically, effects of spinal curvature and posture are ignored in this, next higher level, model. Similar comments apply to the thoraco-lumbar junction models (described later) that have used two-disc and one vertebra unit for injuries and injury mechanisms. These considerations necessitate the use of multilevel functional units/segments so that the effects of curvature and posture can be incorporated. Of course, the most appropriate model would be the use of intact PMHS although issues exist, including difficulties in experimentation, loading, data recording along the spine and data reductions.

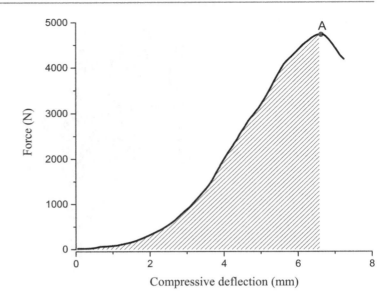

Fig. 15.5 Typical force-deflection from thoracic vertebral body compression loading test. The *shaded area* under the response curve up to the point A, failure load represents the energy stored in the specimen (Adapted from Ref. [13])

In this chapter an attempt is made to provide some insights to these approaches for the thoracic spine.

15.2.1 Component Models

15.2.1.1 Vertebrae

Failure properties of the vertebra have been determined by subjecting isolated vertebral bodies without posterior elements to external forces. Although tensile properties have been reported [8], only compressive loading properties are presented. This is because the vertebral body resists or responds to compression more often than tension, especially in the thoracic spine. An earlier study by Messerer [9] reported failure loads from specimens ranging in age from 25 to 81 years. In a later study, Sonoda [8] tested 26 specimens (age: 22–39 years) in the longitudinal direction under compression and reported failure loads. The strain rates used in these two studies were not dynamic. Kazarian and Graves (1977) conducted one of the earlier dynamic compressive tests on thoracic vertebral bodies [10]. Specimens were prepared from four human cadavers of 26–38 years of age. To ensure uniform load distribution, endplates were potted with dental acrylic. Specimens were tested using a hydraulic testing machine. Each specimen was compressed to 50 % of its original height. The authors reported failure load, displacement to failure, stiffness and energy to failure for three strain rates: 44.4, 0.4 and 0.004 per second. Gozulov et al. (1966) reported compressive failure loads from 530 thoracolumbar vertebrae (age: 19–40 years) using a strain rate of 0.008/s [11]. Yoganandan et al. (1988) subjected thoracic vertebral bodies to axial compressive forces (age: 34–79 years) using an electrohydraulic testing device at a strain rate of 0.13/s [12]. Failure load, displacement at failure, stiffness and energy to failure were reported (Fig. 15.5). The summary data from these studies is included (Table 15.1).

Thoracic spinal column can be divided into lower, middle and upper regions to provide an average-type response for each region. High and low strain rates for this initial analysis were based on rates greater than 1 and lesser than 1 per second. From a failure load carrying perspective, age-dependent strength properties have been documented in literatures wherein vertebral bodies from different regions of the spinal column have been subjected to axial compressive loading to failure. An earlier study reported a bilinear fit as the relation between strength and age. The transition of the bilinear response occurs at 40 years of age with a rapid decrease up to 40 years [14].

Table 15.1 Summary of biomechanical parameters from vertebral bodies for two strain rates

Parameter			Group A						Group B		
			Low strain-rate			High strain-rate			Low strain-rate		
			Upper thoracic	Middle thoracic	Lower thoracic	Upper thoracic	Middle thoracic	Lower thoracic	Upper thoracic	Middle thoracic	Lower thoracic
Area (cm²)		Mean	6.70	9.79	14.21	7.62	8.78	12.47	5.13	8.30	11.87
		SD	0.30	0.44	0.80	0.99	0.76	0.42	0.63	1.08	0.07
Load (kN)		Mean	3.41	4.50	6.30	4.52	7.05	8.32	4.00	3.13	3.55
		SD	0.33	0.36	0.53	0.42	0.68	0.81	1.68	0.34	0.30
Deflection (mm)		Mean	3.20	2.81	2.91	2.10	2.15	2.27	6.83	5.75	7.41
		SD	0.18	0.27	0.25	0.32	0.18	0.20	1.60	0.35	0.54
Failure energy (J)		Mean	4.73	5.10	6.59	3.95	5.55	7.20	19.20		14.40
		SD	0.58	0.63	1.21	0.88	0.76	0.83	8.34		2.02
Stiffness (N/mm)		Mean	2,134	3,626	5,048	4,343	5,574	6,249	540		538
		SD	209	357	406	785	511	617	173		55

SD standard deviation

In contrast, more recent studies have shown the inverse relation to be linear with failure strength and increasing age [15]. These data were found to be true for specimens with intact and excised cortices of lumbar vertebrae. While it is possible to use age as a continuous variable in analyses of data, a dichotomous separation was chosen at 40 years of age as a first step, based on the earlier study.

The mean cross-sectional areas for groups A (less than 40 years) and B (greater than or equal to 40 years) were: from 5.2 to 19.7 cm² and 4.5 to 12.8 cm², respectively [16, 10]. Table 15.1 shows these data for the three regions of the thoracic spine for both age groups. The cross-sectional areas showed an increasing trend from superior to inferior direction for both age groups. Both age group and thoracic region showed statistical significance ($p < 0.05$). However, the cross-sectional area decreased in group B compared to group A.

At high and low strain rates for group A, mean failure loads ranged from 0.8 to 8.6 kN and 3 to 11.1 kN. The low strain rates for group B ranged from 0.29 to 7.0 kN. There were no data for this group at the high strain rate. An increasing trend can be observed in the load carrying capacity from the caudal to the cephalad levels for both age groups at both strain rates, reflecting the increase in the cross-sectional areas at these levels. The linear correlation coefficient between the failure load and cross-sectional area was poor ($R^2 = 0.4$), however. Failure loads at the higher rate were almost double the load at lower strain rates for group A specimens perhaps reflecting the inherent viscous response of the vertebral body at higher rates [10].

At low and high strain rates, group A deflection to failure ranged from 2.2 to 14.2 mm and 1.1 to 3.1 mm. Low strain rates group B specimens ranged from 1.5 to 4.9 mm. The spinal region, age group, and strain rate were found to be statistically significant. The deflection to failure did not significantly vary across spinal region. However, deflections decreased by approximately 33 % for group A specimens at higher strain rates compared to the deflections at low strain rates for all regions. The decrease may be attributed to the viscoelastic behavior of the vertebral body at high strain rates. On average across all regions, deflection increased by 117 % in group B specimens at low strain rates compared to group A specimens. Because specimens in group B were tested without capping the endplates, cross-section areas might have increased contributing to additional deflection.

At low and high strain rates for group A, stiffness ranged from 31 to 1,145 N/mm and 2,353 to 7,861 N/mm. Low strain rates for group B ranged from 1,072 to 6,922 N/mm. The spinal region, age group, and strain rates were statistically significant. An increasing trend in stiffness from the upper to the lower thoracic regions was apparent in group A for both strain rates. At low and high rates, group A specimen stiffness increased by 150 % and 38 % in the lower thoracic spine compared to the upper thoracic spine. Stiffness decreased by 85 % in the group B compared to group A at low strain rates. The secant stiffness used in the definition may have contributed to the low stiffness data from group B.

At low and high strain rates, group A failure energy ranged from 1.4 to 44.2 J and 1.4 and 9.8 J. At low strain rates group B failure energy ranged from 1.7 and 14.2 J. Age group showed statistical significance; however, strain rate and spine region were not significantly different. Failure energy exhibited an increasing trend across spine region for group A at both the strain rates, and the reverse was seen for group B. Energy increased approximately by 20 % in the lower thoracic region compared to the upper thoracic region in group A for both the strain rates. Whereas a 20 % decrease in energy occurred in group B across all spine regions.

15.2.1.2 Intervertebral Discs

Because of its inherent structure, it is difficult to test the entire disc without endplates/adjacent vertebral body attachments, although individual fibers of the annulus or its "coupon" can be loaded using a testing device. Consequently, to test the entire disc, disc segment or one level functional spinal unit is needed. The disc segment consists of the superior and inferior bodies and the disc without the posterior elements of the intervertebral joint. The functional unit includes the posterior elements. Thus, testing the disc segment delineates the responses of the body and disc without contributions from the posterior elements.

During forward bending, anterior region of the discs experience compressive loads, and posterior part exhibits tensile load. Therefore, it is important to determine the strengths of the discs in tension and compression. However, tensile loading is presented as thoracic disc data are sparse for compression. Pintar (1986) conducted tensile tests using ten unembalmed male PMHS specimens [16]. Age of the specimens ranged from 45 to 82 years. The disc segment was fixed at the vertebral ends using Steinman pins and pulled in direct axial tension to failure at a strain rate of 0.5/s using an electro-hydraulic testing machine. The following analysis was performed by dividing the thoracic discs into the upper, middle, and lower regions, representing C7–T5, T5–T8, and T9–L1 levels. Failure loads showed an increasing trend from the upper to lower spine and ranged from 334 to 1,219 N for all regions. There was an approximate 40 % and 60 % increase in the load carrying capacity in the mid-thoracic and lower thoracic regions compared to the upper thoracic regions, reflecting increased cross-sectional areas from upper to lower spine.

Deflection to failure ranged from 4.2 to 19.1 mm for all regions. A decreasing trend occurred in the failure deflection inferiorly. Deflection decreased by 18 % and 27 % in the middle and lower thoracic regions compared to the upper regions. This reduction in deflection may be due to increase in stiffness from upper to lower thoracic region. Stiffness ranged from 32.3 to 226.6 N/mm for all regions. Data did not show any particular trend of variation from upper to lower spine region. Stiffness increased by 100 % and 56 % in the middle and lower thoracic regions compared to upper thoracic discs.

Energy to failure for thoracic discs ranged from 0.94 to 14.2 J from upper to lower thoracic region. An increasing trend was observed in the deformation to failure inferiorly. Energy increased by 50 % and 67 % in the middle and lower thoracic regions compared to the upper thoracic region. Despite decrease in failure deformation toward the lower thoracic region, the increase in stiffness may have attributed to higher energy to failure in lower thoracic discs. These data are shown (Table 15.2).

15.2.1.3 Ligaments

Because of the soft tissue nature of the ligament, it has to be tested by incorporating the connection with the bone, similar to the discs. This leads to models similar to functional unit/disc segment, but in this case, the disc has to be transected leaving only the ligament under test to be loaded in tension. This is termed in situ testing methods. The following analysis was performed by dividing

Table 15.2 Summary of biomechanical parameters from disc tests at low strain rates

Parameter	Statistics	Upper thoracic spine	Middle thoracic spine	Lower thoracic spine
Force (N)	Mean	504.60	716.20	800.75
	SD	61.32	140.00	186.61
Deflection (mm)	Mean	10.44	8.66	8.08
	SD	2.97	2.28	1.83
Energy (J)	Mean	2.98	4.66	4.92
	SD	0.87	2.49	1.98
Stiffness (N/mm)	Mean	85.00	160.68	120.54
	SD	26.48	29.43	3.56

SD standard deviation

the ligaments into the upper, middle, and lower regions, representing C7–T5, T5–T8, and T9–L1 levels.

Yoganandan et al. (1988) tested anterior longitudinal ligament (ALL), posterior longitudinal ligament (PLL), ligamentum flavum (LF), and interspinous ligament (IL) harvested from cadavers aged between 34 and 83 years [12]. The experimental setup is similar to the one used by Pintar et al. (1986) for testing thoracic intervertebral discs. Prepared specimens were loaded at 0.5/s until failure. Because of the paucity of dynamic responses of thoracic ligaments, scaling ratios based on cervical spine ligaments reported by Yoganandan et al. (1989) were used to compute dynamic responses of thoracic ALL and LF in this review [17]. For ALL, dynamic scaling factors of 0.33, 1.0, 0.53, and 0.48 were used. For LF, dynamic scaling factors of 0.40, 0.57, 0.33, and 0.33 corresponding to the force, deflection, energy, and stiffness obtained from this cervical study were used to extrapolate dynamic properties of thoracic ligaments for the three regions.

At low and high strain rates, tensile failure loads ranged from 58 to 939 N and 457 to 1,191 N for ALL and from 80 to 489 N and 427 to 1,193 N for LF. At low strain rates for PLL, failure load ranged from 18 to 343 N. At low strain rates for IL, failure load ranged from 9 to 778 N. An increasing trend was observed in failure load across the three-spine region for ALL and IL. However there was no significant variation in failure load for LF and PLL for the three regions. Approximately, there was a 120 % and 200 % increase in the load carrying capacity in the mid-thoracic and lower thoracic regions compared to the upper thoracic region for the ALL. A 200 % increase in load carrying capacity was observed in ALL and LF at high strain rates compared to low strain rates.

At low and high strain rates for ALL, displacements to failure ranged from 3 to 35 mm and 5 to 14 mm. At low and high strain rates for LF, failure displacements ranged from 2 to 19 mm and 6 to 15 mm. At low strain rates for PLL, failure displacement ranged from 2 to 21 mm. At low strain rates for IL, failure displacement ranged from 2 to 25 mm. An increasing trend was observed in failure displacement across the spine region for ALL, LF, and PL. There was no significant change in the parameter for IL in the three regions. Approximately, there was a 50 % and 63 % increase in the failure displacement in mid-thoracic and lower thoracic regions compared to upper thoracic region for ALL. Increase in strain rate did not alter failure displacement for ALL. Approximately, 13 % and 24 % increase was observed in failure displacement in mid-thoracic and lower thoracic regions compared to upper thoracic LF. A 63 % increase in failure displacement was observed in LF at high strain rates compared to low strain rates. A 50 % and 75 % increase was observed in failure displacements in the mid-thoracic and lower thoracic regions compared to upper thoracic PLL.

At low and high strain rates for ALL, stiffness ranged from 7 to 193 N/mm and 40 to 227 N/mm. At low and high strain rates for LF, stiffness ranged from 9 to 128 N/mm and 35 to 154 N/mm. At low strain rates for PLL, stiffness ranged from

Table 15.3 Summary of biomechanical parameters from tensile ligament tests at low strain rates

Parameter	Statistics	ALL	LF	PLL	IL
Upper thoracic					
Load (N)	Mean	165.06	211.29	102.36	51.36
	SD	35.08	32.12	16.58	9.97
Deflection (mm)	Mean	7.78	8.41	3.59	7.91
	SD	1.48	1.22	0.52	1.64
Failure energy (J)	Mean	1.6	2	0.38	0.41
	SD	0.51	0.45	0.09	0.11
Stiffness (N/mm)	Mean	35.33	28.41	91.8	9.92
	SD	10.55	4.15	61.65	3.83
Middle thoracic					
Load (N)	Mean	357.13	221.19	128.5	57
	SD	61.97	26.51	26.12	12.76
Deflection (mm)	Mean	12.05	8.82	5.41	7.41
	SD	1.68	0.85	0.9	1.39
Failure energy (J)	Mean	4.38	1.94	0.76	0.56
	SD	1	0.27	0.21	0.16
Stiffness (N/mm)	Mean	37.93	30.19	25.95	7.71
	SD	9.01	6.91	3.7	1.19
Lower thoracic					
Load (N)	Mean	415.59	243	99.93	135
	SD	54.99	21.47	23.1	48.48
Deflection (mm)	Mean	13.66	8.67	6.39	9.31
	SD	1.74	0.75	1.51	1.82
Failure energy (J)	Mean	5.95	2.11	0.54	1.77
	SD	1.24	0.24	0.14	0.9
Stiffness (N/mm)	Mean	38.36	31.5	29.91	17.49
	SD	9.33	3.9	9.87	4.06

SD standard deviation

6 to 890 N/mm. At low strain rates for IL, stiffness ranged between 2 and 52 N/mm. An increasing trend was observed in the stiffness across the spine region for ALL while PLL showed a decreasing trend. However, there is no significant variation in stiffness for LF and IL across cranial to caudal. Approximately, there was a 50 % and 100 % increase in stiffness in mid-thoracic and lower thoracic regions compared to upper thoracic ALL. A 50 % decrease in stiffness was observed in lower thoracic PLL compared to upper thoracic PLL. One-hundred and 200 % increases in stiffness were observed in ALL and LF at high strain rates compared to low strain rates.

At low and high strain rates for ALL, failure energy ranged from 2 to 22 J and 4 to 16 J. At low and high strain rates for LF, failure energy ranged from 0.7 to 6.0 J and 2.7 to 16 J. At low strain rates for PLL, failure energy ranged from 0.1 to 2.6 J. At low strain rates for IL, failure energy ranged between 0.2 and 14 J. An increasing trend was observed in failure energy across the spine region for ALL and IL. However there was no significant variation in failure energy for LF and PLL across cranial to caudal regions. Approximately, there was a 300 % and 400 % increase in failure energy in mid-thoracic and lower thoracic regions compared to upper thoracic ALL. A 200 % increase in failure energy was observed in ALL and LF at high strain rates compared to low strain rates. These data are shown (Tables 15.3 and 15.4).

Table 15.4 Summary of biomechanical parameters from tensile ligament tests at high strain rates

Parameter	Statistics	ALL	LF	PLL
Upper thoracic				
Load (N)	Mean	457	515	330.6
Deflection (mm)	Mean	7.7	14	1.7
Failure energy (J)	Mean	3.52	7.21	0.56
Stiffness (N/mm)	Mean	59.35	36.79	194.47
Middle thoracic				
Load (N)	Mean	761.46	721.48	503.8
	SD	95.28	133.9	79.42
Deflection (mm)	Mean	9.94	11.22	4.15
	SD	1.58	1.36	2.15
Failure Energy (J)	Mean	7.47	8.39	2.55
	SD	1.42	2.13	1.62
Stiffness (N/mm)	Mean	87.47	65.37	192.48
	SD	19.32	9.45	51.21
Lower thoracic				
Load (N)	Mean	964.87	731.6	547.2
	SD	169.66	116.6	133.7
Deflection (mm)	Mean	9.33	10.15	2.13
	SD	2.6	4.65	0.13
Failure Energy (J)	Mean	9.14	6.88	1.15
	SD	3.6	2.22	0.22
Stiffness (N/mm)	Mean	126.9	97.89	262.11
	SD	50.84	56.33	78.34

SD standard deviation

15.2.2 Segmented Models: Functional and Multi-segment Units

A multi-segment spine unit (MSU) consists of more than one FSU. Commonly, the vertebral body at the base of the spine unit is potted and loads are applied to the top of the unit. Although tests are conducted with FSU/MSU, results are discussed in terms of individual components of the spinal unit. Maiman et al. (1986) tested T4–L5 spinal columns from 14 unembalmed male cadavers [18]. Specimens were fixed at the ends in 15-cm-diameter aluminum cylinders using Steinmann pins through vertebrae. The preparation was tested in an electrohydraulic testing machine at a strain rate of 1 cm/s (approximately 0.5/s for 50 % compression). Although these tests resulted in thoracic and lumbar injuries, data were extracted from specimens with injuries to the thoracic spine. Kifune et al. (1995) tested ten T11–L1 spinal units (age: 28–71 years) [19]. Specimens were mounted at T11 and L1 in polyester resin and tested on a drop tower impact apparatus. A guided drop mass was allowed to fall freely at a velocity of 5.2 m/s onto an impounder resting on top of the specimen. Panjabi et al. (1995) tested 14 T11–L1 spine units (age: 19–84 years) using the same device at 5.1 m/s [20]. Langrana et al. (2002) quasi-statically tested seven human cadaveric (age: 45–65 years) thoracic motion segments in a MTS testing device [21].

At low and high strain rates, failure loads ranged from 0.6 to 5.6 kN and 2 to 6.8 kN. A 10 % and 30 % increase in failure loads was observed in the mid- thoracic and lower thoracic regions compared to upper spine region. Whereas there was a 43 % increase in the upper spine region compared to mid column at low strain rates. Compression and wedge fractures occurred at low strain rates, while other types of fractures occurred at high rates. At low and high strain rates, mean forces for compression fractures were 1.1 and 3.7 kN; for wedge fractures mean forces were 2.3 and 6.5 kN. At high strain rates, failure loads between wedge and burst fractures were not significantly different. However, failure loads for burst fracture were significantly greater than the other fracture types.

15.2.3 Intact Human Cadaver Studies

Yoganandan et al. (1986) dropped 15 intact male PMHS (age: from 54 to 82 years), at velocities from 4.2, 4.9 and 5.4 m/s onto a force plate [22]. Experiments were done with unrestrained heads, neck pre-flexed, and head/neck complex restrained by special fixtures. X-rays and CT scans were obtained and cryomicrotomy studies were done to document injuries. Both cervical and thoracic spine injuries were identified. Maiman et al. (1986) tested five PMHS to study compressive fractures in thoracolumbar spinal column [18]. Forces at the rate of 1 cm/s were applied at the T2–T3 region with cadavers in the seated posture. Posterior elements were surgically exposed at the proximal end to facilitate axial loading to the torso. Restraints were used to minimize unintended motion of PMHS during loading. The following observations are made from combining data from these studies. Failure force at the midthoracic region occurred at applied peak force of 1.1 kN for the single specimen tested at the low strain rate of 0.5/s. However peak forces varied from 1.1 to 1.6 kN for injuries to the lower thoracic region at the same low strain rate. At higher strain rates, mean failure load ranged from 14.3 to 14.7 kN, 4.5 and 12.4 kN, 4.5 and 7.1 kN for failures in the upper, middle, and lower spinal regions. At low strain rates, only wedge fractures were frequent, similar to those observed in FSU and MSU studies. At low and high strain rates, wedge fractures occurred at a mean peak load of 1.1 and 6.7 kN. At high strain rates, data did not show significant differences in failure loads required to produce burst and compression fractures. The mean load required to produce burst fractures was 9 kN.

15.3 Injury Simulations

15.3.1 Thoracolumbar Junction Injuries

This area is active for clinicians as injuries occur to the thoraco-lumbar (T-L) junction to the dorsal spine in environments such as falls [1]. Generally T11–L2 levels are involved. The two- and three-column definitions of spinal stability are based on the integrity of the different regions of the spine at a segmental level [23, 24] (Fig. 15.6). In order to delineate the role of each column and factors such as spinal canal compromise, researchers have produced fractures (endplate, wedge, burst, etc.) to the T-L junction using two-disc-one-vertebra models using drop-weight techniques. Typically, the distal and proximal vertebrae are fixed such that the superior and inferior discs and the middle vertebra are unconstrained. Postural effects have been simulated using a pre-oriented wedge attached to the fixation to produce compression-flexion type injuries. Although data have been reported on the forces and accelerations required for T-L injuries in some studies, consistent and reliable information can be gleaned only based on the input energy to the two-disc-one-vertebra model. While somewhat surprising, all studies appear to have been based on an arbitrarily chosen energy of 200 Nm (10 kg mass dropped from 2 m), initiating from earlier experiments [25]. Accurate biomechanical thresholds (example, peak axial force or force at the onset of injury) were not determined in these clinical-based experimental investigations (Fig. 15.6).

15.3.2 Upper and Lower Thoracic Spinal Column Studies

In a combined Crash Injury Research Engineering and Network (CIREN) and NASS-CDS investigation, Pintar et al. (2012) analyzed data from motor vehicle crashes and found that injuries are increasing to the dorsal spine despite public awareness for safety, high manual seatbelt use in the United States and advancements in vehicle designs [26].

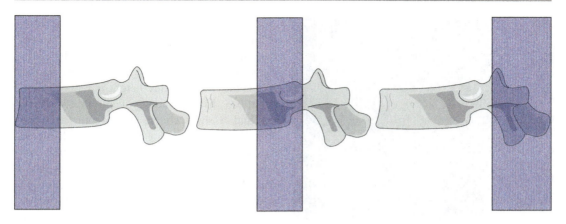

Fig. 15.6 The three-column concept used to define instability

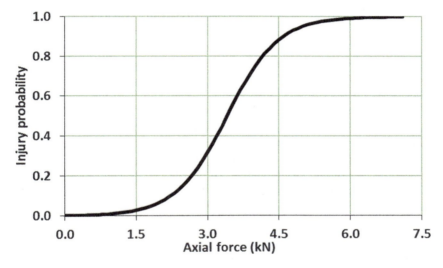

Fig. 15.7 Logistic regression curve for thoracic spinal column axial force

The authors concluded that compression-related thoraco-lumbar spinal column fractures occur in frontal impacts in modern vehicle environments for which "the mechanism of injury is poorly understood." They also underscored the lack of injury assessments for spine fractures in the current motor vehicle crash standards throughout the world [26]. Recently, studies were conducted using upper (T2–T6) and lower (T7–T11) PMHS thoracic spines to determine the biomechanics of impact. Specifically, forces, bending moments and fracture patterns were determined [27]. The superior and inferior ends of the columns were fixed using PMMA, load cells were attached to the ends and the specimen was dropped at velocities of 2.8, 3.4 and 5.4 m/s. The former two velocities served as the low energy drops where unstable thoracic spine fractures did not occur. In contrast, the highest velocity was associated with unstable injuries. Intermittent evaluations consisted of palpations and x-rays. Injuries were assessed using posttest x-rays and computed tomography scans. All data were gathered according to accepted techniques. The mean peak forces ranged from 1.6 to 4.3 and 1.3 to 5.1 kN for the two regions of the thoracic spine. Logistic regression analysis (Fig. 15.7) indicated that 50 % probability of fracture is associated with a peak axial force of 3.4 kN (data

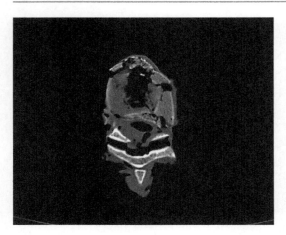

Fig. 15.8 Axial posttest CT scan image showing burst fracture in a thoracic spine specimen (T8)

combined for both regions of the thoracic spine). These results indicate the initial tolerance limits of dorsal spines under vertical loading. A comparison of laboratory-produced thoracic spine fracture with clinical injury is demonstrated (Fig. 15.8). Filtering of acoustic signals in the 70–300 kHz bandwidth from sensors attached to vertebra have shown that the onset of compression fracture is close to the peak force, thus providing confidence in using maximum forces values for describing injury thresholds.

15.4 Clinical Injuries to the Thoracic Spine

There are four major categories of injury to the thoracic spine. These include wedge fractures, burst fractures, flexion-distraction injuries, and fracture-dislocations [28]. A wedge compression fracture is predominantly a flexion injury typically with some compression loading reflecting the degree of eccentricity. A force applied to the back of the head or shoulders results in an exaggeration of the normal thoracic kyphosis [2]. This causes the superior vertebral body to impact the anterior-superior endplate of the vertebral body below, resulting in a fracture of the anterior column, while the middle and posterior columns are often intact [24]. The fracture can involve compression of the trabecular bone, disruption of the cortical mantle, or both. These are seen as a loss of anterior vertebral body height on lateral radiographs causing a characteristic wedge deformity. Neurologic deficit is rare with these injuries unless the interior body height is reduced more than 50 %. Management typically involves bracing and a short course of narcotics for pain control; if the posterior elements are disrupted, fixation is typically necessary. Occasionally, disruption of the posterior ligamentous complex is seen. Persistent pain may require surgical management with vertebroplasty or kyphoplasty.

A pure axial load applied to spine can result in a burst fracture. The axial load can be due to a fall from a height. This results in a failure of the anterior and middle columns [29]. While the posterior neural arch remains intact, widening of the interpedicular distance is seen in almost all cases [30]. As a result of the compressive force, bony fragments from the posterior vertebral body can be retropulsed into the spinal canal, potentially causing neurologic injury. In cases of neurologic deficit or disruption of the posterior ligament complex, surgical management is pursued.

The flexion-distraction injury, commonly referred to as the Chance fracture, is a traumatic disruption of the middle and posterior columns. The most common mechanisms are falls from height and motor vehicle collisions. Prior to the introduction of three-point harnesses in motor vehicles, lap belts were commonly used. The lap belt acts as a fulcrum across which the unconstrained torso can rotate [31]. This can produce a flexion-distraction injury to the bony elements, the disc and ligament complex, or middle and posterior columns. The anterior column including the anterior longitudinal ligament remains intact. Management of these fractures depends on whether the injury is primarily bony or ligamentous. Bracing can sometimes be used with a pure bony Chance fracture, while posterior instrumentation and fusion are needed for ligamentous injuries [31].

Fracture-dislocation injuries can be produced by a variety of mechanisms [24]. The forces involved can be flexion, rotation, shear, either alone or in combination with one another. The injuries these produce are as varied as the mechanisms. For example, an injury with flexion and shear forces can be dislocated the facets bilaterally and produce a listhesis of one vertebral body on another. Neurologic injury is common with such injuries, and these are typically managed with operative decompression and stabilization.

15.5 Summary

Recognizing that relatively very few thoracic spine studies have been conducted in contrast to cervical and lumbar spines in literatures, a brief review of biomechanical models of the thoracic spine was presented in this chapter. Experimental models included the vertebral bodies, ligaments, discs, spinal units and intact PMHS. The peak force, deflection, stiffness and energy parameters were selected for comparing the responses based on two age groups and two strain rates for the upper, middle and lower thoracic regions. Depending on the parameter, group, strain rate and region, different trends were identified. The trends also depended on the selected experimental model. In general, compressive failure loads, stiffness and energies of vertebral bodies increased from the upper to the lower regions and from lower to higher strain rates, while the failure deflections did not vary by region although higher rates resulted in lower deflections. For discs, while the failure loads and stiffness increased from the upper to the lower regions, deflections showed an opposite trend and stiffness was invariant across the three regions. All ligaments showed an increasing trend in the peak tensile force, stiffness and energy from the upper to the lower thoracic regions (except PLL), and the same phenomenon was found to be true for deflection (except LF). FSU and MSU responded with greater failure loads from the upper to the lower regions and from low to high strain rates. The fracture pattern was such that compression and wedge fractures occurred at low strain rates. In contrast, intact PMHS drop tests showed that the applied peak forces are the greatest for injuries to the upper spine, followed by the middle and lower spine regions. As seen in the other experimental models, high rates resulted in greater peak forces for all regions. While wedge fractures were common at low strain rates, higher rates produced compression, burst and wedge fractures.

These analyses clearly show loading/strain rate effects on injuries and injury metrics. However, the rates used in previous studies are lower than those encountered in situations such as underbody blast from land mines or improvised explosive devices. Thus, studies are needed with these parametric inputs to better understand the biomechanics as applied to these environments. Manikins with proven performance in the thoracic region specific to underbody blast scenarios do not exit. Likewise, with the recent recognition of thoracolumbar injuries in automotive frontal impacts with belted and unbelted occupants and rollovers, human thorax studies are also needed to improve occupant safety. This brief chapter described the complexity of the spine to external loading and underscores the need to obtain data from additional samples and by testing in other modes of external loading such as combined compression and flexion wherein combined injury criteria instead of the peak force metric may better describe the injury biomechanics. The suggested studies will facilitate the development and validation of manikins with improved biofidelity for potential use in military and automotive areas.

Acknowledgements This material is the result of work supported with resources and the use of facilities at the Zablocki VA Medical Center, Milwaukee, Wisconsin and the Medical College of Wisconsin. All authors are part-time employees of the Zablocki VA Medical Center, Milwaukee, Wisconsin. Any views expressed in this article are those of the authors and not necessarily representative of the funding organizations.

References

1. Maiman DJ, Pintar FA (1992) Anatomy and clinical biomechanics of the thoracic spine. Clin Neurosurg 38:296–324
2. Kazarian L (1981) Injuries to the human spinal column: biomechanics and injury classification. Exerc Sport Sci Rev 9:297–352
3. Yamada H (1973) Strength of biological materials. Robert E Krieger, Huntington

4. Roaf R (1960) A study of the mechanics of spinal injuries. J Bone Joint Surg 42:810–823
5. Prasad P, King AI, Ewing C (1974) The role of articular facets during+Gz acceleration. J Appl Mech 41:321–326
6. Panjabi MM, Hausfeld JN, White AA 3rd (1981) A biomechanical study of the ligamentous stability of the thoracic spine in man. Acta Orthop Scand 52(3):315–326
7. Sances A Jr, Myklebust JB, Maiman DJ, Larson SJ, Cusick JF, Jodat RW (1984) The biomechanics of spinal injuries. Crit Rev Biomed Eng 11(1):1–76
8. Sonoda T (1962) Studies on the strength for compression, tension and torsion of the human vertebral column. J Kyoto Pref Med Univ 71:13
9. Messerer O (1880) Uber Elaslicitat and Festigkeit der Meuschlichen Knochen. J C Cottaschen Buch handling, Stuttgart
10. Kazarian L, Graves GA (1977) Compressive strength characteristics of the human vertebral centrum. 2(1):1–14. doi:citeulike-article-id:3195063
11. Gozulov SA, Korzheniyantz Y, Skruypnik VG, Sushkov NY (1966) A study of compression strength of vertebrae in man. Arkh Anat Gistol Embryo 51:6
12. Yoganandan N, Pintar F, Anthony Sances J, Maiman D, Myklebust J, Harris G (1988) Biomechanical investigations of the human thoracolumbar spine. Society of Automotive Engineers, Warrendale
13. Yoganandan N, Mykiebust JB, Cusick JF, Wilson CR, Sances A Jr (1988) Functional biomechanics of the thoracolumbar vertebral cortex. Clin Biomech 3(1):11–18. doi:10.1016/0268-0033(88)90119-2
14. Rockoff SD, Sweet E, Bleustein J (1969) The relative contribution of trabecular and cortical bone to the strength of human lumbar vertebrae. Calcif Tissue Res 3(2):163–175
15. Mosekilde L, Mosekilde L (1986) Normal vertebral body size and compressive strength: relations to age and to vertebral and iliac trabecular bone compressive strength. Bone 7(3):207–212, http://dx.doi.org/10.1016/8756-3282(86)90019-0
16. Pintar F (1986) The biomechanics of spinal elements. Marquette University, Milwaukee
17. Yoganandan N, Pintar F, Butler J, Reinartz J, Sances A Jr, Larson SJ (1989) Dynamic response of human cervical spine ligaments. Spine 14(10):1102–1110
18. Maiman DJ, Anthony Sances J, Myklebust JB, Chilbert M, Yoganandan N, Pintar F, Larson SJ (1986) Experimental trauma of the human thoracolumbar spine. In: Sances J, Thomas DJ, Ewing CL, Larson SL, Unterharnscheidt F (eds) Mechanisms of head and spine trauma. Aloray Publishers, New York, p 16
19. Kifune M, Panjabi MM, Arand M, Liu W (1995) Fracture pattern and instability of thoracolumbar injuries. Eur Spine J 4(2):98–103
20. Panjabi MM, Oxland TR, Kifune M, Arand M, Wen L, Chen A (1995) Validity of the three-column theory of thoracolumbar fractures. A biomechanic investigation. Spine 20(10):1122–1127
21. Langrana NA, Harten RR, Lin DC, Reiter MF, Lee CK (2002) Acute thoracolumbar burst fractures: a new view of loading mechanisms. Spine 27(5): 498–508
22. Yoganandan N, Sances A Jr, Maiman DJ, Myklebust JB, Pech P, Larson SJ (1986) Experimental spinal injuries with vertical impact. Spine 11(9):855–860
23. Nicoll EA (1949) Fractures of the dorso-lumbar spine. J Bone Joint Surg 31B(3):376–394
24. Denis F (1983) The three column spine and its significance in the classification of acute thoracolumbar spinal injuries. Spine 8(8):817–831
25. Willen J, Lindahl S, Irstam L, Aldman B, Nordwall A (1984) The thoracolumbar crush fracture. An experimental study on instant axial dynamic loading: the resulting fracture type and its stability. Spine 9(6): 624–631
26. Pintar FA, Yoganandan N, Maiman DJ, Scarboro M, Rudd RW (2012) Thoracolumbar spine fractures in frontal impact crashes. Ann Adv Automot Med 56: 277–283
27. Yoganandan N, Arun MWJ, Stemper BD, Pintar FA, Maiman DJ (2013) Biomechanics of human thoracolumbar spinal column trauma from vertical impact loading. Ann Adv Automot Med 57:155–166
28. Looby S, Flanders A (2011) Spine trauma. Radiol Clin North Am 49(1):129–163. doi:10.1016/j.rcl.2010.07.019
29. Dai LY, Jiang SD, Wang XY, Jiang LS (2007) A review of the management of thoracolumbar burst fractures. Surg Neurol 67(3):221–231; discussion 231. doi:10.1016/j.surneu.2006.08.081
30. Willen J, Anderson J, Toomoka K, Singer K (1990) The natural history of burst fractures at the thoracolumbar junction. J Spinal Disord 3(1):39–46
31. Liu YJ, Chang MC, Wang ST, Yu WK, Liu CL, Chen TH (2003) Flexion-distraction injury of the thoracolumbar spine. Injury 34(12):920–923

Lumbar Spine Injury Biomechanics

16

Brian D. Stemper, Frank A. Pintar, and Jamie L. Baisden

Abstract

The primary biomechanical function of the lumbar spine is to bear the weight of the torso, head-neck, and upper extremities and support physiologic movement. The lumbar spinal column resides vertically between the thoracic spine and sacrum, and consists of five bony vertebrae interconnected by soft tissues including the intervertebral discs, ligaments, and muscles to maintain the integrity of the column under physiologic and traumatic environments. Injuries secondary to excessive deformations or loading resulting from external dynamic forces such as falls, or in military environments, aviator ejections, helicopter crashes or underbody blasts, can result in fracture of the lumbar spine with or without mechanical and clinical instability, and loss of normal function. These types of injuries can have significant consequences for the patient. Mechanically-induced traumas are transmitted to the lumbar spine in a variety of different ways. For example, axial or eccentric compressive forces transmitted to the lumbar spine through a vehicle seat sustaining high-rate vertical acceleration may result in different fracture types (e.g., burst fracture versus anteriorly-oriented wedge fracture), lead to mechanical instability, and impair normal daily activities. These acute consequences are in addition to the chronic effects of lumbar spine trauma including chronic back and lower extremity pain due to spinal degeneration, spinal cord or nerve root injury, or loss of lower limb sensation and function. This chapter outlines lumbar spine injury classification including mechanisms and clinical implication, describes experimental techniques used to understand injury mechanics,

B.D. Stemper, Ph.D. (✉) • F.A. Pintar, Ph.D.
Neuroscience Research Lab, Department of
Neurosurgery, Medical College of Wisconsin,
Milwaukee, WI, USA
e-mail: bstemper@mcw.edu; fpintar@mcw.edu

J.L. Baisden, M.D.
Department of Neurosurgery, Medical College
of Wisconsin, Milwaukee, WI, USA
e-mail: jbaisden@mcw.edu

and provides a listing of biomechanical fracture tolerance and injury criteria from experimental studies incorporating human cadavers. Due to the breadth of literature on lumbar spine injury mechanics, this chapter is not intended to be comprehensive. Rather, the reader will be provided with a overview of concepts relevant to the contemporary understanding of lumbar spine injury mechanics and tolerance.

16.1 Biomechanically Relevant Anatomy

The human spinal column has a sigmoid shape when viewed laterally, or in the sagittal plane. It is composed of 33 vertebrae interconnected by fibro-cartilaginous intervertebral discs in the anterior aspect, articular facet joints postero-laterally, and ligaments spanning adjacent vertebrae in multiple locations across the segment. Typically, there are seven cervical, twelve thoracic, five lumbar (L1–L5, Fig. 16.1), five fused sacral, and four separate vertebrae combined to form the coccyx. Among these, only the cervical, thoracic, and lumbar are flexible. In the sagittal plane, cervical and lumbar regions have an anteriorly convex shape, and the thoracic spine and sacrum are anteriorly concave. The lumbar spine is located in the abdominal body region and is the focus of this chapter.

16.1.1 Vertebrae

The five lumbar vertebrae are aligned with a prominent convex curvature in the lateral or mid-sagittal plane, otherwise known as lordosis. Lumbar vertebrae consist of an outer casing of dense and compact cortical bone. The interior part of the body is composed of cancellous bone aligned in lattice manner to resist axial force while minimizing mass, the effectiveness of which is proportional to its density. Lumbar vertebrae consist of the large body ventrally and posterior elements dorsally (Fig. 16.1). Extending from the vertebral body and moving posteriorly are the pedicles, transverse processes, articular processes forming the facet joints, laminae, and spinous processes. The vertebral foramen is a prominent feature of the lumbar vertebrae and is formed by the posterior aspect of the vertebral body and medial aspects of the pedicles, articular processes, and laminae. Extended from the base of the cranium and formed by all 33 vertebral foramina is the spinal canal, through which the spinal cord traverses. The cord transforms into the cauda equina near the second lumbar vertebra. Nerve roots exit the spinal canal at every vertebral level through the intervertebral foramina, formed by the postero-lateral aspect of the vertebral body, cranial and caudal aspects of opposing pedicles, and the anterior aspect of the articular processes.

The vertebral bodies form the most massive portion of the vertebra, are located anteriorly, and

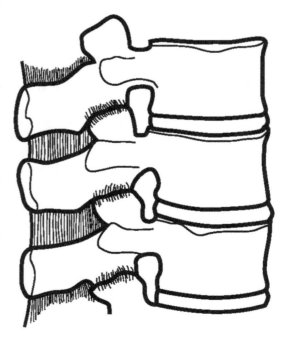

Fig. 16.1 Normal lumbar spine

are largest in the lumbar region, compared to thoracic and cervical regions. Lumbar vertebral bodies have relatively flat and kidney-shaped cranial and caudal surfaces. Transverse processes extend laterally from the pedicles. Facet joints form between opposing surfaces of the articular processes with joint surfaces oriented ventral-laterlly for the cranial processes and dorso-medially for the caudal processes. The pars interarticularis lies between superior and inferior facet joint processes. The large, flat, and vertically oriented spinous processes are located at the dorsal aspect of the vertebrae, are largest in the cranial region and decrease in size at the caudal levels. The five bony vertebrae of the lumbar spine are interconnected by soft tissues and joints, as described below.

16.1.2 Endplates

The endplates are the cranial and caudal surfaces of the vertebral bodies and form a thin cartilaginous interface between the bony vertebral body and the fibrocartilaginous intervertebral disc. The endplates are composed of hyaline cartilage. They are fused to the vertebral body by a calcium layer through which small pores penetrate for the nutrition of the intervertebral disc. The inferior zone of the vertebra remains in contact with the cartilage by the lamina cribrosa, the sieve-like surface. Osmotic diffusion occurs through this layer.

16.1.3 Intervertebral Discs

Intervertebral joints consist of a form of cartilaginous joint known as a symphysis. The primary component of the symphysis is the intervertebral disc, although the endplate (discussed above) is also a component. Lumbar intervertebral discs are located between adjacent vertebral bodies from the thoracolumbar junction (T12–L1) through the lumbo-sacral junction (L5–S1) and are connected to the bodies by the endplates. Their concentrically arranged components are: the outer alternating layer of collagen fibers forming the peripheral rim of the annulus fibrosus; a fibrocartilage component forming a major portion of the annulus fibrosus; the transitional region between the central nucleus pulposus (core), where the annulus fibrosus and the nucleus pulposus merge; and the nucleus pulposus made of a soft, pulpy, highly elastic mucoprotein gel containing various mucopolysaccharides, collagen matrix and water with relatively low collagen fibril. Intervertebral discs attach to the vertebral bodies centrally via the endplates and peripherally via the annulus fibers and ligaments. While some of the annulus fibers blend into the anterior and posterior longitudinal ligaments, others attach to the rim of the vertebra. The disc resists axial compression and tension, lateral and anteroposterior shear, and axial rotation. The shape of the disc in the lumbar spine is such that the lordotic curvature is maintained by the greater height ventrally than dorsally.

16.1.4 Ligaments

Ligaments are multilayered and composed primarily of elastin and collagen in different rations. In very general terms, collagen adds strength to the ligament whereas elastin adds elasticity. Spinal ligaments connect adjacent vertebrae (e.g., interspinous ligament) or extend over several segments (e.g., anterior longitudinal ligament). They are uniaxial structures. As such, they are capable of resisting only tension and buckle under compression. The anterior longitudinal ligament (ALL) runs from the occiput to the sacrum. It consists primarily of long, arranged collagen fibers which are aligned in interdigitating layers. The deep layer extends only to adjacent vertebrae, the middle layer over a few vertebral levels, and the outer over four to five levels. This stratification is of significance in regulating physiologic motion. This ligament functions to prevent hyperextension and excessive distraction; it is functionally active in extension and rotation. The posterior longitudinal ligament (PLL) originates from C2, continues to the coccyx, and is located on the dorsal surface of the vertebral bodies. While this multilayered ligament closely adheres to the disc annulus, attachments to the vertebral bodies are minimal. It is broader in the area of the intervertebral disc and very thin in the area of the vertebral bodies. Deeper layers span only the intervertebral disc and superficial layers can extent across multiple vertebral segments. The cross-section is considerably smaller and the tensile response is weaker than ALL. The ligamentum flava span between adjacent surfaces of laminae and are discontinuous, spanning from the cranial surface of the caudal vertebra to the caudal surface of the cranial vertebra. Ligamentum flava are present from C2–C3 to the sacrum. These ligaments are also termed yellow ligaments due to their appearance. Laterally, it is confluent with the joint capsules. Eighty percent elastin, and under some tension preload at rest, ligamentum flava are quite effective in returning the laminae to their resting positions following flexion. Capsular ligaments are intimate to the facet joint capsule and attach to the vertebrae adjacent to the articular joint. Their fibers are aligned normal to the facets, limiting distraction and sliding of the facet joints and hyperflexion of the segment. The interspinous ligaments connect adjacent spinous processes and are met by the ligamentum flavum anteriorly and the supraspinous ligament posteriorly. The supraspinous ligament is a fibrous ligament running along the distal extent of the spinous processes from seventh cervical vertebra to the sacrum. Interspinous and supraspinous ligaments act to resist flexion bending.

16.1.5 Facet Joints

Facet joints are articular joints that are located postero-laterally to the intervertebral disc at each vertebral segment (T12–L1 through L5–S1). There are two facet joints per segment (right and left sides). The joints are formed by the superior articular process from the caudal vertebra (facing dorso-medially) and the inferior articular process from the cranial vertebra (facing ventro-laterally). Opposing surfaces of the articular processes consist of a smooth and resilient layer of hyaline or articular cartilage. Surrounding the joint is a joint capsule and capsular ligament. Capsular ligaments are composed primarily of collagenous fibers and provide resistance to joint distraction that can occur during a variety of segmental movements. The joint capsule, also known as the synovial membrane or articular capsule, forms a complete envelope around the joint and acts to maintain joint integrity by containing the synovial fluid. The synovial fluid facilitates articulations and allows for 'gliding' of opposing articular surfaces that occurs during physiologic motions including flexion/extension, lateral bending, and axial rotations.

16.2 Injuries to the Spine

Injuries to the lumbar spine result from direct violence to the column or specific vertebrae. Unlike penetrating trauma, violence to the spine can most commonly be attributed to gross motions or acceleration of the body/torso. Examples include anterior bending of the torso resulting in flexion loads on the lumbar spine or vertical acceleration of the pelvis leading to axial compressive loads on the lumbar column. Loads placed on the tissues lead to deformation, with the magnitude, rate, and direction/type of loading responsible for the tissue distortion profile. Injury occurs when deformation exceeds physiologic limits of the tissue. The type and location of tissue deformation is dependent upon the applied load. Pure loads can take the form of linear forces or rotational bending moments. Linear forces can be applied in any direction, but are generally broken down into components based on axial tension/compression perpendicular to the horizontal plane, anterior-posterior shear perpendicular to the frontal plane, or lateral shear perpendicular to the sagittal plane. Likewise, bending moment components include flexion/extension in the sagittal plane, lateral flexion in the coronal plane, and axial twist in the horizontal plane.

16.2.1 Injury Classification

Lumbar spine injuries can be classified according to loading mechanism. Injuries can occur under tension, compression, shear, or bending, although some of these mechanisms are less common due to the inherent characteristics of the in situ lumbar spine. Compression-related injuries are most common type in the thoraco-lumbar spine [1, 2]. Compression injuries occur in any number of high-rate axial loading scenarios including parachuting, falls from height, and motor vehicle crashes [3–8]. These injuries also occur in a variety of military- and sporting-related activities including skiing, snowboarding, and other sporting accidents [9–12], aviator ejection [13–20], helicopter crash [21–23], and underbody blast [24–31]. Injury mechanisms involve the primary component of axial force that can be coupled with varying magnitudes of bending. Injury type and severity are controlled by the biomechanical aspects of the insult. Specifically, axial force applied through the center of rotation of an intervertebral segment results in burst fracture. Axial forces offset from the center of rotation of a segment develop coupled bending moments, with greater offset distances resulting in greater moment magnitudes. Axial forces offset anteriorly, posteriorly, or laterally result in flexion-related injuries, extension-related injuries, or lateral bending-related fractures, respectively.

Tension and shear injuries can also occur in the lumbar spine. Tension (distraction) of the lumbar column is not a common loading scenario for humans. However, localized distraction occurs in different tissues during bending. For example, during forward flexion the anterior structures of the lumbar spine sustain localized compression loading. However, tissues posterior to the axis of rotation are distracted. This is particularly relevant for soft tissues such as ligaments and facet joint components, although bony fractures can also result from localized distraction (see Chance Fracture below). Likewise, shear injuries can affect he lumbar spine in some cases. However, the coupled effect of abdominal tissues most often transforms shear loading applied to the abdomen into bending of the spine. This section outlines different types of lumbar spine fractures, highlighting biomechanical mechanisms and briefly indicating the acute clinical outcome.

Burst fractures result from pure compression transmitted directly along the line of the vertebral bodies (Fig. 16.2). Due to the inherent lordotic

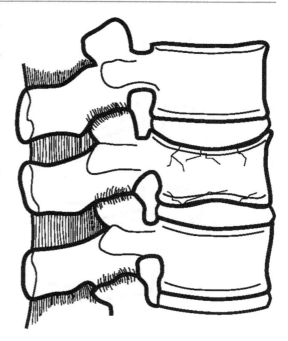

Fig. 16.2 Burst fracture

curvature of the lumbar spine, pre-flexion is necessary to induce a purely compressive state [32]. From a biomechanical perspective, a relatively uniform compressive load is applied across the axial plane of the vertebral body. This results in loss of vertebral body height due to fracture of anterior and posterior cortices. Axial force also leads to fracture of one or both endplates, forcing the intervertebral disc nucleus into the vertebral body, and resulting in a burst pattern [32, 33]. High energy fractures can result in retropulsion of bony fragments into the spinal canal and associated neurological deficit. Posterior element fractures may also be involved, but are not required for this classification. Ligaments commonly remain intact and the spine is mechanically stable. These injuries have been considered to be clinically stable or unstable. Ferguson and Allen reported that these fractures were generally stable, but other clinicians have reported progressive neurological injury or advancing post-injury deformity [33–35].

Anterior wedge fractures result from axial compression combined with flexion [36] or flexion alone [32] (Fig. 16.3). This combination can result from axial loads applied anterior to the

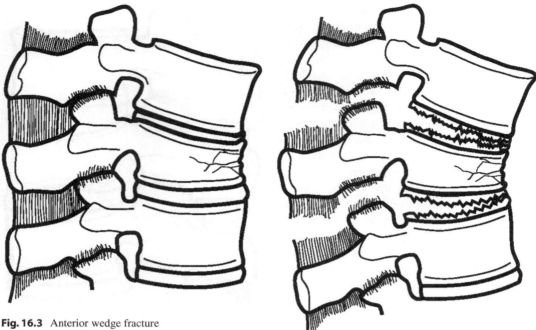

Fig. 16.3 Anterior wedge fracture

Fig. 16.4 Fracture dislocation including body fracture and facet dislocation

center of rotation of the vertebral segment or axial loading combined with anterior bending of the torso. In terms of biomechanics, tissues anterior to the center of rotation (e.g., anterior aspect of the vertebral body) sustain compression whereas middle- and posterior-column tissues (e.g., dorsal to the sagittal plane center of the vertebral body) are subjected to tension. Fractures affect the vertebral body and involve greater loss of body height anteriorly than posteriorly. In many cases, the posterior body height is unaffected. This results in a wedge-shaped profile as seen on lateral X-rays. While dynamic compressions are likely greater, wedge fractures typically present with less than 50 % anterior body height loss due to post-injury restitution. Slight and moderate wedge fractures are stable as the posterior aspect of the vertebral body and posterior ligamentous complex remain primarily intact [37]. Severe wedge fractures can occur with or without disc injury and involve disruption of the posterior ligamentous complex including ligamentous rupture or spinous process fracture. These fractures are unstable and typically referred to as fracture dislocations, which will be discussed below.

Wedge fractures can also occur laterally. Two mechanisms have been proposed to result in lateral wedge fractures. Flexion combined with rotation is the more commonly cited mechanism [2, 36], although Ferguson implicated compression combined with lateral bending [33]. Both mechanisms result in unilateral wedging, with the opposite side remaining intact, attributed to compression on the concave side and tension on the convex side. Radiographically, these injuries are evident in a frontal plane wedge-shaped profile when viewed on anterior-posterior X-rays. The lumbar column may also appear to have a lateral curvature centered about the injured level. Nicoll describes a unilateral wedging combined with transverse process fracture on the convex side and posterior intervertebral joint fracture on the concave side [2]. These injuries are generally considered to be clinically unstable, often associated with prolonged unilateral neurological deficit.

Fracture dislocation is a generic terms that refers to a condition involving fracture of the vertebra coupled with dislocation [36] (Fig. 16.4).

Dislocations can also occur in the absence of any bony fracture. These injuries commonly include rupture of the posterior interspinous ligament [2]. Depending on the status of the capsular ligaments, facet dislocation can occur, resulting in conditions involving upward subluxation, perching, forward dislocation, or forward dislocation with locking [2]. Kaufer and Hayes classified fracture dislocations into five groups based on the presence and type of anterior and posterior fracture or dislocation [38]. The mechanism for these injuries often involves flexion coupled with axial rotation or lateral bending. The coupled bending component is necessary in the lumbar spine region due to the inherent stability of the column, which can be attributed to the large vertebral bodies, wide and flat intervertebral discs, and well developed longitudinal ligaments. In the case of axial rotation, one or both articular processes can fracture causing the upper vertebra to rotate about the lower, breaking off a wedge-shaped section from the anterior region of the inferior vertebral body. A large coupled shear component can also contribute to fracture dislocations [36]. Fracture dislocations are clinically unstable with a propensity toward progressive deformity and acute neurological deterioration [2, 33, 38].

Chance fractures were first described by G.Q. Chance in 1948 [39] (Fig. 16.5). The mechanism of injury involves flexion coupled with distraction. Fractures initiate in the posterior aspect of the neural arch, often including the spinous process, and extend anteriorly into the posterior aspect of the vertebral body, terminating in an upward curve that extends toward the superior endplate just anterior to the neural foramen. These fractures occur in the absence of anterior vertebral body disruption or dislocation of the facet joints. The mechanism of these injuries is attributed to hyperflexion plus distraction. In many cases, these injuries were thought to result from the use of a lap seatbelt in automobile collisions [40, 41]. This mechanism is particularly relevant for improperly positioned lapbelts or pediatric occupants with immature pelvis to support the restraint [40–42], although other authors claim these injuries are rare in children 43. In these cases, the lapbelt would function as a

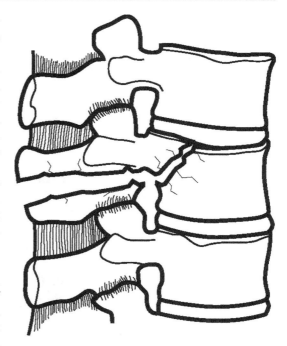

Fig. 16.5 Chance fracture

fulcrum for the spine to rotate about, resulting in tension injuries of the posterior lumbar vertebra. Incorporation of a properly worn shoulder belt to support the torso has minimized the likelihood of these injuries. From a clinical perspective, these injuries are considered to be stable, with little chance of neurological deficit [2, 42].

16.3 Loading Issues

Numerous biomechanical investigations of the lumbar spine have been conducted using whole body cadaveric specimens, lumbar columns, spine segments, and isolated tissues. These investigations have quantified quasi-static and dynamic physiologic, degenerated, and traumatic responses of the lumbar spine under a variety of loading conditions including compression, bending, and shear. A comprehensive review of lumbar spine biomechanical research is beyond the scope of this chapter. Rather, this chapter aims to highlight experimental biomechanical methods and provide tolerance-related information for the lumbar spine. Due to the prominence of axial loading

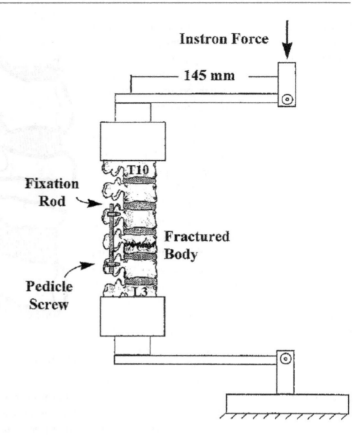

Fig. 16.6 Experimental model incorporating an electro-hydraulic testing device to induce compression-flexion on a lumbar spine column (From Mermelstein et al., Spine 1998 with permission)

on lumbar spine injury mechanics, a significant portion of existing research has focused on this mode. Accordingly, this chapter will focus primarily on experimental studies of axial loading. The following sections provide a description of three experimental protocols that have been used to investigate dynamic axial tolerance of lumbar spine components, and a fourth method incorporating whole body specimens.

16.3.1 Electro-hydraulic Testing Device

A considerable amount of experimental effort has been applied toward the understanding of lumbar spine injury tolerance during dynamic axial loading. Much of that research has been conducted using either electro-hydraulic testing devices or weight-drop apparatuses. Research incorporating electro-hydraulic testing devices have been conducted using whole lumbar columns [44–46], column segments (e.g., two-vertebra motion segments) [47–49], isolated vertebral bodies [50–58], and components including ligaments and annular tissues [59–75]. These devices apply quasi-static or dynamic axial loads using the piston of the electro-hydraulic device. An advantage of electro-hydraulic testing devices is that piston excursion is computer controlled, which leads to a high level of control over the loading versus time pulse, although these devices are somewhat limited in loading rate. Loads can be distributed across the vertebral endplate, applied at a distance from the specimen to induce a bending moment, or locally applied to a specific region of the endplate or intervertebral disc using an indentor. By design, electrohydraulic testing devices are typically uniaxial and compression is the most common loading mode. However, the devices can also impart compression combined with bending through application of axial load at a distance from the center of segmental or column rotation using a moment arm or vertebral arch (Fig. 16.6). For example, compression-flexion loading can be induced through load application to a moment arm extending anteriorly from the cranial vertebra. Likewise, compression-extension and compression-lateral bending are induced using moment arms extending posteriorly or laterally. Oblique loading can be produced through application of the load in the anterior- or posterior-lateral locations. Distance of the load application from the center of rotation controls the ratio of axial load to bending. Load application at a greater distance from the center of rotation results in a higher ratio of bending to axial load. Likewise, load application closer to the center of the vertebra leads to a higher component of compression.

Testing is generally conducted with specimens in neutral position, although some series have applied flexion or extension loads to bias the initial position and study orientation effects [48]. Custom devices have the ability to apply axial tensile or compressive forces at loading rates up to 9 m/s. However, testing of lumbar columns and segments has generally been conducted at quasi-static rates as low as 1.0 mm/min [47] or dynamic rates up to 1.0 m/s [46]. Testing of individual vertebrae has been conducted at rates up to 2.5 m/s [51]. Specimens can be instrumented, with reflective markers, accelerometers, and strain gauges to obtain level-by-level kinematic (displacement and angulation) or localized compressive information. Loads at the impacted and distal ends are recorded using load cells and localized (level-specific) loads can be computed by coupling load cell data with kinematics information. Because of the controlled nature of load application using the piston, dynamic subfailure loading can be used to quantify the physiologic response of lumbar tissues. Likewise, injuries produced during dynamic loading can be correlated with biomechanical measures to derive injury tolerance information. However, it is difficult to achieve a constant velocity during biofidelity testing as the piston has to initiate its travel from rest, and deceleration initiates prior to the point of peak displacement. Constant loading rates can be obtained with piston overshoot, by setting piston displacement to a maximum level well beyond the expected fracture displacement. Inertial effects of the piston require compensation for force measurements from load cells attached to the piston and in-line with the loading vector.

Electro-hydraulic testing devices are also useful for testing of isolated tissues under appropriate loading modes. Although tension of the lumbar spine is rare, specific soft tissues sustain tension during different loading situations. For example, dorsal soft tissues sustain tension during segmental flexion. Likewise, due to Poisson's effect, intervertebral disc annular tissues sustain tension during segmental compression. Similar to compression, tension loading rates can vary from quasi-static to dynamic and maximum distraction can be maintained within the physiologic range or can enter the traumatic realm. A variety of biomechanical studies have been conducted using electro-hydraulic testing devices to quantify tensile response of lumbar spine ligaments [63–65, 67, 68, 70, 72] and intervertebral disc material [60, 61, 66, 76].

16.3.2 Weight-Drop Apparatus

The weight-drop apparatus is another method of load application similar to the electro-hydraulic testing device, with the primary exception that loads are applied by dropping a weight onto the cranial end of the specimen instead of using the piston (Fig. 16.7). Numerous studies have been conducted using the weight-drop apparatus and incorporating 2- and 3-vertebrae segments [77–79] or longer lumbar columns [80–83]. This test setup applies dynamic compressive loads by impacting the spine using a decelerated weight. The weight is accelerated by gravity until impacting the spine and is either guided or allowed to fall without constraint. Similar to the electro-hydraulic testing device, compression is the most common loading mode, although compression combined with bending can be applied through application of axial load at a distance from the center of segmental or column rotation using a moment arm.

Testing is most commonly conducted with specimens in neutral position, although pre-flexion has been applied in some cases [81, 84, 85]. Rate of loading is controlled by mass of the impactor and its closing velocity (i.e., velocity at the time of

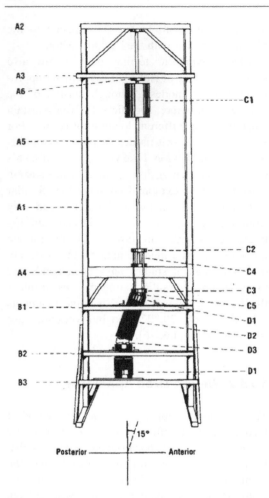

Fig. 16.7 Experimental model incorporating a weight-drop device to induce compression-flexion on a lumbar spine column (From Cotterill et al., J Orthop Res 1987 with permission)

weight from a height that does not induce fracture and is used to quantify the physiologic response of lumbar tissues. Likewise, injuries produced during dynamic loading can be correlated with biomechanical information to derive tolerance information. However, it is difficult to achieve a constant velocity during biofidelity testing as deceleration of the dropped weight initiates immediately upon contact with the specimen.

16.3.3 Drop Tests

Drop tests are used to replicate vertical acceleration conditions in either component or whole body cadaver tests (Fig. 16.8). Benefits of this experimental setup realistic loading and boundary conditions and the ability to relate injury tolerance to external metrics associated with the loading environment (i.e., acceleration of the lumbar spine base). These tests involve a drop tower of varying height, at least one platform connected to a guide rail using linear bearings or a cart mechanism, and pulse-shaping material at the bottom of the tower to modulate characteristics of the deceleration pulse. Testing involves mounting the specimen to the platform, raising the platform to a specific height, release with gravity accelerating the platform downward, and impact to the pulse-shaping material at the base of the drop tower. Characteristics of the deceleration versus time pulse are controlled using initial height, mechanical properties of the pulse-shaping material, and amount of pulse shaping material. Peak accelerations as high as 65 G with rates of onset as high as 2,500 G/s have been achieved using this model [86]. In general, greater peak accelerations can be obtained with drops from greater initial height and steeper rates of onset (i.e., shorter pulses) are obtained with stiffer pulse-shaping material. In the case of isolated components, similar biomechanical information can be collected to that for the weight-drop and electro-hydraulic testing devices including forces at the top and base of the specimen, three-dimensional spinal kinematics (e.g., linear and angular motions and accelerations), localized strains, and fracture information from acoustic sensors.

impact). Control of maximum compression is difficult as this method relies on specimen impact to halt the excursion of the dropped weight. Loading rate can be quantified as the rate of load application (N/s). Specimens can be instrumented, with reflective markers, accelerometers, and strain gauges to obtain level-by-level kinematic (displacement and angulation) or localized compressive information. Loads at the impacted and distal ends are recorded using load cells and localized (level-specific) loads can be computed by coupling load cell data with kinematics information. Subfailure testing is performed by dropping the

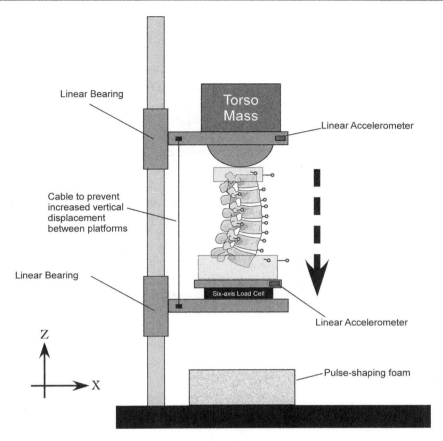

Fig. 16.8 Experimental model incorporating a drop tower apparatus to induce compression-flexion on a lumbar spine column (From Stemper et al., J Biomech Eng 2011 with permission)

A second decoupled platform with variable mass can be added to the cranial aspect of the spine to simulate the torso. The mass can be the same for all specimens tested under a given protocol, for consistency, or can be varied to mimic specimen-specific torso mass. Lumbar spine components can be tested in compression or, since the mass can be attached to the upper platform via a telescoping linkage and load application can be moved anterior-posteriorly, compression combined with flexion, extension, or lateral bending.

16.4 Specimen Details

Different types of experimental models exist to determine the biomechanical properties, replicate real-world injuries, derive injury mechanisms, and determine human tolerance in terms of variables such as forces and risk curves using the above described experimental techniques. Some of the more common experimental models are discussed in this section.

16.4.1 Isolated Components

Testing of isolated components such as vertebral bodies has been performed to quantify the structural or material response of the isolated vertebral body or endplate. Testing of vertebral bodies and endplates is typically performed using electrohydraulic testing device with flat horizontal plate for bodies and an indentor for endplates [51, 87–90]. These tests have been conducted from quasi-static to dynamic rates. Axial force is measured using a load cell, compression displacement is measured using two-dimensional videography or the piston LVDT, and vertebral body strain is measured using strain gauges. These types of testing are ideal for the quantification of high-rate material properties as loading conditions are controlled, repeatable, and applied directly to the tissue of interest.

Isolated soft tissues have also been tested to quantify the structural and material response. Those tests are typically conducted by distracting the specimen in tension using an electro-hydraulic testing device [61, 70]. Testing has quantified the quasi-static, dynamic, and viscoelastic response of isolated ligaments and annular tissues. Test specimens are commonly arranged in an I-shaped mechanical test specimen. However, attachment to the test frame can be difficult for smaller tissues (i.e., fascicles) and test coupons may be required [64].

16.4.2 Segmented Columns

Segmented column models are used to experimentally delineate the gross biomechanical responses of the spine at a macro level and determine tolerance characteristics. Effects of lordotic curvature are incorporated because more than one functional unit is used. Pre-alignment of the spine can be incorporated to account for effects of body posture. However, the degree of inclusion of these factors depends on the number of spinal segments. Consequently, these models tend to be more realistic from injury reproduction perspectives although failure responses of individual components cannot be quantified because the load-path at a segmental level is unknown. Three-dimensional motions of the intervertebral levels have been obtained at high rates of 1,000 samples per second. Two-dimensional motions using high resolution digital cameras can be obtained at much higher rates (~50,000 frames per second). Likewise, local accelerations and strains of individual vertebrae can obtained using accelerometers, strain gauges, and acoustic emission sensors to determine the timing of fracture or spinal instability. Positioning the segmented column on an x–y cross table mounted to the platform of an electro-hydraulic testing device is needed to achieve the intended posture or pre-alignment. The transmitted forces and moments can be recorded at the inferior end using a six-axis load cell. Forces and moments at the segmental level of injury may be estimated although the local dynamics are not known. High-speed video images can be taken to document macroscopic failures, high-speed x-rays can be obtained for bony fractures, and localized segmental motions analyses can be performed using this model. Strict control of the experimental loading conditions can be achieved using segmented columns. Testing can be conducted at injurious levels, or below the threshold for injury to quantify the physiologic response.

Another methodology to apply dynamic loads to the segmented column is using free-fall or drop techniques as described above. This involves fixing the ends of the column, applying preloads (if any), controlling alignment by techniques such as pre-flexing using cables, and dropping on to targets with known stiffness to modulate the pulse. Ensuing motions of the column following initial contact with the target may induce continuing loads and contribute to additional injuries. However, load limiters have been used to prevent this occurrence.

16.5 Biomechanical Data

A number of studies have been performed to characterize lumbar spine fractures and quantify biomechanical tolerance due to axial loading. Specific aims of these studies were to clarify the injury mechanism, observe fracture patterns, measure spinal canal occlusion, compare surgical instrumentation techniques, or understand biomechanics of injury. As mentioned above, in many cases, the electro-hydraulic testing device or weight drop models were employed. However, other studies have incorporated the alternative models described above. Physiologic and injury tolerance information has been derived from these studies. Although not comprehensive, some of the relevant findings are discussed below.

16.5.1 Compressive Load to Failure and Tolerance

Dynamic compression of isolated vertebral bodies using electro-hydraulic test devices has been used to define compressive tolerance of endplates or the body as a whole in the lumbar spine. Those studies have incorporated electro-hydraulic testing devices with indentor (endplate) or flat plate (vertebral body) attachments to the piston to load specimens under quasi-static or dynamic loads. Studies of endplate tolerance have demonstrated significant rate dependence [51] and regional dependence across the surface of the endplate [91]. Strength of the endplate was previously theorized to play a strong role in formation of vertebral burst fractures [32], one of the primary injury types sustained during high-rate dynamic axial loading. Endplate strength was previous correlated to bone mineral density of the vertebral body [92–94]. Because bone mineral density is known to decrease with age, eventually leading to osteoporosis, specimen selection for injury tolerance investigations is critical, and must be performed in light of the population of interest. This can minimize the necessity to scale injury tolerance values obtained from specimens with older ages or osteoporotic spines. Investigations of vertebral body fracture mechanics have generally demonstrated rate of loading effects on tolerance [50–52, 95]. For example, Ochia et al. subjected isolated lumbar vertebral bodies to compressive loading rates of 10 mm/s or 2.5 m/s and demonstrated significantly increased fracture tolerance at the higher loading rate [51]. Kazarian and Graves demonstrated a similar finding for the thoracic spine across loading rates of 0.09 mm/s, 9 mm/s, and 0.9 m/s [50]. A summary of vertebral body testing is provided in Table 16.1. For studies including severely osteoporotic spines, only data from normal and osteoporotic spines are included in the table.

Researchers investigating lumbar spine tolerance have long acknowledged the importance of loading rate as an influencing factor. This fact has particular relevance to the military environment, wherein injuries can occur across a variety of loading rates from relatively low rate (falls) to extremely high rate (underbody blast) [24, 27–30]. The weight-drop method (described above) was one of the first experimental models to impart high-rate axial loading to the lumbar spine, as first described by Hirsch and Nachemson in 1954 [96]. Perey later reported on injury types resulting from experimental modeling of axial compression using the weight-drop method [52]. These experiments produced approximate maximum loads between 10,300 and 13,200 N within 6.0 ms by dropping a mass of 15 kg from a height of 0.5 m. Endplate fractures occurred in 26 % of experiments; wedge-shaped vertebral compression fractures occurred in 8 %. Willen et al. produced more severe compression fractures (i.e., burst fractures) by dropping a 10 kg mass from 2.0 m onto 3-vertebrae thoracolumbar specimens [78]. That study qualitatively demonstrated an age dependence, wherein specimens from cadavers greater than 70 years of age tended to completely collapse in compression and vertebrae from cadavers less than 40 years of age sustained the comminuted fracture pattern characteristic of burst fractures as defined by Denis [97]. These experiments tend to agree with clinical literature that has reported burst fractures generally occurring in younger patients [98, 99]. Subsequent studies have provided confirming results to demonstrate that burst fractures are generally produced under high-rate loading scenarios [52, 78, 100–102]. Testing characteristics and fracture biomechanical data from these studies are summarized in Table 16.2.

Spinal orientation at the time of impact is a factor that has commonly been associated with influencing injury risk during axial loading the lumbar spine. Initial spinal orientation was shown

Table 16.1 Summary of vertebral body (VB) and endplate (EP) testing in literature

Parameter	Unit	Range
Investigated spinal levels	N/A	T12–L5
Testing velocity	m/s	0.01–4.0
VB fracture displacement	mm	2.3–6.5
VB fracture force	kN	4.9–14.9
VB fracture stress	N/mm^2	3.7–7.0
EP fracture force	N	55–170
EP fracture stress	N/mm^2	6.3–7.5

Table 16.2 Summary of short-segment weight-drop literature

Parameter	Unit	Range
Number of spinal levels	N/A	2–3
Impactor mass	kg	2.3–18
Initial impactor height	m	0.5–2.0
Impactor closing velocity	m/s	3.1–6.2
Spine angle	°	0–15
Fracture force	kN	5.3–13.2

Table 16.3 Summary of short-segment electro-hydraulic testing device literature

Parameter	Unit	Range
Number of spinal levels	N/A	2–3
Testing velocity	m/s	0.01–2.5
Spine angle	°	−15–0
Fracture force	kN	2.8–12.4

Table 16.4 Summary of lumbar column electro-hydraulic testing device literature

Parameter	Unit	Range
Number of spinal levels	N/A	6–7
Testing velocity	m/s	1.0
Spine angle	°	0
Fracture force	kN	3.3–5.9

to drastically affect the physiological level-by-level segmental kinematics of the lumbar column during dynamic axial loading applied using a drop tower apparatus [86]. Weight drop studies have incorporated protocols with lumbar spine specimens in neutral or pre-flexed positions. Panjabi et al. quantified the differences in fracture tolerance between 3-vertebrae spinal segments oriented in either neutral or 15° pre-flexed positions [102]. Pre-flexion decreased fracture tolerance by 7 % from 6.7 ± 2.0 kN in neutral position to 6.2 ± 2.3 kN for 15° of pre-flexion. Likewise, comparison between studies highlights decreasing tolerance for lumbar spines in pre-flexed positions. Testing of lumbar spines in neutral position resulted in fracture tolerance between 6.0 and 13.2 kN [52, 78, 102] Fracture tolerance was between 5.3 and 6.6 kN when spines were pre-flexed to 8° [100, 101]. Although the weight-drop model will not be incorporating in testing protocols for this project, effects of initial specimen orientation will be included in the test matrix to demonstrate differing injury tolerance between neutral position and orientations including flexion, extension, or lateral bending.

Fracture tolerance of lumbar segments and columns has been investigated using the electro-hydraulic testing apparatus setup in multiple investigations. Whole lumbar columns have demonstrated fracture tolerance of between 3,303 and 12,535 in one study [45] and 5,009 or 5,911 in another study [46]. However, both fractures were obtained at the cranial level in the second study, which may indicate failures resulting more from fixation artifact than axial loading conditions. Other studies using 3-vertebra segments have demonstrated effects of loading rate or specimen orientation. Langrana et al. investigated effects of specimen orientation on fracture tolerance and demonstrated considerably lower tolerance for specimens tested in neutral position (2.8 ± 0.7 kN) than for specimens tested with 15° of pre-extension (5.8 ± 1.8 kN) [48]. Effects of specimen orientation are important for real-world application as occupants of different vehicles inherently have different seated postures which changes the orientation of the lumbar spine relative to the applied load and influences injury tolerance/risk. Another experimental study incorporating the electrohydraulic testing device model investigated effect of loading rate on fracture tolerance [103]. That study identified increasing fracture tolerance for higher rates of axial loading with specimens positioned in neutral posture. Fracture tolerance was 3.3 ± 1.2 kN for specimens tested at compression rates of 10 mm/s and 4.2 ± 1.7 kN for specimens tested at 2.5 m/s. Understanding rate effects on lumbar spine fracture tolerance has importance for the development of injury mitigation devices. For example, underbody blast is likely to load the lumbar spine at higher rates than aviator ejection, automotive, and fall environments. A summary of human cadaver experiments incorporating the electro-hydraulic testing device setup is provided in Tables 16.3 (short segment) and 16.4 (lumbar columns) below.

While short segment (2- or 3-vertebrae) experimental models provide controlled and repeatable testing protocols, application to the lumbar column mechanics is somewhat limited due to limited incorporation of lordotic curvature and continuous ligamentous structures such as the anterior longitudinal ligament. In 3-vertebrae constructs, only on vertebra is exposed to traumatic loading, which changes the inherent biomechanics including level-by-level load transmission. Therefore, lumbar column testing may be more suited to the understanding of the physiological response and injury tolerance of the lumbar spine associated with axial loading environments.

Testing of whole column specimens was not reported until more recently. Yoganandan and colleagues tested full lumbar columns under quasistatic testing (2.5 mm/s) in the compression-flexion mode using the electro-hydraulic piston model [44]. Fracture occurred at an average load of 3.8 ± 0.5 kN. More recently, Duma et al. subjected whole lumbar columns to dynamic compression loading at a rate of 1.0 m/s [46]. Fracture tolerance in that study was reported as a combination of compression force and bending moment: 5.4 ± 0.5 N and 201 ± 51 Nm. Although the two studies are not directly comparable due to differences in experimental protocol including stress risers [44], it is worth noting that fracture tolerance increase by approximately 40 % under dynamic loading. This highlights a dependence of fracture tolerance on loading rate.

A limited number of investigations have focused on quantifying lumbar column tolerance using a drop tower apparatus [86]. Those studies focused on quantification of military loading rate effects on injury tolerance and location in lumbar columns. Aviator ejection and helicopter crash pulses were simulated with accelerations of 21 and 58 G, respectively. Rates of acceleration onset were 371 and 2,068 G/s. Fracture tolerance increased from 5.7 kN in the lower rate ejection tests to 6.7 kN in the higher rate helicopter crash tests, demonstrating a clear rate dependence. However, also important was that injury locations migrated from primarily upper lumbar spine (e.g., L1, L2) during ejection tests to lower lumbar spine (e.g., L3, L4) during higher rate helicopter crash tests. Unique kinematic data were also collected and used to inform isolated tissue studies.

Table 16.5 Summary of lumbar column drop tower literature

Parameter	Unit	Range
Number of spinal levels	N/A	6
Peak accelerations	G	20.7–65
Rates of onset	G/s	228–2,638
Vertebral body compression rate	m/s	0.5–1.25
Spine angle	°	0
Fracture force	kN	5.2–7.8

For example, two kinematic targets were placed on the anterior of each vertebral body to measure compression rates during fracture. During burst fractures sustained in ejection-simulating tests, vertebral bodies were compressed at rates below 1.0 m/s [95]. That compression rate increased up to 1.25 m/s during helicopter crash tests. A benefit of these studies is the ability to replicate seat, pelvis, or lower lumbar vertical accelerations in the controlled laboratory environment. A summary of these studies is provided in Table 16.5.

16.5.2 Tolerance Criteria

Injury metrics are used to predict occurrence or risk of injury under well-defined dynamic loading scenarios. Although it is more computationally expedient to represent injury tolerance using a single metric (e.g., axial force) and value (e.g., 6 kN), injury risk is often more precisely predicted using a complex computation. Injury metrics are often developed in conjunction with biomechanical testing of cadavers, wherein injuries can be produced under quantifiable and repeatable loading environments. Statistical analysis can then be used to quantify computational metrics that are most predictive of injury. Due to non-homogeneous geometry and material properties of the spine, injury metrics must be developed for specific loading environments and injury types. To date, there remains a gap in the development of a robust injury metric for the lumbar spine. For example, separate injury metrics have been developed in the cervical spine for automotive frontal and rear impacts.

Development of tolerance criteria for the lumbar spine initiated following medical reports of thoraco-lumbar compression fractures sustained by military aviators during ejection from aircraft, which subjected the aviator to high-rate vertical accelerations [104–108]. Initial efforts focused on development of whole body tolerance [109, 110]. While providing useful performance envelopes and design targets for safety engineers, whole body metrics lacked specificity for injury mechanisms and types, such as thoracolumbar compression fractures. Focusing on the spine as a component, Latham developed a mechanical model consisting of lumped parameter elements that was mathematically represented by a second order differential equation accounting for deflection, damping, natural frequency and acceleration of the system [111]. Stech and Payne incorporated the model into an analysis of vertical accelerations, under the assumption that the spine was the primary load-bearing structure in that mode [112]. Their analysis included parameters determined using experimental research and computational analysis [113, 114]. That work was premised on the theory that deflection of the lumbar column, as predicted using their lumped parameter model, was the primary indicator of vertebral fracture. This work resulted in the determination of the Dynamic Response Index (DRI), which is a dimensionless parameter derived to represent the maximum spinal compression experienced during an acceleration event and predict the probability of spinal injury for different age groups [112]. For example, he 50 % probability of spinal injury was estimated to be a DRI value of 21.3 for occupants 27.9 years of age, which was the mean age of the U.S. Air Force aviator population in 1969. Accordingly, the Air Force set a maximum DRI safety level of 18 to represent a 5 % chance of spinal injury during ejection in specification MIL-S-9479B, the military standard for ejection seat systems in aircraft. Further investigations developed spinal injury probabilistic relationships with DRI based on operational data [115].

Where the DRI was developed under the theory that spinal injury is displacement controlled, other researchers have taken an alternate approach by investigating metrics correlating compressive force to injury. For example, Chandler reasoned that acceleration or DRI alone could not account for effects of occupant restraint loading [116]. Accordingly, he suggested that axial force measured between the lumbar spine and pelvis would more accurately account for injury risk. That study referenced work performed at the Federal Aviation Administration Civil Aeromedical Institute (CAMI) to develop a lumbar spine injury metric based on axial compressive force. Those studies measured axial load in the lumbar region of a human manikin subjected to vertical accelerations and demonstrated the influence of occupant positioning. Based on findings from that study, the General Aviation Safety Panel (GASP) recommended a load limit of 6,672 N for axial accelerations. This value corresponded to a DRI value of 19 and a spinal injury risk of approximately 9 %. The Federal Aviation Administration adopted that load limit as a pass/fail condition for their dynamic test procedure for seats within transport category aircraft.

As demonstrated in this brief review, DRI and axial compressive force are the primary lumbar spine injury metrics used to predict injury in vertical accelerative environments. However, these metrics do not fully account for non-uniformity in lumbar spine geometry/material properties and complex loading environments associated with off-axis or out-of-position situations. As demonstrated earlier in this review, these factors are critical in the prediction of injury risk and occurrence. For example, pure lumbar spine axial compressive force alone cannot cause anterior wedge fractures, and consequently lumbar spine injury susceptibility cannot be assessed using compressive forces and ignoring other biomechanical factors. As such, research efforts for the lumbar spine should investigate and develop lumbar spine injury metrics which are sufficiently robust and applicable regardless of the loading environment.

16.6 Summary

The purpose of the review has been to evaluate available biomechanical data relevant to lumbar spine injury tolerance. Research studies using

PHMS were reviewed with a focus on lumbar spine injuries and their mechanisms. Peak injury metrics and human injury risk functions developed from the forces, rotations, and moments measured were included.

The review discussed quasi-static and dynamic tests using lumbar columns, motion segments, and isolated tissues such as vertebral bodies. Dynamic loading protocols have incorporated electro-hydraulic test devices, weight drops, and drop towers. Testing velocity during isolated vertebral body and endplate testing was between 0.25 and 4.0 m/s. Weight drop studies have been conducted using short- and long-segment cadaveric models with mass between 2.3 and 18 kg impacting the spine at velocities between 3.1 and 6.2 m/s. Electro-hydraulic piston tests have been conducted using short- and long-segment models, as well as whole lumbar columns, at rates from quasi-static to 2.5 m/s. Drop tower studies have been conducted using whole lumbar columns at rates approximating military aviator ejection and helicopter crash. These tests have provided information regarding effects of lumbar spine posture at the time of impact on injury tolerance. Injury tolerance in these tests has been somewhat consistent, given the range of experimental conditions. Variation in fracture tolerance can likely be attributed to differing age/BMD of specimens incorporated in the studies, differing loading rates, and differing initial posture.

Acknowledgments The authors gratefully acknowledge the contributions Kim Chapman and the Zablocki VA Medical Center Medical Media for providing many of the figures used in this chapter.

References

1. Griffith HB, Gleave JR, Taylor RG (1966) Changing patterns of fracture in the dorsal and lumbar spine. Br Med J 1:891–894
2. Nicoll EA (1949) Fractures of the dorso-lumbar spine. J Bone Joint Surg Br 31B:376–394
3. Wang MC, Pintar F, Yoganandan N, Maiman DJ (2009) The continued burden of spine fractures after motor vehicle crashes. J Neurosurg Spine 10:86–92
4. Richards D, Carhart M, Raasch C, Pierce J, Steffey D, Ostarello A (2006) Incidence of thoracic and lumbar spine injuries for restrained occupants in frontal collisions. Annu Proc Assoc Adv Automot Med 50:125–139
5. Smith JA, Siegel JH, Siddiqi SQ (2005) Spine and spinal cord injury in motor vehicle crashes: a function of change in velocity and energy dissipation on impact with respect to the direction of crash. J Trauma 59:117–131
6. Richter D, Hahn MP, Ostermann PA, Ekkernkamp A, Muhr G (1996) Vertical deceleration injuries: a comparative study of the injury patterns of 101 patients after accidental and intentional high falls. Injury 27:655–659
7. Inamasu J, Guiot BH (2007) Thoracolumbar junction injuries after motor vehicle collision: are there differences in restrained and nonrestrained front seat occupants? J Neurosurg Spine 7:311–314
8. Hsu JM, Joseph T, Ellis AM (2003) Thoracolumbar fracture in blunt trauma patients: guidelines for diagnosis and imaging. Injury 34:426–433
9. Aleman KB, Meyers MC (2010) Mountain biking injuries in children and adolescents. Sports Med 40:77–90
10. Gertzbein SD, Khoury D, Bullington A, St John TA, Larson AI (2012) Thoracic and lumbar fractures associated with skiing and snowboarding injuries according to the AO Comprehensive Classification. Am J Sports Med 40:1750–1754
11. Alexander MJ (1985) Biomechanical aspects of lumbar spine injuries in athletes: a review. Can J Appl Sport Sci 10:1–20
12. Khan N, Husain S, Haak M (2008) Thoracolumbar injuries in the athlete. Sports Med Arthrosc Rev 16:16–25
13. Smiley JR (1964) Rcaf ejection experience: decade 1952–1961. Aerosp Med 35:125–129
14. Hearon B, Thomas H (1982) Mechanism of vertebral fracture in the F/FB-111 ejection experience. Aviat Space Environ Med 53:440–448
15. Edwards M (1996) Anthropometric measurements and ejection injuries. Aviat Space Environ Med 67:1144–1147
16. Osborne RG, Cook AA (1997) Vertebral fracture after aircraft ejection during Operation Desert Storm. Aviat Space Environ Med 68:337–341
17. Williams CS (1993) F-16 pilot experience with combat ejections during the Persian Gulf War. Aviat Space Environ Med 64:845–847
18. Moreno Vazquez JM, Duran Tejeda MR, Garcia Alcon JL (1999) Report of ejections in the Spanish Air Force, 1979–1995: an epidemiological and comparative study. Aviat Space Environ Med 70:686–691
19. Lewis ME (2006) Survivability and injuries from use of rocket-assisted ejection seats: analysis of 232 cases. Aviat Space Environ Med 77:936–943
20. Nakamura A (2007) Ejection experience 1956–2004 in Japan: an epidemiological study. Aviat Space Environ Med 78:54–58

21. Shanahan DF, Shanahan MO (1989) Injury in U.S. Army helicopter crashes October 1979-September 1985. J Trauma 29:415–422; discussion 423
22. Scullion JE, Heys SD, Page G (1987) Pattern of injuries in survivors of a helicopter crash. Injury 18:13–14
23. Italiano P (1966) Vertebral fractures of pilots in helicopter accidents. Riv Med Aeronaut Spaz 29:577–602
24. Helgeson MD, Lehman RA Jr, Cooper P, Frisch M, Andersen RC, Bellabarba C (2011) Retrospective review of lumbosacral dissociations in blast injuries. Spine 36:E469–E475
25. Ragel BT, Allred CD, Brevard S, Davis RT, Frank EH (2009) Fractures of the thoracolumbar spine sustained by soldiers in vehicles attacked by improvised explosive devices. Spine 34:2400–2405
26. Poopitaya S, Kanchanaroek K (2009) Injuries of the thoracolumbar spine from tertiary blast injury in Thai military personnel during conflict in southern Thailand. J Med Assoc Thai 92(Suppl 1):S129–S134
27. Schoenfeld AJ, Lehman RA Jr, Hsu JR (2012) Evaluation and management of combat-related spinal injuries: a review based on recent experiences. Spine J 12:817–823
28. Schoenfeld AJ, Goodman GP, Belmont PJ Jr (2012) Characterization of combat-related spinal injuries sustained by a US Army Brigade Combat Team during Operation Iraqi Freedom. Spine J 12:771–776
29. Blair JA, Patzkowski JC, Schoenfeld AJ, Cross Rivera JD, Grenier ES, Lehman RA Jr, Hsu JR (2012) Spinal column injuries among Americans in the global war on terrorism. J Bone Joint Surg Am 94:e135(131–139)
30. Lehman RA Jr, Paik H, Eckel TT, Helgeson MD, Cooper PB, Bellabarba C (2012) Low lumbar burst fractures: a unique fracture mechanism sustained in our current overseas conflicts. Spine J 12:784–790
31. Kang DG, Lehman RA Jr, Carragee EJ (2012) Wartime spine injuries: understanding the improvised explosive device and biophysics of blast trauma. Spine J 12:849–857
32. Holdsworth F (1970) Fractures, dislocations, and fracture-dislocations of the spine. J Bone Joint Surg Am 52:1534–1551
33. Ferguson RL, Allen BL Jr (1984) A mechanistic classification of thoracolumbar spine fractures. Clin Orthop Relat Res Oct:77–88
34. Larson SJ, Maiman DJ (1999) Surgery of the lumbar spine. Thieme, New York
35. Davies WE, Morris JH, Hill V (1980) An analysis of conservative (non-surgical) management of thoracolumbar fractures and fracture-dislocations with neural damage. J Bone Joint Surg Am 62:1324–1328
36. White AA, Panjabi MM (2013) Clinical biomechanics of the spine. Williams & Wilkins, Philadelphia
37. Westerborn A, Olsson O (1951) Mechanics, treatment and prognosis of fractures of the dorso-lumbar spine. Acta Chir Scand 102:59–83
38. Kaufer H, Hayes JT (1966) Lumbar fracture-dislocation. A study of twenty-one cases. J Bone Joint Surg Am 48:712–730
39. Chance GQ (1948) Note on a type of flexion fracture of the spine. Br J Radiol 21:452
40. Anderson PA, Rivara FP, Maier RV, Drake C (1991) The epidemiology of seatbelt-associated injuries. J Trauma 31:60–67
41. Howland WJ, Curry JL, Buffington CB (1965) Fulcrum fractures of the lumbar spine. Transverse fracture induced by an improperly placed seat belt. JAMA 193:240–241
42. Raney EM, Bennett JT (1992) Pediatric Chance fracture. Spine 17:1522–1524
43. Gallagher DJ, Heinrich SD (1990) Pediatric Chance fracture. J Orthop Trauma 4:183–187
44. Yoganandan N, Larson SJ, Pintar F, Maiman DJ, Reinartz J, Sances A Jr (1990) Biomechanics of lumbar pedicle screw/plate fixation in trauma. Neurosurgery 27:873–880; discussion 880–871
45. Shono Y, McAfee PC, Cunningham BW (1994) Experimental study of thoracolumbar burst fractures. A radiographic and biomechanical analysis of anterior and posterior instrumentation systems. Spine 19:1711–1722
46. Duma SM, Kemper AR, McNeely DM, Brolinson PG, Matsuoka F (2006) Biomechanical response of the lumbar spine in dynamic compression. Biomed Sci Instrum 42:476–481
47. Hongo M, Abe E, Shimada Y, Murai H, Ishikawa N, Sato K (1999) Surface strain distribution on thoracic and lumbar vertebrae under axial compression. The role in burst fractures. Spine 24:1197–1202
48. Langrana NA, Harten RR, Lin DC, Reiter MF, Lee CK (2002) Acute thoracolumbar burst fractures: a new view of loading mechanisms. Spine 27:498–508
49. Shirado O, Zdeblick TA, McAfee PC, Cunningham BW, DeGroot H, Warden KE (1992) Quantitative histologic study of the influence of anterior spinal instrumentation and biodegradable polymer on lumbar interbody fusion after corpectomy. A canine model. Spine 17:795–803
50. Kazarian L, Graves GA (1977) Compressive strength characteristics of the human vertebral centrum. Spine 2:1–14
51. Ochia RS, Tencer AF, Ching RP (2003) Effect of loading rate on endplate and vertebral body strength in human lumbar vertebrae. J Biomech 36:1875–1881
52. Perey O (1957) Fracture of the vertebral end-plate in the lumbar spine; an experimental biochemical investigation. Acta Orthop Scand Supplementum 25:1–101
53. Alkalay RN, von Stechow D, Torres K, Hassan S, Sommerich R, Zurakowski D (2008) The effect of cement augmentation on the geometry and structural response of recovered osteopenic vertebrae: an anterior-wedge fracture model. Spine 33:1627–1636

54. Belkoff SM, Mathis JM, Jasper LE, Deramond H (2001) The biomechanics of vertebroplasty. The effect of cement volume on mechanical behavior. Spine 26:1537–1541
55. Hansson T, Roos B, Nachemson A (1980) The bone mineral content and ultimate compressive strength of lumbar vertebrae. Spine 5:46–55
56. Steens J, Verdonschot N, Aalsma AM, Hosman AJ (2007) The influence of endplate-to-endplate cement augmentation on vertebral strength and stiffness in vertebroplasty. Spine 32:E419–E422
57. Bartley MH Jr, Arnold JS, Haslam RK, Jee WS (1966) The relationship of bone strength and bone quantity in health, disease, and aging. J Gerontol 21:517–521
58. Bell GH, Dunbar O, Beck JS, Gibb A (1967) Variations in strength of vertebrae with age and their relation to osteoporosis. Calcif Tissue Res 1:75–86
59. Skaggs DL, Weidenbaum M, Iatridis JC, Ratcliffe A, Mow VC (1994) Regional variation in tensile properties and biochemical composition of the human lumbar anulus fibrosus. Spine 19:1310–1319
60. Ebara S, Iatridis JC, Setton LA, Foster RJ, Mow VC, Weidenbaum M (1996) Tensile properties of nondegenerate human lumbar anulus fibrosus. Spine 21:452–461
61. Acaroglu ER, Iatridis JC, Setton LA, Foster RJ, Mow VC, Weidenbaum M (1995) Degeneration and aging affect the tensile behavior of human lumbar anulus fibrosus. Spine 20:2690–2701
62. Green TP, Adams MA, Dolan P (1993) Tensile properties of the annulus fibrosus II. Ultimate tensile strength and fatigue life. Eur Spine J 2:209–214
63. Ambrosetti-Giudici S, Gedet P, Ferguson SJ, Chegini S, Burger J (2010) Viscoelastic properties of the ovine posterior spinal ligaments are strain dependent. Clin Biomech 25:97–102
64. Lucas SR, Bass CR, Crandall JR, Kent RW, Shen FH, Salzar RS (2009) Viscoelastic and failure properties of spine ligament collagen fascicles. Biomech Model Mechanobiol 8:487–498
65. Bass CR, Planchak CJ, Salzar RS, Lucas SR, Rafaels KA, Shender BS, Paskoff G (2007) The temperature-dependent viscoelasticity of porcine lumbar spine ligaments. Spine 32:E436–E442
66. Lu WW, Luk KD, Holmes AD, Cheung KM, Leong JC (2005) Pure shear properties of lumbar spinal joints and the effect of tissue sectioning on load sharing. Spine 30:E204–E209
67. Iida T, Abumi K, Kotani Y, Kaneda K (2002) Effects of aging and spinal degeneration on mechanical properties of lumbar supraspinous and interspinous ligaments. Spine J 2:95–100
68. Neumann P, Keller TS, Ekstrom L, Hansson T (1994) Effect of strain rate and bone mineral on the structural properties of the human anterior longitudinal ligament. Spine 19:205–211
69. Neumann P, Ekstrom LA, Keller TS, Perry L, Hansson TH (1994) Aging, vertebral density, and disc degeneration alter the tensile stress-strain characteristics of the human anterior longitudinal ligament. J Orthop Res 12:103–112
70. Pintar FA, Yoganandan N, Myers T, Elhagediab A, Sances A Jr (1992) Biomechanical properties of human lumbar spine ligaments. J Biomech 25:1351–1356
71. Neumann P, Keller TS, Ekstrom L, Perry L, Hansson TH, Spengler DM (1992) Mechanical properties of the human lumbar anterior longitudinal ligament. J Biomech 25:1185–1194
72. Hukins DW, Kirby MC, Sikoryn TA, Aspden RM, Cox AJ (1990) Comparison of structure, mechanical properties, and functions of lumbar spinal ligaments. Spine 15:787–795
73. Hasberry S, Pearcy MJ (1986) Temperature dependence of the tensile properties of interspinous ligaments of sheep. J Biomed Eng 8:62–66
74. Nachemson AL, Evans JH (1968) Some mechanical properties of the third human lumbar interlaminar ligament (ligamentum flavum). J Biomech 1:211–220
75. Tkaczuk H (1968) Tensile properties of human lumbar longitudinal ligaments. Acta Orthop Scand Suppl 115:111+
76. Zhu D, Gu G, Wu W, Gong H, Zhu W, Jiang T, Cao Z (2008) Micro-structure and mechanical properties of annulus fibrous of the L4-5 and L5-S1 intervertebral discs. Clin Biomech 23(Suppl 1):S74–S82
77. Hirsch C (1955) The reaction of intervertebral discs to compression forces. J Bone joint Surg Am 37-A:1188–1196
78. Willen J, Lindahl S, Irstam L, Aldman B, Nordwall A (1984) The thoracolumbar crush fracture. An experimental study on instant axial dynamic loading: the resulting fracture type and its stability. Spine 9:624–631
79. Oxland TR, Panjabi MM, Lin RM (1994) Axes of motion of thoracolumbar burst fractures. J Spinal Disord 7:130–138
80. Fredrickson BE, Mann KA, Yuan HA, Lubicky JP (1988) Reduction of the intracanal fragment in experimental burst fractures. Spine 13:267–271
81. Mermelstein LE, McLain RF, Yerby SA (1998) Reinforcement of thoracolumbar burst fractures with calcium phosphate cement. A biomechanical study. Spine 23:664–670; discussion 670–661
82. Kallemeier PM, Beaubien BP, Buttermann GR, Polga DJ, Wood KB (2008) In vitro analysis of anterior and posterior fixation in an experimental unstable burst fracture model. J Spinal Disord Tech 21:216–224
83. Jones HL, Crawley AL, Noble PC, Schoenfeld AJ, Weiner BK (2011) A novel method for the reproducible production of thoracolumbar burst fractures in human cadaveric specimens. Spine J 11:447–451
84. Kifune M, Panjabi MM, Liu W, Arand M, Vasavada A, Oxland T (1997) Functional morphology of the spinal canal after endplate, wedge, and burst fractures. J Spinal Disord 10:457–466

85. Panjabi MM, Kifune M, Wen L, Arand M, Oxland TR, Lin RM, Yoon WS, Vasavada A (1995) Dynamic canal encroachment during thoracolumbar burst fractures. J Spinal Disord 8:39–48
86. Stemper BD, Storvik SG, Yoganandan N, Baisden JL, Fijalkowski RJ, Pintar FA, Shender BS, Paskoff GR (2011) A new PMHS model for lumbar spine injuries during vertical acceleration. J Biomech Eng 133:081002
87. Gozulov SA, Korzhen'iants VA, Skrypnik VG, Sushkov Iu N (1966) Study of the durability of the human vertebrae under compression. Arkh Anat Gistol Embriol 51:13–18
88. Ash JH, Kerrigan JR, Arregui-Dalmases C, Del Pozo E, Crandall J (2010) Endplate indentation of the fourth lumbar vertebra – biomed 2010. Biomed Sci Instrum 46:160–165
89. Labrom RD, Tan JS, Reilly CW, Tredwell SJ, Fisher CG, Oxland TR (2005) The effect of interbody cage positioning on lumbosacral vertebral endplate failure in compression. Spine 30:E556–E561
90. Hansson T, Roos B (1980) The influence of age, height, and weight on the bone mineral content of lumbar vertebrae. Spine 5:545–551
91. Hou Y, Yuan W (2012) Influences of disc degeneration and bone mineral density on the structural properties of lumbar end plates. Spine J 12:249–256
92. Closkey RF, Parsons JR, Lee CK, Blacksin MF, Zimmerman MC (1993) Mechanics of interbody spinal fusion. Analysis of critical bone graft area. Spine 18:1011–1015
93. Hollowell JP, Vollmer DG, Wilson CR, Pintar FA, Yoganandan N (1996) Biomechanical analysis of thoracolumbar interbody constructs. How important is the endplate? Spine 21:1032–1036
94. Jost B, Cripton PA, Lund T, Oxland TR, Lippuner K, Jaeger P, Nolte LP (1998) Compressive strength of interbody cages in the lumbar spine: the effect of cage shape, posterior instrumentation and bone density. Eur Spine J 7:132–141
95. Stemper BD, Yoganandan N, Baisden JL, Pintar FA, Shender BS (2012) Rate-dependent failure characteristics of thoraco-lumbar vertebrae: application to the military environment. In: Proceedings of the ASME 2012 summer bioengineering conference, Fajardo
96. Hirsch C, Nachemson A (1954) New observations on the mechanical behavior of lumbar discs. Acta Orthop Scand 23:254–283
97. Denis F (1983) The three column spine and its significance in the classification of acute thoracolumbar spinal injuries. Spine 8:817–831
98. Dai LY, Yao WF, Cui YM, Zhou Q (2004) Thoracolumbar fractures in patients with multiple injuries: diagnosis and treatment-a review of 147 cases. J Trauma 56:348–355
99. Wittenberg RH, Hargus S, Steffen R, Muhr G, Botel U (2002) Noncontiguous unstable spine fractures. Spine 27:254–257
100. Kifune M, Panjabi MM, Arand M, Liu W (1995) Fracture pattern and instability of thoracolumbar injuries. Eur Spine J 4:98–103
101. Panjabi MM, Kifune M, Liu W, Arand M, Vasavada A, Oxland TR (1998) Graded thoracolumbar spinal injuries: development of multidirectional instability. Eur Spine J 7:332–339
102. Panjabi MM, Oxland TR, Kifune M, Arand M, Wen L, Chen A (1995) Validity of the three-column theory of thoracolumbar fractures. A biomechanic investigation. Spine 20:1122–1127
103. Ochia RS, Ching RP (2002) Internal pressure measurements during burst fracture formation in human lumbar vertebrae. Spine 27:1160–1167
104. Ewing CL (1966) Vertebral fracture in jet aircraft accidents: a statistical analysis for the period 1959 through 1963, U. S. Navy. Aerosp Med 37:505–508
105. Harrison WD (1979) Ejection experience in F/FB-111 Aircraft/1967–1978. Safe J
106. Smelsey SO (1970) Study of pilots who have made multiple ejections. Aerosp Med 41:563–566
107. Smiley JR (1965) RCAF ejection experience 1952–1961. RCAF Institute of Aviation Medicine, Toronto, pp 18
108. Sandstedt P (1989) Experiences of rocket seat ejections in the Swedish Air Force: 1967–1987. Aviat Space Environ Med 60:367–373
109. Eiband AM (1959) Human tolerance to rapidly applied accelerations: a summary of the literature. National Aeronautics and Space Administration (NASA), Washington, DC
110. Weiss MS, Matson DL, Mawn SV (1989) Guidelines for safe human exposure to impact acceleration. Naval Biodynamics Laboratory, New Orleans
111. Latham F (1957) A study in body ballistics: seat ejection. Proc R Soc Lond B Biol Sci 147:121–139
112. Stech EI, Payne PR (1969) Dynamic models of the human body. Aerospace Medical Research Laboratory
113. Brown T, Hansen RJ, Yorra AJ (1957) Some mechanical tests on the lumbosacral spine with particular reference to the intervertebral discs; a preliminary report. J Bone Joint Surg Am 39-A:1135–1164
114. Coermann RR (1962) The mechanical impedance of the human body in sitting and standing position at low frequencies. Hum Factors 4:227–253
115. Brinkley JW, Shaffer JT (1971) Dynamic simulation techniques for the design of escape systems: current applications and future air force requirements. Wright-Patterson Air Force Base
116. Chandler R (1985) Human injury criteria relative to civil aircraft seat and restraint systems. Society of Automotive Engineers (SAE), 851847

Knee, Thigh, and Hip Injury Biomechanics

17

Jonathan D. Rupp

Abstract

This chapter summarizes research on the injury mechanisms, tolerances, and responses of the whole knee-thigh-hip (KTH) complex and its components to dynamic loading. The focus is on KTH injury to seated adult vehicle occupants and pedestrians involved in road traffic crashes. Injuries from falls, sports, and other activities are not specifically discussed, although much of the injury biomechanical research described is relevant to these activities.

17.1 Anatomy

Figure 17.1 illustrates the definitions of knee, thigh, and hip used in this chapter, which are based on those currently used in the BioTab method for injury causation assessment [1]. The knee consists of the patella, femoral condyles, and knee ligaments. The thigh consists of the supracondylar, shaft, and subtrochanteric region of the femur, while the hip includes the femoral head, neck, and acetabulum, or hip socket, of the pelvis. Note that parts of the femur are included in the knee, thigh, and hip.

Figure 17.2 illustrates the anatomy of the knee. The patella is a bone located over the anterior knee that primarily functions to increase the extension moment generated by the thigh musculature and protect the knee from anterior impacts. It is roughly triangular in shape, with a slight curve on the superior surface, where the quadriceps tendon inserts, and an apex at its inferior surface where the patellar ligament has its origin. The patella is mostly trabecular bone surrounded by a thin cortex. The majority of the posterior surface of the patella is articular cartilage that is divided in to lateral and medial portions by small central ridge. This ridge tends to translate along the intracondylar notch of the femur during normal knee flexion so the lateral and medial aspects of the retropatellar (i.e., posterior) surface contact the lateral and medial femoral condyles, respectively.

The major ligaments of the knee include the posterior cruciate ligament (PCL), the anterior cruciate ligament (ACL), the medial collateral ligament (MCL) and the lateral collateral ligament (LCL). The PCL is composed of anterior and posterior bundles. The proximal attachment

J.D. Rupp, Ph.D. (✉)
University of Michigan Transportation Research Institute and the Departments of Emergency Medicine and Biomedical Engineering, University of Michigan, Ann Arbor, MI, USA
e-mail: jrupp@umich.edu

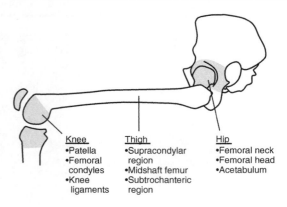

Fig. 17.1 Definitions of the knee, thigh, and hip (Adapted from Rupp et al. (2003). Effects of Hip Posture on the Frontal Impact Tolerance of the Human Hip Joint. Stapp Car Crash J 47:21–37. Used with permission of the Stapp Association)

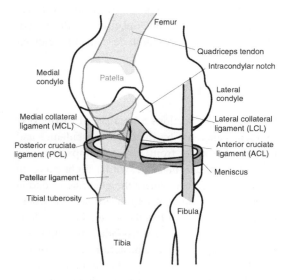

Fig. 17.2 Illustration of knee anatomy

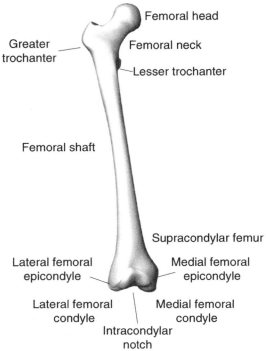

Fig. 17.3 Femur anatomy

of the PCL is on the anterior-lateral portion of the medial femoral condyle, within the intracondylar notch and its distal attachment is on the posterior aspect of the medial tibial plateau. Because of this orientation the PCL provides the primary resistance to posterior translation of the tibia relative to the knee. The ACL is functionally described as having posterior-lateral and anterior-medial bundles, both of which have their proximal attachment in the posterior-medial aspect of the lateral femoral condyle within the intracondylar notch and have their distal attachment in the interspinous area of the anteriomedial tibia plateau (i.e., the superior tibia). Both bundles resist anterior displacement of the tibia relative to the femoral condyles with the anterior bundle providing more resistance when the knee is flexed and the posterior bundle providing more resistance when the knee is extended [2].

The MCL has its proximal attachment on the medial femoral epicondyle, near the adductor tubercle and its distal attachment on the lateral aspect of the tibial metaphysis below the knee joint, under the pes anserinus. The MCL provides primary resistance to valgus (lateral inward) bending of the knee. The LCL has its superior attachment to the lateral femoral epicondyle and its distal attachment to the fibular head. Its primary function is to resist varus (lateral outward) bending of the knee.

The femur is a long bone that, at its distal end, consists of medial and lateral condyles that articulate along the superior tibial plateau to form the knee joint. Proximal to the femoral condyles the femur narrows over the supracondylar region into a quasi cylindrical shaft that has a slight posterior to anterior curvature (Fig. 17.3). The proximal shaft of the femur widens into a subtrochanteric region followed by the trochanteric region, where many of the major muscles of the hip insert. The femoral neck arises from the trochanteric region, oblique to the femoral shaft. The medial femoral neck widens into a spherical femoral head, which rotates within the acetabulum of the pelvis. The acetabulum is the joint surface of the hip on the pelvis, where the illium, ischium, and pubis join. The hip is covered by a fibrous joint capsule that attaches to the rim of the acetabulum and the medial aspect of the trochanter (Fig. 17.4). The ischiofemoral, iliofemoral, and pubofemoral ligaments arise from the pelvis around the acetabulum and insert on the trochanter and act to stabilize the hip.

17.2 Types, Causes, and Mechanisms of Injury

17.2.1 Frontal Impact

KTH injuries to occupants in frontal motor-vehicle crashes typically occur from loading of the anterior surface of the flexed knee either by front-seat knee restraints or by interaction between the knees of rear seat passengers and the seat backs in front of them. This load is transferred from the patella and/or femoral condyles, through the femur and into the posterior acetabular surface of the pelvis. KTH injuries to occupants in frontal impacts with a lateral component may also occur from door or center console loading.

Figure 17.5 shows the distribution of types of KTH fractures and dislocations in frontal crashes. Knee injuries account for 16 % of all

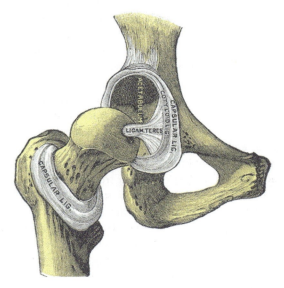

Fig. 17.4 Hip anatomy (From Gray's Anatomy 20th Edition, 1916)

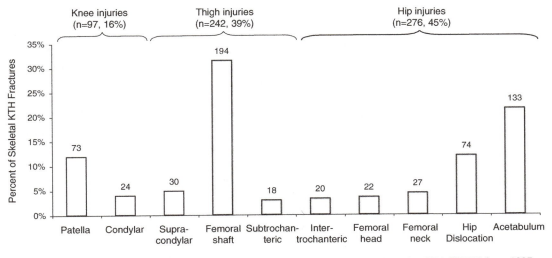

Fig. 17.5 Distribution of knee, thigh, and hip fractures and dislocations in frontal crashes UM CIREN from 1997 to 2003 (Adapted from Chang et al. [3])

KTH injuries [3]. The most common type of knee injury is patellar fracture, which occurs when the patella is compressed against the femoral condyles. The likelihood of patellar fracture may be increased by tension in the patella caused by knee extensor muscle activation from pre-crash braking or bracing [4]. Knee impact with an angled surface such as knee bolster may also produce tension in the patella by causing it to slide superiorly, which places the patellar ligament in tension and is the likely cause of patellar ligament avulsion. Injuries to the posterior cruciate ligament (PCL) and other knee ligaments in frontal crashes are substantially less common than knee fracture, accounting for less than 1 % of moderate severity injuries in frontal crashes [5]. Injuries to the PCL and other knee ligaments may occur if loading is applied to the anterior tibia instead of the patella and femoral condyles [6] or may result from lacerations caused by interaction between the anterior knee and sharp surfaces.

Fractures of the femoral condyles typically occur because the patella is driven into the intracondylar notch [7] and, acting like a wedge, splits the condyles. This wedging action usually also causes a supracondylar femur fracture. Supracondylar femur fractures in the absence of a split condylar fracture are typically associated with a high-energy impact to the knee [8] and are typically highly comminuted.

Fractures to the thigh account for 39 % of KTH injuries [3] and are primarily femoral shaft fractures rather than intertrochanteric or supracondylar femur fractures. Femoral shaft fractures are usually due to axial compression. The curvature of the femoral shaft, with the midshaft lying both anterior and lateral to the line of action between the hip and knee, induces a posterior-to-anterior bending moment in the distal two-thirds of the femoral shaft under axial loading [9–11]. Knee loading that is preferentially applied to the medial femoral condyle rather than the lateral femoral condyle may also cause increased bending moments in the femur [12]. Because the center-of-pressure of the patella on the femoral condyles is below the midline of the femur when the knee is more flexed (i.e., the included angle between the tibia and femur is less than ~90°), loading of the flexed knee may also increase the likelihood of femoral shaft fracture by increasing posterior-to-anterior bending moment.

The mechanisms of trochanteric femur fractures from knee loading have not been extensively studied, because these injuries are not common. However, when these fractures do occur they are thought to involve tension in the region of the greater trochanter generated by axial loading through the femoral shaft, the support of the femoral head in the acetabulum, and the moment arm provided by the femoral neck [13]. Femoral neck fractures occur from a similar mechanism [10].

Hip fractures and dislocations account for 45 % of all KTH injuries in frontal crashes [3]. The most common hip injury is acetabular fracture, and in particular, fracture of the posterior acetabulum, although many different types of acetabular fractures can occur from knee loading [10, 14]. The pre-impact posture of the hip, changes in the posture of the hip during knee loading, and potential secondary loading paths from interaction between the trochanter and components of the seat or side door interior may affect hip fracture patterns in frontal crashes.

Morphologic differences in the acetabulum between men and women may also affect the acetabular fracture patterns and the likelihood of acetabular fracture [15, 16]. Specifically, women tend to have acetabular surfaces that, in a seated posture, cover a greater portion of the femoral head than men, and it has been hypothesized that stresses in the female acetabulum will be lower than those in the male acetabulum assuming similar femoral head size.

Body mass distribution also likely affects load transmission to the hip by changing the effective mass of the KTH complex and the amounts of body mass that become coupled to the KTH proximal and distal to the hip [17]. Body shapes that have a greater proportion of the flesh and torso mass proximal to the hip will likely see high forces transmitted to the hip and a greater likelihood for hip injury.

Body shape and body mass index (BMI) also alter the interaction between the seat belt and the pelvis such that a greater amount of flesh must be deformed before the lap belt can apply a restraint force to the pelvis. As a result, obese belted occupants have greater knee excursion and their knees contact knee restraints at higher velocities than normal BMI occupants [18, 19]. In addition to shifting the belt forward relative to the pelvis, the anterior body contours associated with obesity also tend to raise the belt above the pelvis and onto the abdomen [20].

17.2.2 Lateral Impact

KTH injuries to motor-vehicle occupants involved in lateral impacts occur from loading of the knee, thigh, and/or by the intruding door or side interior components as well as by lateral acceleration of the vehicle that drives the occupant into the door. Lateral impacts with a frontal component may also produce KTH injuries from loading of the anterior knee as described above. Loading of the pelvis by the seat belt and/or center console by the inboard hip are also hypothesized to increase the likelihood of pelvis injury.

Side impacts produce different patterns of lower-extremity injury patterns to occupants on the impacted side of the vehicle (nearside occupants) compared to occupants on the opposite side of the vehicle (farside occupants). For nearside occupants, approximately 70 % of AIS 2+ lower extremity skeletal injuries are to the hip/pelvis, 3 % are to the thigh, and 6 % are to the knee [21]. In contrast, for farside occupants, approximately 38 % of all AIS 2+ lower-extremity injuries are to the hip/pelvis, 5 % are to the thigh, and 17 % are to the knee [21]. Reasons for the differences in KTH injury patterns between farside and nearside occupants include that farside occupants receive less focal pelvic loading from vehicle side structures and components but have greater lateral movements of the leg-ankle-foot complex relative to the KTH. It should also be noted that most of the knee injuries to both farside and nearside occupants were AIS 2 (moderate severity) knee sprains, which have been considered minor injuries by other researchers and excluded from study [22, 23].

17.2.3 Pedestrian Impact

Knee-thigh-hip injuries from lateral impact to the lower extremities in a standing posture (i.e., pedestrian impact) typically occur from loading by the grill or the bumper of the vehicle, but can also result from impact by the hood or contact with the ground. Lower-extremity injuries represent approximately 26 % of all AIS 3+ pedestrian injuries [24] and 37 % of all AIS 2–4 injuries [25]. Approximately half of all AIS 2+ adult lower-extremity injuries in pedestrian impact are to the pelvis, femur, and knee [26]. For adults, approximately 65 % of all AIS 2+ KTH injuries in pedestrian impact are to the femur, 23 % are to the pelvis, and 12 % are to the knee [26]. Most of these injuries are skeletal fractures, although a meaningful proportion of knee injuries involve ligaments, most commonly the medial collateral ligament [27].

Ligamentous knee injuries in lateral impact typically occur from a combination of shear loading of the knee and lateral bending typically caused by impact to the leg, knee, or thigh. Fractures of the thigh are thought to occur from bending, while hip fractures occur from trochanteric loading, similar to lateral impacts to vehicle occupants.

17.3 Injury Tolerance and Biomechanical Response

17.3.1 Whole KTH

17.3.1.1 Injury Tolerance

The first biomechanical study of the tolerance of the KTH to frontal knee impact in an automotive crash environment was conducted by Patrick et al. [28]. In this study, an impact sled was used to decelerate seated whole embalmed cadavers using knee loading by lightly padded knee stops that were instrumented to measure forces applied to each knee. In tests of ten cadavers, fractures to knee, thigh, and hip (in different tests) were produced. The lowest fracture-producing force was 6.2 kN. Patrick et al. [29] later revised this tolerance value to 8.7 kN after an additional series of similar tests did not produce fractures at forces below 8.7 kN, and an analysis of posttest x-rays from their earlier study indicated that several of the subjects that sustained fractures at forces below 8.7 kN were osteoporotic.

Powell et al. [30] also measured the tolerance of the KTH by impacting the flexed knees of eight embalmed and two unembalmed stationary cadavers with a rigid flat-faced 15.6-kg ballistic pendulum. A repetitive test methodology was used where each cadaver knee was impacted at successively increasing impact velocities until fracture was produced. Fracture-producing impact speeds varied from approximately 5–9 m/s and the resulting knee loading rates, where reported, were above 4,000 N/ms. Injuries produced in this study were primarily to the knee and supracondylar femur, and the average fracture force associated with these injuries was 10 kN.

Melvin et al. [31] and Melvin and Stalnaker [8] recognized that much of the earlier KTH tolerance data were from tests on embalmed cadavers and were, therefore, of questionable value because of changes in skeletal mechanical properties caused by embalming. Melvin and Stalnaker therefore performed 31 impacts to the flexed knees of 14 unembalmed cadavers using an 11-kg flat-faced ballistic mass with varying amounts of padding on the impact surface. These tests almost exclusively resulted in knee and supracondylar femur fractures. Fracture-producing impact speeds were generally above 11 m/s and the associated knee loading rates, when published, were above 5,000 N/ms. The average fracture force from these tests was 18.4 kN, and no fractures were observed at forces below 13.4 kN.

The high incidence of knee and distal femur injuries coupled with the low incidence of hip fractures in early biomechanical studies [8, 29–31] were thought to indicate that the tolerance of the hip to anterior knee impact loading is greater than that of the thigh or knee and that the distal femur was the weakest part of the KTH complex. The dynamic tolerance of the distal femur was, therefore, considered to be an appropriate injury criterion for protecting the entire KTH complex. Consequently, distal femur and knee fracture force data collected from the dynamic knee loading performed by Melvin et al., Patrick et al., and Powell et al. [29–31] were used to establish a maximum force criterion of 10 kN for use in frontal crash testing. This criterion was implemented in Federal Motor Vehicle Safety Standard (FMVSS) 208, which currently requires that the peak force measured at the distal shaft femur of a midsize-male Hybrid III ATD not exceed 10 kN in various types of staged crash tests. The 10-kN maximum femur force criterion was later correlated to a 35 % risk of fracture or dislocation to the KTH complex [32].

Responding to concerns that real-world frontal crashes in the late 1970s were resulting in a significant number of injuries to the midshaft femur, proximal femur, and hip while the existing body of KTH tolerance data consisted almost exclusively of distal femur and knee injuries and few hip injuries, Melvin and Nusholtz [33] conducted a series of sled tests in which six whole unembalmed cadavers were decelerated by contact with padded instrumented knee stops. It was hypothesized that the longer knee loading durations and more realistic occupant kinematics provided by decelerating a moving body by applying force to the knees would be more likely to produce proximal femur and hip injuries. Two of the cadavers used in these sled tests were osteoporotic and consequently, data from these tests were not used in their analyses. In the remaining four tests, injuries to the knee, thigh, and hip were produced at peak applied forces that ranged from 10.4 to 25.6 kN. Because this range is similar to the range of peak applied forces associated with distal femur and knee fracture in the pendulum impact testing performed by Melvin and Stalnaker [8], the researchers concluded that the longer loading durations resulting from whole-body impact testing are associated with injuries to the femoral shaft, proximal femur, and hip.

Viano and Stalnaker [9] and Stalnaker et al. [34] impacted the knees of 12 seated unembalmed cadavers using an 11-kg flat-faced rigid ballistic mass. In these tests, femurs were denuded to allow for optical measurement of femoral deformation and knee impact velocities varied from 12 to 14 m/s. Contrary to the results of other pendulum-knee impact studies, these studies produced injuries to the femoral shaft and proximal femur at forces of 6–10 kN in addition to injuries to the distal femur and knee. Based on observations of the deformation of the femur, the primary mode of femur failure for knee impact loading was determined to be femur bending caused by loading of the anterior surface of the knee along with the anatomic curvature of the femoral shaft and the moment arm created by the femoral neck. One possible reason that femoral shaft and proximal femur fractures were produced in the Viano and Stalnaker study [9], but not in other pendulum-impact studies, is that the Viano and Stalnaker tests were conducted with a knee posture that was flexed so that the included angle between the tibia and the femur was approximately 45° (knee highly flexed). By loading the knee in this posture, force was applied to the posterior portion of the femoral condyles (below the midline of the femur), which would increase the inferior-superior bending moment on the femur and thereby increase the likelihood of femur fracture in bending.

Although a substantial amount of research has been done to quantify the tolerance of the whole KTH complex to knee impact loading and the tolerances of components of the KTH complex, few studies have explored load transmission from the knee to the hip and its effect of the risks of knee, thigh, and hip injuries. Horsch and Patrick [35] unilaterally impacted the knees of six whole cadavers, whole lower extremity (femur and all distal anatomy) specimens, and lower extremity specimens that were sectioned and potted at the mid-femur. In each case, impacts were performed using a flat-faced ballistic pendulum of varying mass at different velocities. Knee impact forces and femur and tibia accelerations from these tests have been used to define knee impact response specifications for crash test dummies and have demonstrated that loosely coupled mass in the lower extremities affects knee impact response.

Donnelly and Roberts [36] impacted the flexed knees of nine unembalmed cadavers using a rigid flat-faced 56-kg pendulum. A series of impacts at successively increasing velocity was performed on each subject until a knee, thigh, or hip fracture occurred. Knee impact forces applied to cadaver knees were measured and compared. In addition, a load cell was implanted in the distal femur of one KTH complex so that the axial force internal to the cadaver femur could be measured. Similar to earlier biomechanical studies, the rigid impacts to cadaver knees produced primarily knee and distal femur fractures and few femoral shaft or proximal femur fractures.

Peak axial forces measured by the implanted femur load cells were, on average, 53 % of the peak forces applied to the cadaver knees. The decreases in force between the knee and the location of the femur load cell were attributed to acceleration of the mass between the knee and femur load cell. Because the mass of the femur load cell and associated potting material was substantially larger than the tissue that it replaced, the drop in force between the knee and the femur load cell observed by Donnelly and Roberts was likely larger than that which would be observed in a cadaver without a femur load cell.

Rupp et al. [23] reviewed the results of earlier biomechanical studies and noted that most knee and supracondylar femur fractures were associated with high-rate knee impacts (2,000–5,000 N/ms loading rates) where fracture occurred within 5-ms after the start of impact. Rupp [13] also noted that the limited number of femoral shaft and hip injuries tended to occur in sled tests and under padded knee loading, both of which are associated with longer loading durations and lower loading rates. The association between higher knee loading rates and knee and distal femur fractures, coupled with the finding that there is a 2–5-ms lag between the start of loading of the knee and the start of cadaver pelvis acceleration [34] led Rupp et al. [23] to propose that laxity in the cadaver hip joint, which would prevent the development of a reaction force at the hip until it was removed was the primary reason that high-rate knee loading did not produce hip and proximal femur fractures.

Rupp et al. [17] further explored the response of the body to knee impact and force transmission from the knee to the hip using a combination of physical testing and lumped parameter computational modeling. Multiple knee impact tests were conducted using five unembalmed male cadavers that were approximately midsize in both stature and mass based on U.S. anthropometric data. In these tests, the knees were symmetrically impacted using loading conditions that spanned the 5th–95th percentiles of those observed in FMVSS 208 and NCAP tests (Hybrid III femur forces between 1.3 and 7.3 kN and femur loading rates between 100 and 1,000 N/ms). In addition, each cadaver was impacted after the connection between the thigh flesh and pelvis was cut, after the thigh flesh was removed, and after the torso was removed to provide data on the effects of these components on knee impact response.

Data from these tests and results of other studies were used with data on body segment masses to develop and validate a one-dimensional lumped-parameter model of the seated cadaver. Simulations of the whole body cadaver tests performed with this model predict that approximately 54 % of the peak force applied to the knee is transmitted to the hip and that this percentage is relatively constant over the range of knee impact conditions that are of interest for injury assessment in frontal crashes. Simulation results also confirm the hypothesis that high-rate, short-duration knee loading by a rigid surface is more likely to produce knee/distal femur fractures and less likely to produce hip fractures due to laxity in the hip that delays recruitment of pelvis mass and the development of fracture-level forces at the hip until after the fracture tolerance of the knee/femur has been exceeded.

17.3.1.1 Effects of Muscle Tension on KTH Injury

Existing biomechanical tolerance data and knee impact response studies with cadavers have produced almost exclusively knee and distal femur injuries and hip injuries, and few femoral shaft fractures. In contrast, data from frontal crashes collected by the Crash Injury Research and Engineering Network (CIREN) demonstrate that approximately one third of KTH injuries in frontal crashes are to the midshaft of the femur [13]. Chang et al. [3] hypothesized that the differences between real-world and laboratory KTH injury patterns are in part due to muscle activation in crashes from occupant braking/bracing that cannot be reasonably reproduced in cadaver tests. Specifically, Chang et al. [3] proposed that muscle tension increases flesh-to-skeletal mass coupling in the KTH and applies a pre-load to the femoral shaft and that the combination of these factors changes force transmission from the knee to the hip and increases the bending moment in the femoral shaft such that the risk of femoral shaft fracture is increased relative to the risks of other KTH injuries.

To test this hypothesis, Chang et al. [3] developed and validated a new finite element (FE) model that included detailed lower-extremity musculature and that tied muscle mass to force-generating elements so that the effects of muscle tension on flesh-to-skeletal mass coupling could be simulated. Activation of major thigh flexors and extensors using forces reported in the literature for bracing increased flesh-to-skeletal mass coupling and thereby increased the peak knee impact force relative to that produced without muscle tension for symmetric knee impact loading. Muscle tension also increased the percentage drop in force between the knee and the hip by preferentially increasing the mass coupled to the KTH distal to the hip and increased bending moments at the mid shaft of the femur. These results suggest that lower-extremity muscle tension tends to increase the potential for mid-shaft femur fractures relative to the potential for hip injuries.

Chang et al. [37] addressed some of the shortcomings of their previous work by developing improved estimates of lower extremity muscle activation for maximal braking and bracing exertions and applying these to estimates to the lower extremities of their FE model. Results of symmetric knee-impact simulations with this model using the improved muscle activation data supported the previous finding that muscle activation increases the drop in force between the knee and the hip. However, simulation results also demonstrated the reduction in the magnitude of force transmitted from the knee to the hip for a given knee impact condition was modulated by the increased force at the knee and by increased compressive forces at the hip due to activation of lower-extremity muscles. As a result, activation of lower-extremity muscle tension is not thought to substantially affect the risk of hip injuries. However, maximal muscle activation substantially increases the bending moments in the femoral shaft, thereby increasing the risk of femoral shaft from knee loading fractures from knee loading by 20–40 %.

17.3.1.1 Risk Curves

Injury risk curves for the skeletal knee-thigh-hip (KTH) complex have been developed that describe the probability of KTH fracture as a function of force applied to the anterior surface of the flexed knee. Morgan et al. [32] fit a Weibull distribution to peak knee-impact force data from 34 KTH fracture-producing tests from 6 studies in the National Highway Traffic Safety Administration (NHTSA) biomechanics database [22, 26, 38–41].

Kuppa et al. [22] recognized that Morgan et al. [32] excluded data from tests in which injury was not produced and that this biased the Morgan et al. risk curve. To address this shortcoming, Kuppa et al. used logistic regression to reanalyze peak knee impact force data from injury-producing and non injury-producing knee impacts in the six studies from which Morgan et al. obtained their data. Separate risk curves were developed for AIS 2+ and AIS 3+ injuries (Eqs. 17.1 and 17.2, respectively). However, because the AIS 3+ KTH injuries in the Morgan et al. dataset were almost exclusively distal femur fractures, the Kuppa et al. AIS 3+ risk curve is generally accepted to apply to the distal femur and not other parts of the KTH. The Kuppa et al. [22] risk curves were applied to the Hybrid III midsize male ATD and used to assess injury risk in frontal crash tests by the National Highway Traffic Safety Administration [42]. Injury risk curves for the small female ATD were developed using scaling [42, 43]. Fracture forces analyzed by Kuppa were not normalized to account for differences in response that are expected with varying subject size. Therefore, the Kuppa et al. risk curves are most applicable to vehicle occupants with the mean cadaver age and mass in the dataset Kuppa analyzed, which are 63 years and 69 kg, respectively.

$$P(AIS2+) = \frac{1}{1+e^{(5.7949-0.5196F)}} \quad (17.1)$$

$$P(AIS3+) = \frac{1}{1+e^{(4.9795-0.326F)}} \quad (17.2)$$

where F is peak force (kN) applied to the human knee or peak force at the Hybrid III femur load cell.

Laituri et al. [44] noted that the Kuppa et al. risk curves do not account for subject age and are not based on all available data. Therefore, they developed a new set of KTH risk curves for AIS 2+ injuries (Eq. 17.3). The dataset the Laituri et al. used to develop risk curves was limited to knee impacts performed with padded surfaces. Survival analysis was used to fit a Weibull distribution to the peak force data and to account for censoring (i.e., the occurrence of fracture at an unknown time prior to the time of peak force or the failure of a test to produce injury) in the dataset. The effect of cadaver age on fracture force was accounted for by restricting the analysis to peak forces from tests in which the subjects were older (age 60+) and by scaling the shape parameter of the resulting Weibull distribution (i.e., the risk curve) using relationships between age group and the tolerance of the femoral shaft to axial compression reported by Aoji et al. [45]. Prior to analysis, all peak-force data were normalized to account for the assumed effects of differences in subject size on fracture force using a dimensional analysis technique called equal-stress equal-velocity scaling [46]. Risk curves for females and different sizes of male occupants were developed using similar scaling techniques.

$$p(FX) = 1 - e^{-(\frac{F}{a})^{3.4936}} \quad (17.3)$$

where F is peak force in kN applied to the knee normalized to a midsize male reference mass (75 kg) using equal-stress equal velocity scaling and a is an age scaling parameter that is equal to 15.9499 for ages 15–39, 15.0971 for ages 40–59, and 13.5495 for ages 60+.

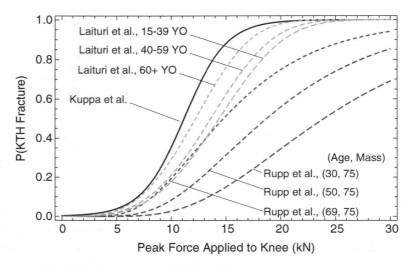

Fig. 17.6 Comparison of KTH risk curves for midsize males from Kuppa et al. [22], Laituri et al. [44] and Rupp et al. [47]

Rupp et al. [47] noted that Laituri et al. [44] used a limited dataset along with scaling to account for age and gender/stature effects rather than using the large amount of empirical data on KTH fracture force to define these effects. Rupp et al. therefore developed a new KTH risk curve for the whole KTH that was parameterized to account for the effects of subject characteristics and test conditions on KTH fracture forces and that appropriately accounted for censoring in peak knee impact force data (Eq. 17.4). The parametric KTH injury risk curve was used to define new KTH injury risk curves that apply to the adult small female and midsize male. Figure 17.6 compares the KTH risk curves developed by Kuppa et al., Laituri et al. and Rupp et al.

$$p(FX) = \Phi\left[\frac{Ln[F] - (0.0081m - 0.0124a + b)}{0.4519}\right] \quad (17.4)$$

where,

- Φ is the cumulative distribution function of the standard normal distribution,
- F is peak force applied to the knee in kN,
- m is mass in kg,
- a is age in years, and
- b is 2.6224 for rigid and 2.9396 for padded knee impacts.

The risk curves described above relate force applied to the knee to the probability of any KTH fracture. However, because all of these risk curves are based primarily on injury data from studies that produced almost exclusively knee and distal femur fractures, these risk curves primarily apply to the knee and distal femur and are not applicable to the hip [10, 13].

Risk curves that apply to the entire KTH have traditionally been applied to peak axial compressive force measurements made by load cells located in the distal third of a crash test dummy femur. However, multiple studies have found that crash test dummies and cadavers produce different knee impact forces under similar knee loading conditions. Early studies found that knee impact forces produced by dummies were substantially higher than those produced by cadavers and it was therefore assumed that evaluating the probability of KTH injury using dummy femur load cell data and injury criteria based on cadaver knee impact force was a conservative approach [35, 36]. However, later work demonstrated that the differences between cadaver and dummy knee impact forces depend on the force-deflection characteristics of the surface loading the knee [13, 17, 48]. For impacts with a surface with constant stiffness, current dummies will over estimate KTH risk while for impacts with surfaces that are force limiting (e.g., crushable aluminum honeycomb) current dummies can underestimate KTH injury risk [13, 17, 48]. Ongoing development of the THOR, which is the next generation of the adult frontal-impact crash test dummy, should address these issues by improving the knee impact force biofidelity of the KTH complex [49].

17.3.2 Hip Injury

17.3.2.1 Frontal Impact

In real-world frontal crashes, the left hip is much more frequently injured than the right hip when the crash dynamics tend to move the occupant forward and to the left, while right hip injuries are more frequent in crashes where the occupant moves forward and to the right [10]. The KTH on the side of the body to which the occupant tends to move in a frontal crash will generally experience higher forces than the contralateral knee and lateral motion of the occupant's pelvis induces adduction of the thigh (i.e., inward rotation of the hip). Hip adduction decreases the contact area between the femoral head and the acetabular surface and thereby decreases the force required to cause hip fracture [14].

Yoganandan et al. [50] first explored the effects of knee loading rate and thigh angle on KTH fracture tolerance using data from impacts to the anterior surfaces of the flexed knees of six whole unembalmed seated cadavers. These tests were performed using a 23.4-kg rigid flat-faced pendulum that impacted a cadaver knee at speeds between 4.3 and 7.6 m/s. A compressive preload was applied to the KTH complex to simulate muscle tension by a loop of seatbelt webbing wrapped around the knees and fastened behind the subject's pelvis. A paired test methodology was used where one KTH complex from each subject was impacted so that the direction of applied force was approximately aligned with the long axis of the femur. The contralateral KTH complex was subsequently tested with the thigh either adducted/abducted and/or flexed from the direction of applied force. Results of these tests suggested that acetabular fractures are more likely with combined flexion and adduction, but no association between thigh angle and the force required to cause acetabular fracture could be ascertained because of the small sample size and large range of postures tested.

Rupp et al. [23] quantified the fracture tolerance of the cadaver KTH complex to loading applied to knee loading using a setup in which the pelvis was fixed to ensure that the force applied to the knee was the same as the force transmitted to the hip. All of these fixed-pelvis tests were conducted with the pelvis and thigh oriented in a typical automotive seated posture [51]. Knee loading was applied through an interface that was custom molded to the shape of each subject's knee to ensure that applied forces were distributed over the entire anterior surface of the knee. All tests produced posterior acetabular fractures and the distributions of fracture types that occurred in these tests matched those observed in real-world frontal crashes from the University of Michigan CIREN database, suggesting that the fixed pelvis boundary condition did not affect test results beyond eliminating the inertially induced decrease in force transmitted from the knee to the hip. The average acetabular fracture tolerance from these tests was 5.7 kN (SD = 1.4 kN).

Rupp et al. [23] also measured the fracture tolerance of the uninjured knee/femur specimens that were used in the fixed-pelvis tests by supporting the femoral head in an "acetabular cup" and applying a force to the knee. All of these knee/femur tolerance tests produced femoral neck fractures, and the average fracture tolerance from these tests was 7.6 kN (SD = 1.6 kN). When considered together with the results of previous studies, the injury patterns and fracture forces measured in the fixed-pelvis and fixed-femoral-head tests demonstrate that the acetabulum is the weakest part of the KTH complex for axial knee impacts in typical automotive postures, and that the femoral neck has the next lowest tolerance to distributed knee impact loading.

Rupp et al. [10] quantified the effects of hip posture on hip tolerance using similar fixed-pelvis test methods as Rupp et al. [23] but with a

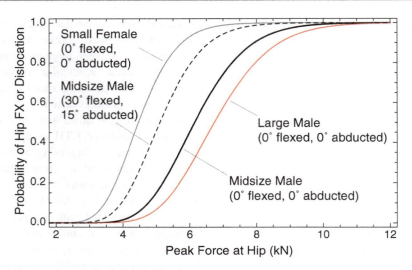

Fig. 17.7 Hip injury risk curve for small female (150 cm), midsize male (176 cm) and large male (186 cm) occupants in a standard automotive seated posture (0° flexion, 0° abduction) and a risk curve for the midsize male in a 30° flexed and 15° abducted hip posture

larger sample size. A paired comparison test methodology was used in which one hip from each subject was tested in a standard automotive seated posture and the contralateral hip was tested in a posture that was either ab-/ad-ducted and/or flexed from this posture. When data from these tests were combined with data from Rupp et al. [23], the average fracture tolerance of the hip in the neutral posture was 6.1 ± 1.5 kN. Hip tolerance decreased by an average of 34 ± 4 % with 30° of flexion from the neutral posture and by 18 ± 8 % with 10° of adduction from the neutral posture. Subsequent fixed-pelvis testing reported by Rupp [13] demonstrated that 15° of hip abduction increased hip tolerance by approximately 15 % and that the effects of hip posture on tolerance are generally linear with posture. Isolated hip tolerance testing by Bass et al. [52] conducted using similar methods to those of Rupp et al. [10] confirm the effects of hip posture on hip injury tolerance.

Rupp et al. [53] applied survival analysis to fracture force data from 27 hip tolerance tests conducted with the hip oriented in a standard automotive seated posture and developed a risk curve that was parametric with specimen stature (which was the only parameter among age, gender, body mass index, bone mineral density, and stature that was a significant predictor of fracture force). Based on data from Rupp et al. [13], this risk curve was modified so the mean shifted depending on the amount of hip flexion or abduction. Equation 17.5 and Fig. 17.7 show this risk curve. Typical values of flexion and abduction used with this risk curve are 30° and 15°, which represent typical amounts of peak hip flexion and hip abduction from a standard automotive seated posture that are present in a frontal crash at the time of peak femur force.

$$p(FX) = \Phi\left[\frac{Ln[F] - \left(Ln\left[\begin{array}{c}Exp[(0.2141 - 0.0114s)] \\ *\left(1 - \dfrac{f-a}{100}\right)\end{array}\right]\right)}{0.1991}\right]$$

(17.5)

where,

f is the cumulative distribution function of the standard normal distribution that describes the relationship between peak force at the hip in kN and the probability of hip fracture,
s is the target stature,
f is the hip flexion angle in degrees, and
a is the hip abduction angle in degrees.

Table 17.1 Loading conditions used in biomechanical studies that considered KTH response and injury in lateral impact

Study	Impactor type	Pelvis and KTH regions loaded	Recommended injury threshold
Cesari and Ramet [54]	Rigid and padded, spherical impact surface (r = 60 cm), 12 kg or 16 kg mass	Greater trochanter	10 kN for midsize males
Nusholtz et al. [55]	Rigid and padded Flat pendulum impact	Greater trochanter and iliac wing	None
Viano [56]	Rigid, 15-cm diameter pendulum	Greater trochanter and iliac wing	27 % Compression = 25 % risk
Cavanaugh et al. [62]	Segmented wall, rigid and padded w/pelvis offset	Greater trochanter and part of thigh	8 kN = 25 % risk for midsize male $V_{max}C_{max} = 2.7$ m/s[a] 36 % compression[a]
Zhu et al. [57]	Segmented wall, rigid w/pelvis offset	Greater trochanter and part of thigh	7.3 kN peak force = 25 % risk, $F_{avg} = 5$ kN[b]
Bouquet et al. [58, 59]	Segmented rigid pendulum impactor	Iliac wing, greater trochanter separately	46 mm half pelvis deflection = 50 % AIS 2+ injury
Pintar et al. [61], Maltese et al. [60]	Segmented wall, rigid and padded w/varying offsets	Iliac wing, greater trochanter and thigh with one impactor	None
Lessley et al. [63]	Segmented flat rigid wall w/KTH offset	Separately loaded knee, thigh, and pelvis	None
Miller and Rupp [64]	Segmented wall, padded w/abdomen offset	Iliac wing, greater trochanter and thigh separately	None
Kallieris et al. [65]	Segmented wall, rigid	Abdomen, iliac wing, greater trochanter and part of thigh	None

[a]Based on normalized pelvis half width deformation calculated from digitized film
[b]F_{avg} = slope of the integral of the applied force history between 10 % and 90 % of peak

17.3.2.1 Lateral Impact

Much of the early data related to the biomechanical response of the KTH complex in lateral impact were collected by either decelerating whole cadavers in seated postures into instrumented segmented load "walls" or by impacting the greater trochanter and iliac wing with a ballistic pendulum [54–61]. Table 17.1 summarizes the loading conditions used in these studies, the resulting injuries, and recommended injury thresholds. Studies that did not produce hip/pelvis injury are excluded from Table 17.1.

Several of the studies listed in Table 17.1 used varying amount of padding over surfaces that interacted with the body and did not characterize the mechanical response of this padding at a level sufficient for tests to be subsequently reproduced [61, 66]. The use of data from these studies for defining injury thresholds for components of the KTH is further complicated by the impactor shapes used in early studies, which tended to be large enough to load either the entire KTH or to simultaneously load the greater trochanter, iliac wing, and/or varying amounts of the thigh flesh [60, 67]. As a result of these limitations, data from earlier studies in Table 17.1 that used padded impact surfaces have limited utility for developing and validating crash test dummies and computational models, and are primarily useful for defining the relationship between mechanical loads applied to the KTH and resulting injury, assuming that such relationships are not affected by padding. Data from studies that used rigid impact surfaces are more useful for defining response and tolerance of the entire KTH or segments of the KTH, as are data from more recent studies (e.g., [60, 61]) for which the mechanical response of the padding is well characterized.

Multiple studies have developed force and deflection-based injury criteria for the hip/pelvis in lateral impact. Based on a series of pendulum impacts to the greater trochanter on 19 cadavers, Cesari and Ramet [54] reported fracture forces between 4.48 and 12.92 kN for male cadavers. Relationships between the Livi index (a measure of obesity) and fracture force were used to identify 10 kN as the force associated with a 50 % probability of pelvis fracture. Nusholtz et al. [55] impacted the iliac wing and trochanter with flat-rigid and padded 50-kg ballistic pendulum impactors and found fracture forces between 3.3 and 10.7 kN, but did not recommend an injury threshold. Viano [56] impacted the trochanter and iliac wing with a flat-faced rigid pendulum. Analyses of these data suggested that pelvic compression is the best predictor of pelvis fracture in lateral impact and that 27 % compression of the whole pelvis breadth is associated with a 25 % risk of fracture. Cavanaugh et al. [62] laterally decelerated whole cadavers into rigid and padded impact walls that contacted the greater trochanter and part of the thigh. Cavanaugh et al. reported that a peak force of 8 kN, a peak velocity*peak compression ($V_{max}C_{max}$) of 2.7 m/s, and 36 % half pelvis compression are all associated with a 25 % risk of pelvis fracture in lateral impact. Using test methods similar to Cavanaugh et al. [62], Zhu et al. [57] found that a 25 % risk of hip fracture was associated with a 7.3 kN peak force and a 5 kN average force, where average force was determined by taking the slope of the integrated force (momentum transfer) history from 10 % to 90 % of peak.

Bouquet et al. [58, 59] impacted the lateral pelves of whole seated cadavers with a pendulum that had a segmented impact surface that was capable of separately measuring loads applied to the iliac wing and greater trochanter. Logistic regression performed on data from fracture and non-fracture producing test suggest that 46 mm of hemi pelvis deflection is associated with a 50 % risk of AIS 2+ hip/pelvis injury.

Kuppa [66] reanalyzed the Bouquet et al. [58, 59] data and developed an injury risk curve for the pelvis in lateral impact by using logistic regression. Equation 17.6 provides the resulting risk curve. Although this risk curve was developed using data from a single study, its predictions are consistent with analyses of data from Cavanaugh et al. [62]. Kuppa et al. [66] also developed relationships between Eq. 17.6 and crash test dummy responses and used these relationships to specify on IARVs for use in lateral impact testing with side impact crash-test dummies.

$$P(AIS2 + pelvis\,injury) = \frac{1}{1 + e^{(6.804 - 0.0089a - 0.7424F)}} \quad (17.6)$$

where F is the sum of forces applied to the trochanter and iliac wing in kN and age is in years.

Although multiple studies have defined the forces and pelvis accelerations produced by lateral impact, most studies of the lateral-impact response of the pelvis have either loaded the iliac wing and trochanter with a single flat plate [60, 61], or have simultaneously loaded the trochanter and thigh with a single plate [62]. The studies that have separated the response of the iliac wing from that of the greater trochanter report that 68 % ± 7 % of the force applied to the pelvis is transmitted to the greater trochanter [58, 59, 64]. Reasons for this include that the greater trochanter is usually (but not always) located more outboard than the iliac wing and the iliac wing is less stiff than the greater trochanter under lateral impact loading. Response corridors for the pelvis in lateral impact loading are defined in ISO/TR9790 [68], Maltese et al. [60], Irwin et al. [67], and Miller and Rupp [64].

17.3.3 Thigh Tolerance, Response, and Risk Curves

17.3.3.1 Frontal Impact

As indicated above, thigh fractures in frontal crashes occur from compression-bending generated by loading of the anterior surface of the flexed knee and the anterior-to-posterior curvature of the femoral shaft as well as medial-lateral bending generated by the eccentricity of the femur caused by the femoral neck. Ivarsson et al. [11] explored the effects of combined axial

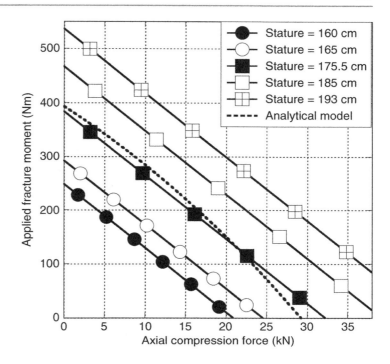

Fig. 17.8 Combinations of axial compressive force and P-A bending moment associated with femoral shaft fracture for different values of stature (From Ivarsson et al. (2009), Stapp Car Crash J, 53: 251–290. Used with permission of the Stapp Association)

compression-bending on the tolerance of the femoral shaft by dynamically applying varying levels of axial compression to isolated, simply supported femoral shaft specimens and then impacting these specimens at midshaft to create three-point bending. Results of these tests indicate that A-P bending increases the compressive force needed to cause fracture, while P-A bending, such as is produced by axial knee loading and the natural curvature of the femoral shaft, decreases tolerance. Figure 17.8 shows, and Eq. 17.7 describes, the relationship between fracture tolerance (50 % probability of fracture) and axial compression and P-A moment at midshaft. This relationship is based on the assumption of a linear relationship between applied axial compression and P-A bending moment, which is not strongly supported by the Ivarsson et al. data but is consistent with analytic models derived from beam-bending theory for linear elastic materials.

$$M = -1148 - 0.0119P + 8.73S \quad (17.7)$$

where M is moment in Nm, P is axial compressive force in N and S is subject stature in cm.

Rupp et al. [69] developed a ±1 SD target response corridor (Fig. 17.9) from force-deflection data obtained by dynamically loading the flexed knees of 20 femur/leg/foot specimens obtained from 11 cadavers [23]. As described above, these tests used a simply supported femoral head and applied force to the knee/femur complex along an axis that was defined by the midpoint between the femoral condyles and the center of the femoral head. Force was applied to the knee through a knee interface that distributed patella and femoral condyles and thereby minimized the effects of patellofemoral compliance on femur force-deflection response.

17.3.3.1 Lateral Impact

The bending tolerance of the midshaft of the femur has been studied by Martens et al. [70] who dynamically loaded femurs from 33 unembalmed cadavers in anterior-posterior four-point bending. Results of these tests indicated that the mean A-P bending moment required to cause fracture is 373 ± 84 N•m.

Kerrigan et al. [71] loaded femoral shaft specimens in three-point bending by applying a dynamic force directed in a lateral-to-medial direction to either the midshaft or a point that was one third of the length of the shaft proximal from the most inferior point on the femoral condyles.

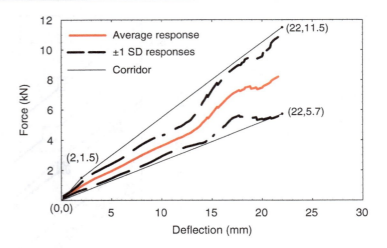

Fig. 17.9 Femur force-deflection response corridor developed from tests in which the femoral head was supported by an acetabular cup and loading was applied to the patella and femoral condyles along an axis defined by the midpoint between the femoral condyles and hip joint center (From Rupp et al. [69])

A subset of tests was performed in which flesh was removed to explore reasons why earlier studies with denuded femurs [72] reported lower fracture forces than tests where the thigh flesh was still present. Results of these tests were combined with similar tests reported by the authors in earlier studies [72, 73] and the combined dataset was normalized to the femur geometry of a midsize male using scaling techniques. A parametric Weibull distribution (Eq. 17.8) fit to the combined dataset describes how the risk of femur fracture for a midsize male varies with impact location, bending moment, and whether flesh is present over the impact site.

$$p(FX) = 1 - e^{-e^{[6.2424*\ln(m)-a-b+36.4101]}} \quad (17.8)$$

where m is the peak moment in the midsize male femoral shaft in Nm, a is 0.9026 if loading is applied through flesh and 0 otherwise, and b is 1.4483 if impact is at the midshaft and zero if impact is to the distal third of the femur.

Kennedy et al. [74] conducted dynamic lateral-to-medial (L-M) and posterior-to-anterior (P-A) three-point bending tests on 45 femurs from 29 cadavers. Femurs were simply supported by cups potted to the ends of the femoral shaft. Bending moments were generating by dropping a cylindrical impactor on to the flesh covering the midshaft. Results of these tests confirmed the bending tolerances reported by Martens et al. [70] and Kerrigan et al. [71]. Results also demonstrate that the tolerance of the femur to three-point bending is not significantly different for loading applied in L-M and P-A directions and that the tensile strain on the external surface of the bone opposite the impact location is 1.3 ± 0.3 % for lateral-medial bending and 1.2 ± 0.3 % for posterior-anterior bending. Equation 17.9 shows the risk curve developed by Kennedy et al. to relate peak femur bending moment and cortical cross-sectional area at midshaft to the probability of femur fracture. For reference, the 5th, 50th, and 95th percentiles of midshaft femur cross sectional area for male subjects from the Kennedy et al. [74] subject population determined from quantitative computed tomography are 389, 467, and 621 mm^2. For female subjects, the 5th, 50th and 95th percentiles of midshaft femur cross sectional area are 251, 324, and 434 mm^2, respectively.

$$p(FX) = 1 - e^{-e^{[9.3704*\ln(m)-(46.3140+0.0216a)]}} \quad (17.9)$$

where m is the peak moment in the femoral shaft in Nm and a is the midshaft cross-sectional area.

Ivarsson et al. [75] characterized the dynamic force-deflection and moment-deflection response of six isolated femoral shaft specimens to L-M three-point bending generating by a force applied through flesh at a location that was 1/3 of the length of the femur proximal to the knee. Figures 17.10 and 17.11 show corridors describing these responses.

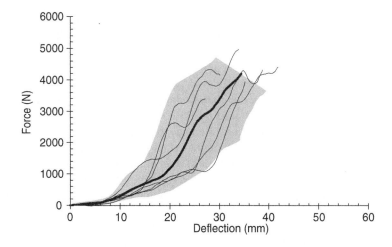

Fig. 17.10 Applied force-deflection response corridor for the isolated midsize male femoral shaft loaded in L-M three-point bending (Adapted from Ivarsson et al. [75]. Used with permission)

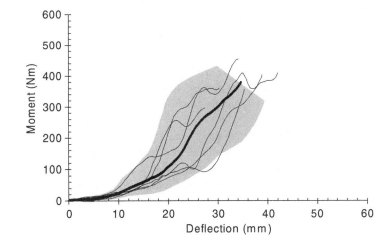

Fig. 17.11 Applied moment-deflection response corridor for the isolated midsize male femoral shaft loaded in L-M three-point bending (Adapted from Ivarsson et al. [75]. Used with permission)

17.3.4 Knee Injury

17.3.4.1 Frontal Impact

As indicated above a substantial number of tests have been conducted in which the knees of whole cadavers have been impacted to characterize the tolerance of the entire KTH complex. Identifying knee fracture tolerance from these tests is difficult because results are censored (i.e., results are reported as peak applied forces but patellar fracture, which was present in almost every knee fracture-producing test, can occur before the time of peak force). However, meta analyses of early whole body knee impact tests indicate that knee impacts with a rigid surface are more likely to produce knee injury and that this injury occurs at a lower force [47, 76].

Studies by Haut [76], Atkinson et al. [4, 77], and Hayashi et al. [78] have explored the effects of the stiffness of the surface that loads the knee on knee fracture force by impacting the patellae of isolated knee specimens with flat, rigid surfaces and surfaces that distribute impact forces over the surface of the patella and the surface of the patella and femoral condyles. Data from Haut [76] suggest that the tolerance of the isolated knee to forces distributed across the patella and femoral condyles at a 90° tibia-to-femur angle is approximately 10.2 kN. The discrepancy between the 10.2 kN reported by Haut and the much larger values reported by Melvin for well-padded impacts may have been due to bottoming out the thinner (1-in.-thick) pad used by Haut et al. For this reason, the 10.2 kN knee tolerance to padded impact reported by Haut is thought to be low and a higher tolerance, like that from Melvin et al., is likely representative of the tolerance of the knee to distributed knee loading, such as is thought to occur in current vehicles [79].

The fracture tolerance of the knee to loading that is directed along the long-axis of the femur and distributed across the anterior patella, but not on to the femoral condyles, has not been well characterized. However, data from impacts to isolated knees performed by Atkinson et al. [77] suggest that it is higher than 7 kN. Atkinson et al. also demonstrate that occult micro fractures in the retropatellar and femoral cartilage and underlying tissues can occur at forces substantially lower than the forces typically associated with gross fracture.

Meyer and Haut [80] demonstrated that the fracture tolerance of the patella is also affected by the angle of the surface loading the flexed knee relative to the long-axis of the femur. For knee loading by a flat rigid surface along an axis that is rotated 15°–30° medial from the long axis of the femur, patellar fracture tolerance to rigid impact decreases by ~40 %. The obliquity of the loading angle reduced the contact area between the impactor and the patellar surface and altered the contact pressure on the retro-patellar surface so that most load is transmitted through the medial aspect of the patella on to the lateral femoral condyle. The patellofemoral joint was also found to be stiffer in axial loading than oblique loading, primarily because the patella slides laterally on the femoral condyles when loading is sufficiently oblique.

Knee angle has also been hypothesized to affect knee fracture force and fracture pattern. Haut [76] also studied the effects of tibia-to-femur angle on knee tolerance and contact stresses on the anterior surface of the patella by impacting isolated distal femur/knee/leg specimens from unembalmed cadavers. One knee from each cadaver was impacted with a flat rigid surface that applied a focal load to the patella. The contralateral knee was impacted with a padded surface that distributed loads across the patella and femoral condyles. Each pair of knees was tested at a tibia-to-femur angle of 60°, 90°, or 110°. Knee fracture patterns also appeared to change with knee angle such that more split condylar fractures were more often produced with smaller knee angles (more flexed knees). However, too few subjects were tested for the effects of knee angle on fracture type and tolerance to be clearly defined.

Changes in knee fracture pattern with increased knee flexion are likely caused by changes in the position of the patella on the femoral condyles. At more flexed knee angles the patella is positioned lower on the femoral condyles where it can act like a "wedge" and drive the femoral condyles apart [30, 76]. The split condylar fractures produced by this wedging action are the primary type of condylar fracture reported in the biomechanical literature. However, the risk of split condylar femur fracture may be exacerbated in the cadaver model because of a lack of tension in the quadriceps tendon, which in a living occupant may help to prevent the patella from migrating into a position lower on the femoral condyles where it would be in a position to produce a greater amount of wedging. On the other hand, high levels of muscle tension may decrease knee tolerance by inducing a tensile preload in the patella [4].

Figure 17.12 illustrates how knee flexion angle is thought to affect knee tolerance for impacts that only load the patella. As knee flexion increases, the patella moves over the surface of the femoral condyles, which changes the support conditions on the posterior patella and consequently affects knee tolerance. As illustrated on the right side of Fig. 17.12, when the knee is flexed past a 90° tibia-to-femur angle, the patella is positioned over the lower part of the femoral condyles and the lateral edges of the patella are supported by the femoral condyles while the middle of the patella is unsupported. In this position,

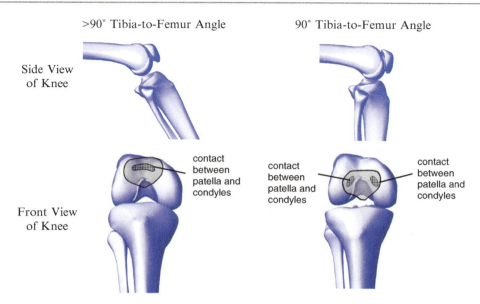

Fig. 17.12 Change in position and contact location of the patella on the femoral condyles with knee flexion (Adapted from Aglietti et al. [81])

loading of the anterior surface of the patella will produce three-point bending of the patella. In contrast, when the knee is more extended (as illustrated on the left side of Fig. 17.12), there is a larger continuous area of the posterior patellar surface supported by the femoral condyles, which is less likely to produce a wedging action to split the femoral condyles and is thought to be less likely to produce patellar fracture.

As indicated above, posterior cruciate ligament (PCL) tears and avulsions are the most common ligamentous knee injury in frontal crashes. PCL injuries primarily occur when the tibia is displaced posteriorly and externally rotated relative to the femur. Several studies of the PCL tolerance to posterior translation of the tibia relative to the femoral condyles have been performed. Viano et al. [6] applied a dynamic posterior displacement to the tibias of five isolated 90° flexed cadaver mid-femur to mid-tibia knee specimens from which the surrounding flesh and the patella had been removed. PCL injury occurred in four of five tests. Force-displacement responses were observed to have a linear region followed by a plateau and a subsequent rise in force. The plateau in force-displacement response was hypothesized to result from failure of collagen fibers in the PCL that occurred prior to rupture. The reported tibia posterior displacement associated with PCL injury was 14.4 ± 4 mm. However, a subsequent reanalysis of the data indicated that the plateau region in force-displacement responses associated with PCL injury was due to analytic errors and that all PCL injuries in this dataset occurred at a peak in applied force [82]. After correcting for these errors, the tibia displacement associated with PCL injury was 16.2 ± 3.9 mm. Analyses of these data were used to establish an injury assessment reference value (IARV) of 15 mm of posterior tibial displacement in the Hybrid III ATD for PCL injury [79].

Balasubramanian et al. [83] conducted additional tests in which the tibias of isolated knee specimens from eight cadavers were displaced posteriorly to augment the Viano et al. [6] data and to explore whether removal of the patella and the tissues surrounding the knee by Viano et al. affected tibia displacements at PCL failure. Force-displacement responses from these tests did not exhibit the rise-plateau-rise behavior in applied force-tibia displacement responses

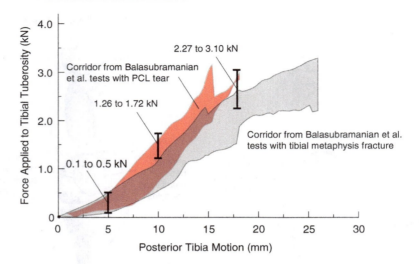

Fig. 17.13 Force-deflection corridors for posterior tibial displacement relative to the femoral condyles from Balasubramanian et al. [83] compared to Hybrid III response specifications (shown as error bars) derived Viano et al. [6] (Adapted from Balasubramanian et al. [83])

observed by Viano et al., even when specimens were tested with the patella and knee flesh removed. Force-displacement responses and tibia displacements at PCL rupture from these tests were similar to those in the reanalyzed Viano data and therefore data from the Balasubramanian et al. and the Viano et al. studies were combined. In this combined dataset, the posterior tibia displacement associated with PCL rupture was 16.8 ± 2.9 mm. It should be noted that neither Balasubramanian et al. nor Viano et al. simulated muscle tension, which will likely alter knee force-deflection characteristics from anterior-to-posterior tibia motion. Both studies also constrained rotation of the tibia, which may affect PCL tolerance.

Figure 17.13 shows force-displacement response corridors developed by Balasubramanian for tests that resulted in tibial metaphysis fracture and tests that resulted in PCL rupture. Specifications for tibia force-posterior tibia displacement for the Hybrid III ATD knee, which were derived from the Viano et al. data, are shown as vertical range bars.

Injuries to the anterior cruciate ligament (ACL) of the flexed knee in frontal motor-vehicle crashes have also been produced from foot loading, which, because of the angle of the tibial plateau, causes the tibia of the flexed knee to translate anteriorly and rotate internally and stretch the ACL [84]. Mean tibial axial compressive force associated with ACL tears from this mechanism was 5.8 ± 2.9 kN. Displacement of the tibia relative to the femoral condyles associated with ACL tear was 9.5 ± 1.6 mm. It should be noted that this scenario for ACL injury is substantially less likely in the presence of anterior constraint of the tibia, because such a constraint increases resistance of the knee to A-P motion [85]. Knee A-P knee stiffness increases with increasing axial compressive force in the tibia, shear force applied to the tibia may not be a reliable injury criterion without consideration of level of axial compressive force in the tibia [85].

17.3.4.1 Lateral Impact

Studies of the tolerance of the fully extended knee to dynamic lateral to medial shear and valgus bending, such as occurs when the lower extremity of a standing/walking pedestrian is impacted by a vehicle bumper, have been performed using whole cadavers, whole cadaver lower-extremity specimens with varying boundary conditions, and isolated knee specimens [72, 83–94].

Early whole body testing used generic vehicle simulators with varying bumper heights and shapes along with varying cadaver knee angle [87, 88]. Results of these studies demonstrated that increasing impact velocity increases the likelihood of lower extremity fracture and decreases the likelihood of ligamentous knee injury primarily because fracture occurs before the displacement necessary to cause injury occurs in higher severity impacts. Tests resulting in ligamentous injury to the struck-side knee tended to produce valgus bending greater than 30° while lower valgus bending angles less than 15° tended to be associated with KTH fracture without ligamentous knee injury.

Studies of the tolerance of the knee to lateral shear and bending have been performed by Kazjer et al. [89, 90] who loaded the knees of 36 cadaver lower extremity and hemi pelvis specimens using a 40 kg impactor at velocities of 15–16 km/h and 20 km/h. All tests used a fully extended knee, a ~400-N downward preload to the hemi pelvis, and supports that restricted lateral motion at the greater trochanter and medial femoral epicondyle. Shear impacts were performed by simultaneously impacting the tibia at the head of the fibula and just above the ankle. Results of these tests indicated that KTH fractures tended to occur earlier in the impact event (~5 ms) while ligament injuries tended to occur later in the event (~15 to 20 ms), after the knee had time to displace inward.

Kazjer et al. [91] conducted additional impacts at higher velocities to the knees of whole cadavers in standing postures that were rigidly supported at the greater trochanter and at the supracondylar region of the medial femur. Methods were otherwise similar to Kazjer et al. [89], except that "shear" impacts were performed by impacting the tibia just below the knee joint in an L-M direction at a much higher velocity (40 km/h) with an impact with much lower mass (6.3 kg). "Bending" impacts were generated by impacting the tibial plafond in medial-to-lateral direction using the same impactor mass and speed. Shear tests produced articular fractures and related ligament damage at 2.6 ± 0.5 kN and 489 ± 141 Nm at the knee joint, respectively.

The large difference in the bending moment at fracture between the Kazjer et al. [90] and Kazjer et al. [91] studies (123 ± 85 Nm vs 489 ± 141 Nm) prompted Kazjer et al. [92] to conduct additional tests to determine if these differences were explained by differences in impact speed, the rigidity of the femoral support, or the generally older cadaver population used in the earlier study. In the Kazjer et al. [92] tests, a younger cadaver population was used with the same methods as Kazjer et al. [91] (rigid support of the femur at the greater trochanter and the medial supracondylar region and impacts with a 6.3 kg impactor just below the tibial plateau and at the ankle). Bending moments at the time of knee fracture in these tests were similar to those of Kazjer et al. [91].

Kerrigan et al. [72], Klinich and Schneider [95] and Bhalla et al. [96] noted multiple issues with the Kazjer et al. studies including that (1) the injuries produced are not representative of pedestrian lower extremity injuries observed in real-world crashes, (2) the bending tolerance of the knee reported in the later Kazjer et al. studies is similar or greater than reported tolerances of the femoral shaft to three-point and four-point bending, and (3) the majority of the injuries in these tests are at the locations where the femur was supported. As a result of these limitations, it has been suggested that Kazjer et al. tests are not appropriate for defining knee ligament tolerance or response.

Subsequent test series have addressed the shortcomings of the Kazjer et al. studies [72, 93, 94]. These tests series eliminated injuries associated with lower-extremity support conditions and provided for improved measurement of forces and moments at the knee joint by testing with isolated full-extended knee-joint preparations and applying displacements and rotations to the fixtures used to mount these preparations. Kerrigan et al. [72] loaded three isolated knees in L-M shear and another three knees using L-M four-point bending. Shear tests produced exclusively ACL injuries and osteochondral defects, which are not representative of pedestrian knee ligament injuries from similar loading in the field [94]. This led subsequent studies to load the knee in

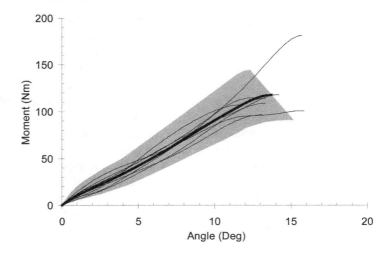

Fig. 17.14 Valgus bending moment-knee angle response corridor developed by Ivarsson et al. (2004) (Adapted from Ivarsson et al. [75])

combined shear and bending, which is thought to be more representative of the knee loading in vehicle bumper to pedestrian impacts [93, 94]. Knee bending tests produced MCL injuries at 134 ± 7 Nm. Results from a subsequent larger series of similar four-point knee bending tests indicate that bending moment at MCL failure is between 90 and 110 Nm [93]. This range is lower than the values reported by Kerrigan et al. because Bose et al. compensated load cell measurements to account for the rotational inertia of the mass between the load cell and knee-joint center, while Kerrigan et al. [72] did not.

Ivarsson et al. [75] developed a response corridor that describes the relationship between valgus bending moment at the knee joint center and knee bending from four-point knee bending tests reported by Bose et al. [93], normalized to midsize male geometry (Fig. 17.14). Ivarsson et al. also analyzed these same data to characterize the probability of knee ligament failure as a function of bending moment at the knee joint center (Eq. 17.10) and knee valgus bending angle (Eq. 17.11). In developing these equations, the bending moment and knee bending angle at failure were identified using acoustic bursts measured by acoustic emission sensors attached to the medial supracondylar femur and the medial tibia at the level of the tibial tuberosity. These relationships produce slightly higher risks for a given bending moment or knee angle than relationships between peak L-M bending moment and peak knee angle and the probability of knee ligament failure also found in Ivarsson et al. [75],

$$p(knee\, ligament\, injury) = 1 - e^{-e^{[4.2297*\ln(m) - 20.4496]}} \quad (17.10)$$

$$p(knee\, ligament\, injury) = 1 - e^{-e^{[10.7804*\ln(\beta) - 28.5412]}} \quad (17.11)$$

where m is valgus bending and moment β is valgus knee bending angle.

Bose et al. [94] performed four-point and three-point L-M bending tests on 40 isolated knee joint preparations in a fully-extended knee posture to explore the effects of bending moment and varying combinations of shear force at the knee on ligament failure. The rates of knee bending in these tests were representative of those produced in 40 km/h impacts. The effect of knee valgus bending angle on the shear displacement required to produce knee failure determined from the Bose et al. experimental data is described by Eq. 17.12.

$$d = \sqrt{1189 - 6851\sin^2(0.28 + 0.0087\alpha)} \quad (17.12)$$

where d is the shear displacement of the knee at failure and α is the valgus bending angle at failure.

17.3.4.1 Isolated Knee Ligament Tolerance

The tolerances and responses of isolated human knee ligaments to dynamic tensile loading have been characterized by multiple sources [e.g., 96–105]. However, only recently have studies used strain rates that are high enough to be relevant to pedestrian impact. Kerrigan et al. [97] loaded bone-ligament-bone MCL and bone-ligament-bone LCL specimens from six cadavers at either 1.6 or 1,600 mm/s to failure. The higher of these loading rates corresponded to strain rates between approximately 30 and 60 s^{-1}, which were thought to be representative of pedestrian impact resulting in injury. An important feature of these tests was that the locations and orientations of the femoral and tibial bone segments were set to reproduce a standing posture with a fully extended knee. Although too few tests were conducted to draw meaningful conclusions on specific failure characteristics, test results suggested differences that the LCL failed at lower forces than the MCL.

van Dommelen et al. [101–103] used the methods developed by Kerrigan et al. [97] to characterize the tolerance and structural response of a larger number of MCL and LCL ligaments as well as the anterior and posterior bundles of both the PCL and ACL over a greater range of strain rates. A substantial rate dependence on ligament material properties was observed with stiffer and more linear responses at higher loading rates. Table 17.2 lists the failure forces and the elongations at failure reported for the knee ligaments. Elongation was defined as the change in the shortest distance between ligament insertions parallel to the long axis of the ligament relative to the initial distance immediately prior to a failure test [95]. All specimens tested by Kerrigan and van Dommelen were from elderly and male cadavers. The tolerance data shown in Table 17.2 may therefore be less applicable to female and/or younger populations, both of which are believed to have different response and/or tolerance [103, 104].

Table 17.2 Failure forces and elongations reported by van Dommelen et al. [101] for isolated ligaments tested at strain rates representative of pedestrian impacts

Ligament	N tests	Failure force (kN)	Elongation at failure
MCL	5	1.40 ± 0.32	0.39 ± 0.09
LCL	4	0.54 ± 0.11	0.18 ± 0.02
aACL	4	0.99 ± 0.56	0.18 ± 0.03
pACL	3	1.00 ± 0.19	0.22 ± 0.03
aPCL	2	0.36–0.94[a]	0.16–0.20[a]
pPCL	2	0.29 ± 0.09	0.14 ± 0.01

[a]Range

References

1. Schneider LW, Rupp JD, Scarboro M, Pintar F, Arbogast KB, Rudd RW, Sochor MR, Stitzel J, Sherwood C, MacWilliams JB, Halloway D, Ridella S, Eppinger R (2011) BioTab—a new method for analyzing and documenting injury causation in motor-vehicle crashes. Traffic Inj Prev 12(3): 256–265
2. Amis AA, Dawkins GP (1991) Functional anatomy of the anterior cruciate ligament: fibre bundle actions related to ligament replacements and injuries. J Bone Joint Surg Br 73(2):260–267
3. Chang CY, Rupp JD, Kikuchi N, Schneider LW (2008) Development of a finite element model to study the effects of muscle forces on knee-thigh-hip injuries in frontal crashes. Stapp Car Crash J 52:475–504
4. Atkinson PJ, Atkinson T, Haut R, Eusebi C, Maripudi V, Hill T, Sambatur K (1998) Development of injury criteria for human surrogates to address current trends in knee-to-instrument panel injuries. In: Proceedings of the 41st Stapp Car Crash Conference, Paper no. 983146. Society of Automotive Engineers, Warrendale, pp 13–20
5. Kuppa S, Fessahaie O (2003) An overview of knee-thigh-hip injuries in frontal crashes in the United States. In: Proceedings of the 18th international technical conference on the enhances safety of vehicles, Paper no. 416. National Highway Traffic Safety Administration, Washington, DC
6. Viano D, Culver C, Haut R, Melvin J, Bender M, Culver R, Levine R (1978) Bolster impacts to the knee and tibia of human cadavers and an anthropomorphic dummy. In: Stapp Car Crash Conference, Paper No. SAE 780896, Proceedings of the 22nd Stapp Car Crash Conference, Society of Automotive Engineers, Warrendale, PA
7. Powell WR, Ojala SJ, Advani SH, Martin RB (1975) Cadaver femur responses to longitudinal impacts. In: Proceedings of the nineteenth Stapp Car Crash Conference, Paper no. 751160. Society of Automotive Engineers, Warrendale, pp 561–579
8. Melvin JW, Stalnaker RL (1976) Tolerance and response of the knee-femur-pelvis complex to axial impact. Report no. UM-HSRI-76-33. The University

of Michigan, Highway Safety Research Institute, Ann Arbor
9. Viano DC, Stalnaker RL (1980) Mechanisms of femoral fracture. J Biomech 13(8):701–715
10. Rupp JD, Reed MP, Jeffreys TJ, Schneider LW (2003) Effects of hip posture on the frontal impact tolerance of the human hip joint. Stapp Car Crash J 47:21–33
11. Ivarsson BJ, Genovese D, Crandall JR, Bolton JR, Untaroiu CD, Bose D (2009) The tolerance of the femoral shaft in combined axial compression and bending loading. Stapp Car Crash J 53:251–290
12. Viano DC, Khalil TB (1976c) Investigation of impact response and fracture of the human femur by finite element modeling. In: Proceedings of the mathematical modeling biodynamic response to impact. SP-412. Society of Automotive Engineers, SAE 760773, pp 53–60
13. Rupp JD (2006) Biomechanics of hip fractures in frontal motor-vehicle crashes. Ph.D. dissertation. The University of Michigan, Ann Arbor
14. Letournel E, Judet R (1993) Fractures of the acetabulum. Springer, New York
15. Wang S, Brede C, Lange D, Poster C, Lange A et al (2004) Gender differences in hip anatomy: possible implications for injury tolerance in frontal collisions. Annu Proc Assoc Adv Automot Med 48:287–301
16. Holcombe S, Kohoyda-Inglis C, Wang L, Goulet J, Wang S, Kent R (2011) Patterns of acetabular femoral head coverage. Stapp Car Crash J 55:479–490
17. Rupp JD, Miller CS, Reed MP, Madura NH, Klinich KD, Schneider LW (2008) Characterization of knee-thigh-hip response in frontal impact using biomechanical testing and computational simulations. Stapp Car Crash J 52:421–474
18. Kent RW, Forman JL, Bolstrom O (2010) Is there really a "cushion effect"?: a biomechanical investigation of crash injury mechanisms in the obese. Obesity 18(4):749–753
19. Turkovich MT (2010) The effects of obesity on occupant injury risk in frontal impact: a computer modeling approach. Dissertation. Available from: http://etd.library.pitt.edu/ETD/available/etd-11222010 140242/unrestricted/Turkovich_Dissertation_Nov23.pdf
20. Reed MP, Ebert-Hamilton SM, Rupp JD (in press) Effects of obesity on seat belt fit. Traffic Inj Prev. doi :10.1080/15389588.2012.659363
21. Banglmaier RF, Rouhana SW, Beillas P, Yang KH (2003) Lower extremity injuries in lateral impact: a retrospective study. In: 47th annual proceedings. AAAM, Lisbon, 22–24 Sept 2003, pp 425–444
22. Kuppa S, Wang J, Haffner M, Eppinger R (2001) Lower extremity injuries and associated injury criteria. In: Proceedings of the 17th international technical conference on the enhanced safety of vehicles, Paper no. 457. National Highway Traffic Safety Administration, Washington, DC
23. Rupp JD, Reed MP, Van Ee CA, Kuppa S, Wang SC, Goulet JA, Schneider LW (2002) The tolerance of the human hip to dynamic knee loading. Stapp Car Crash J 46:211–228
24. Chidester A, Isenberg R (2001) Final report – the pedestrian crash data study. In: Proceedings of the 17th international conference on the enhanced safety of vehicles (ESV), Amsterdam
25. Van Hoof J, de Lange R, Wismans J (2003) Improving pedestrian safety using numerical models. Stapp Car Crash J 47:401–436
26. Mizuno Y, Ishikawa H (2001) Summary of IHRA Pedestrian Safety Working Group activities proposed test methods to evaluate pedestrian protection afforded by passenger cars. In: Proceedings of the 17th International Technical Conference on the Enhanced Safety of Vehicle, Amsterdam, The Netherlands
27. Kerrigan J, Arregui-Dalmases C, Foster J, Crandall J, Rizzo A (2012) Pedestrian injury analysis: field data vs. laboratory experiments, Paper IRC-12-75. In: Proceedings of the 2012 IRCOBI conference Dublin, Ireland
28. Partick LM, Kroell CK, Mertz HM (1965) Forces on the human body in simulated crashes. In: Proceedings of the ninth Stapp Car Crash Conference. University of Minnesota, pp 237–260
29. Partick LM, Mertz HM, Kroell CK (1967) Cadaver knee, chest and head impact loads. In: Proceedings of the eleventh Stapp Car Crash Conference. Society of Automotive Engineers, Warrendale, pp 168–182
30. Powell WR, Ojala SJ, Advani SH, Martin RB (1975) Cadaver femur responses to longitudinal impacts. In: Proceedings of the nineteenth Stapp Car Crash Conference, Paper no. 751160. Society of Automotive Engineers, Warrendale, pp 561–579
31. Melvin JW, Stalnaker RL, Alem NM, Benson JB, Mohan D (1975) Impact response and tolerance of the lower extremities. In: Proceedings of the nineteenth Stapp Car Crash Conference, Paper no. 751159. Society of Automotive Engineers, Warrendale, pp 543–559
32. Morgan R, Eppinger R, Marcus J (1989) Human cadaver patella-femur-pelvis injury due to dynamic loading of the patella. In: Proceedings of the 12th international technical conference on experimental safety vehicles. National Highway Traffic Safety Administration, Washington, DC
33. Melvin JW, Nusholtz GS (1980) Tolerance and response of the knee-femur-pelvis complex to axial impacts—impact sled tests. Final report no. UM-HSRI-80-27. The University of Michigan, Highway Safety Research Institute, Ann Arbor
34. Stalnaker RL, Nusholtz GS, Melvin JW (1977) Femur impact study. Report no. UM-HSRI-77-25. The University of Michigan, Highway Safety Research Institute, Ann Arbor
35. Horsch JD, Partick LM (1976) Cadaver and dummy knee impact response. SAE paper no. 760799. Society of Automotive Engineers, Warrendale
36. Donnelly BR, Roberts DP (1987) Comparison of cadaver and Hybrid III dummy response to axial

impacts of the femur. In: Proceedings of the thirty-first Stapp Car Crash Conference, Paper no. 872204. Society of Automotive Engineers, Warrendale, pp 105–116

37. Chang CY, Rupp JD, Reed MP, Hughes RE, Schneider LW (2009) Predicitng the effects of muscle activation on knee, thigh, and hip injury in frontal crashes using a finite element model with muscle forces from subject testing and musculoskeletal modeling. Stapp Car Crash J 53:475–504

38. Kroell CK, Schnieder DC, Nahum AM (1976) Comparative knee impact response of Part 572 dummy and cadaver subjects. In: Proceedings of the twentieth Stapp Car Crash Conference, Paper no. 760817. Society of Automotive Engineers, Warrendale, pp 585–606

39. Cheng R, Yang KH, Levine RS, King AI (1982) Dynamic impact loading of the femur under passive restrained condition. In: Proceedings of the twenty-sixth Stapp Car Crash Conference, Paper no. 821665. Society of Automotive Engineers, Warrendale, pp 101–118.

40. Cheng R, Yang KH, Levine RS, King AI (1984) Dynamic impact loading of the femur under passive restrained condition. In: Proceedings of the twenty-eighth Stapp Car Crash Conference, Paper no. 841665. Society of Automotive Engineers, Warrendale, pp 101–118

41. Leung YC, Hue B, Fayon A, Tarrière C, Harmon H, Got C, Patel A, Hureau J (1983) Study of "knee-thigh-hip" protection criterion. In: Proceedings of the twenty-seventh Stapp Car Crash Conference, Paper no. 831629. Society of Automotive Engineers, Warrendale, pp 351–364

42. Eppinger RH, Sun E, Bandak F, Haffner M, Khaewpong N, Maltese M, Kuppa S et al (1999) Development of improved injury criteria for the assessment of advanced automotive restraint systems—II. National Highway Traffic Safety Administration, Washington, DC

43. Mertz HJ, Irwin AL, Prasad P (2003) Biomechanical and scaling bases for frontal and side impact injury assessment reference values. Stapp Car Crash J 47:155–188

44. Laituri TR, Henry S, Sullivan K, Prasad P (2006) Derivation and theoretical assessment of a set of biomechanics-based AIS 2+ risk equations for the knee-thigh-hip-complex. Stapp Car Crash J 50:97–130

45. Aoji O, Motoshima T, Bando T (1959) On the effective sectional areas and maximum compressive loads of diaphysis of human long bones. J Kyoto Pref Med Univ 65:979–984

46. Eppinger RH (1978) Prediction of thoracic injury using measurable parameters. In: Proceedings of the 6th international technical conference on experimental safety vehicles. National Highway Traffic Safety Administration, Washington, DC, pp 770–780

47. Rupp JD, Flannagan CA, Kuppa SM (2010) Injury risk curves for the skeletal knee-thigh-hip complex for knee-impact loading. Accid Anal Prev 42(1): 153–158

48. Rupp JD, Reed MP, Miller CS, Madura NH, Klinich KD, Kuppa SM, Schneider LW (2009) Development of new criteria for assessing the risk of knee-thigh-hip injury in frontal impacts using Hybrid III femur force measurements. In: Proceedings of the 21st international technical conference on the enhanced safety of vehicles, Paper 09–0306. National Highway Traffic Safety Administration, Washington, DC

49. Ridella SA, Parent DP (2011) Modifications to improve the durability, usability and biofidelity of the THOR NT dummy. In: Proceedings of the 22nd international technical conference on the enhanced safety of vehicles, Paper 11–0312. National Highway Traffic Safety Administration, Washington, DC

50. Yoganandan NA, Pintar FA, Gennarelli TA, Maltese MR, Eppinger RH (2001) Mechanisms and factors involved in hip injuries during frontal crashes. Stapp Car Crash J 45:437–448

51. Schneider LW, Robbins DH, Pflüg MA, Snyder RG (1983) Development of anthropometrically based design specifications for an advanced adult anthropomorphic dummy family, vol 1. Report no. DOT-HS-806-715. U.S. Department of Transportation, National Highway Traffic Safety Administration, Washington, DC

52. Bass CR, Kent R, Salzar R, Millington S, Davis M et al (2004) Development of injury criteria for pelvic fracture in frontal crashes. In: Proceedings of the 2004 IRCOBI conference. IRCOBI, Bern

53. Rupp JD, Flannagan CA, Kuppa SM (2010) Development of an injury risk curve for the hip for use in frontal impact crash testing. J Biomech 43(3):527–531

54. Cesari D, Ramet M (1982) Pelvic tolerance and protection criteria in side impact. In: Proceedings of the 26th Stapp Car Crash Conference. Society of Automotive Engineers, Warrendale, pp 145–154

55. Nusholtz GS, Alem NM, Melvin JW (1982) Impact response and injury of the pelvis. In: Proceedings of the 26th Stapp Car Crash Conference. Society of Automotive Engineers, Warrendale, pp 103–144

56. Viano D (1989) Biomechanical responses and injuries in blunt lateral impact. In: Proceedings of the thirty-third Stapp Car Crash Conference. Society of Automotive Engineers, Warrendale, pp 113–142

57. Zhu Y, Cavanaugh J, King A (1993) Pelvic biomechanical response and padding benefits in side impact based on a cadaveric test series. In: Proceedings of the 37th Stapp Car Crash Conference. San Antonio, pp 223–233

58. Bouquet R, Ramet M, Bermond F, Cesari D (1994) Thoracic and pelvis human response to impact. In: 14th ESV conference proceedings. Munich, pp 100–109

59. Bouquet R, Ramet M, Bermond F, Caire Y, Talantikite Y, Robin S, Voiglio E (1998) Pelvis human response to lateral impact. In: 16th ESV conference proceedings. Windsor, pp 1665–1686

60. Maltese M, Eppinger R, Rhule H, Donnelly B, Pintar F, Yoganandan N (2002) Response targets of human surrogates in lateral impacts. Stapp Car Crash J Society of Automotive Engineers, Warrendale, PA. 46:321–351
61. Pintar FA, Yoganandan N, Hines MH, Maltese MR, McFadden J, Saul R, Eppinger R, Khaewpong N, Kleinberger M (1997) Chestband analysis of human tolerance to side impact. In: Proceedings of the 41st Stapp Car Crash Conference, Society of Automotive Engineers, Warrendale, PA, pp 63–74
62. Cavanaugh JM, Walilko TJ, Malhotra A, Zhu Y, King AI (1990) Biomechanical response and injury tolerance of the pelvis in twelve sled side impacts. SAE 902305. In: Proceedings of the thirty-fourth Stapp Car Crash Conference, P-236. SAE, Warrendale
63. Lessley D, Shaw G, Parent D, Arregui-Dalmases C, Kindig M et al (2010) Whole-body response to pure lateral impact. Stapp Car Crash J 54:289–336
64. Miller CS, Rupp JD (2011) PMHS impact response in low and high speed near side impacts. UMTRI-2011-10
65. Kallieris D, Mattern R, Schmidt G, Eppinger RH (1981) Quantification of side impact responses and injuries. In: Proceedings of the 25th Stapp Car Crash Conference. SAE, Warrendale, pp 329–366
66. Kuppa S (2004) Injury criteria for side impact dummies. Docket NHTSA-2004-17694. Available from: http://www.regulations.gov/#!documentDetail;D=NHTSA-2004-17694-0004
67. Irwin AI, Sutterfield A, Hsu TP, Kim A, Mertz JJ, Rouhana SW, Scherer R (2006) Side impact response corridors for the rigid flat-wall and offset-wall side impact tests of NHTSA using the ISO method of corridor development. Stapp Car Crash J 49:423–456
68. ISOTR9790 (1999) Road vehicles-lateral impact response requirements to assess the biofidelity of the dummy. Technical report no. 9790. International Standards Organization, American National Standards Institute, New York
69. Rupp JD, Reed MP, Madura NH, Kuppa S, Schneider LW (2003) Comparison of knee/femur force-deflection response of the THOR-NT, Hybrid III, and human cadaver to dynamic frontal-impact knee loading. In: Proceedings of the 18th international technical conference on the enhanced safety of vehicles. National Highway Traffic Safety Administration, Washington, DC
70. Martens M, van Audekercke R, de Meester P, Mulier JC (1986) Mechanical behaviour of femoral bones in bending loading. J Biomech 19(6):443–454
71. Kerrigan JR, Drinkwater DC, Kam CY, Murphy DB, Ivarsson BJ, Crandall JR (2004) Tolerance of the human leg and thigh in dynamic lateral-medial 3-point bending. Technical paper presented at the 2004 international crashworthiness conference, San Francisco
72. Kerrigan JR, Bhalla KS, Madeley NJ, Funk JR, Bose D, Crandall JR (2003) Experiments for establishing pedestrian-impact lower limb injury criteria. Airbags and safety test methodology. SAE no. 2003-01-0895. SAE, Warrendale, pp 191–205
73. Funk JR, Kerrigan JR, Crandall JR (2004) Dynamic bending tolerance and elastic-plastic material properties of the human femur. Annu Proc Assoc Adv Automot Med 48:212–233
74. Kennedy EA, Hurst WJ, Stitzel JD, Cormier JM, Hansen GA, Smith EP, Duma SM (2004) Lateral and posterior dynamic bending of the mid-shaft femur: fracture risk curves for the adult population. Stapp Car Crash J 48:27–52
75. Ivarsson BJ, Lessley DJ, Kerrigan JR, Bhalla KS, Bose D, Crandall JR, Kent RW (2004) Dynamic response corridors and injury thresholds of the pedestrian lower extremities. In: 2004 IRCOBI Conference on the Biomechanics of Impact, Graz, Austria
76. Haut RH (1989) Contact pressures in the patello-femoral joint during impact loading on the human flexed knee. J Biomech 7:272–280
77. Atkinson PJ, Garcia JJ, Altiero NJ, Haut RC (1997) The influence of impact interface on human knee injury: implications for instrument panel design and the lower extremity injury criterion. In: Proceedings of the 41st Stapp Car Crash Conference, Paper no. 973327. Society of Automotive Engineers, Warrendale, pp 167–180
78. Hayashi S, Choi H, Levine R, Yang K, King A (1996) Experimental and analytical study of knee fracture mechanisms in a frontal knee impact. In: Proceedings of the 40th Stapp Car Crash Conference. SAE, Warrendale, pp 161–171
79. Rupp JD, Miller CS, Madura NH, Reed MP, Schneider LW (2007) Characterization of knee impacts in frontal crashes. In: Proceedings of the 20th international technical conference on the enhanced safety of vehicles, Paper 07–0345. National Highway Traffic Safety Administration, Washington, DC
80. Meyer E, Haut R (2003) The effect of impact angle on knee tolerance to rigid impacts. Stapp Car Crash J 47:1–19
81. Aglietti P, Insall JN, Walker PS, Trent P (1975) A new patella prosthesis: design and application. Clin Orthop Relat Res 107:175–187
82. Mertz HJ (1984) Injury assessment values used to evaluate Hybrid III response measurements. NHTSA Docket 74–14, Notice 32, Enclosure 2 of Attachment I of Part III of General Motors Submission USG 2284
83. Balasubramanian S, Beillas P, Belwadi A, Hardy WN, Yang KH, King AI, Masuda M (2004) Below knee impact responses using cadaveric specimens. Stapp Car Crash J 48:71–88
84. Jayaraman VM, Sevensma ET, Kitagawa M, Haut RC (2001) Experimental constraint can affect injury patterns in the human knee during tibial-femoral joint loading. Stapp Car Crash J 45:449–467
85. Meyer EG, Sinnott MT, Haut RH, Jayarama GS, Smith WE (2004) The effect of axial load in the tibia on the response of the 90° flexed knee to blunt

impacts with a deformable interface. Stapp Car Crash J 48:53–70
86. Bunketorp O, Romanus B, Hansson T, Aldman B, Thorngren L, Eppinger RH (1983) Experimental study of a compliant bumper system. In: Proceedings of the twenty-seventh Stapp Car Crash Conference. Report no. SAE 831623. SAE, Warrendale, pp 287–297
87. Cesari D, Cavallero C, Roche H (1989) Mechanisms producing lower extremity injuries in pedestrian accident situations. In: Proceedings of the 33rd annual conference of the Association for the Advancement of Automotive Medicine. AAAM, Des Plaines, pp 415–422
88. Cesari D, Cassan F, Moffatt C (1988) Interaction between human leg and car bumper in pedestrian tests. In: Proceedings of the 1988 international conference on the biomechanics of impacts. IRCOBI, Bron, pp 259–269
89. Kajzer J, Cavallero C, Ghanouchi S, Bonnoit J, Ghorbel A (1990) Response of the knee joint in lateral impact: effect of shearing loads. In: Proceedings of the 1990 international conference on the biomechanics of impacts. IRCOBI, Bron, pp 293–304
90. Kajzer J, Cavallero C, Bonnoit J, Morjane A, Ghanouchi S (1993) Response of the knee joint in lateral impact: effect of bending moment. In: Proceedings of the 1993 international conference on the biomechanics of impacts. IRCOBI, Bron, pp 105–116
91. Kajzer J, Schroeder G, Ishikawa H, Matsui Y, Bosch U (1997) Shearing and bending effects at the knee joint at high speed lateral loading. In: Proceedings of the forty-first Stapp Car Crash Conference. Report no. SAE 973326. SAE, Warrendale, pp 151–265
92. Kajzer J, Matsui Y, Ishikawa H, Schroeder G, Bosch U (1999) Shearing and bending effects at the knee joint at low speed lateral loading. Occupant Protection (SAE-SP-1432). Report no. SAE 1999-01-0712. SAE, Warrendale, pp 129–140
93. Bose D, Bhalla KS, van Rooij L, Millington SA, Studley A, Crandall JR (2004) Response of the Knee joint to the pedestrian impact loading environment, Paper 2004-01-1608. Society of Automotive Engineers
94. Bose D, Bhalla KS, Untaroiu CD, Ivarsson BJ, Crandall JR, Hurwitz S (2008) Injury tolerance and moment response of the knee joint to combined valgus bending and shear loading. J Biomech Eng 130(3) pg. 031008-1 to 031008-8 DOI: 10.1115/1.2907767
95. Klinich KD, Schneider LW (2003) Biomechanics of pedestrian injuries related to lower extremity injury assessment tools: a review of the literature and analysis of pedestrian crash database. Report UMTRI- 2003-25. University of Michigan Transportation Research Institute, Ann Arbor
96. Bhalla K, Bose D, Madeley NJ, Kerrigan J, Crandall J, Longhitano D, Takahashi Y (2003) Evaluation of the response of mechanical pedestrian knee joint impactors in bending and shear loading. In: Proceedings of the 18th international technical conference on the enhanced safety of vehicles. National Highway Traffic Safety Administration, Washington, DC
97. Kerrigan JR, Ivarsson BJ, Bose D, Madeley NJ, Millington SA, Bhalla KS, Crandall JR (2003) Rate sensitive constitutive and failure properties of human knee collateral ligaments. In: 2004 IRCOBI Conference on the Biomechanics of Impact, Lisbon, Portugal
98. Kennedy JC, Hawkins RJ, Willis RB, Danylchuck KD (1976) Tension studies in human knee ligaments. Yield point, ultimate failure, and disruption of the cruciate and tibial collateral ligaments. J Bone Joint Surg Am 58(3):350–355
99. Butler DL, Kay MD, Stouffer DC (1986) Comparison of material properties in fascicle-bone units from human patellar tendon and knee ligaments. J Biomech 19:425–432
100. Arnoux P-J, Cavallero C, Chabrand P, Brunet C (2002) Knee ligament failure under dynamic loadings. Int J Crashworthiness 7(3):255–268
101. van Dommelen J, Minary Jolandan M, Ivarsson BJ, Millington SA, Raut M, Kerrigan JR, Crandall JR, Diduch D (2006) Nonlinear viscoelastic behavior of human knee ligaments subjected to complex loading histories. Ann Biomed Eng 34(6):1008–1018
102. van Dommelen J, Ivarsson BJ, Minary Jolandan M, Millington SA, Raut M, Kerrigan JR, Crandall JR, Diduch D (2005) Characterization of the rate-dependent mechanical properties and failure of human knee ligaments. SAE transactions. J Passeng Cars Mech Syst 114(6):80–90. Based on SAE paper 2005-01-0293
103. van Dommelen J, Minary Jolandan M, Ivarsson BJ, Millington SA, Raut M, Kerrigan JR, Crandall JR, Diduch D (2005) Pedestrian injuries: viscoelastic properties of human knee ligaments at high loading rates. Traffic Inj Prev 6:278–287
104. Chandrashekar N, Mansouri H, Slauterbeck J, Hashemi J (2006) Sex-based differences in the tensile properties of the human anterior cruciate ligament. J Biomech 39(16):2943–2950
105. Woo SLY, Hollis JM, Adams DJ, Lyon RM, Takai S (1991) Tensile properties of the human femur-anterior cruciate ligament-tibia complex. The effects of specimen age and orientation. Am J Sports Med 19(3):217–225

Leg, Foot, and Ankle Injury Biomechanics

18

Robert S. Salzar, W. Brent Lievers, Ann M. Bailey, and Jeff R. Crandall

Abstract

Though rarely life-threatening by themselves, lower extremity injuries are often a debilitating and costly injury affecting the broad populace, and can occur as result of a variety of different injury mechanisms. This chapter details basic lower extremity anthropometry, and reviews the currently available studies and injury criteria available for automotive, sports, and military related injury mechanisms. Detailed analysis and critique of published injury studies are presented.

18.1 Lower Extremity Anatomy

The human lower extremity is comprised of six morphologically distinct regions: the femur head/acetabulum joint, the thigh/femur, the knee/patella, the lower leg (tibia/fibula), the ankle (tibia/fibula/talus joint), and the foot (calcaneous, cuboid, tarsels and metatarsels) (Fig. 18.1). In total, this construct is made up of 29 distinct bones, over 72 articulating surfaces, over 30 synovial joints, over 100 ligaments, and 30 muscle attachments [1]. The tibia and fibula are attached through slightly articulating surfaces at both the proximal and distal end of the bones. The ankle joint is comprised of the talocrural joint, the talocalcaneal joint, the talocalcaneonavicular joint, and the transverse tarsal joint. The talocrural joint contains the distal articular surface of the tibia/fibula and the talus while the talocalcaneal joint is comprised of the three articular facets between the talus and the calcaneus.

The three-dimensional motion of the hindfoot joints (Fig. 18.2) is clinically described as consisting of three anatomical planes of motion. Flexion occurs in the sagittal (X-Z) plane where dorsiflexion moves the superior (i.e. dorsal) surface of the foot towards the axis of the leg and plantarflexion moves the inferior (i.e. plantar) surface of the foot towards the axis of the tibia. Inversion and eversion occurs in the coronal (Y-Z) plane where inversion moves the plantar surface towards the median plane and eversion

R.S. Salzar, Ph.D. (✉)
Center for Applied Biomechanics,
University of Virginia, Charlottesville, VA, USA
e-mail: Salzar@virginia.edu

W.B. Lievers, Ph.D.
Bharti School of Engineering, Laurentian University, Sudbury, ON, Canada
e-mail: blievers@laurentian.ca

A.M. Bailey, BS • J.R. Crandall, Ph.D.
Center for Applied Biomechanics,
University of Virginia, Charlottesville, VA, USA
e-mail: Amb9um@virginia.edu; jrc2h@virginia.edu

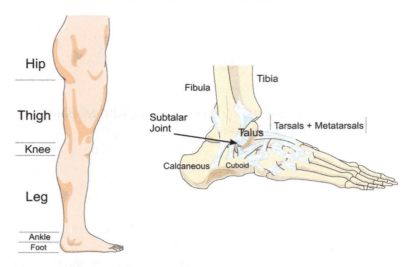

Fig. 18.1 Anatomy of the lower extremities

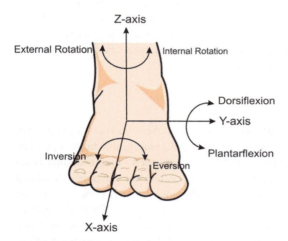

Fig. 18.2 Anatomical motions and coordinate axes of hindfoot joints

moves the plantar surface away from the median plane. Rotation of the foot occurs in the transverse (X–Y) plane where internal rotation moves the toes away from the median plane. During motion of the hindfoot, the joints of the ankle/foot complex work together rather than acting in isolation. While the hindfoot joints are often described as hinge joints, the true axis for this motion is not contained within any anatomical plane and changes orientation over the complete range of motion. In particular, flexion of the hindfoot joints is coupled to both rotation and inversion/eversion of the foot.

Foot and ankle anthropometry was collected in a study by Crandall et al. [1] from detailed measurements of cadavers tested at the University of Virginia (UVA) and Renault, as well as from a volunteer study conducted by Parham et al. [2] for the United States Army, and a compilation of studies by Diffrient et al. [3]. Measurements from the University of Virginia study were based on 39 male cadavers with an average mass of 77.5±18.3 kg and an average stature of 175±7.6 cm, and from Renault, based on 11 male cadavers, with an average stature of 175±12 cm. The study by Parham et al. (n=293) consisted of male soldiers with an average stature of 175.6±7.1 cm and average mass of 75.7±11.4 kg. The Diffrient et al. study was composed of volunteers with an average stature of 174.8 cm. A summary of detailed foot and ankle dimensions is shown in Table 18.1 (Fig. 18.3).

18.2 Automotive-Related Injuries

Lower limb injuries sustained during moderate-to-severe automobile collisions have occurred throughout history of the automobile as a result of toe-pan intrusion alone. Before the advent and wide-spread availability of air-bag equipped vehicles, the occupant was usually killed from head and thoracic trauma, so lower limb fractures were never well documented. As the number of

Table 18.1 Dimensions of the human foot and ankle

Label	Measurement	Study	Dimension (cm)
A	Foot length	UVA [1]	24.4 ± 1.5
B	Ball length (heel to head of 5th metatarsal)	Diffrient [3]	16.3
C	Heel width	Parham [2]	7.0 ± 0.44
D	Ball length (heel to head of first metatarsal)	Diffrient [3]	19.6
E	Foot breadth at metatarsal-phalangeal joints	Parham [2]	10.5 ± 0.56
F	Medial malleolus height form head to floor	UVA [1]	8.3 ± 1.1
	Medial malleolus height from tip to floor	UVA [1]	8.3 ± 1.3
G	Lateral malleolus height form head to floor	Parham [2]	7.2 ± 0.29
	Lateral malleolus height from tip to floor	UVA [1]	6.9 ± 1.2
H	Ankle width at level of medial malleolus	Diffrient [3]	7.6
I	Soft tissue thickness from posterior heel to calcaneous	UVA [1]	0.8 ± 0.4
J	Soft tissue thickness from distal heel to calcaneous	UVA [1]	1.6 ± 0.4
K	Plantar arch height from floor	Parham [2]	3.03 ± 0.60
L	Ankle length from heel to front of ankle (tibia)	Parham [2]	10.8 ± 0.72
M	Heel to head of lateral malleolus	Diffrient [3]	6.6
O	Tibia height from distal heel to tibial medial margin	UVA [1]	47.0 ± 4.07

air-bag equipped vehicles increased in the fleet, the number of collision survivors with serious lower limb injuries has seemingly increased. In fact, as the result of air-bags, injuries to the lower extremities have become the most frequent "non-minor" injury to happen in a frontal collision.

Trauma to the foot and ankle make up approximately 30 % of these non-minor lower extremity injuries, and approximately 10 % of all non-minor injuries. Because of the frequency and severity of these lower limb injuries, significant research has been conducted over the last 20 years to understand the etiology and prevention of these injuries.

Similarly, there are approximately 15,000 hip injuries in the United States each year due to frontal crashes [4]. The study of Kuppa and Fessahaie [5] emphasizes that knee-thigh-hip injuries are the most common form of lower extremity injuries and are also the most life threatening due to the relatively high risk of severing an artery from the broken bones of the femur or pelvis.

Overall, the lower extremities represent the anatomic region most frequently injured in automobile collisions, and 20–30 % of these injuries occur at the knee as a result of contact with the knee bolsters of an automobile. Although less lethal than injuries to the head and neck, knee injuries result in significant recovery time and health care costs. Knee bolsters of varying stiffness are now common among automakers to reduce the load transmitted to the knee during a collision. Though protective of the knee, incorrect design of the knee bolster can shift the injury to the femur or hip.

As outlined in the beginning of this chapter, the biomechanics of the foot, ankle, and lower tibia/fibula involves a complex system of bones, ligaments, and load paths that depend not only on boundary conditions, but load rate and energy delivered to the foot as well.

18.2.1 Tibia/Fibula Response and Failure

The following section provides a review of both the publicly available literature and previous testing on the bending and axial load response and tolerance of the tibia-fibula complex. This section is meant to highlight the existing injury criteria and data available for finite element validation. This review of the leg/tibia/fibula is only concerned

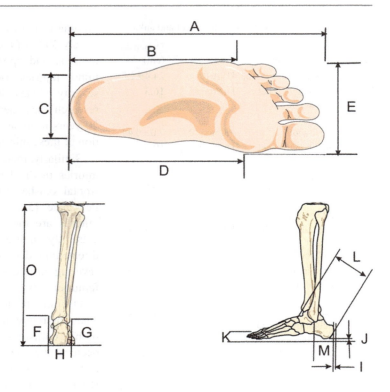

Fig. 18.3 Physical dimensions of the human foot and ankle

with whole bone/segment response to bending and axial load. Testing on the response and failure of the tibial plateau/condyles, the tibial plafond, and lateral and medial malleloli are discussed in the knee and ankle joint validation sections. For human finite element model validation of whole bone tibia/fibula and leg segment models, both the fracture tolerance and the structural bending response are discussed in the following sections.

18.2.1.1 Tibia/Fibula/Leg Fracture Tolerance in Bending

Numerous studies have been conducted to determine the structural fracture tolerance of the human leg in bending (Table 18.2). The studies report a significant variation in fracture tolerance that is likely due to both the wide variety of test conditions applied in the tests as well as variation in the human tissue specimens used. Injury risk functions that account for the applicable covariates are ideal for FE model validation of fracture/failure tolerance since they permit the application of conservative estimates, with statistical bases, to model simulations.

Loading rate is used as a primary evaluation criterion for previous studies. Some of these studies report results from tests conducted at quasi-static loading rates, which may significantly underestimate the bending tolerance of the femur under impact loading [17, 18]. Carter and Hayes [17] estimated that both strength and modulus of cortical bone were proportional to $\dot{\varepsilon}^{.06}$, where $\dot{\varepsilon}$ is the strain rate. Additionally, Mather [8] and Martens [19] suggest that bone exhibits an increase in strength and stiffness at higher strain rates during dynamic loading, and Schreiber et al. [12] showed a dramatic difference in bending tolerance of the human leg between quasi-static tests and dynamic tests. While this suggests bone may experience a strain rate sensitivity of fracture tolerance, this effect may be smaller than inter-specimen variation when strain rates vary by less than an order of magnitude. For an order of magnitude estimate, unpublished data from vehicle-pedestrian impact experiments performed using strain gage instrumented PMHS at the University of Virginia suggest that the tibia can experience strain rates from 100 %/s to as high as

Table 18.2 Previous studies on the failure tolerance in bending of the tibia/fibula/leg

Study	Sample size	Fracture tolerance	Notes
Weber 1859 in Nyquist 1986 [6]	9 (4 M, 5 F)	165/125 Nm (M/F)	Quasi-static, tibia only
Messerer 1880 in Nyquist 1986 [6]	12 (6 M, 6 F)	207/124 Nm (M/F)	Quasi-static, tibia only
Mather 1967 [7]	16	3,244 (+/−783) N	Quasi-static, tibia only
Mather 1968 [8]	318	78.6 J/63.1 J (M/F)	Dynamic, energy required to fracture half of specimens tested
Motoshima 1960 in Yamada 1970 [9]	35	182 Nm	Quasi-static, females 5/6 as strong as males, bending tolerance same in all directions
Nyquist et al. 1985 [10]	10/9 (A-P/L-M bend)	300/317 Nm (AP/LM)	Dynamic (2.1–4.7 m/s), Force data are filtered to cutoff of 100 Hz, L-M direction, with flesh and fibula
Rabl et al. 1996 [11]	16/5/5/6 (Ant/Lat/Med/Pos)	5,787 N/6,430 N/6,117 N/4,556 N	Dynamic (0.3 m/s), moment/span not reported, constrained bone end boundary condition
Schreiber et al. 1997 [12]	10/12 (Q/D)	241/408 Nm (Q/D)	Quasi-static and dynamic (5.6 m/s), P-A direction
Kerrigan et al. 2003a [13]	8 M (4 pairs)	310 Nm	1 Quasi-static, 7 dynamic (1.45 m/s), L-M direction, 1 with flesh and fibula and 7 only bare tibiae
Kerrigan et al. 2003b [14]	4/6 (Q/D)	168/378 Nm (Q/D)	Quasi-static and dynamic (1.6 m/s), L-M direction, with flesh and fibula
Kerrigan et al. 2004 [15]	19	50 % Risk for 50th percentile Male: 312 Nm	Dynamic, L-M direction, with flesh and fibula
Ivarsson et al. 2005 [16]	4/4 (distal/proximal loading)		Dynamic, L-M direction, with flesh and fibula

700 %/s depending on pedestrian stance and vehicle geometry. Thus, tests where specimens were loaded only quasi-statically, will not be considered in the following review.

Mather [8] tested a large number of tibiae in dynamic drop testing with different impact energies to determine which proportion of specimens failed at each level of impact energy. No moment or force calculations were made and thus the results of the study cannot be used in the development of injury risk functions.

In Rabl et al. [11], the authors' goals were to gain basic information on the fracture behavior and fracture morphology of the tibia and to determine the relevance of the results for general practice. Thirty-two human tibiae (16 matched pairs) were loaded in bending, with 8 specimens each loaded on each of the 4 anatomical faces of the bone (posterior, anterior, lateral, and medial). The authors constrained the ends of the bone, which likely influenced the wide range of fracture tolerances presented (2,475–12,206 N).

The authors do not report span length or moment data, nor do they report any anthropometrical measurements of the specimens (age, gender, stature, and weight were provided). The specimens were loaded at 0.3 m/s and data was sampled at 600 Hz from an accelerating load cell that was not inertially compensated. Additionally, only 20 cm long sections of tibiae were used, harvested from the specimen 6 cm distal to the tibial plateau regardless of specimen length. Thus, results from their study cannot be used to develop injury risk functions for bending loading of the tibia/fibula/leg. However, interestingly, despite mid-shaft loading of all specimens, the study reports distal third fractures in almost 60 % of the anterior loading cases, as well as in 20 % of cases with medial loading and 20 % of cases with lateral loading.

Almost all of the remaining studies (other than Nyquist et al. [10]) were performed by the University of Virginia [12–16]. Among these studies, only Schreiber et al. [12] presents data for leg bending in the posterior-anterior (P-A) direction, while the rest present data for bending in the lateral-medial (L-M) direction. Studies presenting data on the tolerance of the leg/tibia/fibula under L-M loading will be discussed first, followed by studies discussing the tolerance in P-A direction.

18.2.1.2 Dynamic Bending Tolerance of the Human Leg/Tibia/Fibula Under L-M Loading

In Kerrigan et al. [13], four matched pairs (all male) of leg specimens were loaded in L-M three point bending (Fig. 18.4). The epiphyses of each specimen were potted in rigid potting cups and attached to rollers. The rollers rested on greased support plates to create a simple support boundary condition at each end. The central loading was provided by a padded aluminum ram attached to the actuator of a servohydraulic materials testing machine able to deliver downward displacement pulses at controllable rates as high as 1.6 m/s. Each of the four matched pair bending tests was used to investigate the influence of a particular parameter on the bending strength of the leg/tibia (Table 18.3). In the first two pair

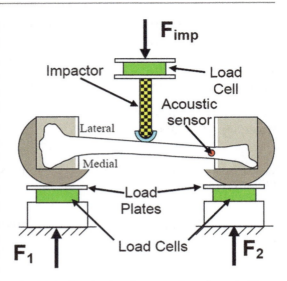

Fig. 18.4 Tibia/leg bending test setup schematic (From the work of Kerrigan et al. 2003a)

of specimens, the influence of location of the central loading point was investigated by comparing the fracture tolerance of the tibia loaded at the mid-shaft versus loading at the proximal third location (first pair), and mid-shaft versus loading at the distal third location (second pair). The distal third and proximal third loading locations were defined by measuring to a point located 33 % of the bone's length measured from the distal tip of the medial malleolus and from the proximal tip of the tibial spine. The fracture tolerance in bending was slightly higher in proximal third loading than in mid-shaft loading in the first pair, and slightly higher in mid-shaft loading than in distal third loading in the second pair. The effect of rate sensitivity was investigated using the third pair of specimens, both loaded at the proximal third location; however, one was loaded at 0.001 m/s while the other was loaded at the dynamic rate. A larger difference in fracture tolerance was reported in the pair loaded at rate than in either of the loading location pairs. In the fourth pair, one denuded tibia was loaded to failure at the mid-shaft and the other specimen with tissue intact (fibula and surrounding flesh were intact, including proximal and distal tibio-fibular joints). The greatest difference in fracture tolerance was noted in this pair: the denuded tibia

Table 18.3 Tibia/leg bending test data [13]

Spec.	Impact location	Tibia length[a] (mm)	Mid shaft circum.[b] (mm)	Span (mm)	Proximal moment arm (mm)	Distal moment arm (mm)	Moment arm ratio[c]	Summed support load[d] (N)	Peak moment (Nm)
122 L	Mid diaph.	385.0	320.0	343.2	172.1	171.1	1.01	−3,842	−333.8
122R	Prox ⅓	385.0	320.0	337.8	102.2	235.6	0.43	−4,644	−342.1
140 L	Mid diaph.	405.0	355.0	363.4	181.7	181.7	1.00	−2,656	−251.7
140R	Dist ⅓	410.0	355.0	325.6	222.3	103.3	2.15	−2,458	−224.7
150 L[e]	Prox ⅓	405.0	390.0	352.5	113.1	239.4	0.47	−3,549	−283.4
150R	Prox ⅓	390.0	390.0	351.1	108.3	242.8	0.45	−4,152	−329.8
134R	Mid diaph.	418.0	290.0	368.4	182.1	186.3	0.98	−3,017	−301.5
134 L	Mid diaph.[f]	426.0	315.0	368.6	183.8	184.8	0.99	−4,276	−416.5
Mean		403.0	342.0	351.3				−3,574.5	−310.4
SD		15.3	36.5	15.3				−796.7	−59.5

[a]Tibia length is defined as the distance from the most proximal point of the tibia spine to the most distal point of the medial malleolus parallel to the longitudinal axis of the bone
[b]Value taken before flesh was removed
[c]Proximal moment arm divided by distal moment arm
[d]Value taken at the time of fracture initiation
[e]Specimen tested with quasi-static loading rate
[f]Bone tested with flesh around bone still intact

failed at a bending moment less than 75 % of the failure moment in the leg segment case. This result highlights the importance of not only validating whole bone bending response of the tibia and fibula, but that the entire leg specimen (tibia, fibula and surrounding flesh) bending tolerance should be as well.

A second study was performed by the University of Virginia [14], where 10 whole leg specimens (tibia, fibula, and surrounding flesh) were loaded in L-M three point bending (Fig. 18.5). This study was performed to highlight the rate sensitivity of the bending tolerance and response of whole leg specimens: six specimens were loaded in dynamic bending at 1.6 m/s, and four specimens were loaded in quasi-static bending. The tests were performed using identical hardware and equipment to those performed in the other University of Virginia study [13]. The bending responses were scaled and used to develop a quasi-static and a dynamic response corridor.

In the third University of Virginia study [15], previously published data were combined with data from new experiments in an effort to create an injury risk function for the dynamic L-M bending tolerance of the leg. The new tests presented in this study included four (no matched pairs) leg

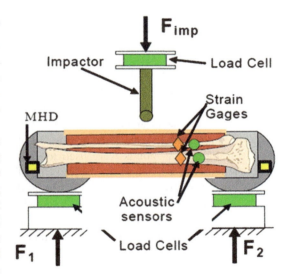

Fig. 18.5 Test diagram for three point bending tests (From the work of Kerrigan et al. 2003b)

specimen bending tests performed identically to those presented in the other two Kerrigan et al. [13, 14] studies. The previously published data that were included in the development of the injury risk function were the single leg specimens from Kerrigan et al. [13], the six dynamically loaded leg specimens from Kerrigan et al. [14], and nine tests published by Nyquist et al. [10].

Nyquist et al. [10] performed a series of leg/tibia bending experiments that included nine dynamic (2.1–4.7 m/s) L-M impacts to simply supported leg specimens (with tibia, fibula and surrounding flesh). Nyquist et al. [10] reported that force data were filtered at 100 Hz to "smooth the traces and slightly attenuate the short duration spikes". Upon examination of the filtered and unfiltered data, it was concluded that the filtered data attenuated the peak forces by approximately 10 %. Nyquist also reported that there was no evidence to suggest that tibial fracture occurs at the peak force, so only filtered values are presented in the paper.

Thus for the purpose of the injury risk function development in Kerrigan et al. [15], all of the data from Nyquist et al. [10] were considered to be right-censored (fracture happens at a force or moment greater than the published values) and data from the other previous University of Virginia studies were to be considered exact, or uncensored (for more information on data censoring or injury risk function development see Kent and Funk [20]). The (moment-based) fracture tolerance of each of the 20 specimens was geometrically scaled to represent the 50th percentile male fracture tolerance using a whole bone length scale factor and 379 mm as the 50th percentile male tibia length. After scaling, 19 of the 20 available data points were used to construct the injury risk function for the leg. One of the Nyquist et al. [10] data points was not used because after scaling the fracture tolerance (634 Nm) was 2.5 standard deviations from the mean of the Nyquist et al. [10] data, 3.2 standard deviations away from the mean of all of the data in Table 18.2, and lastly because it was a right-censored point.

A parametric injury risk curve was developed from a Weibull univariant survival model that was fit to the scaled fracture moment data (8 right censored points and 11 exact points) to determine the relationship between the probability of fracture and peak applied moment. A Weibull model was chosen because it is not necessarily symmetric, permits a zero risk value at zero moment, and has been shown to be no worse than other distributions for modeling biomechanical injury data [20].

The injury risk function for mid shaft L-M bending of the leg is given with 95 % confidence intervals in Fig. 18.6.

In the fourth University of Virginia study [16], the results of eight more dynamic L-M leg (tibia, fibula and surrounding flesh) bending experiments were presented. Four of the specimens were loaded at the distal third location and the remaining four specimens were loaded at the proximal third location. While the scaled (to 50th percentile male) bending tolerance data are presented in this study, no statistical analyses were performed due to the low number of specimens tested in each bending scenario.

In terms of bare bone (tibia or fibula) bending experiments the only available tolerance data for specimens loaded dynamically in the L-M direction were performed by the University of Virginia. However, none of the data have large enough sample sizes to determine injury risk functions with reasonable confidence intervals. Dynamic tolerance for seven bare tibiae loaded in the L-M direction were presented in Kerrigan et al. [13], and the University of Virginia has unpublished data for bending tests performed on the fibulae separated from the tibiae tested for the Kerrigan et al. [13] study. The fibula bending experiments were performed in the L-M direction where matched pairs were used in a manner similar to the tibia tests from Kerrigan et al. [13] to evaluate the influence of loading location and loading rate.

18.2.1.3 Dynamic Bending Tolerance of the Human Leg/Tibia/Fibula Under P-A Loading

Nyquist et al. [10] conducted A-P bending tests on leg specimens in addition to the L-M tests. Nyquist concluded that the tibia failure moment does not depend on the loading direction when comparing A-P bending (avgs. 320 Nm male, 280 Nm female) to L-M bending (avgs. 330 Nm male and 264 Nm female). Despite the shortcomings of the data (see previous section), Nyquist confirmed the earlier finding by Yamada [9] that bending strength was independent of loading direction.

In one study, Schreiber et al. [12] conducted dynamic (5.55 m/s impact velocity) P-A three-point bending tests of the leg complex for 12 intact

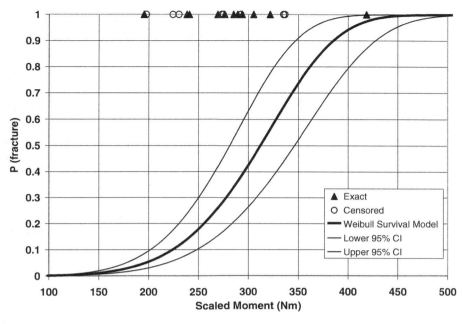

Fig. 18.6 Injury risk function and 95 % confidence interval for the human leg loaded dynamically at the mid shaft in three-point bending (Reproduced with permission from Kerrigan et al. 2004)

leg specimens (Table 18.4). For comparison, six additional tests were conducted in A-P direction using identical test conditions (see Ref. [21] and Table 18.4). A Student's t-test of the P-A and A-P loading scenarios showed no statistically significant differences with a 95 % level of confidence. Crandall et al. [21] concluded that by combining the University of Virginia data with the results of Nyquist et al. [10], the data indicated that the tibia has approximately the same bending moment to failure for all loading directions.

The concept behind the tibia index comes from classical beam theory that involves a prismatic beam subjected to a combination of bending and compression. The form of the tibia index is closely related to the formula for the response of a beam-column:

$$\frac{P}{P_{cr}} + \frac{C_m M}{M_P\left(1 - \dfrac{P}{P_e}\right)} \leq 1 \quad (18.1)$$

where P is the axial compression force, M is the applied bending moment across the beam, C_m is an empirically derived coefficient, M_P is the plastic moment capacity ($M_p = \sigma_y Z_P$) where σ_y is the yield strength of that material and Z_P is the plastic section modulus. P_{cr} is the compression capacity of the beam, or Euler load if the slenderness ratio is sufficiently large.

Following the form of this equation, Mertz [22] developed the tibia index as an injury criterion to aid in the design of injury mitigation strategies for lower limb injuries. In this case, the beam-column relation was reworked as the tibia index relation:

$$\frac{F}{F_{cr}} + \frac{M}{M_{cr}} \leq 1 \quad (18.2)$$

where F is the axial force in the tibia, M is the bending moment in the tibia, and F_{cr} and M_{cr} are the critical axial force and bending moment. If the tibia index exceeds the value of 1.0, a tibia injury is predicted.

Using tibia properties derived by Yamada [9] and scaling them to different anthropometries, Mertz determined the critical compression force and moments for various occupants (Table 18.5).

Since the introduction of the tibia index to the biomechanics community, one study [12] has

Table 18.4 Dynamic three point bending test results [21]

Leg	Energy to failure (J)	Mid-shaft moment (Nm)	Age (years)	Body mass (kg)	Fracture location
Antero-posterior bending [21]					
98-FF-14-RL	6.73	265.28	63	55.8	Mid-shaft
98-FF-7-LL	6.37	341.03	72	74.8	Dist. ⅓
98-FF-10-EXT-LL	12.46	340.88	86	66.2	Mid-shaft
98-FF-8-EXT-LL	11.55	425.96	77	69.4	Mid-shaft
98-FF-14-LL	6.24	260.13	63	55.8	Mid-shaft
98-FM-94-RL	54.27	736.64	65	106.1	Mid-shaft
Average	16.27	394.99	71.00	71.35	
Std. dev.	18.82	178.07	9.23	18.62	
Posteroanterior bending [12]					
48-L	32.4	239	83	61.3	Mid-shaft
1004-R	65.8	535	59	59.9	Mid-shaft
1003-R	67	577	77	69.5	Mid-shaft
1000-R	77.6	458	85	105	Mid-shaft
1005-R	57.9	445	75	73.6	Mid-shaft
1006-R	68.7	372	70	80.4	Mid-shaft
1002-R	55.1	259	70	57.9	Mid-shaft
1010-R	67.3	440	55	73.9	Mid-shaft
50-R	40.2	371	68	46.7	Dist ⅓
68-R	120.4	424	56	72.6	Mid and Dist ⅓
69-R	116.7	534	62	90.3	Mid and Dist ⅓
73-L	54.17	242	61	57.2	Mid-shaft
Average	68.61	408.00	68.42	70.69	
Std. dev.	26.48	115.48	10.13	15.97	

Table 18.5 Tibia index critical forces and moments for different anthropometries

Critical value	Small female	Midsize male	Large male
F_{cr} (kN)	22.9	35.9	42.2
M_{cr} (Nm)	115	225	307

evaluated the validity of the criterion for combined axial and bending loads. This study combined a dynamic bending load and a static compression load and produced a decrease in bending moment by 19 % with an averaged scaled failure moment for a 505 male of 372 Nm. This appears to confirm that the interaction of the bending and compression forces combine to affect the fracture tolerance of the tibia.

Crandall et al. [21] reviewed all subsequent failure tests involving the tibia and offered revised tolerance values for the tibia index. Specifically, this review found that axial and bending tolerance for quasi-static loads should be approximately 10.3 kN and 240 Nm, respectively. Assuming that the fibula can handle 10 % of the static axial load of the tibia, the critical static force for the tibia index increases to 11.3 kN. Review of dynamic bending tests suggest a dynamic bending tolerance of 450 Nm. Using the static-to-dynamic scaling presented by Schreiber et al., the dynamic threshold of 16.6 kN is obtained without the fibula, and 18.3 kN with the fibula.

18.2.2 Ankle Injury

As previously noted, axial loading of the ankle complex produces the most serious injuries, but the most frequent injuries (i.e., malleolar fractures) are produced by a number of different mechanisms. Since we could not eliminate any loading modes, we have included a detailed summary of the literature along with recommended studies to be used for model validation.

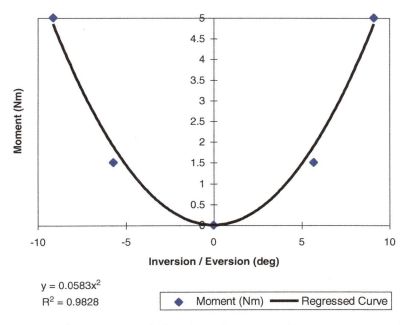

Fig. 18.7 Subtalar joint stiffness curve: inversion/eversion (From the work of Kjaersgaard-Andersen et al. 1988)

18.2.2.1 Xversion Injuries

For simplification purposes, the ankle and subtalar joints have been linked in this chapter for the major motions of the hindfoot. Between the talus and calcaneus lies the subtalar joint, which has the primary function of enabling xversion rotation of the hindfoot. A summary of the response and injury data in both quasi-static and dynamic testing is provided.

Quasi-static Xversion Response

Investigations by the automobile safety community are the only studies with great detail on the quasi-static stiffness of the subtalar joint. Kjaersgaard-Anderson et al. [23] reported the range of motion of the human hindfoot for varying degrees of dorsiflexion. Maximum values for the range of motion used were the result of a 1.5 Nm moment. This moment is much lower than the moment expected in the subtalar joint during toepan intrusion. That being the case, a stiffness function for the subtalar joint was derived that contained the data point of Kjaersgaard-Anderson et al. The actual hindfoot resistance to motion in adduction/abduction and Z-axis rotation was assumed to be a parabolic function of the joint angle about the respective axis. The curve used to represent the resistance to joint motion was calculated as a second order polynomial that intersected the 0° – 0 Nm point and the point measured by Kjaersgaard-Anderson et al. (Fig. 18.7). This stiffness function was assumed for all angles of ankle flexion.

More recently, several studies have been performed by researchers in the automobile safety field for development of response corridors and injury thresholds. Parenteau et al. [24] performed quasi-static inversion and eversion tests on 16 cadaver specimens. The calcaneus of each leg specimen was rigidly fixed with screws and the foot was rotated about a fixed anterior-posterior axis. Petit et al. [25] performed quasi-static inversion and eversion tests on 15 cadaver specimens using a test methodology similar to Parenteau et al. [24], but with 400 N of tension applied to the Achilles tendon to simulate emergency braking. Tests were terminated at 45° of rotation, so not all specimens were injured.

Jaffredo et al. [26] developed a unique test fixture that applied dynamic moments about a prescribed xversion axis. The fixture was developed by the Centre Européen d'Etudes de Sécurité et d'Analyse des Risques (CEESAR) to

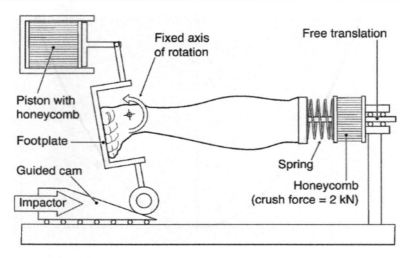

Fig. 18.8 Test apparatus developed by Jaffredo et al. [26] and used subsequently by Funk et al. [33] (Reproduced with permission from the Stapp Association)

dynamically (~1,000°/s) apply pure moments about the inversion-eversion axis of the foot through the subtalar joint center. The test apparatus incorporated an adjustable footplate free to rotate about an offset, fixed vertical axis. Rotation of the footplate was driven by a pneumatic impactor striking a guided cam on one side of the plate and controlled by crushable honeycomb inside a piston on the other side (Fig. 18.8).

The striking side of the test apparatus could be reversed to alter the direction of rotation of the footplate. The amount of footplate rotation was prescribed to either 25° or 48° using different sized cams. In some tests, specimens were initially positioned in 14° of inversion or eversion so that a lower (11°) or higher (62°) final level of rotation could be achieved. In a subsequent study, Funk et al. [27] used the CEESAR apparatus and modified the fixture to include initial dorsiflexion and axial loading of the ankle complex. In a real world car crash severe enough to cause injury, the foot/ankle complex may experience significant axial loading and dorsiflexion [28]. These additional loading factors may influence the response, failure strength and injury mode of the foot/ankle complex in inversion and eversion. Thus, Funk et al. [27] attempted to examine the effects of axial preloading and dorsiflexion on the injury tolerance of the human ankle/subtalar joint in dynamic inversion and eversion.

The combined UVA-CEESAR data set consists of 44 tests, including 29 tests conducted at the University of Virginia and 15 tests conducted at CEESAR. Of the 15 CEESAR tests, the 9 non-injury tests were reported by Jaffredo et al. [26]. Given the comparable test apparatus, the results of Jaffredo et al. and Funk et al. appear similar for the case of no axial load or dorsiflexion.

In an effort to summarize the data for FE modeling, composite response graphs for both inversion and xversion have been developed based on all of the available response data and are shown in Figs. 18.9 and 18.10. Since there was a question in terms of the timing of the fracture and the validity of the response beyond the first acoustic burst (potentially denoting fracture), the response curves have been plotted using both the absolute peak moment as well as the moment-angle response until the time of first acoustic burst.

Dynamic Xversion Injury

The tolerance of the ankle/subtalar joint to inversion and eversion loading has been studied, but limited data is available that is appropriate for use

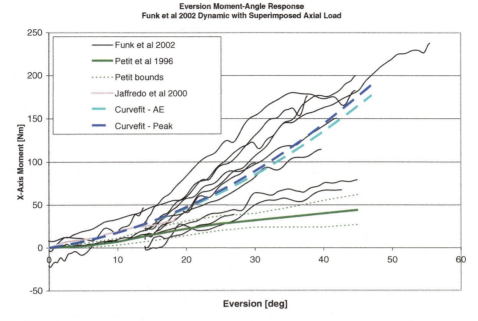

Fig. 18.9 Eversion response showing a comparison of the available quasi-static and dynamic test data

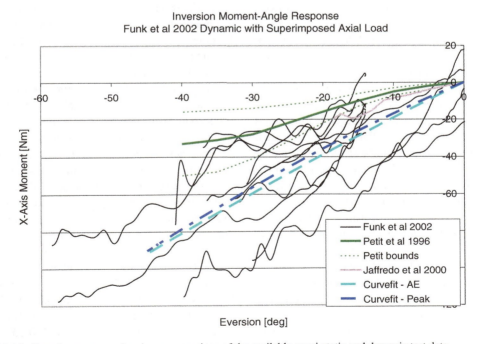

Fig. 18.10 Eversion response showing a comparison of the available quasi-static and dynamic test data

Table 18.6 Summary of Xversion testing for injury prior to Jaffredo et al. [26] and Funk et al. [33]

Investigator	Loading conditions	Directions	Sample size	Failure moment (Nm)	Failure angle
Begeman et al. [32]	Dynamic (~2,000°/s) 700 N axial load	Inversion	n = 6	33 ± 7	61° ± 6°
		Eversion	n = 7	41 ± 14	59° ± 6°
Parenteau et al. [24]	Quasistatic (~6°/s) No axial load	Inversion	n = 8	34 ± 15	34° ± 8°
		Eversion	n = 8	48 ± 12	32° ± 7°
Petit et al. [25]	Quasistatic (~7°/s) 400 N Achilles tension	Inversion	n = 5	40 ± 26	34° ± 8°
		Eversion	n = 5	35 ± 9	32° ± 7°

in model development and validation. Several investigators have generated ankle injuries experimentally in cadaver specimens to study injury patterns, but have not measured failure strength [29–31]. Begeman et al. [32] performed dynamic inversion and eversion tests on 18 cadaveric leg specimens. A plate was attached to the foot of each specimen and struck by an impactor at an off-axis location 50 mm medial or lateral to the ankle center. The impact generated peak axial loads ranging from 270 to 1,300 N (average = 700 N) as well as a moment about the ankle joint. Ligaments were exposed and stained with a dye so that the time of injury could be determined from high-speed video. Injuries were generated in 13 specimens and consisted primarily of collateral ligament tears and avulsions, as well as 3 malleolar fractures. Peak ankle moments ranged from 9 to 120 Nm, with a mean failure moment of 33 ± 7 Nm in inversion and 41 ± 14 Nm in eversion (Table 18.6). High-speed video revealed that ligaments began tearing before fracture in some tests, but that fracture occurred at the time of peak moment. In spite of this finding, Begeman et al. [32] claimed that peak moment was not a good predictor of injury, and suggested that the best threshold of injury was 60° ± 6° degrees of inversion or eversion, as measured by the rotation of the footplate relative to the tibia. For the Parenteau et al. [24] tests, every test produced injury (primarily collateral ligament tears and avulsions, as well as two malleolar fractures). Failure was defined by audible joint damage or a drop in ankle moment. Inversion failure occurred at 34 ± 15 Nm and 34 ± 8°, and eversion failure occurred at 48 ± 12 Nm and 32 ± 7° (Table 18.6). Specimen age, gender, and bone mineral content were not found to be significant predictors of failure moment or failure angle, except in the case of inversion failure moment, which was significantly lower for females than males. For the tests performed by Petit et al. [25] failure was defined as the first significant drop in the ankle moment-angle curve. Inversion failure occurred in five specimens at 40 ± 26 Nm and 34 ± 8° and eversion failure occurred in five specimens at 35 ± 9 Nm and 32 ± 7° (Table 18.6).

Funk et al. [33] performed tests under three different loading conditions: neutral flexion with no axial preload, neutral flexion with 2 kN axial preload, and 30° of dorsiflexion with 2 kN axial preload. Forty-four tests were conducted on cadaveric lower limbs, with injury occurring in 30 specimens. Common injuries included malleolar fractures, osteochondral fractures of the talus, fractures of the lateral process of the talus, and collateral ligament tears, depending on the loading configuration (Fig. 18.11). The time of injury was determined either by the peak ankle moment or by a sudden drop in ankle moment that was accompanied by a burst of acoustic emission.

Characteristic moment-angle curves to injury were generated for each loading configuration. Neutrally flexed ankles with no applied axial preload sustained injury at 21 ± 5 Nm and 38° ± 8° in inversion, and 47 ± 21 Nm and 28° ± 4° in eversion. For ankles tested in neutral flexion with 2 kN of axial preload, inversion failure occurred at 77 ± 27 Nm and 40° ± 12°, and eversion failure occurred at 142 ± 100 Nm and 41° ± 14°. Ankles dorsiflexed 30° and axially preloaded to 2 kN sustained inversion injury at 62 ± 31 Nm and 33° ± 4°, and eversion injury at 140 ± 53 Nm and 40° ± 6°.

Survival analyses were performed to generate injury risk curves in terms of joint moment and rotation angle. Rotation direction, dorsiflexion

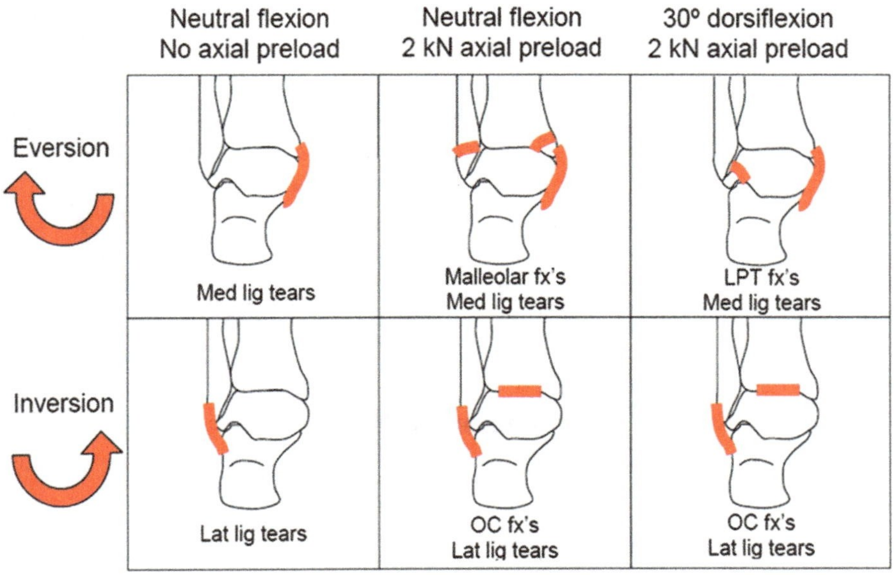

Fig. 18.11 Dependence of xversion injury pattern on the combination of initial position and concomitant axial load [33] (Reproduced with permission from the Stapp Association)

angle, specimen age, gender, mass, height, and BMD were not found to be significant predictors of the angle at injury. However, axial load alone did not model the angle data very well ($R2 = 0.16$). The following closed-form moment survivor function was generated using only angle and axial load:

$$S(\theta \mid xi) = \exp\left\{-\exp\begin{bmatrix} 4.88 \cdot \ln(\theta) - \\ 17.5 - 0.41 \cdot \\ \text{Axialload}(\text{kN}) \end{bmatrix}\right\} \quad (18.3)$$

A graphical depiction of the survival function is shown in Fig. 18.12. Since biomechanists will be developing 5th and 95th male and female models, the analysis performed by Funk et al. [33] examining size and gender dependence are also useful for model validation and evaluation. In particular, Funk et al. found a large sensitivity of the response to the age and gender of the specimen (Fig. 18.13).

18.2.2.2 Dorsiflexion

The more controlled environment of quasi-static tests permitted evaluation of not only the moment-angle characteristics, but also the joint centers of rotation, which were shown to vary slightly with ankle position [24, 25]. Without inertial effects, the results were found to be less variable overall. Petit et al. [25] tested 25 limbs quasi-statically in dorsiflexion with simulated muscle tension applied through the Achilles tendon. The legs were amputated mid-shaft, and the proximal ends of the tibia and fibula were potted together in a cup with cement and Steinmann pins. The calcaneus was potted similarly. In order to apply muscle tension, the Achilles tendon was cut and secured to a cable, which was routed through the test fixture to a counterweight. Dorsiflexion tests were performed with 900 N of tension applied to the Achilles. The prepared specimens were placed in a test fixture which applied rotation to the joint with an electric motor while allowing the specimen to translate in two directions. Of seven legs tested in dorsiflexion, the center of rotation was found to be 76 ± 8 mm proximal from the sole of the foot and 61 ± 6 mm anterior from the posterior face of the heel. The dorsiflexion tests had an average 47 ± 17 N-m and $49° \pm 5°$ failure point. Typical injuries included medial malleolus fractures and calcaneofibular ligament tears.

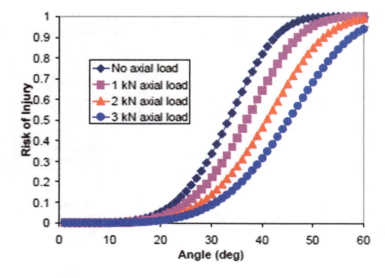

Fig. 18.12 Injury risk as a function of subtalar joint moment for different rotation directions (Ev, Inv), levels of axial preload (1–3 kN) [33] (Reproduced with permission from the Stapp Association)

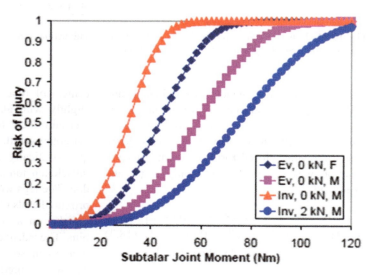

Fig. 18.13 Injury risk as a function of inversion/eversion angle for different levels of axial preload [33] (Reproduced with permission from the Stapp Association)

The quasi-static tests by Parenteau et al. [24] were performed in a fixture similar to the one used by Petit et al. [25], but without the applied muscle tension. Specimen preparation involved cutting the leg mid-shaft and potting the tibia and fibula together in a cup with cement and Steinmann pins. The calcaneus was held with pins and cement in a cup, as well. Torque was applied to the calcaneus cup, which induced rotation in the ankle or subtalar joint, depending on the orientation of the specimen. A motor applied torque at a constant rate of rotation until failure was detected by a drop in moment. Nine tests were performed in dorsiflexion, five of which sustained an injury. Injuries included two lateral malleolus fractures, two calcaneofibular ligament tears and one interosseous talocalcaneal ligament tear. Failure, defined by a drop in moment with increased rotation, occurred at an average moment of 33.1 ± 16.5 N-m and dorsiflexion of 44.0° ± 10.9°, while the average injury moment was 36.2 ± 14.8 N-m with dorsiflexion of 47.0° ± 5.3°. The center of rotation in dorsiflexion was found to be located at the ankle joint, near the center of the talar dome.

Table 18.7 ROM for applied 6 Nm torque with differing boundary conditions

		Fixed femur	Severed knee	Potted tibia-fibula	Severed fibula
Constrained	Dorsiflexion	4.7±2.4°	17.5±4.4°	22.7±1.7°	24.5±1.4°
	Plantarflexion	28.3±4.0°	29.4±4.5°	29.3±5.0°	31.8±5.6°
Unconstrained	Dorsiflexion	4.6±2.5°	21.9±7.1°	30.4±6.9°	30.0±6.4°
	Plantarflexion	32.6±7.3°	32.7±5.5°	33.5±6.5°	34.9±6.1°

Schreiber et al. [34] conducted quasi-static tests in a six-degree of freedom test fixture that was capable of both constrained (rotation about on the dorsiflexion axis) and unconstrained motion. Six cadaveric lower limbs were amputated above the knee and subjected to complete cycles of dorsiflexion and plantarflexion at non-injurious loading levels. For each lower limb, four boundary conditions were used to determine the influence of various below-knee anatomical structures on the ankle response. The ankle test apparatus consisted of a gimbal assembly, with two links centered at the ankle's theoretical center of rotation. A sprocket, driven by a hydraulic piston at a constant rate, produced a moment in the outermost gimbal assembly aligned with the flexion axis of rotation. As the ankle was forced into flexion, the other axes were unconstrained. The tibia was mounted rigidly to a fixed frame while the foot was mounted to a foot plate fixed to the inner gimbal link. Translation was accommodated by roller bearings in the inner gimbal link. In all, six degrees of freedom (three rotational and three translational), were permitted in the ankle's motion. The results suggest that dorsiflexion of the ankle is highly dependent on the triceps surae muscle group, the superior tibiofibular joint, and the relative spacing of the ankle mortise. The insensitivity of plantarflexion motion to varying anatomical conditions suggests that the boney stops, ligaments, and other local anatomical structures determine the extension response.

The overall utility of the tests by Schreiber et al. [34] for FE model validation may lie in the sequential identification of anatomical components to the overall response of the lower limb. Using a range of motion (ROM) defined as the angle achieved when 6 Nm of torque was applied in any direction, Schreiber et al. produced a stiffness mapping with different boundary conditions (Table 18.7).

Colville et al. [35] measured strain in the lateral ligaments of 10 human cadaver ankles while moving the ankle joint quasi-statically from dorsiflexion to plantarflexion. They studied the anterior talofibular, calcaneofibular, posterior talofibular, anterior tibiofibular, and posterior tibiofibular ligaments. Strain measurements in the ligaments were recorded continuously using Mercury-filled Silastic strain gauges 8 mm in length.

The dissected ankle preparations were secured to a Genucom Knee Analysis System which, when coupled with an electrogoniometer and force plate, allowed kinematic and kinetic information. Strain in the anterior talofibular ligament increased when the ankle was moved into greater degrees of plantar flexion while strain in the calcaneofibular ligament increased as the talus was dorsiflexed. These findings provide model validation information regarding the combined function of the anterior talofibular and calcaneofibular ligaments. The authors also measured maximum strain in the posterior talofibular ligament when the ankle was dorsiflexed. The strain in the anterior and posterior tibiofibular ligaments increased when the ankle was dorsiflexed.

Begeman and Prasad [36] performed dynamic testing to investigate the impact response of the ankle joint complex to axial loads with dorsiflexion. Impact loads were applied to the plantar surface of the foot with a 16.5 kg impactor moving at 3.0–8.1 m/s. The legs were potted at the knee, but no mention was made of how the plantarflexor musculature was handled. With the leg horizontal and the foot 90° to the tibia, the impactor head, which was a cylindrical bar 25 mm in diameter covered in padding, struck the foot 62.5 mm above the ankle joint to apply axial load while forcing the foot into dorsiflexion. Reaction loads measured at

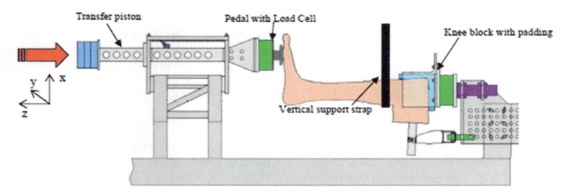

Fig. 18.14 Dorsiflexion test setup of Rudd et al. [39] (Reproduced with permission from the Stapp Association)

the knee potting were used to calculate the moment at the ankle joint. Results were presented for 18 cadaver limbs (9 matched pairs).

A second study by Begeman et al. [37] was similar to the 1,990 work except for improvements made to the test fixture. The point of impact was varied between 0 and 50 mm above the ankle joint to produce different combinations of axial load and dorsiflexion. The tibia and fibula were potted together at the mid-shaft, and no mention was made of the plantarflexor musculature. Twelve matched pairs of human cadaver limbs were tested. They reported range of motion values before and after the tests, and found that injured specimens had a much larger dorsiflexion range of motion after the impact. They concluded that a pre- and post-test comparison of range-of-motion could be used to detect injury, but that pre-test range-of-motion did not indicate susceptibility to injury.

Portier et al. [38] and Crandall et al. [1] also investigated dynamic ankle dorsiflexion response and tolerance. Portier performed component tests with an impactor. The cadaver lower limbs were kept intact, and a six-axis load cell was implanted in the mid-shaft of the tibia. In order to account for the fibula load as well, the fibula was cut and the distal portion was fixed to the distal tibia with a screw. Component tests were performed with an impactor-driven brake pedal positioned at the head of the first metatarsal, which forced the foot into dorsiflexion at velocities of approximately 4 m/s [38]. A total of ten cadaver limbs (five matched pairs) were tested, but four tests were performed with an instrumented heel support that was found to complicate the experiments.

Rudd et al. [39] developed a dynamic dorsiflexion fixture based on the work of Portier et al. [38]. The test rig was constructed such that a simplified, yet realistic pedal load was applied to the forefoot to force the foot into dorsiflexion. Specimen preparation maintained all of the musculature that affects the ankle joint and allowed for accurate measurement of the loads and moments in the leg. The test fixture consisted of a transfer piston, a brake pedal and a knee block, with impact energy provided by a pneumatic impactor (Fig. 18.14). Leg specimens were placed horizontally in the fixture, and the foot was held in a nearly neutral flexion angle. The knee was constrained within padded brackets at the knee block, and the femur was tied back with a load bolt. A strap was hung from above and placed around the leg at the proximal third for support. Unlike the tests performed by Begeman et al. [36, 37], this arrangement permitted the use of an intact specimen with the entire leg in its normal anatomical state [39].

All specimens were sectioned above the knee at mid-femur to preserve the functional anatomy of the knee joint and lower leg musculature. The soft tissue of the proximal thigh was removed and the femur was cut at the mid-shaft such that the bone protruded approximately 10 cm above the point where the soft tissue was removed [39]. Cadaveric specimens were instrumented to measure leg forces

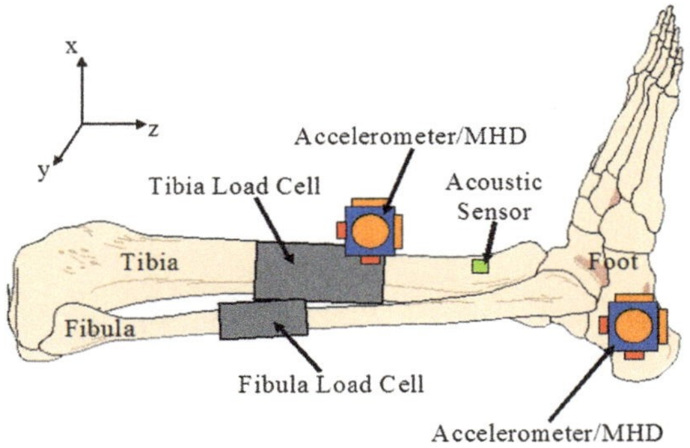

Fig. 18.15 Schematic of instrumented specimen [39] (Reproduced with permission from the Stapp Association)

and moments, leg/foot accelerations and angular rates, and Achilles tension. Changes to the instrumentation were made between the first and second series of cadaver tests, with the main difference being the addition of a fibula load cell. A five-axis tibia load cell was installed by removing a section of the mid-shaft of the tibia [40], and fixing special cups to the distal and proximal sections of the tibia using epoxy (Fig. 18.15). The load cell was secured between the cups with cradles, which were tightened to provide rigid fixation. A jig was used during the preparation process to ensure proper alignment and spacing of all of the components. The six-axis fibula load cell installation was performed with a similar method, except alignment and spacing were not maintained with a jig due to the smaller size of the fibula (see Ref. [27]). Cubes with three accelerometers and three angular rate sensors were fixed to the specimens prior to each test. The tibia cube was affixed directly to the distal tibia load cell cradle, and the foot cube was screwed to the calcaneus. Load cell and accelerometer data were used to calculate moments at the ankle joint. Geometrical terms were measured from x-ray analysis.

Measurement of the Achilles tension was accomplished by use of an arthroscopically implantable force probe (AIFP, Microstrain – Burlington, Vermont), which was inserted into the mid-substance of the Achilles tendon just before each test (see Refs. [41–43]). During leg preparation, incisions were made in the posterior distal leg flesh to expose the Achilles tendon.

Gauze was then wrapped around the proximal part of the tendon and sutured in place to help grip the tendon for the calibration process. A small two-centimeter square area of the anterior distal tibia was denuded and cleaned with ether for mounting an acoustic sensor. The sensor (Nano 30, Physical Acoustics, Princeton Junction, NJ) was glued to the bone with cyanoacrylate adhesive, and had an operating range of 125–750 kHz with a center frequency of 140 kHz.

Moment-angle curves were calculated for each test with the properties inertially compensated and transformed to the ankle joint center in order to produce an unbiased dorsiflexion response curve. Mass-scaled ankle moment-angle curves from the 18 cadaver tests were fit to a second-order polynomial using a least-squares regression routine. The resulting curve was forced to pass through the origin and the peak average mass-scaled fracture moment and angle (62.4 N-m and 40°). Averaging the coefficients yielded the following relationship between dorsiflexion angle and ankle joint moment in the human:

$$M_y = 0.05733\theta + 0.037567\theta^2 \quad (18.4)$$

where θ is in degrees of dorsiflexion and M_y is in N-m.

In their dynamic dorsiflexion tests, Begeman and Prasad [36] typically produced malleolar fractures and torn ligaments (AIS 2). There was considerable variation in the results; for injury tests, ankle moments ranged between 70 and 210 N-m. Based on the lack of knowledge of the state of the plantarflexor musculature during their tests, it is uncertain if the reported moments are ankle joint moments (resistance from the bony structures, ligaments and synovial membrane) or ankle section moments (including the effects of passive musculature). Their data indicated that 45° was the injury threshold for dorsiflexion, and they found no correlation between axial forces or bending moments and injury.

In a second study by Begeman et al. [37], the ankle injuries generated by dorsiflexion were similar to those in the first Begeman study [36], and included malleolar fractures, torn ligaments, a tibia end fracture, a talus fracture and a torn

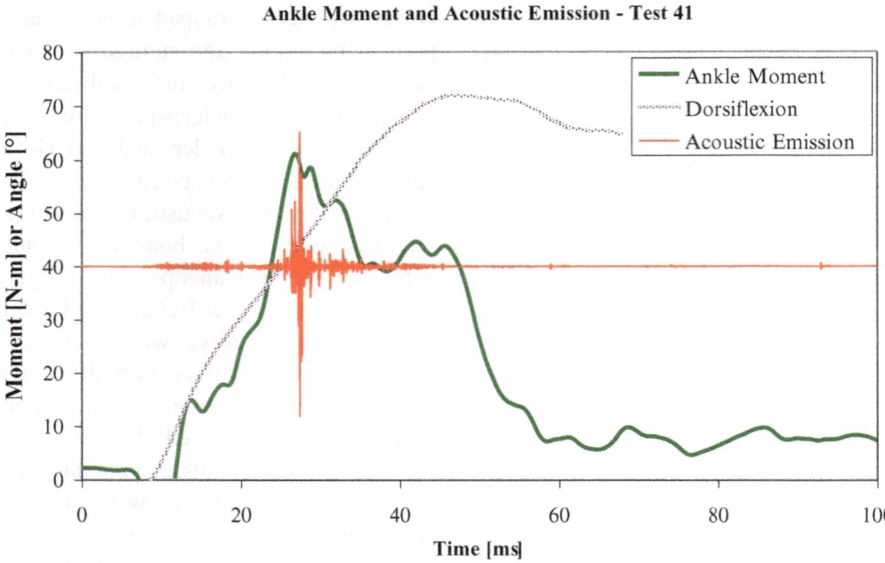

Ankle Moment and Acoustic Emission - Test 41

Fig. 18.16 Ankle moment and acoustic emission from an exemplar test showing distinct acoustic emission burst coincident with peak moment, indicating fracture (From the work of Rudd et al. 2004)

capsule, all of which were classified as AIS 2. Fourteen out of twenty-four specimens sustained a fracture, but Begeman et al. concluded that moment was not a good predictor of injury since it was highly variable among the tests. A better correlation was found for a rotation threshold, which was concluded to be 45° ± 10° of dorsiflexion to cause injury in the dynamic cadaveric tests. Volunteer tests indicated that 45° of dorsiflexion was a tolerable angle, and moment versus angle characteristics were similar to those from cadaver tests.

As previously reported, Portier et al. [38] performed impact tests to the forefoot in order to generate dorsiflexion of the ankle. In the tests with the heel support, severe levels of dorsiflexion created medial malleolus fractures in three of the four specimens. Without the heel support, one of six ankles sustained a medial malleolus fracture and two sustained malleolus avulsions. Portier presented detailed procedures for calculation of the ankle moment, and found that the ankle joint was injured at an average of 60 N-m within the joint and 180 N-m in the ankle section (ankle section moments account for passive musculature). From film analysis of markers on the calcaneus and tibia, fracture occurred at approximately 30° of dorsiflexion.

In the dynamic dorsiflexion tests performed by Rudd et al. [39], 11 of the 20 specimens tested sustained a bony fracture at the ankle joint. In addition to hard tissue injuries, ligament ruptures occurred in four specimens and the majority of the specimens showed some kind of osteochondral or cartilaginous defect during the dissection. It is of interest to note that the onset of fracture, as indicated by the most energetic acoustic event, was not always coincident with the peak ankle moment, a finding that justifies and necessitates the use of acoustic emission to determine fracture time (Fig. 18.16). Rudd et al. provided an injury risk function along with confidence intervals for both moment and angle that can be used for FE model validation.

Crandall et al. [1] conducted both quasi-static and dynamic dorsiflexion tests using a combination of volunteers and PMHS specimens. For the quasi-static tests, ten PMHS and eight volunteers were placed into a multi-degree-of-freedom test device and tested for moment-angle. These results were further compared with contemporary dummy foot/ankle designs. These results are shown in Fig. 18.17.

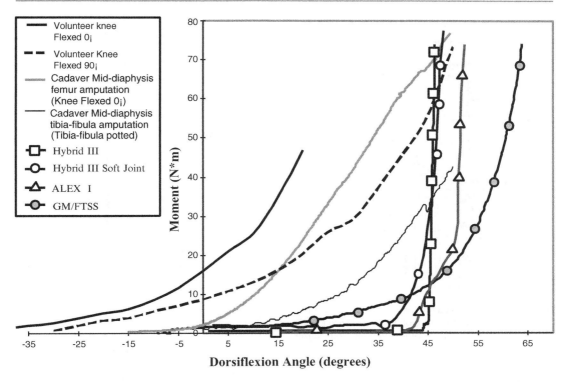

Fig. 18.17 Quasi-static average moment-angle responses in dorsiflexion (From the work of Crandall et al. 1996)

18.2.3 Knee Response and Failure

The following section provides a review of both publicly available literature and previous University of Virginia testing on the response and tolerance of the human knee in shear and bending. This section is meant to highlight what injury criteria, and data are available for FE validation.

18.2.3.1 Lateral Loading

Kajzer et al. [44] conducted a series of tests to determine the tolerance of the human knee joint in lateromedial (LM) shear. Intact lower limbs were harvested from six male and five female preserved (Winckler method) human cadavers in the age range 67–87 years. The limbs were mounted in an upright position and impacted in the lateromedial direction at 15 (n=9) or 20 (n=10) km/h by a 40 kg twin-pronged padded (50 mm of Styrodur©) impactor that contacted at the ankle joint and just below the knee joint (Fig. 18.18). Load cells backing up the impactor padding recorded the individual time histories of impact force at the knee and ankle. Two plates, padded with 25 mm of Styrodur© and backed by load cells, supported the medial side of the limb at the levels of the proximal and distal femur. The foot was positioned on a plate that was free to translate in the direction of impact to minimize the influence of ground friction. An axial preload of 40 kg was applied to each specimen to simulate half the body weight.

Injuries produced in the tests included femoral shaft fracture, femoral condyle fracture, cartilage damage, tibial condyle fracture, tibial shaft fracture, fibula fracture, knee ligament (ACL, PCL, MCL, and LCL) avulsion, rupture, and partial tear. The authors concluded that it was necessary to consider two different injury mechanisms. The first mechanism was said to be directly related to the knee impact force that was responsible for contact injuries including fractures to the fibular head and lateral tibial condyle, extra-articular injuries, and shaft fractures of the tibia and femur.

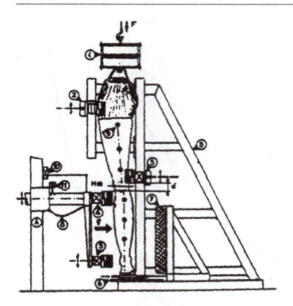

Fig. 18.18 Experimental test set-up used by Kajzer et al. [44] to investigate the tolerance of the human knee joint in LM shear [44] (Reproduced with permission from Kajzer et al. 1990)

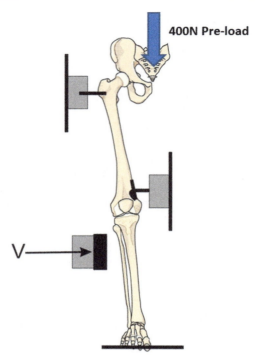

Fig. 18.19 Experimental test set-up used by Kajzer et al. [45, 46] to investigate the tolerance of the human knee joint in LM shear (From the work of Kajzer et al. 1997)

For impact velocities in the range of 15–20 km/h, injuries due to the first injury mechanism were said to occur approximately 5 ms after initial contact between the impactor and limb. The mean ± SD of peak knee impact force associated with injury due to the first injury mechanism was 1.80 ± 0.38 kN and 2.57 ± 0.45 kN at 15 and 20 km/h impact velocity, respectively. The second injury mechanism was said to correlate with the force transferred through the knee joint and resulted in injuries associated with relative displacement between the tibia and femur such as injury to the knee ligaments and femoral cartilage. For impact velocities in the range of 15–20 km/h, injuries due to the second injury mechanism were said to occur approximately 15–20 ms after initial contact between the impactor and limb. The mean ± SD of peak knee reaction force associated with injury due to the second injury mechanism was 2.57 ± 0.37 kN and 3.22 ± 0.46 kN at 15 and 20 km/h impact velocity, respectively.

In a more recent study, Kajzer et al. [45] used a slightly different experimental set-up and a higher impact velocity (40 km/h) to determine the tolerance of the human knee joint in lateromedial (LM) shear (Fig. 18.19). The right lower limb of nine male and one female human cadavers ranging age from 35 to 69 years was tested in which the subject was positioned supine on a bench with the femur through the trochanter and a fixation plate screwed to the femur just superior to the femoral metaphysic. The subject was axially preloaded at 400 N to simulate the influence of half the body weight and impacted by a 6.25 kg Styrodure© padded impactor contacting the specimen just inferior to the tibial plateau. The injuries produced in the tests included ACL avulsion (60 % of the specimens), MCL avulsion (20 % of the specimens), femoral shaft fracture (70 % of the specimens), tibial shaft fracture (10 % of the specimens), tibial articular fracture (30 % of the specimens), and femoral articular fracture (10 % of the specimens). The mean ± SD of the maximum shearing force and bending moment at the knee joint level were 3.0 ± 0.6 kN and 491 ± 128 Nm, respectively. The authors reported that femur fracture at, near, or superior to the knee fixation plate was the initial mode of knee joint failure for 40 % of the tested specimens. These fractures were actually seen in 70 % of the specimens but reported as the source of initial joint failure only in 40 %. In four of the seven specimens that demonstrated this type of fracture, failure occurred at or near the distal end of the knee fixation plate suggesting that they might be associates with knee joint failure. However, the other three fractures were at the proximal end of the knee fixation plate or superior to it.

Using the same set-up as in their 1997 study but only half the impact velocity (20 km/h), Kajzer et al. [46] reinvestigated the tolerance of the human knee joint in lateromedial (LM) shear (Fig. 18.19) using three male and two female human cadavers ranging in age from 42 to 83 years. The injuries produced included ACL avulsion (n = 2 specimens), ACL stretching (n = 1 specimen), femoral shaft fracture (n = 1 specimen). The mean ± SD of the maximum shearing force and bending moment at the knee joint level were 2.5 ± 0.1 kN and 415 ± 60 Nm, respectively. From the summary above, it is clear that one of the most extensive research programs on the response and injury tolerance of the human knee joint in lateromedial bending (valgus bending) that has been undertaken is the multiple test series reported on by Kajzer et al. [45–47]. Although the results from the 1997 test series have served as validation data for the lower leg-form that is currently the prescribed evaluation tool in the New Car Assessment Program (NCAP) pedestrian test protocol in the European Union, Australia, and Japan, the test methodology has been criticized for being unrealistic [48] and the results for being incorrect [49]. Despite that the test set-up used by Kajzer and co-workers in their 1997 and 1999 studies was slightly different from the one used in their 1993 study, Kajzer et al. [46] stated that the only explanation to the failure moments being approximately three times higher in the two later studies was a difference in quality of the cadaveric test specimens. In contrast, Kerrigan et al. [48] pointed to the fact that 70 % of the knee joints that Kajzer et al. [45] tested in bending did actually fail by means of femoral

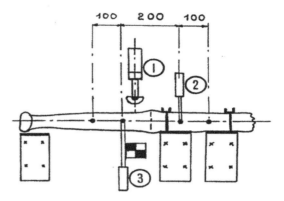

Fig. 18.20 Schematic of the experimental set-up used by Ramet et al. [50] to determine the response and tolerance of the human knee joint in quasi-static LM shear [50]

fracture near, at, or superior to the knee fixation plate and thus suggested that the high failure moments reported by Kajzer et al. could be attributed to the fact that the knee joint was not isolated from the long bones. The most recent explanation to the high bending moments reported by Kajzer et al. [45, 46] is offered by Konosu et al. [49] who showed that Kajzer et al. had made a sign error in their calculation of the failure bending moments. Konosu et al. used the correct formula and recalculated the failure moments from three of Kajzer et al.'s tests for which the originally published failure moments were 329, 450, and 367 Nm and showed that the corresponding correct values were 65, 139, and 127 Nm, respectively.

Ramet et al. [50] investigated the tolerance of the human knee in quasi-static lateral shear. Ten intact lower extremities were harvested from two male and three female PMHS and tested in the set-up shown in Fig. 18.20. The femur was firmly connected to the two underlying supports by a mechanical screw tightening system while the foot rested on an additional support of equal height (tibial rotation was prevented by a mechanical stop at the level of the foot). Using an impactor with load cell (number 1 in Fig. 18.20), the leg was quasi-statically displaced in the lateral direction relative to the femur and the relative displacement between the femur and tibia was measured by displacement transducers (number 2 and 3 in Fig. 18.20) screwed to the tibia and femur. Based on the data from seven of the ten

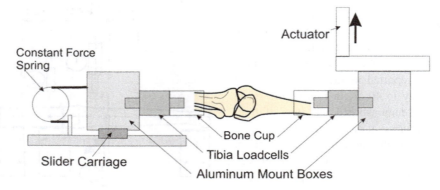

Fig. 18.21 Knee-joint lateral shear test schematic. Constant force springs are used on the femur side to provide the 700 N axial force through the joint [48] (From the work of Kerrigan et al. 2003)

tested specimens (two specimens were not injured and data were missing for one), the authors concluded that initial micro damage occurs at a relative lateral displacement between the femur and tibia of approximately 12 mm while macroscopic lesions start to develop after at least 22 mm of relative displacement. Thus, 20 mm of relative displacement was recommended as an acceptable limit in static shear. The force level associated with injury varied widely (0.75–3 kN) and the authors therefore concluded that force is not a valid predictor of injury for the knee in quasi-static lateral shear.

Kerrigan et al. [48] performed a pilot study specifically designed to eliminate artifact injuries and reduce inertial contribution to the measured knee dynamics. By using a servo-hydraulic test machine to apply loading in displacement control, two cadaveric knee specimens harvested from two males aged 62 and 66 years were loaded in dynamic shear (Fig. 18.21). Unlike past studies, Kerrigan et al. [48] used isolated knee joints to accurately measure the bending moment and shear force close to the knee joint. A compressive axial force equivalent to the body weight was superimposed on the knee joint. The recorded shear force values estimated at the instant of primary ligament injury appear in Table 18.8. The time instant for primary injury was decided based on acoustic signal emission data and drop in bending moment.

The pure shearing of the knee joint specimens in Kerrigan et al.'s tests produced osteo-chondral defects which have never been reported in real world pedestrian crashes or in staged collisions using cadavers. In addition, ACL injuries seen in the study rarely occur as isolated ligamentous injuries. Thus, it was hypothesized that the pure shear loading of the knee in the presence of compressive loads produced in these tests was an extreme loading environment that does not occur in real-world pedestrian impacts. Therefore, to experimentally produce injuries seen in pedestrian impacts, it is necessary to replicate a loading environment with realistic proportions of bending moment and lateral shearing.

A pilot study performed by Bose et al. [51] evaluated the stiffness response of the isolated knee subjected to combined loading with different proportions of bending and shear loading in a three-point bending set up (Fig. 18.22). The three-point configuration provided a combination of bending moment and lateral shear at the knee center. The proportion of superimposed shear force on the bending moment at the knee

Table 18.8 Shear test measured failure values

Specimen	Event time (ms)	Actuator displacement (mm)	Shear force (N)
167 L	31.5	39.6	1,667.3
169R	14.7	15.7	563.2
Mean	23.1	27.7	1,115.3
SD	11.9	16.9	780.7

Note: all values presented are those that occur at the first joint failure as determined by strain data and acoustic emission [48]

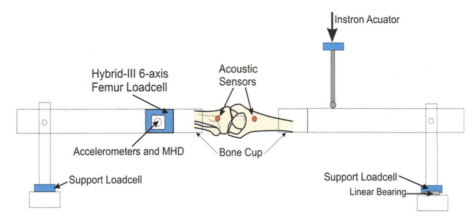

Fig. 18.22 Test schematic for the three-point bending (combined bend-shear loading) test (From the work of Bose et al. 2004)

center described by the M/V ratio [51] was varied by altering the distance between the support and knee center. The dynamic bending rate of the knee was chosen to be 1.5°/ms, and accordingly a rate of 1°/s was chosen to represent the quasi-static bending rate. The combined loading effects of bending and shearing produced mainly injuries to the MCL, which is in agreement with the injury pattern observed in pedestrian victims.

18.2.3.2 Response Testing of the Knee in Lateral Loading

Data on the response of the human knee joint in lateral shear are relatively sparse. From Kajzer et al.'s [44–46], response data are only available from three tests. In their 1990 publication, Kajzer and co-authors only include an example of a "typical" force-time history from a test (Fig. 18.23). Kajzer et al. [45] provide data from one test that resulted in ACL avulsion and another test that produced a comminuted supracondylar fracture.

Ramet et al. [50] provided force-displacement histories for the nine quasi-static lateral shear tests for which data were successfully collected (Fig. 18.24). Based on the data shown in Fig. 18.24, Ramet et al. suggested a response corridor with the upper bound defined by the x, y (displacement (cm), force (N)) coordinates (0, 0) and (3.5, 3,000) and the lower bound defined by (0, 0), (1, 0), and (3.5, 2,350).

Kerrigan et al. [48] provides time histories of bending moment and shear force (Fig. 18.25), and a plot of shear force versus shear displacement (Fig. 18.26).

Bose et al. [51] reported the bending moment-angle history of the eight specimen tests for different proportions of shear loading. The bending moment-angle response shown in Fig. 18.27 has been compensated for inertial effects using an analytical method [51]. An updated version [52] of this work includes more than 40 specimens and provides a more complete analysis of the knee joint under combined bending and shear loading.

18.2.3.3 Bending Loading of the Knee Tolerance Testing and Injury Criteria

Kajzer et al. [47] conducted a series of tests to determine the tolerance of the human knee joint in valgus bending. Intact lower limbs were harvested from nine male preserved (Winckler method) human cadavers with an average age of 77 years. The limbs were mounted in an upright position and impacted in the mediolateral direction at 16 (n = 7) or 20 (n = 10) km/h by a 40 kg padded (50 mm of Styrodur©) impactor that contacted the limb just above the ankle joint (Fig. 18.28). The tests were performed with 40 kg of axial preload applied to the extremity to simulate half the body weight. The authors reported an average failure moment of 101 Nm at 16 km/h

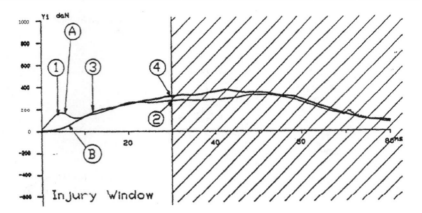

Fig. 18.23 "Typical" time histories of knee impact force (*A*) and knee reaction force (*B*) recorded in the dynamic lateral shear tests reported by Kajzer et al. [44]. According to the authors, the numbers in the graph correspond to local injury (*1*), maximum knee impact force during the injury window (*2*), knee reaction force at the time of knee joint destabilization, and maximum knee reaction force in the injury window (*4*) [44] (Reproduced with permission from Kajzer et al. 1990)

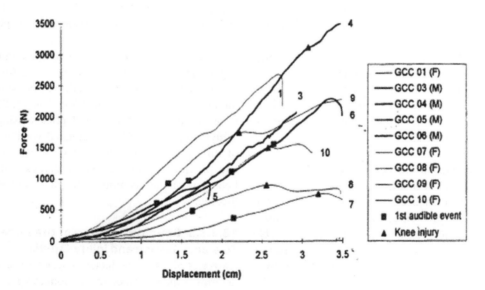

Fig. 18.24 Force-displacement histories recorded in quasi-static lateral shear tests of human knees as reported by Ramet et al. [50]. Note that the specimens used in tests GCC 05 and GCC 06 did not show any sign of injury at the post-test autopsy [50] (Reproduced with permission from Ramet et al. 1995)

impact velocity and 123 Nm at 20 km/h impact velocity. The injuries produced included partial or total rupture of the MCL, MCL avulsion, and fracture of the medial condyle.

In a more recent study, Kajzer et al. [45] used a slightly different experimental set-up and a higher impact velocity (40 km/h) to determine the tolerance of the human knee joint in valgus bending. The right lower limb of nine male and one female human cadavers ranging age from 35 to 69 years was tested in the experimental set-up shown in Fig. 18.29 in which the subject was positioned supine on a bench with the femur fixed by a knee plate and a trochanter fixation

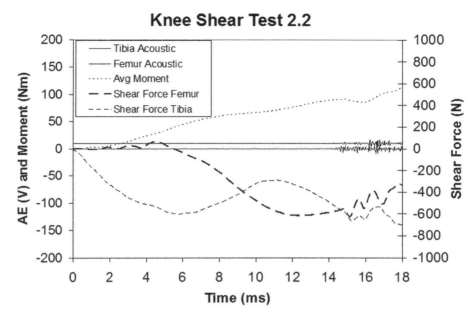

Fig. 18.25 Time histories of bending moment, shear force (recorded at the femur and at the tibia), and acoustic emission as recorded in one of the knee shear tests reported by Kerrigan et al. [48] (From the work of Kerrigan et al. 2003a)

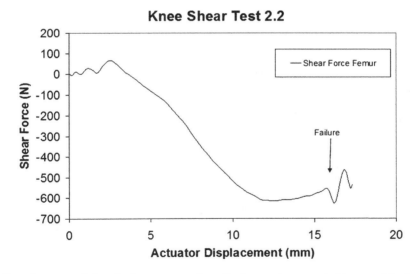

Fig. 18.26 Shear force recorded at the femur versus shear displacement as recorded in one of the knee shear tests reported by Kerrigan et al. [48] (From the work of Kerrigan et al. 2003a)

screw. An axial preload of 400 N was applied to the subject to simulate half the body weight. The leg was impacted at the distal tibial articular surface by a 6.25 kg Styrodure© padded impactor. The two most common types of initial knee joint failure were ligament avulsion (in 30 % of the tests) and femur fracture (in 70 % of the tests). The mean ± SD of the maximum shearing force and bending moment at the knee joint level were 1.6 ± 0.6 kN and 388 ± 89 Nm, respectively. As previously noted, it is clear that one of the most extensive research programs on the response and

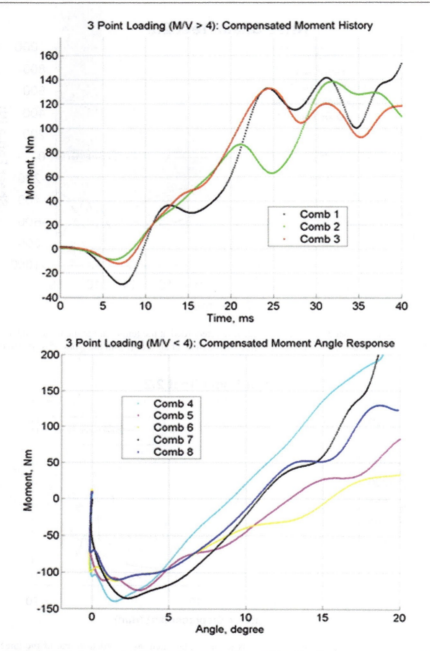

Fig. 18.27 Valgus bending moment–bending angle response for combined loading tests with higher (M/V > 4) and lower (M/V < 4) proportions of lateral shearing [51] (From the work of Bose et al. 2004)

injury tolerance of the human knee joint in lateromedial bending (valgus bending) that has been undertaken is the multiple test series reported on by Kajzer et al. [45–47].

Using the same set-up as in their 1997 study (Fig. 18.29), but only half the impact velocity (20 km/h), Kajzer et al. [46] reinvestigated the tolerance of the human knee joint in valgus bending using three male and two female human cadavers ranging age from 42 to 83 years. The injuries produced included MCL avulsion (n = 1 specimen), MCL stretching (n = 1 specimen), and femoral metaphysis fracture (n = 1 specimen). The mean ± SD of the maximum shearing force

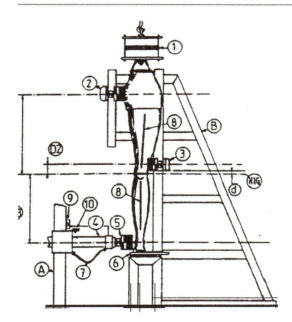

Fig. 18.28 Experimental test set-up used by Kajzer et al. [47] to investigate the tolerance of the human knee joint valgus bending [47] (Reproduced with permission from Kajzer et al. 1993)

and bending moment at the knee joint level were 1.5 ± 0.4 kN and 349 ± 131 Nm, respectively.

Ramet et al. [50] investigated the tolerance of the human knee joint in quasi-static valgus bending. Ten intact lower extremities were harvested from three male and two female cadavers ranging in age from 67 to 81 years and tested in the set-up shown in Fig. 18.30. The femur was firmly connected to the two underlying supports by a mechanical screw while the tibia was attached to a duralumin section of length 1.5 m. The end of the duralumin bar was slowly rotated counter-clockwise to create an increasing valgus bending moment in the knee joint. Five of the ten tested specimens did not show any major injury despite that they were subjected to rotations of 15°, 22°, and 24°. The other five specimens demonstrated MCL rupture (n = 4 specimens), meniscus injury (n = 2), cartilage damage (n = 2), bone avulsion (n = 1), and residual laxity (n = 1). Based on the bending angles and moments at which initial injury appeared to occur, the authors concluded that macroscopic injury in quasi-static valgus bending occurs at valgus angles above 16° and bending moments above 100 Nm.

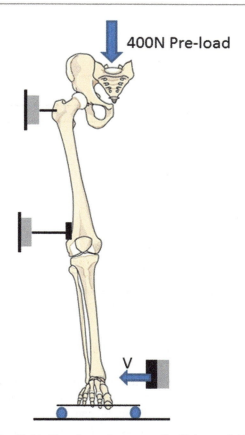

Fig. 18.29 Experimental set-up used by Kajzer et al. [45, 46] in lateral bending tests of the human knee

To account for the shortcomings in evaluating knee injury response using whole limb specimens, Kerrigan et al. [48] performed four-point bending on three isolated knee specimens (Fig. 18.31). In each of the bending tests, MCL injuries were produced. In one test where a larger bending angle was applied to the knee joint, the ACL was also injured. From an analysis of the test data, it was hypothesized that the MCL is damaged first during valgus knee bending, followed by the ACL, which is loaded after MCL failure.

Using a similar four-point bending set-up to the one used by Kerrigan et al. [48], Bose et al. [51] evaluated response and injury of eight isolated knee specimens and predominantly observed MCL injuries. Ivarsson et al. [53] scaled the data from Bose et al. [51] to the 50th percentile adult male and developed univariate Weibull survival models predicting the risk of knee injury in dynamic valgus bending as function of knee

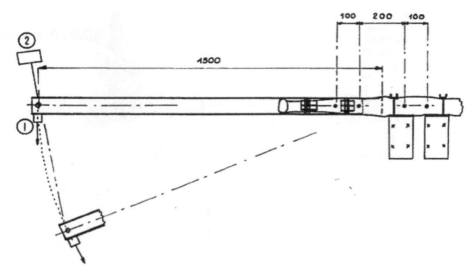

Fig. 18.30 Schematic of the experimental set-up used by Ramet et al. [50] to determine the response and tolerance of the human knee joint in quasi-static valgus bending [50] (Reproduced with permission from Ramet et al. 1995)

moment and knee bending angle (Figs. 18.32 and 18.33).

Combining the test results of experiments done in 2004, Bose et al. [52] reported on the injury tolerance of the knee joint in combined loading using a total of 40 knee specimens. An injury threshold function for the knee in terms of bending angle and shear displacement was determined by regression analysis of the experimental data. The region of no-injury for the knee is described as the area bounded by the injury threshold function $\left(d = \sqrt{1189 - 6850\sin^2(0.2900 + 0.0087\alpha)}\right)$, where d and α are the shear displacement and valgus bending angle respectively), the abscissa axis (dshear=0), and the ordinate axis (αvalgus=0) (Fig. 18.34). The threshold values for the bending angle (16.2°) and shear displacement (25.2 mm) estimated from the injury threshold function are in agreement with previously published knee injury threshold data. The continuous knee injury function expressed in terms of bending angle and shear displacement enabled injury prediction for combined loading conditions such as those observed in pedestrian crashes.

Response Testing of the Knee

From Kajzer et al.'s [45–47] tests, response data are only available from a few tests. In their 1993 publication, Kajzer and co-authors only include an example of a "typical" moment-time history from a 20 km/h test. Kajzer et al. [45] provide data from one test that resulted in ACL, PCL, and MCL avulsion and another test that produced a transverse femur fracture.

Ramet et al. [50] provided bending moment-bending angle histories for the nine quasi-static valgus bending tests for which data were successfully collected (Fig. 18.35). Based on the data shown in Fig. 18.34, Ramet et al. suggested a response corridor with the upper bound defined by the x, y (bending angle (degrees), bending moment (Nm)) coordinates (0, 25) and (25, 180) and the lower bound defined by (0, 0), (5, 6), and (25, 130).

Table 18.9 shows the bending moment and angle recorded by Kerrigan et al. [48] at the first instant of ligament injury as indicated by the acoustic emission signals. Bose et al. [51] reported the bending moment-angle response for the eight specimens loaded in pure bending using a four-point loading set-up (Fig. 18.36). The valgus bending response curves shown in Fig. 18.37 have been compensated for inertial effects using an analytical method [51]. The bending moment-angle response obtained from the experimental tests were further used by Ivarsson et al. [53] to develop a dynamic response corridor for the knee joint in pure bending.

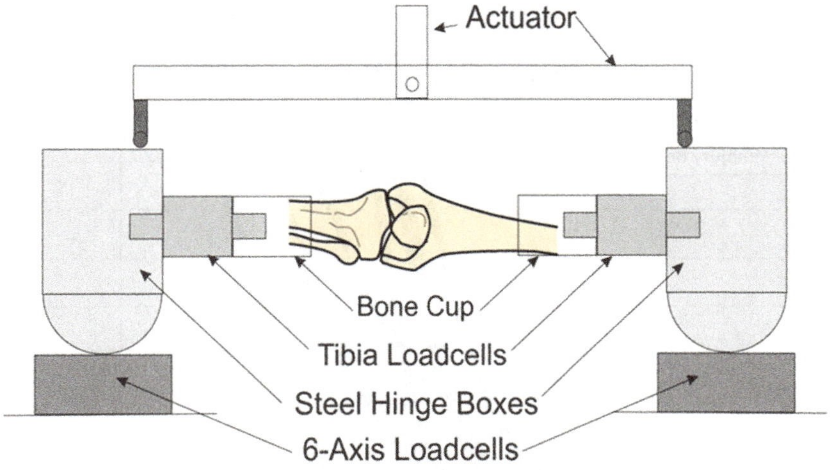

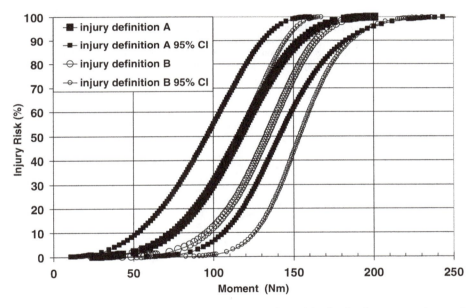

Fig. 18.31 Knee-joint lateral-medial four-point bending test schematic. Guide rails on the side of the contact plate (not shown) prevent knee flexion during bending [48] (From the work of Kerrigan et al. 2003)

Fig. 18.32 Univariate Weibull survival model predicting the risk of knee injury (MCL injury) in dynamic valgus bending of the 50th percentile male knee as function of bending moment based on injury definition A (injury occurs at the time of the first local moment peak occurring within 1–2 ms of significant acoustic emission burst) and injury definition B (injury occurs at the time of maximum moment) [53] (Reproduced with permission from Ivarsson et al. 2004)

18.3 Sports-Related Injuries

The complexity of the human lower extremity, particularly the foot and ankle, results in a wide range of possible injury patterns. This diversity makes it impossible to speak generally about "athletic lower extremity injuries", as an athlete will be predisposed towards certain subsets of injuries based on the sport in which they participate, their level of experience, competitiveness and fitness, their anthropometry, and other variables.

In order to provide some focus for this review, particular emphasis will be given to the injuries associated with American football.

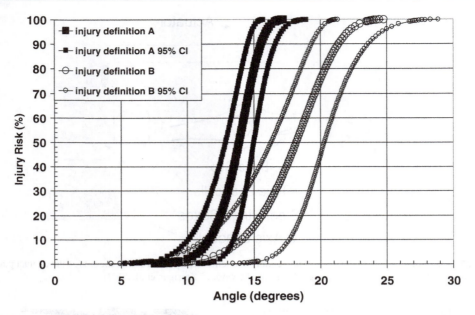

Fig. 18.33 Univariate Weibull survival model predicting the risk of knee injury (MCL injury) in dynamic valgus bending of the 50th percentile male knee as function of bending angle based on injury definition A (injury occurs at the time of the first local moment peak occurring within 1–2 ms of significant acoustic emission burst) and injury definition B (injury occurs at the time of maximum moment) [53] (Reproduced with permission from Ivarsson et al. 2004)

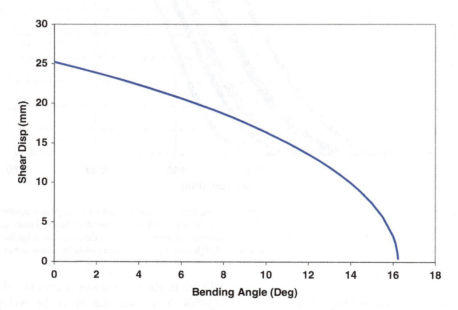

Fig. 18.34 Injury threshold function in terms of knee bending angle and shear displacement [52] (From the work of Bose et al. 2008)

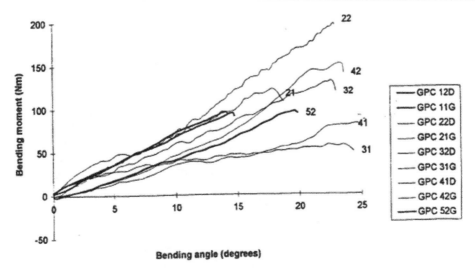

Fig. 18.35 Bending moment-bending angle histories recorded in quasi-static valgus bending tests of human knees as reported by Ramet et al. [50]. Note that the specimens used in tests GPC 11, GPC 12, GPC 31, GPC 32, and GPC 41 did not show any sign of injury at the post-test autopsy [50] (Reproduced with permission from Ramet et al. 1995)

Table 18.9 Bending test failure values [48]

Specimen	Event time (ms)	Impactor displacement (mm)	Bending moment (Nm)	Bending angle (°)
124R	15.1	11.3	142.2	13.5
167 L	13.6	10.4	140.5	12.1
135 L	14.6	8.3	130.4	11.6
Mean	14.3	9.7	136.9	12.0
SD	0.7	1.4	5.5	1.1

A study conducted by Kaplan et al. [54] on collegiate athletes at the National Football League (NFL) Scouting Combine, an event intended to evaluate player performance in preparation for the annual NFL Draft, found that 72 % of the 320 players studied had a history of foot and ankle injuries. Seven injuries were identified as the most prevalent and we will focus on three where recent advances have been made: metatarsophalangeal sprains (turf toe); tarsometatarsal (Lisfranc) dislocations; and, syndesmotic sprains.

18.3.1 Metatarsophalangeal (MTP) Joint Sprains

The hallux is critical to athletic performance, particularly in sports involving large amounts of running and cutting [55]. Partial or total tears of the ligamentous structures of the first metatarsophalangeal (MTP1) joint are referred to colloquially as "turf toe" [56]. Unfortunately, the informality of this term often causes people to dismiss the severity of MTP1 joint sprains. Players with MTP1 sprains may require weeks before returning to full participation, and up to 6 months of recovery when surgery is required [57].

Three mechanisms of injury have been identified and are illustrated in Fig. 18.38. The retrospective study conducted by Rodeo et al. [59] suggests that hyperextension is the dominant injury mechanism (85 %), while hyperflexion (12 %) and valgus/varus loading (4 %) are much less common. Please note that this study was specific to professional football; the distribution of injury mechanisms may vary in different sports [60].

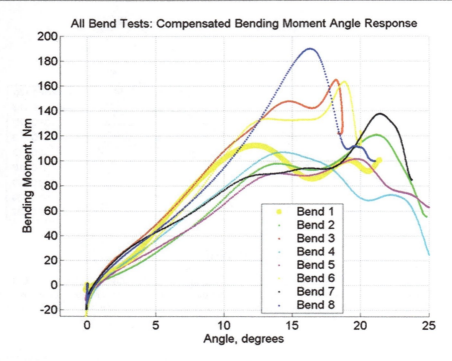

Fig. 18.36 Valgus bending moment–bending angle response from four-point bending tests of cadaveric knee joints [51] (From the work of Bose et al. 2004)

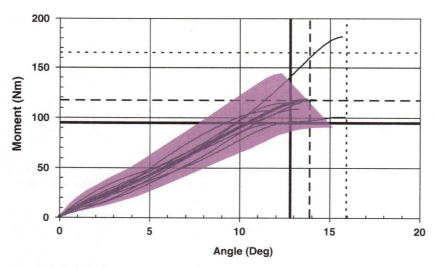

Fig. 18.37 Moment-angle response corridor for the 50th percentile male knee subjected to dynamic four-point valgus bending. The scaled responses from the tests used for developing the corridors along with their characteristic average (in *bold*) appear on the plots. The *solid, dashed,* and *dotted lines* indicate bending moment and bending angle levels associated with 25 %, 50 %, and 95 % risk of knee injury, respectively [53] (Reproduced with permission from Ivarsson et al. 2004)

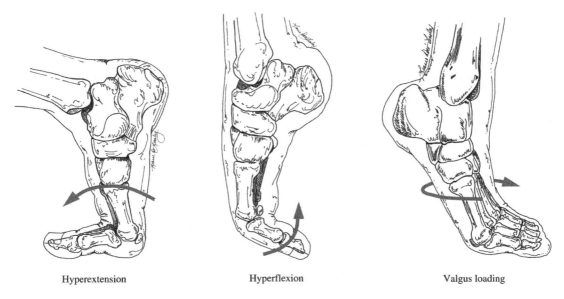

Hyperextension　　　　Hyperflexion　　　　Valgus loading

Fig. 18.38 Three mechanisms of injury for MTP1 sprains [58] (from the work of Frimenko et al. 2012)

Very little cadaveric experimentation has been performed to quantify the tolerance of the MTP1 joint to injurious loading. An early study by Prieskorn et al. [61] performed forced hyperextension of the hallux in a manner that produced relevant injuries. Hyperextension was performed at a rate of 489°/s to a maximum angle of 110°. Loading was applied using an Instron testing system, and a cable and flywheel was used to transform this linear displacement into rotation. Unfortunately, the data is reported in terms of the linear force recorded at the Instron rather than the moment applied to the hallux. Five tests were performed to failure; however, the absence of any sub-catastrophic tests prohibits the development of an angle- or moment-based injury criterion.

Work has recently been performed at the University of Virginia that attempts to overcome the limitations of the Prieskorn study [61] such that a hyperextension injury criterion might be developed [62]. Twenty cadaveric limbs were tested by subjecting the toes to prescribed levels of dorsiflexion. During each test, the angle of the MTP1 joint was calculated in three-dimensions using Vicon motion capture markers rigidly fixed to the first metatarsal and proximal phalanx bones. The maximum hallux dorsiflexion angles (θ_{max}) varied between 44° and 92° and produced both an injured and non-injured population. Injury status was confirmed by post-test necropsies. A two-parameter Weibull survival analysis resulted in a probability of injury given by:

$$P = 1 - \exp\left(-\left(\frac{\theta_{max}}{80.57°}\right)^{9.2}\right).$$

This analysis suggests that a player has a 50 % risk of MTP1 sprain with a hallux dorsiflexion angle of 78° (95 % confidence interval: 21°–88°).

18.3.2 Tarsometatarsal (Lisfranc) Sprains and Dislocations

The tarsometatarsal (TMT) joint is often referred to as the Lisfranc joint in honor of Jacques Lisfranc de Saint-Martin, the nineteenth century French military surgeon. He did not study midfoot sprains and dislocations, but rather described a rapid means of foot amputation through the TMT joint. His name has also been applied to other structures within the midfoot, such as the strong, oblique ligament bundle that connects the medial cuneiform (C1) to the base of the second metatarsal (M2).

The terms Lisfranc sprain or Lisfranc injury are often used as shorthand for either an injury to the TMT joint itself or to an injury specific to the C1/M2 ligament. These injuries are not limited to athletic contexts. A recent review of over 2,000 cases documented in the literature suggests that 43 % occurred in automotive accidents, 24 % were due to falls, jumps and twists, 13 % resulted from crushing injuries, while only 10 % were sports-related [63].

It should be pointed out that the TMT joint is really a series of articulations between the tarsal and metatarsal bones. More detailed descriptions of the anatomy are given elsewhere (e.g. Ref. [64]), but a few important features of this region are worth noting. First, the arrangement and geometry of the bones is integral to the longitudinal and transverse arches of the foot. The wedge-shaped base of M2, in particular, is often seen as acting as the keystone of the transverse arch. Second, the base of M2 is recessed relative to the other metatarsals. This mortise configuration makes the second ray the stiffest and least mobile [65].

The C1/M2 ligament bundle, which stabilizes the base of the second metatarsal, is critical to the integrity of the midfoot. It is generally agreed to consist of a dorsal and an interosseus ligament; however, some authors also identify a third, plantar ligament. Both the geometry and the mechanical behavior of these ligaments have been examined in some detail [66–68], with the latter characterized using component-level tensile testing. Further study is needed to quantify the failure properties of these ligaments under boundary conditions more representative of those sustained during injurious motion.

By comparison, the biomechanics of the TMT joint are much less well understood. Some early exploratory investigations were performed in the 1960s and 1970s to qualitatively describe the effects of various loading patterns [69–71]. For example, forefoot eversion was observed to laterally dislocate the metatarsals, whereas forefoot dorsiflexion did not produce a dislocation [69, 70]. Unfortunately, the experimental approaches used in these early studies are poorly described and documented. This uncertainty makes inter-study comparisons difficult, particularly the lack of quantitative details regarding applied loads and moments.

A further limitation of these early studies is that they produced somewhat contradictory or confusing results. For example, crushing was not observed to produce dislocation [70], despite ample epidemiologic evidence that this is a common cause of injury [63]. Methodological issues, such as loading rate and the location load is applied, are likely responsible for the failure to produce crushing injuries. These results underscore a challenge confronting investigators. The anatomy and biomechanics of the foot generally, and the midfoot specifically, are extremely complicated and the possible patterns of super-imposed loads are extensive. The large number of variables makes it extremely difficult to generate specific midfoot injuries in a controlled, repeatable manner while simultaneously subjecting the foot to kinematics and kinetics that are physiologically relevant.

A second experimental approach has been serial-sectioning: artificially inducing ligamentous disruption, loading the injured foot, and then observing the changes in mechanical behavior [72–75]. While this technique is helpful for understanding the loading required to generate clinically observed dislocation patterns, it does not necessarily follow that those same loads are responsible for creating the ligamentous injuries. Nevertheless, these studies support the importance of the midfoot ligaments for maintaining the integrity of the foot, along with the roles of external and muscle loading in determining the resulting dislocation injury patterns.

More recently, Smith et al. [76, 77] performed pendulum impact tests. These experiments were intended to mimic the effects on the foot of brake-pedal loading or toe-pan intrusion during automotive impacts. While not designed to generate TMT dislocations exclusively, the tests did produce these injuries in some cases. Feet were impacted in either a so-called "plantar normal" (ankle with 0° plantarflexion) or a "plantar flexed" configuration (ankle 30°–50° plantarflexion). The neutral configuration was abandoned after 13 tests as very few injuries were created; however, three tests successfully generated TMT dislocations (23 %). The plantarflexed condition resulted in 11 general TMT dislocations, with 5 injuries specific to the C1/M2 ligaments.

Of course, most of these studies are not specific to the mechanics and loading rates expected in athletes. A common consensus within the literature is that athletic TMT injuries, particularly in American football, are a result of loading along the axis of the foot with some level of superimposed rotation. Recent tests have investigated this loading combination in an Instron test machine [78]. The foot was loaded along the long axis of the foot, with the toes dorsiflexed. While injuries were successfully generated along the first ray in the navicular-first cuneiform and the first cuneiform-first metatarsal ligaments, the very stout C1/M2 ligaments were uninjured.

18.3.3 Syndesmotic Ankle Sprains

Syndesmotic ankle sprains, also referred to as high-ankle sprains, are a common source of injury in professional football players. Kaplan et al. [54] found that they were second only to lateral ankle sprains in the most common foot and ankle injuries reported by collegiate players at the NFL Scouting Combine. Despite being less common than lateral ankle sprains, syndesmotic sprains are much more severe and require more prolonged recovery time [79, 80].

Three structures maintain the tibiofibular syndesmosis: the anterior tibiofibular ligament (ATFL), the posterior tibiofibular ligament (PTFL), and the interosseus ligament (IOL) sheet. Guise [79] hypothesized that supination-internal rotation of the foot and pronation-external rotation of the foot were the primary mechanisms for damaging these ligaments. While rotation is commonly cited as a contributing factor [81], evidence for dorsiflexion of the ankle has also been presented [82]. The latter is significant as Close [83] has shown that dorsiflexion of the talus caused an increase in the distance between the malleoli. This increased separation between the distal tibia and fibula would place greater strain on the syndesmotic ligaments.

Serial-sectioning studies have been performed to document the effects of the various ligaments on the integrity and behavior of the ankle [83–85]. These studies have confirmed that tibiofibular diastasis increases as the ligaments are severed (ATFL, IOL, then PTFL). Since diastasis is a frequent clinical means of diagnosing syndesmotic sprains, this research is certainly of interest to the medical community. Unfortunately, again like the tarsometarsal ligament sectioning experiments described in the previous section, these studies fail to provide any insight into the mechanisms or loads required to generate injuries.

More recently, a great deal of work has been performed at Michigan State University using both computational and experimental modeling of syndesmotic sprains [86–89]. Wei et al. [88] performed experimental studies where cadaveric lower extremities were dissected distal to the knee and then embedded in epoxy. A custom biaxial testing device was then used to rotate the 20° dorsiflexed foot through controlled amounts of external rotation. Failure was diagnosed through changes to the torque-time curves, and resulted in failure torques of 85–90 Nm. Both neutral and highly everted foot positions were also investigated. No statistically significant differences in failure torque were observed; however, the rotational angle of failure for the neutral foot (38°) was significant lower than in the everted foot (47°). The patterns of injury were also notably different. Complete or partial rupture of the ATFL was observed in all five feet tested in the everted condition, whereas the anterior deltoid ligament was the ligamentous structure most often injured in the neutral orientation.

18.3.4 Summary

The study of athletic lower extremity injury lags well behind the state-of-the-art research being performed in other contexts. Only recently has the knowledge gained through automotive and military injury research begun to be applied to the study of sports injuries. Further biomechanical investigations, performed under loading regimes and rate representative of athletic conditions, are still needed.

Even more importantly, this new knowledge must now be applied to improve the safety of sport. Data-based changes to athletic equipment, to athletic playing surfaces, or the rules of competition, must all be considered to protect the bodies and careers of athletes. Translating this fundamental research into the development of appropriate countermeasures will be critical to ensuring that all athletes – professional or amateur – can maximize their performance while maximizing their safety.

18.4 Military-Related Injuries

While overlap exists between lower extremity injuries induced by automobile accidents and military incidents, military-induced injuries encompass a larger range of rates than automotive injuries. Under this extended range of loading conditions, the body and its materials react differently than under the comparatively tame loading environment of automotive accidents. Early lower extremity injury biomechanics efforts focused on the relatively easier to reproduce automotive loading conditions. However, recent military conflicts have lead researchers to change their focus to higher rate loading. Biomechanics research on military-induced lower extremity injury can be divided into two main categories: landmine blast and vehicle underbelly blast.

18.4.1 Landmine Blast

Anti-personnel (AP) landmines are one cause of military lower extremity injuries that biomechanists are investigating. These mines generally consist of less than 250 g of explosive [90] and cause severe injury to the lower extremity as well as injury to other parts of the body from blast air velocities of up to 1,500 m/s [91]. Due to the fact that these landmines remain in position after military conflicts conclude, they continue to threaten the safety of civilians even after the conflict is over. The frequency of injury due to AP landmines has led to the development of various types of personal protective equipment (PPE) to

Table 18.10 Summary of PMHS AP landmine tests

Study	Acceleration	Charge size	Force (kN)
Harris et al. 1999 [92]			3.89–11.0
Makris and Islam 2000 [95]	150–10,300 g	25–100 g C4	7.13–10
Dionne et al. 2001 [94]	500–6,500 g	25–200 g C4	
Chaloner et al. 2002 [96]	565–863 g	25–100 g C4	8.16–20.13
Bass et al. 2012 [97]	75, 100, 200 g C4		8.6 (50 % fracture)
Wolff et al. 2005 [98]		100 g TNT	

decrease the severity of injury to the lower extremity. These PPE boots include the Wellco™ Blast Boot, and Med-Eng™ boot, and the BFR™ boot, which help to shield the foot from forces created by the landmine blast.

Several research efforts have focused on determining the efficacy of these boots and others (Table 18.10). Harris et al. [92] performed a study that used four PMHS with implanted tibia load cells to determine the effectiveness of different boots at shielding the tibia from injurious forces. Injuries sustained from these tests were coded using the Mine Trauma Score. This study revealed that most damage was localized to the distal portions of the lower extremity since femur and tibia strains were below those associated with fracture as reported by Reilly and Burstein in 1975 [93]. In these cases the damage to the distal portion of the lower extremity dissipates energy and decreases the amount of energy transmitted proximally. Forces for these tests ranged from 4 kN to just over 11 kN with time to peaks ranging from less than half a millisecond to just under 2 ms.

Dionne et al. [94] performed similar tests comparing the efficacy of landmine-resistant boots using a "non-biofidelic mechanical leg". For these tests, different C4 charge masses were used and the resulting accelerations were compared for each of the different boots and for different charge masses (Fig. 18.39). From the mechanical leg data, Dionne et al. concluded that the spider boot could reduce accelerations by up to 90% [94]. The effectiveness of the spider boot was then verified using PMHS tests, which reduced the severity of injury when compared to other boots under the same loading conditions. The spider boot's ability to successfully redirect the loading path into the lower extremity proved to be a significant contribution to understanding injury prevention under such intense loading conditions.

Further testing was performed as part of the Lower Extremity Assessment Program (LEAP) that consisted of 38 tests with 20 PMHS to determine the performance of a variety of military footwear [99]. Each specimen was given a Mine Trauma Score according to the severity of injury incurred during the testing, which consisted of an AP mine blast of one of three levels intensity. From this PMHS testing, Bass et al. [97] found that 8.6 kN produced a 50 % risk of AFIS-S 2 injury or greater. This injury risk function is shown in Fig. 18.40. Note that comparing the injury risk function in Fig. 18.40 to the injury risk function developed for automotive rates (Fig. 18.12), determines that a higher force is required for a 50 % risk of injury than for the higher rate loading. This observation contributes to the need for a rate-sensitive injury criterion for military applications that often consist of higher rate loading.

18.4.2 Vehicle Underbelly Blast

Vehicle underbelly blast (UBB) is another prevalent cause of injury in military conflicts that has sparked interest in the field of biomechanics. Medical injury data from the field reveals that 32 % of soldiers wounded in UBB events sustained foot/ankle fractures, while 16 % sustained both foot/ankle and tibia/fibula fractures [101]. For soldiers killed in UBB, the blast's destruction has a larger reach: 28 % sustained foot/ankle fractures, while 27 % sustained both foot/ankle and tibia/fibula fractures. This trend leads to questions of the load and energy transmissibility of the lower extremities during the high-rate, low momentum conditions of UBB events.

High rate axial load has been identified as the main injury mechanism in the lower limb resulting from UBB [107]. There is currently no objective test methodology for determining the transmission of forces, nor the risk of injury to the lower extremities due to foot-well intrusion from UBB; however, the significant body of research on lower extremity injuries resulting from automobile crashes can act as a basis for comparing more recent high-rate loading studies. Lower limb injuries sustained in automobile crashes have been heavily researched due to their frequency and high likelihood of impairment and disability. A review of the literature suggests that lower limb injuries account for roughly one-third of all moderate-to-serious injuries sustained by motor vehicle occupants involved in frontal crashes [108–110]. Since intrusion of the foot-well region is often postulated as the primary mechanism of below-knee lower limb injuries, intrusion characteristics such a s toe-pan displacement, toe-pan acceleration, intrusion onset rate, intrusion duration, and intrusion initiation time have been examined for their potential to produce lower limb trauma. Testing performed at the University of Virginia provides a basis of knowledge on the effects of intrusion on injury risk using both component tests and sled tests [100, 109, 111, 112].

The injury mechanisms of the lower limb associated with intrusion of the foot-well include inertial loading, entrapment, excessive motion of the joints, and subsequent contact with other structures within the occupant compartment [113]. In terms of mechanisms associated with these injuries, the most severe trauma is normally sustained from axial loading of the limb. Biomechanical testing has been conducted to develop basic injury criteria for axial loading of the below-knee structures. For automotive rates of loading, Yoganandan et al. [104] conducted a series of axial impact tests to the human foot-ankle complex and found a mean dynamic force at fracture (calcaneus and distal tibia) to be 15.1 kN. Funk et al. [100] determined injury risk function for axial loads to the foot/ankle complex from a study that included axial loads up to approximately 12 kN.

Other injury mechanisms and criteria for the lower limb structures include ankle dorsiflexion [39], hindfoot xversion [100], bending of the

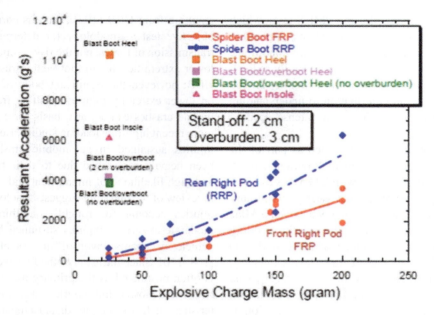

Fig. 18.39 Comparison of acceleration to explosive charge mass [94] (Figure and data courtesy of Allen-Vanguard Corporation, testing conducted at DRDC Suffield, Canada)

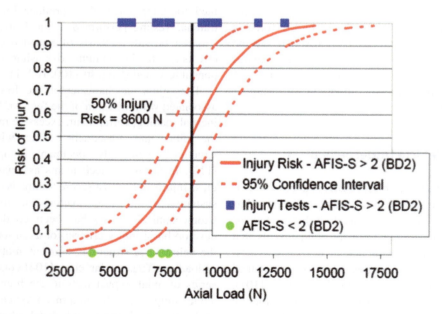

Fig. 18.40 Injury risk function for AFIS-S>2 injury [97] (From the work of Bass et al. 2004)

tibia and fibula [114], and bending of the femur [115, 116]. While these tests provide valid criteria for the automotive environment, they possess several critical limitations for these in the development of injury countermeasures in the UBB environment. These tests have not been developed for rates of loading indicative of the vehicle UBB environment, and they have not included footwear. The applicability of studies and criteria remain unanswered despite the more recent higher-rate research being performed.

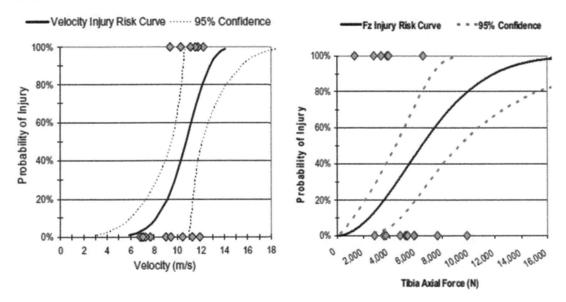

Fig. 18.41 Injury risk curves for higher rate loading using Weibull Regression [103] (Reproduced with permission from the Stapp Association)

The appropriateness of automotive rate tests for UBB applications can be understood by examining the different injury criteria developed from each loading type. For example, in Funk et al. [100], a test series investigating automotive intrusion, produced average toe-pan velocities of 5 m/s with a load duration of 10 ms (approximately 50 g of acceleration, compared to estimates of UBB accelerations of over 500 g). In that study of 43 specimens, this load rate produced 9 talus, 25 calcaneus, 7 pilon, 4 medial malleolus, 8 lateral malleolus, 2 fibula, and 12 tibia plateau fractures. McKay et al. [103] observed similar injuries for higher velocity loading conditions ranging from 7.2 to 11.6 m/s with load durations of 3 ms. However, the injury criteria developed by McKay et al. [103] (Fig. 18.41) differed from the Funk et al. [100] criteria. For the higher velocity loading, the 50% risk of injury occurs at 6.4 kN [103] as opposed to 8.3 kN for 50% injury risk for a 45 year old male [100]. This significant difference in injury criteria necessitates further research on the effect of loading rate on injury to the lower extremities.

In 2001, Wang reported that UBB can produce accelerations averaging 100 g over timespans of 3–10 ms with a floor plate velocity of 12 m/s [117]. For most tests performed, the acceleration is not reported. Others have reported floor plate velocities produced by mine blasts to reach up to 30 m/s in 6 to 10 ms [107]. A summary of previous lower extremity tests with sufficient detail for cross-comparison is shown in Table 18.11. Several efforts to reproduce UBB type loading in a controlled environment have been made. These include linear impactors as well as actual explosions. A summary of the lower extremity tests pertaining to UBB is shown in Tables 18.12 and 18.13.

Anti-vehicular mine blast injuries can be subdivided into four different categories based on the injury mechanism. Ramasamy et al. [118] summarized these categories and their injury mechanisms as presented in Table 18.14. According to statistics collected from various sources, Ramasamy further breaks down these types of injuries as they apply to the bones of the lower extremities, and presents the most prevalent mechanisms of injury present in UBB in Table 18.12.

The high rates at which underbelly blast loads the lower extremities have produced a need for an improved anthropometric test dummy (ATD). The MIL-Lx and the THOR lower extremities were built with the intent to better mimic the characteristics of the human leg under high rate

Table 18.11 Summary of previous lower extremity underbody blast research

	Boundary condition	Hammer mass (kg)	Velocity (m/s)	Max energy (J)	Force (kN)	Acceleration (G)
Schueler et al. 1995 [102]	Whole body	38	12.5	2,968.8	16	250
McKay and Bir 2009 [103]	Femur potted	36.7	7.2–11.8	941–2,494	4–6	Unreported
Yoganandan et al. 1996 [104]	Ballast 16.8 kg	25	7.6	722	4.3–11.4	Unreported
Kitagawa et al. 1998 [105]	Fixed end	18	3.99	143.3	7–9	Unreported
Quenneville et al. 2010 [106]	Free end	3.9	13.9	109.6	15	Unreported

Table 18.12 Common UBB injury mechanisms [107]

Explosion type	Pathophysiology	Fracture characteristics
Primary	Blast wave-mediated fracture	Traumatic amputation; short oblique/transverse fractures
Secondary	Direct impact of fragment	Highly comminuted multi-fragmentary fractures
Primary and secondary	Contact with the seat of explosion, resulting in wave and fragment injury (e.g. anti-personnel landmine explosion)	Traumatic or sub-total amputation with significant soft tissue injury and fragments
Tertiary	Displacement of the casualty or objects near the casualty	Axial loading, three-point bending, spiral fractures

Table 18.13 Summary of PMHS laboratory underbelly blast tests

Study	Velocity (m/s)	Duration (ms)	Energy (J)	Force (kN)	50 % Risk
Roberts 1993 [122]	4.6			12.2	
Schueler et al. 1995 [102]	12.5		2,968.8	16	9.7 m/s
Yoganandan et al. 1996 [104]	2.2–7.6		145–722	4.3–11.4	25 years = 7.0 kN, 45 years = 5.4 kN, 65 years = 3.8 kN
Bir et al. 2008 [119]	3.8–7.1		265–926		
McKay and Bir 2009 [103]	7.2	3	941	5.38	6.429 kN weibull
McKay and Bir 2009 [103]	9.9	3	1,802	4.58	5.931 kN log reg
McKay and Bir 2009 [103]	11.6	3	2,494	4.4	10.8 m/s Weibull

loading. Studies performed using the Hybrid-III ATD leg have shown that the stiffness of the Hybrid-III leg in comparison to PMHS legs produces forces up to three times those present in PMHS specimens under the same loading conditions [119]. With the motivation to produce an ATD lower extremity which can more accurately predict the forces a human leg will experience, a number of studies have been completed in order to develop as well as validate newer ATD limbs such as the THOR and the MIL-Lx [119–121]. These studies are summarized in Tables 18.13 and 18.15.

While many lower extremity studies have focused on axial loading of the foot and leg, the methodology for producing the loading is varied among these studies. Studies performed by Bir and McKay at Wayne State University used 24

18 Leg, Foot, and Ankle Injury Biomechanics

Table 18.14 Classification of AV-mine blast injuries [118]

Blast injury	Mechanism of injury	Clinical effects	Mitigation requirement	Vehicle mitigation
Primary	Blast shock wave	Traumatic amputation, soft tissue deformation, fractures via brisance effect	Reduce blast transfer	Increased standoff and improved gas dynamic characteristics mitigating brisance and the inclusion of blast mitigating materials
Secondary	Fragments from mine products, energized soil ejecta, and vehicle fragments	Penetrating wounds to the lower extremity; extremity fractures from direct impact of fragments	Reduce fragments, or protect against fragments	Improved armor protection of vehicle floor and improved personal protection
Tertiary	Global—vehicle acceleration Local—floor pan deformation	Significant axial loading leading to lower limb (especially calcaneal) injuries	Reduce vehicle acceleration, reduce capture of pressure wave by vehicle, increase resistance to floor plate geometrical changes	Increased standoff. V-shaped hull design. Occupant restraints to prevent collision injuries
Quaternary	Thermal injuries	Burns	Protect against burns	Fire resistant materials in vehicles and fire retardant clothing

Table 18.15 Summary of ATD underbelly blast tests

Study	Velocity	Duration	Impactor mass	Energy	Force
Hybrid-III studies:					
Bir et al. 2008 [119]	4.7 m/s	6 ms	24 kg	265 J	10.017 kN
Bir et al. 2008 [119]	8.3 m/s	NA	37 kg	1,274 J	15–20 kN
Quenneville and Dunning 2012 [121]	2–7 m/s	1.8–3.3 ms	6.8 kg	14–167 J	
THOR studies:					
Bir et al. 2008 [119]	4.7 m/s	12 ms	24 kg	265 J	3.845 kN
Bir et al. 2008 [119]	8.3 m/s	3 ms/10 ms	37 kg	1,274 J	7.316 kN/8.646 kN
MIL-Lx studies:					
Quenneville and Dunning 2012 [121]	2–7 m/s; 3.4 m/s	15.2–20.5 ms	6.8 kg	14–167 J	3 kN
McKay and Bir 2009 [103]	2.6–7.2 m/s	13 ms	36.7 kg	124–951 J	4.5 kN

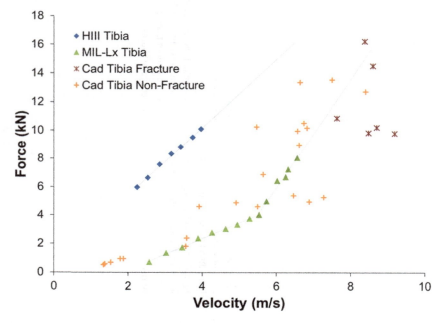

Fig. 18.42 Hybrid-III, MIL-Lx and PMHS axial tibia force comparison for various impact velocities [121] (Reproduced with permission from Quenneville and Dunning 2012)

and 37 kg impact masses at velocities ranging from 4.7 to 11.6 m/s to produce different energy inputs [103, 119]. Quenneville and Dunning [121] used a much smaller impact mass of 6.8 kg at velocities ranging from 2 to 7 m/s. Each of these tests was meant to represent axial UBB loading to the lower extremities, but the test conditions were very different (Fig. 18.42).

A number of tests have performed comparisons of PMHS and ATD lower extremity response. These tests have revealed that at higher rates of loading, the Hybrid-III leg can become up to three times stiffer than the PMHS leg [119]. Thus when using the Hybrid-III ATD at higher rates the current leg injury standard of 5.4 kN for 10 % risk of injury overestimates risk of injury [90]. Other ATD lower extremities such as the Thor-LX and the Mil-LX have been found to produce more realistic results in comparison with PMHS results; similar peak forces were observed for the same loading conditions in the PMHS, Thor-LX, and Mil-LX [119].

While a variety of research has focused on developing injury criteria for the lower extremities under high rate loading conditions, little work has focused on the upper leg. Seven percent of those wounded in a UBB event had femur fractures while 33 % of those killed in UBB events had femur fractures [123] and injury risk statistics derived from live fire ATD testing and pre-existing injury criteria suggest there is a 35 % risk of injuring the upper leg during a UBB event [124], however no PMHS testing currently exists to validate this statistic. Earlier research has produced a variety of information about femur injury criteria for automotive rates, but it is not yet clear whether this research can be applied to military-induced high rate injuries [14–16].

References

1. Crandall J, Portier L, Petit P, Hall G, Klopp G, Bass C, Hurwitz S, Trosseille X, Tarriere C, Pilkey W, Lavaste F, Lassau M (1996) Biomechanical response and physical properties of the leg, foot, and ankle. Paper 962424, Society of Automotive Engineering
2. Parham KR, Gordon CC, Bensel CK (1992) Anthropometry of the foot and lower leg of U.S. army soldiers: Fort Jackson, SC—1985. United States Army Natick Research, Development and Engineering Center, Natick
3. Diffrient N, Tilley AR, Bardagjy JC (1974) Humanscale 1/2/3. The MIT Press, Massachusetts Institute of Technology, Cambridge
4. Rupp J, Reed M, Van Ee C, Kuppa S, Wang S, Goulet J, Schneider L (2002) The tolerance of the human hip to dynamic knee loading. Stapp Car Crash J 46:211–228
5. Kuppa S, Fessahaie O (2003) An overview of knee-thigh-hip injuries in frontal crashes in the United States. Paper 416. In: Proceedings of the 18th international technical conference on the enhanced safety of vehicles
6. Nyquist G (1986) Injury tolerance characteristics of the adult human lower extremities under static and dynamic loading. SAE paper 861925, Society of Automotive Engineers
7. Mather BS (1967) Correlations between strength and other properties of long bones. J Trauma 7(5):633–638
8. Mather SB (1968) Impact tolerance of the human leg. J Trauma 8(6):1084–1088
9. Yamada H (1970) Strength of biological materials. The Williams and Wilkins, Baltimore
10. Nyquist G, Cheng R, El-Bohy A, King A (1985) Tibia bending: strength and response. SAE Paper 851728, The Society of Automotive Engineers
11. Rabl W, Haid C, Krismer M (1996) Biomechanical properties of the human tibia: fracture behavior and morphology. Forensic Sci Int 83:39–49
12. Schreiber P, Crandall J, Micek T, Hurwitz S, Nusholtz G (1997) Static and dynamic bending strength of the leg. In: Proceedings of the IRCOBI conference on the biomechanics of impact. Hanover, 24–26 Sept 1997
13. Kerrigan J, Bhalla K, Madeley NJ, Funk J, Bose D, Crandall J (2003a) Experiments for establishing pedestrian-impact lower limb injury criteria'. SAE 2003-01-0895. Society of Automotive Engineers (SAE) Congress, Detroit
14. Kerrigan J, Bhalla K, Madeley NJ, Crandall J, Deng B (2003b) Response corridors for the human leg in 3-point lateral bending. Paper 1281. The 7th US National Congress on computational mechanics, Albuquerque
15. Kerrigan J, Drinkwater DC, Kam CY, Murphy D, Ivarsson BJ, Crandall J, Patrie J (2004) Tolerance of the human leg and thigh in dynamic latero-medial bending. Int J Crashworthiness 9(6):607–623
16. Ivarsson J, Kerrigan J, Lessley D, Drinkwater C, Kam C, Murphy D, Crandall J, Kent R (2005) Dynamic response corridors of the human thigh and leg in non-midpoint three-point bending. Paper no. 2005-01-0305, Society of Automotive Engineers
17. Carter DR, Hayes WC (1977) The compressive behavior of bone as a two-phase porous structure. J Bone Joint Surg 59(7):954–962
18. McElhaney JH (1966) Dynamic response of bone and muscle tissue. J Appl Physiol 21(4):1231–1236
19. Martens M, van Audekercke R, de Meester P, Mulier J (1980) The mechanical characteristics of the long bones of the lower extremity in torsional loading. J Biomech 13:667–676
20. Kent RW, Funk JR (2004) Data censoring and parametric distribution assignment in the development of injury risk functions from biomechanical data. SAE 2004-01-0317. Society of Automotive Engineers (SAE) Congress, Detroit
21. Crandall J, Funk J, Rudd R, Tourret L (1999) The tibia index: a step in the right direction. In: Proceedings of the Toyota international symposium on human life support biomechanics. Nagoya, Japan
22. Mertz HJ (1993) Antropometric test devices. In: Nahum AM, Melvin JW (eds) Accidental injury: biomechanics and prevention. Springer-Verlage, New York
23. Kjaersgaard-Anderson P, Wethelund JO, Helmig P, Soballe K (1988) The stabilizing effect of the ligamentous structures in the sinus and canalis tarsi on

movements in the hindfoot. Am J Sports Med 16(5): 512–516
24. Parenteau CS, Viano DC, Petit PY (1998) Biomechanical properties of human cadaveric ankle-subtalar joints in quasi-static loading. J Biomech Eng 120:105–111
25. Petit P, Portier L, Foret-Bruno JY, Trosseille X, Parenteau C, Coltat JC, Tarrière C, Lassau JP (1996) Quasistatic characterization of the human foot-ankle joints in a simulated tensed state and updated accidentological data. In: Proceedings of the international research council on the biomechanics of impact. Dublin, Ireland, pp 363–376
26. Jaffredo A, Potier P, Robin S, Le Coz JY, Lassau JP (2000) Cadaver lower limb dynamic response in inversion-eversion. In: Proceedings of the IRCOBI conference on the biomechanics of impact. Montpellier, France
27. Funk JR, Crandall JR, Tourret L, MacMahon C, Bass CR, Khaewpong K, Eppinger R (2002) The axial injury tolerance of the human foot/ankle complex and the effect of Achilles tension. J Biomed Eng 124:750–757
28. Morgan RM, Eppinger RH, Hennessey BC (1991) Ankle joint injury mechanism for adults in frontal automotive impact. Paper 912902, Stapp Car Crash Conference. San Diego, CA
29. Lauge-Hansen N (1950) Fractures of the ankle II: combined experimental-surgical and experimental-roentgenologic investigations. Arch Surg 60:957–985
30. Dias LS (1979) The lateral ankle sprain: an experimental study. J Trauma 19:266–269
31. Rasmussen O, Kromann Andersen C (1983) Experimental ankle injuries analysis of the traumatology of the ankle ligaments. Acta Orthop Scand 54(3):356–362
32. Begeman P, Balakrishnan P, Levine R, King A (1993) Dynamic human ankle response to inversion and eversion. In: Proceedings of the 37th Stapp Car Crash Conference. San Antonio, TX. SAE 933115, pp 83–93
33. Funk JR, Srinivasan SCM, Crandall JR, Khaewpong N, Eppinger RH, Jaffredo AS, Potier P, Petit PY (2002) The effects of axial preload and dorsiflexion on the tolerance of the ankle/subtalar joint to dynamic inversion and eversion. Stapp Car Crash J 46:245–265
34. Schreiber P, Crandall J, Dekel E, Hall G, Pilkey W (1995) The effects of lower extremity boundary conditions on ankle response during joint rotation tests. In: Proceedings of the 23rd international workshop on human subjects for biomechanical research. San Diego, CA
35. Colville MR, Marder RA, Boyle JJ, Zarins B (1990) Strain measurement in lateral ankle ligaments. Am J Sports Med 18(2):196
36. Begeman PC, Prasad P (1990) Human ankle impact response in dorsiflexion. In: Proceedings of the 34th Stapp Car Crash Conference. Society of Automotive Engineers, pp 39–53
37. Begeman P, Balakrishnan P, Levine R, King A (1992) Human ankle response in dorsiflexion. Injury prevention through biomechanics symposium proceedings, Wayne State University
38. Portier L, Petit P, Domont A, Trosseille X, Le Coz J-Y, Tarriere C, Lassau J-P (1997) Dynamic biomechanical dorsiflexion responses and tolerances of the ankle joint complex. SAE paper 973330, pp 207–224
39. Rudd R, Crandall J, Millington S, Hurwitz S (2004) Injury tolerance and response of the ankle joint in dynamic dorsiflexion. Stapp Car Crash J. Nashville, TN, 48:1–26
40. Kennett K, Crandall J, Bass C, Klopp G (1996) In situ measurement of loads in the tibia. In: Proceedings of the 24th international workshop on human subjects for biomechanical research. Albuquerque, NM
41. Fleming BC, Good L, Peura GD, Beynnon BD (1999) Calibration and application of an intra-articular force transducer for the measurement of patellar tendon graft forces: an in situ evaluation. J Biomech Eng 121(4):393–398
42. Hall G, Klopp G, Crandall J, Carmines D, Hale J (1997) Rate independent characteristics of an arthroscopically implantable force probe in the Achilles tendon. In: Proceedings of the 21st annual meeting of the American Society of Biomechanics. Clemson University, SC
43. Fleming BC, Peura GD, Beynnon BD (2000) Factors influencing the output of an implantable force transducer. J Biomech 33:889–893
44. Kajzer J, Cavallero C, Ghanouchi S, Bonnoit J, Ghorbel A (1990) Response of the knee joint in lateral impact-effect of shearing loads. In: Proceedings of the IRCOBI conference on the biomechanics of impact, Berlin, 11–13 Sept 1990
45. Kajzer J, Schroeder G, Ishikawa H, Matsui Y, Bosch U (1997) Shearing and bending effects at the knee joint at high speed lateral loading. SAE paper 973326, Society of Automotive Engineers
46. Kajzer J, Ishikawa H, Matsui Y, Schroeder G, Bosch U (1999) Shearing and bending effects at the knee joint at low speed lateral loading. SAE paper 1999-01-0712, Society of Automotive Engineers
47. Kajzer J, Cavallero C, Bonnoit J, Morjane A, Ghanouchi S (1993) Response of the knee joint in lateral impact-effect of bending moment. In: Proceedings of the international conference on the biomechanics of impact (IRCOBI), Eindhoven, 8–10 Sept 1993
48. Kerrigan JR, Ivarsson BJ, Bose D, Madeley NJ, Millington SA, Bhalla KS, Crandall JR (2003) Rate-sensitive constitutive and failure properties of human collateral knee ligaments. In: Proceedings of the international research council on the biomechanics of impacts (IRCOBI) conference. Lisbon, Portugal
49. Konosu A, Issiki T, Tanahashi M (2005) Development of a biofidelic flexible pedestrian leg-form impactor (Flex-Pli 2004) and evaluation of its biofidelity at the component level and at the assembly level. Paper 2005-01-1879, Society of Automotive Engineers (SAE)

50. Ramet M, Bouquet R, Bermond F, Caire Y (1995) Shearing and bending human knee joint tests in quasi-static lateral load. In: Proceedings of the IRCOBI conference on the biomechanics of impact. Brunnen, Switzerland
51. Bose D, Bhalla K, Rooij L, Millington S, Studley A, Crandall J (2004) Response of the knee joint to the pedestrian impact loading environment. Paper no. 2004-01-1608. World Congress SAE, Detroit
52. Bose D, Bhalla K, Untaroiu C, Ivarsson BJ, Crandall J, Hurwitz S (2008) Injury tolerance and moment response of the knee joint to combined valgus bending and shear loading. J Biomech Eng 130(3):031008. doi: 10.1115/1.2907767
53. Ivarsson J, Lessley D, Kerrigan J, Bhalla K, Bose D, Crandall J, Kent R (2004) Dynamic response corridors and injury thresholds of the pedestrian lower extremities. In: Proceedings of the IRCOBI conference on the biomechanics of impact, Graz
54. Kaplan LD, Jost PW, Honkamp N, Norwig J, West R, Bradley JP (2011) Incidence and variance of foot and ankle injuries in elite college football players. Am J Orthop 40(1):40–44
55. Orendurff MS, Rohr ES, Segal AD, Medley JW, Green JR III, Kadel NJ (2008) Regional foot pressure during running, cutting, jumping, and landing. Am J Sports Med 36(3):566–571. doi:10.1177/0363546507309315
56. Bowers KD Jr, Martin RB (1976) Turf-toe: a shoe-surface related football injury. Med Sci Sports 8(2):81–83
57. Anderson RB (2002) Turf toe injuries of the hallux metatarsophalangeal joint. Tech Foot Ankle Surg1(2):102–111.doi:10.1097/00132587-200212000-00004
58. Frimenko RE, Lievers WB, Coughlin MJ, Anderson RB, Crandall JR, Kent RW (2012) Etiology and biomechanics of first metatarsophalangeal joint sprains (turf toe) in athletes. Crit Rev Biomed Eng 40(1):43–61
59. Rodeo SA, O'Brien S, Warren RF, Barnes R, Wickiewicz TL, Dillingham MF (1990) Turf-toe: an analysis of metatarsophalangeal joint sprains in professional football players. Am J Sports Med 18(3):280–285
60. Frey C, Andersen GD, Feder KS (1996) Plantar-flexion injury to the metatarsophalangeal joint ("sand toe"). Foot Ankle Int 17(9):576–581
61. Prieskorn D, Graves S, Yen M, Ray J Jr, Schultz R (1995) Integrity of the first-metatarsophalangeal joint: a biomechanical analysis. Foot Ankle Int 16(6):357–362
62. Frimenko RE, Lievers WB, Riley PO, Park JS, Hogan MV, Crandall JR, Kent RW (2013) Development of an injury risk function for first metatarsophalangeal joint sprains. Med Sci Sports Exerc. 45(11):2144–2150. doi:10.1249/MSS.0b013e3182994a10
63. Lievers WB, Frimenko RE, Crandall JR, Kent RW, Park JS (2012) Age, sex, causal and injury patterns in tarsometatarsal dislocations: a literature review of over 2000 cases. Foot (Edinb) 22(3):117–124. doi:10.1016/j.foot.2012.03.003
64. de Palma L, Santucci A, Sabetta SP, Rapali S (1997) Anatomy of the Lisfranc joint complex. Foot Ankle Int 18(6):356–364
65. Ouzounian TJ, Shereff MJ (1989) In vitro determination of midfoot motion. Foot Ankle 10(3):140–146
66. Johnson A, Hill K, Ward J, Ficke J (2008) Anatomy of the Lisfranc ligament. Foot Ankle Spec 1(1):19–23. doi:10.1177/1938640007312300
67. Kura H, Luo ZP, Kitaoka HB, Smutz WP, An KN (2001) Mechanical behavior of the Lisfranc and dorsal cuneometatarsal ligaments: in vitro biomechanical study. J Orthop Trauma 15(2):107–110
68. Solan MC, Moorman CT III, Miyamoto RG, Jasper LE, Belko SM (2001) Ligamentous restraints of the second tarsometatarsal joint: a biomechanical evaluation. Foot Ankle Int 22(8):637–641
69. Jeffreys TE (1963) Lisfranc's fracture-dislocation: a clinical and experimental study of tarso-metatarsal dislocations and fracture-dislocations. J Bone Joint Surg 45B(3):546–551
70. Wilson DW (1972) Injuries of the tarso-metatarsal joints: etiology, classification and results of treatment. J Bone Joint Surg 54B(4):677–686
71. Wiley JJ (1971) The mechanism of tarso-metatarsal joint injuries. J Bone Joint Surg 53B(3):474–482
72. Charrois O, Béqué T, Mulier GP, Masquelet AC (1998) Luxation plantair de l'articulation tarso-métatarsienne (articulation de Lisfranc): a propos d'un cas [Lisfranc plantar fracture dislocation: a case report]. Rev Chir Orthop 84(2):197–201
73. Nishi H, Takao M, Uchio Y, Yamagami N (2004) Isolated plantar dislocation of the intermediate cuneiform bone: a case report. J Bone Joint Surg 86A(8):1772–1777
74. Kadel N, Boenisch M, Tietz C, Trepman E (2005) Stability of Lisfranc joint in ballet pointe position. Foot Ankle Int 26(5):394–400
75. Kaar S, Femino J, Morag Y (2007) Lisfranc joint displacement following sequential ligament sectioning. J Bone Joint Surg 89A(10):2225–2232. doi:10.2106/JBJS.F.00958
76. Smith BR, Begeman PC, Leland R, Levine RS, Yang KH, King AI (2003) A mechanism of injury to the forefoot in car crashes. In: Proceedings of the 2003 international IRCOBI conference on the biomechanics of impact, Lisbon, 25 Sept 2003
77. Smith BR, Begeman PC, Leland R, Meehan R, Levine RS, Yang KH, King AI (2005) A mechanism of injury to the forefoot in car crashes. Traffic Inj Prev 6(2):156–169. doi:10.1080/15389580590931635
78. Frimenko RE, Lievers WB, Riley PO, Crandall JR, Kent RW (2012) A method to induce navicular-cuneiform/cuneiform-first metatarsal sprain in athletes. In: Proceedings of the international research council on the biomechanics of injury (IRCOBI) conference, Dublin, 12–14 Sept 2012
79. Guise ER (1976) Rotational ligamentous injuries to the ankle in football. Am J Sports Med 4(1):1–6. doi:10.1177/036354657600400101

80. Boytim MJ, Fischer DA, Neumann L (1991) Syndesmotic ankle sprains. Am J Sports Med 19(3): 294–298. doi:10.1177/036354659101900315
81. Nussbaum ED, Hosea TM, Sieler SD, Incremona BR, Kessler DE (2001) Prospective evaluation of syndesmotic ankle sprains without diastasis. Am J Sports Med 29(1):31–35
82. Hopkinson WJ, St Pierre P, Ryan JB, Wheeler JH (1990) Syndesmosis sprains of the ankle. Foot Ankle 10(6):325–330
83. Close JR (1956) Some applications of the functional anatomy of the ankle joint. J Bone Joint Surg 38A(4):761–781
84. Rasmussen O, Tovborg-Jensen I, Boe S (1982) Distal tibio-fibular ligaments: analysis of function. Acta Orthop Scand 53(4):681–686. doi:10.3109/17453678208992276
85. Xenos JS, Hopkinson WJ, Mulligan ME, Olson EJ, Popovic NA (1995) The tibiofibular syndesmosis: evaluation of the ligamentous structures, methods of fixation, and radiographic assessment. J Bone Joint Surg 77A(6):847–856
86. Wei F, Villwock MR, Meyer EG, Powell JW, Haut RC (2010) A biomechanical investigation of ankle injury under excessive external foot rotation in the human cadaver. J Biomech Eng 132(9):091001. doi:10.1115/1.4002025
87. Wei F, Hunley SC, Powell JW, Haut RC (2011) Development and validation of a computational model to study the effect of foot constraint on ankle injury due to external rotation. Ann Biomed Eng 39(2):756–765. doi:10.1007/s10439-010-0234-9
88. Wei F, Post JM, Braman JE, Meyer EG, Powell JW, Haut RC (2012) Eversion during external rotation of the human cadaver foot produces high ankle sprains. J Orthop Res 30(9):1423–1429. doi:10.1002/jor.22085
89. Meyer EG, Wei F, Button K, Powell JW, Haut RC (2012) Determination of ligament strain during high ankle sprains due to excessive external foot rotation in sports. In: Proceedings of the international research council on the biomechanics of injury (IRCOBI) conference, Dublin, 12–14 Sept 2012
90. NATO Science and Technology Centres (2007) Test methodology for protection of vehicle occupants against anti-vehicular landmine effects. ISBN 978-92-837-0068-5. https://www.cso.nato.int/pubs/rdp.asp?RDP=RTO-TR-HFM-090
91. Bass CR, Hall GW, Crandall JR, Pilkey WD (1996) The influence of padding and shoes on the dynamic response of dummy lower extremities. SAE technical papers series, Detroit
92. Harris RM, Griffin LV, Hayda RA, Rountree MS, Bryant RG, Rossiter ND et al (1999) The effects of antipersonnel blast mines of the lower extremity. In: IRCOBI conference proceedings, Sitges, pp 457–467
93. Reilly DT, Burstein AH (1975) The elastic and ultimate properties of compact bone tissue. J Biomech 8(6):393–405
94. Dionne JP, Makris A, Nerenberg J (2001) Blast evaluation of spider boot foot protection system employing surrogates and biological specimens. In: IRCOBI conference proceedings. Isle of Man, UK
95. Makris A, Islam S (2000) Performance tests of 'spider boot' for demining. World EOD Gazette, pp 33–44
96. Chaloner EJ, McMaster J, Hinsley DE (2002) Principles and problems underlying testing the effectiveness of blast protective footwear. J R Army Med Corps 148(1):38–43
97. Bass CR, Folk B, Salzar RS, Davis M, Donnellan L, Harris R, Rountree M, Gardner M, Harcke T, Rouse E, Oliver W, Sanderson E, Waclawik S, Holthe M, Hauck B (2004) Development of a test methodology to evaluate mine protective footwear. In: Proceedings of the Personal Armor Systems Symposium (PASS), The Hague
98. Wolff K, Prusa A, Wibmer A, Rankl P, Firbas W, Teufelsbauer H (2005) Effect of body armor on simulated landmine blasts to cadaveric legs. J Trauma 59(1):202–208
99. U.S. Army Institute of Surgical Research (2000) Final report of the Lower Extremity Assessment Program (LEAP 99–2). Fort Sam Houston, TX
100. Funk, JR, Tourret, L, Crandall, JR, Pilkey, WD, McMaster, J, Khaewpong, N, Eppinger, R. (2001) The Effect of Active Muscle Tension on the Axial Injury Tolerance of the Lower Extremity. Paper 237, Proceedings of the 17th International Technical Conference on the Enhanced Safety of Vehicles (ESV), Amsterdam, The Netherlands
101. Vasquez K, Logsdon K, Shivers B, Chancey C (2011) Medical injury data 10 Nov 2011. Unclassified//Public Release
102. Schueler F, Mattern R, Zeidler F, Scheunert D (1995) Injuries of the lower legs-foot, ankle joint, tibia. Mechanisms, tolerance limits, injury—criteria evaluation of a recent biomechanics experiment series (impact tests with a pneumatic biomechanic impactor). In: Proceedings of the international research council on the biomechanics of impact, Brunnen
103. McKay BJ, Bir CA (2009) Lower extremity injury criteria for evaluating military vehicle occupant injury in underbelly blast events. Stapp Car Crash J. 2009 Nov;53:229–249
104. Yoganandan, N., Pintar, F., Boynton, M., Begeman, P. et al., Dynamic Axial Tolerance of the Human Foot-Ankle Complex, SAE Technical Paper 962426, 1996, doi:10.4271/962426
105. Kitagawa Y, Ichikawa H, Pal C, King A (1998) Lower leg injuries caused by dynamic axial loading and muscle tensing. In: Proceedings of the international technical conference on the Enhanced Safety of Vehicles (ESV). Windsor, Ontario, Canada
106. Quenneville C, Fraser G, Dunning C (2010) Development of an apparatus to produce fractures from short duration high impulse loading with an application in the lower leg. J Biomech Eng 132(1): 014502. doi: 10.1115/1.4000084

107. Ramasamy A, Masouros S, Newell N, Hill AM, Proud WG, Brown KA et al (2011) In vehicle extremity injuries from improvised explosive devices: current and future foci. Philos Trans Roy Soc Biomech 366:160–170
108. Otte D, von Rheinbaben H, Zwipp H (1992) Biomechanics of injuries to the foot and ankle joint of car drivers and improvements for an optimal car floor development. In: Stapp Car Crash Conference proceedings. Seattle, WA
109. Crandall JR, Martin PG, Kuhlmann T, Klopp GS, Sieveka EM, Pilkey WD et al (1995) The influence of footwell intrusion on lower extremity response and injury in frontal crashes. Annual proceedings of the Association for Advancement of Automotive Medicine. Chicago, IL, pp 269–286
110. Krueger HJ, Heuser G, Kraemer B, Schmitz A (1995) Proceedings of the fourteenth international technical conference on enhanced safety of vehicles. Munich, pp 528–534
111. Funk, JR, Tourret, L, Crandall, JR. (2000) Estimation of Fibula Load-Sharing During Dynamic Axial Loading of the Lower Extremity. Proceedings of the 24th Annual Meeting of the American Society of Biomechanics, Chicago, IL
112. Rudd RW, Crandall JR, Hjerpe E, Haland Y (2001) Evaluation of lower limb injury mitigation from inflatable carpet in sled tests with intrusion using the THOR Lx. In: Proceedings of the 17th international technical conference on the enhanced safety of vehicles. Amsterdam, The Netherlands, pp 4–7
113. Crandall JR, Kuppa SM, Hall GW, Pilkey WD, Hurwitz SR (1998) Injury mechanisms and criteria for the human foot and ankle under axial impacts to the foot. Int J Crashworthiness 3(2):147–162
114. Schreiber P, Crandall JR, Dekel E, Hall GW, Pilkey WD (1995) The effects of lower extremity boundary conditions on ankle response during joint rotation tests. In: Proceedings of the 23rd international workshop on human subjects for biomechanical research. National Highway Traffic and Safety Administration, US DOT
115. Funk, JR, Crandall, JR. (2004) Calculation of Long Bone Loading Using Strain Gauges. Proceedings of the 32nd International Workshop on Human Subjects for Biomechanical Research, National Highway Traffic Safety Administration, U.S. D.O.T., Nashville, TN
116. Ivarsson JB, Genovese D, Crandall JR, Bolton JR, Untaroiu CD, Bose D (2009) The tolerance of the femoral shaft in combined axial compression and bending loading. Stapp Car Crash J. Savannah, GA 53:251
117. Wang JJ, Bird R, Swinton B, Krstic A (2001) Protection of lower limbs against floor impact in army vehicles experiencing landmine explosion. J Battlefield Technol 4(3):8–12
118. Ramasamy A, Hill AM, Hepper AE, Bull AM, Clasper JC (2009) Blast mines: a background for clinicians on physics. Injury mechanisms and vehicle protection. J R Army Med Corps 155: 258–264
119. Bir C, Barbir A, Dosquett F, Wilhelm M, van der Horst M, Wolfe G (2008) Validation of lower limb surrogates as injury assessment tools in floor impacts due to anti-vehicular land mines. Mil Med 173(12): 1180–1184
120. Pandelani T, Reinecke D, Phillippens M, Dosquet F, Beetge F (2010) The practical evaluation of the MII-Lx lower leg when subjected to simulated vehicle under belly blast load conditions. Personal Armour Systems Symposium, Quebec City, Quebec, Canada
121. Quenneville CE, Dunning CE (2012) Evaluation of the biofidelity of the HIII and MIL-Lx lower leg surrogates under axial impact loading. Traffic Inj Prev. 2012;13(1):81–85.doi:10.1080/15389588.2011.623251
122. Roberts, D, Donnelly, B, Severin, C, Medige, J (1993) Injury mechanisms and tolerance of the human ankle joint, Centers for Disease Control, Atlanta, GA
123. Alvarez JG (2011) Injuries of concern & medical research plan for Warrior Injury Assessment Manikin Project (WIAMan). U.S. Army Medical Research and Materiel Command
124. Tegtmeyer M (2011) The WIAMan development program: objectives and rationale. Army Research Laboratory, Aberdeen, MD

Pain Biomechanics

19

Nathan D. Crosby, Jenell R. Smith, and Beth A. Winkelstein

Abstract

Painful injury is a tremendous problem affecting a large proportion of the population during their lifetime. Pain most commonly results from trauma and/or injury, with the spine as the most common injury site for producing chronic pain. This chapter reviews relevant anatomy, mechanics and techniques used to study cervical spine pain in particular, since it is the region most prone to injury and because the largest body of research has been in this region. In vivo, many behavioral assessments exist to functionally measure painful outcomes, with hyperalgesia and allodynia as the two primary classes of behavioral responses. Both provide quantitative measures of pain that relate to the clinical presentation of symptoms. This chapter begins with a brief review of the anatomical tissues that are most at risk for injury and pain generation. This is followed by a review highlighting the cellular and molecular mechanisms of nociception and pain, addressing modifications in the periphery and in the central nervous system (CNS). The relationship between injury biomechanics and the local and spinal cellular responses are presented in the context of nociception and pain, with specific focus on injury to the facet capsule and nerve root because of their common involvement in cervical spine injury. Lastly, a brief review of experimental methodologies used to study pain biomechanics, ranging from the macroscopic to the cellular and molecular scales, is provided along with contextualization of the relative advantages and limitations of each. A brief summary integrates all of the sections to identify future areas of research that will define a more detailed understanding of pain biomechanics.

N.D. Crosby, Ph.D. • J.R. Smith, Ph.D.
Department of Bioengineering, University of Pennsylvania, Philadelphia, PA, USA
e-mail: ncrosby@seas.upenn.edu; jenells@seas.upenn.edu

B.A. Winkelstein, Ph.D. (✉)
Departments of Bioengineering and Neurosurgery, University of Pennsylvania, Philadelphia, PA, USA
e-mail: winkelst@seas.upenn.edu

19.1 Introduction

Painful injury is a tremendous problem affecting a large proportion of the population at some point over the course of a lifetime [1, 2]. In most cases, pain results from trauma and/or injury; and the spine is the site of many of those injuries producing chronic pain. As such, here we review relevant anatomy, mechanics and techniques used to study spine pain biomechanics, with a particular focus on the cervical spine. We focus on cervical spine pain biomechanics because this spinal region is most prone to injury and because the largest body of integrated research has been in this region, providing a good platform for discussion.

When considering the topic of pain, it is important to define that pain is a combination of sensory and emotional experiences and nociception is the set of physiological responses that transmit the pain signals [3]. In vivo, many behavioral assessments have been developed and validated to functionally measure painful outcomes (www.iasp-pain.org). Hyperalgesia and allodynia are two primary classes of behavioral responses. Hyperalgesia is an amplified response to a stimulus (mechanical, temperature or chemical) and suggests that more pain is perceived. Allodynia is characterized as pain that is evoked by a typically non-noxious stimulus. Hyperalgesia and allodynia provide quantitative measures of pain that relate to the clinical presentation of symptoms [4].

We begin with a brief review of the anatomical tissues in the cervical spine that are most at risk for injury and pain generation. In the following section we highlight the cellular and molecular mechanisms of nociception and pain, presenting a general schema for tissue injury in the periphery and discussing responses there and in the central nervous system (CNS). In the fourth section we provide a detailed presentation of the relationship between injury biomechanics and the local and spinal cellular responses in the context of nociception and pain. We specifically focus on facet capsule and nerve root injury because of their common involvement in cervical spine injury and the large body of work in these areas. Lastly, we present a brief review of experimental methodologies used to study pain biomechanics, ranging from the macroscopic to the cellular and molecular scales, contextualizing the relative advantages and limitations of each. We conclude with a brief summary that integrates all of the sections to identify future areas of research that will define a more detailed understanding of pain biomechanics.

19.2 Anatomy of Spinal Tissues with the Potential to Generate Pain

The anatomical components of the spine, and the cervical spine in particular, are diverse in their structure, function, and ability to generate pain. Hard and soft tissues in the cervical spine each have the potential to directly and/or indirectly generate pain. These anatomical structures include the bones and their connecting soft tissues, such as the intervertebral disc and ligaments, as well as neural tissues (i.e. nerve root, DRG and spinal cord) and paraspinal muscles (Fig. 19.1). In addition, in the cervical spine, the vascular system has limited indirect involvement in pain from neck injuries, and so will be addressed only briefly in this chapter. All of these structures are biomechanically important and can be injured and/or generate pain, though to varying degrees, under different physiologic *and* abnormal or injurious biomechanical loading scenarios.

Many of these structures have innate populations of nociceptive fibers [5–8], supporting their ability to generate pain when such fibers are activated during neck and/or tissue loading. Because of their potential for injury and highly likely contribution to pain generation, we briefly review them here, in that context. However, a detailed review of these anatomical structures can be found elsewhere [9, 10]. In addition, for the major tissue components we also provide a brief discussion of the potential loading scenarios that may initiate pain and also relate relevant anatomy to those injury conditions. In a later section we particularly focus on specific tissues

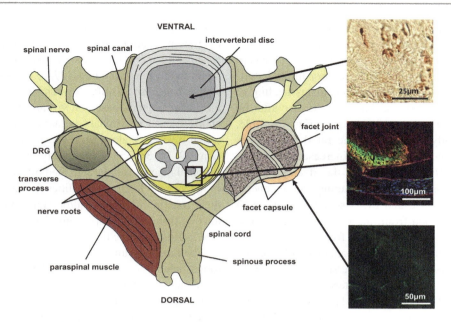

Fig. 19.1 Schematic illustration of an axial view of cervical spine anatomy, including soft and hard tissues with the potential for generating pain. The articulating intervertebral disc and facet joints limit motion and transmit loads, but also have the capacity for pain, due to their innervation. Neural components, such as the spinal cord, nerve roots and dorsal root ganglion (DRG), are also at risk for painful injury. The *insets* provide immunohistochemical images showing the presence of nerve fibers (positive for CGRP) in the intervertebral disc (*brown*), dorsal root (*red*), and facet capsule (*green*)

and the relationship between their injury, pain and the nociceptive mechanisms that are affected by mechanical contributions.

19.2.1 Bone, Intervertebral Disc, and Ligaments

The cervical spine includes seven vertebrae (C1–C7), with C1 and C2 (the atlas and axis, respectively) retaining specialized function since they interface with the base of the skull and each other. Each pair of bony vertebrae of the cervical spine are coupled by the collagenous intervertebral discs anteriorly and the bilateral articulating facet joints posterolaterally (Fig. 19.1). The vertebrae themselves are innervated most extensively in the periosteum that lines the outer surface of each vertebra [5]. Large intraosseus nerves and neurovascular bundles permeate the vertebral body, penetrating to central and peripheral regions including the marrow and the cartilaginous endplates that are adjacent to the intervertebral disc [5, 11].

The intervertebral disc is a collagenous structure that connects to its adjacent vertebral bodies via a cartilaginous endplate (Fig. 19.1). The disc consists of an outer annulus fibrosus made up of concentric rings of stiff fibrocartilage and an inner, gelatinous nucleus pulposus [12]. The primary function of the intervertebral disc is to absorb and transmit axial compressive loading through the spinal column and to control loading by distributing stresses within the dense fibrocartilage of its annulus fibrosus. However, its neuroanatomy suggests that it may also have the capacity for pain generation, owing to the fact that the outer one-third of the annulus fibrosus of the normal asymptomatic disc is innervated by peptidergic fibers and free nerve terminals, which are unencapsulated nerve endings that often respond to noxious stimuli (Fig. 19.1 inset) [7, 11, 13, 14]. These nerve terminals can be activated by a variety of stimuli, including altered disc loading by injury and spinal degeneration, or even from exposure to chemical mediators [13, 15]. Despite evidence of disc innervation by sensory fibers, and a recent focus on the hypothesis of disc-mediated pain from aberrant nerve fiber in-growth in discogenic pain conditions [16–18], there are very limited studies defining the cellular mechanisms leading to persistent pain that arises from the disc after its undergoing abnormal loading. For example, whole body vibration in the rat that produces pain has been shown to also increase neurotrophins (i.e. NGF, BDNF) in the disc [19, 20]. Nonetheless, the effects of aging and degeneration of the disc and the associated anatomical changes in, and from, the vertebrae are a major pathway for initiating pain from this spinal tissue, particularly in the lumbar spine [14, 16, 21].

The cervical vertebrae are stabilized by many ligaments in both the anterior and posterior regions of the spine (Fig. 19.1) that normally contribute to maintaining the head position and providing neck proprioception, which is the sensory input that aids in awareness of relative position and movement of the body during spinal motions. The ligaments in the cervical spine include the anterior and posterior longitudinal ligaments that span the entire vertebral column, ligaments that connect pairs of adjacent vertebrae (ligamentum flavum, interspinous, and supraspinous ligaments), and the bilateral facet capsular ligaments that enclose the apposing surfaces of the articular pillars of adjacent vertebrae [22, 23]. The vertebrae of the upper cervical spine have additional ligaments that facilitate and restrict articulations between the atlas (C1) and axis (C2), as well as with the occiput [22]. The anterior longitudinal, facet capsular, supraspinous, and interspinous ligaments are at particular risk for undergoing excessive strain during injurious loading of the cervical spine, because of their role in bearing tensile loads during spinal motion [24, 25].

These ligaments have a diverse morphology according to their function, but even with their varied roles and structures, they are quite similar in their neuroanatomical features [8]. Within their dense collagenous networks, spinal ligaments are innervated by both proprioceptive and nociceptive bundles, including small diameter sensory fibers with free nerve endings that respond to noxious stimuli (Fig. 19.1 inset) [8, 26, 27]. Numerous immunohistochemical studies have reported the longitudinal, supraspinous, interspinous, and facet capsular ligaments to be innervated by nociceptive fibers and free nerve endings [27–32]. Histological studies have also identified low and high threshold mechanoreceptors, based on visualization of specialized sensory terminals, in many of these same spinal ligaments, along with nociceptive innervation [29, 31, 33–35]. The capacity of nerve fibers in spinal ligaments to be activated by a wide range of mechanical loading modalities further suggests their role in initiating and/or signaling painful loading from cervical spine injuries due to excessive loading of those tissues.

The facet capsular ligaments enclose the facet joints, which are bilateral articulations between each pair of adjacent vertebrae that contain a variety of innervated structures with the potential to generate pain (Fig. 19.1). The medial branches of the dorsal primary ramus innervate the cervical facet joints from the two levels superior and inferior to the joint [36], with mechanoreceptive and nociceptive fibers terminating in the joint and its capsule [6, 34, 37]. The presence of these afferents in the facet joints suggests that this joint has the potential to produce pain under abnormal and/or excessive loading scenarios. Furthermore, clinical studies have demonstrated the involvement of the facet joint in neck pain both by inducing pain through joint provocation (either by distention or injection of chemically irritating agents) and relieving pain through nerve ablation or anesthetic blockade [38–42]. The articulating surfaces of the joint themselves are covered with articular cartilage, and synovial folds exist in the joint space to dissipate stress and maintain stability. Synovial folds in cervical joints are believed to contribute to neck pain and headache following trauma [43, 44], and recent morphological studies have opined several hypotheses for injury mechanisms [45]. Taken together, the complicated biomechanics associated with many types of neck loading, the potential simultaneous activation of pain afferents in facet joint's synovial folds and facet capsule, and the multifaceted clinical evidence all strongly suggest this joint as a primary anatomical source of neck pain due to its, or the spine's, injury.

The bony and soft tissue structures in the cervical spine articulate and interact to support movement and distribute and absorb loading to the various individual tissues. Many biomechanical studies using in vivo and post-mortem human spinal tissues provide a detailed understanding of cervical spine biomechanics under many physiologic and injurious loading conditions [46–50]. During normal loading of the spine, the intervertebral disc and the complement of spinal ligaments guide and restrict motions of the head and neck within physiologic limits. The disc absorbs and distributes the compressive forces that are applied to the anterior column of the spine during flexion or extension. The spinal ligaments serve to resist tensile loading, while their proprioceptive and nociceptive innervation provide sensory feedback. The facet capsular ligament and its role in spinal loading, injury, and pain will be discussed in a later section.

The anatomical structures of the cervical spine can be injured or can fail completely when the compressive, tensile, or shear loading that is experienced by a single tissue exceeds its tolerance during extreme loading scenarios. For example, the disc is particularly susceptible to shear loading and torsion, especially without protection from the facet joints [50]. Likewise, the spinal ligaments have widely varying failure strengths [51, 52], but excessive spinal flexion or extension can exceed those limits, causing their rupture and activating nociceptive signaling pathways. These injuries trigger complex pain cascades and initiate a variety of different cellular mechanisms, both in the injured tissue and in the peripheral nervous system (PNS) and the CNS [53, 54]; relevant pain mechanisms are discussed in greater detail in the later Sect. 19.3.

19.2.2 Spinal Cord, Nerve Root, and Dorsal Root Ganglion (DRG)

The spinal cord is enclosed by a system of three elastic, collagenous membranes making up the meninges: the innermost pia mater, the arachnoid mater, and the outermost dura mater. The dura and arachnoid mater lie directly apposed to each other, separating the vertebral column from the spinal cord and the cerebrospinal fluid within. This dual membranous layer has a longitudinally-oriented collagenous structure that conveys substantial tensile strength along that axis, along with elastic fibers that maintain flexibility [12, 55]. At each spinal level, nerve roots terminate in the spinal cord within the meningeal layers. As they course peripherally, the nerve roots traverse the dura mater and exit the vertebrae, where a separate, albeit less mechanically-robust structure (the epineurium), envelops them [12]. The dura mater contains small, peptidergic nociceptive nerve fibers that express the neuropeptides calcitonin gene-related peptide (CGRP) and substance P [32, 56, 57]. Compared to the cranial meninges, the spinal dura mater has few of these fibers, and so, may play a limited role in the pathogenesis of pain [56]. However, the meninges may have a role in modulating pain through their ability to release pro-inflammatory cytokines into the cerebrospinal fluid that bathes the tissues of the CNS [58].

The spinal cord itself is highly organized and composed of white matter and gray matter (Fig. 19.1). White matter consists largely of myelinated axonal tracts conveying ascending and descending signals to and from the brain [12]. White matter tracts are found at the periphery of the spinal cord and somatotopically organized in the dorsal, ventral, and lateral columns. In contrast, the gray matter includes the neuronal cell bodies, unmyelinated axons, and dendritic structures. The gray matter is divided into the dorsal and ventral horns, with the commissural and intermediate gray matter connecting the horns centrally. The gray matter of the spinal cord is arranged in layers, or laminae, each containing distinct populations of neurons and receiving input from specific groups of fibers [59]. For example, laminae I and II along the superficial most dorsal boundary of the gray matter contain synaptic terminations of exclusively nociceptive fibers [60]. Afferent sensory fibers and descending motor axons enter and exit the spinal cord through the dorsal and ventral rootlets that originate in the gray matter [12].

The spinal cord is involved in the processing and amplification of painful sensory input from peripheral tissues. The innate potential of the spinal cord to generate pain is derived from damage to the nervous tissue itself, since it is comprised of neuronal processes and has no explicit sensory nerve endings. Spinal cord injuries (SCI) are diverse and have many potential causes due to mechanical injury, including partial or complete transection of the cord, contusion from blunt impact or pressure from broken bones or herniated discs, and sharp incision [61–65]. Each of these injuries can lead to varying degrees of cell death, demyelination, and axonal disruption and degeneration due to the initial mechanical insult, accompanied by secondary neuronal, immune and inflammatory responses [12, 66, 67]. Although motor deficits are often a primary outcome of spinal cord injury, widely varying pathogenic pain states often accompany motor deficiency and their cause and mechanisms remain an active area of research [68, 69].

Rootlets come off the spinal cord both anteriorly and posteriorly and combine to form the dorsal and ventral nerve roots that comprise the bilateral peripheral spinal nerves at each level (Fig. 19.1). In general, the posterior dorsal rootlets are afferent sensory fibers and the anterior ventral rootlets are effecter motor fibers, though the ventral roots also contain some afferent fibers [70]. The dorsal and ventral roots come together in the neural foramen to form the spinal nerve as they progress distally to the periphery (Fig. 19.1); but, unlike peripheral nerves the nerve roots are not enclosed by a thick, protective epineurial sheath. The cell bodies of peripheral nerves are located in the dorsal root ganglion (DRG), which is particularly sensitive to mechanical loading, and even slight compression of normal DRGs can produce sustained neuronal activity and pain [71–73]. Direct impingement on the nerve roots and/or DRG can occur by distortion of the normal size and shape of the neural foraminae, which can occur during extreme movements of the cervical spine, stenosis, or slow osteophytic growth, or even by a herniated disc [74, 75]. The mechanics of compression- and tension-based nerve root injuries, i.e. maximum load, rate of loading, and duration of loading, contribute to the extent of pain produced after injury [76, 77]. These and other mechanical parameters that control and modulate nerve root-induced pain and pathology are further discussed in Sect. 19.4.

19.2.3 Paraspinal Musculature

Skeletal muscle constitutes much of the overall volume of the neck (Fig. 19.1), including the superficial and deep layers of muscle with attachments on the skull, shoulders, vertebrae, and spinal ligaments. Neck musculature has a dual role in neck pain – (1) by the potential for its own injury and (2) by altering the kinematics and loading of other structures in the neck that may exacerbate existing cervical spine loading and neck pain. Loading of neck musculature can lead to pain since as many as one-half of the neural units in skeletal muscle have some nociceptive function, likely through Aδ- and C-type fibers with their numerous un-encapsulated free nerve endings [78, 79].

In many neck loading scenarios, the muscles themselves can undergo concomitant contraction and elongation as a result of head or torso movement [25, 80]. For example, during whiplash, loading of the cervical musculature induces strains that exceed injury thresholds, namely in the posterior cervical muscles (splenius capitis, semispinalis capitis, trapezius); these supra-threshold strains have been reported to correlate with reports of pain in the posterior cervical region following whiplash [25, 80]. Further, after such trauma, skeletal muscles also undergo substantial structural and functional changes, including fatty infiltration, tissue degeneration, and altered activation thresholds and contraction timing, all of which can contribute to chronic pain and motor impairment [81–83]. These types of injuries can also lead to the development and maintenance of local inflammatory responses in the injured tissue that are a common peripheral mechanism for muscle tenderness and even further sensitization of nociceptors [84–86].

Though neck muscles are at risk for their own injury, they also pose a threat of injury to the other spinal tissues with which they have insertions or make direct contact. For example, semispinalis and multifidus muscle fibers insert directly on the facet capsule in the cervical spine and present a direct pathway for added loading to the capsule during muscle activation and facet joint extension [87]. Furthermore, neck muscle activation affects head and neck kinematics throughout the cervical spine, thereby also potentially altering the loading to other cervical spine tissues. Reflexive activation of neck musculature during whiplash causes significant variations in muscle extension, head displacement, and head acceleration [88]. Contraction of muscles that have their line of action along the long-axis of the spine increases axial loading in the spine and facet joints and contributes to intervertebral disc compression [25]. These altered loading states could modify the overall spine mechanics, for example potentially increasing tensile loading of spinal ligaments or disc shear, and provoking or exacerbating injury in those tissues.

19.2.4 Vasculature

The vascular components of the spine are not independently viable sources of pain, but their injury in certain loading scenarios can aggravate mechanically-induced injuries to other cervical spine tissues. The vertebral arteries supply blood to the head, brain, and neck tissues as they travel bilaterally along each cervical vertebra [89, 90] (Fig. 19.1). During certain neck loading scenarios outside physiological conditions, the vertebral arteries can become elongated, stretched, or pinched, which can potentially tear the artery [91]. For example, a vehicle occupant with a head-turned posture who undergoes a rear-end collision can undergo neck motions which cause elongation of the vertebral artery that significantly exceeds physiological tolerances [25, 92]. In particular, the vertebral artery is especially vulnerable to mechanical loading and is susceptible to pinching or tearing in the atlanto-axial region of the upper cervical spine where it passes over bony landmarks [90, 93]. Vertebral artery damage can compromise the blood supply to the brain and can also lead to secondary nociceptive pain arising from increased fluid pressure around the spinal cord [25, 94, 95]. These injuries can produce symptoms ranging from headache and neck pain to vertigo and paresthesias [94]. Given the potential mechanisms and outcomes of vertebral artery disruption, this tissue may contribute to the development of pain after injurious mechanical loading of the cervical spine. However, because of its relatively limited involvement compared to other more vulnerable and better-studied tissues, we do not focus on it in this chapter.

19.3 Pain Mechanisms

Mechanical stimulation, such as tissue loading, activates the mechanosensitive receptors that innervate the tissue to transduce the stimulus to an electrophysiological event in primary afferent fibers (Fig. 19.2). For stimuli that are above the threshold for pain sensation, the primary sensory afferents transmit nociceptive signals to the spinal cord. Normally, these nociceptive signals provide feedback to protect the organism from tissue damage or the potential for damage, so they are termed 'physiologic' [96, 97]. Primary sensory afferents in the PNS and higher order pathways in the spinal cord and brain convey, as well as amplify and suppress, painful sensory information [96, 98]. Potentiation of nociceptive processing leads to a state of pain hypersensitivity, or 'sensitization', which can become chronic or pathologic and no longer serves the physiologic purpose of pain [99, 100]. Sensitization can occur peripherally by potentiation of primary afferents, in the dorsal horn of the spinal cord, and/or supraspinally in the brain. Here, we briefly highlight the peripheral and central mechanisms of sensitization to provide a basic discussion of the pathways by which injury can lead to acute and persistent pain states. These mechanisms are reviewed in greater detail elsewhere in this book, as well as in the literature [101, 102].

Nociceptive pain is established in response to stimulus or stimuli (mechanical, chemical, thermal) that exceed the activation threshold

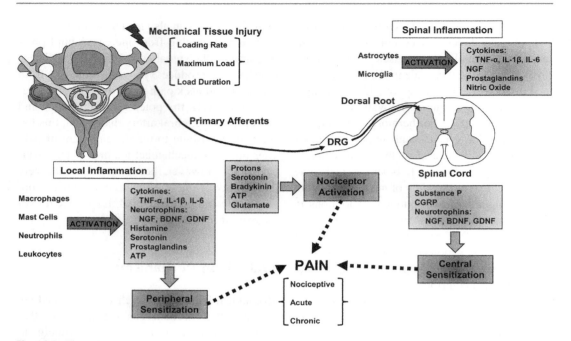

Fig. 19.2 Tissue injury induces a host of cascades in the peripheral tissues where injury occurs that also activates nociceptive afferents and can lead to central sensitization in the spinal cord and lead to pain. Nociceptor activation leads to neurotransmitter and neurotrophin synthesis in the DRG, and their transport to the spinal cord initiates spinal inflammatory cascades. Activated spinal glial cells release inflammatory mediators that further exacerbate neuronal excitability in the dorsal horn of the spinal cord. Specific characteristics of the mechanical inputs, such as loading rate, magnitude and duration, all have been shown to modulate aspects, as well as the collection, of this mechanism

for a given population of nociceptive fibers. Those fibers include small-diameter unmyelinated C-fibers or medium-diameter thinly-myelinated Aδ-fibers that transmit sensory signals from the periphery to the dorsal horn of the spinal cord through cell bodies in the DRG (Fig. 19.2) [78, 101, 103]. C-fibers are further classified as peptidergic or non-peptidergic, based on their characteristic expression of substance P and CGRP in the case of peptidergic fibers, and the purinergic ATP receptor, P2X3, or binding of IB4 in non-peptidergic fibers [104, 105]. Nociceptors also express receptors for a variety of neurotrophins, which are secreted factors important for neuronal growth and survival. Nerve growth factor (NGF) and glial-derived neurotrophic factor (GDNF) are neurotrophins, each of which may play an important role in pain modulation due to the presence of their receptors on peptidergic and non-peptidergic nociceptors, respectively [106].

Nociceptive pain continues only in the presence of sustained noxious stimuli [103], though pain can outlast the stimuli through peripheral sensitization, which is a potentiation of primary sensory fibers leading to lowered firing thresholds and increased firing rates once activated. Peripheral sensitization most commonly leads to decreased mechanical and thermal thresholds in the immediate area of injury [103, 105, 107]. This peripheral sensitization results from the production and peripheral release of neurotransmitters and other inflammatory factors from neurons and infiltrating non-neuronal cells in response to noxious stimuli or tissue damage [105, 108]. Nociceptors are rapidly sensitized directly via peripherally expressed receptors for mediators such as bradykinin, amines, prostanoids, NGF, and protons [108]. After injury, immune cells, including neutrophils, macrophages, and lymphocytes, initiate a cytokine cascade that may indirectly and/or directly sensitize nociceptors, further maintaining hypersensitivity [108, 109]. Peripheral sensitization includes altered nociceptive responses at the site of injury, in part as an adaptive response to promote tissue repair and healing; but, those changes can also contribute to the development of pathological pain, along with sensitization of spinal and supraspinal systems [100]. This collection of systems throughout the PNS and CNS plays a significant role in the transition from physiologic pain to the development and maintenance of persistent pain [103, 107, 110].

Persistent pain, or chronic pain, may long outlast any initiating injury event or tissue damage, and is due in large part to spinal sensitization (Fig. 19.2). Spinal sensitization is characterized by increased spontaneous activity, reduced thresholds for activation by both non-noxious and noxious stimuli, heightened responses to noxious stimuli, and sensitivity at secondary sites with no precipitating tissue damage (Fig. 19.3) [107]. Since many hypotheses have been proposed to explain the exact mechanism by which the spinal cord becomes sensitized or hyperexcitable, we only briefly highlight these theories in the context of injury; more extensive discussions are numerous in the literature [102, 105, 107, 108]. A nociceptive signal is transmitted to the spinal cord through the synaptic release of glutamate and other neurotransmitters (Fig. 19.2). Sustained nociceptive input induces rapid post-translational modifications to post-synaptic receptors that lead to increased excitability, and slower translational and transcriptional changes that promote excitability [105, 107]. Strengthening of spinal and supraspinal (i.e. brain) nociceptive signaling by changes in the ionotropic and metabotropic neurotransmitter receptors is accompanied by structural plasticity (e.g. neurite outgrowth, excitatory synaptogenesis) that may incorporate low-threshold $A\beta$-fibers into nociceptive pathways, causing normally non-noxious stimuli to activate the same nociceptive pathways as supra-threshold stimuli [111–114]. Ultimately, all of these responses contribute to central sensitization, though they are by no means the only mechanisms by which plasticity in the CNS modifies nociceptive pathways.

Trophic, neuroimmune, and neuroinflammatory modulators and integrated cascades also play a role in central sensitization, just as they do in peripheral sensitization. NGF, GDNF, and a third neurotrophin, brain-derived neurotrophic factor (BDNF), each have spinal neuromodulatory roles once transported to, or released from, spinal terminals of nociceptors [115–118]. Peripheral neuroimmune and neuroinflammatory factors also activate glial cells in the CNS, including microglia and astrocytes [54]. Both types of glial cells respond to injury by changing their morphology, proliferating, and releasing several pro-inflammatory cytokines (e.g. IL-1β, TNF-α, IL-6) and mediators that further sensitize both the glial and pre- and post-synaptic spinal neuronal responses [109, 117, 119].

The pain cascades described above can be activated by mechanical loading to individual tissues or combinations of structures in the cervical spine (Fig. 19.2). For example, the primary afferent fibers that innervate the facet capsular ligament are particularly sensitive to mechanical forces generated in the capsule by distraction of the facet joint as a whole [120, 121]. Activation of those mechanosensitive afferents during injury may be sufficient to cause central sensitization by inducing excitatory changes in the neurons of the dorsal horn (Fig. 19.3) [122, 123]. Mechanical loading of the facet joint also modulates spinal glial activation and both peripheral and spinal expression of the neuropeptide, substance P, both of which are involved in the generation and maintenance of pain [124, 125]. Likewise, local injury biomechanics can affect pain by directly loading nervous tissues, like compression of the nerve root, which is known to affect neuronal structure and physiology, and induce microglial and Schwann cell proliferation and astrocytic activation at the site of injury and in the spinal cord [126–128]. The pain states experienced after mechanical injuries to both the musculoskeletal and neural tissues of the cervical spine have the potential to develop both peripherally- and centrally-mediated components and include neuropathic, inflammatory, and neuroimmune characteristics (Fig. 19.2). The relationship between injury biomechanics and pain mechanisms will be elucidated in the next section for two common cervical spine injuries – facet joint distraction and nerve root compression.

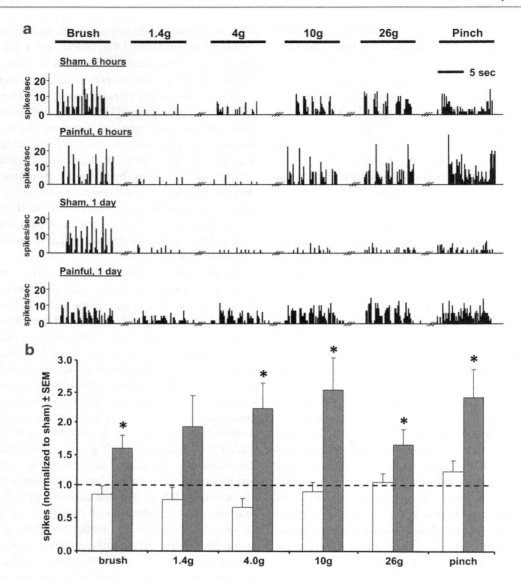

Fig. 19.3 Electrophysiological measurements of dorsal horn neuronal excitability after facet capsule injury. The evoked firing response of neurons in the C6/C7 spinal dorsal horn recorded after a painful C6/C7 facet capsule injury show increased firing over sham surgery by day 1, in response to a train of non-noxious and noxious mechanical stimuli applied to the forepaw, including light brushing, non-noxious and noxious von Frey (1.4 g, 4 g, 10 g, 26 g) filaments, and a noxious 60 g pinch. (**a**) Representative firing rates for neurons at 6 h or 1 day after an injury and sham procedure. (**b**) Spike totals for each stimulus normalized to sham illustrate the effect of injury on neuronal excitability. Evoked firing at 6 h (*white bar*) after a facet capsule injury was not different from sham, while firing was elevated across all stimuli by 1 day (*gray bar*) after sham [122]

19.4 Biomechanics and Pain in Cervical Spine Injuries

As highlighted in Sect. 19.2, there are several anatomic features in the cervical spine which have been recognized as having the highest likelihood of generating pain from neck injury. Among the most vulnerable and most well-studied are the facet capsular ligament and nerve root. Moreover, both such tissues have been well characterized in their response to mechanical loading and provide interesting contrast given one is ligamentous and the other is neural tissue. With growing research efforts focused on both of these tissues, as well as their differences in mechanical responses, cellular composition and potential relevance to different types of neck injuries, we focus on each of them separately in this section. We provide clinical context for each injury modality, local and central responses to loading, and specifically discuss how the different mechanical parameters related to tissue loading/injury modulate aspects of the complicated neuroimmune cascades leading to pain (Fig. 19.2).

19.4.1 Biomechanics and Painful Facet Joint Injury

As described in Sect. 19.2 above, the cervical facet joint and its capsular ligament are among the cervical spinal tissues with great potential to generate pain when injured, particularly during whiplash [36–38]. Using anesthetic nerve blocks or provocative testing clinically, the cervical facet joint is identified as a source of pain in as many as 62 % of neck pain cases [129–131]. Facet joint injury has been hypothesized to result from two potential injury mechanisms in the joint that can lead to pain. The first produces excessive compression of the joint and impingement of the articular surfaces [132]. The articulating surfaces of the facet joints glide across one another during normal flexion, extension, or torsion of the spine [133]. Abnormal extension of the motion segment can force the bony facet of the superior vertebra to compress against the superior facet of the inferior vertebra, pinching the articular cartilage and synovial folds [132, 134, 135]. Extreme articular cartilage compression can lead to degeneration and osteoarthritis in the joint; furthermore, the synovial folds also contain nociceptive fibers and so can be a source of pain during loading [6, 136]. The second mechanism of facet joint injury includes excessive loading of the facet capsular ligament beyond its mechanical tolerance. The facet joint undergoes shear, compression, and tension during different phases of whiplash, each of which can produce tensile strain in the facet capsule [137–141]. Facet capsule strains that exceed the physiologic limit have been shown to activate mechanoreceptors and nociceptive fibers in the joint in a caprine model [120, 121].

The kinematics of the head and cervical spine during whiplash has been well defined by numerous human-volunteer and post-mortem human subject studies simulating rear-impact collisions [135, 137, 139, 142, 143]. Based on volunteer studies using high-speed X-ray imaging during low-speed rear impacts, the spine forms an 'S' curvature in the first 100–120 ms after impact, with the lower cervical spine undergoing extension and the upper cervical spine flexing [134, 136, 144]. The lower cervical spinal levels undergo vertebral retraction and extension, imposing tensile and shear loading to the facet joints and capsular ligaments. In particular, many studies have reported tension across the facet capsule during whiplash that induces strain levels greater than those experienced during normal physiologic motions [135, 145, 146]. In a biomechanical study using cervical spines from post-mortem human subjects, capsular ligament strains were the greatest at C6/C7, peaking at 29.5 ± 25.7 % and 39.9 ± 26.3 % strain for 6.5 g and 8 g accelerations, respectively [135]. Peak capsule strains during whiplash can be as high as five times the peak strains for the same cervical levels during normal neck bending, which have been reported to peak at 6.2 ± 5.6 % [135, 146]. Capsular strains can be further increased by more complicated out-of-plane head postures, such as a turned head [143].

Despite evidence of excessive facet capsule strains being established during whiplash [135, 143, 146], facet capsule ruptures have been reported not to occur during simulated exposures in cervical spine studies [135, 146–150]. In fact, the strains at facet capsule rupture exceed the strains experienced both during whiplash and those required for partial or minor ligament rupture [143, 151]. Furthermore, removing the contribution of the facet capsule precludes the development of behavioral hypersensitivity in animal models [125, 152]. Together, these biomechanical findings suggest that pain from facet joint injury results from *sub-failure* kinematics and kinetics established in the facet capsule, and that input from the afferents in the capsule is required for the establishment of pain after joint loading. This recent hypothesis is based on biomechanical, physiological and behavioral data and is contrary to prior assertions based only on biomechanical data that tissue failure is sufficient to induce such pathology.

Recently, in vivo models of facet injury have been developed that provide useful research platforms to define relationships between facet joint biomechanics, capsular loading, activation of the afferents that innervate the facet capsular ligament, CNS modifications and pain behaviors [122, 123, 153–158]. In a caprine model, distraction of the cervical facet joint produces electrophysiological activity in the afferents that innervate the joint. Specifically, mechanoreceptive fibers innervating the facet capsular ligament are activated during and for a short period (i.e. several minutes) after tensile loading of the capsule [120, 121]. Low-threshold mechanoreceptors are activated at physiologic strain levels (10–15 %) and are hypothesized to be proprioceptive [120]. A separate population of high-threshold mechanoreceptors is activated at facet capsule strains of 25–47 % [120, 121, 157], significantly higher than the strains to activate low-threshold afferents but not different from those during whiplash [135, 146, 159]. Those same magnitudes of capsular strain that activate capsule-innervating fibers are also associated with degenerative changes in the nerve fibers themselves, including localized swelling and disruption of axons, which can lead to spontaneous and neuropathic pain [160, 161]. The demonstration of activation of mechanosensitive afferents in the facet capsule during physiologic and, most notably injurious strain of the capsule, substantiates the hypothesized, but only recently proven, assertion that certain loading scenarios of the facet joint activate peripheral and central pain mechanisms that underlie clinical reports of facet-mediated pain following whiplash.

Facet capsule strains that activate afferent firing in the goat also produce persistent behavioral sensitivity in rats that mimics the primary and secondary hyperalgesia reported by whiplash and neck pain patients [135, 143, 149, 162, 163]. In a rodent model, behavioral hypersensitivity to mechanical stimulation develops in the shoulder, neck and forepaw following tensile loading of the joint that produces strains of up to 42 % in the facet capsule [155, 156, 164]. Behavioral sensitivity begins by 1 day and is sustained for up to 6 weeks after joint loading [165]. The onset of mechanical sensitivity after injury is indicative of both hyperalgesia, or pain hypersensitivity, and allodynia, which is an increased pain response to non-noxious stimuli [3]. Behavioral sensitivity at sites remote from the injury, like the forepaw, is termed 'secondary hyperalgesia' and is a key clinical indication that the pain may be driven at least in part by central sensitization, or the potentiation of nociceptive elements located in the dorsal horn of the spinal cord [107]. Behavioral sensitivity is caused by the same magnitude of facet capsule strains that activate afferent fibers in the capsule; this suggests that the initial biomechanically-induced electrophysiological activity in the facet capsule may activate both peripheral and central pain mechanisms that result in primary and secondary hyperalgesia.

Facet capsule injury induces electrophysiological changes in the dorsal horn of the spinal cord that are consistent with those of central sensitization, including increased spontaneous activity, reduced thresholds for activation by both non-noxious and noxious stimuli, and heightened responses to noxious stimuli. As early as 1 day after injury, neurons in the dorsal horn at the same spinal level as the injury develop hyperexcitability in response to both innocuous and noxious mechanical stimuli at the forepaw (Fig. 19.3) [122]. That hyperexcitability is sustained for at least 7 days after the initial capsule injury [123]. Neurons also exhibit increased spontaneous activity and afterdischarge, as well as a functional shift of secondary neurons in the dorsal horn from low-threshold mechanoreceptors (LTMs) or nociceptive-specific (NS) neurons to wide dynamic range (WDR) neurons [123]. Although LTMs are normally activated only by light mechanical stimulation and NS neurons only by nociceptive stimuli, WDR neurons are responsive to a large range of stimuli, including both innocuous and noxious stimuli [96]. A gain-of-function shift of LTMs to WDR neurons can produce increased amplification of noxious stimuli in the CNS. Likewise, a shift of NS neurons to

WDR neurons can activate nociceptive pathways by normally non-noxious stimuli, leading to decreased pain thresholds and behavioral hypersensitivity [107]. Electrophysiological changes such as those that occur after facet joint injury are likely driven by both peripheral (facet capsule, afferent fiber, DRG) and spinal cellular cascades that ultimately increase neuronal excitability.

Facet joint injury induces a host of cellular-level changes in the peripheral and central nervous systems that contribute to increased neuronal activity and ultimately pain. Many of these changes directly, and indirectly, modulate elements of the glutamatergic system, which is the primary excitatory neurotransmitter system in the spinal cord [154, 158, 166]. For example, facet joint injury upregulates protein kinase C epsilon (PKCε), a regulator of glutamate release and potentiation of ionotropic glutamate receptor NMDA [166–168]. The metabotropic glutamate receptor, mGluR5, which is important for enhanced nociceptive signaling in dorsal horn neurons [169–171], is also upregulated in the DRG and spinal cord as early as 1 day after facet capsule injury [158]. Further, upregulated mGluR5 can contribute to astrocytic activation and enhanced PKCε activity, suggesting another potential mechanism by which spinal neuronal excitability may become increased after painful facet capsule loading [158, 172, 173]. Facet capsule strains induced by whiplash-like joint loading can also disrupt the ability of capsule-innervating afferent fibers to produce substance P mRNA for at least 1 week after injury [124]. At the same time, substance P protein levels are increased in the DRG and decreased in the spinal cord over the same time frame [124], suggesting that joint injury may induce axonal dysfunction that affects protein utilization and retrograde transport along afferent fibers. The effects on cellular function that are induced by painful joint loading extend also to spinal glial cells, including increased spinal astrocytic activation and reactive synthesis and release of pro-inflammatory cytokines in both the DRG and spinal cord [125, 164, 174–176]. The integrated neuronal and astroglial cascades that have been shown to be activated after painful facet capsule injury collectively contribute to a general increase in excitatory signaling in the dorsal horn that can manifest as peripheral and central sensitization observed after mechanical facet joint injury.

Biomechanical evidence supports that a range of capsular ligament sub-failure strains can generate pain and contribute to sustained sensitivity in association with a host of cellular nociceptive responses. For this particular tissue injury, animal models have provided valuable platforms to study the effects of biomechanical loading on the initiation of pain-related electrophysiologic, immunologic, inflammatory, and neurophysiologic changes in both peripheral and central nervous system structures. Further, such systems have allowed the temporal and spatial definition of an integrated schema from the initial injury spanning the development of persistent pain (Fig. 19.2). Continued integrated biomechanical and physiologic research is needed to relate the findings in animal models to injury scenarios in humans, which are currently informed by simulation studies and clinical reports, but which lack the ability to incorporate cellular and molecular level assays. Although injury to the cervical facet joint is a major contributor to whiplash-associated pain, other anatomical structures in the cervical spine are vulnerable to mechanical insult during different loading scenarios. Of note, the cervical nerve roots, which will be addressed in the next section, are highly susceptible to compression- and tension-based loading and contribute to pain based on the magnitude, duration and rate at which they are deformed.

19.4.2 Biomechanics and Painful Nerve Root Injury

As described above, nerve root compression is a common injury in the cervical spine and leads to a host of pain states, which depend on the mechanical profiles of the injury (Figs. 19.1 and 19.2). The clinical syndrome of radiculopathy originates after nerve root injury that can be induced by transient or sustained compression or tension, chemical stimuli, or a combination of both [177–179]. Cervical radiculopathy is characterized clinically as pain, paresthesia or hypersensitivity to mechanical and temperature stimuli radiating down the arm [180–183]. Although both mechanically- and chemically-induced injury to the cervical nerve roots have been confirmed to contribute to the development of chronic radicular pain [184–186], we focus on mechanically-induced neural trauma and its contribution to nerve root-mediated pain in the cervical spine. Mechanical trauma to the nerve roots can occur due to a number of diseases, disorders, and/or spinal loading scenarios. Direct compression, or associated tethering, of the roots can be induced slowly over an extended period of time which is often associated with foraminal narrowing that can occur with the natural progression of spinal aging or degeneration. In contrast, traumatic injury to the nerve root tissue can result from rapid increases in high-magnitude loads or deformations that can be produced by injuries, such as a herniated disc protruding into the intermedullary cavity or spinal trauma causing the intervertebral foramen to become obstructed [180, 187].

Pain induced by nerve root loading can be sustained long after the removal of the tissue insult. The specific mechanics of injury, including the type of load (i.e. compressive or tensile), the loading rate, the maximum load, and the duration of loading all contribute to the resulting behavioral sensitivity (Fig. 19.2) [53]. For example, one such modulator is the duration of root compression, which can be only transient in some injury conditions, including sports or automotive related traumas, or sustained due to disc herniation and spondylosis [180, 188–191]. This discrepancy has also complicated the clinical understanding of such injuries and their associated pathologies because a transient compression may not be visible by conventional imaging. Consequently, the resulting neck pain would not be attributed to a mechanical injury to the nerve root. Animal models of mechanical root injury have been instrumental in enabling the development of a better understanding of the cellular and molecular mechanisms that are initiated after nerve root trauma and how they contribute to the development of acute and/or chronic pain. Although the effects of nerve root compression profiles have been the focus of more investigations than those for tension, both injury modalities have been shown to play a role in pain.

Using animal models, the rate at which tension is applied to the nerve root and the maximum tensile strain have been shown to alter both the mechanical response of the neural tissue and the physiologic function of the neurons that are undergoing the tensile loading [192, 193]. Nerve roots tested in tension to failure exhibit a rate dependence to their mechanical strength at failure; a dynamic loading rate (15 mm/s) results in greater mechanical parameters than those for quasistatic loading (0.01 mm/s): maximum load (13.9 ± 7.5 versus 5.7 ± 2.7 g), maximum stress (624.9 ± 306.8 versus 257.9 ± 111.3 kPa), and elastic modulus (2.9 ± 1.5 versus 1.8 ± 0.8 MPa) [192]. This rate-dependent response is not surprising as it has been commonly demonstrated for other soft tissues [194, 195]. Of interest though, that study also reported that the magnitude of tensile strain applied to the root also affects measures of the tissue's physiologic function, as assessed by electrically-evoked compound action potentials [193]. Tensile strains between 10 % and 20 % were sufficient to decrease the conduction velocity and action potential amplitudes compared to strains less than 10 % [192]. Further, strains greater than 20 % completely blocked action potential conduction irrespective of the strain rate [193]. Although these studies measured structural and functional outcomes in the lumbar nerve roots of the rat during tensile injury, similar tension-based painful injuries occur in the cervical nerve roots and the pain outcomes are hypothesized to have a similar dependence on the mechanics of injury.

As with modeling the facet joint injury, animal models of painful nerve root compression have defined the pathophysiological consequences of root compression and in some cases can relate the tissue injury to those mechanisms in the context of pain. There is extensive literature on the role of nerve root (and cauda equina) compression in mediating low back pain that has formed the basis of most of the work in the cervical spine [4]. More severe nerve root compressions (i.e. higher loads, greater deformation, longer compression times) produce a more-robust behavioral sensitivity response [76, 196–200]. In fact, a transient cervical nerve root compression (applied for 15 min) in the rat using a compressive load of 26 mN elicits mechanical allodynia for only 24 h whereas compressive loads at or above the higher (38 mN) load applied to the nerve root for the same duration produces mechanical allodynia that is sustained for at least 7 days after injury [201]. This suggests a load threshold exists for root compression to induce pain that is sustained. In the same rat injury model, a load of 98.1 mN, which is even further above that load threshold when applied for 15 min, does not initiate mechanical or thermal sensitivity when it is applied for a duration of only 3 min [76, 202, 203]. The degree of tissue impingement or disruption produced by the compression is yet another mechanical parameter that modulates nerve root-induced behavioral sensitivity. For example, greater lumbar nerve root compressions (with an average strain of 39 ± 7 %) also produce more robust mechanical allodynia responses over a 14 day postoperative period compared to those compressions inducing lower strains of 21 ± 5 % [198]. Although that study relates the initial nerve root strain to the pain outcomes, the degree of nerve root deformation that is *maintained* after the removal of a transient compression has also been related to the resulting behavioral hypersensitivity [76]. Cervical roots compressed for 30 s or 3 min recover to over 88 ± 5 % of their original width and do not produce mechanical allodynia. In contrast, nerve roots compressed for 15 min recover to only 72 ± 13 % and *do* induce mechanical allodynia [76]. Together, these studies suggest disruption of the overall structural integrity of the nerve root *during* and *after* its compression may directly contribute to the onset and maintenance of nerve root-mediated pain.

Compression of the nerve root applies direct compression to the afferent fibers within the root and induces a wide array of phenotypic and electrophysiological changes at the location of compression (Fig. 19.2). Biochemical and cellular changes are also initiated at locations remote from the injury site, including in the cell bodies of the compressed afferents (located within the DRG) as well as in neurons and glia within the spinal cord with the release of cytotoxic amounts of neurotransmitters and cytokines at the synapses of the injured afferents [107, 204]. As with nerve root-mediated behavioral sensitivity, cellular changes induced by painful root compression also depend on the mechanics of the primary initial tissue compression [76, 77, 196, 197, 200, 202, 203]. The biomechanical changes that occur within cells at the primary site of injury and at remote locations provide the link between the specific injury mechanics (i.e. maximum load, rate of loading, and load hold time) and nerve root-induced behavioral sensitivity.

The magnitude of root compression contributes to changes in that tissue's structure through the degeneration of its constituent neurons and the associated infiltration of immune cells, and also modulates levels of neuropeptides that are expressed by the compressed and/or injured neurons in the root. For a painful cervical root compression in the rat that lasts 15 min a load of 32 ± 9 mN is sufficient to reduce expression of neurofilaments in the neurons in the injured nerve root by 1 week after injury, indicating their degeneration at this time point [201]. This load threshold is similar to the load threshold that also induces persistent behavioral sensitivity (38 mN) [205], suggesting that disruption of neuronal structure in the nerve root after its compression is associated with the maintenance of pain. In addition the magnitude of compression load also modulates the expression of neuropeptides in the DRG; the number of small diameter nociceptive afferents in the DRG that express substance P decreases by day 7 after a 15 min compression proportionate to the peak compressive load applied to the root [196, 201, 205]. This decrease in substance P in the DRG may be attributed to an increase in the synaptic release and utilization following painful root compression and may also contribute to the subsequent behavioral sensitivity [205, 206]. Yet, a compressive load of 20 ± 10 mN, which is *below* the load thresholds for producing sustained mechanical allodynia and neuronal degeneration, is sufficient to induce macrophage infiltration in the nerve root at this time, which indicates that macrophage infiltration at the root may not directly depend on the load magnitude [201]. Although macrophage infiltration at the root 7 days after its compression does not appear to be directly related to the presence of behavioral sensitivity, by day 14, when pain is still present, infiltrated macrophages exhibit a phagocytotic morphology, aiding in the removal of myelin debris from degenerated neurons [126]. This response suggests that non-quiescent macrophages at the nerve root may be involved in prolonging pain after nerve root compression. Taken together, this collection of studies suggests that afferent responses, including their degeneration and altered transport of substance P, is controlled at least partially by the magnitude of compression and dictates nerve root-mediated pain at this time point.

In addition to the magnitude of the applied load, the duration of compression mediates the disruption of neuronal structure and function. A nerve root compressed for a short period (30 s or 3 min) at 98.1 mN, which is above the load threshold for producing sustained behavioral sensitivity when applied for 15 min [196], exhibits almost full recovery of its original shape (over 88 ± 5 % of the original width of the root) [76]. This suggests that despite the magnitude of load being painful if applied for a certain time period, if the same load is applied for a short duration there is insufficient time for the cytoskeletal structures to become sufficiently deformed and/or undergo injury at the axonal level. That study suggests that a critical duration of compression exists between 3 and 15 min in which the disruption in root structure remains after the compression is removed. That same duration is also necessary to induce immediate changes in electrophysiological responses by the compressed neurons where they synapse in the spinal cord. During a transient 98.1 mN compression, mechanically-evoked action potentials are significantly reduced in the spinal cord of the rat at 6.6 ± 3.0 min of applied compression [207]. Since the duration of nerve root compression required to develop sustained pain for this 98.1 mN load is also between 3 and 10 min [202], the sustained disruption of axonal structure may influence the propagation of action potentials from the nerve root to the spinal cord and contribute to the initiation pain.

The rate at which nerve roots are compressed also mediates the consequences on neuronal function, and therefore, may also contribute to behavioral sensitivity just as load magnitude and duration have modulatory effects. In a porcine model of nerve root compression, Olmarker et al. [208] was among the first to demonstrate that faster rates of root compression (0.05–1 s to reach the maximum load) reduce neuronal conduction velocities in the compressed nerve root compared to slower rates (20 s to reach that same maximum load) [208]. Due to the viscoelastic nature of neural tissue like the nerve root, as the loading rate increases the peak load decreases. This has been demonstrated for rodent nerve root tissue during its compression; a 0.65 mm compression applied at a rate of 0.004 mm/s generates a peak reaction load of 102.3 ± 31.8 mN, which is significantly lower than the peak load of 177.9 ± 63.5 mN when the root is compressed the same amount but at a dynamic rate of 2 mm/s [205]. This study highlights the dependence of the mechanical parameters of nerve root compression on each other, suggesting that the peak load, rate of loading, degree of deformation and hold time must all be considered when defining the role of compression mechanics in nerve root-mediated pain.

As mentioned in Sect. 19.3 on pain mechanisms, the mechanics of nerve root compression also regulate neuronal signaling in the spinal cord where the injured afferents synapse (Fig. 19.2). The fact that evoked action potentials in the spinal cord are reduced (or abolished) at approximately 6 min of compression by a 98.1 mN load [207] indicates that a disruption of neuronal signaling occurs *during* the injury itself. Alterations in neuronal signaling are also present in the spinal cord at 1 week *after* root compression [196]. Spinal expression of the neuropeptides, substance P and CGRP, is decreased on day 7 after a painful compression of 98.1 mN for 15 min [196]. Of the two peptides, only CGRP expression depends on the magnitude of load [196], but the load threshold necessary to induce changes in that peptide was below that necessary to induce persistent behavioral sensitivity (19.5 and 38.2 mN, respectively) [196, 201]. Relatedly, the extent to which the spinal glutamatergic system is modified after nerve root compression in the rat depends on the magnitude of applied load, with increased expression of the glutamate receptor mGluR5 evident on day 7 [203]. However, spinal mGluR5 expression is elevated after a load of 558.6 mN compared to 98.1 mN, despite the absence of different behavioral sensitivity responses at the same time point [203]. These findings indicate that the modulation of biochemical factors other than CGRP and mGluR5 in the spinal cord may play a role in compressive load controlling the production and maintenance of behavioral sensitivity and that different responses may be mediated by different mechanical factors.

The magnitude of compression also controls the extent of spinal astrocyte activation, and does so in parallel with behavioral responses [76, 197, 200]. In contrast to biomarkers such as mGluR5 which exhibit load dependence even when behavioral outcomes do not [203], compressive loads of 98.1 and 588.6 mN, which result in similar levels of persistent mechanical allodynia also induce similar degrees of spinal astrocyte activation [197, 203]. This implicates activated spinal astrocytes in partially contributing to the load dependence seen with persistent behavioral sensitivity after root compression. Spinal astrocyte activation is not observed on day 7 after shorter compressions (30 s, 3 min) that also do not produce mechanical allodynia, but is evident after longer compressions (10 or 15 min) [76, 202]. Interestingly, the longer duration compressions are above that required to reduce neuronal firing in the spinal cord (6.6 ± 3.0 min) [207] and correlate to durations that also do not allow for sufficient structural recovery in the nerve root after the compression is removed [76]. Taken together, this collective group of studies implies that mechanical parameters of compression may immediately modulate axonal firing in the spinal cord and structural recovery of the nerve root which may further be involved in spinal astrocyte activation contributing to the maintenance of nerve root-mediated pain (Fig. 19.2).

19.4.3 Other Anatomic Sources of Mechanically-Induced Neck Pain

This chapter has primarily focused on pain originating from mechanical injury to the cervical facet joint and nerve root because of the strong history of engineering research related to these tissues, as well as the detailed understanding of pain physiology. Nonetheless, mechanical loading of other anatomic sources, such as the spinal cord, intervertebral disc and paraspinal musculature, of the cervical spine does occur and also contributes to neck pain (Fig. 19.1).

Traumatic spinal cord injuries (SCIs) can initiate persistent pain along with the other host of pathologies that are far more debilitating; a review of SCIs is beyond the scope of this chapter and can be found with extensive discussion elsewhere in the literature [209–213]. The mechanical trauma experienced by the spinal cord during incomplete spinal cord injury induces neuronal death and degeneration and glial scarring within the spinal cord, which contribute to persistent neuronal dysfunction and pain [214]. These cellular and molecular changes occur at the same spinal level as the injury, also known as primary or 'at-level' responses, and also at spinal levels above and below the injury, which are known as 'secondary' injury responses [211, 213, 215]. Pain from SCI has an immediate onset and is believed to be caused by the hyperexcitability of injured afferents themselves [211]. Secondary pain develops gradually and may be due to the widespread activation of glial cells that is induced by a host of pathological processes including microvascular hemorrhage, ischemia, edema, free radical-induced oxidative stress and excitotoxicity [211, 214, 216]. The extent of spreading of these secondary responses after SCI depends directly on the energy delivered to the neural tissue during injurious impact and increases the extent of pain [214, 216]. Similar to the pain induced by mechanical injury to the facet capsule and/or nerve root, the cascade of CNS responses that is initiated following spinal cord injury includes neuronal (chemical and anatomical) and inflammatory components that together contribute to pain (Fig. 19.2) [213].

Neck pain can also originate from injury to the intervertebral disc or nearby musculature since these structures are also innervated by nociceptive neurons (Fig. 19.1) [11, 14, 78, 79, 217]. Similar to the vulnerability of the facet joints to injury during certain neck movements, the cervical discs are also at risk for excessive compression, tension and shear during neck motions occurring during traumatic events which only increases as the disc degenerates with age [218, 219]. However, cervical disc injuries can be caused by more severe loading scenarios that involve multiple tissues [220] and are frequently accompanied by concomitant injuries, such as intervertebral canal narrowing and gross anatomical disruption and instability of the cervical spine, which themselves can also contribute to pain as discussed above [221]. Although subfailure muscle injury (e.g. sprains) in the cervical spine can induce pain [80, 222, 223], it is often short-lived and does not become chronic [25, 84]. Yet, as with disc injuries, injuries to the cervical musculature can result in spinal instability which can affect joint loading and amplify pre-existing pain states [83, 86, 88].

19.5 Experimental Techniques to Study Pain Biomechanics

Although most areas of contemporary injury biomechanics require the integration of engineering approaches to define the mechanics with the physiological sequelae of tissue loading, pain biomechanics, in particular, requires experimental techniques which incorporate both such methodologies for a complete understanding of how tissue injury mediates the function or dysfunction of tissues, cells, and molecular cascades. With the advent of more sophisticated technologies, with improved spatial and temporal resolution, it has become possible to make mechanical measurements in different model systems – ranging from the human to the isolated cell – and to relate those mechanics to cellular function/dysfunction beyond those of mechanical function.

Because of the particular importance of these insights for pain biomechanics, in this section we briefly review such experimental techniques which have particular utility for understanding and defining the mechanotransduction of pain. We begin with the human model and then highlight specific methodologies which are useful in animal model systems. Then we present sophisticated biomechanical techniques with increasingly smaller length scales that enable testing of cell/tissue constructs, in particular indentation techniques, and atomic force microscopy (AFM) for cell mechanics.

19.5.1 Human Volunteer and Post-mortem Models

Certainly, the most relevant model system is the human and for the case of pain studies, it would be the human volunteer or clinical patient, in which there is pain. However, ethical codes understandably prohibit both injurious and painful testing on live human subjects. As such, a host of non-invasive approaches have been developed to enable clinical imaging to define tissue geometry, organization, and function in the context of pain. In the laboratory setting, such approaches are being applied and modified to integrate with mechanical testing in order to develop mechanistic understandings of tissue responses, as well as to develop potential biomarkers for pain. Further, investigators continue to apply similar techniques to post-mortem human test subjects to provide fully instrumented simulations of "real-world" injuries. In this section we highlight these techniques briefly to provide context for their application in pain.

Traditional imaging modalities used for clinical and/or human volunteer studies, such as magnetic resonance imaging (MRI) and computed tomography (CT), have been used to identify bony motions and/or bony or soft tissue ruptures or failures [144, 224]. These are particularly helpful for providing real-world metrics of tissue responses and clinical correlates of injuries. Further, these techniques have, and continue to, provide important measurements from arguably *the* most relevant injury model – the human. However, conventional imaging techniques often cannot identify less severe tissue damage that can be sustained during cervical spinal ligament sprains and other painful injuries [141, 225, 226]. Further, in some cases, such techniques do not have the spatial or temporal resolution to localize tissue injury at all [24]. At the same time, non-invasive imaging techniques have been developed to quantify biomechanical and kinematic properties of the fibrous microstructure of collagenous tissues in the laboratory setting. In particular, several techniques, including polarized light imaging, optical coherence tomography, confocal microscopy, and small angle light scattering, have been developed and implemented to define the effects of tissue loading on collagen network structure and alignment [227–232]. These optical imaging techniques have long been associated with mechanical studies of biological tissues. Although macro-scale measurements, including visual tracking of fiduciary markers, have been used to quantify motions and strains in excised specimens and intact human post-mortem subjects [135, 142, 143], there is increasing evidence that the spatial resolution of traditional imaging is not sufficient to capture the kinematics of the complex microstructure, which can be particularly important for the collagenous ligaments (i.e. facet capsule) in the cervical spine.

Quantitative polarized light imaging (QPLI) can be used to acquire microstructural fiber alignment and strain data during tensile, shear, or biaxial loading [229, 232–234]. The natural optical properties of collagen fibers enable the dynamic measurement of both the mean fiber direction and the strength of alignment in that mean direction by transmitting polarized light through the collagenous tissue [227, 235–237]. Changes in collagen fiber alignment may indicate microstructural changes that underlie the altered biomechanics that are observed after subfailure loading, including increased tissue laxity and decreased stiffness [161]. The facet capsular ligament is particularly susceptible to sub-failure loading that can lead to persistent pain, potentially through changes at the microstructural level. Indeed, in both human and rat cervical facet capsular ligaments, fiber realignment was detected at macro-level strains well below those that caused tissue rupture, suggesting that local strains at the microstructural level may differ from those observed at the macro-scale and could lead to activation of mechanoreceptors in the facet capsule prior to gross tissue failure [151, 234, 238]. In the context of pain biomechanics, imaging modalities like QPLI expand our understanding of mechanical loading and its effects on biological tissues across multiple spatial scales and can help inform about the local loading environment of the afferents in the tissue that is being injured. Although not directly informing the pain mechanisms that are initiated by injurious loading to the cervical spine, the macro- and microstructural information gleaned from biomechanical imaging studies can suggest potential mechanical stimuli and mechanical environments for those pain mechanisms.

19.5.2 In Vivo Animal Models

As summarized above in Sect. 19.4, in vivo injury models provide tremendous value in their serving as platforms to directly integrate injury biomechanics to pain outcomes and nociceptive cascades over time and through anatomical regions. The cellular and biochemical responses in these models are analyzed quantitatively using a host of ever-advancing assays such as reverse transcription polymerase chain reaction (RT-PCR) to quantify mRNA expression, and enzyme-linked immunosorbent assay (ELISA), Western Blot and immunohistochemistry to measure protein expression. In the context of pain and neuronal injury, these assays can be supplemented with techniques that enable the analysis of neuronal physiological function, such as electrophysiological recordings and calcium mobilization assays. Here we review such techniques in the context of biomechanics, including those assays that are uniquely relevant to pain and/or injury biomechanics.

Nociceptive cascades can be studied in detail by the electrophysiological response and adaptation to non-noxious or noxious stimuli. Both extracellular and intracellular electrophysiological measurements at the tissue and cellular levels provide insight into the events leading from mechanical injury to nociceptive or pathological pain. This section provides examples of the applications of electrophysiology as they pertain to biomechanical injury in the cervical spine; the literature contains more detailed reviews of the relevant neurobiology and methodologies [59, 239–241].

Extracellular electrophysiology is used to measure the electrical activity in single cells or larger populations by recording membrane potentials with a sensitive electrode that is placed in very close proximity to the body of the cell. These techniques can be applied for making measurements either in vivo or in vitro. But, currently, they provide great benefit to understand neuronal function by their ability to record neuronal activity in vivo in awake or anesthetized subjects, providing measurements in the "closest-to-natural" state of a model, rather than in an artificial environment. Microelectrodes for extracellular recording are designed to include single recording sites or tetrode clusters of electrodes on a single probe for more powerful identification of individual spiking cells in large populations [242]. Sophisticated spike templating and clustering methods have been developed to process data from collections of cells to either analyze local field potentials from populations of cells or isolate potentials from individual neurons to analyze spiking rates, spontaneous or evoked activity, and numerous other features of neuronal activity [243–245].

Extracellular electrophysiological recordings can provide real-time information about mechanical loading of tissues as well as the changes in neuronal excitability and plasticity provoked by injurious tissue loading. Afferents originating in the C5/C6 facet capsule increase their firing rates with tensile stretch of the capsule, and firing becomes saturated at strain levels similar to those experienced during whiplash [120, 121, 153]. Secondary neurons in the dorsal horn have also shown evidence of activation and plasticity after both facet joint injury and nerve root compression [122, 123, 207]. In particular, facet joint distraction that induces painful capsule strain and behavioral hypersensitivity also increases spontaneous firing rates of neurons in the deep laminae of the dorsal horn at the same cervical level as the facet capsule injury [122]. After facet capsule injury, neurons also exhibit increased evoked firing in response to both non-noxious and noxious stimuli; this neuronal hyperexcitability is evident at 1 day and persists to at least 7 days after the injury (Fig. 19.3) [122, 123]. Extracellular electrophysiological data can be integrated with other data such as behavioral sensitivity and tissue assays in animal models to elucidate biomechanical pain mechanisms [122, 123, 246–248].

Single-cell intracellular recording can be more technically demanding than extracellular techniques because the electrode must be inserted into the cell body or nerve fiber without damaging the cell membrane. However, the technique offers precision and resolution that are difficult to achieve with extracellular electrodes. Patch clamp recording is the predominant intracellular technique, using glass micropipettes filled with ionic solution to monitor and control membrane voltages in both in vivo and in vitro settings [239]. Neuropathic and inflammatory pain following mechanical injury often involve the potentiation of primary afferents and, in chronic pain cases, the long-term potentiation of secondary and projection neurons in the dorsal horn of the spinal cord [249, 250]. In both cases, intracellular recording of individual neurons in the dorsal horn in vivo or in organotypic culture can be used to monitor pain-related changes in excitability and the response of neurons to pharmacological agents.

For studies in which larger neuronal population dynamics are of interest (i.e., high throughput recording or studies of interactions between non-adjacent neurons), arrays of up to hundreds of electrodes can be used to record electrophysiological data from widely distributed neurons. Multielectrode arrays (MEAs) capture electrophysiological data from multiple sites simultaneously, increasing output and enabling new types of network physiology experimentation that cannot be performed with individual electrodes [241, 245, 251, 252]. Arrays add dimensionality to recordings, so electrophysiological activity can be monitored both temporally and spatially as signals are passed through networks of cells. As electrode and data processing technology have developed in recent years, studying organotypic and dissociated neuronal cultures on MEAs has become an effective way to monitor a network of neurons in real-time during an applied mechanical insult or treatment stimulus [245, 253, 254]. The recent development of flexible electrode arrays has enabled MEA use in vivo as well [254–257], and this technology could be used in the future to enhance the biomechanical study of the cervical spine in both physiologic and injurious loading conditions. The activity of afferent fibers innervating the facet capsule has been recorded from the dorsal rootlets as they enter the spinal dorsal horn [153]; multi-electrode array technology offers the possibility of recording activity in the peripheral terminals of afferents in the capsule itself, to relate biomechanical data to activity of mechanosensitive afferents both spatially and temporally. In the future, this technology may enable what imaging modalities like QPLI cannot – providing an integrative link between the biomechanical loading of tissues and the activation of nociceptive fibers and pain mechanisms.

19.5.3 Mechanical Loading to Constructs and Isolated Cell Populations

Animal models relating the effects of mechanical loading parameters to the phenotypic and functional responses of neurons and other cells within neural and other tissues are instrumental for defining the mechanisms involved in pain processing. Yet, it is beneficial, and often necessary, to supplement such in vivo injury models with ex vivo tests that allow for a greater degree of control over sample preparation and testing parameters. Mechanical testing of isolated neural tissue is employed for numerous reasons, including to define the change in tissue structural properties after injury and to study the response of certain cell types to direct mechanical insults. This section will briefly discuss its use for measuring mechanical properties of tissue after injury as well as for its use as mimicking injurious loading ex vivo.

Applying mechanical loads to living CNS tissue is frequently performed to measure stiffness and/or viscosity of the specimen after injury. For example, one study used a modified atomic force microscope (AFM) to measure the unrelaxed elastic modulus of spinal cord tissue in anatomical regions that undergo glial scarring after traumatic SCI in rats [258]. The modulus of the spinal cord decreased at both 2 and 8 weeks post-injury at the site of the injury but the decrease was also observed across spinal levels rostral to the site of injury [258]. Since the degree of spreading of the cellular and biochemical changes induced by SCI correlates to the extent of pain (see Sect. 19.4 above) it is possible that the change in mechanical properties of neural tissue after injury correlates with the spreading of pain [205, 211, 216].

Regional stiffness values of isolated tissues also can be obtained by using customized micro-indentation devices (Fig. 19.4) [259]. The micro-indentation device has the capability to measure tissue stiffness of the rat spinal dorsal horn through incremental (50 μm every 15 s) tissue indentation by a blunt-ended cylindrical probe (Fig. 19.4b). The storage modulus, as well as more complicated relationships, can be determined using force-indentation response data (Fig. 19.4c). Techniques such as AFM and micro-indentation that measure the mechanical response of neural tissue can provide specific regional data about the local tissue mechanics after injury, in particular for cases in which there may not be any macroscale injury detected or when the tissue's mechanical milieu may be modified by the cells that reside there and become altered in response to chemical mediators [260, 261].

Mechanical testing of excised living neural tissue is also valuable to provide a model system that mimics tissue injury but enables measurements of cellular consequences of such tissue loading. Gross mechanical loading by both compression and tension has been performed on whole spinal cord samples and individual segments of spinal cords [258, 262]. Ouyang et al. [262] found that compressing white matter samples from guinea pig spinal cord to greater than 50 % of their original thickness at a rate of 0.05 mm/s reduced the amplitude of complex action potentials during the compression proportionally

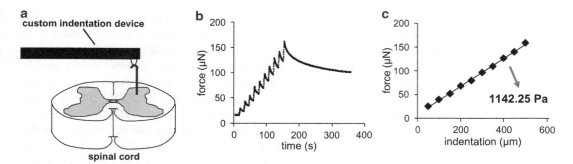

Fig. 19.4 Microindentation devices are used to measure mechanical properties of isolated tissues, such as spinal cord from the rat. (**a**) Schematic indicating spatial resolution of the device with a probe that enables indentation and measurements in gray and white matter of the spinal cord, as well as in different regions of the dorsal horn, where different cell populations reside and functions occur. (**b**) Representative force-time response to an incremental indentation of the probe with a 50 μm indentation applied every 15 s. (**c**) The corresponding force-indentation response to (**b**), with a slope corresponding to the local tissue stiffness

to the applied maximum load. Further, at compressions greater than 70 % of the original thickness, action potential propagation was completely blocked [262]. That immediate change in neuronal function after compression that is dependent on the mechanics of the compression is consistent with in vivo observations using neural tissue injury models. Nerve root compression models demonstrate that the degree of axonal deformation (either magnitude of deformation or time of deformation) directly alters axonal structure and function and may contribute to sustained pain [76, 198, 207]. Combining in vivo and ex vivo studies of mechanical loading provides a better understanding of how mechanical profiles directly alter cellular function in such a way that may contribute to the initiation and maintenance of pain.

Although biomechanical tests of CNS tissue ex vivo can be comparable to in vivo mechanical injury scenarios, it may also be desirable to measure responses of one or more different populations of cells isolated from the responses of other constituent cell types within that tissue. In those cases, cells can be cultured in artificial constructs that mimic or enable one or another feature of the native tissue and undergo mechanical injury and/or testing. For example, purified cultures of neurons and/or astrocytes are grown on both deformable substrates and within polymer scaffolds in order to simulate injurious loading of the brain. They can be exposed to scaffold deformations in order to distribute loads to the cells in a way that simulates the loading environment that exists in vivo; for example, astrocytes on deformable membranes stretched to 25 % strain using a single air impulse undergo cellular activation by 1 day after injury [260]. To simulate ligament injury, neurons are embedded into collagen gel matrices to undergo tensile loads in a three-dimensional controlled environment [263]. Such an approach can be coupled with biochemical assays to define immediate physiological outcomes to controlled neuronal loading (Fig. 19.5). If a dog-bone collagen gel matrix seeded with rat cortical neurons undergoes tension to failure at 0.5 mm/s there is immediate increased expression of phosphorylated extracellular signal-related kinase in the neurons, indicating cellular activation (Fig. 19.5). These types of in vitro experiments and techniques allow for the analysis of how individual aspects of highly controlled mechanical loads activate cells involved in nociceptive signaling. The combination of in vivo, ex vivo and in vitro experiments has greatly strengthened the understanding of the cellular mechanisms that contribute to mechanically-induced pain.

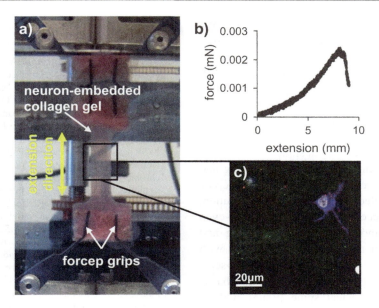

Fig. 19.5 Mechanical testing of collagen gels containing neurons enable defining relationships between macro- and micro-level mechanics, and physiological cellular responses. (**a**) Test set-up of a dog-bone shaped collagen gel containing neurons from rat brain cortexes in an Instron. (**b**) Representative force-extension response of the collagen gel sample loaded to failure under tension at 0.5 mm/s. (**c**) Immunolabeled image of same collagen (*green*) gel containing a neuron (*blue* = MAP2, cell body; *white* = DAPI, nucleus) indicating increased expression and secretion of phosphorylated extracellular signal-related kinase (*red* = pERK) immediately after tensile loading. Purple color indicates the co-localization of neuronal pERK

19.6 Summary

This chapter reviewed the general biomechanics and pain mechanisms in the context of cervical spine biomechanics. Numerous soft and hard tissues in the cervical spine are innervated with nociceptive and mechanosensitive fibers and have the potential to generate pain, including those of neural and ligamentous composition. Biomechanical loading of these tissues can activate tissue-innervating fibers, and loading to a sufficient extent (magnitude, deformation, load, duration) can lead to pain. For both subfailure and failure tissue loading, injury activates complex inflammatory, immune, and electrophysiological cascades that lead to peripheral and central sensitization and pain hypersensitivity.

In particular, the injury mechanisms and pain outcomes for the facet joint and nerve root provide important comparisons between different tissue types, their response to mechanical loading and the cascades that are initiated locally and in the spinal cord. For example, the facet joint is particularly susceptible to injury through excessive strain of the capsular ligament that is characterized by electrophysiological changes in the capsule itself and in the spinal cord indicative of central sensitization, as well as neuronal and glial modifications traditionally associated with pain. The nerve root can be injured by tensile or, more commonly, direct compression and pain outcomes are highly dependent on the mechanical profile of the injury, including rate, duration, and magnitude of compression. Much like facet joint injuries, nerve root compression induces a host of cellular and electrophysiological changes in the DRG and spinal cord that promote pain. For facet joint, nerve root, and other cervical spine injuries, human volunteer and post-mortem human subject studies form a foundation for the kinematic study of isolated tissues and whole cervical spines using traditional and novel imaging modalities. More recently, integrative techniques have sought to combine our knowledge of tissue biomechanics with the complicated physiological cascades associated with tissue injury and pain. The result has been the development of biochemical assays, electrophysiological recording platforms, and techniques for biomechanical testing on ever-smaller spatial scales, all of which have contributed to our understanding of pain biomechanics in the cervical spine.

As the experimental – both engineering and physiological – techniques continue to become more sophisticated and have greater resolution, the advances can continue along the path to even more clearly understanding the initiating and subsequent pain responses. For example, with greater advances in imaging and micro-scale mechanical testing of biologic tissues, mechanotransduction in cervical spine tissues will be studied on increasingly smaller spatial scales relating biomechanics to the cellular, or even molecular, level consequences. Similarly, much of the research that has characterized pain-generating cellular and molecular responses to tissue loading has been performed in animal models that, while valuable as a research tool, have limitations when being interpreted for clinical pain conditions. For example, facet capsular ligament damage at subfailure strains has been visualized in the laboratory setting [161], but such damage is not yet evident using clinical imaging with traditional imaging modalities. The development of techniques that bridge the laboratory and clinical settings will help validate and corroborate findings in both such environments, leveraging the benefits of each and helping to eliminate many potential limitations with each.

Acknowledgments The authors thank Kathryn Lee, Kristen Nicholson, and Sonia Kartha for providing panels for the immunohistochemistry images. This work was supported by funding from the National Institutes of Health/National Institute of Arthritis, Musculoskeletal and Skin Diseases (#AR056288 and BIRT Supplement), the Department of Defense (W81XWH-10-1-1002 and W81XWH-10-2-0140), the National Science Foundation (Grant No. 054745), and the Cervical Spine Research Society, as well as the Catherine D. Sharpe and Ashton Foundations.

References

1. Breivik H, Collett B, Ventafridda V et al (2006) Survey of chronic pain in Europe: prevalence, impact on daily life, and treatment. Eur J Pain 10(4):287–333
2. Hogg-Johnson S, van der Velde G, Carroll LJ et al (2008) The burden and determinants of neck pain in the general population. Eur Spine J 17(S1):S39–S51
3. Loeser JD, Treede RD (2008) The Kyoto protocol of IASP basic pain terminology. Pain 137(3):473–477
4. DeLeo JA, Winkelstein BA (2002) Physiology of chronic spinal pain syndromes: from animal models to biomechanics. Spine 27(22):2526–2537
5. Antonacci MD, Mody DR, Heggeness MH (1998) Innervation of the human vertebral body: a histologic study. J Spinal Disord 1(6):526–531
6. Inami S, Shiga T, Tsujino A et al (2001) Immunohistochemical demonstration of nerve fibers in the synovial fold of the human cervical facet joint. J Orthop Res 19(4):593–596
7. Palmgren T, Gronblad M, Virri J et al (1999) An immunohistochemical study of nerve structures in the anulus fibrosus of human normal lumbar intervertebral discs. Spine 24(20):2075–2079
8. Rhalmi S, Yahia LH, Newman N et al (1993) Immunohistochemical study of nerves in lumbar spine ligaments. Spine 18(2):264–267
9. Bland JH (1990) Anatomy and physiology of the cervical spine. Semin Arthritis Rheum 20(1):1–20
10. Clark CR (ed) (2005) The cervical spine. Lippincott, Philadelphia
11. Fagan A, Moore R, Vernon Roberts B et al (2003) ISSLS prize winner: the innervation of the intervertebral disc: a quantitative analysis. Spine 28(23):2570–2576
12. Watson C, Paxinos G, Kayalioglu G (eds) (2009) The spinal cord. Elsevier, London
13. Ozawa T, Ohtori S, Inoue G et al (2006) The degenerated lumbar intervertebral disc is innervated primarily by peptide-containing sensory nerve fibers in humans. Spine 31(21):2418–2422
14. Yoshizawa H, O'Brien JP, Smith WT et al (1980) The neuropathology of intervertebral discs removed for low-back pain. J Pathol 132(2):95–104
15. Wang DL, Jiang SD, Dai LY (2007) Biologic response of the intervertebral disc to static and dynamic compression in vitro. Spine 32(23):2521–2528
16. Freemont AJ (2009) The cellular pathobiology of the degenerated intervertebral disc and discogenic back pain. Rheumatology 48:5–10
17. Freemont AJ, Peacock TE, Goupille P et al (1997) Nerve ingrowth into diseased intervertebral disc in chronic back pain. Lancet 350:178–181
18. Ohtori S, Takahashi K, Chiba T et al (2002) Substance P and calcitonin gene-related peptide immunoreactive sensory DRG neurons innervating

the lumbar intervertebral discs in rats. Ann Anat 184:235–240
19. Branconi JL, Guarino BB, Baig HA et al (2012) Painful whole body vibration increases NGF and BDNF in cervical intervertebral discs in the rat. Paper presented at the Northeast Bioengineering Conference, Philadelphia
20. Kartha S, Zeeman M, Baig H et al (2014) Upregulation of BDNF and NGF in cervical intervertebral discs exposed to painful whole-body vibration. Spine 39(19):1542–1548
21. Kokubo Y, Uchida K, Kobayashi S et al (2008) Herniated and spondylotic intervertebral discs of the human cervical spine: histological and immunohistological findings in 500 en bloc surgical samples. J Neurosurg Spine 9(3):285–295
22. Panjabi MM, Oxland TR, Parks EH (1991) Quantitative anatomy of cervical spine ligaments part 1: upper cervical spine. J Spinal Disord 4(3):270–276
23. Panjabi MM, Oxland TR, Parks EH (1991) Quantitative anatomy of cervical spine ligaments part 2: middle and lower cervical spine. J Spinal Disord 4(3):277–285
24. Curatolo M, Bogduk N, Ivancic PC et al (2011) The role of tissue damage in whiplash-associated disorders. Spine 36(25S):S309–S315
25. Siegmund GP, Winkelstein BA, Ivancic PC et al (2009) The anatomy and biomechanics of acute and chronic whiplash injury. Traffic Inj Prev 10(2):101–112
26. Higuchi K, Sato T (2002) Anatomical study of lumbar spine innervation. Folia Morphol (Warsz) 61(2):71–79
27. Imai S, Hukuda S, Maeda T (1995) Dually innervating nociceptive networks in the rat lumbar posterior longitudinal ligaments. Spine 20(19):2086–2092
28. el-Bohy A, Cavanaugh JM, Getchell ML et al (1988) Localization of substance P and neurofilament immunoreactive fibers in the lumbar facet joint capsule and supraspinous ligament of the rabbit. Brain Res 460(2):379–382
29. Kallakuri S, Cavanaugh JM, Blagoev DC (1998) An immunohistochemical study of innervation of lumbar spinal dura and longitudinal ligaments. Spine 23(4):403–411
30. Imai S, Konttinen YT, Tokunaga Y et al (1997) An ultrastructural study of calcitonin gene-related peptide-immunoreactive nerve fibers innervating the rat posterior longitudinal ligament. A morphologic basis for their possible efferent actions. Spine 22(17):1941–1947
31. Yahia LH, Newman N, Rivard CH (1988) Neurohistology of lumbar spine ligaments. Acta Orthop Scand 59(5):508–512
32. Yamada H, Honda T, Yaginuma H et al (2001) Comparison of sensory and sympathetic innervation of the dura mater and posterior longitudinal ligament in the cervical spine after removal of the stellate ganglion. J Comp Neurol 434(1):86–100
33. Jiang H, Russell G, Raso JV et al (1995) The nature and distribution of the innervation of human supraspinal and interspinal ligaments. Spine 20(8):869–876
34. McLain RF (1994) Mechanoreceptor endings in human cervical facet joints. Spine 19(5):495–501
35. McClain RF, Raiszadeh KR (1995) Mechanoreceptor endings of the cervical, thoracic, and lumbar spine. Iowa Orthop J 15:147–155
36. Bogduk N, Marsland A (1988) The cervical zygapophysial joints as a source of neck pain. Spine 13(6):610–617
37. Kallakuri S, Singh A, Chen C et al (2004) Demonstration of substance P, calcitonin gene-related peptide, and protein gene product 9.5 containing nerve fibers in human cervical facet joint capsules. Spine 29(11):1182–1186
38. Dwyer A, Aprill C, Bogduk N (1990) Cervical zygapophyseal joint pain patterns. 1: A study in normal volunteers. Spine 15(6):453–457
39. Fukui S, Ohseto K, Shiotani M et al (1996) Referred pain distribution of the cervical zygapophyseal joints and cervical dorsal rami. Pain 68(1):79–83
40. Lord SM, Barnsley L, Bogduk N (1995) Percutaneous radiofrequency neurotomy in the treatment of cervical zygapophysial joint pain: a caution. Neurosurgery 36(4):732–739
41. Lord SM, Barnsley L, Wallis BJ et al (1996) Percutaneous radio-frequency neurotomy for chronic cervical zygapophyseal-joint pain. N Engl J Med 335(23):1721–1726
42. Slipman CW, Lipetz JS, Plastaras CT et al (2001) Therapeutic zygapophyseal joint injections for headaches emanating from the C2-3 joint. Am J Phys Med Rehabil 80(3):182–188
43. Yoganandan N, Knowles SA, Maiman DJ et al (2003) Anatomic study of the morphology of human cervical facet joint. Spine 28(20):2317–2323
44. Yoganandan N, Pintar FA, Cusick JF et al (1998) Head-neck biomechanics in simulated rear impact. Annu Proc Assoc Adv Automot Med 42:209–231
45. Webb AL, Collins P, Rassoulian H et al (2011) Synovial folds – a pain in the neck? Man Ther 16(2):118–124
46. Adams MA, Dolan P (2005) Spine biomechanics. J Biomech 38(10):1972–1983
47. Cusick JF, Yoganandan N (2012) Biomechanics of the cervical spine 4: major injuries. Clin Biomech 17(1):1–20
48. Myers BS, Winkelstein BA (1995) Epidemiology, classification, mechanism, and tolerance of human cervical spine injuries. Crit Rev Biomed Eng 23(5–6):307–409
49. Wen N, Lavaste F, Santin JJ, Lassau JP (1993) Three-dimensional biomechanical properties of the human cervical spine in vitro. I. Analysis of normal motion. Eur Spine J 2(1):2–11
50. White AA, Panjabi MM (1990) Clinical biomechanics of the spine, 2nd edn. Lippincott, New York

51. Myklebust JB, Pintar FA, Yoganandan N et al (1988) Tensile strength of spinal ligaments. Spine 13(5): 526–531
52. Yoganandan N, Pintar F, Butler J et al (1989) Dynamic response of human cervical spine ligaments. Spine 14(10):1102–1110
53. Winkelstein BA (2004) Mechanisms of central sensitization, neuroimmunology & injury biomechanics in persistent pain: implications for musculoskeletal disorders. J Electromyogr Kinesiol 14(1):87–93
54. Winkelstein BA (2011) How can animal models inform on the transition to chronic symptoms in whiplash? Spine 36(25s):S218–S225
55. Fricke B, Andres KH, Von During M (2001) Nerve fibers innervating the cranial and spinal meninges: morphology of nerve fiber terminals and their structural integration. Microsc Res Techn 53(2):96–105
56. Kumar R, Berger RJ, Dunsker SB et al (1996) Innervation of the spinal dura – myth or reality? Spine 21(1):18–25
57. Yamada H, Honda T, Kikuchi S et al (1998) Direct innervation of sensory fibers from the dorsal root ganglion of the cervical dura mater of rats. Spine 23(14):1524–1529
58. Wieseler-Frank J, Jekich BM, Mahoney JH et al (2007) A novel immune-to-CNS communication pathway: cells of the meninges surrounding the spinal cord CSF space produce proinflammatory cytokines in response to an inflammatory stimulus. Brain Behav Immun 21(5):711–718
59. McMahon S, Koltzenburg M (eds) (2005) Wall and Melzack's textbook of pain. Churchill Livingstone, London
60. Rexed B (1952) The cytoarchitectonic organization of the spinal cord in the cat. J Comp Neurol 96: 415–496
61. Basso DM, Beattie MS, Bresnahan JC (1996) Graded histological and locomotor outcomes after spinal cord contusion using the NYU weight-drop device versus transection. Exp Neurol 139:244–256
62. Belanger M, Drew T, Provencher J et al (1996) A comparison of treadmill locomotion in adult cats before and after spinal transection. J Neurophysiol 76:471–491
63. Bresnahan JC, Beattie MS, Todd FD et al (1987) A behavioral and anatomical analysis of spinal cord injury produced by a feedback-controlled impaction device. Exp Neurol 95:548–570
64. Muir GD, Whishaw IQ (2000) Red nucleus lesions impair overground locomotion in rats: a kinetic analysis. Eur J Neurosci 12:1113–1122
65. Profyris C, Cheema SS, Zang D et al (2004) Degenerative and regenerative mechanisms governing spinal cord injury. Neurobiol Dis 15:415–436
66. Hausmann ON (2003) Post-traumatic inflammation following spinal cord injury. Spinal Cord 41:369–378
67. Sroga JM, Jones TB, Kigerl KA et al (2003) Rats and mice exhibit distinct inflammatory reactions after spinal cord injury. J Comp Neurol 462:223–240
68. Beric A (1997) Post-spinal cord injury pain states. Anesthesiol Clin North Am 17(2):445–463
69. Siddall PJ, McClelland JM, Rutkowski SB et al (2003) A longitudinal study of the prevalence and characteristics of pain in the first 5 years following spinal cord injury. Pain 103:249–257
70. Coggeshall RE (1979) Afferent fibers in the ventral root. Neurosurgery 4(5):443–448
71. Hanai F, Matsui N, Hongo N (1996) Changes in responses of wide dynamic range neurons in the spinal dorsal horn after dorsal root or dorsal root ganglion compression. Spine 21(12):1408–1415
72. Hu SJ, Xing JL (1998) An experimental model for chronic compression of dorsal root ganglion produced by intervertebral foramen stenosis in the rat. Pain 77(1):15–23
73. Van Zundert J, Harney D, Joosten EA et al (2006) The role of the dorsal root ganglion in cervical radicular pain: diagnosis, pathophysiology, and rationale for treatment. Reg Anesth Pain Med 31(2):152–167
74. Kitagawa T, Fujiwara A, Kobayashi N et al (2004) Morphologic changes in the cervical neural foramen due to flexion and extension – in vivo imaging study. Spine 29(24):2821–2825
75. Tominaga Y, Maak TG, Ivancic PC et al (2006) Head-turned rear impact causing dynamic cervical intervertebral foramen narrowing: implications for ganglion and nerve root injury. J Neurosurg Spine 4(5):380–387
76. Rothman SM, Nicholson KJ, Winkelstein BA (2010) Time-dependent mechanics and measures of glial activation and behavioral sensitivity in a rodent model of radiculopathy. J Neurotrauma 27(5): 803–814
77. Winkelstein BA, DeLeo JA (2004) Mechanical thresholds for initiation and persistence of pain following nerve root injury: mechanical and chemical contributions at injury. J Biomech Eng 126(2): 258–263
78. Cavanaugh JM (1995) Neural mechanisms of lumbar pain. Spine 20(16):1804–1809
79. Mense S, Meyer H (1985) Different types of slowly conducting afferent units in cat skeletal muscle and tendon. J Physiol 363:403–417
80. Vasavada AN, Brault JR, Siegmund GP (2007) Musculotendon and fascicle strains in anterior and posterior neck muscles during whiplash injury. Spine 32(7):756–765
81. Elliott J, Jull GM, Noteboom JT (2006) Fatty infiltration in the cervical extensor muscles in persistent whiplash-associated disorders: a magnetic resonance imaging analysis. Spine 31(22):E847–E855
82. Elliott J, Pedler A, Kenardy J et al (2011) The temporal development of fatty infiltrates in the neck muscles following whiplash injury: an association with pain and posttraumatic stress. PLoS One 6(6):e21194
83. O'leary S, Falla D, Elliot JM et al (2009) Muscle dysfunction in cervical spine pain: implications for

assessment and management. J Orthop Sports Phys Ther 39(5):324–333
84. Graven-Nielsen T, Mense S (2001) The peripheral apparatus of muscle pain: evidence from animal and human studies. Clin J Pain 17(1):2–10
85. Smith LL (1991) Acute-inflammation – the underlying mechanism in delayed onset muscle soreness. Med Sci Sports Exerc 23(5):542–551
86. Stauber WT (2004) Factors involved in strain-induced injury in skeletal muscles and outcomes of prolonged exposures. J Electromyogr Kinesiol 14(1):61–70
87. Winkelstein BA, McLendon RE, Barbir A et al (2001) An anatomical investigation of the human cervical facet capsule, quantifying muscle insertion area. J Anat 198(Pt 4):455–461
88. Siegmund GP, Sanderson DJ, Myers BS et al (2003) Rapid neck muscle adaptation alters the head kinematics of aware and unaware subjects undergoing multiple whiplash-like perturbations. J Biomech 36(4):473–482
89. Chopard RP, de Miranda Neto MH, Lucas GA et al (1992) The vertebral artery: its relationship with adjoining tissues in its course intra and inter transverse processes in man. Rev Paul Med 110(6):245–250
90. Thiel HW (1991) Gross morphology and pathoanatomy of the vertebral arteries. J Manipulative Physiol Ther 14(2):133–141
91. Carlson EJ, Tominaga Y, Ivancic PC et al (2007) Dynamic vertebral artery elongation during frontal and side impacts. Spine J 7(2):222–228
92. Chung YS, Han DH (2002) Vertebrobasilar dissection: a possible role of whiplash injury in its pathogenesis. Neurol Res 24(2):129–138
93. Kawchuk GN, Wynd S, Anderson T (2004) Defining the effect of cervical manipulation on vertebral artery integrity: establishment of an animal model. J Manipulative Physiol Ther 27(9):539–546
94. Saeed AB, Shuaib A, Al-Sulaiti G et al (2000) Vertebral artery dissection: warning symptoms, clinical features and prognosis in 26 patients. Can J Neurol Sci 27(4):292–296
95. Svensson MY, Aldman B, Bostrom O et al (1998) Transient pressure gradients in the pig spinal canal during experimental whiplash motion causing membrane dysfunction in spinal ganglion nerve cells. Orthopade 27(12):820–826
96. Steeds CE (2009) The anatomy and physiology of pain. Surgery 27(12):507–511
97. Woolf CJ, Salter MW (2000) Neuronal plasticity: increasing the gain in pain. Science 288:1765–1768
98. Woolf CJ (2011) Central sensitization: implications for the diagnosis and treatment of pain. Pain 152:S2–S15
99. Coderre TJ, Katz J, Vaccarino AL et al (1993) Contribution of central neuroplasticity to pathological pain: review of clinical and experimental evidence. Pain 52:259–285
100. Dubner R, Ruda MA (1992) Activity-dependent neuronal plasticity following tissue injury and inflammation. Trends Neurosci 15(3):96–103
101. Markenson JA (1996) Mechanisms of chronic pain. Am J Med 101(1A):6S–18S
102. Millan MJ (1999) The induction of pain: an integrative review. Prog Neurobiol 57(1):1–164
103. Costigan M, Scholz J, Woolf CJ (2009) Neuropathic pain: a maladaptive response of the nervous system to damage. Annu Rev Neurosci 32:1–32
104. Braz JM, Nassar MA, Wood JN et al (2005) Parallel "pain" pathways arise from subpopulations of primary afferent nociceptor. Neuron 47(6):787–793
105. Julius D, Basbaum AI (2001) Molecular mechanisms of nociception. Nature 413(6852):203–210
106. Boucher TJ, McMahon SB (2001) Neurotrophic factors and neuropathic pain. Curr Opin Pharmacol 1(1):66–72
107. Latremoliere A, Woolf CJ (2009) Central sensitization: a generator of pain hypersensitivity by central neural plasticity. J Pain 10(9):895–926
108. Hucho T, Levine JD (2007) Signaling pathways in sensitization: toward a nociceptor cell biology. Neuron 55(3):365–376
109. Verri WA Jr, Cunha TM, Parada CA et al (2006) Hypernociceptive role of cytokines and chemokines: targets for analgesic drug development? Pharmacol Ther 112(1):116–138
110. Ji RR, Kohno T, Moore KA et al (2003) Central sensitization and LTP: do pain and memory share similar mechanisms? Trends Neurosci 26(12):696–705
111. Devor M (2009) Ectopic discharge in Abeta afferents as a source of neuropathic pain. Exp Brain Res 196:115–128
112. Djouhri L, Lawson SN (2004) Abeta-fiber nociceptive primary afferent neurons: a review of incidence and properties in relation to other afferent A-fiber neurons in mammals. Brain Res Rev 46(2):131–145
113. Kohama I, Ishikawa K, Kocsis JD (2000) Synaptic reorganization in the substantia gelatinosa after peripheral nerve neuroma formation: aberrant innervation of lamina II neurons by Abeta afferents. J Neurosci 20(4):1538–1549
114. Woolf CJ, Shortland P, Coggeshall RE (1992) Peripheral nerve injury triggers central sprouting of myelinated afferents. Nature 355:75–78
115. Gould HJ 3rd, Gould TN, England JD et al (2000) A possible role for nerve growth factor in the augmentation of sodium channels in models of chronic pain. Brain Res 854(1–2):19–29
116. Matayoshi S, Jiang N, Katafuchi T et al (2005) Actions of brain-derived neurotrophic factor on spinal nociceptive transmission during inflammation in the rat. J Physiol 569(Pt 2):685–695
117. Moalem G, Tracey DJ (2006) Immune and inflammatory mechanisms in neuropathic pain. Brain Res Rev 51(2):240–264
118. Pezet S, McMahon SB (2006) Neurotrophins: mediators and modulators of pain. Annu Rev Neurosci 29:507–538
119. DeLeo JA, Yezierski RP (2001) The role of neuroinflammation and neuroimmune activation in persistent pain. Pain 90(1–2):1–6

120. Chen C, Lu Y, Kallakuri S et al (2006) Distribution of Adelta and C-fiber receptors in the cervical facet joint capsule and their response to stretch. J Bone Joint Surg 88:1807–1816
121. Lu Y, Chen C, Kallakuri S et al (2005) Neural response of cervical facet joint capsule to stretch: a study of whiplash pain mechanism. Stapp Car Crash J 49:49–65
122. Crosby ND, Weisshaar CL, Winkelstein BA (2013) Spinal neuronal plasticity is evident within 1 day after a painful cervical facet joint injury. Neurosci Lett 542:102–106
123. Quinn KP, Dong L, Golder FJ et al (2010) Neuronal hyperexcitability in the dorsal horn after painful facet joint injury. Pain 151:414–421
124. Lee KE, Winkelstein BA (2009) Joint distraction magnitude is associated with different behavioral outcomes and substance P levels for cervical facet joint loading in the rat. J Pain 10(4):436–445
125. Winkelstein BA, Santos DG (2009) An intact facet capsular ligament modulates behavioral sensitivity and spinal glial activation produced by cervical facet joint tension. Spine 33(8):856–862
126. Chang YW, Winkelstein BA (2011) Schwann cell proliferation and macrophage infiltration are evident at day 14 after painful cervical nerve root compression in the rat. J Neurotrauma 28(12):2429–2438
127. Rothman SM, Huang Z, Lee KE et al (2009) Cytokine mRNA expression in painful radiculopathy. J Pain 10(1):90–99
128. Rothman SM, Winkelstein BA (2010) Cytokine antagonism reduces pain and modulates spinal astrocytic reactivity after cervical nerve root compression. Ann Biomed Eng 38(8):2563–2576
129. Aprill C, Bogduk N (1992) The prevalence of cervical zygapophyseal joint pain: a first approximation. Spine 17(7):744–747
130. Barnsley L, Lord S, Bogduk N (1993) Comparative local anaesthetic blocks in the diagnosis of cervical zygapophysial joint pain. Pain 55(1):99–106
131. Lord SM, Barnsley L, Wallis BJ et al (1996) Chronic cervical zygapophysial joint pain after whiplash: a placebo-controlled prevalence study. Spine 21(15):1737–1744
132. Stemper BD, Yoganandan N, Gennarelli TA et al (2005) Localized cervical facet joint kinematics under physiological and whiplash loading. J Neurosurg Spine 3(6):471–476
133. Jaumard NV, Welch WC, Winkelstein BA (2011) Spinal facet joint biomechanics and mechanotransduction in normal, injury and degenerative conditions. J Biomech Eng 133(7):071010
134. Kaneoka K, Ono K, Inami S et al (1999) Motion analysis of cervical vertebrae during whiplash loading. Spine 24(8):763–769
135. Pearson AM, Ivancic PC, Ito S et al (2004) Facet joint kinematics and injury mechanisms during simulated whiplash. Spine 29(4):390–397
136. Bogduk N, Yoganandan N (2001) Biomechanics of the cervical spine. Part 3: minor injuries. Clin Biomech 16(4):267–275
137. Deng B, Begeman PC, Yang KH et al (2000) Kinematics of human cadaver cervical spine during low speed read-end impacts. Stapp Car Crash J 44:171–188
138. Panjabi MM (1998) Cervical spine models for biomechanical research. Spine 23:2684–2700
139. Panjabi MM, Cholewicki J, Nibu K et al (1998) Simulation of whiplash trauma using whole cervical spine specimens. Spine 23:17–24
140. Ono K, Kaneoka K, Wittek A et al (1997) Cervical injury mechanism based on the analysis of human cervical vertebral motion and head-neck-torso kinematics during low speed rear impacts. Stapp Car Crash J 41:339–356
141. Yoganandan N, Cusick JF, Pintar FA et al (2001) Whiplash injury determination with conventional spine imaging and cryomicrotomy. Spine 26(22):2443–2448
142. Sundararajan S, Prasad P, Demetropoulos CK et al (2004) Effect of head-neck position on cervical facet stretch of post mortem human subjects during low speed rear end impacts. Stapp Car Crash J 48:331–372
143. Winkelstein BA, Nightingale RW, Richardson WJ et al (2000) The cervical facet capsule and its role in whiplash injury: a biomechanical investigation. Spine 25(10):1238–1246
144. Panjabi MM, Cholewicki J, Nibu K et al (1998) Mechanism of whiplash injury. Clin Biomech 13(4):239–249
145. Luan F, Yang KH, Deng B et al (2000) Qualitative analysis of neck kinematics during low-speed rear-end impact. Clin Biomech 15(9):649–657
146. Panjabi MM, Cholewicki J, Nibu K et al (1998) Capsular ligament stretches during in vitro whiplash simulations. J Spinal Disord 11(3):227–232
147. Grauer JN, Panjabi MM, Cholewicki J et al (1997) Whiplash produces an S-shaped curvature of the neck with hyperextension at lower levels. Spine 22(21):2489–2494
148. Siegmund GP, Myers BS, Davis MB et al (2000) Human cervical motion segment flexibility and facet capsular ligament strain under combined posterior shear, extension and axial compression. Stapp Car Crash J 44:159–170
149. Siegmund GP, Myers BS, Davis MB et al (2001) Mechanical evidence of cervical facet capsule injury during whiplash: a cadaveric study using combined shear, compression, and extension loading. Spine 26(19):2095–2101
150. Yoganandan N, Pintar FA (1997) Inertial loading of the human cervical spine. J Biomech Eng 119(3):237–240
151. Quinn KP, Winkelstein BA (2007) Cervical facet capsular ligament yield defines the threshold for injury and persistent joint-mediated neck pain. J Biomech 40(10):2299–2306

152. Lee KE, Franklin AN, Davis MB et al (2006) Tensile cervical facet capsule ligament mechanics: failure and subfailure responses in the rodent. J Biomech 39(7):1256–1264
153. Chen C, Lu Y, Cavanaugh JM et al (2005) Recording of neural activity from goat cervical facet joint capsule using custom-designed miniature electrodes. Spine 30(12):1367–1372
154. Dong L, Winkelstein BA (2010) Simulated whiplash modulates expression of the glutamatergic system in the spinal cord suggesting spinal plasticity is associated with painful dynamic cervical facet loading. J Neurotrauma 27(1):163–174
155. Lee KE, Davis MB, Winkelstein BA (2008) Capsular ligament involvement in the development of mechanical hyperalgesia after facet joint loading: behavioral and inflammatory outcomes in a rodent model of pain. J Neurotrauma 25(11):1383–1393
156. Lee KE, Thinnes JH, Gokhin DS et al (2004) A novel rodent neck pain model of facet-mediated behavioral hypersensitivity: implications for persistent pain and whiplash injury. J Neurosci Methods 137(2):151–159
157. Lu Y, Chen C, Kallakuri S et al (2005) Neurophysiological and biomechanical characterization of goat cervical facet joint capsules. J Orthop Res 23:779–787
158. Weisshaar CL, Dong L, Bowman AS et al (2010) Metabotropic glutamate receptor-5 and protein kinase C-epsilon increase in dorsal root ganglion neurons and spinal glial activation in an adolescent rat model of painful neck injury. J Neurotrauma 27(12):2261–2271
159. Lu Y, Chen C, Kallakuri S et al (2005) Development of an in vivo method to investigate biomechanical and neurophysiological properties of spine facet joint capsules. Eur Spine J 14(6):565–572
160. Kallakuri S, Singh A, Lu Y et al (2008) Tensile stretching of cervical facet joint capsule and related axonal changes. Eur Spine J 17(4):556–563
161. Quinn KP, Lee KE, Ahaghotu CC et al (2007) Structural changes in the cervical facet capsular ligament: potential contributions to pain following subfailure loading. Stapp Car Crash J 51:169–187
162. Curatolo M, Petersen-Felix S, Arendt-Nielsen L et al (2001) Central hypersensitivity in chronic pain after whiplash injury. Clin J Pain 17(4):306–315
163. Herren-Gerber R, Weiss S, Arendt-Nielsen L et al (2004) Modulation of central hypersensitivity by nociceptive input in chronic pain after whiplash injury. Pain 5(4):366–376
164. Lee KE, Davis MB, Mejilla RM et al (2004) In vivo cervical facet capsule distraction: mechanical implications for whiplash & neck pain. Stapp Car Crash J 48:373–396
165. Rothman SM, Hubbard RD, Lee KE et al (2008) Detection, transmission, and perception of pain. In: Slipman C, Simeone F, Derby R (eds) Interventional spine: an algorithmic approach. Saunders, Philadelphia
166. Dong L, Quindlen JC, Lipschutz DE et al (2012) Whiplash-like facet joint loading initiates glutamatergic responses in the DRG and spinal cord associated with behavioral hypersensitivity. Brain Res 1461:51–63
167. Deriemer S, Strong J, Albert K et al (1985) Enhancement of calcium current in Aplysia neurons by phorbol ester and protein kinase C. Nature 313:313–316
168. Xu T, Jiang W, Du D et al (2007) Role of spinal metabotropic glutamate receptor subtype 5 in the development of tolerance to morphine-induced antinociception in rat. Neurosci Lett 420:155–159
169. Azkue JJ, Liu XG, Zimmermann M et al (2003) Induction of long-term potentiation of C fibre-evoked spinal field potentials requires recruitment of group I, but not group II/III metabotropic glutamate receptors. Pain 106:373–379
170. Derjean D, Bertrand S, Le Masson G et al (2003) Dybamic balance of metabotropic inputs causes dorsal horn neurons to switch functional states. Nat Neurosci 6:274–281
171. Young MR, Fleetwood-Walker SM, Dickinson T et al (1997) Behavioral and electrophysiological evidence supporting a role for group I metabotropic glutamate receptors in the mediation of nociceptive inputs to the rat spinal cord. Brain Res 777:161–169
172. Gwak Y, Hulsebosch C (2005) Upregulation of group I metabotropic glutamate receptors in neurons and astrocytes in the dorsal horn following spinal cord injury. Exp Neurol 195:236–243
173. Liang Y, Huang C, Hsu K (2005) Characterization of long-term potentiation of primary afferent transmission at trigeminal synapses of juvenile rats: essential role of subtype 5 metabotropic glutamate receptors. Pain 114:417–428
174. Cao H, Zhang YQ (2008) Spinal glial activation contributes to pathological pain states. Neurosci Biobehav Rev 32(5):972–983
175. DeLeo JA, Tanga FY, Tawfik VL (2004) Neuroimmune activation and neuroinflammation in chronic pain and opioid tolerance/hyperalgesia. Neuroscientist 10(1):40–52
176. Suter MR, Wen YR, Decosterd I et al (2007) Do glial cells control pain? Neuron Glia Biol 3:255–268
177. Frymoyer JW (1998) Back pain and sciatica. N Engl J Med 318(5):291–300
178. Kelly JD, Aliquo D, Sitler MR et al (2000) Association of burners with cervical canal and foraminal stenosis. Am J Sports Med 28(2):214–217
179. Nuckley DJ, Konodi MA, Raynak GC et al (2002) Neural space integrity of the lower cervical spine: effect of normal range of motion. Spine 27(6):587–595
180. Abbed KM, Coumans J (2007) Cervical radiculopathy: pathophysiology, presentation, and clinical evaluation. Neurosurgery 60(1):28–34
181. Caridi JM, Matthias P, Hughes AP (2011) Cervical radiculopathy: a review. HSS J 7:265–272

182. Harden N, Cohen M (2003) Unmet needs in the management of neuropathic pain. J Pain Symptom Manage 25(5):S12–S17
183. Slipman CW, Plastaras CT, Palmitier RA et al (1998) Symptom provocation of fluoroscopically guided cervical nerve root stimulation: are dynatomal maps identical to dermatomal maps? Spine 23(20):2235–2242
184. Ellenberg MR, Honet JC, Treanor WJ (1994) Cervical radiculopathy. Arch Phys Med Rehabil 75(3):342–532
185. Kawakami M, Hashizume H, Nishi H et al (2003) Comparison of neuropathic pain induced by the application of normal and mechanically compressed nucleus pulposus to lumbar nerve roots in the rat. J Orthop Res 21(3):535–539
186. Rutkowski MD, Winkelstein BA, Hickey WF et al (2002) Lumbar nerve root injury induces central nervous system neuroimmune activation and neuroinflammation in the rat: relationship to painful radiculopathy. Spine 27(15):1604–1613
187. Sunderland S (1974) Mechanisms of cervical nerve root avulsion in injuries of the neck and shoulder. J Neurosurg 41:705–714
188. Krivickas LS, Wilbourn AJ (2000) Peripheral nerve injuries in athletes: a case series of over 200 injuries. Semin Neurol 20(2):225–232
189. Panjabi MM, Maak TG, Ivancic PC et al (2006) Dynamic intervertebral foramen narrowing during simulated rear impact. Spine 31(5):128–134
190. Tominaga Y, Maak TG, Ivancic PC et al (2006) Head-turned rear impact causing dynamic cervical intervertebral foramen narrowing: implications for ganglion and nerve root injury. J Neurosurg Spine 26(1):57–62
191. Wainner RS, Gill H (2000) Diagnosis and nonoperative management of cervical radiculopathy. J Orthop Sports Phys Ther 30(12):728–744
192. Singh A, Lu Y, Chaoyang C et al (2006) Mechanical properties of spinal nerve roots subjected to tension at different strain rates. J Biomech 39(9):1669–1676
193. Singh A, Kallakuri A, Chen C et al (2009) Structural and functional changes in nerve roots due to tension at various strains and strain rates: an in-vivo study. J Neurotrauma 26(4):627–640
194. Fung YC (1972) Stress-strain-history relations of soft tissues in simple elongation. In: Fung YC (ed) Biomechanics: its foundations and objectives. Prentice-Hall, Englewood Cliffs
195. Yoganandan N, Kumaresan S, Pintar FA (2001) Biomechanics of the cervical spine part 2. Cervical spine soft tissue responses and biomechanical modeling. Clin Biomech 16(1):1–27
196. Hubbard RD, Chen Z, Winkelstein BA (2008) Transient cervical nerve root compression modulates pain: load thresholds for allodynia and sustained changes in spinal neuropeptide expression. J Biomech 41:677–685
197. Hubbard RD, Winkelstein BA (2005) Transient cervical nerve root compression in the rat induces bilateral forepaw allodynia and spinal glial activation: mechanical factors in painful neck injuries. Spine 30(17):1924–1932
198. Winkelstein BA, DeLeo JA (2002) Nerve root injury severity differentially modulates spinal glial activation in a rat lumbar radiculopathy model: considerations for persistent pain. Brain Res 956(2):294–301
199. Winkelstein BA, Rutkowski MD, Weinstein JN et al (2001) Quantification of neural tissue injury in a rat radiculopathy model: comparison of local deformation, behavioral outcomes, and spinal cytokine mRNA for two surgeons. J Neurosci Methods 111(1):49–57
200. Winkelstein BA, Weinstein JN, DeLeo JA (2002) The role of mechanical deformation in lumbar radiculopathy: an in vivo model. Spine 27(1):27–33
201. Hubbard RD, Winkelstein BA (2008) Dorsal root compression produces myelinated axonal degeneration near the biomechanical thresholds for mechanical behavioral hypersensitivity. Exp Neurol 212(2): 482–489
202. Nicholson KJ, Gilliland TM, Winkelstein BA (2013) Duration of nerve root compressive trauma modulates the subsequent thermal hyperalgesia & spinal expression of the glutamate transporter, GLT1. Paper presented at the Proceedings of the ASME 2013 summer bioengineering conference, Sunriver, Oregon, 26–29 Jun 2013
203. Nicholson KJ, Guarino BB, Winkelstein BA (2012) Transient nerve root compression load and duration differentially mediate behavioral sensitivity and associated spinal astrocyte activation and mGluR5 expression. Neuroscience 209:187–195
204. Scholz J, Woolf CJ (2007) The neuropathic triad: neurons, immune cells and glia. Nat Neurosci 10(11):1361–1368
205. Hubbard RD, Quinn KP, Martinez JJ et al (2008) The role of graded nerve root compression on axonal damage, neuropeptide changes, and pain-related behaviors. Stapp Car Crash J 52:33–58
206. Kobayashi S, Baba H, Uchida K et al (2005) Effect of mechanical compression on the lumbar nerve root: localization and changes of intraradicular inflammatory cytokines, nitric oxide, and cyclooxygenase. Spine 30(15):1699–1705
207. Nicholson KJ, Quindlen JC, Winkelstein BA (2011) Development of a duration threshold for modulating evoked neuronal responses after nerve root compression injury. Stapp Car Crash J 55:1–24
208. Olmarker K, Holm S, Rydevik B (1990) Importance of compression onset rate for the degree of impairment of impulse propagation in experimental compression injury of the porcine cauda equina. Spine 15(5):416–419
209. Burchiel KJ, Hsu FPK (2001) Pain and spasticity after spinal cord injury: mechanisms and treatment. Spine 26(24):S146–S160
210. Finnerup NB, Jensen TS (2004) Spinal cord injury pain: mechanisms and treatment. Eur J Neurol 11(2):73–82

211. Hulsebosch CE, Hains BC, Crown ED et al (2009) Mechanisms of chronic central neuropathic pain after spinal cord injury. Brain Res Rev 60(1):202–213
212. Vierck CJ, Siddall P, Yezierski RP (2000) Pain following spinal cord injury: animal models and mechanistic studies. Pain 89(1):1–5
213. Yezierski RP (2009) Spinal cord injury pain: spinal and supraspinal mechanisms. J Rehabil Res Dev 46(1):95–107
214. Kwon BK, Tetslaff W, Grauer JN et al (2004) Pathophysiology and pharmacologic treatment of acute spinal cord injury. Spine J 4(4):451–464
215. Saab CY, Waxman SG, Hains BC (2008) Alarm or curse? The pain of neuroinflammation. Brain Res Rev 58(1):226–235
216. Hulsebosch CE (2002) Recent advances in pathophysiology and treatment of spinal cord injury. Adv Physiol Educ 26:238–255
217. Yamashita T, Cavanaugh JM, el-Bohy AA et al (1990) Mechanosensitive afferent units in the lumbar facet joint. J Bone Joint Surg Am 72(6):865–870
218. Adams MA, Roughley PJ (2006) What is intervertebral disc degeneration, and what causes it? Spine 31(18):2151–2161
219. Setton LA, Chen J (2006) Mechanobiology of the intervertebral disc and relevance to disc degeneration. J Bone Joint Surg Am 88(Suppl 2):52–57
220. Dai LY, Jia LS (2000) Central cord injury complicating acute cervical disc herniation in trauma. Spine 25(3):331–335
221. Berrington NR, van Staden JF, Willers JG et al (1993) Cervical intervertebral disc prolapse associated with traumatic facet dislocations. Surg Neurol 40(5):395–399
222. Brault JR, Siegmund GP, Wheeler JB (2000) Cervical muscle response during whiplash: evidence of a lengthening muscle contraction. Clin Biomech 15(6):426–435
223. Siegmund GP, Blouin JS, Carpenter MG et al (2008) Are cervical multifidus muscles active during whiplash and startle? An initial experimental study. BMC Musculoskelet Disord 9:80
224. Davis SJ, Teresi LM, Bradley WG et al (1991) Cervical spine hyperextension injuries: MR findings. Radiology 180(1):245–251
225. Kliewer MA, Gray L, Paver J et al (1993) Acute spinal ligament disruption: MR imaging with anatomic correlation. J Magn Reson Imaging 3(6):855–861
226. McMahon PJ, Dheer S, Raikin SM et al (2009) MRI of injuries to the first interosseous cuneometatarsal (Lisfranc) ligament. Skeletal Radiol 38(3):255–260
227. Billiar KL, Sacks MS (1997) A method to quantify the fiber kinematics of planar tissues under biaxial stretch. J Biomech 30(7):753–756
228. Hansen KA, Weiss JA, Barton JK (2002) Recruitment of tendon crimp with applied tensile strain. J Biomech Eng 124(1):72–77
229. Robinson PS, Tranquillo RT (2009) Planar biaxial behavior of fibrin-based tissue-engineered heart valve leaflets. Tissue Eng Part A 15(10):2763–2772
230. Snedeker JG, Pelled G, Zilberman Y et al (2006) Endoscopic cellular microscopy for in vivo biomechanical assessment of tendon function. J Biomed Opt 11(6):064010
231. Snedeker JG, Pelled G, Zilberman Y et al (2008) An analytical model for elucidating tendon tissue structure and biomechanical function from in vivo cellular confocal microscopy images. Cells Tissues Organs 190:111–119
232. Tower TT, Neidert MR, Tranquillo RT (2002) Fiber alignment imaging during mechanical testing of soft tissues. Ann Biomed Eng 30(10):1221–1233
233. Sander EA, Barocas VH (2009) Comparison of 2D fiber network orientation measurement methods. J Biomed Mater Res A 88(2):322–331
234. Quinn KP, Bauman JA, Crosby ND et al (2010) Anomalous fiber realignment during tensile loading of the rat facet capsular ligament identifies mechanically induced damage and physiological dysfunction. J Biomech 43(10):1870–1875
235. Dickey JP, Hewlett BR, Dumas GA et al (1998) Measuring collagen fiber orientation: a two-dimensional quantitative macroscopic technique. J Biomech Eng 120(4):537–540
236. Tower TT, Tranquillo RT (2001) Alignment maps of tissues: I. Microscopic elliptical polarimetry. Biophys J 81(5):2954–2963
237. Whittaker P, Canham PB (1991) Demonstration of quantitative fabric analysis of tendon collagen using two-dimensional polarized light microscopy. Matrix 11(1):56–62
238. Quinn KP, Winkelstein BA (2010) Full field strain measurements of collagenous tissue by tracking fiber alignment through vector correlation. J Biomech 43(13):2637–2640
239. Liem LK, Simard JM, Song Y et al (1995) The patch clamp technique. Neurosurgery 36(2):382–392
240. Stanfa LC, Dickenson AH (2004) In vivo electrophysiology of dorsal horn neurons. In: Luo ZD (ed) Methods in molecular medicine. Humana, Totowa
241. Zhao Y, Inayat S, Dikin DA et al (2009) Patch clamp technique: review of the current state of the art and potential contributions from nanoengineering. Proc IMechE 222:1–11
242. Pine J (2006) A history of MEA development. In: Taketani M, Baudry M (eds) Advances in network electrophysiology: using multi-electrode arrays. Springer, New York
243. Lewicki MS (1998) A review of methods for spike sorting: the detection and classification of neural action potentials. Comput Neural Syst 9(4):R53–R78
244. Schouenberg J (1984) Functional and topographical properties of field potentials evoked in rat dorsal horn neurons by cutaneous C-fiber stimulation. J Physiol 356:169–192
245. Stett A, Egert U, Guenther E et al (2003) Biological applications of microelectrode arrays in drug discovery and basic research. Anal Bioanal Chem 377:486–495
246. Hains BC, Willis WD, Hulsebosch CE (2003) Temporal plasticity of dorsal horn somatosensory

neurons after acute and chronic spinal cord hemisection in rat. Brain Res 970:238–241
247. Martindale JC, Wilson AW, Reeve AJ et al (2007) Chronic secondary hypersensitivity of dorsal horn neurons following inflammation of the knee joint. Pain 133:79–86
248. Vernon H, Sun K, Zhang T et al (2009) Central sensitization induced in trigeminal and upper cervical dorsal horn neurons by noxious stimulation of deep cervical paraspinal tissues in rats with minimal surgical trauma. J Manipulative Physiol Ther 32(7):506–514
249. Ikeda H, Heinke B, Ruscheweyh R et al (2003) Synaptic plasticity in spinal lamina I projection neurons that mediate hyperalgesia. Science 299: 1237–1240
250. Randic M, Jiang MC, Cerne R (1993) Long-term potentiation and long-term depression of primary afferent neurotransmission in the rat spinal cord. J Neurosci 13(12):5228–5241
251. Eytan D, Marom S (2006) Dynamics and effective topology underlying synchronization in networks of cortical neurons. J Neurosci 26(33):8465–8476
252. Hofmann F, Bading H (2006) Long term recordings with multi-electrode arrays: studies of transcription-dependent neuronal plasticity and axonal regeneration. J Physiol Paris 99(2–3):125–132
253. Kutzing MK, Luo V, Firestein BL (2011) Measurement of synchronous activity by microelectrode arrays uncovers differential effects of sublethal and lethal glutamate concentrations on cortical neurons. Ann Biomed Eng 39(8):2252–2262
254. Yu Z, Graudejus O, Tsay C et al (2009) Monitoring hippocampus electrical activity in vitro on an elastically deformable microelectrode array. J Neurotrauma 26:1135–1145
255. Cheung KC, Renaud P, Tanila H et al (2007) Flexible polyimide microelectrode array for in vivo recordings and current source density analysis. Biosens Bioelectron 22(8):1783–1790
256. Timko BP, Cohen-Karni T, Yu G et al (2009) Electrical recordings from hearts with flexible nanowire device arrays. Nano Lett 9(2):914–918
257. Viventi J, Kim DH, Moss JD et al (2010) A conformal, bio-interfaced class of silicon electronics for mapping cardiac electrophysiology. Sci Transl Med 2(24):24ra22
258. Saxena T, Gilbert J, Stelzner D et al (2012) Mechanical characterization of the injured spinal cord after lateral spinal hemisection injury in the rat. J Neurotrauma 29(9):1747–1757
259. Levental I, Levental KR, Klein EA et al (2010) A simple indentation device for measuring micrometer-scale tissue stiffness. J Phys Condens Matter 22(19): 194120
260. Miller WJ, Leventhal I, Scarcella D et al (2009) Mechanically induced reactive gliosis causes ATP-mediated alterations in astrocyte stiffness. J Neurotrauma 26(5):789–797
261. Vergara D, Martignago R, Leporatti S et al (2009) Biomechanical and proteomic analysis of IFN-β-treated astrocytes. Nanotechnology 20:244106–455115
262. Ouyang H, Galle B, Li J et al (2008) Biomechanics of spinal cord injury: a multimodal investigation using ex vivo guinea pig spinal cord white matter. J Neurotrauma 25(1):19–29
263. East E, de Oliveira DB, Golding JP et al (2010) Alignment of astrocytes increases neuronal growth in three-dimensional collagen gels as is maintained following plastic compression to form a spinal cord repair conduit. Tissue Eng Part A 16(10):3173–3184

Thoracolumbar Pain: Neural Mechanisms and Biomechanics

20

John M. Cavanaugh, Chaoyang Chen, and Srinivasu Kallakuri

Abstract

Based on extensive research the anatomic components that have been considered to cause back pain include the intervertebral disks, paraspinal muscles, spinal ligaments, facet joint capsules, dorsal roots and dorsal root ganglia. While mechanical strain is an important causative factor in the initiation of spinal pain, biochemical irritation plays an important role in sensitizing nerve endings in the injured tissue. In addition, central sensitization at the spinal cord level can be an important contributor to persistent spinal pain. The mechanical input that results in back pain includes both acute mechanical loading and repetitive cyclic loading. The latter is particularly important in disc injury and degeneration and in muscle fatigue to the lumbar spine. In this chapter the biomechanics and neural mechanisms of thoracolumbar pain are described in the following order: (1) mechanisms of pain initiation and maintenance (nociceptive pain vs. neuropathic pain, acute vs. chronic pain, peripheral sensitization, central sensitization), (2) biomechanics of noxious loading of the thoraco-lumbar spine, (3) joint, ligament and skeletal pain (involving intervertebral disc, anterior and posterior longitudinal ligaments, ligamentum flavum, dura, facet joints, bone and periosteum), (4) neuropathic pain (involving nerve roots and dorsal root ganglia) and (5) muscle pain. Further research is required to better characterize the relationship between measured mechanical input to tissues of the spine and the molecular, histological and neurophysiologic outcomes that result in spinal pain. This understanding can help in the development of preventive strategies to reduce the risk of spinal injury and pain.

J.M. Cavanaugh, M.D. (✉) • C. Chen, M.D.
S. Kallakuri, Ph.D.
Department of Biomedical Engineering, Wayne State
University, Detroit, MI, USA
e-mail: jmc@wayne.edu; cchen@wayne.edu;
skallakuri@wayne.edu

20.1 Mechanisms of Pain Initiation and Maintenance

Chronic pain affects about 100 million adults in the United States, with an estimated annual cost of $560-$635 billion [1]. Musculoskeletal pain, especially joint and back pain, is the most common type of chronic pain [1]. Activity-limiting low back pain is estimated to have a world-wide lifetime prevalence of about 39% [2]. Sources of pain generation include the facet joint capsules and ligaments, the annulus of the disc and adjacent posterior longitudinal ligaments. The muscles of the back may also be a source of pain from spasm, fatigue or from strain during overloading.

All these tissues contain nerves called nociceptors that are the signals of pain. When tissue is strained beyond its elastic limits minor trauma occurs and can progress as the strain increases. Nociceptors are the body's natural stress transducers that the strain has moved beyond the physiologic into the tissue damaging realm. If the strain is only at the physiologic limit, reducing the strain will relieve pain, if the strain moves into the range of tissue damage, pain will persist after the tissue is unloaded.

Inflammation and tissue repair occurs at the damaged tissue site. Inflammation is part of the repair process and also makes the nociceptors more sensitive. This sensitization helps keep the damaged tissue from being overloaded to allow proper healing. Chronic pain persists beyond normal healing, and lasts more than several months (variously defined as 3-6 months) [1]. The reasons for chronic pain are often unclear but can include: poor healing, nerve injury, sensitization of the nerve pathways in the spinal cord and sympathetically maintained pain.

Pain is an unpleasant sensory and emotional experience associated with actual or potential tissue damage, or described in terms of such damage [3]. Types of pain include nociceptive and neuropathic pain. Nociceptive pain results from direct activation of nociceptors. Neuropathic pain results from direct injury to nerves in the peripheral or central nervous systems and may have a burning or electric sensation [4].

Nociceptors exist as free nerve endings and they are activated by mechanical, thermal, or chemical stimuli. They appear as fine branches and do not appear to have peripheral structures that transduce or filter stimuli. Thermal nociceptors are small diameter, thinly myelinated A-delta fibers that conduct at about 5–30 meters per second (m/s). They are activated by noxious temperatures greater than 45 °C or less than 5 °C. Mechanical nociceptors are activated by high mechanical stress that signals impending tissue damage or actual tissue damage. These nociceptors are A-delta fibers that conduct at 5–30 m/s. Activation of the A-delta nociceptors is associated with sharp, pricking pain. Polymodal nociceptors are activated by a variety high intensity mechanical and chemical stimuli and hot or cold stimuli. These are small diameter, unmyelinated C-fibers that conduct slowly at 0.5–2.5 m/s. They are associated with aching and burning pain. Both A-delta and C fibers are widely distributed in skin and in deep tissues. Viscera and joints contain silent nociceptors. They typically do not fire unless their threshold is reduced by inflammation and chemical insults [4].

Pain modulation includes peripheral sensitization, central sensitization and inhibition. In peripheral sensitization, tissue injury and inflammation cause release of various chemicals from non-neuronal cells, including mast cells, macrophages, platelets, immune cells, endothelial cells, Schwann cells, keratinocytes and fibroblasts. These chemicals include protons (H+), purines (adenosine, adenosine triphosphate), nerve growth factor (NGF), cytokines such as tumor necrosis factor (TNF-α) and interleukins (IL-1β, IL-6), leukemia inhibitory factor (LIF), prostaglandin E2 (PGE2), bradykinin, histamine, serotonin (5-HT), platelet activating factor (PAF), and endothelin. In addition, neurogenic mediators, such substance P (SP) and calcitonin gene related peptide (CGRP), are released from peripheral nociceptor terminals. These various mediators can sensitize or can activate nociceptors [4, 5] (Fig. 20.1).

Central Sensitization: Central sensitization occurs in spinal cord dorsal horn neurons as a result of sustained activation of nociceptors. Neurotransmitters in the dorsal horn include glutamate, N-methyl-D-asparate (NMDA) and

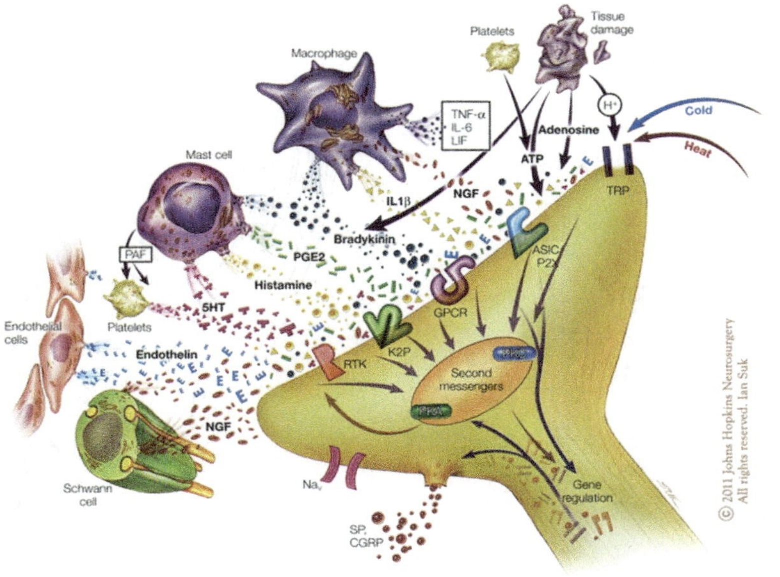

Fig. 20.1 The figure illustrates potential mediators of peripheral sensitization after inflammation. Tissue injury and inflammation lead to the release of various chemicals from non-neuronal cells, including mast cells, macrophages, platelets, immune cells, endothelial cells, Schwann cells, keratinocytes and fibroblasts. These chemicals include protons (H+), purines (adenosine, adenosine triphosphate), nerve growth factor (*NGF*), cytokines such as tumor necrosis factor (*TNF-α*) and interleukins (*IL-1β*, *IL-6*), leukemia inhibitory factor (*LIF*), prostaglandin E2 (*PGE2*), bradykinin, histamine, serotonin (*5-HT*), platelet activating factor (*PAF*), and endothelin. Neurogenic mediators, such as SP and CGRP, are released from peripheral nociceptor terminals. The released chemicals may act directly to alter the sensitivity of peripheral nociceptors or indirectly via coupling to one or more peripheral membrane-bound receptors, including transient receptor potential (*TRP*) channels, acid-sensitive ion channels (*ASICs*), purinergic (*P2X*) receptors, G protein–coupled receptors (*GPCRs*), two-pore potassium channels (*K2P*), and receptor tyrosine kinase (*RTK*). Binding of the ligands to these receptors can initiate a cascade of events that includes activation of second-messenger systems (protein kinase A [*PKA*] and C [*PKC*]) and alteration of gene regulation (Reproduced with permission from Ringkamp et al. [5])

SP (a neuropeptide). Neuropeptides can diffuse considerable distances from their site of release because there is no specific reuptake mechanism. Thus, many postsynaptic terminals can be influenced. Peptide levels are significantly increased in persistent pain. Peptides, including substance P, contribute to the excitability of dorsal horn neurons and to the poor localization of some painful conditions [4]. Central sensitization cane be both short term and long term [6] (Fig. 20.2).

Pain generators in the spine include the intervertebral discs, paraspinal muscles, facet joint capsules, dorsal root ganglia, nerve roots and dorsal horn [7]. Disc, muscle and facet joint capsules all contain nociceptors that can be a source of peripheral sensitization if these tissues are injured or inflamed. The dorsal roots and dorsal root ganglion are sources of neuropathic pain which is described later in this chapter [7] (Fig. 20.3).

20.2 Biomechanics of Noxious Loading of Thoraco-lumbar Spine

20.2.1 Injury Tolerance of the Thoraco-lumbar Spine

+Gz acceleration (upward vertical acceleration to the torso and spine) occurs in ejection from aircraft seats, falling from a height onto the buttocks or vehicle occupant loading of the seat pan

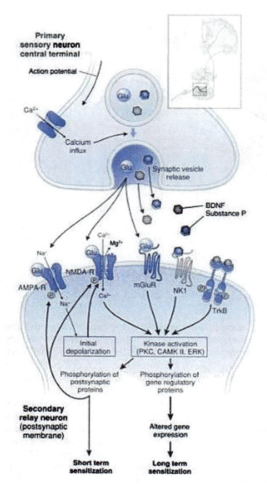

Fig. 20.2 Flow chart displaying the possible outcomes of central sensitization. An action potential from a primary sensory neuron central terminal can lead to both short term and long term sensitization. *AMPA-R* alpha amino-3hydroxy-5-methyl-4isoxazolepropionic acid receptor; *BDNF* brain-derived neurotrophic factor; *CAMK II* calmodulin dependent preotein kinase II; *ERK* extracellular signal-regulated kinase; *NMDA-R* N-methyl-D-aspartate receptor; *PKC* protein kinase C (Reproduced with permission from Golan et al. [6])

during vertical or frontal impact of the vehicle. Vertical loading occurs when explosive devices deploy under a vehicle and cause the vehicle to accelerate skyward. Such is the case when improvised explosive devices (IEDs) were deployed against coalition forces in the recent wars in Iraq and Afghanistan.

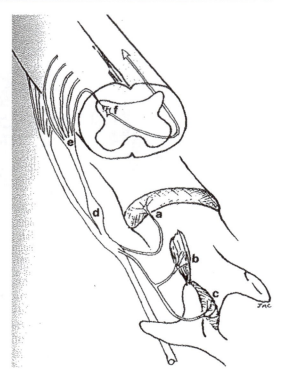

Fig. 20.3 Diagram showing various pain generators in the spine. These include: intervertebral discs (*a*); paraspinal muscles (*b*); facet joint capsules (*c*); dorsal root ganglia (*d*); nerve roots (*e*); dorsal horn (*f*) (Reproduced with permission from Cavanaugh [7])

20.2.1.1 Volunteer Data from Servicemen

Eiband reported that with lap, shoulder straps and face curtain in use, 16 g's for 0.04 s has been endured in catapult seat experiments. A 20 g median value has been assumed as a safe limit for ejection seat performance since World War II at pulse duration of 0.005–0.5 s. Hogs subjected to 110 g's for 0.002 s recovered within a few days. Onset rate to g level (g's per second) is termed jolt. 500 to 1300 g's per second has jolted the subject severely but without permanent injury. A jolt of 115–180 g's per second is considered preferable [8].

The type of torso support is critical in regards to thoraco-lumbar spine injury tolerance. The minimum support required to avoid this was lap and shoulder belts in the review by Eiband [8]. Use of the armrest increases vertical loading tolerance further by reducing the load on lumbar vertebrae. Providing support for the head and neck reduces the severe neck flexion that can occur in sudden vertical loading [8].

20.2.1.2 Cadaver Studies
Ewing et al. tested embalmed cadavers in hyperextended, erect and flexed modes in +Gz acceleration. In the hyperextended mode g-level to fracture averaged 17.6 g, compared to 10.4 g in erect and 9.0 g in flexed mode [9].

Vulcan et al. reported that the CG of the body lies 0.75–1.45 in. anterior to the centerline of the T9 vertebra. In four cadavers vertebral bodies were strain gauged at T11 to L4. In cases where the shoulder was allowed to rotate forward there was a peak in the vertebral strain levels that corresponded to peak shoulder strap loading. If the shoulders were held back with tight straps from flexing forward the large vertebral strains were avoided. Thus, upper body flexion plays a key role in stresses of the lower vertebrae under sudden vertical loading [10].

Prasad et al. showed that the facets share part of the loading during hyperextension. In the erect mode the facet capsular ligaments go into tension, causing the vertebral bodies to sustain more compressive load than the total spine load [11]. In the hyperextended mode the facets relieve the vertebral bodies of some of the compressive load. Yang and King showed that the mechanism of load transmission is by the bottoming of the tip of the inferior facet onto the lamina below [12].

In summary, thoraco-lumbar spine injury tolerance increases as the spine posture changes from flexion to hyperextension. As the body becomes more flexed and trunk support is removed, this tolerance level is reduced. Vertebral wedge fractures can occur when there is a flexion component in addition to the Gz loading. In severe cases of compressive loading, burst fractures with retropulsion of vertebral body fragments into the spinal canal can occur, causing spinal cord and/or nerve root injury.

20.2.1.3 Gx Acceleration (Sternum to Spine Acceleration as Occurs in Frontal Impacts)
Using a 2D math model Prasad and King reported that the lumbar vertebrae could undergo compression in a frontal impact. It was hypothesized that the spine attempted to straighten out during -Gx acceleration, loading the lumbar spine [13]. This was supported in a study by Begeman et al who measured large seat pan loads in cadaver tests in which the upper torso was restrained [14].

Miniaci and McLaren attributed four anterolateral wedge fractures to the wearing of lap-shoulder belts in motor vehicle accidents [15].

20.3 Joint, Ligament and Skeletal Pain

A diagram of the anatomy of thoraco-lumbar motion segments is illustrated in Fig. 20.4 [16]. Mechanisms by which individual structures of the lumbar spine can be sources of pain are described below.

20.3.1 Intervertebral Discs

20.3.1.1 Disc Herniation
The herniated (ruptured) intervertebral disc can be a source of pain. The disc outer annulus contains nerve endings that can transmit signals that lead to pain. Degenerated discs have pain fibers that extend further into the disc than in healthy discs. In addition, the nucleus pulposus, the gel-like material that can extrude out of a ruptured disc can cause inflammation that sensitizes nerve endings to cause low back pain and nerve roots to cause sciatica.

The disc is stronger in vertical loading than are the vertebral bodies. The vertebral bodies will usually fail first. Disc ruptures from a single loading event are highly unlikely based on biomechanical studies. Henzel et al. observed that early researchers noted that the vertebral body broke before the adjacent disc incurred any dam-

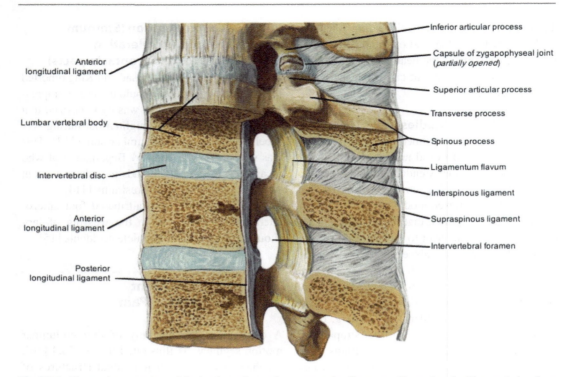

Fig. 20.4 The motion segments of the lumbar spine and accompanying ligaments (Reproduced with permission from Freemont [16])

age [17]. Yoganandan et al. tested individual discs and vertebral bodies and noted that thoracic and lumbar vertebral bodies had failure loads of 2,642–4,590 N, healthy discs 11,030 N and degenerated discs 5,300 N on average [18]. Brinckmann performed a study in which 25 PMHS lumbar intervertebral discs had the annular fibers from the inside out, preserving ½ to 1 mm of the outer annulus. The motion segment was then loaded to 1,000 N (225 lb). There were no disc ruptures, bulging was minimal and there was never extrusion of disc material. Disc ruptures did not occur even after the vertebral body was loaded to failure [19].

Adams and Hutton were able to produce disc ruptures in the case of extreme flexion combined with high compressive loads [20]. Sixty-one joints were tested to failure. Twenty-six failed by prolapse (rupture) of the intervertebral disc. The average compressive force to produce rupture was 5,448 N and the average flexion angle 12.8° between two vertebrae. This angle would represent a flexion of over 60° in five vertebrae, an extreme condition beyond the physiologic range. Thus, hyperflexion between lumbar vertebrae and high compressive loading are conditions that have produced disc rupture experimentally. However, other studies indicate the vertebrae will fracture at these loads before the disc ruptures.

Gordon et al. produced disc rupture by loading 14 human cadaveric lumbar motion segments (two vertebrae plus the disc) at 1.5 Hz for an average of 6.9 h, in a combination of 7° of flexion, 3° of rotation and 1,334 N of compression. Ten discs failed through annular protrusions. The average number of cycles to failure was 36,750. This study supports repetitive loading over time as a primary cause of disc herniation [21].

Callaghan ad McGill noted that while intervertebral disc herniations are observed clinically, consistent reproduction of this injury in the laboratory has been elusive [22]. They performed a study designed to examine the biomechanical response and failure mechanics of spine motion segments to highly repetitive low magnitude complex loading. They tested porcine cervical spine motion segments (C3–C4) in applied axial compressive loads with pure flexion/extension moments. Dynamic testing was conducted to a maximum of 86,400 bending cycles at a rate of 1 Hz with simultaneous torques, angular rotation and axial deformations recorded for the duration of the test. The results supported the notion that intervertebral disc herniation may be more linked to repeated flexion extension motions than applied joint compression, at least with younger, non-degenerated specimens.

In summary, herniation of a lumbar disc from a single accident is not impossible but highly unusual. It has been replicated in an extreme combination of compression and flexion, as shown by Adams and Hutton. Disc rupture is typically a wear-and-tear phenomenon that occurs over thousands of loading cycles and is not the result of a single loading event. Loads high enough to rupture a disc will more likely fracture a vertebral body first.

20.3.1.2 Disc Innervation and Pain

Clinically the role of degenerating intervertebral discs and the consequences of ruptured disc in the ensuing symptoms of low back pain have been well documented. A recent review by Bogduk et al. suggests that discogenic pain can occur and be diagnosed under strict operational criteria [23]. For an anatomical structure such as intervertebral disc (IVD) to be a source of pain, a nerve supply is required. One of the earliest citations on IVD innervation in relationship to back pain dates back to a report by Wiberg [24]. However interest in disc innervation predates this as indicated by studies of others in support of IVD innervation [25, 26] with one of the prior studies of the time reporting no nerve elements in humans discs but in the surrounding ligaments [27]. This lack of nerve fibers in disc annular region was again reported [28]. During later years, Mulligan reported disc annulus innervation using a canine thoracic vertebral column [29]. A previous study by Stilwell reported unmyelinated nerve fibres with free nerve endings being confined to the superficial lamina of loose connective tissue of the annulus [30]. However,

Yoshizawa showed profuse non-myelinated axonal networks and free nerve terminals in the outer (lateral) half of the annulus of IVDs removed for low back pain [31]. In one of the subsequent anatomical studies Bogduk and colleagues offered irrefutable evidence of nerve supply to human lumbar IVD. They described that branches of sinuvertebral nerve innervate the posterior aspect of the IVDs and that the posterolateral aspects of the discs receive branches from adjacent ventral primary rami and from the grey rami communicantes near their junction with the ventral primary rami [32, 33] (Fig. 20.5A). Subsequently, Kojima et al. described that sinuvertebral innervation to posterolateral portion of the annulus fibrosus (AF) in rat lumbar IVD and indicated no nerve fibers and terminals in deep layer of the AF or the nucleus pulposus [34]. Using a retrograde transport method, Morinaga et al. showed IVD afferent innervation in rat specimens. They demonstrated that the anterior portion of the L5–L6 lumbar IVD was innervated from L1 or L2 spinal nerves and suggested this as a possible explanation for why patients with lower lumbar disc lesions sometimes complain of inguinal pain corresponding to the L1–L2 dermatome [35]. Using a similar approach, Ohtori et al. showed that that the dorsal portion of the L1–L6 discs of rats was shown to be multi-segmentally innervated by the T11 through L6 dorsal root ganglia using paravertebral sympathetic trunks and sinuvertebral nerves as nerve pathways [36, 37]. Nevertheless, the data on disc innervation from rodent studies needs to be interpreted cautiously in cases of human discogenic pain.

Immunohistochemical evidence for IVD innervation was supported using antibodies to protein gene product (PGP 9.5), a general neuronal marker. Nerve fibers containing PGP 9.5 were observed in the outer annulus running between and across the collagenous lamellae in human disc specimens [38]. Palmgren et al. on the other hand showed PGP 9.5 reactive nerve fibers penetrating 3.5 and 1.1 mm into the annulus in normal human IVD specimens, whereas Willenegger et al. showed PGP 9.5 immunoreactive nerve fibers in the periphery of the IVD in sections from dogs [39, 40]. In an earlier study Cavanaugh

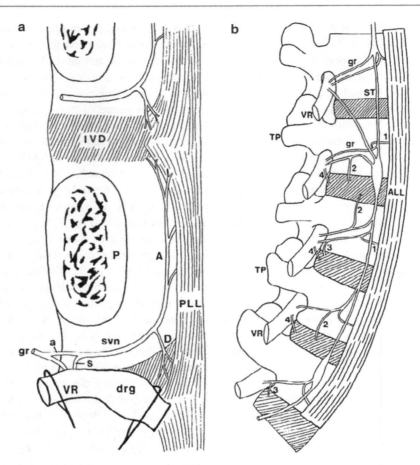

Fig. 20.5 (a) Diagram of a left lumbar sinuvertebral nerve (*svn*) with pedicle of the vertebra (*P*) transected and the neural arch and the dural sac removed. The ventral ramus (*VR*) and dorsal root ganglion (*drg*) are retracted to reveal the origins of the sinuvertebral nerve from a somatic root (*s*) from the ventral ramus, and an autonomic root (*a*) from the grey ramus communicans (*gr*). Ascending (*A*) and descending (*D*) branches innervate the posterior longitudinal ligament (*PLL*) and intervertebral discs (*IVD*) at the level of the foraminal entry of the nerve and at the next level above. (b) A diagram of nerves innervating the anterior and lateral aspects of the lumbar vertebral column. *TP* transverse process, *ALL* anterior longitudinal ligament, *VR* ventral ramus, *ST* sympathetic trunk, *gr* grey rami communicantes. *1* are branches to anterior longitudinal ligament, *2* are branches to lateral aspects of intervertebral disc, *3* are branches to intervertebral disc from grey rami, *4* are branches to intervertebral disc from ventral rami (Reproduced with permission from Bogduk [33])

et al., using a silver impregnation method, showed numerous fine profiles of nerve fibers limited to the superficial anulus in rabbit lumbar IVD [41].

The IVDs have also been reported to have various types of nerve endings. Roofe reported naked nerve endings terminating in annulus and posterior longitudinal ligament [25]. Stilwell reported unmyelinated nerve fibers with free nerve endings [30]. Simple free nerve endings were also reported by Hirsch et al. and Jackson et al. [42–44]. Malinsky found five types of free nerve endings in the outer annulus layer: lone, simple, free; branching; shrubby; mesh-like loops, and clusters running in parallel. Malinsky also found encapsulated and partially encapsulated nerve endings on the external lateral annulus surface [45]. The presence of nerve endings was further confirmed by Roberts et al. who reported mechanoreceptors resembling Pacinian corpuscles, Ruffini endings, and, most frequently, Golgi tendon organs in the outer 2–3 lamellae of the IVD and associated anterior longitudinal ligament (ALL) from human and bovine specimens [46]. Ruffini, Golgi type and free nerve endings were also reported by others in the superficial layers of the annulus and they suggested that the anterior part of the disc has a greater frequency of encapsulated receptors [42].

Mooney reviewed much of the current knowledge on low back pain and concluded that the disc may be primary source in the production of low back pain, but the mechanisms of pain production are uncertain [47]. Hisrch et al. reported that 0.3 ml (11 %) of hypertonic saline injected into the disc produced severe pain, 'identical to a real lumbago', that could not be localized but that was described as a deep aching across the low back [43]. Kuslich et al. demonstrated that of 144 patients, whose central lateral annulus was stimulated, 71 % had pain by the procedure and 30 % experienced significant pain. In this study, 18 spine tissue sites were stimulated with blunt surgical instruments or electrical current of low voltage. Significant pain was reported by 90 % of patients upon stimulation of central lateral annulus, and by 15 % upon central annulus stimulation [48]. In a retrospective study of a cohort of 28 consecutive patients, De Palma et al. concluded that outer annular fissures were a source of discogenic pain attributed to sensitized nociceptors in annular tears. The subjects underwent provocation diskography and analgesic diskography utilizing a balloon-tipped intradiskal catheter allowing intradiskal injection of anesthetic. Eighty percent of painful intervertebral disks as detected by provocation diskography were sufficiently anesthetized, resulting in >50 % reduction in low-back pain during analgesic diskography [49].

The disc is made up of an inner gelatinous nucleus pulposus which is confined above and below by the end plates of the vertebral bodies and circumferentially by the fibres of the thick annulus fibrosus (AF) [16] (Fig. 20.4). The cells of the annulus are fibroblast-like and the cells of the nucleus pulposus (NP) are chondrocyte–like [50]. The Chondrocyte-like cells of the NP synthesize type II collagen, proteoglycans, and non-collagenous proteins that form the matrix of the nucleus pulposus and the cartilage endplate. Fibroblast-like cells synthesize type I and type II collagen for the annulus fibrosus [51]. The disc material undergoes constant turnover by getting degraded by matrix metalloproteinases (MMPs) secreted by the chondrocytes [52, 53]. Furthermore, disc degeneration and a persistent state of inflammation may arise from synthesis of abnormal disc components, or an increase in synthesis of mediators such as nitric oxide, interleukins, prostaglandin E2, and matrix metalloproteinases [54–58] that can lead to matrix degradation. It has also been suggested that the number of nerve fibers increases following disc degeneration that innervate not only the outer AF and also the inner nucleus pulposus. Also increased are the number of mechanoreceptors in the superficial layers of the disc [59]. Furthermore, a recent investigation reported that increased transcription growth factor beta (TGF-β) and SP levels were predominantly found in chronic degenerative disc disease (DDD) with MMP3 increased in acute herniated disc material suggesting variations in the expression of mediators between acute and chronic processes [60]. Another investigation by Richardson et al. showed that nerve growth factor (NGF) significantly correlated with both MMP-10 and substance P mRNA in degenerate NP samples and suggested that MMP-10 expression increases in the symptomatic degenerated IVD [61].

Experimentally, the potential role for disc in pain generation comes from some recent studies that showed animals with epidural presence of nucleus pulposus or disc puncture showing behavior of pain [62, 63]. Behavioral and histological signs of alleviation by pharmacological intervention by using tumor necrosis factor (TNF) inhibitor [64–66] were also reported. Neurophysiologically, the released disc material is capable of inducing functional changes in adjacent exposed nerve roots as evidenced by reduced conduction velocity [67, 68]. It has been suggested that the leakage of disc material from annular tears to adjacent nerve roots is one pathophysiologic mechanism of low back pain following disc degeneration. The role of other factors such as

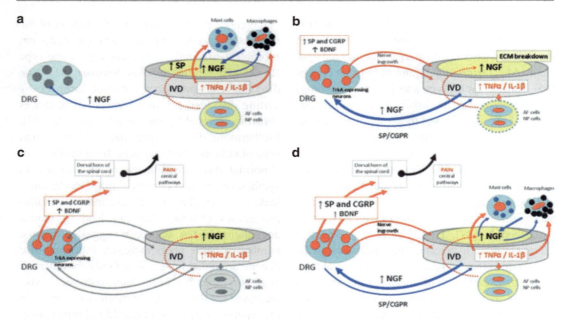

Fig. 20.6 Schematic representation of the potential mechanisms involved in the genesis of the discogenic pain. (**a**) Inflammation causes release of proinflammatory cytokines in the intervertebral disc (*IVD*), which act on mast cells and macrophages to trigger secretion of *NGF*. Cells in the IVD upregulate expression of NGF and substance P (*SP*). Increased levels of NGF can be retrogradely transported to dorsal root ganglia (*DRGs*) or stimulate mastocytes and macrophages locally initiating a positive feedback loop. (**b**) Increased levels of NGF reaching the DRGs act on TrkA-expressing neurons inducing expression of peptides that mediate pain [SP and calcitonin gene-related peptide (*CGRP*)]. The increased levels of NGF in the IVD, as well as the breakdown of the IVD aggrecans, result in ingrowth of nociceptive nerve fibers. Anterograde transport of SP and CGRP to the IVD also may occur to maintain pain. (**c**) Synaptic transmission in lamina I and II of the dorsal horn of the spinal cord is mediated by SP and CGRP. In addition, brain-derived neurotrophic factor (*BDNF*) produced in DRG is released to these laminas and modulates pain transmission. (**d**) It is proposed that these complex networks are able to originate and maintain pain of IVD origin (Reproduced with permission from Garcia-Cosamalon et al. [59])

ageing, genes, nutrition, toxic factors, metabolic disorders, and various mechanical factors has also been reviewed by others [51, 16]. It is suggested that pressure and chemical irritation of nociceptive nerves is dependent on degenerated discs or their material exciting sensory neural elements, especially in the posterior longitudinal ligament and possibly also in the peripheral parts of the annulus fibrosus [69]. It was also suggested that interleukin 1 beta (IL-1β) is generated during IVD degeneration, which stimulates the expression of vascular endothelial growth factor (VEGF), nerve growth factor (NGF), and brain derived growth factor (BDNF), resulting in angiogenesis and innervation [70]. It is also plausible that the released trophic factors can act on nociceptive nerve fibers and also contribute to further release of inflammatory mediators [59, 71] (Fig. 20.6).

20.3.2 Anterior Longitudinal Ligament (ALL)

The anterior longitudinal ligament (ALL) runs on the anterior (ventral) surface of the spine traversing the vertebral bodies and intervertebral discs (Fig. 20.4) [16]. Similar to the IVD, nerve supply to ALL is controversial with very limited studies supporting a less developed innervation of ALL. Bogduk performed one of the earliest investigation of ALL innervation and reported that they were innervated by the branches of the grey rami communicantes [33, 32] (Fig. 20.5B). In subsequent work, Sato et al. performed a detailed investigation of the nerve supply to ALL. They reported that the ALL receives nerve branches from the sympathetic trunk and splanchnic nerves non-segmentally and suggested that the sympathetic nerves may be involved in proprioception of the spinal column [72]. The presence of nerve fibers in ALL was further supported by immunohistochemical studies of Kallakuri et al. who reported dense network of PGP 9.5 reactive nerve fibers. They also reported SP and calcitonin gene related peptide (CGRP) reactive nerve fibers in ALL supporting a putative nociceptive role for this tissue [73]. The presence of mechanoreceptors in ALL was reported by Roberts et al. and suggested that they may play a role in sensing posture, movement and nociception [46]. Clinically by a rare case of heterotopic ossification (HO) in the plane of ALL, symptoms of low back pain due to a fracture in HO was also reported [74] suggesting a role for ALL in mediating LBP.

20.3.3 Posterior Longitudinal Ligament

The posterior longitudinal ligament (PLL) is located within the vertebral canal and extends along the posterior aspect of the vertebral bodies from the body of axis to the sacrum. It is narrow at the level of vertebral body and is wider at intervertebral discs (IVDs) [16] (Fig. 20.4). The nerve supply to the longitudinal ligaments has been reported previously by several investigators [25, 29, 44, 75–76]. Bogduk in one of the earliest neuroanatomic investigations revealed that the PLL is also innervated by branches of sinuvertebral nerves and grey rami communicantes [32, 33] (Fig. 20.5A). These findings were further supported by Higuchi and Sato who also reported PLL innervation from sinuvertebral nerve via the branches of the deep transverse rami of the rami communicantes [72]. Kojima using the acetylcholinesterase (AchE) enzyme histochemistry offered further details on PLL innervation in rat. They reported that the meningeal branch of the spinal nerve (the sinuvertebral nerve) enters the vertebral canal and divides into ascending and descending branches which fuse with those from adjacent vertebrae. They give off transverse branches, connecting with those from the opposite side to form the superficial nerve fiber network in the intervertebral segment, which spreads to the vertebral segment of the PLL. Apart from this superficial nervous network, many nerve fibres enter through the posterolateral portion of the annulus fibrosus (AF) and form a dense, fine nerve fiber network in the deep layer of the intervertebral portion of the PLL and the superficial layer of the disc annulus [34]. Cavanaugh et al. using a silver impregnation technique further supported PLL innervation in close association with AF of IVD in rabbits with occasional encapsulated endings in the annular surface and PLL [41].

One of the earliest evidence of PLL neuropeptide innervation comes from a study by Korkala et al. who showed SP reactive nerve fibers in human PLL samples collected during surgery [78]. Later, Konttinen et al. showed SP and CGRP reactive fibers that often co-localized with cytoskeletal neurofilaments in the PLL of human lumbar disc specimens [69]. Ahmed et al. reported peptidergic (neuropeptide Y, NPY and vasoactive intestinal polypeptide, VIP) and noradrenergic (tyrosine hydroxylase, TH) immunoreactive nerve fibers in the spinal ligaments in preparations of rat lumbar spine [79]. Imai et al. using CGRP and TH as markers reported that the lumbar PLL was dually innervated with one system as polysegmentally innervated and closely associated with autonomic innervation and the other one as unisegmentally innervated and not associated with autonomic fibers [80]. von During and colleagues showed PGP 9.5 reactive nerve fibers with the PLL in the region of IVD being rich in capillaries forming a dense plexus within the ventral part and extending to the outer AF [81]. Kallakuri et al. show further evidence of PGP 9.5 reactive network of large and medium sized bundles and small diameter fibers. They also reported PLL nerve fibers reactive to SP, CGRP and some showing nicotinamide adenine dinucleotide phosphate (NADPH)-diaphorase activity [73]. Yahia and colleagues in their scanning electron microscopy (SEM) study reported PLL innervation both in the superficial and deep layers using neurofilament protein (NFP) immunohistochemistry and suggested that most of the nerve fibers terminated as simple free endings that can act as nociceptors [82]. In neurophysiology studies in the cat, mechanosensitive afferent units in the lumbar posterior longitudinal ligament were reported. The majority of the units were located around the intervertebral disc level with conduction velocities of the units in the range of Group III (0.5–2.5 m/s) and Group IV (2.5–20 m/s) with putative nociceptive function [83].

20.3.4 Dural Innervation

The current literature on dural innervation is very limited, with variations on the extent of innervation of dorsal and ventral aspects [79, 84–87]. The dorsal aspect of the dura is considered to be less densely innervated than the ventral aspect. Groen et al. offered some detailed observations on dorsal dural innervation in studies conducted on human fetal tissue using acetylcholinesterase staining. They described the nerves on the dorsal dura as small in number without forming a plexus and not reaching the medial region of the dura [86]. However, immunohistochemical studies in rabbits by Kallakuri et al. have shown that dorsal dura has small bundles of PGP 9.5 reactive nerve fibers forming connections with adjacent nerve fibers. Many nerves ran longitudinally and perpendicularly and extended large distances on the surface, with nerve fibers reaching midline occasionally in rabbit lumbar dural preparations. The authors further showed prominent SP, CGRP and TH immunoreactive as well as NADPH diaphorase reactive nerve fibers on the dorsal aspect of the dura. The CGRP, TH, and NADPH diaphorase reactive fibers were described as being restricted to dural sleeve on the lateral margins [73]. Furthermore, Ahmed et al. have also reported the presence of NPY-, TH- and VIP-positive fibers in the dorsal aspect [79]. SP and CGRP immunoreactive fibers on the lumbar dorsal dura was further confirmed by a more recent investigation which also suggested that the density of SP and CGRP nerve fibers increases following laminectomy [88]. Thus, pain originating in the dura may be related to the sensitization of various nociceptive nerve fibers. Intradural disc herniation and dural erosion are themselves rare [89].

Innervation of the ventral aspect of the dura has been more extensively reported [30, 77, 86, 87, 90–92]. Groen et al. through their work on human fetuses using acetylcholinesterase staining described that ventral dura receives innervation from the sinuvertebral nerves, the nerve plexus of the posterior longitudinal ligament as well as the plexus of the radicular branches of segmental arteries. They also reported the nerves may extend up to eight segments with overlap between adjacent nerves and attributed this as possible explanation for extrasegmentally referred dural pain [86]. Kallakuri et al. reported PGP 9.5 reactive nerve fibers extending throughout the rabbit ventral dura matrix and suggested a multisegmental innervation on the ventral aspect. Similarly, they also reported prominent SP, CGRP, TH reactive and NADPH diaphorase reactive nerve fibers on the ventral dura [73] (Fig. 20.7). Neuropeptide (NPY, TH and VIP) reactive nerve fibers in the dura were also reported by others [79, 84]. Kumar et al. concluded the SP and CGRP innervation of the lumbar spinal dura was not dense [85]. Furthermore, an extensive network of acetylcholinesterase containing nerve fibers has been reported in rat spinal dura [93].

20.3.5 Ligamentum Flavum

There is very little description on the nerve supply to the ligamentum flavum. Ligmentum flava connect the lamina of adjacent vertebra. On a macroscopic level, this tissue has superficial and deep layers whose fibers are opposite and form close connections with the tendons of some spinal erector muscles. Microscopically, the ligamentum flavum reveals prominent elastic fibers [94]. Although studies by Vandenabeele et al. [95] and Ashton et al. [96], offer no support for the presence of nerve fibers, Bucknill and colleagues in a recent immunohistochemical investigation offered definitive evidence of nerve fibres in ligamentum flavum as shown by PGP 9.5 immunoreactivity in samples harvested from low back pain patients. They further showed the presence of sodium channel (SNS/PN3 and NaN/SNS2) immunoreactivity in a subset of nerve fibers of the ligamentum flavum [97]. The presence of ligamentum flavum innervation is consistent with

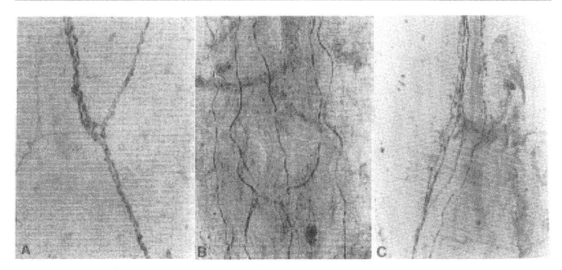

Fig. 20.7 Innervation of dura and the longitudinal ligaments. Immunoreactive fibers to protein gene product 9.5 (PGP 9.5), a general neuronal marker were demonstrated in dorsal dura (**a**) and posterior longitudinal ligament (**b**). PGP 9.5 immunoreactive fibers were also shown in ventral dura and anterior longitudinal ligament. Figure (**c**) shows SP reactive nerve fibers in PLL. Such nerve fibers were also shown in the dura and ALL (Reproduced with permission from Kallakuri et al. [73])

the previous observations of Rhalmi and colleagues who showed neurofilament immunoreactive nerve fibers close to blood vessels and fat globules [98]. It has been suggested that the bulging of the ligamentum flavum contributes to narrowing of the spinal canal which may also have potential implications in the genesis of painful spine conditions [99]. Furthermore, it has also been suggested that a hypertrophic ligamentum flavum may contribute to nerve root compression at the level of the lateral spinal recess and these hypertrophied parts needs to be removed completely during surgical decompression [100]. Also, ligamentum flavum hematoma can compress the adjacent nerve root and in turn contribute to painful conditions [101–103].

Summary: The above review offers extensive evidence in support of nociceptive neuropeptide innervation of the longitudinal ligaments (anterior and posterior), the spinal dura mater and the ligamentum flavum. Considering the important and well-established roles these neuropeptides (SP and CGRP) play in inflammation and pain, it plausible that these structures can be a source of pain. This may particularly be crucial in the case of PLL and ventral dura, considering their close proximity to the IVD. Release of inflammatory mediators from degenerated discs [67, 68] can potentially affect the neural elements in these tissues, leading to altered sensation and pain.

20.3.6 Facet Joints

The morphology of the lumbar facet capsules was studied by Yamashita et al. Via microscopic examination of cadaveric specimens the facet capsules were found to have an outer tough ligamentous layer and a softer inner elastic layer [104].

Nerve supply to lumbar facets occurs via the medial branches of the dorsal rami [33] (Fig. 20.8). The L1–L4 dorsal rami divide into medial and lateral branches near the transverse process. The medial branch travels medially around the base of the superior articular process and under the mamillo-accessory ligament. The medial branch divides into articular branches, supplying the facet joints above and below. Medial branches also supply multifidus and

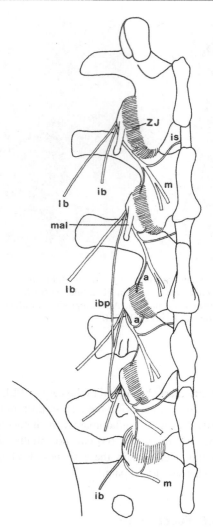

Fig. 20.8 A diagram of the dorsal view of the branches of the left lumbar dorsal rami. Mamillo-accessory ligaments (*mal*) have been lifted in-situ covering the L1 and L2 medial branches. *ZJ* zygapophysial joint, *m* medial branch, *lb* lateral branch, *ib* intermediate branch of the dorsal rami, *ibp* intermediate branch plexus, *is* interspinous branch, *a* articular branches of the dorsal rami (Reproduced with permission from Bogduk [33])

interspinous muscles and the interspinous ligament. The L5 medial articular branch supplies the L5–S1 facet joint and multifidus muscle [33].

To treat facet pain, facet injections as well as medial branch denervation have been carried out and reported by several investigators. Schwarzer et al. injected lumbar facet joints of 146 patients with non-specific low back pain and no definitive radiologic findings. Fifteen percent of these patients had pain relief with a short acting anesthetic (lidocaine) and longer pain relief with a longer acting anesthetic (bupivacaine) [105].

Kuslich et al. reported on 193 awake patients undergoing surgery for disc herniation and/or spinal stenosis. On these patients progressive local anesthesia was performed with 1 % lidocaine. Fifty-four percent of patients had some sensation when the facet capsule was stimulated and 20 % had significant pain [48].

Manchikanti et al. carried out a retrospective analysis of diagnostic facet joint injections in 424 spinal pain patients, divided into 6 age groups. The prevalence of cervical facet joint-related pain was 33–42 % and similar between age groups. The prevalence of lumbar facet joint pain ranged from 18% to 44 % and was higher in older age groups [106].

Eubanks et al. examined a total of 647 cadaveric lumbar spines for evidence of lumbar facet arthrosis. Grading ranged from 0 to 4 (from no arthritis to complete ankylosis). Facet arthrosis was present at the following vertebral levels: 53 % (L1–L2), 66 % (L2–L3), 72 % (L3–L4), 79 % (L4–L5), and 59 % (L5–S1). By decade, facet arthrosis was present in 57 % of 20- to 29-year-olds, 82 % of 30- to 39-year-olds, 93 % of 40- to 49-year-olds, 97 % in 50- to 59-year-olds, and 100 % in those >60 years old [107].

Beaman et al. examined 14 facet joints from patients with low back pain. All facets exhibited cartilage surface irregularity and fibrillation. Erosion channels had progressed into subchondral bone and substance P positive fibers were observed within these erosion channels. Beaman et al. proposed that high pressure at facet contact points may lead to pain at degenerated surfaces, that hypertrophy of the arthritic facet may also cause capsular stretch and pain and that this dual effect may play an important role in low back pain [108].

The objective in a study by Igarashi et al. was to quantify various inflammatory cytokines in facet joint tissue in surgical cases of stenosis and lumbar disc herniation. IL-1beta was detected in joint cartilage and synovium in both groups and TNF-alpha in the synovium in stenosis. IL-6 was high in joint cartilage and synovium in both groups. The concentration was significantly higher in stenosis

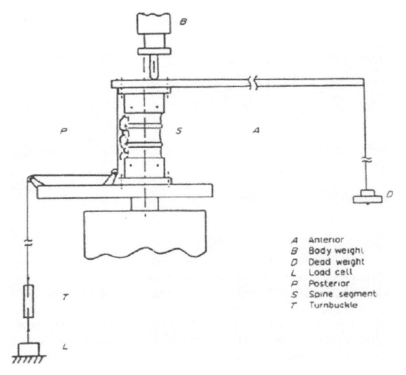

Fig. 20.9 Schematic diagram of the test set-up for measuring facet contact pressure in human cadaveric lumbar motion segments (Reproduced with permission from El-Bohy et al. [110])

than in LDH. The authors suggested that inflammatory cytokines in degenerated facet joints may have some relation to the cause of pain in degenerative lumbar disorders [109].

In a biomechanical study Yang and King reported on the effect of the eccentricity of the applied load on lumbar facet loads. They observed that as facet loads increased, the facets bottomed out on the laminae below, causing the facets to pivot and stretch the superior portion of the joint capsule [12]. El-bohy et al. included extensor muscle action to overcome moments due to eccentric loads which represented body weight and an external load acting 340 mm anterior to the center of the disc. Facet pressure was measured in all cases when muscle load was applied to counteract body weight [110] (Fig. 20.9). This pressure increased when more muscle force was applied to balance the externally applied flexion moment. When the anterior load was released suddenly, there was a large increase in facet pressure with a concomitant decrease in disc pressure [110].

El-Bohy et al. studied lumbar facet capsule strains with specimens loaded in flexion and extension. In extension, the facets bottomed out on the lamina below. This resulted in large capsular strains in the superior portion of the facet capsules which was quantified via video racking of targets placed on the capsule [110].

20.3.6.1 Facet Joint and Muscle Neurophysiology in Control Animals

The biomechanical studies of King, Yang and El-Bohy formed the biomechanical basis for a hypothesis that the facet joint capsule is a source of low back pain and that the pain may arise from large strains in the joint capsule that cause pain receptors to fire. To test this hypothesis, the following neuroanatomic and neurophysiologic studies were performed.

Yamashita et al. performed neurophysiologic studies of sensory neurons originating from the L5–L6 facet joint and surrounding muscle by recording from split L5 lumbar nerve roots in

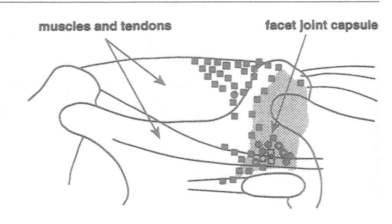

Fig. 20.10 Diagram localizing the receptive fields of the units that were identified at the facet joint and adjacent tissues in anesthetized New Zealand White rabbits. Square = units with thresholds of <6.0 g, circle = units with thresholds of >6.0 g. If the unit responded to stretch, the symbol is hollow (Reproduced with permission from Yamashita et al. [112])

anesthetized New Zealand White rabbits [111, 112]. Thirty mechanosensitive units were identified at the facet joint and 27 others in the muscles and tendons near their insertion into the facet [112] (Fig. 20.10). Of the 30 units at the facet joint, 13 were in the capsule, 15 in the border regions between capsule and muscle or tendon, and two in the ligamentum flavum (Fig. 20.10). Of the 30 units in the facet joint, two units had CVs <2.5 m/s (group IV), 17 units had CVs of 2.5–20 m/s (group III), and 11 units had CVs >20 m/s (group II). Nine units had thresholds >6.0 g, 17 units had thresholds <6.0 g, and four units were not examined. Eight units responded to joint movement caused by pulling on the isolated L5 lamina. Five of these were in the medial aspect of the facet joint. These units were most responsive to 1–2 mm of caudal-to-rostral stretch. The two units in the ligamentum flavum were most responsive to ventral-to-dorsal stretch. Of the 27 units in muscle and tendon, 1 unit had a CV < 2.5 m/s (group IV), 4 units had CVs of 2.5–20 m/s (group III), and 22 units had CVs > 20 m/s (group II). One unit had a threshold >6.0 g and 26 units had thresholds <6.0 g. Thus, the facet joint contained a much higher proportion of high-threshold, low conduction velocity units than muscle. It is these units that are likely to serve as a pain function. These studies demonstrated that some units in the joint capsule were indeed responsive to stretch. However, the stretch was applied through a somewhat non-physiologic means (pulling on a surgically isolated joint) so the next step was to determine whether spine loading would cause capsular stretch that would cause facet joint units to fire.

20.3.6.2 Spine Loading Studies

Twenty-four in vitro experiments were performed in order to determine if facet joint receptors would respond to spine loading [113]. The abdominal aorta of the anesthetized New Zealand White rabbit was cannulated and perfused with Krebs solution oxygenated with 95 % O_2, 5 % CO_2. The lumbar spine was removed with bone rongeurs and placed in a chamber perfused with the same solution. Receptive fields were characterized and nerve recordings made for 2–3 h as the spine was loaded in tension and compression. A 100 lb load cell (Entran Devices ELF-100, Fairfield, NJ) recorded tensile and compressive loads. In these experiments a typical response was a vigorous multi-unit discharge during loading. During compressive loading the spine underwent dorsal-lateral bending that caused the facet joints to articulate along the joint plane, stretching the facet joint capsule. Loading excited both units with spontaneous activity, as well as units that were silent before loading was initiated. Three patterns were observed: (1) short duration bursts during onset or change in loading (phasic-type responses), (2) prolonged discharges that began during low levels of loads of 300–500 g (slowly adapting, low-threshold mechanoreceptors), and (3) prolonged discharges activated by higher loads of 3–5.5 kg (slowly adapting, high-threshold mechanoreceptors). It is this latter response which strongly suggested activation of pain fibers. Fourteen small myelinated or unmyelinated fibers (groups III and IV) responded around the facet joint. It was often not possible to track the response of these individual units to spine loading as this loading naturally elicited firing from many nerves.

20.3.6.3 The Effects of Facet Joint Inflammation on Nerve Activity

In both the *in vivo* and *in vitro* experiments described above, nerves typically stopped firing after the mechanical stimulus was removed or shortly thereafter, even if the stimulus was noxious. Thus, these results demonstrated that capsular stretch could cause the onset of facet pain but does not explain its persistence. It is likely that obvious tissue damage (i.e. tearing of muscle or capsule) would result in persistent discharge. The possible role of inflammatory mediators in facet joint pain was studied as a step in understanding the persistence of low back pain. These studies focused on neurogenic and non-neurogenic inflammatory mediators. Kaolin (a silica product) and carrageenan (a seaweed product) are commonly used to produce an acute tissue inflammation which results in a release of histamine, bradykinin and prostaglandins into extracellular tissue. These algesic (pain-producing) chemicals can excite and/or sensitize nerve endings [114–116].

Ozaktay et al. demonstrated the effects of inflammation in rabbit lumbar facet joint capsule and adjacent tissue. Type II carrageenan was injected into the receptive fields of mechanosensitive sensory neurons. Background discharge rate increased in two phases over a time period of 150 min: the first phase (0–30 min) and the second phase (45–150 min). Silent unit discharge rates increased in the first 15 min and persisted beyond 75 min. Histological examination revealed inflammation in carrageenan injected tissues. In normal saline injected controls there were no changes observed. The electrophysiological results showed that inflammation of the facet joint and deep back muscles caused (1) increases in multi-unit discharge rate, (2) sensitization to mechanical stimuli and (3) recruitment of previously silent units [117].

20.3.6.4 The Effects of Substance P on Unit Activity

The carrageenan studies demonstrated that tissue inflammation had a profound effect on the mechanical sensitivity and discharge rate of facet joint units. Another aspect of tissue inflammation is neurogenic inflammation and is the result of inflammatory chemicals released from the nerves themselves. One of the most studied neurogenic mediators is substance P, an eleven-amino-acid neuropeptide released from small nerve fibers. It has been shown to be released from nerve endings in the knee joint [118]. Substance P causes vasodilatation, plasma extravasation and release of histamine from mast cells [116]. These are important in the inflammatory cascade which can prolong pain. Substance P is also released at nerve synapses in the dorsal horn of the spinal cord to facilitate action potential transmission in spinal cord pain pathways [119, 116]. Substance P has been demonstrated in nerves of lumbar spinal tissues [78] and in small nerve fibers of the facet joint [120]. However, its effect on lumbar spinal tissue nerve activity had never been studied. In 15 rabbits a study was undertaken to examine how substance P affects the mechanosensitive afferent units identified in the lumbar facet joint and adjacent tissues of the rabbit [112]. Substance P was applied to the receptive fields of the units by means of microinjection and afferent activity of the units was recorded from dorsal root filaments. Changes of afferent discharge rates and von Frey thresholds were measured for 30 min. Most of the units (83.3 %) showed an increase in spontaneous discharge rate after the injection of 10 ug *of* substance P into the receptive field; 54.2 % of the units showed immediate onset and 29.2 % of the units showed slow onset of the excitation [112] (Fig. 20.11). One-third of the units showed decreased von Frey thresholds after the application of substance P (evidence of sensitization). Substance P had an excitatory effect on 81.8 % of the units with thresholds >5.0 g and conduction velocities <30 m/s that

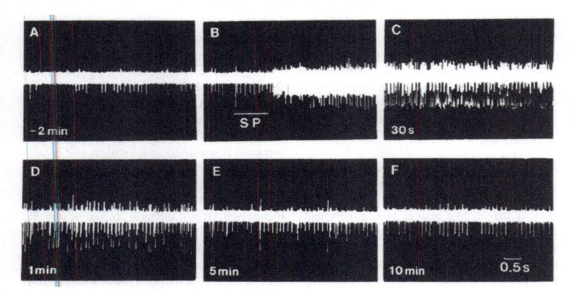

Fig. 20.11 Illustration of afferent discharge rates in lumbar facet joints of rabbits due to injection of 10 μg of substance P (*SP*) into receptive field. Discharges are shown before injection, at injection, and 30 s, 1 min, 5 min, and 10 min after SP injection (Reproduced with permission from Yamashita et al. [112])

may serve as nociceptors, and 84.6 % of the units with thresholds <2.0 g which may serve as proprioceptors. These studies suggest that release of substance P may contribute to transmission of both nociceptive and proprioceptive sensations. Substance P may have had a direct effect on the nerve endings in paraspinal tissue, or acted indirectly through its influence on vasodilatation, plasma extravasation and histamine release [112].

In summary the region of the lumbar facet joints contain an abundance of proprioceptors and nociceptors in the capsule and adjacent multifidus muscle. Biomechanically, the capsule can undergo high strains in flexion and extension. The nerve endings in the capsule are activated and sensitized by inflammatory mediators. Clinical studies indicate the lumbar facet joints may be the source of 15–45 % of chronic axial spinal pain.

20.4 Neuropathic Pain

Neuropathic pain results from direct injury to nerves and often includes a burning or electric sensation [4]. Substantial sensory functional loss, motor dysfunction, and neuropathic pain can result from nerve Injuries. These have major social consequences in terms of health care and long periods of sick-leave. Spinal instability is an important cause of low back pain. Clinical studies indicate that the application of an external fixator to the painful segment of the spine can significantly reduce spinal pain [121].

Anatomic components of the spinal column include nerve tissue, including peripheral nerves, spinal nerves, nerve roots and dorsal root ganglia (DRG). Injury to spinal nerve tissues causes sensory and motor dysfunction, as well as neuropathic pain. Lumbar disk herniation can cause low back pain and sciatic pain. This sciatic pain is also referred to as radicular pain that radiates along the sciatic nerve and its peripheral innervations. Sciatic pain results from lumbar nerve root and/or dorsal root ganglion (DRG) irritation, including mechanical compression and chemical stimulation from herniated discs or from lumbar osteophytes. Mechanical and chemical irritation can cause nerve root or DRG injury, leading to the interruption of axon continuity, degeneration of nerve fibers distal to the lesion and eventual death of somatosensory neurons located in the DRG [122–124].

Acute exposure of rat nerve root to nucleus pulposus (a component from ruptured intervertebral disk) resulted in increased number of axons with neuropathy and higher intensity of prolonged ectopic discharges on compression [125]. The ectopic discharges from nerve roots and DRG are sources of neuropathic pain and paresthesias in low back pain patients. At the nerve injury site, neuromas may also be formed. Neuromas are an important sources for ectopic discharge at sites of nerve injury where voltage-sensitive Na+ channels are increased on remodeling axon membranes [126]. These sites of hyperexcitability are the sources of neuropathic pain.

Neural plasticity after peripheral nerve injury and regeneration response to injury can become maladaptive [127], leading to sensitization of the nociceptive somatosensory neurons (located in DRG and capable of encoding noxious stimuli) in the peripheral nerve system (PNS) and in the central nervous system (CNS), resulting in persistent and debilitating pain [128]. Symptoms include hyperalgesia (increased pain from a stimulus that normally provokes pain), allodynia (pain due to a stimulus that does not normally provoke pain) and even "spontaneous pain" (if the nature of the stimulus is uncertain) [128].

In the somatosensory system, the intensity of a sensation is signaled by the frequency of action potentials ("spikes") in afferent nerve fibers [129]. It has been suggested that in chronic pain states there is an increased spike frequency in nerve root and/or DRG nociceptive neurons [130]. Animal experiments demonstrated that compression of the nerve root and DRG elicited prolonged discharges of spikes [125, 131]. The herniated nucleus pulpous (NP) is inflammatory. Clinically, in low back pain patients undergoing lumbar surgery for disc herniation or spinal stenosis, nerve roots under epidural or local anesthesia have been shown to be very sensitive to compression and mechanical manipulation. Uncomfortable sensations and sciatic pain can be produced in response to mechanical manipulation on the nerve roots in patients with a herniated disc during spinal surgery [48, 132]. Compression on nerve roots with chronic constriction also produces ectopic discharges including nociceptive activity and changes of DRG mechanosensitivity [133]. Threshold pressure to the nerve root for ectopic discharges can be changed after exposure of nerve root to phospholipase A_2 (a chemical found in herniated nucleus pulposus) [134]. Furthermore, exposure of nerve roots to NP also results in changes of threshold pressure required to evoke ectopic discharges, decreased from 2.2 to 0.9 g/mm^2, indicative of nerve root mechanical-sensitization [125]. In addition, the duration of ectopic discharge was longer in acute phospholipase A_2 exposed nerve roots (7.86 s) than normal nerve roots (0.09 s) upon mechanical stimulation [134].

Animal studies have shown that transient compression to the normal spinal dorsal nerve root only elicited a transient ectopic discharge lasting for 0.09 s, while the same stimulation to the inflammatory nerve caused prolong ectopic discharge lasting for more than 7 s [134, 125]. On the other hand, small transient pressures to the dorsal root ganglion elicit afterdischarges [7] (Fig. 20.12). Sodium channels play an important role in the neural activity including action potential generation and conduction. Considerable attention has been focused on the role of sodium channels in the pathophysiology of neuropathic pain and as therapeutic targets [135]. Increased expression of sodium channel occurs at the neuroma forming sites where nerve is injured [136]. Although there is no clear evidence to support neuroma formation in the herniated disk compressed nerve root, one study did find that exposure of nerve root to NP leads to abnormal sodium channel expression, including accumulation and redistribution in the dorsal root and DRG [125]. Exposure of nerve root to NP and chronic mild compression for 7 days causes nerve conduction dysfunction and neuropathology [137, 125]. These studies suggest that exposure of nerve roots to autografted nucleus pulposus without mechanical compression causes nerve root functional and structure changes. In addition, after application of nucleus pulposus to the nerve root, the corresponding dorsal root ganglion demonstrated increased excitability and mechanical hypersensitivity, suggestive of NP-induced excitatory changes in the dorsal root ganglion [131].

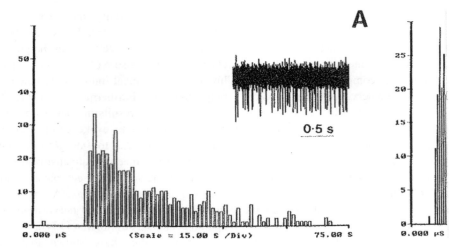

Fig. 20.12 Discharge rate versus time histogram, illustrating the response of a lumbar dorsal root ganglion (DRG) to probing with a nylon filament rated at 1.3 g of force. Nerve discharge continued for approximately 1 min after the stimulus was removed (bin width=1 s). The *inset* shows a sample of the multiunit response (Reproduced with permission from Cavanaugh [7])

Spinal nerves can also be subjected to stretch. In the "nerve tension sign" for lumbar disk herniation patients cannot raise their leg up to the normal level of 70°. The result is termed as positive Lasègue's sign [138]. Lasègue's sign is determined in the straight leg raise (SLR) test; with the patient lying down on his or her back on an examination table or exam floor, the examiner lifts the patient's leg while the knee is straight. The spinal nerve is pulled distally during this maneuver. Lumbar spinal nerves are relatively mobile. The range of excursion is 1.5 mm for the L4 nerve root, 3.0 mm for the L5 lumbar nerve root, and 4 mm for the S1 nerve root [139]. SLR may draw the spinal nerve roots distally by 2–8 mm and firmly fix the root to the anterior wall of the spinal canal [140]. Under the circumstance of lumbar disk herniation, the mobile range of the nerve root decreases, leading to increased injury risk from spinal nerve overstretch.

Injury to lumbar nerve roots has been investigated under various strain magnitudes and stretch speeds in a rat model. In this model L5 nerve roots were stretched by an actuator and recordings of nerve strain were made by tracking targets on the nerve root with a high speed video camera [141–144]. Nerve conduction dysfunction was used as a criterion for nerve injury by measuring the changes of magnitude of compound action potential (CAP). CAPs are the summation of action potentials from many nerve fibers provoked by electrical stimulus pulses. Decrease of CAP magnitude indicates a portion of nerve fibers in the nerve trunk have lost function. At low strains (<10 %) and low stretch rate (20 mm/s), nerves showed recovery over 6 h post-stretch, based on CAP amplitude percentage change compared to the baseline. The strain that caused 50 % probability of *mild* nerve dysfunction was 8.68 % strain in 20 mm/s group; 7.32 % strain in 200 mm/s group; and 6.27 % strain in 800 mm/s group. The strain that caused 50 % probability of *moderate* nerve dysfunction was 31.25 % strain in the 20 mm/s stretch speed group; 22.14 % strain in the 200 mm/s group; and 9.55 % strain in the 800 mm/s group [141]. These studies and other similar studies [142–144] focused on establishing a relationship between strain and displacement rate to spinal nerves, and the resulting functional loss and probability of recovery. In summary Chen et al. found that higher stretch rates and higher strain produced greater deficits in nerve conduction function. At low strains (<10 %) nerves showed partial to full recovery over 6 h. At stretch rates of 20 and 200 mm/s nerves showed partial recovery over 6 h. The stretch rate of 800 mm/s produced severe deficits in nerve function out to 6 h after stretch, without functional recovery at strains >10 % [141].

20.5 Muscle Pain

Muscle pain is a very common experience and can range from mild to excruciating. Most occurrences of acute muscle pain are self-limited strains. Persistent, chronic, or recurrent symptoms of muscle pain are frequently associated with degenerative lumbar disc disease or spondylolytic stress lesions. Most instances of myalgia result from overloaded stress, tension, or overusing the muscle during physical activity [145]. Eccentric contractions, accumulated muscle damage and/or a single injurious event can lead to muscle injury [146]. Accumulated muscle damage may result from the neuromuscular inhibition following muscle injury. Neuromuscular inhibition may impede the rehabilitation process and subsequently lead to maladaptive muscle structure and function, including eccentric weakness and atrophy of the previously injured muscles [146]. Certain muscles are susceptible to strain injury (e.g. lower limb muscles that cross multiple joints or have complex architecture). These muscles have a strain threshold for both passive and active injury. Strains are the result of excessive stretch or stretch while the muscle is being activated. When the muscle tears, the damage is commonly localized near the muscle-tendon junction [147]. Excitatory and inhibitory reflex arcs (the ligamento-muscular reflex) exist between the ligaments/tendons and recruited muscles in maintaining joint stability. These reflex arcs control the quality of movement and safety/stability of the joint, hence impact the safety and health of workers and athletes [148].

Mechanoreceptors exist in the ligaments of the major joints and include four major types of nerve endings: Golgi, Pacinian corpuscles, Ruffini endings, and free nerve endings [149]. Physiologic behaviors of the mechanoreceptors have been investigated extensively [115, 150–152]. Stretch of cervical facet joint capsule elicited corresponding paraspinal muscle contraction [153, 154] indicating that the cervical ligamento-muscular reflex pathways are activated via tensile facet joint capsule stretch and extend to superficial and deep musculature on the anterior and posterior aspects of the neck, ipsilateral and contralateral to the side of FJC stretch. Understanding the injury strain thresholds will aid in understanding the limits of facet joint movement required to prevent muscle spasm and facet capsule ligament injury.

Occupational factors plus muscle response can significantly contribute to the generation of low back pain. For example of long-term truck drivers, it has been reported that 40 % suffer from chronic back problems, the figure jumps to over 65 % for professional drivers with 12–15 years behind the wheel [155]. Many physical factors contribute to the development of low back disorders (LBD) [156]. Among such physical exposures encountered in working conditions, whole-body vibration (WBV) has repeatedly been identified as a risk factor for low back pain [157, 158]. Sitting with awkward postures plus WBV increases low back pain fourfold [155]. Paraspinal muscles play an important role in back pain generation. Epidemiological studies in helicopter pilots have also suggested that LBP is associated with WBV [159–161]. These studies are suggestive that the cyclic response of the erector spinae (ES) muscles due to WBV lead to muscle fatigue, causing LBP in pilots.

Researchers have demonstrated that mechanical vibration (namely between 30 and 120 Hz) directly applied to a muscle belly or tendon elicits the tonic vibration reflex (TVR) [162, 163]. This is a neuromuscular response caused by excitation of muscle spindles leading to enhanced muscle activity. The TVR has also been suggested to occur in the back muscles at frequencies between 1 and 5 Hz [158, 164]. This increased muscle activity that is necessary to dampen the vibratory waves [165] could lead to muscle fatigue, which has been shown to affect neuromuscular coordination [166] and proprioception [167]. These leave the spine at increased risk of injury. It has been reported that exposure to vibration causes muscle-skeletal pain of the spine [159, 168, 169] with radiological pathologic findings occurring earlier in the spine of individuals exposed to long periods of vibration [168]. The relationship between whole-body vibration (WBV) and low back pain (LBP) has also been investigated in other studies [170–172].

Work/rest ratios have an effect on muscle fatigue associated with the amount of rest allowance provided following muscle contraction. Experiments in rats demonstrated that the higher duty cycle and shorter cycle time group resulted in significantly greater fatigue than the lower duty cycle and longer cycle time group, as measured by increased M-wave amplitude and area. A longer M-wave duration was observed in the high duty cycle, long cycle time group, suggesting that the combination of low duty cycle and long cycle times leads to less fatigue. In high duty cycle scenarios, short cycle times result in less fatigue [173].

Muscle pain differs in several aspects from cutaneous and visceral pain [174]. Subjectively, muscle pain is perceived as aching and cramping; it is difficult to localize and exhibits referral to other deep somatic tissues such as fascia, muscle, and joints [174]. Objectively, the information from muscle nociceptors is processed differently in the central nervous system. Muscle pain has a special relay in the mesencephalon [175] and is inhibited more strongly by the descending pain-modulating pathways than is cutaneous pain [176]. In addition, cortical imaging data have shown that muscle pain activates areas in the human cortex that differ in location from those activated by cutaneous pain [177].

Nociceptive nerve endings in muscles and other tissues are equipped with a multitude of receptor molecules for endogenous pain-producing and sensitizing agents. Particularly interesting molecules are the purinergic receptors, which can be activated by adenosine triphosphate (ATP), and the vanilloid receptor, which is sensitive to protons (low pH) [178]. The purinergic receptors are activated by tissue damage because cell necrosis is associated with the release of ATP. A low pH is present in many pathologic conditions such as ischemia and inflammation [179]. At the spinal and medullar level, painful muscle lesions induce marked neuroplastic changes that result in hyperexcitability and hyperactivity of nociceptive neurons. This central sensitization is the basis for the spontaneous pain and hyperalgesia of patients [180]. The transition from acute to chronic muscle pain is complete when the initially functional changes are transformed into structural ones. Patients with morphologic alterations in their nociceptive system are difficult to treat because the changes require time to normalize [174].

For muscular pain, the relationship between whole-body vibration and corresponding muscle activity relationship in vehicle drivers has been studied. Root-mean-square (RMS) is most frequently used as a predictor for muscle activity. Recently, root-mean-quad (RMQ) [181] and power (V2/Hz and EMG V2/Hz) [182] have been showed to be predictors of muscle activity in response to mechanical stimulation. Electromyography (EMG) measures the electrical activity of muscles at rest and during contraction. Correlation between muscle contraction force and EMG signals has been characterized and mathematical equations have been developed to depict the magnitude of contraction force based on EMG signals [183, 184]. EMG characteristics associated with muscle pain have not been identified, although efforts have been devoted to detect muscular-source discomfort based on surface electromyography [185]. Waveform recognition and template matching (WRTM) [150, 186–188] may be a method to detect muscular nociceptor response to noxious muscle stretch or prolonged contraction. WRTM methods have been successfully used to monitor mechanoreceptor activity, including nociceptor activity, in the facet joint capsule in response to physiologic and noxious facet capsule stretche [152, 189, 190]. Classification of the sensory receptors is based on conduction velocity, activation threshold, and adaptation patterns [150, 191]. It is possible that by combining EMG and WRTM methods, EMG characteristics associated with muscle pain may be identified when nociceptors are activated during noxious stretch, prolonged contraction, or spastic contraction.

20.6 Summary

Based on extensive research the anatomic components that have been considered to cause back pain include the intervertebral disks, paraspinal muscles, spinal ligaments, facet joint capsules, dorsal roots and dorsal root ganglia. While mechanical strain is an important causative factor in the initiation of spinal pain, biochemical irritation of peripheral tissues is important in sensitizing the injured tissue. In addition, central sensitization at the spinal cord level has been shown to be an important contributor to persistent spinal pain. The mechanical input that results in spinal pain includes both acute mechanical loading and repetitive cyclic loading. The latter is particularly important in disc injury and degeneration and in muscle fatigue to the lumbar spine. While much research has been performed on the neuroanatomy, neurophysiology, molecular biology and biomechanics of thoraco-lumbar spine injury and pain, relatively little research has combined these specialties to address the relationship between measured mechanical input to tissues of the spine and the molecular, histological and neurophysiologic outcomes that result in spinal pain. Further work is required in this area to fully understand the etiologies of thoraco-lumbar pain and to develop preventive strategies to prevent such pain in the workplace and other environments.

Acknowledgements The authors thank Preston Lemanski for his significant efforts in the organization of references and figures in this chapter.

References

1. Relieving pain in America: a blueprint for transforming prevention, care, education and research. Committee on Advancing Pain Research, Care and Education; Insititute of Medicine, The National Academies; The National Academies Press; Washington, DC, 2011
2. Hoy D, Bain C, Williams G, et al. A systematic review of the global prevalence of low back pain. Arthritis Rheum 2012; 64:2028–2037
3. Merskey H, Lindblom U, Mumford JM, et al (1994) Part III: Pain terms, a current list with definitions and notes on usage. In: Merskey H, Bogduk N (eds) Classification of chronic pain, 2nd edn. IASP Task Force on Taxonomy, Seattle, pp 209–214
4. Kandel E, Schwartz J, Jessel T, Siegelbaum S, Hudspeth A (2013) Principles of neural science, 5th edn. McGraw-Hill, New York
5. Ringkamp MRS, Campbell JN, Meyer RA (2013) Peripheral mechanisms of cutaneous nociception. Wall and Melzack's textbook of pain, 6th edn. Elsevier/Saunders, Philadelphia
6. Golan DE, Tashjian AH, Armstrong EF (2007) Principles of pharmacology: the pharmacologic basis of drug therapy. 2nd edn. Wolters Kluwer Health, Baltimore
7. Cavanaugh JM (1995) Neural mechanisms of lumbar pain. Spine 20(16):1804–1809
8. Eiband AM (1959) Human tolerance to rapidly applied accelerations: a summary of the literature. National Aeronautics and Space Administration, Washington (Memorandum 5-19-59E)
9. Ewing CL, King AI, Prasad P (1972) Structural considerations of the human vertebral column under +gz impact acceleration. J Aircraft 9:84–90
10. Vulcan AP, King AI, Nakamura GS (1970) Effects of bending on the vertebral column during +gz acceleration. Aerosp Med 41(3):294–300
11. Prasad P, King AI, Ewing CL (1974) The role of articular facets during +Gz acceleration. J Appl Mech 41(2):321–326. doi:10.1115/1.3423284
12. Yang KH, King AI (1984) Mechanism of facet load transmission as a hypothesis for low back pain. Spine 9(6):557–565
13. Prasad P, King AI (1974) An experimentally validated dynamic model of the spine. J Appl Mech 41:546–550
14. Begeman P, King AI, and Prasad P (1973) Spinal loads resulting from -gx acceleration. SAE Technical Paper 730977, Society of Automotive Engineers, Warrendale, PA
15. Miniaci A, McLaren AC (1989) Anterolateral compression fracture of the thoracolumbar spine. A seat belt injury. Clin Orthop Relat Res 240:153–156
16. Freemont AJ (2009) The cellular pathobiology of the degenerate intervertebral disc and discogenic back pain. Rheumatology 48(1):5–10. doi:10.1093/rheumatology/ken396
17. Henzel JH, Mohr GC, von Gierke HE (1968) Reappraisal of biodynamic implications of human ejections. Aerosp Med 39(3):231–240
18. Yoganandan N, Pintar F, Sances A, Maiman D, Myklebust J, Harris G, Ray G (1988) Biomechanical investigations of the human thoracolumbar spine. SAE technical paper (881331)
19. Brinckmann P (1986) Injury of the annulus fibrosus and disc protrusions. An in vitro investigation on human lumbar discs. Spine (Phila Pa 1976) 11(2):149–153
20. Adams MA, Hutton WC (1982) Prolapsed intervertebral disc: a hyperflexion injury. Spine 7(3):184–191
21. Gordon SJ, Yang KH, Mayer PJ, Mace AH Jr, Kish VL, Radin EL (1991) Mechanism of disc rupture. A preliminary report. Spine (Phila Pa 1976) 16(4):450–456
22. Callaghan JP, McGill SM (2001) Intervertebral disc herniation: studies on a porcine model exposed to highly repetitive flexion/extension motion with compressive force. Clin Biomech (Bristol, Avon) 16(1):28–37
23. Bogduk N, Aprill C, Derby R (2013) Lumbar discogenic pain: state-of-the-art review. Pain Med. doi:10.1111/pme.12082

24. Wiberg G (1949) Back pain in relation to the nerve supply of the intervertebral disc. Acta Orthop Scand 19(2):211–221, illust
25. Roofe PG (1940) Innervation of annulus fibrosus and posterior longitudinal ligament. Arch Neurol Psychiat 44:100
26. Ehrenhaft JL (1943) Development of the vertebral column as related to certain congenital and pathological changes. Surg Gynecol Obst 76:282–292
27. Jung A, Brunschwig A (1932) Recherches histologiques sur l'innervation des articulations des corps vertebraux. Presse medicale 1:316–317
28. Kumar S, Davis PR (1973) Lumbar vertebral innervation and intra-abdominal pressure. J Anat 114(Pt 1):47–53
29. Mulligan JH (1957) The innervation of the ligaments attached to the bodies of the vertebrae. J Anat 91(4):455–465
30. Stilwell DL (1956) The nerve supply of the vertebral column and its associated structures in the monkey. Anat Rec 125(139–169)
31. Yoshizawa H, O'Brien JP, Smith WT, Trumper M (1980) The neuropathology of intervertebral discs removed for low-back pain. J Pathol 132(2):95–104. doi:10.1002/path.1711320202
32. Bogduk N, Tynan W, Wilson AS (1981) The nerve supply to the human lumbar intervertebral discs. J Anat 132(Pt 1):39–56
33. Bogduk N (1983) The innervation of the lumbar spine. Spine 8(3):286–293
34. Kojima Y, Maeda T, Arai R, Shichikawa K (1990) Nerve supply to the posterior longitudinal ligament and the intervertebral disc of the rat vertebral column as studied by acetylcholinesterase histochemistry. I. Distribution in the lumbar region. J Anat 169:237–246
35. Morinaga T, Takahashi K, Yamagata M, Chiba T, Tanaka K, Takahashi Y, Nakamura S, Suseki K, Moriya H (1996) Sensory innervation to the anterior portion of lumbar intervertebral disc. Spine 21(16):1848–1851
36. Ohtori S, Takahashi Y, Takahashi K, Yamagata M, Chiba T, Tanaka K, Hirayama J, Moriya H (1999) Sensory innervation of the dorsal portion of the lumbar intervertebral disc in rats. Spine 24(22):2295–2299
37. Ohtori S, Takahashi K, Chiba T, Yamagata M, Sameda H, Moriya H (2001) Sensory innervation of the dorsal portion of the lumbar intervertebral discs in rats. Spine 26(8):946–950
38. Ashton IK, Roberts S, Jaffray DC, Polak JM, Eisenstein SM (1994) Neuropeptides in the human intervertebral disc. Journal Orthop Res 12(2):186–192. doi:10.1002/jor.1100120206
39. Willenegger S, Friess AE, Lang J, Stoffel MH (2005) Immunohistochemical demonstration of lumbar intervertebral disc innervation in the dog. Anat Histol Embryol 34(2):123–128. doi:10.1111/j.1439-0264.2004.00593.x
40. Palmgren T, Gronblad M, Virri J, Kaapa E, Karaharju E (1999) An immunohistochemical study of nerve structures in the anulus fibrosus of human normal lumbar intervertebral discs. Spine 24(20):2075–2079
41. Cavanaugh JM, Kallakuri S, Ozaktay AC (1995) Innervation of the rabbit lumbar intervertebral disc and posterior longitudinal ligament. Spine 20(19):2080–2085
42. Dimitroulias A, Tsonidis C, Natsis K, Venizelos I, Djau SN, Tsitsopoulos P (2010) An immunohistochemical study of mechanoreceptors in lumbar spine intervertebral discs. J Clin Neurosci 17(6):742–745. doi:10.1016/j.jocn.2009.09.032
43. Hirsch C, Ingelmark BE, Miller M (1963) The anatomical basis for low back pain. Studies on the presence of sensory nerve endings in ligamentous, capsular and intervertebral disc structures in the human lumbar spine. Acta Orthop Scand 33:1–17
44. Jackson HC 2nd, Winkelmann RK, Bickel WH (1966) Nerve endings in the human lumbar spinal column and related structures. J Bone Joint Surg Am 48(7):1272–1281
45. Malinsky J (1959) The ontogenetic development of nerve terminations in the intervertebral discs of man. (Histology of intervertebral discs, 11th communication). Acta Anat (Basel) 38:96–113
46. Roberts S, Eisenstein SM, Menage J, Evans EH, Ashton IK (1995) Mechanoreceptors in intervertebral discs. Morphology, distribution, and neuropeptides. Spine 20(24):2645–2651
47. Mooney V (1987) Presidential address. International Society for the Study of the Lumbar Spine. Dallas, 1986. Where is the pain coming from? Spine 12(8):754–759
48. Kuslich SD, Ulstrom CL, Michael CJ (1991) The tissue origin of low back pain and sciatica: a report of pain response to tissue stimulation during operations on the lumbar spine using local anesthesia. Orthop Clin North Am 22(2):181–187
49. DePalma MJ, Lee JE, Peterson L, Wolfer L, Ketchum JM, Derby R (2009) Are outer annular fissures stimulated during diskography the source of diskogenic low-back pain? An analysis of analgesic diskography data. Pain Med 10(3):488–494. doi:10.1111/j.1526-4637.2009.00602.x
50. Peng BG (2013) Pathophysiology, diagnosis, and treatment of discogenic low back pain. World J Orthop 4(2):42–52. doi:10.5312/wjo.v4.i2.42
51. Hadjipavlou AG, Tzermiadianos MN, Bogduk N, Zindrick MR (2008) The pathophysiology of disc degeneration: a critical review. J Bone Joint Surg 90(10):1261–1270. doi:10.1302/0301-620X.90B10.20910
52. Matrisian LM (1990) Metalloproteinases and their inhibitors in matrix remodeling. Trends Genet 6(4):121–125
53. Goupille P, Jayson MI, Valat JP, Freemont AJ (1998) Matrix metalloproteinases: the clue to intervertebral disc degeneration? Spine 23(14):1612–1626
54. Kang JD, Stefanovic-Racic M, McIntyre LA, Georgescu HI, Evans CH (1997) Toward a biochemical understanding of human intervertebral disc

degeneration and herniation. Contributions of nitric oxide, interleukins, prostaglandin E2, and matrix metalloproteinases. Spine 22(10):1065–1073

55. Kang JD, Georgescu HI, McIntyre-Larkin L, Stefanovic-Racic M, Donaldson WF 3rd, Evans CH (1996) Herniated lumbar intervertebral discs spontaneously produce matrix metalloproteinases, nitric oxide, interleukin-6, and prostaglandin E2. Spine 21(3):271–277

56. Kanemoto M, Hukuda S, Komiya Y, Katsuura A, Nishioka J (1996) Immunohistochemical study of matrix metalloproteinase-3 and tissue inhibitor of metalloproteinase-1 human intervertebral discs. Spine 21(1):1–8

57. Doita M, Kanatani T, Harada T, Mizuno K (1996) Immunohistologic study of the ruptured intervertebral disc of the lumbar spine. Spine 21(2):235–241

58. Podichetty VK (2007) The aging spine: the role of inflammatory mediators in intervertebral disc degeneration. Cell Mol Biol 53(5):4–18

59. Garcia-Cosamalon J, del Valle ME, Calavia MG, Garcia-Suarez O, Lopez-Muniz A, Otero J, Vega JA (2010) Intervertebral disc, sensory nerves and neurotrophins: who is who in discogenic pain? J Anat 217(1):1–15.doi:10.1111/j.1469-7580.2010.01227.x

60. Schroeder M, Viezens L, Schaefer C, Friedrichs B, Algenstaedt P, Ruther W, Wiesner L, Hansen-Algenstaedt N (2013) Chemokine profile of disc degeneration with acute or chronic pain. J Neurosurg Spine18(5):496–503.doi:10.3171/2013.1.SPINE12483

61. Richardson SM, Doyle P, Minogue BM, Gnanalingham K, Hoyland JA (2009) Increased expression of matrix metalloproteinase-10, nerve growth factor and substance P in the painful degenerate intervertebral disc. Arthritis Res Ther 11(4): R126. doi:10.1186/ar2793

62. Nilsson E, Nakamae T, Olmarker K (2011) Pain behavior changes following disc puncture relate to nucleus pulposus rather than to the disc injury per se: an experimental study in rats. Open Orthop J 5:72–77. doi:10.2174/1874325001105010072

63. Olmarker K (2008) Puncture of a lumbar intervertebral disc induces changes in spontaneous pain behavior: an experimental study in rats. Spine 33(8): 850–855. doi:10.1097/BRS.0b013e31816b46ca

64. Nakamae T, Ochi M, Olmarker K (2011) Pharmacological inhibition of tumor necrosis factor may reduce pain behavior changes induced by experimental disc puncture in the rat: an experimental study in rats. Spine 36(4):E232–E236. doi:10.1097/BRS.0b013e3181d8bef3

65. Murata Y, Onda A, Rydevik B, Takahashi K, Olmarker K (2004) Selective inhibition of tumor necrosis factor-alpha prevents nucleus pulposus-induced histologic changes in the dorsal root ganglion. Spine 29(22):2477–2484

66. Olmarker K, Nutu M, Storkson R (2003) Changes in spontaneous behavior in rats exposed to experimental disc herniation are blocked by selective TNF-alpha inhibition. Spine 28(15):1635–1641; discussion 1642. doi:10.1097/01.BRS.0000083162.35476.FF

67. Kayama S, Konno S, Olmarker K, Yabuki S, Kikuchi S (1996) Incision of the anulus fibrosus induces nerve root morphologic, vascular, and functional changes. An experimental study. Spine 21(22): 2539–2543

68. Olmarker K, Brisby H, Yabuki S, Nordborg C, Rydevik B (1997) The effects of normal, frozen, and hyaluronidase-digested nucleus pulposus on nerve root structure and function. Spine 22(5):471–475; discussion 476

69. Konttinen YT, Gronblad M, Antti-Poika I, Seitsalo S, Santavirta S, Hukkanen M, Polak JM (1990) Neuroimmunohistochemical analysis of peridiscal nociceptive neural elements. Spine 15(5):383–386

70. Lee JM, Song JY, Baek M, Jung HY, Kang H, Han IB, Kwon YD, Shin DE (2011) Interleukin-1beta induces angiogenesis and innervation in human intervertebral disc degeneration. J Orthop Res 29(2):265–269. doi:10.1002/jor.21210

71. Wallach D, Arumugam TU, Boldin MP, Cantarella G, Ganesh KA, Goltsev Y, Goncharov TM, Kovalenko AV, Rajput A, Varfolomeev EE, Zhang SQ (2002) How are the regulators regulated? The search for mechanisms that impose specificity on induction of cell death and NF-kappaB activation by members of the TNF/NGF receptor family. Arthritis Res 4(Suppl 3):S189–S196

72. Higuchi K, Sato T (2002) Anatomical study of lumbar spine innervation. Folia Morphol 61(2):71–79

73. Kallakuri S, Cavanaugh JM, Blagoev DC (1998) An immunohistochemical study of innervation of lumbar spinal dura and longitudinal ligaments. Spine 23(4):403–411

74. Czyrny JJ, Glasauer FE (1995) An unusual location for heterotopic ossification: lumbar anterior longitudinal ligament. J Spinal Cord Med 18(3):194–199

75. Groen GJ, Baljet B, Drukker J (1990) Nerves and nerve plexuses of the human vertebral column. Am J Anat 188(3):282–296. doi:10.1002/aja.1001880307

76. Herlihy WF (1949) The sinu-vertebral nerve. N Z Med J 48(264):214–216

77. Pedersen HE, Blunck CF, Gardner E (1956) The anatomy of lumbosacral posterior rami and meningeal branches of spinal nerve (sinu-vertebral nerves); with an experimental study of their functions. J Bone Joint Surg Am 38-A(2):377–391

78. Korkala O, Gronblad M, Liesi P, Karaharju E (1985) Immunohistochemical demonstration of nociceptors in the ligamentous structures of the lumbar spine. Spine 10(2):156–157

79. Ahmed M, Bjurholm A, Kreicbergs A, Schultzberg M (1993) Neuropeptide Y, tyrosine hydroxylase and vasoactive intestinal polypeptide-immunoreactive nerve fibers in the vertebral bodies, discs, dura mater, and spinal ligaments of the rat lumbar spine. Spine 18(2):268–273

80. Imai S, Hukuda S, Maeda T (1995) Dually innervating nociceptive networks in the rat lumbar posterior longitudinal ligaments. Spine 20(19):2086–2092

81. von During M, Fricke B, Dahlmann A (1995) Topography and distribution of nerve fibers in the

posterior longitudinal ligament of the rat: an immunocytochemical and electron-microscopical study. Cell Tissue Res 281(2):325–338
82. Yahia L, Newman N (1993) A scanning electron microscopic and immunohistochemical study of spinal ligaments innervation. Ann Anat 175(2):111–114
83. Sekine M, Yamashita T, Takebayashi T, Sakamoto N, Minaki Y, Ishii S (2001) Mechanosensitive afferent units in the lumbar posterior longitudinal ligament. Spine 26(14):1516–1521
84. Ahmed M, Bjurholm A, Kreicbergs A, Schultzberg M (1991) SP- and CGRP-immunoreactive nerve fibers in the rat lumbar spine. Neuroorthopedics 12:19–28
85. Kumar R, Berger RJ, Dunsker SB, Keller JT (1996) Innervation of the spinal dura. Myth or reality? Spine 21(1):18–26
86. Groen GJ, Baljet B, Drukker J (1988) The innervation of the spinal dura mater: anatomy and clinical implications. Acta Neurochir 92(1–4):39–46
87. Konnai Y, Honda T, Sekiguchi Y, Kikuchi S, Sugiura Y (2000) Sensory innervation of the lumbar dura mater passing through the sympathetic trunk in rats. Spine 25(7):776–782
88. Saxler G, Brankamp J, von Knoch M, Loer F, Hilken G, Hanesch U (2008) The density of nociceptive SP- and CGRP-immunopositive nerve fibers in the dura mater lumbalis of rats is enhanced after laminectomy, even after application of autologous fat grafts. Eur Spine J 17(10):1362–1372. doi:10.1007/s00586-008-0741-7
89. Floeth F, Herdmann J (2012) Chronic dura erosion and intradural lumbar disc herniation: CT and MR imaging and intraoperative photographs of a transdural sequestrectomy. Eur Spine J 21(Suppl 4):S453–S457. doi:10.1007/s00586-011-2073-2
90. Edgar MA, Ghadially JA (1976) Innervation of the lumbar spine. Clin Orthop Relat Res 115:35–41
91. Edgar MA, Nundy S (1966) Innervation of the spinal dura mater. J Neurol Neurosurg Psychiatry 29:530–534
92. Kimmel DL (1961) Innervation of spinal dura mater and dura mater of the posterior cranial fossa. Neurology 11:800–809
93. Artico M, Cavallotti C (2001) Catecholaminergic and acetylcholine esterase containing nerves of cranial and spinal dura mater in humans and rodents. Microsc Res Tech 53(3):212–220. doi:10.1002/jemt.1085
94. Viejo-Fuertes D, Liguoro D, Rivel J, Midy D, Guerin J (1998) Morphologic and histologic study of the ligamentum flavum in the thoraco-lumbar region. Surg Radiol Anat 20(3):171–176
95. Vandenabeele F, Creemers J, Lambrichts I, Robberechts W (1995) Fine structure of vesiculated nerve profiles in the human lumbar facet joint. J Anat 187(Pt 3):681–692
96. Ashton IK, Ashton BA, Gibson SJ, Polak JM, Jaffray DC, Eisenstein SM (1992) Morphological basis for back pain: the demonstration of nerve fibers and neuropeptides in the lumbar facet joint capsule but not in ligamentum flavum. J Orthop Res 10(1):72–78. doi:10.1002/jor.1100100109
97. Bucknill AT, Coward K, Plumpton C, Tate S, Bountra C, Birch R, Sandison A, Hughes SP, Anand P (2002) Nerve fibers in lumbar spine structures and injured spinal roots express the sensory neuron-specific sodium channels SNS/PN3 and NaN/SNS2. Spine 27(2):135–140
98. Rhalmi S, Yahia LH, Newman N, Isler M (1993) Immunohistochemical study of nerves in lumbar spine ligaments. Spine 18(2):264–267
99. Hansson T, Suzuki N, Hebelka H, Gaulitz A (2009) The narrowing of the lumbar spinal canal during loaded MRI: the effects of the disc and ligamentum flavum. Eur Spine J 18(5):679–686. doi:10.1007/s00586-009-0919-7
100. Winkler PA, Zausinger S, Milz S, Buettner A, Wiesmann M, Tonn JC (2007) Morphometric studies of the ligamentum flavum: a correlative microanatomical and MRI study of the lumbar spine. Zentralbl Neurochir 68(4):200–204. doi:10.1055/s-2007-985853
101. Keynan O, Smorgick Y, Schwartz AJ, Ashkenazi E, Floman Y (2006) Spontaneous ligamentum flavum hematoma in the lumbar spine. Skeletal Radiol 35(9):687–689. doi:10.1007/s00256-005-0945-4
102. Shimada Y, Kasukawa Y, Miyakoshi N, Hongo M, Ando S, Itoi E (2006) Chronic subdural hematoma coexisting with ligamentum flavum hematoma in the lumbar spine: a case report. Tohoku J Exp Med 210(1):83–89
103. Minamide A, Yoshida M, Tamaki T, Natsumi K (1999) Ligamentum flavum hematoma in the lumbar spine. J Orthop Sci 4(5):376–379
104. Yamashita T, Minaki Y, Ozaktay AC, Cavanaugh JM, King AI (1996) A morphological study of the fibrous capsule of the human lumbar facet joint. Spine (Phila Pa 1976) 21(5):538–543
105. Schwarzer AC, Aprill CN, Derby R, Fortin J, Kine G, Bogduk N (1994) Clinical features of patients with pain stemming from the lumbar zygapophysial joints. Is the lumbar facet syndrome a clinical entity? Spine (Phila Pa 1976) 19(10):1132–1137
106. Manchikanti L, Manchikanti KN, Cash KA, Singh V, Giordano J (2008) Age-related prevalence of facet-joint involvement in chronic neck and low back pain. Pain Physician 11(1):67–75
107. Eubanks JD, Lee MJ, Cassinelli E, Ahn NU (2007) Prevalence of lumbar facet arthrosis and its relationship to age, sex, and race: an anatomic study of cadaveric specimens. Spine (Phila Pa 1976) 32(19):2058–2062. doi:10.1097/BRS.0b013e318145a3a9
108. Beaman DN, Graziano GP, Glover RA, Wojtys EM, Chang V (1993) Substance P innervation of lumbar spine facet joints. Spine (Phila Pa 1976) 18(8):1044–1049
109. Igarashi A, Kikuchi S, Konno S, Olmarker K (2004) Inflammatory cytokines released from the facet joint tissue in degenerative lumbar spinal disorders. Spine (Phila Pa 1976) 29(19):2091–2095

110. El-Bohy AA, Yang KH, King AI (1989) Experimental verification of facet load transmission by direct measurement of facet lamina contact pressure. J Biomech 22(8–9):931–941
111. Yamashita T, Cavanaugh JM, El-Bohy A, Getchell TV, King AI (1990) Mechanosensitive afferent units in the lumbar facet joint. J Bone Joint Surg Am 72A(6):865–880
112. Yamashita T, Cavanaugh JM, Ozaktay AC, Avramov A, Getchell TV, King AI (1993) Effects of substance P on the mechanosensitive units in the lumbar facet joint and adjacent tissue. J Orthop Res 11:205–214
113. Avramov AI, Cavanaugh JM, Ozaktay AC, Getchell TV, King AI (1992) Effects of controlled mechanical loading on group III and IV afferents from the lumbar facet joint: an in vitro study. J Bone Joint Surg Am 74-A(10):1464–1471
114. Berberich P, Hoheisel U, Mense S (1988) Effects of a carrageenan-induced myositis on the discharge properties of group III and IV muscle receptors in the cat. J Neurophysiol 59(5):1395–1409
115. Grigg P, Schaible HG, Schmidt RF (1986) Mechanical sensitivity of group III and IV afferents from posterior articular nerve in normal and inflamed cat knee. J Neurophysiol 55(4):635–643
116. Rang HP, Bevan S, Dray A (1991) Chemical activation of nociceptive peripheral neurones. Br Med Bull 47(3):534–548
117. Ozaktay AC, Cavanaugh JM, Blagoev D, Getchell TV, King AI (1994) Effects of carrageenan induced inflammation in rabbit lumbar facet joint capsule and adjacent tissue. Neurosci Res 20(4):355–364
118. Yaksh TL (1988) Substance P release from knee joint afferent terminals: modulation by opioids. Brain Res 458(2):319–324
119. Coderre TJ, Katz J, Vaccarino AL, Melzack R (1993) Contribution of central neuroplasticity to pathological pain: review of clinical and experimental evidence. Pain 52(3):259–285
120. El-Bohy A, Cavanaugh JM, Getchell ML, Bulas T, Getchell TV, King AI (1988) Localization of substance P and neurofilament immunoreactive fibers in the lumbar facet joint capsule and supraspinous ligament of the rabbit. Brain Res 460(2):379–382
121. Panjabi MM (2003) Clinical spinal instability and low back pain. J Electromyogr Kinesiol 13(4):371–379
122. Atlasi MA, Mehdizadeh M, Bahadori MH, Joghataei MT (2009) Morphological identification of cell death in dorsal root ganglion neurons following peripheral nerve injury and repair in adult rat. Iran Biomed J 13(2):65–72
123. Gladman SJ, Ward RE, Michael-Titus AT, Knight MM, Priestley JV (2010) The effect of mechanical strain or hypoxia on cell death in subpopulations of rat dorsal root ganglion neurons in vitro. Neuroscience 171(2):577–587. doi:10.1016/j.neuroscience.2010.07.009
124. Ma J, Novikov LN, Wiberg M, Kellerth JO (2001) Delayed loss of spinal motoneurons after peripheral nerve injury in adult rats: a quantitative morphological study. Exp Brain Res 139(2):216–223
125. Chen C, Cavanaugh JM, Song Z, Takebayashi T, Kallakuri S, Wooley PH (2004) Effects of nucleus pulposus on nerve root neural activity, mechanosensitivity, axonal morphology, and sodium channel expression. Spine (Phila Pa 1976) 29(1):17–25. doi:10.1097/01.brs.0000096675.01484.87
126. Matzner O, Devor M (1994) Hyperexcitability at sites of nerve injury depends on voltage-sensitive Na+ channels. J Neurophysiol 72(1):349–359
127. Navarro X, Vivo M, Valero-Cabre A (2007) Neural plasticity after peripheral nerve injury and regeneration. Prog Neurobiol 82(4):163–201. doi:10.1016/j.pneurobio.2007.06.005
128. Kuner R (2010) Central mechanisms of pathological pain. Nat Med 16(11):1258–1266. doi:10.1038/nm.2231
129. Adrian ED, Zotterman Y (1926) The impulses produced by sensory nerve-endings: Part II. The response of a single end-organ. J Physiol 61:151–171
130. Cuellar JM, Montesano PX, Antognini JF, Carstens E (2005) Application of nucleus pulposus to L5 dorsal root ganglion in rats enhances nociceptive dorsal horn neuronal windup. J Neurophysiol 94(1):35–48. doi:10.1152/jn.00762.2004
131. Takebayashi T, Cavanaugh JM, Cuneyt Ozaktay A, Kallakuri S, Chen C (2001) Effect of nucleus pulposus on the neural activity of dorsal root ganglion. Spine (Phila Pa 1976) 26(8):940–945
132. Smyth MJ, Wright V (1958) Sciatica and the intervertebral disc; an experimental study. J Bone Joint Surg Am 40-A(6):1401–1418
133. Howe JF, Loeser JD, Calvin WH (1977) Mechanosensitivity of dorsal root ganglia and chronically injured axons: a physiological basis for the radicular pain of nerve root compression. Pain 3(1):25–41
134. Chen C, Cavanaugh JM, Ozaktay AC, Kallakuri S, King AI (1997) Effects of phospholipase A2 on lumbar nerve root structure and function. Spine (Phila Pa 1976) 22(10):1057–1064
135. Moldovan M, Alvarez S, Romer Rosberg M, Krarup C (2013) Axonal voltage-gated ion channels as pharmacological targets for pain. Eur J Pharmacol 708 (1–3):105–112. doi:10.1016/j.ejphar.2013.03.001
136. Devor M, Govrin-Lippmann R, Angelides K (1993) Na+ channel immunolocalization in peripheral mammalian axons and changes following nerve injury and neuroma formation. J Neurosci 13(5):1976–1992
137. Olmarker K, Nordborg C, Larsson K, Rydevik B (1996) Ultrastructural changes in spinal nerve roots induced by autologous nucleus pulposus. Spine (Phila Pa 1976) 21(4):411–414
138. Rabin A, Gerszten PC, Karausky P, Bunker CH, Potter DM, Welch WC (2007) The sensitivity of the seated straight-leg raise test compared with the

supine straight-leg raise test in patients presenting with magnetic resonance imaging evidence of lumbar nerve root compression. Arch Phys Med Rehabil 88(7):840–843. doi:10.1016/j.apmr.2007.04.016
139. Goddard MD, Reid JD (1965) Movements induced by straight leg raising in the lumbo-sacral roots, nerves and plexus, and in the intrapelvic section of the sciatic nerve. J Neurol Neurosurg Psychiatry 28(1):12–18
140. Charnley J (1951) Orthopaedic signs in the diagnosis of disc protrusion. With special reference to the straight-leg-raising test. Lancet 1(6648):186–192
141. Chen C, Virk G, Yaldo J, Tanimoto K, Kallakuri S, Cavanaugh JM (2012) Determination of spinal nerve injury tolerance to stretch. J Neurotrauma 29:A79–A79
142. Singh A, Kallakuri S, Chen C, Cavanaugh JM (2009) Structural and functional changes in nerve roots due to tension at various strains and strain rates: an in-vivo study. J Neurotrauma 26(4):627–640. doi:10.1089/neu.2008.0621
143. Singh A, Lu Y, Chen C, Cavanaugh JM (2006) Mechanical properties of spinal nerve roots subjected to tension at different strain rates. J Biomech 39(9):1669–1676. doi:10.1016/j.jbiomech.2005.04.023
144. Singh A, Lu Y, Chen C, Kallakuri S, Cavanaugh JM (2006) A new model of traumatic axonal injury to determine the effects of strain and displacement rates. Stapp Car Crash J 50:601–623
145. Bono CM (2004) Low-back pain in athletes. J Bone Joint Surg Am 86-A(2):382–396
146. Opar DA, Williams MD, Shield AJ (2012) Hamstring strain injuries: factors that lead to injury and re-injury. Sports Med 42(3):209–226. doi:10.2165/11594800-000000000-00000
147. Garrett WE Jr (1996) Muscle strain injuries. Am J Sports Med 24(6 Suppl):S2–S8
148. Solomonow M (2006) Sensory-motor control of ligaments and associated neuromuscular disorders. J Electromyogr Kinesiol 16(6):549–567. doi:10.1016/j.jelekin.2006.08.004
149. Petrie S, Collins JG, Solomonow M, Wink C, Chuinard R, D'Ambrosia R (1998) Mechanoreceptors in the human elbow ligaments. J Hand Surg Am 23(3):512–518. doi:10.1016/s0363-5023(05)80470-8
150. Chen C, Lu Y, Kallakuri S, Patwardhan A, Cavanaugh JM (2006) Distribution of A-delta and C-fiber receptors in the cervical facet joint capsule and their response to stretch. J Bone Joint Surg Am 88(8):1807–1816. doi:10.2106/jbjs.e.00880
151. Schaible HG, Schmidt RF (1983) Activation of groups III and IV sensory units in medial articular nerve by local mechanical stimulation of knee joint. J Neurophysiol 49(1):35–44
152. Lu Y, Chen C, Kallakuri S, Patwardhan A, Cavanaugh JM (2005) Neural response of cervical facet joint capsule to stretch: a study of whiplash pain mechanism. Stapp Car Crash J 49:49–65
153. Azar NR, Kallakuri S, Chen C, Lu Y, Cavanaugh JM (2009) Strain and load thresholds for cervical muscle recruitment in response to quasi-static tensile stretch of the caprine C5-C6 facet joint capsule. J Electromyogr Kinesiol 19(6):e387–e394. doi:10.1016/j.jelekin.2009.01.002
154. Azar NR, Kallakuri S, Chen C, Cavanaugh JM (2011) Muscular response to physiologic tensile stretch of the caprine c5/6 facet joint capsule: dynamic recruitment thresholds and latencies. Stapp Car Crash J 55:441–460
155. Lis AM, Black KM, Korn H, Nordin M (2007) Association between sitting and occupational LBP. Eur Spine J 16(2):283–298
156. Jonsson E, Nachemson A (2000) Collected knowledge about back pain and neck pain. What we know–and what we don't know. Lakartidningen 97(44):4974–4980
157. Bovenzi M (1996) Low back pain disorders and exposure to whole-body vibration in the workplace. Semin Perinatol 20(1):38–53
158. Seidel H (1993) Selected health risks caused by long-term, whole-body vibration. Am J Ind Med 23(4):589–604
159. Bongers PM, Hulshof CT, Dijkstra L, Boshuizen HC, Groenhout HJ, Valken E (1990) Back pain and exposure to whole body vibration in helicopter pilots. Ergonomics 33(8):1007–1026. doi:10.1080/00140139008925309
160. Shanahan DF, Reading TE (1984) Helicopter pilot back pain: a preliminary study. Aviat Space Environ Med 55(2):117–121
161. Froom P, Hanegbi R, Ribak J, Gross M (1987) Low back pain in the AH-1 Cobra helicopter. Aviat Space Environ Med 58(4):315–318
162. Desmedt JE (1983) Mechanisms of vibration-induced inhibition or potentiation: tonic vibration reflex and vibration paradox in man. Adv Neurol 39:671–683
163. Vermeersch D, Vermeersch L, Vermeersch G (1986) The tonic vibration reflex of the musculus quadriceps femoris can be used to measure the change in tonus of the postural type. Electromyogr Clin Neurophysiol 26(7):481–487
164. Richter J, Meister A, Bluethner R, Seidel H (1988) Subjective evaluation of isolated and combined exposure to whole-body vibration and noise by means of cross-modality matching. Act Nerv Super (Praha) 30(1):47–51
165. Wakeling JM, Nigg BM (2001) Modification of soft tissue vibrations in the leg by muscular activity. J Appl Physiol 90(2):412–420
166. Ng GY, Cheng JM (2002) The effects of patellar taping on pain and neuromuscular performance in subjects with patellofemoral pain syndrome. Clin Rehabil 16(8):821–827
167. Taimela S, Kankaanpaa M, Luoto S (1999) The effect of lumbar fatigue on the ability to sense a change in lumbar position. A controlled study. Spine (Phila Pa 1976) 24(13):1322–1327

168. Dupuis H, Zerlett G (1987) Whole-body vibration and disorders of the spine. Int Arch Occup Environ Health 59(4):323–336
169. Dupuis H (1994) Medical and occupational preconditions for vibration-induced spinal disorders: occupational disease no. 2110 in Germany. Int Arch Occup Environ Health 66(5):303–308
170. Tiemessen IJ, Hulshof CT, Frings-Dresen MH (2008) Low back pain in drivers exposed to whole body vibration: analysis of a dose-response pattern. Occup Environ Med 65(10):667–675. doi:10.1136/oem.2007.035147
171. Frymoyer JW, Pope MH, Costanza MC, Rosen JC, Goggin JE, Wilder DG (1980) Epidemiologic studies of low-back pain. Spine (Phila Pa 1976) 5(5):419–423
172. Lings S, Leboeuf-Yde C (2000) Whole-body vibration and low back pain: a systematic, critical review of the epidemiological literature 1992–1999. Int Arch Occup Environ Health 73(5):290–297
173. Wawrow PT, Jakobi JM, Cavanaugh JM (2011) Fatigue response of rat medial longissimus muscles induced with electrical stimulation at various work/rest ratios. J Electromyogr Kinesiol 21(6):939–946. doi:10.1016/j.jelekin.2011.08.007
174. Mense S (2003) The pathogenesis of muscle pain. Curr Pain Headache Rep 7(6):419–425
175. Keay KA, Bandler R (1993) Deep and superficial noxious stimulation increases Fos-like immunoreactivity in different regions of the midbrain periaqueductal grey of the rat. Neurosci Lett 154(1–2):23–26
176. Berkowitz S (1978) The induction of II-III translocations by tris-(2,3-dibromopropyl) phosphate in Drosophila. Mutat Res 57(3):385–387
177. Svensson P, Minoshima S, Beydoun A, Morrow TJ, Casey KL (1997) Cerebral processing of acute skin and muscle pain in humans. J Neurophysiol 78(1):450–460
178. Ding Y, Cesare P, Drew L, Nikitaki D, Wood JN (2000) ATP, P2X receptors and pain pathways. J Auton Nerv Syst 81(1–3):289–294
179. Sluka KA, Kalra A, Moore SA (2001) Unilateral intramuscular injections of acidic saline produce a bilateral, long-lasting hyperalgesia. Muscle Nerve 24(1):37–46
180. Woolf CJ, Salter MW (2000) Neuronal plasticity: increasing the gain in pain. Science 288(5472):1765–1769
181. Bovenzi M (2009) Metrics of whole-body vibration and exposure-response relationship for low back pain in professional drivers: a prospective cohort study. Int Arch Occup Environ Health 82(7):893–917. doi:10.1007/s00420-008-0376-3
182. Chen C, Cheng B, Wang Z, Chen D, Tao X, Cavanaugh JM (2013) Back muscle activity while operating a vehicle. In: Proceedings of the FISITA 2012 world automotive congress, vol 197. Lecture notes in electrical engineering. Springer, Berlin/Heidelberg, pp 801–811. doi:10.1007/978-3-642-33805-2_65
183. Ramirez A, Grasa J, Alonso A, Soteras F, Osta R, Munoz MJ, Calvo B (2010) Active response of skeletal muscle: in vivo experimental results and model formulation. J Theor Biol 267(4):546–553. doi:10.1016/j.jtbi.2010.09.018
184. Tao X, Chen C, Cheng B, Wang Z, Wang W, Cavanaugh JM (2012) Characterization of muscle contraction force, electromyogram and fatigue in response to electrical stimuli – a preliminary study. Paper presented at The 2012 meeting of the Biomedical Engineering Society, Atlanta
185. Tao X, Cheng B, Wang W, Zhang F, Li G, Chen C, Cavanaugh JM (2012) SEMG based prediction for lumbar muscle fatigue during prolonged driving. Paper presented at The 34th FISITA world automotive congress
186. Lewicki MS (1998) A review of methods for spike sorting: the detection and classification of neural action potentials. Network 9(4):R53–R78
187. Snider RK, Bonds AB (1998) Classification of non-stationary neural signals. J Neurosci Methods 84(1–2):155–166
188. Chen C, Lu Y, Cavanaugh JM, Kallakuri S, Patwardhan A (2005) Recording of neural activity from goat cervical facet joint capsule using custom-designed miniature electrodes. Spine (Phila Pa 1976) 30(12):1367–1372
189. Lu Y, Chen C, Kallakuri S, Patwardhan A, Cavanaugh JM (2005) Development of an in vivo method to investigate biomechanical and neurophysiological properties of spine facet joint capsules. Eur Spine J 14(6):565–572. doi:10.1007/s00586-004-0835-9
190. Lu Y, Chen C, Kallakuri S, Patwardhan A, Cavanaugh JM (2005) Neurophysiological and biomechanical characterization of goat cervical facet joint capsules. J Orthop Res 23(4):779–787. doi:10.1016/j.orthres.2005.01.002
191. Wyke B (1979) Neurology of the cervical spinal joints. Physiotherapy 65(3):72–76

Role of Muscles in Accidental Injury

21

Gunter P. Siegmund, Dennis D. Chimich, and Benjamin S. Elkin

Abstract

Skeletal muscle is the most abundant tissue in the human body and can play various roles in the context of accidental injuries. First, muscles make up 38 ± 5 % of male total body mass and 31 ± 6 % of female total body mass (Janssen et al., J Appl Physiol 89(1):81–88, 2000), and thus represent a considerable proportion of the body's inertia. Second, muscles provide padding to many bones and other tissues, and thus can attenuate impacts to the body. Third, muscles generate forces within the body that alter the load state of other tissues during an impact. And finally, muscles themselves can be injured by impacts to the body. Despite these varied roles, muscles are often ignored in the study of accidental injury. For some types of accidental injury, muscles indeed contribute little or nothing to the injury mechanism. For other types of injury, however, muscle forces can exacerbate, mitigate and sometimes even cause specific injuries.

The goal of this chapter is to review our current understanding of how skeletal muscles affect accidental injury. Our focus is on traumatic injuries, but we address chronic or overuse injuries where they contribute to the understanding of traumatic injuries. We begin with a brief overview of muscle mechanics and then examine the role of muscles on injuries to various anatomic regions, including the head, spine, upper extremity and lower extremity. We close with a consideration of how muscle activation affects whole body motion and traumatic injury patterns in general.

G.P. Siegmund, Ph.D. (✉)
MEA Forensic Engineers & Scientists, Richmond, BC, Canada

School of Kinesiology, University of British Columbia, Vancouver, BC, Canada
e-mail: gunter.siegmund@meaforensic.com

D.D. Chimich, M.Sc. • B.S. Elkin, Ph.D.
MEA Forensic Engineers & Scientists, Richmond, BC, Canada
e-mail: dennis.chimich@meaforensic.com; benjamin.elkin@meaforensic.com

21.1 Muscle Mechanics

21.1.1 Structure and Function of Skeletal Muscle

Muscles move the body by pulling on bones to generate moments about joints. At a microscopic level, skeletal muscle, or voluntary muscle, consists of an interdigitated matrix of myosin and actin filaments arranged into contractile units called sarcomeres. A string of sarcomeres form a myofibril, and a bundle of myofibrils form a muscle fiber, which is considered a single cell with multiple nuclei. A bundle of muscle fibers then form a fascicle, and multiple fascicles form a muscle. Human sarcomeres have an optimal length (L_0) of 2.65 μm and on average operate over lengths varying from 0.71 L_0 to 1.24 L_0 [1]. Fiber diameters vary widely (e.g. 20–100 μm) even within the same muscle [2] and large muscles can contain hundreds of thousands of muscle fibers. For example, there are 200,000–900,000 fibers (depending on age) in the human vastus lateralis muscle [3], and a million or more fibers in the medial gastrocnemius muscle [4].

Muscle activation is controlled by motor neurons that reside in the spinal cord for most muscles or in the brain stem for facial, eye and some neck muscles. A single motor neuron controls between 5 fibers (eye muscles) [5] and about 2,000 fibers (gastrocnemius muscle) [4]. The collection of all muscle fibers controlled by a single motor neuron is called a motor unit. A motor unit is the smallest functional unit of muscle controlled by the nervous system. Based on data from non-human primates, whole muscles contain from 10 to 1,500 motor units [6], with the number of motor units and the number of fibers within each motor unit determining the degree of control within a specific muscle.

When a motor neuron discharges, all fibers within the motor unit receive an electrical signal called an action potential. This electrical current is propagated to each sarcomere in the motor unit and causes it to shorten. Shortening across many sarcomeres causes the muscle to contract, which shortens its length and increases its diameter. The nervous system controls the magnitude of the force a muscle generates by varying both the discharge frequency of a motor neuron and the number of motor units it recruits. Small motor units are recruited first and offer fine motor control at low force levels, whereas large motor units are recruited later and offer large force generation capabilities [7]. This orderly recruitment of small to large motor units appears to be maintained during both voluntary and reflex muscle activations [8].

Muscle fibers and fascicles are held together by connective tissues that transfer the muscle forces to tendons that then attach to bone. Bones connect to each other at joints, and muscle forces generate moments about these joints that then lead to rotations and translations of adjacent body segments.

21.1.2 Force Generation

A muscle's architecture, i.e. the arrangement of its fibers relative to the axis of force generation, determines the force a muscle can produce [9]. Longitudinally arranged muscles have fibers that run parallel to its force-generating axis, whereas uni-pennate and multi-pennate muscles have fibers that run at one or more angles relative to its force-generating axis. The angle between a muscle's fibers and its force-generating axis is called the pennation angle. Pennation angles are generally less than 10° in the limb muscles, but there are exceptions (e.g. ~15° in semimembranosus, ~17° in medial gastrocnemius, and ~25° in soleus) [9–12]. Pennation angles vary from 0° to 30° in neck muscles [13].

The maximum contraction force a muscle can generate is a function of the number and diameter of its muscle fibers, the pennation angle (θ), the activation level (a varying between 0 and 1), and the specific tension (σ) of the muscle. The number of fibers, their diameter and the muscle pennation angle are typically combined into an effective area known as the physiologic cross-sectional area (PCSA) of the muscle (Eqs. 21.1 and 21.2). PCSA is not a real cross sectional area of the muscle, but rather the effective area of a theoretical cylindrical muscle with a length equal to the fiber length. Practically, PCSA is determined by measuring a muscle's mass (M), fiber length (l) and pennation angle, and assuming a density (ρ) of 1.06 g/cm^3 for mammalian muscle [14]. The specific tension of muscles is typically estimated to be between 0.33 and 0.50 MPa [12, 15], although more recent in vivo estimates place it between 0.55 and 0.60 MPa [16].

$$PCSA = \frac{M\cos\theta}{\rho l} \quad (21.1)$$

$$F = a \times \sigma \times PCSA \quad (21.2)$$

Three other factors affect the force a muscle functionally produces. First, the interdigitated structure of sarcomeres means that there is a varying degree of overlap between the actin and myosin molecules. At its optimum length (L_0), all of the available binding sites are available and the sarcomere can generate its maximum force. At both shorter and longer lengths, progressively fewer binding sites are available and a sarcomere's force-generating capacity progressively diminishes (see active tension curve in Fig. 21.1a).

Second, the connective tissues within the muscle generate passive resistance to lengthening similar to any tissue placed in tension (see passive tension curve in Fig. 21.1a). The force generating properties depicted in Fig. 21.1a are for isometric conditions, i.e. when muscle fiber length remains constant. Note that muscle fibers in a contracting muscle spanning a fixed joint angle actually shorten due to lengthening of the tendon, and therefore a fixed joint angle is not truly isometric from the muscle fiber's perspective.

Third, the speed and direction of a muscle's contraction can attenuate or amplify a muscle's force. The chemical processes at the binding sites limit the speed that a muscle can shorten, with the force varying inversely with shortening velocity (Fig. 21.1b). The maximum shortening velocity (V_{max}) is often taken to be 10 L_0/s [17], however others have reported maximum shortening velocities between 1 and 2 L_0/s [18] and variations in maximum shortening velocity with activation level [19]. In contrast to shortening, a lengthening muscle can generate a force greater than its peak isometric force (Fig. 21.1b). This amplification occurs because the force to physically break the bond between the myosin and actin binding site is greater than the force the myosin molecule can generate during the conformational change it undergoes while actively contracting.

The net effect of these length and velocity dependencies is the Force-Length-Velocity (FLV) curve shown in Fig. 21.1c. This normalized surface is used to scale the force values obtained from Eq. 21.1. Techniques for implementing these properties of muscle into computational models can be found in Zajac [17].

Separation of muscle force into active and passive components is a simplification, and more recent work suggests that the passive force component depends on the level of muscle activation [20] and on the muscle contractile history [21].

21.1.3 Muscle Activation Timing

Muscles must be active to generate large forces over their normal operating lengths. In the context of their potential to affect accidental injuries, this means that muscles must be either active before impact, i.e. pre-impact bracing, or activated rapidly during or immediately following impact. There is no debate that a braced muscle is active during impact and can thus alter the forces applied to other tissues during impact; however, there has been debate whether an initially relaxed muscle can be reflexively activated by an impact and then generate sufficient force to affect an injury caused by the impact. Also encapsulated in this latter debate is whether a braced muscle's force can further increase early enough to affect injury.

Reflex muscle activation can be viewed as three sequential steps (Fig. 21.2). The first step precedes muscle activation and is the interval between the onset of a stimulus and the onset of electrical activity in a muscle. This time interval, called the reflex time, is governed by properties of the stimulus, the sensory organs and the sensorimotor pathways involved. The second step is the time between the onset of electrical activity in the muscle and the onset of muscle force. This time interval, called the electromechanical delay (EMD), is a function of muscle and tendon properties, and is sensitive to how it is measured. The final step is the time between the onset of muscle force and peak muscle force.

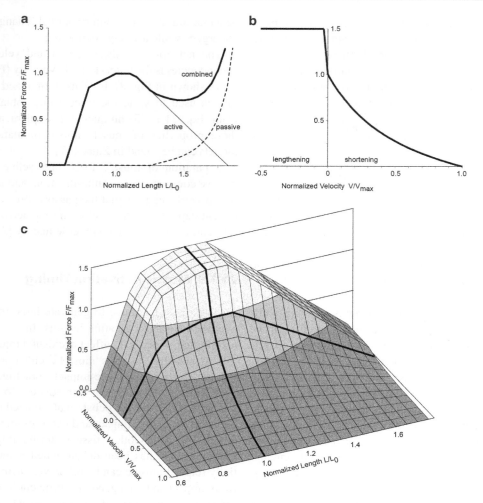

Fig. 21.1 Force, length and velocity relationships for muscle. (**a**) Normalized force curve showing active (*thin line*), passive (*dashed line*) and combined force (*thick line*) in maximally activated muscle as a function of the normalized length. (**b**) Normalized force as a function of the normalized shortening/lengthening velocity in maximally activated muscle. (**c**) Surface representing the normalized force as a function of both normalized length and normalized velocity for maximally activated muscle. Lower levels of activation will produce forces below the surface. F_{max}, maximum force; L_0, optimum length; and V_{max}, maximum shortening velocity

The reflex times for many axial and appendicular muscles vary from 55 to 100 ms for stimuli varying from a rear-end car crash, supine free fall and acoustic startle [22–25]. These reflex times are shorter than voluntary onset latencies, which, for comparison, are about 105–110 ms in the sternocleidomastoid muscle [26, 27] and 128–130 ms in the tibialis anterior and soleus muscles [28].

Reflex times can shorten slightly with increasing stimulus intensity [23], but there is a minimum reflex time below which increases in stimulus intensity have no further shortening effect. This floor is set by physical constraints like nerve conduction velocities and synapse times. Reflex times can also be shorter when multiple sensory modalities are stimulated. For instance, neck muscle onset latencies during a rear-end collision without noise were about 15 ms longer than when the same collision pulse was presented simultaneously with a loud tone [29]. Onset latencies in the neck muscles are also slightly shorter in females than males [23, 25].

EMD times between 13 and 95 ms have been reported in the literature [30–32]. The large range

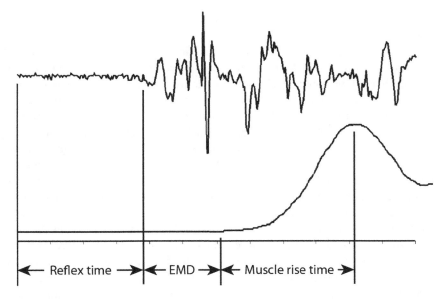

Fig. 21.2 Schematics showing the relationship between electrical activity in a muscle (*upper trace*) and the production of force by that muscle (*lower trace*). The stimulus triggering the muscle response occurs at the start of the traces

stems from different ways of measuring its endpoint. Some authors used the onset of movement (video or contact release), which results in long EMD times, whereas others have used the onset of acceleration, which can be detected earlier than the onset of movement and thus results in shorter EMD times. Corcos et al. [33] showed that minimizing measurement-induced artifact by using an accelerometer mounted over a bony prominence yielded EMD times of 13.2 ± 1.1 ms.

Muscle rise times are governed by the intensity of the contraction and the inertia, damping and stiffness of the system the muscle is attempting to move or react against. As a result, there is no estimate of muscle rise time that can be universally applied. However, if the delay between peak electromyographic activity and peak force is similar to the delay between the onset of electromyographic activity and the onset of acceleration, then the rise time of the EMG signal may be a good surrogate for the muscle rise time in a particular situation. During low severity rear-end impacts, the time between sternocleidomastoid (SCM) muscle onset and peak activity (measured using a root mean square of the electromyographic signal) is about 35–43 ms [25, 34].

Using SCM as an example, the total time between bumper contact and peak force in the muscles during a low-speed rear-end impact is about 117–132 ms [25, 30]. As we will see later, this is likely fast enough to affect a whiplash injury, but not fast enough to affect other types of injury.

21.1.4 Muscle Mechanical Properties

Muscle is a difficult tissue to characterize from a mechanical perspective. In addition to being sensitive to loading rates like other soft tissues in the human body, muscle's highly organized structure creates large anisotropy and its contractile ability results in different properties at different levels of contraction.

Numerous experiments have shown that passive muscle is an anisotropic, highly non-linear viscoelastic material [35–37]. At low strain rates of 0.0005 s^{-1} in fresh porcine muscle, compression across the fiber direction yielded stresses that were 2–2.5 times higher than those along the fiber direction at 30 % strain [36]. As strain rates increased from 0.0005 to 0.10 s^{-1}, the stress at 30 % strain along the fibers increased more rapidly than across the fibers, ultimately reaching 1.3 times the stress across the fibers at 0.10 s^{-1} [37].

At compressive strain rates of 700–3,700 s^{-1}, passive bovine muscle was again highly rate sensitive [38–40]. Van Sligtenhorst et al. [40] reported no significant difference in the response between loading along and transverse to the fiber direction at rates of 700–1,250 s^{-1}, although greater difficulty in producing the transverse specimens and greater variability in their results may have contributed to this finding. Song et al. [39] on the other hand reported greater strain rate sensitivity along the fiber direction than across it, particularly at the very high strain rates.

Intermediate strain rates are perhaps more relevant to many typical impact scenarios. For instance, about half of all pedestrian impacts occur at car speeds below 25 km/h (6.9 m/s) and about 80 % occur at car speeds below 40 km/h (11.2 m/s) [41, 42]. Muscle tissue overlying the lateral and posterior surfaces of the lower limb is 25–30 mm and 50–60 mm thick respectively [43] and thus yields strain rates of 115–448 s^{-1}. Compression tests on fresh human muscle at strain rates of 136–262 s^{-1} showed significant rate effects across even this narrow range of strain rates [44].

Fewer studies have examined the three dimensional properties of muscle in tension. In passive rabbit muscles lengthened quasi-statically at 0.0005 s^{-1}, higher linear moduli and lower strains to failure were seen along the fiber direction than across it [45]. This relative stiffness was opposite to that observed in compression, where the modulus was greater across the fiber direction than along the fiber direction at similarly low strain rates [36, 37] and is consistent with a transversely isotropic incompressible material.

Much of these data are generated from ex-vivo tissue, and therefore the effect of pressurized vasculature was not included. In vivo data is generally limited to low strain rates and unidirectional loading [35, 46]. Muscle tissue properties also change considerably during rigor, however, post-rigor properties were not significantly different from pre-rigor properties at both low strain rates [47] and high strain rates [40].

21.1.5 Muscle Injury

Functional injury to muscle-tendon units occurs over a wide range of strains. Direct injury to the sarcomeres occurs from eccentric contractions; i.e. imposed lengthening during active contraction. Lengthening strains of 5–20 % in active muscle have been shown to cause injury in animal studies [48, 49]. These injuries consisted of a loss of actin and myosin interdigitation at the microscopic level. More macroscopically, damage at or near the muscle tendon junctions has been observed at about 25 % strain in passive muscle [50]. These injuries result in small focal areas of muscle fiber rupture and hemorrhage near the distal myotendinous junction. Complete rupture of passively stretched muscle also occurs at the distal myotendinous junction at strains varying from 73 % to 225 % depending on the specific muscle [51]. Increasing complexity in muscle architecture (fusiform, uni-pennate, bi-pennate and multi-pennate) was associated with increasing strains to rupture. In active muscle, the rupture force increased by about 15 %, but the strain to failure and failure site did not vary [52]. These data indicate that a muscle's passive force rather than its active force dominates its failure [53]. All of the tests described above were performed at strain rates at or below 0.7 s^{-1}. At higher strain rates (~6 s^{-1}), ruptures in passive muscle have also been reported in the muscle belly [54].

Direct injury to muscle can also occur from blunt trauma. Tensed muscle distributes the impact force more broadly than relaxed muscle [55]. In simulated bumper impacts into the lower leg at 2.5 m/s using ~1.8 kg impactors, larger muscle compression occurred in relaxed muscle than in tensed muscle for posterior impacts to the gastrocnemius/soleus muscles, whereas no difference was seen between relaxed and tensed muscle in lateral impacts over the thinner peroneus longus and/or extensor digitorum longus muscles [43]. These findings suggest that muscle could have a protective effect on the underlying bone, but this protective effect varies with muscle thickness and level of activity.

21.2 Muscle's Effect on Accidental Injury

Since peak muscle forces are less than the peak inertial or contact forces that can develop in many impact situations, we expect muscle forces, whether from pre-impact bracing or reflex activation, to have a greater effect on injury in low severity events than in high severity events. Moreover, given the large proportion of total body mass made up of muscle and the fact that muscle is not rigidly coupled to the skeleton, the amount and activation level of muscle can alter the effective mass of the body during an impact. This behavior differs from fat and other tissues whose effective mass cannot be voluntarily changed.

21.2.1 Head and Brain Injuries

Head and brain injuries generally occur as a result of head impacts. Increasing injury frequency and severity are generally related to increasing impact force and the induced kinematics. Neck muscle forces can alter the head kinematics, but the magnitude of the neck muscle forces in relation to the head mass (~3.5–4.5 kg) suggest that neck muscles will have the greatest effect at the least severe end of the injury spectrum, i.e., concussions. Moreover, since head impacts typically have durations (5–15 ms) shorter than neck muscle reflex latencies, neck muscles can play a role only if tensed before impact.

Sport-related concussion occurs more frequently in females than males [56, 57]. One potential explanation for this phenomenon is the greater neck muscle strength in males than females [58]. This hypothesis is supported by a model showing that increased neck stiffness (achieved by increased neck muscle strength and activation) can attenuate head kinematics, which could reduce the risk of concussion [59].

If strong, active neck muscles can attenuate peak head kinematics, then players with strong necks who are anticipating an impact might be expected to experience lower severity head impacts than players with weak necks who are not anticipating the impact. Static neck strength measured in flexion, extension, anterolateral flexion, posterolateral flexion and axial rotation did not correlate with peak head linear or angular acceleration in youth hockey players [60]. Anticipation, however, decreased angular head acceleration by about 250 rad/s^2 in medium severity head impacts (50th to 75th percentile severity) in male youth ice hockey (14 years old), but a similar reduction was not observed in the most severe (upper quartile) head impacts [61].

Increased neck strength brought about by resistance training has also failed to reduce peak head acceleration. No changes in displacement or angular acceleration to forced flexion and extension of the head was observed in male and female soccer players after an 8 week training program that increased neck strength [62]. Other work in college-aged males showed that 7–10 % increases in neck strength from an 8-week training program has no effect on linear or angular head acceleration when tackling a standard football tackling dummy [63]. Acceleration data in both studies were obtained by double differentiating motion data acquired at 120 Hz, and therefore may not accurately reflect actual accelerations. Nonetheless, acceleration levels in both studies were low compared to sport-related head impacts that have caused concussion, and it is at these low impact levels that neck muscles would likely have the largest effect.

Based on current data, the postulated benefit of strong neck muscles on reducing the risk of concussion has not been detected. No studies were found that showed a role for neck muscles in reducing moderate or severe head/brain injuries.

21.2.2 Spine Injuries

Muscles are an important functional part of the spine. Muscles insert on or originate from every spinal level, with some deep muscles spanning only adjacent vertebrae and some superficial muscles spanning many vertebrae. Muscle activity is needed to achieve the dynamic equilibrium required for upright activities [64] and to stabilize the spine [65].

Muscle can in some cases cause spinal fractures in the absence of external forces. Vertebral fractures have been reported in up to 16 % of epileptic seizures [66], and although it remains unclear whether some fractures are due to the muscle contraction or a related fall, the circumstances of some cases indicate that spinal muscles alone can cause compression and burst fractures [67].

21.2.2.1 Cervical Spine

Muscles comprise the majority of the neck's volume (Fig. 21.3). The superficial muscles, such as sternocleidomastoid (SCM) or trapezius, attach to the skull, shoulder girdle, and ligamentum nuchae but do not generally attach directly to the cervical vertebrae. Deeper muscles, such as splenius, semispinalis, longissimus, scalenes, and longus, attach on multiple cervical vertebrae. The deepest neck muscles, the multifidus muscles, also insert directly on the facet capsule of cervical vertebrae and may be relevant to injury of the capsular ligaments [68, 69]. Most neck muscles have complex architecture, with extensive internal tendons [13] and a high density of muscle spindles [70].

Neck muscles can potentially affect the genesis of whiplash injury in three ways. First, direct attachment of the multifidus muscles to the capsular ligament [68, 69] combined with early activation of these muscles in some subjects during a rear-end collision may increase the peak strain in the capsular ligaments [71]. Second, neck muscles are oriented primarily vertically and therefore their activation produces axial compression of the cervical spine. This increases the loads on intervertebral discs and facet joints. And third, reflex muscle activation alters the kinematic response of the head and neck during whiplash-induced motion. In subjects exposed to a series of identical perturbations without a head restraint, habituation of the muscle response amplitude by about 50 % was accompanied by a 20 % increase in neck extension and a 30 % decrease in head acceleration [72]. Whether the higher muscle activation levels are harmful or protective to other neck tissues remains unclear.

At higher impact severities, muscles also appear to affect the kinematics and the type and

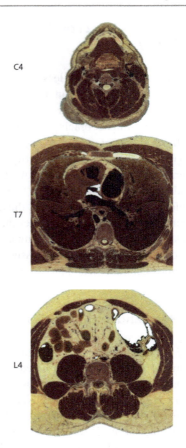

Fig. 21.3 Horizontal cross sections of the Visible Human Male at the *C4*, *T7* and *L4* vertebral levels. Note the greater muscle area surrounding the neck and lumbar spine compared to the thoracic spine

severity of injury. Maximum muscle activation with a reflex time of 25 ms generated better agreement between a comprehensive head/neck model and volunteer data in high severity frontal and lateral collisions [73]. Moreover, a comparison of passive and active muscles in frontal (60 km/h) and lateral (25 km/h) impacts showed that active muscles reduced the strain in many spinal ligaments [74].

Under tensile loading, maximally activated muscles in an LS-DYNA model increased the tolerance of the cervical spine from 1,800 to 4,160 N by providing an alternate load path [47]. More refined activation schemes mimicking relaxed and tensed neck muscles yielded tensile tolerances of 3,100–3,700 N [75]. In both studies, muscle activation shifted the site of injury from the lower cervical spine to the upper cervical spine, an injury site the authors reported was more consistent with those observed clinically.

Under compressive loads, simulated muscle forces also alter the injury behavior of the neck. Axial loading can buckle the spine and cause different types of vertebral fractures and dislocations depending on the line of action [76]. These injuries were produced in inverted drop tests using ligamentous spines without musculature and in a neutral lordotic posture [77]. While neck muscles cannot activate reflexively early enough to affect an injury that occurs within 20 ms of impact, the neck muscles are active and the neck posture may not be neutral following the prolonged vehicle motion leading to a diving head impact in a rollover collision [78]. Earlier inverted drop tests found that cadavers with their necks restrained by cables generating 140–200 N of pre-compression had less head rotation (unquantified) and higher head contact forces (3,000–7,100 N versus 9,800–14,700 N) [79]. More cervical spine fractures occurred in the restrained cadavers than in the unrestrained cadavers, a pattern that suggested muscle restraint could adversely affect injury potential in compressive neck injuries. Head pocketing in padding, which also constrains neck motion, has also been shown to increase cervical spine injuries [80].

In numerical simulations of diving head impacts in rollover collisions, maximally active neck muscles nearly doubled the vertebral fracture risk compared to passive neck muscles [81]. In simulations of sagittal-plane impacts from below (from 50° to 70° below horizontal), maximal neck muscle activation virtually eliminated head flexion at 5 g and reduced it by 50 % and 36 % at 13.5 g and 22 g respectively [82]. Ligament strains increased with full muscle activation during the 5 g impact, likely because the muscle induced ligament strains exceeded the impact induced strains. At the two higher impact severities, maximal muscle activation reduced peak ligament strains, likely by providing an alternate load path.

21.2.2.2 Thoracolumbar Spine

The thoracolumbar spine provides the main load path for compression and shear related to supporting the upper body. While less muscle is present in the portion of the thoracic spine that gains some of its stability from the rib cage, the thoracolumbar spine is again surrounded by considerable muscle (Fig. 21.3). Without muscle, the ligamentous lumbar spine buckles at compressive loads of 80–100 N [33], yet it supports compressive loads of 440–1,000 N during relaxed standing and about 4,000–5,500 N when lifting 10–33 kg at various speeds [83–86]. Simulating the compressive loads applied by muscles between adjacent vertebrae increased the capacity of the ligamentous spine to support a compressive load without buckling to about 1,200 N [87]. Together, these studies indicate that spinal musculature is integral to thoracolumbar spine stability, and that relatively small stabilizing forces between adjacent vertebra, possibly generated by the deep multifidus and rotatores muscles, can increase the stability and load-carrying capacity of the lumbar spine significantly [88, 89].

Cholewicki and McGill [83] showed that the lumbar spine's stability index, which they described as the root average spine stiffness slope in all directions, increased with muscle activation, meaning that the lumbar spine became more stable with increasing external loading. This implied, somewhat paradoxically, that the lumber spine was most vulnerable to instability-induced injuries at low loads; a finding they proposed explains why some lumbar spine injuries occur during seemingly trivial tasks such as picking up a pencil from the floor.

Despite the importance of muscles in lumbar spine function and stability, there is relatively little research on the role of muscles in lumbar spine injuries. Much of the research has instead focused on occupational injuries, and in particular lifting. For lifting injuries, posture, load positioning, and expectation have been shown to affect the loads in the lumbar spine [90], and fatigue and improper muscle activation strategies have been shown to affect the risk of ligament or disc injury [83, 91].

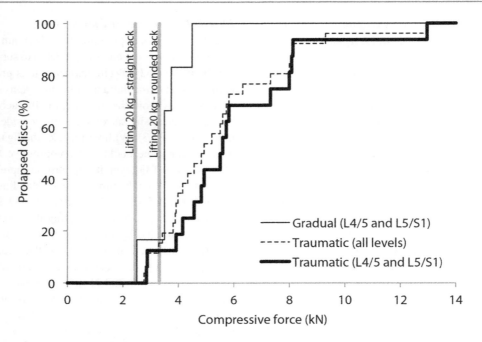

Fig. 21.4 Summary of the compressive forces generating gradual and traumatic disc herniations/prolapses at various lumbar levels [92, 93]. Also shown are the estimated compressive loads at the L4/5 disc level for lifting 20 kg with proper lifting technique (bent legs and straight back) and improper lifting technique (rounded back)

Lumbar disc herniations, protrusions or extrusions can occur gradually or traumatically when the spine is flexed and compressed [92, 93]. Cyclical loading has generated gradual prolapses at loads varying from 2,500 to 4,500 N (Fig. 21.4) [93]. Despite testing vertebrae from all levels, these injuries only occurred at the L4/5 and L5/S1 levels. Traumatic disc herniations occurred at a compression load of 2,760–12,968 N (mean 5,448 ± 2,366 N) [92]. These injuries occurred at all five levels, though most (16 of 26) occurred at the L4/5 and L5/S1 levels (Fig. 21.4). As a result, many researchers have focused on the lumbar compressive loads.

Compressive loads in the lumbar spine can be inferred from intradiscal pressure measurements. Based on stress profilometry measurements [94], the intradiscal pressure is relatively constant through the middle of the disc but falls off over the outer 3–10 mm. Assuming a linear pressure drop over the outer 5 mm of a circular disc of 22.5 mm radius [95], the effective area of the disc over which the mid-disc pressure acts is about 0.80 times its actual area. Using this effective area and intradiscal pressure data for a 70 kg male with an L4/5 disc area of 1,800 mm^2, lumbar compressive loads were 144 N when lying supine, 720 N when standing relaxed, 1,584 N when holding 20 kg close to the body, 2,592 N when holding the same mass 60 cm in front of the chest, 2,448 N when lifting 20 kg properly, and 3,312 N when lifting the same mass with a rounded back [96]. Proper lifting technique involves bending the knees and lifting with a straight back. The difference between the two holding conditions and the difference between the two lifting conditions are due to different levels of back muscle contraction needed to perform each task. These muscle-induced differences are consistent with Takahashi et al. [97], who found that actual disc compression forces when standing upright and flexing forward with a rounded back were 1.43–1.74 times higher than predicted by a theoretical model without muscles. When compared to the lifting loads described above, we see that lifting improperly can generate lumbar loads in a flexed spine within the range shown to traumatically herniate some lumbar discs (Fig. 21.4).

Epidemiologically, acute lumbar disc herniations are not associated with car crashes, though they are associated occupationally with driving a vehicle [98]. Nonetheless, lumbar disc herniations are occasionally attributed to car crashes, particularly relatively minor rear-end crashes. When seated normally in an automobile, the lumbar spine is flexed, though not maximally [99]. The added flexion that arises from a slouched seated posture has not been quantified. Intradiscal pressures at L4/5 and L5/S1 in an 83 kg male were 0.5 bar (0.05 MPa) in an "ideal" automobile seat position, 0.95 bar (0.095 MPa) when the seat pan was adjusted to increase the weight borne by the thighs, and 1.5 bar (0.15 MPa) when the seat pan was adjusted to increase the weight borne by the buttocks [100]. These pressures are low compared to those measured by Wilke et al. [96], who reported 0.10 MPa lying supine, 0.33 MPa when sitting relaxed but erect in an armchair, and 0.27 MPa when "strongly" slouching in the armchair. If the pressures reported by Zenk et al. [100] are correct, they suggest a compressive load in the lumbar spine as low as 72 N (using the same assumptions as above to calculate force from pressure). Inertial lumbar spine compression in a rear-end impact is reportedly less than 870 N over a range of speed changes from 8 to 24 km/h (Gates et al. 2010) [101], although the biofidelity of the lumbar spine of Hydrid III and BioRID II dummies in low and moderate severity rear-end impact remains unproven.

During a rear-end collision, muscle activity is also present in the paralumbar muscles [102] and can compress the lumbar spine. While relaxed and seated in the vehicle, these researchers measured root mean squared (RMS) EMG levels between 1 % and 7 % (mean 2.3 ± 1.6 %) of those observed during a maximal voluntary contraction (MVC). During rear-end impacts with a speed change of 7.5–10 km/h, paralumbar EMG levels varied from 6 % to 67 % of MVC (mean 20 ± 16 %).

Estimating force from EMG amplitude is problematic, but as a first approximation, it can yield some insight into the scale of the problem. The erector spinae has a unilateral PCSA of about 11.3 cm^2 (corrected for pennation angle) [103], which for a specific tension of 0.5 MPa equates to a maximum bilateral tension of 1,128 N. Assuming no contribution from other muscles, the net compressive load on the lumbar spine from the static intradiscal measurements (72 N), the dynamic dummy loads (870 N) and 20 % of the maximum erector spinae force (226 N) sums to 1,168 N. This compression level is below both the lowest compressive load causing a traumatic disc herniation (2,760 N) [92] and the lowest compressive load causing a gradual disc prolapse from cyclical loading (2,500 N) [93]. This simplified analysis suggests that low-speed rear-end collisions are not a likely cause of traumatic lumbar disc herniations, however further work is needed to better understand how an initially slouched posture in a car seat affects the static, dynamic and muscle-related loads during a rear-end collision.

The role of muscles in lumbar spine fractures is not well studied. Transverse process fractures occur in up to 29 % of patients with lumbar fractures and can be caused by either direct trauma or avulsion by excessive contraction of the psoas or quadratus lumborum muscles [104]. Beyond this relatively minor injury and the seizure-related fractures described earlier, there is little research examining the contribution of muscles to lumbar spine fractures.

21.2.3 Upper Extremity Injuries

A chief function of the upper extremity is to move and position the arm and hand. Larger shoulder muscles produce movement of the whole upper extremity while smaller wrist muscles are sufficient to move the relatively lighter hand. Major muscles of the upper extremity include: the rotator cuff muscles which provide stability and assist other shoulder muscles like the deltoid in shoulder motion; the biceps and triceps muscle groups which flex and extend the elbow; and the flexor and extensor muscle groups of the forearm that work individually or together to produce wrist flexion and extension, ulnar and radial deviation, and forearm pronation and supination. The muscles of the upper extremities are innervated by efferent nerve fibers emerging from levels C4 to T1 of the spinal cord.

Traumatic upper extremity injuries commonly include fractures, dislocations, ligament tears, and/

or tendon ruptures. These injuries can occur from direct contact or indirectly through an extended or braced arm and can occur in high- and low-energy events. The upper extremity is also at risk for repetitive or overuse injures such as tendinitis, tendinosis or impingement syndromes.

21.2.3.1 Shoulder

The glenohumeral joint is the primary joint of the shoulder and has the greatest range of motion of any joint in the body [105, 106]. This range of motion comes at the cost of joint stability and injury; the glenohumeral joint is the most frequently dislocated major joint of the body [107, 108]. Glenohumeral joint stability is primarily provided by the complex interaction of static soft tissue stabilizers, like ligaments and capsular structures, and dynamic soft tissue stabilizers, like the surrounding muscles. The ligaments and capsule primarily restrain the joint at the end of the shoulder's motion range and are lax in the mid-range of motion. In contrast, muscles provide dynamic stability throughout the range of shoulder motion but their effect is greatest in the mid-range of motion [109].

Muscles are thought to help stabilize the shoulder joint by five mechanisms: passive muscle tension; joint motion that secondarily tightens the passive ligamentous constraints; activation causing compression of the humeral head into the glenoid concavity (i.e. concavity compression); barrier effect of contracted muscles; and coordinated activation to redirect the joint force through the center of the glenoid surface [110]. Depending on joint position and the direction of injurious loading, different shoulder muscles affect joint stability to different degrees [111].

Shoulder dislocation can occur from direct trauma to the shoulder but more commonly occurs from indirect loading through the humerus [112]. In direct impact to the shoulder, the muscles and soft tissue about the shoulder absorb some of the impact energy and lessen the force applied to the shoulder. However, in lateral shoulder impacts simulating side motor vehicle collisions, the soft tissue of the cadaver shoulders (averaging 6.4 ± 4.2 mm thick) was not sufficient to prevent shoulder injuries like clavicle fracture and sternoclavicular laxity at impact speeds ranging from 13 to 25 km/h [113].

The vast majority of shoulder dislocations occur in the anterior direction from a combination of abduction, extension, and external rotation forces to the arm. These arm motions and forces indirectly load the anterior capsule and ligaments, glenoid rim, and rotator cuff [112]. Shoulder muscle activation stabilizes the joint and resists the anteriorly directed forces. The effectiveness of a shoulder muscle as a stabilizer depends on the magnitude of the muscle force and its line of action relative to the joint center [109]. Shoulder muscle forces can be broken down into joint compression and shear components. Shoulder joint compression increases joint stability by bringing the articular surfaces together whereas shear forces decrease stability by acting to displace the surfaces. For example, in a cadaveric model of anterior shoulder dislocation, passive tension of 609 ± 244 N is generated in the pectoralis major muscle which applies an anterior shear force critical to dislocation [114].

Of particular importance to shoulder stability and injury are the rotator cuff muscles: subscapularis, supraspinatus, infraspinatus, and teres minor [115]. These muscles, together with the intra-articular long head of the biceps muscle, compress the humeral head into the glenoid cavity (Fig. 21.5) [106, 110]. In addition to the cuff muscles, the outer sleeve of shoulder muscles, like the deltoid, pectoralis major and the latissimus dorsi, can also contribute to joint compression in certain shoulder positions [106, 110] although the cuff muscles are more active than these peripherally located muscles [105].

The rotator cuff and deltoid muscles generate about 337 ± 88 N of glenohumeral joint compression in cadaveric shoulders at 90° of abduction [116, 117] and 569 ± 141 N of compression in a forced apprehension position, i.e. forced abduction and external rotation [114]. Increased rotator cuff muscle activation increases the concavity compression in the shoulder joint and the stability of the joint to external translating forces [106, 118]. Increasing shoulder compression pre-load from 50 to 100 N increases the translational force required to cause anterior dislocation from 17 ± 6 N to 29 ± 5 N [106]. Conversely, a decrease scapulohumeral balance and depends on coordinated muscle action. This balance is affected by numerous factors including injury, instability, muscle fatigue, degeneration and altered joint mechanics.

The shoulder muscle activation level, and the resulting stability or protection of the joint, are affected by the upper extremity action being performed. In an elevated arm position, increasing handgrip force to 50 % of maximum increased rotator cuff muscle activity [119]. While the arm positions tested do not replicate a driving posture, these data suggest that gripping the steering wheel in anticipation of a crash may increase rotator cuff muscle activity and in turn increase shoulder joint stability. This has potential implications in the relative shoulder injury risk for drivers versus passengers.

To increase stability, shoulder muscles must be active before subluxation or dislocation occurs. In relaxed subjects, muscle onset times (defined as 5 % of MVC) varied from 110 to 220 ms in the anterior deltoid, pectoralis major, upper subscapuularis, biceps long head, teres minor, latissimus dorsi, lower subscapularis, infraspinatus, and supraspinatus muscles when an unexpected anterior translation force was applied to the humeral head with the shoulder in the apprehension position [105]. Anterior/inferior dislocations have been produced traumatically in-vitro in this shoulder position [120]. Although anterior muscles activated before posterior muscles, these reflex latencies are likely too long to prevent an anterior traumatic instability episode in this arm position.

Prior shoulder muscle activation shortens these latencies from 80 to 133 ms at 0 % muscle contraction (relaxed) to 64–81 ms (10–52 % faster) at 20 % of MVC and to 70–89 ms (3–41 % faster) at 50 % of MVC. Initial muscle activity increases the sensitivity of the muscle spindles, which detect the perturbation-induced muscle stretch sooner and thus provides a faster reflexive response [121]. EMD and muscle rise times increase these delays. Since the time to dislocate or sublux a shoulder is currently unknown, it is not known whether reflex activation occurs quickly enough to protect the joint.

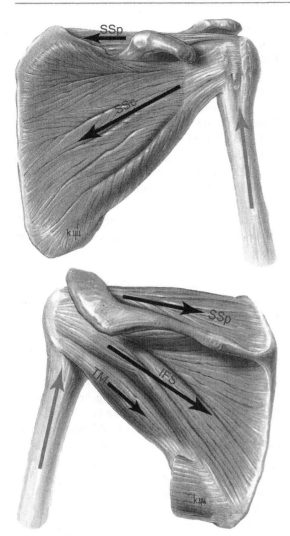

Fig. 21.5 Approximate lines of action (*black arrows*) of the left rotator cuff muscles compressing the humeral head into the glenoid. The deltoid muscle (not shown) can pull the humerus superiorly (*gray arrows*) and apply a destabilizing shear force. *SSp* supraspinatus, *SSc* subscapularis, *IFS* infraspinatus, *TM* teres minor (LifeART image copyright (2000) Wolters Kluwer Health, Inc.- Lippincott Williams & Wilkins. All rights reserved)

in concavity compression through simulated inactivity of the supraspinatus and subscapularis muscles resulted in an 18 % and 17 % decrease in the force to anteriorly dislocate the shoulder, respectively [115]. In order for concavity compression to be most effective in terms of achieving stability, the joint reaction force must be kept within the glenoid fossa; this is referred to as

The rotator cuff muscles are stressed and at risk for injury during many possible shoulder motions. Numerous rotator cuff injury patterns exist but usually the supraspinatus muscle is involved [122]. Rotator cuff tears typically occur in the over-40 age group due to extrinsic and intrinsic factors and are usually the end result of an ongoing process [123]. Extrinsic factors include impingement of the rotator cuff against the anteroinferior aspect of the acromion [124]. Intrinsic factors involve pathology within the tendon, usually as a result of rotator cuff overuse and overloading, and include changes in vascularity and degeneration that can lead to decreased rotator cuff function and altered shoulder mechanics [124].

In patients under 40 years old, rotator cuff lesions are more often related to activity and repetitive trauma [125]. Repetitive overhead throwing involves forceful muscular movements of the arm and shoulder and can lead to minor inflammation, subtle injury, altered mechanics and altered loading of the rotator cuff. A fall onto an outstretched arm is a common traumatic mechanism of rotator cuff injury [122], perhaps due to the shoulder experiencing the greatest deflection and absorbing the majority of the impact energy of a fall [126].

Superior glenoid labrum injuries occur at or near the tendinous insertion of the long head of the biceps onto the glenoid rim. Given the anatomic proximity and inter-connection between the labrum and the bicep tendon, the long head of the biceps muscle and tendon play a role in some labral injury mechanisms. Although the precise pathogenesis of these lesions has not yet been established, mechanisms of injury include falling or a direct blow to the shoulder, glenohumeral subluxation or dislocation, heavy lifting, and overhead racquet sports and throwing [127]. In elite-level throwing athletes, a proposed mechanism of injury is repetitive labral traction applied by the long head of the biceps tendon during the deceleration phase of throwing [128]. This is supported by EMG findings that indicate tension in the biceps and brachialis muscles is greatest at the start of the deceleration phase [129]. In a cadaveric model of lifting a large load with the arm at the side, rapidly applied traction to the long head of the biceps tendon produces SLAP (superior labrum anterior posterior) lesions at failure loads of about 550 N [130]. Introducing an inferior subluxation prior to applying biceps tension increases the incidence of SLAP lesion occurrence and is consistent with the association between labral tears and joint subluxation or dislocation [128].

Cadaveric testing has been performed to assess a traumatic labral injury mechanism in forward and backward falls onto an outstretched hand [131]. All five simulated forward falls resulted in labral injury but only two backward falls result in injury. The SLAP lesions were not created by bicep tendon tension (mean peak tension $= 82.5 \pm 12.1$ N) but rather by shearing forces caused by impact between the humerus and the glenoid. The shearing force is affected by the rotator cuff muscles, which along with other shoulder muscles, will be highly activated in a fall. Dynamic muscle activity data during a standing height fall have not yet been published, so the effectiveness of rotator cuff muscle activity in limiting shear forces and limiting or preventing these labral lesions is not known [131].

The tensile loads applied to the bony attachments of contracting shoulder muscles can also lead to avulsion fractures, and in extreme cases, even joint dislocations [132]. For example, the superior and lateral borders as well as the inferior angle of the scapula suffer avulsion fractures at the attachment sites of the omohyoid and supraspinatus, the teres major and serratus anterior, and the teres minor muscles, respectively [133, 134]. Mechanisms for scapular avulsion fractures include: uncoordinated muscle contraction due to electroconvulsive therapy, electric shocks, or epileptic seizures; and, muscle contraction against a resisted force as a result of trauma or excessive exertion [135, 136].

21.2.3.2 Elbow

The elbow consists of the humeroulnar, humeroradial and radioulnar articulations. The radial and ulnar collateral ligament complexes of the elbow help stabilize the joint in response to varus and valgus loading. The muscles associated with the joint are the brachialis, biceps and brachioradialis muscles anteriorly; the triceps and anconeus muscles posteriorly; the supinator muscle and common extensor tendon laterally; and the common flexor tendon and flexor carpi ulnaris muscle medially [137]. The triceps and aconeus muscles extend the forearm, while the brachialis, biceps and brachioradialis muscles flex the forearm. Supination is achieved by the supinator and biceps brachii muscles while pronation is achieved by the pronator quadratus and pronator teres muscles.

Like the shoulder, the elbow achieves considerable dynamic joint stability through compression by the muscles crossing the joint, particularly the aconeus, triceps, and biceps muscles [138–141]. This is particularly true in elbow flexion where there is less bony contact [142, 143].

Elbow injury can occur from single traumatic events like a fall onto an outstretched arm, or from repetitive loading in overhead athletes like baseball, tennis, or volleyball players [144]. Common elbow injuries include tendinitis, bursitis, ligamentous strain or rupture, bony fracture and dislocation. The majority of acute elbow dislocations are posterior or posterolateral [145, 146]. Posterior elbow dislocation is often the result of a fall onto an extended and outstretched arm and hand. Even low height (6 cm) falls can generate relatively high axial compressive loads, up to 50 % body weight, at the elbow [147]. Based on the typical associated soft tissue injury patterns, axial compression, hyperextension, and valgus forces are applied at the elbow during posterior dislocation [145]. Elbow muscle activation patterns in response to this mechanism for posterior elbow dislocation have not been reported.

Although strong muscles are thought to protect the elbow, flexor muscles of the elbow provide little resistance to joint dislocation at loads up to 22 N in human subjects [148]. The limited muscle stabilization observed in this study may be partly due to specific study design factors like knowledge of loading timing, dislocation load direction, low load magnitudes and low load application rates that were tolerated by the subjects without difficulty. Therefore, the absence of effective muscle response in this study may have been due to the non-traumatic non-injurious loads applied to the elbow.

In contrast, human subjects exposed to an expected elbow extension perturbation demonstrate muscle co-contraction prior to the perturbation, suggesting muscular contribution to elbow stability may reduce the injury risk caused by sudden elbow joint loading [149]. In tests where the perturbation is unexpected, there was an increase in reflex muscle activity (defined as 25–150 ms post-perturbation onset). Better quantification of elbow muscle contribution to joint stability is currently needed to understand injury risk during sudden elbow loading [149].

Forward falls are a common source of traumatic upper extremity injury, including fractures of the distal forearm/wrist, the supracondylar region of the elbow, and the humeral neck. Peak force is highest at the wrist in experimental falls with the elbow locked in extension [126], but elbow flexion beyond 12° provides a muscle damping effect that reduces axial force to the upper extremity and delays the maximum ground reaction force [147, 150]. The effect of elbow flexion on ground reaction force at the hand varies between studies, with one study showing no effect on peak hand force [147] and others showing that fall arrest strategies like elbow flexion and reducing hand velocity can substantially reduce the peak force applied to the distal forearm during hand-to-ground impact [150, 151]. This variation may be related to the timing of elbow flexion relative to ground impact.

While elbow flexion and muscle activation mitigate injury in some upper extremity structures, it can exacerbate injury in others. Eccentric loading of the contracted triceps during a forward fall generates a tensile force at the triceps insertion onto the olecranon process [152, 153]. Pre-impact muscle activity and the stretch reflex further increase the potential for tendon/muscle rupture or even avulsion fracture, particularly in osteopenic bone [133, 154].

Muscle activation also stiffens the extremities in response to impact loading. Changes in limb stiffness may increase transmission of impact shock in the lower extremity [155], a premise that has been shown to have an injurious effect [156]. In simulated forward falls, increasing forearm muscle activation (from 12 % to 48 % MVC) stiffens the forearm and increases the rate at which the reaction force travels up the forearm [157]. These increased loading rates suggest a stiffer pathway for load transmission and an associated increase in bone injury risk [158]. Given the potentially high loading rates in motor vehicle collisions, forearm muscle activation may increase load transmission in drivers who are holding the steering wheel at impact.

Forearm fractures also commonly occur from direct contact with the steering wheel or airbag components [159], particularly over the ulna where there is little soft tissue. The typical injury mechanism for drivers is transverse loading of the forearm during the initial punch-out phase of airbag deployment. Thicker subcutaneous tissue on the underside of the forearm may attenuate the force applied by the deploying airbag reducing force transmission directly to the forearm [160]. In these tests, two cadavers with thicker subcutaneous tissue over the forearm did not sustain fractures. A similar soft tissue cushioning may also protect the humerus against injury in motor vehicle collisions [161].

Repetitive loads to the elbow can result in overuse injuries like joint laxity (from excessive ligament strain or even rupture) and tendonitis. Overhead throwing and the associated valgus extension overload can lead to elbow injury, particularly in elite pitchers. Elite pitchers generate varus elbow torques of up to 64 ± 12 Nm, which is above the 32 ± 10 Nm reported for ulnar collateral ligament rupture [162]. This suggests that muscles carry some of the load and reduce the forces on the medial passive structures of the elbow. Electromyographic studies have shown maximal activity in the flexor-pronator muscle group during the acceleration phase of throwing [163]. Simulated contraction of flexor-pronator muscles in cadavers significantly decreased elbow valgus angle and decreased medial collateral ligament strain [164, 165]. This activation may help stabilize the elbow during this motion and reduce or at least share the applied forces with the medial ulnar collateral ligament.

Excessive muscle forces or repetitive muscle contractions can also result in elbow avulsion fractures at the tendinous insertion into the bone. Although occurring infrequently overall, elbow avulsion injuries occur most commonly at the medial epicondyle in adolescents and may be acute or chronic [166]. "Little League elbow" is associated with a forceful throwing motion and recurrent or isolated contraction of the flexor-pronator muscles during the acceleration phase of throwing. These muscles attach to the medial epicondyle growth plate in adolescents and can pull the growth plate away from the bone. Fracture-separation of the medial epicondyle also occurs in adolescents during arm wrestling when one wrestler tries to force the end of the match or counteracts a pinning move [167]. These actions represent a shift from concentric to eccentric muscle contraction, which generates peak flexor forces about 37 % greater than the forces generated by concentric contraction [168] and can change a non-injurious muscle load to an injurious one.

21.2.3.3 Wrist

The wrist consists of the distal radius and ulna, a proximal and distal row of carpal bones, and the proximal end of the metacarpal bones. The bones form a series of joints between the forearm and hand including the radiocarpal joint (commonly referred to as the wrist joint), distal radioulnar joint (DRUJ), and the midcarpal joints [169].

Most of the muscles that move the wrist are in the forearm and originate at the elbow. The wrist extensor tendons travel over the dorsal aspect of the wrist and include: abductor pollicis longis (radial wrist abductor); extensor carpi radialis longis/extensor carpi radialis brevis (radial wrist extensors); and, extensor carpi ulnaris (ulnar wrist extensor) [169]. The main wrist flexor muscles are the flexor carpi radialis and the flexor carpi ulnaris, the most powerful wrist muscle due to its multiple short muscle fibers.

Most investigations into the stability of the wrist in response to injurious forces focus on boney geometry and interaction as well as the restraint provided by ligamentous and capsular structures. While numerous tendons cross the wrist joint, relatively few studies have addressed the potential stabilizing effect of the muscles of the wrist joint [170]. In order for relaxed muscles to contribute to the mechanical stability of the joint in traumatic situations, they must be able to sense and respond quickly enough to injurious loading conditions. Several different mechanoreceptors, which sense transient and continuous events and relay pain from excessive deformation or damage to the tissue, have been identified in palmar ligaments and suggest a protective ligamentomuscular reflex in the wrist [171].

Pre-activation of wrist muscles during a fall also influences joint stability and injury potential. In forward falls onto an outstretched hand, the palmar surface of the hand contacts the ground first and the impact force is transmitted through the scaphoid/lunate into the radius. This loading tends to rotate the scaphoid into flexion and pronation, and stretch the scapholunate ligament. Simultaneous activation of the extensor and flexor carpi muscles and the abductor pollicis longus muscle result in flexion and supination of the scaphoid [172]. Supination of the scaphoid counteracts its tendency to pronate under axial loading and maintains or moves the scaphoid to a position in which the dorsal scapholunate ligament is better protected [172].

In motor vehicle collisions, occupants aware of an impending impact brace for the collision. Bracing affects how the body interacts with the vehicle interior, the loads applied to the body and the resulting injury risk. Wrist fracture risk in side airbag deployments depends on interaction of the hand with the door handgrip and grip strength [173, 174]. In seat-mounted side airbag deployments, the airbag strikes the back of the elbow and applies an axial load through the forearm into the hand against the door handgrip. Grip strength on the handgrip affects hand, wrist and elbow kinematics and in turn the peak forearm force. In a simulation with a weak grip (10 N grip force), the upper extremity maintains contact with the airbag during deployment and the elbow is forced into full extension, which results in a high compressive load (4,760 N) to the forearm [173]. In a simulation of a strong grip (418 N grip force), the elbow slips inboard of the deploying airbag prior to full elbow extension and thus does not undergo direct airbag contact or prolonged loading. The altered forearm load path associated with the strong grip results in a peak axial forearm load about 40 % less than with the weak grip.

The scaphoid bone is the most commonly fractured carpal bone and occurs in motor vehicle collisions, sports, and in forward falls from standing height onto an extended wrist. In a fall, the soft tissue directly over the palm can play a significant role in energy absorption and affect injury risk to the scaphoid and distal radius. About 30–55 % of the total impact energy is absorbed by the skin and subcutaneous tissue, 25–40 % is absorbed by muscle and tendinous structures, and 10–15 % is absorbed by the radius [175]. As well, increased soft tissue thickness over the palmar surface, particularly between the palm and the scaphoid, directs the impact force away from the scaphoid thus potentially reducing fracture risk in forward falls [176]. As noted earlier, wrist fracture can also be affected by the muscles of the upper arm and arm position. Absorption of impact energy through a flexed elbow and muscle action can reduce the loads to the wrist and the risk of fracture from a fall compared to landing with the elbow locked in extension.

Pre-existing injury or instability of the wrist can temper the ability of muscle action to protect the wrist. Co-contraction of the wrist muscles reorients the carpal bones from their relaxed positions, however, the orientation that results from an imbalance in the pronator-supinator muscles can stretch some carpal ligaments and increase their risk for injury. In addition, contraction of the extensor carpi ulnaris (a pronator) may increase wrist instability in the presence of an injured or torn scapholunate ligament [172].

21.2.4 Lower Extremity Injuries

The lower extremities include the pelvis, hip, upper and lower leg, knee, ankle and foot. They are primarily responsible for carrying the load of the body, propelling it through space, and resisting landing forces following a jump. The lower extremities must regularly counteract ground reaction forces ranging from one to five times body weight (in some events up to ten times) while maintaining stability under the additional application of torque and other externally applied loads [177, 178]. Injuries of the lower extremities include fracture of bony structures, ligament and tendon strains and tears, and injuries to the muscles.

Active and passive muscles can mitigate lower extremity injury in several ways. In low-energy injuries, such as falls and sports-related injuries, muscle activation and recruitment can adjust posture and internally distribute loads in ways that protect from injuries. Soft tissues, composed of skin, fat, and muscle, can also act as a cushion to blunt impacts to the lower extremities but their contribution in this way is minimal. In high-energy injuries such as motor vehicle crashes, bracing by the lower extremities can protect other regions of the body by reducing peak occupant acceleration and excursion [179].

Active muscle contraction can modify the loading pattern and increase load magnitude on bony structures and ligaments of the lower extremity. Studies of long bone fractures in the leg generally do not consider the effect of active muscle contraction. However, in 3-point bending tests, thresholds for fracture are affected by axial preloading [180]. These preloads can be caused by muscle contraction [181]. During frontal motor vehicle crashes, pre-impact bracing of the lower extremity muscles often occurs [182]. Studies on cadaveric legs have shown that axial forces are amplified by simulated pre-impact muscle contraction, increasing the risk of tibia fracture [183, 184]. Compressive loading of the femur due to muscle bracing has also been used to explain femur fracture in real world frontal motor vehicle collisions where femur loads were otherwise predicted to be below injury thresholds based on external loading alone [185]. In some cases, muscle contraction can cause the traumatic injury itself. Case reports of avulsion injury at the tibial tuberosity demonstrate that muscle-generated forces alone are capable of causing traumatic injury to the lower extremities [186]. Injury risk can also be modified by acute muscle fatigue [187] and chronic differences in strength ratio between opposing muscle groups [188].

Major muscles of the hip and knee include gluteal, adductor, hamstring group, and quadriceps group muscles that control flexion/extension, abduction/adduction, and rotation of the hip and knee in addition to maintaining joint stability. Major muscles of the ankle and foot include the tibialis, peroneus, gastrocnemius, flexor, and extensor muscles that control dorsiflexion and plantarflexion of the feet in addition to other more complex stabilizing motions. The muscles of the lower extremities are innervated by efferent nerve fibers emerging from levels L2 to S3 of the spinal cord.

21.2.4.1 Pelvis and Hip

The pelvis is comprised of the two hip bones, sacrum, and coccyx. The proximal femur, which rests within the acetabulum of the hip bones, is composed of the head, neck, and trochanteric regions. These structures are the only transmission path to the ground for the weight of the head, arms, and torso. Pelvic injuries include avulsions of muscle insertions, isolated fracture of the pelvic ring, and fractures of the sacrum and coccyx. Hip injuries include traumatic hip dislocations, fractures of the acetabulum, and fractures of the neck of the femur. The greater and lesser trochanters are also susceptible to avulsion injury during vigorous athletics [189].

The pelvis and hip are held together by strong ligaments and thick surrounding muscle mass, so large forces are required to dislocate (luxate) the hip. These large forces often lead to acetabular or proximal femur fracture with hip dislocation. These injuries can occur in high-energy events such as motor vehicle crashes or low-energy events such as falls or skiing incidents. In frontal motor vehicle crashes, posterior hip dislocation often occurs through unrestrained knee impact with the dash. Other mechanisms of dislocation have been proposed where active muscle contraction is required to transmit forces into dislocating the hip. Monma and Sugita [190] proposed a mechanism of hip dislocation in frontal motor vehicle crashes requiring active bracing of the right leg against the brake prior to impact leading to traumatic posterior dislocation of the hip. Active muscle contraction forces alone, without external application of force, are capable of injuring the hip. While most acetabular fractures are due to direct impact to the hip [191], a review of seizure-induced acetabular fractures found that in some cases, seizure alone was capable of causing acetabular fracture [192].

In falls, the role of muscles in mitigating pelvic and hip injury is two-fold: (1) muscles can alter fall kinematics by initiating protective posturing and (2) muscles contribute to hip protection through cushioning of falls. In a study of six young and healthy individuals (22–35 years old), Hsiao and Robinovitch [193] found that wrist contact and pelvic contact with the ground occurred at an average of 680 ± 116 ms and 715 ± 160 ms, respectively, following initiation of the fall. This time window provides enough time for voluntary muscle activation to adjust posture during the fall. In fact, a later study by Robinovitch et al. [194] found that this time window was sufficient time for young and elderly women to break a fall with a hand (except in the case of an elderly woman falling laterally). Postural movements in the lower extremities can also reduce forces on the hip at impact. In backwards falls, a squatting motion can reduce hip impact velocity by 18 % [195]. However, these protective strategies are sensitive to reaction times and a delay of 300 ms can significantly reduce the protective effect of postural responses [151].

Impact forces can also be mediated by the soft tissues (including muscles) covering the hip. A study of hip impacts in volunteers using a rapid pelvis release methodology found that both muscle thickness and a relaxed state reduced force in direct impacts to the hip [196]. Paradoxically, this study suggested that hip muscle contraction, of the sort required to reduce the kinetic energy of the fall, could lead to increased forces transmitted to the hip at impact. A later study demonstrated a reduced effect of contraction on force transmission but a significant dependence on configuration with increased force to the femur when falling with the trunk upright versus recumbent [197]. Ultimately, however, reduction of hip impact force from muscle cushioning alone is generally not sufficient to reduce the forces below fracture thresholds in falls [198].

In motor vehicle crashes, injury to the pelvis is common. The pelvis is particularly susceptible to fracture in lateral impacts and hip dislocations are common in severe frontal impacts due to unrestrained knee bolster impacts. Fracture tolerances for the pelvis in lateral impacts range from ~3 to 10 kN with trochanteric soft tissue thickness having a small but significant effect on tolerance and bone mineral density having a large effect [199, 200]. The effect of bracing or active muscle contraction on pelvic fracture mechanisms during lateral impacts has not been investigated. In frontal impacts, it has been suggested that bracing, especially by the driver against the brake pedal, can lead to increased risk of hip dislocation [190]. Chang et al. [201] used EMG data normalized to MVC to estimate muscle forces in the lower extremities during simulated maximum braking. Incorporation of muscle activation into simulations of the lower extremity in frontal impact suggested that muscle activation due to braking increases the effective mass of the body coupled to the knee, increasing knee impact force with the knee bolster, and increasing the risk of femur fracture. However, muscle activation did not have an effect on the likelihood of hip fracture. Bracing can also affect excursion of the lower extremities during frontal impacts [179], potentially reducing the force with which the knee contacts the knee bolster or preventing contact altogether. However, a human volunteer study of body kinematics during low (2.5 g) and medium severity (5 g) frontal impacts with and without bracing found that while bracing reduced the forward excursion of the knees and hips by ~50–60 % in low severity impacts, it did not have a significant effect on their excursion in medium severity impacts [202]. This suggests that bracing will not reduce the risk of pelvic injury in high severity frontal impacts.

21.2.4.2 Knee

The knee is classified as a double condyloid joint, meaning that it supports flexion/extension of the leg and rotation while in flexion. It relies on ligaments, muscles and tendons to remain stable while bearing load. The medial collateral ligament (MCL) and lateral collateral ligament (LCL) generally resist varus and valgus loading. The anterior cruciate ligament (ACL) and posterior cruciate ligament (PCL) primarily restrict the anterior and posterior movement of the tibia relative to the femur. Secondarily, the cruciate ligaments provide resistance to valgus, varus and tibial rotation. The patella is located between the quadriceps tendon and the attachment point on the tibial turberosity, and articulates with the femoral condyles to form the patellofemoral joint. This joint experiences large forces, especially when the knee is flexed and the quadriceps muscle is active. Muscles in the knee resist applied loads but reflexive muscle contractions, requiring about 220 ms in response to a stimulus, are too slow to protect the knee during rapid loading [203]. Still, preparatory contractions and stretch reflexes may play a role in knee joint protection. PCL injury, sometimes referred to as 'dashboard knee', requires a posteriorly directed force on the tibia. Muscles do not play a significant role in PCL injuries other than to place the knee in a flexed position prior to application of posterior loading on the tibia in frontal crashes.

Noncontact injuries of the ACL often occur in sporting activities that require rapid changes in direction (cutting), decelerating from running, jump landings with the leg extended, and pivoting around a planted foot [204]. Muscles play a significant role in non-contact ACL injury through neuromuscular activation, relative muscle strength and recruitment, and muscle fatigue. Stabilization of the knee during dynamic activity relies on the neuromuscular control system to coordinate muscle contractions in a timely manner [205]. For example, in preparation for large forces at the knee, co-activation of the hamstrings with the quadriceps is critical to protecting the knee joint from forces that can lead to ACL injury [206]. Quadriceps activation is much higher than hamstring activation, ranging from an average of 64–87 % of MVC higher, for sidestep cuts, crosscuts, stopping and landing [207]. Especially when the knee is close to full extension, excessive quadriceps activation can cause significant shear force on the tibia in the anterior direction which can lead to increased strain on the ACL [208] (Fig. 21.6). DeMorat et al. [209] found that a 4,500 N simulated quadriceps contraction in cadaver knees at 20° flexion produced significant anterior displacement of the tibia and ACL injury. To protect against ACL injury, hamstring co-contraction can counteract this strain at knee flexion angles of 15–120° [208, 210]. Gender differences in quadriceps-to-hamstring strength ratio have been observed, with mature females having a sig-

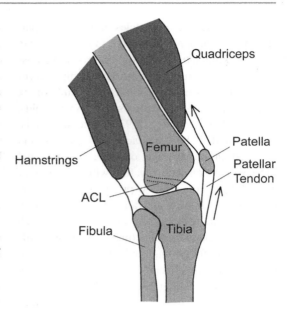

Fig. 21.6 Lateral view of the knee joint. Quadriceps contraction can lead to anterior displacement of the tibia, increasing tensile strain on the ACL. Hamstring co-contraction stabilizes the knee and counteracts the tibial displacement caused by quadriceps contraction

nificantly higher ratio than immature girls, immature boys, and mature boys [211]. In addition, differences in quadriceps activation and knee/hip motion between females and males have been observed in tasks that mimic common ACL injury mechanisms [212, 213]. These gender differences in relative hamstring strength and neuromuscular control may explain the increased incidence of ACL injury in female athletes [188, 189, 214].

Muscle fatigue may also increase the risk of ACL injury by affecting neuromuscular control and coordination of muscle contraction that could lead to joint laxity [215] and altered knee and hip mechanics [187, 216]. Training and conditioning exercises have been proposed to address the increased risk of ACL injury due to factors such as muscle fatigue and quadriceps-to-hamstring strength ratio [217–219].

ACL and other knee ligament injuries caused by direct loading such as in pedestrian impacts can also be affected by muscle contraction. However, mechanisms of knee injury and tolerances to injury during pedestrian impacts have generally been investigated using finite element models validated only against cadaver experiments which do not include the effect of muscle activation [220]. A recent series of finite element studies incorporating muscle contraction in low speed pedestrian impacts to the leg suggest that reflex muscle activation is protective to the ligaments of the knee, reducing knee ligament forces in below knee impacts [221]. Simulated lateral impacts with muscles activated by stretch reflex predicted a two-fold and greater decrease in ligament loading compared to passive muscles [222]. As in non-contact loading, simulations suggest that hamstring force in particular can reduce ACL and PCL strains while the gastrocnemius generally affects MCL strain in lateral impacts at 25 km/h [223]. Simulated impacts at, above, and below the knee in frontal, posterior, and various lateral directions found that peak ligament strains were lower in unaware pedestrians with stretch reflex implemented in the model compared to the cadaver and braced aware pedestrian models [224]. Risk of ligament injury in real world crashes may therefore be lower than the risk predicted from cadaver studies alone.

Patellar avulsion, where the patellar tendon tears away from the tibia, is a rare injury of the patellofemoral joint that is caused directly by muscle contraction [225]. It usually occurs in young athletes with the knee in a flexed position coupled with violent contraction of the quadriceps [226]. A real life case of patellar tendon rupture during a weightlifting competition was caught on video and analyzed to determine the forces and loading rates on the tendon during knee flexion that led to rupture [227]. Tensile loads in the patellar tendon were calculated by summing estimates of the net forces and moments at each joint in a rigid body model of the lift. Knee flexion angle was 89.2° when the tendon failed at a knee extensor moment of 550–560 Nm and patellar tendon tension of approximately 14.5 kN. Time to rupture from movement initiation was 380 ms.

21.2.4.3 Ankle and Foot

The ankle and foot are capable of complex multiplanar and multiaxial motions that provide the body with support and balance during standing and while in motion [228]. They are composed of multiple bony structures that provide rigidity and lever arm mechanisms, multiple joints which provide several degrees of freedom for motion, and muscles and tendons which respond rapidly through proprioceptive feedback to control foot movement and stability. Muscles that control the eversion and inversion of the foot such as the peroneal muscles play a significant role in ankle stability and therefore contribute to mechanisms of ankle sprain but can also interact with fracture mechanisms during axial loading such as in frontal motor vehicle crashes. Muscle activation can also contribute to calcaneal tendon rupture.

Lateral ankle sprain is one of the most common ankle injuries, often occurring in sport or from walking on uneven surfaces. Lateral ankle sprains occur with rapid inversion of the foot (rolling over the lateral aspect foot) leading to lateral ligament strain [229], often implicating the anterior talofibular and calcaneofibular ligaments [230, 231]. Several case studies of ankle inversion injuries caught on video have been used to determine the kinematics and kinetics of ankle rotation during a lateral sprain [232, 233]. In general, maximum ankle rotation occurs between 80 and 180 ms following ground contact with inversion and internal rotation. One case report captured a lateral ankle sprain during cutting maneuvers performed in a laboratory environment while muscle activity was recorded (Fig. 21.7) [234]. Bursts of tibialis anterior and peroneus longus muscle activation began at 40–45 ms with the first peaks in activation occurring at 62 ms (tibialis anterior) and 74 ms (peroneus longus) after ground contact. Maximum ankle rotation occurred at 150 ms after ground contact suggesting that muscular stretch reflexes may play a role in the mechanisms of ankle injury.

Ankle instability may be affected by motor response to rapid supination or eversion of the foot which can be delayed in individuals complaining of ankle instability relative to healthy controls. Specifically, the peroneus longus, peroneus brevis and tibialis anterior muscles demonstrate reaction times about 10 ms slower on average in human volunteers with unstable ankles compared to healthy volunteers [235].

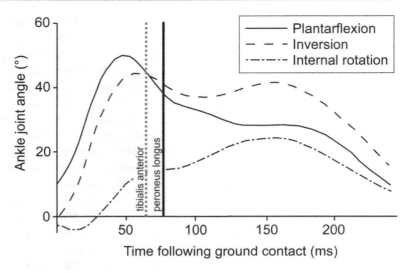

Fig. 21.7 Ankle joint kinematics (plantarflexion, inversion, and internal rotation) measured in a single case of ankle sprain (Adapted from Gehring, Wissler, Mornieux, & Gollhofer. How to sprain your ankle – a biomechanical case report of an inversion trauma, *J Biomech*, 46, Pages 175–178, Copyright (2013), with permission from Elsevier [235]. Time points for the first peak in muscle activity recorded during the injury event are shown for the tibialis anterior (*vertical grey dashed line*) and the peroneus longus (*vertical black line*)

Eversion-to-inversion ankle strength has also been explored as a risk factor for ankle injury with a ratio >1 being correlated with an increased incidence of ankle injury [236]. Other intrinsic risk factors have also been identified that may contribute to risk of ankle sprain such as endurance, balance, coordination, and muscle strength in dorsiflexion [237]. Fatigue of the peroneus longus in particular may lead to increased ankle instability and possibly increased risk of injury [238].

The foot can also undergo injury moderated by muscle activity. Overuse and chronic injuries to the foot can be caused by abnormal muscle control during gait due to muscle fatigue. In a study of prolonged marches in military recruits, peroneus longus muscle fatigue was associated with increased calcaneal and metatarsal contact stress that could serve as a mechanism for stress fractures in these regions [238]. Calcaneal (Achilles) tendon rupture is another injury of the foot that occurs mainly in sporting activities during sudden acceleration or jumping [239]. Achilles tendon rupture occurs when forces exceed the tensile strength of the tendon or its insertion point on the calcaneus due to muscle contraction, rapid foot dorsiflexion or plantarflexion, or direct impact to the tendon. It has been suggested that non-uniform stress through the cross sectional area of the Achilles tendon can be caused by differences in individual muscle forces from the soleus and gastrocnemius muscles [240]. The non-uniform stress distribution can result in stress concentrations within the tendon which may represent a mechanism of rupture.

Axial loading of the foot and ankle in frontal motor vehicle crashes often results in injury to the midfoot, forefoot, malleoli, calcaneus and tibia. Active bracing occurs in majority of occupants during frontal impact [182] and this bracing can lead to additional axial loading [179] or redistribution of the internal loads through the lower extremity at impact [184]. In drivers braking before simulated impact, the Achilles tendon requires about 1.5 kN of force to support maximum braking [241]. Kitagawa et al. [183] and McMaster et al. [242] found that pre-load at the Achilles tendon of at least 1.5 kN during pendulum impact to the midfoot amplified axial forces in the tibia and foot/ankle complex, leading to calcaneal and tibial pilon fractures. However, these studies did not compare to cases without

Achilles tendon pretension. Funk et al. [243] used a full plantar plate to impact cadaver legs, half of which included Achilles tendon pretension and half of which did not. Achilles tension increased the axial tibial force and the number of tibial pilon fractures. Thus bracing before impact can increase the risk of lower extremity injury during frontal crashes.

21.2.5 Whole Body

In frontal crashes, pre-impact leg bracing can absorb considerable energy and alter the distribution of forces applied to the body. Based on data figures presented by Armstrong et al. [179], the bracing pre-load applied through the floor pan was $1,180 \pm 600$ N in 30 sled impacts with speed changes of 2.7–5.8 m/s and accelerations of 3.7–15.5 g. There was no difference in preloads in the lap and lap + torso belt configurations. These researchers estimated that pre-impact leg bracing could absorb 44–55 % of the body's initial kinetic energy. Chandler and Christian [245] observed similar horizontal preloads ($1,222 \pm 410$ N) for lap belts, but lower preloads (819 ± 334 N) for lap + torso belts in frontal impacts with a speed change of 6.5 ± 0.5 m/s with $12.1 \pm 1.6\,g$ acceleration. Bracing pre-loads were about 20–30 % of peak floor pan loads.

Thoracic muscle tensing increases the stiffness of the chest to blunt impacts at low severity levels, but this stiffening essentially disappears at impact levels that generate chest deflections large enough to cause injuries [245, 246].

Choi et al. [247] attempted to quantify bracing in the upper and lower extremities, however they neglected to report their impact severity and whether the brake pedal forces were from one or both lower limbs. Given their gravity sled dropped a maximum of 1 m, we can assume their impact speed was less than 4.4 m/s. Peak steering wheel forces were 151 ± 79 N and peak brake pedal forces were 274 ± 95 N. Maximum brake pedal forces in stationary configurations have been measured to be 529 ± 242 N and in pre-crash simulations as 245 ± 123 N [248]. Others have measured maximum braking forces of 750 N [249].

Choi et al. [247] then used a finite element model with 16 muscles to simulate their volunteer experiments. Using a whole body computational model also based on the volunteer data for Choi et al. [247], lower head and sternum excursions and slightly higher knee excursions were observed with pre-impact bracing in a frontal impact with a 57 km/h speed change [250, 251]. When evaluating injury metrics, these authors found peak femur loads increased 46–55 % due to increased dash interaction. At the relatively low acceleration rates that occur during emergency braking (0.6–0.8 g), pre-braking bracing had a large effect on the peak occupant motion [252, 253].

These and other whole body simulations suffer from a lack of usable activation data for all of the muscles that respond in these braking and collision situations. The validity of extrapolating muscle bracing levels acquired at low levels to impact severities three to four times more severe is also unknown.

21.3 Summary

Muscles, whether passive or active, can both mitigate and exacerbate accidental injury. Muscles not only move the body, they stiffen joints, cushion direct impacts, and transfer impact loads to other body regions. The role muscles play in a particular injury depends on their geometry, morphology and activation level in relation to the injurious loads and tissues being loaded. Active muscle can increase the effective mass of the body and thus increase the impact forces applied to some tissues. Active muscle can also distribute loads and attenuate the impact forces in other tissues. Muscle forces appear to play a proportionally larger role in lower severity events.

Our understanding of the role of muscle in accidental injury is incomplete. Studies that accurately probe the action and effect of muscles during actual injury events are challenging to design. In vitro studies provide an avenue to examine loading at and above injury thresholds, but they suffer from the lack of a real physiologic environment and intact neuromuscular control. In vivo studies, on the other hand, can be illustrative of the importance of an intact neuromuscular system, but cannot be performed at injury thresholds. Scaling of in vivo data acquired at sub-injurious loads to events occurring near injury threshold requires assumptions that may not hold at increased loads. Computational simulations that explore the effects of muscles on injury (both passive and active) can be useful in examining the effects of different muscle properties, but again caution should be used when interpreting results of these studies as they are often based on volunteer data at sub-injurious loading and validated against cadaver or dummy experiments. Although further study of muscles and injury is needed, this review nevertheless highlights the importance of including the role of muscles in the study of accidental injury.

References

1. Burkholder TJ, Lieber RL (2001) Sarcomere length operating range of vertebrate muscles during movement. J Exp Biol 204(Pt 9):1529–1536
2. Hegarty PV, Hooper AC (1971) Sarcomere length and fibre diameter distributions in four different mouse skeletal muscles. J Anat 110(Pt 2):249–257
3. Lexell J (1995) Human aging, muscle mass, and fiber type composition. J Gerontol A: Biol Med Sci 50(Spec No):11–16
4. Feinstein B, Lindegard B, Nyman E, Wohlfart G (1955) Morphologic studies of motor units in normal human muscles. Acta Anat 23(2):127–142
5. Torre M (1953) [Number and dimensions of the motor units of the extrinsic eye muscles and, in general, of skeletal muscles connected with the sensory organs]. Schweizer Archiv fur Neurologie und Psychiatrie Archives suisses de neurologie et de psychiatrie Archivio svizzero di neurologia e psichiatria 72(1–2):362–376
6. Jenny AB, Inukai J (1983) Principles of motor organization of the monkey cervical spinal cord. J Neurosci 3(3):567–575
7. Denny-Brown D, Pennybacker JB (1938) Fibrillation and fasciculation in voluntary muscle. Brain 61: 311–341
8. Bawa P, Binder MD, Ruenzel P, Henneman E (1984) Recruitment order of motoneurons in stretch reflexes is highly correlated with their axonal conduction velocity. J Neurophysiol 52(3):410–420
9. Lieber RL, Jacobson MD, Fazeli BM, Abrams RA, Botte MJ (1992) Architecture of selected muscles of the arm and forearm: anatomy and implications for tendon transfer. J Hand Surg 17(5):787–798
10. Lieber RL, Fazeli BM, Botte MJ (1990) Architecture of selected wrist flexor and extensor muscles. J Hand Surg 15(2):244–250
11. Wickiewicz TL, Roy RR, Powell PL, Edgerton VR (1983) Muscle architecture of the human lower limb. Clin Orthop Relat Res 179:275–283
12. Winters JM, Stark L (1988) Estimated mechanical properties of synergistic muscles involved in movements of a variety of human joints. J Biomech 21(12):1027–1041
13. Kamibayashi LK, Richmond FJ (1998) Morphometry of human neck muscles. Spine 23(12):1314–1323
14. Mendez J, Keys A (1960) Density and composition of mammalian muscle. Metab Clin Exp 9:184–188
15. Myers BS, Woolley CT, Slotter TL, Garrett WE, Best TM (1998) The influence of strain rate on the passive and stimulated engineering stress – large strain behavior of the rabbit tibialis anterior muscle. J Biomech Eng 120(1):126–132
16. O'Brien TD, Reeves ND, Baltzopoulos V, Jones DA, Maganaris CN (2010) In vivo measurements of muscle specific tension in adults and children. Exp Physiol 95(1):202–210. doi:10.1113/expphysiol.2009.048967
17. Zajac FE (1989) Muscle and tendon: properties, models, scaling, and application to biomechanics and motor control. Crit Rev Biomed Eng 17(4):359–411
18. Srinivasan RC, Lungren MP, Langenderfer JE, Hughes RE (2007) Fiber type composition and maximum shortening velocity of muscles crossing the human shoulder. Clin Anat 20(2):144–149. doi:10.1002/ca.20349
19. Chow JW, Darling WG (1999) The maximum shortening velocity of muscle should be scaled with activation. J Appl Physiol 86(3):1025–1031
20. Ehret AE, Bol M, Itskov M (2011) A continuum constitutive model for the active behavior of skeletal muscle. J Mech Phys Solids 59:625–636
21. Herzog W, Leonard TR (2005) The role of passive structures in force enhancement of skeletal muscles following active stretch. J Biomech 38(3):409–415. doi:10.1016/j.jbiomech.2004.05.001
22. Bisdorff AR, Bronstein AM, Gresty MA (1994) Responses in neck and facial muscles to sudden free fall and a startling auditory stimulus. Electroencephalogr Clin Neurophysiol 93(6): 409–416
23. Brault JR, Siegmund GP, Wheeler JB (2000) Cervical muscle response during whiplash: evidence of a lengthening muscle contraction. Clin Biomech 15(6):426–435
24. Brown P, Rothwell JC, Thompson PD, Britton TC, Day BL, Marsden CD (1991) New observations on the normal auditory startle reflex in man. Brain 114(Pt 4):1891–1902
25. Siegmund GP, Sanderson DJ, Myers BS, Inglis JT (2003) Awareness affects the response of human subjects exposed to a single whiplash-like perturba-

tion. Spine 28(7):671–679. doi:10.1097/01. BRS.0000051911.45505.D3
26. Mazzini L, Schieppati M (1992) Preferential activation of the sternocleidomastoid muscles by the ipsilateral motor cortex during voluntary rapid head rotations in humans. In: Berthoz A, Vidal P, Graf W (eds) The head-neck sensory motor system. Oxford University Press, Oxford, UK
27. Siegmund GP, Inglis JT, Sanderson DJ (2001) Startle response of human neck muscles sculpted by readiness to perform ballistic head movements. J Physiol 535(Pt 1):289–300
28. Geertsen SS, Zuur AT, Nielsen JB (2010) Voluntary activation of ankle muscles is accompanied by subcortical facilitation of their antagonists. J Physiol 588(Pt 13):2391–2402. doi:10.1113/jphysiol.2010.190678
29. Blouin JS, Siegmund GP, Timothy Inglis J (2007) Interaction between acoustic startle and habituated neck postural responses in seated subjects. J Appl Physiol 102(4):1574–1586. doi:10.1152/japplphysiol.00703.2006
30. Corcos DM, Gottlieb GL, Latash ML, Almeida GL, Agarwal GC (1992) Electromechanical delay: an experimental artifact. J Electromyogr Kinesiol 2(2):59–68. doi:10.1016/1050-6411(92)90017-D
31. Nilsson J, Tesch P, Thorstensson A (1977) Fatigue and EMG of repeated fast voluntary contractions in man. Acta Physiol Scand 101(2):194–198
32. Winter EM, Brookes FB (1991) Electromechanical response times and muscle elasticity in men and women. Eur J Appl Physiol Occup Physiol 63(2):124–128
33. Crisco JJ, Panjabi MM, Yamamoto I, Oxland TR (1992) Euler stability of the human liga-mentous lumbar spine. Part II: experiment. Clin Biomech 7:27–32
34. Magnusson ML, Pope MH, Hasselquist L, Bolte KM, Ross M, Goel VK, Lee JS, Spratt K, Clark CR, Wilder DG (1999) Cervical electromyographic activity during low-speed rear impact. Eur Spine J 8(2):118–125
35. Bosboom EM, Hesselink MK, Oomens CW, Bouten CV, Drost MR, Baaijens FP (2001) Passive transverse mechanical properties of skeletal muscle under in vivo compression. J Biomech 34(10):1365–1368
36. Van Loocke M, Lyons CG, Simms CK (2006) A validated model of passive muscle in compression. J Biomech 39(16):2999–3009. doi:10.1016/j.jbiomech.2005.10.016
37. Van Loocke M, Lyons CG, Simms CK (2008) Viscoelastic properties of passive skeletal muscle in compression: stress-relaxation behaviour and constitutive modelling. J Biomech 41(7):1555–1566. doi:10.1016/j.jbiomech.2008.02.007
38. McElhaney JH (1966) Dynamic response of bone and muscle tissue. J Appl Physiol 21(4):1231–1236
39. Song B, Chen W, Ge Y, Weerasooriya T (2007) Dynamic and quasi-static compressive response of porcine muscle. J Biomech 40(13):2999–3005. doi:10.1016/j.jbiomech.2007.02.001
40. Van Sligtenhorst C, Cronin DS, Wayne Brodland G (2006) High strain rate compressive properties of bovine muscle tissue determined using a split Hopkinson bar apparatus. J Biomech 39(10):1852–1858. doi:10.1016/j.jbiomech.2005.05.015
41. Rosen E, Sander U (2009) Pedestrian fatality risk as a function of car impact speed. Accid Anal Prev 41(3):536–542. doi:10.1016/j.aap.2009.02.002
42. Tefft BC (2013) Impact speed and a pedestrian's risk of severe injury or death. Accid Anal Prev 50:871–878. doi:10.1016/j.aap.2012.07.022
43. Dhaliwal TS, Beillas P, Chou CC, Prasad P, Yang KH, King AI (2002) Structural response of lower leg muscles in compression: a low impact energy study employing volunteers, cadavers and the hybrid III. Stapp Car Crash J 46:229–243
44. Balaraman K, Mukherjee S, Chawla A, Malhotra R (2006) Inverse finite element characterization of soft tissues using impact experiments and Taguchi methods (2006-01-0252). SAE International, Warrendale
45. Morrow DA, Haut Donahue TL, Odegard GM, Kaufman KR (2010) Transversely isotropic tensile material properties of skeletal muscle tissue. J Mech Behav Biomed Mater 3(1):124–129. doi:10.1016/j.jmbbm.2009.03.004
46. Best TM, McElhaney J, Garrett WE Jr, Myers BS (1994) Characterization of the passive responses of live skeletal muscle using the quasi-linear theory of viscoelasticity. J Biomech 27(4):413–419
47. Van Ee CA, Chasse AL, Myers BS (2000) Quantifying skeletal muscle properties in cadaveric test specimens: effects of mechanical loading, post-mortem time, and freezer storage. J Biomech Eng 122(1):9–14
48. Macpherson PC, Schork MA, Faulkner JA (1996) Contraction-induced injury to single fiber segments from fast and slow muscles of rats by single stretches. Am J Physiol 271(5 Pt 1):C1438–C1446
49. McCully KK, Faulkner JA (1985) Injury to skeletal muscle fibers of mice following lengthening contractions. J Appl Physiol 59(1):119–126
50. Noonan TJ, Best TM, Seaber AV, Garrett WE Jr (1994) Identification of a threshold for skeletal muscle injury. Am J Sports Med 22(2):257–261
51. Garrett WE Jr, Nikolaou PK, Ribbeck BM, Glisson RR, Seaber AV (1988) The effect of muscle architecture on the biomechanical failure properties of skeletal muscle under passive extension. Am J Sports Med 16(1):7–12
52. Garrett WE Jr, Safran MR, Seaber AV, Glisson RR, Ribbeck BM (1987) Biomechanical comparison of stimulated and nonstimulated skeletal muscle pulled to failure. Am J Sports Med 15(5):448–454
53. Hang YS, Tsuang YH, Sun JS, Cheng CK, Liu TK (1996) Failure of stimulated skeletal muscle mainly contributed by passive force: an in vivo rabbit model. Clin Biomech 11:343–347

54. Lin R, Chang G, Chang L (1999) Biomechanical properties of muscle-tendon unit under high-speed passive stretch. Clin Biomech 14(6):412–417
55. Crisco JJ, Hentel KD, Jackson WO, Goehner K, Jokl P (1996) Maximal contraction lessens impact response in a muscle contusion model. J Biomech 29(10):1291–1296
56. Covassin T, Swanik CB, Sachs ML (2003) Sex differences and the incidence of concussions among collegiate athletes. J Athl Train 38(3):238–244
57. Dick RW (2009) Is there a gender difference in concussion incidence and outcomes? Br J Sports Med 43(Suppl 1):i46–i50. doi:10.1136/bjsm.2009.058172
58. Tierney RT, Sitler MR, Swanik CB, Swanik KA, Higgins M, Torg J (2005) Gender differences in head-neck segment dynamic stabilization during head acceleration. Med Sci Sports Exerc 37(2):272–279
59. Viano DC, Casson IR, Pellman EJ (2007) Concussion in professional football: biomechanics of the struck player – part 14. Neurosurgery 61(2):313–327. doi:10.1227/01.NEU.0000279969.02685.D0, discussion 327–318
60. Mihalik JP, Guskiewicz KM, Marshall SW, Greenwald RM, Blackburn JT, Cantu RC (2011) Does cervical muscle strength in youth ice hockey players affect head impact biomechanics? Clin J Sport Med 21(5):416–421. doi:10.1097/JSM.0B013E31822C8A5C
61. Mihalik JP, Blackburn JT, Greenwald RM, Cantu RC, Marshall SW, Guskiewicz KM (2010) Collision type and player anticipation affect head impact severity among youth ice hockey players. Pediatrics 125(6):e1394–e1401. doi:10.1542/peds.2009-2849
62. Mansell J, Tierney RT, Sitler MR, Swanik KA, Stearne D (2005) Resistance training and head-neck segment dynamic stabilization in male and female collegiate soccer players. J Athl Train 40(4):310–319
63. Lisman P, Signorile JF, Del Rossi G, Asfour S, Eltoukhy M, Stambolian D, Jacobs KA (2012) Investigation of the effects of cervical strength training on neck strength, EMG, and head kinematics during a football tackle. Int J Sports Sci Eng 6:131–140
64. Bergmark A (1989) Stability of the lumbar spine. A study in mechanical engineering. Acta Orthop Scand Suppl 230:1–54
65. Kettler A, Hartwig E, Schultheiss M, Claes L, Wilke HJ (2002) Mechanically simulated muscle forces strongly stabilize intact and injured upper cervical spine specimens. J Biomech 35(3):339–346
66. Pedersen KK, Christiansen C, Ahlgren P, Lund M (1976) Incidence of fractures of the vertebral spine in epileptic patients. Acta Neurol Scand 54(2):200–203
67. Mehlhorn AT, Strohm PC, Hausschildt O, Schmal H, Sudkamp NP (2007) Seizure-induced muscle force can caused lumbar spine fracture. Acta Chir Orthop Traumatol Cech 74(3):202–205
68. Anderson JS, Hsu AW, Vasavada AN (2005) Morphology, architecture, and biomechanics of human cervical multifidus. Spine 30(4):E86–E91
69. Winkelstein BA, McLendon RE, Barbir A, Myers BS (2001) An anatomical investigation of the human cervical facet capsule, quantifying muscle insertion area. J Anat 198(Pt 4):455–461
70. Boyd-Clark LC, Briggs CA, Galea MP (2002) Muscle spindle distribution, morphology, and density in longus colli and multifidus muscles of the cervical spine. Spine 27(7):694–701
71. Siegmund GP, Blouin JS, Carpenter MG, Brault JR, Inglis JT (2008) Are cervical multifidus muscles active during whiplash and startle? An initial experimental study. BMC Musculoskelet Disord 9:80. doi:10.1186/1471-2474-9-80
72. Siegmund GP, Sanderson DJ, Myers BS, Inglis JT (2003) Rapid neck muscle adaptation alters the head kinematics of aware and unaware subjects undergoing multiple whiplash-like perturbations. J Biomech 36(4):473–482
73. Van der Horst MJ, Thunnissen JGM, Happee R, Van Haaster RMHP, Wismans JSHM (1997) The influence of muscle activity on head-neck response during impact (973346). SAE International, Warrendale
74. Brolin K, Halldin P, Leijonhufvud I (2005) The effect of muscle activation on neck response. Traffic Inj Prev 6(1):67–76. doi:10.1080/15389580590903203
75. Chancey VC, Nightingale RW, Van Ee CA, Knaub KE, Myers BS (2003) Improved estimation of human neck tensile tolerance: reducing the range of reported tolerance using anthropometrically correct muscles and optimized physiologic initial conditions. Stapp Car Crash J 47:135–153
76. Myers BS, Winkelstein BA (1995) Epidemiology, classification, mechanism, and tolerance of human cervical spine injuries. Crit Rev Biomed Eng 23(5–6):307–409
77. Nightingale RW, McElhaney JH, Richardson WJ, Myers BS (1996) Dynamic responses of the head and cervical spine to axial impact loading. J Biomech 29(3):307–318
78. Yamaguchi GT, Carhart MR, Larson R, Richards D, Pierce J, Raasch CC, Scher I, Corrigan CF (2005) Electromyographic activity and posturing of the human neck during rollover tests (2005-01-0302). SAE International, Warrendale
79. Yoganandan N, Sances A Jr, Maiman DJ, Myklebust JB, Pech P, Larson SJ (1986) Experimental spinal injuries with vertical impact. Spine 11(9):855–860
80. Nightingale RW, Richardson WJ, Myers BS (1997) The effects of padded surfaces on the risk for cervical spine injury. Spine 22(20):2380–2387
81. Hu J, Yang KH, Chou CC, King AI (2008) A numerical investigation of factors affecting cervical spine injuries during rollover crashes. Spine 33(23):2529–2535. doi:10.1097/BRS.0b013e318184aca0
82. Brolin K, Hedenstierna S, Halldin P, Bass C, Alem N (2008) The importance of muscle tension on the outcome of impacts with a major vertical component. Int J Crashworthiness 13:487–498
83. Cholewicki J, McGill SM (1996) Mechanical stability of the in vivo lumbar spine: implications for injury and chronic low back pain. Clin Biomech 11(1):1–15
84. Dolan P, Adams MA (1993) The relationship between EMG activity and extensor moment genera-

tion in the erector spinae muscles during bending and lifting activities. J Biomech 26(4–5):513–522
85. Nachemson AL (1981) Disc pressure measurements. Spine 6(1):93–97
86. Schultz A, Andersson G, Ortengren R, Haderspeck K, Nachemson A (1982) Loads on the lumbar spine. Validation of a biomechanical analysis by measurements of intradiscal pressures and myoelectric signals. J Bone Joint Surg (Am Vol) 64(5):713–720
87. Patwardhan AG, Havey RM, Meade KP, Lee B, Dunlap B (1999) A follower load increases the load-carrying capacity of the lumbar spine in compression. Spine 24(10):1003–1009
88. Patwardhan AG, Meade KP, Lee B (2001) A frontal plane model of the lumbar spine subjected to a follower load: implications for the role of muscles. J Biomech Eng 123(3):212–217
89. Wilke HJ, Wolf S, Claes LE, Arand M, Wiesend A (1995) Stability increase of the lumbar spine with different muscle groups. A biomechanical in vitro study. Spine 20(2):192–198
90. Mannion AF, Adams MA, Dolan P (2000) Sudden and unexpected loading generates high forces on the lumbar spine. Spine 25(7):842–852
91. Solomonow M, Zhou BH, Baratta RV, Lu Y, Harris M (1999) Biomechanics of increased exposure to lumbar injury caused by cyclic loading: part 1. Loss of reflexive muscular stabilization. Spine 24(23):2426–2434
92. Adams MA, Hutton WC (1982) Prolapsed intervertebral disc. A hyperflexion injury 1981 Volvo Award in Basic Science. Spine 7(3):184–191
93. Adams MA, Hutton WC (1985) Gradual disc prolapse. Spine 10(6):524–531
94. McNally DS, Adams MA (1992) Internal intervertebral disc mechanics as revealed by stress profilometry. Spine 17(1):66–73
95. Sato K, Kikuchi S, Yonezawa T (1999) In vivo intradiscal pressure measurement in healthy individuals and in patients with ongoing back problems. Spine 24(23):2468–2474
96. Wilke HJ, Neef P, Caimi M, Hoogland T, Claes LE (1999) New in vivo measurements of pressures in the intervertebral disc in daily life. Spine 24(8):755–762
97. Takahashi I, Kikuchi S, Sato K, Sato N (2006) Mechanical load of the lumbar spine during forward bending motion of the trunk-a biomechanical study. Spine 31(1):18–23
98. Kelsey JL (1975) An epidemiological study of acute herniated lumbar intervertebral discs. Rheumatol Rehabil 14(3):144–159
99. Banks R, Martini J, Smith H, Bowles A, McNish T, Howard R (2000) Alignment of the lumbar vertebrae in a driving posture. J Crash Prev Inj Control 2:123–130
100. Zenk R, Franz M, Bubb H, Vink P (2012) Technical note: spine loading in automotive seating. Appl Ergon 43(2):290–295. doi:10.1016/j.apergo.2011.06.004
101. Gates D, Bridges A, Welch TDJ, Lam TM, Scher I, Yamaguchi G (2010) Lumbar loads in low to moderate speed rear impacts (2010-01-0141). SAE International, Warrendale
102. Szabo TJ, Welcher JB (1996) Human subject kinematics and electromyographic activity during low speed rear impacts (962432). SAE International, Warrendale
103. Delp SL, Suryanarayanan S, Murray WM, Uhlir J, Triolo RJ (2001) Architecture of the rectus abdominis, quadratus lumborum, and erector spinae. J Biomech 34(3):371–375
104. Miller CD, Blyth P, Civil ID (2000) Lumbar transverse process fractures – a sentinel marker of abdominal organ injuries. Injury 31(10):773–776
105. Latimer JA, Tibone JE, Pink MM, Mohr KJ, Perry J (1998) Shoulder reaction time and muscle-firing patterns in response to an anterior translation force. J Shoulder Elb Surg 7:610–615
106. Lippitt S, Matsen F (1993) Mechanisms of glenohumeral joint stability. Clin Orthop Relat Res 291:20–28
107. Blasier RB, Soslowsky LJ, Malicky DM, Palmer ML (1997) Posterior glenohumeral subluxation: active and passive stabilization in a biomechanical model. J Bone Joint Surg (Am Vol) 79(3):433–440
108. Hindle P, Davidson EK, Biant LC, Court-Brown CM (2013) Appendicular joint dislocations. Injury. doi:10.1016/j.injury.2013.01.043
109. Ackland DC, Pandy MG (2009) Lines of action and stabilizing potential of the shoulder musculature. J Anat 215(2):184–197. doi:10.1111/j.1469-7580.2009.01090.x
110. Abboud JA, Soslowsky LJ (2002) Interplay of the static and dynamic restraints in glenohumeral instability. Clin Orthop Relat Res 400:48–57
111. Labriola JE, Lee TQ, Debski RE, McMahon PJ (2005) Stability and instability of the glenohumeral joint: the role of shoulder muscles. J Shoulder Elbow Surg 14(1 Suppl S):32S–38S. doi:10.1016/j.jse.2004.09.014
112. Rockwood CA Jr, Wirth MA (1996) Subluxations and dislocations about the glenohumeral joint. In: Rockwood and Green's fractures in adults, 4th edn. Lippincott-Raven, Philadelphia
113. Bolte JH, Hines MH, McFadden JD, Saul RA (2000) Shoulder response characteristics and injury due to lateral glenohumeral joint impacts. Stapp Car Crash J 44:261–280
114. McMahon PJ, Lee TQ (2002) Muscles may contribute to shoulder dislocation and stability. Clin Orthop Relat Res (403 Suppl):S18–S25
115. Blasier RB, Guldberg RE, Rothman ED (1992) Anterior shoulder stability: contributions of rotator cuff forces and the capsular ligaments in a cadaver model. J Shoulder Elb Surg 1(3):140–150. doi:10.1016/1058-2746(92)90091-G
116. Apreleva M, Parsons IM, Warner JJ, Fu FH, Woo SL (2000) Experimental investigation of reaction forces at the glenohumeral joint during active abduction. J Shoulder Elb Surg 9(5):409–417. doi:10.1067/mse.2000.106321

117. van der Helm FC (1994) Analysis of the kinematic and dynamic behavior of the shoulder mechanism. J Biomech 27(5):527–550
118. Parsons IM, Apreleva M, Fu FH, Woo SL (2002) The effect of rotator cuff tears on reaction forces at the glenohumeral joint. J Orthop Res 20(3):439–446. doi:10.1016/S0736-0266(01)00137-1
119. Sporrong H, Palmerud G, Herberts P (1996) Hand grip increases shoulder muscle activity, An EMG analysis with static hand contractions in 9 subjects. Acta Orthop Scand 67(5):485–490
120. McMahon PJ, Chow S, Sciaroni L, Yang BY, Lee TQ (2003) A novel cadaveric model for anterior-inferior shoulder dislocation using forcible apprehension positioning. J Rehabil Res Dev 40(4):349–359
121. Myers JB, Riemann BL, Ju YY, Hwang JH, McMahon PJ, Lephart SM (2003) Shoulder muscle reflex latencies under various levels of muscle contraction. Clin Orthop Relat Res 407:92–101
122. Mall NA, Lee AS, Chahal J, Sherman SL, Romeo AA, Verma NN, Cole BJ (2013) An evidenced-based examination of the epidemiology and outcomes of traumatic rotator cuff tears. Arthroscopy 29(2):366–376. doi:10.1016/j.arthro.2012.06.024
123. Ko JY, Huang CC, Chen WJ, Chen CE, Chen SH, Wang CJ (2006) Pathogenesis of partial tear of the rotator cuff: a clinical and pathologic study. J Shoulder Elb Surg 15(3):271–278. doi:10.1016/j.jse.2005.10.013
124. Lewis JS (2009) Rotator cuff tendinopathy. Br J Sports Med 43(4):236–241. doi:10.1136/bjsm.2008.052175
125. Blevins FT, Hayes WM, Warren RF (1996) Rotator cuff injury in contact athletes. Am J Sports Med 24(3):263–267
126. Chiu J, Robinovitch SN (1998) Prediction of upper extremity impact forces during falls on the outstretched hand. J Biomech 31(12):1169–1176
127. Snyder SJ, Banas MP, Karzel RP (1995) An analysis of 140 injuries to the superior glenoid labrum. J Shoulder Elb Surg 4(4):243–248
128. Andrews JR, Carson WG Jr, McLeod WD (1985) Glenoid labrum tears related to the long head of the biceps. Am J Sports Med 13(5):337–341
129. Jobe FW, Moynes DR, Tibone JE, Perry J (1984) An EMG analysis of the shoulder in pitching. A second report. Am J Sports Med 12(3):218–220
130. Bey MJ, Elders GJ, Huston LJ, Kuhn JE, Blasier RB, Soslowsky LJ (1998) The mechanism of creation of superior labrum, anterior, and posterior lesions in a dynamic biomechanical model of the shoulder: the role of inferior subluxation. J Shoulder Elb Surg 7(4):397–401
131. Clavert P, Bonnomet F, Kempf JF, Boutemy P, Braun M, Kahn JL (2004) Contribution to the study of the pathogenesis of type II superior labrum anterior-posterior lesions: a cadaveric model of a fall on the outstretched hand. J Shoulder Elb Surg 13(1):45–50. doi:10.1016/S1058274603002519
132. Stueland DT, Stamas P Jr, Welter TM, Cleveland DA (1989) Bilateral humeral fractures from electrically induced muscular spasm. J Emerg Med 7(5):457–459
133. Tehranzadeh J (1987) The spectrum of avulsion and avulsion-like injuries of the musculoskeletal system. Radiographics 7(5):945–974
134. Vochteloo AJ, Henket M, Vincken PW, Nagels J (2012) Bony avulsion of the supraspinatus origin from the scapular spine. J Orthop Traumatol 13(1):51–53. doi:10.1007/s10195-011-0173-8
135. Heyse-Moore GH, Stoker DJ (1982) Avulsion fractures of the scapula. Skelet Radiol 9(1):27–32
136. Kelly JP (1954) Fractures complicating electroconvulsive therapy and chronic epilepsy. J Bone Joint Surg Br Vol 36-B(1):70–79
137. Salmons S (1995) Muscle. In: Williams PL (ed) Gray's anatomy, 38th edn. Churchill Livingstone, New York
138. Alcid JG, Ahmad CS, Lee TQ (2004) Elbow anatomy and structural biomechanics. Clin Sports Med 23(4):503–517. doi:10.1016/j.csm.2004.06.008, vii
139. de Haan J, Schep NW, Eygendaal D, Kleinrensink GJ, Tuinebreijer WE, den Hartog D (2011) Stability of the elbow joint: relevant anatomy and clinical implications of in vitro biomechanical studies. Open Orthop J 5:168–176. doi:10.2174/1874325001105010168
140. Fornalski S, Gupta R, Lee TQ (2003) Anatomy and biomechanics of the elbow joint. Tech Hand Up Extrem Surg 7(4):168–178
141. Zimmerman NB (2002) Clinical application of advances in elbow and forearm anatomy and biomechanics. Hand Clin 18(1):1–19
142. Bryce CD, Armstrong AD (2008) Anatomy and biomechanics of the elbow. Orthop Clin N Am 39(2):141–154. doi:10.1016/j.ocl.2007.12.001, v
143. Morrey BF, An KN (1983) Articular and ligamentous contributions to the stability of the elbow joint. Am J Sports Med 11(5):315–319
144. Safran MR, Baillargeon D (2005) Soft-tissue stabilizers of the elbow. J Shoulder Elbow Surg 14(1 Suppl S):179S–185S. doi:10.1016/j.jse.2004.09.032
145. Hotchkiss RN (1996) Fractures and dislocations of the elbow. In: Rockwood and Green's fractures in adults, 4th edn. Lippincott-Raven, Philadelphia
146. Linscheid RL, Wheeler DK (1965) Elbow dislocations. JAMA 194(11):1171–1176
147. Chou PH, Chou YL, Lin CJ, Su FC, Lou SZ, Lin CF, Huang GF (2001) Effect of elbow flexion on upper extremity impact forces during a fall. Clin Biomech 16(10):888–894
148. Hanson CT, Joslow B, Danoff JV, Alon G (1981) Electromyographic response of the elbow flexors to a changing, dislocating force. Arch Phys Med Rehabil 62(12):631–634
149. Holmes MW, Keir PJ (2012) Posture and hand load alter muscular response to sudden elbow perturbations. J Electromyogr Kinesiol 22(2):191–198. doi:10.1016/j.jelekin.2011.11.006
150. DeGoede KM, Ashton-Miller JA (2002) Fall arrest strategy affects peak hand impact force in a forward fall. J Biomech 35(6):843–848
151. Lo J, Ashton-Miller JA (2008) Effect of upper and lower extremity control strategies on predicted injury risk during simulated forward falls: a study in

healthy young adults. J Biomech Eng 130(4):041015. doi:10.1115/1.2947275
152. Levy M, Goldberg I, Meir I (1982) Fracture of the head of the radius with a tear or avulsion of the triceps tendon. A new syndrome? J Bone Joint Surg (Br) 64(1):70–72
153. Yazdi HR, Qomashi I, Ghorban Hoseini M (2012) Neglected triceps tendon avulsion: case report, literature review, and a new repair method. Am J Orthop 41(7):E96–E99
154. Dietz V, Noth J, Schmidtbleicher D (1981) Interaction between pre-activity and stretch reflex in human triceps brachii during landing from forward falls. J Physiol 311:113–125
155. Holmes AM, Andrews DM (2006) The effect of leg muscle activation state and localized muscle fatigue on tibial response during impact. J Appl Biomech 22(4):275–284
156. Milner CE, Ferber R, Pollard CD, Hamill J, Davis IS (2006) Biomechanical factors associated with tibial stress fracture in female runners. Med Sci Sports Exerc 38(2):323–328. doi:10.1249/01.mss.0000183477.75808.92
157. Burkhart TA, Andrews DM (2010) Activation level of extensor carpi ulnaris affects wrist and elbow acceleration responses following simulated forward falls. J Electromyogr Kinesiol 20(6):1203–1210. doi:10.1016/j.jelekin.2010.07.008
158. Hansen U, Zioupos P, Simpson R, Currey JD, Hynd D (2008) The effect of strain rate on the mechanical properties of human cortical bone. J Biomech Eng 130(1):011011. doi:10.1115/1.2838032
159. Conroy C, Schwartz A, Hoyt DB, Brent Eastman A, Pacyna S, Holbrook TL, Vaughan T, Sise M, Kennedy F, Velky T, Erwin S (2007) Upper extremity fracture patterns following motor vehicle crashes differ for drivers and passengers. Injury 38(3):350–357. doi:10.1016/j.injury.2006.03.017
160. McKendrew C, Hines MH, Litsky A, Saul RA (1998) Assessment of forearm injury due to a deploying driver-side airbag (98-S5-O-09). In: Proceedings of the 16th ESV conference, Windsor
161. Chong M, Broome G, Mahadeva D, Wang S (2011) Upper extremity injuries in restrained front-seat occupants after motor vehicle crashes. J Trauma 70(4):838–844. doi:10.1097/TA.0b013e3181df6848
162. Fleisig GS, Andrews JR, Dillman CJ, Escamilla RF (1995) Kinetics of baseball pitching with implications about injury mechanisms. Am J Sports Med 23(2):233–239
163. Park MC, Ahmad CS (2004) Dynamic contributions of the flexor-pronator mass to elbow valgus stability. J Bone Joint Surg (Am Vol) 86-A(10):2268–2274
164. Lin F, Kohli N, Perlmutter S, Lim D, Nuber GW, Makhsous M (2007) Muscle contribution to elbow joint valgus stability. J Shoulder Elb Surg 16(6):795–802. doi:10.1016/j.jse.2007.03.024
165. Udall JH, Fitzpatrick MJ, McGarry MH, Leba TB, Lee TQ (2009) Effects of flexor-pronator muscle loading on valgus stability of the elbow with an intact, stretched, and resected medial ulnar collateral ligament. J Shoulder Elb Surg 18(5):773–778. doi:10.1016/j.jse.2009.03.008
166. Stevens MA, El-Khoury GY, Kathol MH, Brandser EA, Chow S (1999) Imaging features of avulsion injuries. Radiographics 19(3):655–672
167. Ogawa K, Ui M (1996) Fracture-separation of the medial humeral epicondyle caused by arm wrestling. J Trauma 41(3):494–497
168. Doss WS, Karpovich PV (1965) A comparison of concentric, eccentric, and isometric strength of elbow flexors. J Appl Physiol 20:351–353
169. Cooney WP, Linscheid RL, Dobyns JH (1996) Fractures and dislocations of the wrist. In: Rockwood and Green's fractures in adults, 4th edn. Lippincott-Ravel, Philadelphia
170. Ruby LK (1992) Wrist Biomechanics. Instr Course Lect 41:25–32
171. Petrie S, Collins J, Solomonow M, Wink C, Chuinard R (1997) Mechanoreceptors in the palmar wrist ligaments. J Bone Joint Surg Br Vol 79(3):494–496
172. Salva-Coll G, Garcia-Elias M, Leon-Lopez MT, Llusa-Perez M, Rodriguez-Baeza A (2011) Effects of forearm muscles on carpal stability. J Hand Surg Eur Vol 36(7):553–559. doi:10.1177/1753193411407671
173. Boggess BM, Sieveka EM, Crandall JR, Pilkey WD, Duma SM (2001) Interaction of the hand and wrist with a door handgrip during static side air bag deployment: simulation study using the CVS/ATB multi-body program (2001-01-0170). SAE International, Warrendale
174. Sokol JA, Potier P, Robin S, Le Coz JY, Lassau JP (1998) Upper extremity interaction with side impact airbag. IRCOBI, Göteborg
175. Nikolic V, Hancevic J, Hudec M, Banovie B (1975) Absorption of the impact energy in the palmar soft tissues. Anat Embryol 148(2):215–221
176. Choi WJ, Robinovitch SN (2011) Pressure distribution over the palm region during forward falls on the outstretched hands. J Biomech 44(3):532–539. doi:10.1016/j.jbiomech.2010.09.011
177. McNair PJ, Prapavessis H (1999) Normative data of vertical ground reaction forces during landing from a jump. J Sci Med Sport 2(1):86–88
178. Ortega DR, Bies ECR, de la Rosa FJB (2010) Analysis of the vertical ground reaction forces and temporal factors in the landing phase of a countermovement jump. J Sports Sci and Med 9:282–287
179. Armstrong RW, Waters HP, Stapp JP (1968) Human muscular restraint during sled deceleration (680793). SAE International, Warrendale
180. Schreiber P, Crandall JR, Hurwitz S, Nusholtz GS (1998) Static and dynamic bending strength of the leg. Int J Crashworthiness 3:295–308
181. Nordsletten L, Ekeland A (1993) Muscle contraction increases the structural capacity of the lower leg: an in vivo study in the rat. J Orthop Res 11(2):299–304. doi:10.1002/jor.1100110218
182. Morris R, Cross G (2005) Improved understanding of passenger behaviour during pre-impact events to aid

smart restraint development (05-0320). In: Proceedings of the 19th ESV conference, Washington, DC
183. Kitagawa Y, Ichikawa H, Pal C, King AI, Levine RS (1998) Lower leg injuries caused by dynamic axial loading and muscle testing (98-S7-O-09). In: Proceedings of the 16th ESV conference, Windsor
184. Klopp GS, Crandall JR, Sieveka EM, Pilkey WD (1995) Simulation of muscle tensing in pre-impact bracing. IRCOBI, Brunnen
185. Tencer AF, Kaufman R, Ryan K, Grossman DC, Henley BM, Mann F, Mock C, Rivara F, Wang S, Augenstein J, Hoyt D, Eastman B, Crash Injury Research and Engineering Network (CIREN) (2002) Femur fractures in relatively low speed frontal crashes: the possible role of muscle forces. Accid Anal Prev 34(1):1–11
186. Maffulli N, Grewal R (1997) Avulsion of the tibial tuberosity: muscles too strong for a growth plate. Clin J Sport Med 7(2):129–132, discussion 132–133
187. Nyland JA, Shapiro R, Stine RL, Horn TS, Ireland ML (1994) Relationship of fatigued run and rapid stop to ground reaction forces, lower extremity kinematics, and muscle activation. J Orthop Sports Phys Ther 20(3):132–137
188. Myer GD, Ford KR, Barber Foss KD, Liu C, Nick TG, Hewett TE (2009) The relationship of hamstrings and quadriceps strength to anterior cruciate ligament injury in female athletes. Clin J Sport Med 19(1):3–8. doi:10.1097/JSM.0b013e318190bddb
189. Anderson K, Strickland SM, Warren R (2001) Hip and groin injuries in athletes. Am J Sports Med 29(4):521–533
190. Monma H, Sugita T (2001) Is the mechanism of traumatic posterior dislocation of the hip a brake pedal injury rather than a dashboard injury? Injury 32(3):221–222
191. Pearson JR, Hargadon EJ (1962) Fractures of the pelvis involving the floor of the acetabulum. J Bone Joint Surg Br Vol 44-B:550–561
192. Mader TJ, Booth J, Gaudet C, Hynds-Decoteau R (2006) Seizure-induced acetabular fractures: 5-year experience and literature review. Am J Emerg Med 24(2):230–232. doi:10.1016/j.ajem.2005.10.011
193. Hsiao ET, Robinovitch SN (1998) Common protective movements govern unexpected falls from standing height. J Biomech 31(1):1–9
194. Robinovitch SN, Normandin SC, Stotz P, Maurer JD (2005) Time requirement for young and elderly women to move into a position for breaking a fall with outstretched hands. J Gerontol A: Biol Med Sci 60(12):1553–1557
195. Robinovitch SN, Brumer R, Maurer J (2004) Effect of the "squat protective response" on impact velocity during backward falls. J Biomech 37(9):1329–1337. doi:10.1016/j.jbiomech.2003.12.015
196. Robinovitch SN, Hayes WC, McMahon TA (1991) Prediction of femoral impact forces in falls on the hip. J Biomech Eng 113(4):366–374
197. Robinovitch SN, Hayes WC, McMahon TA (1997) Distribution of contact force during impact to the hip. Ann Biomed Eng 25(3):499–508
198. Robinovitch SN, McMahon TA, Hayes WC (1995) Force attenuation in trochanteric soft tissues during impact from a fall. J Orthop Res 13(6):956–962. doi:10.1002/jor.1100130621
199. Etheridge BS, Beason DP, Lopez RR, Alonso JE, McGwin G, Eberhardt AW (2005) Effects of trochanteric soft tissues and bone density on fracture of the female pelvis in experimental side impacts. Ann Biomed Eng 33(2):248–254
200. Song E, Fontaine L, Troseille X, Guillemot H (2005) Pelvis bone fracture modeling in lateral impact. In: Proceedings of the 19th ESV conference, Washington, DC
201. Chang CY, Rupp JD, Reed MP, Hughes RE, Schneider LW (2009) Predicting the effects of muscle activation on knee, thigh, and hip injuries in frontal crashes using a finite-element model with muscle forces from subject testing and musculoskeletal modeling. Stapp Car Crash J 53:291–328
202. Beeman SM, Kemper AR, Madigan ML, Duma SM (2011) Effects of bracing on human kinematics in low-speed frontal sled tests. Ann Biomed Eng 39(12):2998–3010. doi:10.1007/s10439-011-0379-1
203. Pope MH, Johnson RJ, Brown DW, Tighe C (1979) The role of the musculature in injuries to the medial collateral ligament. J Bone Joint Surg (Am Vol) 61(3):398–402
204. Alentorn-Geli E, Myer GD, Silvers HJ, Samitier G, Romero D, Lazaro-Haro C, Cugat R (2009) Prevention of non-contact anterior cruciate ligament injuries in soccer players. Part 1: mechanisms of injury and underlying risk factors. Knee Surg Sports Traumatol Arthrosc 17(7):705–729. doi:10.1007/s00167-009-0813-1
205. Lephart SM, Abt JP, Ferris CM (2002) Neuromuscular contributions to anterior cruciate ligament injuries in females. Curr Opin Rheumatol 14(2):168–173
206. Shimokochi Y, Shultz SJ (2008) Mechanisms of non-contact anterior cruciate ligament injury. J Athl Train 43(4):396–408. doi:10.4085/1062-6050-43.4.396
207. Colby S, Francisco A, Yu B, Kirkendall D, Finch M, Garrett W Jr (2000) Electromyographic and kinematic analysis of cutting maneuvers. Implications for anterior cruciate ligament injury. Am J Sports Med 28(2):234–240
208. Li G, Rudy TW, Sakane M, Kanamori A, Ma CB, Woo SL (1999) The importance of quadriceps and hamstring muscle loading on knee kinematics and in-situ forces in the ACL. J Biomech 32(4):395–400
209. DeMorat G, Weinhold P, Blackburn T, Chudik S, Garrett W (2004) Aggressive quadriceps loading can induce noncontact anterior cruciate ligament injury. Am J Sports Med 32(2):477–483
210. Fujiya H, Kousa P, Fleming BC, Churchill DL, Beynnon BD (2011) Effect of muscle loads and

torque applied to the tibia on the strain behavior of the anterior cruciate ligament: an in vitro investigation. Clin Biomech 26:1005–1011
211. Ahmad CS, Clark AM, Heilmann N, Schoeb JS, Gardner TR, Levine WN (2006) Effect of gender and maturity on quadriceps-to-hamstring strength ratio and anterior cruciate ligament laxity. Am J Sports Med 34(3):370–374. doi:10.1177/0363546505280426
212. Chappell JD, Creighton RA, Giuliani C, Yu B, Garrett WE (2007) Kinematics and electromyography of landing preparation in vertical stop-jump: risks for noncontact anterior cruciate ligament injury. Am J Sports Med 35(2):235–241. doi:10.1177/0363546506294077
213. Myer GD, Ford KR, Hewett TE (2005) The effects of gender on quadriceps muscle activation strategies during a maneuver that mimics a high ACL injury risk position. J Electromyogr Kinesiol 15(2):181–189. doi:10.1016/j.jelekin.2004.08.006
214. Arendt EA, Agel J, Dick R (1999) Anterior cruciate ligament injury patterns among collegiate men and women. J Athl Train 34(2):86–92
215. Rozzi SL, Lephart SM, Fu FH (1999) Effects of muscular fatigue on knee joint laxity and neuromuscular characteristics of male and female athletes. J Athl Train 34(2):106–114
216. Thomas AC, McLean SG, Palmieri-Smith RM (2010) Quadriceps and hamstrings fatigue alters hip and knee mechanics. J Appl Biomech 26(2):159–170
217. Cowling EJ, Steele JR, McNair PJ (2003) Effect of verbal instructions on muscle activity and risk of injury to the anterior cruciate ligament during landing. Br J Sports Med 37(2):126–130
218. Gilchrist J, Mandelbaum BR, Melancon H, Ryan GW, Silvers HJ, Griffin LY, Watanabe DS, Dick RW, Dvorak J (2008) A randomized controlled trial to prevent noncontact anterior cruciate ligament injury in female collegiate soccer players. Am J Sports Med 36(8):1476–1483. doi:10.1177/0363546508318188
219. Holcomb WR, Rubley MD, Lee HJ, Guadagnoli MA (2007) Effect of hamstring-emphasized resistance training on hamstring:quadriceps strength ratios. J Strength Cond Res 21(1):41–47. doi:10.1519/R-18795.1
220. Ruan JS, El-Jawahri R, Barbat S, Rouhana SW, Prasad P (2008) Impact response and biomechanical analysis of the knee-thigh-hip complex in frontal impacts with a full human body finite element model. Stapp Car Crash J 52:505–526
221. Mukherjee S, Chawla A, Karthikeyan B, Soni A (2007) Finite element crash simulations of the human body: passive and active muscle modelling. Sadhana 32:409–426
222. Soni A, Chawla A, Mukherjee S (2006) Effect of active muscle forces on the response of knee joint at low-speed lateral impacts (2006-01-0460). SAE International, Warrendale
223. Soni A, Chawla A, Mukherjee S, Malhotra R (2009) Sensitivity analysis of muscle parameters and identification of effective muscles in low speed lateral impact at just below the knee (2009-01-1211). SAE International, Warrendale
224. Chawla A, Mukherjee S, Soni A, Malhotra R (2008) Effect of active muscle forces on knee injury risks for pedestrian standing posture at low-speed impacts. Traffic Inj Prev 9(6):544–551. doi:10.1080/15389580802338228
225. Hand WL, Hand CR, Dunn AW (1971) Avulsion fractures of the tibial tubercle. J Bone Joint Surg (Am Vol) 53(8):1579–1583
226. Shields CL, Ashby ME (1975) Diagnosis in patellar tendon avulsion. J Natl Med Assoc 67(3):231–232
227. Zernicke RF, Garhammer J, Jobe FW (1977) Human patellar-tendon rupture. J Bone Joint Surg (Am Vol) 59(2):179–183
228. Abboud J (2002) Relevant foot biomechanics. Curr Orthop 16:165–179
229. Andersen TE, Floerenes TW, Arnason A, Bahr R (2004) Video analysis of the mechanisms for ankle injuries in football. Am J Sports Med 32(1 Suppl):69S–79S
230. Bahr R, Krosshaug T (2005) Understanding injury mechanisms: a key component of preventing injuries in sport. Br J Sports Med 39(6):324–329. doi:10.1136/bjsm.2005.018341
231. de Asla RJ, Kozanek M, Wan L, Rubash HE, Li G (2009) Function of anterior talofibular and calcaneofibular ligaments during in-vivo motion of the ankle joint complex. J Orthop Surg Res 4:7. doi:10.1186/1749-799X-4-7
232. Fong DT, Ha SC, Mok KM, Chan CW, Chan KM (2012) Kinematics analysis of ankle inversion ligamentous sprain injuries in sports: five cases from televised tennis competitions. Am J Sports Med 40(11):2627–2632. doi:10.1177/0363546512458259
233. Kristianslund E, Bahr R, Krosshaug T (2011) Kinematics and kinetics of an accidental lateral ankle sprain. J Biomech 44(14):2576–2578. doi:10.1016/j.jbiomech.2011.07.014
234. Gehring D, Wissler S, Mornieux G, Gollhofer A (2013) How to sprain your ankle – a biomechanical case report of an inversion trauma. J Biomech 46(1):175–178. doi:10.1016/j.jbiomech.2012.09.016
235. Mitchell A, Dyson R, Hale T, Abraham C (2008) Biomechanics of ankle instability. Part 1: reaction time to simulated ankle sprain. Med Sci Sports Exerc 40(8):1515–1521
236. Baumhauer JF, Alosa DM, Renstrom AF, Trevino S, Beynnon B (1995) A prospective study of ankle injury risk factors. Am J Sports Med 23(5):564–570
237. Willems TM, Witvrouw E, Delbaere K, Mahieu N, De Bourdeaudhuij I, De Clercq D (2005) Intrinsic risk factors for inversion ankle sprains in male subjects: a prospective study. Am J Sports Med 33(3):415–423
238. Gefen A (2002) Biomechanical analysis of fatigue-related foot injury mechanisms in athletes and recruits during intensive marching. Med Biol Eng Comput 40:302–310

239. Leppilahti J, Orava S (1998) Total Achilles tendon rupture. A review. Sports Med 25(2):79–100
240. Arndt AN, Komi PV, Bruggemann GP, Lukkariniemi J (1998) Individual muscle contributions to the in vivo Achilles tendon force. Clin Biomech 13(7):532–541
241. Manning PW (1998) Dynamic response and injury mechanism in the human foot and ankle and an analysis of dummy biofidelity. In: Proceedings of the 16th ESV conference, Windsor
242. McMaster J, Parry M, Wallace WA, Wheeler L, Owen C, Lowne R, Oakley C, Roberts AK (2000) Biomechanics of ankle and hindfoot injuries in dynamic axial loading. Stapp Car Crash J 44:357–377
243. Funk JR, Crandall JR, Tourret LJ, MacMahon CB, Bass CR, Khaewpong N, Eppinger RH (2001) The effect of active muscle tension on the axial injury tolerance of the human foot/ankle complex (237). In: Proceedings of the 17th ESV conference, Amsterdam
244. Chandler RF, Christian RA (1970) Crash testing of humans in automotive seats (700361). SAE International, Warrendale
245. Shaw G, Lessley D, Crandall J, Kent R, Kitis L (2005) Elimination of thoracic muscle tensing effects for frontal crash dummies. SAE, Warrendale
246. Kent R, Bass C, Woods W, Salzar R, Melvin J (2004) The role of muscle tensing on the force deflection response of the thorax and a reassessment of frontal impact biofidelity corridors. In: IRCOBI conference on the biomechanics, Graz, Austria
247. Choi HY, Sah SJ, Lee B, Cho HS, Kang SJ, Mun MS, Lee I, Lee J (2005) Experimental and numerical studies of muscular activations of bracing occupants (05-0139). In: Proceedings of the 19th ESV conference, Washington, DC
248. Hault-Dubrulle A, Robache F, Drazetic P, Morvan H (2009) Pre-crash phase analysis using a driving simulator. Influence of atypical position on injuries and airbag adaptation (09-0534). In: Proceedings of the 21st ESV conference, Stuttgart
249. Iwamoto M, Nakahira Y, Sugiyama T (2011) Investigation of pre-impact bracing effects for injury outcome using an active human FE model with 3D geometry of muscles (11-0150). In: Proceedings of the 22nd ESV conference, Washington, DC
250. Bose D, Crandall JR (2008) Influence of active muscle contribution on the injury response of restrained car occupants. Ann Adv Automot Med 52:61–72
251. Bose D, Crandall JR, Untaroiu CD, Maslen EH (2010) Influence of pre-collision occupant parameters on injury outcome in a frontal collision. Accid Anal Prev 42(4):1398–1407. doi:10.1016/j.aap.2010.03.004
252. Ejima S, Zama Y, Ono K, Kaneoka K, Shiina I, Asada H (2009) Prediction of pre-impact occupant kinematic behavior based on the muscle activity during frontal collision (09-0913). In: Proceedings of the 21st ESV conference, Stuttgart
253. Osth J, Brolin K, Carlsson S, Wismans J, Davidsson J (2012) The occupant response to autonomous braking: a modeling approach that accounts for active musculature. Traffic Inj Prev 13(3):265–277. doi:10.1080/15389588.2011.649437

Pediatric Biomechanics

Kristy B. Arbogast and Matthew R. Maltese

Abstract

During the human postnatal developmental process, extensive tissue and morphological changes occur. Many take place in the first few years of life but substantial development for several body regions continues well into young adulthood. Along with overall change in size, these material and structural changes influence the biomechanical response of child such that they respond differently to traumatic load than an adult. Understanding the unique biomechanical response of the child is challenging, as compared to the wealth of biomechanical data on the adult response to trauma, pediatric biomechanical data are relatively sparse. As a result, quantitative scaling relationships based on anatomical and material differences have been historically used to understand the biomechanics of the child. The last decade, however, has seen a tremendous increase in contributions to the biomechanics literature based upon pediatric subjects – volunteers, postmortem human subjects, and animal models – thus increasing our knowledge of how to design injury mitigation systems to protect the young.

In this chapter, aspects of developmental anatomy and biomechanical knowledge are reviewed to provide context for pediatric human injury prediction. Emphasis is initially placed on the head and brain as this body region represents the most common seriously injured body region for children in virtually all unintentional injury modes. Specifically, head injuries are particularly relevant clinically as the developing brain is difficult to evaluate and treat, and even mild brain injuries in childhood can lead to deficits that remain long after the injury. Discussion follows on the cervical spine and thorax as these body regions are not only important from an injury mitigation standpoint but they govern the kinematics of the head during traumatic loading and therefore play a role in head injury protection. A brief description follows for the other body regions: the abdomen

K.B. Arbogast, Ph.D. (✉) • M.R. Maltese, Ph.D.
Children's Hospital of Philadelphia, University
of Pennsylvania, Philadelphia, PA, USA
e-mail: arbogast@email.chop.edu;
maltese@email.chop.edu

and extremities as well as an outline of the scaling theory used by many researchers to scale adult biomechanical data to the child. The biomechanics data contained in this chapter may assist in improving the accuracy of pediatric injury criteria and the biofidelity of child anthropometric test devices (ATD) and human body computer models. Because of space limitations, this chapter does not serve as an inclusive data repository for all pediatric material property and biomechanical response data but rather summarizes the seminal publications in the field and directs the reader to other resources for more detailed data.

22.1 Head

Much like the adult, the pediatric brain is highly incompressible and extremely soft relative to the skull, with heterogeneity in stiffness and degree of anisotropy. In general, the vault can be thought of as rigid and thus subject to no shape change when subjected to a direct blunt blow. This condition holds true for most helmeted and padded impact conditions. There are two exceptions to the rigid skull assumption, that being in the case of the infant, where the incomplete development of skull bones and deformability of the sutures leads to change in shape from direct impact. The second is when skull fracture occurs. In both of these conditions, brain deformation results from both inertial forces and shape change of the brain case. To develop mathematical relationships between direct or inertial forces that are experienced by the brain during a traumatic brain injury (TBI) event and the resulting injury, one must understand the structure and material properties of the skull and brain.

22.1.1 Developmental Anatomy

22.1.1.1 The Skull

The skull consists of the neurocranium and viscerocranium. The term skull generally refers to the entire bony structure and cranium refers only to the fused regions, i.e. without the mandible. The adult neurocranium is a series of irregularly shaped fused flat bones. Eight bones make up the neurocranium: frontal, left and right parietal, left and right temporal, occipital, sphenoid and ethmoid. The sphenoid and ethmoid bones provide junction with the face and anterior base of the skull. The calvarium consists of the frontal, parietal, and occipital bones the form the bulk of the convexity on the top. Figure 22.1 illustrates the superior aspect of the adult and newborn skull. In the newborn, the occipital condyles are elongated and flat instead of the curved shape seen in the adult. The tympanic rings form the prominent features of the base of the skull and provide attachment for the tympanic membranes. The calvarium extends laterally and posteriorly beyond the base of the skull. The viscerocranium consists of the bones that support the face.

Unlike the mature adult human cranial bone, which consists of two layers of rigid cortical bone (inner and outer tables) housing the relatively deformable cancellous (diploe) layer, in the newborn the cranial bone is primarily cortical, with no diploe component. Around 3–6 months of age, the structure gradually transforms into the sandwich composition of the adult skull. At birth, the calvarial bones are thin and ossification does not extend into the suture lines of the skull. The frontal bone is divided along the midline by the metopic suture and outlined posteriorly by the coronal suture. The parietal bones are two symmetric plates covering the lateral aspect of the skull. They are separated by the sagittal suture in the midline. Anteriorly, they are abutted by the coronal suture, posteriorly by the lambdoidal suture, and inferiorly but the squamosal suture. The occipital bone, or more specifically the squamous portion of the occipital bone, is formed by the lambdoidal suture that traverses

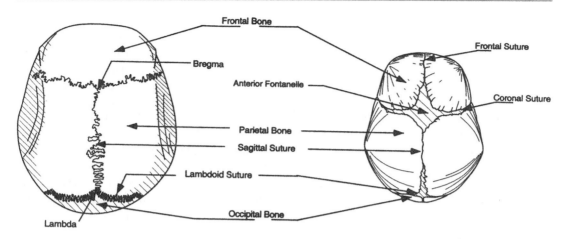

Fig. 22.1 Superior aspect of the adult (*left*) and newborn (*right*) skull. Various bones and sutures are indicated (Adapted from Williams [1])

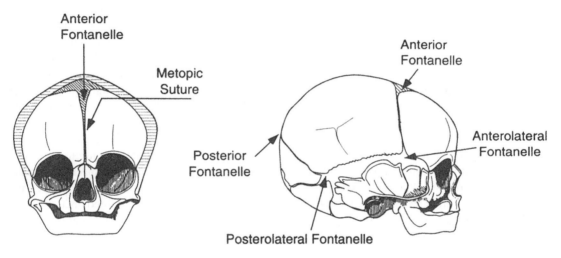

Fig. 22.2 Front and side views illustrating the fontanelles (Redrawn from Agur and Lee [2], Tindall et al. [4], Williams [1], and Youman [5])

symmetrically from the lambdoid and merges into the posteriolateral fontanelle. With age, the sutures continue to diminish in size, but closure occurs at different intervals. The metopic suture closes during the sixth to eighth year of life. Complete fusion of the bony plates with obliteration of the sutures occurs around 20 years of age when the skull has reaches is full definitive size.

The soft spots or fontanelles of a newborn head are large patches of fibrous tissue located between the boney plates of the skull [1–5]. Fontanelles can be compared to the synchondrosis of the cervical spine (Fig. 22.2). They are not fused in the newborn. There are six major fontanelles with variable smaller or accessory fontanelles usually located along the sagittal suture. The six fontanelles are the anterior, posterior, and the paired anterolateral and posteriolateral fontanelles. The anterior fontanelle is located along the bregma of the intersection of the coronal, sagittal, and metopic sutures. This fontanelle is the largest at birth with an average diameter of 25 mm. The posterior fontanelle is located along the lambda, which is the intersection of the sagittal and lambdoidal sutures. The anterolateral fontanelles are located along the pterion, which is the intersection of the coronal and squamosal sutures. Closure of the fontanelles occurs at various times. The posterior and anterolateral fontanelles close within 2–3 months after birth, and the posteriolateral fontanelle closes approximately at 1 year. In contrast, the anterior fontanelle closes at around 18–24 months of age.

22.1.1.2 Brain

In utero, a specialized region of cells forms the neural plate from which the structures of the nervous system are formed by cellular division. As the cells along the outer most edge proliferate more rapidly, the neural groove is created. This process continues with the margins growing up and around to form the neural tube, with closure initially at the area of the neck, and proceeding in the rostral and caudal directions. The various components of the central nervous system develop from the neural tube. Once closed, the neural tube represents all components of the nervous system with the rostral portion forming the cerebral hemispheres and brainstem, and the caudal portion forming the spinal cord. These events form the basic blueprint of the human nervous system from which growth in utero will proceed until birth.

Post-natal, the brain water content decreases and there is general increase in the volume the brain. By age 2, the brain has reached nearly 80 % of its adult volume and by age 5 the brain is adult-size. From age 4 to 20 brain white matter volume increases linearly, but these increases are spatially inhomogeneous with white matter increases in the dorsal prefrontal cortex, but not in the ventral prefrontal regions during this time period. Cortical gray matter volume increases non-linearly to maximum at age 12 and then decreases to age 20 [1, 6–8].

22.1.2 Biomechanical Studies

22.1.2.1 Skull Material Properties

While studies describing the mechanical properties of the adult human cranial bone under compression, tension, bending and shear are available, a paucity of such information exists for the developing pediatric population [9–12]. Specific to the growing population, in an early study, cranial bone specimens from two full-term newborns (gestation age 40 weeks) and one 6-year-old human cadaver were tested using three-point bending technique at a quasi-static rate of 0.5 mm/min [13–15]. In the two newborn specimens, frontal and parietal bone samples were tested. In the 6-year-old specimen, parietal bone samples were used. Tests were conducted along an axis parallel and perpendicular to the long axis of the specimen. Data from the 6-year-old parietal specimen indicated differences in the modulus between the parallel (7.38 ± 0.84 GPa) and perpendicular (5.86 ± 0.69 GPa) orientations. Significant differences in the elastic modulus were also found between parallel (3.88 ± 0.78 GPa) and perpendicular (0.951 ± 0.572 GPa) fiber orientation for the newborn (Fig. 22.3). In addition to these data, tests were conducted on four fetal cadavers with gestational age ranging from 25 to 40 weeks (Table 22.1). Significant differences in the elastic modulus were found between parallel (1.65 ± 1.17 GPa) and perpendicular (0.145 ± 0.062 GPa) orientations.

In a later study, cranial bone specimens from 10 human cadavers with age ranging from 20 to 42 gestational weeks were tested using three-point bending techniques at a loading rate of 0.0083 mm/s [14]. The bending stiffness ranged from 1.4 to 25.1 N/mm. The elastic modulus for the perpendicular orientation (1.044–1.996 GPa) was lower than the modulus for the parallel orientation (3.02–7.46 GPa). These data, on a specimen-by-specimen basis, are included in Table 22.2.

Margulies and Thibault obtained human infant cranial bone specimens from four cadavers [16]. Subjects ranged in age from 25 weeks gestation to 6 months of age. From each cadaver, bilateral strips of parietal bone adjacent to the sagittal suture were excised. All strips were cut parallel to sagittal suture. For each test specimen, the rupture modulus (σ_{rupt}), elastic modulus (E), and energy per unit volume absorbed to failure (U_o) were calculated from the measured force and centerline deflection data. Tests were conducted

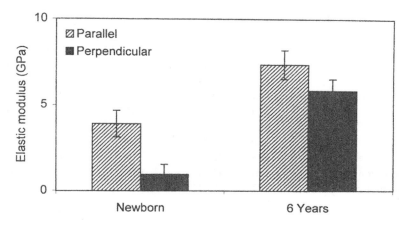

Fig. 22.3 Comparison of the elastic modulus between the newborn and 6-year-old human cadavers in three-point bending under parallel and perpendicular orientations (Adapted from McPherson and Kriewall [15])

Table 22.1 Elastic modulus from three-point bending tests on six human cadavers (Adapted from McPherson and Kriewall [15])

Gestational age (weeks)	Elastic modulus (GPa)							
	Parallel orientation				Perpendicular orientation			
	Parietal bone	n	Frontal bone	n	Parietal bone	n	Frontal bone	n
25 ± 2	1.30 ± 0.6	8	–		0.12 ± 0.01	8		
27 ± 2	0.94 ± 0.41	10	–		0.18 ± 0.03	3		
28 ± 2	3.62 ± 0.46	5	–		0.14 ± 0.08	5		
38 ± 2	4.24 ± 0.73	9	–		0.84 ± 0.19	8		
40 ± 2	4.01 ± 1.28	3	3.05 ± 0.88	2	1.74 ± 0.59	3	1.70 ± 0.79	2
40 ± 2	3.51 ± 0.50	10	3.06 ± 0.84	10	0.57 ± 0.14	5		

n number of samples tested in each cadaver

Table 22.2 Biomechanical properties of human fetal parietal bone under three-point bending at loading rate of 0.0083 mm/s. Data obtained from 10 cadavers (Adapted from Kriewall [14])

Gestational age (weeks)	Stiffness (N/mm)	Elastic modulus (MPa)	
		Parallel orientation	Perpendicular orientation
20	4.1	4,232	1,238
30	3.3	3,478	1,044
30	1.4	3,634	1,626
33	2.3	3,029	1,996
34	4.7	3,390	1,646
39	17.5	4,301	–
40/newborn	25.1	5,167	1,434
40/newborn	20.4	2,961	–
40/newborn	6.8	3,594	1,684
2-week-old	14.1	7,360	–

at slow (0.042 mm/s) and fast (42.33 mm/s) rates. These data are included in Table 22.3.

Coats and Margulies tested human infant cranial bone in three point bending at high rates similar to those experienced in a fall [17]. Human pediatric cranial bone and suture were collected at autopsy (46 specimens from 21 infant calveria). Two cranial specimens were removed and frozen from each subject: one occipital bone and one parietal parasagittal sample. A drop test apparatus was designed and validated to test samples in three-point bending and tension at high rates. Bending modulus, ultimate stress, and ultimate strain were calculated (Table 22.4). The authors emphasized that the ultimate strain reported should be considered an underestimation

Table 22.3 Mechanical properties of human infant parietal bone in three-point bending (Adapted from Margulies and Thibault [16])

Cadaver	Age	Sample	Side	Rate (mm/s)	σ_{Rupt} (MPa)	E (MPa)	U_o (N·mm/mm³)
1	25 weeks gest.	1	Left	0.042	4.5	71.6	0.0312
		2	Left	42.33	4.0	43.8	
2	30 week gest.	1	Left	0.042	3.1	95.3	
		2	Right	42.33	11.2	444.5	0.0624
		3	Right	0.042	14.9	618.8	
		4	Right	42.33	8.9	407.7	0.0575
		5	Right	42.33	17.0	455.4	0.0675
3	1 week term	1	Left	42.33	10.6	820.9	0.0607
4	6 months term	1	Left	0.042	42.1	2,111.7	0.1392
		2	Right	0.042	44.6	2,199.4	0.1834
		3	Left	42.33		2,671.9	
		4	Right	42.33	71.7	3,582.2	0.4361

of the actual strain of the material. The authors conducted a three-way ANOVA of specimen location, donor age, and strain rate and found a significant influence of both location (parietal/occipital) and donor age on bending modulus and ultimate stress. Parietal bone ultimate stress and modulus were larger than occipital bone. Ultimate stress and modulus increased with the age of the donor. Ultimate stress was significantly affected by the three-way interaction between donor age, location, and strain rate. Ultimate strain was the only parameter that was sensitive to strain rate, and it increased with strain rate.

Davis et al. harvested the calvarium from a 6 year old cadaver, and conducted 4 point bending experiments on 47 samples [18]. Samples were classified by morphology as either "cortical only" (n=7), "sandwiched" defined as layered cortical and trabecular (n=17), "transitional" defined as between cortical and sandwiched (n=5), and "across closed suture" (n=18). All suture specimens failed at the suture. Cortical-only specimens (1.88±0.16 mm) were significantly thinner than sandwich structure specimens (3.38±0.45 mm) or suture specimens (3.60±0.71 mm), but sandwich and suture specimens were not significantly different from one another. Average strain rates spanned two orders of magnitude (0.045, 0.44, and 2.2 s^{-1}) but did not affect modulus, ultimate strain or ultimate stress. The effective modulus of elasticity of cortical (9.87±1.24 GPa), sandwich (3.69±0.92 GPa), and suture (1.10±0.53 GPa) specimens were all different from one another. Cortical specimens had the highest ultimate strength (184.49±25.19 MPa), followed by sandwich specimens, (82.87±22.00 MPa), followed by the suture specimens (27.18±9.23 MPa). Ultimate strain was not dependent on morphology. Plots of ultimate stress (Fig. 22.4) and elastic modulus (Fig. 22.5) versus specimen thickness are provided. Despite large differences in the elastic modulus or the ultimate stress between specimens of differing morphology, the authors noted that the bending stiffness was uniform between cortical and suture specimen morphologies. This suggests that the skull utilizes thickness and structure as compensatory mechanisms for the relative weakness and lower elastic modulus of the sutures at this age.

Although investigations have been conducted delineating the characteristics of the adult sutures [9, 19], biomechanical studies of sutures of the human skull in the pediatric age group have received less attention. For pediatric specimens, Davis et al. described the properties of suture in the 6 year old, and these data have been discussed above [18]. For the infant, where the suture is quite soft, Coats and Margulies harvested human infant pediatric cranial bone-suture-bone specimens from the coronal suture (n=14 specimens from 11 calveria) and tested them in tension to rupture at rates of 1.2 and 2.38 m/s [17]. Rupture

Table 22.4 High rate three point bending data for infant cranial bone (Adapted from Coats and Margulies [17])

Cranium	Age	Region	Bending modulus (MPa)	Ultimate stress (MPa)	Ultimate strain (mm/mm)
1	21 weeks gest	Occipital	181.1	12.5	0.0627
2	21 weeks gest	Occipital	45.3	8.8	0.0071
3	21 weeks gest	Occipital	89.4	9.8	0.0027
		Occipital	132.9	12.3	0.002
		Parietal	50.2	5.5	0.0037
		Parietal	120.1	5.6	0.0089
4	32 weeks gest	Occipital	58.7	3.3	0.0416
5	34 weeks gest	Parietal	552.9	81.1	0.0045
6	35 weeks gest	Parietal	97	7	0.0026
7	38 weeks gest	Occipital	290	12.6	0.0467
		Occipital	448.1	31.4	0.0735
8	39 weeks gest	Occipital	211.1	6.7	0.0501
		Occipital	229.3	7.4	0.0346
		Parietal	253.9	7.6	0.0264
		Parietal	933.1	31.8	0.0431
10	19 days old	Parietal	336.8	37.8	0.149
12	21 days old	Occipital	550.7	5.8	0.0125
		Occipital	516.2	4.6	0.0068
		Parietal	182.7	8.4	0.045
13	1 month old	Occipital	449.2	18.5	0.0465
		Parietal	815.5	53.7	0.0753
14	1.5 months old	Occipital	28.6	8.7	0.0068
		Occipital	57.7	13.5	0.0039
15	1.5 months old	Parietal	372.4	19.7	0.07
		Parietal	518.2	29.6	0.0533
		Parietal	581.3	25.6	0.0639
16	1 months, 23 days old	Occipital	421.4	15.1	0.0314
17	2 months old	Parietal	297.4	14.2	0.0515
		Parietal	522.4	27.1	0.0765
18	2 months, 9 days old	Occipital	186.4	3.1	0.0259
		Occipital	186.1	5.7	0.0268
19	3 months old	Occipital	1,317.6	43.4	0.0254
		Occipital	463.5	26.1	0.0456
		Parietal	1,155.2	69.7	0.0807
20	4.5 months old	Occipital	317.7	16.4	0.0542
		Occipital	392.8	19.5	0.0538
		Parietal	552.4	23.7	0.05
21	11 months old	Occipital	602.9	37.6	0.003
		Occipital	322.1	17.3	0.0031
		Parietal	783.8	52.1	0.0032
		Parietal	573	48.8	0.0034
22	12 months old	Occipital	104.2	6.2	0.0538
		Occipital	621.8	21.2	0.1613
		Parietal	200.8	19.3	0.1217
		Parietal	566.5	51.5	0.0936
23	13 months old	Parietal	216.8	15.1	0.0885

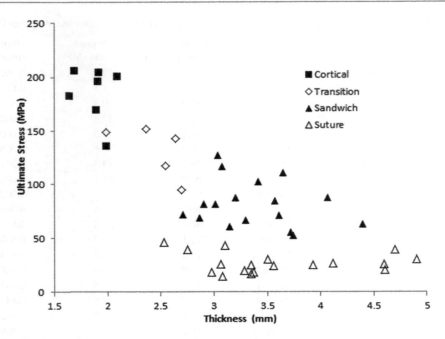

Fig. 22.4 Ultimate stress vs. specimen thickness from 4-point bending tests of samples from a 6 year old cranium (Adapted from Davis et al. [18])

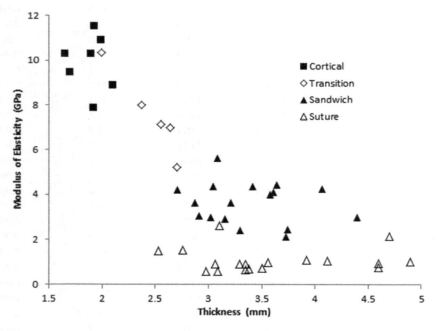

Fig. 22.5 Modulus of elasticity vs. specimen thickness from 4-point bending tests of samples from a 6 year old cranium (Adapted from Davis et al. [18])

Table 22.5 Material properties for infant cranial suture tested in tension (Adapted from Coats and Margulies [17])

Specimen number	Age	Elastic modulus (MPa)	Ultimate stress (MPa)	Ultimate strain (mm/mm)
1	21 weeks gest	N/A	3.5	N/A
2	28 weeks gest	14.2	6.7	0.6943
		13.2	6.3	4.7029
4	32 weeks gest	4.3	4.5	2.8320
5	34 weeks gest	6.9	5.7	1.2776
6	35 weeks gest	8.1	4.2	1.1234
9	2 days old	3.8	3.7	1.2776
11	2 weeks old	6.4	2.2	1.0930
		3.8	3.1	1.2014
16	1 month, 23 days old	N/A	4.6	N/A
		N/A	7.2	N/A
18	2 months, 9 days old	N/A	6.8	N/A
21	11 months old	4.2	4.1	1.2511
22	12 months old	16.2	3.5	0.3324

consistently occurred at the bone-suture interface. The authors found no significant effect of donor age or strain rate on the elastic modulus, ultimate stress, or ultimate strain of pediatric suture. Elastic modulus, ultimate stress, and ultimate strain for infant human suture are shown in Table 22.5.

These data can be used to scale injury criteria [20] or as material properties in finite element models [21, 22], for the purpose of predicting skull fracture and estimating skull deformation during direct impact and the resulting interaction between the brain and skull.

22.1.2.2 Brain Material Properties

Substantial evidence links cell death with deformation of nervous system tissue [23–27]. Thus, for the purposes of predicting brain injury, one must predict brain deformation. From an engineering mechanics perspective, a constitutive relationship links the tractions and body (inertial) forces experienced by the brain during a TBI event with the resulting deformation of the brain tissue. Thus, many have sought to elucidate the material properties of the brain to inform scaling of whole head kinematic adult TBI tolerance between species [28] and between ages within the same species [29], and for the development of pediatric finite element head/brain models [21]. (For a comprehensive review of brain material properties, see Chatelin et al. [30].)

Brain tissue is non-linear and viscoelastic. Thus, when determining the mechanical properties of brain tissue, the experimental protocol must deform the tissue samples at strain magnitudes and rates similar to that which is observed in TBI events. Direct observation of brain deformation for the calculation of strain during impact events is extremely difficult. Several studies have quantified brain-skull displacements in animals and human cadavers [31–33], but this technique currently lacks the spatial resolution to capture strain. Tagged magnetic resonance imaging of low amplitude head accelerations in humans that produce no injury, strains range from 2 % to 5 % [34, 35]. In hemisection rapid rotation experiments of immature pig cadavers exposed to angular accelerations that would produce mild TBI, strains range from 9 % to 39 % [36]. Thus, we can expect that the relevant strains for more severe events would exceed 40 %, and thus mechanical tissue property tests should be in this range.

Determination of brain material properties generally requires removing tissue material from a live subject or fresh cadaver. It should be noted that difficulties exist with regard to characterization of the mechanical properties of soft tissue material ex vivo because of its quick deterioration postmortem [37–39]. It is well know that mechanical properties such as stress-relaxation characteristics differ significantly between live and dead brain tissue, and the variations maybe due to alterations in the parenchyma and vasculature pressure. Consequently, if one were to use autopsied brain tissue for the determination of the mechanical properties, time would be a critical factor. Porcine studies have been conducted to determine the mechanical response of the brain (to shear loads) within 3 h postmortem [40, 41]. However, practical and logistic constraints may limit the use of human cadaver brain to directly determine mechanical properties. Although it is possible to obtain tissues from operating rooms during surgery (with appropriate institutional approval and adherence to government regulations), the tissue may not all be fully usable for biomechanical material property testing because of the presence of pathologies. With careful attention to the aforementioned issues, limited material property testing of pediatric human [42] and animal brain tissue [40, 43–45] has been conducted. We present below a summary of the mechanical tissue testing that has been performed on both pediatric animal and human subjects.

Prange and Margulies [40] tested fresh swine white and gray matter at the infant and toddler ages, and adult swine and human specimens (surgical discards), using a parallel plate shear device at strains up to 50 %. The data were fit to a first order Ogden hyperelastic material model.

By examining brain tissue properties at specific strains across a broad strain spectrum (2.5–50 %), the authors found that the infant is approximately twice as stiff as either the toddler or the adult porcine animals at strains relevant to injurious TBI (Table 22.6). In addition, the authors noted significant heterogeneity (up to 50 %) in material stiffness between corpus callosum and gray matter, and between two white matter regions (corona radiata and corpus callosum.)

It should be noted, that at small strains (approximately less than 5–10 %) which are relevant to sub-injurious or possibly mild injury head impact events, the aforementioned trend reverses and the infant brain is softer than the

Table 22.6 Ogden material property parameters for pediatric and adult tissue (Adapted from Prange and Margulies [40])

Species	Human	Porcine	
Region	Gray matter	Mixed gray/white matter	
Age	Adult	5 day old	4 week old
# samples	6	6	5
α	0.0323	0.0100	0.0326
μ_0 (Pa)	295.7	526.9	216.5
C_1	0.335	0.332	0.316
τ_1 (s)	2.40	2.96	3.00
C_2	0.461	0.389	0.428
τ_2 (s)	0.146	0.181	0.190

adult. Thibault and Margulies [41] collected samples of frontal cerebrum from neonatal (2–3 days) and adult pigs and tested them within 3 h post-mortem. The complex shear modulus of the samples was measured in a custom-designed oscillatory shear testing device at engineering shear strain amplitudes of 2.5 % or 5 % from 20 to 200 Hz. In this range, the elastic and viscous components of the complex shear modulus increased significantly with the age-based development of the cerebral region of the brain.

These trends are also observed in the rat. Finan et al. used microindentation to determine viscoelastic properties of different anatomical structures in sagittal slices of juvenile and adult rat brain [43]. Strains were limited to 10 %. The authors found that the brain becomes stiffer as the animal matures. Similarly, Gefen et al. conducted controlled 1 mm indentation of a force probe into the intact rat brain, both inside the braincase and with the brain removed from the skull [44]. Rats were either 13, 17, 43 or 90 days old, and test samples were either preconditioned (PC) or non-preconditioned (NPC), the former to study the effect of repeated blows to the head. The instantaneous (Gi) and long-term (G') shear moduli of brain tissue were determined. The immature rat brain (Gi = 3,336 Pa NPC, 1,754 Pa PC; G' = 786 Pa NPC, 626 Pa PC) was significantly stiffer ($p < 0.05$) than the mature brains (Gi = 1,721 Pa NPC, 1,232 Pa PC; G' = 508 Pa NPC, 398 Pa PC).

These results at small strains are also reflected in human pediatric tissue studies. Chatelin et al.

(2012) conducted a rheological study on autopsy brain tissue from seven subjects aged 2 months old to 55 years old. Three tissue morphologies were extracted – gray matter (thalamus), white matter (corona radiata), and brain stem. Samples were tested 24–48 h post-mortem in oscillatory shear to 0.5 % engineering strain, at frequencies ranging from 0.5 to 10 Hz. Results show a two to three fold increase in storage or loss modulus between the three infant subjects (2, 2 and 5 months old) and either the two toddler subjects (22 and 23 months old), or the two adult subjects (50 and 55 year old).

The compendium of these data highlights the importance of test conditions on the stiffness of brain material across the child to adult age spectrum. These results suggest that at small strains the infant brain is softer than the adult, but at large strains most relevant for moderate and severe TBI, the adult is softer than the infant. It is also important to note that the small strain studies have demonstrated an increase in storage or loss modulus with increasing strain rate [46, 47], and at this writing we are unaware of any brain tissue testing conducted at both high strain *and* high strain rates comparable to those observed in injurious TBI events. More research is required to find the appropriate tissue properties for brain material at both high strain and high strain rates.

Selection of appropriate material properties of the brain can have significant effect on the predicted outcome in finite element models of the head. The coefficients of the material model for the brain will directly affect the resulting strains, with stiffer brain materials leading to lower strains, given the same input conditions. In the infant, where the vault is softer due to the pliable sutures and incomplete skull bone formation, variations in brain material stiffness may influence the response of the skull. For example, when simulating falls in the infant, varying brain tissue material properties within the ranges reported in the literature can increase by approximately one-third the maximum principal stress in the skull [21], and thus selection of FE brain material properties may affect not only the strain in the brain, but also the prediction of skull fracture in the young.

22.1.3 Brain and Skull as a System

Response of the whole pediatric head has been confined to only one study of neonates. Prange et al. [48] conducted compression and drop tests on three fresh frozen isolated human head specimens ages 1, 3, and 11 days post-natal. First, each head was exposed to compression between two rigid flat plates. The maximum head deformation during the compression was limited to 5 % of the total gauge length of the head to ensure no damage or fracture. Following preconditioning, the head was compressed at constant velocity using four different rates – 0.05 mm/s (quasi-static), 1.0, 10 and 50 mm/s. Compression tests were conducted in the anterior-posterior (AP) and right-left (RL) directions. Following the compression tests, each head was dropped on a rigid flat surface from 15 or 30 cm. At each drop height, the head was impacted once in each of the five locations: vertex, occiput, forehead, right parietal bone, and left parietal bone. For biofidelity evaluation of the infant ATD, similar experiments were conducted with the CRABI 6 month old ATD head.

The force deflection data for all tests exhibited an initial toe region with increasing stiffness at higher displacements. The linear stiffness was dependent on rate of displacement. The average stiffness values for the 0.05, 1.0, 10, and 50 mm/s constant velocity compression tests were 7.45, 23.3, 29.9, and 29.5 N/mm respectively. The quasi-static stiffness was significantly less than the three higher rates (Fig. 22.6). Peak acceleration and HIC varied with drop height but did not change between impact locations. Maximum head acceleration by drop height and impact location is shown in Fig. 22.7. Average peak acceleration and HIC during the 30 cm drop height impacts were 55.3 g and 84.1 respectively. Both these average values were significantly greater than the average peak acceleration and HIC values of 38.9 g and 32.9 respectively for the 15 cm drop height impacts. The acceleration pulse durations were not significantly different between the 15 and 30 cm drop height impacts with an average pulse duration of 18.3 ms. At both 15 and 30 cm drop heights, CRABI acceleration was similar to the cadaver

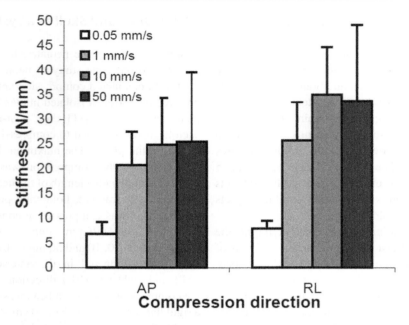

Fig. 22.6 Stiffness values determined from head compression tests at different constant velocities in two different directions from three neonate human cadavers. AP is anterior-posterior and RL is right-left (Reprinted with permission. The Stapp Association from Prange et al. [48])

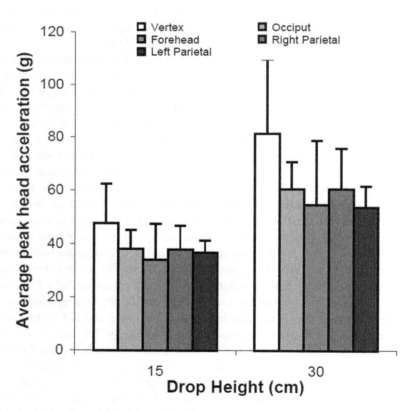

Fig. 22.7 Average peak head acceleration from drop tests of three neonate human cadavers, stratified by drop height and impact location (Reprinted with permission. The Stapp Association from Prange et al. [48])

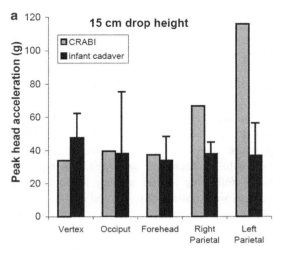

Fig. 22.8 CRABI and infant cadaver peak head acceleration for five impact locations from 15 cm (**a**) and 30 cm (**b**) drop heights. Error bars indicate standard deviation (Reprinted with permission. The Stapp Association from Prange et al. [48])

data for the vertex, occiput, and forehead impacts but significantly greater for the lateral impact (Fig. 22.8).

22.1.4 Injury Scaling Methods in Pediatric TBI Animals Models

Studies of human patient populations suggest that young children who sustain severe TBI in early childhood or moderate or severe TBI in infancy may be particularly vulnerable to significant residual cognitive impairment [49–53]. However, it is unclear in these studies if the youngest patients have poorer outcomes because they experienced more severe TBI events than their older counterparts, or if there is a biochemical or physiological vulnerability in the young. To separate TBI severity from vulnerability, researchers turn to animal models of human pediatric and adult TBI. By imparting the same scaled TBI insult to the adult and pediatric animal, the biochemical or physiological vulnerability between ages can be quantified. Several methods of such scaling are described below.

22.1.4.1 Mass-Acceleration Scaling

Ibrahim et al. conducted rapid rotation TBI experiments on pigs representing the infant (3–5 day old pig) and toddler (4 week old) humans [36]. To scale loading conditions between animal ages, Holbourn's scaling law for acceleration was utilized [54],

$$\alpha_i = \alpha_t \left[\frac{M_t}{M_i} \right]^{\frac{2}{3}}$$

where the subscripts i and t refer to the infant and toddler, M is the mass of the brain, and α is the angular acceleration of the head. The infant animals were exposed to higher angular accelerations than the toddler pigs, such that both age groups experience similar theoretical brain tissue strain. The authors found no difference in the percentage volume of axonal injury or subarachnoid hemorrhage score between these two animals of differing age that received similar scaled TBI insults. While these data suggest that scaling is an accurate method to transform injury thresholds between the old and young, an assumption is that the brain tissue material modulus is similar across the age spectrum.

22.1.4.2 Scaled Cortical Impact

Controlled cortical impact is a laboratory TBI injury method where an indentor is rapidly driven into the brain. Unlike the rapid rotation mechanism described above which leads to diffuse brain injury, the most severe location of tissue damage is typically at the site of indentation. Depending on the severity of injury and time since the injury occurred, injured tissue may extend to the contralateral hemisphere. When studying brain injury across the child-adult age spectrum, *scaled* cortical impact is necessary to ensure experiments across groups of animals of different ages receive the same impact severity relative to their brain size. One such method is to ensure a uniform percent volume of displaced brain by the impactor across animal ages, as performed by Missios et al. [55]. Those authors performed scaled controlled cortical impacts in the infant, toddler and early adolescent age groups of pig. Cortical indentation was performed with a stainless steel impactor applied directly to the exposed brain (some protocols will leave the dura intact). For each animal age group, a specific indentor diameter and penetration depth was used such that the indentor displaced 1 % of the total brain volume. In total, 67 (30 males) animals were tested. At 7 days post-indentation, the animals were sacrificed and five coronal brain sections centered around the site of indentation were the basis for injury quantification. Injury was defined as any tissue with the presence of necrosis with or without hemorrhage, neuronal dropout or damage, and/or reactive gliosis, all identified by light microscopy in each of the five coronal brain sections. Injury was evaluated 7 days following TBI. The percent area of injured tissue was determined and compared across ages.

Immediately following indentation, the degree of cortical surface hemorrhage at the site indentation was uniform across all age groups, suggesting that the scaling approach led to uniform mechanical insult across age. Seven days after indentation, the percent area of injured tissue increased as a function of age. Infants had smaller lesion sizes than toddlers (3.0 ± 0.8 percent of cortical surface area versus 8.5 ± 1.6) and toddlers had smaller lesion sizes than adolescents (8.5 ± 1.6 versus 25.1 ± 2.7).

22.1.4.3 Pressure Scaling

Bittingau et al. exposed 3, 10, 14 and 30 day old rats to weight drop impacts proximal to the sensorimotor cortex [56]. To ensure severity of injury was not influenced by skull thickness, the skull at the location of impact was mechanically thinned prior to injury in animals in the 10, 14 and 30 day old rats, such that each had a similar skull thickness to the 3 day old rat. To ensure animals in each age group received a mechanically similar insult, the height from which the weight was dropped was adjusted to yield equal contact pressure across all ages. Results suggest that vulnerability decreases with age. Twenty-four hours following scaled impact, apoptotic cell damage was substantially elevated in 3 and 7 day old animals, compared to the 10 and 14 day rats, and injury was almost undetectable at 30 days. Interestingly, apoptosis was worst for the 7 day old animals – the second youngest animal. This non-linearity in the vulnerability-age relationship may be due to biomechanical factors – such as age-dependent changes in the stiffness of rat brain [43, 44] – or biochemical factors, such as documented N-methyl-D-asparate (NMDA) receptor hypersensitivity in the 7 day old rat leading to excitotoxicity [57, 58].

22.1.4.4 Scaled Acute Subdural Hematoma

Durham and Duhaime induced experimental subdural hematoma (SDH) in 15 pigs of age 5 days, 1 month, and 4 months old [59]. To induce SDH, a volume of blood equal to 10 % of the intracranial volume of each pig was injected through a right frontal burr hole. Following injection, the dura was sealed with cyanoacrylate glue and scalp sutured. At 7 days post-injury, the animals were euthanized. A 1.5 cm thick coronal section centered under the injection site was analyzed for injury via histology including hematoxylin and eosin staining and TUNEL staining. A significant difference in percentage of injured hemisphere was noted between the 5-day old group (0 %) and the 1 month old (7 %) and 4-month old (9 %) animals.

22.1.4.5 Acute Injury Severity Scaling

Another approach to comparing the outcome following brain injury between adult and pediatric animals exposed to the same scaled insult is to adjust the mechanical insult until the acute injury response is equal across age groups, and then examine long term outcome. For example, Cernak et al. examined the response to TBI in rats across the pediatric to adult age spectrum [60]. Rats were subjected to a single TBI event at either 7 days old, 14 days old, 21 days old or 3 months old. To induce TBI, animals were positioned prone while a pneumatically-driven impactor struck the bare skull. A metal disc was affixed to the skull to eliminate injury differences cause by skull maturity, and the animal head was supported by a gel pillow that allowed the head to rotate as a result of the impactor strike. To provide uniform impact intensity across the age groups, the authors tuned the impactor stroke in each age group such that a similar (20–25 %) acute (<5 min) mortality rate was observed across all age groups. This technique allowed the researchers to study the short- and long-term consequences of the same injury severity induced across a broad age range.

Responses between young and old animals show several important trends. While all animals regardless of age exhibited signs of vasogenic edema compared to uninjured age-match shams at 20 min post-injury, pediatric animals resolved their edema at 4 and 24 h following injury while edema persisted in adults at these later time points. In addition, younger animals showed persistent and substantial deficits in motor function at multiple time points up to 90 days following TBI, compared to age-matched shams. Conversely, adults had only modest deficits that were resolved within 7 days post TBI. Interestingly, cognitive testing showed an opposite trend, with pediatric animals showing little difference from shams in Morris water maze platform tests, and adults showing consistently poorer performance as compared to shams.

Scaling of the TBI event across animals of the same age within the same species is an emerging area of TBI research, with all the aforementioned publications occurring in the last 6 years. These methods of ensuring proportionality of TBI insult across age are essential to scaling well validated and accepted adult injury criteria to children. Clearly, this emerging branch of TBI research has not reached any conclusion, with some studies showing equivalent vulnerability between the young and old, some showing the eldest to be more vulnerable than the youngest, and some showing the young to be more vulnerable than the adult. These contradictory conclusions are fueled by differences in species chosen for research, the time point(s) post-injury when deficit is assessed, and the histological or behavioral methods used to quantify deficit. Additional research is this area will elucidate the answers to these questions.

In addition to recognizing the role of impact severity on outcome, the direction of impact also plays an important role. Eucker et al. conducted non-impact rapid rotation TBI experiments in three anatomic planes of head rotation that produced mild to severe TBI, using 3–5 day old pigs that represent approximately the 1 year old human [61]. The authors found that regional cerebral blood flow, primary brain injury (extra-axial bleeding, diffuse axonal injury), and secondary brain injury (ischemia and infarct) differed by head rotation direction, with sagittal and axial rotations resulting in moderate to severe TBI, and coronal injuries resulting in mild to no TBI. When considering the motion of the brain in these experiments, it is important to note that the coronal rotation in the biped human is most like an axial rotation in the pig.

22.1.5 Injury Criteria

The widely used Head Injury Criterion (HIC) was developed in the 1970s. The basis of the HIC lies in the Wayne State University tolerance curve which was developed by dropping embalmed human cadaver foreheads onto unyielding flat surfaces [62–64]. In its final form, the tolerance curve was developed by combining the result from a wide variety of pulse shapes, cadavers, animals, human volunteers, clinical research, and injury mechanisms. Skull fracture and/or concussion was used as the failure criterion except for long-duration human volunteer tests wherein no apparent injuries were reported. The criterion is given by the following equation:

$$HIC = max\left[\{1/(t_2 - t_1)\} \int_{t_1}^{t_2} a(t) dt \right]^{2.5}$$

where t_1 and t_2 are arbitrary times during the acceleration phase, and a(t) is the resultant acceleration response. Attempts have been made to establish pediatric head injury tolerance using adult data for the HIC and angular thresholds. A widely used procedure for scaling the angular/rotational acceleration between species is based on the mass of the brain as described above for scaling across age [54, 65].

$$\alpha_H = \alpha_A [M_A / M_H]^{\frac{2}{3}}$$

where subscripts H and A refer to the human and animal, respectively, M is the mass of the brain, and α is the angular acceleration of the head. The underlying assumption is that the density, material property, and brain shape between the animal and the human are equal. This equation results in a value of 1,700 rad/s² for the angular acceleration threshold for the adult human. Using this approach and assuming the mass of the infant brain to be 0.5 kg, pediatric head injury tolerances have been obtained [66]. Because developmental and anatomic differences exist between the pediatric and adult skull and brain, this equation was modified in later studies [29, 41] to incorporate the age-related material property changes in the brain tissues,

$$\alpha_C = \alpha_{Ad} [M_{Ad} / M_C]^{\frac{2}{3}} \{G'_C / G'_{Ad}\}$$

where the subscripts Ad and C refer to the adult and child, respectively, and G' is the elastic modulus of the brain. In the case where the infant brain is stiffer than the adult [40], the acceleration tolerance of the child would be higher than that which was predicted by considering brain mass alone. However, in the case where the infant brain is softer than the adult [40], the acceleration tolerance of the child would be lower than that which was predicted by considering brain mass alone. Additional research in this area is required to elucidate these relationships.

Recognizing the need to account for both material and geometric variations, as an initial step, injury assessment reference values for the 6 month old infant ATD were obtained by combining geometric and material scaling data between the adult and pediatric groups [67]. The skull bone modulus of elasticity from literature was used to determine the modulus of elasticity ratio for the skull of the 6 month old infant ATD [15]. The mechanical properties of the human skull between the newborn and the 6-year-old were approximated by a linear fit. Using the principles of scaling discussed in the earlier section and detailed in the Appendix, injury assessment reference values developed for the adult midsize male (Hybrid III) and scaled to estimate the corresponding values for the HIC and head acceleration for the child dummies. Injury assessment reference values were developed for the CRABI 6-month-old ATD with deploying passenger airbags [67]. The recommended values are peak resultant head acceleration of 50 g associated with a 22-ms HIC of 390. It was stated that the 390 value for HIC should not be permitted for shorter time duration pulses with values above 50 g. Using skull bone modulus as a governing parameter for material scaling, HIC values of 121, 275, and 525 were obtained for the 12 month old, 3 year old, and 6 year old dummies, respectively [68].

The advent of the finite element modeling in biomechanics research affords the opportunity for determining head injury thresholds for children. Klinich et al. developed a finite element model (FEM) of a 6-month-old infant head using available material properties and humanlike geometry [22]. The infant head FEM was used to simulate different injury and no-injury loading conditions based on child restraint response data from reconstruction tests. Logistic regression analysis was used to estimate threshold stresses associated with skull fracture. The acceleration responses of the infant head FEM and the CRABI ATD were compared for the no-injury and injury producing conditions. Provisional injury assessment reference values were estimated for the current CRABI ATD (Table 22.7).

22.2 Neck

22.2.1 Developmental Anatomy

The major structural components of the neck include the cervical spinal column and soft tissues. The cervical spine consists of seven vertebrae (C1–C7). The intervertebral discs start

Table 22.7 Tolerance limits for various head injury criteria, derived from finite element model reconstructions of rear-facing infant seat airbag fatalities (Adapted from Klinich et al. [35])

	Head	Tolerance for **injury** lies between		Tolerance for **severe injury** lies between	
Peak resultant accel (g)	Infant head	46	152	152	217
	CRABI FEM	42	143	N/A	N/A
	CRABI ATD	46	128	N/A	N/A
3 ms clip resultant accel (g)	Infant head	42	107	107	108
	CRABI FEM	40	110	N/A	N/A
	CRABI ATD	45	77	N/A	N/A
Mean resultant accel (g)	Infant head	15	79	79	85
	CRABI FEM	12	63	N/A	N/A
	CRABI ATD	17	49	N/A	N/A
HIC (unlimited)	Infant head	207	881	881	1,477
	CRABI FEM	143	758	N/A	N/A
	CRABI ATD	239	376	N/A	N/A
HIC (15 ms)	Infant head	146	881	881	1,477
	CRABI FEM	98	758	N/A	N/A
	CRABI ATD	161	376	N/A	N/A

inferiorly from C2. Spinal ligaments interconnect from the base of the skull (occiput) to C7 and proceed distally. The spinal cord that originates from the foramen magnum is housed in the osseous-ligament anatomy of the spinal column. Muscles originate and insert at various locations along the neck and proceed along the cranial and caudal directions. A brief description of the developmental anatomy of the various components is given below.

22.2.1.1 Vertebrae

The primary developmental characteristic of the vertebral column is the presence of cartilaginous tissue, termed synchondroses, which convert to bone as the individual ages.

Atlas

The first vertebra (C1) is called the atlas. It is formed from three primary ossification centers; two bilaterally in the posterior neural arches, present at birth and one anteriorly, developing several months to 2 years postnatally. The junctions between the anterior and bilateral posterior centers are called the neurocentral synchondroses. The two neural arch centers meet dorsally at the posterior synchondrosis. Fusion of the posterior synchondrosis occurs first – in the fourth or fifth year of life, followed by the neurocentral synchondrosis approximately 1 year later. The spinal canal fully forms and attains the mature adult size following complete fusion of the primary ossification centers [69–71].

Axis

The second vertebra (C2) is called the axis. It is formed from five primary ossification centers; one in the centrum (vertebral body location), two bilaterally in the posterior neural arches, and two in the odontoid process [72–75]. The two centers in the odontoid are separate in utero and join to form a single ossification center by the seventh month prenatally. The posterior synchondrosis forms at the junction of the neural arches and fuses in postnatal year 3 to 4. The cartilaginous regions between the odontoid and the neural arches, the dentoneural synchondroses, ossify at a similar time. The odontoid process is connected to the body of C2 by dentocentral synchondrosis and fuses by postnatal year 4 to 6. Paired neurocentral synchondroses connect the two posterior arches and centrum and fuse at a similar time. The synchondrosis that joins the tip of the odointoid process to its body forms between ages 3 and 6 years and remains cartilaginous until 12 years of age and in certain cases never ossifies. As in the case of the atlas, the spinal canal reaches its mature size following closure of the posterior and neurocentral synchondroses [69–71].

Typical Cervical Vertebrae

Each of the five vertebrae of the lower cervical spine, C3–C7, is formed from three primary ossification centers, one in the anterior centrum and two in the posterior neural arches [72–77]. The neurocentral synchondroses are the joining element between the neural arches and centrum bilaterally. The two neural arches are connected to each other by the posterior synchondrosis which fuses in postnatal year 2. Joining of the anterior and posterior ossification centers occurs in postnatal year 3 or 4. Similar to C1 and C2, the spinal canal size attains the adult dimension following completion of these primary ossifications [69, 70]. Figure 22.9 illustrates the characteristics

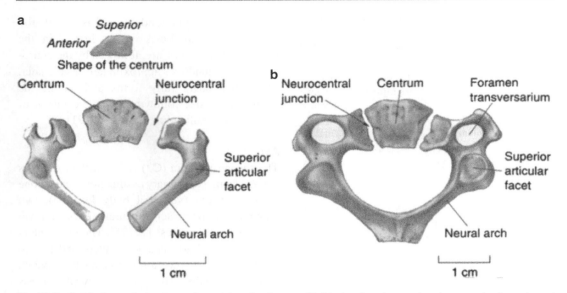

Fig. 22.9 Cervical vertebrae of a perinate (**a**) and a 3 year old (**b**) showing the synchondroses and other relevant vertebral anatomy (Reprinted from Scheuer and Black [78])

of ossification in the atlas, axis, and a typical cervical vertebra.

During development, the overall dimensions and shape of the vertebrae change. With advancing age, vertebral bodies attain a more rectangular shape and increase in height as the superior and inferior growth plates develop [79–85]. This growth continues until puberty. The growth plates contribute to the longitudinal development of the vertebral body. The latitudinal growth occurs by the expansion of the neural arch ossification centers that develops into three independent growth zones bilaterally: the pedicle, lamina, and transverse processes. The uncovertebral joints of C3–C7, formed between the lateral edges of the superior surfaces of one vertebral body and the inferior surface of the vertebral body directly above, do not form until age 6 years [78, 86–90]. These joints are responsible for the saddle shape of the cervical vertebrae and coupled motion of the spine [91, 92].

Facet Joints

Bilateral facet joints form the posterior connection between the inferior articular surface of the superior vertebra and the superior articular surface of the inferior vertebra (Fig. 22.10). The two surfaces are connected by the synovial joint consisting of the synovial fluid, synovial membrane, articular cartilage, and capsular ligaments. Developmentally, orientations of the facet joints change and the variations depend on the vertebral level. The joints in the upper spine are more horizontal than those of the lower cervical spine and become more upright with advancing age (Fig. 22.11) [72–74, 83, 93, 94]. The more horizontal orientation of the upper spine facet joints, together with the softer intervertebral components, appears as pseudosubluxation in younger ages [95–99]. Pseudosubluxation, initially defined through lateral x-rays of the pediatric cervical spine of the neck in extension by Cattell and Filtzer in 1965, refers to normal anterior displacement of a vertebra relative to its inferior adjacent vertebra that is so pronounced that it resembles an injurious condition [97]. Approximately 40 % of children demonstrate non-injurious pseudosubluxation at C2–C3 and 20 % at C3–C4. The changing facet joint orientation likely contributes to variations in the distribution of external load with increasing age (i.e. the relative contribution of axial versus shear loading) [83, 100].

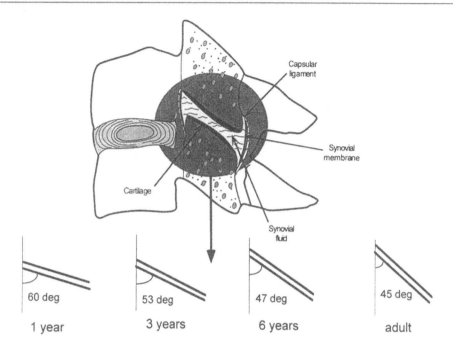

Fig. 22.10 Exploded facet joint illustrating its components (*top*). Variation of the facet joint orientation as a function of age is shown (*bottom*)

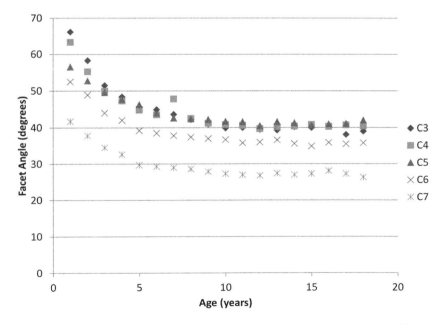

Fig. 22.11 Variation of the facet joint angle as a function of age (Data adapted from Kasai et al. [83])

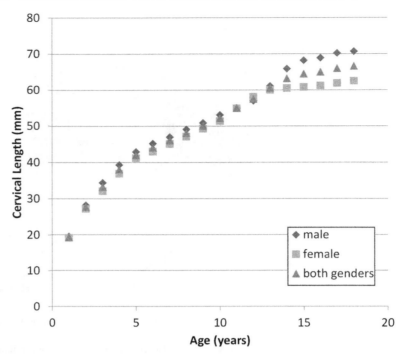

Fig. 22.12 Variation of the summated height of the typical cervical vertebral bodies denoted as cervical length from C3 to C7 as a function of age and gender (Data adapted from Kasai et al. [83])

Quantitative Description of Vertebral Growth

The most comprehensive quantitative study of vertebral developmental anatomy was performed by Kasai et al. via radiographs of approximately 350 children from birth to 18 years [83]. The authors quantified numerous metrics including vertebral diameter, vertebral height, facet joint angle, lordosis angle from C3 to C7 and cervical length. Selected developmental observations include: an increase in vertebral body diameter and height with age (Fig. 22.12), an increase in the initially horizontal facet angle through age 10 years, and decrease in the cervical lordosis angle through age 9 years and then increase through the ages of puberty. In addition to dimensional changes, these authors quantified the available sliding motion of adjacent vertebra, defined as the relative horizontal displacement of adjacent vertebrae during flexion and extension as a percent of intervertebral disc diameter. As an example, for C2–C3 the maximum sliding motion ranged from 34 % for 1 year olds to 13 % for 18 year olds. Other cervical levels demonstrated similar observations. This reference provides a wealth of additional quantitative information.

Age-Related Groupings

Based on vertebral growth and characteristic distributions in the constituents between the cartilaginous structures of the synchondrosis and the bone itself, pediatric structures can be broadly categorized into four groups [78, 86–90]. In general, the age group between newborn and 1 year can be skeletally represented by the presence of three primary ossification centers. The age group between 1 and 3 years can be represented by the fusion of the posterior synchondrosis. The age group between 3 and 6 years is marked by fusion of the bilateral neurocentral synchondroses, and the 11- to 14- year (approximately puberty) group corresponds to secondary ossification and initiation of the development of the uncinate and uncovertebral anatomy (Fig. 22.13). Skeletally mature adult vertebral anatomy in the human occurs during the second decade of life.

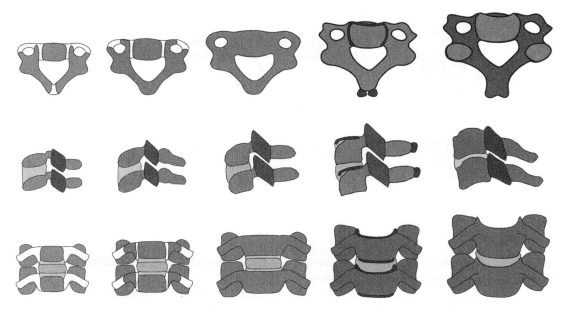

Fig. 22.13 Typical cervical vertebrae (*top row*) as viewed from the *top* and functional units as viewed from the side (*middle row*) and front (*bottom row*). Typical growth patterns are illustrated. Starting from the *left*, patterns represent groups I, II, III, IV and adult spines. Cartilage is shown in lighter color and bony components are shown in darker color. Secondary ossifications are shown in the darkest color for group IV (all three rows). In the *middle row*, the facet joints are expanded to illustrate changes in their orientation with age

22.2.1.2 Ligaments

Cervical vertebrae are connected to each other and the base of the skull by ligaments. Ligaments are unique to the upper cervical region, i.e., the occiput-atlas-axis complex [101, 102]. Beginning anteriorly, the anterior atlanto-occipital membrane connects the occiput to C1; it is renamed the anterior longitudinal ligament below C1. The apical ligament attaches from the occiput to the tip of the odontoid process of C2. The alar ligaments attach from the superior-lateral aspect of the odontoid process and run obliquely to the occiput. The cruciate ligament has a strong transverse portion that runs laterally around the odontoid process and attaches at both ends to the medial aspects of the arch of the atlas. The vertical cruciate ligament attaches from the occiput, just posterior to the apical ligament, intertwines with its transverse portion, and attaches again to the posterior-inferior aspect of the vertebral body of the axis. The tectorial membrane attaches to the anterior one-third of the occiput just posterior to the vertical cruciate ligament. This ligament tapers inferiorly to become continuous with the posterior longitudinal ligament. The posterior atlanto-occipital membrane connects the superior aspect of the posterior arch of the atlas to the occiput [103]. Proceeding inferiorly from C2, the anterior and posterior longitudinal ligament, ligamentum flavum, bilateral capsular, and interspinous ligaments connect one vertebra to its immediate adjacent vertebra (Fig. 22.14).

22.2.1.3 Intervertebral Discs

Intervertebral discs are present between the vertebrae and are composed of a nucleus pulposus, annulus fibers, and ground matrix. These components undergo considerable developmental changes [73–75, 104–110]. Typically, around 1 year of age, the disc is characterized by a large nucleus with loosely embedded annular fibers resulting in lack of clear distinction between the nucleus and annulus. At approximately 3 years of age, the annulus fibers become better organized and a clearer division from the nucleus pulposus can be observed. Around 6 years of age, fibers increase in stiffness and density, further accentuating

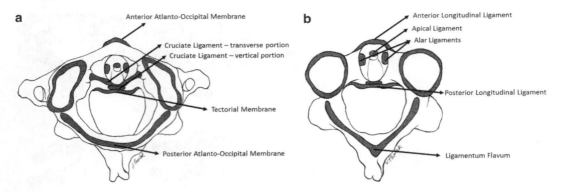

Fig. 22.14 Schematic diagrams of the atlanto-axial complex. (**a**) Superior view of C1 and C2 showing the location of the major ligaments. (**b**) Superior view of C2 emphasizing the ligaments

the demarcation between the nucleus and annulus boundaries. The geometry of the disc changes with age as well; initially it is biconvex until approximately age 10 years when the convexity begins to become less pronounced.

22.2.1.4 Muscles

Neck muscles connect the ligamentous cervical spinal column with the head and torso. Although the neck muscles are numerous, they are classified based on the motions they produce. Muscles active during flexion are classified as flexors (e.g., sternocleidomastoid, longus colli). Similarly, muscles active during extension are termed extensors (e.g., splenius capitis, trapezius). During the developmental process, cross-sectional area of the flexors and extensor muscles increase [111–113]. However, the difference between pediatric and adult muscle cross-sectional area varies by muscle type (Fig. 22.15) [111]. Additional studies have quantified that neck muscle maximum voluntary contraction as measured by peak force and muscle endurance as measured by the ability to sustain that force increases with age [112]. Differences between the genders do not appear until adolescence. These muscular changes lead to overall increases in neck anthropometry. Specifically, neck circumference and lateral breadth increase linearly as a function of age. A detailed quantification of anthropometric variations with age can be found in Snyder 1977 [113].

22.2.2 Biomechanical Studies

22.2.2.1 Structural Response: Human Volunteers

Human volunteer experiments have a long established history in biomechanics research. Early researchers used themselves as test specimens [114] or enrolled adult human volunteers to define the response of neck to dynamic loading environments [115–117]. Limited studies however have been done using pediatric human volunteers.

Arbogast et al. through non-injurious frontal sled tests on human volunteers quantified the kinematic response of the restrained child's head and spine and compared pediatric kinematics to those of the adult [118]. Normalized forward and downward excursion of the spine significantly decreased with age (Fig. 22.16). The majority of the spine flexion occurred at the base of the neck not in the upper cervical spine and the magnitude of flexion was greatest for the youngest subjects. Additional flexion occurred in the thoracic spine as well. These findings pointed to decreasing head-to-neck girth ratio with increasing age as the primary factor governing the age-based differences in spinal trajectories. Other factors, such as muscle response and cervical vertebral structural properties, may also contribute to the differences, but were not evaluated in this study. Using inverse dynamics, the same authors quantified the forces and moments at the upper cervical spine [100] (Fig. 22.17). Axial

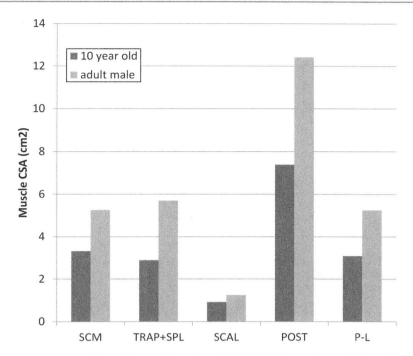

Fig. 22.15 Muscle cross-sectional area for several cervical spine muscles for an average 10 year old and an average adult male. Muscles include sternocleidomastoid (*SCM*), trapezius (*TRAP*), splenius capitis (*SPL*), scalene muscles (*SCAL*), the posterior neck muscles (*POST*) which included trapezius, splenius capitis, splenius cervicis, semispinaliscapitis, semispinaliscervicis, and multifidus; and the postero-lateral muscles (*P-L*) which included longissimus and levator scapula (Data adapted from Dawson et al. [111])

force decreased with increasing age while flexion moment increased with increasing age. Increases in shear force with age were explained statistically by age-based changes in the relative size of the head and neck. These data represent the only dynamic tests to date of pediatric human volunteers evaluating the kinematics and kinetics of the cervical spine.

Human volunteers tests have also be used to define cervical spine range of motion (ROM). Several studies quantified the ROM for flexion/extension, lateral bending and axial rotation across various relevant pediatric age ranges (Table 22.8). In interpreting the data in the table, in particular the age effects, it is important to consider the specific age range evaluated in each study.

22.2.2.2 Structural Response: Human Cadaveric Specimens

Whole Body

The first study on cervical spine structural response was conducted in 1874 using four newborn whole-body cadavers and one 2-week-old child cadaver [125]. Force was quasistatically applied to the lower extremities in increasing increments until the cervical spinal column failed. Severed vertebrae detached from adjacent segments as ligaments failed. The force to complete failure increased from the newborn (471 ± 79 N) to the 2-week old specimen (654 N).

There has been limited dynamic study of whole body pediatric PMHS. To the authors' knowledge, 15 pediatric whole body PMHS sled tests have been conducted to date. Subjects range in age from 2 to 13 years at speeds from 31 to 50 km/h [126–130]. These studies examine a diverse set of restraint conditions and the instrumentation, documentation and injury coding reflect the age in which the studies were done. The focus of these studies was on the kinematics of the head and torso rather than the cervical spine. As a result, little data can be extracted from these efforts on the cervical spine structural response.

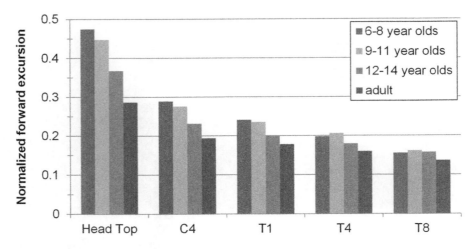

Fig. 22.16 Age-based differences in normalized forward excursion for restrained human volunteers in a low-speed frontal sled test. Skeletal landmarks include head top (HT), C4, T1, T4, and T8 (Reprinted with permission. The Stapp Association from Arbogast et al. [118])

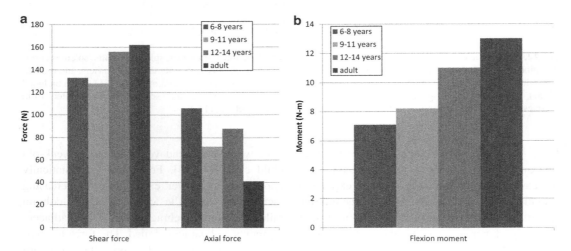

Fig. 22.17 Age-based differences in (**a**) shear force and axial force and (**b**) flexion moment for restrained human volunteers in a low-speed frontal sled test (Data adapted from Seacrist et al. [119])

Segments

McGowan et al. tested several functional spinal units from all spinal levels from a fresh frozen 8-h-old human PMHS spine in axial tension at 1.25 mm/s [131]. Fixation failures occurred in the cervical and thoracic specimens. Force-displacement curves indicated a peak force of 216 N at a displacement of approximately 4 mm and a distraction stiffness of 94 N/mm.

More recently, whole pediatric PMHS cervical spines (age 2–12 years; n = 10) were subject to sagittal plane bending and tensile load to failure [132]. Results are summarized in Table 22.9. Data were compared between subjects 2–4 years and 5–12 years. Only tensile force at failure showed a statistically significant relationship with age with increasing failure load with increasing age (Fig. 22.18). Of note, most failures occurred at the site of fixation and thus, the reported data is likely influenced by this artifact of the testing protocol.

Lastly, a comprehensive study on the tensile properties of the pediatric spine was performed by Luck et al. [133, 134]. Eighteen PMHS spines

Table 22.8 Pediatric cervical spine range of motion studies

Reference	Age group	Measures	Statistical difference between genders	Statistical difference across age
Ohman and Beckung [120]	2–10 months	Axial rotation Lateral bending	Not evaluated	Increased rotation with age
Lewandowski and Szulc [121]	3–7 years	Flexion/extension Lateral bending Axial rotation	None	Not evaluated
Arbogast et al. [122]	3–12 years	Flexion/extension Lateral bending Axial rotation	None	Increased flexion with age Increased rotation with age
Lynch-Caris et al. [123]	8–10 years	Flexion/extension Lateral bending Axial rotation	Not evaluated	No consistent trends
Seacrist et al. [119]	6–12 years	Passive flexion	Increased flexion by females	Increased flexion compared to adults
Greaves et al. [124]	4–17 years	Helical axis of motion (HAM) in: Flexion/extension Lateral bending Axial rotation	More anterior HAM location in females for flexion/extension	Less anterior HAM location with age for flexion/extension and axial rotation. Age relationships varied by gender.

Table 22.9 Average values from PMHS cervical spine testing, 2–12 years [132]

Bending stiffness (N-m/degree)	0.041
Tensile failure load (N)	725 ± 171
Displacement at failure (mm)	20.2 ± 3.2
Tensile stiffness (N/mm)	35 ± 6

ranging in age from 20 weeks gestation to 14 years were tested. Whole spines were tested as well as the following spinal segments: O–C2, C4–C5, and C6–C7. Tensile load to failure sharply increased with increasing age for all spinal levels (Fig. 22.19). Data from a single 14 year old PMHS showed ultimate strength 120 % of the adult average (mean age 58 years) suggesting that age related changes continue throughout the lifespan. For the younger cohort (24 weeks gestation to 24 days old) for which there is the most data, tensile stiffness increased as spinal level became more caudal however no differences in tensile strength were found across spinal level (Table 22.10). For the older specimens (>1 month old), the upper cervical spine demonstrated increased strength compared to the lower cervical spine – a finding similar to data in the adult [135–137].

One additional study on human pediatric cervical spine specimens (age 2–28 years) has been conducted [138]. Segment (C1–C2, C3–C5, and C6–C7) flexibility was assessed in tension, compression, flexion, extension, lateral bending, and axial rotation and then failure mechanics in tension (C1–C2), compression (C3–C5), and extension (C5–C6) were quantified. Both the functional and failure properties demonstrated significant relationships with age.

22.2.2.3 Structural Response: Animal Cadaveric Specimens

Post mortem animal subjects, including porcine, baboon, and caprine specimens, have also been utilized to understand the structural response of the spine. In the context of airbag loading, porcine (mean age: 10 weeks) and baboon specimens (mean age 5.2 years) were subject to crash pulse of either 14.5 g (56 km/h delta v) or 8.4 g (33.6 km/h delta v) [139]. The animal was placed on a sled and held vertically by a tether that was released prior to airbag deployment. Seven

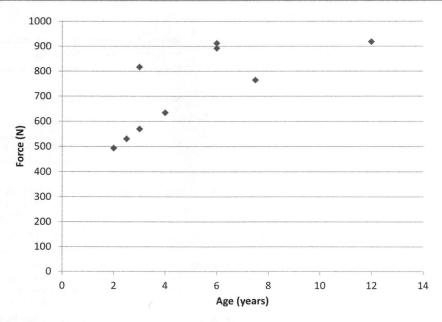

Fig. 22.18 Tensile load at failure of whole pediatric PMHS cervical spines as a function of age (Data adapted from Ouyang et al. [132])

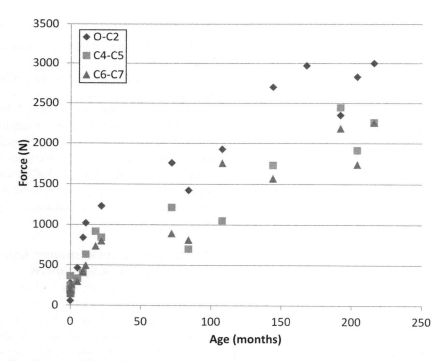

Fig. 22.19 Tensile load at failure of functional spinal units of pediatric PMHS as a function of age (Data adapted from Luck et al. [134])

Table 22.10 Average values from PMHS cervical spine testing, 24 weeks gestation-24 days old [133]

	Whole spine	Upper C-spine	C4–C5	C6–C7
Tensile strength (N)	n/a	230.9 + 38.0	212.8 + 60.9	187.1 + 39.4
Tensile stiffness (N/mm)	6.8 + 1.0	9.9 + 2.1	47.5 + 8.4	43.5 + 8.5

different animal positions and 10 types of airbag inflators were used in combination with eight types of airbags, four types of folds, and three types of covers. The 10 week old porcine was chosen to mimic a 3 year old child based on weight (porcine: 15.7 kg; human: 14.9 kg), thoracic (porcine: 143 mm; human: 165 mm) and abdominal breadth (porcine: 144 mm; human: 165 mm). Despite these similarities, the pig has no chin structure for air bag interaction because its neck attaches to the dorsal region of the skull, resulting in its snout being somewhat aligned with the cervical column. The fore-aft range of motion of the head-neck structure of the pig is less than that of the child. This results in a smaller rearward motion to produce extension neck trauma in the pig. The neck circumference of the pig is approximately twice that of the child because of its large dorsal neck musculature. Other deficiencies include chest anthropometry (depth to- width ratio of the pig is the inverse of that of the child) and head shape.

Twenty-four of the 46 animals experienced significant neck injuries. Two animals suffered fatalities secondary to neck trauma. The most frequent neck injury (20 of the 24 with significant neck injury) was hemorrhage in the occipito-atlantal joint capsules. Three animals sustained posterior element fractures of the C5, C6, or C7 vertebra. Pathologic changes in the spinal cord occurred in 4 of the 24 animals with significant neck trauma. To determine the neck kinetics, parallel testing was conducted using a Hybrid III 3-year-old ATD and its kinematics were matched with animal kinematics. Statistical analysis indicated that tension in the neck of the ATD had the strongest statistical relationship with injury. An AIS3+ neck injury risk curve based on neck tension estimated a force of 1,160 N corresponding to a 50 % probability of injury. Little improvement in injury predictability was found when moment was added to the analysis [139].

Prasad and Daniel built on this work and conducted 15 paired dynamic airbag tests using 12- to 15-week-old piglets and a 3-year-old child ATD [140]. The child ATD was a Ford-built version of the General Motors child ATD, but with a different head to reduce ringing. Neck injuries occurred in 7 of the 15 cases; 4 were fatal. Injuries were primarily concentrated at the occipito-atlanto-axial complex and were attributed to high tensile and bending stresses at the occipito-atlantal and atlanto-axial joints. For the piglet, tensile loads in the range of 1,500 N for 11 ms or 2,100 N for 3–6 ms corresponded to fatal or severe neck injuries. These authors were the first to suggest a composite metric for indication of neck injury by combining axial tension with bending moment. The constant stress line had values of 2,000 N tension and 34 N-m extension-moment (Fig. 22.20).

The caprine has been commonly used as an animal surrogate for cervical spine structural testing. Pintar et al. initially evaluated tensile tolerance and tensile and bending stiffness across age and defined scaling ratios [141]. Hilker et al. further refined these estimates by incorporating additional adult data – see Table 22.11 [142]. Clarke et al. evaluated bending response in a caprine specimen and quantified increases in bending stiffness with age [145].

Ching and Nuckley utilized an immature baboon model to evaluate cervical spine biomechanics [143, 146–148]. Isolated functional spinal units from specimens ranging in age from 2 to 26 human equivalent years were tested in tension [143]. Scaling ratios of tensile failure and tensile stiffness relative to the adult were developed (Table 22.11). A decreasing trend in tensile failure load was observed when moving from the upper to lower cervical spine except for the youngest age group which demonstrated no difference across spinal level. The effect of loading rate on these findings was further examined using specimens of 10 human

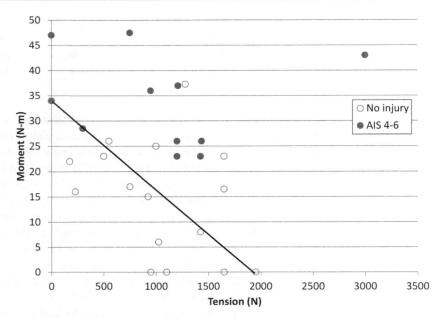

Fig. 22.20 Moment versus axial force relationship for no injury and injury (AIS 4–6) cases from paired dynamic airbag tests conducted using 12–15 week old piglets and a 3-year-old child test ATD. The constant stress line is indicated (Redrawn from Prasad and Daniel [140] with permission. The Stapp Association)

Table 22.11 Tensile scale factors developed from several biomechanical studies

Age group	Tension failure – caprine (N) [142]	Tension failure –baboon (N) [143]	Tension failure – [144]	Tension failure – FMVSS 208 (N)	Tension stiffness – caprine (N-m) [142]	Tension stiffness – baboon (N-m) [143]
1 year old	0.10	–	0.25	0.24	0.13	–
3 year old	0.16	0.33	0.30	0.34	0.18	0.54
6 year old	0.30	0.55	0.37	0.45	0.38	0.71
12 year old	0.62	0.66	–	–	0.66	0.76
Adult male	1.00	1.00	1.00	1.00	1.00	1.00

Table 22.12 Peak limits and N_{ij} critical intercepts incorporated into USFMVSS 208

	Peak limits		N_{ij} critical intercepts			
Age	Tension (N)	Compression (N)	Tension (N)	Compression (N)	Flexion (N-m)	Extension (N-m)
1 year	780	960	1,460	1,460	43	17
3 year	1,130	1,380	2,120	2,120	68	27
6 year	1,490	1,820	2,800	2,800	93	37
Adult	4,170	4,000	6,806	6,160	310	135

equivalent years subjected to tensile loading at rates from 0.5 to 5,000 mm/s [147]. Metrics of tensile mechanics (stiffness: 2×, failure load: 4×) increased with loading rate according to a power relationship.

Nuckley et al. evaluated compressive tolerance using baboon functional spinal unit segments and observed increases in compressive failure load with age [148]. They also observed differences across spine level that varied with age. For the youngest specimens, the upper cervical spine had the lowest tolerance; while for those specimens greater than 8 human equivalent years, the lower cervical spine was the weakest. These compressive studies were extended to additional mechanical parameters such as stiffness, bulk elastic modulus and bulk strength and similar increases with age were observed. Vertebral stiffness increased from 1,218 N/mm at 1 year to 3,534 N/mm at 30 years via a second order relationship. Elias et al. built on this work and explored the role of loading rate on compression mechanics [149]. Stiffness and failure load demonstrated an increase with loading rate.

These baboon studies were brought to completion in 2006 with a study of functional spinal units (O–C2 to C7–T) tested non-destructively in tension and compression then failed in tension [146]. Regression relationships were developed to quantify the increase in stiffness and failure load with age. Differences in biomechanical response across spinal levels were also observed.

Biomechanical studies using animal surrogates offer an advantage of plentiful specimen supply which leads to a more comprehensive data set; however they are limited by the need to properly scale responses to the human. As such, these data need to be examined in the context of the reliability of the human age-equivalency estimation which is often based on matching development anatomy.

22.2.2.4 Injury Assessment Reference Values and Scaling of Stiffness and Failure Loads

Using the principles of scaling that includes geometric similitude and mechanical property differences, several attempts have been made to scale human adult and pediatric animal thresholds to estimate human pediatric tolerance and injury assessment reference values. A more thorough discussion of the methods is contained later in this chapter. For the cervical spine specifically, initial scaling efforts used strength data from the calcaneal tendon to determine the mechanical property scale factor for the tensile strength of the neck [67, 150]. Implicit in this is the assumption that the developmental characteristics leading to the mechanical response are similar between the calcaneal tendon and the neck ligamentous tissues.

Efforts to include neck injury criteria as part of the United States Federal Motor Vehicle Safety Standards required the development of a specific criterion as well as methods for scaling that criterion across age. From this, a combined metric neck injury criterion, Nij, was defined based on a combination of axial forces and moments [68]. Nij is defined as the sum of the normalized forces and moments.

$$Nij = F / F_{int} + M / M_{int}$$

where F represents axial force, M represents bending moment, and the subscript "int" denotes the critical intercept value used for normalization. A Nij limit of 1.0 corresponds to a 15 % risk of serious neck injury; a limit of 1.4 corresponds to a 30 % injury risk. The critical intercepts were initially determined from reanalysis of the porcine airbag test data described above to determine the level of tension and extension moment that best predicted the injury outcomes for the 3 year old and then further modified by Eppinger et al. based on comments from industry and other stakeholders [151]. Critical intercept values for flexion moment were set by maintaining the previously reported ratio of 2.5 between flexion and extension. Compression tolerance was chosen to be the same as tension. To obtain values for the 1-year-old and 6-year-old, intercepts for force and moment were scaled according to the second and third powers, respectively, of the neck length which was estimated by neck circumference. It was suggested that since material stiffness variations were already incorporated into the ATD neck design, neck injury criteria needed to only be scaled using geometric factors. Critical intercept values incorporated in the FMVSS 208 based on this work are shown in Table 22.12.

Other cervical spine biomechanical scaling efforts used biomechanical properties of spine components obtained from adult post mortem human subjects in combination with material and neck geometry data to determine scale factors for 1-, 3- and 6-year-olds in tension, compression, flexion and extension [144]. As these efforts as well as the caprine and baboon studies described above were conducted after the regulatory development of *Nij*, tensile scale factors developed from those studies differ slightly and are summarized below (Table 22.11).

22.2.2.5 Material Properties

Yoganandan et al. synthesized mechanical property data of components such as the vertebra, ligaments, cartilage, spinal cord, muscles, and disks; information was obtained from studies in literature and in-house tests [152]. Most these data were based on adult specimens; however Kumaresan et al. scaled these properties to the 1-, 3-, and 6-year-old for the development of pediatric cervical spine finite element models [153, 154]. Specific material properties used and the source of those data are outlined in detail in those references.

22.3 Thorax

The thorax is a key structure in restraint design. In frontal crash protection of the seat belt restrained occupant, the thorax interfaces directly with the seat belt webbing and, if available, with the frontal airbag and as such absorbs much of the inertial load of body in an effort to prevent occupant ejection and mitigate head injury. In a side impact, door and child restraint padding and/or torso side airbags will also interact directly with the thorax. As such, detailed knowledge of pediatric thoracic biomechanics is required, though at this writing the body of literature contains mainly tissue testing from coupon or whole rib samples, and a limited number of tests on whole intact thoraces in pediatric post-mortem subjects or cardiopulmonary resuscitation recipients. These data are reviewed herein, to provide a resource for those who will design restraints and/or tools for restraint designers (crash test dummies and human body computer models, for example). A comprehensive review of the available literature on pediatric material properties, including those for the tissues of the thorax, has recently been published by Franklyn et al. [155]. More recently, Kent et al. [156] provided a comprehensive review of thoracic biomechanics, drawing on the data summarized by Franklyn, and also adding more recent research. Here, we review all of the relevant literature for pediatric restraint design, and where confined by space restrictions we direct the reader to the original publications for more details.

22.3.1 Developmental Anatomy

The boney thorax consists of the 12 left and right ribs of the rib cage, the sternum, and the 12 thoracic vertebrae comprising the thoracic spine. The superior 10 ribs are connected to the sternum via the costal cartilage at the anterior, and each at a corresponding thoracic vertebra in the posterior. The two most inferior ribs, the so-called floating ribs, are connected only at the vertebra. The visceral contents include all soft tissue superior to the diaphragm, most notably the heart and lungs, the great vessels of the cardiovascular system including the aorta, vena cava and pulmonary vessels, and the trachea and esophagus. In addition, the thymus gland, thoracic duct, thoracic lymph nodes and a network of nerves also are contained in the thorax. The sternum consists of 6 main bones – the manubrium superiorly, followed by sternebrae 1 through 4 and the xiphoid process. The fourth sternebra appears at age 12 months, while the xiphoid process appears at 3–6 years (Fig. 22.21). Fusing between sternebrae begins at age 4 years and continues through age 20 years. The sternum as a whole descends with respect to spine from birth up until age 2–3 years, causing the ribs to angle downward when viewed laterally, and the shaft of the rib to show signs of axial twist deformation. The costal cartilage also calcifies with age, likely influencing its flexibility [90].

22.3.2 Biomechanical Studies: Anatomic Components

22.3.2.1 Ribs

Kemper et al. conducted dynamic tensile tests at a nominal strain rate of 0.5 s^{-1} on cortical rib bone coupons harvested from anterior, lateral,

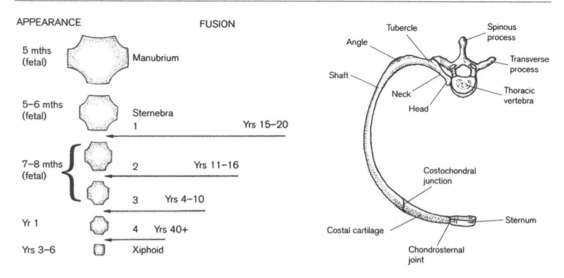

Fig. 22.21 Sternum (*left*) and rib and thoracic vertebra anatomy (*right*) (Reprinted from Scheuer and Black [90])

and posterior locations on ribs 1 through 12 of an 18 year old male cadaver [157]. While the inclusion of an 18 year old does not strictly categorize this study as pediatric, it does describe the changes in cortical bone elasticity across a broad age range when compared with the entire test population including males and females up to age 67. The samples harvested from the 18-year-old donor had significantly lower elastic modulus and ultimate stress than the older subjects, and significantly greater ultimate strain and strain energy density at failure. The mean elastic modulus was approximately 50 % greater in the pool of older samples as compared to the 18 year old (15.1 MPa vs. 10.0 MPa) (Fig. 22.22) and the ultimate stress was approximately 20 % greater (129.3 MPa vs. 106.3 MPa). Perhaps most striking was the difference in ultimate strain, with a mean value almost a factor of two greater for the younger samples (4.24 % vs. 2.25 %). The mean strain energy density at failure was approximately 60 % greater in the 18-year-old samples (3.5 MPa vs. 2.2 MPa).

Theis (as cited in Stürtz [158]) reported on results from quasi-static bending tests on ribs 6 and 7 from children under the age of 14 years (n = 3) and dynamic bending tests on ribs 6 and 7 from children under the age of 14 years (n = 17, bending rate = 2.9 m/s), persons 15–64 years (n = 65, bending rate = 2.0 m/s) and persons older than 65 years (n = 31, bending rate = 1.7 m/s). The average breaking load for the 3 pediatric ribs that were loaded quasi-statically was 240 N, whereas the average dynamic breaking loads for the ribs from the children under the age of 14 years, persons 15–64 years and persons older than 65 years were 234 N, 372 N and 173 N, respectively.

22.3.3 Vertebrae

As described by Kent et al. [156], Weaver and Chalmers reported compression tests of trabecular bone samples harvested from the third lumbar vertebrae of human cadavers, 6 of which were under the age of 20 years [159]. The compressive strength of those pediatric samples was found to be similar to the group of middle-aged cadavers (around age 40), but lower than the group of samples age 20–30. The compressive strength of the pediatric samples ranged from 2.4 to 3.7 MPa (see Franklyn et al. [155]). Mosekilde et al. [160, 161] and Mosekilde and Moskilde [162] reported compression tests on trabecular bone samples from several thoracic and lumbar vertebrae of human cadavers, including two age 15 and 19 years. Maximum compressive stress, energy absorption, and other properties were measured.

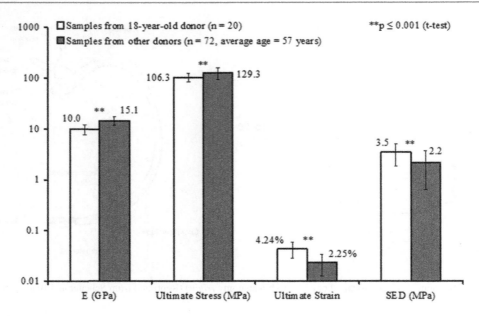

Fig. 22.22 Mean elastic modulus of rib cortical bone samples from an 18-year-old donor compared to samples from older donors. Error bars indicate standard deviation. Note logarithmic scale on ordinate (Reprinted with permission. The Stapp Association from Kemper et al. [157])

The maximum compressive stress ranged from 4.2 to 4.7 MPa for trabecular bone cored samples, and was 5.85 MPa for a whole vertebra tested from the 15-year-old. Energy absorption ranged from 0.64 to 0.85 mJ/mm^2. Kleinberger et al. [68] summarized data originally collected by McElhaney et al. [163] and concluded that the ultimate compressive strength of human lumbar vertebral cancellous bone increased throughout pediatric development and decreased with age above approximately 20 years (Fig. 22.23).

22.3.4 Costal Cartilage

As described by Kent et al. [156], Yamada reported the results of tensile tests of samples of costal cartilages from 28 subjects grouped into age categories [150]. The ultimate tensile strength was reported to be 0.46 MPa and invariant with age up to age 19, while the ultimate percent elongation was reported to decrease slightly during pediatric development, from 31.2 % in the samples age 0–9 years to 28.2 % in the samples age 10–19 years. Oyen et al. [164] presented the results of spherical indentation experiments on juvenile porcine costal cartilage samples and developed a linear viscoelastic model of the tissue using a correspondence technique developed by Mattice et al. [165]. That study found that the cartilage midsubstance exhibited a time-zero elastic modulus, E0, of 5.34 ± 1.59 MPa, a long-time modulus, E∞, of 0.73 ± 0.27 MPa and that the assumption of linear viscoelasticity was reasonable over the strain magnitudes considered. The authors also concluded that those modulus values were similar to those seen for adult humans.

22.3.5 Vascular and Cardiac Tissue

Yamada included the tensile breaking load (TBL), ultimate percent elongation (UPE), and ultimate tensile stress from samples of thoracic aorta and coronary artery [150]. These data were summarized by Franklyn et al. in a table that is partially reproduced here (Table 22.13) [155]. The tensile breaking load was found to increase with age, but this appears to be primarily a structural phenomenon related to the increase in size with age.

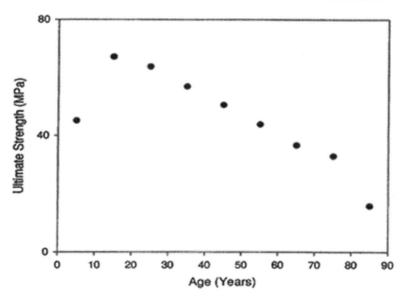

Fig. 22.23 Ultimate compressive strength of human lumbar vertebral trabecular bone as a function of age (From McElhaney et al. [163] as reported by Kleinberger et al. [68])

Khamin reported significant decreases with aging (after age 11 years) in the ultimate tensile stress of the aortic wall [166]. Interestingly, Yamada and Khamin separately tested the ultimate tensile strength (UTS) of aorta, and as shown in Table 22.13 showed dramatic differences in results. That is, UTS of aorta in the longitudinal directions was 0.92 ± 0.06 MPa as reported by Yamada, but 2.44 ± 0.14 MPa reported by Khamin, and similar differences were observed in transverse loading. This variation between research labs could be a result of specimen preparation, location of thoracic aorta sample (Yamada, for example, averaged the values from ascending and descending aorta), and grip type.

22.3.6 Whole Thorax Biomechanical Experiments

Very few published studies have addressed the issues of the strength and blunt impact tolerance of the intact pediatric thorax. Ouyang et al. [167] tested 9 pediatric cadavers aged 2–12 years, subjecting them to Kroell-style [168] frontal sternal impact by means of an impactor weighing 2.5 kg (cadavers aged 2–3 years at time of death) or 3.5 kg (cadavers aged 5–12 years at time of death) at impact speeds ranging from 5.9 to 6.5 m/s (Table 22.14). Following impact, six of the impacted subjects had pneumothoraces, one had a bleeding thymus gland and two had no signs of thoracic injury (Table 22.15). None of the subjects sustained a rib fracture. Time histories of the viscous criterion, acceleration, force and deformation were recorded in all the tests. The subjects were divided into two age cohorts and force-deformation cross-plots were generated and reduced to impact response corridors. The authors reported good qualitative agreement between the thoracic force and deflection data of the older pediatric PMHS cohort and a scaled biofidelity corridor developed from adult PMHS impacts, but quantitatively the pediatric PMHS cohort exhibited a lower force-deflection onset rate, higher maximum forces, and lower maximum deflection than the scaled corridor.

As a result of experimental and analytical limitations in the Ouyang study, Parent reinterpreted the pediatric PMHS blunt thoracic impact response data and corridor development [169]. The external deflection for each subject was

Table 22.13 Mechanical properties of thoracic aorta (Adapted from Franklyn et al. [155])

Property	Author	Age group			
		Fetal[a]	Infant/child	Adolescent	Young adult
TBL longitudinal (g/mm)	Yamada	35–80	116.5 ± 14	137 ± 7	146.5 ± 7
TBL transverse (g/mm)	Yamada	58–105	157 ± 19	204 ± 15	213 ± 15
UPE longitudinal (%)	Yamada	109	67.5 ± 4	98 ± 5	90 ± 8
UPE transverse (%)	Yamada	99	69.5 ± 5	104.5 ± 8	101 ± 8
UTS longitudinal (MPa)	Yamada	0.78–1.14	1.02 ± 0.08	0.92 ± 0.06	0.87 ± 0.06
	Khamin (n = 109)			2.44 ± 0.14	1.79 ± 0.04
UTS transverse (MPa)	Yamada	1.27–1.46	1.40 ± 0.10	1.31 ± 0.10	1.22 ± 0.05
	Khamin (n = 107)			4.40 ± 0.12	3.57 ± 0.06

[a]Graphs only were presented in the paper, so data points were measured from the graphs

Table 22.14 Test matrix from Ouyang et al. [167] blunt hub thoracic impact tests

Subject	Impactor mass (kg)	Impactor diam. (cm)	Impact speed (m/s)	Age/gender	Wt. (kg)	Ht. (cm)	Cause of death
1	2.5	5.0	5.9	2/F	13.0	97.0	Fluorocetamide poisoning
2	2.5	5.0	5.9	2.5/M	10.5	87.5	Cerebral edema
3	2.5	5.0	6.0	3/M	13.5	93.0	Brain tumor
4	2.5	5.0	6.0	3/F	10.5	85.0	Heart disease
5	3.5	7.5	6.4	5/M	13.0	101.0	Cerebritis
6	3.5	7.5	5.9	6/M	16.5	108.0	Leukemia
7	3.5	7.5	6.5	6/M	20.0	109.0	Mediterranean anemia
8	3.5	7.5	6.0	7.5/F	17.0	117.0	Acute urinemia
9	3.5	7.5	6.2	12/F	29.0	142.5	Leukemia

Table 22.15 Test results from Ouyang et al. [167] blunt hub thoracic impact tests

Subject	Max. ext. chest def. (mm, % of undeformed)	Max. viscous criterion (m/s)	Max. accel. at T4 (g)	Max. applied force (N)	Observed trauma
1	57.7, 46.1 %	1.9	73.2	740	Rt. pneumothorax
2	45.0, 40.9 %	2.7	63.9	790	Rt. pneumothorax
3	44.8, 40.7 %	2.1	63.9	825	None
4	44.9, 35.9 %	1.5	124.8	750	None
5	55.7, 46.4 %	2.1	62.4	900	Pneumothorax
6	44.5, 37.1 %	0.7	71.2	1,200	Pneumothorax
7	31.5, 24.2 %	1.0	91.7	1,560	Rt. pneumothorax
8	48.7, 40.6 %	1.6	64.0	900	Bleeding thymus gland
9	72.3, 48.2 %	4.5	35.6	1,130	Lt. pneumothorax

determined by measuring the distance between the impactor and the spine of the subject using high-speed video captured during the event. The deflection was then synchronized with the mass-compensated force by minimizing the error between the impactor motion measured from the high-speed video and the impactor motion calculated by double-integration of the impactor accelerometer. This process was intended to correct for the potential error of misjudging impact time when using high-speed video alone to determine external deflection and thereby to improve synchronization of the deflection and force time-histories. These results were then scaled to create

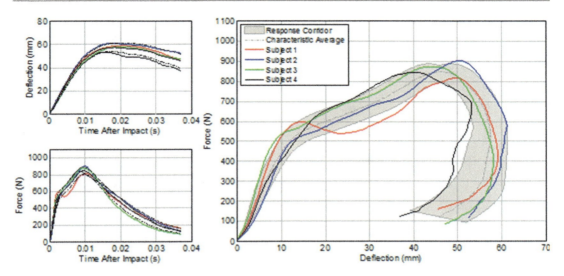

Fig. 22.24 Three year old response to blunt sternal impact of a 2.5 kg impactor at 6 m/s (Reprinted from Parent [169] with permission)

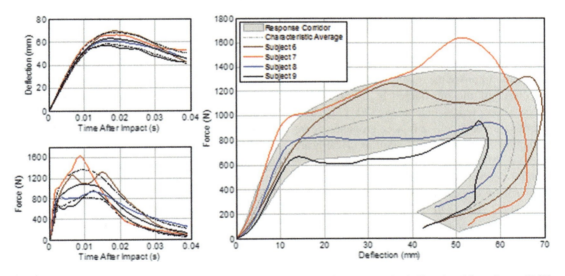

Fig. 22.25 Six year old response to blunt sternal impact of a 3.5 kg impactor at 6 m/s (Reprinted from Parent [169] with permission)

performance corridors for blunt sternal impact for the 3 year old (Fig. 22.24) and 6 year old (Fig. 22.25).

Given the small number of pediatric PMHS subjects available for impact testing, alternative methods for obtaining pediatric thoracic force and deflection data are necessary to further quantify any potential differences between actual pediatric thoracic response and the scaled pediatric thoracic biofidelity corridors. In particular, cardiopulmonary resuscitation (CPR) involves the deflection of the sternum toward the spine in a manner not dissimilar to the PMHS experiments previously described. Various electro-mechanical

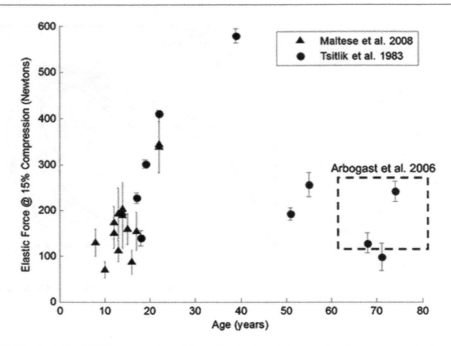

Fig. 22.26 Elastic force at 15 % compression of the pediatric chest during cardiopulmonary resuscitation (Reprinted with permission. The Stapp Association from Maltese et al. [173])

devices have been developed over the past three decades to improve the quality of CPR and study the effect of the mechanics of thoracic compression on clinical outcomes [170–172]. Most recently, a load cell and accelerometer force-deflection Sensor (FDS) (Philips Healthcare, Andover, MA) has been integrated into a patient monitor-defibrillator to provide visual and audio feedback on the quality of CPR chest compressions [173]. The FDS is interposed between the hand(s) of the person administering CPR and the sternum of the patient, and it records force and acceleration data during each compression cycle. Maltese et al. reported on 18 subjects (11 females) ages 8–22 years who received CPR chest compressions [173]. At the time this study was conducted, clinical resuscitation guidelines prescribed targets for CPR chest compressions: 38–51 mm of chest compression for the adult, and one-third to one-half the anterior-posterior (AP) chest depth for the child [174]. In terms of absolute deflection, the CPR compression target for the 6 year old child is 47–72 mm, assuming an AP chest depth of 143 mm [20]. For comparison, chest deflections in hub impact testing with PMHS range from approximately 50–70 mm in adults [168] and from 31.5 to 73 mm in children [167].

These data become most useful when compared with other similar studies of adults in CPR. In particular, Tsitlik et al. instrumented a mechanical chest compressor device used in clinical chest compressions on 10 patients age 17–72 receiving CPR chest compressions [171]. Of note, subjects of similar age in the Tsitlik and Maltese cohorts have comparable elastic forces at 15 % compression. In addition, Arbogast et al. [175] analyzed force-deflection data from 91 adult subjects who received out-of-hospital CPR chest compressions using force-accelerometer technology similar to that which was used by Maltese et al. [173]. Individual subject age was not available in this study, though the interquartile range (IQR) and median age was 70 (IQR 61–81). Elastic force at 15 % compression for this study was 188 N (SD 80). When all three CPR cohorts are taken together, the elastic stiffness of chest climbs from age 8 to age 40, but falls after age 40 to force levels comparable to pediatric patients (Fig. 22.26). This behavior suggests a complexity to scaling the thoracic

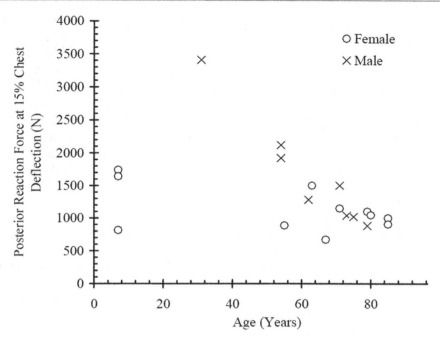

Fig. 22.27 Elastic force at 15 % compression of the pediatric chest during dynamic belt loading (Reprinted with permission. The Stapp Association from Kent et al. [177])

force-displacement between adult and child that is not captured in current scaling techniques [20].

Even more compelling is when the CPR data is compared to dynamic seat belt loading data. Kent et al. reported a series of diagonal belt and distributed loading tabletop experiments on three cadavers aged 6, 7 and 15 years at time of death [176]. In this protocol, the cadaveric subject lies prone on a flat rigid table, and a length of belt webbing is drawn across the chest with each belt end falling below the height of the tabletop. Via high-speed hydraulic pistons, the belt ends are pulled downward such that the belt compresses the chest in a manner similar to a frontal motor vehicle crash with a shoulder-belted occupant. Of note, when plotted with similar experiments with 15 adult cadavers (Fig. 22.27), the overall elastic force vs. age plot takes a similar shape as was presented by Maltese et al. [173].

Aging bone shows a decrease in elastic modulus beyond adult middle age, and ribs alone in bending demonstrate decreased breaking strength with increased age [157], but no studies have sufficient data to determine age-based changes in the stiffness of the whole thorax. The CPR and dynamic belt testing data suggests a reduction in thoracic stiffness beyond middle age that supports the material and rib strength findings. Indeed, it is likely that the rise in stiffness to age 40 occurs in the absence of rib fractures, and reduction in stiffness after age 40 is related to the increased incidence of rib fracture. For example, in the young population, three of the patients reported by Maltese et al. [173] received autopsies, and no rib or sternal fractures were found, whereas Tsitlik et al. [171] found rib fractures in the 55- and 68-year-old subjects in Fig. 22.26. In general, rib fractures are rare in children secondary to CPR (0–2 %), but are more common in adult (13–97 %) [178]. This pattern extends to the impact environment – no rib fractures were found in a recent series of blunt impacts into the thoraces of nine PMHS ages 2–12 years [167], yet the same type of test performed on adults produced rib fractures in 18 of 22 subjects [168]. This pattern is likely due to several aspects of pediatric development in the rib cage, including the large sections of cartilaginous tissue in the costal and intra-sternebrae space, which ossify with maturity and thus provide greater flexibility of the pediatric chest compared to that of the adult.

The pronounced differences in thoracic injury mechanics between children and adults are reflected in the results of Ouyang et al. [167], who reported that six of the nine subjects received pneumothoraces in the absence of rib fracture. Clinically, children frequently receive pulmonary injury in the absence of rib fracture (which is rare in adults, [179]), and when rib fractures are present in children they are associated with severe trauma [180]. This has implications for development of thoracic injury criteria for children. Criteria and thresholds developed from adult PMHS impact data and scaled to the child will predict best the injuries present in those experiments – rib fractures with the occasional soft tissue injury [151, 168, 181], but thoracic injury criteria for children must do the opposite – predict primarily soft tissue injury, with the occasional rib fracture. It is not clear that the current methods of developing thoracic injury criteria for children, which involve scaling adult-based criteria that are often based on rib fracture severity in adult cadaver tests, is a valid method for arriving at the appropriate injury criteria or thresholds for children.

22.4 Abdomen

22.4.1 Developmental Anatomy

The abdomen is bounded superiorly by diaphragm and inferiorly by the pelvis. Its organs can be separated into solid and hollow. The solid organs are the liver, spleen, kidneys, adrenal glands, and ovaries. The hollow organs are the stomach, intestines, urinary bladder, and uterus. Both the size of the abdomen relative to the size of the body and the relative size of the organs change with development. Stocker and Dehner summarized the average abdominal organ weights for children by year of age from birth through age 19 years [182] (Fig. 22.28). The liver is the largest abdominal organ at birth due to the many functions it performs in utero and continues to grow increasing by almost ten times in weight by age 10. The stomach which is small at birth rapidly increases in size. The kidneys occupy a relatively large space in the abdomen at birth. The small intestine is short at birth and constantly grows until puberty. The boney thorax is more superior in the younger child and as a result, does not provide as much protection for the solid organs that underlie it. As a result, the abdomen of the young child appears rather protruded until the pelvis widens, especially during puberty, and the abdominal organs move inferiorly.

22.4.2 Biomechanical Studies

22.4.2.1 Material Properties of Isolated Abdominal Organs

Material property studies of pediatric abdominal organs are extremely limited in the literature. The most comprehensive studies have been conducted on the liver and spleen. Fazekas et al. as summarized by Schmidt quantified the compressive load until rupture in intact human livers and spleens from specimens age 10–86 years [183, 184]. The force required to rupture was greatest in the single 10-year-old specimen indicating that tolerance decreases from childhood to adulthood. Stingl et al. measured the ultimate tensile stress of the liver and spleen in human PMHS specimens – 4 of which were less than 20 years old – and found no relationship with age [185]. Seki and Iwamoto tested hepatic and splenic tissue from immature swine representing a 10 year old human. They quantified the breaking stress and observed that it was significantly greater when the serous membrane was present than when it had been removed [186]. Viscoelastic properties of the pediatric liver [187] and kidney [188] in shear were determined from the bovine and porcine respectively. Mattice conducted stress relaxation tests on immature porcine kidney and observed that trends with age were small compared to biological variability among the specimens [189].

Tests on hollow organs are even more limited. Zhao et al. defined the uniaxial tensile properties of the immature rodent and rabbit stomach and quantified variations by location, direction and species [190]. Based on fetal and pediatric human

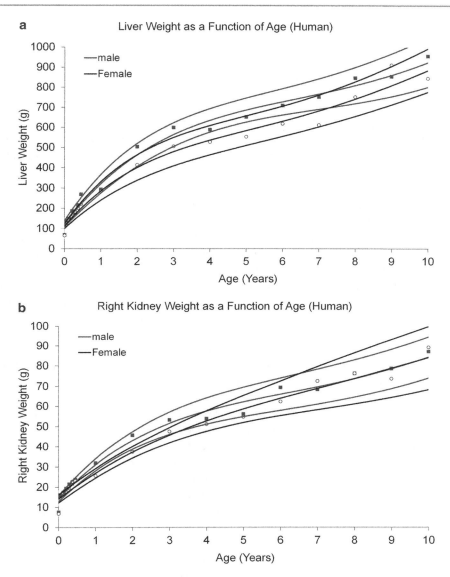

Fig. 22.28 Age trends in abdominal organ mass for males (*thick lines*) and females (*thin lines*) with 95th percentile corridors: (**a**) liver (**b**) right kidney (Adapted from Stocker and Dehner 2002)

tissue, Yamada demonstrated that the strength of the stomach increases with age with the peak being achieved between 10 and 19 years of age [150]. The tolerance of the immature porcine bowel to pinching was determined by Heijnsdijk et al. [191] Pediatric porcine tissue showed no difference in tolerance compared to adult human tissue and the perforation force of the large bowel was greater than that of the small bowel.

In general, these data on isolated abdominal organs provide some estimates of material properties that can be used in computer simulations such as finite element modeling, but lack the methodological detail and broad sample sizes to fully quantify age-based trends. For specific values, the reader is referred to Kent et al. and Franklyn et al. [155, 156].

22.4.2.2 Whole Body Animal Studies

As with many other body regions, the pig has been used to describe the whole body response of the abdomen. In one of the earlier tests, mini-pigs corresponding to an 8–12 year old child were accelerated into the rear end of a vehicle with several variations of a rear spoiler [192]. From these tests, a tolerance of 981 N for AIS3 abdominal injury was suggested.

More recently, Kent et al. developed a porcine model of the 6-year-old abdomen by quantifying anthropometric dimensions and organ masses and comparing them to equivalent published data from the human [193]. From this effort, they were able to identify the specific age porcine – 77 days and 21.4 kg – that corresponded best to the target human age. Quasistatic seat belt loading of the porcine abdomen compared well to similar data from human volunteers described below [194]. A detailed test matrix was conducted using a transversely oriented lap belt (5 cm wide) and varying impact location (upper versus lower abdomen), penetration depth (up to two-thirds of abdominal depth), muscle tensing (yes/no), and penetration rate (quasistatic and up to 7.8 m/s). The upper abdomen was found to be stiffer quasistatically and showed more variation to loading rate with stiffness increasing with increasing rate. The lower abdomen was rate insensitive. Muscle stimulation had a small effect in the quasistatic tests but was negligible at higher rates. The injuries sustained in these porcine tests were then described and injury risk curves developed [195]. Maximum penetration and belt tension were the best predictors of both AIS 2+ and AIS 3+ abdominal injury (Fig. 22.29). It is important to note, that over the range of rates tested, penetration rate did not influence tolerance. These tolerance values were much lower than that proposed by Gögler et al. above suggesting that the mode of loading (impact versus non-impact and loading surface) influence injury thresholds.

22.4.2.3 Post-mortem Human Subject Studies

Limited data exist using post mortem human subjects to quantify abdominal response as changes post mortem make assessment of abdominal injury in a cadaveric model challenging. Sled tests of four PMHS (age 2.5–11 years) restrained by a lap belt provided some insight into tolerable values of belt loading [128]. The tests were conducted at sled velocities of 30–40 km/h, deceleration over 70 ms and belt forces were between 5.7 and 6.6 kN. No abdominal organ injuries were sustained. There are several other studies of full-body pediatric PMHS sled tests [126, 127, 130]; only one subject – a 6-year-old in a five-point child restraint harness sustained any injury to the abdominal organs (liver contusion of unknown severity). These tests are described in more detail elsewhere in this chapter.

In order to more comprehensively assess the validity of the porcine corridors developed above, Kent et al. conducted similar belt-loading tests on three pediatric PMHS subjects of ages 6, 7, and 15 years [176]. The pediatric PMHS exhibited a response very similar to the porcine data in that the lower abdomen was stiffer than the upper abdomen and the overall response was stiffer with increasing rates. The upper abdomen of the human was slightly stiffer than that of the porcine while the lower abdomen was similar across species (Fig. 22.30). The conclusions from this work were that the porcine corridors appear to be a good representation of the pediatric abdominal response.

The same pediatric PMHS studies reported by Ouyang for the spine and thorax [132, 167] also included frontal impacts to the mid abdomen. While these data are not recorded in archival/peer-review journals, they were published in Kent et al. which stated that they were obtained from personal communications from Ouyang et al. who performed the tests [156]. The impacts were located vertically one-third of the distance from the umbilicus to the bottom of the sternum. Two different impactors were used – 2.5 kg, 5.0 cm diameter or 3.5 kg, 7.5 cm diameter; impacts were conducted at 6.3 m/s. Abdominal deformation of approximately 10 cm (53–83 % of abdominal depth) resulted in peak force of 414–1,070 N. One subject sustained no abdominal injuries but the remaining subjects sustained perforations of the hollow organs and hemorrhages of the solid organs. There were no significant differences across age.

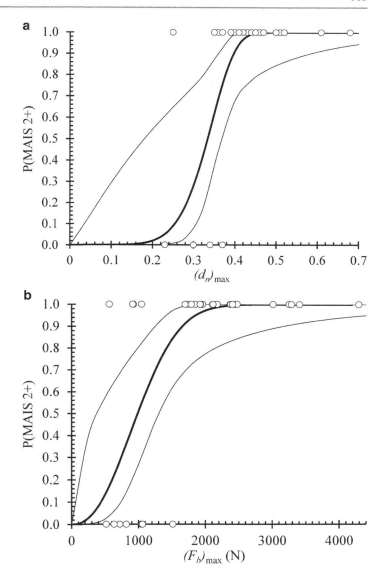

Fig. 22.29 AIS2+ abdominal injury risk functions and 95 % confidence limits for (**a**) maximum penetration and (**b**) belt force as determined by Kent et al. [195] (Reprinted with permission. The Stapp Association)

22.4.2.4 Human Volunteer Studies

To the authors' knowledge, there is only one pediatric human volunteer study on abdominal response. Chamouard et al. conducted a series of quasi-static tests with six child volunteers (mean age: 6.1 years) using a 2-point seatbelt oriented laterally across the lower abdomen [194]. Corridors representing the upper and lower bounds of the force vs. belt penetration data were created. These data were used in comparison to the porcine corridors described above.

22.5 Extremities

22.5.1 Developmental Anatomy

Like other body regions, extremities develop with advancing age. Because of space limitations, the authors chose to highlight only key aspects of the development process. The reader is referred to the literature for a more detailed description [78, 90, 196].

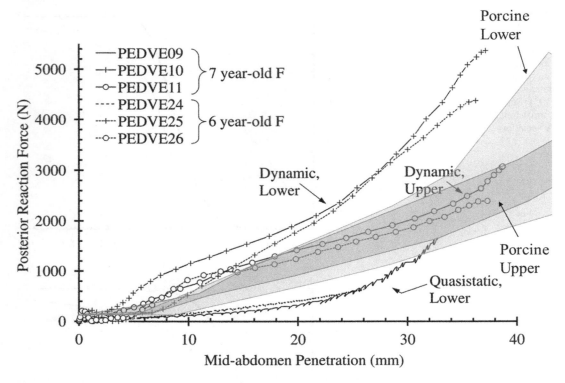

Fig. 22.30 Abdominal force-penetration response of two pediatric PMHS compared to envelopes bracketing the porcine response reported by Kent et al. [193]. PMHS data adapted from Kent et al. [176]

22.5.1.1 Growth Plate and Fracture Patterns: Long Bones

Pediatric long bones are characterized by a growth plate, or region of cartilage, between the epiphysis and metaphysis that serves as the source of longitudinal growth. The presence of this cartilaginous tissue makes radiological interpretation of fractures difficult and results in injury patterns that are unique to the child – termed Salter Harris fractures [197]. Extremity fractures that involve the growth plate can raise concerns about long term growth disturbances.

Fractures in children not involving the growth plate also have unique patterns specific to the immature skeleton [196, 198]. Due to differences in material properties discussed below, pediatric long bones often do not fracture completely in transverse, spiral or oblique patterns but rather fracture incompletely to form greenstick or torus/buckle fractures. A greenstick fracture, common in the forearm, occurs when the bone bends but does not fracture completely and the compression side of the fracture still has an intact cortex. The name is derived from the breakage pattern of a living stick. A torus or buckle fracture is a compression fracture of the metaphysis of the bone where the bone buckles under axial load.

22.5.1.2 Pelvis Development and Injury Patterns

The pelvis undergoes tremendous multi-dimensional structural change over the pediatric age range and is inherently a different mechanical structure in children as compared to adults. In a young child, the three bones of the pelvis (the ilium, the ischium, and the pubis) are loosely connected by cartilaginous tissue. These three bones come together to form the acetabulum and interface with the sacrum (Fig. 22.31). Growth occurs in all three dimensions and ossification occurs throughout childhood and adolescence and is not complete until early adulthood (18–22 years of age), much later than other bones of the body. Benchmarks of ossification occur at 7–8

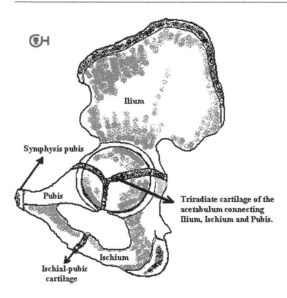

Fig. 22.31 Lateral view of the anatomy of the developing pelvis. The ischial-pubic cartilage fuses at approximately 7 years of age while the triradiate cartilage fuses during puberty (Adapted from Gray and Clemente [101])

years of age (ischio-pubic cartilage) and 12–14 years of age (triradiate cartilage) [101]. Before these two ossification milestones, the individual bones of the pelvis are connected by cartilage and therefore have relative motion. The ligaments are more elastic than those in adults leading to large displacements without fracture [199].

These developmental differences influence the pattern of injury: (1) below age 7 or 8 years where pelvic fractures are extremely rare, (2) between 8 years and puberty where isolated pubic rami fractures occur, and (3) post-pubescent adolescents whose fracture pattern of multiple pelvic bones is similar to that of adults [200, 201]. Due to the lack of ossification between the pelvic bones in children, energy applied to the pelvis is absorbed by the flexible cartilaginous pelvic structure and the stress concentrations are minimized in the other pelvic bones. In contrast, in adolescents and adults, the bones are fused and a contiguous path exists for the forces to travel throughout the pelvis.

22.5.2 Biomechanical Studies

Many studies have quantified the material properties of the lower and upper extremities; however due to lack of space, only those with human tissue are briefly described here. The most common bone studied is the femur. Quasistatic and dynamic tensile and bending tests of human femoral cortical bone have determined that the femur responds with increasing strength and stiffness from youth to the third or fourth decade and then decreases [106, 150, 202–206]. Specifically, tensile tests have revealed that pediatric femur tissue has a low modulus of elasticity, lower ash content, and absorbs more energy to fracture than the adult femurs [207]. This is in part due to changes in bone mineral density which increases rapidly during the first two decades of life [208]. Bending tests conducted by Currey et al. confirmed many of the tensile findings extend to this loading mode [209]. Variation of femur bending modulus with age is depicted in Fig. 22.32. There has been very little study of human trabecular bone however a study in immature sheep matched results from Currey and Butler [207] except that energy absorption increased with age [210]. With regard to the shear strength tolerance of the femur, isolated femur testing from child cadavers ranging from 5 days to 15 years has been conducted and demonstrated shear strength increases with age [211].

Quasistatic and dynamic three-point bending tests have been conducted on intact thighs of pediatric PMHS age 1–6 years by Miltner and Kallieris [212]. Quasistatic fracture moment increased as a function of age from 7.1 to 110 N. More recently, Ouyang et al. conducted impact tests of pediatric (2–12 years) PMHS pelves and upper and lower extremities. Lateral pelvic impact with compression exceeding 50 % of the pelvis width resulted in no injuries [213]. This response is in contrast to similar tests on adult pelvic bones where a 25 % risk of injury was seen with approximately 30 % of compression [214, 215]. Quasistatic and three-point bending tests of the long bones of the upper and lower extremities (stripped of soft tissues) demonstrated fracture moment generally increases with age [216].

Soft tissues of the lower extremities including tendons and ligaments have also demonstrated age-related biomechanical changes. Data on calcaneal tendon, summarized by Yamada, showed an increase in ultimate tensile strength and strain

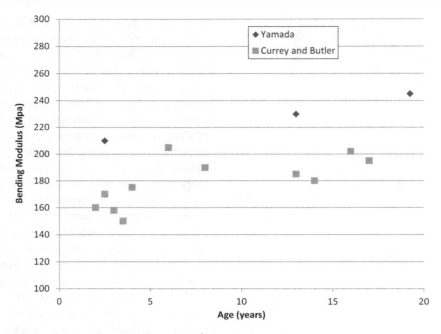

Fig. 22.32 Variation of femur bending modulus with age (Data adapted from Currey and Butler [207] and Yamada [150])

with age [150]. These data were used in the scaling relationships described elsewhere in this chapter. Kubo et al. in a study of human volunteers age 10–25 years conducting isometric knee joint extension demonstrated that tensile stiffness of the knee ligaments increase with age [217]. The most extensive work on immature ligaments has been conducted in a rabbit model and show increase in strength and stiffness with age [218–222]. These authors highlight that the weakness in the lower extremity structure of the child is not the tendons or the ligaments themselves but rather where they insert on the bone. This biomechanical property is confirmed via lower extremity injury patterns in young children [196].

22.6 Summary

The pediatric structural components of the human body develop and attain maturity throughout diverse physiologic processes that proceed at varied pace well into young adulthood. These developmental changes translate into differences in how the child's body withstands and responds to load. The focus of this chapter was on three primary body regions: the head, the neck and the thorax. This emphasis is based on epidemiological data which highlights the importance of head injury mitigation as the primary prevention priority for children. Injuries to the brain and skull are among the most common injuries sustained by children through a variety of mechanisms. Spine and thoracic biomechanics are not only important from an injury mitigation standpoint but they govern the kinematics of the head during traumatic loading and therefore play a role in head injury protection.

Research is underway to develop child-specific brain injury criteria across the spectrum of severity. Mild traumatic brain injuries deserve particular attention as concerns are mounting as to the toll of concussion on the developing brain. The pediatric neck is characterized by the substantial flexibility and relative motion of the vertebrae compared to the adult. As children grow, the spine's structure changes in ways that influence flexibility and lead to increasing stiffness with age. Research has shown that the active range of motion, passive flexion, and dynamic flexion of the spine decrease with age. It is particularly important that these biomechanical properties be incorporated into ATDs and computational models so that the injury predictive ability of these tools is improved.

As children age, the ribs and the sternum structurally change; the skeletal components fuse and the ribs angle downward and twist. Researchers have found that the chest stiffness increases through middle age and then decreases in the elderly. This is manifested at the organ level as greater failure strain in pediatric rib tissue than in the mature skeleton. As a result, rib fractures are uncommon in the youngest children and loads to the thoracic cage are often transferred to the underlying thoracic soft tissues. These findings have implications for age-based thoracic injury criteria suggesting that that different metrics may be needed for different age groups.

Due to space constraints, limited attention was paid to the biomechanics of the pediatric abdomen and the lower extremity. Injuries to the abdomen are a common injury sustained by children especially to those in motor vehicle crashes restrained by seat belts. Recent data has utilized an animal model to quantify the tolerance of the abdomen to blunt loading and has identified force and penetration depth as key metrics in predicting injury. Lower extremity injuries are common in children involved in car crashes as passengers or pedestrians. Though not typically life-threatening, lower extremity injuries are disabling and can directly impair normal orthopedic growth if the fracture involves the growth plate. Limited data currently exist on the biomechanical and physiological properties of the pediatric lower extremities and as a result, pediatric ATD and computational models have limited detail in this body region.

The last decade has witnessed a dramatic increase in pediatric-specific biomechanical knowledge. Such data will facilitate development of better child ATD and computer models and therefore improve the tools available to design better interventions to keep children safe. It is important to recognize that even within the pediatric age range, structure and tissue properties vary with age. Biomechanics research must focus on defining anthropometrics, age-dependent injury tolerance and material and whole body response across the entire pediatric age. These data should be quantified at loading rates relevant to injurious conditions, and across all impact directions. Effort should be placed on validation or further development of the scaling methodologies – both across age and across species – in order to benefit from the richness of animal model, PMHS and human volunteer data. Lastly, as is clear from the references in this chapter, many of the fundamental data are rather dated and as a result, lack the experimental and data collection detail afforded modern studies. Independent confirmation of these data with improved experimental methods and documentation would be beneficial.

22.7 Appendix

The pediatric Anthropomorphic Test Device (ATD) and associated injury criteria are key tools for the evaluation and optimization of automotive restraint systems for child occupants. In addition, the advent of finite element modeling has brought to bear pediatric human body models (HBM) for computational simulation of crashes. The ATD/HBM interacts with the restraints or interior of the vehicle, and must do so in a biofidelic manner to ensure restraint designs protect humans. Injury criteria as well must accurately predict injury in the would-be human occupant.

For adults, post-mortem human subject (PMHS) and volunteer testing has become the mainstay of defining biofidelity requirements for ATD/HBM, and also for developing injury criteria. For children, the paucity of pediatric PMHS and volunteer tests to define injury limits has led to scaling techniques to translate adult-based data to children [151, 169, 223–225]. While aforementioned studies all use different techniques, here we describe the basics of scaling using dimensional analysis. As new data from pediatric subjects become available, there is evidence that scaling is inaccurate in some instances [29, 177]. Thus, the scaling techniques we describe below should be used as a first order approximation and interim solution until actual data with pediatric subjects can be obtained. Our derivation is based upon methods of Langhaar [226].

As an example, we scale angular acceleration between two deformable bodies that represent adult and child human head. Scaling angular accelerations of the two heads in time and magnitude according to the equations below, we expect that brain material contained in head will experience the same strain time-history. These techniques are not unique to the head as they have been applied, for example, to the thorax [227].

The basis of Langhaar's approach is the assumption of scaling ratios for three system properties wherein the fundamental quantities of mass, length and time are represented. The three system properties we chose here are length, density, and modulus of elasticity. We define scaling ratios (λ) as follows,

$$\lambda_l = \frac{l_a}{l_c} \quad \lambda_\rho = \frac{\rho_a}{\rho_c} \quad \lambda_G = \frac{G_a}{G_c} \quad (22.1)$$

where l is a length dimension of the whole brain, ρ and G are the density and elastic modulus, respectively, of the brain material, and the subscripts a and c refer to adult and child, respectively. Scaling ratios for other kinematic quantities are found by dimensional analysis. Heretofore we use the symbol $\Rightarrow$ to signify "has units of", and T, L, and M to denote the units of time, length, and mass, respectively. Thus, angular acceleration (α) has units of radians per second squared,

$$\alpha \Rightarrow \frac{radians}{T^2} \quad (22.2)$$

where T is the unit of time. Angle, measured in radians, is strictly defined as the arc length divided by the radius of curvature, thus

$$\alpha \Rightarrow \frac{L/L}{T^2} = \frac{1}{T^2} \quad (22.3)$$

Performing dimensional analysis, we know that

$$l \Rightarrow L \quad \rho \Rightarrow \frac{M}{L^3} \quad G \Rightarrow \frac{M}{LT^2} \quad (22.4)$$

and thus we can define λ_α in terms of λ_l, λ_ρ and λ_G,

Table 22.16 Scaling equations for various mechanical quantities

Quantity	Symbol	Scaling relationship
Length	L	λ_L
Density	ρ	λ_ρ
Modulus	G	λ_G
Area	A	λ_L^2
Volume	V	λ_L^3
Displacement	u	λ_L
Velocity	v	$\lambda_G^{1/2} \lambda_\rho^{-1/2}$
Acceleration	a	$\lambda_G \lambda_\rho^{-1} \lambda_L^{-1}$
Mass	m	$\lambda_L^3 \lambda_\rho$
Frequency	F	$\lambda_L^3 \lambda_\rho$
Time	t	$\lambda_L^3 \lambda_\rho$
Stiffness	k	$\lambda_L \lambda_G$
Pressure	P	λ_G
Force	F	$\lambda_L^2 \lambda_G$
Stress	σ	λ_G
Strain	ε	1
Energy	E	$\lambda_L^3 \lambda_G$
Angular displacement	θ	1
Angular velocity	ω	$\lambda_G^{1/2} \lambda_\rho^{-1/2} \lambda_L^{-1}$
Angular acceleration	α	$\lambda_G \lambda_\rho^{-1} \lambda_L^{-2}$

$$\lambda_\alpha = \lambda_G \lambda_\rho^{-1} \lambda_l^{-2} \quad (22.5)$$

We assume that a single length scale factor is sufficient to scale the geometry between the adult and child, and thus,

$$\lambda_l = \lambda_m^{1/3} \quad (22.6)$$

and thus,

$$\lambda_\alpha = \lambda_G \lambda_\rho^{-1} \lambda_m^{-2/3} \quad (22.7)$$

Equation 22.6 is the "shape equivalence" assumption, and effectively means that the adult and child scale equally in all dimensions for the body region of interest. Inspection of human developmental anthropometry reveals that the shape equivalence assumption is often violated [113].

By similar dimensional analysis, we can derive scaling equations for angular velocity (ω), angular displacement (θ), time (t), linear acceleration (a), velocity (v), and displacement (u)

$$\lambda_\omega = \lambda_G^{1/2} \lambda_\rho^{-1/2} \lambda_M^{-1/3} \quad \lambda_\theta = 1 \quad (22.8)$$

$$\lambda_t = \lambda_G^{-1/2} \lambda_\rho^{1/2} \lambda_M^{1/3}$$

$$\lambda_a = \lambda_G \lambda_\rho^{-1} \lambda_m^{-1/3} \quad \lambda_v = \lambda_G^{1/2} \lambda_\rho^{-1/2}$$

$$\lambda_u = \lambda_L$$

Using the above methods, a complete set of scaling equations is shown in Table 22.16.

References

1. Williams P (1995) Gray's anatomy. Churchill Livingstone, New York
2. Agur A, Lee M (1991) Grant's atlas of anatomy, 9th edn. Williams & Wilkins, Baltimore
3. Tanner J (1962) Growth at adolescence. Blackwell Scientific, Oxford
4. Tindall G, Cooper P, Barrow D (1996) The practice of neurosurgery. Williams & Wilkins, Baltimore. http://www.worldcat.org/title/practice-of-neurosurgery/oclc/30894343
5. Youman J (1996) Neurological surgery. WB Saunders, Philadelphia
6. Behrman R, Vaughan VI (1987) Developmental pediatrics: growth and development. In: Nelson textbook of pediatrics, 13th edn. Philadelphia
7. Casey BJ, Giedd JN, Thomas KM (2000) Structural and functional brain development and its relation to cognitive development. Biol Psychol 54:241–257
8. Giedd JN, Blumenthal J, Jeffries NO, Castellanos FX, Liu H, Zijdenbos A, Paus T, Evans AC, Rapoport JL (1999) Brain development during childhood and adolescence: a longitudinal MRI study. Nat Neurosci 2:861–863
9. Hubbard R (1971) Flexure of layered cranial bone. J Biomech 4:251–263
10. McElhaney JH, Fogle JL, Melvin JW, Haynes RR, Roberts VL, Alemt NM, Alem NM (1970) Mechanical properties of cranial bone. J Biomech 3:495–511
11. Melvin J, Evans F (1971) A strain energy approach to the mechanics of skull fracture. In: Proceedings of the 15th Stapp car crash conference, Coronado
12. Yoganandan N, Pintar FA, Sances A, Walsh PR, Ewing CL, Thomas DJ, Snyder RG (1995) Biomechanics of skull fracture. J Neurotrauma 12:659–668
13. Kriewall J, Mcpherson GK, Tsai AC (1981) Bending properties and ash content of fetal cranial bone. J Biomech 14:73–79
14. Kriewall T (1982) Structural, mechanical, and material properties of fetal cranial bone. Am J Obstet Gynecol 143:707–714
15. McPherson GK, Kriewall TJ (1980) The elastic modulus of fetal cranial bone: a first step towards an understanding of the biomechanics of fetal head molding. J Biomech 13:9–16
16. Margulies SS, Thibault KL (2000) Infant skull and suture properties: measurements and implications for mechanisms of pediatric brain injury. J Biomech Eng 122:364–371
17. Coats B, Margulies SS (2006) Material properties of human infant skull and suture at high rates. J Neurotrauma 23:1222–1232
18. Davis MT, Loyd AM, Shen H-YH, Mulroy MH, Nightingale RW, Myers BS, Bass CD (2012) The mechanical and morphological properties of 6 year-old cranial bone. J Biomech 45:2493–2498
19. Jaslow C (1990) Mechanical properties of cranial sutures. J Biomech 23:313–321
20. Irwin A, Mertz HJ (1997) Biomechanical basis for the CRABI and hybrid III child dummies. In: Proceedings of the 41st Stapp car crash conference. Orlando
21. Coats B, Margulies SS, Ji S (2007) Parametric study of head impact in the infant. Stapp Car Crash J 1–15
22. Klinich KD, Hulbert GM, Schneider LW, Beach PV (2002) Estimating infant head injury criteria and impact response using crash reconstruction and finite element modeling. In: Proceedings of the 46th Stapp car crash J. 46:165–194
23. Bain AC, Meaney DF (2000) Tissue-level thresholds for axonal damage in an experimental model of central nervous system white matter injury. J Biomech Eng 122:615–622
24. Geddes DM, Cargill RS 2nd, LaPlaca MC (2003) Mechanical stretch to neurons results in a strain rate and magnitude-dependent increase in plasma membrane permeability. J Neurotrauma 20:1039–1049
25. Morrison B III, Cater HL, Wang CC-B, Thomas FC, Hung CT, Ateshian GA, Sundstrom LE (2003) A tissue level tolerance criterion for living brain developed with an in vitro model of traumatic mechanical loading. Stapp Car Crash J 47:93–105
26. Singh A, Kallakuri S, Chen C, Cavanaugh JM (2009) Structural and functional changes in nerve roots due to tension at various strains and strain rates: an in-vivo study. J Neurotrauma 26:627–640
27. Smith DH, Wolf J, Lusardi TA, Lee VM-Y, Meaney DF (1999) High tolerance and delayed elastic response of cultured axons to dynamic stretch injury. J Neurosci 19:4263–4269
28. Margulies SS, Thibault LE, Gennarelli TA (1990) Physical model simulations of brain injury in the primate. J Biomech 23:823–836
29. Ibrahim NGN, Ralston J, Smith C, Margulies SS (2010) Physiological and pathological responses to head rotations in toddler piglets. J Neurotrauma 1035:1021–1035
30. Chatelin S, Constantinesco AA, Willinger RR (2010) Fifty years of brain tissue mechanical testing: from in vitro to in vivo investigations. Biorheology 47:255–276

31. Gurdjian E (1970) Movements of the brain and brain stem from impact induced linear and angular acceleration. Trans Am Neurol Soc 95:248–249
32. Shelden CH, Pudenz RH, Restarski JS, Craig WM (1943) The lucite calvarium-A method for direct observation of the brain. J Neurosurg 3:487–505
33. Zou H, Schmiedeler JP, Hardy WN (2007) Separating brain motion into rigid body displacement and deformation under low-severity impacts. J Biomech 40:1183–1191
34. Bayly PV, Cohen T, Leister E, Ajo D, Leuthardt E, Genin G (2005) Deformation of the human brain induced by mild acceleration. J Neurotrauma 22: 845–856
35. Sabet AA, Christoforou E, Zatlin B, Genin GM, Philip V, Bayly PV (2008) Deformation of the human brain induced by mild angular head acceleration. J Biomech 41:307–315
36. Ibrahim NGNN, Szczesny SES, Eucker S, Margulies SS, Hall S, Natesh R, Szczesny SES, Ryall K, Eucker S, Coats B, Margulies SS (2010) In situ deformations in the immature brain during rapid rotations. J Biomech Eng 132:044501
37. Fitzgerald E, Freeland A (1970) Viscoelastic response of intervertebral discs at audio-frequencies. Med Biol Eng 9:459–478
38. Hayes W, Bodine A (1978) Flow-independent viscoelastic properties of articular cartilage matrix. J Biomech 11:407–419
39. Metz H, McElhaney J, Ommaya A (1970) A comparison of the elasticity of live, dead, and fixed brain tissue. J Biomech 3:453–458
40. Prange MT, Margulies SS (2002) Regional, directional, and age-dependent properties of the brain undergoing large deformation. J Biomech Eng 124: 244–252
41. Thibault KL, Margulies SS (1998) Age-dependent material properties of the porcine cerebrum: effect on pediatric inertial head injury criteria. J Biomech 31:1119–1126
42. Chatelin S, Vappou J, Roth S, Raul JS, Willinger R (2012) Towards child versus adult brain mechanical properties. J Mech Behav Biomed Mater 6:166–173
43. Finan JD, Elkin BS, Pearson EM, Kalbian IL, Morrison B (2012) Viscoelastic properties of the rat brain in the sagittal plane: effects of anatomical structure and age. Ann Biomed Eng 40:70–78
44. Gefen A, Gefen N, Zhu Q, Raghupathi R, Margulies SS (2003) Age-dependent changes in material properties of the brain and braincase of the rat. J Neurotrauma 20:1163–1177
45. Gefen A, Margulies SS (2004) Are in vivo and in situ brain tissues mechanically similar? J Biomech 37:1339–1352
46. Arbogast KB, Margulies SS (1998) Material characterization of the brainstem from oscillatory shear tests. J Biomech 31:801–807
47. Nicolle S, Lounis M, Willinger R, Palierne J-F (2005) Shear linear behavior of brain tissue over a large frequency range. Biorheology 42:209–223
48. Prange MT, Luck JF, Dibb A, Van Ee C, Nightingale RW, Myers BS (2004) Mechanical properties and anthropometry of the human infant head. Stapp Car Crash J 48:279–299
49. Yeates KO, Kaizar E, Rusin J, Bangert B, Dietrich A, Nuss K, Wright M, Taylor HG (2012) Reliable change in postconcussive symptoms and its functional consequences among children with mild traumatic brain injury. Arch Pediatr Adolesc Med 166: 615–622
50. Anderson V, Catroppa C, Morse S, Haritou F, Rosenfeld J (2005) Functional plasticity or vulnerability after early brain injury? Pediatrics 116: 1374–1382
51. Beauchamp MH, Ditchfield M, Maller JJ, Catroppa C, Godfrey C, Rosenfeld JV, Kean MJ, Anderson V (2011) Hippocampus, amygdala and global brain changes 10 years after childhood traumatic brain injury. Int J Dev Neurosci 29:137–143
52. McKinlay A (2010) Controversies and outcomes associated with mild traumatic brain injury in childhood and adolescences. Child Care Health Dev 36:3–21
53. Wade SL, Gerry Taylor H, Yeates KO, Drotar D, Stancin T, Minich NM, Schluchter M (2006) Long-term parental and family adaptation following pediatric brain injury. J Pediatr Psychol 31:1072–1083
54. Holbourn AHS (1956) Private communication to Strich
55. Missios S, Harris BT, Dodge CP, Simoni MK, Costine BA, Lee Y-L, Quebada PB, Hillier SC, Adams LB, Duhaime A-C (2009) Scaled cortical impact in immature swine. J Neurotrauma 26:1943–1951
56. Bittigau P, Sifringer M, Pohl D, Stadthaus D, Ishimaru M, Shimizu H, Ikeda M, Lang D, Speer A, Olney JW, Ikonomidou C (1999) Apoptotic neurodegeneration following trauma is markedly enhanced in the immature brain. Ann Neurol 45:724–735
57. Ikonomidou C, Mosinger JL, Salles KS, Labruyere J, Olney JW (1989) Sensitivity of the developing rat brain to hypobaric/ischemic damage parallels sensitivity to N-methyl-aspartate neurotoxicity. J Neurosci 9:2809–2818
58. McDonald JW, Silverstein FS, Johnston MV (1988) Neurotoxicity of N-methyl-D-aspartate is markedly enhanced in developing rat central nervous system. Brain Res 459:200–203
59. Durham SR, Duhaime A-C (2007) Maturation-dependent response of the immature brain to experimental subdural hematoma. J Neurotrauma 24:5–14
60. Cernak I, Chang T, Ahmed F, Cruz MI, Vink R, Stoica B, Faden AI (2010) Pathophysiological response to experimental diffuse brain trauma differs as a function of developmental age. Dev Neurosci 32:442–453
61. Eucker S, Smith C, Ralston J, Friess SH, Margulies SS (2011) Physiological and histopathological responses following closed rotation head injury depend on direction of head motion. Exp Neurol 227:79–88

62. Backaitis S, Mertz H (1995) Hybird III: the first human-like crash test ATD, vol PT44. Society of Automotive Engineers, Warrendale
63. Sances AJ, Yoganandan N (1986) Human head injury tolerance. Mechanisms of head and spine trauma. Aloray, Goshen
64. Versace J (1971) A review of the severity index. In: Proceedings of the 15th Stapp car crash conference. Society of Automotive Engineers, Warrendale, Coronado, pp 771–796
65. Ommaya A, Yarnell P, Hirsch A, Harris E (1967) Scaling of experimental data on cerebral concussion in sub-human primates to concussion threshold for man. In: Proceedings of the 11th Stapp car crash conference, Anaheim, pp 73–80
66. Duhaime A, Gennarelli T, Thibault L, Bruce D, Margulies S, Wiser R (1987) The shaken baby syndrome. J Neurosurg 66:409–415
67. Melvin J (1995) Injury assessment reference values for the CRABI 6-month infant ATD in a rear-facing infant restraint with airbag deployment. In: SAE World Congress, Detroit, MI
68. Kleinberger M, Sun E, Eppinger RH, Kuppa S, Saul R (1998) Development of improved injury criteria for the assessment of automotive restraint systems. National Highway Traffic Safety Administration
69. Hinck VC, Hopkins CE, Savara BS (1962) Sagittal diameter of the cervical spinal canal in children. Radiology 79:97–108
70. Tulsi RS (1971) Growth of the human vertebral column. An osteological study. Acta Anat (Basel) 79(4):570–580
71. Yousefzadeh DK, El-Khoury GY, Smith WL (1982) Normal sagittal diameter and variation in the pediatric cervical spine. Radiology 144(2):319–325
72. Bailey DK (1952) The normal cervical spine in infants and children. Radiology 59(5):712–719
73. Ogden J, Grogan D, Light T (1987) Postnatal development and growth of musculoskeletal system. In: Albright J, Brand R (eds) The scientific basis of orthopedics. Appleton and Lange, Norwalk
74. O'Rahilly R, Benson D (1985) Development of vertebral column. In: Bradford D, Hensinger RN (eds) The pediatric spine. Thieme, New York, pp 3–17
75. Verbout AJ (1985) The development of the vertebral column. Adv Anat Embryol Cell Biol 90:1–122
76. Chandraraj S, Briggs CA (1991) Multiple growth cartilages in the neural arch. Anat Rec 230(1): 114–120
77. Ford DM, McFadden KD, Bagnall KM (1982) Sequence of ossification in human vertebral neural arch centers. Anat Rec 203(1):175–178
78. Scheuer L, Black SM (2004) The juvenile skeleton. Elsevier Academic Press, London/San Diego
79. Bick EM, Copel JW (1950) Longitudinal growth of the human vertebra; a contribution to human osteogeny. J Bone Joint Surg Am 32(A:4):803–814
80. Carpenter EB (1961) Normal and abnormal growth of the spine. Clin Orthop 21:49–55
81. Gooding CA, Neuhauser EB (1965) Growth and development of the vertebral body in the presence and absence of normal stress. Am J Roentgenol Radium Ther Nucl Med 93:388–394
82. Haas S (1939) Growth in length of vertebrae. Arch Surg 38:245–249
83. Kasai T, Ikata T, Katoh S, Miyake R, Tsubo M (1996) Growth of the cervical spine with special reference to its lordosis and mobility. Spine (Phila Pa 1976) 21(18):2067–2073
84. Ogden J, Ganey T, Sasse J, Neame P, Hilbelink D (1994) Development and maturation of the axial skeleton. In: Weinsten D (ed) The pediatric spine: principles and practice. Raven, New York
85. Roaf R (1960) Vertebral growth and its mechanical control. J Bone Joint Surg (Br) 42-B:40–59
86. Boreadis AG, Gershon-Cohen J (1956) Luschka joints of the cervical spine. Radiology 66(2): 181–187
87. Compere E, Tachdjian M, Kernahan W (1959) Luschka joints: their anatomy, physiology and pathology. Orthopedics 1:159–168
88. Hayashi K, Yabuki T (1985) Origin of the uncus and of Luschka's joint in the cervical spine. J Bone Joint Surg Am 67(5):788–791
89. Kumaresan S, Yoganandan N, Pintar F (1997) Methodology to quantify the uncovertebral joint in the human cervical spine. J Musculoskeletal Res 1(2):1–9
90. Scheuer L, Black SM (2000) Developmental juvenile osteology. Academic, San Diego
91. Maiman D, Yoganandan N (1991) Biomechanica of cervical spine trauma. In: Black P (ed) Clinical neurosurgery, vol 37. Williams and Wilkins, Baltimore, pp 543–570
92. Yoganandan N, Pintar F, Larson SJ, Sances A Jr (1996) Frontiers in head and neck trauma: clinical and biomechanical. IOS Press, Amsterdam
93. Kumaresan S, Yoganandan N, Pintar F (1997) Age-specific pediatric cervical spine biomechanical responses: three-dimensional nonlinear finite element models. In: Proceedings of the Stapp car crash conference, Orlando
94. Kumaresan S, Yoganandan N, Pintar FA (1998) Finite element modeling approaches of human cervical spine facet joint capsule. J Biomech 31(4): 371–376
95. Bonadio WA (1993) Cervical spine trauma in children: Part II. Mechanisms and manifestations of injury, therapeutic considerations. Am J Emerg Med 11(3):256–278
96. Bonadio WA (1993) Cervical spine trauma in children: Part I. General concepts, normal anatomy, radiographic evaluation. Am J Emerg Med 11(2): 158–165
97. Cattell HS, Filtzer DL (1965) Pseudosubluxation and other normal variations in the cervical spine in children. A study of one hundred and sixty children. J Bone Joint Surg Am 47(7):1295–1309

98. Evans D, Bethem D (1985) Cervical spine injuries in children. J Pediatr Orthop 9:563–568
99. Hadley MN, Zabramski JM, Browner CM, Rekate H, Sonntag VK (1988) Pediatric spinal trauma. Review of 122 cases of spinal cord and vertebral column injuries. J Neurosurg 68(1):18–24
100. Seacrist T, Arbogast KB, Maltese MR, Garcia-Espana JF, Lopez-Valdes FJ, Kent RW, Tanji H, Higuchi K, Balasubramanian S (2012) Kinetics of the cervical spine in pediatric and adult volunteers during low speed frontal impacts. J Biomech 45(1): 99–106
101. Gray H, Clemente C (1984) Gray's anatomy of the human body. Lea and Febinger, New York
102. Myklebust JB, Pintar F, Yoganandan N, Cusick JF, Maiman D, Myers TJ, Sances A Jr (1988) Tensile strength of spinal ligaments. Spine (Phila Pa 1976) 13(5):526–531
103. Yoganandan N, Pintar F (1999) Biomechanics of the cranio-cervical junction. In: Boeker D (ed) Cranio-cervical junction – anatomy, physiology, therapy. Bierman Verlag, Koln, pp 2–14
104. Coventry M, Ghormley R, Kernohan J (1945) Intervetebral disc: its microscopic anatomy and pathology. Part I: Anatomy, development, and physiology. J Bone Joint Surg 27(1):105–112
105. Hallen A (1962) Collagen and ground substance of human intervertebral disc at different ages. Acta Chem Scand 16(3):705–710
106. Hirsch C, Evans F (1965) Studies on some physical properties of infant compact bone. Acta Orthop Scand 35:300–313
107. Oda J, Tanaka H, Tsuzuki N (1988) Intervertebral disc changes with aging of human cervical vertebra. From the neonate to the eighties. Spine (Phila Pa 1976) 13(11):1205–1211
108. Peacock A (1956) Observations on postnatal structure of invertebral disc in man. J Anat 86(2):162–179
109. Taylor JR (1970) Growth of human intervertebral disc. J Anat 107(Pt 1):183–184
110. Walmsley R (1953) The development and growth of the intervertebral disc. Edinb Med J 60(8):341–364
111. Dawson RM, Latif Z, Haacke EM, Cavanaugh JM (2013) Magnetic resonance imaging-based relationships between neck muscle cross-sectional area and neck circumference for adults and children. Eur Spine J 22(2):446–452
112. Lavallee AV, Ching RP, Nuckley DJ (2013) Developmental biomechanics of neck musculature. J Biomech 46(3):527–534
113. Snyder R (1977) Anthropometry of infants, children and youths to age 18 for product safety design. University of Michigan, Ann Arbor
114. Stapp J (1949) Human response to linear deceleration. Air Force report
115. Ewing CL, Thomas D, Beeler G, Patrick L, GIllis D (1968) Dynamic response of head and neck of the living human to gravity impact acceleration. In: Proceedings of the Stapp car crash conference, Detroit
116. Mertz H, Patrick L (1971) Strength and response of the human neck. In: Proceedings of the Stapp car crash conference, Coronado
117. Wismans J, Philippens M, van Oorschot D, Kallieris D, Mattern R (1987) Comparison of human volunteer and cadaver head-neck response in frontal flexion. In: Proceedings of the Stapp car crash conference, New Orleans
118. Arbogast KB, Balasubramanian S, Seacrist T, Maltese MR, Garcia-Espana JF, Hopely T, Constans E, Lopez-Valdes FJ, Kent RW, Tanji H, Higuchi K (2009) Comparison of kinematic responses of the head and spine for children and adults in low-speed frontal sled tests. Stapp Car Crash J 53:329–372
119. Seacrist T, Saffioti J, Balasubramanian S, Kadlowec J, Sterner R, Garcia-Espana JF, Arbogast KB, Maltese MR (2012) Passive cervical spine flexion: the effect of age and gender. Clin Biomech (Bristol, Avon) 27(4):326–333
120. Ohman AM, Beckung ER (2008) Reference values for range of motion and muscle function of the neck in infants. Pediatr Phys Ther 20(1):53–58
121. Lewandowski J, Szulc P (2003) The range of motion of the cervical spine in children aged from 3 to 7 years – an electrogoniometric study. Folia Morphol (Warsz) 62(4):459–461
122. Arbogast KB, Gholve PA, Friedman JE, Maltese MR, Tomasello MF, Dormans JP (2007) Normal cervical spine range of motion in children 3–12 years old. Spine (Phila Pa 1976) 32(10):E309–E315
123. Lynch-Caris T, Brelin-Fornari J, Pelt C (2006) Cervical range of motion data in children. In: SAE World Congress, Detroit, SAE Tech Paper #2006-01-1140
124. Greaves LL, Van Toen C, Melnyk A, Koenig L, Zhu Q, Tredwell S, Mulpuri K, Cripton PA (2009) Pediatric and adult three-dimensional cervical spine kinematics: effect of age and sex through overall motion. Spine (Phila Pa 1976) 34(16):1650–1657
125. Duncan JM (1874) Laboratory note: on the tensile strength of the fresh adult foetus. Br Med J 2(729):763–764
126. Brun-Cassan F, Page M, Pincemaille Y (1993) Comparative study of restrained child dummies and cadavers in experimental crashes. In: Child occupant protection symposium, San Antonio
127. Dejammes M, Tarriere C, Thomas C (1984) Exploration of biomechanical data towards a better evaluation of tolerance for children involved in automotive accidents. In: Proceedings of the Stapp car crash conference, Chicago
128. Kallieris D, Barz J, Schmidt G (1976) Comparison between child cadavers and child ATD by using child restraint systems in simulated collisions. In: Proceedings of the Stapp car crash conference, Dearborn, pp 511–542
129. Mattern R, Kallieris D, Riedl H, von Wiren B (2002) Reanalysis of two child PMHS-tests. Final report. University of Heidelberg, Heidelberg

130. Wismans J, Maltha J, Melvin J (1979) Child restraint evaluation by experimental and mathematical simulation. In: Proceedings of the Stapp car crash conference, San Diego
131. McGowan DA, Voo L, Liu Y (1993) Distraction failure of immature spine. ASME Adv Bioeng 24:24–25
132. Ouyang J, Zhu Q, Zhao W, Xu Y, Chen W, Zhong S (2005) Biomechanical assessment of the pediatric cervical spine under bending and tensile loading. Spine (Phila Pa 1976) 30(24):E716–E723
133. Luck JF, Nightingale RW, Loyd AM, Prange MT, Dibb AT, Song Y, Fronheiser L, Myers BS (2008) Tensile mechanical properties of the perinatal and pediatric PMHS osteoligamentous cervical spine. Stapp Car Crash J 52:107–134
134. Luck JF, Nightingale RW, Song Y, Kait JR, Loyd AM, Myers BS, Bass CR (2013) Tensile failure properties of the perinatal, neonatal, and pediatric cadaveric cervical spine. Spine (Phila Pa 1976) 38(1):E1–E12
135. Dibb AT, Nightingale RW, Luck JF, Chancey VC, Fronheiser LE, Myers BS (2009) Tension and combined tension-extension structural response and tolerance properties of the human male ligamentous cervical spine. J Biomech Eng 131(8): 081008
136. Nightingale RW, Carol Chancey V, Ottaviano D, Luck JF, Tran L, Prange M, Myers BS (2007) Flexion and extension structural properties and strengths for male cervical spine segments. J Biomech 40(3):535–542
137. Van Ee CA, Nightingale RW, Camacho DL, Chancey VC, Knaub KE, Sun EA, Myers BS (2000) Tensile properties of the human muscular and ligamentous cervical spine. Stapp Car Crash J 44:85–102
138. Nuckley DJ, Linders DR, Ching RP (2013) Developmental biomechanics of the human cervical spine. J Biomech 46:1147–1154
139. Mertz H, Driscoll G, Lenox J, Nyquist G, Weber D (1982) Responses of animals exposed to deployment of carious passenger inflatable restraint system concepts for a variety of collision severities and animal positions. In: Proceedings of the International Technical Conference on Experimental Safety Vehicles, Kyoto
140. Prasad P, Daniel R (1984) Biomechanical analysis of head, neck, and torso injuries to child surrogates due to sudden torso acceleration. In: Proceedings of the Stapp car crash conference, Chicago
141. Pintar FA, Mayer RG, Yoganandan N, Sun E (2000) Child neck strength characteristics using an animal model. Stapp Car Crash J 44:77–83
142. Hilker CE, Yoganandan N, Pintar FA (2002) Experimental determination of adult and pediatric neck scale factors. Stapp Car Crash J 46:417–429
143. Ching RP, Nuckley DJ, Hertsted SM, Eck MP, Mann FA, Sun EA (2001) Tensile mechanics of the developing cervical spine. Stapp Car Crash J 45:329–336
144. Yoganandan N, Kumaresan S, Pintar FA (2000) Geometric and mechanical properties of human cervical spine ligaments. J Biomech Eng 122(6): 623–629
145. Clarke EC, Appleyard RC, Bilston LE (2007) Immature sheep spines are more flexible than mature spines: an in vitro biomechanical study. Spine (Phila Pa 1976) 32(26):2970–2979
146. Nuckley DJ, Ching RP (2006) Developmental biomechanics of the cervical spine: tension and compression. J Biomech 39(16):3045–3054
147. Nuckley DJ, Hertsted SM, Eck MP, Ching RP (2005) Effect of displacement rate on the tensile mechanics of pediatric cervical functional spinal units. J Biomech 38(11):2266–2275
148. Nuckley DJ, Hertsted SM, Ku GS, Eck MP, Ching RP (2002) Compressive tolerance of the maturing cervical spine. Stapp Car Crash J 46:431–440
149. Elias PZ, Nuckley DJ, Ching RP (2006) Effect of loading rate on the compressive mechanics of the immature baboon cervical spine. J Biomech Eng 128(1):18–23
150. Yamada H, Evans FG (1970) Strength of biological materials. Williams & Wilkins, Baltimore
151. Eppinger RH, Sun E, Bandak F, Haffner M, Khaewpong N, Maltese MR, Kuppa S, Nguyen T, Takhounts E, Tannous R, Zhang A, Saul R (1999) Development of improved injury criteria for the assessment of advanced automotive restraint systems – II: supplement to NHTSA docket. National Highway Traffic Safety Administration
152. Yoganandan N, Kumaresan S, Pintar FA (2001) Biomechanics of the cervical spine. Part 2. Cervical spine soft tissue responses and biomechanical modeling. Clin Biomech (Bristol, Avon) 16(1):1–27
153. Kumaresan S, Yoganandan N, Pintar F, Mueller W (2000) Biomechanics of pediatric cervical spine: compression, flexion and extension responses. J Crash Prev Inj Control 2:87–101
154. Kumaresan S, Yoganandan N, Pintar FA, Maiman DJ, Kuppa S (2000) Biomechanical study of pediatric human cervical spine: a finite element approach. J Biomech Eng 122(1):60–71
155. Franklyn M, Peiris S, Huber C, Yang KH (2007) Pediatric material properties: a review of human child and animal surrogates. Crit Rev Biomed Eng 35(3–4):197–342
156. Kent R, Ivarsson J, Maltese M (2013) Experimental injury biomechanics of the pediatric thorax and abdomen. In: Crandall JR, Myers BS, Meaney DF, Schmidke D (eds) Pediatric injury biomechanics. Springer, New York, pp 221–286
157. Kemper AR, McNally C, Kennedy EA, Manoogian SJ, Rath AL, Ng TP, Stitzel JD, Smith EP, Duma SM, Matsuoka F (2005) Material properties of human rib cortical bone from dynamic tension coupon testing. Stapp Car Crash J 49:199–230
158. Sturtz G (1980) Biomechanical data of children. In: Proceedings of the Stapp car crash conference.

Society of Automotive Engineers, Warrendale. Orlando, SAE Paper No. 801313
159. Weaver JK, Chalmers J (1966) Cancellous bone: its strength and changes with aging and an evaluation of some methods for measuring its mineral content I. Age changes in cancellous bone. J Bone Joint Surg 48:289–299
160. Mosekilde L, Danielsen CC (1987) Biomechanical competence of vertebral trabecular bone in relation to ash density and age in normal individuals. Bone 8:79–85
161. Mosekilde L, Viidik A, Mosekilde L (1985) Correlation between the compressive strength of iliac and vertebral trabecular bone in normal individuals. Bone 6:291–295
162. Mosekilde L, Mosekilde L (1986) Normal vertebral body size and compressive strength: relations to age and to vertebral and iliac trabecular bone compressive strength. Bone 7:207–212
163. McElhaney J, Alem NM, Roberts VL (1970) A porous block model for cancellous bones. Proceedings of the Winter Annual Meeting of the American Society of Mechanical Engineers, New York
164. Oyen ML, Lau AG, Kindig MW, Stacey SC, Kent RW (2006) Mechanical properties of structural tissues of the pediatric thorax. J Biomech 39:S156
165. Mattice JM, Lau AG, Oyen ML, Kent RW (2006) Spherical indentation load-relaxation of soft biological tissues. J Mater Res 21:2003–2010
166. Khamin NS (1977) Strength properties of the human aorta and their variation with age. Polym Mech 13:100–104
167. Ouyang J, Zhao W, Xu Y, Chen W, Zhong S (2006) Thoracic impact testing of pediatric cadaveric subjects. J Trauma 61(6):1492–1500
168. Kroell C, Nahum A (1974) Impact tolerance and response of the human thorax II. In: Proceedings of the 18th Stapp car crash conference. Society of Automotive Engineers, Warrendale, Ann Arbor
169. Parent DP (2009) Scaling and optimization of thoracic impact response in pediatric subjects. University of Virginia, Charlottesville
170. Gruben KG, Guerci AD, Halperin HR, Popel AS, Tsitlik JE (1993) Sternal force-displacement relationship during cardiopulmonary resuscitation. J Biomech Eng 115:195–201
171. Tsitlik JE, Weisfeldt ML, Chandra N, Effron MB, Halperin HR, Levin HR (1983) Elastic properties of the human chest during cardiopulmonary resuscitation. Crit Care Med 11:685–692
172. Vallis CJ, Mackenzie I, Lucas BG (1979) The force necessary for external cardiac compression. Practitioner 223:268–270
173. Maltese MR, Castner T, Niles D, Nishisaki A, Balasubramanian S, Nysaether J, Sutton R, Nadkarni V, Arbogast KB (2008) Methods for determining pediatric thoracic force-deflection characteristics from cardiopulmonary resuscitation. Stapp Car Crash J 52:83–106
174. American Heart Association (2006) 2005 American Heart Association guidelines for cardiopulmonary resuscitation and emergency cardiovascular care. Circulation 112(Suppl)
175. Arbogast KB, Maltese MR, Nadkarni VM, Steen PA, Nysaether JB (2006) Anterior-posterior thoracic force–deflection characteristics measured during cardiopulmonary resuscitation: comparison to post-mortem human subject data. Stapp Car Crash J 50:131–145
176. Kent R, Lopez-Valdes FJ, Lamp J, Lau S, Parent D, Kerrigan J, Lessley D, Salzar R, Sochor M, Bass D, Maltese MR (2012) Biomechanical response targets for physical and computational models of the pediatric trunk. Traffic Inj Prev 13(5):499–506
177. Kent R, Salzar R, Kerrigan J, Parent D, Lessley D, Sochor M, Luck JF, Loyd A, Song Y, Nightingale R, Bass CRD, Maltese MR (2009) Pediatric thoracoabdominal biomechanics. Stapp Car Crash J 53:373–401
178. Hoke RS, Chamberlain D (2004) Skeletal chest injuries secondary to cardiopulmonary resuscitation. Resuscitation 63:327–338
179. Holmes JF, Sokolove PE, Brant WE, Kuppermann N (2002) A clinical decision rule for identifying children with thoracic injuries after blunt torso trauma [see comment]. Ann Emerg Med 39:492–499
180. Garcia VF, Gotschall CS, Eichelberger MR, Bowman LM (1990) Rib fractures in children: a marker of severe trauma. J Trauma Inj Infect Crit Care 30:695–700
181. Kallieris D, Mattern R, Schmidt G, Eppinger R (1981) Quantification of side impact responses and injuries. In: Proceedings of the Stapp car crash conference. Society of Automotive Engineers, Warrendale, San Francisco, pp 329–366
182. Stocker JT, Dehner LP (1992) Pediatric pathology. Lippincott, Philadelphia
183. Fazekas JG, Kosa F, Jobba G, Meszaros E (1972) Compression strength of the human spleen under the action of blunt force. Arch Kriminol 149(5):158–174
184. Schmidt G (1979) The age as a factor influencing soft tissue injuries. In: Proceedings of the International Conference on the Biomechanics of Impact
185. Stingl J, Baca V, Cech P, Kovanda J, Kovandova H, Mandys V, Rejmontova J, Sosna B (2002) Morphology and some biomechanical properties of human liver and spleen. Surg Radiol Anat 24(5):285–289
186. Seki S, Iwamoto H (1998) Disruptive forces for swine heart, liver, and spleen: their breaking stresses. J Trauma 45(6):1079–1083
187. Liu Z, Bilston L (2000) On the viscoelastic character of liver tissue: experiments and modelling of the linear behaviour. Biorheology 37(3):191–201
188. Nasseri S, Bilston L, Phan-Thien N (2002) Viscoelastic properties of pig kidney in shear, experimental results and modeling. Rheol Acta 41:180–192
189. Mattice J (2006) Age-dependent changes in the viscoelastic response of the porcine kidney parenchyma using spherical indentation and finite element analysis. University of Virginia, Charlottesville

190. Zhao DL, Shi JS (2005) [Bone marrow metastasis of stomach cancer: a case report]. Zhonghua Zhong Liu Za Zhi 27(12):712
191. Heijnsdijk EA, van der Voort M, de Visser H, Dankelman J, Gouma DJ (2003) Inter- and intraindividual variabilities of perforation forces of human and pig bowel tissue. Surg Endosc 17(12):1923–1926
192. Gögler E, Best A, Braess H (1977) Biomechanical experiments with animals on abdominal tolerance levels. In: Stapp car crash conference, New Orleans, pp 713–751
193. Kent R, Stacey S, Kindig M, Forman J, Woods W, Rouhana SW, Higuchi K, Tanji H, Lawrence SS, Arbogast KB (2006) Biomechanical response of the pediatric abdomen, part 1: development of an experimental model and quantification of structural response to dynamic belt loading. Stapp Car Crash J 50:1–26
194. Chamouard F, Tarriere C, Baudrit P (1996) Protection of children on board vehicles: influence of pelvis design and thigh and abdomen stiffness on the submarining risk for dummies installed on a booster. In: Proceedings of the 15th Enhanced Safety of Vehicles Conference
195. Kent R, Stacey S, Kindig M, Woods W, Evans J, Rouhana SW, Higuchi K, Tanji H, St Lawrence S, Arbogast KB (2008) Biomechanical response of the pediatric abdomen, Part 2: injuries and their correlation with engineering parameters. Stapp Car Crash J 52:135–166
196. Rockwood CA, Beaty JH, Kasser JR (2010) Rockwood and Wilkins' fractures in children, 7th edn. Wolters Kluwer/Lippincott/Williams & Wilkins, Philadelphia
197. Salter RB, Harris WR (1963) Injuries involving the epiphyseal plate. J Bone Joint Surg 45:587–622
198. Crandall J, Myers B, Meaney D, Schmidtke S (2013) Pediatric injury biomechanics. Springer, New York
199. Canale S, Daugherty K, Jones L (1998) Campbell's operative orthopedics. Mosby, New York
200. Arbogast KB, Mari-Gowda S, Kallan MJ, Durbin DR, Winston FK (2002) Pediatric pelvic fractures in side impact collisions. Stapp Car Crash J 46:285–296
201. Silber JS, Flynn JM (2002) Changing patterns of pediatric pelvic fractures with skeletal maturation: implications for classification and management. J Pediatr Orthop 22(1):22–26
202. Vinz H (1969) Mechanical principles of typical fractures in children. Zentralbl Chir 94(45):1509–1514
203. Vinz H (1970) Change in the resistance properties of compact bone tissue in the course of aging. Gegenbaurs Morphol Jahrb 115(2):257–272
204. Vinz H (1972) Firmness of pure bone substance. Approximation method for the determination of bone tissue firmness related to the cavity-free cross section. Gegenbaurs Morphol Jahrb 117(4):453–460
205. Wall JC (1974) The effects of age and strain rate on the mechanical properties of bone. In: International Meeting on Biomechanics of Trauma in Children, Lyon, pp 185–193
206. Wall JC, Chatterji SK, Jeffery JW (1979) Age-related changes in the density and tensile strength of human femoral cortical bone. Calcif Tissue Int 27(2):105–108
207. Currey JD, Butler G (1975) The mechanical properties of bone tissue in children. J Bone Joint Surg Am 57(6):810–814
208. Currey JD (1979) Changes in the impact energy absorption of bone with age. J Biomech 12(6):459–469
209. Currey JD, Brear K, Zioupos P (1996) The effects of ageing and changes in mineral content in degrading the toughness of human femora. J Biomech 29(2):257–260
210. Nafei A, Danielsen CC, Linde F, Hvid I (2000) Properties of growing trabecular ovine bone. Part I: mechanical and physical properties. J Bone Joint Surg (Br) 82(6):910–920
211. Chung SM, Batterman SC, Brighton CT (1976) Shear strength of the human femoral capital epiphyseal plate. J Bone Joint Surg Am 58(1):94–103
212. Miltner E, Kallieris D (1989) Quasi-static and dynamic bending stress of the pediatric femur for producing a femoral fracture. Z Rechtsmed 102(8):535–544
213. Ouyang J, Zhu QA, Zhao WD, Xu YQ, Chen WS, Zhong SZ (2003) Experimental cadaveric study of lateral impact of the pelvis in children. Di Yi Jun Yi Da Xue Xue Bao 23(5):397–401, 408
214. Cavanaugh JM, Waliko T, Malthora A (1990) Biomechanical response and injury of the pelvis in twelve sled side impacts. Stapp Car Crash J, Orlando, 1–12
215. Viano D (1989) Biomechanical responses and injuries in blunt lateral impacts. In: Proceedings of the Stapp car crash conference, Washington DC, pp 113–142
216. Ouyang J, Zhu QA, Zhao WD, Xu YQ, Chen WS, Zhong SZ (2003) Biomechanical character of extremity long bones in children. Chin J Clin Anat 21:620–623
217. Kubo K, Kanehisa H, Kawakami Y, Fukanaga T (2001) Growth changes in the elastic properties of human tendon structures. Int J Sports Med 22(2):138–143
218. Lam TC, Frank CB, Shrive NG (1993) Changes in the cyclic and static relaxations of the rabbit medial collateral ligament complex during maturation. J Biomech 26(1):9–17
219. Woo SL, Hollis JM, Adams DJ, Lyon RM, Takai S (1991) Tensile properties of the human femur-anterior cruciate ligament-tibia complex. The effects of specimen age and orientation. Am J Sports Med 19(3):217–225
220. Woo SL, Ohland KJ, Weiss JA (1990) Aging and sex-related changes in the biomechanical properties of the rabbit medial collateral ligament. Mech Ageing Dev 56(2):129–142

221. Woo SL, Orlando CA, Gomez MA, Frank CB, Akeson WH (1986) Tensile properties of the medial collateral ligament as a function of age. J Orthop Res 4(2):133–141
222. Woo SL, Peterson RH, Ohland KJ, Sites TJ, Danto MI (1990) The effects of strain rate on the properties of the medial collateral ligament in skeletally immature and mature rabbits: a biomechanical and histological study. J Orthop Res 8(5):712–721
223. Ivarsson BJ, Crandall JR, Longhitano D, Okamoto M (2004) Lateral injury criteria for the 6-year-old pedestrian – part II: criteria for the upper and lower extremities. SAE 2004 World Congress and exposition. Society of Automotive Engineers, Detroit. SAE Paper No. 2004-01-1755
224. Maltese MR, Arbogast KB, Wang Z, Craig M (2011) Scaling methods applied to thoracic force displacement characteristics derived from cardiopulmonary resuscitation. In: Proceedings of the 22nd Enhanced Safety of Vehicles Conference, Washington, DC
225. Van Ratingen MR, Twisk D, Schrooten M, Beusenberg MC (1997) Biomechanically based design and performance targets for a 3-year old child crash ATD for frontal and side impact. In: Proceedings of the 41st Stapp car crash conference, Orlando
226. Langhaar H (1951) Dimensional analysis and theory of models. Wiley, New York
227. Eppinger RH, Marcus JH, Morgan RM (1984) Development of ATD and injury index for NHTSA's thoracic side impact protection research program. SAE Government Industry Meeting and Exposition. Paper No. 840885. The Society of Automotive Engineers, Warrendale

Best Practice Recommendations for Protecting Child Occupants

Kathleen D. Klinich and Miriam A. Manary

Abstract

Pediatric occupants of motor vehicles need the specialized protection provided by child restraint systems because of their immature and developing body structures. The best child restraint system for a child depends on the size, age, and development level. As children proceed through the four phases of restraints systems (rear-facing harnessed restraints, forward-facing harnessed restraints, belt-positioning booster seats, and seat belts), their level of protection decreases, so graduation to the next type of product should be delayed as long as possible given the specifications for each product. The protection afforded children in crashes also depends on their seating position and safety features in the vehicles. Even when the most appropriate child restraint is selected, its ability to provide protection in a crash depends on how it is installed and used. This review describes how the basic principles of occupant protection are applied to the design of child restraint systems. Best practice recommendations for child occupant protection (primarily from the US perspective) are outlined, together with the research basis that supports them.

23.1 Effectiveness of Child Restraint Systems

The number of children killed in motor-vehicle crashes has dropped by 41 % from 2000 to 2009, and it is no longer the leading cause of death for children aged 1–4 [1]. These reductions in fatalities have, in part, resulted from improvements in child restraint design and increased use of proper restraint prompted by education and strengthened child restraint use laws. The American Academy of Pediatrics [2, 3] and the National Highway Traffic Safety Administration [4] recommend four sequential steps to properly protect child passengers in motor-vehicle crashes:

1. Rear-facing restraint with harness
2. Forward-facing restraint with harness
3. Belt-positioning booster seat used with a three-point belt restraint
4. Vehicle seat belt

K.D. Klinich, Ph.D. (✉) • M.A. Manary, M.Eng.
Transportation Research Institute, University of Michigan, Ann Arbor, MI, USA
e-mail: kklinich@umich.edu; mmanary@umich.edu

This sequence of recommendations was based on the effectiveness of child restraint systems estimated using crash databases and laboratory data. Compared to unrestrained children, the risk of death and serious injury is reduced by 71 % for rear-facing child restraints and 54 % for forward-facing harnessed child restraints [5]. Zaloshnji et al. [6] and Arbogast et al. [7] report that the odds of AIS2+ injury are 78–82 % lower for children using forward-facing child restraints rather than vehicle three-point belt systems. Recommendations to remain rear-facing longer are based on a study by Henary et al. [8], which showed that injury rates of children aged 0–2 in forward-facing restraint systems are 1.76 times higher than those in rear-facing restraints.

Elliot et al. [9] analyzed FARS data to demonstrate that for children aged 2–6 years, using a child restraint (without gross misuse) resulted in a 28 % lower risk of death compared to children using a seatbelt. For children aged 4–8 years, using a booster seat lowers fatality risk (relative to unrestrained occupants) by 55–67 %, while using a seatbelt has a similar level in fatality reduction of 52 % [10]. Arbogast et al. [11] report that booster seats reduce the risk of AIS2+ injury by 55 % relative to seatbelts among children aged 4–8.

The percentage of children traveling unrestrained in motorvehicles has dropped from 90 % in 1976 [12] to 13 % in 2008 [13]. Though they make up a small proportion of occupants, unrestrained children aged 0–15 make up 54 % of child fatalities in motor-vehicle crashes. Thus reducing the rate of nonuse from 13 % to 0 % would lead to substantial reductions in fatalities.

Misuse of child restraints continues to be a problem. Several studies have documented that between 63 % and 90 % of child restraints have at least one installation error [14–18]. However, the studies generally do not classify errors by their severity; some errors have greater consequences than others and/or may only be a problem in more severe crashes. But some laboratory and computational studies have demonstrated that several small errors can have a similar result as one large error [19–24]. While misuse remains a challenge, the relatively high effectiveness rates for child restraints are based on their frequent rate of misuse. Child restraints would likely be even more effective if their misuse rates were lower.

23.2 Occupant Restraint Systems

23.2.1 Design Concepts

When a child restraint harness or vehicle seat belt is initially positioned snugly across the occupant, the restraint systems can begin restraining the occupant earlier in the impact event, resulting in the lowest amount of loading to restrain the occupant. Padding and airbags provide supplemental restraint to absorb energy that would otherwise load the occupant. With loose harness or belt webbing, restraint cannot begin until the occupant moves forward into the webbing, which increases the force needed to stop the occupant.

Design of restraint systems involves distributing loads across the occupants' most robust components. Seatbelts for adults and older children are designed to restrain through the shoulder and pelvis bones. For children, rearward-facing orientation distributes load across the child's back. Those using forward-facing harnessed restraints have straps that provide 5-points of restraint to distribute load. Belts must be properly positioned to provide effective restraint and avoid causing injury. Lap belts positioned over the abdomen on a child not using a booster, or a loose belt that allows the occupant to "submarine" under the lap belt, can load the abdomen and cause serious injuries [25, 26]. Using only a lap belt allows excessive loading of the abdomen and lumbar spine [27, 28]. Children and adults sometimes misuse their lap-and-shoulder belt by placing the shoulder portion of the belt behind their back or under the arm. This eliminates torso restraint and can cause the same problems seen with only lap belt use [29, 30].

Although belt-restraint systems have the potential to result in injuries, they can provide occupant protection in different types of crashes, including rollovers and those involving multiple impacts. With belt restraints, the loads are proportional to occupant mass. A belt system restraining an adult weighing 80 kg will apply higher restraining forces than a child weighing 30 kg. Although pediatric tissues may not be as strong as an adult's, the injury potential from belt loading is lower because of the child's lower mass. Some of the more recent advancements in airbag design also provide variable restraint force based on occupant mass, but the first-generation of frontal airbags produced the same deployment force for all occupants. Since they were designed to stop an unrestrained adult in a crash, this level of force proved to be lethal for some child occupants.

The design of a child restraint varies with the child's size and rear- or forward-facing direction, as well as the method of installing the child restraint and securing the child within the restraint. However, all child restraints follow the design principle of tightly coupling the child to the vehicle because it increases the time the restraint can work, which decreases the maximum force applied the child. The vehicle seatbelt has typically been used to attach the child restraint to the vehicle. Another option available to attach the child restraint to the vehicle is the LATCH system, which stands for Lower Anchors and Tethers for CHildren. This is a dedicated system consisting of vehicle and child restraint hardware that was first required in the US in 2002. Once the child restraint is installed in the vehicle, a separate harness secures the child, leading to two links between the vehicle and the child. With either attachment method (LATCH or seatbelt) the installation must be tight and the harness must be snug to provide effective restraint.

23.2.2 Seating Positions, Airbags, and Children

Because it is farthest from the vehicle exterior, the center second row seat is considered the safest seating position. Field data confirm this presumption [31–33]. However, before airbags were provided for right-front passengers, caregivers often placed infants in the right-front position to allow the driver to monitor the infant during travel [34]. Children also rode in the front seat to take advantage of the lap-and-shoulder belt restraint, because until 1989, they were not required in the rear seat.

The introduction of frontal-impact airbags in the early 1990s in the United States had the unintended consequence of producing restraint systems with the potential to mortally injure children [35, 36]. In the United States, airbags have been designated the cause of fatal injury to over 28 infants and 152 older children [37]. In vehicles with first-generation airbags, the risk of fatal injury to children in the right-front seating position is estimated to increase by 34–63 % [38]. Most of the fatalities involved direct contact between the airbag and/or airbag housing that produced head or neck injury in either rear-facing children or older children who were unrestrained or out-of-position and too close to the airbag during deployment [39].

The initial response to the danger posed by first-generation airbags was to recommend that children under 13 years of age use the rear seat. Educational campaigns and airbag warning labels have increased the rate of rear seat occupancy to 83 % of children under age 7, including almost all rear-facing children and half of children aged 8–12 [40]. Federal regulations also require occupant detection systems and allow airbag designs that deploy with less force. Compared to first-generation airbags, second-generation airbags reduce fatal injury risk by 65 % for children aged 0–4, 46 % for children aged 5–9, and 32 % for children aged 10–12 who were seated in the front seat [32]. Another study identified a 34 % risk reduction for children under 6 between second- and first-generation airbags [41].

Despite technological airbag improvements, the recommendation for children under age 13 to use the rear seat for travel remains, because many vehicles in the fleet still have first-generation airbags [2, 4]. Also, several studies have demonstrated that the rear seat is safer for belted occupants [42, 43], partly because the occupants are further from intrusion that can occur during frontal crashes [44].

Test procedures have been developed for side airbags to ensure that their introduction would not have the dangerous unintended consequences seen with frontal-impact airbags. Vehicle manufacturers voluntarily perform tests to check that side airbags do not pose a hazard to "out-of-position" children seated next to airbags [45]. During side impacts and rollovers, curtain airbags deploy from the vehicle roofline to protect the head. Examination of crash data show almost no injuries to children from side or curtain airbags, demonstrating that the currently used test procedures are effective [46, 47].

23.3 Child Restraint Installation

23.3.1 Seatbelt

New occupant restraint laws and their enforcement increased the rate of seatbelt use in the US to 85 % in 2010 from 11 % in 1982 [48, 49]. To encourage belt use by adult occupants, vehicle manufacturers improved comfort and ease-of-use of vehicle belt systems as they also improved belt effectiveness. However, some changes that improved belt restraint for adults made child restraint installation more difficult. As an example, child restraints can be secured more easily if the seatbelt anchorage is located at the vehicle seat bight (the intersection of the seatback and seat cushion.) But a seatbelt anchorage located forward of the bight often provides a better restraint angle for adults. As a result of incompatibilities, FMVSS 208 was amended with a seatbelt lockability requirement in 1996, and SAE Recommended Practice J1819 [50] was developed to provide design guidelines for vehicle and child restraint manufacturers.

One of the most common child restraint installation errors is the failure to lock the seatbelt. To improve comfort for adults, nearly all seatbelts include an emergency locking retractor. This feature allows the occupant to move during normal driving, but locks during a crash to limit forward motion of the occupant. Seatbelts with emergency locking retractors permit too much movement of child restraints during typical driving. To allow a tight child restraint installation, most emergency locking retractors are switchable, and can be adjusted to an automatic locking retractor that allows tightening of the belt through the child restraint belt path. The automatic locking retractor is usually activated by pulling the webbing all the way out of the retractor, then reeling it in as the belt is tightened.

Another type of vehicle hardware available to allow tight child restraint installation is a locking latchplate. This feature is usually built into the latch plate and keeps belt webbing from feeding into the lap belt.

Belt lockoffs are provided on some child restraints to clamp down on seatbelt webbing to lock it. If a child restraint does not have belt lockoffs, the child restraint manufacturer must provide a locking clip. This device prevents webbing from moving from the shoulder portion to the lap portion of the belt. However, locking clips are often misused. If they are not properly positioned within one inch of the latchplate, they can allow belt slack, deform, or become a projectile during a crash.

23.3.2 LATCH

The frequent difficulty of installing child restraints with seatbelts led to implementation of the Child Restraint Anchorage System in 1999 by the National Highway Traffic Safety Administration (NHTSA). FMVSS 225 defines the vehicle requirements [51], while revisions to FMVSS 213 define the child restraint requirements [52]. In the US, the system is commonly known as LATCH, which stands for Lower Anchors and Tethers for CHildren. The concept is based on the ISOFix [53] system developed by International Standards Organization (ISO). Since 2002, all vehicles and child restraints sold in the United States are equipped with LATCH hardware.

In the United States, the vehicle hardware consists of two lower anchors located near the bight of the vehicle seat (Fig. 23.1) and a tether anchor (Fig. 23.2) located rearward of the seating position on the rear filler panel, seatback, floor, or vehicle roof. The child restraint has lower connectors to provide the main attachment to the lower anchors. Forward-facing products have a top tether strap that limits head excursion when attached appropriately to the tether anchor. The ISOFix approach also has two lower attachment points, but allows a variety of methods to limit forward rotation of the child restraint, including tether or floor support.

For US child restraints, most products attach to the lower anchors using a length of webbing with connectors and adjustment hardware on each end. CRS lower connectors are usually hook-on or push-on styles (Fig. 23.3). Many products route the LATCH strap through the vehicle belt path on the child restraint rather than using a seatbelt (Fig. 23.4a), or attach them to each side of the child restraint (Fig. 23.4b). A few US products use rigid lower LATCH connectors (Fig. 23.4c), which are required by ISOFix in Europe. Although products must meet FMVSS 213 excursion requirements when tested without a tether, the tether improves occupant protection in forward-facing restraints by reducing child head excursion when used with either the vehicle seat belt or the LATCH strap. Also, the tether often helps to obtain a better installation.

LATCH does make child restraint installation easier in some vehicles compared to using the seatbelt. But in others, the characteristics of the LATCH hardware can make them difficult to use [54, 55]. Instead of reducing misuse, LATCH often results in new types of misuse. Notably, only about half of caregivers use the top tether with forward-facing products [15, 56]. In a study of vehicle LATCH hardware characteristics, lower anchors more than 2 cm within the vehicle seat bight, high attachment forces, and low clearance around the lower anchors increase the likelihood of installation errors [57].

Another issue with LATCH relates to the strength of the anchorages. FMVSS 225 specifies a quasi-static pull test to evaluate the lower and tether anchorages. Since most harnessed child restraints were designed for children weighing up to 18 kg (40 lb) when LATCH was first introduced, some vehicle manufacturers have stated that LATCH hardware could only be used for children up to this weight. However, child restraints have become heavier as they incorporate more ease-of-use and side impact protection features. They have also been designed to accommodate larger children using a harness, with some products allowing harness use up to 39 kg (85 lb). NHTSA issued regulation changes indicating that anchors designed to meet the quasi-static strength

Fig. 23.1 Lower anchor bars embedded in vehicle seat

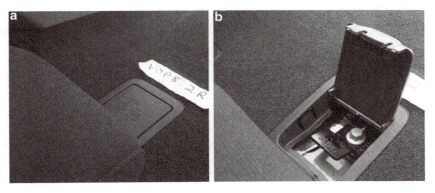

Fig. 23.2 Tether anchor hardware

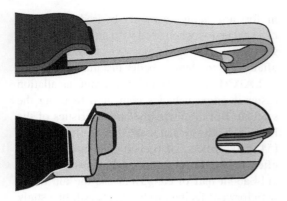

Fig. 23.3 Hook-on (*top*) and push-on (*bottom*) LATCH strap connectors

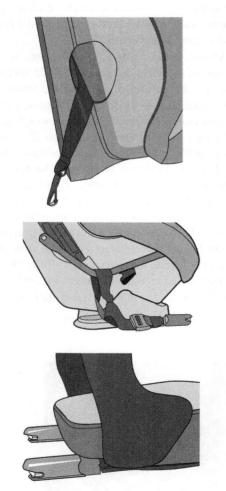

Fig. 23.4 Implementations of LATCH on US child restraints: LATCH strap routed through belt path (*top*), attached to bar on each side (*center*), rigid LATCH (*bottom*)

requirements should be strong enough to secure a combined weight of 29 kg (65 lb) for the child and child restraint [58]; child restraint manufacturers must label the acceptable child weight range based on the weight of the product. With some child restraint weights exceeding 11 kg (25 lb), this limits the ability to use lower anchors to secure the product over its usable lifetime. All child restraints can still be installed with the seatbelt as the primary connection, but there are concerns that caregivers will not realize the need to transition from LATCH to seatbelt securement and failures may occur.

In some cases, seatbelts may provide a better install than LATCH for any weight of child because of incompatibilities. Best practice recommends that a child restraint should be installed using the method that most easily provides a tight installation. All forward-facing installations should use a tether. In their statement regarding limits for lower anchors, NHTSA did not specify limits for tether anchors. Since the top tether supplements the main securement method, and provides a safety benefit by reducing head excursion, they should be used consistently despite a small risk of injury from a hypothetical tether anchorage failure in a severe crash.

23.3.3 Usability and Vehicle/Child Restraint Compatibility

Although vehicle LATCH hardware designs and child restraints have often improved, incompatibility between some vehicles and some child restraints continue to cause installation issues. Misuse of child restraints continues to be a problem. Several different approaches have been implemented to make child restraints easier to use; FMVSS 213 only deals with usability peripherally relative to requirements for labeling and manuals.

A consumer information program, the Ease-of-Use (EOU) Rating system [59], has been developed by NHTSA to publish information about different features available on child restraints. The rating system has promoted positive changes in child restraint products, leading to development of innovative features. To address vehicle/child restraint compatibility, NHTSA has proposed an additional consumer information program where vehicle manufacturers would list child restraints that "fit" in particular vehicles based on a specific set of key installation factors [60].

The Child Restraints Group of the ISO has developed a three-part evaluation system that assesses the usability of ISOFIX (LATCH-type) hardware. The system addresses child restraints, the vehicle environment, and the interaction between specific pairings of child restraints and vehicles [61, 62]. Several of the vehicle features assessed include elements of the vehicle owner's manual, labeling and visibility of anchors, the ability to confuse the anchors with other vehicle hardware, the actions required to prepare the anchors, and incompatibilities between anchors and seatbelts.

A recommended practice has been drafted by the SAE Children's Restraint Systems Standards Committee with the goal of improving compatibility when child restraints are installed in vehicles using LATCH [63]. The document describes procedures and tools to assess vehicle and child restraint hardware. Vehicle factors include measuring the clearance around lower anchorages, and measuring the force required to attach lower connectors, while child restraint factors include maximum and maximum sizes for the dimensions of LATCH connector hardware.

23.4 Child Restraint Systems

Best practice recommendations for proper child restraint include four main stages: rear-facing harnessed restraint, forward-facing harnessed restraint, booster seat with a lap-and-shoulder belt restraint, and seat belts. Both the child's maturity level and physical size influence the proper choice of restraint for a child. Because each step in the sequence reduces the level of occupant protection, children should delay moving to the next level as long as possible.

23.4.1 Rear-facing Child Restraints

Since 2011, both the National Highway Traffic Safety Administration and the American Academy of Pediatrics recommend that children travel rear-facing until they no longer fit in their restraint. Based on available products and typical child sizes, most children in the US should be able to travel rear-facing through the age of 2 years. Henary et al. [8] analyzed NASS-CDS data and found that children aged 0–2 years in rear-facing restraints are 5.53 times safer in side impacts than children in forward-facing restraints; the comparable ratio is 1.23 in frontal impacts. These limited US data agree with results from Sweden (where best practice keeps children rear-facing through age 4 years) showing that the risk of AIS2+ injury is 90 % lower for rear-facing children than unrestrained children [64, 65]. Older educational materials may include text stating that children only should remain rear-facing through age 1 or 10 kg (20 lb), but this is no longer considered a safe practice.

Two styles of rear-facing child restraints are currently sold in the US. Infant or rear-facing only restraints typically have a shell portion (with carrying handle) to secure the child that are attached to the vehicle using a separate base to simplify repeated installation (Fig. 23.5a). However, most can be secured with the seatbelt without the base (Fig. 23.5b). Either the seatbelt or lower anchorages can be used to attach to the vehicle. These products are designed to restrain children between 9 and 16 kg (20–35 lb). Rear-facing convertible products (Fig. 23.5c) can be used rear-facing for children weighing up to 13–20 kg (30–45 lb), and are then converted to use in forward-facing mode. Best practice for rear-facing use dictates that the top of the child's head should have at least 2.5 cm (1 in.) clearance to the top edge of the child restraint. Many children will outgrow their rear-facing restraint by length before they reach the maximum weight.

The internal harness of a rear-facing restraint secures the child within the shell. However, the main restraining forces in a frontal impact are provided by the back of the child restraint and

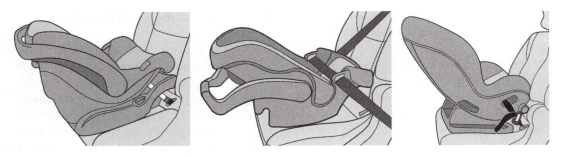

Fig. 23.5 Rear-facing child restraints: rear-facing only with base (*left*), rear-facing only without base (*middle*), and convertible installed rear-facing (*right*)

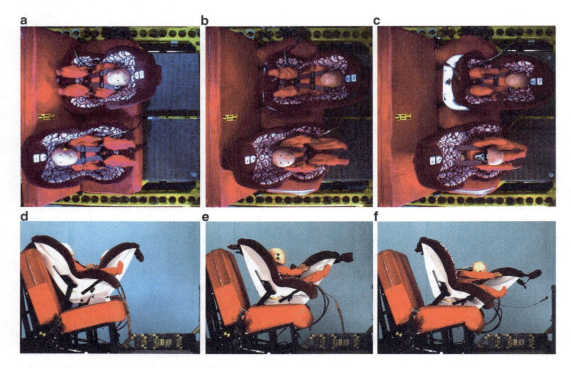

Fig. 23.6 Sequence showing advantage of using rear-facing child restraint compared to forward-facing child restraint

distributed across the head and back of the child. The head and neck are supported in the same way as the torso, which reduces the tension and flexion forces that can occur on the neck with forward-facing children. Figure 23.6 shows a sequence illustrating the difference in head kinematics during a simulated frontal impact when the same child restraint is used rear-facing and forward-facing. The peak axial neck forces with the forward-facing restraint are four times higher than the dummy restrained in the rear-facing restraint.

Misuse of rear-facing restraints include forward-facing use, which could result in injurious loading or ejection because the belt path was not designed for this loading direction. If a convertible restraint is installed rear-facing using the forward-facing belt path (or vice versa), similar negative consequences could occur. Allowing a child to travel forward-facing too early increases spinal cord injury risk; the youngest children's heads comprise a larger proportion of their body mass, resulting in greater tensile load to the cervical spine to stop forward head motion.

When installing rear-facing child restraints, achieving the correct angle is critical. A rear-facing restraint installed too upright may cause the small infant's head to tip forward and impair breathing. A rear-facing restraint installed too reclined will not allow the back of the restraint to effectively restrain the child.

In a frontal crash, the ability of the restraint to restrain the child decreases as the angle of the back support is reclined more than 45°, because the force projecting the baby upwards along the seatback becomes larger than the reaction force provided by the back of the child restraint. For larger rear-facing children, placing the restraint more upright would allow better crash protection. For newborns, the child restraint angle must balance between crash protection and keeping the head from hanging forward and potentially cutting off the airway. A back support angle of 45° (shown in Fig. 23.7) from vertical is considered the best compromise between these two opposing issues.

Because installation angle varies with the geometry of the vehicle seat, child restraint manufacturers usually provide some sort of angle indicator to provide guidance during installation. However, indicator placement is often based on where the child restraint performs best in FMVSS 213 testing. Some child restraint products now have different angle indicators for smaller and larger children that account for the changing needs of rear-facing children, but not all manufacturers do so. For newborn children, if the required installation angle based on manufacturer instructions places the child too upright, parents should switch to another rear-facing child restraint or car bed.

When rear-facing child restraints are installed in the rear vehicle seat, allowing the back of the restraint to initially contact the back of the front seat will limit forward rotation and improve protection during a crash. Forces can be higher if there is an initial gap and the child restraint then suddenly contacts the seatback [66, 67]. This initial configuration is not allowed by some vehicle manufacturers because the contact can interfere with the occupant sensing systems used with advanced airbags. Some child restraint manufacturers also prohibit the initial contact because of concerns regarding potential adverse interaction in rear impacts between the child restraint and front seatback.

Some child restraint manufacturers recommend using a tether with rear-facing child restraints. This practice is not covered by FMVSS 213, which specifies that child restraints meet requirements without using a tether. FMVSS 225 also does not specify tether anchors located below and forward of a vehicle seating position that can be used with a rear-facing child restraint. Tethering rear-facing child restraints is more common in Scandinavia and Australia. The Australian method (Fig. 23.8a) routes the tether towards the back of the vehicle and uses the same tether anchorage used with forward-facing child restraints. This tethering strategy prevents forward movement and rotation. The Swedish method (Fig. 23.8b) routes the tether to a point forward and below the vehicle seat to help control the initial restraint angle and limit rebound rotation towards the back of the vehicle [68].

When evaluating the effect of tethering rear-facing during laboratory testing, any type of rear-facing restraint, with or without tether, provided better protection than a forward-facing restraint with a tether. Among the rear-facing conditions,

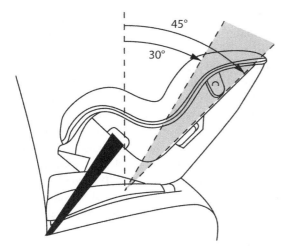

Fig. 23.7 Angles for rear-facing child restraints

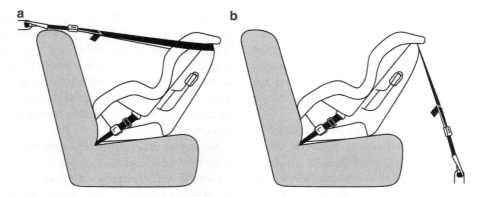

Fig. 23.8 Tethering rear-facing child restraints: Australian method (*left*) and Swedish method (*right*)

the Australian tether method led to the lowest head and chest accelerations, as well as the lowest forward head excursions. During simulated rear-impact testing, none of the tethered rear-facing conditions produced excessive neck loads relative to the injury reference values listed in FMVSS 208 [69].

As in other crash modes, the first priority of a restraint system during side impact is to prevent head contact. If preventing head contact is not possible, the object that the head contacts should be designed to absorb energy. Because of the differences in belt path locations and the child's center of gravity between rear-facing and forward-facing child restraints, a rear-facing restraint will move more towards the side of impact relative to a forward-facing child restraint. Despite this greater motion, analysis of crash data indicates that children in forward-facing child restraints are five times more likely to sustain serious injury in side impacts compared to children in rear-facing restraints [8]. Because most side impacts have a frontal deceleration component, occupants will move toward the struck side and towards the front of the vehicle. This frontal deceleration component will cause the head of a rear-facing child to move further into the shell within the side wings of the restraint, while the head of a forward-facing child moves forward out of the protective sides of the restraint, exposing it to potential contact from intruding vehicle components. Child restraint systems equipped with rigid LATCH connectors reduce movement of both the child restraint and child in side impacts, improving the ability to prevent head contact [70].

Rear-facing child restraints can rotate towards the rear vehicle seatback during rear impacts and rollovers. In these situations, the harness provides the main occupant restraint. When rear-facing child restraints were first introduced, the designers envisioned this "cocooning" movement as a benefit to shield infants from debris [71]. With the development of larger rear-facing restraints designed to accommodate larger children, there is greater potential for head contact with interior vehicle components during a rear impact or rollover. However, no crash investigations have documented head injuries from this injury mechanism to date.

Most rear-facing child restraints employ a five-point harness, although previous designs used a three-point harness that did not have straps routed over the hips. To provide child restraints that can accommodate children weighing from 2 up to 18 kg (4 to up to 40 lb) rear-facing, child restraint manufacturers have provided more adjustable harnesses, a greater number of shoulder slots, and padded inserts. Padding should not be used if it pushes the infant's head toward the chest. If needed for lateral support, additional firm padding such as a rolled blanket or towel can be placed between the infant and the side of the restraint. To prevent slouching, caregivers can place rolled towels between the infant and crotch strap [72].

However, thick soft padding should never be placed between the back of the restraint and the child or between the child and the shoulder straps. Unless tested and provided by the child restraint manufacturer, this type of padding would create harness slack as it compresses during dynamic loading, potentially leading to ejection of the child or excessive loading during a crash.

To fit the harness to a rear-facing child, use the shell harness slots that are at or below the child's shoulder height. If higher harness slots are used, the child could slide up the seatback during a crash, possibly allowing the child's head to move beyond the protective shell of the child restraint. With smaller babies, the head would likely be retained within the shell if higher slots are incorrectly used, but using higher slots could increase loading through the shoulders. Loose harnesses increase the level of loading and the likelihood of ejection. During a crash, chest clips help keep the harness positioned on the shoulders but cannot compensate for a loose harness.

Despite advances in frontal impact airbag systems, the right-front passenger seating position should never be used for rear-facing child restraint installation. A rear-facing child restraint installed in the right-front passenger seating position places the child at high risk of injury or death during a crash. For a rear-facing child installed in the right-front position, the child's head is placed very close to the airbag module. To provide an adult occupant with energy-absorbing cushion of restraint during a crash, the airbag must deploy at high speed from a small folded package, as much as 300 km/h (186 mi/h). The force of an airbag contacting the back of a rear-facing child restraint during inflation produces accelerations in infant dummy heads from 100 to 200 g [73, 74], with 50 g considered the injury threshold for children the size of a 6-month-old dummy [74, 75]. Figure 23.9 shows an airbag contacting a rear-facing child restraint. Laboratory measurements indicate that head injuries occur from the initial contact, even though the airbag can also rotate the infant and restraint into the vehicle seatback.

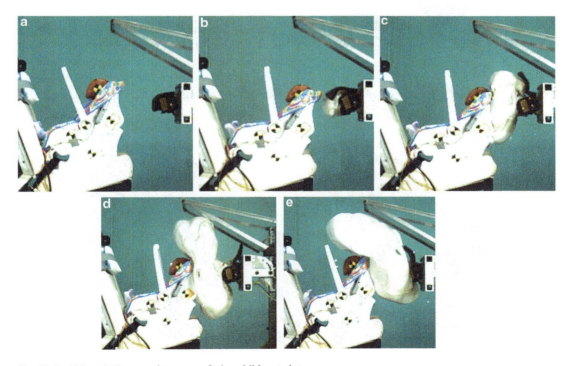

Fig. 23.9 Airbag deployment into a rear-facing child restraint

23.4.2 Car Bed Restraints

A car bed (Fig. 23.10) is a suitable choice for infants with medical issues that prevent them from using the semi-reclined position provided by a rear-facing restraint system. Car-bed restraints allow the infant to lie flat, usually on its back. A car bed is installed with its length perpendicular to the fore-aft direction in the vehicle, with the baby's head placed nearest the vehicle center.

During frontal crashes, the side of the car bed restraint distributes restraining forces along the length of the infant's body. During rebound or rollover, the harness provides the restraint. Although there are no documented field data, infants in car beds during side impacts may be more likely to sustain head or neck injury than those in rear-facing restraints, particularly if the head is closest to the struck side of the vehicle [76].

Rear-facing restraints are the first choice of the American Academy of Pediatrics for infants, but they acknowledge that positional apnea may require use of a car bed [77–79]. Current best practice for infants born at less than 37 weeks gestation is to evaluate them before hospital discharge to detect possible apnea, bradycardia, or oxygen desaturation that might occur during use of a rear-facing child restraint [80].

23.4.3 Forward-Facing Child Restraints

Children traveling forward-facing in a harnessed restraint system can either use a convertible child restraint in the forward-facing mode (Fig. 23.11a) or a combination child restraint equipped with a harness that can be reconfigured into booster mode (Fig. 23.11b). Some products are available that are designed for only forward-facing harnessed use.

While the maximum child weight limit of some forward-facing restraints is up to 22 kg (40 lb), many child restraints allow children up to 40 kg (90 lb) to use the harness system. All forward-facing child restraints sold in the US

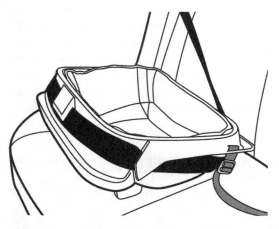

Fig. 23.10 Example of a car-bed restraint

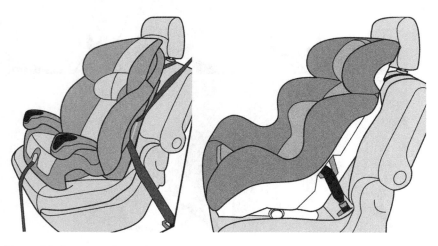

Fig. 23.11 Forward-facing convertible (*left*) and combination restraint used in forward-facing harnessed mode (*right*)

can be installed with either a seatbelt or LATCH lower attachments. In addition, all child restraint manufacturers currently recommend that the tether be used with forward-facing child restraints to minimize forward head movement during a crash.

The best protection of a forward-facing child requires a snug fit of the harness and tight installation of the child restraint to couple it to the vehicle. Most current child restraints use a five-point harness (Fig. 23.12a) because they permit the best snug fit around the child. However, a few child restraint models use a tray shield, shoulder straps, and crotch strap (Fig. 23.12b) to secure the child.

The five-point strap harness design is based on restraint systems used by race car drivers and the military. The five "points" are a strap over each shoulder, a strap over each hip, and a crotch strap between the legs. Many products use a harness adjuster that tightens the harness with a single pull or twist of a knob. If the harness is loose rather than snug, the child can move closer to vehicle components at the onset of the impact and experience higher acceleration levels when they eventually load the harness. Ejection or injury to the thorax or abdomen can occur with a misrouted or improperly buckled harness.

With forward-facing restraints, the shoulder straps should be placed through harness slots that are at or above the child shoulders. If the harness is routed to slots below the shoulders, it essentially introduces slack to the system, since the child can move forward before the harnesses start to restrain the shoulders. Sometimes the lower slots on convertible restraints are not strengthened for use in forward-facing mode; using these slots for forward-facing could lead to failure of the child restraint shell.

Switching a child from rearward-facing to forward-facing should be delayed as long as possible, because rear-facing is the safest mode of child restraint. When forward-facing in frontal crashes, the cervical spine experiences significant loading as the harness holds back the shoulders and the head moves forward and is stopped by the neck. In a frontal 48 km/h (30 mph crash) with a vehicle compartment deceleration of 25 g, the occupant's head stops later and more abruptly than the vehicle and can experience deceleration levels of up to 60 or 70 g. The strength of the vertebrae and connecting ligaments determines whether the spinal cord will be injured [81, 82]. In adults, the cervical spine is fully ossified and can experience tensile forces corresponding to decelerations of up to 100 g [83], so spinal cord injury is almost always accompanied by vertebral fracture. With children, the flexible characteristics of the cervical spine that allow growth permits

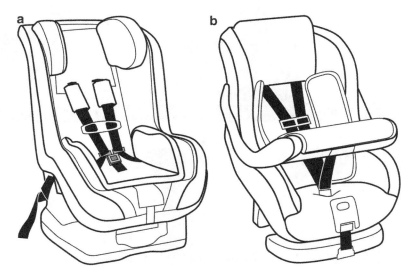

Fig. 23.12 Forward-facing restraints with 5-point harness (**a**-*left*) and tray shield (**b**-*right*)

children to sustain injury to the spinal cord even if the vertebrae remain intact [84–86]. Spinal cord injury risk decreases with age and increases with crash severity [82]. Although serious cervical spine injuries are rare among forward-facing children using proper restraint, crash investigations provide evidence of serious long-term negative consequences from spinal cord injury [73, 87–90]. Because of the potential for negative consequences from cervical spine injuries, best practice recommends keeping smaller children rear-facing as long as possible.

To reduce forward rotation of the child in frontal crashes, caregivers should always use top tethers with forward-facing child restraints [91, 92]. The kinematics of a forward-facing harnessed 3-year-old dummy restrained with a tether (closer to camera) and untethered (farther from camera) in a frontal 48 km/h crash are shown in Fig. 23.13. With a tethered child restraint, head excursion is an average of 150 mm (6 in.) less than the untethered child restraint. Less head excursion during a crash translates to lower likelihood of interior head contact. Head and face injuries are most common among children injured in forward-facing child restraints [93–95].

Although rare, head contact can also lead to vertebral fractures and dislocations, as well as spinal cord injury [82, 83], so reducing head excursion and contact can also reduce risk of neck injury. Using the top tether can also improve coupling between the vehicle and child restraint. But the tether must be tight, as slack degrades tether effectiveness. About half of caregivers fail to use tethers with forward-facing installations, making it a leading type of misuse [56].

While frontal impacts occur more frequently, risk of serious or fatal injury is higher in side impacts [96]. The transition from rear- to forward-facing restraints should be delayed because the risk of injury in side impacts is so much lower for rear-facing children [8]. Forward-facing restraints with larger padded sidewings may offer improved protection when attempting to prevent injuries to the most commonly injured regions of the head and face [97–99]. In addition, child restraints with rigid LATCH connectors were better at preventing rotation towards the struck side of the vehicle during laboratory testing [70]. However, the same series of tests showed negligible benefit of the tether on lateral head excursion during side impact. Some child

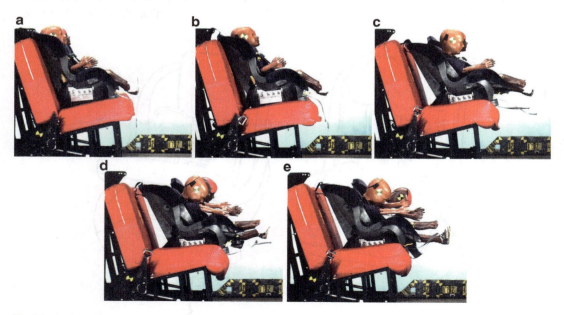

Fig. 23.13 Sequence of photos showing reduced excursion when tether is used (nearest to camera)

restraints now include additional energy absorbing elements, such as a side air cushion, which show potential for improved side impact protection [100].

Like all children under age 13, those in forward-facing child restraints should use a rear seating position. However, if the rear seat is fully occupied with other children, a properly restrained child in a forward-facing harnessed restraint would be the best choice for the right-front seat. Risk of injury from the airbag for such a child should be similar to the risk of a belted adult. Compared to a child in a booster seat or seatbelt, the harness prevents the child from leaning forward and being too close to the airbag. None of the children who sustained fatal injuries from deploying airbags were properly using a forward-facing child restraint [37]. If a child in a forward-facing restraint is using the right-front seating position, the vehicle seat should be placed as rearward as possible while allowing use of the rear seat.

23.4.4 Boosters

Children who can no longer fit in a harnessed forward-facing restraint should switch to a belt-positioning booster seat used with a vehicle lap-and-shoulder belt. Because the risk of injury is lower for forward-facing children in harnessed restraints compared to children using boosters, the transition should be delayed as long as possible.

Booster seats reposition the child and redirect vehicle belts to improve their routing relative to the child's body, but do not actually restrain the child. According to both the NHTSA and the AAP, children should use booster seats until they fit in seat belts alone. This means most children should use boosters through age 8–12 years [2]. As a result of public education programs, state laws requiring boosters, and more available booster products, use of booster seats among children aged 4–8 years has risen from 15 % in 2000 to 63 % in 2007 [13].

Analysis of children in crashes indicates that children aged 4–8 using boosters are 45 % less likely to sustain injury compared to children using seatbelts alone [11, 101]. In particular, boosters prevent abdominal injury: the ratio of abdomen injury risk for children without and with boosters and the vehicle seatbelt is 8 [102]. Figure 23.14 shows how the lap belt can load the abdomen when a booster is not used.

Studies performed with child volunteers, vehicle seats, and seat belt geometries over the past decade have improved understanding of how different booster seat designs improve belt fit [103–105]. As shown Fig. 23.15, use of a booster repositions the child upwards relative to the vehicle belt. Even with backless boosters, the higher elevation places the shoulder belt further away from the neck (so it is comfortable to use) and restrains the child through the shoulder in a crash.

As the child booster elevates the child, the lap belt angle also increases relative to the lap belt

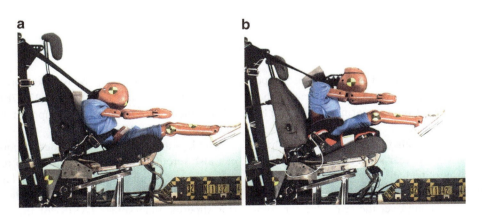

Fig. 23.14 Without a booster (*left*), seatbelt can load the abdomen

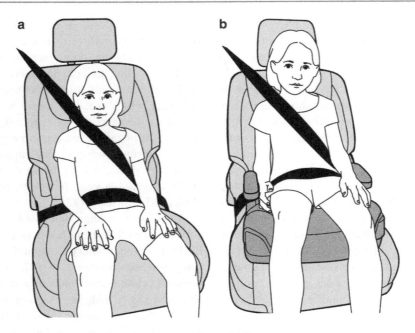

Fig. 23.15 Booster improves belt fit by raising the child upwards relative to the vehicle belt system

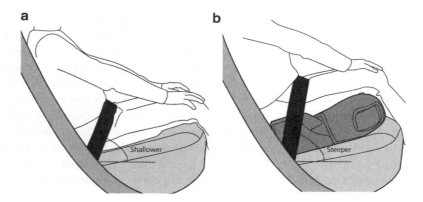

Fig. 23.16 Using a booster (even without a belt guide) increases lap belt angle, better allowing the lap belt to restrain the pelvis

anchor point as illustrated in Fig. 23.16. Steeper lap belt angles improve restraint because they prevent the child from submarining under the lap belt during a crash.

Using boosters also improves childrens' posture. Most vehicle rear seats are longer than the buttock-to-popliteal distance of older children, so children will tend to move forward to allow their knees to hang over the seat edge for comfort, which also shifts their pelves forward (Fig. 23.17a) [105–107]. A booster seat provides a compatible cushion length (Fig. 23.17b) and allows them to sit comfortably upright as an older child would (Fig. 23.17c).

The lap- and shoulder-belt guides of boosters can also reroute and improve belt fit. To reduce the chance of submarining in a crash, the lap belt should be positioned completely below the anterior superior iliac spines (ASIS) of the pelvis. Some boosters have well-designed lap belt guides

Fig. 23.17 Sitting on the vehicle seat, a 7-year-old child slouches to provide a comfortable angle over the front of the seat (*left*). Using a booster (*center*) provides a better match between cushion length and thigh length, allowing them to achieve an upright posture similar to a larger child (*right*)

that not only keep the belt low, touching the thighs during travel, but also resist upward motion of the belt during dynamic loading. The most effective shoulder belt guides place the shoulder belt near the clavicle's midline (Fig. 23.18). Some improperly designed shoulder belt guides place the belt off the child's shoulder, or can allow slack to accumulate in the belt because it prevents retraction of the belt after the child leans forward.

The US market currently has four types of belt-positioning boosters: backless boosters (Fig. 23.19a), removable-back boosters (Fig. 23.19b), highback boosters (Fig. 23.19c), and vehicle-integrated boosters. When the vehicle seat and head restraint are tall enough to support the child's head at the level of the tops of the ears, children can use backless boosters.

Field crash data indicate similar low levels of injury risk in boosters with and without backs [11]. It seems as though boosters with backs would have a greater potential to offer better side impact protection and keep the children in a better lateral position relative to the belt, especially while sleeping. However, backless boosters typically place children further rearward and away from forward vehicle structures. In addition, because the lap-and-shoulder belt allows the child to move during travel, they often do so. One study showed that children seated in boosters with less prominent wings leaned forward for 25 % of a journey, while those seated in boosters with larger (more protective) sidewings leaned forward 55 % of the time [108].

Fig. 23.18 Belt-positioning boosters with a shoulder belt guide can also improve belt fit

Although many boosters have shoulder-belt guides that improve belt fit statically, they often do not keep the belt on the dummy shoulder during dynamic testing [109, 110]. These data indicate

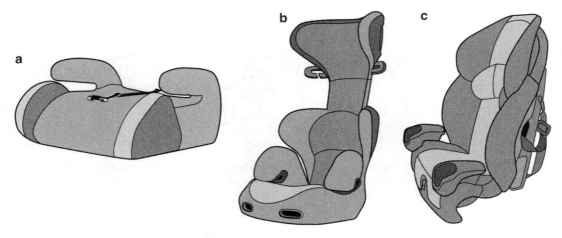

Fig. 23.19 Three styles of belt-positioning booster seats: backless (*left*), removable back (*center*), and highback (*right*)

that the vehicle seating position requiring the least amount of shoulder belt routing by the booster seat guide should be selected. Several studies demonstrate that belt fit provided by boosters depends on the vehicle geometry [111–113] and the Insurance Institute for Highway Safety rates booster seats for their effectiveness at providing good belt fit over a range of vehicle belt geometries [104]. Despite the range of improvements in belt fit provided by boosters, the effectiveness of boosters at reducing injury risk in crashes suggests that using any booster is likely to provide better seat belt fit than the vehicle belt alone.

23.4.5 Seatbelts for Children

The occupant posture and size, vehicle seat size and shape, and belt system geometry and features all affect the belt fit of an occupant [114]. Although a child may have good belt fit in one vehicle, belt fit may be poor in another vehicle. To achieve good lap belt fit, the belt should be as low as possible on the pelvis, either touching or lying flat over the thighs. The child pelvis is shorter and less prominent than an adult pelvis. Thus it is even more critical for children that the lap belt can engage the pelvis during a crash. To check lap belt fit, a child can locate the top of their own pelvis by finding the bony points at the top front of the pelvis and make sure it is above the lap belt (Fig. 23.20). For good shoulder belt fit, it should lie about halfway between the neck and arm, flat on the shoulder, and should cross the middle of the sternum.

Early studies of belt fit with boosters led to the common recommendation that children should use a booster until they are 148 cm (58 in.) tall with a clothed weight of 37 kg (81 lb) [106]. This size represents a 90th percentile 9-year-old, a 50th percentile 11-year-old, and a 5th percentile 13-year-old. Although a straightforward limit on height, weight, or age is easiest to use when crafting legislation or education campaigns, many children larger than this size have better belt fit with a booster.

To obtain the best protection from the vehicle seat belt, the child should sit up straight, with the buttocks against the seatback and with the feet touching the floor. This posture helps prevent the belt from sliding up over the anterior-superior iliac spines into the abdomen. If the child moves forward to bend their knees over the seat edge, this position will make the child slouch and rotate the pelvis rearward. This posture can lead the lap belt to be positioned over the abdomen, preventing the belt from restraining the child through the pelvis. The child should use a booster if they cannot sit upright on the vehicle seat or if the shoulder belt crosses the throat.

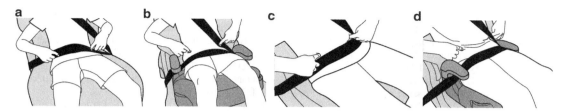

Fig. 23.20 On a 7YO child seated on a vehicle seat (**a**), the vehicle belt lies on the abdomen above the pelvis. When using a booster (**b**), the lap belt lies below the top of the pelvis. For a 10YO child on the vehicle seat, the lap belt lies just below the pelvis (**c**), providing acceptable fit. Even so, the 10YO using the booster experiences much better fit over the hips (**d**)

Though shoulder belts touching the neck are more likely to cause discomfort than injury (unless they are very loose) [115–117], the irritation often causes the child to misroute the shoulder belt behind the back or under the arm. If the shoulder belt is routed behind the back, the potential for injurious belt loading increases because there is no torso restraint and the inboard portion of the lap belt can be pulled up [118]. Misrouting the belt behind the back can also eliminate early loading of the shoulder belt that snugs the lap belt and may cause the retractor to lock the belt. Routing the shoulder belt under the arm can also produce serious injuries to the thorax in a crash [119, 120].

Shoulder belt positioners, though unregulated, have been advertised as minimizing discomfort from the shoulder belt touching the occupant's neck. Several of these products shift the shoulder belt by linking it to the lap belt, which improves shoulder belt fit by pulling up the lap belt and making its fit worse [111, 121]. The potential negative effects of shoulder belt positioners cannot be quantified, as child dummies cannot currently quantify abdominal loading. Shoulder belt positioners should not be used in place of belt-positioning boosters, which are proven in the field to reduce injury, particularly to the abdomen.

References

1. Gilchrist J, Ballesteros MF, Parker EM (2012) Vital signs: unintentional injury deaths among persons aged 0–19 years – United States, 2000–2009, vol 61. Centers for Disease Control and Prevention. http://www.cdc.gov/mmwr/preview/mmwrhtml/mm6115a5.htm?s_cid=mm6115a5_w
2. American Academy of Pediatrics, Committee on Injury, Violence, and Poison Prevention (2011) Technical report: child passenger safety. Pediatrics 127:1060–1068
3. Durbin DR, American Academy of Pediatrics, Committee on Injury, Violence, and Poison Prevention (2011) Policy statement: child passenger safety. Pediatrics 127(4):788–793. doi:10.1542/peds.2011-0213
4. National Highway Traffic Safety Administration (2011) Car seat recommendations for children. 4StpsFlyer. http://www.nhtsa.gov/ChildSafety/Guidance
5. Kahane CJ (1986) An evaluation of child passenger safety-the effectiveness and benefits of safety seats. DOT HS 806 890
6. Zaloshnja E, Miller TR, Hendrie D (2007) Effectiveness of child safety seats vs safety belts for children aged 2 to 3 years. Arch Pediatr Adolesc Med 161(1):65–68. doi:10.1001/archpedi.161.1.65
7. Arbogast KB, Durbin DR, Cornejo RA, Kallan MJ, Winston FK (2004) An evaluation of the effectiveness of forward facing child restraint systems. Accid Anal Prev 36(4):585–589
8. Henary B, Sherwood CP, Crandall JR, Kent RW, Vaca FE, Arbogast KB, Bull MJ (2007) Car safety seats for children: rear facing for best protection. Inj Prev 13(6):398–402. doi:10.1136/ip.2006.015115
9. Elliott MR, Kallan MJ, Durbin DR, Winston FK (2006) Effectiveness of child safety seats vs seat belts in reducing risk for death in children in passenger vehicle crashes. Arch Pediatr Adolesc Med 160(6):617–621. doi:10.1001/archpedi.160.6.617
10. Morgan C (1999) Effectiveness of lap/shoulder belts in the back outboard seating positions. DOT HS 808 945
11. Arbogast KB, Jermakian JS, Kallan MJ, Durbin DR (2009) Effectiveness of belt positioning booster seats: an updated assessment. Pediatrics 124(5):1281–1286. doi:10.1542/peds.2009-0908

12. Williams AF (1976) Observed child restraint use in automobiles. Am J Dis Child 130(12):1311–1317
13. National Highway Traffic Safety Administration (2009) Child restraint use in 2008-overall results. Traffic Safety Facts DOT HS 811 135
14. Decina LE, Lococo KH (2005) Child restraint system use and misuse in six states. Accid Anal Prev 37(3):583–590. doi:10.1016/j.aap.2005.01.006
15. Decina LE, Lococo KH (2007) Observed LATCH use and misuse characteristics of child restraint systems in seven states. J Saf Res 38(3):273–281. doi:10.1016/j.jsr.2006.08.009
16. Eby DW, Kostyniuk LP (1999) A statewide analysis of child safety seat use and misuse in Michigan. Accid Anal Prev 31(5):555–566
17. Dukehart JG, Walker L, Lococo KH, Decina LE, Staplin L (2007) Safe kids checkup events: a national study. SafeKids Worldwide, Washington, DC
18. O'Neil J, Daniels DM, Talty JL, Bull MJ (2009) Seat belt misuse among children transported in belt-positioning booster seats. Accid Anal Prev 41(3):425–429
19. Tai A, Bilston LE, Brown J (2011) The cumulative effect of multiple forms of minor incorrect use in forward facing child restraints on head injury risk. ESV Paper 11-0118
20. Lesire P, Cuny S, Alonzo F, Tejera G, Cataldi M (2007) Misuse of child restraint systems in crash situations – danger and possible consequences. Annu Proc Assoc Adv Automot Med 51:207–222
21. Weber K, Melvin JW (1983) Injury potential with misused child restraint systems. SAE 831604. Paper presented at the 27th Stapp car crash conference. Society of Automotive Engineers, Warrendale
22. Kapoor T, Altenhof W, Snowdon A, Howard A, Rasico J, Zhu F, Baggio D (2011) A numerical investigation into the effect of CRS misuse on the injury potential of children in frontal and side impact crashes. Accid Anal Prev 43(4):1438–1450. doi:10.1016/j.aap.2011.02.022
23. Lalonde S, Legault F, Pedder J (2003) Relative degradation of safety to children when automotive restraint systems are misused. ESV Paper 03-0085
24. Bilston LE, Yuen M, Brown J (2007) Reconstruction of crashes involving injured child occupants: the risk of serious injuries associated with sub-optimal restraint use may be reduced by better controlling occupant kinematics. Traffic Inj Prev 8(1):47–61. doi:10.1080/15389580600990352
25. Rouhana SW (1993) Nahum AM, Melvin JW, (eds) Biomechanics of abdominal trauma. In: Accidental injury biomechanics and prevention. Springer, New York, pp 391–428
26. Rutledge R, Thomason M, Oller D, Meredith W, Moylan J, Clancy T, Cunningham P, Baker C (1991) The spectrum of abdominal injuries associated with the use of seat belts. J Trauma 31(6):820–825, discussion 825–826
27. Johnson DL, Falci S (1990) The diagnosis and treatment of pediatric lumbar spine injuries caused by rear seat lap belts. Neurosurgery 26(3):434–441
28. King AI, (1993) Nahum AM, Melvin JW (eds) Injury to the thoraco-lumbar spine and pelvis. In: Accidental injury biomechanics and prevention. Springer, New York, pp 429–459
29. McGrath N, Fitzpatrick P, Okafor I, Ryan S, Hensey O, Nicholson AJ (2010) Lap belt injuries in children. Ir Med J 103(7):216–218
30. Louman-Gardiner K, Mulpuri K, Perdios A, Tredwell S, Cripton PA (2008) Pediatric lumbar chance fractures in British Columbia: chart review and analysis of the use of shoulder restraints in MVAs. Accid Anal Prev 40(4):1424–1429. doi:10.1016/j.aap.2008.03.007
31. Kallan MJ, Durbin DR, Arbogast KB (2008) Seating patterns and corresponding risk of injury among 0- to 3-year-old children in child safety seats. Pediatrics 121(5):e1342–e1347. doi:10.1542/peds.2007-1512
32. Braver ER, Scerbo M, Kufera JA, Alexander MT, Volpini K, Lloyd JP (2008) Deaths among drivers and right-front passengers in frontal collisions: redesigned air bags relative to first-generation air bags. Traffic Inj Prev 9(1):48–58. doi:10.1080/15389580701722787
33. Mayrose J, Priya A (2008) The safest seat: effect of seating position on occupant mortality. J Saf Res 39(4):433–436. doi:10.1016/j.jsr.2008.06.003
34. Edwards J, Sullivan K (1997) Where are all the children seated and when are they restrained? SAE 971550. Reprinted in child occupant protection 2nd symposium. Paper presented at the Society of Automotive Engineers, Warrendale
35. Quinones-Hinojosa A, Jun P, Manley GT, Knudson MM, Gupta N (2005) Airbag deployment and improperly restrained children: a lethal combination. J Trauma 59(3):729–733
36. Braver ER, Ferguson SA, Greene MA, Lund AK (1997) Reductions in deaths in frontal crashes among right front passengers in vehicles equipped with passenger air bags. JAMA 278(17):1437–1439
37. National Highway Traffic Safety Administration (2009) Traffic Safety Facts 2008 DOT HS 811-170
38. Braver ER, Whitfield R, Ferguson SA (1998) Seating positions and children's risk of dying in motor vehicle crashes. Inj Prev 4(3):181–187
39. National Highway Traffic Safety Administration (2007) Counts of frontal air bag related fatalities and seriously injured persons. AB0707
40. Greenspan AI, Dellinger AM, Chen J (2010) Restraint use and seating position among children less than 13 years of age: Is it still a problem? J Saf Res 41(2):183–185. doi:10.1016/j.jsr.2010.03.001
41. Olson CM, Cummings P, Rivara FP (2006) Association of first- and second-generation air bags with front occupant death in car crashes: a matched cohort study. Am J Epidemiol 164(2):161–169. doi:10.1093/aje/kwj167
42. Berg MD, Cook L, Corneli HM, Vernon DD, Dean JM (2000) Effect of seating position and restraint use on injuries to children in motor vehicle crashes. Pediatrics 105(4 Pt 1):831–835

43. Durbin DR, Chen I, Smith R, Elliott MR, Winston FK (2005) Effects of seating position and appropriate restraint use on the risk of injury to children in motor vehicle crashes. Pediatrics 115(3):e305–e309. doi:10.1542/peds.2004-1522
44. Evans SL, Nance ML, Arbogast KB, Elliott MR, Winston FK (2009) Passenger compartment intrusion as a predictor of significant injury for children in motor vehicle crashes. J Trauma 66(2):504–507. doi:10.1097/TA.0b013e318166d295
45. Side Airbag Out-of-Position Injury Technical Working Group (2003) Recommended procedures for evaluating occupant injury risk from deploying side airbags
46. Hallman JJ, Brasel KJ, Yoganandan N, Pintar FA (2009) Splenic trauma as an adverse effect of torso-protecting side airbags: biomechanical and case evidence. Ann Adv Automot Med 53:13–24
47. Arbogast KB, Kallan MJ (2007) The exposure of children to deploying side air bags: an initial field assessment. Annu Proc Assoc Adv Automot Med 51:245–259
48. Lund AK (1986) Voluntary seat belt use among U.S. drivers: geographic, socioeconomic and demographic variation. Accid Anal Prev 18(1):43–50
49. National Highway Traffic Safety Administration (2010) Seat belt use in 2009-use rates in the states and territories. Traffic Safety Facts DOT HS 811 324
50. Society of Automotive Engineers (1994) Securing child restraint systems in motor vehicles. SAE J1819 (R). Society of Automotive Engineers, Warrendale
51. Code of Federal Regulations (2006) Title 49, transportation, part 571.225; child restraint anchorage systems US Government Printing Office
52. Code of Federal Regulations (2011) Title 49, transportation, part 571.213; child restraint systems US Government Printing Office
53. International Standards Organization (1999) Road vehicles – anchorages in vehicles and attachments to anchorages for child restraint systems – part 1: seat bight anchorages and attachments. ISO/IS 13216-1. ISO, Geneva
54. Insurance Institute for Highway Safety (2003) Status report, vol 38, number 5
55. SafeRide News (2011) The LATCH manual 2011. SafeRide News Publications, Edmonds
56. Jermakian JS, Wells JK (2011) Observed use of tethers in forward-facing child restraint systems. Inj Prev 17:317–374
57. Klinich KD, Flannagan CAC, Jermakian JS, McCartt AT, Manary MA, Moore JL, Wells JK (2012) Vehicle LATCH system features associated with correct child restraint installations. Traffic Inj Prev. doi:10.1080/15389588.2012.701030
58. National Highway Traffic Safety Administration (2012) 49 CFR part 571 10-year-old final rule docket no. NHTSA-2011-0176
59. National Highway Traffic Safety Administration (2006) Consumer information rating program for child restraints docket. NHTSA-2006-25344
60. National Highway Traffic Safety Administration (2011) Consumer information: program for child restraint systems, request for comments. Federal Register 25 Feb 2011, 76 FR 10637
61. International Standards Organization (2010) Road vehicles – methods and criteria for usability evaluation of child restraint systems and their interface with vehicle anchorage systems – part 1 vehicles and child restraints equipped with ISOFIX anchorages and attachments. CD 29061-1, Geneva
62. Pedder J, Hillebrandt D (2007) The development and application of a child restraint usability rating system. ESV Paper 07-0509
63. Society of Automotive Engineers (2009) Guidelines for implementation of the child restraint anchorage system or LATCH system in motor vehicles and child restraint systems. Child Restraints Systems Committee, Draft recommended practice #J2893, Versions #1 and #2
64. Isaksson-Hellman I, Jakobsson L, Gustafsson C, Norin H (1997) Trends and effects of child restraint systems based on Volvo's Swedish accident database. SAE 973299. Paper presented at the child occupant protection 2nd symposium. Society of Automotive Engineers, Warrendale, p 84
65. Jakobsson L, Wiberg H, Isaksson-Hellman I, Gustafsson J (2007) Rear seat safety for the growing child – a new 2-state integrated booster cushion. ESV Paper 07-0322
66. Tylko S (2011) Interactions of rear facing child restraints with the vehicle interior during frontal crash tests. ESV Paper 11-0406
67. Sherwood CP, Gopalan Y, Abdelilah RJ Crandall JR (2005) Vehicle interior interactions and kinematics of rear facing child restraints in frontal crashes. In: Annual proceedings of the Association for Advanced Automotive Medicine, pp 215–228
68. Sherwood C, Abdelilah Y, Crandall J (2005) The effect of Swedish tethers on the performance of rear-facing child restraints in frontal crashes. ESV Paper 05-0346
69. Manary MA, Reed MP, Klinich KD, Ritchie NL, Schneider LW (2006) The effects of tethering rear-facing child restraint systems on ATD responses. Annu Proc Annu Proc Assoc Adv Automot Med 50:397–410
70. Klinich KD, Ritchie NL, Manary MA, Reed MP, Tamborra N, Schneider LW (2005) Kinematics of the Q3S ATD in a child restraint under far-side impact loading. NHTSA, Washington, DC. Paper presented at the nineteenth ESV conference proceedings, Washington, DC. Paper Number 05-0262
71. Feles N (1970) Design and development of the General Motors infant safety carrier. SAE 700042. Paper presented at the Society of Automotive Engineers, New York
72. American Academy of Pediatrics (2012) Car safety seats: information for families for 2012, p 30
73. Weber K, Dalmatas D, Hendrick B (1993) Investigation of dummy response and restraint con-

figuration factors associated with upper spinal cord injury in a forward-facing child restraint. SAE 933101. Child occupant protection SAE International
74. Klinich KD, Hulbert GM, Schneider LW (2002) Estimating infant head injury criteria and impact response using crash reconstruction and finite element modeling. Stapp Car Crash J 46:165–194
75. Melvin JW (1995) Injury assessment reference values for the CRABI 6-month dummy in a rear-facing infant restraint with airbag deployment. SAE 950872. Society of Automotive Engineers, Warrendale
76. Weber K (1990) Comparison of car-bed and rearfacing infant restraint systems. Paper presented at the 12th international technical conference on experimental safety vehicles. National Highway Traffic Safety Administration, Washington, DC
77. Bull M, Agran P, Laraque D, Pollack SH, Smith GA, Spivak HR, Tenenbein M, Tully SB, Brenner RA, Bryn S, Neverman C, Schieber RA, Stanwick R, Tinsworth D, Tully WP, Garcia V, Katcher ML (1999) American Academy of Pediatrics, Committee on Injury and Poison Prevention. Safe transportation of newborns at hospital discharge. Pediatrics 104(4 Pt 1):986–987
78. Degrazia M, Guo CY, Wilkinson AA, Rhein L (2010) Weight and age as predictors for passing the infant car seat challenge. Pediatrics 125(3):526–531. doi:10.1542/peds.2009-1715
79. Nagase H, Yonetani M, Uetani Y, Nakamura H (2002) Effects of child seats on the cardiorespiratory function of newborns. Pediatr Int 44(1):60–63
80. Bull MJ, Engle WA, Committee on Injury, Violence, and Poison Prevention and Committee on Fetus and Newborn, American Academy of Pediatrics (2009) Safe transportation of preterm and low birth weight infants at hospital discharge. Pediatrics 123(5):1424–1429. doi:10.1542/peds.2009-0559
81. Huelke DF, Mackay GM, Morris A, Bradford M (1992) Car crashes and non-head impact cervical spine injuries in infants and children. SAE 920563. Society of Automotive Engineers, Warrendale
82. Stalnaker RL (1993) Spinal cord injuries to children in real world accidents. SAE 933100. Child occupant protection. Society of Automotive Engineers, Warrendale
83. McElhaney JH, Myers BS, (1993) Nahum AM, Melvin JW (eds) Biomechanical aspects of cervical trauma. In: Accidental injury biomechanics and prevention. Springer, New York, pp 311–361
84. Kumaresan S, Yoganandan N, Pintar FA, Maiman DJ, Kuppa S (2000) Biomechanical study of pediatric human cervical spine: a finite element approach. J Biomech Eng 122(1):60–71
85. Kumaresan S, Pintar FA, Mueller W (1998) Yoganandan N (ed) One, three and six year old pediatric cervical spine finite element models. In: Frontiers in head and neck trauma. IOS Press, Amsterdam, pp 509–523
86. Myers BS, Winkelstein BA (1995) Epidemiology, classification, mechanism, and tolerance of human cervical spine injuries. Crit Rev Biomed Eng 23(5–6):307–409
87. Fuchs S, Barthel MJ, Flannery AM, Christoffel KK (1989) Cervical spine fractures sustained by young children in forward-facing car seats. Pediatrics 84(2):348–354
88. Hoy GA, Cole WG (1993) The paediatric cervical seat belt syndrome. Injury 24(5):297–299
89. Langwieder K, Hummel T, Felsch B, Klanner W (1990) Injury risks of children in cars – epidemiology and effect of child restraint systems. Paper presented at the XXIII FISITA Congress, vol 1. Associazione Tecnica dell'Automobile, Turin
90. Trosseille X, Tarriere C (1993) Neck injury criteria for children from real crash reconstructions. SAE 933103. Child occupant protection. Society of Automotive Engineers, Warrendale
91. Brown J, Kelly P, Griffiths M, Tong S, Pak R, Gibson T (1995) The performance of tethered and untethered forward facing child restraints. Paper presented at the international IRCOBI conference on the biomechanics of impact, Bron
92. Legault F, Gardner W, Vincent A (1997) The effect of top tether strap configurations on child restraint performance. SAE 973304. Paper presented at the child occupant protection 2nd symposium. Society of Automotive Engineers, Warrendale
93. Nance ML, Kallan MJ, Arbogast KB, Park MS, Durbin DR, Winston FK (2010) Factors associated with clinically significant head injury in children involved in motor vehicle crashes. Traffic Inj Prev 11(6):600–605. doi:10.1080/15389588.2010.513072
94. Arbogast KB, Durbin DR, Kallan MJ, Menon RA, Lincoln AE, Winston FK (2002) The role of restraint and seat position in pediatric facial fractures. J Trauma 52(4):693–698
95. Arbogast KB, Wozniak S, Locey CM, Maltese MR, Zonfrillo MR (2012) Head impact contact points for restrained child occupants. Traffic Inj Prev 13(2):172–181. doi:10.1080/15389588.2011.642834
96. Viano DC, Parenteau CS (2008) Fatalities of children 0-7 years old in the second row. Traffic Inj Prev 9(3):231–237
97. Orzechowski KM, Edgerton EA, Bulas DI, McLaughlin PM, Eichelberger MR (2003) Patterns of injury to restrained children in side impact motor vehicle crashes: the side impact syndrome. J Trauma 54(6):1094–1101. doi:10.1097/01.TA.0000067288.11456.98
98. Arbogast KB, Locey CM, Zonfrillo MR, Maltese MR (2010) Protection of children restrained in child safety seats in side impact crashes. J Trauma 69(4):913–923. doi:10.1097/TA.0b013e3181e883f9
99. Maltese MR, Locey CM, Jermakian JS, Nance ML, Arbogast KB (2007) Injury causation scenarios in belt-restrained nearside child occupants. Stapp Car Crash J 51:299–311
100. Bendjellal F, Sciclune G, Frank R, Grospietsch M, Whiteway A, Flood W, Marsilio W (2011) Applying side impact cushion technology to child restraint systems. ESV Paper 11-0138-O

101. Durbin DR, Elliott MR, Winston FK (2003) Belt-positioning booster seats and reduction in risk of injury among children in vehicle crashes. JAMA 289(21):2835–2840. doi:10.1001/jama.289.21.2835
102. Jermakian JS, Kallan MJ, Arbogast KB (2007) Abdominal injury risk for children seated in belt positioning booster seats. ESV Paper 07-0441
103. Reed MP, Ebert SM, Klinich KD, Manary MA (2008) Assessing child belt fit. Effects of vehicle seat and belt geometry on belt fit for children with and without belt-positioning booster seats, vol I. UM 2008-49-1
104. Reed MP, Ebert SM, Sherwood CP, Klinich KD, Manary MA (2009) Evaluation of the static belt fit provided by belt-positioning booster seats. Accid Anal Prev 41(3):598–607. doi:10.1016/j.aap.2009.02.009
105. Bilston LE, Sagar N (2007) Geometry of rear seats and child restraints compared to child anthropometry. Stapp Car Crash J 51:275–298
106. Klinich KD, Pritz HB, Beebe MS, Welty K, Burton RW (1994) Study of older child restraint/booster seat fit and NASS injury analysis. DOT/HS 808:248
107. Huang S, Reed M (2006) Comparison of child body dimensions with rear seat geometry. SAE Technical Paper 2006-01-1142. doi:10.4271/2006-01-1142
108. Andersson M, Bohman K, Osvalder AL (2010) Effect of booster seat design on children's choice of seating positions during naturalistic riding. Ann Adv Automot Med 54:171–180
109. Tylko S, Dalmotas D (2005) Protection of rear seat occupants in frontal crashes. ESV Paper 05-258
110. Klinich KD, Reed MP, Ritchie NL, Manary MA, Schneider LW, Rupp JD (2008) Assessing child belt fit. Effect of restraint configuration, booster seat designs, seating procedure, and belt fit on the dynamic response of the hybrid III 10YO ATD in sled tests, vol II. UM-2008-49-2
111. Brown J, Kelly P, Suratno B, Paine M, Griffiths M (2009) The need for enhanced protocols for assessing the dynamic performance of booster seats in frontal impacts. Traffic Inj Prev 10(1):58–69. doi:10.1080/15389580802493155
112. McDougall A, Brown J, Beck B, Bilston LE (2011) The effect of varied seat belt anchorage locations on booster seat sash guide effectiveness ESV Paper Number 11-0135
113. Klinich KD, Reed MP, Manary MA, Orton NR, Rupp JD (2011) Optimizing protection for rear seat occupants: assessing booster performance with realistic belt geometry using the hybrid III 6YO ATD. UM-2011-40
114. Klinich KD, Reed MP, Ebert SM, Rupp JD (2011) Effects of realistic vehicle seats, cushion length, and lap belt geometry on child ATD kinematics. UM-2011-20
115. Kortchinsky T, Meyer P, Blanot S, Orliaguet G, Puget S, Carli P (2008) Misuse of an adult seat belt in a 7-year-old child: a source of dramatic injuries and a plea for booster seat use. Pediatr Emerg Care 24(3):161–163. doi:10.1097/PEC.0b013e3181668b18
116. Appleton I (1983) Young children and adult seat belts: is it a good idea to put children in adult belts? New Zealand Ministry of Transport, Road Transport Division, Wellington, p 33. doi:10.1016/S0001-4575(03)00065-4
117. Corben CW, Herbert DC (1981) Children wearing approved restraints and adult's belts in crashes. New South Wales, Traffic Accident Research Unit, Sydney
118. Brown J, Bilston L (2007) Spinal injuries in rear seated child occupants aged 8–16 years. ESV Paper 07-0461
119. Gotschall CS, Better AI, Bulas DI, Eichelberger MR, Bents F, Warner M (1998) Injuries to children restrained in 2- and 3-point belts. Paper presented at the Association for the Advancement of Automotive Medicine 42nd conference. AAAM, Des Plaines
120. States JD, Huelke DF, Dance M, Green RN (1987) Fatal injuries caused by underarm use of shoulder belts. J Trauma 27(7):740–745
121. Sullivan LK, Chambers FK (1994) Evaluation of devices to improve shoulder belt fit. DOT/HS 808 383. National Highway Traffic Safety Administration, Vehicle Research and Test Center, East Liberty, Ohio

Pedestrian Injury Biomechanics and Protection

24

Ciaran Knut Simms, Denis Wood, and Rikard Fredriksson

Abstract

Pedestrians account for about one third of road accident fatalities worldwide, but there are large regional variations. In general, in highly motorized countries pedestrians account for around 10–20 % of fatalities, but in less motorized countries, pedestrians can account for over 50 % of fatalities. Pedestrians are frequently classed as vulnerable road users as they have a higher fatality rate than vehicle occupants. Protecting pedestrians from vehicle collisions requires a combination of road engineering, vehicle design, legislation/enforcement and accident avoidance technology. The separation of pedestrians from fast-moving motorized vehicles is preferable and pre-crash sensing methods combined with autonomous braking technology can greatly reduce the occurrence and severity of pedestrian accidents. However, these approaches cannot prevent all accidents, and vehicle/pedestrian collisions remain a real and frequent problem.

Pedestrian kinematics during the vehicle contact phase are strongly correlated to the vehicle speed and the height of the vehicle front-end structures relative to the pedestrian height. However, the subsequent ground contact is a highly variable event, which nonetheless accounts for a significant proportion of head injuries.

C.K. Simms, Ph.D. (✉)
Department of Mechanical and Manufacturing Engineering, Trinity College, Dublin, Ireland
e-mail: csimms@tcd.ie

D. Wood, Ph.D.
Denis Wood Associates, Dublin, Ireland
e-mail: info@deniswood.com

R. Fredriksson, Ph.D.
Biomechanics & Restraints, Autoliv Research, Vargarda, Sweden
e-mail: Rikard.fredriksson@autoliv.com

Vehicle impact speed is the main determinant in pedestrian injury outcome. However, despite a popular view that pedestrian safety cannot be significantly improved due to the mass and stiffness disparity between unprotected humans and motorized vehicles, it is now well established that vehicle design has a significant effect on the severity and distribution of pedestrian injuries arising from vehicle impact. In particular, the height of the bonnet leading-edge relative to the pedestrian's centre of gravity is significant for the kinematics and subsequent injuries, and the stiffness of the main contact surfaces on the vehicle also plays a significant role.

Many modern cars feature active pedestrian safety devices such as warnings, autonomous braking, external airbags and pop-up hoods. In the future, the combination of improved vehicle shapes with reduced critical stiffness and auto-braking technology are likely to yield further substantial decreases in pedestrian injuries and fatalities.

24.1 Introduction

The proportion of road fatalities who are pedestrians varies significantly in different countries, but overall the protection of pedestrians is the most important road traffic safety priority [1]. This requires a combination of road engineering, vehicle design, legislation/enforcement and accident avoidance technology. The separation of pedestrians from fast-moving motorized vehicles is preferable and pre-crash sensing methods combined with brake-assist technology can greatly reduce the occurrence and severity of pedestrian accidents [2, 3]. However, these approaches cannot prevent all accidents, and vehicle/pedestrian collisions remain a real and frequent problem.

Despite a popular view that pedestrian safety cannot be significantly improved due to the mass and stiffness disparity between unprotected humans and motorized vehicles [4, 5], researchers have known since the 1970s that vehicle design has a significant effect on the severity and distribution of pedestrian injuries arising from vehicle impact [6–9].

This chapter provides an overview of the magnitude of the pedestrian injury problem, the main factors influencing pedestrian injuries in the event of a collision, and a summary of the current and future vehicle based countermeasures for pedestrian protection.

24.1.1 Pedestrian Injury Statistics

It is estimated that over one million people die from road accidents and that about one third of these are pedestrians [10]. The proportion of road accident fatalities who are pedestrians varies substantially throughout the world, with large divisions evident between low and high income countries. In general, in highly motorized countries such as the US and in the EU, pedestrians account for around 10–20 % of fatalities, but in less motorized countries such as India and China pedestrians account for a significantly greater percentage of road fatalities [11]. An explanation of these variations lies in the cultural as well as socio-economic differences between these regions.

The International Harmonized Research Activities (IHRA) working group has pooled pedestrian injury data from Australia, Europe Japan and the USA to give the most extensive pedestrian injury database, and a summary of their findings is presented in Table 24.1. The head and leg regions each account for about one third while the thorax region accounts for one fifth of AIS2–6 injuries (The AIS is an Abbreviated Injury Score ranging from minor injury (AIS1), to serious (AIS3) to unsurvivable (AIS6) [12]). In highly motorized countries, children account for nearly one third of the IHRA dataset [13]. The accident severity is also age-dependent, with older people having a greater likelihood of high severity injuries [14, 15].

The strong correlation between vehicle impact speed and pedestrian injury severity is well documented [5, 16–20]. However, absolute risk levels are disputed and the newer data from Rosen and Sander [20] shown in Fig. 24.1 may be the most reliable to date. It shows a steep increase in fatality risk for impacts above about 40 km/h.

24.1.2 Injuries from Vehicle and Ground Contact

In most cases, after being impacted by a vehicle, the pedestrian strikes the ground and significant injuries can be attributed to both the vehicle and the ground impacts [21]. Ground contact is especially important for child pedestrians [14].

Table 24.1 The main IHRA body regions injured and their frequency [14]

Body region	% AIS2–6 injuries
Head	31.4 %
Face	4.2 %
Neck	1.4 %
Chest	10.3 %
Abdomen	5.4 %
Pelvis	6.3 %
Arms	8.2 %
Legs	32.6 %

24.1.3 Injury Risk as a Function of Vehicle Type

Impacts with larger passenger cars result in more serious pedestrian injuries than smaller cars [22–24]. However, there is an even stronger relationship between vehicle type (cars, trucks, SUVs etc.) and pedestrian injury/fatality risk. Lefler et al. [25] used a combination of the Pedestrian Crash Data Study (PCDS) data, the General Estimates Data and the Fatal Accident Reporting System data from the US to show that 11.5 % of pedestrians struck by large SUVs are killed, compared with 4.5 % for pedestrians struck by cars. A more recent analysis concluded that the odds-ratio for pedestrian fatalities from LTVs impacts compared to cars is closer to 1.5 [26].

Longhitano et al. [27, 28] used the Pedestrian Crash Data System to analyze the influence of vehicle type on pedestrian injury distribution, see Fig. 24.2. Ground impact injuries were not included. They found AIS3+ head injuries in 71 % of cases for car impacts compared to 81 % of cases for LTVs. AIS3+ injuries of the mid-body regions were found in car impacts in only 25 % of cases compared to 60 % of cases for LTVs. They concluded that the head was the most frequently injured body region for both vehicle categories, but the next most important body region is the lower extremity for impacts with

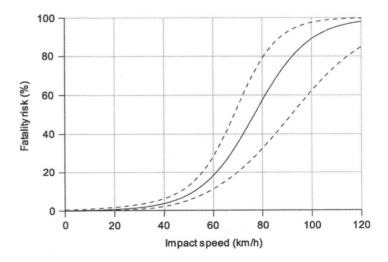

Fig. 24.1 Probability of pedestrian fatality as a function of vehicle impact speed (km/h) [20]

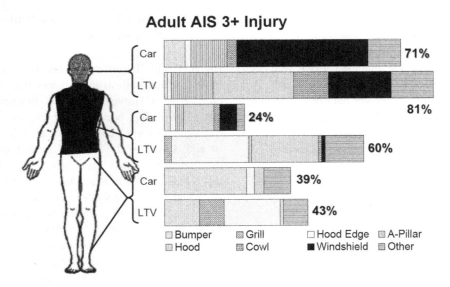

Fig. 24.2 Distribution of AIS3+ pedestrian injuries as a function of vehicle type (Adapted from [27])

cars, whereas for impacts with LTVs it is the torso. In less motorized countries, the most severe pedestrian injuries (AIS4-5) occur in accidents which involve either buses or trucks [29].

A recent breakdown of body regions injured and the vehicle components involved for European data is shown in Table 24.2. The data show that the overwhelming majority of lower extremity injuries are caused by the bumper. For head injuries, the windshield and A-Pillar are mainly responsible, but a significant proportion of head injuries comes from subsequent contact with the ground.

24.1.4 Pedestrian Impact Kinematics

The movement of pedestrians during and after vehicle impact is key to understanding the resulting injury mechanisms and for informing improved vehicle design. The most important factors determining pedestrian kinematics during impact are the shape and speed of the vehicle and the height, stance and speed of the body relative to the vehicle at the instant of impact. Film evidence of pedestrian accidents shows that significant variations in the impact configurations and the resulting kinematics frequently occur. This chapter focuses on the most common impact sequences, but individual accident cases are unlikely to be fully described by the descriptions in this chapter.

About 80 % of pedestrians were standing up and moving across the road when struck from the side by the fronts of passenger cars [31, 32], and about 60 % of pedestrians made no avoidance maneuver prior to impact such as jumping, accelerating, turning away or stopping [33]. Therefore the most important pedestrian accident scenario is a side-struck pedestrian moving at walking pace across the vehicle line of travel. However, this configuration may not be representative of children, as one analysis found that nearly half of child pedestrians struck by vehicles were running when struck [34].

24.1.5 Wrap Projection

Pedestrian wrap projection occurs when the principal vehicle impact force is applied below the centre of gravity. This is the most common sequence for pedestrians struck by vehicle fronts and can occur regardless of the orientation of the pedestrian at impact. In very low speed impacts

Table 24.2 Body regions injured and the vehicle components involved [30], reprinted with permission from the Association for the Advancement of Automotive Medicine

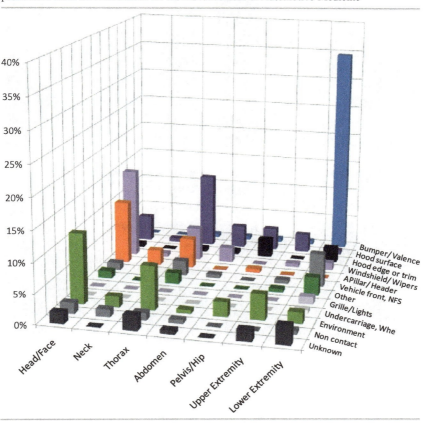

the pedestrian may be able to prevent head impact by active bracing using the arms. For moderate to high severity collisions, primary impact is followed by a further contact where the head and/or arms/shoulders are struck by the bonnet/windscreen area on the vehicle, and the pedestrian may then continue to rotate towards the roof. When the front of a passenger car strikes a side-facing adult pedestrian, the first contact is between the bumper and the knee region, see Fig. 24.3, which shows a 40 km/h staged pedestrian impact with a cadaver [35]. The following timings relate to this staged impact. Initially the struck-side leg is accelerated in the direction of vehicle travel. Due to inertia effects, the movements of the non-contacted body regions lag behind those of the struck region, facilitated by a combination of articulation in the joints, soft tissue deformation and bending of the long bones. The lower leg and foot on the struck side fold underneath the bumper and the non-struck leg remains still for at least 30 ms after initial vehicle contact. Similarly, the pelvis and the whole upper body remain almost horizontal for 40 ms, see Fig. 24.3.

By about 20 ms, there is contact between the bonnet leading edge and the pedestrian upper leg/pelvis on the struck side, the severity of which depends on the vehicle shape. Rotation of the upper body onto the bonnet then proceeds so that 120 ms after primary impact the torso has rotated sufficiently for the arm and shoulder on the struck side to contact the bonnet. The motion of the head initially lags behind the torso as its inertia causes lateral bending in the neck. However, the resulting lateral bending moment in the neck causes the head to "catch up" with the torso rotation and by

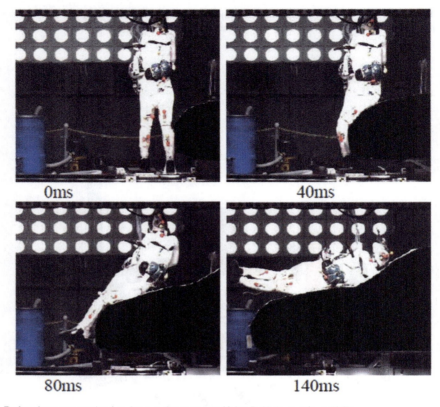

Fig. 24.3 Pedestrian wrap projection for a cadaver test at 40 km/h (Adapted from [35])

about 120 ms the head contacts the vehicle in the region of the bonnet/windscreen/grille.

The length measured along the vehicle front profile from the ground to the pedestrian head impact location is the Wrap Around Distance WAD [36], which provides a useful measure of which vehicle front-end components contact the head in pedestrian/cyclist impacts. Taller pedestrians are twice as likely to have a head impact on the windscreen, while shorter ones are 50 % more likely to have a head impact on the bonnet [37]. In general, a low bonnet leading edge height results in the pedestrian being carried further onto the car and closer to the windshield [38].

24.1.6 Forward Projection

Forward projection occurs when the pedestrian centre of gravity is within the primary impact zone. This typically occurs where an adult is struck by a minivan or a truck, a small adult is struck by a large SUV or a child is struck by a passenger car, van, truck or an SUV. The location of first contact depends on the size of the struck body relative to the vehicle front, but subsequent contacts follow in very quick succession since the pedestrian is struck by an almost flat and vertical surface, see Fig. 24.4 [39]. As a result, the mid-body region is rapidly accelerated in the direction of vehicle travel with very little rotation so that the mid-body region and the vehicle attain an almost common post-impact velocity.

Forward projection impacts occur provided the height of the vehicle's front leading edge is above the pedestrian's centre of gravity height. In such cases, the impact will be effectively concentric and the mid-body region and the vehicle front attain a common post-impact velocity. When the bonnet leading edge height is just below the body centre of gravity height, the resulting kinematics appear as a combination of wrap and forward projection.

Fig. 24.4 Pedestrian forward projection kinematics: vehicle impact speed of 16.9 km/h against a stationary dummy [39]

24.1.7 Post Head Impact Kinematics for Forward and Wrap Projection Cases

Figure 24.5 shows the results of a staged wrap projection impact of a stationary adult pedestrian dummy at 58 km/h [40]. Only the post head impact movement is shown. Evidence from this and other staged tests shows little restitution in the head impact [41], which therefore attains a common linear velocity with the bonnet/windscreen impact location immediately after head impact. The body then pivots about the head, either lifting the legs towards the roof in higher severity collisions or dropping the body back towards the bonnet due to gravity in lower severity collisions. The kinematic sequence depends mostly on the vehicle shape and impact speed, and the impact angle and angular velocity of the body at head impact.

Separation of the pedestrian from the vehicle following head impact in wrap projections marks the beginning of the airborne trajectory (Fig. 24.5), and the occurrence of this depends substantially on vehicle braking. In cases where the body does not drop back onto the bonnet, separation will occur if the vehicle is braking. However, in the subsequent somersault type movement of the body further contacts with the bonnet are possible if vehicle braking is insufficient. For forward projection, following separation from the vehicle due to braking, the body falls over and contacts the ground. In cases where the vehicle does not brake sufficiently following impact, the vehicle may run over the pedestrian. The contact with the ground is highly variable for both wrap and forward projection pedestrian impacts, and prediction of ground contact kinematics is very difficult. Nonetheless, high fronted vehicles lead to more severe head ground contact conditions [42, 43].

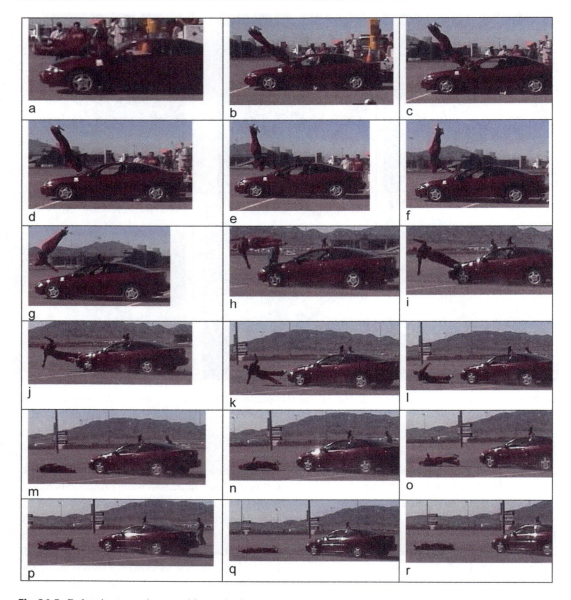

Fig. 24.5 Pedestrian ground contact kinematics in a staged test (Adapted from [40])

24.1.8 Vehicle Design Standards for Pedestrian Protection

The safety of pedestrians was not a serious consideration in vehicle design until the 1980s. This followed the belief that little could be done to protect pedestrians in the event of a vehicle impact [4, 7] but also by vehicle manufacturers' reluctance to develop an area not governed by legislation and not considered to provide sufficient added value to the vehicle. In consequence, existing standards are relatively new and subject to updates and legal implementation is evolving. However, there is now substantial public appetite in many countries for the regulation of vehicle design for pedestrian safety, as evidenced by the introduction of pedestrian safety testing by consumer driven safety organizations such as the

New Car Assessment Programmes operational in Europe, Japan and Australia.

The main bodies who have developed pedestrian safety standards are Working Groups of the European Enhanced Vehicle Safety Committee (EEVC), the International Organization for Standardisation (ISO) and the International Harmonised Research Activities/United Nations Economic Commission for Europe (UNECE). The IHRA has now largely superseded the work of the ISO and their work has been followed by the development of a Global Technical Regulation (GTR) for pedestrian protection [44]. The tests proposed by the EEVC, ISO and the IHRA are all similar and are aimed at protecting pedestrians struck from the side by a vehicle at speeds up to 40 km/h, since this was judged to be the upper limit at which pedestrian protection could be achieved and over 80 % of pedestrian accidents are below this speed [15, 45, 46]. In all cases pedestrian subsystem impactor tests were developed in favour of pedestrian dummy tests, see Fig. 24.6. This is quite different to the full scale dummy tests stipulated in frontal impact vehicle occupant protection standards (e.g. FMVSS 208 [47]) and several arguments have been proposed to justify this, mostly relating to repeatability and dummy biofidelity. The principal goal of a subsystem impactor test is to reproduce the force interaction between the corresponding vehicle and pedestrian body parts, and provide an assessment of the resulting injury likelihood by comparison with accident or cadaver data. However, the choice of impactors in lieu of a whole-body dummy makes it almost impossible to determine the net change in pedestrian risk resulting from a specific vehicle design alteration [48].

The current EEVC legform, developed by the UK Transport Research Laboratory [50, 51], consists of two foam-covered rigid segments representing the upper and lower leg of a 50 percentile adult male connected by a simulated knee joint. It was designed to be propelled in free flight at 40 km/h against the vehicle bumper, and the knee lateral bending angle and shear displacement and upper tibia lateral acceleration are compared to tolerance thresholds derived from biomechanical testing, see Table 24.3.

An alternative flexible legform impactor (FlexPLI) with a more biofidelic knee joint [52] (see Fig. 24.7) has been developed by the Japanese Automobile Manufacturers Association (JAMA) in collaboration with the Japanese Automobile Research Institute (JARI) [53]. The rigid TRL rather than the flexible FlexPLI legform has been chosen for the first phase of the GTR, but it is planned to introduce the FlexPLI in future. For the GTR a higher tolerance of 19° for lateral knee bending for the TRL impactor was adopted [44], see Table 24.3.

The EEVC has considered the isolated legform impactor test representative of a whole-body dummy test if the bumper impact occurs at or below the knee joint level (i.e. 500 mm or less above the ground) [55]. For vehicles with high bumpers, the bumper impacts the femur part of the legform impactor and the absence of the hip joint constraint and upper body mass can lead to rigid body rotation of the legform impactor without lateral bending of the knee [56, 57]. To account for this, the EEVC and GTR have allowed for an optional alternative horizontal upper legform to bumper test in cases where the

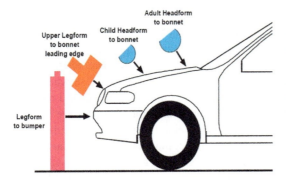

Fig. 24.6 Pedestrian subsystem tests (Adapted from [49])

Table 24.3 EURO-NCAP and GTR biomechanical thresholds for 40 km/h free flight adult headform impact to vehicle bumper

	Knee bending angle (°)	Knee shear displacement (mm)	Tibia head acceleration (g)
EEVC	15	6	150
GTR	19	6	170

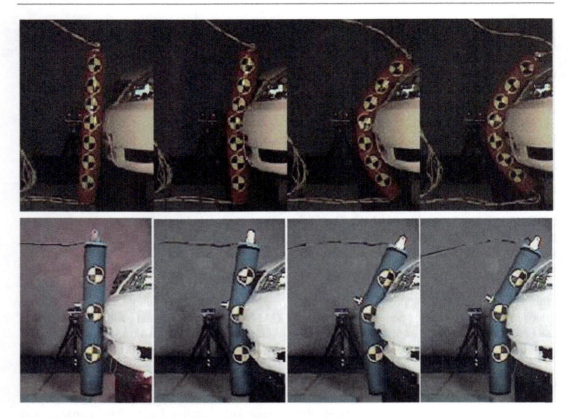

Fig. 24.7 Comparison of TRL and JAMA-JARI legforms in contact with a vehicle bumper showing increased flexibility of the JAMA-JARI prototype (Adapted from [54])

bumper is located high and close to the bonnet leading edge [55]. However, the acceptance level for the EEVC upper legform test in this configuration is demanding (5 kN force and 300 Nm bending moment) and the EEVC have used this to discourage a design trend towards higher bumpers [58]. A comparison between the EEVC legform impactor test and staged tests using cadavers struck by a Peugeot 206 and a Renault Twingo showed that the critical bending moment proposed by the EEVC for the legform impactor is appropriate, but that the shear displacement threshold is not [59].

24.1.9 Upper Legform to Bonnet Leading Edge

The bonnet leading edge contact with an adult pedestrian is mostly in the region of the femur [60], and the EEVC upper legform impactor represents a (shortened) segment of an adult femur [55], see Fig. 24.8. There are load cells at the upper and lower supports of the upper legform, and strain gauges measure the peak bending moment in the middle of the impactor. Staged tests using an adult dummy struck by instrumented car fronts [61] and multibody pedestrian models [62] showed that the effective mass, impact velocity and impact angle for the upper leg depend in a predictable manner on the shape of the vehicle front. This resulted in input parameters for the EEVC upper legform impactor test where the vehicle's bonnet leading edge height and bumper lead are measured and look-up tables are used to define impactor velocity, impact angle and impact energy. The impact mass is then determined from the impact velocity and energy.

The biomechanical tolerance thresholds proposed for the impactor are intended to predict a

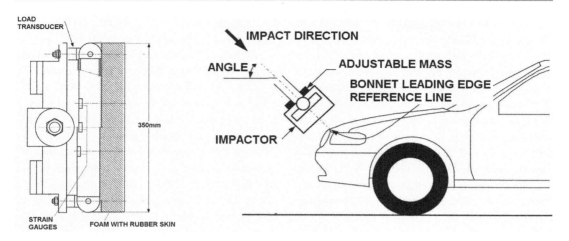

Fig. 24.8 The EEVC upper legform impactor and its application (Adapted from [58])

20 % risk of fractures of the femur and pelvis and the EEVC threshold values are 5 kN for the sum of impactor forces and a 300 Nm bending moment limit. Cadaver studies have shown that femur and pelvis injuries for impact energies below 200 J are insignificant and therefore no EEVC bonnet leading edge test is proposed for vehicles where the upper legform impact energy is less than 200 J. A recent comparison between the EEVC upper legform impactor test and staged tests using cadavers and multibody simulations showed that the upper leg impact angle and velocity in the EEVC specification are reasonably representative of real accidents for both a Peugeot 206 and a Renault Twingo [59]. However, the high incidence of spinal, thoracic and abdominal injuries in a sample of 80 pedestrian accidents highlights the types of serious injuries that are not addressed by the current subsystem testing protocols [63].

There is currently no GTR upper legform test for the bonnet leading edge, partly due to difficulties encountered with this test by the EEVC [58], where the streamlined shape of modern passenger cars has resulted in a substantial reduction in bonnet leading edge injuries [58]. However, data from the USA shows a large number of midbody injuries have been found due to the prevalence of light trucks and vans, and a means to evaluate these is important [64].

24.1.10 Headform to Bonnet Top: Adult and Child

The EEVC and the ISO developed a child headform to assess child head injury risk from contact with the forward sections of the bonnet and an adult headform test to assess adult pedestrian injury risk from contact with the rear part of the bonnet, wings and scuttle. The EEVC headform tests are limited to the front of the vehicle up to and including the base of the windscreen, because these were the only areas that were thought feasible to improve [65]. Both the EEVC and ISO headforms are spherical and featureless to increase test repeatability and both have the same construction: a semi-rigid sphere, covered in a deformable rubber or foam "skin". The impactors are designed to be propelled in free flight at 40 km/h at a predetermined angle against various portions of the bonnet top. The IHRA headform specifications are the same as for the ISO [66].

The EEVC adult headform has shown good correlation of the resultant acceleration time history by comparison with some cadaver tests [67]. In cadaver tests and computer simulations with a vehicle impact speed of 40 km/h, head impact velocities to the bonnet were reportedly 43–50 km/h, with the direction of impact approximately 65° to the horizontal [67]. Computer simulations also showed the head impact angle for adults to be circa 65° and 40–55° for children. However, the EEVC have recognized that head impact velocity varies substantially as a function of vehicle shape [55, 56] and the simplification of a fixed impact velocity for the headform test eliminates the influence of overall vehicle shape on head impact conditions. A recent comparison between the EEVC headform impactor test and cadaver tests with a Peugeot 206 and a Renault Twingo showed that the cadaver head impact velocity and the impact angle are generally lower than the EEVC specification [59]. Other cadaver tests have shown the velocity of head impact on the vehicle can be both higher and lower than the original vehicle impact velocity on the pedestrian [68–70], and these findings may be a function of pedestrian stature and vehicle shape.

The ISO/IHRA have proposed slightly different impactor specifications and impact conditions than the EEVC. The differences in the impactor masses are small, but there are substantial differences in the vehicle regions covered and in the prescribed impact conditions. For the EEVC tests, the child headform test area is bounded by Wrap Around Distances of 1,000–1,500 mm, while the adult headform test area is bounded by a WAD of 1,500–2,100 mm, unless the latter extends beyond the windscreen, in which case the lower windscreen frame forms the rear boundary of the test area. The EEVC test does not include the windscreen and only includes the windscreen frame if it lies within the specified WAD, even though the EEVC have accepted that most adult head impact locations are outside the zone considered by their test method [58]. Other tests, such as those of EuroNCAP, define different zones for testing on the vehicle [71].

The acceptance level for the EuroNCAP headform is now a Head Injury Criterion (HIC) score of 650. The HIC equation derives from power regression analysis of the Wayne State Tolerance Curve [72, 73] and the HIC is iteratively calculated from the resultant head centre of gravity acceleration (a) time history in g units:

$$HIC = max\left[[t_2 - t_1] \left[\frac{1}{t_2 - t_1} \right] \int_{t_1}^{t_2} a dt \right]^{2.5}$$

$$t_2 - t_1 \leq 15 ms$$

All proposed tests only account for the linear acceleration of the headform via the HIC, and do not include rotational acceleration effects which are known to be significant for brain trauma such as diffuse axonal injuries [74]. However, a HIC value of 1,000 has previously been found to be a good indicator of serious head injury for real pedestrian cases involving adults and children [75] and it was concluded that there was insufficient data available to propose a separate robust HIC acceptance level for child protection [55].

For the GTR, child and adult headform tests at 35 km/h are defined. The threshold HIC score for the bonnet or windscreen is 1,000 and the windscreen frame and A-pillars are excluded from the test zone [64]. The reason for this exclusion is that it was considered that the A-pillar stiffness requirements for vehicle rollover protection preclude reduction of HIC scores for pedestrian head impact against the A-pillars. For the windscreen, it has been argued that the low HIC scores for mid-windscreen contacts mean that the advantages of including the windscreen in the test would be minimal [76]. However, the windscreen is responsible for many head injuries and the potential for improved windscreen design will be addressed later in this chapter.

While current tests have the capacity to assess the effectiveness of active features such as pop-up hoods and bonnet airbags (see later sections), they cannot assess the effect active features such as brake assist and autonomous braking. In the near future, tests will therefore need to be designed to assess the effectiveness of integrated active and passive safety features for pedestrian safety.

24.1.11 Implementation into Legislation

Pedestrian safety legislation varies considerably in different jurisdictions and developments are ongoing. In the United States, there is no legislation. In the EU, the EEVC pedestrian legform and headform impactor tests have been included in a European Directive [77] which is in place since 2005. Japan introduced pedestrian regulations also in 2005 for head protection. Japan and Korea will introduce the GTR in 2013.

24.1.12 The Influence of Vehicle Design on Pedestrian Injuries

The principles governing the individual influences of vehicle shape and stiffness on pedestrian injuries are relatively straightforward and can be summarized by Ashton's statement in 1979 that a smooth and compliant front profile may reduce overall pedestrian injury severity [6]. The three main vehicle related considerations for pedestrian and cyclist injuries are mass, shape and stiffness. Vehicle mass will first be shown not to be very significant. In contrast, vehicle stiffness will be seen from first principles to be a fundamental parameter for pedestrian and cyclist injury and the importance of vehicle shape will also be demonstrated since this determines the impact locations of the vehicle surfaces on the body. Furthermore, it will be seen that shape influences the energy associated with each impact, while stiffness determines the corresponding force. Therefore shape and stiffness combine to determine injury likelihood for pedestrians when struck by a particular vehicle type.

There are hundreds of vehicle types which can be potentially involved in a pedestrian accident. These include cars, buses, vans, sport utility vehicles and trucks and significant variations are found even within these categories, see Fig. 24.9.

Snedeker et al. [78] published average bumper height, bonnet leading edge height and bumper lead for cars, SUVs and vans, see Table 24.4. Great variations in commercial truck sizes are found and these are therefore not included here. There are also significant differences in pedestrian stature with age, gender and ethnicity, nutrition etc., see Fig. 24.10 [79].

The bumper typically strikes the lower limbs, while the bonnet leading edge strikes in the region of the hip and head impact subsequently occurs on with either the bonnet top or the windscreen area, see Fig. 24.11.

Table 24.4 Typical vehicle front end dimensions for different classes [78]

Height above ground	Car (mm)	Van (mm)	SUV (mm)
Bumper height	500	580	640
Bumper lead	140	160	140
Bonnet leading edge	740	860	1,020

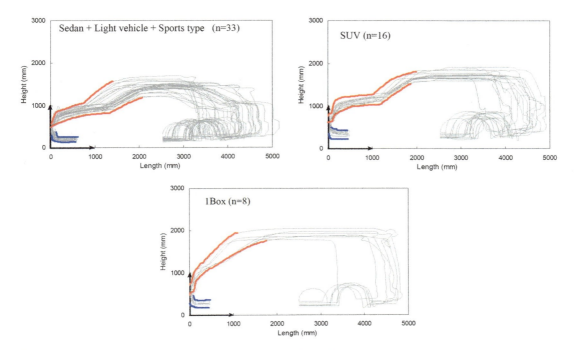

Fig. 24.9 Variation in vehicle shape as characterized by the IHRA [14]

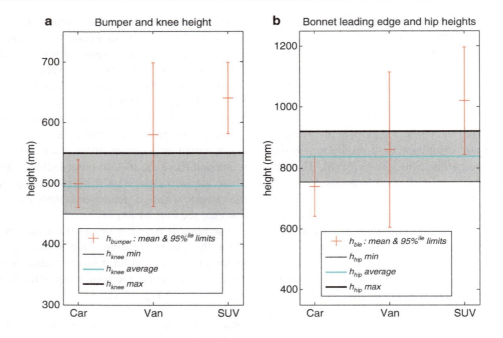

Fig. 24.10 (a) Bumper height (h_{bumper}) and adult knee height (h_{knee}) and (b) bonnet leading edge (h_{ble}) and adult hip heights (h_{hip}) for different vehicle classes [79, 78]

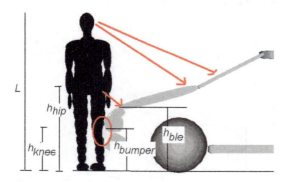

Fig. 24.11 Typical contact locations in vehicle pedestrian accidents [11]

Therefore, a comparison of bumper height (h_{bumper}) with knee height (h_{knee}) and bonnet leading edge height (h_{ble}) with hip height (h_{hip}) for the different vehicle classes is instructive, see Fig. 24.10. The average and 95 percentile limits are given for the adult knee and hip heights as well as for the bumper and bonnet leading edge for each vehicle type [78]. The variability in bonnet leading edge height for vans means that there are cases where contact occurs both above and below the hip height, depending on pedestrian stature.

24.1.13 Influence of Vehicle Mass

It can be shown from fundamental considerations that the mass difference between most motorized vehicles and pedestrians is sufficiently large to make the considerable mass disparity between different vehicle types striking pedestrians and cyclists largely insignificant. For example, consider a pedestrian modeled as a rigid rod struck by the bonnet leading-edge (ignoring the bumper contact) of vehicles of the same shape but ranging in mass from 800 to 5,000 kg. Assuming a common post-impact velocity at the impact location, and a vertical height difference (h) between the pedestrian cg and the bonnet leading edge of 0.15 m, the ratio of pedestrian centre of gravity velocity change (Δv_{ped_cg}) to vehicle impact

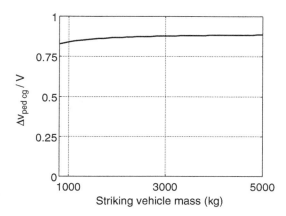

Fig. 24.12 Ratio of pedestrian centre of gravity velocity change (Δv_{ped_cg}) to collision velocity (V) estimated using momentum conservation and a single segment pedestrian representation [11]

velocity (V) can be used as a crude measure of impact severity normalised with respect to the vehicle impact speed [11], and Fig. 24.12 shows the very low sensitivity of this ratio to striking vehicle mass over the large mass range considered.

This indicates that the substantial mass difference between passenger cars (mean mass 1,275 kg) and heavier vehicles such as SUVs (mean mass 1,625 kg [80]) and even larger trucks does not greatly influence pedestrian injury risk. A similar result was obtained with detailed Madymo multibody modeling [81] and analysis of accident data [24, 25], and it can be concluded that mass variation alone is not a very important vehicle design parameter for pedestrian protection.

24.1.14 Influence of Vehicle Stiffness

The stiffness of the vehicle structure significantly influences pedestrian injury severity. This is obvious from the fundamental relationship between maximum crush (d_{max}) and mean acceleration ($\bar{a}$) for the idealised case of a plastic, concentric impact between a vehicle mass M striking a pedestrian mass m at a velocity V:

$$\bar{a} = \frac{1}{2} \frac{V^2}{d_{max}} \left[\frac{M}{M+m} \right]$$

This equation shows that impact severity assessed by mean acceleration $\bar{a}$ increases with the square of the impact velocity for a fixed crush depth, and is inversely proportional to deformation for a fixed impact velocity. Therefore, maximizing crush depth and minimizing impact speed are essential for pedestrian and cyclist protection.

The distribution of vehicle stiffness has recently been determined from analysis of EuroNCAP pedestrian impactor tests [82]. Results from 425 tests were divided into categories according to headform HIC score and legform shear force and bending moment. The categories were coded green for good, red for poor and yellow for intermediate, and the mean and ±1 standard deviation force versus deformation corridors were derived. The results for the bumper, bonnet (front, middle and rear) and windscreen base are shown in Fig. 24.13. Significant differences between vehicle regions and vehicle types are present, and therefore the current vehicle fleet presents a broad range of pedestrian and cyclist injury risk based on stiffness alone. Not surprisingly, therefore, adjusting the stiffness of a small production car without influencing the styling showed that substantial improvements could be made to the EEVC test results scores [83].

24.1.15 Influence of Overall Vehicle Shape

Vehicle shape relative to pedestrian height determines which body regions are struck and the impact energy associated with each body region impact. We have seen that there are major differences in injury patterns for different classes of vehicle. The bonnet surface and windshield are the most important injury inducing contact regions for passenger vehicles, while for light trucks and vans (including SUVs) the main injury sources are the bonnet surface and the bonnet leading edge. Vehicle shape is substantially dependent on vehicle class and the majority of vehicles striking pedestrians can be broadly classified as cars, SUVs and trucks/vans. Using the mean dimensions for cars, SUVs and vans ([78]), the conservation of momentum applied to a single

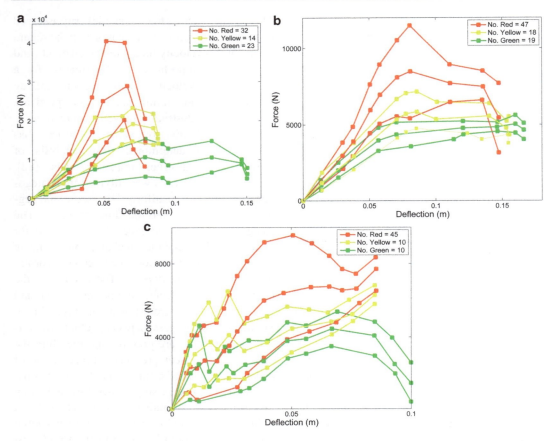

Fig. 24.13 Mean and ±1 standard deviation stiffness plots for (**a**) the bumper, (**b**) bonnet front and (**c**) windscreen base derived from EuroNCAP pedestrian impactor tests (Adapted from [82]): colours represent vehicle rating: *green* = good, *red* = poor, *yellow* is intermediate

segment pedestrian model can again be used to predict the generalized momentum transfer in the primary vehicle impact associated with different vehicle shapes. As a first approximation when assessing overall pedestrian kinematics as a function of vehicle class, it is reasonable to consider the bumper and bonnet leading edge contacts as a single contact occurring at the average height of the bumper and bonnet. Assuming as before a 50th percentile male mass and height, a mean car mass of 1,265 kg and mean SUV/van masses of 1,625 kg, the relative severity of the primary impact for each of the vehicle types can be assessed by considering the linear and angular post impact velocities of the pedestrian segment. The centre of gravity linear velocity (Δv_{ped_cg}) normalised by vehicle impact speed (V) and the normalized body angular velocity change ($\Delta \omega$) can then be found using conservation of momentum [11] and these kinematic quantities are a measure of the linear and angular impulse transmitted to the struck pedestrian and their variation for the average impact height of cars, vans and SUVs are shown in Fig. 24.14. This approach clearly shows that the lower impact height of cars results in a significantly lower linear impulse in the primary impact compared to SUVs, but this is accompanied by a significantly higher angular impulse for car impacts compared to SUVs.

We will see that the velocity of the subsequent head contact with the bonnet or windscreen is substantially influenced by whole body rotation

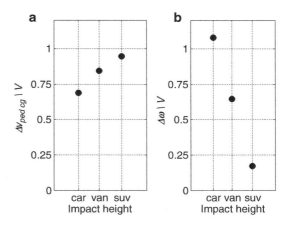

Fig. 24.14 Predictions for the influence of vehicle type (car, van or SUV based on dimensions from [78]) on pedestrian kinematics in primary impact, from [11]: (**a**) normalized linear velocity change of pedestrian body centre of gravity ($\Delta v_{ped\_cg}/V$) and (**b**) normalized angular velocity change of the pedestrian body ($\Delta \omega/V$)

following primary impact and therefore it can be concluded from this simple analysis that the generalized shapes associated with cars, vans and SUVs result in substantial variations in head and midbody loading for pedestrians. This highly simplified analysis shows why pedestrian impacts from high fronted vehicles result in more severe midbody loading and lower head impact velocity on the vehicle due to reduced rotation.

24.1.16 Bumper Shape

Bumper shape is defined by three main measures: height from the ground relative to the pedestrian height, protrusion from the vehicle (or bumper lead), and the vertical spread of the bumper (i.e. distance from upper to lower bumper edges). The relative bumper height determines the initial contact location on the body, and the relative bumper height, shape and stiffness are the main determinants of the magnitude and distribution of the resulting bumper force. Apart from soft tissue cuts and abrasions, the principal injuries from bumper impact are fracture of the tibia, fibula and femur and tearing of the ligaments in the knee. These injuries occur due to high stresses in the contact region and the distribution of shear force and bending moment due to the inertia of the leg/thigh and the hip, knee and ankle joint constraint forces and moments. The bumper height relative to knee height is significant and results in different loading patterns of the knee joint and the long bones of the leg.

Analysis of accident data, mathematical models and cadaver and dummy tests has shown that, for vehicle bumpers striking adult pedestrians, the maximum bending moment in the leg coincides approximately with the height of the bumper, and the knee shear force and bending moment are highest for impact close to the knee level due to the inertia of the leg below the knee. As a result, a bumper striking close to the knee causes large lateral deflections in the knee joint, posing a significant risk of knee ligament injury. Conversely, for bumper impacts in the mid-tibia region, the bending moment in the knee is minimized as the upper and lower leg segments rotate together and lateral bending of the knee joint is prevented. This implies an optimum bumper contact height with the leg for minimizing knee injuries at about 35 and 24 cm above the ground for adults and children respectively. Furthermore, lowering the impact height below the knee not only reduces knee injuries, but limits involvement of the critical pelvic and abdominal areas. Bumpers that strike above the knee level, common in off-road vehicles and multipurpose vehicles, cause the upper body and leg to rotate in opposite directions and can result in severe knee trauma. These vehicles have a requirement for high ground clearance which make it difficult to introduce methods of pushing the pedestrian legs forward [4, 6, 8, 61, 84–94].

The bumper height also influences the magnitude of the bumper contact force. Cadaver tests have shown that the peak bumper force was larger for a relative bumper height of 0.9 compared to 0.6 when the bumper was almost rigid, but this was not seen for a compliant bumper and no clear correlation between bumper force and injuries in the leg area was observed [8, 87]. Similarly, multibody modeling has shown that raising the relative bumper height from 0.5 to 0.9 increases the bumper force [95].

The bumper height also influences which bones in the leg are injured. Analyses of light trucks and vans compared to cars have shown a greater risk of injuries above compared to below the knee [27, 80]. A detailed review of leg injuries from high fronted vehicles in the Pedestrian Crash Data Study showed that a higher bumper is more likely to cause femur fracture while a lower bumper is likely to cause tibia fracture and/or knee ligament injury [57]. The average relative bumper lower heights for femur fracture, ligament injury and tibia fracture were 0.95, 0.79 and 0.78 respectively. Furthermore, the impact velocity was statistically different for the femur fracture group compared to the knee ligament injury group [57, 96]. For bumper impacts above the knee at about 40 km/h femur fracture is normal, but at 20–30 km/h ligament injuries usually occur instead. For bumper impact below the knee at close to 40 km/h tibia fracture is normal but ligament injuries usually occur at impact velocities of 20–30 km/h. These and other accident data and experimental tests therefore show that in impacts with sufficient energy to cause bone fracture a protective effect for the knee ligaments is found [96–100].

24.1.17 Secondary Bumper

A significant reduction in knee injuries can be achieved by the provision of a secondary bumper close to the ground which supports the lower leg, see Fig. 24.15, and vehicle designs have developed accordingly, see Fig. 24.16 [4, 101–106]. For the double bumper system compared to the standard bumper, dummy tests showed a 20 % lower knee bending moment, and the shear force and lateral bending angle at the knee and the tibia acceleration were more than 50 % lower [91], and only simple long bone injuries were reported [104].

For vehicles such as SUVs designed for off road usage, the secondary lower bumper for pedestrian protection frequently conflicts with ground clearance requirements, and an inflating bumper airbag has been proposed to provide leg protection by reducing lateral knee loading [107], see Fig. 24.17.

Simplified modeling has shown that for bumper contacts well below the knee – for example from a car with a secondary bumper – the femur shear and bending loads are low and this explains why the femur is rarely fractured in these cases [11]. For bumper contact close to the knee, tibia and knee ligament injuries frequently occur together. For bumper contact above the knee – i.e. from a high fronted vehicle without a secondary

 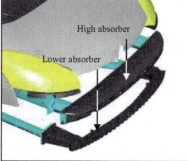

Fig. 24.15 Upper and lower bumper in the Citroen C6 (Adapted from [106])

Fig. 24.16 Development of vehicle front end style (Adapted from [105])

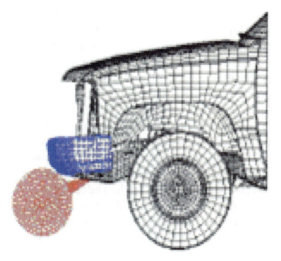

Fig. 24.17 Possibility to reduce the lateral knee loading by an inflating bumper airbag [107]

bumper – the shear and bending in the knee and particularly the femur are high. When the femur fractures, the knee bending moment is relieved and this explains why femur fracture does not generally occur together with tibia and ligament injuries [11].

24.1.18 Bumper Lead

Bumper lead is a measure of how far the bumper protrudes in front of the vehicle, and traditionally bumper lead was used as a means to protect the vehicle. Bumper lead largely determines the time delay between bumper and bonnet leading edge contact with the pedestrian. A short bumper lead has been recommended to minimize the injuries to the lower leg and knee [6], and modern cars generally feature very short or entirely absent bumper leads, see Fig. 24.16. Dummy tests show that the influence of bumper lead on the knee bending moment is slight, but increasing the bumper lead from 100 to 200 mm resulted in a 10 % increase in knee shear force [91].

24.1.19 Bumper Stiffness

We have seen that average acceleration is inversely proportional to deformation and early investigation of real-world pedestrian accidents quickly led to the recommendation of compliant vehicle front profiles [6, 88]. The ideal vehicle bumper design for pedestrian protection allows sufficient compliance to cushion the impact while limiting knee lateral bending [102]. Dummy tests with a softer bumper showed highly decreased knee shear force and tibia acceleration but no significant reduction in the knee bending moment [91]. Cadaver tests showed that stiff bumpers cause more severe tibia fractures, but reduced injuries to the knee and ankle due to the protective effect of bone fracture at the impact site [104].

However, a significant problem for the optimization of vehicle front bumper stiffness is caused by bumper regulations designed to minimize vehicle repair costs [1, 103, 108]. To comply with bumper standards in many countries, the bumper structure must be able to absorb considerable

Fig. 24.18 Honda Civic bumper tests (Adapted from [65])

energy, and this prevents large decreases in bumper stiffness [101]. It has been shown that to provide adequate leg protection, free space between the bumper skin and bumper structure is required [109, 110], and the European Traffic Safety Council recommend that the front face of the bumper needs to crush by 5–7.5 cm to protect the leg [110]. The deep bumper and the integrated air dam or spoiler common in modern cars are beneficial in distributing the contact force in legform impacts. Honda uses a combination of deforming loop and crush cans at the front face of the bumper armature and a two-stage energy management system has been implemented in the bumper which helps absorb leg impact loading, without compromising bumper energy absorption in low speed vehicle impacts, see Fig. 24.18. Opel have used low density pedestrian protection foam in front of the stiff aluminium bumper crossmember to absorb the impact energy together with sufficient deformation space to avoid the impactor hitting the stiff aluminium bumper crossmember or the foam bottoming out [111]. A recent study using the rigid TRL legform impactor in tests found some modern bumpers feature a significant protective effect from the foam around the rigid front cross member, but less effectiveness was observed near the sides of these bumpers where the foam covering is thinner [112].

24.1.20 Shape of Bonnet and Bonnet Leading Edge

The height and shape of the bonnet leading edge are important parameters for vehicle/pedestrian collisions because they strongly influence injuries caused directly by the bonnet edge contact as well as the subsequent head kinematics. Adult midbody injuries from bumper or bonnet leading edge contact are significant because pelvis and upper thigh trauma often cause permanent disablement [15] and a review of pedestrian accidents has found that up to 17 % of pedestrian fatalities involved only a chest, spine or abdomen injury [18]. Similarly, the kinematic motion of the head is important because this determines the location and velocity of the head impact on the bonnet or windscreen. For children, head contact can occur directly with the bonnet leading edge [113].

Analysis of accidents, and tests with dummies and cadavers and detailed multibody/finite element models have all shown that pelvic loading is much greater for higher profile vehicles and therefore bonnet leading edge height is a significant factor for femur and pelvis injuries [8, 37, 78, 81, 114–117]. Partly in response to these findings, the bonnet leading edge of most modern cars is substantially lower and more rounded than for older cars (see Fig. 24.16), where high bonnet leading edges resulted in abdominal injuries accounting for almost 20 % of all injuries and 7 % of serious/fatal injuries [15]. Conversely, no serious injuries to the pelvic area were produced in cadaver tests with a lower vehicle profile [8]. The bonnet leading edge contact force increases substantially with bonnet leading edge height [61, 91] because of the increased effective mass of the pedestrian as the bonnet leading edge strikes closer to the pedestrian centre of gravity.

Pelvis height largely coincides with the pedestrian centre of gravity height and these results are therefore doubly significant for pedestrian injury: not only does the impulse increase with relative bonnet leading edge height, the relative bonnet leading edge height also determines whether contact is made with the femur or directly with the critical pelvis/abdomen region. Therefore, bonnet height has a direct influence on vehicle aggressivity for pedestrians.

Analysis of pedestrian accidents showed that pelvic fractures predominantly occur when there is a direct impact with the bonnet leading edge [9]. This is particularly important in the context of the proportion of vehicles classified as light trucks or vans (including SUVs), most of which have a high bonnet leading edge. Analysis of accident data showed that pedestrian fatality risk in collisions with these vehicles is substantially higher than for conventional passenger cars [25, 43, 80]. Multibody modeling has shown that the primary reason for the increased hazard to pedestrians from high fronted vehicles is the increased height of the bonnet leading edge for these vehicles compared to conventional cars. The location of the bumper and bonnet edge contacts is such that the midbody region is directly struck in an SUV/pedestrian collision, allowing less rotation of the body. This means that for pedestrians struck by high fronted vehicles there is the combination of a harder primary impact which occurs directly with a critical midbody region. Lowering the bumper and bonnet and reducing the stiffness for SUVs would help to reduce injuries to these midbody regions [81].

Analysis using the THUMS finite element model has shown that the bonnet leading edge radius influences pelvis and femur injury risk since it alters the effective closing speed of contact of the bonnet edge with the midbody regions. The closing speed of contact between the thigh and the car bonnet can be made significantly smaller than the vehicle impact speed, depending on the roundness of the bonnet leading edge. In an ideal design, acceleration of the distal femur by the bumper and rolling motion imparted to the thigh by the bonnet radius can greatly reduce the energy of the hip impact [78, 116]. It was concluded that a car with a sufficiently low bonnet leading edge height, large bonnet edge radius, moderate bumper lead and high bumper edge height would practically exclude the possibility of femur fracture in primary lateral impact of a 50th percentile male at impact speeds less than 40 km/h [78].

Real-world injuries and dummy and cadaver tests show a strong interactive effect between the bumper and the bonnet leading edge [61, 100, 116, 118, 119]. Since the pedestrian generally attains a substantially common velocity with the vehicle during the impact process, bonnet leading edge impulse is greatest for a short bumper lead and a high bonnet. Conversely, the bonnet leading edge impulse is reduced when the bumper lead is increased and the bonnet edge is lowered. In dummy tests, it was found that the bumper tends to shield low bonnet leading edges from impact, and conversely, the bumper has little effect on bonnet leading-edge force for high bonnet leading edges. In modern cars, the risk of chest injury due to passenger car impact has been reported as extremely low, but this was partly because the elbow to bonnet impact provides cushioning [120].

We have seen that the height of the bonnet leading edge significantly influences the subsequent rate of upper body rotation for both adults and children in vehicle pedestrian impacts, and this affects the speed of the subsequent head impact on the bonnet/windscreen. This has been shown experimentally using dummy tests, where the HIC score strongly depended on the head impact speed, see Fig. 24.19. It is evident from this data that the overall effective stiffness of the windscreen is lower than for the bonnet (hood).

Accident analysis and tests using cadavers and dummies and multibody models showed that increasing bonnet leading-edge height for cars reduces the subsequent head impact speed on the bonnet or windscreen [8, 37, 61, 95, 116, 117, 122, 123]. The dependence of head impact speed (normalized by vehicle impact speed) on bonnet leading edge height for different vehicle types has been reported using multibody modeling and dummy tests [120, 123], see Table 24.5. The multibody simulation results show higher speed ratios than the dummy tests, but both modeling approaches indicate increased head impact velocities for cars compared to SUVs. However, considerable disparity between the multibody modeling and dummy test results are evident. A more recent comparison between the Madymo multibody pedestrian model and staged cadaver and dummy tests showed relative variations in predicted head impact speed to be 16 % of the vehicle impact speed on average [124], but it is clear that head impact speed is sensitive to initial stance and body compliance [125].

24.1.21 Bonnet Leading Edge Stiffness

Multibody simulations have reported no clear relationship between stiffness of the bonnet leading edge and HIC score [126–128] but have shown that the thigh impact force is almost linearly dependent on the bonnet leading edge stiffness [123]. This is particularly relevant for impact with children, where a direct head impact with the bonnet leading edge can occur [113], and HIC scores of over 3,000 have been found for headform contact with the bonnet latch [121]. Headlamps and their housing can be very stiff and should be designed as energy absorbing components [4, 106, 111]. In general, the bonnet edges near the wings are stiffer than the central region and accident data showed the risk of AIS2+ femur or pelvis injury was higher at the leading edge of the wing than at the leading edge of the bonnet [114].

A bonnet leading edge airbag for pedestrian protection in SUV impacts has been developed [129], see Fig. 24.20. This design passed the EuroNCAP upper legform impactor test and in full scale tests the airbag decreased the risk of chest pelvis injuries considerably, especially in the chest and abdomen area. Furthermore, simulations of the device showed that it was largely insensitive to the underlying bonnet leading edge stiffness.

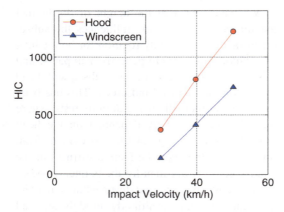

Fig. 24.19 Relationship between HIC score and head impact velocity from dummy tests for impact with the bonnet (hood) and windscreen [121]

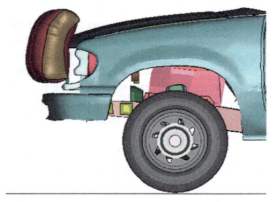

Fig. 24.20 SUV bonnet leading edge airbag (Adapted from [129])

Table 24.5 Head impact speed normalized by vehicle impact speed for 40 km/h dummy tests with different vehicle type impact for dummy tests and multibody simulations [120, 123]

	Compact car	Midsized car	Large car	Minivan	SUV/light truck
Dummy tests [120]	0.80	0.88	–	–	0.72
Multibody simulations [123]	1.18	–	0.92	0.97	0.9

24.1.22 Head Impact on the Bonnet/Windscreen

The bonnet leading edge height and the bonnet length relative to pedestrian height strongly influence where on the vehicle the head contact occurs. For adult pedestrians, high fronted vehicles or passenger cars with long bonnets usually result in a head contact on the bonnet top [43]. In contrast, for adults struck by passenger cars with short bonnets and/or lower bonnet leading edges, head contact usually occurs in the windscreen region [15, 89]. For some combinations of pedestrian height and vehicle shape, head contact with the cowl/windscreen or with the roof can occur. For children, head contact is generally with the bonnet top or the bonnet leading edge.

Head injury likelihood is strongly influenced by the stiffness of the vehicle surface, and the available deformation space before almost rigid components are contacted. The latter is particularly relevant for head impact on the bonnet top, which is a relatively compliant structure, but bottoming out can lead to head contact with stiff engine components [4, 130]. Plastic bonnets are thicker and therefore effectively stiffer than sheet metal bonnets and have a higher risk of head injury [131]. Aluminium bonnets are less stiff than steel, but they can bottom out and this results in a similar HIC score [132]. Collapsible bonnet arresters have been developed [106] and bonnet design has evolved to create a more uniform stiffness profile [111, 133, 134], see Fig. 24.21.

Headform impacts on the car body and windscreen [121] have shown a clear dependency of HIC score on dynamic deformation, see Fig. 24.22. The variation in HIC score for various head impact locations on the bonnet/windscreen of Japanese cars determined using the EEVC headform is shown in Fig. 24.23 [121].

The detailed design of the vehicle strongly influences the HIC score, with softer parts immediately adjoining stiffer elements. The bonnet/fender seam, bonnet hinge and stopper, cowl, corner of the windscreen frame and bottom of the A-pillar are very stiff and lead to high HIC scores [132, 135]. A full cover bonnet has been proposed to reduce head injury risks at the edges [136] and for the cowl area a bonnet overhang has recently shown good protective capability, see Fig. 24.24 [137]. For the fender area the Honda CR-V has deformable flanges under the fender-hood junction and crush space in adjacent areas of the bonnet reinforcement, see Fig. 24.25 [137]. The stiffness of the A-pillar of newer vehicles for occupant compartment protection can result in

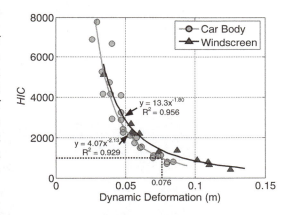

Fig. 24.22 Dependency of HIC score on dynamic deformation from 40 km/h headform tests (Adapted from [121])

Fig. 24.21 Design of bonnets: (**a**) traditional, (**b**) multicone design (Adapted from [133]) and (**c**) hybrid wireframe design [134]

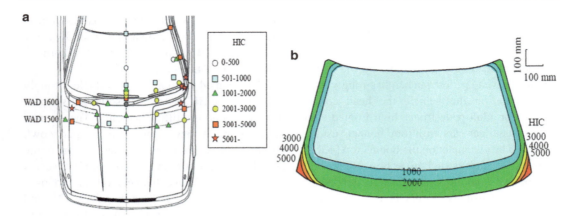

Fig. 24.23 HIC scores using EEVC adult headform for Japanese passenger cars at different levels of WAD (Adapted from [121])

Fig. 24.24 Exposed cowl and covered cowl (Adapted from [137])

Fig. 24.25 Honda CR-V fender countermeasures (Adapted from [137])

very high HIC scores unless specific pedestrian protection features are implemented.

Although the stiffness of the windscreen is lower than surrounding structures, due to the number of head contacts which occur with the windscreen [14], see Table 24.2, its design is important for pedestrian injuries. Analysis of accident data in the 1990s indicated that more horizontal windscreens reduce injuries [138], but recent multibody modeling [139] has reported conflicting evidence with lower head angular and linear accelerations for more vertical windscreens.

Recent headform testing with production windscreens and finite element modeling has shown a significant influence of the windscreen material on head acceleration [140, 141].

24.1.23 Primary Countermeasures: Warning and Auto-braking

Primary safety systems have been introduced to either aid the driver in reducing speed or to automatically reduce the speed of the impacting car in a pedestrian crash, see Fig. 24.26. Infra-red systems detecting living creatures such as animals or pedestrians and displaying the image on a screen to the driver were introduced in the early 2000s (e.g. Cadillac, Lexus) and were later followed by systems which additionally warned the driver (e.g. BMW, Audi, Honda, Mercedes, Toyota). The "brake assist" system in the brake pedal senses the braking intention of the driver and automatically optimizes braking performance. The brake assist systems were mandated in new vehicles in Europe in 2008. Since these depend on driver action they were estimated to be activated only in 50 % of accidents [142]. It is then natural to develop this system into an automatic system without driver intervention. A system was introduced in 2009 that detected pedestrians and gently applied the brakes if no driver action was noticed after a warning [143]. Recently, an auto-brake system was introduced that detects pedestrians and other vehicles/objects and automatically applies full braking before an imminent impact. This system has been claimed to be able to brake to a full stop from 25 km/h and thereby completely avoid low-speed pedestrian crashes [144]. At higher speeds, crash energy can be substantially reduced. This system is using radar and mono-vision camera. Other sensors under development for activating automatic braking include stereo-vision and infra-red cameras. A typical sensor field of view is around 40°. Systems with gentle braking will typically apply a brake force of 0.4–0.5 g while systems with full braking can apply brake forces up to 0.8–1.0 g. Rosén et al. [145] estimated that an ideal auto-brake system, activated for all visible pedestrians within a forward-looking angle of 40° one second prior to impact, would reduce fatalities (when struck by car fronts) by 40 % and seriously injured by 27 %.

24.1.24 Secondary Safety Countermeasures: Deployable Hoods and Airbags

Deformation space is crucial for head protection and this is traditionally achieved through under bonnet clearance. However, modern cars are densely packed under the bonnet and several designs for a lifting bonnet in the event of a pedestrian impact have been proposed [106, 146–148]. These are also called active hoods, pop-up hoods or deployable hoods, and are now in production in a large number of vehicles, where Jaguar and Citroën were the first in 2005. They lift the rear hood part between 50 and 120 mm to enable energy absorption of the head

Fig. 24.26 Primary pedestrian safety systems; (driver display of) pedestrian warning system (*left*, with permission from Autoliv Inc.), auto-brake system detecting pedestrians at danger (*right*, with permission from Volvo Car Corp.)

Fig. 24.27 Deployable hood (*left*, with permission from Autoliv Inc.) and windshield airbag (*right*, with permission from Volvo Car Corp.)

impact preventing a second "bottoming out" impact to structures underneath the hood in the engine compartment. Staged tests with dummies have indicated their ability to maintain HIC scores in head bonnet impacts below 1,000 for a 40 km/h impact [146, 149]. Fredriksson et al. [150] showed, in a combined experimental and finite element study, that an under-hood distance of 100 mm reduced both skull fracture-related and brain-related injury criteria to acceptable levels in 40 km/h headform impacts. The same study with dummy tests using Polar II and a real vehicle showed a large reduction in head loading by a deployable hood system compared to a standard hood.

For deployable hoods to be activated in accidents they are connected to a sensor and an actuator, which must make the decision and perform the lifting motion within a short time period, see Fig. 24.27 left. For the lower part of the windshield and the A-pillars, airbags have been proposed to enhance head protection [4, 147, 151, 152]. It has been shown that this is the area that produces most severe and fatal injuries in the windshield area [153]. This design is also beneficial since it reduces the problem that the driver sight is blocked if triggered falsely. The first airbag of this kind was put into production in 2012 by Volvo in the V40 model called Pedestrian Airbag Technology (PAT), similar to Fig. 24.27 right. Dummy tests and simulations have shown that the deployed hood for a standard sedan-type passenger car needs to be in position within less than 60 ms after the first leg impact to the front of a typical bonnet-type car at a crash speed of 40 km/h. A windshield airbag typically needs 90–100 ms for the same type of car, while a car design with a more vertical hood and front (e.g. multi-purpose vehicles MPV or vans) requires positioning significantly earlier of both the deployable hood and airbag. A pedestrian airbag needs more energy absorption distance than a deployable hood, due to the limited energy absorption capability in the first phase of the airbag impact. A typical airbag thickness is 200 mm and the volume can vary between 80 and 140 l depending on the car size, roughly similar to a passenger side occupant airbag.

While mitigation of leg injuries is quite well known and implemented, only limited solutions to protect the upper body have been implemented thus far. No design solutions have been developed to mitigate chest injury which is a real-life concern with many fatal injuries.

24.1.25 Integrated Systems of Combined Primary and Secondary Safety

The pre-crash, or primary, safety measures and the in-crash, or secondary, safety measures can be combined into integrated systems. Integrated pedestrian systems have not been introduced as standard equipment in production cars. Fredriksson and Rosén [154] showed in a theoretical study that an ideal integrated system would be more effective than an ideal single primary system such as autonomous braking. They studied 54 representative, severely head injured (AIS3+) pedestrians in detail to estimate the potential of theoretical primary and secondary systems and the potential of combining them into an integrated system. The study concluded that if combining the systems into an integrated system it protected a significantly higher number, than the individual systems, from severe (AIS3+) head injury. Although the Fredriksson and Rosén study showed theoretically that primary and secondary systems complement each other to increase the protection potential, there is a need to further study the potential of integrated systems including information from real tests or simulations with the countermeasures.

24.1.26 Evaluation of Production Vehicles via Pedestrian Safety Regulations

A 1998 programme of 269 tests using the EEVC impactors applied to five passenger cars, an MPV and an SUV showed that most vehicles failed the EEVC Working Group 10 tests [155]. The upper legform and legform impactor performances were particularly poor, and it was concluded that a major redesign of vehicle s would be required to pass the tests. The best results were achieved with the head impactor tests, where about one quarter of head locations achieved compliance. By 2001, most cars tested according to the EEVC tests performed badly in the pedestrian tests [110]. The vehicles that did best were Japanese manufacturers, but even they did not pass by 2001.

In a subsequent evaluation of EuroNCAP pedestrian tests (which use the EEVC impactors), it was found that although over 54 % of the child headform impacts passed, only 25 % of adult headforms tests were compliant. For the legform tests, only 14 % of cases passed and only 1 % of the upper legform tests were compliant. More recently, Mallory et al. [137] evaluated 11 vehicles selected to represent the US car fleet with a focus on large passenger vehicles. They performed head impact tests according to the draft Global Technical Regulation (GTR) and found that the peripheral areas of the head impact test zone produced the most severe impact, especially the hinges, cowl, wiper spindles and the wings.

The results of the GTR testing of US vehicles do not show a clear connection between vehicle size and head injury risk [137]. HIC scores measured centrally were lower than those measured at the sides, rear and front of the test area. The bonnet hinge location had the highest average HIC value, in particular where there was an exposed hinge with no bonnet covering to dissipate energy. The area adjacent to the cowl, including the wiper spindles, is challenging for pedestrian safety design. The worst performers did not have bonnet overhang over the cowl and one vehicle tested allowed direct contact of the headform with the cowl. The area of the test zone adjacent to the fender was also problematic for many vehicles: none of the four fender area impacts had HIC scores below 1,000. The best performer in the fender area impact was the Honda CR-V, which has deformable flanges under the fender-hood junction, as well built-in crush space in adjacent areas of the bonnet reinforcement.

A recent comparison of EURONCAP scores with pedestrian accident outcomes in Sweden showed a significant reduction of injury severity for cars with better pedestrian scoring, although cars with a high score could not be studied due to lack of cases. However, for cars with Brake Assist, no significant injury reductions were found [156], possibly due to fairly small sample size. A study of nearly 700 pedestrian accidents from the German In-Depth Accident Study found a correlation between the EuroNCAP score and moderate pedestrian injuries, but reported significant injury ranges within a single EuroNCAP score and even vehicles with good ratings were responsible for significant injuries [157].

24.2 Concluding Remarks

This chapter has shown that the mechanisms of pedestrian injuries from vehicle impact are now largely understood. Fundamental considerations, accident data and a variety of modeling approaches have shown that the combination of vehicle shape and stiffness have a significant influence on pedestrian injury outcome, and it has been shown how detailed front-end design and active safety measures can be used to reduce injury risks for pedestrians.

Although the basic principles governing the effects of vehicle shape and stiffness on injury outcome are relatively straightforward, there are many confounding factors which complicate the implementation of vehicle design for pedestrian safety, and isolating the individual influences of shape and stiffness is a difficult task. Nonetheless, vehicle designers now have the scientific and technical knowledge required to alter most vehicle designs and implement active safety measures to radically reduce the likelihood and severity of pedestrian injury in the event of a vehicle collision. This can be achieved through a combination of vehicle shape configuration, structural stiffness reductions in the probable impact zones and the use of safety devices such as autonomous braking, lifting bonnets and airbags in regions where sufficient structural compliance cannot be achieved. However, it is clear from the variety of vehicle shapes and stiffness properties present in current production vehicles that there is a large variability in the aggressivity of the current vehicle fleet for pedestrians, and an optimum design has not yet been achieved.

References

1. Mackay G (1992) Mechanisms of injury and biomechanics: vehicle design and crash performance. World J Surg 16:420–427
2. Marchal P, Gavrila D, Letellier L, Meinecke M, Morris R, Mathias M (2003) SAVE-U: an innovative sensor platform for vulnerable road user protection. In: Proceedings of World Congress on intelligent transport systems and services, Madrid
3. Mlekusch B, Wilfling C, Groger U, Dukart A, Mark F (2004) Active pedestrian protection system development. In: Vehicle aggressivity and compatibility, structural crashworthiness and pedestrian safety, Detroit. SAE paper no 2004-01-1604
4. Crandall J, Bhalla K, Madeley N (2002) Designing road vehicles for pedestrian protection. Br Med J 324:1145–1148
5. Wakefield H (1961) Systematic automobile design for pedestrian injury prevention. In: Stapp car crash conference, Twin Cities, Minnesota, pp 193–218
6. Ashton S, Mackay G (1979) Car design for pedestrian injury minimisation. In: Experimental safety vehicles conference, Paris, pp 630–640
7. Fisher A, Hall R (1972) The influence of car frontal design on pedestrian accident trauma. Accid Anal Prev 4:47–58
8. Pritz H, Hassler C, Herridge J, Weis EJ (1975) Experimental study of pedestrian injury minimisation through vehicle design. Society of Automotive Engineers. SAE technical paper 751166, Warrendale, Pennsylvania
9. Ashton S (1981) Factors associated with pelvic and knee injuries in pedestrians struck by the front of cars. Society of Automotive Engineers, Illinois. SAE paper no 811026
10. WHO (2009) Global status report on road safety. World Health Organisation (WHO), Geneva
11. Simms C, Wood D (2009) Pedestrian and cyclist impact – a biomechanical perspective. Springer, Dordrecht
12. AAAM (2008) Abbreviated injury scale 2008. American Association for Automotive Medicine, Des Plaines, San Diego, California
13. Mizuno Y (2003) Summary of IHRA pedestrian safety working group activities – proposed test methods to evaluate pedestrian protection offered by passenger cars. In: Experimental safety vehicles conference, Nagoya. ESV paper no 580
14. Mizuno Y (2005) Summary of IHRA pedestrian safety working group activities – proposed test methods to evaluate pedestrian protection offered by passenger cars. In: Administration UDoTNHTS (ed) Experimental safety vehicles conference, Washington, DC. ESV paper no 05-0138-O
15. Danner M, Langwieder K, Wachter W (1979) Injuries to pedestrians in real accidents and their relation to collision and car characteristics. Society of Automotive Engineers. SAE paper no 791008
16. Neal-Sturgess C, Coley G, De Olivera P (2002) Pedestrian injuries: effects in impact speed and contact stiffness. In: Vehicle safety. IMechE, London, pp 311–322
17. Walz F, Hoefliger M, Fehlmann W (1983) Speed limit reduction from 60 to 50 km per hour and pedestrian injuries. In: Stapp car crash conference. SAE paper no 831625, Warendale, Pennsylvania

18. Fildes B, Gabler HC, Otte D, Linder A, Sparke L (2004) Pedestrian impact priorities using real-world crash data and harm. In: IRCOBI conference, Graz, Austria, pp 167–177
19. Wood D (1991) Pedestrian impact, injury, and injury causation. In: Peters B, Peters G (eds) Automotive engineering and litigation, vol 4. Wiley Law Publications, New York, pp 41–72
20. Rosen E, Sander U (2009) Pedestrian fatality risk as a function of car impact speed. Accid Anal Prev 41(3):536–542. doi:10.1016/j.aap.2009.02.002
21. Foret-Bruno J, Faverjon G, Le Coz J (1998) Injury pattern of pedestrians hit by cars of recent design. In: Experimental safety vehicles conference. ESV paper no 980-S10-O-02, Windsor, Ontario
22. Galloway D, Patel A (1982) The pedestrian problem: a 12 month review of pedestrian accidents. Injury 13(4):294–298
23. Atkins R, Turner W, Duthie R, Wilde B (1988) Injuries to pedestrians in road traffic accidents. Br Med J 297(6661):1431–1434
24. Mizuno K, Kajzer J (1999) Compatibility problems in frontal, side, single car collisions and car to pedestrian accidents in Japan. Accid Anal Prev 31(4):381–391
25. Lefler DE, Gabler HC (2004) The fatality and injury risk of light truck impacts with pedestrians in the United States. Accid Anal Prev 36(2):295–304
26. Desapriya E, Subzwari S, Sasges D, Basic A, Alidina A, Turcotte K, Pike I (2010) Do light truck vehicles (LTV) impose greater risk of pedestrian injury than passenger cars? a meta-analysis and systematic review. Traffic Inj Prev 11(1):48–56
27. Longhitano D, Ivarsson J, Henary B, Crandall J (2005) Torso injury trends for pedestrians struck by cars and LTVs. In: Experimental safety vehicles conference. ESV paper no 05-0411, Washington, DC
28. Longhitano D, Henary B, Bhalla K, Ivarsson J, Crandall J (2005) Influence of vehicle body type on pedestrian injury distribution. Society of Automotive Engineers. SAE paper no 2005-01-1876
29. Malini E, Victor D (1990) Measures to improve pedestrian safety: lessons from experience in Madras. J Traffic Med 18(4):266
30. Mallory A, Fredriksson R, Rosen E, Donnelly B (2012) Pedestrian injuries by source: serious and disabling injuries in us and European cases. Ann Adv Automot Med 56:13–24
31. Neal-Sturgess C, Carter E, Hardy R, Cuerden R, Guerra L, Yang J (2007) APROSYS European in-depth pedestrian database. In: Experimental safety vehicles conference. ESV paper no 07-0177-O, Lyon, France
32. Yang J (2005) Review of injury biomechanics in car-pedestrian collisions. Int J Vehic Saf 1(1/2/3):100–117
33. Jarrett K, Saul R (1998) Pedestrian injury – analysis of the PCDS field collision data. In: Experimental safety vehicles conference, Windsor, Ontario
34. Yao J, Yang J, Otte D (2007) Head injuries in child pedestrian accidents – in-depth case analysis and reconstructions. Traffic Inj Prev 8:94–100
35. Kerrigan J, Murphy DB, Drinkwater D, Kam C, Bose D, Crandall J (2005) Kinematic corridors for PMHS tested in full-scale pedestrian impact tests. In: Experimental safety vehicles conference. ESV paper no 05-0394, Washington, DC
36. Ashton S, Pedder J, Mackay G (1978) Pedestrian head injuries. In: AAAM, Ann Arbor, pp 237–244
37. Otte D (1994) Influence of the fronthood length for the safety of pedestrians in car accidents and demands to the safety of small vehicles. In: Society of Automotive Engineers. SAE paper no 942232, Detroit, Michigan
38. Wood D (1995) Determination of speed from throw. In: Bohan TL, Damask AC (eds) Forensic accident investigation. Lexis Law Publishing, Charlottesville
39. CrashTestServices (2007) www.crashtest-service.com. Accessed 2007
40. EVU-CTS (2005) crashconferences.com. http://www.crashconferences.com/arccsi/2005Conference.html. Accessed 3 June 2009
41. Wood D (1988) Impact and movement of pedestrians in frontal collisions with vehicles. Proc Inst Mech Eng D Automob Eng 202:101–110
42. Simms C, Ormond T, Wood D (2011) The influence of vehicle shape on pedestrian ground contact mechanisms. In: IRCOBI (ed) Proceedings of IRCOBI conference, Poland
43. Roudsari B, Mock C, Kaufmann R (2005) An evaluation of the association between vehicle type and the source and severity of pedestrian injuries. Traffic Inj Prev 6:185–192
44. UNECE (2009) Global technical regulation no. 9 pedestrian safety. UNECE
45. Otte D (1997) Pedestrian impact at front end of car. Accident Research Unit, Medical University of Hannover
46. Otte D (1997) Injuries to pedestrians caused by impacts with the front edge of car bonnets. EEVC WG17, Brussels, UNECE
47. FMVSS (1991) FMVSS 208: occupant crash protection. NHTSA
48. Mizuno K, Ishikawa H (2001) Summary of IHRA pedestrian safety working group activities – proposed test methods to evaluate pedestrian protection offered by passenger cars. Society of Automotive Engineers. SAE paper no 2001-06-0136
49. Lawrence G, Hardy B, Caroll J, Donaldson W, Visvikis C, Peel D (2006) A study on the feasibility of measures relating to the protection of pedestrians and other vulnerable Road users – final report. Transport Research Laboratory
50. TRL (2006) Factors influencing pedestrian safety: a literature review. Transport Research Laboratory
51. Lawrence G, Thornton S (1996) The development and evaluation of the TRL legform impactor. Transport Research Laboratory

52. Konosu A, Tanahashi M (2003) Development of a biofidelity pedestrian legform impactor – introduction of JAMA JARI legform impactor version 2002. In: Experimental safety vehicles conference. ESV paper no 378, Gothenburg, Sweden
53. Wittek A, Konosu A, Matsui Y, Ishikawa H, Shams T, McDonald J (2001) A new legform impactor for evaluation of car aggressiveness in car – pedestrian accidents. In: Experimental safety vehicles conference. ESV paper no 184, Amsterdam, The Netherlands
54. Notsu M, Nishimoto T, Konosu A, Ishikawa H (2005) J-MLIT research into a pedestrian lower extremity protection – evaluation tests for pedestrian legform impactors. In: Experimental safety vehicles conference. Paper number 05-0193, Washington, DC
55. EEVC (1994) Proposals to evaluate pedestrian protection for passenger cars – EEVC working group 10 report. European Experimental Vehicles Committee
56. Matsui Y, Wittek A, Konosu A (2002) Comparison of pedestrian subsystem safety tests using impactors and full-scale dummy tests. Society of Automotive Engineers. SAE paper no 2002-01-1021
57. Matsui Y (2004) Evaluation of pedestrian subsystem test method using legform and upper legform impactors for assessment of high bumper of vehicle aggressiveness. Traffic Inj Prev 5:76–86
58. EEVC (2002) EEVC Working Group 17 report: improved test methods to evaluate pedestrian protection afforded by passenger cars (December 1998 with September 2002 updates). EEVC
59. Chalandon S, Serre T, Masson C, Arnoux P, Perrin C, Borde P, Cotte C, Brunet C, Cesari D (2007) Comparative study between subsystem and global approaches for the Pedestrian impact. In: Experimental safety vehicles conference. ESV paper no 07-0429, Seoul, Korea
60. Lawrence J, Hardy B, Harris J (1993) Bonnet leading edge subsystems test for cars to assess protection for pedestrians. In: Proceedings of 13th international technical conference on experimental safety vehicles, Eindhoven, The Netherlands, pp 402–413
61. Lawrence G (1989) The influence of car shape on pedestrian impact energies and its application to subsystem tests. In: Experimental safety vehicles conference, Gothenburg, Sweden, pp 1253–1265
62. Janssen E, Nieboer J (1992) Subsystem tests for assessing pedestrian protection based on computer simulations. In: IRCOBI conference, Verona, Italy, pp 263–280
63. Anderson R, McClean A, Streeter L, Ponte G, Sommariva M, Londsay T, Wundersitz L (2002) Severity and type of pedestrian injuries related to vehicle impact locations and results of subsystem impact reconstruction. In: IRCOBI conference, Munich, pp 289–302
64. UNECE (2006) Proposal for a global technical regulation on uniform provisions concerning the approval of vehicles with regard to their construction in order to improve the protection and mitigate the severity of injuries to pedestrians and other vulnerable road users in the event of a collision, Brussels, UNECE
65. Hardy B, Lawrence G, Carroll J, Donaldson W, Visvikis C, Peel D (2006) A study on the feasibility of measures relating to the protection of pedestrians and other vulnerable road users. Transport Research Laboratory, Berks
66. McClean A (2005) Vehicle design for pedestrian protection. The University of Adelaide, Adelaide
67. Glaeser K (1991) Development of head impact procedure for pedestrian protection. In: Experimental safety vehicles conference, Paris, France
68. Kerrigan J, Arregui C, Crandall J (2009) Pedestrian head impact dynamics: comparison of dummy and PMHS in small sedan and large SUV impacts. In: Experimental safety vehicles conference
69. Kerrigan J, Crandall J (2008) A comparative analysis of the pedestrian injury risk predicted by mechanical impactors and post mortem human surrogates. Stapp Car Crash J 52:527–567
70. Masson C, Serre T, Cesari D (2007) Pedestrian-vehicle accident: analysis of 4 full scale tests with PMHS. In: Experimental safety vehicles conference. ESV paper no 07-0428, Lyon, France
71. EURO-NCAP (2012) European New Car Assessment programme, Brussels, EuroNCAP
72. Gadd C (1966) Use of a weighted impulse criterion for estimating injury hazard. In: SAE conference. SAE paper no 660793
73. Versace J (1971) A review of the severity index. In: Stapp car crash conference, Colorado, pp 771–796
74. Margulies S, Thibault L (1990) A proposed tolerance criterion for diffuse axonal injury in man. J Biomech 25(8):917–923
75. MacLaughlin T, Kessler J (1990) Pedestrian head impact against the central hood of motor vehicles – test procedure and results. In: 34th Stapp car crash conference, Orlando. SAE paper 902315
76. UNECE (2008) Global technical regulation no. 9 pedestrian safety. UNECE
77. EC (2003) Directive 2003/102/EC of the European Parliament and of the Council of the 17th November 2003. Protection of pedestrians and other vulnerable Road users before and in the event of a collision with a motor vehicle and amending directive 70/156/EEC, vol OJ L 321, Brussels, EU
78. Snedeker J, Walz F, Muser M, Lanz C (2005) Assessing femur and pelvis injury risk in current pedestrian collisions: comparison of full-bodied PMTO impacts, and a human body finite element model. In: Experimental safety vehicles conference. ESV paper no 05-103, Washington, DC
79. Pheasant S, Haslegrave C (2006) Bodyspace – anthropometry, ergonomics and the design of work. Taylor and Francis, London
80. Ballesteros M, Dischinger P, Langenberg P (2004) Pedestrian injuries and vehicle type in Maryland, 1995–1999. Accid Anal Prev 36(1):73–81
81. Simms C, Wood D (2006) Pedestrian risk from cars and sport utility vehicles – a comparative analytical study. IMechE J Automob Eng 220:1085–1100

82. Martinez L, Guerra L, Ferichola G, Garcia A, Yang J (2007) Stiffness corridors of the European fleet for pedestrian simulation. In: Experimental safety vehicles conference. ESV paper no 07-0267, Lyon, France
83. Kalliske I, Friesen F (2001) Improvements to pedestrian protection as exemplified on a standard sized car. In: Experimental safety vehicles conference. ESV paper no 283, Amsterdam, The Netherlands
84. Konosu A, Takahiro I, Tanahashi M (2005) Development of the pedestrian lower extremity protection car using a biofidelic flexible pedestrian legform impactor. In: Experimental safety vehicles conference. ESV paper no 05-0106
85. Bacon D, Wilson M (1976) Bumper characteristics for improved pedestrian safety. Society of Automotive Engineers. SAE paper no 760812, Warrendale, Pennsylvania
86. Stuertz G, Suren E, Gotzen L, Behrens S, Richter K (1976) Biomechanics of real child pedestrian accidents. Society of Automotive Engineers. SAE paper no 760814, Warrendale, Pennsylvania
87. Bunketorp O, Romanus B, Hansson T, Aldman B, Thorngren L, Eppinger R (1983) Experimental study of a compliant bumper system. Society of Automotive Engineers. SAE paper no 831623, Warrendale, Pennsylvania
88. Aldman B, Thorngren L, Bunketorp O, Romanus B (1980) An experimental model for the study of the lower leg in car pedestrian impacts. In: IRCOBI conference, Birmingham, UK, pp 180–193
89. Stcherbarcheff G, Tarriere C, Duclos P, Fayon A, Got C, Patel A (1975) Simulation of collisions between pedestrians and vehicles using adult and child dummies. Society of Automotive Engineers. SAE paper no 751167, Detroit, Michigan. Warrendale, Pennsylvania
90. Harris J (1976) Research and development towards improved protection for pedestrian struck by cars. In: Experimental safety vehicles conference, Arlington, Virginia
91. Ishikawa H, Kajzer J, Ono K, Sakurai M (1992) Simulation of car impact to pedestrian lower extremity: influence of different car front shapes and dummy parameters on test results. In: IRCOBI conference, Verona, Italy, pp 1–12
92. Aldman B, Lundell B, Thorngren L, Bunketorp O, Romanus B (1979) Physical simulation of human leg bumper impacts. In: IRCOBI conference, pp 180–193
93. Cesari D (1988) Interaction between human leg and car bumper in pedestrian tests. In: IRCOBI conference, Germany, pp 259–269
94. Yang J, Lovsund P, Cavallero C, Bonnoit J (2000) A human body 3-D mathematical model for simulation of car-pedestrian impacts. J Crash Prev Inj Control 2(2):131–149
95. Ishikawa H, Yamazaki K (1991) Sasaki A Current situation of pedestrian accidents and research into pedestrian protection in Japan. In: Experimental safety vehicles conference, Paris, France, pp 281–291
96. Matsui Y (2005) Effects of vehicle bumper height and impact velocity on type of lower extremity injury in vehicle pedestrian accidents. Accid Anal Prev 37:633–640
97. Kajzer J, Schroeder G, Ishikawa H, Matsui Y, Bosch U (1997) Shearing and bending effects at the knee joint at low speed lateral loading. In: SAE transactions. SAE paper no 1997-01-0712
98. Kajzer J, Ishikawa H, Schroeder G, Matsui Y, Bosch U (1999) Shear and bending effects at the knee joint at high speed lateral loading. Society of Automotive Engineers. SAE paper no 973326, Warrendale, Pennsylvania
99. Matsui Y, Schroeder G, Bosch U (2004) Injury pattern and response of human thigh under lateral loading simulating car-pedestrian impact. Society of Automotive Engineers. SAE paper no 2004-01-1603, Warrendale, Pennsylvania
100. Edwards KJ, Green JF (1999) Analysis of the inter-relationship of pedestrian leg and pelvis injuries. In: IRCOBI conference, Bron, pp 355–369
101. Kajzer J, Schroeder G (1992) Examination of different bumper system using hybrid II, RSPD subsystem and cadavers. Society of Automotive Engineers. SAE paper no 922519, Warrendale, Pennsylvania, pp 119–127
102. Schuster P (2006) Current trends in bumper design for pedestrian impact. Society of Automotive Engineers. SAE paper no 2006-01-0464
103. Schuster P, Staines B (1998) Determination of bumper styling and engineering parameters to reduce pedestrian leg injuries. Society of Automotive Engineers, Warrendale, Pennsylvania
104. Groesch L, Heiss W (1989) Bumper configurations for conflicting requirements: existing performance versus pedestrian protection. In: Experimental safety vehicles conference, Gothenburg, Sweden, pp 1266–1273
105. Otte D, Haasper C (2005) Technical parameters and mechanisms for the injury risk of the knee joint or vulnerable road users impacted by cars and road traffic accidents. In: IRCOBI conference, Prague, pp 281–298
106. Pinecki C, Zeitouni R (2007) Technical solutions for enhancing the pedestrian protection. In: Experimental safety vehicles conference. ESV paper no 07-0307, Lyon, France
107. Pipkorn B, Fredriksson R, Olsson J (2007) Bumper bag for SUV to passenger vehicle compatibility and pedestrian protection. In: Experimental safety vehicles conference. ESV paper no 07-0056, Lyon, France
108. Kajzer J (1991) The biomechanics of knee injuries. Chalmers Technical University, Gothenburg
109. Bosma F, Gaalman H, Souren W (2001) Closure and trim design for pedestrian impact. In: Experimental safety vehicles. ESV paper no 322, Warrendale, Pennsylvania
110. ETSC (2001) Priorities for EU motor vehicle safety design – pedestrian safety. European traffic safety Council, Brussels
111. Wanke T, Thompson G, Kerkeling C 2005) Pedestrian measures for the Opel Zafira II. In:

Experimental safety vehicles conference. ESV paper no 05-0237, Washington, DC
112. Matsui Y, Hitosugi M, Mizuno K (2011) Severity of vehicle bumper location in vehicle-to-pedestrian impact accidents. Forensic Sci Int 212(3):205–209. doi:10.1016/j.forsciint.2011.06.012
113. Ashton S (1979) Some factors influencing the injuries sustained by child pedestrians struck by the fronts of cars. Society Automotive Engineers, Warrendale, Pennsylvania, pp 353–380
114. Matsui Y, Ishikawa H, Sasaki A (1999) Pedestrian injuries induced by the bonnet leading edge in current car – pedestrian accidents. Society of Automotive Engineers. SAE paper 1999-01-0713, Warrendale, Pennsylvania
115. Kramer M (1975) Pedestrian vehicle accident simulation through dummy tests. Society of Automotive Engineers. SAE paper no 751165, Warrendale, Pennsylvania
116. Pritz H, Pereira J (1983) Pedestrian hip impact simulator development and hood edge location consideration on injury severity. Society of Automotive Engineers. SAE paper no 831627, Warrendale, Pennsylvania
117. Niederer P, Schlumpf MR (1984) Influence of vehicle front geometry on impacted pedestrian kinematics. Society of Automotive Engineers. SAE paper no 841663, Warrendale, Pennsylvania
118. Lucchini E, Weissner R (1978) Influence of bumper adjustment on the kinematics of an impacted pedestrian. In: IRCOBI conference, Lyon, France, pp 172–182
119. Ashton S, Mackay G (1979) A review of real-world studies of pedestrian injury. Jahrestagung der Deutschen Gesellschaft fuer Verkehrsmedizin, Cologne
120. Matsui Y, Wittek A, Tanahashi M (2005) Pedestrian kinematics due to impact by various passenger cars using full-scale dummy. J Vehic Saf Res, pp 64–84
121. Mizuno K, Yonezawa H, Kajzer J (2001) Pedestrian headform impact tests for various vehicle locations. In: Experimental safety vehicles conference. ESV paper no 278, Amsterdam, The Netherlands
122. Mackay G (1972) Injury to pedestrians. Report to Committee on Pedestrian Safety, Committee on the Challenges of Modern Society, NATO, Brussels.
123. Liu X, Yang J, Lovsund P (2002) A study of influences of vehicle speed and front structure on pedestrian impact responses using mathematical models. Traffic Inj Prev 3:31–42
124. Elliot J, Lyons M, Kerrigan J, Wood D, Simms C (2012) Predictive capabilities of the MADYMO multibody pedestrian model: three-dimensional head translation and rotation, head impact time and head impact velocity. IMechE J Multibody Dyn 226(3):266–277
125. Simms CK, Wood DP (2006) Effects of pre-impact pedestrian position and motion on kinematics and injuries from vehicle and ground contact. Int J Crashworthiness, pp 345–356
126. Higuchi K, Akiyama A (1989) The effect of vehicle structure's characteristics on pedestrian behaviour. In: Experimental safety vehicles conference, Gothenburg, Sweden, pp 323–329
127. Liu X, Yang J (2002) Effects of vehicle impact velocity and front-end structure on the dynamic responses of child pedestrians. In: IRCOBI conference, Munich conference, pp 19–30
128. Liu X, Yang J (2003) Effects of vehicle impact velocity and front-end structure on dynamic responses of child pedestrians. Traffic Inj Prev 4:337–344
129. Fredriksson R, Flink E, Bostrom O, Backman K (2007) Injury mitigation in SUV-to-pedestrian impacts. In: Experimental safety vehicles conference. ESV paper no 07-0380
130. Schwarz D, Bachem H, Opbroek E (2004) Comparison of steel and aluminium hood with same design in view of pedestrian head impact. Society of Automotive Engineers. 2004-01-1605, Warrendale, Pennsylvania
131. Kessler J (1987) Development of countermeasures to reduce pedestrian head injury. In: Experimental safety vehicles conference, Washington, DC, pp 784–796
132. Pritz H (1983) Experimental investigation of pedestrian head impact on hoods and fenders of production vehicles. Society of Automotive Engineers. SAE paper no 830055, Warrendale, Pennsylvania
133. Kerkeling C, Schaefer J, Thompson G (2005) Structural hood and hinge concepts for pedestrian protection. In: Experimental safety vehicles conference. ESV paper no 05-0304
134. Belingardi G, Scattina A (2009) Development of an hybrid hood to improve pedestrian safety in case of vehicle impact. Paper presented at the experimental safety vehicles paper no 09-0026, Stuttgart
135. Mizuno K, Aiba T, Kajzer J (1999) Influences of vehicle from shape on injuries in vehicle – pedestrian impact. In: Japanese Society of Automotive Engineers, Warrendale, Pennsylvania, pp 55–60
136. Kessler J, Monk M (1991) NHTSA pedestrian head injury mitigation research program -status report. In: Experiment safety vehicles conference, Paris, France, pp 1226–1236
137. Mallory A, Stammen J, Meyerson S (2007) Pedestrian GTR testing of current vehicles. In: Experimental safety vehicles conference. ESV paper no 07-0313, Lyon, France
138. Otte D (1994) Design and structure of the windscreen as part of injury reduction for car occupants, pedestrians and bicycles. Society of Automotive Engineers. SAE paper no 942231, Warrendale, Pennsylvania
139. Lyons M, Simms C (2012) Predicting the influence of windscreen design on pedestrian head injuries. In: IRCOBI, Dublin
140. Pinecki C, Fontaine L, Adalian C, Jeanneau C, Zeitouni Z (2011) Pedestrian protection – physical and numerical analysis of the protection offered by

the windscreen. ESV paper number 11-0432, Washington, DC
141. Xu J, Li Y, Ge D, Liu B, Zhu M, Park T et al (2011) Automotive windshield – pedestrian head impact: energy absorption capability of interlayer material. Int J Automot Technol 12(5):687–695
142. Hannawald L, Kauer F (2004) Equal effectiveness study on pedestrian protection. Technische Universität Dresden, Dresden
143. Lexus (2011) Advanced pre-collision system (APCS) with driver attention monitor.http://www.lexus.com/models/LS/features/safety/advanced_precollision_system_apcs_with_driver_attention_monitor.html. Accessed 13 Jan 2011
144. VolvoCars (2010) A revolution in pedestrian safety – Volvo's automatic braking system now reacts to people as well as vehicles. www.volvocars.com/za/top/about/news-events/pages/default.aspx?itemid=24. Accessed 22 Aug 2010
145. Rosén E, Källhammer J-E, Eriksson D, Nentwich M, Fredriksson R, Smith K (2010) Pedestrian injury mitigation by autonomous braking. Accid Anal Prev 42(6):1949–1957
146. Fredriksson R, Haland Y, Yang J (2001) Evaluation of a new pedestrian head injury protection system for the sensor in the bumper and lifting of the bonnets rear part. In: Experimental safety vehicles conference. ESV paper no 131, Amsterdam, The Netherlands
147. Maki T, Asai T, Kajzer J (2003) Development of future pedestrian protection technologies. In: Experimental safety vehicles conference. ESV paper no 165, Nagoya, Japan
148. Krenn M, Mlekusch B, Wilfling C, Dobida F, Deutscher E (2003) Development and evaluation of a kinematic hood for pedestrian protection. Society of Automotive Engineers. SAE paper no 2003-01-0897, Warrendale, Pennsylvania
149. Nagatomi K, Hanayama K, Ishizaki T, Sasaki A, Matsuda K (2005) Development and full-scale dummy tests of a pop-up hood system for pedestrian protection. In: Experimental safety vehicles conference. ESV paper no 05-0113, Washington, DC
150. Fredriksson R, Zhang L, Boström O (2009) Influence of deployable hood systems on finite element modelled brain response for vulnerable road users. Int J Vehic Saf 4(1):29–44
151. Autoliv (2002) Annual report 2001. Stockholm
152. Autoliv (2010) Annual report 2009. Stockholm
153. Fredriksson R, Rosén E, Kullgren A (2010) Priorities of pedestrian protection – a real-life study of severe injuries and car sources. Accid Anal Prev 42(6):1672–1681
154. Fredriksson R, Rosén E (2012) Integrated pedestrian countermeasures – potential of head injury reduction combining passive and active countermeasures. Saf Sci 50(3):400–407. doi:10.1016/j.ssci.2011.09.019
155. Green J (1998) A technical evaluation of the EEVC proposal on pedestrian protection test methodology. In: Experimental safety vehicles conference. ESV paper no 98-S10-O-04, Windsor, Ontario
156. Strandroth J, Rizzi M, Sternlund S, Lie A, Tingvall C (2011) The correlation between pedestrian injury severity in real-life crashes and Euro NCAP pedestrian test results. Traffic Inj Prev 12(6):604–613
157. Liers H, Hannawald L (2011) Benefit estimation of secondary safety measures in realworld Pedestrian accidents. ESV paper no 11-0300, Washington, DC

Design and Testing of Sports Helmets: Biomechanical and Practical Considerations

James A. Newman

Abstract

The first few sections of this chapter chronicle the development of the modern sports helmet and the evolution of their performance standards in the context of research in the biomechanics of head trauma. How this research has been employed in performance testing protocols and how this may have influenced helmet design is reviewed. The design of the testing protocols, the biofidelity of the test devices, the injury assessment functions available and the failure limits employed in recognized standards are all issues that bear on the application of biomechanical fundamentals. The second half of this chapter discusses practical considerations in helmet design including the specification of performance requirements and material and design constraints. Recent advances in helmet designs and test methods are also briefly reviewed.

25.1 Biomechanical Considerations

The main way by which biomechanics has influenced helmet design is not so much in our understanding of different head injury mechanisms, but rather in a better appreciation of the biophysical characteristics of the head, the development of kinematic head injury assessment functions and the evolution of testing methods. Melvin and Lighthall [1] have provided an exceptional overview of our current understanding of brain injury mechanisms. This insight has provided better ways to assess the capabilities of a helmet without first placing it on a human being.

At the outset, it is clear that a broad understanding of what a helmet should do preceded any clear understanding of head injury mechanisms. The basic principals have always been to provide a hard outer shell to ward off external agents and padding to help cushion the blow. That the shell distributed the applied force and thereby reduced localization of loading, and thus the propensity for skull fracture, may not have occurred to early inventors. That padding served to absorb impact energy thereby reducing the

J.A. Newman, Ph.D. (✉)
Newman Biomechanical Engineering Consulting Inc., Edmonton, AB, Canada
e-mail: newman@biomechanical.engineering.com

inertial loading on the head and thence reducing acceleration-induced injuries was also not likely given much thought [2].

Beginning in the early 1940s, in what were the first of several attempts to try to relate external loading to brain injury, Gurdjian, Webster and Lissner at Wayne State University [3, 4], impacted living dog's heads. Perhaps not surprisingly now, they observed that the harder the dogs were struck, the more likely or the more serious a head injury would be. In later experiments, they applied jets of air directly to animals' exposed brain. They noted a correlation between the force of the air and the duration of exposure to the blast and the severity of concussive effects.

A few years later, Lissner and coworkers continued this line of enquiry with a different model [5]. They dropped cadavers onto their heads from different heights onto a flat steel plate until they could produce a skull fracture. In these experiments they monitored the intracranial pressure during impact as well as the cadaver head's acceleration. By examining acceleration and relating it to those tests when a fracture occurred, they devised a crude relationship between the probability of concussion, the average head acceleration in Gs and the time duration of the impact. That this should relate to brain injury, particularly concussion, was imbedded in their hypothesis that nearly everyone who sustains a skull fracture sustains a concussion. An adequate description of the head response, i.e., its movement, was felt to be contained entirely within the linear acceleration-time history during the impact.[1] By combining all this data, along with other estimates from accidental free falls, they developed what became and what is still referred to as the Wayne State Concussion Tolerance curve [6]. This function attempted to relate the level of tolerable linear acceleration to how long that acceleration lasted.

By 1953, the helmet industry was coming to grips with some of these biomechanical concepts as evidenced in the following text from an American football helmet patent (Turner and Harvey US2,634,415). Something of an oversimplification perhaps, but the inventors recognized that severe movement of the head can be injurious even if the skull does not fracture.

> A head jolt properly may be defined as a sudden and/or severe change in the direction in which the head is moving, or the velocity with which it is moving, or both. The avoidance of sharp and/or severe head jolts is of vital importance. The human brain 'floats' in the skull much as the yolk of an egg has floating suspension in its associated egg white. A sharp or severe jolt can rupture an egg yoke without fracturing the egg shell. Similarly, a sharp or severe jolt can cause fatal injury to a human brain without fracturing the skull which houses it, and often with only minor, if any evidence of injury at the outside of the head. There have been many such fatal injuries in the playing of football.

Previously Holbourn [7] had hypothesized that the predominant mechanism of brain injury was not linear motion at all but rather was due to rotation of the head. Holbourn maintained that you could disturb/distort/disrupt the contents of the skull much more readily by rotating the head than by accelerating it linearly.

Over 20 years later, researchers in the US, such as Gurdjian, Lissner, Hodgson et al. [8], continued to support the concept that translational (linear) acceleration was the most important mechanism. By the 70s, other US researchers, notably Ommaya and coworkers [9] who had subjected live monkeys to linear and angular impact motions, had concluded that *"no convincing evidence has to this date been presented which relates brain injury and concussion to translational motion of the head..."*.

Nowadays, it is generally acknowledged that deformation of the brain and associated injury might be understood only by knowing the full three dimensional history of the head's motion following impact. Distortion of brain tissue as a result of head impact is being studied with finite element mathematical models that can predict the extent and distribution of the deformation from different types of impact [10]. One such model is

---

[1] One of the many problems with this approach is that most people who sustain a concussion, or many of the other kinds of brain injury, do so without having their skulls fractured. Another problem that followed, and that has lingered for years, was the idea that brain injury was *caused* by acceleration.

the Simulated Injury Monitor (SIMon) finite element head model developed by NHTSA [11]. This model allows one to compute brain distortion pattern histories but more importantly, generates several criterion functions that have been (more or less) correlated to the severity of various brain injury types in humans.[2] SIMon has not been validated for every conceivable impact scenario and no doubt will be refined as more human head injury tolerance data becomes available.

Some investigators have gone so far as to put mathematical helmets on the mathematical heads to see what the mathematically predicted effects are [12]. To do so, in addition to modeling the skull, its contents and the helmet's geometric and physical properties, the complete characterization of the head motion in time (in all three dimensions) including all linear and rotational components is required.

As will be discussed further on, the evaluation of protective headgear requires the use of a surrogate or dummy head upon which a helmet may be worn and impacted. Since test headforms have no *brains*, it has been necessary to develop empirical relations between the way the head moves when struck and the likelihood of brain injury. Among the head injury assessment functions that have found their way into helmet standards are:

- The resultant linear acceleration of the center of gravity of the test headform. Usually expressed in gravitational units, it is used extensively in the evaluation of helmet impact performance.
- The Severity Index. Developed by Gadd [13], it is the function employed in the NOCSAE helmet standards.
- NHTSA's Head Injury Criterion HIC. First referenced in 1974, it is the current assessment function for the evaluation of closed head injury probability in automotive frontal crash testing and certification. It relies solely on a portion of the resultant linear acceleration history of the ATD head following impact. It appears in the European motorcycle helmet standard EC 22 and the FIA standard for auto racing headgear.

One area that continues to frustrate the evolution of better helmet performance standards, and thus perhaps better helmets, is the rotation issue. Biomechanically, there is little argument today that Holbourn [7] was right and it is rotational motion that dominates the nature and extent of brain injury not simply linear translational acceleration. Indeed, over 20 years ago, tolerance curves relating various types of brain injuries to rotational motion of the head had been proposed [14]. As far back as 1986, this author introduced a head injury assessment function that endeavored to take into account both translational and rotational acceleration [15]. This function, GAMBIT, a weighted sum of maximum linear and angular acceleration, was employed in an effort to better understand the nature of protective headgear for soccer [16]. Several years ago, the author with colleagues Shewchenko and Welbourne [17] introduced a more general head injury assessment function, the Head Impact Power HIP. This function, which considers the maximum rate of translational and rotational energy transfer to be the controlling element in inertially induced brain injury, was successfully used in the development of a new North American football helmet (The Riddell Revolution). HIP was also employed to help quantify head injury threats in soccer [18]. The most recent rotation based criterion function is the Kinematic Brain Injury Criterion BRIC [19]. This empirical function is a weighted sum of the maximum angular acceleration and maximum angular velocity and has been correlated quite well to cumulative brain strain.

To this day however, the failure criteria for every published helmet performance standard in the World continue to be based solely on linear acceleration of a test headform. How we will reconcile this conundrum remains to be seen.

---

[2] In particular;
- The cumulative strain damage measure CSDM for diffuse axonal injury DAI
- The relative motion damage measure RMDM for acute subdural hematoma ASDH and
- The dilatational damage measure DDM for contusion and focal lesions.

25.2 The Evolution of Helmet Performance Standards

Performance standards for sporting helmets have their origins in the first of all such standards – that of the British Standards Institute BS 1869-1952 Crash Helmets for Racing Motorcyclists. This was followed in 1954 by BS 2826-1954, Protective Helmets and Peaks for Racing Car Drivers. Both standards provided tests to determine the energy dissipation capability of a helmet when subjected to a known impact. Standards for non-racing motorcyclists soon followed.

Biomechanically speaking not much was known or utilized in these early standards. The importance of proper anatomical shape of the head was appreciated and a table documenting the polar coordinates of a test headform, for a medium sized head, was developed. These same dimensional characteristics are employed to this day though they have been extrapolated to a variety of sizes. Their origin remains something of a mystery but likely lies in the art of British haberdashery.[3]

From the point of view of test headform biofidelity, nothing was implemented. In fact the original test headforms were constructed of wood – wood with rather precise characteristics vis "*...a wooden headform shall be made up of horizontal laminations of birch having a density of 40-45 pounds per cubic foot at a moisture content of 12 percent. The wood shall be straight in grain, free from defects and free from dote*" (BS 1869, 1952).

In terms of a helmet test failure criterion, today there is an appreciation that it should somehow reflect human tolerance values. In the 50s, it was known that when certain helmets were tested according to the BSI protocol – which involved dropping a wooden block onto the helmeted headform, some helmets transmitted higher forces than did others. On the assumption that lower would be better, the failure level was set at 5,000 lb force.

In the US, the first foray into the sports helmet standard issue was pioneered by Snively. In 1959 he virtually single handedly published the first North American standard for racing crash helmets (Snell Memorial Foundation 1959). His approach was to measure test headform acceleration directly using a freely moving headform. The test headform that he employed was not made of wood but rather was described as being of K-1A magnesium alloy. Dimensional characteristics were not specified. Other than weighing 12 lb, there were no other biofidelity requirements and there was no consideration regarding different sizes of heads and helmets. Using the BSI standard as a guide, the failure criterion was set at 400G (equivalent to 4,800 lb force).

Later, substantiating what would now be considered a high failure criterion, Snively and Chichester [20] supplemented the data with a groundbreaking analysis of accident-involved helmets. By *replicating* the damage to 10 helmets worn by racing drivers who had survived a crash, they deduced that; "*Survival limits of localized head acceleration of brief duration in man have been shown to exceed 450G.*"[4]

By 1961, the American Standards Association ASA had established a helmet standard committee and by 1966 had published its first standard for "Vehicular Users".[5] The test procedures were essentially as Snively had first devised but some additional consideration of biomechanical tolerance levels was given. The WSU tolerance curve had now been published [6] and it pointed out that there should perhaps be time limits on acceleration exposure.[6] The standard retained the 400G maximum limit but set time duration limits at lower levels. These so-called *dwell times*, required that acceleration in excess of 200G not persist for longer than 2 msec and at 150Gs for

---

[3] As there was no need to have much dimensional precision below the hat band region, only the top of the head was modeled.

[4] Nearly 30 years later, Hopes and Chinn [21] using a similar helmet damage replication technique, determined that "*Current helmets are too strong, and their design is optimized for an impact severity that gives little chance of survival.*"

[5] This standard was the forerunner to today's US DoT standard 218 for motorcycle helmets which to this day, retains these same failure criteria.

[6] This was intended to limit velocity change following impact.

no longer than 4 ms. The headform meanwhile remained the rather un-anthropomorphic K-1A alloy. In 1970, Snell upgraded its standard permitting only 300Gs maximum but with no dwell times. Snell has continued to upgrade its standard over the years. Its most recent standard for motorcycle helmets [22] employs six different size headforms each with a different mass and failure level that range from 243 to 275Gs depending on headform size.

In Europe, the first country to take head injury in sports seriously, particularly ice hockey, was Sweden where the use of helmets began in the mid-1950s and became mandatory for all players taking part in the game in 1963. All players were required to wear Swedish Hockey Association *approved* helmets.

Two years later in 1965 the Canadian Amateur Hockey Association made helmet wearing mandatory for all non-adults playing as amateurs. Subsequently the Amateur Hockey Association of the United States established a new rule requiring players to wear head protection as well. Unfortunately at that time, except for the rather ill-defined standard in Sweden, there were no accepted performance specifications for hockey helmets.

In 1968, two teenage hockey players in Canada died from closed head injuries they sustained even though both were wearing *helmets*. This led to a request by the CAHA to the Canadian Standards Association to form a technical committee to develop a set of realistic performance specifications for hockey headgear and for a procedure to certify product to that standard. In 1975 the first CSA standard for hockey helmets was published and that same year all CAHA players were required to wear CSA certified helmets.

Football helmets took a somewhat different approach. In American football, concern for injuries dated back to the early twentieth century. In 1905, for example, helmets were not worn and there were 18 deaths and 129 serious injuries. Perhaps not surprisingly, later studies determined that most fatalities in football were due to head injury [23].

Encouraged by the ongoing research at Wayne State University, athletic products manufacturers decided to try to implement some of the research findings to improve helmet performance. To this end, the National Operating Committee for Athletic Equipment NOCSAE was formed in 1969 with the intent of developing a better standard for football helmets. Hodgson et al. [8] recognized the possible importance of headform biofidelity and developed a test headform to better model the anthropometrics and dynamic response of the human head. Based upon cadaver data the new test headform modeled both the geometric, inertial and frequency response characteristics of the head quite well.[7] Originally in one size only, several sizes were subsequently developed.

The second important development was to employ the Severity Index [13] as a measure of helmet performance. The failure criterion, in keeping with Gadd's view was initially set to SI = 1,000. This turned out to be too severe for helmets being produced at the time and the criterion value was moved upward to 1,500. In 1973 NOCSAE published its Standard Method of Impact Test and Performance Requirements for Football Helmets. Since about 1980, virtually every football helmet sold and used in the United States has had to meet the NOCSAE standard.[8] The introduction of the NOCSAE standard had a profound effect on head injuries in football in the US. Since its implementation, serious head injury rates dropped of the order of tenfold.

Recent work with football players in the National Football League, has found the concussion threshold at a much lower value of SI than had previously been accepted [24]. Furthermore, replication of actual incidents where players were concussed has led to the suggestion that the NOCSAE helmet test protocol could use updating. To this end NOCSAE introduced improved headforms and working with the NFL is currently examining the manner by which its impact test methods and failure criteria might be improved [25].

---

[7] Like the human skull, and unlike any headform before it, the NOCSAE headform could fracture if impacts were too severe – not a particularly good feature of a piece of test equipment.

[8] The failure criterion has since been lowered to SI = 1,200.

In regard to motorcycle helmets, huge strides were made in Europe with the publication of the COST 327 report [26]. The ensuing ECE Reg 22-05 motorcycle helmet standard now represents the state-of-the-art in performance specifications in this area; partly because of biomechanical considerations. Importantly, though from a test repeatability perspective, contentious, the test allows the headform to freely fall without any of the usual customary (guided free fall) constraints of contemporary protocols. This, it is argued, allows the head to respond in a more human-like fashion better permitting the implementation of human based criteria. Halldin et al. [27] have proposed an oblique impact test to address these same concerns. Unlike many other standards (NOCSAE excepted), each of the five ECE test headforms comprise the entire head rather than the partial headform employed by others. The anthropometrics are considered good and various sizes of appropriate mass and mass distribution are used. No attempt has been made to have biodynamic fidelity as this is now generally considered to be relatively unimportant in helmeted impacts.[9] This standard also sets the maximum allowable headform acceleration at 275Gs and (though the US Department of Transportation proposed and then revoked the idea of incorporating HIC in its performance standard in 1974) ECE Reg. 22-05 does in fact require that HIC not exceed 2,400.

A recently published standard for Formula 1 race car drivers (FIA 8860-2004) [28], also employs maximum acceleration and HIC as failure criteria even though it employs the Snell-like guided free fall methodology.

25.3 The Evolution of the Sports Helmet

The physics and certain material and design considerations regarding protective headgear have previously been outlined by Newman [29].

The remainder of this chapter is intended to supplement that prior publication.

Initially helmets, for whatever sporting application, were no more than leather bonnets. In auto and motorcycle racing, these designs, usually worn with goggles, were borrowed from earlier aviators and served primarily to keep the head warm and the hair in place. In the late 1800s and early 1900s, American football players and the occasional ice hockey player also wore the equivalent of a soft leather hat. Some employed a fleece or felt lining or were padded somewhat with cotton batting.

The concept of a hard shell, dating back to medieval times, tacitly acknowledged that distribution of the force to the head would reduce the probability of skull fracture – now a biomechanical tenet. Or perhaps it was seen as simply a better way to deflect objects from the head. However, no hard shell appeared on these early *helmets*.

As demands on the leather football helmet design increased, the outer leather was treated to make it hard (in a similar fashion to that developed many years prior for firefighters' headwear). Individual hard leather pieces were usually sewn to a hard fiber material crown section. Initially, they were simply lined with felt, fleece or some other padding but a few years later, with the introduction of a rudimentary inner suspension, something of a breakthrough in football headgear design had occurred. This new device had the capacity to absorb and distribute blows to the head somewhat more effectively than the floppy leather caps previously in use. But there was still a long way to go.

It would not be until the middle of the twentieth century that it was recognized that there were at least two types of sporting headgear. One dealing with the one-time life threatening blow that could occur in certain sports (e.g. auto racing), the other to deal with repeated low level blows associated with the game, (e.g. American football, ice hockey, etc.).

The need for a *crash helmet* was born shortly after the widespread introduction of the motorcycle in the 1900s. By the 1930s the use of hard shell helmets in international and grand prix auto racing had become standard gear. The situation

---

[9] Insofar as the failure criteria are based simply on linear acceleration (or functions thereof), it is assumed that the headform moves as a rigid body.

was similar in motorcycle racing. The very first of the modern hard *crash* helmet shells was not constructed of rigid leather but made up of several layers of cardboard held together with glue. Later, linen or other fine cloth impregnated with varnish resins was used. The sheets were laid up in a simple inverted dome-shaped mold and the composite was then allowed to cure to a solid shape. With a shape familiar to certain kitchenware, the name *pudding bowl* was often applied to these early helmets [30].

The football helmet industry was making strides of its own. In 1939, Riddell introduced what appears to have been the first helmet to employ a molded plastic shell. Unfortunately, early production methods were not that good and fracturing of the plastic shells during game play gave a bad reputation to plastic helmets for the first few years.

In 1941, Riddell patented an ingenious arrangement of fabric straps designed to keep the rigid plastic shell off the wearer's head (US patent 2,250,275) but, far more importantly, provided a means for absorbing impacts to the head. Sounding like a fundamental understanding of certain biomechanical principles, the patent itself states;

> ....a shock at any point about the surface of the helmet is not transmitted directly to the wearer's head in the vicinity of the blow but is transmitted by slings, and thus spread over a large area of the wearer's head.

The hard shell Riddell suspension helmet debuted in 1949 in the NFL. Other sporting applications would, in time, pick up on this design concept. Many other football helmet manufacturers however took a different approach.

In US patent 2,634,415 (Turner and Harvey 1953), the first padded hard shell football helmet was described. Claiming that "*...tape or strap suspensions have been sadly inadequate to the avoidance of severe head jolts*", they proposed a resilient closed cell rubber-like foam be placed throughout the shell interior. Cavities in the liner aligned with holes in the shell were provided for ventilation. Often cased in leather, this liner inside the hard molded plastic shell proved to be reasonably effective in dissipating impact energy and was the model for the current modern athletic helmet.

However, the design was very hot and heavy and not well ventilated and as a result, the web suspension design continued to prevail for many years – until more rigorous helmet performance specifications came along.

Meanwhile, Lombard et al. [31] working in the military aviation field, had also deduced that the Riddell-like suspension system, no matter how finely tuned, used space between the wearer's head and the inside of the helmet shell that could be better utilized for impact management. Filed in 1947 and issued in 1953, their patent (US 2,625,683), would alter crash helmet design in ways that have basically not since changed. Their idea was to fill, as completely as possible the gap between the head and the shell with crushable, energy absorbing material such as polyurethane foam. Clearly, though energy absorption was improved, this was not a repetitive impact kind of design as performance degraded significantly with subsequent impacts.

In order that their concept might achieve acceptance beyond the aviation community, Lombard and Roth formed Toptex Corporation in 1954. The plan was to produce motorcycle and auto racing helmets with crushable energy absorbing polyurethane liners. Importantly, at some point in time, a non-polyurethane liner material, expanded polystyrene bead EPSB foam, was selected for these helmets. This foam was cheap, readily available, relatively easy to manufacture, light weight, its mechanical properties could be fine-tuned but most importantly, it crushed more or less completely upon impact – and stayed crushed. The motorcycle, auto racing helmet as we know it was born. To this day, virtually all helmets of the *accidental* genre (i.e. crash helmets) employ this very same material.

During 1954, in what would become (for the last part of the twentieth century at least) the preeminent helmet producer in the world, some auto racing enthusiasts began to manufacture helmets in a garage behind the Bell Auto Parts store in Southern California. The initial helmet shells were hand laminated fiberglass resin composites with a thick semi-rigid foam polyurethane liner. They were the first to extend the pudding bowl to cover more of the head in a *jet* helmet style.

Along with their extended coverage of the head, they were believed at the time to be among the most protective race helmets ever designed.

In 1957, Snively conducted a biomechanics of head injury study that, though crude by today's standards, had a profound impact on modern helmet design and performance [32]. In this article, he discussed his testing of helmets then currently available to the racing community. He had helmets placed on the head of cadavers and subjected each to a massive impact. Six cadaver/helmet experiments in all were conducted on six different brands of helmets then available. In every case but one, the helmets failed to prevent the cadaver head from sustaining what would, in a living human, be a life threatening skull fracture. The only helmet to not result in a skull fracture was the helmet being made by Lombard and Roth's company Toptex. Though all helmets tested had a hard outer shell, this was the only one with a non-resilient foam liner and it was the one that would change the face of auto and motorcycle racing, cycling and equestrian helmet design, over the next 50 years.

Today, there is probably little disagreement that the fundamental objective of good helmet design, whether athletic or crash type, is to distribute impact loading over as large an area of the head as possible and to reduce the total force on the wearer's head as much as possible.

25.4 Requirements of the Modern Sports Helmet

25.4.1 General Considerations

As discussed above, there are two basic types of sports helmets; the so-called crash helmet and the athletic helmet.[10]

The crash helmet is intended to deal with the rare single life-threatening impact. This includes all those cases where impact to the head is accidental and possibly fatal such as motorcycling, bicycling, skiing, auto racing, and equestrian activities. Like all modern helmets of this type, once it's been involved in an accidental impact it is usually good practice to discard it. It's designed to take one reasonably sized hit at any one site and that's it. It cannot be used again since its capacity to absorb impact has been depleted (or at least substantially reduced).

The athletic helmet category comprises mainly sports such as amateur boxing, football, hockey, lacrosse and so on. The impacts to the head are usually not of a lethal nature and, more or less, are expected. Most importantly, they are expected over and over again. Obviously one would not wish to discard the helmet following every hit to the head. However sports impacts are not usually as severe as would be encountered in, for example, a racing car accident. At these lower impact levels, the multi-impact helmet will often perform better than the one time unit. However the reverse is not true. At energy levels encountered in for example a harness racing accident, the multi-impact helmet would likely be incapable of dealing with the high energy blows.

Design specifications, in general, can usually be broken down into two categories: functional and user requirements. The former makes the device *work*, the latter insures that it is acquired and used.[11] As with most forms of personal protective devices, the functional requirements of a helmet can be broken down additionally into two distinct categories.

1. Performance requirements, which stipulate the nature and level of protection the helmet is expected to provide and
2. Operational requirements which are specific to the application but for which protection is the secondary function.

---

[10] Other types of helmets include those intended to provide ballistic protection such as employed by military personnel and police SWAT teams and those for ancillary operations where the helmet serves mainly as a platform for other devices (e.g. sky diving).

[11] User requirements have to do principally with personal factors related to whether or not the helmet is considered acceptable by the wearer. These factors might include; value, comfort and appearance, and will not be considered further here.

25.4.2 Performance Requirements

In the case of safety helmet design, the performance requirements are those which are generally addressed by performance standards.[12] The three most significant of such requirements are shock attenuation, penetration resistance and helmet retention.[13] These directly address the threats to the head and the need to keep the helmet in place when required. It is the environment in which specific head gear needs to provide this kind of protection that makes helmet designs different. Helmets for hockey and bicycling for example, have many of the same basic performance requirements as, for example, auto racing. They just function in different environments and will differ substantially depending on the specific nature of the threats with which they must deal.

25.4.2.1 Shock Attenuating Capability

Probably the most important performance feature no matter what the application is the capacity of any helmet to deal with the energy associated with an impact to the head. That impact might be due to striking ones head on the roadway after falling from a motorcycle at speed, or, striking ones helmeted head against that of an opposing player in a football game. The primary purpose of the helmet is to distribute and reduce the forces that would otherwise develop when impact occurs. *Distributing* the force over a larger area reduces localized loading on the skull. This reduces the likelihood of skull fracture and brain damage immediately below the impact site. *Reducing* the magnitude of the force reduces the extent to which the head accelerates away from the impact and thereby reduces internal brain injuries associated with head motion.

Typically, as discussed earlier, standard test methods for this requirement entail the use of a dummy headform. The headform is fitted with an accelerometer cluster mounted at its center of gravity. The helmeted headform is dropped from some predetermined height onto a specified anvil and the ability of the helmet to reduce the impact loading to the headform is assessed each time the helmet is impacted. Different standards of course have different test conditions and different pass/fail criteria.[14] As discussed earlier, it is here where the field of head injury biomechanics has and can make a contribution to helmet design and performance.

However, virtually every sports helmet standard test method employs a guided headform that is constrained to respond only in an essentially linear manner.[15] Thus the application of any rotationally based injury assessment functions is a mute point.

25.4.2.2 Penetration Resistance

The ability of a helmet to prevent intrusion by some sharp object can be important. Some race car drivers have been killed as a result of errant vehicle parts piercing their helmet during a crash. In ice hockey for example, penetration by a skate blade could be an important consideration. In motorcycling, penetration resistance is of a secondary concern. However occasionally, a motorcyclist's head may encounter an object such as a guard rail support bolt or some other aggressive roadside object.

Motorcycle and auto racing helmets can, depending on the standard, be subjected to a test whereby a heavy sharp penetrator is dropped onto the stationary test helmeted headform from a predetermined height. If penetration of the shell occurs, or if the headform is touched by the penetrator, the helmet fails the test.

Equestrian helmet standards are one of the few that measure and require shock attenuation when struck with a penetrating object.[16]

---

[12] There appears to be no shortage of organizations that prepare and publish sports helmet standards. Among these are; ASTM, BSI, CPSC, CSA, DOT, and NOCSAE to name several.

[13] Less-often referenced requirements are for helmet stability, abrasion resistance, overall stiffness and for shell friction and conspicuity.

[14] These are thoroughly reviewed in Newman (2007) [33].

[15] ECE reg 22-04 is the only exception. It allows the headform to rebound freely after impact.

[16] A very aggressive anvil intended to represent the threat from a horse's shod hoof.

25.4.2.3 Retention Capability

Any protective headwear is of little value if it doesn't remain properly in place on the wearer's head. Dynamic stability of the helmet on the head is an integral part of retention capability and is addressed in certain standards. There was a time not long ago, when many motorcycle helmet models could be easily removed from the head of the wearer without disengaging the retention system [34, 35]. It's still possible to find such helmets in other applications.

Retention system testing at its simplest (and most common) measures the elasticity of the chin strap assembly. The chin strap is fitted around a set of rollers simulating the jaw and a force is applied to the fastened strap. The force may be statically or dynamically applied (depending on the standard). The strap must not break and must not stretch more than a certain amount. Another more rigorous test involves trying to pull the fastened helmet off a headform by rolling it forward and backward on the head. If the helmet shifts too much or comes off the headform completely, the helmet fails the test.

Curiously, perhaps because the designs tend to be historically good, there is no standard requirement for retention capability for football helmets. Though there is a retention system requirement for hockey helmets, it is apparent that players often wear the strap so loose that the helmet can easily be removed without unfastening it.

25.4.3 Operational Requirements

A list of operational requirements could be quite extensive and could include the adaptation of communication systems,[17] ventilation,[18] fitting mechanisms[19] and face shields.[20]

---

[17] Radio to permit coach to quarterback communications in football, for example.

[18] Forced air induction into helmet in closed cockpit race cars, for example.

[19] Internal adjustment devices to accommodate different head sizes in bicycling helmets, for example.

[20] Off road or dirt track racers, for example need to be able to install and use "tear-offs" when their visor becomes too dirty to see through.

None of these are *comfort* issues. They are essential operational requirements that must be addressed in order that the wearer do whatever it is he has to do while wearing the helmet.

25.4.4 Conflicting Requirements

Invariably, it seems, these sets of requirements are at conflict with one another and improper consideration of each can, at one extreme, lead to a "hot", heavy, expensive unattractive helmet that provides excellent protection to, at the other extreme, a cool, light, attractive, inexpensive device that offers virtually no protection. In both cases, the design would be a failure, since in one case it would "work" but would not be used and in the second would be used but not "work". The optimum clearly therefore lies somewhere between these two extremes. Similarly a helmet design would be a failure if, in spite of being excellent from a protection and user standpoint, it did not permit the wearer to perform the function he is expected to undertake (i.e. to not "operate").

Restrictions of the above type, whether imposed by standards, operational requirements or consumer attitudes, clearly limit the extent to which the designer can manipulate the various material and geometry alternatives available to him.

25.5 The Design of the Modern Sports Helmet

We shall not go into detail here. Readers who wish to know more about helmet design are referred to Newman (2007) [33] from which this section is adapted. Suffice it to say a safety helmet usually consists of four primary elements. They are the outer shell, an energy absorbing liner, a comfort liner and a retention system. We shall consider here just the shell and the liner.

25.5.1 Shell Design

From a functional point of view, the object of the shell is to provide a hard, smooth outer surface which serves to distribute the impact load over a large area. It also provides a penetration shield against high-speed objects and in addition, serves to protect both the wearer and the underlying liner of the helmet from abrasion. By virtue of its *smoothness* it also minimizes rotational effects when subject to glancing blows.

Common alternatives for the shell are fiber reinforced plastic (FRP) composites and thermoplastics. The former include woven fiberglass, polyaramid and/or carbon fiber bonded with a resin matrix. Thermoplastics typically are polycarbonate, acrylonitrile butadiene styrene ABS, or alloys of the two. Others, whose attractiveness depends on the particular application, include high density polyethylene, nylon and even metal. The modern bicycling helmet is a significant departure from these general rules as the shell is largely decorative. It is usually a very thin vacuum-formed polyvinyl sheet that is bonded to the liner during forming.

In an effort to address rotational injuries, several novel shell designs have recently been introduced. In 2012, Halldin was awarded a patent for the MIPS® system whereby the shell can slide somewhat relative to the underlying liner and head [36]. At least one company, Scott Sports has integrated this technology into certain mountain bike helmets. In another example, Lazer Helmets of Belgium has introduced a motorcycle helmet for which a secondary *skin* on the outside of the shell permits relative sliding between it and the shell (much like skin will *slide* on bone. (See http://www.lazerhelmets.com/innovations/superskin/).

25.5.2 Impact Liner

Interior to the shell is the liner of the helmet. It is this element that, by way of its deformation and in some cases, its partial destruction, is largely responsible for absorbing the energy of impact.

In order to perform its function effectively, the liner material must deform at stress levels below that which would cause head injury. Its strength should be largely insensitive to impact velocity and, to maximize net energy absorption, it should have slow (or no) recovery characteristics. These requirements dictate that the liner should have a well-defined, relatively constant low crushing/yield strength and be relatively strain-rate insensitive.

There are two basic types of foams used in helmet liner applications; resilient and crushable. Resilient liners are employed in repetitive athletic applications. Crushable materials are used in crash helmets for that *one-time* accidental impact.

A perfect resilient liner system is one that deforms upon impact, does not rebound but, before the next blow to the head occurs, recovers fully its shape and its physical properties. Many rubber-like types of foam have this general characteristic and have appeared in many different football and hockey-like helmets over the years. Some manufacturers have gone to great lengths to design energy absorbing systems exhibiting these properties by combining various fluid and gas filled bladder structures with foam materials. Pliable, vinyl nitride foam is the basic material of modern choice though some helmets continue to use polyurethane, polyethylene and polypropylene foams.

Material choices for a crushable liner have over the years included everything from rigid urethane foam to paper and aluminum honeycomb structures. Today, without question, the most popular, perhaps the only *crushable* liner used in crash helmets today, is made of expanded polystyrene bead foam (EPSB). The one governing factor that determines the crushing properties of the material, aside from the exact bead chemistry itself, is the bulk density of the molded part [37]. This can be controlled quite accurately by controlling the degree of pre-expansion before the beads are inserted into the mold. It is not uncommon nowadays to find bicycle and motorcycle helmets in which the EPSB liner assembly comprises components of different densities. This is often done to optimize the impact attenuation at different sites on the helmet and to accommodate ventilation slots.

In an effort to combine the best of both types of materials, some manufacturers have employed expanded polypropylene (EPP) bead foams. The manufacturing process (and the appearance) is not unlike that of EPS liners but the resultant product has interestingly different properties. In particular, following an impact it will regain most of its structural properties. Such a liner is often thought to be suitable in some racing accident situations where the wearer may have his head exposed to multiple impacts during the course of his accident.[21] The downside is that as the material recovers, it may increase the effective impact velocity of the head.

There are modern contenders to the crushable-resilient energy absorbing materials currently used in most helmets. One such alternative is Skydex®. This structure is akin to a group of plastic egg cartons stacked on top of each other maller. Skydex is custom made and the size and spacing of the *egg cup* depressions can be varied as can the thickness of the plastic material itself. The result is a highly *tunable* energy absorbing material that has already found application in some US navy helmets and in the Schutt DNA football helmet. Another possible contender is called Brock®. Currently being considered to absorb energy in a variety of forms of athletic padding, it's a lot like EPPB foam – without the E. Rather than being expanded and thereby having no gaps between them, individual polypropylene beads are essentially glued together. The PP beads remain more or less spherical with gaps between through which air can pass. It has excellent energy absorbing capabilities that can be tuned to a specific application.

There is little doubt that newer materials will be developed and touted as the greatest energy attenuating material ever. However the laws of physics as discussed in Newman [29] will prevail and there will always be limits as to what can be reasonably achieved.

25.6 Summary and Discussion

This chapter has provided a brief overview of some of the biomechanical engineering and practical considerations in protective headgear design and performance. There are of course many issues which have not been considered and many applications that have been given but passing attention.

The discussion is on operational and user requirements but does not, for example, take into account the special needs of children. For years, many helmet manufacturers seemed to have regarded children simply as small adults. Their anthropometry, injury tolerances and their use of protective headgear are quite different from adults.[22]

No consideration has been shown here as to the merits of these various standards. Do the designs which we have available reflect the true hazards to which we might be exposed? Or do they reflect some arbitrary set of laboratory tests devised by a committee somewhere who may or may not know how to adequately interpret the needs of the user? And what exactly constitutes a *pass* on any one of these tests?

Many other issues have been omitted from this brief. Among these is consideration of the different kinds of head injuries that can occur, the variability among humans and the influence of head size, shape and gender on such matters. Virtually no consideration has been given to the trade-offs that may have to be made between brain injury and neck and face injury.

Many problems in helmet design still have not been adequately dealt with. Dynamic retention of the helmet on the head is an example which, since first commented upon [35], has shown some improvement but which still needs work. Another is fogging or other visual disturbances of faceshields. A problem for some time in certain situations, this has been only partially addressed in some helmet designs today. Long term durability of contemporary helmets continues not to be assured. However, more recently manufacturers are at least providing information about when to replace a helmet.

Probably the most important concern at this time is the ongoing uncertainty about the extent to which helmets can deal with brain injury associated with rotational motion of the head.

---

[21] Skateboarding and BMX are examples.

[22] The Snell Foundation with FIA did publish a standard for children's motorsports [38] but as of this writing, no manufacturer produces such helmets.

In spite of recent design innovations, test standards have not found a consistent way to measure rotational motion of a test headform. In fact most common helmet test methods preclude any rotation of the headform. Even if this was not the case, reliable rotational injury assessment functions have yet to be fully developed. Even though it would be possible to measure rotational acceleration, a meaningful failure criterion has not yet been established.

Helmet design has progressed significantly in the past several 1,000 years and somewhat marginally during the past 25. Further progress will, relatively speaking, be even more marginal. One thing is certain, helmets are better than they were 50 years ago and this is partially due to the recognition of certain basic biomechanical concepts. However, little improvement can be expected for the next 50 years until we determine ways to implement more of what little we know about the biomechanics of head injury.

References

1. Melvin JW, Lighthall JW (2002) Brain-injury biomechanics. In: Nahum AM, Melvin J (eds) Accidental injury – biomechanics and prevention, 2nd edn. Springer, New York
2. Tenner E (2003) Our own devices. Alfred A. Knopf, New York
3. Gurdjian ES, Lissner HR (1944) Mechanism of head injury as studied by the cathode ray oscilloscope, preliminary report. J Neurosurg 1:393–399
4. Gurdjian ES, Webster JE (1943) Experimental head injury with special reference to the mechanical factors in acute trauma. Surgery, Gynecology and Obstetrics 76:622–634
5. Lissner HR, Lebow M, Evans FG (1960) Experimental studies on the relation between acceleration and intracranial pressure changes in man. Surgery, gynecology & obstetrics 111:329–338
6. Gurdjian ES, Roberts VL, Thomas LM (1966) Tolerance curves of acceleration and intracranial pressure and protective index in experimental head injury. J Trauma 6(5):600–604
7. Holbourn AH (1943) Mechanics of head injury. Lancet 2:438–441
8. Hodgson VR, Mason MW, Thomas LM (1972) Head model for impact. In: Proceedings of the 16th Stapp car crash conference, Detroit, MI
9. Gennarelli TA, Thibault LE, Ommaya AK (1972) Pathophysiologic responses to rotational and translational accelerations of the head. In: Proceedings of the 16th Stapp car crash conference, Detroit, MI
10. King AI, Yang KH, Zhang L, Hardy W (2003) Is head injury caused by linear or angular acceleration? In: Proceedings of the IRCOBI conference, Lisbon
11. Takhounts EG, Ridella SA, Hasija V, Tannous RE, Campbell JQ, Malone D, Danelson K, Stitzel J, Rowson S, Duma S (2008) Investigation of traumatic brain injuries using the next generation of simulated injury monitor (SIMon) finite element head model. Stapp car crash journal 52:1–31
12. Aare M, Kleiven S, Halldin P (2003) Injury criteria for oblique helmet impacts. In: Proceedings of the IRCOBI conference, Lisbon
13. Gadd CW (1966) Use of a weighted impulse criterion for estimating injury hazard. In: Proceedings of the 10th Stapp car crash conference, Hollomon Air Force Base, NM
14. Margulies SS, Thibault LE (1992) A proposed tolerance criterion for diffuse axonal injury in man. Journal of biomechanics 25(8):917–923
15. Newman JA (1986) A generalized model for brain injury threshold (GAMBIT). In: International IRCOBI conference on the biomechanics of impact, Zurich
16. Smith T (2005) Personal communication, re tests of Full 90 headgear
17. Newman JA, Schewchenko N, Welbourne E (2000) A proposed new biomechanical head injury assessment function – the maximum power index. In: Proceedings of the Stapp car crash conference, Atlanta
18. Shewchenko N, Withnall C, Keown M, Gittens R, Dvorak J (2005) Heading in football. Part 1: development of biomechanical methods to investigate head response. British Journal of Sports Medicine 39(Suppl 1):i10–i25. doi:10.1136/bjsm.2005.019034
19. Takhounts EG, Hasija V, Ridella SA, Rowson S, Duma SM (2011) Kinematic rotational brain injury criterion (BRIC). In: Proceedings of the ESV conference, Washington, DC
20. Snively GC, Chichester CO (1961) Impact survival levels of head acceleration in man. J Aerosp Med; 32: 316–320
21. Hopes PD, Chinn BP Helmets (1989) A new look at design and possible protection. In: Proceedings of the IRCOBI conference, Stockholm, Sweden
22. Snell Memorial Foundation (2015) 2010 Helmet standard for use in motorcycling
23. Halstead PD, Alexander CF, Cook EM, Drew RC (2004) Historical evolution of football headgear. Submitted for publication to JATA
24. Newman JA, Barr C, Beusenberg MC, Fournier E, Shewchenko N, Welbourne E, Withnall C (2000) A new biomechanical assessment of mild traumatic brain injury – part 2 – results and conclusions. In: Proceedings of the IRCOBI conference, Montpellier
25. Withnall C, Shewchenko N (2005) Personal communication. Biokinetics and Associates, Ottawa
26. Chinn, Canaple, Derpler, Doyle, Otte, Schuller, Willinger, Cost 327, Motorcycle Safety Helmets, Final report of the action, European Commission, Directorate General for Energy and Transport, Belgium, 327 pp
27. Halldin P, Gilchrist A, Mills NJ (2001) A new oblique impact test for motorcycle helmets. Int J Crashworthiness 6

28. Federation Internationale de l'Automobile (2005) FIA 8860-2004 Advanced Helmet Test Specification
29. Newman JA (2002) Biomechanics of head trauma: head protection. In: Nahum AM, Melvin J (eds) Accidental injury – biomechanics and prevention, 2nd edn. Springer, New York
30. Wagar IJ, Fisher D, Newman JA (1980) Head protection in racing and road traffic. In: Proceedings of the 24th conference of the American Association for Automotive Medicine, pp 446–455, Rochester, NY
31. Lombard CF, Ames SW, Roth HP, Rosenfeld S (1951) Voluntary tolerance of the human to impact accelerations of the head. The Journal of aviation medicine 22(2):109–116
32. Snively GC (1957) Skull busting for safety. Sports Car Illustrated
33. Newman JA (2007) Modern sports helmets. Their history, science and art. Schiffer Publishing, Atglen
34. Mills NJ, Ward R (1985) The biomechanics of motorcycle helmet retention. In: Proceedings of the international council of the biomechanics of injury, Goteborg, Sweden
35. Newman JA, Woodroffe J (1979) The dynamic retention characteristics of motorcycle helmets. Biokinetics Contract Report to Science Centre, Supply and Services Canada, Ottawa
36. Halldin P (2012) Helmet with sliding facilitator arranged at energy absorbing layer EP2440082A1 (European patent), filing data 20110503, priority date 20120418
37. Mills NJ, Gilchrist A (1990) The effectiveness of foams in bicycle and motorcycle helmets. In: Association for the Advancement of Automotive Medicine conference, Scottsdale
38. Snell Memorial Foundation and FIA (2007) 2007 Standard for protective headgear for use in children's motorsports

Normalization and Scaling for Human Response Corridors and Development of Injury Risk Curves

26

Audrey Petitjean, Xavier Trosseille, Narayan Yoganandan, and Frank A. Pintar

Abstract

To 'collate' impact response data from a group of samples from tests using unembalmed human cadavers (Post-Mortem Human Subjects, PMHS) for crashworthiness applications, it is important to transform the fundamental measured variable such as acceleration, force and deflection to a standard or reference, termed as normalization. Scaling can be defined as a process by which normalized data can be transformed from one Standard to another (example, mid-size adult male to large-male and small-size female adults, and pediatric populations). This chapter examines the pros and cons of approaches used in normalization/scaling processes for over four decades in impact biomechanics, with a focus on automotive applications. Specifically, the equal stress equal velocity and impulse momentum methods are discussed. The process of corridor development is critical to ensure that all anthropomorphic test devices mimic human impact responses. The chapter also examines methods used in developing human corridors, ranging from subjective to objective approaches in terms of incorporating individual variations or spread in the measured metrics. The variability in the response should also be taken into account in the determination of the human tolerance to impact, routinely used as a tool to improve occupant protection in crashes. Human tolerance can be expressed in terms of a human injury risk curve. The different empirical and statistical methods

A. Petitjean, M.Sc. (✉)
CEESAR, Nanterre, France
e-mail: Audrey.petitjean@ceesar.asso.fr

X. Trosseille, Ph.D.
LAB PSA Peugeot Citroen Renault, Nanterre, France
e-mail: Xavier.trosseille@lab-france.com

N. Yoganandan, Ph.D.
Department of Neurosurgery, Medical College of Wisconsin, Milwaukee, WI, USA
e-mail: yoga@mcw.edu

F.A. Pintar, Ph.D.
Neuroscience Research Lab, Department of Neurosurgery, Medical College of Wisconsin, Milwaukee, WI, USA
e-mail: fpintar@mcw.edu

historically used in the biomechanical field to develop injury risk curves are presented in this chapter. The normalization group ISO/TC22/SC12/WG6 recommended a procedure to construct injury risk curves aiming at harmonizing the development of injury risk curves. This procedure is presented in detail. It includes several steps taking into account the specific censored status of the biomechanical data and including some checks related to the statistical modeling of the injury risk. Topics allowing future improvement of the procedure are discussed.

26.1 Introduction

Outcomes from experiments using human volunteers and unembalmed human cadavers (termed as Post-Mortem Human Subjects, PMHS), are used to develop biofidelity corridors and derive human injury criteria/risk curves applicable to automotive, sports, aviation and military environments. Human injury criteria are termed as injury reference values in automotive literature [1]. Sled and electrohydraulic piston equipment, an impacting mass attached to a device such as pendulum, and drop tests of the human surrogate itself are commonly used to apply the impact force to components, sub-systems or, whole body PMHS [2–7]. Responses from individual tests can be grouped to determine the mean and standard deviations if inter-specimen variability can be ignored or minimal. This process has been used in automotive impact biomechanics studies [8–10]. However, the inherent biological variability of the human can limit applicability of this approach to derive corridors for assessing dummy biofidelity. Physical, geometrical, material and inertial properties affect responses. It is difficult to control subject selection such that all variables are confined to a narrow range and minimize effects of biological variability. Therefore, for the outcomes of an experiment to be applicable to a specific anthropometry, it is necessary to normalize individual subject responses to a predetermined standard/reference such as mid-size crash test dummy. Subsequent to the conversion of measured specimen-specific data to standard responses, means, standard deviations and human response corridors can be derived, leading to the establishment of dummy biofidelity corridors. The first part of this chapter focuses on some normalization procedures used in impact biomechanical studies with specific reference to the automotive literature. Scaling from one reference to another is discussed next. The relative merits and demerits of these approaches are discussed along with recent advancements in developing biofidelity corridors.

Knowledge on human injury tolerance is important to improve occupant safety. Because of variability, human tolerance to impact is better expressed using a risk curve. It is the result of statistical modeling of the occurrence of injury as a function of a biomechanical metric(s) in response to external mechanical load. The second part of this chapter focuses on methods used to develop human injury probability curves in impact biomechanics literatures. One example is used to demonstrate different approaches. Taken together, the two parts, i.e., normalization and development of risk curves from experiments with human surrogates provide insights into analytical methods of advancing human safety secondary to external mechanical loads.

26.2 Part I: Normalization and Scaling

26.2.1 Equal Stress Equal Velocity Approach

26.2.1.1 Description

This simple approach assumes linear relationships between the fundamental parameters of length, mass and time; and identical density and modulus

Table 26.1 Normalizing factors for the equal stress equal velocity method

Variable	Normalizing factor
Force	Two-thirds root of the ratio of mass of PMHS to reference mass
Deflection	Cube root of the ratio of mass of PMHS to reference mass
Acceleration	Cube root of the ratio of reference mass to mass of PMHS mass
Time	Cube root of the ratio of mass of PMHS to reference mass
Moment	Ratio of mass of PMHS to reference mass

of elasticity between the mass and reference, i.e., dummy for automotive applications [11]. Table 26.1 shows the normalization factors for the force, displacement, time, acceleration, and bending moment to convert specimen-specific data to the reference. Normalization factor is unity for the angulation parameter.

26.2.1.2 Applications

Early applications of this method involved normalization of seatbelt forces from frontal impact tests wherein the total body mass ranged from 36 to 102 kg, thus making a strong case for normalization of data [12]. A later application consisted of normalizing data from 49 PMHS side impact tests with a range of velocities and boundary/contact conditions, and these studies formed a basis for the Thoracic Trauma Index metric, used in the first US side impact regulations, promulgated in 1990 [13]. The equal stress equal velocity method has also been used in recent side impact studies for applications to the mid-size ES-2re and small-size SID-IIs dummies in the United States [14]. Forty-two PMHS sled tests conducted at the Medical College of Wisconsin were used to normalize data to the standard mass of the two different dummies, representing the mid-size male and small female [4, 5, 15]. The latest update of the FMVSS-214 standards incorporates analyses of data from these recent experiments [16]. Other applications include normalizing side impact sled test results to small-size female dummy mass to extract acceleration, force and deflection responses in an attempt to evaluate the biofidelity of fifth-percentile dummies [17]. Yoganandan et al. adopted this approach to determine the force-time histories from oblique side impact PMHS sled tests wherein data were obtained from load cells attached to anthropometry-specific modular load wall.

26.2.2 Impulse Momentum Approach

To determine region-specific normalizing factors, this approach uses specific body region/segmental characteristics and it also accounts for the type of the impact test [18]. Mass and stiffness ratios are used along with assumptions of lumped mass and spring models. Sled tests and whole body free fall/drop tests are treated as one degree of freedom and pendulum tests are treated as two-mass spring systems.

26.2.2.1 One Degree of Freedom Model

This method is explained using lateral impact PMHS sled tests, described in detail in the chapter on thorax injury biomechanics. The purpose of segmentation of load-walls in a side impact sled test is to isolate the loads and kinematics of body regions. Tests with a segmented wall accommodating thorax, abdomen and pelvic load plates enable a portion of the total body mass of the PMHS to load the respective plate. Body-region specific effective mass properties are computed as the ratio of the impulse of the respective load plates to the impact velocity. The impulse is computed from time-zero to the time at which the velocity reaches the sled velocity. The effective body region-specific mass ratio is defined as the ratio of the effective mass to the total body mass of each subject. The average of the effective mass ratios for the entire ensemble tested using the same experimental design is determined for each body region. To determine the standard effective mass of each PMHS for the specific body region, the average effective mass ratio is multiplied by the mass of the reference. The mass ratio factor for each subject is defined as the ratio of the reference effective mass of the body region to the

Table 26.2 Normalizing factors for the impulse momentum method using the one degree of freedom model

Variable	Normalizing factor
Force	Square root of the product of mass and stiffness factors
Deflection	Square root of the ratio of mass to the stiffness factor
Acceleration	Square root of the ratio of stiffness to the mass factor
Time	Square root of the ratio of mass to the stiffness factor

Table 26.3 Normalizing factors for the impulse momentum method using the two degree of freedom model

Variable	Normalizing factor
Force	Square root of the product of mass and stiffness factors and weighting factor
Deflection	Square root of the ratio of mass to the stiffness factor multiplied by the square root of the weighting factor
Acceleration	Square root of the ratio of stiffness to the mass factor multiplied by the square root of the weighting factor
Time	Square root of the ratio of mass to the stiffness factor multiplied by the square root of the weighting factor

effective body-region specific mass of each subject, determined above. The stiffness ratio factor for each subject can be obtained as the ratio of a characteristic length of the body region of the reference (example, right to left lateral distance of the thorax of the mid-size male dummy) to the same measurement from each PMHS. This is termed as the characteristic length approach. If the length is unknown, extending the assumption of geometric similitude to the total body, the stiffness ratio may be estimated using the cube root of the body mass ratios of the standard surrogate and each PMHS. The normalized force, deflection, acceleration and time are obtained using the standard stiffness and mass ratio factors (Table 26.2). Another method of determining the stiffness factor includes the use of the force and deflection from each PMHS test and this is described later.

26.2.2.2 Two Degrees of Freedom Model

Data from tests with a pendulum wherein an impactor is attached to the forward facing end of the device and the impactor delivers the dynamic load to a specific region of the human body (example, shoulder) can be normalized using this model. The mass and acceleration of the impactor are used to determine the force applied to the PMHS [19]. The body-region specific effective mass properties are computed as the ratio of the impulse to the impact velocity. The impulse is computed as the time-integral of the mass times the acceleration of the pendulum impactor divided by the time-integral of the acceleration of the PMHS. The normalizing factors shown (Table 26.2) for the single degree of freedom model is modulated by the weighting factor, defined as the ratio of the sum of the mass of the pendulum and effective mass computed above to the sum of the mass of the pendulum and effective mass of the reference. In other words, all quantities shown in Table 26.2 are multiplied (within the square root function) by this weighting factor (Table 26.3).

26.2.2.3 Normalizing Factor for Velocity

Renormalization to a common velocity is needed if considerable differences exist in the insults between different PMHS tests. This is likely more of an issue with pendulum tests and PMHS with widely varying total body mass. This can also occur from the device aspect stemming from the use of energy absorbing materials or due to the inherent biological variability in the surrogate. Increased body mass index or adipose tissue may also be a factor. A standard impact velocity representative of all tests in the ensemble should be initially chosen for this purpose. The afore determined force and deflection responses are further modulated (multiplied) by the velocity factor, defined as the ratio of the standard velocity to the velocity sustained by the tested PMHS, although the time factor does not need renormalization.

26.2.2.4 Another Method for the Determination of the Stiffness Ratio

Another approach has been suggested to determine the stiffness ratio [9, 18–21]. Using the force and deflection data, the effective stiffness of each PMHS is determined by dividing twice the integral of the force-deflection response by the square of the peak deflection of the PMHS. The stiffness ratio is the stiffness of the reference to the effective stiffness of the PMHS. A limitation is the need to obtain force and deflection data from each PMHS test. This necessitates use of devices such as a chestband or accelerometer in the instrumentation list. The numerator of the stiffness ratio cannot be obtained from geometrical measurements for the reference subject (dummy) and it is test specific. Because not all intact PMHS tests have used chestbands to record deformation, these two features render the application of this method to existing data difficult. Double integration of acceleration data is also not fully accurate to obtain deflections. Although this method appears to be more accurate than using the length as a surrogate for stiffness, experimental design should accommodate the determination of kinetics in the biological surrogate. This may be adopted in future course of experiments if this method proves to be more efficacious, i.e., reduce the spread in the resulting corridors.

26.2.2.5 Applications

This method has been adopted for developing human response corridors from different PMHS experiments [19, 22–29]. Forces and deflection responses were normalized to adult mid-size male and data were obtained from 14 PMHS subjected to 44 blunt lateral impacts at three velocities using a pendulum device [19]. The International Standards Organization for side impact has used this approach to develop regional biofidelity response corridors. Yoganandan et al., used this approach to determine the deflection-time histories from oblique side impact sled tests wherein data were obtained from chestbands wrapped on the thorax and abdomen [21]. The stiffness factors based on the force and deflection responses of each PMHS in lateral impacts was used to compare the results with the characteristic length method used in the impulse momentum approach [5, 15, 20, 30]. The use of PMHS-specific stiffness from the force-deflection responses improved coefficients of variations compared to the characteristic length method.

26.2.3 Mean and Standard Deviation Responses for Corridors

In the beginning subjective methods were used to establish surrogate response corridors [31]. Mathematical processes include the equal stress equal velocity and impulse momentum methods. The mean and plus and minus one standard deviations in the temporal domain for fundamental responses (force, acceleration and deflection) constitute human response corridors. This also applies to the cross-variables, i.e., force-deflection responses.

26.2.3.1 Fundamental Time-Varying Responses

It is important to align the signals of the variables from the impact test. One of the simplest methods is to align the maximum magnitude of the ordinate and shift the signal in time such that each peak occurs at the same time for all PMHS tests [32]. As this method can lead to inaccurate alignments, another method would be to align the signals based on the initiation of the acceleration/deceleration pulse. Although this method appears to be straightforward, difficulties exist if different body regions absorb the impact loading initiating at different times during the loading phase of the PMHS. The cumulative variance method is used [33]. In recent side impact tests, forces and deflections were isolated for different body regions: forces for the thorax, abdomen and pelvis and deflections of the thorax and abdomen [15]. The following methods were established to determine the mean and standard deviation responses and corridors from a variety of test conditions. For flat wall tests, t_0 was determined by initiation of arm contact on the thoracic load plate. In pelvic, thoracic and abdominal offset tests, t_0 was coincident with specimen contact with the offset load plate. Contact with the load plate was determined by finding the first point in time where the force exceeded 200 N and then incrementing backward to find the point in time where the force-time history crossed zero force. The time of occurrence of the zero-crossing load was taken to be the start of the impact event for all signals. While signals appeared to have the same shape, their time of occurrence varied between PMHS and it was attributed to local geometry. To preserve relative timing information between sensors attached to different body regions, the following methods were developed to establish the time of occurrence of each signal.

A characteristic time for each signal was defined to quantify when a particular signal occurred relative to t_0 (Fig. 26.1). Initiating from the peak amplitude of the force, the signal was traced back temporally to determine one-fifth of the peak amplitude. The characteristic time was defined as the time of occurrence of the one-fifth of the peak force. The characteristic time for acceleration signals was determined in the same manner, except that the signal was first integrated and then the peak and one-fifth point of the integrated acceleration were determined, after determining characteristic time. The integrated acceleration curve was discarded and the characteristic time was associated with the original acceleration curve. Average characteristic time for each signal group was determined to quantify average time at which signals in the group occurred relative to t_0. The signals in each signal group were aligned by minimizing the cumulative variance.

$$V_{ab} = \sum_{t}^{t^i} (a_i - b_i)^2 \qquad (26.1)$$

Terms a and b are the magnitudes of the two signals at the ith time step; t is the greater of the start times of signal a and b; and t^i is the lesser of the end times of signals a and b. Signal start and end times were defined as follows. For force and

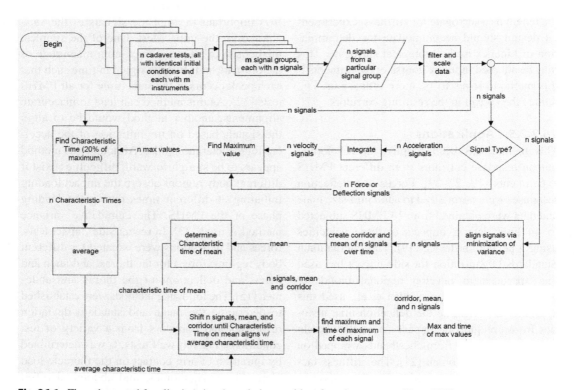

Fig. 26.1 Flow chart used for aligning signals to derive corridors from impact tests (From [15])

deflection signals, starting from the peak of the signal decrement in time until the signal amplitude reached one-fifth of the peak value. This defined the starting of the signal. Again starting from the peak of the signal, increment in time until its amplitude reached one-fifth of the peak value, and this is the end time of the signal. For acceleration signals, the curve is first integrated and the maximum value is determined. Beginning at the peak value and decrementing backward, the end time is the time of the first point that reaches four-fifths of the maximum value, and the start time is the time of the first point that reaches one-fifth of the maximum value.

Two signals can be aligned by time-shifting the signals relative to one another until a minimum variance (Eq. 26.1) would be obtained. To align a set of n signals, one signal from each signal group was chosen as the standard for alignment. This alignment standard was the one that appeared to have the most typical shape of all signals. The cumulative variance between the alignment standard signal (S) and the second signal in the group, V_{s2}, was minimized as follows. The second signal was shifted backward in time by an amount equal to one-third of its duration (end time minus start time), and the variance V_{s2} calculated according to the above equation. The second signal was shifted forward by one time step and the cumulative variance calculated. The process of shifting and calculating the variance continued until the second signal had been forward time-shifted by an amount equal to two-thirds of its total duration; the variance was recorded at each shift step. The shift step with the lowest cumulative variance was considered to have the optimal alignment of the alignment standard and second signals. The process was repeated for all signals in the group, one at a time, optimally aligning them with the standard signal. The mean PMHS response is obtained using the following equation subsequent to the described alignment process.

$$\bar{x}(t) = \frac{\sum_{i=1}^{n} x_i(t)}{n} \quad (26.2)$$

Where the variable $x_i(t)$ represents the magnitude of the signal at time t for the ith PMHS, n is the number signals and $\bar{x}(t)$ represents the mean PMHS response at time t. The standard deviation (SD_t) at each time is obtained from the Eq. 26.3.

$$SD_t = \sqrt{\sum_{i=1}^{n} \frac{(x_{i,t} - x_t)^2}{(n-1)}} \quad (26.3)$$

The corridor expressed as the mean and one standard deviation limits of the response is derived by adding and subtracting the standard deviation from the mean in the temporal domain.

$$\text{Plus / minus one SD corridor} : SD_t +/- \bar{x}(t) \quad (26.4)$$

26.2.3.2 Cross Variable Responses

Responses such as force-deflection curves are obtained by eliminating the abscissa between the two parameters in the fundamental force-time and deflection-time responses. In an early study on the determination of the mechanical properties of PMHS lumbar spine ligaments, Pintar et al. used measured force-time responses from the load cell and deflection-time responses from the linear variable differential transducer attached to the electrohydraulic piston to determine tensile force-deflection properties [34]. This method was later used by Yoganandan et al. to determine the force-deflection properties of PMHS cervical spine ligaments [35]. Force-time and deformation-time signals were normalized with respect to the corresponding failure force and deformation magnitudes. Using the deformation as the governing variable for each force-deformation curve, at every normalized deformation data point, normalized force values were computed. The mean of these values for each ligament resulted in the mean normalized force at the corresponding normalized deformation points. These mean normalized force-deformation curves were transformed into engineering units by suitably multiplying the abscissa and ordinate with the mean failure deformation and mean failure force values, respectively, calculated for that particular ligament.

At each abscissa point on the mean force-deflection curve, the authors computed plus and minus one standard deviations, describing the corridors based on the ordinate [34, 36]. Joining the minus one standard deviation and upper one standard deviation points yields the plus minus one standard deviation response corridor. Although these normalization procedures were derived from static loading, the method can be used in dynamic loadings. The described method does not account for variations in deflections because the deflection was the controlled input from the testing device. However, its standard deviations can be obtained using the same method. This process results in plus and minus one standard deviation at each data point on the mean curve, instead of the deviation for only the ordinate. Two empirical methods, termed as the box and elliptical methods, are proposed for corridor construction when both deviations are considered [37, 38].

Box Method
The box method represents the four corners of the plus and minus standard deviations in the abscissa and the ordinate the following steps are followed. Normalize deflection by dividing deflections for each test (raw curve) by the peak deflection. Determine forces for all raw curves at common normalized deflections (e.g., at every 1 % of peak deflection), and determine the average force at each deflection step. Multiply the normalized deflections by the average peak deflection of the raw curves in the entire ensemble. This results in the characteristic mean response curve. Determine the standard deviation in the force for each point on the characteristic mean response curve. Normalize force by dividing the peak force for each raw curve with its associated peak deflection. Obtain deflections for all raw curves at common normalized force values using an interpolation technique (e.g., 1 % of peak force). Determine the standard deviation in the deflection for each point on the characteristic mean response curve. Determine the "extreme" variation values associated with each point on the characteristic mean response curve, plot the four variation "extreme" curves, and enclose the region bounded by the four extreme curves.

Ellipsoid Method
This is a slight deviation from the box method. This approach treats the four corners of plus and minus standard deviations in the abscissa and ordinate (deflection and force in the cited example above) as the end points of the major and minor axes of an ellipse at each point on the characteristic mean response curve.

26.2.4 Scaling

A distinction can be made between the terms normalization and scaling based on the following considerations. While normalization can be defined as the process by which the measured/derived responses from individual PMHS tests with varying properties are brought into a reference, scaling is the process by which normalized responses can be transformed from one reference to another. For example, responses normalized and applicable to a mid-size male anthropometry can be scaled to pediatric and adult small-size and large-size anthropometries using scaling factors. This process has been used in the United States Federal Motor Vehicle Safety Standards [39]. However, it is possible to directly normalize the original data to such anthropometries by choosing them as reference. This approach has been used in the analysis of side impact response corridors wherein the small-size female total body mass was used to develop fifth percentile female dummy-specific acceleration, force and deflection corridors [17]. The equal stress equal velocity and impulse momentum approaches can be used for either scenario. Scaling factors other than geometry for pediatric applications (bone modulus and calcaneal tendon strength as a surrogate for scaling head injury criteria) have been used [39–42].

26.2.5 Discussion on Normalization and Scaling

The equal stress equal velocity approach relies only on the mass ratio between the reference subject and the tested PMHS in the ensemble on a specimen-by-specimen basis. This is applicable to all types of impact tests: drop, pendulum and sled. The impulse momentum approach takes into account the type of impact test and the effective of mass of the specific body region, regardless of methods used to determine stiffness ratios, based on characteristic length- or force-deflection properties specific to the type of impact test. This approach should lead tighter corridors in principle. Studies evaluating the tightness are limited. An analysis of maximum regional forces from lateral impacts with abdomen offset using the equal stress equal velocity and the impulse momentum approaches are presented in Figs. 26.2 and 26.3 [43, 44]. The mean and standard deviation data for different body regions show no significant bias.

Factors such as loading type and the region of impact to the human body can influence results. Abdomen offset loading has the potential to load the less stiff and more viscous abdominal region compared to the flat wall or thoracic offset loading as these impacts tend to engage the skeletal regions initially [45]. The impulse momentum method may not be completely sensitive to local responses in all testing conditions. The choice of the PMHS may govern the outcome. A well-controlled experimental design such as using anthropometry-specific and segmented load-walls to accurately engage the specific body region, confining the subject selection based on bone mineral density of skeletal structures to minimize variability and ensuring alignment of the subject prior to impact with the load-wall, will likely have a greater influence on the outcomes, regardless of the analysis procedure used to normalize and scale recorded measurements. However, this hypothesis remains further research.

For the corridors and cross variable plots, extensions of the originally proposed analysis, based on PMHS cervical and lumbar spine ligament data, by accounting for the spread in the force and deflection by treating the four corners of the standard deviations as the extreme boundaries of an ellipse or a box are advancements. Studies are limited delineating to determine the superiority of the two box and ellipse methods. The use of either approach may have to account for the following issue: four extreme curves created by plotting the four extreme variation values for each point (plus and minus one standard deviation) on the characteristic average curve, and then enclosing the region bounded by the four extreme curves to create the corridor, may include areas beyond the one standard deviation for the ordinate and abscissa. It should be noted that both variables are treated as dependent parameters, not fully complying with the conventional definitions

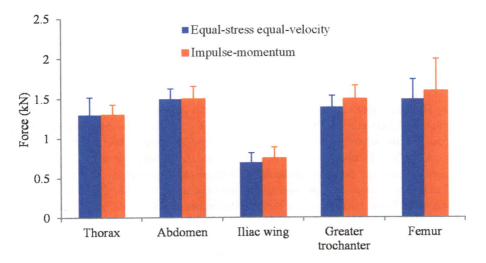

Fig. 26.2 Comparison of the two approaches at 3 m/s velocity in lateral abdominal offset PMHS (From [43, 44])

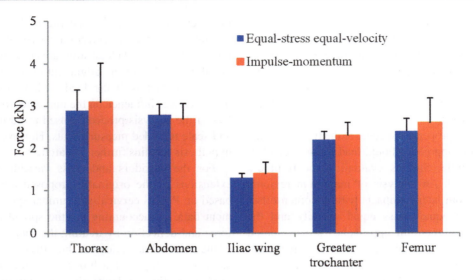

Fig. 26.3 Comparison of the two approaches at 8 m/s velocity in lateral abdominal offset PMHS (From [43, 44])

of insult-response depictions. Additional studies to evaluate the efficacy of these empirical methods with controlled datasets from different impacting conditions are therefore necessary.

26.3 Part II: Human Injury Risk Curve

26.3.1 Rationale for Human Injury Risk Curve Development

In the automotive field, knowledge on human injury tolerance to impact is necessary to design restraint systems for improving occupant protection. It is also necessary to set limits in test protocols for regulatory bodies and consumer-testing organizations. Human tolerance is investigated through relationships between mechanical load (such as a stress or a strain) and injury due to the external load. As explained earlier and elsewhere, there is variability among humans such that injury tolerance to impact does not correspond to one magnitude [46]. Some humans will tolerate more than others. Human injury tolerance limit is therefore better expressed using a distribution instead of one value. The cumulative distribution of the tolerance corresponds to the injury risk curve. A probability of 50 % of injury risk means that one-half of the considered population has a tolerance limit lower than the corresponding injury criterion value and will sustain injury if subjected to conditions corresponding to this level or higher. The injury risk curve is built as a function of an injury criterion. It represents the parameter allowing the prediction of the risk. It can correspond to a strain or a stress (derived metrics) or to more global parameters such as deflection, force and acceleration. The injury risk curve is the result of statistical modeling of the occurrence of the injury as a function of injury criterion sustained by the human body when impacted with an external load.

As described before, normalization processes attempt to reduce variability among the humans as much as possible, taking into account parameters for which the influence on the human response is known. The weight and size of the human are usually considered [12, 18]. Despite normalization processes, injury tolerance among humans are still scattered due to inter-individual differences and the definition of an injury risk curve rather than a single injury tolerance value is necessary. Injury tolerance limits are derived from the injury risk curves and can be used by regulatory bodies, automotive manufacturers and suppliers to ensure that the design of a restraint system or a vehicle component is efficient to protect the occupant during a crash.

The final goal of injury risk curves is to provide injury criteria to be used to ensure protection of the motor vehicle occupant. An efficient way to do it would be to collect information from crashes to develop injury risk curves. Crash investigations provide information on vehicle and occupant conditions and sometimes on the loading of the occupant (such as the load applied on the thorax), although information on the responses of the occupant are usually not available. An alternative solution is to reproduce some crash conditions in laboratory. The most representative surrogates for the occupant would be volunteers, although they are tested at low severity, i.e., sub-injurious loads. Some volunteer tests have been conducted in the biomechanical field but they are mainly used to study human kinematics [47, 48]. The fact that tests are non-injurious is a clear limitation in the development of an injury risk curve. Animal tests have also been considered as some of them allow the testing of a living surrogate at high severity [49, 50]. However, the applicability of resulting curves to the human will depend on the correspondence in terms of anatomy, geometry and material properties between the animal and human. A possibility to test a surrogate closer to the car occupant, allowing the performance of injurious and non-injurious tests and avoiding any correspondence due to difference of species is to use PMHS in laboratory tests. The PMHS are generally older than the mean age of the driving population. This is a well acknowledged aspect in automotive tests for many decades. In some tests, physiological conditions of the living human (re-pressurization of the arterial system, re-insufflation of the lungs, etc.) are reproduced as much as possible to limit the influence of differences between living human and PMHS and their possible influence on the mechanical responses and injuries.

During these tests, when an injury occurs, the time of injury is not always known and when the test is non-injurious, it is not known up to which level no injury would occur. In other words, data are censored. There are different types of censoring.

Data is left or right censored depending on if the measured injury criterion value is greater than or less than the injury tolerance limit (injury tolerance limit being defined as the injury criterion value associated with the occurrence of an injury). It means that when a test is performed and no injury is observed after the test, the measured injury criterion is right censored. When a test is performed and an injury is observed after the test without knowing at which input it occurred, the measured injury criterion is left censored. For example, the thoracic deflection can be measured during a human thoracic impactor test. If no injury is found after the test, the thoracic deflection is right censored because the deflection at which the injury would occur is unknown. If an injury is found after the impact test, the thoracic deflection is left censored because the deflection at which the injury occurred is unknown.

In some configurations, the surrogate is impacted at increased severity until an injury occurs. In that case, the injury tolerance limit is bounded within a known range of values (input of the latest test without injury and of the first injurious test) so the measured injury criterion is interval censored. The measured injury criterion is exact or uncensored when the value of the injury criterion is known at the time of injury. Several attempts have been made to determine the input, i.e., the time at which injury occurs. This has been done using strain gages for rib fractures or acoustic sensors for ankle injuries [51, 52]. Literatures also exist on other regions of the human body. For example, spinal columns have been instrumented with these types of sensors in recent studies [53].

26.3.2 Methods Used to Define Human Injury Risk Curves

26.3.2.1 Mertz-Weber Technique

This method was an earliest approach used in the development of injury risk curves and is a modification of the median rank method [54]. The hypothesis underlying this approach is that the weakest and strongest subjects are closer to their individual injury tolerance limits than the others subjects of the ensemble. Therefore, a normal distribution is fitted between the injury criterion value of the injured subject with the lowest injury criterion value and the one of the non-injured subject with the highest injury criterion value. The mean of the distribution is the mean of the injury criterion values of these two tests. The standard deviation is calculated based on the injury criterion values of these two tests and the number of tests between these two values. This method was used to determine the injury risk curve for the risk of skull fracture and brain injury and for the risk of AIS3+ (Abbreviated Injury Scale) neck injury, in biomechanical studies [55–57]. This method is empirical and has little statistical basis. It cannot account for exact data. The injury risk curve is highly influenced by the strongest and weakest specimens in the ensemble sample, as it is based mainly on these two data points.

26.3.2.2 The Certainty Method

The certainty method was developed to analyze censored data [57]. An assumption of this method is that a correlation exists between the injury criterion and risk, i.e., risk increases when the injury criterion increases. The principle is that if a subject was injured at a given injury criterion value, that subject would have been injured at a higher value. On the contrary, if a subject was not injured at a given injury criterion value, that subject would not have been injured neither at a lower value. For each criterion value, the number of cases is determined based on the above mentioned principle. And then include cases actually tested at this value as well as the cases tested at lower or higher values. Therefore, the same test is accounted for several times in the development of the injury risk curve. The probability of risk corresponds to the ratio of the injured cases at an injury criterion value divided by the total of cases for this value. A fit with a normal distribution is usually done once the non-parametric probability of risk is determined. As an example, the certainty method was used to develop an injury risk curve for the risk of skull fracture and brain injury [57]. The method is based on the hypothesis that a correlation exists between a type of injury and injury criterion. This hypothesis must be checked before the development of the injury risk curve because a physical measurement which has no relationship with the risk of injury will yield an increasing injury risk curve. The method cannot take into account exact data.

26.3.2.3 Logistic Regression

The logistic regression has been widely used in various fields and detailed elsewhere [58]. The logistic regression is a parametric statistical method, modeling the relationship between a binary variable (injured or non-injured, Abbreviate injury score AIS < 3 or ≥3) and one or several explanatory variables (AIS). The logistic distribution of the tolerance is assumed in this parametric method and the probability of risk is given by the following equation.

$$Probability\ of\ risk = \frac{1}{1+e^{-(\beta_0+\beta_1 x_1+\ldots+\beta_i x_i)}} \quad (26.5)$$

Where $x_1, \ldots, x_i$ are explanatory variables (such as thoracic deflection, PMHS age or gender), and $b_0, b_1, \ldots, b_i$ are coefficients of the statistical model associated to these variables. The coefficients of the model are estimated based on the maximum likelihood. The maximum likelihood method allows determining the best fit of the data maximizing the probability of injury for the injured tests and minimizing the probability of injury of the non-injured tests. The logistic regression has been implemented in numerous statistical software and it has been used recently in the biomechanical field [4, 59, 60]. As the logistic distribution is assumed, the probability of risk for an injury criterion value equal to zero will always be greater than zero. A modified logistic regression that imposes a probability of risk equal to zero for an injury criterion value equal to zero was proposed in order to avoid negative tolerance limits for low probability of risk [61]. The logistic regression cannot take into account exact data.

26.3.2.4 The Consistent Threshold and the Extended Consistent Threshold

The Consistent Threshold (CT) and the Extended Consistent Threshold (ECT) are non-parametric statistical methods, developed in order to take into account the censoring of the data and not to make any assumption on the distribution of the probability of the risk of injury. The probability of injury is estimated by the maximum likelihood. The consistent threshold algorithms differ for a sample of censored and mixing of censored and exact data [62, 63]. As in any non-parametric method, no covariates are taken into account and the resulting injury risk curve is a step function. Large steps could indicate that the dataset is small and data are lacking or that there may be several injury mechanisms included in the sample. The CT was used in order to demonstrate that the parametric survival models would fail to identify if a sample includes different injury mechanisms [64]. The CT was also recommended to be used to check the validity of the distribution assumption made in the parametric methods.

26.3.2.5 Survival Analysis

The survival analysis is commonly used in clinical studies to determine the survival time after the onset of a fatal disease. The time of survival is censored and survival analysis has been specifically developed to take this aspect into account. In the biomechanical field, the time variable considered in clinical studies is replaced by injury criterion such as stress, strain, deflection or force. The survival analysis can be non-parametric (Kaplan-Meier), semi-parametric (Cox model) or parametric [65]. The probability of risk is estimated using the maximum likelihood. Survival analysis can handle censored and exact data. The parametric survival analysis has been widely used in biomechanics [65–70]. The probability of risk shown in the following equation is expressed in terms of the hazard function h, where h is the instantaneous risk of injury and $h(u)du$ is the probability for a subject of being injured between u and $u + \Delta u$, given the subject is not injured until u (where u is the injury criterion value).

$$Probability\_of\_risk = 1 - \exp\left(-\int_0^x h(u)du\right) \quad (26.6)$$

Several distributions of the probability of risk can be assumed in the parametric survival analysis: Weibull, log-normal, log-logistic are some examples.

The hazard function is constant or continuously increasing/decreasing for a Weibull distribution, shown below.

$$Probability\_of\_risk = 1 - \exp\left(-\int_0^t h(u)du\right)$$
$$= 1 - \exp\left(-\left(\frac{x}{\lambda}\right)^k\right) \quad (26.7)$$

As an example, for a biomechanical sample for which the $k < 1$, it means that the instantaneous risk of injury decreases with the loading severity. For $k > 1$, the instantaneous injury risk increases with the loading severity. The log-normal and the log-logistic distributions allow a continuously decreasing instantaneous risk of injury or an initial increase of instantaneous risk of injury until a maximum followed by a decrease. The assumption of the distribution assumption should be checked when modeling data with a parametric survival analysis. The parametric survival analysis is usually performed using statistical software. The injury risk curve is easily reported using the formula of the chosen distribution and the values of the coefficients.

26.3.2.6 Comparison of Different Methods

Different methods presented above mentioned have been used in the biomechanical field to propose injury risk curves. Several studies have been conducted to evaluate the influence of the method on the resulting injury risk curve based on the biomechanical samples [57, 71, 72]. Results show that resulting injury risk curves depend on the choice of the method. However, it is not possible to determine which method is the most appropriate as the actual distribution of the risk is unknown. Different methods have also been compared among and to an actual distribution of injury tolerance using statistical simulations [64]. Results show that the parametric survival analysis (Weibull distribution) is the most appropriate method to fit actual distributions evaluated in this study (normal, Weibull). For censored injury criterion, the logistic regression yields similar results. Only the CT and the parametric survival analyses take exact injury criterion into account among the evaluated methods. The probability of risk is shown to be better estimated when increasing the size of the sample and when including exact data in the sample (especially for small sample sizes). For a given sample size, low probabilities of risk are better estimated when injury criterion value is available for low injury severity levels while probabilities over the entire range are better estimated when injury criteria values are widely distributed across the range of the actual tolerance limit.

26.3.3 Recommendations for Development of Injury Risk Curves

The group of normalization ISO/TC22/SC12/WG6 dealing with performance criteria expressed in biomechanical terms has provided recommendations for injury risk curve development in its Technical Specification TS18506 [73]. The group divided the development of the injury risk curve into several steps. Each will be illustrated with the analysis of human injury risk of the lower limb in axial loading based on the tests described in [74].

26.3.3.1 Step 1: Collect Relevant Data

The collection of relevant data is the first step. It should include injury types and severities, as well as injury criteria values measured during the tests. An injury risk curve can be built for any injury severity ($AIS \geq 2$, $AIS \geq 3$...), or for any type of injuries (fracture, soft tissue trauma or contusion...). Co-variables can also be included in the dataset, such as PMHS characteristics such as age, bone mineral content and gender. Relevant variables for the prediction of the injury risk should be properly selected by the biomechanical expert(s), according to the knowledge on the injury mechanisms and factors influencing the tolerance.

Application

The lower limb sample includes 52 PMHS tests from the dataset described above. The injury to be addressed is the fracture of the foot. The authors chose the impact force as the main explanatory variable and included the PMHS age as a co-variable.

26.3.3.2 Step 2: Assign the Censoring Status (Exact, Left, Right, Interval Censored)

It is important to take the censoring status of the injury criterion into account, as exact data brings in more information than censored data. The censoring status can be exact, left, right, or interval censored. The coding of the censoring status should be checked and made according to the statistical software used.

Application

In the lower limb sample considered for this presentation, some PMHS lower legs did not result in fracture. The impact force was coded right censored. Some PMHS lower legs sustained injuries after sustaining the mechanical load. The peak impact force was coded left censored. Some PMHS lower legs were impacted once at sub-injurious level and then at an injurious level. The impact force was coded interval censored and the lower bound of the interval being the force of the non-injurious test and the upper bound being the force of the injurious test.

26.3.3.3 Step 3: Check for Multiple Injury Mechanisms

The complexity of the human body may lead to different injury mechanisms [71]. Before applying a parametric method and making the assumption of a distribution of the probability of injury, it is important to check that the injury risk curve built with a non-parametric statistical method (such as CT) does not show discontinuities or change of slope in the injury risk curve. Such characteristics of the injury risk curve might indicate that several injury mechanisms are included in the sample. This check can only be done for one variable at a time because it uses a non-parametric statistical method. If several explanatory variables are used in the development of the injury risk curves, the sample should be divided in sub-samples dividing the range of the explanatory variable in several classes and building the injury risk curve with the CT for each class.

26.3.3.4 Step 4: Separate Data Sample by Injury Mechanism

If there is no evidence of multiple injury mechanisms, data can be fitted with a parametric survival analysis and coefficients of the model can be estimated (Step 5). Otherwise, the biomechanically tested sample should be separated into samples with a single injury mechanism before proceeding with the following steps.

Application

The sample was divided into four age classes. The size of the samples for the age classes from 20–40 and 80–100 years were too small to have adequate data (5 and 1 respectively). Injury risk curves derived with the CT method for the class of age 40–60 year-old (15 points) and 60–80 year-old (22 points) are presented (Fig. 26.4). The large steps could be a sign for multiple injury mechanisms, but given the size of the samples, it more likely gives a proof of the limited amount of data in each class.

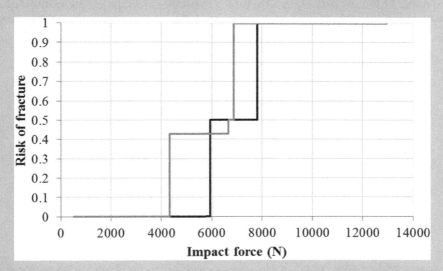

Fig. 26.4 Fracture risk curve as a function of the impact force built with the CT for the class of age 40–60 (*black*) and 60–80 (*grey*), in the illustration

26.3.3.5 Step 5: Estimate the Distribution Parameters

Variables of the parametric survival analysis are estimated with the maximum likelihood method [75]. This calculation is usually implemented in commercial statistical software. The Weibull, log-normal and log-logistic distributions are recommended to be investigated because they ensure zero risk of injury for zero stimuli (contrary to the normal or logistic distributions). The resulting injury risk curves will be compared later in the process of development of the injury risk curves in order to recommend the one that best predicts the injury risk function.

26.3.3.6 Step 6: Identify Overly Influential Observations

Some tests in the sample might influence the estimation of the coefficient of the parametric survival analysis than others. They are overly influential. It is recommended to identify them for calculating the dfbeta's statistics. This statistic indicates the change of each parameter estimate when deleting one observation of the sample at a time. The cut-off value recommended in the literature over which the test is considered as overly influential is $2/\sqrt{\text{sample size}}$ [76]. For the size of the samples usually available in the biomechanical field (assumed to be in the 40–50 range), the cut-off value is approximately 0.3. All the overlying influential tests should be checked for any specificity in terms of test measurements and PMHS characteristics or injuries. These tests should be kept in the analysis in case there is no evidence of errors concerning these tests.

26.3.3.7 Step 7: Check the Distribution Assumption

The distribution assumption should be checked. It can be done graphically using a Q-Q plot ("Q" stands for quantile). The percentiles of the distribution are plotted against the corresponding percentiles of the biomechanical sample. The fact that the

Application

Five out of the 52 tests were considered to belong to overly influential experiments based on the age coefficient. Figure 26.5 shows that the overly influential tests correspond to very young PMHS compared to the others and to the lowest impact force among the injured PMHS.

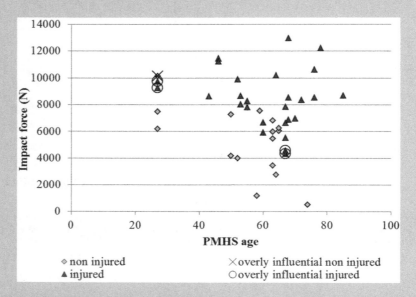

Fig. 26.5 Impact force as a function of the PMHS age for the injured and non-injured subjects and for the overly influential tests

plot follows a line through the origin with a slope equal to unity indicates that the chosen distribution is appropriate. It is also possible to graphically plot the cumulative risk calculated with the survival analysis with a given distribution against the cumulative risk calculated with a non-parametric such as the CT/ECT method. The fact that the cumulative risks lay close one to the other indicates that the chosen distribution is appropriate. This check can only be done for one variable at a time because it uses a non-parametric statistical method. Therefore it is only applicable for injury risk curve as a function of one variable.

26.3.3.8 Step 8: Choose the Distribution

It is recommended to choose the best fit among several distributions evaluated assessing the Akaike information criterion (AIC). The AIC is based on the likelihood of the model but is also an indicator of parsimony as it takes into account the number of variables used in the model (AIC=−2*log likelihood+2*number of variables). The AIC value decreases with the quality of the fit. The distribution with the lowest AIC is then selected.

Application

The AIC values for the Weibull, log-normal and log-logistic distributions were 54, 57 and 57 for this dataset. The Weibull distribution is more appropriate than the other two because the difference of AIC is greater than 2. As a consequence, the Weibull distribution is chosen to develop the injury risk curve.

Application
The parametric survival injury risk curves (for 50 and 70 year-old age specimens) are superimposed to the non-parametric injury risk curves (Fig. 26.6). The number of steps of the non-parametric curves is too limited to assess in a robust manner the chosen distribution. However, given the availability of data, there is no evidence that the chosen distribution is not compatible with the underlying data.

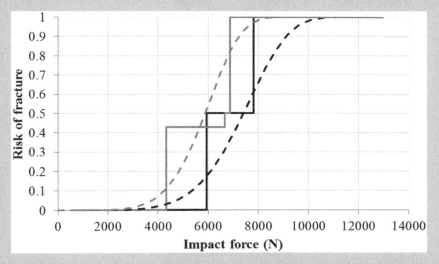

Fig. 26.6 Fracture risk curve as a function of the impact force built with the CT for the class of age 40–60 (*black*) and 60–80 (*grey*) and built with the parametric survival analysis

26.3.3.9 Step 9: Check the Validity of the Predictions Against Existing Results

Injury risk curves based on results from laboratory tests are aimed at predicting the risk of injury in the real-world situations. When possible, it is then essential to check the validity of the predicted injury risk curves against real world results, such as accident observations.

26.3.3.10 Step 10: Calculate the 95 % Confidence Interval

The injury risk curve is the best estimate of the probability of risk. However, depending on the sample size, the censoring status of the tests, or the distribution of the tests relative to the actual injury risk curve, the confidence on the tolerance limits can vary. It is recommended to provide an indicator of the confidence together with the tolerance limit provided. A quality index was defined for that purpose.

Step 10.1

First, the 95 % confidence interval of the injury risk curve is calculated with the normal approximation of the error.

Step 10.2

It is then used to calculate the relative size of the confidence interval. It is defined as the width of the 95 % confidence interval at a given injury risk relative to the value of the stimulus at this same injury risk. It was chosen to be determined at 5 %, 25 % and 50 % risks of injury in order to provide confidence indicators for levels of risk of injury used in the regulatory protocols as well as in consumer-testing protocols.

Application

The entire dataset was used to calculate 95 % confidence intervals of the probability of fracture risk as a function of the impact force for the median age of the PMHS (Fig. 26.7).

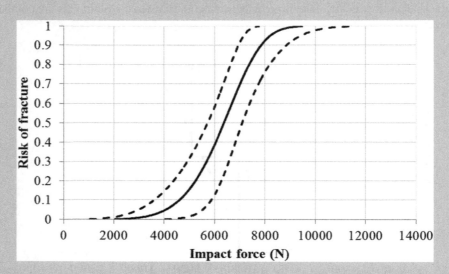

Fig. 26.7 Fracture risk curve as a function of the impact force for the median age of the PMHS derived with the parametric survival analysis (*solid line*) and 95 % confidence intervals

Table 26.4 Categories of the quality index based on the relative size of the 95 % confidence interval

Quality index	Relative size of the 95 % confidence interval
Good	From 0 to 0.5
Fair	From 0.5 to 1
Marginal	From 1 to 1.5
Unacceptable	Over 1.5

26.3.3.11 Step 11: Assess the Quality Index of the Injury Risk Curves

As a first approach, biomechanical samples were used to construct injury risk curves according to the above mentioned process. The quality indices were calculated for each sample and were used to propose four categories of quality index (Table 26.4). The four categories were determined to have curves in each of them. Once these four categories were defined, they were used to assess the quality of any injury risk curve. It is recommended to provide the quality index along with each injury risk curve. This has not been done consistently in previous impact biomechanical studies.

Application

The quality index were determined to be fair at 5 % and good at 25 % and 50 % of fracture risk for the median age of the PMHS (63 year-old) from this dataset.

26.3.3.12 Step 12: Recommend One Curve per Body Region, Injury Type, and Injury Severity

Several injury risk curves could be derived for a given injury mechanism and for a given body region. The prediction of the risk could then be contradictory between the injury risk curves. It is then recommended to select only one injury risk curve per body region, injury type and injury severity. The best predictor of the risk should be selected in such instances. Nevertheless, several injury risks curves are necessary to address several injury mechanisms for a given body region.

Step 12.1

Injury risk curves are compared using their AIC. A difference of at least two in terms of AIC illustrates a difference in the fit of the data between the different injury risk curves [77]. It is recommended to select the injury risk curve with the lowest AIC with a difference of two in terms of AIC over the others. The absolute value of AIC does not have practical meaning, by itself. However, AIC is useful when comparing several models. As AIC takes into account the number of explanatory variables, models with different number of explanatory variables can be compared, as long as the sample of tests on which the injury risk curves are based is the same in size and in terms of injured/non-injured cases.

Step 12.2

An injury risk curve can be better than another but still present poor ability to predict the injury risk properly. It was chosen not to recommend an injury risk curve with an "unacceptable" quality index.

Step 12.3

Several injury risk curves might be equivalent in terms of AIC and all with quality indices might be from good to marginal. In addition of these statistical tools, engineering judgment should be used to recommend one curve over the others. Once the process of selection of the injury risk curve is completed, the injury risk curve should be recommended with its associated quality indices in order for the final user to have a good idea of the quality of the provided tool. Depending on the biomechanical sample, the parametric survival analysis might not converge in some cases (censoring, small size, distribution of the tests, etc.). In those cases, a non-parametric method such as CT would allow to provide a tolerance limit. However, the fact that the parametric survival does not converge clearly indicates a lack of information. No tolerance limit based on CT should therefore be recommended in that case. The injury risk curve based on the CT would show large steps anyway.

26.4 Discussion

The choice of explanatory variables to be associated to a given injury mechanism is a major important step. If the relevant explanatory variables are not selected, the injury risk curve developed will be of poor quality and will not allow the prediction of the injury risk. In order to improve the predictive ability of the injury risk curves, exact data should be used as much as possible. Increasing the number of tests available to develop the injury risk curve should also help this process. Moreover, human injury risk curves in biomechanics are often developed based on PMHS tests. Resulting injury risk curves normally apply to a population having the same characteristics as the PMHS included in the sample. Tolerance limits determined from PMHS tests should normally be somehow corrected before being applied to a population with different characteristics (example, age). If it is not possible, tolerance limits should be considered cautiously. Furthermore, human tolerance to impact is influenced by age, as it is often an explanatory variable considered when developing human injury risk curves. The range of age of the PMHS considered in the biomechanical samples is generally around 60–80 years. The effect of age on the probability of injury is then determined on a relative narrow range of age and might not be very accurate and robust for all ages, even in the adult range. Moreover, when proposing an injury risk curve for an age for which the number of cases is scarce, little information is often available to develop the risk curve from these biomechanical tests. The confidence in the tolerance limits associated to this age will be lower than for ages for which the number of cases is more important and the quality index of the risk curve will be poorer. This should be recognized and underscored in any recommendation. In addition, probabilities of risk at which tolerance limits are considered differ depending of the type of application. The confidence in a particular level of risk is improved if test results are available for these levels of risk. If low probabilities of risk are of interest, results from tests corresponding to low probabilities of risk will increase the confidence of the tolerance limit in this area. If the 50 % of risk is the main focus, results from tests corresponding to this level help providing a tolerance limit with an improved confidence. Finally, acknowledging that human injury risk curves are proposed based on laboratory tests, it is important to underscore that the purpose of the human injury risk curve is to improve occupant protection, regardless of the application, civilian (automotive for example) or military. Ensuring the appropriateness of laboratory-based injury risk curves to properly evaluate the actual injury risk is critical.

26.5 Summary

Common methods are described in this chapter for converting responses recorded during PMHS experiments to reference responses. Generally the mid-size male total body mass is used as the initial standard to normalize biomechanical responses in automotive crashworthiness studies. The equal stress equal velocity and impulse momentum methods, and methods used in developing corridors were explored. The impulse momentum approach is the method of choice in recent studies because of its potential to accommodate different variables in the normalization process. Studies are needed to show tightness in the corridors with the use of this method. While the mean and plus and minus standard deviation responses have been the choice for describing human response corridors, it should be noted that for crashworthiness applications such as anthropometric test dummies, smoothening of the responses are needed as physical materials do not have the same fidelity as human tissues/body region. Some recommendations are available from the ISO normalization group dealing with human tolerance to develop an injury risk curve from a given dataset. They include the use of the survival analysis statistical method specific to the censored status of the biomechanical data, several checks necessary when statistical modeling is performed, as well as the recommendation of the injury risk curve along with an indicator of the quality of the curve. However, there is no recommendation to assist the biomechanical researcher to elaborate the biomechanical dataset to develop the curve. The required characteristics of the dataset allowing to develop an injury risk curve of good quality depends on the censoring status (a high proportion of exact data improves the quality), size of the sample (a high number of cases improves the quality), and distribution of the cases relative to the actual injury distribution which is unknown. General recommendations have been provided in this chapter based on some statistical simulations. Applying parametric survival analysis requires checking some hypothesis such that the chosen distribution is appropriate. When several variables are used in the analysis, the dataset should be divided in subgroup to evaluate the appropriateness of the chosen distributions. The dataset available in the impact biomechanics field are often limited in size such that it is not possible to generate sub-groups for which the distribution can be checked. A semi-parametric approach such as Cox regression would be an alternative as no assumption on the form of the distribution is made. Its applicability as well as the acceptability of its assumptions in the biomechanical field could be investigated in the future. Finally, data used in biomechanics are most of the time censored. The type of censoring is taken into account using the survival analysis. However, how much the data is censored is not always known. When developing an AIS ≥ 3 injury risk curve, a test with AIS1 or AIS2 is coded the same way while the level of censoring is different. Including this type of information in the modeling of the data will improve the quality of the injury risk curve. These studies should be designed in the future.

Acknowledgments This material is the result of work supported with resources and the use of facilities at the Zablocki VA Medical Center, Milwaukee, Wisconsin and the Medical College of Wisconsin. Narayan Yoganandan and Frank. A. Pintar are part-time employees of the Zablocki VA Medical Center, Milwaukee, Wisconsin. Any views expressed in this chapter are those of the authors and not necessarily representative of the funding organizations.

References

1. Mertz HJ (2002) Injury risk assessment based on dummy responses. In: Nahum AM, Melvin JW (eds) Accidental injury: biomechanics and prevention, 2nd edn. Springer, New York, pp 89–102
2. Cavanaugh JM, Walilko TJ, Malhotra A, Zhu Y, King AI (1990) Biomechanical response and injury tolerance of the thorax in twelve sled side impacts. In: Proceedings of the 34th Stapp car crash conference, Orlando, Florida, United States, pp 23–38
3. Kallieris D, Mattern R, Schmidt G, Eppinger RH (1981) Quantification of side impact responses and injuries. In: Proceedings of the 25th Stapp car crash conference, San Francisco, 28–30 Sept 1981. Society of Automotive Engineers, pp 329–368
4. Kuppa S, Eppinger RH, McKoy F, Nguyen T, Pintar FA, Yoganandan N (2003) Development of side impact thoracic injury criteria and their application to the modified ES-2 dummy with rib extensions (ES-2re). Stapp Car Crash J 47:189–210
5. Pintar FA, Yoganandan N, Hines MH, Maltese MR, McFadden J, Saul R, Eppinger RH, Khaewpong N, Kleinberger M (1997) Chestband analysis of human tolerance to side impact. In: Proceedings of the 41st Stapp car crash conference, Lake Buena Vista, pp 63–74
6. Yoganandan N, Morgan RM, Eppinger RH, Pintar FA, Sances A Jr, Williams A (1996) Mechanisms of thoracic injury in frontal impact. J Biomech Eng 118(4):595–597
7. Yoganandan N, Pintar FA, Stemper BD, Gennarelli TA, Weigelt JA (2007) Biomechanics of side impact: injury criteria, aging occupants, and airbag technology. J Biomech 40(2):227–243. doi:10.1016/j.jbiomech.2006.01.002
8. Stemper BD, Yoganandan N, Pintar FA (2004) Response corridors of the human head-neck complex in rear impact. Annu Proc Assoc Adv Automot Med 48:149–163
9. Lessley D, Shaw G, Parent D, Arregui-Dalmases C, Kindig M, Riley P, Purtsezov S, Sochor M, Gochenour T, Bolton J, Subit D, Crandall J, Takayama S, Ono K, Kamiji K, Yasuki T (2010) Whole-body response to pure lateral impact. Stapp Car Crash J 54:289–336
10. Yoganandan N, Zhang J, Pintar F (2004) Force and acceleration corridors from lateral head impact. Traffic Inj Prev 5(4):368–373. doi:10.1080/15389580490510336
11. Eppinger R (1976) Prediction of thoracic injury using measurable experimental parameters. In: Proceedings of the 6th international conference on experimental safety vehicles, Washington, DC. NHTSA, pp 770–779
12. Eppinger R, Marcus J, Morgan R (1984) Development of dummy and injury index for NHTSA's thoracic side impact protection research program. In: Government/industrial meeting and exposition, Washington, DC, 21–24 May 1984. US DOT NHTSA, pp 1–29
13. FMVSS-214 (1990) Federal Register, Docket No. 88–06. 49 CFR; Parts 571. side impact protection, vol 55. US Government Printing Office, Washington, DC

14. Kuppa S (2004) Injury criteria for side impact dummies. Washington, DC
15. Maltese MR, Eppinger RH, Rhule HH, Donnelly BR, Pintar FA, Yoganandan N (2002) Response corridors of human surrogates in lateral impacts. Stapp Car Crash J 46:321–351
16. FMVSS-214 (2008) 49Code of federal regulations: 571.214, vol 55. US Government Printing Office, Washington, DC
17. Yoganandan N, Pintar FA (2005) Deflection, acceleration, and force corridors for small females in side impacts. Traffic Inj Prev 6(4):379–386. doi:10.1080/15389580500256888
18. Mertz HJ (1984) A procedure for normalizing impact response data. SAE 840884
19. Viano D (1989) Biomechanical responses and injuries in blunt lateral impact. In: Stapp car crash conference, Washington D.C, United States, pp 113–142
20. Moorhouse KM (2013) An improved normalization methodology for developing mean human response curves. In: Proceedings of the enhanced safety of vehicles conference, Seoul
21. Yoganandan N, Humm J, Arun MWJ, Pintar FA (2013) Oblique lateral impact biofidelity deflection corridors from post mortem human surrogates. Stapp Car Crash J 58:427–440
22. Eppinger R, Augenstyn K, Robbins DH (1978) Development of a promising universal thoracic trauma prediction methodology. In: Stapp car crash conference, Warrendale
23. Irwin A, Walilko T, Cavanaugh J, Zhu Y, King A (1993) Displacement responses of the shoulder and thorax in lateral sled impacts. In: Stapp car crash conference, San Antonio, pp 166–173
24. Marcus JH, Morgan RM, Eppinger R (1978) Human Response to and Injury from lateral Impact. In: Stapp car crash conference, 1983, Warrendale
25. Walfisch G, Fayon A, Tarriere C, Rosey J, Guillon F, Got C, Patel A, Stalnaker R (1980) Designing of a dummy's abdomen for detecting injuries in side impact collisons. In: IRCOBI, Birmingham, United Kingdom, pp 149–164
26. Cesari D, Ramet M (1980) Evaluation of pelvic fracture tolerance in side impact. In: Stapp car crash conference, Troy, 20–21 Oct 1980
27. Cesari D, Ramet M (1982) Pelvic tolerance and protection criteria in side impact. In: Stapp car crash conference, Ann Arbor, 20–21 Oct 1982, pp 145–154
28. Cesari D, Ramet M, Bouquet R (1983) Tolerance of human pelvis to fracture and proposed pelvic protection criterion to be measured on side ipact dummies. In: Proceedings of the 9th international technical conference on experimental safety vehicles, Kyoto, pp 261–269
29. Tarriere C, Walfisch G, Fayon A, Got C, Guillon F (1979) Synthesis of human impact tolerance obtained from lateral impact simulations. In: 7th international technical conference on experimental safety vehicles, Washington, DC, pp 359–373
30. Shaw JM, Herriott RG, McFadden JD, Donnelly BR, Bolte JH (2006) Oblique and lateral impact response of the PMHS thorax. Stapp Car Crash J 50:147–167
31. Lobdell TE, Kroell CK, Schneider DC, Hering WE, Nahum AM (1973) Impact response of the human thorax. In: King WF, Mertz HJ (eds) Human impact response: measurement and simulation. Plenum Press, New York, pp 201–245
32. Morgan RM, Marcus J, Eppinger R (1981) Correlation of side impact dummy/cadaver tests. In: Stapp car crash conference, San Francisco, California, United States, pp 301–326
33. McAdams H (1973) The repeatability of dummy performance. In: Kwam HJ (ed) Human impact response. Plenum Press, New York, pp 35–67
34. Pintar FA, Yoganandan N, Myers T, Elhagediab A, Sances A Jr (1992) Biomechanical properties of human lumbar spine ligaments. J Biomech 25(11): 1351–1356
35. Yoganandan N, Pintar FA, Kumaresan S (1998) Biomechanical assessment of human cervical spine ligaments. In: Stapp car crash conference, Tempe, pp 223–236
36. Yoganandan N, Kumaresan S, Pintar FA (2000) Geometric and mechanical properties of human cervical spine ligaments. J Biomech Eng 122(6): 623–629
37. Ash J, Lessley D, Forman JL, Zhand C, Shaw JM, Crandall J (2013) Whole-body kinematics: response corridors for restrained PMHS in frontal impacts. In: International conference on the biomechanics of injury, Dublin, 16–18 Sept 2013
38. Lessley D, Crandall J, Shaw G, Kent RW, Funk J (2004) A normalization technique for developing corridors from individual subject responses. In: SAE World Congress, Detroit, 8–11 Mar 2004
39. Kleinberger M, Sun E, Eppinger R, Kuppa S, Saul R (1998) Development of improved injury criteria for the assessment of advanced automotive restraint systems. NHTSA, Washington, DC
40. Kleinberger M, Yoganandan N, Kumaresan S (1998) Biomechanical considerations for child occupant protection. In: AAAM, Charlottesville, pp 115–136
41. Melvin JW (1995) Injury assessment reference values for the CRABI 6-month infant dummy in a rear-facing infant restraint with airbag deployment. In: SAE Congress and exposition, Detroit. SAE, pp 1–12
42. Yoganandan N, Kumaresan S, Pintar F, Gennarelli T (2000) Pediatric biomechanics. In: Nahum A, Melvin J (eds) Accidental injury: biomechanics and prevention. Springer, New York, pp 550–587
43. Miller C, Rupp J (2011) PMHS impact response in low and high-speed nearside impacts. US DOT, Washington, DC
44. Miller C, Rupp J (2013) PMHS Impact response in 3 m/s and 8 m/s nearside impacts with abdomen offset. Stapp Car Crash J 58:387–425
45. Yoganandan N, Pintar FA, Maltese MR (2001) Biomechanics of abdominal injuries. Crit Rev Biomed Eng 29(2):173–246

46. Praxl N (2011) How reliable are injury risk curves? In: 22nd ESV conference proceedings, Washington, DC, paper n°11-0089
47. Arbogast KB, Balasubramanian S, Seacrist T, Maltese MR, Garcia-Espana JF, Hopely T, Constans E, Lopez-Valdes FJ, Kent RW, Tanji H, Higuchi K (2009) Comparison of kinematic responses of the head and spine for children and adults in low-speed frontal sled tests. Stapp Car Crash J 53:329–372
48. Ewing CL, Thomas DJ, Majewski PL, Black R, Lustik L (1977) Measurement of head, T1 and pelvic response to Gx impact acceleration. In: Proceedings of the 21st Stapp car crash conference, Society of Automotive Engineers, Warrendale
49. Mertz HJ, Driscoll GD, Lenox JB, Nyquist GW, Weber DA (1982) Responses of animals exposed to deployment of various passenger inflatable restraint system concepts for a variety of collision severities and animal positions. In: Proceedings of the 9th international technical conference on experimental safety vehicles, Kyoto, pp 352–368
50. Prasad P, Daniel R (1984) Biomechanical analysis of head, neck, and torso injuries to child surrogates due to sudden torso acceleration. In: Stapp car crash conference, Chicago, Nov 1984. Society of Automotive Engineering, pp 25–40
51. Leport T, Baudrit P, Potier P, Trosseille X, Lecuyer E, Vallancien G (2011) Study of rib fracture mechanisms based on the rib strain profiles in side and forward oblique impact. Stapp Car Crash J 55:199–250
52. Rudd R, Crandall J, Millington S, Hurwitz S, Hoglund N (2004) Injury tolerance and response of the ankle joint in dynamic dorsiflexion. Stapp Car Crash J 48:1–26
53. Yoganandan N, Arun MWJ, Stemper BD, Pintar FA, Maiman DJ (2013) Biomechanics of human thoracolumbarpinal column trauma from vertical impact loading. Ann Adv Automot Med 57:155–166
54. Mertz HJ, Weber DA (1982) Interpretations of the impact responses of a 3-year-old child dummy relative to child injury potential, SAE 826048, SP-736. Automatic Occupant Protection Systems, Feb 1988
55. AIS (1990) The abbreviated injury scale. The Association for the Advancement of Medicine, Chicago
56. Mertz HJ, Prasad P, Irwin AL (1997) Injury risk curves for children and adults in frontal and rear collisions. In: Proceedings of the 41st Stapp car crash conference. Society of Automotive Engineers, Warrendale, pp 13–30
57. Mertz HJ, Prasad P, Nusholtz G (1996) Head injury risk assessment for forehead impacts. SAE Paper 960099. Society of Automotive Engineers, Warrendale
58. Hosmer DW, Lemeshow S (2000) Applied logistic regression, 2nd edn. Wiley, New York
59. Eppinger R, Sun E, Bandak F, Haffner M, Khaewpong N, Maltese M, Kuppa S, Nguyen T, Takhounts E, Tannous R, Zhang A, Saul R (1999) Development of improved injury criteria for the assessment of advanced automotive restraint systems – II. National Highway Traffic Safety Administration, U.S. Department of Transportation, Docket No. NHTSA-99-6407, Nov 1999
60. Laituri TR, Prasad P, Kachowski BP, Sullivan K, Przybylo PA (2003) Predictions of AIS3+ thoracic risks for belted occupants in full-engagement, real-world frontal impacts: sensitivity to various theoretical risk curves. SAE Paper 2003-01-1355, Society of Automotive Engineers, Warrendale
61. Nakahira Y, Furrkawa K, Niimi H, Ishihara T, Miki K (2000) A combined evaluation method and a modified maximum likelihood method for injury risk curves. In: Proceedings of the IRCOBI, international conference of the biomechanics of impact, Montpellier, pp 147–155
62. Di Domenico L, Nusholtz G (2005) Risk curve boundaries. Traffic Inj Prev 6(1):86–94. doi:10.1080/15389580590903212
63. Nusholtz G, Mosier R (1999) Consistent threshold estimate for doubly censored biomechanical data. SAE Paper 1999-01-0714. Society of Automotive Engineers, Warrendale
64. Petitjean A, Trosseille X (2011) Statistical simulations to evaluate the methods of the construction of injury risk curves. Stapp Car Crash J 55:411–440
65. Pintar FA, Yoganandan N, Voo L (1998) Effect of age and loading rate on human cervical spine injury threshold. Spine 23(18):1957–1962
66. Funk JR, Srinivasan SC, Crandall JR, Khaewpong N, Eppinger RH, Jaffredo AS, Potier P, Petit PY (2002) The effects of axial preload and dorsiflexion on the tolerance of the ankle/subtalar joint to dynamic inversion and eversion. Stapp Car Crash J 46:245–265
67. Johannsen H, Trosseille X, Lesire P, Beillas P (2012) Estimating Q-dummy injury criteria using the CASPER project results and scaling adult reference values. In: Proceedings of the IRCOBI, international conference of the biomechanics of impact, Dublin
68. Kent RW, Funk JR (2004) Data censoring and parametric distribution assignment in the development of injury risk functions from biomechanical data. SAE Paper N°2004-01-0317. Society of Automotive Engineers, Warrendale
69. Petitjean A, Trosseille X, Praxl N, Hynd D, Irwin A (2012) Injury risk curves for the WorldSID 50th male dummy. Stapp Car Crash J 56:323–347
70. Takhounts E, Craig M, Moorhouse K, McFadden J, Hasija V (2013) Development of brain injury criteria (BrIC). Stapp Car Crash J 57:243–266
71. Di Domenico L, Nusholtz GN (2003) Comparison of parametric and non-parametric methods for determining injury risk. SAE Paper 2003-01-1362. Society of Automotive Engineers, Warrendale
72. Wang L, Banglmaier R, Prasad P (2003) Injury risk assessment of several crash data sets. SAE Paper 2003-01-1214. Society of Automotive Engineers, Warrendale
73. ISO TS18506 (2013) Road vehicles – procedure to construct injury risk curves for the evaluation of road

users protection in crash tests. International Standards Organization, American National Standards Institute, New York

74. Yoganandan N, Pintar FA, Boynton M, Begeman P, Prasad P, Kuppa SM, Morgan RM, Eppinger RH (1996) Dynamic axial tolerance of the human foot-ankle complex. In: Stapp car crash conference, Albuquerque

75. Hosmer DW, Lemeshow S (1999) Applied survival analysis: regression modeling of time to event data. Wiley, New York

76. Belsley DA, Kuh E, Welsch RE (1980) Regression diagnostics: identifying influential data and sources of collinearity. Wiley, New York

77. Burnham KP, Anderson D (2010) Model selection and multi-model inference. Springer, New York

Injury Criteria and Motor Vehicle Regulations

27

Priyaranjan Prasad

Abstract

A historical review of frontal and side impact regulations in the U.S.A. and Europe has been conducted. The biomechanical bases of injury criteria utilized in these regulations have been examined. It is shown that the biomechanical database utilized in the current impact regulations are based on research conducted before the mid-1980s. Research conducted since then has not had much of an impact on regulatory injury criteria, although upgrades in the US FMVSS 208 and the FMVSS 214 have attempted to synthesize all available data to develop tolerance levels for various size dummies.

27.1 Introduction

The influence of impact biomechanics on automotive safety regulations worldwide has been profound over the years. Since the promulgation of the Safety Act in the USA in 1966, passenger vehicles have been subjected to many safety regulations that dictated the way they were designed. Early safety regulations controlled the energy absorption designs of vehicle interiors through component tests, steering column movement in frontal barrier impacts, strength of doors for side impact and dynamic fuel system integrity of vehicles in frontal, side and rear crashes. With the compulsory introduction of three-point seat belts starting around 1974, the need to dynamically test the seat belts in vehicles subjected to frontal crashes against rigid barriers became a necessity without any existing regulations to do so. The need became more acute as frontal airbags were proposed to satisfy the "passive" requirements of the Federal Motor Vehicle Safety Standard 208 (FMVSS 208). Dynamic testing of the vehicle and its restraint system required repeatable and reproducible crash test procedures with objective test devices- crash test dummies (also known as Anthropomorphic Test Devices, ATD's)- and acceptance criteria based on the responses of the dummy. It is in the development of a suitable ATD whose response to impact was humanlike ("biofidelic") where impact biomechanics played a major role. Biomechanics research was aimed at developing the impact response of various parts of the human body, mechanisms of injury as they related to ATD response- like accelerations,

P. Prasad, Ph.D. (✉)
Prasad Consulting, LLC, Plymouth, MI, USA
e-mail: priya@prasadengg.com

forces, moments, deflections, etc. that needed to be controlled during crash tests to assure a level of safety commensurate with the severity of crash being simulated. Even before the adoption of the FMVSS 208, a New Car Assessment Program was introduced in 1976. This involved frontal crash testing of cars with belted dummies in the front outboard seating positions against a rigid barrier at 35 mph. The dummy used was called the Hybrid II dummy- certified as an objective test device- representing a mid-sized American male in mass and size. This dummy had limited response measurement capabilities- head and chest accelerations and femur forces. Therefore the rating system was based on these measurements.

The Hybrid II dummy was developed by General Motors in 1972 to assess the integrity of lap and shoulder belt restraint system. Whereas, this dummy had the size, shape and mass distribution of a 50th percentile adult male, its biofidelic responses in various parts (e.g. head, neck, and chest) were suspect. Due to lack of biofidelity and limited instrumentation, the injury criteria specified in the FMVSS208 were based on resultant accelerations of the head and chest, and axial femur loads.

While compliance to the FMVSS208 and NCAP rating tests continued with the Hybrid II dummy for many years, substantial biomechanical impact research was initiated to establish mechanical responses and associated injury criteria for a new generation of crash test dummies for frontal impact evaluation of vehicles. The results of several more years of research led to the development of the Hybrid III dummy representing a mid-sized American male in size and weight. It had substantially more biofidelity than that of the Hybrid II in all its components. Along with improved biofidelity, instrumentation capabilities were substantially added to measure more meaningful impact responses like neck loads, chest deflections, forces and moments in the femur and tibia. In 1984, this dummy was adopted as an alternate dummy to demonstrate compliance with the FMVSS208. Subsequently, the Hybrid II dummy was phased out of the FMVSS208, and the Hybrid III dummy became the only dummy that could be used for frontal compliance purposes.

With the introduction of the Hybrid III dummy in 1984, Mertz [1] proposed a set of Injury Assessment Reference Values associated with several injury indicating responses being measured in the dummy. Not all of these responses were chosen for this initial regulation. Head Injury Criteria, spinal acceleration, chest deflections, and femur loads were the only responses regulated by the FMVSS208. The FMVSS208 was further modified in 1998 to include neck forces and moments as regulated responses. Frontal Crash Regulations in Canada, Europe, Japan, and Australia have also accepted the Hybrid III dummy as the test device with controls over the measured dummy responses. The limits specified on the dummy responses vary somewhat from the FMVSS208. Whereas the FMVSS208 specifies controls over head, neck, chest, and femur responses, the European directive (96/79 ECE Directive) has additional controls placed on the tibia. The ECE directive for frontal impact basically adopted all the Injury Assessment Reference Values (IARV's) proposed by Mertz [1, 2] with some minor modifications. The biomechanical origins and basis of these IARV's will be discussed later in this paper. It should be mentioned that these IARV's as suggested by Mertz were based on the state of the biomechanical knowledge in 1984, and spurred considerable re-examination and basic research in the biomechanics of the head, neck, thoracic and lower limb responses and injury criteria.

In the US, the Hybrid III family of dummies was incorporated into the Advanced Airbag Rule (Part of the FMVSS208). A small-female dummy (generally referred to as the 5th percentile dummy), a 12-month infant dummy (CRABI), a 3 year and a 6 year child dummies were incorporated into the FMVSS208. The goal of this rule is to minimize the potentially adverse effects of airbag inflation while maximizing its benefits to a range of the human population. The frontal crash regulation in the US is more extensive than in any other country. Figures 27.1, 27.2 and 27.3 show the test conditions in the FMVSS208 and Fig. 27.4 shows the US Regulation and its comparison with the frontal regulations in other regions of the world. For completeness, the frontal

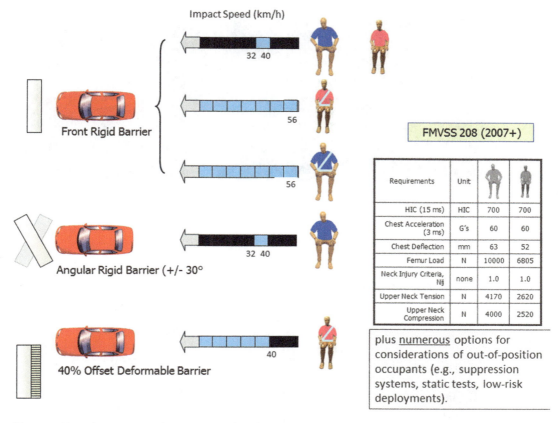

Fig. 27.1 Frontal crash test configurations in US FMVSS208 (Advanced Airbag Rule)

Fig. 27.2 Driver side airbag out-of-position tests in FMVSS208

NCAP tests are also shown in Fig. 27.4. The FMVSS208 involves conducting frontal barrier tests with belted and unbelted dummies, two adult dummy sizes and tests with Out-of-Position occupants. With increase in testing with various size dummies, injury criteria to be utilized with the various dummies had to be developed and will be discussed next.

Tables 27.1 and 27.2 show the injury criteria that have to be satisfied for Head, Chest, Femur

Fig. 27.3 Passenger side OOP tests in FMVSS208 with child dummies. Minor setup differences between the ISO positions and FMVSS208 positions exist

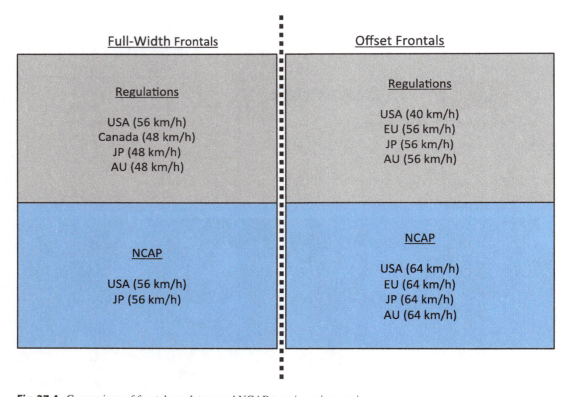

Fig. 27.4 Comparison of frontal regulatory and NCAP tests in various regions

and Neck for the various size dummies being tested in the FMVSS208. It should be noted that in all other regions of the world, only the Hybrid III Mid-Size male dummy is in the regulatory frontal test. The injury criteria specified in the ECE R94 are also shown in Table 27.1. Notably different from the FMVSS208 is the inclusion of the Viscous Criteria in the ECE R94 and the continued use of HIC 36 ms. The femur loads in the ECE R94 are time dependent. Table 27.2 shows

Table 27.1 Head, chest and femur injury limits in FMVSS208 and ECE R94

	Mid-male	Small female	6 years old child	3 years old child	12 month infant
Head HIC 15 ms	700	700	700	570	390
Chest (G) 3 ms cumdur	60	60	60	55	50
Chest defl. (mm)	63	52	40	34	No reg.
Femur load (kN)	10	6.805	No reg.	No reg.	No reg.

ECE R94 Limits for Mid-Size Male Hybrid III
Head: HIC 36 ms 1,000. Peak resultant 3 ms head acceleration not to exceed 80 G's
Chest deflection: 50 mm Chest V*C: 1
Femur load limit: Duration of loading over given force 9.07 kN for 0 ms
Duration of loading over given force 7.58 kN for $\geq$ 10 ms

Table 27.2 Neck injury limits in FMVSS208

Nij intercept	Mid-male	Small female	6 years old child	3 years old child	12 month infant
Nij = F/Fc + M/Mc, where Fc and Mc intercepts are in the table below:					
Tension (N)	6,806	4,287	2,800	2,120	1,460
Compression (N)	6,160	3,880	2,800	2,120	1,460
Flexion (Nm)	310	155	93	68	43
Extension (Nm)	135	67	37	27	17
Limit	1	1	1	1	1
	Mid-male	Small female	6 years old child	3 years old child	12 month infant
Additional limits on tension and compression (N):					
Tension	4,170	2,620	1,490	1,130	780
Compression	4,000	2,520	1,820	1,380	960

Note: The child intercept values are for OOP Testing, and the Adult values are for in-position testing

Table 27.3 Neck injury limits in ECE R94

Dur. of load	0 ms	25 ms	35 ms	60+ ms
Tension (kN)	3.3	Linear 25–35	2.9	1.1
Shear (kN)	3.1	1.5	1.5/1.1@45 ms	1.1
Neck ext. (Nm)	57	57	57	57

the newly developed Nij for the FMVSS208 Advanced Airbag Rule. Table 27.3 shows the neck injury criteria (NIC) in the ECE R94. The functional equivalence between the FMVSS208 and ECE R94 has not been established, however there is no known difference in neck injury risk between Europe and US.

Side Impact regulations were first introduced in the US in what is known as the FMVSS214. It was followed by the regulation in Europe- ECE R95. The crash test configurations in the two regions are different as shown in Fig. 27.5, and initially even the ATD's were different. The earliest dummy designed for side impact testing is commonly known as the USSID and became the regulatory dummy in the US FMVSS214 standard in 1984. At the time of the regulation, repeatability and reproducibility of the dummy design was established and the associated injury criteria were already researched. A parallel dummy development with additional measurement capability was ongoing in Europe resulting in the development and inclusion of this dummy, Eurosid I, in the European side impact directive followed a few years after the introduction of the US FMVSS214. Since the Eurosid I had biofidelity for lateral impacts in the chest and abdomen and load measurement capability in the pelvis, it is not surprising that these responses were regulated in Europe. In terms of biomechanical basis for injury criteria used with the two dummies, the USSID used acceleration-based

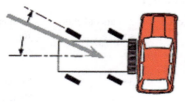 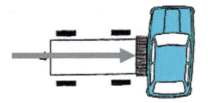

USA
- 1367 kg impactor
- Square and Uniform Stiffness Deformable Face
- 27° Crabbed impact
- 54 km/h impact speed
- 50th Percentile DOT SID in Front and Rear seats (changing now)

Europe and Japan
- 950 kg impactor
- Square and Variable Stiffness Deformable Face
- Perpendicular impact
- 50 km/h impact speed
- 50th Percentile EuroSID II in

- Australia and Canada accept either regulatory side impact test
- US now has an Angle Pole Test that is being phased-in

Fig. 27.5 Side impact regulatory tests in USA, Europe and Japan

Table 27.4 Side impact injury limits in FMVSS214 and ECE R95

Regulation	FMVSS214	FMVSS214	ECE R95
Dummy	ES-2re 50th male	SID-IIs 5th female	ES-2 50th male
Head HIC 36 ms	1,000	1,000	1,000
Neck	No requirement	No requirement	No requirement
Chest	44 mm rib defln.	82 g lower spine	42 mm rib defln.
Abdomen force	2.5 kN	No requirement	2.5 kN
Pelvis	6.0 kN	5.525 kN	6.0 kN

criteria, Thoracic Trauma Index (TTI), and pelvic acceleration for which the dummy was designed. The Eurosid I was designed for biofidelity in chest deflection, abdominal and pelvic loads. The injury criteria adopted were deflection and load based. The USSID has now been replaced by ES-IIre that is similar to the European ES-II dummy (currently used in the European testing). The TTI is no longer used to demonstrate compliance to the FMVSS214. Table 27.4 lists the injury criteria currently being used in the FMVSS214 and the ECE R94. Notable difference between the two regulations is the lack of the small female dummy in the ECE tests. The injury criteria used by the two regulations for the Mid-Male dummies are essentially the same.

Once again, the side impact regulation in the US has gained in complexity with the inclusion of a small female dummy and an additional crash condition- an oblique pole test. The justification for the upgrades in the FMVSS214 have been reported by Samaha and Elliott in 2003 [3]. The addition of a small female dummy, the SID IIs, injury criteria applicable to this dummy had to be developed. The Insurance Institute for Highway Safety in the US has been conducting side impact ratings of light vehicles using the SID IIs dummy for several years. One would have expected that the FMVSS214 would have adopted functionally equivalent injury criteria used by the IIHS. However, NHTSA decided to develop different injury criteria for the regulation.

The following parts of this paper discuss the biomechanical basis of the injury criteria currently incorporated in frontal and side impact regulations in the US and Europe. Note that other regions of the world like Australia, Japan and China have incorporated parts of both the US and the European regulations in terms of test procedures, dummies and injury criteria. Although regulations for rear impacts and rollovers do not include dummy related responses, potential injury criteria that might enter future regulations will be discussed at the end of this paper.

27.2 The Biomechanical Basis of Regulatory Injury Criteria

The biomechanical basis for injury criteria is established by tests performed on animals and human cadaveric specimens. The tests generally include measurement of dynamic responses to known stimulus, e.g. forces, accelerations, etc., and failures, if any, in the specimens to the stimulus. In general the mechanisms of injuries are established through dose/response type of analysis. Threshold levels of injury producing stimulus are established, and the latest trend is to develop risk curves that would predict the percent of population likely to be injured versus the stimulus. In many cases voluntary exposures to known stimulus establish non-injury-producing levels. If large amounts of well documented accident data are available, it is possible to reconstruct accidents of varying severity with anthropomorphic test devices in crash environments. These test results can be compared to injury outcomes of accident victims to establish the levels of stimulus that produce certain levels of injuries in the real world. Further biomechanical basis for injury criteria can be provided by theoretical, mathematical modeling studies. These studies can establish the relevance of the utilized injury criteria on an engineering basis.

The following parts of this paper examine the research- laboratory testing and theoretical simulations- that formed the basis of injury criteria currently being used in governmental crash regulations and crashworthiness assessment testing worldwide.

27.2.1 Head Injury Criteria

The currently used worldwide regulatory criteria for controlling head injuries is commonly known as the HIC. Ever since its adoption in the FMVSS208 in 1972, this criteria has been controversial [4]. It is interesting to trace the development of HIC to evaluate its biomechanical basis. The earliest tolerance criterion for head injuries was introduced by Lissner et al. [5] in 1960 and is known as the Wayne State Tolerance Curve (WSTC). This tolerance curve was based on human cadaver skull fracture data, pressure application directly to the exposed brain of animals, and impacts to animals, and human volunteer data. Further examination of the WSTC was conducted in Japan over several years of research with impact experiments on animals and human cadaveric skulls [6]. Once again, scaling techniques were used to derive threshold of concussion curve, and turned out to be very similar to that originally proposed by Lissner et al. [5] and modified in the long duration regime by Patrick [7]. Theoretical justification of this curve was provided by finite element analysis of the skull and the brain by Ruan and Prasad [8]. In this study, iso-stress curves for the skull and the brain were developed as a function of average head acceleration and its time duration.

As early as in 1966, Gadd analyzed the basic biomechanical data supporting the WSTC and other animal human exposure data to propose the concept of severity index for evaluating head injury potential. The severity index was calculated by integrating the head acceleration raised to the power 2.5 ($a^{2.5}$) over the entire duration of the pulse with 1,000 being the critical value. Early evaluation of the severity index was conducted by Hodgson et al. [9] who concluded that a good correlation existed between the severity index and degree of concussion in 29 stumptail monkeys subjected to impact, and with frontal bone fractures in cadaveric specimens. It was soon recognized that time duration of the head acceleration pulse affects the S.I., to the point that under a 1G environment, SI of 1,000 is exceeded every 1,000 s. Therefore, limitation of the pulse duration for calculation of S.I. was important. Versace [10] proposed an alternative formulation of SI which subsequently was adopted in the FMVSS208 in 1972. The new formulation is commonly known as the Head Injury Criteria (HIC). Ever since its introduction, HIC has been controversial [4].

Within the deliberations of the International Standards Organization (ISO) working group on biomechanics, it was suggested that the critical value of HIC be raised to 1,500 in frontal impact regulatory tests at 48 kph. The U.S. delegation rejected this proposal based on existing cadaveric data in which HIC's were measured along with skull fractures and brain injuries. The rationale supporting the position taken by the U.S. delegation was explained in a paper by Prasad and Mertz [11]. The investigation also showed that human volunteers had undergone HIC exposures in the 1,000 range with airbags without any brain injuries. These HIC durations were long (30–36 ms) suggesting that limitations of the HIC duration were essential to evaluate the risk of head injuries. They suggested that calculation of HIC should be restricted to a maximum of 15 ms. They also developed a head injury risk curve that associated the probability of head injury versus HIC. A HIC level around 1,450 corresponded to a 50 % risk of skull fracture, while a HIC level of 700 corresponded to a 5 % risk of skull fracture. The head injury risk curve was later shown to predict the efficacy of reconditioned helmets against injury producing head impact in High School football [12].

Instead of limiting HIC durations to 15 ms, NHTSA chose to adopt 36 ms duration in the FMVSS208. However, Transport Canada adopted 15 ms HIC duration, but set the HIC limit at 700. This level of HIC corresponded to a 5 % risk of skull fracture or serious head injury as predicted by the Prasad and Mertz risk curve [11]. Recognizing that a 700, 15 ms HIC was more stringent than the 1,000, 36 ms HIC, NHTSA also adopted the Canadian Regulation in the 2003 phase-in of the Advanced Airbag Rule (FMVSS208). Even though, scaling techniques showed that the HIC for the 5th percentile dummy head should be 779, a limit of 700 was adopted for the 5th percentile dummy and the 6 years old dummy. Scaling was used to establish the limit of 500 for the 3 years old dummy and 390 for the CRABI dummy. Details of the scaling techniques can be found in an Alliance Submission [13] and NHTSA's report by Eppinger et al. [14].

In the European regulation the HIC (renamed as the Head Performance Criteria (HPC)) is calculated only if contact between the occupant head and a vehicle component occurs. In these situations, HIC calculation is limited to the duration of contact with a maximum of 36 ms. An HPC of 1,000 is the limit. Additionally, the resultant head acceleration cannot exceed 80 g's for more than 3 ms cumulatively. The HPC is considered to be satisfied if during the test no contact occurs between the head and any vehicle component. This last requirement is consistent with field data observations in the past– non-contact head injuries are extremely rare if any- and supported by more recent analysis of real world data by Yoganandan et al. [15].

As far as any theoretical basis for HIC is concerned, controversy continues. Many reasons are advanced, the main one being that HIC is based solely on the measurement of linear accelerations at the center of gravity of the head and its unit is "seconds". Obviously, angular accelerations also cause shear deformations in the brain. However, limits of angular accelerations have not yet been introduced in any regulation in the world. If the limit is near 13,600 rad/s^2 [16], such high angular accelerations are hardly ever seen in regulatory testing of restraint systems in frontal impacts, and may be the reason why angular acceleration limits have not been specified in any regulation so far. It is currently believed that the effect of combined linear and angular acceleration on brain responses can be accounted for through detailed mathematical models of the human head.

27.2.2 Neck Injury Criteria

The European frontal impact regulation has adopted neck injury criteria proposed by Mertz [1] in 1984. Mertz's proposal consisted of specifying limits on peak extension and flexion moments at the occipital condyles of the dummy (head/neck junction) and time varying tension, compression and shear forces as measured by load cells in the Hybrid III dummy head. The peak extension and flexion moment limits were based on cadaver and human volunteer tests and accident reconstructions. The time varying tension forces were derived from accident reconstruction data using Volvo vehicles and the Hybrid III dummy for which real world injury data were available. The long duration end of the force-time curve (>45 ms) is based on muscle strength of a male volunteer. In terms of peak tensile force, the limit was suggested to be 3.3 kN which is in close agreement, to 3.1 kN, which was proposed by McElhaney and Myers [17] who based this conclusion on available biomechanical data. Compressive tolerance of the neck proposed by Mertz was based on reconstruction of injury-producing high school American football "spearing maneuvers" against a mechanical tackling device with the Hybrid III dummy [18]. It was an indirect way of determining dynamic axial compression tolerance limit. The peak tolerable load was 4.0 kN that dropped down to 1.1 kN if the duration of loading was 30 ms or longer. The peak load of 4.0 kN compares to 4.8 kN ± 1.3 kN mean axial load to failure in six cadaveric rotation constrained specimens reported-by McElhaney and Myers [17]. It should be noted that Mertz's proposal called the neck injury tolerance levels as Injury Assessment Reference Values. Further, it was stated that being below these IARV's meant that significant neck injury, due to the loading condition considered, was unlikely. With further biomechanical research conducted since the introduction of the Hybrid III 50th percentile male dummy and the development of a family of Hybrid III dummies, the determination of Injury Assessment Reference Values for these new dummies took on heightened importance. The Advanced Airbag Rule required changes in the FMVSS208 to further include child and 5th percentile dummies, NHTSA [19].

Scaling techniques have been used by NHTSA and the American Automobile Manufacturers Association to derive IARV's for various sized dummies and limit values in the regulatory tests have been suggested. Readers interested in following the development of these regulatory values are referred to the Docket of FMVSS208 [20]. The main departure from previous regulatory limits on peak neck responses is to consider the effects of combined loading of the neck as suggested by Prasad and Daniel in 1984 [21]. It was shown that higher axial forces could be tolerated with an absence of bending moments in the neck, and higher bending moments were tolerable with an absence of axial forces in the neck. This data was augmented by similar, previously unpublished, data developed by GM. The combined data was made available to NHTSA for simultaneous analysis by their experts and by Ford/GM experts. An analysis of the data was published by Mertz and Prasad [22] in 2000. In this paper, scaling factors to predict failure loads for different sized dummies were presented along with failure tensile loads, bending moments and critical tensile forces and extension moments for a combined injury criteria. In the same time frame, the results of animal studies conducted under the sponsorship of NHTSA were published by Yoganandan et al. [23] and Pintar et al. [24]. These studies concluded that the scaling factors proposed by NHTSA and the AAM were reasonable.

Chest Injury Criteria: The earliest chest injury criteria to enter a crash test regulation were specified in the FMVSS208 in the U.S.. Recognizing that the test device in the regulation was the Hybrid II dummy, which had a very stiff and non-biofidelic thorax design, chest acceleration was the only possible injury response that was meaningful to control. The chest acceleration limit placed in 30-mph rigid barrier tests was 60G's for 3 ms-time duration.

This acceleration limit was based on voluntary exposures in early rocket-sled tests conducted by Col. J.P. Stapp. Stapp himself was exposed to 40G's for 100 ms without injury. Another subject had undergone 45G's for 44 ms. It was assumed that for shorter durations higher accelerations could be tolerated. Considerable controversy ensued with the adoption of the acceleration based criteria. Mertz and Gadd [25] reported that there was no evidence that "even a 60G chest acceleration level would not be tolerable with an adequate restraint system" for pulse durations less than 100 ms. They preferred to specify thoracic compression limit. Early biomechanical basis was examined by Walfisch et al. [26] who analyzed their cadaver database and concluded that "50G to 70G measured on a part 572 dummy seems to correspond to an acceptable level of injury for the restrained population exposed to risk of accident." However, chest acceleration was not generally accepted by the biomechanical community, and with the development of a more biofidelic dummy, the Hybrid III, chest deflection became the measurement of choice to be controlled in the Canadian and European regulations. However the resultant chest acceleration continues to be part of the Advanced Airbag Rule (FMVSS208). The limit is set at 60 G's with a 3 ms cumulative duration for the 50th, 5th and the 6 years old dummies. The limit is lowered to 55 G's for the 3 years old dummy and to 50 G's for the 12 month (CRABI) dummy.

Substantial research was conducted by General Motors in establishing impact response of the chest in dynamic blunt loading conditions. Analysis of pendulum impacts to cadaveric specimens was conducted by Kroell, Nahum, Schneider, Neathery and Mertz [27, 28] who established 75 mm sternal compression as being the 50 % probability of AIS 3 injury in blunt loading. Mertz [1, 2] suggested this limit for airbag loading, but suggested 50 mm sternum deflection as the limit for belt loading in the FMVSS208 tests. Accident reconstruction studies were conducted by Mertz, Horsch, Melvin, Horn, and Viano [29, 30] with Renault and Volvo restraint systems. These studies concluded that 50 mm of sternal deflection produced by belt loading of the Hybrid III dummy corresponded to a 50 % risk of AIS3+ thoracic injury for the population at risk in real world frontal crashes. The European and Canadian frontal impact regulations accepted this limit, but the U.S. FMVSS208 continued with 75 mm of sternal deflection as the limit regardless of the restraint system utilized till 2003. From 2003 to 2006, the Advanced Airbag Rule (FMVSS208) was phased in. In this regulation, a 63 mm chest deflection limit was adopted to be functionally equivalent to the Canadian and the European Regulations. The functional equivalency comes about due to the difference between the compliance methods between US and Europe or Canada. The US requires self-certification whereas in other countries, witness testing is required for certification of a vehicle. The design goal utilized by most manufacturers account for repeatability and reproducibility of the test results. To ensure meeting the regulation, the design goals are set substantially below the regulatory limits. Essentially, a compliance margin of 20 % was added to 50 mm – the European and Canadian regulation with the understanding that the average of the fleet will be designed at 50 mm to ensure the vast majority would meet the regulatory limits.

Whereas early biomechanical studies concentrated on sternal rib deflections as injury producing responses, studies utilizing anesthetized animals and human cadaveric specimens in early to mid-1980s showed that rate of compression was also an important injury producing agent. These studies were conducted by researchers at the G.M. Research Laboratories. The early results of these studies were published by Kroell et al. [31] and Lau and Viano [32] in 1986. A chest injury criteria called the Viscous Criteria was proposed by Lau and Viano [32]. This criteria explained heart, lung, liver and soft tissue injuries. It also accounted for strain rate effects in soft tissues subjected to impact. The Viscous Criteria consisted of the product of deformation velocity (m/s) and the maximum normalized chest deflection. Numerical value of 1 for this product corresponds to a 20 % risk of AIS ≥ 4 injury and seemed to be independent of animal species. A Viscous Criteria, now commonly referred as V*C, of 1 has been adopted in several regulations for frontal and side impacts, but has not been adopted in the US regulations.

It should be mentioned that measurement of deflection in tests with cadaveric specimens had been problematic in the 1970s and 1980s. The development of a device that would yield deflection and the shape of the thorax in impact conditions was pursued by NHTSA. One such device, commonly known as the "Chest Band," was reported by Eppinger [33] in 1989. Several cadaveric tests in frontal and side impact conditions have since been conducted. This device has made it possible to monitor deflections and accelerations simultaneously in test specimens subjected to impact. Kuppa and Eppinger [34] published a collection of PMHS test data generated in several laboratories in US and Europe. The reported data has been a subject of analysis by many over the years. One such analysis by Prasad [35] showed that deflection was a better predictor of rib fractures than spinal accelerations. All, but one of the PMHS, had AIS 3+ chest injuries above 30 % maximum normalized chest deflection. This level of chest compression was associated with 52 mm of internal sternum deflection. Logistic analysis of the data indicated this level of sternum deflection as being associated with a 50 % probability of AIS 3+ chest injury for the PMHS tested whose average age was 58 years. Although spinal acceleration continues to be an injury indicator in the FMVSS208, chest deflection is the response utilized in the NCAP rating system.

Chest injury criteria for side impact: The earliest side impact regulation utilizing dynamic test of vehicles with dummy occupants is the FMVSS214 in the US. In its earlier form, the FMVSS 214 utilized a side impact dummy and the Thoracic Trauma Index (TTI) as the injury criteria. It has been mentioned earlier that thoracic deflection measurements were problematic in the 70s and the 80s. As a result, early biomechanical testing of animal or human cadaveric specimen utilized acceleration measurements at the thoracic spine and the rib cage with a 12-accelerometer array reported by Robbins et al. [36]. It was also assumed that a dummy designed to meet biofidelity corridors based on accelerations would have adequate biofidelity for side impact testing. Such a dummy, the SID, was designed and subsequently adopted in the FMVSS214.

Statistical correlation between acceleration responses of T-12 and the ribs were used to predict AIS4+ injuries. Eppinger [37] found that the average of Peak T-12 acceleration and peak rib acceleration correlated with chest injuries. Additionally age was a major contributor to injuries and was included in the formulation of the Thoracic Trauma Index. It should be noted that acceleration based criteria was the only meaningful criteria use with this dummy since the biofidelity in deflection was doubtful and not measured in the cadaver tests. Considerable controversy continues with the use of TTI. To further examine the validity of TTI or other injury measures 26 cadaver tests using "chest bands" and accelerations have been conducted and reported by Pintar et al. [38]. Analysis of the data indicated that normalized chest deflections, V*C and the TTI have merit as injury criteria. However, a combination of TTI and maximum normalized chest deflection was the best statistical predictor of chest injuries in lateral impacts. Obviously, there is then a need for a test device that has biofidelity for both acceleration and deflection responses. The FMVSS214 has been upgraded and a slightly modified form of the ES-II dummy has replaced the SID, and the TTI is no longer used as an injury criterion in the regulation. The process of the selection of injury criteria in the FMVSS214 has been described by Kuppa [39] in a NHTSA document.

The current European chest injury criterion is based solely on maximum deflection of the thorax. The supporting biomechanical data came from drop tests conducted at the APR laboratories and reported by Stalnaker et al. [40]. Interestingly, the authors found that 30 % normalized compression of the thorax based on whole chest width was the tolerance for AIS or less. This is very similar to what was discussed earlier for frontal impacts.

Femur Injury Criteria: The earliest proposal for limits on axial compression of the femur was based on sled tests with cadavers conducted by Patrick, Kroell and Mertz in 1966 [41]. This evaluation of cadaver test data suggested a conservative threshold level of 6.23 kN for protection of the femur and hip complex. Experiments on cadaveric specimens with impactors of various sizes and mass were conducted by Powell et al. [42] and Melvin et al. [43] in 1974 and 1975. When all these tests were analyzed, it became clear that tolerable compressive loads depended on the duration of loading. Viano and Khalil [44] utilized a finite element model of the femur to show that time duration of impact was an important parameter. The FMVSS208 adopted a peak compressive force of 10 kN as the injury criteria without any time duration limitation. However the European Regulation specified a time dependent peak force criteria. The allowable peak compressive force is 9.07 kN, but the allowable peak force drops down to 7.58 kN for durations greater than 10 ms. Between 0 to 10 ms durations, a linear interpolation between 9.07 kN and 7.58 kN is used. These numbers are the same or close enough to those suggested by Mertz [2].

Tibia Injury Criteria: The tibia injury criteria currently adopted in European frontal impact regulations is based on research conducted by Mertz who suggested the Tibia Index and tolerable axial compressive loads in the tibia. The origin of the currently used 8 kN maximum compressive load is based on accident reconstruction of airbag crashes in which fractures of tibial condyles were observed. These injuries were independent of crash severity and were attributed to airbag-knee interactions. Mertz noted that if the load on the medial or lateral condyles exceeded 4 kN fractures occurred. Since the maximum load per condyle could be only 4 kN, the maximum allowable tibia compressive force could not exceed 8 kN. For the fracture of the tibial shaft Mertz's suggestion was to use the Tibia Index as defined below:

$$TI = (M/M_c) + (F/F_c)$$

Where $M_c = 225$ N-m and $F_c = 35.9$ kN and the TI could not exceed 1.0.

The critical values for compression and bending were based on earlier strength data from Yamada [45]. Since, the maximum force could not exceed 8 kN, the Tibia Index acted as a control on the maximum bending moment. Further examination of bending moment to failure data published by Nyquist [46] indicated that the critical moment of 225 Nm was too conservative for a 50th percentile male. The average failure bending moment for a 50th percentile male was closer to 300 Nm. As a result, the acceptable Tibia Index adopted in the European regulation was raised to 1.3.

27.3 Injury Criteria for Rollover Crashes

Currently, regulations for rollover protection do not include dummy responses that have to be controlled as in frontal and side impacts. In fact, several dynamic test procedures have been suggested, but none has been incorporated in any rule anywhere in the world. This has a lot to do with the fact that repeatability and reproducibility of test results with dummies has been an issue in the past. The vast majority of rollover research tests conducted in the past with dummy occupants have been with the Hybrid III mid-size male dummy and the dummy responses reported have been compared to the injury criteria proposed by Mertz [1, 2]. The subject of the biofidelity of the Hybrid III dummy for rollover testing has been controversial over the years, especially for the compressive stiffness of the neck. A review of the available test data, and their various interpretations has been conducted by Piziali et al. [47] who concluded that the "neck stiffness of the Hybrid III dummy and the human cadaver is reasonably similar for both fully constrained and rotationally constrained compressive loading." Regardless of the controversy, the neck injury criteria currently utilized with the Hybrid III family of dummies have been introduced in the US and European regulations as discussed earlier in this paper. If another dummy is introduced in the future, many of the biomechanical tests used to establish the injury criteria for the Hybrid III dummy family will have to be repeated and/or reanalyzed for developing dummy specific injury criteria.

27.4 Injury Criteria for Rear Impacts

Currently, a low speed sled test is specified in the Head Restraint Standard (FMVSS202a) in US, and is the only dynamic regulation in place to evaluate head restraints. The dynamic test is an alternative to satisfy the static requirements of the head restraint standard. Note that the FMVSS202a is a regulation aimed at reducing the likelihood of Soft Tissue Neck Injuries (STNI) in low speed rear impacts. These are AIS1 type of injuries that predominantly occur in rear impacts and the after effects of this injury are commonly known as "Whiplash Associated Disorders, i.e. WAD." Several hypotheses as to the mechanism of the injury have been advanced, but none have been proven out. Based on the several hypotheses, several injury criteria have been advanced. In this paper, all possible injury criteria that have been proposed to address low speed rear impacts will not be discussed as it is an evolving field of research. Injury criteria in regulations and public domain tests will be discussed briefly. For a broad coverage of the subject, readers are referred to Yoganandan and Pintar [48] and Kuppa [49].

The earliest hypothesis was that whiplash injuries occurred due to excessive rotation of the head relative to the torso- hyperextension of the neck, Macnab [50]. This led to the introduction of head restraints in vehicles whose minimum height above the Seating Reference Point was specified in the FMVSS202. The standard was upgraded in 2004 with higher head restraints and a dynamic test was added in what is known as the FMVSS202a. Rating agencies like the EuroNCAP and the IIHS also set up head restraint height and backset requirements. The goal of the geometry standards is to reduce head displacements relative to the torso.

Motion of the head relative to the torso causes shear between adjacent cervical vertebrae. Particularly important is the shear in the facet capsules which have been shown to be the site for pain by a team led by Cavanaugh at Wayne State University, Lu et al. [51] and Chen et al. [52]. As early as 1997, Yang et al. [53] studied the role of cervical facet joints in rear impacts and advanced stretch of the capsule as an injury mechanism. Two following studies by Deng et al. [54] and Sundararajan et al. [55] have estimated stretch in the facet capsules of PMHS's in low speed rear impacts through high speed x-ray imaging. Sundararajan's study has shown that maximum stretch in the facet capsules along the cervical spine occurred substantially after head-to-head restraint contact. What cannot be deduced from tests conducted with PMHS's is the amount of stretch in living humans in similar severity impacts. As far as low speed rear impacts are concerned, the subject of facet stretch in humans versus in PMHS's in similar severity impacts will continue to be controversial as the effect of muscle tone and strength are difficult to estimate.

A third hypothesis has been advanced by Svensson et al. [56] in 1993 who postulated that pressure changes in the spinal canal occurred during whiplash extension motion of the head relative to the torso. The pressure changes possibly caused injury to the spinal ganglia. This was followed up by Bostrom et al. [57] who proposed a new Neck Injury Criterion (NIC).

27.5 Low Speed Rear Impact Regulation

NHTSA has adopted the Hybrid III dummy for compliance testing in the FMVSS202a after considering alternative dummies like the BioRID II and the RID 2. The choice of the Hybrid III dummy for low speed rear impact testing has been justified by Kuppa [49]. Among some reasons was the fact that a self-aligning head restraint and seat was designed by Saab using the Hybrid III dummy with substantial effectiveness in the field, Viano and Olsen [58]. Viano and Davidsson [59] also showed that the displacement of the head relative to T-1 of the Hybrid III was similar to those observed in low speed rear impacts of volunteers and utilized the Neck Displacement Criteria (NDC) to design the Saab seat and head restraint. Although little noticed in the scientific literature, the 2000 Ford Taurus front outboard seats and head restraints were designed using the Hybrid III dummy and controlling the corrected lower neck moments. An IIHS report [60] has reported a reduction in whiplash injuries, especially for females who have been traditionally overrepresented in the incidence of whiplash injuries. A study by Prasad et al. [61] had shown that estimated moments at the base of the Hybrid III neck (corrected Lower Neck Moments) ranked three seats according to insurance claims for whiplash. Kuppa [49] has shown a strong correlation between neck displacement relative to T1 and corrected lower neck moments ($R^2 = 0.96$), justifying NHTSA's selection of injury criteria based on the rotation of the head relative to the torso. One would assume a similarly strong correlation between corrected lower neck moments and the NDC.

27.6 Injury Criteria for Whiplash Used in Public Domain Tests

The EuroNCAP and the IIHS have been rating seats/head restraints for whiplash protection. Both organizations use static and dynamic assessments. The static assessment includes head restraint geometry like height and backset. The dynamic assessment consists of sled tests- three in Europe and one by IIHS- with the BioRID II dummy. Injury criterion considered for ratings include NIC, Nkm, Rebound velocity, upper neck shear and tension, T1 Accelerations, and head to head restraint contact time.

Neck Injury Criteria (NIC): A postulated injury mechanism, the hydrodynamic pressure changes in the spinal canal, was advanced by Bostrom et al. [57] and NIC was proposed as an indicator of such pressure changes. Low speed rear impact tests reported by Heitzplatz et al. [62] had shown negative correlation between NIC and insurance claims in Europe. Additionally, NIC maximized just prior to head contact with the head restraint. The hypothesis that whiplash is related to pressure changes in the spinal canal is controversial, King [63]. Reportedly, under minor non-head contact situations, the threshold limit of NIC has been crossed in volunteers who have not had any neck injuries, Funk [64]. NIC may be a surrogate for some other phenomenon that may cause nerve root damage. Even if correct, it does not account for the potential of neck distortions after head-to-head restraint contact during which other forces and moments in the neck seem to reach their highest values.

Nkm: Nkm is a formulation that is similar to the currently used Nij formulation discussed earlier in this paper, except that it uses a combination of shear force and flexion/extension moments, Schmitt et al. [65]. Since the Nkm combines shear force and moments at the head/neck junction of the dummy, it leads to the acceptance of an earlier hypothesis advanced by Yang et al. [53]. However, upper neck moments, in the presence of head restraints, have been shown to be relatively insensitive to crash severity and type of seat, Prasad et al. [66]. Therefore, injury criteria based solely on upper neck moments or its combination with other upper neck forces may be misleading in predicting whiplash injuries.

Upper neck shear force and tension: Forces in the upper neck are also being used in the rating schemes in Europe and by the IIHS. Though these forces, especially shear, are indications of the distortions in the neck, their relevance in predicting whiplash in seats equipped with head restraints is tentative. In a series of tests of three production seats in rear impacts ranging at 8 and 16 kph Delta-V's, upper neck forces and moments as measured with Hybrid III dummies were relatively insensitive to type of seats (three different insurance claims frequency), Prasad et al. [61]. Their results showed that forces measured at the lower neck location, especially the extension moment, were more predictive of the claims frequency associated with the tested seats.

Injury criteria in high speed rear impacts: There are no regulations currently controlling forces/moments in the dummy in high-speed rear impacts. However, the most research tests are conducted with the Hybrid III family of dummies and utilizing existing injury criteria for all body regions. The trend is to also measure lower neck loads- forces and moments whose Injury Assessment Reference Values have been proposed by Prasad et al. [61].

27.7 Summary and Conclusions

An attempt at examining the biomechanical basis of injury criteria currently utilized in regulations worldwide for frontal and side impacts has been made. This historical review, although not an exhaustive one, shows that most of the injury criteria utilized in regulations today have a biomechanical basis in animal and human cadaveric testing, and accident reconstruction. Although most of the studies upon which the regulatory criteria are currently based were conducted before the mid-eighties, subsequent research has further confirmed the choice of the injury criteria. A new neck injury criterion, the Nij, has been added in the FMVSS 208 to be effective from 2003. However, the data on which it is based was developed in the early 1980s. By tracing the history of research to Regulation, it should be obvious that a long time transpires before regulations catch up with research. In most cases the regulation reflects the state of biofidelity of the dummy, and dummy changes would require further adjustments to the acceptance criteria.

A substantial amount of relevant research has followed the introduction of regulations and the biomechanical knowledge base has increased substantially. More advanced and refined test devices and measurement capabilities are under development. It is hoped that in future regulatory activities a synthesis of all existing data will be carried out and refined injury criteria will be adapted. Biomechanical researchers and regulators should note the large time lag between research and regulations.

References

1. Mertz HJ (1984) Injury assessment reference values used to evaluate Hybrid III response measurements. NHTSA docket 74-14. Notice 32. Enclosure 2 of attachment I of part III of general motors submission USG2284
2. Mertz HJ (1993) Anthropomorphic test devices. In: Nahum AM, Melvin JW (eds) Accidental injury. Springer, New York
3. Samaha RR, Elliott DS (2003) NHTSA side impact research: motivation for upgraded test procedures. In: Proceedings of the 18th enhanced safety of vehicles conference, DOT, NHTSA
4. Newman JA (1980) Head injury criteria in automotive crash testing. SAE 801317. In: Proceedings of the 20th Stapp car crash conference, SAE, Warrendale, Pennsylvania, USA
5. Lissner HR, Lebow M, Evans FG (1960) Experimental studies on the relation between acceleration and intracranial pressure changes in man. Surg Gynecol Obstet 111:329–338
6. Ono K, Kikuchi A, Nakamura M, Kobayashi H, Nakamura N (1980) Human head tolerance to sagittal impact reliable estimation deduced from experimental head injury using sub-human primates and human cadaver skulls. SAE 801303. In: Proceedings of the 24th Stapp car crash conference, SAE, Warrendale, Pennsylvania, USA
7. Patrick LM, Gurdjian ES, Lissner HR (1965) Survival by design: head protection. In: Proceedings of the 7th Stapp car crash conference, SAE, Warrendale, Pennsylvania, USA
8. Ruan JS, Prasad P (1995) Coupling of a finite element human head model with a lumped parameter Hybrid III dummy model: preliminary results. J Neurotrauma 12(4):725–734
9. Hodgson VR, Thomas LM, Prasad P (1970) Testing the validity and limitations of the severity index. SAE 700901. In: Proceedings of the 14th Stapp car crash conference, SAE, Warrendale, Pennsylvania, USA
10. Versace J (1971) A review of the severity index. SAE 710881. In: Proceedings of the 15th Stapp car crash conference, SAE, Warrendale, Pennsylvania, USA
11. Prasad P, Mertz HJ (1985) The position of the United States delegation to the ISO Working Group 6 on the use of HIC in the automotive environment. SAE 851246. SAE, Warrendale, Pennsylvania, USA
12. Mertz HJ, Prasad P, Nusholtz GS (1996) Head injury risk assessment for forehead impacts. SAE 960099. SAE, Warrendale, Pennsylvania, USA
13. Alliance (1999) Dummy response limits for FMVSS 208 compliance testing. Docket No. NHTSA-99-6407, Noyice 1, Annex 2 of Alliance Submission
14. Eppinger R, Sun E, Bamndak F, Haffner M, Khaewpong N, Maltese M, Kuppa S, Nguyen T, Takhounts E, Tannous R, Zhang A, Saul R (1999) Development of improved injury criteria for the assessment of advanced automotive restraint systems-II. National Highway Traffic Safety Administration, Washington, DC
15. Yoganandan N, Gennarelli TA, Pintar FA (2007) Characterizing diffuse brain injuries from real world motor vehicle impacts. In: Proceedings of the 20th ESV conference, NHTSA, Washington, DC, USA
16. Pincemaille Y, Trosseille X, Mack P, Tarriere C, Breton F, Renault B (1989) Some new data related to human tolerance obtained from volunteer boxers. SAE 892435. In: Proceedings of the 33rd Stapp car crash conference, SAE, Warrendale, Pennsylvania, USA
17. McElhaney JH, Myers BS (1993) Biomechanical aspects of cervical trauma. In: Nahum AM, Melvin JW (eds) Accidental injury. Springer, New York
18. Mertz HJ, Hodgson VR, Thomas LM, Nyquist GW (1978) An assessment of compressive neck loads under injury-producing conditions. Phys Sportsmed 6:95–106
19. Federal Register, USA (1998) Advanced Technology Air Bags, docket no. NHTSA 98-4405, notice 1, 18 Sept 1998
20. AAMA Comments to docket no. NHTSA 98-4405, no. 48448, DOT docket section (1998)
21. Prasad P, Daniel RP (1984) A biomechanical analysis of head, neck, and torso injuries to child surrogates

due to sudden torso acceleration. SAE 841656. In: Proceedings of the 28th Stapp car crash conference, SAE, Warrendale, Pennsylvania, USA
22. Mertz HJ, Prasad P (2000) Improved neck injury risk curves for tension and extension moment measurements of crash dummies. Stapp Car Crash J 44:59–75
23. Yoganandan N, Eppinger R, Maltese M, Kuppa S, Sun E, Gennarelli TA, Kumaresan S, Pintar FA (2000) Pediatric and small female neck injury scale factors and tolerance based on human spine biomechanical characteristics. In: IRCOBI conference, Lyon, France
24. Pintar FA, Mayer RG, Yoganandan N, Sun E (2000) Child neck strength characteristics using an animal model. Stapp Car Crash J 44:77–83
25. Mertz HJ, Gadd CW (1971) Thoracic tolerance to whole-body deceleration. SAE 710852. In: Proceedings of the 15th Stapp car crash conference, SAE, Warrendale, Pennsylvania, USA
26. Walfisch G, Chamouard F, Lesterlin D, Tarriere C, Brun Cassan F, Mack P, Got C, Guillon F, Patel A, Hureau J (1985) Predictive functions for thoracic injuries to belt wearers in frontal collisions and their conversion into protection criteria. In: Proceedings of the 29th Stapp car crash conference, SAE, Warrendale, Pennsylvania, USA
27. Kroell CK, Schneider DC, Nahum AM (1974) Impact tolerance and response to the human thorax II. SAE 741187. In: Proceedings of the 18th Stapp car crash conference, SAE, Warrendale, Pennsylvania, USA
28. Neathery RF, Kroell CK, Mertz HJ (1975) Prediction of thoracic injury from dummy responses. SAE 751151. In: Proceedings of the 19th Stapp car crash conference, SAE, Warrendale, Pennsylvania, USA
29. Horsch JD, Melvin JW, Viano DC, Mertz HJ (1991) Thoracic injury assessment of belt restraint systems based on Hybrid III chest compression. SAE 912895. In: Proceedings of the 35th Stapp car crash conference, SAE, Warrendale, Pennsylvania, USA
30. Mertz HJ, Horsch JD, Horn G, Lowne RW (1991) Hybrid III sternal deflection associated with thoracic injury severity of occupants restrained with force-limiting shoulder belts. SAE 910812. SAE, Warrendale, Pennsylvania, USA
31. Kroell CK, Allen SD, Warner CY, Perl TR (1986) Interrelationship of velocity and chest compression in blunt thoracic impact to Swine II. SAE 861881. In: Proceedings of the 30th Stapp car crash conference, SAE, Warrendale, Pennsylvania, USA
32. Lau IV, Viano DC (1986) The viscous criterion – bases and applications of an injury severity index for soft tissues. SAE 861882. In: Proceedings of the 30th Stapp car crash conference. SAE, Warrendale, Pennsylvania, USA
33. Eppinger RH (1989) On the development of a deformation measurement system and its application toward developing mechanically based injury indices. SAE 892426. In: Proceedings of the 33rd Stapp car crash conference, SAE, Warrendale, Pennsylvania, USA
34. Kuppa SM, Eppinger RH (1998) Development of an improved thoracic injury criterion. In: Proceedings of the 42nd Stapp car crash conference, SAE, Warrendale, Pennsylvania, USA
35. Prasad P (1999) Biomechanical basis for injury criteria used in crashworthiness regulations. In: IRCOBI conference, Lyon, France
36. Robbins DH, Melvin JW, Stalnaker RL (1976) The prediction of thoracic impact injuries. SAE 760822. In: Proceedings of the 20th Stapp car crash conference, SAE, Warrendale, Pennsylvania, USA
37. Eppinger RH, Marcus JH, Morgan RM (1984) Development of dummy and injury index for NHTSA's thoracic side impact protection research program. SAE 840885. Government/Industry Meeting and Exposition, Washington, DC
38. Pintar FA, Yoganandan N, Hines MH, Maltese MR, McFadden J, Saul R, Eppinger RH, Khaewpong N, Kleinberger M (1997) Chestband analysis of human tolerance to side impact. SAE 973320. In: Proceedings of the 41st Stapp car crash conference, SAE, Warrendale, Pennsylvania, USA
39. Kuppa S (2004) Injury criteria for side impact dummies. DOT/National Highway Traffic Safety Administration, Washington, DC. FMVSS-214 NPRM docket: NHTSA-2004-17694
40. Stalnaker RL, Tarriere C, Fayon A, Walfisch G, Balthazard M, Masset J, Got C, Patel A (1979) Modification of part 572 dummy for lateral impact according to biomechanical data. SAE 791031. In: Proceedings of the 23rd Stapp car crash conference, SAE, Warrendale, Pennsylvania, USA
41. Patrick LM, Kroell CK, Mertz HJ (1965) Forces on the human body in simulated crashes. In: Proceedings of the 9th Stapp car crash conference, SAE, Warrendale, Pennsylvania, USA
42. Powell WR, Advani SH, Clark RN, Ojala SJ, Holt DJ (1974) Investigation of femur response to longitudinal impact. In: Proceedings of the 18th Stapp car crash conference, SAE, Warrendale, Pennsylvania, USA
43. Melvin JW, Stalnaker RL, Alem NM, Benson JB, Mohan D (1975) Impact response and tolerance of the lower extremities. SAE 751159. In: Proceedings of the 19th Stapp car crash conference, SAE, Warrendale, Pennsylvania, USA
44. Viano DC, Khalil TB (1976) Plane strain analysis of a femur midsection. In: Proceedings of the 4th New England bioengineering conference, Pergamma
45. Yamada H (1970) Strength of biological materials. In: Evans FG (ed) Strength of biological materials. Williams and Wilkins, Baltimore
46. Nyquist GW, Cheng R, Elbohy AR, King AI (1985) Tibia bending: strength and response. In: Proceedings of the 29th Stapp car crash conference, SAE, Warrendale, Pennsylvania, USA
47. Piziali R, Hopper R, Girvan D, Merala R (1998) Injury causation in rollover accidents and the biofidelity of hybrid III data in rollover tests. SAE 980362. Mechanics of Protection (SP-1355)

48. Yoganandan N, Pintar FA (2000) Frontiers in whiplash trauma. IOS Press, Amsterdam
49. Kuppa SM (2004) Injury criteria and anthropomorphic test devices for whiplash injury assessment. NHTSA and NTBRC report
50. Macnab I (1964) Acceleration Injuries of the Cervical Spine. J Bone Joint Surg Am 46:1797–1799
51. Lu Y, Chen C, Kallakuri S, Patwardhan A, Cavanaugh JM (2005) Neural response of cervical facet joint capsule to stretch: a study of whiplash pain mechanism. Stapp Car Crash J 49:49–65
52. Chen C, Lu Y, Kallakuri S, Patwardhan A, Cavanaugh JM (2006) Distribution of A-delta and C-fiber receptors in the cervical facet joint capsule and their response to stretch. J Bone Joint Surg Am 88(8):1807–1816. doi:10.2106/JBJS.E.00880
53. Yang KH, Begeman PC, Muser M, Niederer P, Walz F (1997) On the role of cervical facet joints in rear end impact neck injury mechanisms. SAE 970497. SAE, Warrendale, Pennsylvania, USA
54. Deng B, Begeman PC, Yang KH, Tashman S, King AI (2000) Kinematics of human cadaver cervical spine during low speed rear-end impacts. Stapp Car Crash J 44:171–188
55. Sundararajan S, Prasad P, Demetropoulos CK, Tashman S, Begeman PC, Yang KH, King AI (2004) Effect of head-neck position on cervical facet stretch of post mortem human subjects during low speed rear end impacts. Stapp Car Crash J 48:331–372
56. Svensson M (1993) Pressure effects in the spinal canal during whiplash extension motion- a possible cause of injury to the cervical spinal ganglia. In: IRCOBI conference, Lyon, France
57. Bostrom O, Svensson MY, Aldman B, Hansson HA, Haland Y, Lovsund P, Seeman T, Suneson A, Saljo A, Ortengren T (1996) A new neck injury criterion candidate based on injury findings in the cervical spinal ganglia after experimental neck extension trauma. In: IRCOBI conference, Lyon, France
58. Viano DC, Olsen S (2001) The effectiveness of active head restraint in preventing whiplash. J Trauma 51(5):959–969
59. Viano D, Davidsson J (2002) Neck displacements of volunteers, BioRID P3 and hybrid III in rear impacts: implications to whiplash assessment by a Neck Displacement Criterion (NDC). Traffic Inj Prev 3:105–116
60. IIHS (2002) New vehicle seat & head restraint designs are reducing neck injuries in rear-end crashes, news release. Insurance Institute of Highway Safety Website
61. Prasad P, Kim A, Weerappuli D (1997) Biofidelity of anthropomorphic test devices for rear impact. SAE 973342. In: Proceedings of the 41st Stapp car crash conference, SAE, Warrendale, Pennsylvania, USA
62. Heitzplatz F, Sferco R, Fay P, Reim J, Kim A, Prasad P (2003) An evaluation of existing and proposed injury criteria with various dummies to determine their ability to predict the levels of soft tissue neck injury seen in real world accidents. In: Proceedings of the 18th international technical conf on the enhanced safety of vehicles
63. King AI (2012) A retrospective look at my contributions to Stapp. Stapp Car Crash J 56:V–XV
64. Funk JR, Cormier JM, Bain CE, Guzman H, Bonugli EB (2007) An evaluation of various neck injury criteria in vigorous activities. In: IRCOBI conference, Lyon, France
65. Schmitt K, Muser M, Niederer P (2001) A new neck injury criterion candidate for rear-end collisions taking into account shear forces and bending moments. In: Proceeding of the 17th ESV conference, Washington, DC, USA
66. Prasad P, Kim A, Weerappuli D, Roberts V, Schneider D (1997) Relationship between passenger car seat back strength and occupant injury severity in rear end collisions: field and laboratory studies. SAE Paper No. 973343. In: Proceedings of the 41st Stapp car crash conference, SAE, Warrendale, Pennsylvania, USA

Civil Aviation Crash Injury Protection

28

Richard L. DeWeese, David M. Moorcroft, and Joseph A. Pellettiere

Abstract

The Federal Aviation Administration (FAA) has adopted safety requirements intended to protect aircraft occupants during survivable crash scenarios. These requirements include design specifications, static strength tests and dynamic impact tests of seats and restraint systems using instrumented Anthropomorphic Test Devices (ATD). Two orientations of impact test are cited: a combined longitudinal/vertical test with the impact vector 60° from horizontal, and a longitudinal test with the impact vector yawed 10° from the centerline of the aircraft. Injury potential is assessed during the dynamic tests by comparing test results to a set of injury criteria. The static and dynamic test requirements vary by aircraft type due to the differences in energy transmitted to the seats. However, the injury criteria evaluated during these tests are very similar for all aircraft types. The criteria cited in the regulations are: the Head Injury Criteria (HIC), lumbar spine compressive load, shoulder strap load, femur compressive load (for passengers of transport aircraft only), a requirement that the seat belt not bear on the abdomen, and that the shoulder belts (if used) bear on the shoulder.

Side facing seats have unique injury risks. Therefore, tests using an ATD that can measure those risks (the ES-2re) are conducted to ensure that side facing seats provide the same level of safety as forward or aft-facing ones. The injury criteria originally developed to evaluate side impacts in autos have been cited in the aviation requirements, and include: HIC, rib lateral deflection, abdomen force, and pubic symphysis force.

R.L. DeWeese, B.S. (✉) • D.M. Moorcroft, M.S.
Civil Aerospace Medical Institute, Federal Aviation Administration, Oklahoma City, OK, USA
e-mail: Rick.deweese@faa.gov; David.moorcroft@faa.gov

J.A. Pellettiere, Ph.D.
Aviation Safety, Federal Aviation Administration, Washington, DC, USA
e-mail: Joseph.pellettiere@faa.gov

Because aircraft side-facing seats do not always provide full support for the occupant, additional criteria were developed to limit the risk of injuries caused by excessive excursion of the head, torso and legs. These criteria place limits on upper neck loads, femur twist angle, and torso flail, and prohibit significant contact between occupants of multi-place seats.

As new seat configurations and restraint technologies are introduced, additional criteria may be needed to ensure that these new systems provide the same level of safety as conventional seats and restraints. Advancements in biomechanics and injury mitigation technology have the potential to increase the level of safety for all aircraft occupants.

Crashes occur in all modes of transportation, but the risk of their occurrence and the chances of being killed or injured during them vary widely between modes. Commercial air carriers have the lowest accident rate in aviation. From 2002 to 2009, the annual rate of accidents has been less than 5 per million departures, less than 0.3 per 100,000 flight hours, and less than 0.75 per 100 million miles flown [1]. And while public perception may differ, many aircraft accidents are survivable and warrant measures to minimize the number of fatalities and injuries occurring during them. In air carrier accidents from 1983 to 2000, 96 % of occupants survived the accident [2]. General aviation operations (small private aircraft), which include fixed wing as well as rotorcraft, have an accident rate approximately 20 times higher than the commercial air carrier rate, with roughly six accidents per 100,000 flight hours occurring from 2000 to 2009. There is about 1 fatal accident per 100,000 flight hours for this type of aircraft [1]. While not a direct comparison, but as a point of reference, there have been 1.5 fatalities or less per 100 million vehicle miles traveled in U.S. motor vehicle crashes since 2001 [3].

Most aircraft accidents occur during the take-off and landing phases of flight [1]. Typical crash scenarios for fixed wing aircraft include ground-to-ground, such as takeoff abort and landing overrun, and air-to-ground, such as hard landing, undershoot, and stall [4]. Typical crash scenarios for rotorcraft include air-to-ground with high vertical or horizontal impact velocities or with high yaw rates caused by loss of tail rotor effectiveness, ground-to-ground, such as a rollover, air-to-water ditching, and in-flight incidents such as wire impacts [5]. Regardless of aircraft type, a survivable crash requires three components: a livable volume is maintained around the occupant, loads transmitted to the occupants are within human limits, and there is adequate time to evacuate before a post-crash fire reaches the occupants [2]. The ability to quickly egress the aircraft after a crash is crucial to survival due to the likelihood and rapid propagation of post-crash fires.

For certification purposes, the Federal Aviation Administration (FAA) has defined three main categories of aircraft in Title 14 of the Code of Federal Regulations (CFR). These are general aviation (Part 23), transport airplanes (Part 25), and rotorcraft (Part 27 and Part 29). Part 23 aircraft include normal, utility, acrobatic, and commuter category airplanes. They generally consist of planes seating nine or less, not including the pilot seats, and have a maximum takeoff weight of 12,500 lb. The commuter category extends these numbers to 19 passengers and 19,000 lb. Part 25 aircraft are those transport airplanes which generally exceed the weight and/or passenger carrying capacity of Part 23. Part 25 aircraft make up the bulk of the commercial transport aviation on which most passengers travel. Rotorcraft comprises two categories, normal and transport. Part 27 is normal category rotorcraft that seat nine or less passengers and has a maximum takeoff weight of 9,000 lb. The Part 29 transport category are those rotorcraft having higher weight and/or passenger capacity.

The FAA has a number of standards and regulations that are intended to protect occupants in the event of a crash. Static strength requirements for seats and other items of mass and occupant protection provisions are defined in 14 CFR, Part 25, Subparts 561 and 785, along with similar regulations in Parts 23, 27, and 29 [6]. Although they have been revised somewhat over the years, many of the basic requirements have been in place since 1946. The general requirements (Part xx.561) state that the aircraft structure must be designed to protect the occupant when loaded in any direction (upward, downward, forward, rearward, and sideward) during a minor crash landing (Table 28.1). For seats, this is demonstrated by pulling with a specified force on a body block belted into the seat and showing that the load path from the body block to the seat attachment points is maintained. The force applied is based on the specified load factor, occupant weight, and in some cases an additional fitting factor to account for wear on the seat attachment points. The specific occupant protection requirements (Part xx.785) are often referred to as delethalization requirements. They address occupant protection by, among other things, specifying the type of safety belts to be provided and by requiring that head contact with any object that could cause serious injury be prevented, or that those objects be padded.

Table 28.1 Static load requirements

Direction relative to aircraft	Part 23 Small airplanes Load factor	Part 25 Transport airplanes Load factor	Part 27 and 29 Rotorcraft Load factor
Forward	9.0	9.0	16.0
Sideward	1.5	4.0	8.0
Upward	3.0[a]	3.0	4.0
Downward	3.0[b]	6.0	20.0
Rearward	None	1.5	1.5
Occupant weight	215 lb[c]	170 lb	170 lb
Fitting factor	1.33	1.33	1.33

[a]4.5 load factor for aerobatic category
[b]6.0 load factor for commuter category
[c]170 lb occupant weight for commuter and normal category

Accident investigations and research utilizing dynamic impact tests revealed that crash survival could be greatly improved if a means of quantitatively evaluating injury potential of the seating systems was developed and incorporated into the seat safety standards [4]. These new injury criteria, assessed during dynamic tests, would complement, rather than replace, the existing static strength and occupant protection requirements. In the early 1980s, a combined industry and government group known as the General Aviation Safety Panel (GASP) utilized the results of accident investigations, full-scale impact test data, military crashworthiness standards, and automotive injury criteria to formulate recommendations for dynamic test requirements and injury criteria for evaluation of general aviation seating systems. The GASP recommendations served as the basis for additional FAA regulations pertaining to general aviation. The injury criteria cited in the recommendations were also used to develop similar regulations covering rotorcraft and transport category aircraft [7].

Dynamic testing and occupant injury assessment have been required for seats in transport category airplanes certified since the adoption of 14 CFR 25.562 in 1988, and similar regulations in Parts 23, 27, and 29 [8]. The FAA later extended the safety benefit of these regulations to older airliner designs (referred to as derivatives) that are continuing to be built. This regulation required that all newly built aircraft intended for Part 121 (regularly scheduled airline) service delivered after October 2009 also meet the dynamic testing requirements [9]. The regulations require two basic tests (Fig. 28.1 and Table 28.2). Test 1 is a primarily vertical impact test that assesses the structural strength of the seat, and potential spinal injuries. Test 2 is primarily a horizontal test that assesses the structural strength of the seat, with an emphasis on the restraint system's performance. Both tests define the impact in terms of a total velocity change, peak acceleration, and a maximum time to achieve that peak. These impact requirements reflect the forces that could be transmitted to the aircraft cabin during the most likely, survivable impact scenarios.

Illustration shows a forward facing seat

Inertial load shown by arrow

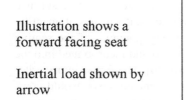
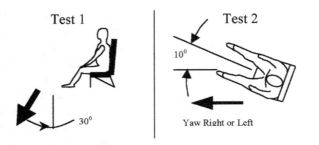

Test Pulse simulating Aircraft Floor Deceleration -Time History

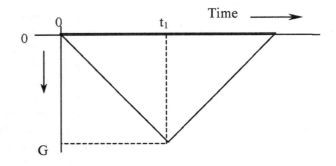

t_1 = Rise time
V_1 = Impact velocity

The ideal pulse is a symmetrical isosceles triangle

Fig. 28.1 Dynamic test requirements- seat orientation and test pulse shape

Table 28.2 Dynamic test requirements-summary

Dynamic test requirements	Part 23	Part 25	Part 27	Part 29
Test 1				
Velocity change (ft/s)	31	35	30	30
Seat pitch (degree)	60	60	60	60
Seat yaw (degree)	0	0	0	0
Peak acceleration (G)	19/15[a]	14	30	30
Time to peak (s)	0.05/0.06[a]	0.08	0.031	0.031
Floor deformation (degree)	None	None	10 pitch	10 pitch
			10 roll	10 roll
Test 2				
Velocity change (ft/s)	42	44	42	42
Seat pitch (degree)	0	0	0	0
Seat yaw (degree)	10	10	10	10
Peak acceleration (G)	26/21[a]	16	18.4	18.4
Time to peak (s)	0.05/0.06[a]	0.09	0.071	0.071
Floor deformation (degree)	10 pitch	10 pitch	10 pitch	10 pitch
	10 roll	10 roll	10 roll	10 roll

[a]Crew/passenger seat requirements

28 Civil Aviation Crash Injury Protection

Table 28.3 Dynamic test requirements-injury criteria pass/fail limits

Injury criteria	Part 23	Part 25	Part 27	Part 29
Max HIC	1,000	1,000	1,000	1,000
Max lumbar load (lb)	1,500	1,500	1,500	1,500
Max shoulder strap load (lb)	1,750/2,000[a]	1,750/2,000[a]	1,750/2,000[a]	1,750/2,000[a]
Max femur load (lb)	N/A	2,250	N/A	N/A

[a]Single/dual strap criteria

Small airplanes and rotorcraft have more severe test requirements than transport category aircraft. For the vertical test, this increase in severity reflects the additional load transmitted to the aircraft cabin due to the lack of crushable space between the bottom of the fuselage and the floor of the small aircraft. The horizontal test has higher severity because the lower momentum of these smaller and lighter aircraft results in a shorter stopping distance. For a given impact velocity, a shorter stopping distance produces a higher level of acceleration. Both of these test conditions have associated injury metrics (Table 28.3) that must be met before a test is considered a pass and the seat is certified for use in aircraft. These injury metrics include limits on lumbar spine and leg loads, limits on head acceleration, and limits on shoulder strap loads. The pass/fail requirements also include qualitative injury criteria that are based on observed occupant interaction rather than a specific measurement. These criteria specifically require that belts remain in place on the pelvis and shoulder. While not explicitly limited, these occupant injury criteria are primarily focused on protecting the occupant during forward and vertical impacts. This chapter will explain the origin and application of the criteria and how their intent has been applied to new seating configurations to maintain the level of safety established in the regulations.

28.1 Injury Criteria Cited in the Aviation Regulations

28.1.1 Criteria Selection

The adoption of the dynamic seat requirements (Part xx.562) included the first occupant injury criteria in aviation regulations. The criteria were selected to protect occupants during expected impact scenarios (primarily longitudinal and vertical impact vectors relative to the aircraft), when seated in the most common types of seats (forward and aft facing). The criteria focused on controlling the most serious injuries that were likely to result from those combinations of seat types and impact scenarios. Some of the criteria selected were originally developed for automotive application but were found to also be applicable to the aviation environment for cases where the loading conditions were similar. However, some loading conditions, such as the high vertical loads that can be experienced in an aviation crash, required unique injury criteria.

28.1.2 Head

Injuries to the head during an aviation accident are possible if the head comes into contact with any of the various structures in the aircraft. These include locations on the seat, such as the seat back, arm rest, tray table, or any equipment mounted on the seat. The head could also contact nearby structures such as a partition, galley wall, instrument panel, glare shield, or flight control. Contacts with hard surfaces can result in skull fractures and multiple types of brain injuries. Brain injuries are also possible even without head contact if inertial forces are high enough [10]. While serious head trauma is an immediate threat to life, in an aviation accident, lesser head injuries that render the occupant unconscious can result in fatality since rapid egress is important.

The head injury criterion (HIC) is used to evaluate head injury risk and is calculated according to the following equation:

$$HIC = \left\{ (t_2 - t_1) \left[\frac{1}{(t_2 - t_1)} \int_{t_1}^{t_2} a(t) dt \right]^{2.5} \right\}_{max}$$

Where:

t_1 is the initial integration time, t_2 is the final integration time, and a(t) is the total acceleration vs. time curve for the head strike, and where (t) is in seconds, and (a) is in units of gravity (g).

Note:

The values of t_1 and t_2 are selected such that the HIC value is the maximum possible for the time period being evaluated.

At the time of its adoption, HIC was considered the best available head injury indicator, and is still the most widely accepted method. The HIC calculation, however, tends to overestimate the injury risk for impacts that include a long period of low G acceleration, as can happen during occupant flailing prior to head contact with a hard surface or during contact with soft surfaces such as air bags. To address this issue, the aviation regulations specify that HIC is calculated using head acceleration data gathered during contact with the airplane interior. In addition to this requirement, the rotorcraft regulations limited the HIC time interval to a maximum of 50 ms, which in the course of the rule's development, had been suggested as another means of addressing HIC's tendency to overestimate the injury risk [11].

A value of 1,000 is cited as a pass/fail limit for HIC in the aviation regulations. Applying a cumulative normal distribution function (Fig. 28.2) to the available injury data indicates that a HIC = 1,000 is a 23 % risk of an AIS-3 or greater (serious) head injury or a 47 % risk of an AIS-2 or greater (moderate) head injury. The Abbreviated Industry Scale (AIS) developed by the Association for the Advancement of Automotive Medicine provides a means of quantifying the severity (or threat to life) of a specific injury [12, 13]. The relationship between HIC and injury is based on impact cases that are of short duration (less than 12 milliseconds), so it is valid for any HIC calculation method that disregards low-G, long duration acceleration by limiting HIC interval duration or excluding pre-impact acceleration.

In U.S. automotive regulations [14], rather than limit the calculation period to only the duration of head contact, the pre- and post-contact periods of low-g acceleration were initially addressed by limiting the HIC time interval to 36 ms (a method referred to as HIC36), with a pass/fail value of 1,000. While this was an effective method for that impact environment, it was not adopted for aviation use because of the longer duration of some aviation impacts. The flail envelope of lap belt-restrained occupants can permit head contact with multiple interior features during a single impact event. Excluding time periods between contacts or calculating HIC separately for each period of contact would not comply with the original purpose of limiting HIC to the duration of contact, which was to exclude pre-contact, low-g acceleration. Calculating the HIC for each individual contact may also lead to allowing some designs to pass, where the method

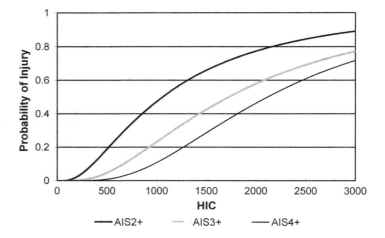

Fig. 28.2 Probability of AIS 2+, 3+, and 4+ head injury as a function of HIC

of continuously calculating HIC would show their non-compliance. Recent research indicates that some types of brain injuries are cumulative, not only during a specific impact event, but over a period of months or years [15, 16]. Beginning the HIC calculation at the time of initial contact and continuing through the end of the head impact event improves the validity of the injury risk prediction by excluding the pre-contact accelerations that can skew the prediction, while at the same time including all of the acceleration that could contribute to injury.

The HIC calculation method in U.S. auto regulations was later revised to limit the maximum time interval to 15 ms and reduce the pass/fail limit to 700 (referred to as HIC15). This change in calculation method was one of several new or revised injury criteria adopted to better assess the safety of autos incorporating airbags. The shorter time interval excludes more of the low-G acceleration produced during typical airbag contact. The lower limit was necessary for the HIC15 calculation to provide an equally stringent evaluation of long duration events. While HIC15 has not been adopted for aviation use at this time, it may be useful for evaluation of aviation airbag installations that produce occupant loading similar to automotive airbags.

28.1.3 Lumbar Spine

Many survivable aircraft crashes have a significant vertical component of acceleration (relative to the occupant). The performance of seats and restraint systems in this impact scenario is evaluated by the combined loading condition (Test 1) defined for each of the aircraft types in the regulations. This test includes a velocity change between 30 and 35 ft/s with a peak acceleration between 14 and 30 G. While the impact vector of this test is inclined 30° off vertical, this still results in a pulse that is primarily directed along the spinal column. In this type of orientation and loading, the injuries expected are those resulting from spinal compression and include: vertebral disc disruptions and herniations, avulsions, and wedge type of fractures to the spinal bodies. While some of these may be moderate, some can be serious and cause extensive complications, including impeding the ability to quickly evacuate the aircraft. Clearly, some means of quantifying the potential for injury due to vertical loading is needed in order to ensure the safety of aircraft occupants.

Over the years, the criteria for assessing the safety of a system to vertical impacts have evolved. Eiband developed the earliest criterion in the 1950s [17]. Using human volunteer and animal data, exposure limits for uninjured, moderately injured, and seriously injured occupants were developed. For vertical impacts, he reported that human volunteers tolerated 10 G for 0.1 s and 15 G for 0.05 s (Fig. 28.3).

Application of the Eiband curve had several limitations. It primarily characterized the response to whole-body acceleration and did not break out injuries by body region. It also was not sensitive to changes in the pulse shape or mitigation methods that may have been developed. Initial ejection seat designs had acceleration limits in the 20-G range. This range falls at the boundary of moderate injury in the Eiband criteria (Fig. 28.3). It was found that spinal fractures frequently occurred during ejection seat incidents and that improved seat designs were needed [18]. As part of a revised ejection seat development program, a new criterion was also developed that is known as the Dynamic Response Index (DRI) [19]. The DRI model represents the spinal column of the human occupant as a lumped mass-spring-damper model. Input to the model consisted of seat pan accelerations, and model output consisted of the deflection time history of the DRI system. The maximum value of the DRI response was the parameter of interest. This value was then correlated with experimental and operational injury data (Fig. 28.4), and a value of 18 was selected as a design goal to limit spinal injury risk [20, 11].

The primary limitation of the DRI model is that it was derived for seats with nearly rigid seat pans with only a small amount of firm cushioning and a typical ejection seat restraint system.

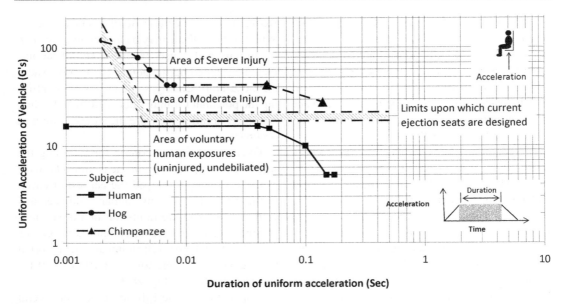

Fig. 28.3 Duration and magnitude of headward acceleration endured by various subjects, Eiband injury tolerance curve for vertical acceleration

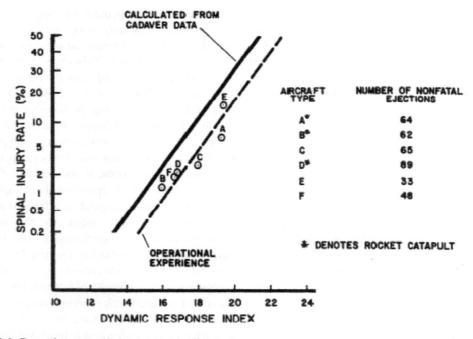

Fig. 28.4 Dynamic response index versus spinal injury rate

Therefore, it is not directly applicable to typical aircraft seats that have compliant seat pans and cushions. To address these issues, the FAA developed a lumbar load tolerance value. Since load in the lumbar region is the primary factor causing injuries, it was surmised that a criterion based directly on measured lumbar load response was prudent. To determine the threshold, the FAA

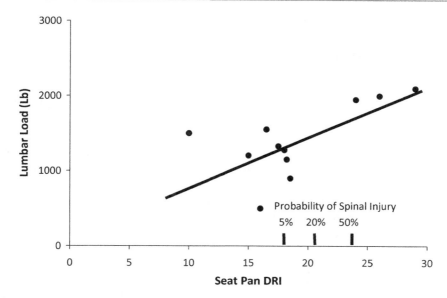

Fig. 28.5 Dynamic response index and associated spinal injury risk versus lumbar load

conducted a series of dynamic impact tests using aviation-specific pulses [11]. These tests included an energy-absorbing seat with a rigid seat pan and an Anthropomorphic Test Device (ATD) that was modified to collect lumbar loads. For each test, a lumbar load was measured, and the DRI of the test condition was calculated (Fig. 28.5). Based upon this correlation, a lumbar load of 1,500-lb measured in the Hybrid II ATD was correlated to a DRI of 19, or approximately a 9 % risk of a detectable spinal injury. This 1,500 lb value was cited in the GASP recommendations that formed the basis of current FAA regulations.

The 1,500 lb lumbar load requirement is applicable to all types of aircraft and must be measured with either the 50th Percentile Male-size Hybrid II ATD or the 50th Percentile Male-size FAA Hybrid III ATD [21].

Providing protection to the lumbar spine may also mitigate heart and aortic injuries. An analysis of military rotorcraft crash data revealed that these soft tissue injuries seldom occurred without an associated head/neck or thoracic injury [22]. Thus, a separate criterion to protect the internal organs during vertical impacts does not appear necessary.

28.1.4 Chest

Chest injuries can result from direct contact with structure or restraint systems and from inertial forces resulting from rapid deceleration of the thorax. Automotive chest injury criteria relating chest acceleration or compression to injury risk were derived for contacts with flat impact surfaces approximating auto steering wheels [11]. In the aviation impact environment, there are fewer direct contact risks (no fixed steering wheel), but the chest can be loaded directly by the upper torso restraint system, and restraint system forces can decelerate the torso at a rate sufficient to cause injuries.

To evaluate the risk of chest injury for occupants wearing upper torso restraint systems, an injury criterion that related restraint tension load to chest injury risk was developed. This criterion was based on a study of auto crashes for which both the strap load and resulting injuries were available for comparison. In seat qualification tests, a single torso strap load must remain below 1,750 lb., and the combined load on dual torso straps must remain below 2,000 lb. These criteria correspond to a 50 % risk of an AIS-3 or greater chest injury [11]. Dynamic tests of seats employing a single torso strap are typically conducted with the seat yawed in the direction that places the strap over the leading shoulder to maximize the load produced.

28.1.5 Leg

Protection of the lower extremities during an aircraft crash is important since an injury to this body region would severely hinder egress. Two examples of impact scenarios that could result in leg injuries are: flailing forward to contact a seat or structure in front of the occupant, and direct loading by the floor or surrounding structure if crash forces cause them to intrude into the cabin. These scenarios can produce combined loading on both the femur and tibia, as well as transmitting loads into the pelvis.

At the time the seat dynamic rules were adopted, the most widely accepted leg injury criterion was the femur axial load limit of 2,250 lb., cited in the Federal Motor Vehicle Safety Standard for Occupant Crash Protection, (FMVSS) 208 [14]. This load limit corresponds to a 35 % risk of an AIS-2 or greater injury or a 15 % risk of an AIS-3 or greater injury to the knee-thigh-hip complex [23]. Meeting this criterion was only required for Part 25 (transport category) aircraft since these larger aircraft present the greatest evacuation challenges. In transport aircraft, passengers must be mobile enough after a crash to negotiate the relatively narrow aisle leading to the exit.

28.1.6 Abdomen

Direct loading of the abdomen by belt systems can cause serious injuries at relatively low applied loads. Improper lap belt geometry can result in the belt sliding up over the iliac crests of the pelvis and into the abdomen as the occupant loads the belt during a forward impact. This effect is often referred to as "submarining," since the occupant essentially slides *under* the lap belt. This tendency can be minimized if the lap belt angle is sufficiently vertical. The vertical component of force reacted by the lap belt prevents it from sliding upward as the occupant translates forward. For restraint systems with upper torso straps attached near the mid-point of the lap belt, the vertical component of force reacted by the lap belt tends to keep the lap belt in place by counteracting the upward-acting tension forces in the shoulder straps. Otherwise, the shoulder strap tension forces can pull the lap belt upward into the abdomen. Some restraint systems use a vertical tie-down strap attached at the mid-point of the lap belt to ensure the lap belt remains in place. As part of a U.S Army program to improve the crashworthiness of their aircraft, recommendations were developed (Fig. 28.6) for lap belt geometry that minimizes the chance of submarining in a forward impact [24].

Because of the serious consequences of abdominal belt intrusion, aviation regulations (Part xx.562) require that the belt remain on the pelvis during the specified dynamic tests. Compliance with this criterion is usually verified by examination of high-speed video and examination of lap belt tension force vs. time history data. The point in time when the lap belt rides up and over the iliac crest is usually indicated by a sudden drop in belt tension load.

28.1.7 Flail Injuries

All dynamically qualified aircraft seats, other than Part 25 (transport category) forward or aft-facing passenger seats, are required to incorporate an upper torso restraint. The primary purpose of this portion of the restraint system is to control the motion of the upper torso well enough that head contact with nearby structure is prevented or reduced in severity. For aerobatic aircraft, the shoulder straps also serve to keep the occupant in the seat during negative G maneuvers. The potential for injuries related to the forces imparted by upper torso restraint loads are addressed by the tension limits previously discussed. Effectiveness of the shoulder strap(s) in controlling the motion of the occupant is evaluated indirectly by the requirement to limit the HIC resulting from any impact. The regulations also

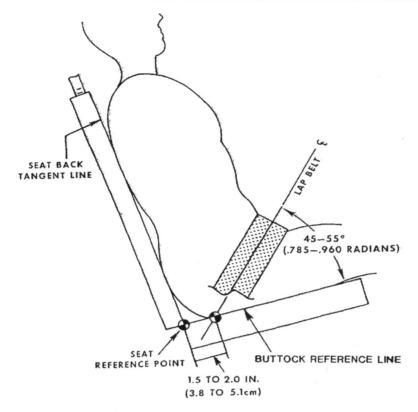

Fig. 28.6 Recommended lap belt anchor geometry

directly evaluate shoulder strap effectiveness during the dynamic tests by requiring that the strap remain on the shoulder.

The location of the shoulder strap anchor (or strap guide) and the point at which it is attached to the lap belt are the primary factors determining whether the shoulder strap will remain in place when dynamically loaded. An anchor point that is too far from the seat centerline can place the strap over the top of the humerus, rather than the clavicle. With a single diagonal shoulder strap system, this placement can tend to force the upper torso to rotate about its Z axis during a forward impact, and slip out of (or roll-out) of the restraint, allowing the occupant to fail forward much further than if the belt had remained in place. This potential to cause injury is evaluated by conducting one of the dynamic qualification tests with the seat yawed 10° in the direction that is most likely to produce occupant roll-out (as is the case when the shoulder belt is over the trailing shoulder).

Compliance with the criterion that the shoulder strap remains on the shoulder is usually verified by examination of high-speed video.

28.2 Injury Criteria for Side-Facing Seats

28.2.1 Special Conditions

The occupant protection provisions for small aircraft (Part 23.785) require that other seat orientations provide the same level of occupant protection as a forward- or aft-facing seat but do not cite specific means of providing that protection. The provisions for large aircraft (Part 25.785) require seats oriented greater than 18° from forward to provide an energy-absorbing rest to support the arms, shoulder, head, and spine, or provide a shoulder harness to prevent head contact with injurious objects. The provisions for rotorcraft (Parts 27 and 29) do not specifically address passenger seat orientation. During an aircraft crash with a longitudinal impact vector, the occupant of a side-facing seat is loaded in the lateral direction (with respect to the occupant), which is the same direction of loading that occurs during an automotive side impact. In auto impacts, Marcus et al. found that lateral loading has a different set of injury risks than frontal loading [25]. Therefore, additional airworthiness standards for transport category airplanes, in the form of special conditions, were adopted to ensure that side-facing seats could provide a level of safety equivalent to that afforded occupants of forward- and aft-facing seats [26]. These special conditions include testing methods and injury criteria that are applied in addition to the existing dynamic requirements. The injury criteria are a combination of those found in current aviation and automotive regulations and ones derived from research quantifying aviation-specific lateral injuries [27]. As of this book's publication, compliance with all of these tests and injury criteria have not been required to certify side-facing seats in other types of aircraft (Parts 23, 27 and 29). For side-facing seats on Part 23 aircraft, special conditions are developed specifically for each installation. The evaluation criteria cited in the special conditions reflect a combination of those automotive side-facing and aviation-specific criteria deemed appropriate for injury evaluation of each seat configuration.

28.2.2 Neck

Unless the occupant of a side-facing seat is seated just aft of a supporting wall, a forward impact can cause large lateral excursions of the torso and head that produce significant neck loading due to inertial forces. Contact of the head with nearby structure can also generate neck loads. To address this potential for injury, the FAA sponsored research that included tests with Post Mortem Human Subjects (PMHS) to derive appropriate neck injury criteria for side-facing seats. The injury criteria developed (outlined below) were incorporated in the Special Conditions for side-facing seats [28].

- Neck Tension: A strong correlation was found between neck tension applied by inertial forces during lateral impacts and neck injury. An injury risk curve was developed for tension forces measured by the ES-2 ATD. An axial neck tension limit of 405 lb represents a 25 % risk of an AIS-3 or greater neck injury. This limit is considerably less than the 937 lb limit cited in FMVSS-208 for upper neck tension resulting from frontal impacts. This reduction in tolerance is likely due to the spinal misalignment that occurs as the neck bends laterally.
- Neck Bending Moment: Initial PMHS tests did not produce high lateral moments in the absence of tension. This loading condition was of interest because in some seat tests, the potential injury mitigation technology (inflatable restraint systems) has produced relatively high bending loads with little tension [29]. The PMHS test associated with the highest moment recorded during the initial research did not produce significant neck injury. This indicated that the onset of injury was likely greater than the 673 in.-lb measured by an ES-2 ATD subjected to the same test conditions. To investigate further, a follow-on project was conducted to assess the load case consisting of low tension and high lateral moments. The highest bending moment produced at the occipital condyles of the PMHS during that follow-on research was 651 in.-lb. This load did not result in a detectable injury. A comparison between ES-2 and PMHS response when loaded in the same manner and at the same severity indicates that this load corresponds to a 1,018 in.-lb moment measured by the ES-2 [30]. Therefore, a value of 1,018 in.-lb lateral bending moment (as measured at the occipital condyle location of the ES-2) can be considered a threshold, below which neck injury is not expected.
- Neck Compression: This research focused on loading conditions that were most likely to occur in typical side-facing seat installations. The investigated conditions did not produce neck compressive loads; however, this load case cannot be ignored since some seat configurations that have nearby structure could

produce neck compression during lateral impacts. The tension and compression limits cited in U.S. auto regulations (49 CFR Part 571.208) for use in forward tests of automobiles are very similar (tension = 937 lb. and compression = 899 lb). If the lateral bending of the neck reduces its tolerance to compression loading in the same manner and to the same degree that it reduces the tolerance to tension loading, then applying the same limit specified for tension loading during side impact (405 lb) to compression loading should provide a similar level of safety for this load case.

- Neck Shear force: The correlation between neck shear and injury indicated by this research was not as strong as the one for tension loading. This is because, as with neck lateral moment, the research method did not produce high shear forces independent of tension loading. So, for the serious injury cases, it is unknown whether the injury was caused by the shear force, the tension force, or the combined effects of both. On the other hand, an upper limit at which severe injury would occur obviously exists. A conservative load limit can be based on the injury risk curve that was developed for resultant shear (Fxy) forces measured by the ES-2. A limit on the neck shear load of 185 lb represents a 25 % risk of an AIS-3 or greater neck injury. While this limit is conservative due to the inclusion of some tension force in the corresponding injury cases, the degree of such conservatism is unknown.

28.2.3 Thorax and Pelvis

The injury criteria cited in U.S. auto safety regulations (49 CFR 571.214) for use with the ES-2re ATD were reviewed for applicability in aviation impact scenarios. Aircraft side-facing seats that provide a supporting wall just forward of the occupant can load the ATD during a forward impact similar to a side impact with an automobile door. Since the loading case is similar for both conditions, all of the ES-2re's injury criteria were found to be useful for evaluating the safety of aircraft side-facing seats and were cited in the Special Conditions [26]. For each of these criteria, the relationship between the probability of injury and a value measureable by the ES-2re has been defined. These criteria and the associated risk of sustaining a specific level of injury are summarized below [12]:

- Ribs: Lateral deflection of any of the three rib modules is limited to 1.73 in. This limits the injury risk to a 50 % chance of an AIS-3+ chest injury.
- Abdomen Force: Limited to a combined load (from all three load cells) of 562 lb. This limits the injury risk to a 33 % chance of an AIS-3+ abdominal injury.
- Pubic Symphysis force: Limited to 1,350 lb. This limits the injury risk to a 25 % chance of an AIS-3+ pelvis injury.

28.2.4 Leg

FAA-sponsored research found that side-facing seat configurations that support the upper leg with an armrest or wall, but do not support the lower legs, can produce serious leg injuries (femur fractures) during a forward impact. The nature of these injuries indicates that they were caused by torque applied to the femur at the knee. The inertia of the lower legs, as they flailed laterally, was evidently sufficient to twist the femur beyond its normal range of rotational motion and fracture it. To ensure side-facing seats provide an equivalent level of safety, a means of assessing this unique injury risk during seat certification tests was needed. The biofidelity of the ES-2re's leg lateral flail and resulting femur torsion has not been validated, so an injury criteria based on measured femur torsion was not feasible. Since the injuries observed were caused by upper leg axial rotation, one way to limit the injury risk is to limit the amount of that rotation. A relationship between rotation angle and specific injury risk is not yet available, but the normal static range of motion (below which injury is likely) has been defined [31]. Based on this data, a limit of 35° of upper leg axial rotation was selected for inclusion in the Special Conditions to ensure that the risk of leg injury is low. The documented range of motion is related to flexibility and therefore varies widely between individuals. The angle chosen is approximately the 50 % percentile range of motion for both genders.

28.2.5 Occupant Support Criteria

Aircraft side-facing seat configurations do not necessarily include end closures or supporting surfaces between the occupants (for seats that accommodate multiple occupants). These seats can permit occupant excursions, articulations, and contacts of the type that would not occur in forward- or aft-facing seats. To address these unique injury risks, the Special Conditions for side-facing seats include limitations on occupant excursion, occupant articulation, and contacts between occupants.

28.2.5.1 Excursion Limits

A side-facing seat without a forward end closure can permit the occupant's pelvis to translate off (or partially off) the end of the seat during a forward impact unless the lap belt is configured to restrain the pelvis laterally. In a similar manner, a seat without an aft end closure could permit the occupant's pelvis to translate off the aft end of the seat if the occupant rebounds with sufficient energy. During a forward impact, the geometry of the lap belt tends to produce significant vertical load on the bottom of the pelvis. Excursion off of the seating surface could allow concentrated asymmetrical loading of the bottom of the pelvis as it slides over the edge of the seat pan. This type of potentially injurious loading does not occur in a forward- or aft-facing seat. Therefore, to provide the same level of safety, the Special Conditions require that the excursion of the pelvis is limited, such that the load path is maintained between the seat pan and the entire load bearing area of the pelvis. The area under and around the ischial tuberosities on the bottom of the pelvis bear much of a seated occupant's vertical load [32]. For the purposes of this evaluation, the load bearing area of the pelvis is defined as a 5-in. deep by 8-in. wide rectangular area with edges 3 in. forward, 2 in. rearward, and 2 in. sideward of each buttock reference point [33].

Another consideration in evaluating occupant excursion is that the general occupant protection requirements of 14 CFR 25.561 (a) and (b), and the similar regulations for Parts 23, 27, and 29, require the occupant to be protected from serious injury when inertia forces are applied in any of the six orthogonal directions (Table 28.1). If application of those loads could be expected to move the occupant in such a way that it was not supported by the seat bottom or back, or that the belts would interact with the occupant in an injurious manner prohibited by the Part xx.562 criteria, then the seat would not be in compliance with the intent of Part xx.561. Essentially, a seat and restraint system that only restrains the occupant safely when loaded in the forward or vertical directions defined in Part xx.562 would not meet all of the requirements of Part xx.561.

28.2.5.2 Articulation Limits

Unless the occupant of a side-facing seat is seated just aft of a supporting surface, a forward impact can produce significant lateral flexion of the upper torso. Conventional upper torso restraints employed in many seat designs can allow significant lateral flexion. This articulation is of concern because lateral flailing over an armrest has been observed to produce serious injuries, including spinal fractures [34]. The ES-2re's abdominal force measurement has been shown to correspond to injuries resulting from horizontal impact on that area of the body. Limiting the loading in this area may prevent some of the injuries produced when occupants flail over an armrest structure, but this criterion was not intended to evaluate spinal or internal injuries caused by excessive lateral bending of the occupant, as those types of injuries were not observed in the typical automotive side impacts that formed the basis of the criterion. In those automotive side impact tests, the occupants are typically fully supported by the vehicle door and window.

At the time of the Special Condition's development, there were no criteria available relating the amount of lateral flail to a specific risk of injury. However, if lateral flexion is limited to the normal static range of motion, then the risk of injury should be low. This range of motion is approximately 40° from the upright position [31]. Ensuring that lateral flexion does not create a significant injury risk is consistent with the goal of providing an equivalent level of safety to a forward- or aft-facing seat, since that type of articulation does not occur to occupants of those seats during forward impacts.

28.2.5.3 Contact Limits

Multiple-place, side-facing couches create a unique injury risk due to the potential for adjacent occupants to flail onto one another during a forward impact. The specific injury risks for these contacts are unknown. Evaluating these contacts during dynamic tests using existing criteria such as HIC is problematic, since the biofidelity of the contact between various parts of the ATD has not been validated. When the Special Conditions were developed, there was no standard means to assess the specific injury risk for the contact between occupants. Since these types of contact do not occur to occupants in forward- or aft-facing seats, one means to ensure the same level of safety is to prohibit contact during the loading phase between the head, pelvis, torso, or shoulder area of one ATD with the adjacent seated ATD's head, pelvis, torso, or shoulder area. Contact during the rebound phase is permitted since it would be unlikely to cause significant injury due to the much lower energy of those contacts.

28.3 Maintaining the Level of Safety

The injury criteria contained in the aviation regulations establish a specific level of safety for aircraft occupants and represent a huge step forward in occupant safety compared to the general requirements that had been historically used. In the crashes that have occurred involving airliners equipped with the improved seats, there have been many passengers whose survival has been attributed, in part, to the performance of the improved seats [35].

By use of Special Conditions, additional criteria were adopted to ensure the level of safety provided to occupants of side-facing seats was the same as for occupants of forward- and aft facing seats. As seat manufacturers continue to innovate, new seat configurations must be evaluated to determine if they produce serious injury modes not addressed by the current criteria. Development of new criteria may be needed to address any new injury modes resulting from future seating innovations.

One of the innovations becoming more common on aircraft is airbags. Just as in autos, airbags have the potential to significantly reduce the risk of injury, when compared to conventional restraint performance. Special Conditions addressing restraint-mounted airbags have been adopted. These requirements do not include new injury criteria since the manner of deployment (away from the occupant) is inherently low risk, and the occupant interaction with the bag during the impact is less severe than the direct contact with surrounding structure that would otherwise be permitted by conventional restraints. However, additional requirements may be necessary to address structure-mounted airbags. Because these systems typically deploy toward the occupant, a comprehensive injury assessment may require injury criteria and test methods similar to those used to assess airbag systems in automobiles.

One of the latest innovations, the obliquely oriented "pod" seats currently being installed in many aircraft, have the potential to load the occupant in unique ways. These seats are a good example of an innovation where advancements in injury biomechanics are necessary to accomplish a comprehensive injury risk assessment.

While the established level of safety has been shown to reduce the number of immediate fatalities in survivable crashes, further research may provide a better means of predicting and mitigating life-threatening injuries. The current head injury limit (HIC = 1,000) can still result in a period of unconsciousness from 1 to 6 h, which would prevent independent egress from the aircraft [36]. As advancements are made in both biomechanical knowledge of head injury and in head impact mitigation technology, it may eventually be possible to align the intended level of safety with what can be measured during a test to increase the likelihood that occupants would not only survive the impact but be conscious and able to evacuate quickly [37].

References

1. National Transportation Safety Board (2011) Review of U.S. Civil Aviation Accidents, NTSB/ARA-11/01
2. National Transportation Safety Board (2001) Survivability of accidents involving part 121 U.S. Air Carrier Operations, 1983 through 2000, NTSB SR01/01
3. NHTSA Fatality Analysis Reporting System Encyclopedia (2013) FARS Data Tables, Trends, Fatalities and Fatality Rates 1994–2012 - State : USA, Washington, DC. www-fars.nhtsa.dot.gov
4. Soltis S, Nissley W (1990) The development of dynamic performance standards for civil aircraft seats. Paper presented at the National Institute of Aviation Research, 1990 Aircraft Interiors Conference, Wichita State University, Kansas
5. Coltman J, Bolukbasi A, Laananen D (1985) Analysis of rotorcraft crash dynamics for development of improved crashworthiness design criteria, DOT/FAA/CT-85/11
6. U.S. Code of Federal Regulations, Title 14, Parts 23.561, 23.785, 25.561, 25.785, 27.561, 27.785, 29.561, 29.785. U.S. GPO, Washington, DC
7. Chandler RF (1993) Development of crash injury protection in civil aviation, Chapter 7. In: Nahum A, Melvin J (eds) Accidental injury, biomechanics and prevention, 1st edn. Springer, New York
8. U.S. Code of Federal Regulations, Title 14, Parts 23.562, 25.562, 27.562, 29.562. U.S. GPO, Washington, DC
9. U.S. Code of Federal Regulations, Title 14, Part 121. Improved seats in air Carrier Transport Category Airplanes. U.S. GPO, Washington, DC; Final Rule; Federal Register 70(186):56542 (2005)
10. Bandak F, Eppinger R (1994) A three-dimensional finite element analysis of the human brain under combined rotational and translational accelerations. Stapp Car Crash J 38, Paper No. 942215
11. Chandler R (1985) Human injury criteria relative to civil aircraft seat and restraint systems. SAE International, Warrendale, SAE Paper 851847
12. Kuppa S (2004) Injury criteria for side impact dummies. DOT/National Highway Traffic Safety Administration, Washington, DC; FMVSS-214 NPRM docket: NHTSA-2004-17694
13. (2008) The abbreviated injury scale 2005 – update 2008. Association for the Advancement of Automotive Medicine, Barrington
14. U.S. Code of Federal Regulations, Title 49, Part 571.208. U.S. GPO, Washington, DC
15. Takhounts E, Eppinger R et al (2003) On the development of the SIMon finite element head model. Stapp Car Crash J 47, Paper No. 03S-04
16. Iverson G, Gaetz M et al (2004) Cumulative effects of concussion in amateur athletes. Brain Inj 18(5): 433–443
17. Eiband A (1959) Human tolerance to rapidly applied accelerations: a summary of the literature. NASA Memorandum 5-19-59E. NASA Lewis Research Center, Cleveland
18. Henzel J (1967) The human spinal column and upward ejection acceleration: an appraisal of biodynamic implications. USAF Aerospace Medical Research Laboratories, AMRL-TR-66-233
19. Stech E, Payne P (1969) Dynamic models of the human body, AMRL-TR-66-157. Aerospace Medical Research Laboratory, Wright-Patterson AFB, Ohio
20. Brinkley J, Shaffer J (1971) Dynamic simulation techniques for the design of escape systems: current applications and future Air Force requirements. USAF Aerospace Medical Research Laboratories, AMRL-TR-71-29
21. Gowdy R, Beebe M, Kelly R et al (1999) A lumbar spine modification to the Hybrid III ATD for aircraft seat tests. SAE International, Warrendale, SAE Paper 1999-01-1609
22. Barth T, Balcena P (2010) Comparison of heart and aortic injuries to the head, neck, and spine injuries in US Army Aircraft accidents from 1983 to 2005. In: Proceedings of the American Helicopter Society forum 66, Phoenix, 10–13 May 2010
23. Kuppa S, Wang J, Haffner M, Eppinger R (2007) Lower extremity injuries and associated injury criteria. DOT/National Highway Traffic Safety Administration, Washington, DC. NHTSA paper no. 457
24. US Army (1989) US Army crash survival design guide, USAAVSCOM TR 89-D-22D Volume IV, Section 7.3.3
25. Marcus JH, Morgan RM, Eppinger RH, Kallieris D, Mattern R, Schmidt G (1983) Human response to and injury from lateral impact, SAE Paper No. 83163424
26. FAA (2012) FAA policy statement, PS-ANM-25-03-R1, Technical criteria for approving side-facing seats. DOT/Federal Aviation Administration, Washington, DC
27. DeWeese R, Moorcroft D, Abramowitz A, Pellettiere J (2012) Civil aircraft side-facing seat research summary. DOT/Federal Aviation Administration, Washington, DC. FAA report no. DOT/FAA/AM-12/18
28. Philippens M, Forbes P, Wismans J, DeWeese R, Moorcroft D (2011) Neck injury criteria for side-facing aircraft seats. DOT/Federal Aviation Administration, Washington, DC. FAA report no. DOT/FAA/AR-09/41

29. DeWeese R, Moorcroft D, Green T, Philippens M (2007) Assessment of injury potential in aircraft side-facing seats using the ES-2 anthropomorphic test dummy. DOT/Federal Aviation Administration, Washington, DC. FAA report no. DOT/FAA/AM-07/13
30. Yoganandan N, Pintar F, Humm J et al (2013) Neck injury criteria for side-facing aircraft seats – phase II. DOT/Federal Aviation Administration, Washington, DC. FAA report no. DOT/FAA/TC-12/44
31. Henry Dreyfuss Associates (2002) The measure of man and woman. Wiley, New York
32. Rosen J, Arcan M (2003) Modeling the human body/seat system in a vibration environment. J Biomech Eng 125:223–231
33. FAA (2006) FAA advisory circular 25.562-1B, dynamic evaluation of seat restraint systems and occupant protection on transport airplanes. U.S. GPO, Washington, DC
34. Pintar F, Yoganandan N, Stemper D et al (2007) Comparison of PMHS, WorldSID, and THOR-NT responses in simulated far side impact. Stapp Car Crash J 51:313–360, Paper No. 2007-22-0014
35. Trimble E (1990) Report on the accident to Boeing 737-400 G-OBME near Kegworth, Leicestershire on 8 January 1989. Aircraft accident report 4/90. HMSO, London
36. Yoganandan N, Pintar F et al (2005) Biomechanical aspects of blunt and penetrating head injuries, IUTAM Paper 011405
37. Grierson A, Jones L (2001) Recommendations for injury prevention in transport aviation accidents. SAE International, Warrendale, SAE Paper 2001-01-2658

Ballistic Injury Biomechanics

Cynthia Bir

Abstract

Penetrating and non-penetrating injuries can result from a ballistic impact. Typical gun shot wounds are the penetrating type. However, when a bullet designed to penetrate hits a piece of personnel protective equipment the result can be a blunt impact injury. Blunt impact injuries can also occur with less-lethal kinetic energy devices. Injuries that result from these high rate impacts are dependent on many factors including: the energy imparted to the body, the surface area that this energy is focused upon as well as the region of the body impacted. Research in the areas of gun shot wounds, direct and indirect impacts to the long bones, behind armor blunt trauma and blunt ballistic impacts will be discussed.

Ballistic injury can be the result of either penetrating or non-penetrating impacts. Typical gun shot wounds are of the penetrating type. However, when the impact occurs over personnel protective equipment or is the result of a round that is design specifically not to penetrate, such as a less-lethal kinetic energy device, non-penetrating, blunt impact injuries can occur.

C. Bir, Ph.D. (✉)
Keck School of Medicine, University of Southern California, Los Angeles, CA, USA
e-mail: cbir@usc.edu

29.1 Gun Shot Wounds

There are several variables that affect the type of wounds caused by bullets. From the ballistics side, key characteristics of the bullet such as caliber, velocity, mass, orientation on impact and potential fragmentation all contribute to the resulting injury. From the human side, both the soft tissues (skin, muscle) and underlying bone and the proximity of each will dictate the damage observed. There are two major mechanisms of injury: crushing and tearing. Crushing is often caused directly by the bullet, while tearing is the result of the expansion of gases as the bullet transverses through the tissues. The first causes a permanent cavity in the tissues, as opposed to the second, which results in a temporary cavity.

Gunshot wounds have been described as either penetrating or perforating [1]. Penetrating wounds are when an entrance wound is present, but there is no exit. Perforating wounds have both an entrance and exit wound. In the case where the bullet grazes the body, both soft and boney tissue damage can occur, but the bullet does not actually penetrate into the body. If the wound is shallow and does not demonstrate injury to the deeper layers of the fascia, then it is considered a graze. If deeper layers of tissue are involved, then the wound is labeled as a guttering wound. These wounds are often easier to detect directionality due to the tearing that occurs on the end of the wound where the bullet leaves the skin.

29.1.1 Entrance Versus Exit

One of the first steps of a forensic investigation of gunshot wounds is to determine entrance versus exit wounds. Depending on the region of the body, caliber of round and distance traveled, these wounds may or may not be easily distinguished. Determining entrance and exit wounds will determine directionality of the bullet, which is an important factor in the overall crime scene investigation.

29.1.1.1 Soft Tissue

The amount of soft tissue damage can help distinguish an entrance from an exit wound. Entrance wounds will generally have less tissue damage than exit wounds if the bullet enters perpendicular to the tissue and has not hit an intermediate object. They are circular in nature often with an abraded ring around the circumference. If the bullet enters at a slight angle, the wound will be more oval in shape with a more abraded region found on the entry side of the wound.

If the bullet starts to tumble or becomes unstable in flight, the entrance wound may be irregular in shape or have tearing at the margins. Bullets that ricochet or hit an intermediate target will also have irregular entrance wounds. Depending on the distance, shots that occur over a bony region such as the skull can result in a stellate wound. This is due to the expansion of gases between the skin and bony layers. As the gases expand the soft tissues cannot stretch fast enough to accommodate the increase volume, therefore tearing will occur. It is important not to interpret these types of wounds as exit wounds despite the large amount of tissue damage. This same gas expansion may cause a muzzle imprint in areas where the tissue is free to expand. The skin will expand as the gases enter the underlying tissues and impact upon the muzzle of the firearm. This imprint can often be matched to the weapon if known and available.

Exit wounds can have varying shapes and sizes. Although typically larger than entrance wounds, this may not always be the case. The larger size is due to the bullet tumbling and deforming as it enters into the body. The amount of tumbling will be dependent on the initial velocity when the bullet enters the body and the amount of distance traveled within the body. High-powered rifles may cause stellate wounds upon exit, which resemble those seen the entrance wounds over flat bony regions as described above.

The lack of abraded margins can help distinguish an entrance wound from an exit. However, in the case where the skin is pressed up against a wall or flat surface, shoring will occur which results in abrasion of the skin caused by contact with the flat surface as the bullet exits. Heavy clothing may also create a shored exit wound.

29.1.1.2 Bony Structures

Due to the dipole nature of the skull, entrance and exit wounds in this region can often be distinguished based on the injury patterns to the skull itself. Damage to the inner and outer tables can be different due to the bending that occurs as the bullet strikes the bone and the resulting stresses that arise in the tissues. This discrepancy causes beveling (Fig. 29.1), which is represented by a smaller diameter on the impacted surface versus a larger diameter on the far side of the impact. Entrance wounds will typically have internal

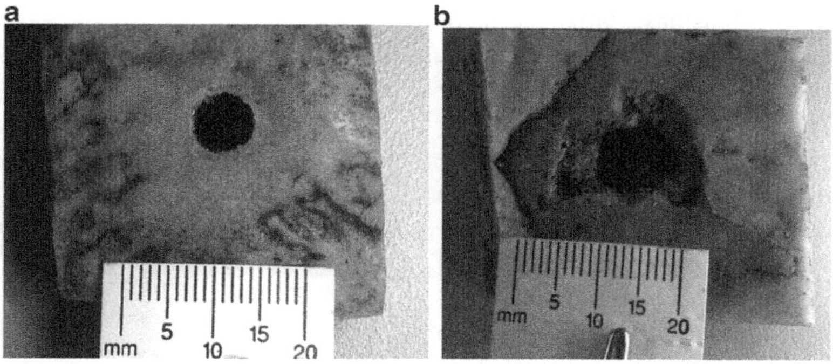

Fig. 29.1 Gunshot wound to the skull. (**a**) Entrance wound on exterior of skull (**b**) exit wound on interior of skull with beveling

beveling and exit wounds will have external beveling. However, in cases where the bullet has started to yaw, or with high power contact wounds, external beveling can be present on an entrance wounds, but is most likely paired with internal beveling to a greater degree.

Beveling can be found evenly distributed around the wound, indicating a more or less perpendicular entrance. In cases of tangential impacts, the amount of beveling may be asymmetrical with greater amount opposite the entrance of the bullet. Tangential bullets may also produce what are known as "keyhole defects" [2]. As the bullet enters the bone tangentially, chipping will occur on the entrance side and the stresses on the opposite side will cause fracture propagation in the outer table causing a large area of beveling. Internal beveling will also be present which can confirm the wound as entrance. Directionality of the bullet through the skull can also be determined based on the presence of this finding.

A bullet may pass through one portion of the body and reenter into another. The most common occurrence is when an arm is struck and the bullet perforates through and then enters the thorax [1]. This phenomenon can result in both an atypical entrance wound on the thorax as well as an atypical shoring exit wound on the arm if pressed against the body. The reentry wound is often oval or crescent in shape and may not have a ring of abrasion.

29.1.2 Distance

The range of fire, or muzzle to target distance, is often the next determination made in gun shot wound cases. Approximate distances can be established based on the presence or absence of debris from the muzzle. If close enough, this debris can be deposited into the skin and clothing of the victim. Typically four ranges have been delineated: contact, near contact, close and distant.

Contact shots will show the presence of gun shot residue (GSR) and soot on the margins of the wound. The hot gases expelled by the gun will burn the soot into the tissues and it cannot be washed away. This should not be confused with bullet wipe. Bullet wipe is the presence of debris that is transferred to the clothing or skin as the bullet passes through. This can happen in both close and distant range firings.

When the muzzle has been placed close to the skin but not in hard contact, loose or near contact wounds may be seen. Loose contact wounds are represented by a deposition of soot around the entrance wound, which can be wiped away. Near contact wounds have a larger ring of soot and with a larger inner ring of seared tissue, due to the dispersion of powder, that is not easily wiped away.

The presence of stippling or tattooing is an indication that the distance to muzzle was of close range. This distance is from 0.012 m (5 in) to approximately 0.457 m (18 in). Stippling is cause by the gunpowder particles striking the skin and causing punctate abrasions. Unlike the deposition of soot, stippling cannot be washed away. If the muzzle is fired at an angle to the skin, there will be a greater dispersion of stippling in the direction of the firing.

When the muzzle of the gun is greater than 0.457 m (18 in) away, stippling will not be present. However, the lack of stippling alone should not determine distance since clothing and intermediate objects may mask its presence. The size and shape of the entrance wound in distant shots will be more representative of the bullet.

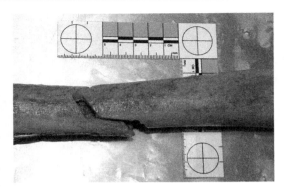

Fig. 29.2 Spiral fracture caused by temporary cavity formation

29.1.3 Direct Versus Indirect Fractures

Skeletal injuries to long bones in the extremities can be caused by both direct impact and indirect impact [3]. Indirect fractures are the result of a projectile passing close to a bone but not directly striking it. They are characterized by a simpler fracture pattern and less medullary contamination when compared with direct fractures [4]. Approximately 10 % of all long bone fractures are indirect. Direct fractures occur when a bullet strikes bone with enough force to cause a fracture [5]. In contrast, these fractures are more complex and generally have more comminution.

Between 1968 and 1970, Huelke et al. authored three papers to describe the "bioballistics" of direct fractures of the femur. Using stainless steel spheres, the authors conducted direct impacts to both unembalmed and embalmed femoral specimens. One of the basic findings was that both the unembalmed and embalmed specimens responded the same in terms of energy loss versus impact velocity. The authors reported that given the same velocity, the diameter and not the mass of the sphere was the main factor determining damage. Cavitation and shock waves were discussed as the reason for this finding. Although, it was reported that cavitation did not appear until the velocity was 600 and 800 ft/s for spheres measuring .406 and .250 in. respectively [6].

More recent work has studied the pathophysiology of indirect fractures of long bones to isolate variables involved with the production of a fracture. By using both strain gage technology and high-speed video, the temporal relationship between the passage of a bullet and occurrence of long bone fracture has been determined [7]. Dougherty et al. (2011) further analyzed the parameters associated with the production of long bone fracture. Using cadaveric specimens, it was determined that indirect fractures were of a simple (oblique or spiral) pattern (Fig. 29.2). For those specimens with fractures, the average wound track to bone distance was 9.68 mm. In contrast, the non-fractured specimens demonstrated a significantly shorter wound track to bone distance of 15.15 mm (p=0.036). In addition, there were no fractures reported when 9-mm bullets with an average velocity of 263 m/s (862 ft/s) were used or when the M995 bullet velocity was below 975 m/s (3,200 ft/s) [8].

29.2 Behind Armor Blunt Trauma

29.2.1 Body Armor: Torso

Of approximately 1,200 officers killed in the line of duty since 1980, it is estimated that more than 30 % could have been saved by body armor [9]. According to the James Guelff Body Armor Act, the risk of dying from gunfire is 14 times higher for an officer not wearing a vest [9]. In addition, the US Department of Justice estimates that 25 % of state, local, and tribal law enforcement officers

Table 29.1 NIJ P-BFS performance test summary

Armor type	Test round	Test bullet	Bullet mass	Conditioned armor test velocity	New armor test velocity
IIA	1	9 mm FMJ RN	8.0 g (124 gr)	355 m/s (1,165 ft/s)	373 m/s (1,225 ft/s)
	2	.40 S&W FMJ	11.7 g (180 gr)	325 m/s (1,065 ft/s)	352 m/s (1,155 ft/s)
II	1	9 mm FMJ RN	8.0 g (124 gr)	379 m/s (1,245 ft/s)	398 m/s (1,305 ft/s)
	2	.357 Magnum JSP	10.2 g (158 gr)	408 m/s (1,340 ft/s)	436 m/s (1,430 ft/s)
IIIA	1	.357 SIG FMJ FN	8.1 g (125 gr)	430 m/s (1,410 ft/s)	448 m/s (1,470 ft/s)
	2	.44 Magnum SJHP	15.6 g (240 gr)	408 m/s (1,340 ft/s)	436 m/s (1,430 ft/s)
III	1	7.62 mm NATO FMJ	9.6 g (147 gr)	847 m/s (2,780 ft/s)	–
IV	1	.30 Caliber M2 AP	10.8 g (166 gr)	878 m/s (2,880 ft/s)	–
Special	–	Each test threat to be specified by armor manufacturer or procuring organization			

are not issued body armor. Since establishing the IACP (International Association of Chiefs of Police)/DuPont™ Kevlar® Survivors' Club® in 1987; over 3,000 law enforcement personnel have survived both ballistic and non-ballistic incidents because they were wearing body armor [10].

Body armor is comprised of fibers that have been woven together into sheets. Numerous sheets are used to make up one ballistic panel. The sheets work individually and together to help prevent the penetration of the bullet. Some materials that are used include: Kevlar®, Spectra® Fiber, Aramid Fiber, and Dyneema. The material fibers work to absorb and spread the energy over the entire torso so all of the energy from the impact is not focused on one area of the body, resulting in serious injury.

The current standard for certifying the protective ability of the thoracic armor (NIJ Standard-0101.06) was released in 2008 by the National Institute of Justice. This Ballistic Resistance of Body Armor standard has been revised several times since the original version was released in 1972. Part of the standard since its inception is the measurement of the deformation behind a backing material, also known as Backface Signature (BFS). This testing method provides a discrete value by which the failure of the vest can be determined. A failure is indicated by a deformation of the backing material greater than 44 mm or a penetration of the vest. A penetration of the vest can be due to the projectile itself, a fragment of the projectile or a fragment from the vest.

There are five types of armor that are covered in the current standard (IIA, II, IIIA, III, IV). Each type has a specified threat level that must be used during the certification testing (Table 29.1). Shot patterns on the vest are also specified.

The current 0101.06 standard also includes a ballistic limit determination test. This test involves altering the velocity of the ammunition in order to determine a velocity at which there is a 50 % chance of penetration. The number of minimum shots required depends on the level of armor being tested. Type II through Type IIIA requires a minimum of 12 shots, while there are a minimum of 6 shots required for Types III and IV.

Although often criticized, the BFS test is based on research performed within the military in the 1970s [11–14]. This research involved a biomedical assessment of injuries resulting from specified ammunition striking a selected armor or a blunt impactor striking the body directly designed to mimic the bullet/armor/body interaction. Key areas of the torso were targeted including the lung and heart. The surrogate often used was an Angora goat, however some studies cite the use of a porcine or canine surrogate. Lung and heart injury as well as non-lethality were key parameters studied.

An extension of this research involved parametric modeling of blunt trauma lethality [11, 14]. Input parameters from the projectile such as mass, velocity, diameter and those of the armor; mass per unit area, were combined with the characteristics (mass and body wall thickness) of the subject being impacted. These input parameters were used then used to predict a level of lethality for the impact.

Newer models to predict the risk of injury as a result of BABT have been developed including biomechanical [15] and finite element models [16, 17]. These models have been developed to address the current limitations of the clay standard, however none have been incorporated into the current standard.

29.2.2 Case Studies: Torso

In an effort to determine the types of BABT injuries being sustained by law enforcement agents, a subset of data was collected from the IACP/DuPont™ Kevlar® Survivors' Club® database. This database is maintained by IACP and represents those officers who have survived due to the fact they were wearing body armor. It includes not only the ballistic cases but also stab/slash and blunt force trauma. Survivors are asked to complete a questionnaire about their incident and the resulting injuries.

As part of a recent ongoing study, cases involving ballistic impacts to the torso were studied. Additional information was collected specific to their injuries including the medical record. Of the 77 cases collected, 71 had adequate data available. Data for each case were obtained through phone interviews, medical records, and police reports. Injured officers were between the ages of 22 and 54, with the average being 34 ± 8 years of age. From the 71 cases that were collected for this study, there were a total of 90 shots stopped by personal body armor. The majority of the shots impacted the anterior chest (74 %, n = 65), followed by abdomen (16 %, n = 12), posterior upper torso (9 %, n = 8), and posterior lower torso (6 %, n = 5).

The level of protection offered by a ballistic vest is an important element of officer safety. The NIJ body armor performance standard has been revised several times since its inception to reflect the growing threats faced by officers. As previously discussed, vests are certified using a combination of weapons and rounds that are commonly used by law enforcement and criminals. The most current standard classifies armor into five types: IIA, II, IIIA, III, and IV. The first three levels are soft armor that protect against various handgun and shotgun ammunitions, while the last two levels are hard plates used in conjunction with soft armor to protect against rifle and armor piercing rounds. The majority of officers in this study wore either a threat level IIA or II vest (Fig. 29.3). From all of the cases, 25 % of the officers had additional protection from a trauma plate or pack.

The injuries that occurred from the 90 shots stopped by the armor have been classified as blunt trauma and backface signature injuries (Table 29.2). The majority of the injuries resulting from impacts to the vest were categorized as blunt trauma injuries (60 %, n = 54). These include less

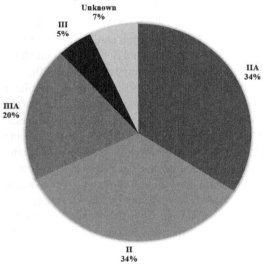

Fig. 29.3 Percentage of body armor threat level based on case studies

Table 29.2 Summary of case study data

Injuries	Body armor threat level				
	IIA	II	IIIA	III	Unknown
Chest blunt trauma	11	18	6	1	3
Chest BFS	7	2	2	1	2
Abdominal blunt trauma	5	1	3	–	–
Abdominal BFS	–	1	2	–	–
Posterior torso blunt trauma	1	3	–	1	1
Posterior torso BFS	1	–	1	–	–
Edge shot – GSW	2	1	–	1	–
No injuries	5	3	3	–	–
Unknown	1	–	1	–	–
Total	33	29	18	4	6

severe injuries involving contusions, abrasions, and rib fractures. Backface signature (BFS) injuries were the next most common injury among the sample population (21 %, n = 19). BFS injuries occur when there is a penetrating injury to the chest even though the bullet is captured within the armor [18]. These injuries have become more prevalent in recent years due to the desired increased in flexibility of the armor systems.

Eleven of the impacts to the vest resulted in no notable injury (12 %). Four impacts struck the edge of the vest and resulted in a gunshot wound (4 %) and the remaining two impacts generated injuries that were unknown (2 %).

Typically, the blunt trauma noted was a mild to severe contusions and abrasions. In addition, rib fractures, liver lacerations, and lung contusions were noted. Four officers sustained rib fractures, one officer experienced a lacerated liver, and one officer was found to have micro-fractures of the ribs and a lung contusion from impacts stopped by their protective vest.

29.2.3 Ballistic Helmets

Not unlike the body armor standard, the NIJ standard 0101.06 for Ballistic Helmets provides guidelines for the certification of ballistic helmets by law enforcement officers. A specialized headform with witness plate and accelerometer at the center of gravity is used. Specific threat levels are referenced for levels of protection: I, IIA, and II (see Table 29.3). Four fair hits without penetration and acceleration levels not to exceed 400 g's are required for certification. This standard is currently being revised by the NIJ.

In an effort to determine the prediction of skull fracture during ballistic loading of the helmet, Bass et al. [19] conducted a series of PMHS tests. Nine (9) helmeted PMHS were impacted with 9 mm ammunition and an additional four (4) were impacted with compliant direct impacts. The authors produced fractures in 5 out of the 9 ballistic impacts. These fractures were reported to be both simple linear and complex linear/depressed in nature. A 50 % risk of skull fracture was reported to occur at peak pressures of 51,200 kPa. It was also determined, that the acceleration based Head Injury Criterion (HIC) was not a good predictor.

Table 29.3 Standard threat levels for NIJ level of protection for ballistic helmets

Helmet	Test round	Test bullet	Bullet mass	Velocity
I	1	22 LRHV Lead	2.6 g (50 gr)	320 ± 12 m/s (1,050 ± 40 ft/s)
	2	38 special RN lead	10.2 g (158 gr)	259 ± 15 m/s (850 ± 50 ft/s)
IIA	1	.357 Magnum JSP	10.2 g (158 gr)	381 ± 15 m/s (1,250 ± 50 ft/s)
	2	9 mm FMJ	8.0 g (124 gr)	332 ± 15 m/s (1,090 ± 50 ft/s)
II	1	.357 Magnum JSP	10.2 g (158 gr)	425 ± m/s (1,395 ± 50 ft/s)
	2	9 mm FMJ	8.0 g (124 gr)	358 ± 15 m/s (1,175 ± 50 ft/s)

Using a finite element model, Pintar et al. [20] explored pressures and strains within the brain of a helmeted head from standardized ballistic threats. An instrumented headform with seven uni-axial load cells was used to determine the force-time histories used as input to the model. It was determined that the volume that exceeded the pressure tolerance level of 40 MPa was dependent on the impact direction: left (5 %), right (24 %), rear (47 %) and front (93 %). None of the elements exceeded the strain limits of .2 % with the standard helmet.

Hisley et al. [21] explored the use of digital image correlation to determine the energy available on the backface of standard helmets. It was determined that the loads were mechanically similar to those seen with blunt ballistic impacts with less-lethal projectiles.

29.3 Blunt Ballistic Impacts

The utilization of less-lethal force in both law enforcement and military operations has increased over the past several years. These less-lethal techniques allow for the use of force that is designed to incapacitate or subdue individuals with a low risk of lethality. Less-lethal arsenals include contact weapons, chemical

Table 29.4 Specifications for less-lethal munitions being manufactured

Caliber of munition	Mass of submunition	Muzzle Velocity	Materials
12 gauge	20–50 g	76–243 m/s	Rubber fin
			Bean bag
			Rubber ball
37/40 mm	20–140 g	50–137 m/s	Foam
			Wood
			Bean bag
			Rubber ball

agents, directed energy devices, capture devices and projectiles. Each type of technology is designed to inflict enough force to deter the situation without causing a fatal outcome.

Projectiles or kinetic energy (KE) rounds give a wide range of applicability with a relative ease of deployment. Often times, they are designed to deploy from weapon systems that are already in the controlling force's possession. Manufacturers have developed a variety of munitions to meet various situational needs. Single fire munitions allow for encounters with a single individual. Multiple projectile rounds are designed to control large crowds and potential riot situations. The proximity of the disturbance has been addressed in the development of both close-range and standard-range rounds.

These munitions differ from the normal ballistic weapons in that they have an increased mass but are deployed at a decreased velocity. A variety of munitions are being manufactured that include both 12 gauge and 37/40 mm rounds. These munitions vary in terms of the number and type of submunition deployed and the specified muzzle velocity (Table 29.4). The deployment of these KE rounds can result in blunt ballistic impacts, which has been defined as those impacts with a mass of 20–200 g and an impact velocity of 20–250 m/s.

Although KE munitions address the largest spectrum of situations while affording the greatest protection for the officer, the frequency of use and the types of injuries being inflicted are not regularly monitored. With no standardized reporting system, case reports are currently the only source of documentation [22–25].

Injury data from primarily Northern Ireland is available to describe the types and severity of injuries associated with the deployment of one type of non-lethal kinetic energy rounds. "Plastic bullets" were first used in Northern Ireland in 1973 as a means for riot control. They were intended to replace a less accurate rubber bullet previously used by security forces [23]. These "safer" bullets are cylindrical in shape measuring 100 mm in length with a 37-mm flat diameter. They are reported to weigh 135 g and are made from a polyvinyl chloride [22]. The muzzle velocity at which these munitions are fired range from 71.5 m/s (160 mph) to 89.4 m/s (200 mph) [24].

In 1987, Metress and Metress reported on 13 deaths resulting from impacts with plastic batons in Northern Ireland. It was reported that these fatalities were the result of impacts to either the head (nine impacts) or chest (four impacts). Deployment of the fatal impacts were reported to be less than 20 yards which contradicts the recommended 'rules of engagement' [24]. An additional death was reported in 1989 [22] for a total of 14 deaths since 1973.

Ritchie (1992) collected data from 1975 to 1989 by means of a retrospective chart review at a district general hospital in Belfast. A total of 123 patients were treated in emergency rooms with 38 being admitted for further care. Of the 126 injuries sustained by these patients, 19 (15 %) were to the head, 22 (17.5 %) were to the chest, 10 (7.9 %) were to the maxillofacial region, 17 (13.5 %) were to the abdomen, 24 (19 %) were to the upper limb, 33 (26.2 %) were to the lower limb and 1 (.8 %) was to the groin. The only death reported was due to ventricular fibrillation of heart function as a result of a blunt chest injury. Ritchie (1992) classified the injuries as either serious or non-serious but did not state a criterion for determining what qualified as a serious injury.

In 2004, Hubbs and Klinger released the first large dataset based on injuries related to less-lethal KE munitions in the United States. The authors investigated 969 firings of less-lethal KE munitions with 867 (92 %) striking the human body. It was reported that the abdomen was the area of the body most commonly impacted at 33 % (n = 263). The chest (n = 152), back (n = 85), leg (n = 119) and arm (n = 115) were the other most commonly impacted regions. Bruising and abrasions account for 81.5 % of the injuries, however 10 deaths were reported. Two (2) of these fatalities were due to the "miss-loads" with the law enforcement agents mistaking breaching rounds as a less-lethal ammunition. Four (4) of the cases involved KE munitions that penetrated into the body causing a fatal wound.

29.3.1 Thoracic Impacts

As part of the ongoing effort to evaluate, predict and prevent injuries caused by KE munitions, several studies have explored the biomechanics of blunt ballistic impacts to various body regions. One of the first regions studied was the thorax. Three impact conditions were established using a rigid impactor: (a) 140 g mass at 20 m/s (b) 140 g mass at 40 m/s and (c) 30 g mass at 60 m/s. Force-time, deflection-time and force-deflection curves characterizing the response of the body to blunt ballistic impacts were established using cadaveric specimens [26]. The determination of a valid injury criterion for assessing blunt ballistic impacts involved the review and re-analysis of existing data. Injury analysis was also conducted on the cadaveric specimens used to establish the biomechanical corridors. From these analyses, the injury criterion of Viscous Criterion (VC) was determined to adequately predict the risk of injury from blunt ballistic impacts. The reanalysis of data first collected by Cooper and Maynard (1986) demonstrated a VC of 2.8 predicted a 25 % risk of severe lung injury. A less severe skeletal injury of Abbreviated Injury Scale (AIS) = 2 is predicted by a VC of .8, based on data from cadaveric specimens [27].

29.3.2 Abdominal Impacts

Similar work was conducted for blunt ballistic impacts to the abdomen [28]. The impact condition was a 45 g rigid projectile with a targeted impact velocity of 60 m/s. Both cadaveric and sus scorfa specimens were used to determine biomechanical response corridors and injury criteria respectively. Six cadaveric specimens were impacted in the epigastric region to create a biomechanical corridor with an average peak force of 4,741 ± 553 N and an average peak deflection of 22 mm. The Blunt Criterion (BC) is based on natural log of the impact energy divided by the product of the specimen mass to the one-third power, wall thickness of the specimen and the area of impact. This criterion was found to be predictive of a 50 % risk of AIS two to three injuries to the liver (BC of .51) and bowel (BC of 1.32).

29.3.3 Head Impacts

While the head is never intended to be the region of impact with these munitions, impacts to the face and skull have been reported. Therefore, a determination of the biomechanical characteristics of the skull/facial structures and resulting fracture tolerance level was made [29]. Thirty-two (32) blunt ballistic impacts were performed to 11 unembalmed, post-mortem human subject heads. Impact locations included the temporal parietal, frontal and zygoma bones. An impactor with a mass of 103.3 g and diameter of 38.1 mm was launched using an air cannon system at velocities between 5.1 and 37.7 m/s. Impacts were paired to achieve a fracture/no fracture impact on each specimen.

For the temporal parietal impacts (n = 14), peak forces for impact condition A (18.8 g ± 2.1 m/s) were 3,211 ± 429 N with deformations at peak force of 7.9 ± 1.6. Peak forces for impact condition B (33.5 g ± 1.5 m/s) were 5,189 ± 992 N with deformations at peak force of 10.5 ± 2.3 mm. Three key injury criteria were found to be predictive: BC, force and strain [30]. Based on logistic curves, a 50 % risk of skull fracture is represented by a BC of 1.61, force of 5,970 N and strain of 5062 με or 0.51 %.

Of the five specimens tested for frontal and zygomatic impacts, there were a total of 20 impacts. There were 4 fractures out of the 10 impacts for both the frontal and zygomatic bones. Frontal bone fractures ranged from linear fractures to comminuted depressed fractures.

In one case, there was a severe linear fracture than ran continuously from the impact site down across the sphenoid, temporal, parietal and occipital bones and ending approximately 30 mm from the external occipital protuberance. For the zygomatic region, there were two tripod fractures in this test series [29]. In addition to the tripod fractures, there were many linear and comminuted fractures to the zygoma, maxilla and bones within the orbit of the eye. The range for frontal bone fractures was from 4,413 N to 9,438 N while non-fracture values were from 2,630 N to 6,623 N. The range for zygomatic fractures was 575 N to 2,746 N. Non-fracture values were from 468 N to 3,711 N.

29.4 Summary

The study of terminal ballistics and target effects is an important process in the treatment and prevention of injury. Understanding the mechanisms of injury will guide medical providers on how to treat these injuries in a trauma setting. The development and revision of standards to evaluate personnel protective equipment rely heavily on injury biomechanics. Ongoing research to determine both the biomechanical responses of the body to such impacts and ways to predict the resulting injuries using surrogates (both biomechanical and computer based), are key steps in the injury prevention process.

References

1. DiMaio VJM (2002) Gunshot wounds: practical aspects of firearms, ballistics, and forensic techniques, 2nd edn. Taylor & Francis, Boca Raton
2. Spitz WU, Spitz DJ, Fisher RS (2006) Spitz and Fisher's medicolegal investigation of death: guidelines for the application of pathology to crime investigation. Charles C. Thomas, Springfield
3. Dougherty PJ, Vaidya R, Silverton CD, Bartlett C, Najibi S (2009) Joint and long-bone gunshot injuries. J Bone Joint Surg Am 91(4):980–997
4. Clasper JC, Hill PF, Watkins PE (2002) Contamination of ballistic fractures: an in vitro model. Injury 33(2):157–160
5. Huelke DF, Harger JH, Buege LJ, Dingman HG (1968) An experimental study in bio-ballistics: femoral fractures produced by projectiles–II. Shaft impacts. J Biomech 1(4):313–321
6. Huelke DF, Harger JH, Buege LJ, Dingman HG, Harger DR (1968) An experimental study in bio-ballistics femoral fractures produced by projectiles. J Biomech 1(2):97–105
7. Sherman D, Dougherty P (2007) Indirect fractures to bones by ballistic injury. Paper presented at the International Society of Biomechanics, Taipei
8. Dougherty PJ, Sherman D, Dau N, Bir C (2011) Ballistic fractures: indirect fracture to bone. J Trauma 71(5):1381–1384. doi:10.1097/TA.0b013e3182117ed9
9. Guelff J, McCurley C (2001) Body Armor Act of 2001. House report 107-193
10. IACP (2013) IACP/DuPont Kevlar Survivors' Club. http://www2.dupont.com/Kevlar/en_US/uses_apps/law_enforcement/survivors_club.html. Accessed 8 Oct 2013
11. Prather RN, Swann CL, Hawkins CE (1977) Backface signatures of soft body armors and the associated trauma effects. Technical report no. ARCSL-TR-77055. US Army Aberdeen Proving Ground, Chemical Systems Laboratory, Aberdeen
12. Goldfarb MA, Ciurej TF, Weinstein MA, LeRoy WA (1975) Method for soft body armor evaluation: medical assessment. Technical report no. EB-TR-74073. Edgewood Arsenal, Aberdeen Proving Ground
13. Metker LW, Prather RN, Johnson EM (1975) A method for determining backface signatures of soft body armor. Technical report no. EB-TR-75029. US Army Armament Research and Development Command, Aberdeen Proving Ground
14. Clare VR, Lewis JH, Mickiewicz AP, Alexander P, Sturdivan LM (1975) Blunt trauma data correlation. Technical report no. EB-TR-75016. Edgewood Arsenal, Aberdeen Proving Ground
15. Hewins K, Anctil B, Stojsih S, Bir C (2012) Ballistic blunt trauma assessment methodology validation. In: Personal armour systems symposium, Nuremberg
16. Raftenberg M (2003) Response of the Wayne State Thorax Model with fabric vest to a 9-mm bullet. ARL-TR-2897. US Army Research Laboratory, Aberdeen
17. Roberts JC, Ward EE, Merkle AC, O'Connor JV (2007) Assessing behind armor blunt trauma in accordance with the National Institute of Justice Standard for Personal Body Armor Protection using finite element modeling. J Trauma 62(5):1127–1133. doi:10.1097/01.ta.0000231779.99416.ee
18. Wilhelm M, Bir C (2008) Injuries to law enforcement officers: the backface signature injury. Forensic Sci Int 174(1):6–11. doi:10.1016/j.forsciint.2007.02.028
19. Bass C, Boggess B, Bush B, Davis M, Harris R, Campman S, Monacci W, Eklund J, Ling G, Sanderson E, Waclawik S (2003) Helmet behind armor blunt trauma. In: NATO/PFP (ed) RTO AVT/HFM specialists' meeting on "Equipment for Personal Protection (AVT-097)" and "Personal Protection: Bio-Mechanical

Issues and Associated Physio-Pathological Risks (HFM-102)", Koblenz
20. Pintar FA, Philippens MM, Zhang J, Yoganandan N (2013) Methodology to determine skull bone and brain responses from ballistic helmet-to-head contact loading using experiments and finite element analysis. Med Eng Phys 35(11):1682–1687. doi:10.1016/j.medengphy.2013.04.015
21. Hisley D, Gurganus J, Lee J, Williams S, Drysdale A (2010) Experimental methodology using Digital Image Correlation (DIC) to assess ballistic helmet blunt trauma. Army Research Laboratory, ARL-TR-0000
22. Steele JA, McBride SJ, Kelly J, Dearden CH, Rocke LG (1999) Plastic bullet injuries in Northern Ireland: experiences during a week of civil disturbance. J Trauma 46(4):711–714
23. Rocke L (1983) Injuries caused by plastic bullets compared with those caused by rubber bullets. Lancet 1(8330):919–920
24. Metress EK, Metress SP (1987) The anatomy of plastic bullet damage and crowd control. Int J Health Serv 17(2):333–342
25. Hubbs K, Klinger D (2004) Impact munitions database of use and effects. Final report 204433
26. Bir C, Viano D, King A (2004) Development of biomechanical response corridors of the thorax to blunt ballistic impacts. J Biomech 37(1):73–79
27. Cooper GJ, Maynard RL (1986) An experimental investigation of the biokinetic principles governing non-penetrating impact to the chest and the influence of the rate of body wall distortion upon the severity of lung injury. In: Proceedings of IRCOBI European impact biomechanics conference, Zurich
28. Eck J (2006) Biomechanical response and abdominal injury due to blunt ballistic impacts. Wayne State University, Detroit
29. Raymond D, Bir C, Crawford G (2007) Biomechanics of the skull and facial bone to blunt ballisitc impacts. Final report grant number: M67854-06-1-5016
30. Raymond D, Van Ee C, Crawford G, Bir C (2009) Tolerance of the skull to blunt ballistic temporoparietal impact. J Biomech 42(15):2479–2485. doi:10.1016/j.jbiomech.2009.07.018

Index

A
Abbreviated Injury Scale (AIS), 33–34, 334–336
Abdomen
  anatomy
    hollow/membranous abdominal organs, 374–377
    solid abdominal organs, 377–378
    vasculature, 378
  civil aviation crash injury protection, 820
  clinical data, 383
  field accident data, 379–383
  injury mechanisms
    anthropomorphic test device abdomen development, 424–428
    general considerations, 408–409
    injury predictors, 409–424
    submarining assessment, 424–428
  material properties testing
    liver, in compression, 389–390
    liver, in tension, 386–388
    membranous tissues, 384–386
    spleen, in compression, 390–391
    spleen, in tension, 390
  pediatric biomechanics
    biomechanical studies, 680–683
    developmental anatomy, 680
  surrogate testing
    restraint testing, 401–407
    rigid impact, 392–401
  trauma biomechanics
    hollow viscus, 72–73
    incidentalomas, 74–75
    injury grading, 75
    solid organs, 70–72
    vascular, 73–74
Accidental injury, muscles role
  muscle mechanics
    force generation, 612–613
    muscle activation timing, 613–615
    muscle injury, 616
    muscle mechanical properties, 615–616
    structure and function, 612
  muscle's effect
    head and brain injuries, 617
    lower extremity injuries, 627–633
    spine injuries, 617–621
    upper extremity injuries, 621–627
    whole body, 633
Activity-related injuries, neck injury biomechanics
  automobile accidents, 294–295
  diving, 297–300
  football, 296–297
  motorcycle accidents, 295
Acute subdural hematoma (ASDH), 7–14
Aging, restraint system biomechanics, 134–136
AIS. *See* Abbreviated Injury Scale (AIS)
Ankle injury, automotive-related injuries
  axial loading, 508
  dorsiflexion
    Achilles tension measurement, 517
    dynamic testing, 515, 518
    900 N tension, 513
    quasi-static tests, 514, 518, 519
    ROM, 6 Nm torque, 515
    subtalar joint stiffness curve, 509
  xversion injuries
    dynamic xversion injury, 510–513
    quasi-static xversion response, 509–510
Anterior cruciate ligament (ACL), 471
Anterior longitudinal ligament (ALL), 590–591
Anterior wedge fractures, 455, 456
Anthropomorphic test device (ATD), 424–428, 687, 799
  frontal crash testing
    child dummies, 88
    CRABI infant dummies, 87
    early ATDs, 84
    hybrid II dummy family, 84–85
    hybrid III dummy family, 85–87
    Q series of infant, 88
    THOR, 87–88
  future ATD, 98–108
  injury assessment reference values, 93, 98–108
  injury risk curves, 93–98
  for rear impact testing
    BioRID II and RID2 dummies, 92–93
    hybrid III family of dummies, 92
  side impact testing
    biofidelic side impact dummy (BioSID), 90
    EUROSID and EUROSID-1, 89–90
    Q3s, 91
    side impact dummy (SID) and SID-HIII, 88–89

Anthropomorphic test device (ATD) (cont.)
   SID-IIs, 90–91
   WorldSID small female, 91
Anti-vehicular mine blast injuries, 539
Aortic trauma
   anatomical location of, 340
   aortic injury mechanisms, 340–341
   epidemiology, 339–340
Arachnoids, 8–9
ASDH. *See* Acute subdural hematoma (ASDH)
ATDs. *See* Anthropomorphic test device (ATD)
Automotive-related injuries. *See also* Injury biomechanics
   ankle injury
      axial loading, 508
      dorsiflexion, 513–519
      subtalar joint stiffness curve, 509
      xversion injuries, 509–513
   human foot and ankle dimensions, 501
   knee response and failure
      bending loading, 523–529
      lateral loading, 519–523
      response testing, 523
   non-minor injury, 500–501
   tibia/fibula response and failure
      dynamic bending tolerance, 504–508 (*see also* Bending tolerance)
      fracture tolerance in bending, 502–504
Aviation regulations, injury criteria
   abdomen, 820
   chest, 819–820
   criteria selection, 815
   flail injuries, 820–821
   head, 815–817
   leg, 820
   lumbar spine, 817–819

B

Ballistic injury biomechanics
   behind armor blunt trauma
      ballistic helmets, 835
      body armor, torso, 832–834
      torso, 834–835
   blunt ballistic impacts
      abdominal impacts, 837
      head impacts, 837–838
      thoracic impacts, 837
   gun shot wounds
      direct *vs.* indirect fractures, 832
      distance, 831–832
      entrance *vs.* exit, 830–831
Behind armor blunt trauma
   ballistic helmets, 835
   body armor, torso, 832–834
   torso, 834–835
Bending tolerance
   dynamic, human leg/tibia/fibula
      under L-M loading, 504–506
      under P-A loading, 506–508

   knee, 491
   midshaft, femur, 485
   test diagram, three point bending tests, 505
   tibia/leg bending test data, 505
Blunt ballistic impacts
   abdominal impacts, 837
   head impacts, 837–838
   thoracic impacts, 837
Blunt carotid injury grading scale, 60
Bone fracture, computational analysis of
   BMD, 185
   bone damage, 192–193
   bone geometry, 185
   bone quality, 185
   computational model verification and validation (V&V), 196–197
   continuum damage mechanics modeling, 193–194
   cortical bone, probabilistic continuum damage modeling of
      FEA, 195–196
      trabecular bone damaging behavior, 196
      uniaxial material modeling, 194–195
   engineering models, 185–186
   fracture risk assessment, 184–185
   probabilistic modeling accounts, 189–192
   skeletal fragility, 184
   statistical shape and density model (SSDM), 187–189
   statistical shape models, 187
Brain injury
   anatomy of head
      central nervous system (CNS), 223
      cerebellum, 224
      cerebrum, 223
      medulla oblongata, 223–224
      meninges, 222
      midbrain, 223
      pons, 223
      scalp, 222
      skull, 222
   biomechanical mechanisms
      biomechanical factors, 235
      developments, 239–242
      functional changes, 253
      in silico simulations, 253–254
      in vitro, salient traumatic loading scenarios, 249–250
      mechanical loading, complex and critical aspects of, 248–249
      mechanical state for individual cells, 250–251
      mechanosensors, 251–253
      multiscale questions, 254
      rotational brain-injury studies, 236–238
      translational motion studies, 238–239
   direct brain deformation models, 232–234
   experimental brain-injury models, 227–228
   head acceleration models, 230–232
   head-impact models, 228–230
   mechanisms, 7
   traumatic brain injury
      diffuse injuries, 224–225

Index

focal injuries, 225–226
skull fractures, 226–227
types of, 224
Brain motion, blunt impact, 4
Burst fracture, 455

C

Car bed restraints, 708
Central nervous system (CNS), 223
Cerebellum, anatomy, 224
Cerebrum, anatomy, 223
Chance fractures, 457
Chest
  civil aviation crash injury protection, 819–820
  trauma biomechanics
    chest wall, 64–65
    diaphragm, 68–69
    heart, 63–64
    injury grading, 69
    mediastinum and thoracic aorta, 66–68
    pulmonary, 65–66
Child Advanced Safety Project for European Roads (CASPER) project, 427
Child Restraint Air Bag Interactions (CRABI), 87
Child restraint system (CRS), 88, 103
  effectiveness, 697–698
  installation
    LATCH, 700–702
    seatbelt, 700
    usability and vehicle/child restraint compatibility, 702–703
  occupant restraint systems
    design concepts, 698–699
    seating positions, airbags, and children, 699–700
  recommendations
    boosters, 711–714
    car bed restraints, 708
    forward-facing child restraints, 708–711
    rear-facing child restraints, 703–707
    seatbelts for children, 714–715
CIREN. *See* Crash Injury Research And Engineering Network (CIREN)
Civil aviation crash injury protection
  injury criteria, aviation regulations
    abdomen, 820
    chest, 819–820
    criteria selection, 815
    flail injuries, 820–821
    head, 815–817
    leg, 820
    lumbar spine, 817–819
  leg, 823–824
  level of safety, 825–826
  occupant support criteria
    articulation limits, 824–825
    contact limits, 825
    excursion limits, 824
  side-facing seats
    neck, 822–823
    special conditions, 821–822
    thorax and pelvis, 823
Clay Shoveler's fracture, 270
CNS. *See* Central nervous system (CNS)
CODES. *See* Crash Outcome Data Evaluation System (CODES)
Component models
  advantage, 439
  compression loading test, 440
  intervertebral discs, 442
  ligaments, 442–445
  vertebrae, 440–442
Co-operative Crash Injury Study (CCIS), 339
Cortical vessels, anatomy of, 9–10
CRABI. *See* Child Restraint Air Bag Interactions (CRABI)
Crash helmet, 760
Crash Injury Research And Engineering Network (CIREN), 39–40, 478
Crash Outcome Data Evaluation System (CODES), 38–39
CRS. *See* Child restraint system (CRS)

D

Diffuse axonal injuries (DAI), 54
Direct brain deformation models, 232–234
Dorsal root ganglion (DRG), 553–554
Drop tests, loading issues, 460–461
Dural innervation, 592
Dura mater, 8
Dynamic Response Index (DRI), 18

E

EDRs. *See* Event Data Recorders (EDRs)
EEVC. *See* European Experimental Vehicle Committee (EEVC)
Electro-hydraulic testing device, 458–459
Electronic Numerical Integrator and Computer (ENIAC), 143
Endplates, biomechanically relevant anatomy, 453
European Experimental Vehicle Committee (EEVC), 89
Event Data Recorders (EDRs), 42–44
Experimental brain-injury models, 227–228
Extremities
  pediatric biomechanics
    biomechanical studies, 685–686
    developmental anatomy, 683–685
  trauma biomechanics
    imaging of, 77–78
    injury grading, 78
    pelvis, 76–77

F

Facet joint injury, 559–561
  biomechanically relevant anatomy, 454
  biomechanical study, 595
  inflammation, nerve activity, 597

Facet joint injury (*cont.*)
    morphology, 593
    and muscle neurophysiology in control animals, 595–596
    nerve supply, 593
    prevalence, 594
    retrospective analysis, 594
    spine loading studies, 596–597
    substance P, unit activity, 597–598
Face, trauma biomechanics, 61–62
Fatality Analysis Reporting System (FARS), 36
FE-based models
    CAE tools and critical aspects in
        material model verifications, 152–153
        meeting safety regulations, 153–156
        mesh convergence, 151–152
    frontal impact, 157–159
    restraint system modeling, 167–170
    rollover, 167
    side impact, 159–167
Femoral condyles fractures, 474
Femur anatomy, 472
Field accident data, 379–383
Flail injuries, 820–821
Force generation, accidental injury, 612–613
Forward-facing child restraints, 708–711
Fracture dislocation, 456–457
Frontal impact
    body mass distribution, 474
    body shape and body mass index, 474–475
    crash testing, ATDs
        child dummies, 88
        CRABI infant dummies, 87
        early ATDs, 84
        hybrid II dummy family, 84–85
        hybrid III dummy family, 85–87
        Q series of infant, 88
        THOR, 87–88
    dislocations, frontal crashes, 473–474
    femoral condyles fractures, 474
    hip fractures and dislocations, 474
    knee loading, 474
    men *vs.* women, acetabulum, 474
    thigh fractures, 474
    trochanteric femur fractures, 474
    types, 473

G
General Estimates System (GES), 37
Glasgow coma scale, 53
Gun shot wounds
    direct *vs.* indirect fractures, 832
    distance, 831–832
    entrance *vs.* exit, 830–831

H
Head
    acceleration models, 230–232
    accidental injury, muscle's effect, 617
    anatomy of
        central nervous system (CNS), 223
        cerebellum, 224
        cerebrum, 223
        medulla oblongata, 223–224
        meninges, 222
        midbrain, 223
        pons, 223
        scalp, 222
        skull, 222
    civil aviation crash injury protection, 815–817
    injury biomechanics
        acute subdural hematoma (ASDH), 7–14
        brain injury mechanisms, 7
    motor vehicle regulations, injury criteria, 799–800
    pediatric biomechanics
        acute injury severity scaling, 656–657
        brain anatomy, 646
        brain and skull, 653–655
        brain material properties, 651–653
        injury criteria, 657–658
        mass-acceleration scaling, 655
        pressure scaling, 656
        scaled acute subdural hematoma, 656
        scaled cortical impact, 655–656
        skull, developmental anatomy, 644–646
        skull material properties, 646–651
    trauma biomechanics, 52–55
Head-impact models, 228–230
Hip
    anatomy, 472, 473
    fractures
        and dislocations, 474
        in elderly, 24–25
    injury tolerance, 481–484
Hollow abdominal organs
    diaphragm, 377
    gallbladder, 376
    large intestine, 376
    small intestine, 376
    stomach, 375–376
    urinary bladder, 376–377
    uterus, 377
Human injury risk curve
    certainty method, 780
    consistent threshold and the extended consistent threshold, 781
    different methods comparison, 781–782
    logistic regression, 780
    Mertz-Weber technique, 779–780
    rationale, 778–779
    recommendations, 782–787
    survival analysis, 781
Human response corridors
    human injury risk curve
        certainty method, 780
        consistent threshold and the extended consistent threshold, 781
        different methods comparison, 781–782
        logistic regression, 780
        Mertz-Weber technique, 779–780

Index

  rationale, 778–779
  recommendations, 782–787
  survival analysis, 781
 normalization and scaling
  applications, 771
  discussion on, 776–778
  equal stress equal velocity approach, 770–771
  impulse momentum approach, 771–773
  mean and standard deviation responses, 773–776
  scaling, 776

I
ICD-9. *See* International Classification of Diseases version 9 (ICD-9)
Improvised explosive devices (IED's), 17
Inflatable restraint biomechanics
 head and face contact mitigation, 120
 load sharing, force distribution, and work, 120–125
 restraint system design flexibility, 125
Injury assessment reference value (IARV), 144
Injury biomechanics
 aims of
  human tolerance, 5–6
  identification and explanation, 3
  mechanical response, quantification of, 3–5
  safety devices and techniques, 6
 data sources
  Crash Injury Research And Engineering Network (CIREN), 39–40
  Crash Outcome Data Evaluation System (CODES), 38–39
  Event Data Recorders (EDRs), 42–44
  Fatality Analysis Reporting System (FARS), 36
  General Estimates System (GES), 37
  National Automotive Sampling System/Crashworthiness Data System (NASS/CDS), 36–37
  National Inpatient Sample (NIS), 41
  National Motor Vehicle Crash Causation Survey (NMVCCS), 37–38
  National Trauma Data Bank (NTDB), 40
  Pedestrian Crash Data Study (PCDS), 40
  Pre-hospital Databases (North Carolina PreMIS), 41–42
  Special Crash Investigations (SCI), 39
  State Crash Data, 38
 definition of, 2
 head injury mechanisms
  acute subdural hematoma (ASDH), 7–14
  brain injury mechanisms, 7
 hip fractures, in elderly, 24–25
 injury metrics
  Abbreviated Injury Scale (AIS), 33–34
  International Classification of Diseases version 9 (ICD-9), 34
  KABCO scale, 34
 methods, 6–7
 mid-foot fracture, mechanism of, 25–26
 motivation to study, 2–3

 neck injury mechanisms, 14–17
 pilon fracture mechanism, 25
 real world crash data, applications of
  injury risk estimation, 46–47
  mortality, 44
  societal cost, 45
 thoracolumbar spine
  acute disc rupture, 22–23
  horizontal acceleration/deceleration, 20–21
  vertical (caudocephalad) accelerations, 17–20
 thorax, 17
Injury grading
 blunt carotid, 60
 chest, 69
 extremity, 79
 head, 55
 spine, 59
Injury predictors
 acceleration, 409–410
 compression, 410–412
 force, 417–418
 maximum force and maximum abdominal compression, 419–422
 pressure, 422–423
 rate effects and viscous injury, 412–416
 wave phenomena, 423–424
Injury risk assessments. *See* Anthropomorphic test device (ATD)
Injury Severity Scale (ISS), 336
Injury simulations
 thoracolumbar junction injuries, 446
 upper and lower thoracic spinal column studies, 446–448
Injury tolerance
 biomechanical study, 475
 Federal Motor Vehicle Safety Standard (FMVSS), 476
 hip, 481–484
 incidence, 476
 knee, 487–493
 multiple knee impact tests, 478
 muscle tension effects, 478–481
 peak axial forces, 477
 repetitive test methodology, 476
 thigh, 484–486
INRETS. *See* National Institute for Research in Transportation and Safety
In silico simulations, brain injury, 253–254
Intact human cadaver studies, 446
International Classification of Diseases version 9 (ICD-9), 34
Intervertebral discs
 anatomy, 437–438
 biomechanically relevant anatomy, 453
 component models, 442
 disc herniation, 585–587
 disc innervation and pain, 587–590
Intervertebral load cell (IVL), 18
Isolated components, specimen details, 461–462
ISS. *See* Injury Severity Scale (ISS)

J
Joint, ligament and skeletal pain. *See* Thoracolumbar pain

K
KABCO scale, 34
Knee
   anatomy, 472
   injury tolerance, 487–493
   response and failure, automotive-related injuries
      bending loading, 523–529
      lateral loading, 519–523
      response testing, 523
Knee-thigh-hip (KTH) complex
   anatomy
      anterior cruciate ligament, 471
      femur, 472
      hip, 472, 473
      knee, 472
      lateral collateral ligament, 471
      medial collateral ligament, 471
      patella location, 471
      posterior cruciate ligament, 471
   definitions, 471, 472
   fractures and dislocations, 473
   frontal impact
      body mass distribution, 474
      body shape and body mass index, 474–475
      dislocations, frontal crashes, 473–474
      femoral condyles fractures, 474
      hip fractures and dislocations, 474
      knee loading, 474
      men *vs.* women, acetabulum, 474
      thigh fractures, 474
      trochanteric femur fractures, 474
      types, 473
   injury tolerance
      biomechanical study, 475
      Federal Motor Vehicle Safety Standard (FMVSS), 476
      hip, 481–484
      incidence, 476
      knee, 487–493
      multiple knee impact tests, 478
      muscle tension effects, 478–481
      peak axial forces, 477
      repetitive test methodology, 476
      thigh, 484–486
   lateral impact, 475
   pedestrian impact, 475

L
Landmine blast
   Lower Extremity Assessment Program (LEAP), 537
   non-biofidelic mechanical leg, 536
   PMHS AP landmine tests, 536
Langhaar's approach, 688
Lateral collateral ligament (LCL), 314, 315, 471
Lateral impact, 475

LCS. *See* Lower cervical spine (LCS)
Leg injuries. *See* Lower extremity injuries
Ligaments
   anatomy, 438–439
   biomechanically relevant anatomy, 453–454
   component models, 442–445
Ligamentum flavum, 592–593
Lisfranc sprain/lisfranc injury. *See* Tarsometatarsal (lisfranc) sprains and dislocations
Liver
   in compression, 389–390
   in tension, 386–388
Lordosis and preflexion, 289–290
Lower cervical spine (LCS), 266–268
Lower extremity injuries
   accidental injury, muscle's effect, 627–633
   anatomy, 499–500
   automotive-related injuries
      ankle injury, 508–519
      human foot and ankle dimensions, 501
      knee response and failure, 519–529
      non-minor injury, 500–501
      tibia/fibula response and failure, 501–508
   civil aviation crash injury protection, 820, 823–824
   military-related injuries
      landmine blast, 536–537
      vehicle underbelly blast, 537–543
   sports-related injuries
      metatarsophalangeal joint sprains, 531–533
      syndesmotic ankle sprains, 535
      tarsometatarsal (lisfranc) sprains and dislocations, 533–535
Low speed rear impact regulation, 805–806
Lumbar spine injury
   anatomy, biomechanically relevant
      endplates, 453
      facet joints, 454
      intervertebral discs, 453
      ligaments, 453–454
      vertebrae, 452–453
   biomechanical data
      compressive load to failure and tolerance, 463–465
      physiologic and injury tolerance, 462
      tolerance criteria, 465–466
   civil aviation crash injury protection, 817–819
   classification, 454–457
   loading issues
      axial loading, 458
      drop tests, 460–461
      electro-hydraulic testing device, 458–459
      weight-drop apparatus, 459–460
   specimen details
      isolated components, 461–462
      segmented columns, 462

M
MAthematical DYnamic MOdel (MADYMO)
   based human models, 149–151
   based rollover models, 147–149

Mechanosensors, 251–253
Medial collateral ligaments (MCL), 314, 471
Medulla oblongata, 223–224
Membranous abdominal organs. *See* Hollow abdominal organs
Meninges, 222
Mertz–Weber technique, 779–780
Metatarsophalangeal joint sprains
  hyperextension, 533
  mechanisms of injury, 531, 533
  Weibull survival analysis, 533
Midbrain anatomy, 223
Mid-foot fracture, mechanism of, 25–26
Military-related injuries
  landmine blast
    Lower Extremity Assessment Program (LEAP), 537
    non-biofidelic mechanical leg, 536
    PMHS AP landmine tests, 536
  vehicle underbelly blast
    acceleration comparison, 538
    anti-vehicular mine blast injuries, 539
    ATD underbelly blast tests, 542
    automotive rate tests, 539
    classification, 541
    injury mechanisms, 537, 540
    injury risk curves, 539
    PMHS laboratory underbelly blast tests, 540
Model validation, importance of
  rigorous approaches, 147
  simplistic approaches, 145–146
Motor vehicle regulations, injury criteria
  biomechanical basis
    head injury criteria, 799–800
    neck injury criteria, 800–804
  low speed rear impact regulation, 805–806
  rear impacts, injury criteria, 804–805
  rollover crashes, injury criteria, 804
  whiplash used, public domain tests, 806
Multi-segment spine unit (MSU), 445–446
Muscle activation timing, accidental injury, 613–615
Muscle injury, accidental injury, 616
Muscle mechanical properties, accidental injury, 615–616
Muscle pain, 601–602
Muscle tension effects
  biomechanical tolerance data, 478
  Crash Injury Research and Engineering Network (CIREN), 478
  injury tolerance, 478–481
  risk curves, 479
  Weibull distribution, 479

N
National Automotive Sampling System/Crashworthiness Data System (NASS/CDS), 36–37
National Center for Statistics and Analysis (NCSA), 39
National Inpatient Sample (NIS), 41
National Institute for Research in Transportation and Safety (INRETS), 89
National Motor Vehicle Crash Causation Survey (NMVCCS), 37–38
National Trauma Data Bank (NTDB), 40
Near-side collisions, 125–128
Neck injury
  activity-related injuries
    automobile accidents, 294–295
    diving, 297–300
    football, 296–297
    motorcycle accidents, 295
  anatomy, 262–265
  experimental studies
    cadavers, 271–272
    cervical spines, 272–275
    sagittal plane bending, 280–282
    tension, 276–278
    torsion, 279–280
  fractures and dislocations
    Clay Shoveler's fracture, 270
    clinical classification, 270–271
    lower cervical spine, 266–268
    upper cervical spine, 268–269
  incidence, 260–262
  lateral bending and lateral shear, 284–285
  mechanisms, 14–17
  motor vehicle regulations, injury criteria, 800–804
  side-facing seats, aviation crash injury, 822–823
  spine injury mechanics and mechanisms
    anatomical variation and asymmetry, 290
    buckling and buckled deformation, 286–287
    cervical musculature, 285–286
    end condition, 291
    head motion, 288–289
    impact attitude and head inertia, 291–292
    impact surface stiffness and padding characteristics, 292–293
    loading rate, 287–288
    lordosis and preflexion, 289–290
  trauma biomechanics, 59–61
  upper cervical spine (UCS), 282–284
Neck, pediatric biomechanics
  biomechanical studies
    injury assessment reference values, 671–672
    material properties, 672
    structural response, 664–671
  developmental anatomy
    age-related groupings, 662–663
    atlas, 659
    axis, 659
    facet joints, 660
    intervertebral discs, 663–664
    ligaments, 663
    muscles, 664
    quantitative description, vertebral growth, 662
    typical cervical vertebrae, 659–660
    vertebrae, 659–663
Nerve root injury, 561–565
Neuropathic pain, 598–600
NIS. *See* National Inpatient Sample (NIS)

N

NMVCCS. *See* National Motor Vehicle Crash Causation Survey (NMVCCS)
Normalized Integral Square Error (NISE) method, 145
Noxious loading biomechanics
    cadaver studies, 585
    Gx acceleration, 585
    injury tolerance, 583–585
NTDB. *See* National Trauma Data Bank (NTDB)

O

Obesity, restraint system biomechanics, 132–134
Occupant FE models
    FE-based ATD modeling, 171–172
    human modeling, 172–177
Occupant support criteria
    articulation limits, 824–825
    civil aviation crash injury protection
        articulation limits, 824–825
        contact limits, 825
        excursion limits, 824
    contact limits, 825
    excursion limits, 824

P

Pain biomechanics
    anatomy
        bone, intervertebral disc, and ligaments, 551–553
        DRG, 553–554
        paraspinal musculature, 554–555
        spinal tissues, 550–551
        vasculature, 555
    cervical spine injuries
        neck pain, 565–566
        and painful facet joint injury, 559–561
        and painful nerve root injury, 561–565
    experimental techniques
        human volunteer and post-mortem models, 566–567
        mechanical loading, 569–571
        in vivo animal models, 567–569
    mechanical stimulation, 555–558
Patella location, 471
PCDS. *See* Pedestrian Crash Data Study (PCDS)
Pedestrian Crash Data Study (PCDS), 40
Pedestrian injury biomechanics and protection, 475
    bonnet shape, 740–742
    bumper
        lead, 739
        leading edge stiffness, 742
        shape, 737–738
        stiffness, 739–740
    combined primary and secondary safety, 747
    deployable hoods and airbags, 745–746
    forward projection, 726
    headform to bonnet top, 731–732
    head impact, bonnet/windscreen, 743–745
    implementation into legislation, 732
    injury risk, function of vehicle type, 723–724
    overall vehicle shape influence, 735–737
    pedestrian impact kinematics, 724
    post head impact kinematics, 727
    production vehicles via pedestrian safety regulations, 747–748
    secondary bumper, 738–739
    statistics, 722–723
    upper legform to bonnet leading edge, 730–731
    vehicle
        design, 733–734
        design standards, pedestrian protection, 728–730
        and ground contact, 723
        mass influence, 734–735
        stiffness influence, 735
    warning and auto-braking, 745
    wrap projection, 724–726
Pediatric biomechanics
    abdomen
        biomechanical studies, 680–683
        developmental anatomy, 680
    extremities
        biomechanical studies, 685–686
        developmental anatomy, 683–685
    head
        biomechanical studies, 646–653
        brain and skull, 653–655
        developmental anatomy, 644–646
        injury criteria, 657–658
        injury scaling methods, animal models, 655–657
    Langhaar's approach, 688
    neck
        biomechanical studies, 664–672
        developmental anatomy, 658–664
    pediatric anthropomorphic test device (ATD), 687
    scaling equations, 688
    thorax
        biomechanical studies, 672–673
        costal cartilage, 674
        developmental anatomy, 672
        vascular and cardiac tissue, 674–675
        vertebrae, 673–674
        whole thorax biomechanical experiments, 675–680
Pelvis, aviation crash injury, 823
Physiologic and injury tolerance, 462
Pilon fracture mechanism, 25
PMHS. *See* Post mortem human subjects (PMHS)
PODS. *See* Probability of Death Score (PODS)
Pons anatomy, 223
Posterior cruciate ligament (PCL), 471
Posterior longitudinal ligament (PLL), 591–592
Post mortem human subjects (PMHS)
    airbag deployment, 323–324
    guided impactors, 320–322
    Sled, 322–323
Pre-hospital Databases (North Carolina PreMIS), 41–42
Probability of Death Score (PODS), 336

R

Range of motion (ROM), 515
Rear-facing child restraints, 703–707

Rear impacts, injury criteria, 804–805
Restraint system biomechanics
    aging, 134–136
    airbag tests, 406–407
    far-side collisions, 128–129
    general considerations, 114–115
    general restraint maxims, 116–120
    inflatable restraint biomechanics
        head and face contact mitigation, 120
        load sharing, force distribution, and work, 120–125
        restraint system design flexibility, 125
    near-side collisions, 125–128
    obesity, 132–134
    rear seat occupants, 131–132
    rollovers, 130–131
    seat belt biomechanics, 116–120
    seatbelt tests, 401–406
Reusable Rate-Sensitive Abdomen (RRSA), 425
Rib fractures and flail chest, 337
Rollover crashes, injury criteria, 804

S

Scalp, anatomy of, 222
SCI. *See* Special Crash Investigations (SCI)
Seat belt biomechanics, 116–120
Segmented columns, 445–446, 462
Side-facing seats, aviation crash injury
    neck, 822–823
    special conditions, 821–822
    thorax and pelvis, 823
Side impact testing, ATDs
    biofidelic side impact dummy (BioSID), 90
    EUROSID and EUROSID-1, 89–90
    Q3s, 91
    side impact dummy (SID) and SID-HIII, 88–89
    SID-IIs, 90–91
    WorldSID, 91
    WorldSID small female, 91
Skull and facial bone injury biomechanics
    anatomy, 203–204, 222
    cranial vault, 206–210
    facial bones, 210–215
    injury classification, 205–206
    severity scales, 205–206
    skull (calvarium) fracture, 216–128
Solid abdominal organs
    kidneys, 378
    liver, 377
    pancreas, 378
    spleen, 377–378
Special Crash Investigations (SCI), 39
Spinal cord trauma, 58–59
Spine injury
    accidental injury, muscle's effect, 617–621
    mechanics and mechanisms
        anatomical variation and asymmetry, 290
        buckling and buckled deformation, 286–287
        cervical musculature, 285–286
        end condition, 291
        head motion, 288–289
        impact attitude and head inertia, 291–292
        impact surface stiffness and padding characteristics, 292–293
        loading rate, 287–288
    trauma biomechanics
        compression and burst fractures, 57
        flexion distractions and fracture dislocations, 57–58
        injury grading, 59
        screening, 56
        spinal cord trauma, 58–59
        transverse process (TP) fractures, 58
        unique cervical spine injuries, 56–57
Spleen
    in compression, 390–391
    in tension, 390
Sports helmets
    biomechanical considerations, 755–757
    evolution, 760–762
    modern design
        impact liner, 765–766
        shell type, 764–765
    performance standard evolution, 758–760
    requirements
        conflicting, 764
        design, 762
        operational, 764
        penetration resistance, 763
        performance, 763
        retention capability, 764
        shock attenuating capability, 763
Sports-related injuries
    metatarsophalangeal joint sprains
        hyperextension, 533
        mechanisms of injury, 531, 533
        Weibull survival analysis, 533
    syndesmotic ankle sprains, 535
    tarsometatarsal (lisfranc) sprains and dislocations
        C1/M2 ligament bundle, 534
        oblique ligament bundle, 533
        plantarflexed condition, 535
        serial-sectioning, 534
SSDM. *See* Statistical shape and density model (SSDM)
Statistical shape and density model (SSDM), 187–189
Subdural hematoma (SDH), 10
Syndesmotic ankle sprains, 535

T

Tarsometatarsal (lisfranc) sprains and dislocations
    C1/M2 ligament bundle, 534
    oblique ligament bundle, 533
    plantarflexed condition, 535
    serial-sectioning, 534
Thigh
    fractures, 474
    injury tolerance, 484–486
Thoracic response, frontal impact, 4

Thoracic spine injury biomechanics
  anatomy
    intervertebral discs, 437–438
    kyphotic (backward curve), 435–436
    ligaments, 438–439
    vertebrae, 436–437
  clinical injuries, 448–449
  component models
    advantage, 439
    compression loading test, 440
    intervertebral discs, 442
    ligaments, 442–445
    vertebrae, 440–442
  injury simulations
    thoracolumbar junction injuries, 446
    upper and lower thoracic spinal column studies, 446–448
  intact human cadaver studies, 446
  segmented models, 445–446
Thoracolumbar pain
  anterior longitudinal ligament, 590–591
  dural innervation, 592
  facet joints
    biomechanical study, 595
    inflammation, nerve activity, 597
    morphology, 593
    and muscle neurophysiology in control animals, 595–596
    nerve supply, 593
    prevalence, 594
    retrospective analysis, 594
    spine loading studies, 596–597
    substance P, unit activity, 597–598
  intervertebral discs
    disc herniation, 585–587
    disc innervation and pain, 587–590
  junction injuries, 446
  ligamentum flavum, 592–593
  muscle pain, 601–602
  neuropathic pain, 598–600
  noxious loading biomechanics
    cadaver studies, 585
    Gx acceleration, 585
    injury tolerance, 583–585
  pain initiation mechanisms and maintenance, 582–583
  posterior longitudinal ligament, 591–592
  spine, injury biomechanics
    acute disc rupture, 22–23
    horizontal acceleration/deceleration, 20–21
    vertical (caudocephalad) accelerations, 17–20
Thorax injury
  age, injury tolerance, 338–339
  aortic trauma
    anatomical location of, 340
    aortic injury mechanisms, 340–341
    epidemiology, 339–340
  assessment and development, 364–366
  epidemiology
    heart and great vessels, 332–334
    lungs and mediastinum, 332
    rib cage, 332
  frontal impact, biomechanical response of
    belt loading, 342–343
    quasi-static tests, 344–345
    sternum, pendulum impacts, 341–342
    two-point belt loading plus knee bolster, 343–344
  frontal impact, injury tolerance of
    acceleration criteria, 345
    belt loading and combined restraint loading, 347–350
    combined compression and acceleration criterion, 347
    compression criteria, 346
    energy criterion, 345–346
    force criteria, 345
    viscous criterion, 346
  general, 336–337
  heart injuries, 338
  hemothorax and pneumothorax, 338
  injury scaling methods
    Abbreviated Injury Scale (AIS), 334–336
    Injury Severity Scale (ISS), 336
    Probability of Death Score (PODS), 336
  lateral impact
    airbag tests, 357
    drop tests, 352
    impactor tests, 356–357
    side impact punch, 357
    sled tests, 352–356
  lung contusions, 337–338
  oblique lateral impact, 357–364
  pediatric biomechanics
    biomechanical studies, 672–673
    costal cartilage, 674
    developmental anatomy, 672
    vascular and cardiac tissue, 674–675
    vertebrae, 673–674
    whole thorax biomechanical experiments, 675–680
  rib fractures and flail chest, 337
  side-facing seats, aviation crash injury, 823
  traumatic rupture of aorta (TRA), 17
Tibia/fibula response and failure, automotive-related injuries
  dynamic bending tolerance, 504–508 (*see also* Bending tolerance)
  fracture tolerance in bending, 502–504
TRA. *See* Traumatic rupture of aorta (TRA)
Transport Research Laboratory (TRL), 89
Trauma biomechanics
  abdomen
    hollow viscus, 72–73
    incidentalomas, 74–75
    injury grading, 75
    solid organs, 70–72
    vascular, 73–74
  background, 51–52
  chest
    chest wall, 64–65
    diaphragm, 68–69
    heart, 63–64

injury grading, 69
mediastinum and thoracic aorta, 66–68
pulmonary, 65–66
external, 79
extremity
imaging of, 77–78
injury grading, 78
pelvis, 76–77
face, 61–62
head, 52–55
neck, 59–61
spine
compression and burst fractures, 57
flexion distractions and fracture dislocations, 57–58
injury grading, 59
screening, 56
spinal cord trauma, 58–59
transverse process (TP) fractures, 58
unique cervical spine injuries, 56–57
Traumatic brain injury
diffuse injuries, 224–225
focal injuries, 225–226
skull fractures, 226–227
types of, 224
Traumatic rupture of aorta (TRA), 17, 339
TRL. *See* Transport Research Laboratory (TRL)
Trochanteric femur fractures, 474

U

Upper cervical spine (UCS), 268–269, 282–284
Upper extremity injuries
accidental injury, muscle's effect, 621–627
biomechanically relevant anatomy
bones, 310–311
joints, 311–313
ligaments, 313–315
muscles, 315
component biomechanics and tolerance
clavicle, 316–317
forearm and elbow, 318–320
humerus, 317

PMHS
airbag deployment, 323–324
guided impactors, 320–322
Sled, 322–323
side impacts
impact types and velocities, 325–326
implications of injuries, 326–327
injury scores, 326
occupants and injury types, 325
other body regions, 326
vehicle factors, 325

V

Vasculature
anatomy, 555
arteries, 378
veins, 378
Vehicle underbelly blast
acceleration comparison, 538
anti-vehicular mine blast injuries, 539
ATD underbelly blast tests, 542
automotive rate tests, 539
classification, 541
injury mechanisms, 537, 540
injury risk curves, 539
PMHS laboratory underbelly blast tests, 540
Vertebrae
anatomy, 436–437
biomechanically relevant anatomy, 452–453
component models, 440–442

W

Wedge fractures, 455, 456
Weibull distribution, 479
Weight-drop apparatus, 459–460

X

Xversion injuries
dynamic xversion injury, 510–513
quasi-static xversion response, 509–510

CPSIA information can be obtained at www.ICGtesting.com
Printed in the USA
BVOW11*1716180115
383850BV00001B/1/P